THE ENCYCLOPEDIA OF MARINE RESOURCES

The

ENCYCLOPEDIA
of
MARINE RESOURCES

EDITED BY

Frank E. Firth

*New England Marine Resources Information Program
University of Rhode Island, Kingston, Rhode Island*

VAN NOSTRAND REINHOLD COMPANY
New York   Cincinnati   Toronto   London   Melbourne

SH
201
.F56
1969

LIBRARY
LOS ANGELES COUNTY MUSEUM OF NATURAL HISTORY

Van Nostrand Reinhold Company Regional Offices:
  Cincinnati, New York, Chicago, Millbrae, Dallas
Van Nostrand Reinhold Company Foreign Offices:
  London, Toronto, Melbourne

Copyright © 1969 by Reinhold Book Corporation

Library of Congress Catalog Number: 70-78014

All rights reserved. No part of this work covered by the
copyrights hereon may be reproduced or used in any form or
by any means—graphic, electronic, or mechanical, including
photocopying, recording, taping, or information storage and
retrieval systems—without written permission of the publisher.
Manufactured in the United States of America.

Published by Van Nostrand Reinhold Company
450 West 33rd Street, New York, N.Y. 10001
Published simultaneously in Canada by
D. Van Nostrand Company (Canada), Ltd.

15 14 13 12 11 10 9 8 7 6 5 4 3 2 1

CONTRIBUTORS

ROBERT B. ABLE, Office of Sea Grant Programs, National Science Foundation, Washington, D.C. *National Sea Grant Program.*

FREDERICK, A. ALDRICH, Marine Sciences Research Laboratory, Memorial University of Newfoundland, St. John's, Newfoundland. *Cephalopoda.*

JOSEPH W. ANGELOVIC, U.S. Bureau of Commercial Fisheries, Beaufort, North Carolina. *Radioactivity in the Sea: Effects on Fisheries.*

CLINTON E. ATKINSON, American Embassy, Tokyo, Japan. *Japanese Fisheries.*

KJELL BAALSRUD, Norsk Institutt For Vannforskning Blindern, Oslo, Norway. *Pollution of Seawater.*

PATRICIA R. BERGQUIST, The University, Auckland, New Zealand. *Sponge Industry.*

ROBERT BIERI, Antioch College, Yellow Springs, Ohio. *Invertebrates as Food; Pelagic Distribution.*

MAURICE BLACKBURN, Scripps Institution of Oceanography, La Jolla, California. *Peru Current.*

E. BOOTH, Institute of Seaweed Research, Inveresk, Midlothian, Scotland. *Seaweeds of Commerce.*

KENNETH J. BOSS, Museum of Comparative Zoology, Harvard University, Cambridge, Massachusetts. *Conchs.*

MARGARETHA BRONGERSMA-SANDERS, Rijksmuseum v Naturlijke Historie, Leiden, Netherlands. *Mortality in the Sea.*

FRANK A. BROWN, JR., Department of Biological Sciences, Northwestern University, Evanston, Illinois. *Biological Clocks in Marine Animals.*

HARVEY R. BULLIS, U.S. Bureau of Commercial Fisheries, Pascagoula, Mississippi. *Fishing Gear and Methods.*

P. R. BURKHOLDER, Lamont Geological Observatory of Columbia University, Palisades, New York. *Antimicrobial Substances.*

WAYNE V. BURT, Department of Oceanography, Oregon State University, Cornvallis, Oregon. *Oceanography: Exploration and Exploitation.*

PHILIP A. BUTLER, U.S. Bureau of Commercial Fisheries, Gulf Breeze, Florida. *Pesticides in the Sea.*

STANLEY A. CAIN, Department of Biological Sciences, Central Washington State College, Ellensburg, Washington. *Government (U.S.) Development of Marine Resources.*

DAVID K. CALDWELL, Marineland Research Laboratory, St. Augustine, Florida. *Porpoises; Sea Turtles.*

MELBA C. CALDWELL, Marineland Research Laboratory, St. Augustine, Florida. *Porpoises; Sea Turtles.*

CANADIAN DEPARTMENT OF FISHERIES, Information Service, Ottawa, Canada. *Canadian Fisheries: General.*

FRANCIS J. CAPTIVA, U.S. Bureau of Commercial Fisheries, Pascagoula, Mississippi. *Fishing Gear and Methods.*

J. LOCKWOOD CHAMBERLIN, U.S. Bureau of Commercial Fisheries, Washington, D.C. *Clams.*

GEORGE M. CLARKE, Manchester, Massachusetts. *Mackerel Fishery, Atlantic; Redfish Fishery.*

SALVATORE COMITINI, Department of Economics, University of Hawaii, Honolulu, Hawaii. *Exploitation of Marine Resources; Japanese Fisheries.*

JOHN A. DASSOW, U.S. Bureau of Commercial Fisheries, Seattle, Washington. *Crab Industry: Pacific Coast.*

G. DAVIES, Ministry of Agriculture, Conway, Caernarvonshire, England. *Mussels As a World Food Resource.*

LOLA T. DEES, U.S. Bureau of Commercial Fisheries, Washington, D.C. *Sea Lamprey.*

C. B. ENRIQUE M. DEL SOLAR, Biologo y Tecnologo Pesquero, Lima, Peru. *Peru Fishery.*

HORACE W. DIAMOND, Morton Salt Co., Chicago, Illinois. *Salt, Solar.*

v

CONTRIBUTORS

ROBERT L. DOW, Department of Sea and Shore Fisheries, Augusta, Maine. *Maine Marine Worm Fishery; Sea Scallop Fishery, Shrimp Fishery: Maine Shrimp Fishery.*

ALEXANDER DRAGOVICH, U.S. Bureau of Commercial Fisheries, Miami, Florida. *Dinoflagellates.*

T. W. DUKE, U.S. Bureau of Commercial Fisheries, Beaufort, North Carolina. *Radioactivity in the Sea: General.*

NORMAN W. DURRANT, U.S. Bureau of Commercial Fisheries, Washington, D.C. *Irish Moss Industry.*

S. DUTT, Andhra University Post-Graduate Centre, Guntur, India. *Indian Fisheries.*

EARL E. EBERT, Department of Fish and Game, Menlo Park, California. *Abalone.*

R. ENDEAN, University of Queensland, Brisbane, Australia. *Biotoxins, Marine.*

RHODES W. FAIRBRIDGE, Department of Geology, Columbia University, New York, New York. *Coral Reefs.*

J. W. FELL, Institute of Marine Science, University of Miami, Miami, Florida. *Fungi, Marine.*

HOWARD W. FIEDELMAN, Morton Salt Co., Chicago, Illinois. *Salt, Solar.*

FRANK E. FIRTH, University of Rhode Island, Kingston, Rhode Island. *Editor; Longline Fishing for Swordfish; Norwegian Fisheries.*

JAMES F. FISH, Graduate School of Oceanography, University of Rhode Island, Kingston, Rhode Island. *Sounds of Marine Animals.*

ALLISTER M. FLEMING, Fisheries Research Board of Canada, St. John's, Newfoundland. *Newfoundland Fisheries.*

M. FORMELA, Sea Fisheries Institute, Gdynia, Poland. *Poland Fisheries.*

JAMES H. FRASER, Department of Agriculture and Fisheries, Torry, Aberdeen, Scotland. *Plankton Resources.*

R. W. GEORGE, Western Australian Museum, Perth, Australia. *Lobsters: Spiny Lobsters.*

PERRY W. GILBERT, Mote Marine Laboratory, Sarasota, Florida. *Sharks: Behavior and Attack Patterns.*

MALVERN GILMARTIN, Hopkins Marine Station, Stanford University, Pacific Grove, California. *Fertility of the Sea.*

RAYMOND M. GILMORE, Natural History Museum, San Diego, California. *Ambergris.*

HAROLD LELAND GOODWIN, Office of Sea Grant Programs, National Science Foundation, Washington, D.C. *National Sea Grant Program.*

WILLIAM L. GRIFFIN, Attorney at Law, Washington, D.C. *Law of Ocean Space.*

J. GRINDLEY, Hydraulics Research Station, Wallingford, Berkshire, England. *Estuarine Environment.*

WILLIAM H. GROSS, The Dow Chemical Co., Midland, Michigan. *Magnesium From the Sea.*

GORDON GUNTER, Gulf Coast Research Laboratory, Ocean Springs, Mississippi. *Gulf Fisheries.*

R. G. HAINES, Kelvin Hughes America Corp., Annapolis, Maryland. *Electronics in Developing Marine Resources.*

BRUCE W. HALSTEAD, M.D., World Life Research Institute, Colton, California. *Toxicology of Marine Animals.*

ROBERT W. HANKS, U.S. Bureau of Commercial Fisheries, Oxford, Maryland. *Clams: Soft-Shell Clams.*

GEORGE Y. HARRY, JR., U.S. Bureau of Commercial Fisheries, Ann Arbor, Michigan. *Salmon.*

JOEL W. HEDGPETH, Marine Science Center, Oregon State University, Newport, Oregon. *Ecology: Marine.*

OLE J. HEGGEM, Economic Affairs Department, U.S. Civil Administration of the Ryukyu Islands. *Ryukyuan Fisheries.*

KENNETH A. HENRY, U.S. Bureau of Commercial Fisheries, Beaufort, North Carolina. *Menhaden Fisheries.*

E. HIGHLEY, Commonwealth Scientific and Industrial Research Organization, Cronulla, New South Wales, Australia. *Pearling Industry.*

C. R. HITZ, U.S. Bureau of Comercial Fisheries, Seattle, Washington. *Fishing Vessels and Support Ships.*

CONTRIBUTORS

ANDREAS HOLMSEN, College of Agriculture, University of Rhode Island, Kingston, Rhode Island. *Red Crab Fishery.*

D. I. D. HOWIE, Trinity College, Dublin, Ireland. *Breeding in Marine Animals.*

DWIGHT L. HOY, U.S. Bureau of Commercial Fisheries, Gloucester, Massachusetts. *Mackerel Fishery, Atlantic.*

DAVID R. IDLER, Fisheries Research Board of Canada, Halifax, Nova Scotia, Canada. *Cod Fishery.*

C. P. IDYLL, Institute of Marine Science, University of Miami, Miami, Florida. *Shrimp Fisheries.*

COLUMBUS O'D. ISELIN, Woods Hole Oceanographic Institution, Woods Hole, Massachusetts. *Ocean: Potential Resources.*

P. M. JANGAARD, Fisheries Research Board of Canada, Halifax, Nova Scotia, Canada. *Cod Fishery.*

KENNETH R. JOHN, Department of Biology, Franklin and Marshall College, Lancaster, Pennsylvania. *Schooling Behavior.*

T. G. KAILIS, Ross Fisheries, Pty. Ltd., West Perth, Australia. *Lobsters: Spiny Lobsters.*

NEVA L. KARRICK, U.S. Bureau of Commercial Fisheries, Seattle, Washington. *Vitamins From Marine Sources.*

J. L. KASK, Inter-American Tropical Tuna Commission, Scripps Institution of Oceanography, La Jolla, California. *Tuna Fisheries.*

JOHN D. KAYLOR, U.S. Bureau of Commercial Fisheries, Gloucester, Massachusetts. *Radiation Preservation of Marine Foods.*

HASTINGS KEITH, U.S. House of Representatives, Washington, D.C. *U.S.S.R. Fisheries.*

BOSTWICK H. KETCHUM, National Science Foundation, Washington, D.C. *Productivity of Marine Communities.*

ROBERT R. KIFER, U.S. Bureau of Commercial Fisheries, College Park, Maryland. *Animal Feeds.*

S. A. KULM, Department of Oceanography, Oregon State University, Cornvallis, Oregon. *Oceanography: Exploration and Exploitation.*

ROBERT J. LEARSON, U.S. Bureau of Commercial Fisheries, Gloucester, Massachusetts. *Breaded Fishery Products.*

IAN LE FEVRE, Sydney, Australia. *Australian Fisheries.*

JOHN LISTON, College of Fisheries, University of Washington, Seattle, Washington. *Microbiology, Marine.*

ALFRED R. LOEBLICH, JR., Chevron Research Co., La Habra, California. *Foraminifera.*

TRAVIS D. LOVE, U.S. Bureau of Commercial Fisheries, Pascagoula, Mississippi. *Nutritive Value of Seafoods.*

CHARLES H. LYLES, U.S. Bureau of Commercial Fisheries, Arlington, Virginia. *United States and World Fishery Statistics.*

FRED T. MACKENZIE, Department of Geology, Northwestern University, Evanston, Illinois. *Chemistry of Seawater.*

F. JENSENIUS MADSEN, Universitetets Zoologiske Museum, Copenhagen, Denmark. *Abyssal Zone.*

THOMAS A. MANAR, U.S. Bureau of Commercial Fisheries, Honolulu, Hawaii. *Hawaiian Fisheries; Pacific Fisheries: Tropical and Subtropical.*

CLARO MARTIN, Quezon City, Philippines. *Philippine Fishery.*

J. MAUCHLINE, The Dunstaffnage Marine Laboratory, Oban, Argyll, Scotland. *Radioactive Waste in Seawater.*

A. W. MAY, Fisheries Research Board of Canada, St. John's, Newfoundland. *Labrador Fisheries.*

GUY C. MCLEOD, New England Aquarium, Boston, Massachusetts. *Sea Farming.*

JOHN L. MERO, Ocean Resources, Inc., La Jolla, California. *Minerals of the Ocean: Potential Resources.*

ARTHUR S. MERRILL, U.S. Bureau of Commercial Fisheries, Oxford, Maryland. *Clams: Surf-Clam Fishery; Quahog Fishery.*

DANIEL MERRIMAN, Sears Foundation for Marine Research, Yale University, New Haven, Connecticut. *Icthyology.*

ALTON B MOODY, Federal Aviation Administration, Washington, D.C. *Navigation.*

CONTRIBUTORS

MELVAN E. MORRIS, JR., U.S. Department of Fish and Game, Kodiak, Alaska. *Crab Industry: King Crab Industry.*

A. G. MOURAD, Battelle Memorial Institute, Columbus, Ohio. *Marine Geodesy.*

M. P. MOYER, The Dow Chemical Co., Midland, Michigan. *Bromine From the Sea.*

ROSS F. NIGRELLI, New York Aquarium and Osborne Laboratories of Marine Sciences, Brooklyn, New York. *Parasites and Diseases.*

ROBERT E. NORLAND, Norland Products, Inc., New Brunswick, New Jersey. *Fish Glue.*

RICHARD E. NORRIS, Department of Botany, University of Washington, Seattle, Washington. *Phytoplankton.*

ARTHUR F. NOVAK, Department of Food Science and Technology, Louisiana State University, Baton Rouge, Louisiana. *Polysaccharide Coatings.*

HIDEO OMURA, The Whales Research Institute, Tokyo, Japan. *Whales and Whaling.*

DONALD F. OTHMER, Department of Chemical Engineering, Polytechnic Institute of Brooklyn, Brooklyn, New York. *Desalination of Seawater; Heat and Power From Seawater.*

E. R. PARISER, Avco Corp., Wilmington, Massachusetts. *Fish Protein Concentrate.*

BASIL A. PARKES, Boston Deep Sea Fisheries, Ltd., Boston, England. *Fish Harvesting.*

GLENN W. PATTERSON, Department of Botany, University of Maryland, College Park, Maryland. *Algae, Marine.*

J. M. PERES, Station Marine d'Endoume et Centre d'Oceanographie, Marseille, France. *Benthonic Domain.*

LEO PINKAS, Department of Fish and Game, Terminal Island, California. *Ecology: Fish.*

A. T. PRUTER, U.S. Bureau of Commercial Fisheries, Seattle, Washington. *Canadian Fisheries: British Columbia; Pacific Fisheries: U.S. West Coast Fisheries.*

M. R. RAMACHANDRA RAO, Department of Food Science and Technology, Louisiana State University, Baton Rouge, Louisiana. *Polysaccharide Coatings.*

W. M. REES, Marine Colloids, Inc., Springfield, New Jersey. *Phycocolloids and Marine Colloids.*

T. R. RICE, U.S. Bureau of Commercial Fisheries, Beaufort, North Carolina. *Radioactivity in the Sea: General; Effects on Fisheries.*

BEATRICE R. RICHARDS, Battelle Memorial Institute, Duxbury, Massachusetts. *Marine Borers.*

JULIUS ROCKWELL, JR., U.S. Bureau of Commercial Fisheries, Arlington, Virginia. *Satellite Sensing of Marine Phenomena.*

JOHN W. ROPES, U.S. Bureau of Commercial Fisheries, Oxford, Maryland. *Clams: Surf Clam Fishery; Quahog Fishery.*

GEORGE A. ROUNSEFELL, Marine Sciences Institute, University of Alabama. *Marking of Fish.*

A. W. SADDINGTON, Kelco Co., San Diego, California. *Alginates From Kelp.*

SAUL B. SAILA, Graduate School of Oceanography, University of Rhode Island, Kingston, Rhode Island. *Lobsters: Offshore Lobsters.*

JOHN C. SAINSBURY, University of Rhode Island, Kingston, Rhode Island. *Education for the Commercial Fisheries.*

F. BRUCE SANFORD, U.S. Bureau of Commercial Fisheries, Seattle, Washington. *Irish Moss Industry.*

LESLIE W. SCATTERGOOD, U.S. Bureau of Commercial Fisheries, Washington, D.C. *Lobsters: General.*

MILNER B. SCHAEFER, Institute of Marine Resources, University of California (San Diego), La Jolla, California. *Oceanography: Oceanography and Marine Fisheries.*

EDWARD A. SCHAEFERS, U.S. Bureau of Commercial Fisheries, Washington, D.C. *Fishery Engineering.*

H. H. SELBY, American Agar and Chemical Co., San Diego, California. *Agar.*

CONTRIBUTORS

WILLIAM N. SHAW, U.S. Bureau of Commercial Fisheries, Oxford, Maryland. *Oysters.*

JOHN MCN. SIEBURTH, Graduate School of Oceanography, University of Rhode Island, Kingston, Rhode Island. *Bacteria in the Sea.*

JOHN B. SKERRY, U.S. Bureau of Commercial Fisheries, Gloucester, Massachusetts. *Atlantic Northwest Fisheries.*

BERNARD EINAR SKUD, U.S. Bureau of Commercial Fisheries, Boothbay Harbor, Maine. *Lobsters: Giant Lobsters.*

JOSEPH W. SLAVIN, U.S. Bureau of Commercial Fisheries, Washington, D.C. *Freezing Fish at Sea.*

DONALD G. SNYDER, U.S. Bureau of Commercial Fisheries, College Park, Maryland. *Fish Meal.*

STEWART SPRINGER, U.S. Bureau of Commercial Fisheries, Washington, D.C. *Sharks: Biological Characteristics.*

G. H. STANDER, Division of Sea Fisheries, Capetown, South Africa. *South Africa, Marine Resources.*

MAURICE E. STANSBY, U.S. Bureau of Commercial Fisheries, Seattle, Washington. *Fats and Oils.*

CHARLES R. STEPHAN, Department of Ocean Engineering, Florida Atlantic University, Boca Raton, Florida. *Ocean: Ocean Engineering.*

RICHARD H. STROUD, Sport Fishing Institute, Washington, D.C. *Sport Fishing.*

BERTIL SWEDMARK, Kristinebergs Zoologiska Station, Fiskebackskil, Sweden. *Interstitial Fauna.*

OTOHIKO TANAKA, Ocean Research Institute, Tokyo, Japan. *Copepods.*

HELEN TAPPAN, Department of Geology, University of California (Los Angeles), Los Angeles, California. *Foraminifera.*

H. L. A. TARR, Fisheries Research Board of Canada, Vancouver, British Columbia, Canada. *Antibiotic Preservation of Fish at Sea.*

MARY H. THOMPSON, U.S. Bureau of Commercial Fisheries, Pascagoula, Mississippi. *Biochemical Composition of Fish.*

SETON H. THOMPSON, U.S. Bureau of Commercial Fisheries, St. Petersburg, Florida. *Marine Fisheries Commissions.*

G. THORSTEINSSON, Marine Research Institute, Reykjavik, Iceland. *Iceland Fisheries.*

CATHERINE A. TIZARD, The University, Auckland, New Zealand. *Sponge Industry.*

CHARLES H. TURNER, Department of Fish and Game, Terminal Island, California. *Artificial Reefs.*

VADIM D. VLADYKOV, Department of Biology, University of Ottawa, Ottawa, Ontario, Canada. *Eels.*

J. J. WATERMAN, Torry Research Station, Ministry of Technology, Aberdeen, Scotland. *British Fisheries.*

J. G. WATKINSON, Marine Department, Wellington, New Zealand. *New Zealand Fisheries.*

PETER T. WILSON, Fisheries Management, Trust Territory of the Pacific Islands, Saipan, Mariana Islands. *Trust Territory of the Pacific Islands.*

HOWARD E. WINN, Graduate School of Oceanography, University of Rhode Island, Kingston, Rhode Island. *Sounds of Marine Animals.*

GEORGE H. WINTERS, Fisheries Research Board of Canada, St. John's, Newfoundland. *Capelin.*

F. F. WRIGHT, Institute of Marine Science, University of Alaska, College, Alaska. *Mineral Exploration Technology.*

PREFACE

The oceans of the world hold the key to the most persistent long range problem of the entire human race, namely, how to provide adequate food for the always increasing population. This key can be used in two ways: the ocean is the greatest potential source not only of essential nutrients, but also of an almost unlimited supply of fresh water. The nutrient possibilities can be exploited by the development of such products as fish protein concentrate, as well as by intensive cultivation of high-protein seafoods through sea farming methods. The fresh water, which can be extracted by future expansion of desalination technology, can be used to irrigate the deserts, thereby converting many of the earth's waste places to fertile, food-producing regions capable of supporting millions of people. Potable water achieved through desalination is already a reality in dozens of locations, and the process holds vast promise for vital agricultural projects.

The purpose of this book is to help serve these needs by presenting the most significant aspects of the ocean's resources, together with summaries of a few closely related topics considered necessary to round out the treatment. At the outset, the editor found himself subject to stringent space limitations, which necessitated shortening some of the articles and eliminating many of the tables and illustrations. The temptation to wander into such fields as oceanography and marine engineering proved almost too great to resist. These areas—vast in themselves—are given brief treatment. Detailed coverage of oceanography can be found in *The Encyclopedia of Oceanography,* edited by R. W. Fairbridge.

The editor has made every effort to obtain contributions from the best qualified authorities from many nations; both he and the publisher wish to express sincere gratitude for their cooperation, truly indispensable for the publication of this volume. As a result of their collaboration, the book ought to be useful to all those who have any contact with, or interest in the oceans and their almost infinite resources that can promote the welfare of mankind. Among those who will find it valuable are biologists, chemists, engineers, food technologists, metallurgists, fishery management specialists, teachers, students in marine-oriented fields, and thoughtful laymen concerned about their world, its problems and solutions.

The editor is happy to have this opportunity to acknowledge the help of everyone who shared in the preparation of this encyclopedia. As all contact with the contributors was conducted by mail, the correspondence was, indeed, a monumental task. For this, the efforts of Mr. and Mrs. Ralph T. Winslow of Lexington, Massachusetts were essential, and are greatly appreciated. The editor is also indebted to Mr. Leslie A. Scattergood and Mr. A. T. Pruter of the U.S. Bureau of Commercial Fisheries for their unfailing assistance in providing introductions to contributors and their helpful interest in all phases of the undertaking.

Finally, I wish to acknowledge the assistance of the most articulate editor and writer I know, Mr. Gessner G. Hawley of Newton Center, Massachusetts, who gave liberally of his time and experience when, at times, the going was rough.

FRANK E. FIRTH

Kingston, Rhode Island
June, 1969

A

ABALONE

Abalones are prosobranch marine snails possessing an auriform shell that surrounds and protects their soft body parts dorsally, and a large muscular foot for locomotion and for firm ventral attachment to rocky substratum.

Primitive mollusks, they are in the phylum Mollusca, class Gastropoda, order Archaeogastropoda, and family Haliotidae. Gastropods, or snails, form the largest and most successful class of mollusks and over 80,000 living species have been described. All gastropods undergo a twisting, or torsion, early in their development wherein the visceral mass rotates 180° counterclockwise and the ctenidia (gills), mantle cavity, and anus, which originally faced to the rear, come to lie behind the head. Abalone development typically proceeds from a short (3–6 days) pelagic free-swimming stage to a benthonic creeping phase and finally to a more sedentary and retiring adult existence.

Geographical Distribution

Abalones inhabit marine waters along the rocky shores of all the continents and among many of the islands in the Pacific, Atlantic, and Indian Oceans. The greatest concentrations, both in numbers of species and individuals, are off the coasts of Australia, Japan, and western North America. The largest species, *Haliotis refescens*, occurs on the California coast; mature individuals average between 7 and 9 in. and some exceed 11 in. in diameter. The next largest, *H. gigantea*, occurs near Japan and attains a length of 10 in. Other large haliotids live off South Australia, New Zealand, and South Africa. The greatest number of species is found in the South and Central Pacific and parts of the Indian Ocean, but no large ones occur in these areas. On the Pacific coast of North America haliotids are found from Sitka, Alaska, south to Cape San Lucas, lower California. Substantial populations also exist at the offshore islands along southern California and lower California.

Abalones are not found in the Gulf of California nor along the west coast of Mexico. Records from these areas have been based on empty shells probably left there by tourists. No haliotids have been reported from the Gulf of Mexico, Central America, or the west coast of South America except at the Galapagos Islands, approximately 700 miles off Ecuador. The only abalone known from the eastern seaboard of North America is *H. pourtalesii*, a small deepwater form that rarely has been collected.

In the eastern hemisphere, the northernmost abalones occur on the outer coast of the Kamchatka Peninsula. From there, they range southward along the Asiatic mainland and are found in the coastal waters of Korea, China, Vietnam, Cambodia, Thailand, and the Malay Peninsula. They extend through the islands of Japan, the Ryukyus, Formosa, the Philippines, the Indonesian Archipelago, Borneo, Ceram, the Moluccas, and southeastward through New Guinea, the Bismarck Archipelago, the Solomons, New Hebrides, New Caledonia, Australia, New Zealand, and Tasmania. Their southernmost limit is the sub-antarctic Macquarie Island, approximately 1000 miles southeast of New Zealand.

Although found throughout most of the Pacific Islands, haliotids are quite rare in some areas, and their range is not completely known. They are present among many of the islands of the Trust Territories (Marianas, Marshalls, Carolines) but are scarce in the western part. They have been reported from many of the islands of Melanesia, and undoubtedly further investigation would reveal new localities in this region. Haliotids are present among most of the Polynesian group, including the Marquesas and the Tuamotus; however, they are not in the Hawaiian Islands and attempts to transplant young black abalones, *H. cracherodii*, from California to Oahu have not been successful. The eastern boundary of the Tuamotus apparently is the southeasternmost limit for *Haliotis* in the Pacific. None is found on Pitcairn, Henderson, Easter, or Sala y Gomez Islands, or on any of the other small, isolated islands in this region.

In southwest Asia, abalones are found on the islands in the Bay of Bengal (Andamans and Nicobars), along the coasts of India and Ceylon, and off the islands of the Indian Ocean. They occur on the shores of the Arabian Sea, the Persian Gulf, and the Red Sea.

Numerous species of *Haliotis* are found on the African east coast extending south along Tanganyika, Mozambique, and Natal, offshore to Madagascar and its surrounding islands (Mauritius, Réunion, etc.). Their range continues southward along the shores of South Africa, around the Cape of Good Hope, and north along the coast of southwest Africa, the Gold Coast on the Gulf of Guinea, the coast of Senegal, and offshore to the Cape Verde, Canary, Madeira, and Azores Islands.

A single species, *Y. lamellosa*, is native to the Mediterranean shores of France, Spain, Italy, Yugoslavia, Greece, Syria, and Egypt. In the northeastern Atlantic, a single species, *H. tuberculata*, lives as far north as Cherbourg, France, and among the Channel Islands.

Frequently, abalone reports mention only the name common in the geographical area concerned. As many as two dozen common names (Table 1) in over 15 languages may be encountered in world publications dealing with abalones.

TABLE 1. COMMON NAMES OF WORLD HALIOTIDS

Abalone	United States
Orielle de Mer	France
Si-ieu	France
Ormer, Ormier, Omar	England
Venus ear	England
Norman shell	Old English
Lapa Burra	Portugal
Senorinas	Spain
Orecchiale	Italy
Patella Reale	Sicily
Ohrsnecke	Germany
Meerohren	Germany
Venus' ear	Greece
Orechio de San Pietro	Adriatic, Dalmatia (Yugoslavia)
Mutton fish	Australia
Paua	New Zealand, Tasmania (Maori)
Karariwha	New Zealand, Tasmania (Maori)
Perlemoen	South Africa
Cholburi	Thailand
Holley	Amboina (Molluccas) Ceram
Telinga Maloli	Malaya
Ria Scatsjo	Malaya
Awabi	Japan
Aulon, Aulone	Spanish American
Aulone	Mexican

Fisheries

Ethnological and anthropological evidence indicates that abalone fisheries flourished for many years in Japan and among the Indians of the coastal regions and Channel Islands of southern California. Commercial fisheries exist in Japan, Korea, Australia, South Africa, on the islands in the English Channel, the United States and lower California. The fisheries in Lower California, Japan, Australia, and the United States are extensive; those of Korea and South Africa moderate, and intermittent in the English Channel. The English Channel haliotids have been subject to heavy exploitation. Several studies have been made and fishery regulations were put into effect in anticipation of producing a sustained yield fishery.

One species of abalone, *H. midae*, is commercially important in the South African fishery. Commercial fishing grounds are found along the south and southwest coast of the republic, the center of the industry being located at Hermanus. Commerical exploitation of *H. midae* began in 1950; landings were 300 metric tons by 1953; and by 1962 an annual production of 1,700 metric tons was achieved.

The Australian fishery is in the development and expansion stage at present, and the potential of this resource has yet to be assessed. Important commerical species include the Australian red abalone, *Notohaliotis ruber*, and white, *Schismotis laevigata*. Red abalones found in N.S.W., eastern Victoria, and Tasmania may reach shell lengths of 6 in. Australian abalones are frozen whole, canned, or dried, and largely exported to overseas markets.

Nine species of abalones are reported from Japan. They occur wherever brown algae flourish in open-sea situations. Commercially important species are *H. gigantea, H. sieboldii, H. discus,* and *H. kamtschatkana*. Abalone meats are utilized fresh, or as canned or dried products.

The Mexican fishery, like that in California, was started in the 1860's in Lower California by the Chinese, who used San Diego as a base of operations. Five species have commercial value, but the pink abalone, *H. corrugata* (called yellow abalone by the Mexicans) and green, *H. fulgens*, predominate in the catch. Principal fishing areas along the west coast of Lower California are between Santa Rosalia and San Juanico Point.

Abalones are marketed fresh, frozen, or canned. Values exported to the United States during 1964 and 1965 were as follows:

Product	1964	1965
Abalone, canned	$3,131,800	$2,876,500
Abalone, fresh and frozen	367,400	336,000

The California abalone fishery started by the Chinese in the early 1850's developed rapidly. By 1879, landings exceeded 4.1 million pounds. This early fishery, centered in the San Diego region, primarily harvested green and black abalones. Since the Chinese were not divers, they harvested the intertidal zone, or fished the shallow waters from skiffs. Because there were no restrictions on the take of abalones, stocks soon became depleted, and ordinances were subsequently passed making it unlawful to fish commercially for abalones except in deeper water.

The present California abalone fishery is centered in the Morro Bay and Santa Barbara regions. Principal commercial species are red, *H. rufescens*, and pink, *H. corrugata*, abalones. The commercial fishery has an annual production between 4 and 5 million pounds and is fairly stable. Present regulations governing the commercial fishery include minimum size limits, a seasonal closure (January 14 to March 16), and a depth restriction (no abalones may be harvested in waters less than 20 ft deep).

Fishing Methods

Several methods are employed for recovering abalones. In many instances fishery regulations or the habitat of the abalone dictate fishing practices. In Japan, abalones are either fished from a skiff or by diving. In the skiff fishery of northern Japan, the fisherman holds in one hand a long pole with a hook on the end. In the other, he holds a look-box. Peering into the box, he manipulates the hook under the shell of the attached abalone, gives a quick jerk on the pole, thus dislodging the abalone and pulls it into the boat on the hook.

The Japanese developed the use of helmet diving for recovering abalones. The helmet diver has a surface air supply via a hose from a boat with an air compressor and walks along the ocean floor. Abalones are pried from the substrate with a short, narrow, flat iron bar, and then placed in a basket for dispatch to the surface.

The most colorful and interesting method for recovering abalones is practiced by the ama divers of southern Japan and Korea. All the Korean ama divers are women. Male ama divers practice in Japan but females predominate. Using no special equipment other than a face mask or goggles ama divers may descend to depths of 80 ft, but generally they dive in 29–40-ft depths.

In Australia, abalones are fished commercially by free swimming, scuba, or hookah divers. Free divers use a face mask and swim fins and may use a rubber diving suit in cold water. Scuba divers carry their air supply in cylinders on their backs and are free from any surface attachment. Hookah divers use lightweight type diving equipment, have a surface-connected air supply via a hose, but swim rather than walk over the sea-floor. Air to the hookah diver is supplied to the mask (full-face mask) or directly to the mouth (via a mouthpiece). The later method then incorporates a smaller face mask not encompassing the mouth. All the divers use short, flat, iron bars to pry the abalones from the rocks. Fishing is normally carried out from small boats and dinghies powered by outboard engines. More distant fishing grounds are reached by larger fishing vessels up to 50 ft in length.

The Chinese used a long pole with a wedge at one end to dislodge abalones from the rocks. The Japanese introduced helmet diving equipment into the California fishery, and the fishery moved from shallow to deeper waters. The technique of helmet diving for abalones was jealously guarded by the Japanese, and Caucasian divers did not achieve any measure of success until 1929.

In the 1950's, some California divers started using the lightweight, hookah diving equipment that now predominates in the fishery. This permits a more rapid bottom coverage and is less expensive in equipment and operating costs. California law prohibits the taking of abalones commercially by free diving, and requires surface-connected hoses be a minimum of 150 ft in length.

A typical fishing boat in the California fishery is 26 ft long, has an inboard engine, and cruises at 15 knots. The crew generally consists of a diver, a boat operator, and a line tender. However, some of the lightweight divers operate independently from an anchored boat, although this is considered unsafe.

TABLE 2. COMMERCIAL ABALONE LANDINGS (000 metric tons)[a]

	1954	1955	1956	1957	1958	1959	1960	1961	1962	1963	1964	1965	1966
Baja California, Mexico	4.6	7.0	10.4	8.1	17.2	6.7	6.0	6.4	7.1	8.3	7.6	7.8	6.7
Japan	3.9	3.9	3.6	3.9	4.7	4.9	4.4	3.8	4.9	4.9	4.6	4.3	5.6
United States	2.0	2.1	2.1	2.7	2.1	2.3	2.1	2.3	2.1	2.2	2.0	2.3	2.5
South Africa	0.7	0.3	0.2	0.3	0.5	0.5	1.0	1.4	1.7	1.7	1.7	1.7	0.7
Republic of Korea	0.4	0.6	0.6	0.8	0.1	0.5	0.5	0.4	0.4	0.4	1.0	0.4	0.5
Australia											0.1	0.4	1.3

[a] Compiled from yearbook of fishery statistics, Food and Agriculture Organization of the United Nations. U.S. landings are from the California Department of Fish and Game.

Processing

Market operators report the catch in dozens and these figures are converted into poundages for statistical and biological records. All published figures and records of the California Department of Fish and Game are based on weight in the round and for a dozen abalones the following conversions are used:

 reds (minimum size $7\frac{3}{4}$ in.) = 50 lb
 pinks (minimum size 6 in.) = 25 lb
 greens (minimum size $6\frac{1}{4}$ in.) = 35 lb

Prior to 1959, the conversion factor for pinks was 35 lb per dozen.

The early Chinese fishermen in California at first gathered abalones for their own use or for local markets. Later they dried the meats and exported them to the Orient, but this practice was prohibited in 1915. The first processing of fresh abalones—slicing into steaks—was started in Monterey in 1913 by the Japanese. After laws were passed to prohibit drying abalones, there was some difficulty educating Californians to eat the fresh product.

Abalone canning was started at Cuyacos in 1905, and soon other canneries were established at San Diego, San Pedro, and Point Lobos near Monterey. Although five canning plants were operating in 1917, by 1928 there were only three. There was never much demand for canned abalones because not only were they expensive to can, but the fresh product was sufficient for the market. When export of abalones from California was prohibited, the last cannery at Point Lobos closed in 1931. Since then, the entire product has been marketed either fresh or frozen. A 1967 enactment again permits the export of abalone meats from California.

Abalones are usually delivered to the plants in the afternoon when the divers are forced to return to port because of winds. The abalones are put in boxes and unloaded from the boats at the processor's dock. Inside the plant, they are placed either on the cement floor or on tables where they remain until the next morning or until they have relaxed sufficiently for easier handling by the workers. The processing begins when a shucker inserts a flat, semicurved iron bar between the shell and flesh and forces it forward to where the muscle is attached. He then exerts a strong quick force to the bar and pops the animal cleanly from the shell.

Most of the viscera is removed and discarded; however, in some plants the entrails and gonads are frozen for fish bait.

After shucking, the abalones are washed either by hand or in a modified cement mixer. After washing, they are spread in a single layer on a cement table and allowed to stand for several hours to let the muscle tissues relax. When the abalones no longer curl around the edges, the epipodium, which contains a black pigment, is trimmed off either with a knife or with a motor driven circular blade. Usually some of the black pigment remains, but when the tough fascia around the columellar muscle is peeled off, all residual epipodial pigment is also removed. Almost the entire processing operation is done by hand so it is necessary to recover a maximum of saleable meat from each abalone. The trimmers and peelers must be quite skillful to keep waste at a minimum and assure the processor of a profit for his product. Shucking, trimming, and peeling reduces an abalone to about one-third of its round weight. The peeled abalones are sliced into steaks approximately $\frac{1}{2}$ in. thick, usually with a slicing machine.

The steaks have to be pounded to break down their connective fibers. If this is not done, the meat retains its extremely tough consistency, making chewing difficult. Although many methods have been tried, no satisfactory substitute has been found for hand-pounding the steaks. This operation, too, calls for skill; the slices must be struck hard enough to break their fibers but not so hard the tissues are crushed and the slice shattered.

The finished product is sorted according to steak size and color, packed in 5 or 10 lb cartons, and placed in a freezer. Premium prices are received for large, white-meat steaks; therefore, abalones which produce the whitest meat and the largest steaks are eagerly sought by both diver and processor. Occasionally abalones will have greyish-colored meat. They taste every bit as good as white-meated abalones, but because they are dark, they bring a lower price.

Most of the processed catch is quick-frozen and distributed to restaurants. There is always a market for a first-quality product; lower grade abalones (darks) may not sell as readily, but usually all red abalones can be marketed.

Some processing plants grind dark abalone meat with trimmings and produce abalone patties. These are packaged, frozen, and distributed to markets.

Abalone shells have been fashioned into many kinds of ornaments. The more decorative shells are sold, intact, to collectors. Recently a market developed in California for crushed abalone shell. The crushed shell is polished and dispersed in casting resin to form extraordinary table tops and innumerable artifacts.

Earl E. Ebert

ABYSSAL ZONE

Definition. The term abyssal originally meant the entire depth area beyond the reach of fishermen. But after the 1860's when explorations of

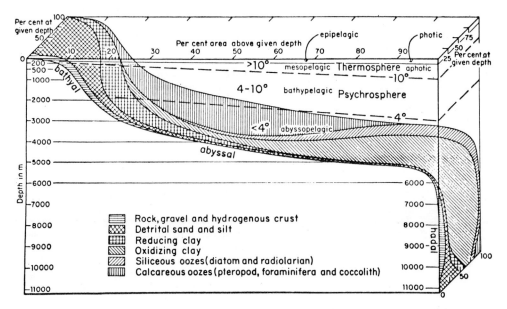

Fig. 1. Ecologic zonation of the deep sea (from *Galathea* Report, Vol. 1, Fig. 6, 1959).

the oceanic deep sea began, culminating with the cruise of the *Challenger* (1872–1876), it became evident that two main faunistic regions were to be distinguished: (1) a lower region with a fairly uniform fauna and a low temperature (to which the term abyssal became restricted) and (2) an upper region, now called the bathyal (or archibenthal) zone (*q.v.*), with a more varied and richer fauna and relatively high temperatures. The upper boundary of the abyssal zone toward the bathyal lies at about 2000 m, varying in different areas from 1000–3000 m, in accordance with the different positions of the 4°C isotherm which roughly delimits the distribution of the endemic elements and its fauna. The distinction between an abyssal and a bathyal fauna is thus obscured in the polar regions; and a true abyssal fauna is not found in the Mediterranean seas with warmer bottom waters. The recent deep-sea investigations have further led to the conception that the trenches and deeps, with depths exceeding 6000–7000 m, constitute a special faunistic region, the hadal.

Area. The abyssal is the world's largest ecological unit, occupying more than three-quarters of the total area of the oceans and adjacent seas, and slightly more than half of the total area of the globe (see Table 1).

Life Conditions

Environmental Factors. Life conditions are very uniform, except for food. The temperature over the greater part of the oceanic deep sea is between 0°–2°C, constant within the geographical

TABLE 1. AREAS OF ABYSSAL ZONE

	Area (million km²)
Total area of the globe	510
The oceans and adjacent seas	361
Depths exceeding 2000 m	305
Depths exceeding 3000 m	278
Depths exceeding 6000 m	5
Calcareous oozes	128
Abyssal clay	102
Siliceous oozes	38

areas and with no seasonal variation. The low temperature is conditioned by the cold-water masses sinking down in the polar regions and from here streaming over the bottom towards the equator. Enclosed deep-sea basins without direct communication with the oceanic abyssal, such as the Mediterranean, the Japan Sea, and the Sulu Sea, have higher bottom temperatures and do not show true abyssal conditions.

The salinity is about 3.5‰. The concentrations of phosphate, carbon dioxide, and silicate, and the alkalinity vary only slightly. The oxygen content is entirely dependent on the continuous supply of oxygenated water masses, as oxygen can be added only in the upper water layers, by the photosynthesis of the pelagic plants, and by absorption of air at the surface. Generally, however, the bottom water contains sufficient oxygen for the support of animal life, i.e., on the average 5–6 cm³/liter in the Atlantic and 3.5–4 cm³/liter in the Pacific. Close to the bottom there may be a decrease in oxygen content which is evidently

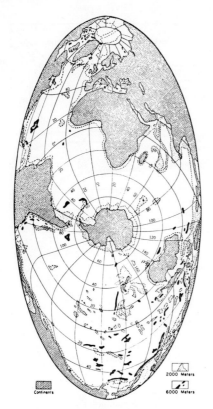

Fig. 2. Abyssal region of the earth, showing 2,000 m contour and depths greater than 6,000 m (Hammer-Aitoff projection; from Ref. 1).

connected with the biological processes. The pressure in the abyssal ranges from 200–700 atm. How far it directly affects the abyssal animals has not yet been discovered.

Substratum. The abyssal is normally a calm milieu where very small particles may settle. The bottom, apart from the mountain ranges and sea mounts, is also mainly covered by a soft clayey ooze of very considerable thickness, perhaps hundreds of meters. Near land, the bottom sediments are derived to a varying degree from terrigenous mineral and organic material, as well as from planktonic organisms. Farther from land, the deposits are predominantly of pelagic origin. At the greatest distance from land, the sedimentation is very slow, only a fraction of a centimeter per 1000 years. The change from primarily terrigenous to pelagic deposits generally takes place at an approximate depth of 2000 m, thus coinciding with the upper limit of the abyssal fauna. Down to depths of 4000 m, in mid and low latitudes, the sediment is primarily a grayish calcareous ooze, largely consisting of the sunken shells of pelagic foraminifera (e.g., *Globigerina*), and skeletons of pelagic flagellates, especially Coccolithidae. Globigerina ooze oc-cupies most of the ocean floor of the Atlantic and the Indian Ocean, and parts of the southern Pacific. Pteropod shells are dominant in the calcareous ooze in certain areas. In depths exceeding 4000–5000 m, the calcareous deposits have largely been dissolved, and the sediment is the brownish abyssal (red) clay composed mainly of aluminium silicate. The siliceous shells of radiolarians are a main constituent of the sediment in the deeper areas of the tropical Pacific and Indian Oceans, where these animals are abundant in the surface waters. In the Antarctic and northernmost Pacific, the sediment is mainly a siliceous ooze formed by the accumulated shells of diatoms living in the cold surface waters here.

Slow currents occur along the bottom, varying in velocity to around 5 cm/sec, and deep-sea photographs showing ripple marks indicate that at least to depths about 3000 m, and around sea mounts, the currents may occasionally increase to 16 cm/sec.

Food Supply. Ultimately, the food supply for deep-sea life is totally dependent upon the supply of organic matter from land and from the upper water layers where it is basically produced by the photoautotrophic phytoplankton, the first link in the oceanic food chain. When the epipelagic organisms sink, they are utilized by the bathypelagic fauna. However, the remains of plants, animals, and excrements which sink below a depth of 2000 m will reach the bottom having been little influenced by the depths, as the abyssopelagic life is very poor. The common assertion that the greater the depth, the less the supply of food is true only in connection with distance from land or from areas with a high productivity at the surface. Abyssal heterotrophic bacteria assimilate the organic matter and are considered to constitute a main source of food, either directly or as the second link in a new food chain. Many kinds of invertebrate animals can also utilize the dissolved organic materials found in the environment. Terrigenous deposits which support a much richer animal life than the eupelagic deposits may be transported far into the ocean by turbidity currents. Plant debris are very abundant on the abyssal bottom in certain tropical areas and cause a concentration of life. The sunken dead bodies of large animals, e.g., whales, form temporary local food supplies and are mentioned as being one explanation of the patchy distribution of many abyssal species.

Fauna

Morphological Characteristics. The environmental conditions are, apart from the pressure, no different from any that can be found in shallower regions, and the abyssal fauna show no fundamental differences from the fauna of these regions. Characteristics shown by the abyssal animals are correlated with their life in darkness, in a calm milieu with a soft bottom. They are uniformly colored, most often grayish, and often very delicately built. The mobile animals have long slender legs, and the sessile ones are often stalked (which may in part be an adaptation for raising them above the oxygen-poor water layer close to the bottom). Many fish and crustaceans are blind, and in some fish the pelvic fins are developed as tactile organs.

Composition. Life in the abyssal region is concentrated on or close to the bottom. All the major groups of marine invertebrates are represented, in addition to several benthic types of fish. Examples are: protozoa, porifera, coelenterata, polychaeta, echiuroidea, crustacea, mollusca, echinoderma, tunicata, and vertebrata.

Infauna and Epifauna. The abyssal fauna, like the cold-water faunas and infaunas of the shallower depths, is to a large extent composed of comparatively few widespread species with a relative abundance of individuals. In addition, there are many species, especially in the epifauna, which show patchy distributions. The epifauna will also be concentrated on and around the sea mounts where increased currents may expose a firm surface or a coarser grade of bottom. Deep-sea photographs have shown, e.g., that crinoids may dominate on one sea mount and poriferans on another. The epifauna is otherwise dependent on the scattered pieces of pumice, volcanic rocks, and manganese nodules, and on other animals with a firm surface to which they may be attached.

Feeding Types. With increasing depth, the number of suspension feeders (feeding from the suspended detritus and dissolved organic particles) and deposit feeders (feeding on the deposited organic matter), increase in relation to that of carnivores and scavengers. The Porcellanasteridae, a family of mud eaters within the otherwise carnivorous sea stars, are unknown from depths less than 1000 m, but constitute a fourth of the species of sea stars recorded from 4000 m, and half of those from 6000 m. The mud-eating holothurians, the Elasipoda, are characteristic animals in the abyssal. Also, some of the abyssal fishes are deposit feeders. The abyssal polychaetes, which are among the dominant animals, are predatory, however.

Density of Life. The density of life in the abyssal zone is very low compared to that of the moderate depths. Expressed in biomass (i.e., quantity of substance in live organisms in grams per square meter) even in the richer abyssal regions (the Antarctic and northern Boreal) it is no more than 1% of that in the sublittoral, and some eupelagic deposits may be almost barren. Also, the productivity may be extremely poor.

Reproduction. The rate of reproduction seems slow, and a considerable longevity is also postulated for abyssal animals in general. It is yet unknown whether they show seasonal life rhythms, but there are indications that for the true abyssal species this is not the case and the reproduction may take place at any time of the year, with only a small number of eggs ripening simultaneously. A non-pelagic development seems to be usual for the true abyssal benthic species and for the majority of those extending into the zone, though some may be expected to have free-swimming larval stages of some duration. Some species possess pelagic larvae, and some of the

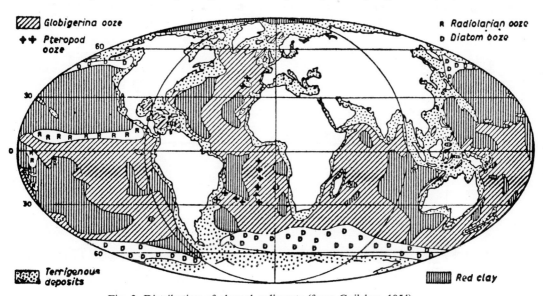

Fig. 3. Distribution of abyssal sediments (from Guilcher, 1954).

TABLE 2. FIGURES FOR BIOMASS IN NORTHWESTERN PACIFIC IN GRAMS PER SQUARE METER

(Birstein and Belyaev, 1955; Birstein 1959)

Coastal zone	1000–5000
50–200 m	200
ca. 4000 m	ca. 5
Kuril–Kamchatka trench	
ca. 6000 m	1.2
ca. 8500 m	0.3
Central part of the ocean floor	0.01
Tonga trench	
10,500 m	0.001

benthic-abyssopelagic fishes, the Synaphobranchidae and Ceratiidae, have larvae living in the upper pelagial.

Subdivisions

Vertical Subdivision. The abyssal zone may be divided into an upper and a lower zone at a depth of about 4500 m, where a change in the composition of the species occurs. A number of the worldwide distributed species belong exclusively to the lower abyssal. At about 4500 m, the water becomes undersaturated with calcite, and this may account in part for the faunal change.

Regional Distribution and Subdivisions. The uniform life conditions and the absence of definite topographical barriers account for the wide distribution of many of the endemic forms in the abyssal. The genera are typically cosmopolitan; a circumtropic, almost worldwide distribution seems to be a general rule also for species of benthic fishes, anthozoa, echinoderms, and tunicates, whereas crustaceans show restricted distributions. The mollusk *Neopilina* is exceptional in its restricted distribution in the east Pacific. The greatest difference in the composition of the abyssal fauna is found between the faunas on either side of the American continent (this area also constitutes the most marked barrier in the abyssal region). With respect to the echinoderms, another barrier for the distribution would seem to be the mid-Pacific deep sea with its very sparse food resources. The extremely

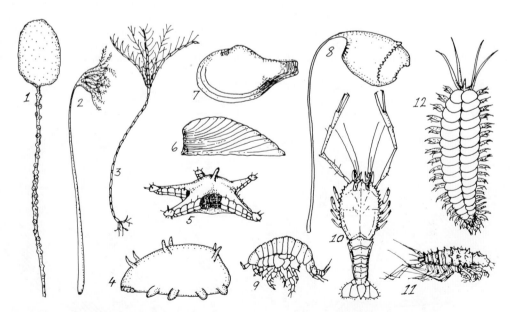

Fig. 4. Examples of abyssal benthic invertebrates: 1, the glass-sponge *Hyalonema* with a colonial sea anemone, *Epizoanthus*, attached on the stalk; 2, the sea-pen *Umbellula*; 3, a stalked crinoid (*Bathycrinidae*); 4, the elasipid sea-cucumber *Elpidia*; 5, the sea-star *Porcellanaster*; 6, the mollusk *Neopilina*; 7, the bivalve *Cuspidaria*; 8, the tunicate *Culeolus*; 9, the amphipod *Harpinia*; 10, a blind-lobster (*Eryonidea*); 11, the isopod *Storthynqura*; 12, a polychaete (*Aphroditidae*).

cold water (<0°C) of the Antarctic deep sea probably accounts for the special composition of the abyssal fauna there. Three main zoogeographical subdivisions of the oceanic abyssal region thus may be distinguished: (1) the Atlantic and Indian Ocean inclusive of the southwestern Pacific, (2) the east Pacific, and (3) the Antarctic deep sea. A more detailed zoogeographical subdivision will, for the greater part, reflect the zoogeography of the bathyal region.

Secondary Abyssal Species. Numerous species extend from the bathyal into the neighboring abyssal and constitute a very considerable part of the actual abyssal fauna. Such species have been called secondary abyssal species or guests, since they, though able to live in the abyssal habitat, may be unable to produce a constant series of generations there. It is probable also that many species may be prevented from spreading into the distant oceanic deep sea because of the decreasing food supply.

Origin of the Fauna

Endemic Forms. Only few groups of higher taxonomic categories are peculiar to the deep sea. A family of sea anemones is as yet known solely from hadal depths (*q.v.*). The sea stars, Porcellanasteridae, are almost exclusively abyssal (1000–7600 m). Two-thirds of the species of the holothurian order, the Elasipoda, are abyssal, and one-third bathyal. The mollusks of the class Monoplacophora (*Neopilina*) are exclusively bathyal-abyssal and represent the only abyssal group with a paleozoic fossil record. The most dominant abyssal groups, such as the Elasipoda, are without fossil records, and only few of the other families represented can be traced as far back as Mesozoic.

Nonendemic Forms. It is believed that the low temperature in the recent abyssal deep sea is a character only acquired in rather recent geological time and that during Late Mesozoic and Early Tertiary the abyssal temperature was about 10°C. The gradual Late Tertiary cooling, culminating during the Quaternary glaciations, may have had a very selective influence on the fauna. Stenothermic abyssal forms adapted to a temperature of about 10°C would become extinct, although they might possibly persist in the bathyal region. Taxonomic evaluations of different genera and species of echinoderms, polychaetes, and crustaceans occurring at various levels of depths have led to the conviction that they become more advanced with increasing depth, and thus indicate a progressive distribution into the abyssal zone. The region may, since Late Mesozoic or Early Tertiary, have been repopulated gradually from the bathyal depths. The endemic deep-sea families and genera, e.g., hexactinellids, crinoids, cidaroids, and elasipods, probably constitute the oldest element. Other groups, gorgonarians, astropectinids, and the macrourids, with shallow-water relatives in the tropical sublittoral, may have extended their range into the abyssal during Early Tertiary, before the fall in temperature had begun to have an effect. *Neopilina* also may not have invaded the abyssal until Early Tertiary. A Late Tertiary invasion may be assumed for the groups which are extensively distributed in the abyssal and have their shallow-water relatives in the temperate regions. A Late Quaternary (postglacial) invasion must have taken place, and is still going on, with regard to those species which have their main occurrence in the polar regions and only a restricted distribution in the neighboring deep sea.

F. JENSENIUS MADSEN

References

1. Bruun, A. F., 1957, "Deep sea and abyssal depths," *Tr. Mar. Ecol. Paleocol. I., Geol. Soc. Am. Mem.,* 67.
2. Ekman, S., 1953, "Zoogeography of the Sea," London, Sidgwick and Jackson.
3. Madsen, F. Jensenius, 1961, "On the zoogeography and origin of the abyssal fauna," *Galathea Report,* 4.
4. Marshall, N. B., 1954, "Aspects on Deep-Sea Biology," New York, Philosophical Library.
5. Murray, J., 1895, "A summary of the scientific results," *Challenger Report, Summary* 2.
6. Zenkevitch, L., 1963, "Biology of the Seas of the U.S.S.R.," London, Allen & Unwin.

Cross reference: *Benthonic Domain.*

AGAR

The terms agar and agar-agar are currently applied interchangeably throughout the world to any gum of seaweed origin which can be dissolved in boiling water to make a liquid that will form a jelly when cooled. The word agar is used also to denote any culture medium containing sufficient colloidal material to permit its use as a solid or semi-solid gel. This article will discuss the seaweed gum alone. Information on agar culture media can be found in the literature devoted to the organisms studied with the help of such mixtures.

Definition

Agar is not a substance of definite and unvarying composition, and a precise, succinct definition is probably impossible. The material obtainable varies in purity and composition to such a degree that the most practical definition can utilize only its important properties. Such a definition has been official in *The U.S. Pharmacopeia*[1] for the past two decades:

"Agar is the colloid extracted from Gelidium cartilagineum, Gracilaria confervoides and related Rhodophyceae. It is insoluble in cold water but soluble in boiling water. A hot 1.5% sol is clear and congeals at 32°–39°C. to a firm, resilient gel which becomes liquid above but not below 85°C.

Total ash	6.5% max
Acid-insoluble ash	0.5% ,,
Foreign organic matter	1.0% ,,
Foreign insoluble matter	1.0% ,,
Foreign starch	absent
Gelatin	absent
Water absorption	5 × min
Moisture	20% max"

The Food Chemicals Codex carries a similar definition[2] and specifies that arsenic, heavy metals or lead shall not exceed 3, 40, or 10 ppm, respectively. A few somewhat different galactan phycocolloids are known which do not comply with the U.S.P. definition. These will here be referred to as agaroids. They are of commercial importance and some are called agar, particularly abroad. Furcellaran and Gracilaria gum of high gelation temperature are examples.

The first commercial agar production occurred in Japan about 1670 and was a uniquely Japanese activity for nearly two centuries until Chinese and Malay production began. Except for brief flurries of activity in Australia, India, and Ceylon, no new agar or agaroid producers appeared until 1915–

Fig. 2. Rope bag used to haul seaweed to surface.

1925, during which period small firms established themselves in the East Indies and California. World War II deprived much of the world of Japanese agar; thus intensive efforts were made in Australia, New Zealand, South Africa, Russia, Denmark, and the United States to fill the gap. Some 20 firms were active at the end of the war. About 10 of these survive today and have been joined in the interval by about 20 producers operating in Argentina, Chile, France, Korea, Mexico, Morocco, Portugal, and Spain. Japan remains the largest supplier of agar, with over 400 establishments. The United States has had only one producer since 1953 and its output is used almost entirely in laboratories.

Collection

Weed is universally gathered by hand, mechanization being difficult on rocky bottoms.

Fig. 1. Hard-hat diver collecting seaweed by hand.

Waders and boatmen with rakes are employed in shallows, while skin diving to 10 m is widely practiced in Japan by trained young women. Hard-hat divers in pressure suits are used in deeper waters in Japan and at all depths in Mexico. However, there is a controversy over the best technique. Tearing the plant from its holdfast and shaking it to disseminate spores is favored by many. Cutting above any stolons followed by shaking is advocated by others. All agree, however, that any holdfast must be returned to the ocean floor. When available, broken concrete, stone, or other rubble is sometimes placed in and around weed beds for anchoring new plants. The seaweed is sent to the surface in rope bags. Cultivation on rope fences and on broken stone has been practiced with some success in Japanese waters for decades but is not general elsewhere.

Where fresh water is available, it is used to wash the collected algae to remove detritus and to reduce the amount of adhering hygroscopic seawater salts. Such washed weed is of slightly higher value and withstands storage better than does unwashed weed, especially in humid surroundings.

Collected weed is air dried as promptly as possible by spreading on the beach or on racks and turning until drying is complete. Drying with the aid of flue-heated air is advantageous in wet climates and is harmless if shallow layers are dried with air at 50°C. Residual moisture should be less than 17% if prolonged storage is anticipated. After drying, weed is compressed to a density of about 400 kg/M^3 (25 lb/ft^3) and baled for transport. Weed for export is covered with cloth or matting and strapped.

Processing

Because agar and agaroids are soluble in hot water, all commercial extraction is done with water at 90–150°C. The extracted sol is dried directly, treated with alcohol, and the precipitate dried, or the sol is congealed, frozen, thawed, and dried. The last is the process most commonly used, because gel freezing causes migration of impurities into the ice phase.

The traditional Japanese process[3] involves the following steps, and is a winter operation, often performed by a single family which has spent the summer fishing and obtaining seaweed:

(1) Pounding and washing to clean and soften the seaweed.
(2) Extraction in open cauldrons over a wood fire with boiling water. Tough seaweeds (Gelidium, etc.) are added first, soft algae (Gracilaria, etc.) last. The pH is adjusted to 5–6 with sulfuric acid and extraction continues for 4–10 hours, after which a second extract from the day before is added and a hypochlorite, hydrosulfite, or bisulfite bleaching agent is introduced. Extraction is ended after 12–15 hours.
(3) Straining. The cooker contents are strained through wire mesh or cloth, the cake is returned to the cauldron for the second extraction and the liquor, containing about 1% agar and agaroids, is allowed to stand for sedimentation.
(4) Gelation. The supernatant liquor is held in wooden trays 1 cm deep until it solidifies.
(5) Freezing. The gel is cut into strips, placed outdoors on straw mats and allowed to freeze. Each day the sunshine melts some of the ice phase and the drainings carry away some salt, color bodies, and other impurities. Sprinkling restores some of the lost moisture and the cycle is repeated for as long as 30 days. Sprinkling is then stopped and the agar strips are allowed to dry. Much care and some luck are required to produce agar of top grade.
(6) Packing. The dried strips are made into bundles and paper wrapped for grading and sale.
(7) Grading. Strip agar is assigned one of five grades at an inspection center according to color, luster, shape, uniformity, gel strength protein, insolubles, moisture, and boron content.

More modern technology is employed by most of the producers who have entered the field since 1925, but since each tends to consider his technique as confidential information, complete descriptions are not available. Some of the improvements introduced are:

(1) Mechanization of raw material pretreatment.
(2) The use of cellulase,[4] pectinases, and bases before extraction.
(3) Continuous, automated countercurrent extraction and precipitation.
(4) Chemical impurities removal by precipitation and ion exchange.
(5) Avoidance and removal of microbiological and metallic contamination.
(6) Adoption of Wood's[5] alcohol-treatment technique with modern equipment.
(7) Drying by continuous roll, vacuum spray, and regenerative air columns.
(8) Protection from atmospheric recontamination.

Consumption

The United States consumption of agar approached 1,000,000 lb in 1966. In addition, about 100,000 lb of furcellaran and other agaroids were imported. A reasonable breakdown of American agar consumption by fields would be:

Microbiology	300,000 lb
Baked goods	200,000 lb
Confectionery	100,000 lb
Meat and poultry	100,000 lb
Desserts and beverages	100,000 lb
Laxatives and health foods	50,000 lb
Pet foods	50,000 lb
Moulages, including dental	30,000 lb
Pharmaceuticals	20,000 lb
Miscellaneous	50,000 lb

From 1956 to 1966, the consumption of commercial agar plus agaroids in the United States doubled, and the use of agar for microbiological purposes tripled. Extrapolation from available data suggests a similar increase by 1978. Similar increases in demand have occurred in Europe, and the prospects are for an even more rapid increase.

Although demand will surely increase, supply may not keep pace in the future unless new weed beds are found and unless all available harvesting areas are exploited efficiently and with due regard for the conservation of this natural resource. Eventually intensive cultivation including protection from enemies such as sea urchins may be practiced.

Uses

In microbiology agar functions as a solidifying or threshold gelation agent in a host of nutrient culture media into which microorganisms can be introduced for cultivation and subsequent enumeration, specific differentiation, and mass cultivation. This use is widespread in food, drug, clinical, and public health laboratories. Bakers and confectioners find agar and agaroids useful as texture-improving antistaling agents, emulsifiers, and stabilizers for fillings, icings, piping gels, and candies. Canners of meats, fish, and poultry use agar and agaroids as gelation agents to make attractive jelly packs of softer tissues which will withstand the stresses of retorting and transportation. For this purpose, the high gelation temperature of Gracilaria agaroids is sometimes an advantage. The packaged foods industries use agar and agaroids in the mixes which have become increasingly popular with the housewife in recent years. Flavored milk gels or puddings called flan are widely consumed in Europe and account for very large sales of agaroids, particularly furcellaran.

Agar is an excellent laxative when used as a gel or in the form of liquid-softened flakes. In the gastrointestinal tract (with ample fluid intake) it seems to insure smooth fecal bulk, to improve peristalsis, and to be non-habituative. Because it is of vegetable origin, agar supplants gelatin in the diets of those to whom the idea of using animal products is distasteful and it is widely available in health food outlets. For some years, agar and agaroid sales to English pet-food makers have increased steadily, and some agar is used in American pet foods.

Although alginates and elastomers have become popular here and abroad for making dental impression materials and moulage compositions, agar is still used in dental prosthesis, criminology, and art where duplicating work of high accuracy is required. Pharmaceutical manufacturers use agar and agaroids as tablet excipients, emulsifiers, capsule ingredients, pill coatings, texturizers, and expanding agents. Some vintners use agar for fining premium wines. Miscellaneous fields include photography, explosives manufacture, dairy products, detergents, decalcomania, metal plating, lithography, cosmetics, pesticides, metallurgy, metallography, analytical chemistry, and clinical laboratory diagnosis. In the last two, agar and agarose gels are finding wide use in gel bead filtration, gel electrophoresis, and immunodiffusion techniques, by which otherwise obscure amino acid and protein anomalies associated with neoplastic and other disease conditions can be detected. In immunoelectrophoretic work, highly purified agarose is the material of choice, although pure agar is advantageous in separating lipoproteins.

Sources

Agar is obtained from certain of the Rhodophyceae of the genera Acanthopeltis, Gelidium, Gracilaria, and Pterocladia. Some species of the genus Gracilaria appear to contain agar at all periods of growth, e.g., South African G. verrucosa, whereas in other species physical properties vary during growth to such a degree that purified extracts must be classified as agaroid at one time and as agar at another, due to seasonal gelation temperature differences. This phenomenon was noted by DeLoach, et al.[6] Gelidiella acerosa may exhibit a similar seasonal change.

TABLE 1. MAJOR SOURCES

Agar	Agaroids
Acanthopeltis japonica	Ahnfeltia plicata
Gelidium amansii	Ceramium, 3 species
G. cartilagineum	Eucheuma, 6 species
G. corneum	Furcellaria fastigiata
G. japonicum	Gracilaria confervoides
G. pristoides	G. lichenoides
Gracilaria verrucosa (?)	Phyllophora nervosa
Pterocladia capillacea	

It should be noted that the above names are subject to change. The taxonomy of the Rhodophyceae, according to the Schmitz–Kylin system, has enjoyed long acceptance, but Dixon[7] and others suggest extensive revision.[8]

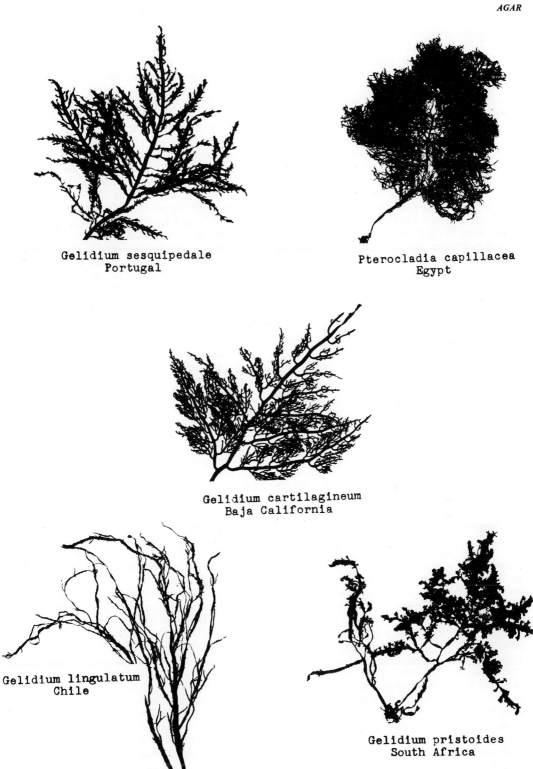

Fig. 3. Five agar-bearing seaweeds.

The plants grow at depths as great as 65 m, but the greatest concentration is found in the region between low tide and 15 m in most species. Porous rock or broken shell is the most favorable bottom, and water turbulence appears to be desirable for good growth and yield. Reproduction is by spores and by regrowth from stolons. One annual harvest is normal but two are sometimes possible in lower latitudes where water temperature exceeds 25°C for only brief periods.

The agar content of harvested and dried Gelidium species varies from 10–70%, dry basis, and is highest in specimens gathered in the warmest months. Approximately equal percentages are found in the thallus and the fronds of a single plant. In the plant, agar could be present as a precursor rather than as agar. This proto-agar may be a cross-linked complex of straight-chain agar units acting as an osmotic and ion-exchange membrane serving to limit the ionic content of the interior of the cell. When Gelidium cartilagineum is reduced to a 5 μ powder, no agar is extracted from the alga by water in a period of 30 minutes at 100°C. However, extraction for 2 hours at 110°C gives normal yields of agar.[9] This suggests the above hypothesis since, in a 5 μ particle, any agar present would be amply exposed to the solvent. Thus diffusion through a membrane would not be a factor.

Chemistry

Agars and agaroids of commerce are linear polysaccharide complexes containing up to 1000 sugar units per molecule of component. The major sugars are β-D-galactopyranose and 3,6-anhydro-α-L-galactopyranose, followed by 6-O-methyl-D-galactose, 4-O-methyl-L-galactose, L-galactose, D-xylose and O-methyl pentose in approximate order of abundance. Small amounts of pyruvic and uronic acids are occasionally found and ethereal sulfate is present, varying from 0.2–6% as SO_4^{--}. In general, the D-galactose accounts for half of the total sugars in agars and a third of the total in agaroids. The L-galactose amounts to about 40% in all.

Araki found agar to be a mixture of two polygalactoses which he named agarose and agaropectin.[10] The agarose component is low in ash and organic sulfate, has high gel strength, and its gels have low resilience. Agaropectin has high ash and ethereal sulfate, low gel strength, high resilience, and high ion-exchange properties. The agarose:agaropectin ratio in agars varies with species, extraction method, and technique of separation. Values of 1.1:1–8.1:1 have been found. In agaroids, the ratio varies from 0.4:1–5:1. However, Tauchiya and Hong report 20:1 in a Gracilaria agaroid.[11]

Agaroses appear to be chains of alternating 3,6-anhydro-α-L-galactopyranose (1→4) units separated from each other by (1→3) rings of β-D-galactopyranose and O-methyl-galactoses. Agaropectins seem to have a main structure similar to their corresponding agaroses but with linked glucuronic, galacturonic, and pyruvic acids and sulfate. Much of the present knowledge of the structure of agar comes from decades of work by many investigators, notably Araki, Arai, and Hirase.[12]

Agarose and agaropectin can be obtained from agar by acetylation, chloroform partition, and deacetylation[10]; by cationic quaternary ammonium compound precipitation[13]; by the use of polyethylene glycol[14]; by the action of selective solvents[15]; and by precipitations with ammonium sulfate[16] and aluminum compounds. The agarose fractions prepared by the above methods from a given alga are similar, but differences among them are revealed, particularly by electrophoresis. This suggests further study.

Agar and agaroids can be considered to be salts of polygalactanic acids. As such, their natural calcium and magnesium cations can be replaced with others. For ashless applications, ammonium and alkanolamine agars and agaroses have been made, for example. However, no derivatives are commercially available.

If kept dry and cool, the agar of commerce is reasonably stable. In the worst case studied, only mild discoloration and a 20% decrease in gel strength have been noted after a year at 25°C. Highly purified agar has been stored for 10 years before equal deterioration has been found. Sterile gels show similar behavior. Sols are thermolabile and are rapidly degraded at very low and very high pH. Generally, a pure agar sol will lose only 5% of its gel strength if autoclaved an hour at 120°C at pH 6.5–7.5.

Very few terrestrial microorganisms attack agar and, until 1945, marine agarolytic bacteria were thought to be rare. Humm[17] however, reported 20 in 1946, and others are occasionally encountered. Pseudomonas atlantica, e.g., yields an agarase[18] which may be useful in differentiating agars and agaroses.

Impurities

For cosmetic, moulage, and technical uses, agar and agaroids are sufficiently pure as offered on the market. For food and pharmaceutical purposes, occasional lots may be unsuitable as judged from their compliance with the specifications of the Food Chemicals Codex.[2] Microbiologists, chemists, and medical technologists prefer and often require agar of the highest possible purity because the work done by each can be adversely affected by the fatty acids and alcohols, sugars, guanylurea compounds, vitamins, and metals present in commercial agar.

Although agar is the most nearly perfect gel-

TABLE 2. COMPOSITION OF AGARS

		Commercial agars	Bacteriological agars	Commercial agaroses
Aluminum	ppm.	50–1000	0–50	0–230
Arsenic	,,	0–2	0	N.D.[a]
Barium	,,	3–50	0–3	0–13
Bismuth	,,	0–5	0	N.D.
Boron	,,	10–300	10–110	N.D.
Cadmium	,,	0–2.1	0–0.5	0.2–1.3
Chromium	,,	0–0.7	0	0–10
Cobalt	,,	0–10	0–1	0–2
Copper	,,	2–66	0.4–53	1–10
Iron	,,	24–800	0–200	25–140
Lead	,,	0.3–100	0.1–10	2–8
Magnesium	,,	200–4200	100–6200	20–200
Manganese	,,	3–200	0.1–2	1–2
Nickel	,,	0–2	0–2	0–1
Rubidium	,,	0–40	0	N.D.
Strontium	,,	5–400	0	N.D.
Tin	,,	0–50	0	N.D.
Titanium	,,	0–200	0–10	N.D.
Zinc	,,	10–2000	10–60	4–20
Free sugars	%	0–32	0–1	0.5–3
Protein	%	0.1–13	0–0.9	0.1–0.4
Sulfate (org.)	%	0.9–4.2	0.2–1.5	0.1–0.4
Sulfate (inorg.).	%	0.6–2.7	0.8–3	0.1–0.3
Insolubles	%	0.2–1.5	0.0–0.3	0.2–0.8
Sol clarity, 2%, mm		20–350	300–1000	200–600
Spores per gram		0–2500	0–10	0–2

[a] N.D. = not determined.

forming agent for microbiological use, it is well known that many microorganisms are sensitive to substances present in commercial agars. The growth-enhancing and -inhibiting effects of cationic impurities are receiving increasing attention[19] and inhibition by agaropectin may prove to be important, thus suggesting its binding in media by diamino ethyl dextran,[20] protamine, or other cationic agent to which the organism is indifferent. The substitution of pure agarose for agar is occasionally justified in these cases.

The following compilation of data obtained in analyzing over 100 agar and agarose samples from all the major producing countries may be of value in showing the range to be anticipated in the case of an individual impurity or constituent. For information on the physical properties of agar and agaroids and for other data not presented above, general works on gums (21–22) may be consulted.

H. H. SELBY

References

1. "The United States Pharmacopeia," 17th Revision, 1965, pp. 17–18, New York.
2. "Food Chemicals Codex," 1966, pp. 17–18, Pub. 1406, Washington, D.C., National Academy of Sciences—National Research Council.
3. Watanabe, K., *J. Agr. Chem. Soc. Japan*, 17, A69 (1941).
4. Hachiga, M., and Hayashi, K., *Hakko Kogaku Zasshi*, 42, No. 4, 207 (1964) (Japan).
5. Wood, E. J. F., "Agar in Australia," p. 33, Council for Scientific and Industrial Research, Bull. 203, Melbourne, 1946.
6. DeLoach, W. S., Wilton, O. Christine, McCaskill, Jean, Humm, H. J., and Wolf, F. A., *Duke Univ. Marine Sta. Bull.* No. 3, 28 (1946).
7. Dixon, P., "The Rhodophyta," *Oceanog. Marine Biol. Ann. Rev.*, I, 177 (1963).
8. Dawson, E. Y., "Marine Botany," 1966, p. 175, New York, Holt, Rinehart and Winston.
9. Selby, Thelma, unpublished, Amer. Agar & Chem. Co. (1946).
10. Araki, C., *Bull. Chem. Soc. Japan*, 29, 543 (1956).
11. Tsuchiya, Y. and Hong, K., "Vth International Symposium on Seaweed," 1966, p. 315, Long Island City, Pergamon Press.
12. Araki, C., Arai, K., and Hirase, S., *Bull. Chem. Soc. Japan*, 40, No. 4, 959 (1967).
13. Hjertén, S., *Biochim. Biophys. Acta*, 62, 445 (1962).
14. Russell, B., Mead, T. H., and Polson, A., *Biochim. Biophys. Acta*, 86, 169 (1964).
15. Katsuura, K., Fuse, T., and Kano, K., *Kogyo Kagaku Zasshi*, 68 (1), 205 (1965) (Japan).
16. Azhitsky, G. Iu. and Kobozev, G. V., *Laboratornoe Delo*, No. 3, 143 (1967) (Russ).
17. Humm, H. J., *Duke Univ. Marine Sta. Bull.* No. 3, 43 (1946).
18. Yaphe, W., "Vth International Symposium on Seaweed," 1966, p. 333, Long Island City, Pergamon Press.
19. Burns, Catherine, "Metals and Microorganisms" (Bibliographies), 1966, 1967, San Diego, American Agar & Chemical Co.

20. Tauraso, N. M., *J. Bacteriol.*, **93**, 1559 (1967).
21. Whistler, R. L., "Industrial Gums," 1959, pp. 15–54; 155–167, New York, Academic Press.
22. "Natural Plant Hydrocolloids," Advances in Chemistry Series, No. 11, pp. 16–19; 101–103, 1954, Washington, D.C., American Chemical Society.

ALBACORE—See **TUNA FISHERIES**

ALGINIC ACID—See **ALGINATES FROM KELP**

ALGAE, MARINE

Most fresh-water algae are distributed world wide and their occurrence is limited primarily by light, temperature, and nutrient conditions. Marine algal species on the other hand have a much more restricted distribution. Thus there is a characteristic flora for the Pacific coast and the Atlantic coast; there is a Japanese flora and a South African flora, etc. These algae occupy two main regions: the tidal zone near land where most algae are usually attached to rocks, sandy bottoms, or other stationary objects; and the surface of the water in the open sea. Most of the algae near land are larger (macroscopic) algae and those of the open sea are usually microscopic forms but there are notable exceptions in both cases.

Nearly all marine algae are photosynthetic plants and as such are the primary producers of food in the oceans. Algal productivity is known in Arctic waters as well as temperate and tropical regions. Since water covers nearly three-fourths of the surface of the earth, it is apparent that there is a tremendous potential for food production from marine algae. It has been estimated that as much as 90% of the photosynthetic carbon reduction on this planet each year is accomplished by algae.* Much of this algal growth is consumed by protozoans and in turn by larger marine animals. Through this food chain some of this algal productivity becomes available to man by eating fish or other seafood. However, as far as man is concerned, this food chain is inefficient. It would be more efficient if he could utilize some of the smaller animals that fed on the algae or, better yet, if he could directly use algae as food. The current use of marine algae for human consumption is more widespread than is generally recognized in the Western Hemisphere. Regular use of algae for food is common in China, Japan, and the tropical islands of the Pacific. The genera most frequently used in the Far East are *Porphyra*,

---

* This is considered too high by many authorities, who place the figure at a more conservative 65–70%. —Ed.

Laminaria, *Sargassum*, and *Monostroma*. Some species of the red algae, green algae and brown algae are chopped and eaten raw in green salads or other dishes. In South America, species of *Ulva* and *Durvillea* are eaten after they have been dried and salted. Many of the larger forms of algae are used as food even though they may not necessarily be cultivated. These algae are collected from the beaches or from rocks in the surf at low tide.

Cultivation for Food

In Japan in particular, *Porphyra* is cultivated and harvested regularly to produce "Asakusa Nori," a preparation which is very popular. The total area used for cultivating *Porphyra* is estimated to be 155,000 acres from which is produced 4000–5000 metric tons of dry algae per year. Sale of this product in Japan brings the equivalent of 28 million U.S. dollars per year. Nets are placed in shallow water to catch the spores of *Porphyra* in the autumn. The young plants are later transferred to an area of lower salinity (where the plants grow best), until they are ready for harvest.

Most of the algae used for food are macroscopic since they are easily collected. However, in recent years there has been much research on microscopic algae to determine the feasibility of their use as food. These experiments show that unicellular algae can be processed into food. However, the problems of obtaining food from the unicellular forms are varied and complex. Although unicellular algae such as *Chlorella* grow rapidly, they are not cultivated in natural bodies of water like *Porphyra*.

A primary problem faced in production of *Chlorella* is that of harvesting, since a process such as centrifugation or filtration must be used for unicellular algae. This is likely to be more expensive than the costs incurred in harvesting the larger algal forms. Production at the pilot-plant level has indicated that on a large scale *Chlorella* can be produced for 17–25 ¢/lb of dry material. More experience and refinement of culture techniques will probably allow a reduction of this price, although it still could not compete with other inexpensive protein sources such as dried alfalfa (3 ¢/lb) and soybeans (6 ¢/lb). However, if a popular preparation of *Chlorella* were developed, expense would not be such a problem since dry *Porphyra* is very popular in Japan at over $4.00/lb of dried algae.

Mineral requirements for an alga such as *Chlorella* are not radically different from those of terrestrial plants. The elements nitrogen, potassium, magnesium, sulfur, and phosphorus are added in relatively large quantities to the nutrient medium while calcium, iron, manganese, zinc, copper, and molybdenum are required only in trace quantities. Some unicellular species such as diatoms require silicon and boron, and blue-green algae have been shown to need supplements of cobalt. For a typical algal culture to reach a maximum rate of production, the requirements are: (1) light for the photosynthetic process, (2) carbon dioxide for photosynthesis (CO_2 is usually mixed with air and bubbled into the culture), (3) inorganic nutrients, (4) favorable culture temperature characteristic of the species being grown, and (5) a means of stirring the culture to keep the cells from settling out.

Nutritional Value

One of the most important factors in the selection of any plant as a food source is the nutritional quality of the plant itself. Marine algae are a rich source of many of the minerals required in human nutrition. Some marine algae have been found to contain iodine in amounts up to 1% of their dry matter. The level of iodine varies from season to season in most algae but is high enough at all times to make algae our richest biological source of iodine. Potassium, sodium, and chlorine are also found in abundance in marine algae, which also contain in somewhat smaller amounts the important minerals iron, copper, manganese, zinc, calcium, and phosphorus. When algae are cultivated artificially the mineral content of the algae is strongly dependent upon the nutrient content of the culture medium. For instance, artificially cultured algae are not normally grown with iodine added to the nutrient medium, so they contain only negligible quantities of iodine.

Algae are considered to contain an abundance of nearly all vitamins. Some algae rank among our best known sources of certain vitamins. For example, most plants do not contain vitamin B_{12}, but some seaweeds contain $1 \mu g$ of vitamin B_{12} per gram of dry weight. This is the approximate amount contained by liver, which is recognized as a good source of vitamin B_{12}. Certain species of *Sargassum* have been reported to contain more vitamin C than is present in the juice of a lime, which is one of our best sources of this vitamin. Provitamin D is found in large quantities in some algae and is apparently quite low in others. It has been concluded that ¼ lb of dried *Chlorella* would provide more than the minimum daily requirement of all vitamins except vitamin C. Large quantities of vitamin C are actually found in *Chlorella* but most of it is lost in drying. The vitamin C content can possibly be preserved by improvements in the processing.

Although the vitamin content of algae is impressive, interest in algae as a source of food is centered around their protein content. In many countries of the world where insufficiency of food is a problem, usually it is not lack of calories, but lack of sufficient protein that is the primary problem. Unicellular algae such as those in the genus *Chlorella*, which embraces both marine and fresh-water species, are recognized to be the richest plant source of protein. As was the case with minerals in algae, protein composition varies tremendously depending primarily upon the nitrogen content of the culture medium. *Chlorella* can be easily produced to contain 60% or more protein on a dry weight basis. As a result of widespread culturing of *Chlorella* in laboratories all over the world, more data is available on the nutritional quality of this alga than any other. Many of the larger seaweeds are also good sources of protein, but their protein content is usually significantly lower than that of *Chlorella*. In order for algal protein to be of use it must be digestable. Limited studies of digestability of *Chlorella* protein indicates that it is similar to the protein of terrestrial plants although the digestability varies according to the method of processing of the algal material. Studies with rats, chicks, and humans show that *Chlorella* is an acceptable food when added to wheat flour or another source of carbohydrates.

Another factor of great nutritional importance is the content of essential amino acids in algal protein. It can be tested for nutritional value by feeding tests with experimental animals or by chemical assay for the individual essential amino acids. Both assay methods indicate that *Chlorella* contains all of the amino acids recognized as essential in human nutrition. One of these amino acids, methionine, is recognized to be present in lower quantities than is nutritionally desirable. However, methionine is found to be low in nearly all plant proteins. Fortunately this problem can be easily overcome since synthetic methionine is readily available. When we remember that protein is the most limiting item in the average diet in many parts of the world, the possibilities of growing algae for food become even more impressive. In Japan production of protein from *Chlorella* cultures has been calculated at 14,000 lb of *protein* per acre per year. A high-protein terrestrial plant such as beans yields only about 400 lb of protein per acre per year.

The lipid fraction algae is thought to be similar to that of most terrestrial plants. The oils of algae are apparently slightly more unsaturated than those of most food plants. The larger marine algae such as *Laminaria* generally have a very low fat content (but the unicellular algae may contain large quantities of fat). When algae are grown in the presence of a sufficient nitrogen source they produce large quantities of protein. Under these conditions the lipid content is normally 10–20%. In the absence of sufficient nitrogen, algae have been produced which contained over 85% of their dry weight as lipid material. Large globules of oil are microscopically visible inside these cells. Other important materials present in the algal lipid fraction are chlorophylls, carotenoids, and sterols. *Chlorella* has been reported to contain 6% chlorophyll which is considerably more than alfalfa (0.2%), which is presently a commercial source of chlorophyll. The content of beta-carotene, an excellent source of vitamin A, is approximately equal in carrots and *Chlorella*. All algal classes with the exception of the blue-green algae have been found to contain sterols. Some of these are important to animals as a precursor in the synthesis of vitamin D. Other sterols have been considered as a possible starting material in the commercial synthesis of drugs such as cortisone, which has a similar chemical structure, but as yet it appears that the amounts of sterols produced in most algae are too small to be of commercial value.

Although the nutritive value of algae is undeniable, with the exception of the Far East, algae have not been utilized to any extent as human food. In many maritime districts over the world, however, algae have been used directly for animal fodder. Small industries have been developed to process algae (chiefly *Ascophyllum*, *Laminaria*, and *Fucus*) into seaweed meal. Cattle fed this meal are reported to have increased butterfat content in their milk and hens fed seaweed meal lay eggs with a higher iodine content. Since marine algae contain a high ash content, some have been burned and the ash used for fertilizer. More often they are added to the soil and plowed under as an organic fertilizer. It has been claimed that this can promote good soil structure. In some coastal areas containing corralline algae, which contain large amounts of calcium, these algae are ground up and applied to the soil in place of lime.

Commercial Importance

Probably the best known industrial use of algae is in the making of agar. Agar is a non-nitrogenous carbohydrate extract from certain Red algae, especially *Gelidium*, *Gracilaria*, *Pterocladia*, and *Campylaephora*. Although dry agar is nearly insoluble in cold water, it dissolves in hot water forming a *sol* which upon cooling becomes a gel. This gel is used as a growth medium for many bacteria, fungi, and algae. Agar is also used in many industrial processes. Carrageenin is another commercially important product obtained from the red algae, primarily *Chondrus crispus*. Over 17 million pounds of *C. crispus* was harvested from Nova Scotia in 1958 for the extraction of carrageenin, which has properties similar to agar. Carrageenin is primarily used to stabilize emulsions and suspend solids in the food, pharmaceutical, textile, and brewing industries.

The most important commercial product from the brown algae is alginic acid or its derivatives. Alginic acid is extracted from brown algae harvested off the coasts of Japan, South Africa, Australia, California, and South America. The alginic acid derivatives are used in industry as emulsifiers, gelling agents, thickeners in the food industry, in cosmetics, and as surface films in the paper industry.

The commercial contribution of the diatoms is called diatomite. It is composed of the siliceous cell walls of diatoms which flourished in certain marine environments millions of years ago. At the death of the diatoms, the cell walls accumulated as sediments in many marine and freshwater basins. Many of these deposits contain silica in a rather pure state and are worked commercially. The finished product is chemically inert and is used primarily as an aid in filtration processes. It is also used as a filler in paints, varnishes, and paper products and also as an insulation material particularly for use at high temperatures.

Although antibiotics have not been chemically identified in all of the algae in which they have been found, these organisms show some promise as a source of antibiotics. Antibacterial extracts have been found in *Ascophyllum*, *Chlorella*, *Laminaria*, *Nitzschia*, and *Rhodomela*. It is thought that these products may serve to limit the bacterial population in nature.

Water "blooms" of certain algae can under certain circumstances cause injury or even death to animals. One type of "bloom" found in the ocean and known as the "red tide" is so called because of the red coloration of a species of Dinoflagellate which is dense enough to make the water appear to be red. These organisms produce a very strong poison which is taken up by marine animals. It has been shown that mussels containing these toxins, when eaten by man, can cause paralytic shellfish poisoning. The greatest outbreaks of "red tide" occur in tropical and subtropical regions. Although these blooms do occur occasionally in northern latitudes of Iceland and Norway, they usually do not have the drastic effects that characterize them in the tropical and subtropical regions.

GLENN W. PATTERSON

References

1. Burlew, John S., Ed., "Algal Culture from Laboratory to Pilot Plant," 1953, Washington, Carnegie Institute.
2. Kachroo, P., Ed., "Proceedings of the Symposium on Algalogy," 1960, Indian Council of Agricultural Research, New Delhi.
3. Krauss, R. W., "Mass culture of algae for food and other compounds," *Am. J. Bot.* **49**, 425 (1962).
4. Lewin, R. A., Ed., "Physiology and Biochemistry of Algae," 1962, New York, Academic Press.
5. Oppenheimer, Carl H., "Marine Biology II, Phytoplankton," 1966, New York, New York Academy of Sciences.

6. Round, F. E., "The Biology of the Algae," 1965, Edward Arnold, Ltd.

Cross references: *Agar; Alginates from Kelp; Dinoflagellates; Irish Moss Industry; Phycocolloids and Marine Colloids; Phytoplankton; Seaweeds of Commerce.*

ALGINATES FROM KELP

Except for the fishing industry, man through the ages has paid little attention to the world's oceanic resources. In particular, he has ignored marine plant life almost completely even though in amount it is the equal of the land plants. Most of these primitive life forms belong to a class called algae. Although many algae consist of only a single, free-floating cell, this family also includes multicellular members referred to as seaweeds. These grow in a narrow, restricted belt of relatively shallow coastal waters about a mile in width. Here, they attach themselves more or less permanently to hard, rocky seabottoms. Growing luxuriantly in cooler waters and varying greatly in size and structure, the kelps are prominent among these plants.

Botanists have classified algae into several groups which differ only in the dominant pigment —green, red, or brown—used by the plant to intercept solar energy. The kelps belong to the brown algae, which are characterized by yellow-brown coloring materials called xanthophyll, carotene, and fucoxanthin. These effectively mask the green chlorophyll which is also present.

Algin is the generic name for the water-soluble polysaccharide gum extracted from brown seaweeds, where it occurs as the main structural constituent of the cell walls. Sodium, potassium, ammonium, and calcium salts of alginic acid, along with propylene glycol alginate, are generally available. All these materials yield viscous solutions which are able to thicken and stabilize many formulations used in the food, pharmaceutical, and chemical industries. These solutions also give useful gels, fibers, and films when precipitated under the right conditions.

Kelp

Algin may be produced economically only in those areas which support an adequate and permanent source of the necessary kelp. The giants among seaweeds, these plants form underwater forests of great size under the proper conditions. In some places, these occur as conspicuous offshore surface growths; on other coasts, these beds become evident only after heavy storms which cast up great quantities of underwater growth on adjacent beaches. If large casts of weed occur regularly in accessible places, an economical harvesting operation becomes possible. Areas such as this, which are the first requirement of a successful algin industry, are rare.

At the present time only two seaweeds are harvested mechanically from self-propelled barges —*Macrocystis pyrifera* and, to a lesser extent, *Ascophyllum nodosum*. Growing in large beds from 50 ft to a mile in width and several miles in length, off the coast of California, the former plant is by far the most important and has formed the basis for the first, and still the largest, algin industry. Similar beds have been reported off the coasts of Mexico, Chile, Australia, and New Zealand. Of these only the Australian material has been harvested in a limited way.

A perennial plant, *Macrocystis pyrifera* grows in water 20–100 ft deep where adequate light can penetrate.[1–6] Anchored to the bottom by a holdfast which may reach a diameter of 4 ft, it sends up as many as 100 individual stipes or stems, each representing a plant in various stages of development. Along each stipe grows a series of alternating fronds or leaves, which are supported at their base by a hollow float bulb. At the surface these fronds form a dense and continuous umbrella of plant material. Varying in length from 50–200 ft, individual stipes have a diameter of $\frac{1}{4}$–$\frac{3}{4}$ in. and possess great strength. Floats are approximately 1–2 in. long by $\frac{1}{2}$–1 in. in diameter, while the fronds vary from 4–6 in. in width and 1–2 ft in length.

Depending on the season, each stipe reaches maturity in 4–6 months. During this period food is synthesized by the floating fronds, which are exposed to sunlight by day while being continuously bathed in nutrient-containing water. At maturity spores are produced in a series of special leaves called sporophylls which grow at the base of the plant. After discharge of the spores, the entire stipe weakens, dies, and is eventually sloughed off by the main plant. Its place in the sun is taken by one of the younger stipes growing out of the original holdfast. Unless uprooted in a storm, the plants will continue to send up new stipes for as long as 12 years. During the summer months a young stipe will grow at rates up to 12 in. per day.

Since this giant kelp occurs in relatively deep water and since its foliage is concentrated in a dense umbrella on the surface, ocean-going barges are able to harvest it economically. Equipped with underwater cutting blades, they shear off the kelp stipes approximately 3 ft below the surface of the water. A moving conveyor picks up the cut kelp, which continues to float, and automatically unloads it onto the barge. Loads of fresh kelp are moved quickly to the plant site for immediate processing into algin without the necessity of prolonged storage. Since *Macrocystis pyrifera* is a perennial and grows very rapidly, harvesters may operate in these beds 2–3 times per year. The removal of mature plants from the surface is beneficial to the kelp bed since increased light penetration stimulates the development of younger stipes reaching toward the surface. Sloughed material does not accumulate in the kelp canopy, where it can greatly increase damage during storms. This is reflected in a reduced amount of drift kelp on nearby beaches. The carefully controlled harvesting of kelp apparently has no effect on fish life in the area.

As the algin market expanded, producers started to use a smaller seaweed, *Ascophyllum nodosum*, which grows in the very shallow intertidal waters of Canada, Great Britain, Norway, and France.[2,5,7] Since these plants are uncovered at low tide, men working from small boats or on land are able to cut and collect this weed by hand. Recently, however, in Nova Scotia, a small barge similar in basic design to the Macrocystis boats has commenced harvesting this plant as it floats at high tide.

Growing almost entirely in breaking surf on rocky shores, Ascophyllum is totally immersed at high tide and completely exposed at low. The severity of the surf in which it grows determines the size of the individual plant. In relatively calm water, it reaches a length of 10–12 ft, but in more exposed areas, breakage reduces this to about 5 ft.

These seaweeds possess a frond which is almost cylindrical in shape and which averages $\frac{1}{4}$ in. in diameter. Anchored to the rocks by a disk and a series of fingerlike attachments, the base of the plant is seldom dislodged. To resist surf action, the stems possess considerable tensile strength. Because the fronds are extensively forked and the forks in turn bear short lateral branches, the plant has a bushy appearance. During spring, certain reproductive branches develop and shed spores. All the stems carry a series of hollow enlargements which accumulate gas to give the plant buoyancy.

In spite of extensive damage in rough surf, Ascophyllum plants live as long as 20 years. Breakage is not necessarily fatal since the plant has the power of initiating new growth at the site of injury. To permit regeneration of the plant after harvesting, it is cut about 6 in. above the holdfast. Growth, however, is slow, about 6 in. per year. To conserve these beds, each annual cutting is limited to 20% of the plants in a given area.

In its beginning, the seaweed industry of Great Britain utilized the bottom kelp, *Laminaria cloustoni*, almost entirely.[2,5-7] Although it is a relatively small perennial attaining a maximum size of 16 ft, this plant is still an important source of algin. Possessing no air bladders, it grows only on hard, rocky ocean bottoms in 5–60 ft of water under conditions which make mechanical harvesting extremely difficult. A small crampon or holdfast attaches the plant to the bottom by gradually surrounding stones of considerable size. A rigid, tapering stipe 5–10 ft high holds the plant upright and terminates in a fan-shaped frond that reaches a length and breadth of 5–6 ft. This frond is split lengthwise into many segments which are sloughed off after the spores are released from their surfaces. Growth occurs at the transition zone between the stipe and frond, new growth appearing just below the old. As the zone grows broader, splits appear which are held together at their tips until the old frond is completely removed. This constant sloughing process results in regular casts of weed on nearby beaches during the late spring and keeps the size of individual plants essentially the same throughout their lives.

In the areas where they grow, *Laminaria cloustoni* plants form dense masses of enormous bulk. Winter storms cause great damage in these beds and produce heavy casts of weed on neighboring beaches. These accumulations are preserved by drying as quickly as possible before storage and later use in the algin factory.

The extensive submarine forests of *Laminaria cloustoni* provide a great part of the raw material for algin production in Great Britain and Norway. In France a second closely related fan kelp forms the basis of an algin industry. Called *Laminaria digita*, this kelp grows in a very narrow zone often located just inside the *Laminaria cloustoni* beds.[3,5-7] About the same size, it differs only in having a smooth, more flexible stipe that allows the plant to lie prone on the ocean floor. Otherwise, its appearance and life history are very similar. Growth of the plant also occurs at the tip of the stipe, old tissue being pushed forward as new forms just below it. Because of expansion of the newly formed tissue, the frond grows wider and becomes segmented at the same time. Here also, plants reach a maximum size as the accumulation of new tissue is balanced by the loss of dead material in rough seas. Because it grows in shallow water, men are able to harvest this seaweed from small boats. However, this operation kills the individual plant, and if the kelp bed is to be conserved, only a quarter of the plants in a given area may be taken each year.

At various times algin producers have purchased kelp meal from other parts of the world to supplement their supply of the native materials. Thus they absorb a considerable harvest of sun-dried Ecklonia maxima that grows along the seacoasts of South Africa.[8] The French also import appreciable quantities of dried Laminaria from Morocco.[9,10]

For many years the Japanese have used all readily available Brown algae in a food product called Kombu. However, several companies produce small quantities of algin from seaweeds such as Ecklonia, Eisenia, Alaria, and Sargassum.

Algin Process

Regardless of the type of seaweed, all manufacturers use essentially the same process for the extraction of algin.[11-14] After a preliminary milling operation, a water wash removes excess salt and any soluble organic impurities that may be present. Digestion of the washed kelp with soda ash at elevated temperatures converts the algin to its soluble sodium salt. Filters or centrifuges free the resulting dilute solution of cellulose and other insolubles. Algin is then precipitated from this clear solution either as its calcium salt or as alginic acid.

If alginic acid is the desired end product, the wet fibers from the above process are dried, milled, and screened to the desired mesh size. Since this material is insoluble, its calcium, sodium, ammonium, and potassium salts are usually produced. To accomplish this, alginic acid is reacted with the corresponding base at a concentration such that a paste is formed before being dried, milled, and packaged in a variety of particle sizes as required by the ultimate consumer. In a very similar process, the reaction of alginic acid with propylene oxide under controlled conditions yields propylene glycol alginate.

Physical Properties. The theoretical equivalent weight of commercial alginic acid is 176.[15] In actual practice this value is not reached, apparently because of the algin molecule's ability to bind additional water very tightly. When efforts are made to remove this water under rigorous conditions, the polymer degrades, often with the introduction of additional acidic groups.

As it exists in kelp, algin possesses a very high molecular weight. Commercial extraction processes greatly reduce this value, resulting, under varying conditions, in materials having a molecular weight of between 30,000 and 200,000. These figures correspond to a degree polymerization of 180 to 930.

Optical rotation measurements reported in the literature for sodium alginate solutions indicate a value between $-113°$ and $-148°$ for $(\alpha)_D^{20°}$. In the presence of normal potassium chloride, the dissociation constant pk for alginic acid has a value of 2.95 between a pH of 2.8 and 6.0.

Chemical Properties. In the dry state, algin picks up moisture readily from the air, reaching an equilibrium water content that depends upon the humidity. Under average conditions, this figure runs close to 10%.

Alginic acid of high molecular weight changes to one of low molecular weight very readily. However, these lower molecular weight products are fairly stable, and complete breakdown to uronic acid occurs only under extremely severe conditions that involve the use of hot acid solutions. Salts of alginic acid are stable, and most commercial products change very little after years of storage. Some depolymerization may occur, however, when high-viscosity materials are stored at elevated temperatures, especially in the presence of moisture.

In the absence of bacterial attack, the stability of alginate solutions is very similar to the above. Thus, solutions of high-viscosity alginates decrease somewhat in viscosity with time at ordinary temperatures, although low-viscosity materials are quite stable.

Although alginic acid as such is not soluble in water, it forms water-soluble salts with monovalent cations such as sodium, potassium, and ammonium and low molecular weight amines. Since it is a linear polymer of high molecular weight, these salts, when dissolved, hold a large amount of water and give very high viscosities at low concentrations. This thickening action, along with their colloidal properties and negative charge, makes them excellent suspension agents.

Since bivalent and trivalent cations have the ability to combine with two or three carboxyl groups, they are able to cross link several algin molecules to form aggregates which are insoluble in water. Depending on the formulation used, the addition of such materials to alginate solutions forms algin gels, films, or fibers as desired.

When suitably activated, each glycuronic acid group of the algin molecule has two hydroxyl groups available for esterification. Diacetyl products of this kind have been made.[16-18] However, the required reaction conditions cause considerable depolymerization of the algin molecule, and the end products possess both low viscosity and poor stability. These derivatives possess the unique property of being soluble in many organic solvents.

Esters involving reaction at the carboxyl group may be prepared readily, and such compounds are available commercially.[19] Since these products are resistant to precipitation by acids and salts, they greatly extend the range of algin's usefulness. Because of hydrolysis, an alkaline medium reconverts this ester to its corresponding salt.

Applications of Algin

The solution properties of monovalent alginate salts are responsible for most commercial algin applications. These salts consist of macromolecules which, in solution, carry a negative charge due to their electrolytic dissociation. Since they are hydrophilic colloids, these solutions are much more viscous than those of ordinary materials.

Because of their great affinity for water, commercial alginates are readily soluble in hot or cold

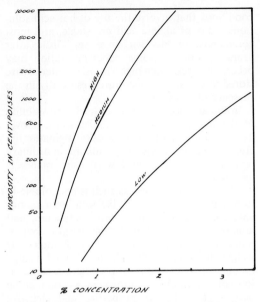

Fig. 1. Viscosity vs concentration at 25°C of representative sodium alginates.

water. The viscosity of such a solution depends upon the concentration used, its molecular weight or degree of polymerization, temperature, pH of the solution, and the effect of other materials that may be present. The various commercial alginates are sold in a variety of closely controlled viscosity ranges that run from 10 to 2000 centipoises for a 1% solution under ordinary conditions. The curves of Fig. 1 illustrate the effect of concentration on aqueous solutions of representative low, medium, and high viscosity sodium alginates at 20°C. An increase or decrease of 1°C in temperature will change the viscosity of these solutions by approximately 2.5%. Heating for short periods results in no coagulation or other change in the properties of these solutions. Since bacteria attack all polysaccharides, including algin, a preservative should be added if these solutions are to be stored for long periods.[7,14,15]

As the pH of an alginate solution is changed from 4 to 10, its viscosity increases slightly at the neutral point. Below a pH of 4, solution viscosity also rises because of the formation of insoluble alginic acid. If the pH is lowered to a range between 3.0 and 3.5, the solution will gel.[7]

In general, alginate solutions are compatible with low concentrations of most alkali, magnesium, and ammonium salts. At high concentrations, salting out of the algin sometimes increases the solution viscosity. Most carbohydrates do not affect alginate solutions which may be used in conjunction with materials such as starch, dextrin, sugar, and soluble cellulose derivatives. They are compatible also with most of the dyes (vats, rapidogens, indigosols, acetates, etc.), resins of the phenolic, urea-formaldehyde, and melamineformaldehyde types, proteins, polyhydric alcohols, and water-soluble gums. In higher concentrations of most organic solvents, such as alcohol or acetone, precipitation takes place often in the form of fibers.[7,15]

By far the greater number of the clear algin gels used in industry are based on the addition of a calcium salt, an acid, or some combination of these to an alginate solution. Different formulations permit variations in the textures of these gels from those which can be described as tender to others which are tough in nature. Usually the formulation gives a slow but uniform release of the precipitating ions into the alginate solution. Sequestering agents such as polyphosphates are added to control setting time. Varying the algin concentration from 0.5% to 2.0% controls the rigidity of the gel and also its setting time. These gels do not melt on heating and are stable up to the boiling point of water. They can be returned to their liquid form by reconverting the algin to its soluble salt.[7,15]

The soluble alginates also lend themselves to the preparation of soluble and insoluble films and filaments. After being cast on a smooth surface, wet algin films are sometimes dried by heating to remove water or ammonia. In other processes these wet films are rendered insoluble by a precipitating solution such as calcium chloride. Filaments are readily prepared by extruding the alginate into a similar precipitating bath. Films consisting of sodium alginate are water-soluble, but those converted to a polyvalent salt are insoluble. Most are clear, tough, and flexible. Allowing the passage of water vapor, they resist penetration by oils, solvents, fats, and other materials. To increase the flexibility of the final film, glycerol and sorbitol are often added to the formulation as plasticizers. To make the film water-resistant, a urea-formaldehyde resin may be included and the mixture put through a final heating step. Soluble algin films are occasionally treated with zinc or aluminum salts for the same purpose.[7,15]

Algin has found acceptance by industry in uses which involve thickening, suspending, emulsifying, stabilizing, gelling, and film-forming. They are particularly useful in the food, pharmaceutical, textile, cosmetic, and paper industries.

A. W. SADDINGTON

References

1. Dept. of Fish and Game, "The Seaweed Story," 1954, State of California.
2. Newton, Lily, "Seaweed Utilization," 1951, London, Sampson Low.

3. Steiner, A. B., "Encyclopedia Americana," 1952, New York, Americana Corp., p. 397.
4. Scofield, W. L., "The harvesting of kelp in California," 1934, Contribution No. 141, California State Fisheries Lab.
5. Fritsch, F. E., "The Structure and Reproduction of the Algae," 1952, Cambridge, Univ. Press.
6. Milne, L. J., and Milne, M., "The Biotic World and Man," 1965, 3rd Ed., Englewood Cliffs, N.J., Prentice-Hall, p. 276.
7. McNeely, W. H., in "Industrial Gums," 1959, New York, Academic Press, p. 55.
8. *Chem. Trade J.*, **133**, 500 (1953).
9. *Chem. Tade J.*, **126**, 1466 (1950).
10. *Nature*, **173**, 766 (1954).
11. Bonniksen, C. W., *Chem. Trade J.*, **128**, 377 (1951).
12. Bashford, L. A., Thomas, R. S., and Woodward, F. N., *J. Soc. Chem. Ind.* (London) **69**, 337 (1950).
13. Tressler, D. K., and Lemon, J. M., "Marine Products of Commerce," 1951, 2nd Ed., New York, Reinhold, p. 94.
14. Black, W. A. P., and Woodward, F. N., in "Natural Plant Hydrocolloids," 1954, Washington, D.C., American Chemical Society, p. 83.
15. McDowell, R. H., "Properties of the Alginates," 1955, London, Alginate Ind., Ltd., 1955.
16. Cefoil, Ltd., Brit. Patent 573,591.
17. Wyeth, Inc., Brit. Patent 676,564.
18. Schweiger, R. G., Can. Patent 715,568 (1965).
19. Steiner, A. B., and McNeely, W. H., U.S. Patent 2,426,125 (1947); 2,463,824 (1949); 2,494,911 (1950); 2,494,912 (1950); Brit. Patent 676,618 (1952).

AMBERGRIS

Ambergris, as known to whalers, is a lumpy substance, resembling fresh fecal boluses in the normally liquid feces of the rectum or descending colon of the sperm whale, where it has been found many times *in situ*, and where it evidently forms. It has also been found floating at sea or beached, in both cases near sperm-whale grounds, usually in small fragments, after ejection by the whale and attrition by the waves. Beach fragments are usually harder and more friable. All recently documented cases of discovery of ambergris *in situ* have also been in males only, usually adults.

Formerly, ambergris was supposed to come only from sick, diseased, and emaciated sperm whales, and there were probably some finds in such animals. Recently however, well-documented finds *in situ*, and even some equally well-documented finds of former hand whaling days, have been in healthy, male sperm whales in good condition.

It was once thought that irritation from the beaks of large squid, upon which the sperm whale feeds in large quantity, caused such impaction as a natural defense mechanism to wall off the irritation by copious secretions of "bile," and some fatty, waxy material.

Formerly, ambergris was valuable in medicine, ritual drinks, and votive offerings. But its only value today is as a fixative in expensive perfumes, to which it also contributes its own peculiar, musky odor. Its medicinal value is nil, and it has lost much value because of the synthesis of other fixatives and even of the essential fixing component of ambergris itself.

Bayne-Cope examined some ambergris from the huge sample collected by R. Clarke in the sub-Antarctic and stated that the "two typifying substances . . . [were] ambreine and epicoprostanol . . .", but he did not give an empirical formula for either, nor any analytical constants. In six small samples from the mass of 926 lb, he analyzed the constituents as follows:

Ether-soluble material	79.8–97.5%
Ambreine	13.6–39.8%
Epicoprostanol	3.4–28.0%
Ash	0.75–8.6%
P_2O_5 insoluble in ether	0.1–2.8%

A spectrographic analysis of ash disclosed Mg and Ca, with small amounts of Na, Mg, Cu, Fe, and Si, with traces of Zn, Al, and Ni.

The differences in composition among the six samples were results of difference in location—on the periphery or within the body of the large original piece. The core, or axis, was "richer" (more ambreine and epicoprostanol?) than at the periphery; and the narrow (posterior?) end was richer than the wider end.

Two samples from another large piece of 340 lb collected by Clarke showed:

Ether-soluble material	95.5, 97.2%
Ambreine	18.1, 28.0%
Epicoprostanol	26.6, 11.4%
Ash	5.0, 1.7%

These two samples came from the core. A third small sample came from the periphery and showed such a low amount of ambreine and epicoprostanol that it was obviously impure.

Figueiredo noted that it was natural to think of ambergris as a pheromone, because such a communicant was common in mammals. But he rejected the idea because of the possibility, as he then thought, that females might produce ambergris. Tomelin noted that "other authors [not mentioned] believe that ambergris is a product of normal secretion of the rectal glands periodically ejected together with feces." But he gave no description of the rectal glands.

The description, however, by J. E. Hamilton in 1915 (Rep. 84th Meeting *Brit. Assoc. Adv. Sci. for 1914*, 139–140) of a heavy cuticular lining "for the last four to six feet" of the intestine [meaning rectum only or rectum *and* descending colon] with crypts, in male sperm whales taken at

Belmullet, Ireland, may have indicated special secretory processes there.

All this discussion of the function of ambergris is surmise as of now, and if ambergris is a pheromone, it is probably normally a liquid secretion passed with the liquid feces, and disseminated into the water, and accumulates only as an impaction after excessive secretion over a short period of time, perhaps during breeding.

RAYMOND M. GILMORE

ANCHOVY—*See* **PACIFIC FISHERIES; PERU CURRENT; PERU FISHERY**

ANIMAL FEEDS

Fish or their products are used in large quantities in animal feeds. For instance, 54% of the total quantity of fish landed in the United States during 1966 was used for industrial products, practically all of which was consumed by animals. The form in which fish are presented to animals varies from raw whole fish to cooked, pressed, and dehydrated products. Whether fish are fed in the raw or processed state, research studies have indicated that fish contain nutrients (protein, fat, vitamins, and minerals) in sufficient quantity and quality to support excellent growth and reproductive performance of animals.

Certain methods concerning the handling of fish products prior to their use in feeding should be discussed. Raw whole fish should be fresh and free from spoilage such as fat rancidity or protein putrefaction, both of which lower the acceptability and nutritive value of the product. In the production of the processed products, poor operating practices may also lower the acceptability and nutritive value of the product. The application of heat at the various stages of production should be well controlled, since excess heat damages the protein and fat. Excess heat may also arise from the oxidation of fat during storage. Usually, however, additions of various antioxidants (ethoxyquin, butylated hydroxytoluene, and butylated hydroxyanisol) coupled with careful handling stabilizes the fat so that any loss of nutritive value and acceptability is minimized.

Farm Animals

Fishery products for animals are used mainly on the farm. Farm animals—particularly poultry (chickens and turkeys) as well as pigs, fish (in pond farming), and cattle—consume nearly 95% of the total industrial fishery products such as fish meal and fish solubles, and in the case of pigs and pond fish, some raw material.

Poultry. Poultry producers have found that fish meal and solubles possess a high nutritive value that cannot be obtained from other feedstuffs. For this reason, fish meal is incorporated in starter, growing–finishing, and breeding diets of chicken and turkey poults. Three factors, however, limit the upper level at which fish meal is used: cost, mineral content, and residual fat content. Excessive amounts of fat may cause off-flavor. Levels of incorporation for fish meal usually range from 2–10% of the total diet; level of incorporation for fish solubles usually range from 1–4%. Chickens and turkeys use about 85% of all the fish meal available in the U.S.

Pigs. In the U.S., the use of fish meal and solubles in pig diets is rather limited in both the total quantity of product consumed and the percentage used in the diet. These two products are used mostly in pig starter diets (that is, they are fed to pigs from about 2–8 weeks of age) and to a lesser extent in sow diets. Levels of use range from 1–5% for fish meal and 1–3% for fish solubles. Little or no fish meal or solubles is used in growing–finishing diets (that is, in diets of pigs weighing from 50–200 lb). The net use of fish meal for pigs in the U.S. is about 5–10% of the total amount of fish meal available. In Europe, fish meal is used to a greater extent in the diet of pigs being incorporated at from 2–10% of the diet. Fish solubles are fed at a level of from 1–4%.

Fish. Recently a new animal industry in this country, fish farming, has evolved. The confined rearing of fish has led to the development of diets high in nutritive value to permit optimum growth. Consequently, fish meal and fish solubles are being used in a rather high proportion in the diet. The amount of fish meal used in the diet ranges from 1–30%, and the amount of solubles used ranges from 1–5%, depending upon the species of fish being fed. For instance, trout feeds may contain from 10–30% fish meal, whereas catfish feeds may contain only from 1–15%. A small quantity of salmon viscera from canning plants is used as food for the confined rearing of fish. These viscera are usually stored and fed in the frozen state. However, of the total quantity of fishery industrial products available, less than 1% is used in fish diets.

Cattle. Cattle can and will consume fish meal and fish solubles. However, in the U.S., essentially none of these products is fed to cattle. Two basic reasons explain this situation: One, cattle, as ruminants, have stomachs that are essentially fermentation vats, which contain bacteria and protozoa flora. This flora is capable of synthesizing protein, beginning with simple non-protein nitrogen compounds such as urea, or with protein of inferior quality. Thus, this fact places fish products of high protein quality in competition with those of relatively low protein quality. Since a price differential exists between products of high and low protein quality, the nature of the ruminant therefore precludes the use of the fish products. However, in European countries, where feedstuffs of lower protein quality are in shorter supply, the relatively expensive fish products become more competitive, and the economically feasible ones are utilized in the diets of both beef and dairy cattle. Levels of use range from 1–5% of the diet.

Pets

Pet foods are another outlet for fishery products. This is especially true of canned food for cats. Canned cat foods may contain from 25–100% fish and from 10–20% dehydrated products, such as fish meal. Dog foods, in general, contain little or no fish products. A few manufacturers, however, may incorporate from 1–5% of fish in either their canned or dry products.

Pet birds thrive on fishery products. Commercial formulas usually contain mostly fish-meal-type products manufactured from crab or shrimp waste, or from both. These products may comprise from 2–10% of the diet.

However, only from 2–5% of the total fishery industrial products available are used in pet foods.

Fur-Bearing Animals

Fish and fish products, owing to their high-quality protein and high fat level, are practically indispensable ingredients in the diets of fur-bearing animals, especially those of mink. The value of fish, however, varies with respect to a number of factors. Some of these factors are kind of fish, levels of fish used, seasonal effects on composition of a given species, spoilage, and the care used in the handling of the fishery material. The use of fish for reproduction and lactation ranges between 40–60% of the diet and for growing and furring ranges between 40–70% of the diet.

No one ration appears to be best for mink. The use of fish will therefore depend upon availability, cost, and potential nutritional problems. In using fish, especially in the raw form, mink farmers must be cognizant of a number of potential problems. Certain species of fish contain the enzyme thiaminase, which destroys vitamin B_1 (thiamine) in the ration. The lack of vitamin B_1 then may result in a neurological condition called "Chastek" paralysis. Another important fact is that fish oils contain polyunsaturated fatty acids, which are highly susceptible to oxidative rancidity. The disorder resulting from the consumption of rancid fat material is termed steatitis. It is due to the destruction of vitamin E or other natural antioxidants. Vitamins A and D, which are fat soluble, may also be destroyed by oxidation.

However, mink ranchers follow a number of rules of thumb, which effectively eliminate the problems. Such precautions are as follows: (1) the varieties of fish and fish products being fed and their approximate composition should be known, (2) several kinds of fish should be blended together—a high level of single species should not be fed (the rule of thumb is that, in general, no more than 15% of any one species should be used), (3) fish should not be stored more than 6 months, (4) additional vitamin E should be fed, particularly during the growing period, if fish are of questionable quality (that is, if the fat may be oxidized), and (5) the fish should be checked for thiaminase activity. If the fish contain thiaminase, they should be cooked or be fed on alternate days with non-thiaminase-containing fish, or fish should be omitted on alternate days.

Fish meal or solvent-extracted fish-meal products are fed to a limited extent in commercial mixed dry feeds. These products, being dehydrated, contain about five times the nutrients (except fat) per pound contained in raw fish. Thus, a level of 5% of meal in the feed is equivalent to a level of 25% of raw or frozen fish in the feed.

Zoo Animals

To some extent, fish similarly are used in the diets of zoo animals. This use is not unexpected, as many zoo animals are carnivorous. The source of flesh for many of the zoo animals in the natural state consists of fish as their entire diet. Ordinarily, over 90% of the diets of sea mammals (seals, sea lions, and walrus) is raw

TABLE 1. SUPPLY OF U.S. FISHERY PRODUCTS, 1955–1966 (ROUND-WEIGHT BASIS)

Year	Domestic catch		Imports[a]		Total
	Billion pounds	%	Billion pounds	%	Billion pounds
1955	4.8	68	2.3	33	7.1
1956	5.3	70	2.3	30	7.6
1957	4.8	67	2.4	33	7.2
1958	4.7	63	2.8	37	7.5
1959	5.1	61	3.3	40	8.5
1960	4.9	60	3.3	40	8.2
1961	5.2	54	4.4	46	9.6
1962	5.4[b]	51	5.1	49	10.4
1963	4.9	42	6.6	58	11.4
1964	4.5	38	7.5	62	12.0
1965	4.8	45	5.8	55	10.5
1966	4.3	35	8.1[b]	65	12.4[b]

[a] Excludes imports of cured cod into Puerto Rico, but includes landings of foreign-caught tuna in American Samoa.
[b] Record.

TABLE 2. SUPPLY OF U.S. INDUSTRIAL FISHERY PRODUCTS, 1955–1966 (ROUND-WEIGHT BASIS)

Year	Domestic catch		Imports		Total
	Billion pounds	%	Billion pounds	%	Billion pounds
1955	2.2	70	1.0	31	3.2
1956	2.6	74	.9	26	3.5
1957	2.3	74	.8	26	3.1
1958	2.1	65	1.1	35	3.2
1959	2.8	66	1.4	34	4.2
1960	2.4	62	1.5	38	4.0
1961	2.7	52	2.5	49	5.2
1962	2.8[a]	49	2.9	51	5.8
1963	2.3	35	4.3	66	6.6
1964	2.0	29	5.1	71	7.2[a]
1965	2.2	41	3.2	59	5.4
1966	1.8	25	5.2[a]	75	7.0

[a] Record.

Note: The weights of the domestic catch and imports represent the live (round) weight of all items except univalve and bivalve mollusks (conchs, clams, oysters, scallops, etc.), which are shown in the weight of meats, excluding the shell.

TABLE 4. SUPPLY OF U.S. FISH SOLUBLES, 1955–1966

Year	U.S. production[a]		Imports		Total
	Thousand tons	%	Thousand tons	%	Thousand tons
1955	111	97	3.3	3	114
1956	129	98	3.0	2	132
1957	122	93	9.7	7	132
1958	130	90	14.6	10	145
1959	165[b]	86	26.6[b]	14	192[b]
1960	99	97	3.2	3	102
1961	112	94	6.7	6	119
1962	125	95	6.3	5	131
1963	107	94	7.1	6	115
1964	93	95	4.5	5	98
1965	95	95	5.1	5	100
1966	85	95	4.3	5	89

[a] Includes homogenized condensed fish for 1955–63; no production in 1964, 1965, and 1966.
[b] Record.

Note: Imports of solubles are understood to be on a wet-weight basis except for those from the Republic of South Africa, which are believed to be on a dry-weight basis.

fish. Reptiles—such as alligators, crocodiles, and turtles—naturally consume fish as a major part of their diets. Snakes, being carnivores, can also live well on a fish diet.

Birds, like the domestic fowl, need and consume fish and fishery industrial products. The form consumed may be raw frozen fish or crustaceans, by penguins, for instance, or fish meal by pheasants. The large and small cats found in the zoo do not differ from domesticated cats in their dietary habits. Similarly, they thrive on diets containing from 50–100% of fish.

Economic Importance

The consumption of fish or fishery industrial products is of international economic importance. These avenues provide a large outlet for the waste from factories canning edible products and for species of fish considered unusable by humans. Nutritionally, these products are very valuable; often, they supply the entire dietary requirements of certain animals. A number of tables (from C.F.S. 4400. Fisheries of the United States 1966) are included to illustrate the economic importance of the fishery products consumed in animal feeds.

Table 1 illustrates the total quantity of fishery products available as domestic catch and import.

TABLE 3. SUPPLY OF U.S. FISH MEAL, 1955–1966

Year	U.S. production		Imports		Total
	Thousand tons	%	Thousand tons	%	Thousand tons
1955	265	73	98	27	363
1956	296	77	90	23	386
1957	264	77	81	24	345
1958	248	71	100	29	349
1959	307	70	133	30	440
1960	290	69	132	31	422
1961	311	59	218	41	529
1962	312[a]	55	252	45	565
1963	256	41	376	60	632
1964	235	35	439	65	674[a]
1965	254	48	271	52	525
1966	224	33	448[a]	67	672

[a] Record.

TABLE 5. SUPPLY OF U.S. FISH OILS, 1950–1966 (EXCLUDING LIVER, WHALE, AND SPERM OIL)

Year	U.S. production	Imports	Total supply	Exports	Available for U.S. consumption
	Million pounds	Million pounds	Million pounds	Million pounds	Million pounds
1950	161	15	176	76	100
1951	134	17	151	50	101
1952	119	16	134	44	90
1953	151	12	163	108	54
1954	162	13	175	142	34
1955	185	10	196	143	53
1956	197	11	208	141	67
1957	148	7	155	115	41
1958	162	5	167	94	73
1959	183	7	190	145	45
1960	206	9	215	144	71
1961	255	8	263	123	141
1962	247	11	258	123	135
1963	184	9	192	262	−70
1964	177	12	189	152	37
1965	193	6	199	104	95
1966	164	13	177	77	100

Table 6. Value of Processed U.S. Fishery Products at Processor's Level, 1965–1966
(Processed from Domestic Catch and Imported Products)

Item	1965		1966[a]	
	Million dollars	Percent of total	Million dollars	Percent of total
Canned	495	44.3	551	44.8
Packaged:				
Fillets and steaks	66	6.0	70	5.7
Fish sticks	36	3.2	36	2.9
Fish portions	56	5.0	58	4.7
Breaded shrimp	77	6.9	94	7.6
Other (fish and shellfish)	251	22.4	293	23.9
Cured	54	4.8	54	4.4
Industrial products	83	7.4	73	6.0
Total	1118	100.0	1229	100.0

[a] Preliminary.

Table 2 shows the quantity of the total used in industrial fishery products, nearly all of which is consumed by animals. Thus, of the approximately 12 billion pounds of total fish landed in 1966, 7 billion pounds or 55% of the total was consumed by animals.

Over 90% of the 7 billion pounds or 3.5 million tons (Table 2) of fish are converted into fish meal, solubles, and fish oil. Tables 3, 4, and 5 give the supply of these three products. In the manufacture of these three products, about four units of fish will be converted, after removal of most of the water, into one unit collectively of meal, solubles, and oil.

Although the industrial products constitute 54% of the total fish landed, the ratio of their economical value to the value of the total catch is comparatively low. Table 6 illustrates this comparative economical value.

Nevertheless, the approximately 73 million-dollar return in 1966 for industrial products is by no means an insignificant amount of money.

Robert R. Kifer

Cross reference: *Fish Meal.*

ANTIBIOTIC PRESERVATION OF FISH AT SEA

In spite of the tremendous advances that have been made in recent years in freezing and otherwise processing fish at sea, it is improbable, for economic and other reasons, that simple chilling will be discarded as one of the principal methods of holding this commodity on vessels in the forseeable future. There are several reasons for this. Many fisheries are located at comparatively short distances from shore bases, and there is still a strong preference by many for unfrozen fish. Freezing on vessels is somewhat costly since the equipment, especially if processing is involved, is somewhat expensive and some skilled workers are required.

For many years chilling and holding fish in crushed ice was the only method employed at sea for fish preservation. However, especially during the past decade, there has been an increasing tendency to introduce other methods. Thus, fish may be held in refrigerated seawater, sprayed with recirculated chilled seawater, iced with ice made from seawater, and iced in either well-insulated holds or in holds the temperature of which has been reduced by means of refrigerated pipes.

Fish are naturally contaminated with bacteria that grow comparatively well under cool conditions, and experiments have shown that the spoilage rate is about doubled if the storage temperature is raised from 30° to 37°F. The critical importance of rapid chilling and constant low-temperature storage is therefore obvious. Though proper chilling delays bacterial spoilage considerably, and also delays visceral autolysis due to enzymes, preservation by this means is purely transient, and varies with species of fish, the initial bacterial contamination, and probably with the nutritional status and state of maturation of the fish.

There are considerable variations in the rate of spoilage of different varieties of fish. Thus, comparatively soft-fleshed fish such as cod, with large visceral cavities, spoil much more rapidly in ice than do halibut, which are firm-fleshed and have small visceral cavities. While it is still desirable that fish be landed soon after capture, there are many instances where this is not possible, or where the fish must be shipped in ice to distant inland destinations after they are landed. It is such circumstances that have prompted the search for a simple means of prolonging the period that fish may be held in chilled condition without serious loss of quality.

A simple treatment with a harmless preservative was sought for many years. Substances which were permitted additives for many foods proved of little or no value for preserving chilled fish. Nitrite salts, permitted as additives in meat curing, gave considerable promise as preservatives for use with fish, and were indeed used for many years in the Canadian maritime provinces for fillet preservation.

It was first reported in 1950 that antibiotics of the tetracycline group were very effective for preserving chilled fish when used in low concentrations. Since that time there has been a considerable volume of research dealing with use of chlortetracycline (CTC or Aureomycin) and oxytetracycline (OTC or Terramycin) in fish preservation, and there have been a number of industrial applications (see references). These antibiotics have been used on shore for dipping fish fillets and at sea for preservation of eviscerated or non-eviscerated fish. This article is restricted to a discussion of the results of experimental and industrial applications of antibiotics at sea. The results of small-scale experiments are usually published, while the results of industrial applications are rarely made available in this manner. Consequently it has proved very difficult to obtain much reliable information regarding the actual extent of commercial applications.

The first use of antibiotics on fishing boats was made in Canada in 1953 where the keeping quality of spring (king) salmon, stored 5 days on a trolling boat at 30°F in refrigerated sea water containing 2 parts per million (ppm) of CTC, was compared with fish stored in ice in the usual manner. A very striking improvement in quality of the fish stored in the CTC-containing sea water was observed. Subsequent trials at sea and on shore confirmed and extended these findings. Thus, during 1954, about 4 tons of flaked ice containing approximately 1.25 ppm of CTC were manufactured daily at one fishing plant for several months. The CTC was stabilized by addition of citric acid, since it proved to be unstable in the hard water which was used.

A number of trials were carried out in which trolling boats carried ordinary ice or CTC-containing ice. Eviscerated coho salmon were stored in these ices for up to 12 days, and in all cases the bacteriological quality of the CTC-treated fish was superior. Large-scale trials were carried out subsequently on trawlers, non-eviscerated "ground fish" (cod, flounders, and "ocean perch") being iced with crushed ice with and without 2.5 ppm of CTC. Bacteriological studies indicated that the CTC-iced fish were of better quality than those iced with ordinary ice. Further studies showed that excellent preservative effects resulted when fish were immersed in 25, 50, or 100 ppm solutions of CTC in sea water before icing them in ordinary ice. These various trials showed that antibiotics may be applied on fishing vessels in several different ways with excellent results so far as inhibition of spoilage is concerned. These results have been confirmed and extended in several different countries.

Between about 1954 and 1956 fairly extensive trials were made with antibiotics in Great Britain. Initial laboratory studies indicated the superiority of antibiotics over several other promising antibacterial agents investigated. Subsequently, work was carried out at sea with various species of eviscerated trawl-caught fish including cod, haddock, whiting, and lemon sole. The comparative quality of fish iced with ordinary ice and with ice containing 5 ppm of CTC was assessed by both subjective and objective procedures, most of which showed the superiority of the antibiotic-treated fish.

Further trials were made in 1963. These, while also indicating the superiority of CTC-containing ice over ordinary ice, did not give such outstanding differences as the earlier trials. Factors such as instability of CTC in the water used to make the ice may have been responsible, or the technique used may not have been as exacting.

While a good deal of comparatively small-scale exploratory work on use of antibiotics in fish and shellfish preservation was carried out in university laboratories in the U.S. many of the vessel experiments were conducted by pharmaceutical companies interested in furthering the applications of antibiotics. The

Fig. 1. Live crabs are transferred to chilled seawater in which they may be held in excellent condition for some days.

keeping quality of eviscerated Atlantic groundfish such as cod and haddock was thoroughly investigated by one group. Fish were treated on fishing vessels by different methods including use of ice containing 5 ppm of CTC and a dip in a 10 ppm solution of CTC followed by icing in CTC-ice. In all instances the antibiotic treatments gave very marked improvements in keeping quality. It was also demonstrated that west coast salmon and halibut could be shipped to New York in excellent condition when CTC-ice was used for preservation. Most of the work on shellfish preservation with antibiotics was carried out in the U.S. or in Canada. The results showed that, perhaps because of the rather alkaline nature of the material, or the comparatively high content of divalent cations (magnesium and calcium salts), quite high concentrations of antibiotics were required for successful preservation. In general it would seem doubtful if the use of antibiotics for this purpose can be recommended under most conditions.

Two nations that are dependent on marine resources for much of the animal protein requirement of their populations, namely Japan and Russia, have conducted fairly extensive trials in use of antibiotics in fish preservation. In Japan the results of certain experiments were first reported about 1955, and since that time a number of large-scale tests have been carried out at sea in several different fisheries. In an early experiment non-eviscerated mackerel were immersed 1 hour in a solution containing 10 ppm of CTC plus 5% sodium chloride on a fishing vessel. Mackerel so treated kept 1.7 times as long as untreated fish. When eviscerated mackerel were treated in this way the factor was increased to 2.6 times. In later work, also carried out at sea, sardines were immersed 30 minutes in solutions containing 10–20 ppm of CTC plus 5% sodium chloride. The fish were held without refrigeration (18°C) and their keeping quality was much better than that of untreated fish. With non-eviscerated herring, storage in chilled seawater containing 10 ppm of CTC, in ice containing 5 ppm, or preservation by application of both methods, caused marked improvements in keeping quality.

Thus, the maximum increase in storage life for the antibiotic-treated fish was 90% at 20°C and 40% at −1–2°C. Groundfish also responded well to antibiotic treatment. An 8-day improvement in keeping quality resulted when yellow croaker were dipped on fishing vessels in a 10 ppm solution of CTC and were subsequently iced with ice containing 10 ppm of the antibiotic. It appears that CTC is used in Japanese fisheries where fish is intended for use with cooked products such as those known collectively as "kamoboko." The extent of these applications is difficult to determine, but one prominent Scottish fisheries scientist stated, in 1962, that 40,000 tons of CTC ice were used annually in Japan.

Several large-scale trials on use of CTC on fishing vessels have been made in Russia. During the maiden voyage of the trawler "Minsk" in 1959, crushed ice containing 3–5 ppm was compared with ordinary ice for icing fish caught in the Barents Sea. The fish were dressed as usual. Cod were eviscerated, beheaded, the gills removed, and the fish then iced in a hold, the air temperature of which was maintained at 2–5°C by mechanical refrigeration. The amount of ice used was 30–50% of the weight of the iced fish. The maximum holding period of the vessel was 18 days. The results of careful grading of the landed fish showed that after the 18-day period 18% of the fish held in the antibiotic ice were still edible (Grades I and II), while those held in ordinary ice were inedible. Later a similar trial was made by another investigator with favorable results. The present author has not been able to find information concerning applications of antibiotics in the Russian fishing industry.

The above, and experiments carried out in several different countries, have shown without question that antibiotics can be used to advantage in retarding bacterial spoilage of chilled fish, especially where prolonged storage periods on vessels or on shore must be observed. The comparative success of such treatment with fish

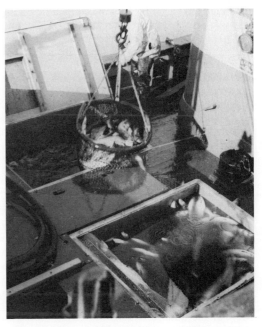

Fig. 2. Salmon are brailed from refrigerated seawater containing antibiotic.

prompted investigations on use in preservation of shellfish and whale carcasses.

The first experiment on shellfish preservation was carried out with cooked crabmeat where it was found that 10–20 ppm of CTC or OTC was required to obtain reasonable extension of shelf life. Later work confirmed the need for these comparatively high concentrations in studies of preservation of shucked oysters, cooked blue crab meat, and raw shrimp. Trials carried out at sea in preserving shrimp with antibiotics in ice or in chilled sea water indicated that only quite high concentrations caused significant preservation. The comparative ineffectiveness of CTC and OTC has been explained by the alkaline reaction of shellfish coupled with their high calcium and magnesium salt content. Ethylenediamine tetraacetic acid, which removes magnesium and calcium salts, greatly improves effectiveness of the antibiotics. For various reasons the use of antibiotics in shellfish preservation has never been adopted commercially.

Whale carcasses are very susceptible to rapid putrefactive spoilage, which is accelerated by the slow cooling of the warm flesh and by the fact that they are naturally contaminated with strictly anaerobic sporeforming bacteria. Considerable success in arresting this putrefaction was attained in Canadian experiments carried out between 1955 and 1956. The antibiotic (50–100 g) was dissolved in seawater and injected into several locations in the visceral cavity by means of the compressed-air device which is normally used to float the carcasses. Subsequent investigations carried out on the North Sea, on the Atlantic near Iceland, off the coast of Chile, and in Russia confirmed these experiments. In general, antibiotic-treated carcasses could be held 2 or even 3 days compared with about 1 day with untreated carcasses before serious spoilage occurred.

A number of permissions to use CTC or OTC in preservation of fish have been granted in different countries. Thus, in 1956 the Canadian Pure Food and Drug Act was amended to permit, with declaration, not over 5 ppm of CTC or OTC on fish for preservative purposes. In 1957 the use of ice containing not over 5 ppm of CTC was authorized for storing fish aboard trawlers operating in the East China Sea which catch fish for "fish cake products" (kamoboko). In 1959 the U.S. Food and Drug Administration also gave limited permission to use CTC on certain non-processed fish or shellfish, but has recently signified its intent to withdraw this permission on the assumption that use of such antibiotics might involve hazards to health. In view of the quite extensive use of antibiotics in certain fisheries for about 10 years without any reported ill effects, this assumption would appear to have no factual basis. Also, research has shown that pathogenic organisms do not tend to gain ascendency with the very low concentration of antibiotic involved. There are several reports that indicate that antibiotics are used in other countries for fish preservation.

In conclusion it cannot be too strongly emphasized that antibiotics must not be used to cover up unsanitary procedures or to extend the shelf life of fish so that the quality of the landed product is not improved. As with the advocated methods of fresh-fish preservation, the goal of the investigator has been to improve the quality of fish that reaches the consumer, not merely to extend keeping quality so that larger loads of fish of average or poor quality can be landed. This method of preservation is comparatively inexpensive and the application methods are not complicated. So far no other preservation method, except radiation pasteurization which is initially very expensive, offers more promise for at-sea use with chilled fish.

H. L. A. TARR

References

1. Anon., "Chilling of fish," Fish Processing Technologists Meeting (FAO), Edited by E. Hess and G. N. Subba Rao, Eds., 1960, The Hague, Netherlands, Govt. Printing Office, 276 pp.
2. Anon., "Antibiotics and the preservation of wet fish." Torry Advisory Note No. 19, 1964, Dept. Sci. Ind. Res., Gt. Britain, 7 pp.
3. Castell, C. H., and Dale, J.," Antibiotic dips for preserving fish fillets," Fisheries Res. Bd. Canada Bull. No. 138, 70 pp. 1963.
4. Dubrova, G. B., "The use of antibiotics in the preservation of fish products," Fisheries Res Bd. Canada Transl. No. 275, 1961, Moscow, Gostorgizdat, 88 pp.
5. Farber, L., "Antibiotics in food preservation," Ann. Rev. Microbiol, 13, 125 (1959).
6. Shewan, J. M., "The use of antibiotics in the preservation of fish," in "Antibiotics in Agriculture," Butterworths, 1962, London, pp. 289–314.
7. Tarr, H. L. A., "Antibiotics in fish preservation," Fish. Res. Bd. Canada Bull. No. 124, 1960, 24 pp.
8. Tarr, H. L. A., "Chemical control of microbiological deterioration" in "Fish as Food," G. Borgstrom, Ed., 1961, New York, Academic Press, Vol. I, pp. 639–680.
9. Tomiyama, T. et al., "The effectiveness and safety of the use of antibiotics in keeping freshness of fish. A symposium," Bull. Japan Soc. Sci. Fisheries, 28, 85 (1962).
10. Wrenshall, C. L., in "Antibiotics, Their Chemistry and Non-Medical Uses," H. S. Goldberg, Ed., 1959, Princeton, N.J., D. Van Nostrand, pp. 449–527.
11. Zaitsev, V. P., "Preservation of fish products by refrigeration," translated from Russian, 1962, Springfield, Virginia, U.S. Dept. of Commerce Clearinghouse for Federal and Scientific Information, pp. 89–92.

ANTIMICROBIAL SUBSTANCES

The antagonistic action of certain organisms in causing growth inhibition of other organisms has been known since the latter part of the 19th century. The word antibiotic was first used by Vuillemin in 1889 to describe the opposition of one organism to the life of another.[1] The concept of antibiotics (*anti*: against; *bios*: life) as powerful chemical substances produced by microorganisms growing under laboratory conditions and having the capacity in low concentrations to inhibit growth or to destroy various bacteria and other microorganisms was clearly set forth by Waksman.[2] The scope of antibiotics has greatly expanded to include antimicrobial substances derived from microscopic organisms growing in culture media, and also diverse kinds of substances produced in nature by fleshy fungi, lichens, plankton, seaweeds, and various animals. Antimicrobial substances produced by marine fauna and flora are here considered from the viewpoint of their nature and origin, and their importance as potential biomedical agents.

Algae

Antibacterial activity was demonstrated by Pratt and co-workers[3,4] in the green alga, *Chlorella vulgaris* and in brown and red seaweeds. The properties of "chlorellin," a kind of fatty acid, were described by Pratt,[5] and by Spoehr, et al.[6] Apparently biosynthesized fatty acids are photooxidized to yield an antibiotic compound.[7] Inhibition of the sensitive alga *Haematococcus* by fatty substances produced by *Chlamydomonas* was demonstrated by Proctor.[8] The antibacterial action of *Rhodomela larix* and *Symphocladia gracilis* may be due to brominated phenolic substances.[9] Saito and colleagues[10,11] demonstrated antibiotic red and brown seaweeds. Kamimoto[12] showed that certain seaweeds inhibit growth of pathogenic bacteria. Chesters and Stott[13] found species of *Halidrys, Pelvetia, Laminaria,* and *Polysiphonia* especially active against bacteria. Roos[14] reported 27 antimicrobial species of marine algae at Kiel. In a survey of 150 species of tropical marine algae growing in Puerto Rico, it was reported that 66 kinds inhibit various bacteria and yeast.[15] Activity of some of these algae against fungi was demonstrated by Welch.[16] Other contributions to knowledge of antibacterial activity in algae have been published.[17-20]

In recent years, many references have appeared concerning antiviral activity of seaweed materials.[21] Gerber et al.,[22] reported antiviral properties in *Chondrus* and *Gelidium*. Substances in *Cladophora* and *Lyngbya* are antibacterial and antiviral,[23] and kelp inhibits influenza viruses.[24] Martinez Nadal, et al.,[25-28] have published papers concerning antibacterial and antiviral properties of *Sargassum* and *Cymopolia*.

The chemical nature of seaweed antimicrobial agents has been studied by several investigators. Methanol extracts of marine algae showed diverse patterns suggesting at least four different substances.[29] Acrylic acid appears to be a common antibacterial substance in seaweeds and phytoplankton[30-32]. The occurrence of B-dimethylpropiothetin in the red alga *Polysiphonia fastigiata* was demonstrated by Challenger and Simpson.[33] Hydrolysis of the dimethylpropiothetin by enzymatic cleavage yields dimethylsulfide and acrylic acid in red and green algae.[34,35] As long ago as 1935 Haas[36] demonstrated evolution of dimethylsulfide by passing a stream of air over *Polysiphonia* heated to 30°C.

Seaweeds contain sesquiterpenes,[37-39] phenolic substances, alcohols, hydrocarbons, and other kinds of compounds,[31] some of which may account for the antimicrobial activity of the plants. Terpenoids are toxic, inhibitory, or pharmacologically active for annelids, ascarids, sea urchins, and fish.[31,40] *Digenea simplex*, a red alga, has long been used in treating ascariasis. The anthelmintic agent in this red seaweed is kainic acid, now sold under the trade mark "Digesan" by Takeda Chemical Industries. A new compound, domoic acid, derived from *Chondria armata*, has been studied for its ability to exterminate parasitic worms. Chlorophyll derivatives may be antibiotic.[41-43] Algal phenols have been shown to be antibacterial.[9-11] A novel brominated phenolic derivative, 2, 3-dibromo benzylalcohol-4, 5-disulfate dipotassium salt, has been isolated from *Polysiphonia lanosa*.[44] Further references to brominated phenols of algae may be found in a paper by Craigie and Gruenig.[45] Tannins produced by brown algae retard growth of fouling animals and bacteria.[46] General discussions of antibacterial substances in seaweeds can be found in papers by Sieburth[47] and by Burkholder.[48]

Antagonism has long been known among the various kinds of microalgae.[49,50] It is claimed that actively growing *Scenedesmus* and *Chlorella* are highly bactericidal to coliforms and *Salmonella*. Anticoliform activity of the diatom *Skeletonema costatum* has been demonstrated by Sieburth and Pratt.[51] Accorinti[52] showed that *Scenedesmus* and *Coelastrum* produce fatty acids inhibitory for *Staphylococcus*. Klein[53] discovered that the green alga *Pythophora* inhibits *Mycobacterium tuberculosis*. *Phaeocystis pouchetii* produces acrylic acid in Antarctic waters, and this substance then passes from the algae through the food chain to penguins.[32,54,55] Antibacterial concentrates have been prepared from the diatom *Asterionella japonica* by Aubert and Gauthier.[56] Many other species of Mediterranean phytoplankton have also been found active against bacteria. The marine blue-green *Trichodesmium erythraeum* is also antibacterial in ocean waters.[57]

Heterotrophic Marine Microorganisms

Many marine bacteria produce antibiotics.[58,59] Antibacterial substances produced by marine bacteria have been studied by Burkholder[60] and by Burkholder, Pfister, and Leitz.[61] From the marine bacterium *Pseudomonas bromoutilis*, a new antibiotic compound, 2-(3, 5-dibromophenyl) 3, 4, 5-tribromopyrrole, has been isolated. This bromine-rich compound was synthesized by Hanessian and Kaltenbronn.[62] The presence of antiyeast and antifungal microbes in the sea has been amply demonstrated by Buck, Meyers and Liefsen, and by Buck and Meyers.[63-65] The antibiotic cephalothin was derived from a product of the fungus, *Cephalosporium*, isolated from the sea off Sardinia;[66] the first successful cephalosporin agent is now marketed under the trade mark "Keflin."[67]

Sponges

Sponges are known to produce antimicrobial substances.[68] The antibiotic "ectyonin" was derived from the red-beard sponge *Microciona prolifera*. Substances with varying degrees of antimicrobial activity have been obtained from temperate and tropical sponges.[69] Extracts of the sponge *Ianthella* sp. inhibited a pink yeast.[63] Extracts from the antibacterial *Haliclona viridis* proved to be toxic to fish and mice.[70] Effects of sponge compounds on viruses have been discussed by Nigrelli et al.[71] Many crystalline compounds have been prepared from diverse species of marine sponges by Sharma and Burkholder.[72-74] One of these antibiotic substances is 2, 6,-dibromo-4-acetamido-4-hydroxycyclohexadienone.

Various Invertebrates

Antimicrobial activity is scattered throughout the marine invertebrates. A steroid glycoside, holothurin, isolated from a sea cucumber,[75] shows antiprotozooal activity.[76] Holothurin suppresses growth of tumor cells.[77] Some species of horny corals are antibacterial.[78] Ciereszko et al.,[79] reported that terpenoid substances isolated from gorgonians may account for their activity against bacteria. Crassin and eunicin, from the gorgonids *Plexaura crassa* and *Eunicea mammosa*, are antiprotozooal and antibacterial.[40]

Extracts of peanut worms, *Bonellia* sp., are toxic to protozoa.[80] Various marine shellfish are antiviral and antibacterial.[81-83] "Paolin," derived from shellfish juice, has some activity against polio and influenza viruses.[84] Abalone juice and clam extracts are antiviral.[85] Mollusk compounds also show activity against sarcoma 180 in mice[86,87] and HeLa cells.[88] The active subtance, "mercenene," is probably a glyco-peptide.[89]

Protochordates and Chordates

Several different kinds of vertebrates have been found to produce antimicrobial materials. The fish, *Grammistes sexlineatus*, forms a toxic substance active against *E. coli*.[90] The acorn worm, *Balanoglossus biminiensis*, produces 2, 6-dibromophenol.[91] The brominated phenol appears to act as an antiseptic in the immediate environment of the worm.[48]

Excluded from this discussion are many marine substances having biomedical significance; they belong in other categories such as venoms, toxins, coagulants, hormones, etc. References to literature on various aspects of marine pharmacognosy and pharmacology may be found in references 71, 92–95.

P. R. BURKHOLDER

References

1. Florey, H. W., et al., "Antibiotics," 1949, London, Oxford Univ. Press, Vol. I, pp. 14–15.
2. Waksman, S. A., "Microbial antagonisms and antibiotic substances," 1945, New York, Commonwealth Fund.
3. Pratt, R., Daniels, T. C., Eiler, J. J., Gunnison, J. B., Kummler, W. D., Oneto, J. F., Spoehr, H. H., Hardin, G. J., Milner, H. W., Smith, J. H. C., and Strain, H. H., "Chlorellin, an antibacterial substance from *Chlorella*," *Science*, **99**, 351 (1944).
4. Pratt, R., Mautner, R. H., Gardner, G. M., Sha, Y., and Dufrenoy, J., "Report on antibiotic activity of seaweed extracts," *J. Am. Pharm. Assoc.*, **40**, 575 (1951).
5. Pratt, R., "Studies on *Chlorella vulgaris*. V. Some properties of the growth inhibitor formed by *Chlorella* cells," *Am. J. Bot.*, **29**, 142 (1942).
6. Spoehr, H. A., Smith, J., Strain, H., Milner, H., and Hardin, G. J., "Fatty acid antibacterials from plants," Carnegie Inst., 1949, Washington, Publ. No. 586, pp. 1–67.
7. Krzywicka, A., "Influence of time exposure to light on the activity of the antibacterial substance of *Chlorella vulgaris*," *Bull. Acad. Pol. Sci.*, Ser. Sci. Biol., **14**, 509 (1966).
8. Proctor, V. W., "Studies on algal antibiosis using *Haematoccus* and *Chlamydomonas*." *Limnol. Oceanogr.*, **2**, 125 (1957).
9. Mautner, H. C., Gardner, G. M., and Pratt, R., "Antibiotic activity of seaweed extracts. II. *Rhodomela larix*," *J. Am. Pharm. Assoc.*, **42**, 294 (1953).
10. Saito, K., and Nakamura, Y., "Sarganin and related phenols from marine algae and their medical functions," *J. Chem. Soc. Japan, Pure Chem. Sect.*, **72**, 992 (1951).
11. Saito, K., and Sameshima, J., "Studies on antibiotic action of algae extracts," *J. Agr. Chem. Soc. Japan*, **29**, 427 (1955).
12. Kamimoto, K., "Studies on the antibacterial substances extracted from seaweeds on the growth of some pathogenic bacteria," *Nippon Saikunaku Zasshi*, **10**, 897 (1955).

13. Chesters, C. G., and Stott, J. A., "The production of antibiotic substances by seaweeds," 1966, Second Intern. Seaweed Symp. held in Trondheim, July 1966, New York, Pergamon Press, pp. 49–54.
14. Roos, H., "Untersuchungen über das Vorkommen antimikrobieller Substanzen in Meeresalgen," *Kiel. Meeresforsch.*, **13**, 41 (1957).
15. Burkholder, P. R., Burkholder, L. M., and Almodovar, L. R., "Antibiotic activity of some marine algae of Puerto Rico," *Botanica Marina*, **2**, 149, (1960).
16. Welch, A. M., "Preliminary survey of fungistatic properties of marine algae." *J. Bact.*, **83**, 97 (1962).
17. Fassina, G., "Recherches sur les propriétés antibiotiques des algues de la côte venitienne," *Arch. Ital. Sci. Farmacol.*, **12**, 238 (1962).
18. Feller, B., "Contribution à l'étude des plaies traitées par un antibiotic derivé des algues. Thèse vétérinaire," 1948, Paris (Alfort).
19. Maurer, C. C., "Antibacterial substances in Chilean marine algae," *Anales Fac. Quim. Farm., Univ. Chile*, **16**, 114 (1965).
20. Walters, B., "Antibiotisch und toxisch wirkende Substanzen aus Algen und Moosen," *Planta Medica*, **12**, 85 (1964).
21. Takemoto, K. K., and Spicer, S. S., "Effects of natural and synthetic sulfated polysaccharides on viruses and cells," *Ann. N.Y. Acad. Sci.*, **130**, 365 (1965).
22. Gerber, P., Dutcher, J. D., Adams, E. V., and Sherman, J. H., "Protective effect of seaweed extracts for chicken embryos infected with influenza B or mumps virus," *Proc. Soc. Exp. Biol. Med.*, **99**, 590 (1958).
23. Starr, T. J., Deig, E. F., Church, K. K., and Allen, M. B., "Antibacterial and antiviral activities of algae extracts studied by acridine orange staining," *Texas Repr. Biol. Med.*, **20**, 271 (1962).
24. Kathan, R. H., "Kelp extracts as antiviral substances," *Ann N.Y. Acad. Sci.*, **130**, 390 (1965).
25. Martinez Nadal, N. G., Casillas Chapel, C. M., Rodriguez, L. V., Rodriguez Perazza, J. R., and Vera, L. T., "Antibiotic properties of marine algae III. *Cymopolia barbata*," *Bot. Marina*, **9**, 21 (1966).
26. Martinez Nadal, N. G., Rodriguez, L. V., and Casillas, C., "Sarganin and chonalgin, new antibiotic substances from marine algae from Puerto Rico," in "Antimicrobial Agents and Chemotherapy, 1963," Publ. by Am. Soc. for Microbiol., pp. 68–72.
27. Martinez Nadal, N. G., Rodriguez, L. V., and Casillas, C., "Isolation and characterization of sarganin complex, a new broad-spectrum antibiotic isolated from marine algae," in "Antimicrobial Agents and Chemotherapy, 1964," Publ. by Am. Soc. for Microbiol., pp. 131–134.
28. Martinez Nadal, N. G., Rodriguez, L. V., and Dolagaray, J. I., "Low toxic effect of antimicrobial substances from marine algae," *Bot. Marina*, **9**, 62 (1966).
29. Allen, M. B., and Dawson, Y., "Production of antibacterial substances by benthic tropical marine algae," *J. Bact.*, **3**, 459 (1960).
30. Challenger, F., Bywood, R., Thomas, P., and Hayward, B. J., "Studies on biological methylation. XVII. The natural occurrence and chemical reactions of some thetins," *Arch. Biochem. Biophys.*, **69**, 514 (1957).
31. Katayama, T., "Volatile constituents," in "Physiology and Biochemistry of Algae," 1962, R. A. Lewin, Ed., New York, Academic Press.
32. Sieburth, J. M., "Acrylic acid, an antibiotic principle in *Phaeocystis* blooms in Antarctic waters," *Science*, **132**, 676 (1960).
33. Challenger, R., and Simpson, M. I., "Studies on biological methylation. XII. A precursor of dimethyl sulfide evolved by *Polysiphonia fastigiata*. Dimethyl-2-carboxyethyl sulphonium hydroxide and its salt," *J. Chem. Soc.*, **3**, 1591 (1948).
34. Bywood, R., and Challenger, R., "The evolution of dimethyl sulfide by *Enteromorpha inestinalis*," *Biochem. J.*, **53**, 26 (1953).
35. Cantoni, G. L., and Anderson, D. G., "Enzymatic cleavage of dimethyl propiothetin by *Polysiphonia lanosa*," *J. Biol. Chem.*, **222**, 171 (1956).
36. Haas, P., "The liberation of methylsulfide by seaweed," *Biochem. J.*, **29**, 1297 (1935).
37. Takaoka, M., and Audo Y., "Studies on essential oil of seaweed I," *J. Chem. Soc. Japan*, **72**, 999 (1961).
38. Katayama, T., "Chemical studies on volatile constituents of seaweeds. X. Antibacterial action of *Enteromorpha sp*.," *Bull. Japan Soc. Sci. Fisheries*, **22**, 248 (1956).
39. Katayama, T., "Structure and antibacterial activity of terpenes," *Bull. Japan Soc. Sci. Fisheries*, **26**, 29 (1960).
40. Ciereszko, L. S., "Chemistry of coelenterates. III. Occurrence of antimicrobial terpenoid compounds in the zooxanthellae of alcyonarians," *Trans. N.Y. Acad. Sci.*, Ser. II, **24**, 502 (1962).
41. Jorgensen, E. G., "Antibiotic substances from cells and culture solutions of unicellular algae with special reference to some chlorophyll derivatives," *Phys. Plant.*, **15**, 530 (1962).
42. Aubert, M., Aubert, J., Gauthier, M., et Daniel, S., "Origine et nature des substances antibiotiques présentes dans le milieu marin. Ière partie. Etude bibliographique et analyse des travaux antérieurs." *Rev. Intern. Oceanogr. Med.*, I, 9 (1966).
43. Aubert, M., Aubert, J., Gauthier, M., Pesando, D., et Daniel, S., "Origine et nature des substances antibiotiques présentes dans le milieu marin. VIe partie. Etude biochimique des substances antibacteriennes extraits d'*Asterionella japonica* (Cleve)," *Rev. Intern. Oceanogr. Med.*, IV, 23 (1966).
44. Hodgkin, J. H., Craigie, J. S., and McInnes, A. G., "A novel brominated phenolic derivative from *Polysiphonia lanosa*," 1965, Proc. Fifth Int. Seaweed Symposium, New York, Pergamon Press, p. 279.
45. Craigie, S. S., and Gruenig, D. E., "Bromophenols from red algae," *Science*, **157**, 1058 (1967).
46. Sieburth, J. M., and Conover, J. T., "Sargassum tannin, an antibiotic which retards fouling," *Nature*, **208**, 52 (1965).
47. Sieburth, J. M., "Antibacterial substances produced by marine algae," *Develop. Indust. Microbiol.*, **5**, 124 (1964).

48. Burkholder, P. R., "Antimicrobial substances from the sea." Conference on Drugs from the Sea, Marine Technology Society, August 28–29, Univ. of Rhode Island, Kingston, R.I., 1967.
49. Flint, L. H., and Moreland, C. F., "Antibiosis in the blue-green algae," *Am. J. Bot.*, **33**, 218 (1946).
50. Lefevre, M., Jakob, H., et Nisbet, M., "Auto- et heteroantagonism chez les algues d'eau naturelles," *Ann. de la Station Centrale d'Hydrobiologie Appl.*, **4**, 5 (1952).
51. Sieburth, J. M., and Pratt, D. M., "Anticoliform activity of sea water associated with the termination of *Skeletonema costatum* blooms," *Trans. N.Y. Acad. Sci., Ser. II*, **24**, 498 (1962).
52. Accorinti, J., "Inhibidores antibacterianos de *Scenedesmus obliquus* y *Coelastrum microporum*. Relaciones con acidos grasos," *Rev. Mus. Arg. Cs. Nat. Bs. As.* No. 4, *Hidrobiologia*, **1**, 137 (1963).
53. Klein, S., "A new anti-tubercular substance," *Bot. Mar.*, **6**, (1964).
54. Sieburth, J. M., "Antibacterial activity of Antarctic marine phytoplankton," *Limnol. Oceanogr*, **4**, 419 (1959).
55. Sieburth, J. M., "Antibiotic properties of acrylic acid, a factor in the gastrointestinal antibiosis of polar marine animals," *J. Bacteriol.*, **82**, 72 (1961).
56. Aubert, M., Gauthier, M., et Daniel, S., "Origine et nature des substances antibiotiques présentes dans le milieu marin. IIIème partie. Activité antibacterienne d'une diatomée marine, *Asterionella japonica* (Cleve)," *Rev. Intern. Oceanogr. Med.*, **I**, 35 (1966).
57. Ramamurthy, V. D., and Krishnamurthy, S., "The antibacterial properties of the marine blue-green alga Trichodesmium erythraeum," *Current Sci.*, **36**, 524 (1967).
58. Rosenfeld, W. D., and Zobell, C. E., "Antibiotic production by marine microorganisms," *J. Bact.*, **54**, 393 (1947).
59. Krasil'nikova, E. N., "Antibiotic properties of microorganisms isolated from various depths of world oceans," *Microbiol.*, **30**, 545 (1961).
60. Burkholder, P. R., "Some nutritional relationships among microbes of sea sediments and waters," in "Symposium on Marine Microbiology," ed. by Carl H. Oppenheimer, 1963, Springfield, Ill., C. C. Thomas, pp. 133–150.
61. Burkholder, P. R., Pfister, R. M., and Leitz, F. M., "Production of a pyrrole antibiotic by a marine bacterium," *Appl. Microbiol.*, **14**, 649 (1966).
62. Hanessian, S., and Kaltenbronn, J. S., "Synthesis of a bromine-rich marine antibiotic," *J. Am. Chem. Soc.*, **88**, 4509 (1966).
63. Buck, J. D., and Meyers, S. P., "Antiyeast activity in the marine environment. I. Ecological considerations," *Limnol. Oceanogr.*, **10**, 385 (1965).
64. Buck, J. D., and Meyers, S. P., "*In vitro* inhibition of *Rhodotorula minuta* by a variant of the marine bacterium, *Pseudomonas piscicida*," *Helgolander wiss. Meeresunters.*, **13**, 171 (1966).
65. Buck, J. D., and Meyers, S. P., "Growth and phosphate requirements of *Pseudomonas piscicida* and related antiyeast pseudomonads, *Bull. Marine Sci.*, **16**, 93 (1966).
66. Brotzu, G., "Richerche su di un nuovo antibiotico," Lav. Inst. Ig. Univ. Cagliari, 1948.
67. Newell, R. W., "Healers from the Sea," 1964, Indianapolis, Indiana, Eli Lilly Co., 32 pp.
68. Nigrelli, R. F., Jakowska, S., and Calventi I., "Ectyonin, an antimicrobial agent from the sponge *Microciona prolifera*, Verrill," *Zoologica*, **44**, 173 (1959).
69. Jakowska, S., and Nigrelli, R. F., "Antimicrobial substances from sponges," *Ann. N.Y. Acad. Sci.*, **90**, 913 (1960).
70. Nigrelli, R. F., Baslow, M., and Jakowska, S., "Further characterization of antimicrobial substances from the Bahamian sponges *Haliclona viridis* and *Tedania ignis*," First Intersci. Conf. on Antimicrobial Agents and Chemotherapy., Am. Soc. for Microbiol., pp. 83–84, 1961.
71. Nigrelli, R. F., Stempien, M. F., Ruggieri, G. D., Liguori, V. R., and Cecil, J. T., "Substances of potential biomedical importance from marine organisms," *Fed. Proc.*, **26**, 1197 (1967).
72. Sharma, G. M., and Burkholder, P. R., "Studies on the antimicrobial substances of sponges I. Isolation, purification and properties of a new bromine-containing antibacterial substance," *J. Antibiotics, Ser. A. (Japan)*, **20**, 200 (1967).
73. Sharma, G. M., and Burkholder, P. R., "Studies on the antimicrobial substances of sponges II. Structure and synthesis of a bromine-containing antibacterial compound from a marine sponge," *Tetrahedron Letters*, No. 42, 4147 (1967).
74. Sharma, G. M., Vig, B., and Burkholder, P. R., "Studies on the antimicrobial substances of marine sponges III. Studies on the chemistry of antibacterial compounds from different types of sponges," Conference on Drugs from the Sea, Marine Technology Society, Washington, D.C.
75. Chanley, J., Ledean, R., Wax, J., Nigrelli. R. F., and Sobotka, H., "Holothurin I. The isolation, properties and sugar components of holothurin A," *J. Am. Chem. Soc.*, **81**, 5180 (1959).
76. Nigrelli, R. F., and Zahl, P. A., "Some biological characteristics of holothurin," *Proc. Soc. Exp. Biol. Med.*, **81**, 379 (1952).
77. Sullivan, T. D., and Nigrelli, R. F., "The antitumorous action of biologics of marine origin I. Survival of Swiss mice inoculated with Krebs 2 ascites tumor and treated with holothurin, a steroid saponin from the sea cucumber, *Actinopyga agassizi*," *Proc. Am. Assoc. Cancer Res.*, **2**, 151 (1956).
78. Burkholder, P. R., and Burkholder, L. M., "Antimicrobial activity of horny corals," *Science*, **127**, 1174 (1958).
79. Ciereszko, L. S., Gifford, D. H., and Weinheimer, A. J., "Chemistry of coelenterates. I. Occurrence of terpenoid compounds in gorgonians," *Ann. N.Y. Acad. Sci.*, **90**, 917 (1960).
80. Ruggieri, G. S., and Nigrelli, R. F., "Effects of Bonellin, a water soluble extract from the proboscis of *Bonellia viridis* on sea urchin development," *Am. Zoologist* **2**, No. 365 (1962).
81. Li, C. P., "Antimicrobial effect of *Abalone* juice," *Proc. Soc. Exp. Biol. Med.*, **103**, 522 (1960).
82. Li, C. P., "Antimicrobial activity of certain

83. Li, C. P., Eddy, B., Prescott, B., Caldes, G., Green, W. R., Martino, E. C., and Young, A. M., "Antiviral activities of paolins from clams," *Ann. N.Y. Acad. Sci.*, **130**, 374 (1965).
84. Li, C. P., Prescott, B., and Jahnes, W. B., "Antiviral activity of a fraction of abalone juice," *Proc. Soc. Exp. Biol. Med.*, **109**, 534 (1962).
85. Li, C. P., Prescott, B., Jahnes, W. C., and Martino, E. C., "Antimicrobial agents from mollusks," *Trans. N.Y. Acad. Sci., Ser. II*, **24**, 504 (1962).
86. Schmeer, M. R., "Growth-inhibiting agents from *Mercenaria* extract. Chemical and biological characteristics," *Science* **144**, 413 (1964).
87. Schmeer, M. R. and Huala, C. V., "Mercenene: in vivo effects of mollusk extracts on sarcoma 180," *Ann. N.Y. Acad. Sci.*, **118**, 603 (1965).
88. Schmeer, M. R., Horton, D., and Tanimura, A., "Mercenene, a tumor inhibitor from *Mercenaria*; purification and characterization studies," *Life Sci.*, **5**, 1169 (1966).
89. Schmeer, M. R., "Mercenene: Growth-inhibiting agent of *Mercenaria* extracts. Further chemical and biological characterization," *Ann. N.Y. Acad. Sci.*, **136**, 211 (1966).
90. Liguori, V. R., Ruggieri, G. D., Baslow, M. H., Stempien, M. F., and Nigrelli, R. F., "Antibiotic and toxic activity of the mucus of the Pacific golden striped bass *Grammistes sexlineatus*," *Am. Zool.*, **3**, Abs. No. 302 (1963).
91. Ashworth, R. B., and Cormier, M. J., "Isolation of 2,6-dibromophenol from the marine hemichordate, *Balanoglossus biminiensis*," *Science*, **155**, 1558 (1967).
92. Jackson, D. F., "Algae and Man," 1964, New York, Plenum Press, 434 pp.
93. Halstead, B. W., "Poisonous and Venomous Marine Animals of the World," 1965, Washington, D.C., U.S. Govt. Printing Office, Vol. I., pp. 994.
94. Russell, F. E., "Comparative pharmacology of some animal toxins," *Federation Proc.*, **26**, 1206 (1967).
95. Russell, F. E., and Saunders, P. R., "Animal Toxins," 1967, New York, Pergamon Press, 428 pp.

Cross references: *Algae, Marine; Bacteria in the Sea; Sponge Industry.*

ARTIFICIAL REEFS

Modification of the ocean environment to favor desirable marine plants and animals is the best way to improve the harvest. One such modification is the construction of artificial (man-made) reefs. It has long been known that certain fish are consistently found around reefs, banks, and floating objects. The ancients anchored bundles of reeds offshore with stones, and caught the dolphinfish attracted to these floating objects. The Japanese have sought to increase their harvest of sea plants and animals by constructing special frames, pilings, and reefs. These have been designed to attract and harbor the desired species and to concentrate them for easier and more productive harvest. However, only in recent years have efforts been so diligently and universally directed toward the scientific study of why these methods succeed—toward determining what structures are optimal in attractiveness and for which species, and what is the dollar return on the investment.

Artificial, as well as natural, reef areas allow for increased carrying capacity of the "land." An ocean floor of sand or mud supports at most a rather limited and specialized grouping of animals. Many of the fish orienting to these bottoms are wanderers which visit a given area for short periods. By contrast, high-relief areas are permanent residence for numerous animals and plants which form biotic assemblages of myriad complexity and beauty. Further, isolated reefs, those located on extensive areas of flat sand or mud bottom, not only concentrate the reef dwellers, and attract the migratory species normally frequenting the area, but are home to a third group of fish which orient to both reef and flat-bottom substrates. Combined, these various animals increase the area's productivity. For example, Dr. John E. Randall, in 1963, observed that a small isolated man-made reef supported nearly 10 times the weight of fish (standing crop) as is supported by an equal-sized portion of a natural reef.

An increase in the number and size of such man-made reefs will help to assure the world's fishermen of increased success. In the developing countries improved fishing conditions can increase the take for food and commerce. Man-made reefs can be placed at the periphery of fishing piers (typically built along the open coast in sand-bottom areas) to attract fish and thus increase the catch. Such reefs are accessible to all sportsmen and thus give a high return (in terms of use) for their cost. However, their effect upon nearshore sand transport and currents must be considered.

Japanese workers have investigated a variety or materials for use in creating additional surfaces for the attachment of seaweeds and for providing "homes" for fish and invertebrates. For example, in the cultivation of nori (seaweed) they now employ special nets, made with ropes suitable for both spore attachment and hand harvesting of the algae. These are strung between thousands of bamboo poles driven into the bay muds. For fish and invertebrates, they have constructed two distinct types of reefs. The first (*tsukiiso*) is low in profile and generally placed in shallow water. These "constructed beach" reefs are considered good for algae attachment, and if constructed with numerous openings they are optimum for invertebrates (e.g., sea cucumber, octopus, spiny lobster, and sea urchin). They have been made of various materials including rocks, timbers, oil drums, and concrete pipes. By contrast, their "fish reefs" (*gyosho*) are of high relief, placed in deeper water, and are primarily built to attract and concentrate edible fish. They are frequently constructed of concrete or timbers in a box-like design, open on all sides. They incorporate extensive openings with relatively little structural material and make an attractive "home" for fish, but are unsuitable for spiny lobster and other invertebrates requiring protective hiding places. Data on cost/return considerations are fragmentary but one report, considering only the sale of the increased spiny lobster catch, indicates that it may take 12 years to amortize a shallow-water Japanese reef.

In the U.S., the first artificial fishing reef was constructed in 1935. Composed of about 1400 butter tubs ballasted with concrete, it was constructed inside Fire Island Inlet, New York, by local party boat skippers, to form a new fishing ground for their passengers. Following this lead other skippers constructed two additional reefs, or "fishing preserves" off the south New Jersey coast. They used automobile bodies, cement-filled drums, old boat hulls, concrete rubble, and the like. Fishermen were brought from as far afield as Philadelphia on special trains to fish these highly productive areas. These excursions were halted by World War II and have not been reinstated.

In subsequent years, fishermen on the Atlantic coast harvested quantities of fish in the vicinity of ships sunk during the war. These harvests caused renewed interest in constructing ocean reefs as fishing areas. At first, and to some extent even now, reefs were constructed without any but the most general biological considerations: to make them convenient for anglers and to compose them of materials at hand.

Although agencies constructing the reefs reported greatly increased catches, few had well-staffed programs to observe and evaluate routinely the results of their reef-building efforts. Volunteer diving groups who periodically examined these reefs usually reported, "the reef abounds with life," but this phrase, no matter how descriptive, does not record the changes taking place. A systematic investigation and evaluation of existing reefs should be a prerequisite to building new ones.

With this in mind, the California Department of Fish and Game, aided financially by Federal Aid to Fish Restoration Funds (Dingell–Johnson program), instigated an Ocean Fish Habitat Development project in 1958. Its purpose was to determine if man-made reefs were practicable off California, what materials should be used, and what would be the return to the fishermen. For this study, the Department employed biologists who were competent scuba divers.

These biologists were to observe the reefs visually, to evaluate their findings, and to make recommendations for improving reef design. Since fish congregate around natural reefs and sunken ships, there was little doubt that they would be attracted to the man-made reefs, but answers were needed for many questions. How many fish? What sizes and kinds? Would they remain and provide fishing? Was there a preference for particular reef materials? Were these fish feeding at the reef or seeking shelter. Answers to these and similar questions were forthcoming during the 6-year study that followed.

Background surveys were conducted to identify the biota prior to reef installation. Transect lines were swum, animals and plants within range of vision were recorded and bottom sediments were examined for suitability to support the reef. This factor is most important. Reefs have failed in some areas because they sank into the bottom muds or were buried by current-carried sediments. In soft-bottom areas (e.g., bays or estuaries) this sinking may be combated by building a foundation for the reef. Both steel-mill slag and landing-strip steel mesh have been suggested as suitable "pad" material, although neither has been tried. Another method might be to drive pilings and festoon them with numerous automobile tires.

Preliminary background studies revealed that small artificial reefs could not compete well with natural reefs. Therefore only flat areas of firm sand or mud, widely separated ($\frac{1}{2}$ mile or more) from existing reefs were considered as reef construction sites.

Prior to construction, it was necessary to obtain permits from the U.S. Army Corps of Engineers and the California State Lands Commission or the city having title to the area. Clearance was also obtained from the U.S. Navy, and fishing interests and other interested groups and agencies were consulted and informed of the plans for a fishing reef. Objections from any group were considered to avoid conflicts of use in these nearshore waters. Once constructed, the reefs were marked in compliance with U.S. Coast Guard permits; both to prevent the reef from becoming a navigational hazard and to assist local fishermen in finding it.

After reef construction, monthly checks were made to record the numbers and species of fish and invertebrates present, and to observe the condition of the reef materials. Temperatures were taken for correlation with seasonal fluctuations in abundance of animals and plants. Visibility was recorded and its effect on the estimates of fish numbers was assessed. The substrates were closely watched for scouring or sanding-in of the reef materials.

As fish populations increased and local sportsmen began fishing the reefs, "creel-checks" were made to determine the catch per angling hour. These often showed as much as a six-fold increase over catches from nearby areas. It was also noted that fishing was successful only when conducted directly over the reef. Fishing a few feet away, on the flat sand, resulted in a smaller catch.

The first reefs were constructed of available material—old automobiles and junked streetcars. Within a few hours, fish were orienting to these reefs and after 6 months over 4000 were tallied during one survey dive around a 20 automobile-body reef. The number of fish recorded on any given survey varied widely. It was found that these variations were related to seasonal movements, changes in water clarity, and a preference of the fish for a particular reef material. To evaluate these preferences, a replication experiment was designed. In 1960, California's Wildlife Conservation Board allocated $18,000 to cover the cost of constructing three multicomponent experimental reefs at 60-ft depths in Santa Monica Bay.

Four different materials, of approximately equal cubic footage, were used in these widely separated reefs. Each reef was composed of 14 automobile bodies, 44 concrete blocks, 300 tons of quarry rock and a streetcar, and each component was between 100 and 200 ft from its nearest neighbor. It had been presumed that these distances were sufficient for separate fish

Fig. 2. Concrete blocks, prefabricated fish "homes," being lowered into the sea off Hermosa Beach, California, August 1960.

populations to form and orient to the preferred material. Some movement of fish between materials occurred when the materials were only 100 ft apart.

The number of fish present at a given material ranged widely, averaging about 846 for all materials and observations. The majority were attracted to the concrete blocks, averaging 1053 per survey, the lone streetcar attracting the fewest, 635. Between these extremes were the automobile bodies, 826, and the rock, 870.

Although the large ($8 \times 5 \times 2\frac{1}{2}$-ft) concrete blocks attracted nearly 18% more fish than the quarry rock, rock reefs are recommended because they cost less and are easier to construct. A 1000-ton rock reef costs approximately $6000 compared to $11,000 for a concrete-block reef of comparable size. Automobile bodies, a favored material in many areas, are not recommended by the California scientists, since the metal rusts completely within 3–5 years and must be replaced at a cost of about $3000. Further, of the four materials studied, only the quarry rock showed no signs of sanding-in or scouring-out the bottom.

Species diversity was greatest around the reefs placed where the waters were relatively clean and the currents active, indicating the importance of reef location. Fish congregated along the reefs in "pot-hole" areas, where 3 or 4 automobiles formed a circle or where the rocks bordered the open sand. Fish were often 3–4 times as numerous here as in other areas of equal size. Apparently these are ecotones, or transitional zones, which are characteristically rich in animal and plant life. They appeared suitable for both the rock- and sand-bottom fishes with both groups freely intermingling. Reefs should be designed to

Fig. 1. The last of six streetcars being placed off Redondo Beach, California, to form an artificial reef, September 25, 1958.

Fig. 3. Old automobile bodies being lowered into the sea off Hermosa Beach, California, August 1960, to form part of an experimental multicomponent artificial reef in 60 ft of water.

maximize this effect. Their configuration should be in circular or square patterns with centralized open areas not exceeding 50–60 ft across.

By the third year, a 60-ft deep southern California reef, of approximately 25,000 ft^2, was found to support some 3400 fish. Included were some 1900 embiotocid perches (56%) and 1100 serranids (32%). The remainder were gobies, damselfishes, cottids, and rockfish, including a few hundred sculpin. Although many of these, particularly the sport species, are semi-resident forms (species which leave the reef for short periods), the reef adds to the acceptable habitat available to them and their progeny. By supplying this additional habitat the reefs not only concentrate fish but, in the long term, allow their total numbers to increase.

In southern California, organisms encrusting the artificial reefs were also studied. At the 60-ft depths, true succession occurred. First observed was a barnacle-hydroid phase followed by less distinct mollusk-polychaete, ascidian-sponge, and encrusting ectoproct stages. Subsequent to these first-year changes, aggregate anemones became important (in the second and third year), then gorgonians (in the third and fourth year) and finally stony corals (in the fifth), at which time the reefs began to resemble mature natural reef areas.

Again, reef location and the proximity of other man-made environmental modifications are important. For example, one artificial reef in close proximity to a large sewage outfall supported an invertebrate population of less diversity and numbers than was observed on similar man-made reefs in other locations. Similarly, natural reefs bathed with the wastes of a complex civilization are often unproductive. Thus, a primary need is to clean up the waters in existing reef areas before constructing new reefs where no natural attractants are found.

The first fish observed around the southern California reefs were semi-resident sand bass, kelp bass, and embiotocid perches. They oriented to the reefs but often left for food or breeding. Next to drift in were resident forms, the gobies, cottids, damselfish and some of the rockfish, who spend their entire life on the reef. As the reef "aged," resident forms outnumbered the larger semi-residents. By the third year of reef life, nearly 80 fish species had been observed. Resident fish ate the organisms encrusting the reef materials and in turn became prey for the larger semi-residents. In addition, schools of predators, such as Pacific bonito, California yellowtail and California barracuda, frequented these high-relief areas to feed upon the reef residents or the schools of bait fish that occasionally aggregated over the reefs.

Observations around offshore oil drilling towers in California (effective man-made reef areas) often revealed schools of forage fish, Pacific sardine, jack mackerel, and Pacific mackerel swimming in the shade of these structures. Whenever a portion of a school ventured away from a tower it was immediately set upon by predators circling just outside the platform supports. By using the smaller species for bait,

Fig. 4. California spiny lobster crowding onto an artificial reef offshore from Hermosa Beach, California, October 1963.

local recreational fishermen sometimes took 10 large California yellowtail in an hour.

Results of studying offshore drilling platforms and their effect upon the marine environment are reported in Fish Bulletin 124, California Department of Fish and Game, entitled "Artificial Habitat in the Marine Environment." This publication also includes the preliminary results of a 6-year reef study. A second report, "Artificial Reef Ecology," is being prepared by the California marine biologists for probable publication in 1969.

The immediate and obvious return from these artificial reefs and reef studies has been increased fish catch. A less tangible but more valuable return is the knowledge gained concerning the nearshore marine environment, which is becoming the most productive and heavily utilized area of the ocean. From this coastal area of the continental shelf, various groundfish, pelagic fish (tunas, engraulids, and clupeids), as well as mollusks, crustaceans, and algae must be harvested in ever-increasing quantities to supply the food and recreational needs of the worlds populace. The employment of artificial (man-made) fishing reefs is one method of increasing this area's productivity.

In summary, artificial reefs can increase the productivity of a nearshore area and produce a sizable return for the cost of construction. However, reef construction must be supervised by local conservation agencies and based upon sound biological findings. Man-made reefs first concentrate reef-dwelling species and then by providing additional habitat, permit increases in their total numbers.

Construction units should be of moderately large bulk, with numerous holes and crevices, rising several feet above the bottom. The over-all reef should cover a broad area and incorporate in its configuration large open spaces less than 60 ft in diameter.

Reef materials should be durable, inexpensive, and easily handled at sea. Thin steel quickly disintegrates by rusting, wooden objects are subject to rapid destruction by teredos, and small, low-profile materials are susceptible to sanding-in or silting-over.

Reefs should be constructed on relatively flat barren sand or mud bottoms, at considerable distances ($\frac{1}{2}$ mile or more) from natural reefs, and in areas free from polluting substances. They must be well marked, perhaps with large floats or lighted buoys.

The water depth over the reef will vary considerably with the construction site and local considerations. Productive man-made reefs have already been placed in water 10–80 ft deep. Reefs constructed around fishing piers are by necessity in shallow water (less than 30 ft deep), but, because of their ready accessibility to the recreational fisherman, they are important.

In areas of soft bottoms (e.g., bays and estuaries) where reef materials may sink, pilings driven into the muds and festooned with old tires may prove useful. Another approach is to build a firm base for the reef materials by placing them upon "pads" of steel-mill slag or landing-strip steel mesh. In areas of strong currents or heavy wave action the materials must be heavy enough to prevent movement.

Reefs must not be installed without adequate biological and physical data for the area in question. These data are necessary to determine the need for such a reef and the proper material and design for its construction.

CHARLES H. TURNER

ATLANTIC NORTHWEST FISHERIES

Discovery of the rich codfish resources off Newfoundland by John Cabot led to the rapid expansion of that fishery more than 450 years ago. Upon his return from a voyage to the northwest Atlantic in 1497, Cabot related, "The sea there is full of fish to such a point that one takes them not only by means of a net, but also with baskets to which one attaches a stone to sink them in the water."

The Newfoundland fishery, from its very beginning in the 1500's, was an international fishery. Portuguese, Spanish, French, and English left the Icelandic fishing grounds for the new ones to the west. The French were foremost in the development of the new fishing grounds; the English continued their fishery along the southwest coast of Iceland. Of the 128 vessels sailing to Newfoundland prior to 1550, 11 were English, 12 Portuguese, 3 joint English and Portuguese, 9 Spanish, and 93 French. Following the decline of the fishery off the Irish coast in 1545, the Spanish moved westward to join in the Newfoundland fishery. The war with England saw the Spanish fishery on Newfoundland's inshore fishery come to an end. The move of the English westward from Icelandic waters to Newfoundland occurred in 1580 when Denmark commenced to enforce payment of license fees to fish in the waters around Iceland. The rise of the English fishery forced the French away from the coast of Newfoundland. During their search for new grounds, they located a fishing bank about 150 miles south of their former fishing grounds. This bank is now known as the Grand Bank of Newfoundland.

The continued expansion of English fishing interests in the early 1600's led to the exploration and finally the settlement of new areas, especially the New England coast. Prior to the establishment of permanent settlements in New England, the fishery had been carried out only in the summer months. Settlement in Newfoundland was discouraging due to difficulties encountered in developing agriculture needed to supply food to the winter residents. With the advantages of settlement, the development of agriculture, lumbering, and shipping, New England rapidly rose as an important base for fishing operations.

The successful operation of the fleets in the Atlantic coast fisheries required access to shore facilities for bait, supplies, and for curing the catch. This requirement was a source of conflict among the countries participating in the fisheries, especially between the English Colonies on the one hand and France and the American Colonies on the other. The Spanish and Portuguese having been pushed from the Newfoundland shore fishery, fished the offshore banks and so were not engaged in the wars and treaties of the northwest Atlantic which were to plague the fisheries until the early 1900's. Between 1790 and 1900, the fisheries went through almost continual economic and competitive readjustments. The history of the struggles by the various European countries in the fisheries of the northwest Atlantic is well documented by Professor Harold A. Innis in his book *The Cod Fisheries*. Dr. C. L. Cutting's book, *Fish Saving*, describes the methods used by the fishermen of the northwest Atlantic to preserve their fish from the days of salting to the modern freezing methods.

The beginning of the 20th century saw some very important modifications in the northwest Atlantic fisheries. These included the change from sail to steam or gasoline for propulsion, the introduction of the otter trawl to largely supplant the hand line and the long line, the development of new fisheries and improvements in fish handling and preservation both at sea and ashore.

The appearance of the factory stern trawler in the 1950's, developed by eastern Atlantic countries for distant-water fishing, has probably been one of the most far-reaching developments in fish catching and processing in the past 50 years. The post-war development of distant-water fisheries by several European nations, and the expansion of the fisheries of the U.S. and Canada, resulted in an associated increase in fishing intensity on the stocks of fish of the northwest Atlantic.

Graham states, "Many changes have taken place in the fisheries of the northwest Atlantic since those early days, but the quantity of fish removed annually has steadily increased ... As cod failed to keep up with the increasing demand, effort was turned to other species such as haddock and redfish, and then herring and hakes."

Early recognition of the problem of reduced abundance and potential depletion of the fisheries of the northwest Atlantic prompted the U.S. to convene a conference in Washington in January, 1959. Eleven countries having an interest in the fisheries were in attendance (Canada, Newfoundland, Iceland, the United Kingdom, the U.S., Denmark, France, Italy, Norway, Portugal, and Spain). The work of this conference resulted in the formation of a conservation treaty and the International Commission for the Northwest Atlantic Fisheries (ICNAF). It was opened for signature on February 8, 1959. The Convention, which went into force July 3, 1950, is designed to provide for international cooperation in the coordination, correlation, and dissemination of information concerning the fisheries of the northwest Atlantic Ocean, and a procedure for cooperative action by the contracting governments regarding measures deemed necessary to maintain a maximum sustained yield from these fisheries. By 1967, the number of countries signatory to the Convention reached 14. In addition to the 10 original signers (Newfoundland is now represented by Canada), the Federal Republic of Germany, Poland, U.S.S.R., and Romania have signified adherence to the Convention.

In the early days of the Commission in the 1950's, the scientists stressed mesh regulation as a means toward protecting the fish stocks against intensive fishing and to insure a maximum sustainable yield. Such regulations, which are now in effect in the Convention area, protect the small fish and allow them to survive a while longer and increase in weight before they are caught to contribute to the total biomass of landings.

A 1-year-old haddock, for example, which weighs less than $\frac{1}{2}$ lb, grows rapidly, and, if permitted to remain in the water another year, would weigh about 1 lb. The natural mortality of these haddock is such that if a given number of 1-year-old haddock, say 10,000, which weigh in the aggregate 3000 lb, were permitted to escape from trawl nets and take their chances with nature for one more year, although a number of them would die from various causes, the total weight which survived would be about 6500 lb. Before 1952, the cod and haddock fleet caught large numbers of immature, unmarketable fish. The bulk of these small fish were destroyed. If it could be managed that fish smaller than a certain size could be released unharmed, then the destruction of the small fish could be ended. The most practical method of saving the small, immature haddock appeared to be that of regulating the size of mesh in trawl nets used on the otter trawlers.

An investigation carried out by the U.S. indicated that the optimum size and age at which haddock should enter the fishery is approximately 40 cm (16 in.), at about $2\frac{1}{2}$ years of age. To insure that fish under 40 cm would escape, it was determined it would be necessary to prohibit the use of trawl nets having a mesh size less than $4\frac{1}{2}$ in. The $4\frac{1}{2}$-in. mesh size is still required in the southern portion of the Convention area. Mesh sizes for the entire Convention area have been recommended to all Governments, but they are in force in only three of the five subareas.

The mesh-size regulation entered into effect June 13, 1953, for Subarea 5, the Georges Bank part of the Convention area. The regulation was considered to be fairly successful until about 1963, at which time, due to increasing numbers of foreign fishing vessels, the effort reached the point that it became evident that mesh regulation alone would not insure continued high landings. The scientists of ICNAF have advised the Commission that cod and haddock are now producing at or near their maximum and that further increases in effort will result in decreased landings rather than increases.

The Commission is now seeking additional conservation measures to prevent overfishing of those stocks of fish presently being heavily exploited.

JOHN B. SKERRY

Cross references: *Cod Fishery; Crab Industry; Lobsters; Menhaden Fisheries; Tuna Fisheries.*

AUSTRALIAN FISHERIES

The continental shelf is a submerged border to the Australian continent, sloping gently from low-water mark to 100 fathoms; beyond this the waters deepen markedly. On the shelf the zooplankton is uniform and moderately abundant; beyond, there is little zooplankton. In 1953, the Australian Commonwealth Government proclaimed sovereignty over the waters of the continental shelf by passing the Pearl Fisheries Act.

The taxonomic range of fish caught in the northern hemisphere is not found in the Australian waters. In Australia, there is a total lack of ganoids, few teleostean flat fish, and the whitebait is the only commercial fish representative of the salmon family.

Ports. There are few natural harbors for fishing boats on the Australian coast. Accordingly, a number of man-made harbors have been developed. Since 1954, the state governments have gradually improved the fishing ports. Estuaries also offer protection, and when in some cases sandbars impeded the entrances, these were dredged and protecting groynes (breakwaters) were constructed. Some small ports remain open to the ocean and the vessels using them are beached for safety. Maintenance and repair facilities are available at many ports, and most fishing boats use two-way radios.

Marketing procedure varies according to the state, but the open auction system is common. Most of the fish produced is sold fresh; it is transported by rail or refrigerated trucks from the port to the major markets. Fisherman's Cooperatives also transport and market the catch; the South Australian Fisherman's Cooperative Ltd. (SAFCOL) in addition to handling, also processes. It now has cannery and freezing operations in Victoria, New South Wales, and Tasmania in addition to South Australia. Fish for canning is sold directly to the different canneries by the fishermen.

The Fleet. The Australian fleet is, with few exceptions, Australian built. In 1965 it comprised 10,900 vessels, valued at $37,526,000 Australian (excluding the pearling fleet), manned by about 13,000 licensed professional fishermen. The total population of Australia in 1965 was 11,478,703. Of the fleet 50% of the boats were 20 ft or less, 25% from 20 to 29 ft, and only a very small number were more than 50 ft.

Longline Fisheries for Shark. Shark fishing forms the basis of an important localized commercial fishery. At least 25% of wet-fish landings in Victoria are shark, where they are popular with the public, and are sold under the name of "flake." Shark are also landed in considerable quantities in Tasmania and South Australia; these catches are mainly marketed in Melbourne. A large part of all fish retailed in Melbourne is flake. Some edible shark are also caught by longliners, off the coast of New South Wales; the amount is comparatively small, and it is mainly

TABLE 1. THE PRINCIPAL FISH PRODUCTS OF AUSTRALIA

West Australia	South Australia	Queensland	New South Wales	Victoria	Tasmania
Crayfish	Crayfish	Prawns	Prawns	Crayfish	Crayfish
Prawns	Tuna	Mullet	Crayfish	Scallops	Scallops
Aust. salmon	Shark	Spanish Mackerel	Tuna	Shark	Barracouta or snoek
Mullet		Cultured pearls	Shark	Snoek	Australian salmon
Snapper		Pearls shells	Oysters	Flathead	
Cultured pearls			Mullet		
Pearl shell			Morwong		
			Flathead		

marketed in Sydney. The two varieties of edible shark most commonly found are the most importantly commercially. They are:

School, or Snapper Shark (*Galeorhinus australis*);
Gummy Shark (*Mustelus antarticus*).

Although both of these species rarely exceed 6 ft in length, have smallish teeth, and are harmless to men, they are closely related to the ferocious tiger shark.

Apart from restrictions on the minimum size to be taken, there are few regulations controlling this aspect of the fishing industry. The fishing is done in deep water miles out to sea, in extra-territorial waters. The minimum over-all length permitted in the catching of school shark is 36 in. for all states, and for the gummy shark, 30 in. in Tasmania and 24 in. in Victoria. There are closed seasons which vary and are of short duration.

The principal method employed in catching shark is longlining, similar in technique to that used in New South Wales for snapper. The shark longline may be either clip type gear or "fixed rig." Usually, the snoods, whether fixed or clip type, are of 72 thread cotton or laid kuralon seine twine. Because of the sharp teeth of the school shark, it is necessary to have a short monel wire leader (about 4 in.) tied to the hook. The latter is usually a long shanked design, size 10/0. The main line is usually a 1-in. sisal or manilla rope. The trend today, however, is to change to synthetic kuralon or laid nylon seine rope.

The methods employed in rigging the gear vary considerably according to the ideas of the individual fisherman, and local conditions. Leaders or snoods vary from 3 to 7 ft in length, and are spaced from $2\frac{1}{2}$ to $3\frac{1}{2}$ fathoms apart. It is generally considered true that a longline rigged with the leaders $3\frac{1}{2}$ fathoms apart, fishes better than one with the leaders grouped more closely together. With the former, the ground is more extensively covered. With the latter, uneven distribution of the fish may lead to better results as, by this arrangement, more baits are presented to the fish, although the risk of tangling the gear is greater.

With fixed gear, the leaders are tied permanently to the main line; when not in use they are coiled into a basket or box, and examined for replacement of hooks or damaged or missing leaders.

Clip gear involves the use of a snap-type stainless spring steel clip which is attached to the mainline end of the leader. When the gear is hauled, all the leaders are unclipped and overhauled if necessary and stacked separately from the mainline, which is wound on to a drum without having to be handled at all. Some use has been made of wire rope as the mainline, and on it, lead stops are molded at appropriate distances apart, so that the leaders can be clipped to the line, but cannot slide along and tangle.

The annual catch, which is now variable between 7.5–10.5 million lb annually, was at a maximum during 1943–50. The saw shark (*Pristiophorus undipinnis*), dog shark (*Squalus megalops*), and edible sharks of other species are also captured and marketed.

Spanish Mackerel Fishing in Queensland. The narrow-banded Spanish mackerel (*Cybium commersoni*) is the basis of a very active fishery along the Queensland coast.

Although some mackerel are taken by beach seining in Wide Bay, and are also occasionally hand-lined, the major portion of the catch is landed by spinning with either live or artificial bait from the rear of a dory or launch. The larger refrigerated vessels may operate up to six small dories, and make long trips, while the smaller boats return daily to port. When a school of mackerel is located, the vessel is usually steered continuously to port. Outriggers are often used, and the lines on one side are longer than on the other, and normally weighted with lead to prevent their meeting and tangling. They are usually 15 fathoms in length and of soft laid cotton twine, followed by a wire leader and a pair of 10/0 barbed hooks which are coupled together. Since fishermen's ideas and preferences and fish behavior vary, many modifications in gear are found along the coast. The favorite bait is fresh garfish, but when the mackerel are really "on the bite" they will strike at small fish, prawns, squid, artificial lures, and feather jigs.

As a food, the Spanish mackerel ranks among the best in Australian waters. It always commands a high price in the Queensland markets, because its flesh, although a trifle dry, has a splendid consistency and flavor, and owing to its "meatiness," the bones cause no trouble. When smoked, there is probably no better fish, in Australia. Spanish mackerel are captured throughout the year and sold fresh. The major part of the catch comes from the northern part of the state. The Queensland product comprises 80% of the Australian mackerel annual catch (over 2 million lb); the remainder is caught off New South Wales, West Australia, and Tasmania.

Traps. Trapping is a more efficient method of catching snapper (especially squire), and far less work than setlining. The fish trap is a wire cage built from 2 in. mesh on a wooden or iron frame. Generally, it is rectangular in shape, about 5×2 ft, with a funnel entrance for the fish; it is filled with bait, weighted by bricks, and lowered to the ocean floor. The traps are buoyed to the surface with glass floats and a flag. In the strong currents they are often swept miles from the place in which they were originally set. More commonly, the buoys dive down in the current, but when it finally ceases they reappear. Under such conditions the fishermen cannot haul their traps to the surface for days on end.

Traps are a favorite method of catching snapper and leather jackets, especially to the north of New South Wales and South Queensland. Mullet is considered the best bait, but at times, when it is unobtainable, good results are obtained with salmon, tuna, bonito, or fish scraps such as leather jacket heads.

The same methods and gear are employed to catch the yellow jacket (*Nelusetta cittate*) which commands a high price at the markets. This fish is of excellent quality, and as a food is white, tender, and well flavored. Another important commercial food fish which at times is caught by trapping, is the black bream, known also as the silver bream or bream (*Mylio Australis*). It is highly prized as a food, and considered by some to be Australia's best table fish. The procedure of catching is the same as for snapper, but the cages are laid in locations around river mouths and estuary training walls and are generally baited with mullet.

The Luderick (*Girella tricuspidata*), also a species of commercial interest, is at times taken in large amounts in the same conventional type of trap, without any bait at all. The traps are laid near rocks or training walls where netting is not practicable, and in some estuaries on the north coast of New South Wales large hauls are made. No one knows why these fish enter the trap unless it is to escape predators, or, one fish in wandering in, acts as a decoy for others.

Estuary Fisheries. The estuaries of Australia have been sources of great quantities of food fish for the home market.

One of the most common and oldest methods of fishing employed in the estuaries is the seine or hauling net, used to encircle a school of fish and haul them from the water. The fish either swim into the net mesh, or, because they have been disturbed, rush into the net. The net is usually made of nylon. It is supported on one end by floats of cork, glass, plastic, or nylon. This is the cork or head line. The bottom of the net is made to sink by attaching lead weights. In some cases, the net may be required to sink to the bottom. It is then referred to as a diver; and for this, the weighting of the lead line must exceed the floatation of the cork line. If the net is required to float off the bottom, appropriate adjustments in the weighting have to be made. Such a net is used to capture pelagic school fish such as garfish (*Hemirhamphus spp.*). This net is particularly suited to shallow waters which have a gradually sloping bottom, free from obstructions.

The fish caught in the Estuarine fisheries include mullet (*Mugil cephalus*), yellow-eye mullet (*Aldrichetta forsteri*) and flat-tail mullet (*Liza argentea*); 90% of this catch comes from New South Wales and Queensland and varies between 10–22 million lb annually. Other estuarine fish are: bream (*Acanthopagrus spp.*), King George whiting (*Sillaginodes punctatus*), school whiting (*Sillago bassensis*), flathead (*Neoplatycephalus richardsoni*), tailor (*Pomatomus saltator*), and different crabs. The methods of capture, in addition to set nets, trawl nets, and gill nets, also include handlines, traps, and dredges.

Beach Seining for Salmon. The Australian salmon (*Aripis trutta*) is not related to the true salmon of the Pacific and Atlantic Oceans of the northern hemisphere. It is a member of the perch family. It occurs seasonally in large shoals on grounds extending along the east coast as far north as Sydney, and up the west coast as far as Perth. It is netted on any suitable ocean beach or in bays or inlets, by means of a hauling seine net. These nets are large and may be as long as 600 fathoms. Many of the larger fishing trawlers turn from their regular fishing, to participate in

TABLE 2. PUBLISHED DATA FOR EXPORTED MARINE PRODUCTS, 1964–1965

Seafoods	Exported to—	Quantity ('000 lb)	Value ($A'000)
Fresh and frozen			
Fish	U.S.A.	2,800	352
Crayfish tails	U.S.A.	7,802	13,601
Whole crayfish	U.S.A.	540	453
Prawns	Japan, U.K., and S. Africa	2,420	1,996
Scallops	U.S.A. and France	1,886	883
Other crustaceans and molluscs	U.S.A. and France	811	561
Total seafoods		17,042	18,206
Other marine products			
Marine animal oil	U.K. and the Netherlands	15,651	864
Marine shells	Germany Fed. Rep., U.S.A.	1,343	451
Pearls	U.S.A.	...	10

netting salmon, while the fish are making their migratory "run," at times working in conjunction with a spotter plane. These vessels generally act as carriers for the net boat, net, and crew. Under these circumstances, the fishermen use a net with a detachable cod end, which is removed when full of fish, sealed off, and taken to the carrier vessel. There the fish are either manually or mechanically brailed aboard. Along the west Australian coastline, the salmon fishermen are highly mechanized. The nets are shot from a net boat and hauled by tractors and four-wheel-drive vehicles. They also use a portable heading and gutting machine, and mobile motor-driven elevators to load fish into trucks for transport up the beach and to the cannery. In the eastern states canneries buy salmon in the round, and the cleaning is done at the factory.

Salmon heads are in great demand in the western Australian crayfishery and sell at about the same price as the fish.

The edible qualities of this fish are not of a high standard while fresh, but when canned improve in flavor and texture. Practically all the catch goes to the local fish canneries. The catch has varied annually from 6 to 12 million lb, and the season also varies from port to port depending on the migratory habits of the salmon.

Export Industries

Australia exports many of her marine products, the most valuable of which are frozen crayfish and crayfish tails. During 1964–65, exports of prawns, scallops, and frozen whole tuna increased significantly, and are continuing to do so. In addition, shell oysters and frozen oyster meat, scallop meat, frozen and canned abalone, pearl shell, cultured pearls, trochus shell, and whale oil, all find overseas markets. The value of the export markets is fast approaching the sum expended annually on marine imports. For 1964–65, the export value was $A20,859,000 as compared with $A28,105,000 for imports.

Western Crayfishery. The western crayfish (*Panulirus cygnus*) is the largest single fishery in the whole of Australia. It is carried on along the west Australian coast, primarily for export. This fish is easily distinguished by a uniform red coloring and continuous simple grooves across the segments of its abdomen. The grounds which are fished off west Australia extend from Bunburry to Shark Bay in inshore and offshore waters, including the reefs of the Houtman Abrolhos, off Geraldton. They cover approximately 8000 square miles within the 45 fathom line and extend between 24°S and 34°S.

The boats used in this fishing vary considerably, according to the exact location. Only 5% of the entire fleet are over 45 ft in length in the northern waters. Much of the "potting" is done in very shallow reefs in which a large vessel with a deep draft is useless. This applies in particular north of 30°S. Nearly half the fleet are specially designed high-speed hulls 15–25 ft in length, called "scooter boats." They work among rocks and nearly submerged bomboras interspersed with many islands. In the south, only 28% of the boats are under 25 ft, while 10% are over 45 ft. Here the main fishing regions are exposed off-shore reefs and banks; consequently, larger boats are needed. Practically all the larger vessels are equipped with pothauling gear and a tipper. Most of the boats also have echo sounders and two-way radios.

The catch is held in a variety of ways. In most of the larger vessels wells are built into the hull. These have seawater constantly circulating through them. The crayfish swim in the wells and so the fishermen can stay out until a load is taken on board.

Some of the smaller boats hold the crayfish in submerged "swimmers," or crates, at their mooring to await transport to the mainland processing works, or, they may bag the crustaceans and unload them daily.

The crayfish is taken by using the conventional lobster pot. However, these vary greatly in both method and material of construction. Cane, batten, wire, stick, and steel are all used to make both the beehive and square traps. The pots are baited with fish, fish heads, sheep heads, cattle hocks, horse meat, or kangaroo meat.

The regulations governing the industry are extensive and strictly enforced. Breaches are severely penalized. The industry is considered as being fished to the maximum.

The number of pots used is controlled by the regulations, and with the exception of large freezer boats, the crew varies from 1 to 4 men. The main base ports are Fremantle, Lancelin, Jurien Bay, Geraldton, and Dongara. The production figures for the past 3 years show a gradual decline from approximately 20 million lb in 1963–1964 to under 18 million lb in 1965–1966. The catch is mainly exported frozen. Some are also exported "whole cooked." This fishery is the most valuable Australian commercial fishery. In 1965–1966 it was valued at about $A11 million, representing 60% of the total crayfish production in Australia, or about 24% of the total value of Australian edible marine products.

Southern Crayfishery. The other important Australian crayfishery is found to the south of Australia, around the Tasmanian coast, the coasts of Victoria and S. Australia, and the islands and rock formations in Bass Strait.

The species of crayfish is the southern crayfish (*Jasus lalandei*) which is found between latitudes 21°–48°S. Fishing is usually restricted to within 40 miles from the coast and waters of less than

50 fathoms. This area is extremely dangerous for fishing craft, and many boats have been wrecked on the reefs and partly submerged bomboras. Furthermore, the weather conditions are at times appalling.

The methods employed are much the same as those of the Westralian industry. Pots are baited, set by a boat, and subsequently hauled. If an area proves fruitful, a vessel will set the maximum legal number of pots permitted and fish the ground for 3–4 days. If the fish are in shallow water, the large vessel will stand off while the gear is set from a small skiff. Conditions must be right. Pots set in a rough or "making" sea can easily be lost. The experience and skill of the individual fisherman are at a premium.

The boats used are mostly built of wood and are diesel powered. They vary between 15–100 ft. All the larger vessels which cover great distances are equipped with echo sounders, two-way radios, and some, automatic pilots. The same methods of keeping the crays alive until marketed or processed, are used, as described for the west Australian industry. The domestic market is supplied principally by whole cooked eastern and southern crayfish. Frozen tails are exported to overseas markets usually through the agency of processing concerns. In 1965–66, South Australia produced over 6 million lb, Tasmania slightly less than 4 million lb, and Victoria 1,648,000 lb.

The season in southern waters is from November to June, but in some areas there is no closed season. The beehive type of crayfish pot is used, and, as in west Australia since the war, pot hauling has now become mechanized.

Abalone Diving. One of Australia's newest fisheries, abalone diving, is growing in importance. The grounds at present being worked are the southern waters of New South Wales, Victoria, Tasmania, South Australia, and those of west Australia to within about 30 miles from Perth.

Prawn Trawling on the Eastern Coastline. Prawns are caught by trawling along both the coastlines of New South Wales and Queensland. Many of the estuaries are also worked to a limited extent, commonly by smaller trawlers. This inshore fishery is strictly controlled by legislation.

The main commercial species are, in order of importance, the school prawn (*Metapenaeus macleayi*) caught during December-April, the eastern king prawn (*Penaeus plebejus*) January–May, and the greentail or inshore greasy back prawn (*Metapenaeus bennettae*).

The prawns are caught by trawling off-shore grounds and estuaries. In the latter, if trawling is prohibited, prawns are captured by beach seine, cast nets, pocket nets, and pocketed scoopnets.

The procedure with the trawl remains the same throughout the prawning industry. Although the fish trawler has sweeplines between the net and boards, to herd the fish into the net, the prawn trawler does not. The cone-shaped net is opened by the trawl boards or doors which shear outwards when the towing pressure is increased. Nets may be as large as 25 fathoms along the headline, although most of the trawlers use 12–20 fathom nets in deep waters. In the estuaries, the size of the net is limited, by regulation, to a maximum of 6 fathoms. Again, in this fishery, although the method remains the same, the gear may vary from port to port depending on local conditions.

The prawns are usually cooked on board, cooled by hosing with fresh sea water, and boxed with plenty of crushed ice. Some vessels are refrigerated. Nearly all carry ice and often brine tanks. To the north along the Queensland coast, and on new grounds off west Australia, prawns are produced for export. The four main species exported are the king prawn (*P. Latisulcatus*) 76%, the tiger prawn (*P. esculentus*) 20%, the banana prawn (*P. merguiensis*) 3%, and the school prawn (*M. macleayi*) 1%. Most of these are exported raw, headless, or cooked with the head on. King prawns go mainly to Japan, South Africa, and the Pacific Islands; tiger prawns to Japan; school prawns to the Pacific Islands; banana prawns to the U.S.A. and United Kingdom. Both local and overseas markets are expanding; the catch has increased from 2 to 13 million lb between 1952–64.

Scallops. Scallops are caught by the dredging method first used commercially in 1919, in Tasmanian waters. They are also found in Port Phillip Bay, Victoria. The main species sought (*Pecten meridionalis*) has suffered a decline in stocks on the Tasmanian grounds and these areas are now yielding poor results. There has been a considerable exploitation of doughboy scallops (*Mimachlamys asperrimus*) in areas now closed, e.g., the D'Entrecasteaux Channel. The Department of Agriculture's Sea Fisheries Division is at present (1967) sampling the area, with a view to the possibility of opening the channel under a permit system which would limit the number of boats taking scallops commercially.

Scallops are also caught in Queensland by prawn trawlers. The trawling is done with otter gear and a slightly modified otter prawning trawl when prawns are scarce. This is normally a secondary industry except for short periods of the year. The grounds worked are the coastal waters from Wide Bay to Yeppoon.

The scallop caught in this region is called the saucer scallop (*Amusium balloti*). There are no regulations controlling this industry at present, and the catch is disposed of, both fresh and frozen, through the Queensland Fish Board and processing firms.

Many suitable areas for scallop remain to be explored and this industry could expand considerably.

Oysters. The Sydney rock oyster (*Crassostrea commercialis*) is one of the finest edible oysters in the world. It is one of the few marine organisms which can be cultivated commercially. It was marketed as soon as the settlement of Sydney was established; the natural beds were stripped and left to recover by natural spatfall. Thus the industry tended to fluctuate according to the rate of growth of the oysters.

1868 saw the introduction of legislation for the regulating of the industry, and the encouragement of oyster farming. The oyster farms or culture areas require suitable culch and space for growth and control of pests, in order to grow the stock productively; the latter results from natural spatting. The farms are found along the coastline of New South Wales, and two main areas in Queensland, Hervey Bay, and Moreton Bay. The production figures for the last few years have been about 13 million lb and of these 90% were grown in New South Wales. The value of the total 1966 New South Wales production from oyster farming is given as $A2,747,550.

Whales. Whaling is the oldest organized fishing industry in Australia beginning in 1803. Originally, this industry depended on the capture of the humpback whale (*Megaptera novaeangliae*) but, by 1962, the population of this species had become so depleted that the International Whaling Commission in 1963 prohibited its capture. The sperm whale is now the only species to be caught in this fishery, and all the whaling stations, with the exception of one at Albany, have ceased operations. During 1965, three whalers caught 668 sperm whales as compared with 711 during 1964. The export value of whale products fell from $A3,052,000 to $A368,000 between 1961–1965.

Southern Bluefin Tuna. Southern bluefin tuna (*Thunnus thynnus maccoyii*) are caught by the live bait and pole method off the coastlines of southern New South Wales, Victoria, and South Australia. With the exception of the bluefin of the northern hemisphere, all the known major species and some of the minor, are found in Australian waters. Commercial fishing for the bluefin began in Australia in 1949.

The boats used in this industry are mostly of a minimum size of 40 ft but they also range in length up to a maximum of 100 ft. The tuna fleet contains some of Australia's finest fishing craft. All the vessels are diesel powered and most have echo sounders, two-way radios, recording thermographs, and automatic pilots. Some of the better vessels are fitted with radar, sonar gear, and are refrigerated. Live-bait tanks and fishing racks are standard equipment.

The live bait and pole method employed to catch the fish is very exciting to watch, and is very arduous for the participating fishermen. While looking for fish, a man is kept in the "look out." He is armed with binoculars and stays aloft for 6–8 hours scanning the ocean for fish. The entire ship's crew depends on this man for their livelihood.

Once a school of fish is sighted the vessel comes alongside the school. The "churner" then throws some of the small fish, which were stored in the bait tanks, among the tuna in an endeavor to make them bite. Once excited, the fish strike wildly and boldly. Immediately, the fishermen standing on the fishing racks, flick a barbless hook with colored feathers attached at the end of an 8 ft leader fixed on a 10 ft Japanese bamboo pole in among the chopping tuna. As soon as a fish strikes, it is jerked from the water with an expert flick of the pole. It flies over the top of the operator and as it reaches the end of the wire, another flick of the pole at precisely the right moment frees the fish from the lure. The fish is thus landed, and the freed lure flies back to the water for another fish.

Sometimes a large run of fish is encountered and then two and sometimes three men will each work a pole to a single communal line and one lure. A heavy leather pad is worn by the operator to take the weight on the end of the pole.

The fish stocks, worked by the Australian fleet, breed in the Indian Ocean, remain in west Australian waters until 2–3 years old, move on to the South Australian and New South Wales grounds where they remain until 5 years old, when they move into deeper waters off the continental shelf. There they are worked quite intensively by Japanese longliners.

In 1963, an investigation into the Japanese method of longlining revealed that, in view of the current cost/price relationship, it would be uneconomical for Australian vessels to take tuna by this means. This is unfortunate, because the Japanese are thus catching tuna which are on the Australian doorstep.

Some Australian boats have designed a modified longline. While the Japanese vessel sets 2000–2500 hooks and then hauls the gear, the Australian longline normally uses only 150–200 hooks. The line is allowed to drift in the current and is patrolled by the boat. When a float is seen to bob up and down, or be pulled under, the vessel maneuvers into position and the relevant portion of the line is pulled to the vessel by a grapple. The fish is taken from the line, the hook rebaited and returned to the water.

Trolling is another method of catching this fish, which is used by boats 20–50 ft long. Many of these vessels operate for tuna only during the "run," and between, return to their normal work.

The gear consists of a number of lines with varying lures and hooks attached. These are towed from the stem of the boat and from outrigger booms. Up to 10 lines are used, varying in length from 3 to 25 fathoms. The lures are towed to where fish are observed, and if possible, around the edge of a "working" school. The fish when striking, hook themselves and are hauled aboard.

Although there have been several attempts to use purse seines, this method has not been successful in Australian waters during the last few years. Lack of suitable gear and lack of knowledge and skill of the method of operating the large net were contributing factors.

Realizing the possibilities, the South Australian Fisherman's Cooperative Ltd. bought an American purse seine vessel, the *Esperito Santo* in November 1965. She weighs 450 tons, is 130 ft in length, and is equipped with a power block and a synthetic net. She is the first suitable vessel to operate here, and keen interest is being shown in the results this vessel attains.

Canned tuna is becoming increasingly popular on the home market, also small quantities of smoked tuna hams and fish meal are consumed locally.

SAFCOL operates canneries on South Australia, Victoria, and New South Wales. Kraft also own a cannery in New South Wales. Approximately ⅔ of the weight of the fish is lost during the canning process. The offal is converted to fish meal. Some frozen tuna is exported to foreign countries. The catching of yellowfin tuna (*Neothunnus macropterus*) is also being developed in New South Wales.

The figures for the 1963–64 tuna catch for Australia (in 1000 lb live weight) total 17,932 and were caught in Queensland; New South Wales, 5689; Victoria, 74; south Australia, 12,085; west Australia, 49; Northern Territory, 5; Tasmania, 29.

The demand for this fish in Australia is indicated by the import figures for this fish during 1964–65, viz., 316,000, value $A90,000.

The Future of the Industry

While those fisheries which have suffered from overexploitation may recover, they can not be expected to develop further. At times, however, there are swarms of pelagic fish in the oceans surrounding Australia. In other countries, pilchards, sardines, and other small varieties of the mackerel family support fleets of vessels engaged in their capture. The further development of the industry may well depend on the long overdue study of these pelagic fish.

In April 1966, funds were provided by the New South Wales Government for the building of a modern steel-hulled fisheries research vessel, at a cost of about $A160,000, for the purpose of scientific investigation of the waters off the New South Wales coast. The findings could lead to the discovery of new grounds and new fisheries being opened up.

The new prawn grounds discovered in the Gulf of Carpenteria and along the coast of west Australia, are already being worked and could be the forerunners of many further discoveries. If this should happen, Australia might well develop a large and important export prawn industry.

The introduction of fish culture in Australia is directed toward the production of fish for sport, and high-priced varieties such as trout and salmon. A farm in Bridport, Tasmania, rear rainbow trout for the luxury market.

There is too the possibility of extending oyster farming into states other than New South Wales. The pearl industry is already showing remarkable growth, which it would appear need only be limited by the availability of naturally grown live pearls.

Finally, the growth of marine flora has recently been commercialized. A factory, built in 1964 in Tasmania, extracts alginates from the seaweed (*Macrocystis pyrifera*). Further similar developments elsewhere are a possibility.

IAN LE FEVRE

Cross references: *Abalone; Sea Scallop Fishery.*

B

BACTERIA IN THE SEA

The study of marine bacteria had its beginnings in the 1880's and has slowly developed during scientific explorations and as marine stations came into existence. However, the body of usable knowledge is still in its infancy compared to the main branches of bacteriology such as those devoted to food, soil, and pathogenic bacteria and the more specialized branches of nutrition, physiology, genetics, and taxonomy. Marine isolates are beginning to be used in these specialized branches, but the field of marine bacteriology is really the microbial ecology of marine habitats. We are interested in the relationships of bacteria to other micro-organisms as well as larger plants and animals and how they influence their microenvironments, and also how their environments, both chemical and physical, affect them. Fundamental questions are: What are marine bacteria? How did they develop in the sea? What are they doing there? What are their relationships with other marine organisms? How does this knowledge help us to wisely develop and utilize our vast resources of the sea as the populations of man continue to develop along the margins of the worlds oceans? How can we keep this environment healthy for both marine organisms and for man? The purpose of this article is to answer these questions to the extent that present knowledge will permit.

All available evidence today suggests that the first forms of life were primitive bacteria which developed in the ancient seas of the developing earth. The atmosphere was not oxidizing as it is today but was reducing and contained hydrogen, helium, methane, ammonia, water, and carbon dioxide, while the forming soil contained inorganic elements which dissolved in the condensing moisture to yield the ancient seas. The energy from electrical discharges, and to a lesser extent solar radiation, in such an atmosphere caused the spontaneous formation of organic compounds, the building blocks of life such as amino acids and organic acids. This material accumulated in the seas to form rich organic soups. Such processes can be duplicated in the laboratory today. Since the earth was sterile or free from life until the first forms developed, there was nothing to compete with the random assemblying of the organic building blocks into self-duplicating living organisms which developed in the oxygen-free atmosphere. As these primitive forms of life consumed and depleted the organic matter of the ancient seas, more complicated micro-organisms developed which could harness the energy from the sun and synthesize organic matter to replace that being consumed by the primitive bacteria. As the photosynthetic populations developed, the fixation of carbon dioxide with the formation of free oxygen helped to change the composition of the atmosphere as did the slow escape of hydrogen and helium. As the earth changed from an oxygen-free and reducing atmosphere to one containing oxygen, the bacteria and subsequent forms of life up the scale living directly or indirectly on preformed organic matter from the green plants also took on aerobic modes of metabolism to yield the type of organisms we know today.

The diversity of bacterial forms and metabolic types appear infinite, filling virtually every environmental niche possible. Metabolizing bacteria are found growing at temperatures of $-2°C$ in seawater and brines between ice crystals to $92°C$ water in hot springs at Yellowstone Park. Oxygen requirements vary from strict aerobes at the water–air interface and in oxygen-saturated water to facultative organisms to obligate anaerobes found only in anoxic basins and sediments. Hydrostatic pressures of 1000 atm (15,000 psi) found at the bottom of the deepest ocean trenches can be tolerated and are apparently required by organisms isolated from these depths. Nutrient requirements range from complex mixtures of organic compounds for the metabolically simpler organisms, which resemble the ancient microbial ancestors, to just the contaminants found in distilled water for the more complex synthetic forms. Others obtain their energy from the oxidation of inorganic compounds or solar energy through bacterial photosynthesis. An organism can be found which will virtually decompose or attack any organic compound. Motile organisms can be seen in such unlikely substrates as acid mine waters and formalin solutions from pickled specimens. Many bacteria are free-living, others are ectocommensals requiring living surfaces, while others are obligate parasites, one type even being dependent upon certain host species of bacteria. This great diversity of bacteria is also the great stumbling block in assessing the true role of bacteria in the marine environment. Bacteria have often been assumed, without any direct evidence, to be responsible for many processes which occur on land. Since we are terrestrial organisms, and the study of microbial processes in soil has been vital to our heritage of living essentially off the land, it is only natural to rely upon this knowledge when we delve into the sea. However, as marine specialists try to work out the food chains, nutrient cycles, and energy flow in the sea they are often at a loss to explain apparent discrepancies. Too many have not stopped to think of the primary differences between land and oceanic regimes.

The most obvious difference between the land and the sea is that in the former the concentration of salts is absent or relatively low and of variable composition, while that of the sea is appreciable and has a very stable composition. It appears that as micro-organisms developed in the sea they developed specific ionic requirements for growth and metabolism. These ancestral forms apparently gave rise to terrestrial organisms which lost these requirements and are at the most halotolerant, while organisms which populated hypersaline bodies increased their tolerance to saturation levels. The morphological, biochemical, and taxonomic types of bacteria found on land and in the sea are in other respects similar or identical. It therefore seems quite logical to assume that the types of bacteria involved in the carbon, nitrogen, and sulfur cycles and other processes in the sea are similar or identical to those on land. This is largely true. However, the relative importance of the processes in the two systems is apparently quite different. This is due to the fact that the processes of organic matter production and consumption in the two environments are basically quite different.

Organic matter production on land is essentially a batch process. After dormancy during winter, seeds or plants sprout and leaf out in spring, grow in summer and ripen in the fall. Although there is some consumption during the growing process, an appreciable part is stored for consumption during winter. Unharvested material slowly decays. In other words consumption on land is largely a detrital economy where everything from bacteria to cows lives largely on dead plant material. Organic matter production in the sea, on the other hand, is essentially an open or continuous system. Production and consumption go hand in hand throughout the year. Winter, especially in temperate waters, is usually equally or more productive than summer. There is no direct storage of organized plant material. It is an economy of animals living on developing plant populations. There is no great seasonal die-off of leaves, grasses, and plants to support a detrital microflora such as is found in soil, except in marshes and intertidal areas. Therefore, the sea is not as dependent upon bacterial or microbial decomposition and regeneration as is the land.

This is also reflected by the concentrations of both organic matter and bacteria. The open oceans support a small but detectable bacterial flora possibly equivalent to that of the desert sand. There are exceptions where localized blooms and productions occur just as oases occur in the desert. As one approaches the shore across the continental shelves, both production and bacterial populations increase until one reaches the bays, estuaries, and marshes where the economies and bacterial populations approach those on land. To understand the role of bacteria in the sea we should look closely at the food chain in the sea.

A simplified food chain is shown in Fig. 1. The primary production of organic matter upon which all other forms of life are dependent, directly or indirectly, is carried on by the phytoplankton and the seaweeds. The latter is usually disregarded as it is largely confined to the rocky littoral and sub-littoral areas and few organisms live directly on seaweeds, However, the seaweeds

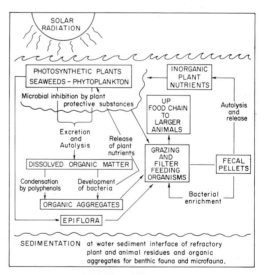

Fig. 1. The hypothetical role of bacteria in the food chain; the transformation of dissolved organic matter to aggregated bacterial cells, the development of an epiflora on solid surfaces, and the enrichment of fecal pellets, which are all potential food for grazing and filter feeding organisms. The source of organic matter for bacterial synthesis is largely from autolysed and excreted organic matter, while the primary breakdown of living cells occurs during ingestion and digestion by predaceous organisms.

as well as the free-floating microscopic plants exude a considerable portion of their photosynthate into the water. This plus organic matter released from defecated and dying plant material by autolysis make up the dissolved organic matter of seawater, the largest fraction of organic matter in the sea. This heterogeneous disorganized reservoir is the closest analogy to the harvested plant stores on land. Even this material, however, is in an open system as it is continually being replenished and constantly utilized to form non-living particulate organic matter for grazing and filter-feeding organisms.

Some of the "organic aggregates" are formed abiogenically by the condensation of soluble polysaccharides and proteinaceous material by polyphenolic substances. These polyphenolic materials are exuded by brown seaweeds which are scattered over the surfaces of the world's oceans as autolyzing bits of matter. This yellow-colored condensing material attains sufficient proportions so that it precipitates from solution as fluffy amorphous material. In addition to these condensates an appreciable fraction of the "organic aggregates" are bacterial in nature. Certain bacteria are floc formers, and bacterial aggregates are common in seawater. This may be due to either the gummy nature of their cell envelope or to the liberation of ammonia during deamination of proteinaceous material, which increases the pH microzonally to force calcium and magnesium salts out of solution.

These fluffy precipitates concentrate dissolved organic matter which supports further bacterial growth. Many organisms including shellfish, various worms, larval fish, and crustaceans may derive an appreciable part of their nutrition from such organic aggregates. The seaweeds themselves, especially on the older parts of the plant, support an appreciable epiflora of filamentous and stalked bacteria. This fauna may be grazed by protozoa or other organisms. This synthesis of particulate organic matter in the form of bacterial cells may be thought of as secondary plant production and is undoubtedly a much overlooked area of organic matter production for higher organisms. The phytoplankton, often called the grass of the sea, is restricted to the photic zone and far outweighs the biomass of bacteria in these areas. However, when one considers that this small bacterial biomass is distributed throughout the entire water column, then even conservative estimates indicate that this source of organic matter may equal that of the phytoplankton.

The importance of bacteria has been ascribed to plant-nutrient regeneration on land and a similar role has been assumed to occur in the sea. On land, nutrient regeneration is largely microbial in nature. The cellulose-supporting structures made refractory by lignification are slowly decomposed by fungi, permitting the released proteins and carbohydrates to be decomposed by bacteria. To a lesser degree this is also true in inshore waters where marsh grasses, sea grasses, and annual seaweed species have an annual die-away. The major plant communities of phytoplankton in the sea, however, do not have cellulose-supporting structures. Although many have silicified or calcified shells, after consumption by filter-feeding organisms the cell contents largely disappear. The fecal material is ejected as bacterially enriched fecal pellets which are subsequently consumed by the same or different species of filter feeders. Soluble nitrogen and phosphorus, the limiting plant nutrients in the sea, are excreted by these grazing and filter-feeding organisms. Attempts to show bacterial regeneration of phosphorus have been futile. Phosphorus liberated from plant and animal material is immediately bound up as polyphosphate in the bacterial cell wall and is only released after autolysis—an unlikely mechanism of regeneration in an open system. Grazing and filter-feeders such as protozoa and zooplankters, however, excrete soluble phosphorus after the ingestion of phosphorus-rich bacteria. Therefore, the assumed role of bacteria in phosphate regeneration was overstated and overlooked a necessary segment of the marine biota. The rapid release of soluble proteinaceous materials by autolysis from senescent plants and dead animals indicates that this may be the primary mechanism of nitrogen regeneration in the sea. Bacteria undoubtedly contribute their share by the depolymerization of polymeric materials and the deamination of amino acids. But this probably occurs mainly from soluble substances released from plants and animals during cell autolysis or digestive processes after consumption.

In addition to the relative absences of seasonal surpluses of decaying materials in the sea, other factors which limit bacterial populations are grazing animals and inhibition by plant-protective substances. In terrestrial plants the presence of protective substances is indicated by the selective feeding habits of insects and the resistance of certain species or varieties to certain diseases. These phenomena are due to secondary plant substances which occur at low but significant concentrations sporadically throughout the plant kingdom. Analogous substances are also produced by many marine plants including the micro- and macro-algae. Snails and sea urchins show detectable preferences for certain seaweed species and the young growing areas of seaweeds are relatively free of a fouling epibiota. Such branch tips of brown seaweeds can be squeezed and the expressed juices are toxic to naked animal forms at dilutions as great as 1:1000 in seawater. Terpenoid hydrocarbons, organic acids, and polyphenolic substances have been shown to be toxic or inhibitory for bacteria, phytoplankters, and larval animal forms. The toxicity of these compounds is minimal compared to the biotoxins of the red-tide dinoflagellates. However, in restricted habitats the less toxic substances accumulate and can reach toxic concentrations. At the plant–water interface they are inhibitory and appear to check the extent of fouling. The type of bacteria which foul seaweeds appear to be uniquely suited for this mode of life. They are either filamentous forms such as *Leucothrix mucor* or are stalked such as *Hyphomicrobium* and *Caulobacter* species. These morphological features remove the organisms from the plant surface but keep them within reach of the excreted algal substances upon which they depend for growth. The more common bacterial forms appear to be held in check.

The land, which occupies less than 30% of the earth's surface, at present yields a thousand-fold the food harvested from the sea. As the cultivation of marine plants and the herding of marine animals is undertaken to utilize the potential of the sea, it becomes necessary to first understand the intricacies of the food web of the sea. Observations that oysters prefer and successfully utilize microscopic non-living plant material rich in carbohydrate indicate that attention should be focused on the mechanism of "organic aggregate" formation, especially from algal polysaccharides, and on ways of duplicating or improving natural processes. Polyphenols which denature immature larval forms are liberated at toxic levels from brown seaweeds at approximately the same time that larval oyster spat are liberated; this fact suggests possible natural limiting factors that can be removed during artificial oyster propagation. The importance of an understanding of microbial ecology for successful mariculture is obvious.

Crowding, which is essential to domestication, is always conducive to disease. This is already evident in a bacterial disease of lobsters that has become endemic in lobster holding tanks. Microbial diseases in natural populations must go largely unnoticed due to the consumption of the weakened members of a population by predators. Overcrowding and the absence of predators in artificial rearing tanks will undoubtedly reveal pathogenic marine forms hitherto unrecognized. This as well as the increased utilization of tropical species which can be highly toxic will force development of the marine health sciences. But marine health sciences will have to go further than this. As man continues to populate and crowd the margins of the world's oceans he not only threatens his own health but that of the marine organisms upon which he will become increasingly dependent. Domestic, agricultural, and industrial pollution is and will continue to adversely affect the estuaries, salt ponds, and marshes, the nurseries of our natural marine populations. The development of mariculture will mean little if pollution abatement is neglected. The beneficial as well as harmful roles of bacteria in the sea will have to be studied in context with other fields in order to more fully understand and utilize our vast marine resources.

JOHN McN. SIEBURTH

Cross references: *Antimicrobial Substances; Microbiology, Marine.*

BENTHONIC DOMAIN

The benthonic domain includes all the bottoms of the oceans from the highest levels of seawater influence, by sprays or waves, down to the deepest part of the oceanic trenches. The marine benthos is the assemblage of organisms that live on or in these bottoms. Some, especially fish and crustaceans of commercial value, live not on the bottom itself but in its vicinity; for this reason they are called nectobenthonic. A given species may be by turns nectobenthonic and truly pelagic; for instance, the European mackerels, although typical pelagic species, spend the winter close to the bottom like a nectobenthonic species; the hakes, which are true nectobenthonic species, swim by night in the pelagos far from the bottom hunting food.

Biomass and Production

As regards the wealth of the benthonic domain, one must consider two concepts—biomass and production.

The biomass, or standing crop, is the amount of living matter existing in a given segment of the bottom, including both the epifauna and infauna. Generally, the biomasses are related to one square meter and concern the fresh (or wet) weight of all plants and animals existing in the community. This is only an approximation because: (a) Except for the pumping systems and the corers, the bottom samplers do not work on the same area on the sediment surface, and at 10–15 cm (or more) of depth, the number of specimens and weight of collected material may be correct for the surface, but less and less accurate with increasing depth of sediment, owing to the shape of the cut made by the jaws of the sampling device. (b) The fresh weight has no value as far as the food chain is concerned; the benthonic species, both vegetal and animal, have very different contents in water and limestone, which are not nutritive materials. It would be better to express the biomasses in dry weight after decalcification; unfortunately this process is much more complex and the data are rarely given in this form. (c) In the biomass, which is only a weight (wet or dry) one generally cannot separate the different trophic levels, which are much too intricate; for example, the standing crop has a very different value for herbivorous animals than for carnivorous.

However, the biomass considered alone is not sufficient to express the richness of a given bottom. Much more important is the concept of production, which is the increment of biomass for a definite time, generally 1 year in the benthonic domain. In other words, biomass is the principal and production is the interest; for a good exploitation of any trophic level the appropriation should not surpass the level of the production. The production of a given species depends on many factors; the most important are the speed, of growth, the age of the first breeding period and the duration of the whole life cycle. Generally speaking, the short-lived species have a more important production in relation to the biomass.

Food Chain in the Benthonic Domain

In the pelagic domain the conditions are rather better than in the benthonic because: (a) The planktonic plants are unicellular and their reproduction rate is generally very high; this means that a very important primary production takes place for a rather small standing crop. (b) The planktonic fauna include a great deal of true herbivorous species (mainly small crustaceans like copepods, ostracods, and many larval stages of benthonic invertebrates) which convert vegetal organic matter into animal (secondary production). Many nectonic species, like the clupeid fish (herring, sardines) feed upon the herbivorous microplankton and represent the third link of the food chain; in their turn these plankton-eating fish are eaten by predators (tuna, marlin, etc.). (c) This food chain of the pelagic domain is rather direct and without much loss; of course, a small herbivorous invertebrate may be eaten, not by a fish but by another invertebrate.

In fact the energy corresponding to one predator, is almost never quite lost from the food chain but only results in a small decrease of the final efficiency in comparison of the planktonic primary production. It seems that this efficiency is about the following: 1.000 g of phytoplankton results in an increase in weight of about 250 g for the herbivorous population; for the third step of the food pyramid a smaller efficiency of about 15% leaves an increase of planktonophagous fish of 37.5 g; with a new decrease of the conversion efficiency (10%) the last step leaves only an increase of about 3.75 g of the stock of carnivorous fish.

In the benthonic domain the efficiency of the food chain, as far as its terminal for human consumption is concerned, is much lower because: (a) The multicellular algae and the phanerogams do not exist on every kind of bottom, even in the Phytal system (euphotic). (b) There are only a few true herbivorous species feeding upon these fresh metaphyts; about 90% of the non-carnivorous invertebrates feed upon "detritus" from such plants both lying on the sediment surface (detritus-feeding animals) and floating in the sea water (suspension-feeding animals). (c) It seems that these microphagous animals cannot utilize the dead and decaying vegetal organic matter itself but the bacteria living upon them; from the point of view of efficiency, these ways are not equivalent because a new decrease of efficiency results from the intercalation of this bacterial link. (d) There is insufficient knowledge of the unicellular algae living on the bottoms, mainly in soft substrata; nevertheless it seems that in some sandy or muddy biotops of the shelf, which apparently are wanting in plants, these algae (chiefly diatoms), which are eaten in fresh condition, play an important part in the diet of the so-called detritus feeders and suspension feeders; the primary production from unicellular benthonic algae varies with the characteristics of the bottom. (e) Unlike the pelagic domain, the tertiary level of the food pyramid in the benthos includes many animals which feed upon the same prey (herbivorous and detritophagous invertebrates) as the fish; these predator invertebrates not only compete with the fish for available food, but also generally cannot be eaten by the carnivorous fish. For instance, a small bivalve or worm may be eaten by a plaice, a gurnard, or another benthonic fish, but may be also eaten by a crab or an octopus. As most of these great invertebrates are of no value for human consumption nor as food for fish, an important part of the secondary production is deflected from the food chain carrying the energy up to fish consumption by man.

Quantitative considerations are gerite hypothetical. Unfortunately, knowledge of the food chains is insufficient, even in the most general features, especially in the benthos, as to the part played by microphytobenthos (diatoms) and bacteria. This latter peculiarity, and the many "lateral escapes" of energy from the direct food chain leading to the fish of commercial value, explains why the estimation of the wealth of a given area of the bottom cannot be found by primary-production measurements, as in the pelagos. Moreover, the planktonic production may interfere with the food chain of the benthos itself. We may only assert that the food chain in the benthonic domain is less efficient than in the pelagic, and that this inequality is reflected in the yields of the respective fisheries.

Biomasses and Production of Different Bottoms

Generally speaking, the biomass of the macrobenthos is more important on hard bottoms than on (and in) soft ones. On the shelf, one often observes a decrease of the biomass with increasing depth; the vegetal part of the biomass is important only in the most shallow waters, but some metaphyts and many unicellular photoautrophic algae (benthonic diatoms) may exist down to 200 m in some places.

On shallow-water hard bottoms the biomass of the belts of brown seaweeds (*Fucus, Laminaria*) may reach 15–20 kg/m^2 (wet weight) and their annual primary production may be about 3–4 kg/m^2. The mussel community, which is very common on the rocky coasts on both the sides of the north Atlantic, has generally a biomass (wet weight) of about 20–25 kg/m^2 (sometimes up to 80 kg/m^2). Nevertheless, when the biomasses are in dry weight after decalcification, the values for the mussel beds and for the communities of brown seaweeds are not very different (1.5–2.5 kg/m^2) owing to the high content of limestone.

Rather important biomasses are also found on rocky substrata of the deeper part of the shelf (circalittoral zone); for instance, in the Mediterranean the population of red coral (*Corallium rubrum*) or some gorgonarians (*Muricaea chameleon*) have biomasses of 500–1000 g/m^2 (dry weight after decalcification).

On the soft bottoms of the shelf, the biomass is less important. Some bottoms of the shelf of the Okhotsk Sea have a biomass of about 1000 g/m^2, and some communities of the North Sea, with a great deal of small bivalve mollusks (*Spisula*), may reach a biomass of more than 3 kg/m^2.

With increasing depth the biomass decreases very quickly, especially when the shelf is rather wide. The values for the communities living on hard bottom are unknown, but, as far as soft bottoms are concerned the average biomass is 1–10 g/m^2 (net weight). On the slope and abyssal plains, except perhaps in some privileged areas, like the abyssal plain of the Atlantic offshore of the United States, the biomass may diminish to a few centigrams, or even milligrams in the deepest trenches. Owing to the lower temperature of the deep waters, the secondary and tertiary production of the abyssal and hadal zone are probably very small and no fisheries may take place in such depths.

Present Status of Benthonic Marine Resources

The statistics of the world sea fisheries are very approximate, not only because some fisheries give irregular results from one year to another, due to weather conditions, commercial difficulties, overfishing, etc., but also because many fishermen, especially in tropical countries, do not want to present their catches to a governmental fisheries officer even if he exists. Probably for 1965 the total catches of marine products are about 50 million metric tons. This amount includes the following components:

Fish	42.5
Mollusks	3.0
Crustaceans	1.2
Other invertebrates	0.2
Seaweeds	0.6
Whales, etc.	2.5

As far as the fish are concerned, about 50% of the total amount are the clupeids (herring, sardines, anchovies, etc.) which are typically planktonophagous and pelagic, and about 8% another natural group of pelagic carnivorous, the scombroids (tuna, mackerel). The group of the flatfish (plaice, sole, flounder, halibut) with only 3% concerns only typically benthonic species. In the 22% of "various" there are some important benthonic groups like the "red fish" (*Sebastes, Sebastodes*). The gadoids, with about 15% of the total world catch, constitute the single natural group of benthonic fish of great commercial value. In the *Gadidae* family, the most important species is the cod (mainly the Atlantic one) which is really a nectobenthonic species, eating many swimming animals, but sometimes also bottom invertebrates, and the haddock, which is a typical benthonic species whose diet almost exclusively consists of bottom invertebrates. The closely related family of *Merlucciidae* includes the different species of hake, which live on the deeper parts of the continental shelves and on the upper levels of the slope; it seems that they are on the bottom only by day and have a true nectonic behavior by night.

As far as the Mollusks are concerned, about 40% of the total catch is cephalopods, mainly squids, which are pelagic; the catches of the benthonic octopuses are less important. The gastropods only represent 1% (abalone and some small snails). The bivalves are far the most important resource of the mollusks with 59% of the total; in fact, most of the landings result from farming (oysters, mussels, clams), but the natural banks are also important, for instance, for scallops.

In the Crustaceans the catches of shrimp represent about 62% of the total and the landings are more and more important due to their high commercial value, and because they are easily preserved by freezing, canning, etc. The main areas of important shrimp fisheries are in the subtropical shallow waters (sometimes brackish), Gulf of Mexico, north-eastern and central Atlantic coast of South America, India, etc. There are also many interesting species of the same family (*Peneidae*) on some muddy bottoms of the upper part of the continental slope

(Mediterranean, western coasts of Europe and North Africa, etc.). Among the "crabs" which represents about 27% of the crustacean landings, the main species are "false-crabs" of the *Lithodidae* family belonging to the group of anomuran, which are closely related to the pagurids; these "King crabs" are cold-loving species, rather rare in the north Atlantic areas, but support very large fisheries in the north Pacific, especially in western areas; some other species exist on the coldest areas of the Chilean coast.

Among other marine products of invertebrates there are some catches of low and local interest like ascidians (some species of *Microcosmus* in Mediterranean and *Pyura* along Japanese coasts), sea-urchins (chiefly in Mediterranean for the species *Paracentrotus lividus* and Japan), holothurians (the Chinese and Viet Namian sea cucumbers "trepang").

The Future of Benthonic Marine Resources

Although in the total landings of marine products the pelagic species represent about 75% and the benthonic products only 25%, it is necessary to promote the demersal fisheries, inasmuch as they often are of a greater and more direct benefit for the populations of some countries where the hunger problem is crucial; indeed, a great increase of landings of pelagic species, generally needs well-equipped ships (echo sounding, freezing) which are limited to advanced countries or governmental offices, because of the capital involved.

Of course, for the fish themselves which represent the most important part of catches of benthonic origin, over-fishing already occurs in the shelf areas where intensive trawling takes place, and the mean value of landings of the commercial species does not rise with the fishing effort; generally, in the same time the average size of specimens of a given species is steadily decreasing. This low efficiency of the trawling originates from insufficient knowledge of the productivity of benthonic communities. When the first sign of overfishing on the shelf populations of the north-eastern Atlantic appeared, some trawlers turned their activities to slope bottom communities; but these, which live in colder waters than the shelf communities, certainly have a lower productivity and the overfishing will occur sooner. Fortunately, many shelf areas do not yet support fisheries, mainly in the intertropical zone, but we ought to obtain an approach of benthonic production values of different bottoms and commercial species before engaging new fishing efforts. In every way, due to the sharp decrease of benthonic standing crop and production with increasing depth, fisheries of demersal species probably cannot take place below 1000 m.

The problem of farming is very difficult except for mollusks; for the sessile or sedentary bivalves the methods of farming for oysters and mussels are rather well known and some may be put in practice for the oysters of mangrove swamps of tropical muddy shores which are now almost unexploited. Farming clams, which are not sessile but sedentary in soft bottoms, is more difficult but encouraging results have been obtained recently in the U.S., and the most common species (*Mercenaria mercenaria*) has been acclimatized on British and French coasts.

As far as the crustaceans are concerned the improvement of the biological cycle and ecology of some shallow-water shrimps is a good beginning, but, for lobster and spiny lobster, the long larval stage will probably prevent any true farming. Recently Russian scientists tried to acclimatize the far-east cold-loving species of king crabs (*Paralithodes*) to the European northern seas.

Farming demersal marine fish is perhaps the most difficult, but also the most attractive. More than 50 years ago C. G. J. Petersen succeeded in obtaining plaice (*Pleuronectes platessa*) of commercial size in some Danish fjords; the young specimens, caught in the North Sea and transported to these shallow-water areas, find there a great deal of small prey and grow more quickly than in natural conditions; recently British scientists obtained the whole development of the same species, but this promising result is not yet of practical value.

It seems that efforts to obtain the whole development of a given species from egg fertilization up to commercial size are more realistic than attempts to increase the production of an important area of bottom. Experiments in discharging nutrient salts in some well-delimited coastal areas, like Scottish "lochs," gave poor results; this is not surprising due to the features of the food chain in the benthonic domain correlated with the low energy transformation efficiency from one link to the next, and the great number of "lateral losses" of the food chain.

J. M. Peres

Cross reference: *Abyssal Zone*

BIOCHEMICAL COMPOSITION OF FISH

Although in recent years there have been an increasing number of research projects devoted to the study of the biochemical composition of fish, the subject is not exhausted. Several complicating factors tend to make specifics impossible, not the least of which is the variation in composition found from species to species. In addition to species variation, several subfactors produce variations within a species. These subfactors are: (1) season of catch; (2) geographic location of catch; (3) size and maturity; (4) sex of fish; (5) food; (6) state in the reproductive cycle; and (7) condition of the fish. Nevertheless, it is possible to make some generalizations regarding the kind and relative amount of the various biochemical components of fish.

Proximate Composition. One of the more extensively studied aspects of the biochemical composition of the fish is their proximate composition. Proximate composition is a measure of the oil, moisture, protein, and ash content of the edible portion of a species (or, in some cases, of the total fish). The moisture content may range from 66–84%; the oil from 0.4–25%; the protein from 13–24%; and the ash from 0.5–2%. The edible portion of fish contains little carbohydrate.

The greatest variation, both within a single species and between species, is found in the oil and moisture content. These two components bear an inverse relation to each other, with the sum total usually approximating 80% of the total weight of the fish. Within a single species, the oil content varies throughout the year, usually reaching its peak just prior to the entry of the fish into its reproductive cycle. Nevertheless, it is possible to categorize fish into two groups: (1) lean fish, the oil content of the edible portion never exceeding 5% (e.g., cod, haddock); and (2) fatty fish, the oil content ranging between 5 and 25% (e.g., herring, menhaden).

Although there is considerable variation in the protein and ash content among various species of fish, these two components tend to remain stable in amount within a single species. The protein content of a species is dependent upon the condition of a fish at the time of capture. That is, if a fish has been relatively active while at the same time feeding relatively little, the protein component will decrease to a certain extent. In this respect, one of the more striking species is the salmon. At the beginning of the spawning run of salmon, its protein content is approximately 20% of its total weight; after the strenuous trip up the river to the spawning grounds, its protein content may fall as low as 10%. For most other species, however, fluctuations in protein content throughout the year are more nearly in the order of 2%.

The crude protein content of fish is usually reported as the percent of nitrogen multiplied by the factor 6.25. This is the accepted method of converting an easily measurable quantity (nitrogen) to a quantity measurable with more difficulty. It is in error by that part of the nitrogen measured that is not bound into a protein component. The nonprotein components consist of free amino acids of the protein type, volatile nitrogen bases (ammonia), non-volatile bases (trimethylamine oxide), creatin, taurine, the betaines, uric acid, anserine, carnosine, purine derivatives, and histamine. Not all of these, with the exception of the free amino acids and ammonia, are present in all species of fish. The amount of the non-protein nitrogen components will also vary with the season, the food of the fish, the condition of the fish, and the part of the reproductive cycle in which the fish is when captured.

Fish muscle is composed of the same fractions of protein as is mammalian muscle. Fillets are composed of segments of muscular tissue separated by fiber septae (myosepts). Extracellular protein (collagen or elastin) is present in a lesser amount than in mammals. Intracellular protein consists of actin, myosin, globulin, and myoalbumin. Present also are compounds of the adenosine triphosphate series plus creatine phosphate. The flesh of lean fish is notably low in purine and pyrimidine nitrogen. The distribution of ribonucleic acid and deoxyribonucleic acid is similar to that of mammals. The milt of various species contains nucleoproteins, which when hydrolyzed, give rise to nucleic acids and protamine of a species-specific type.

Amino Acid Composition. The free amino acid content of the fillets of various species shows patterns differing sufficiently among the species that a species can often be identified by its own particular amino acid pattern. The protein amino acid content, however, varies to a lesser extent. By and large, fish contain amino acids, essential to human nutrition, in amounts comparable to or greater than mammalian muscle. These amino acids include histidine, isoleucine, leucine, lysine, methionine, threonine, tryptophan, and valine. The amounts of lysine and histidine are somewhat higher than are those found in mammals, whereas the amounts of methionine and trytophan are sometimes lower. Of the amounts of non-essential amino acids, the valine content is generally somewhat higher, and the phenylalanine content somewhat lower, than the corresponding amounts found in mammalian tissue. The hydroxyproline content of fish tissues is much lower than that of corresponding mammalian tissues, as is the proline content; on the other hand, arginine values are greater. A typical amino acid analysis of ocean perch fillets—irrespective of season, sex, and maturity (all of which can cause variations)—is shown in Table 1.

Mineral Composition. The mineral content of fish flesh is also subject to the same variations as are the other components. The sodium content of fish not subjected to brine treatment ranges from 37 to 90 mg/100 g. The potassium content

TABLE 1. AMINO ACID CONTENT OF OCEAN PERCH FILLETS WITHOUT REGARD TO SEASON OF CATCH, SIZE, OR SEX

Amino acid	Concentration of amino acid μ moles/mg of nitrogen
Hydroxyproline	0.22
Aspartic acid	4.88
Threonine	2.41
Serine	2.60
Proline	1.86
Glutamic acid	6.30
Glycine	4.39
Alanine	4.42
Valine	2.67
Cystine/2	0.30
Methionine	1.39
Isoleucine	2.77
Leucine	3.85
Tyrosine	1.43
Phenylalanine	1.89
Hydroxylysine	0.03
Lysine	4.75
Histidine	0.83
Arginine	2.10
Tryptophan	0.52

is somewhat higher (230–500 mg/100 g) and is usually about 5 times that of the sodium content of a particular species. Fish flesh is considered to be relatively rich in phosphorus (142–301 mg/100 g), calcium (5–200 mg/100 g), sulfur (58–272 mg/100 g), chlorine (70–325 mg/100 g), and magnesium (28–60 mg/100 g). These latter minerals are subject to considerable variation, however, with respect to both season of catch and species of fish.

Iron content of fish flesh is 0.4–5 mg/100 g, with the dark meat of certain species containing as much as 20 mg/100 g. The copper content of most fish lies within the range of 0.04–0.6 mg/100 g. Manganese, zinc, lead, arsenic, nickel, cadmium, chromium, lithium, cobalt, molybdenum, and selenium are also found in trace amounts. Fish muscle is relatively rich in iodine and fluorine. The range of iodine content is 0.01–0.5 mg/100 g—a much higher value than that reported for mammalians. The range of fluorin is 0.5–1 mg/100 g.

Lipid Composition. The most striking biochemical difference between fish and mammals lies in the area of the fatty acid composition of the oil. Marine and freshwater animals possess a common characteristic—a high degree of unsaturation in the fatty acids composing the oil of the species. Here, again, variability exists due to the species, the diet, the temperature, and the salinity of their habitat, and the period of the reproductive cycle in which the fish is caught. In general, however, marine oils contain large amounts of fatty acids of the polyunsaturated type (more than two double bonds). A large proportion of the polyunsaturated fatty acids of saltwater fish contain more than 20 carbons and from four to six degrees of unsaturation. Relatively little, usually less than 20%, of the fatty acid component consists of the saturated acids such as palmitic acid and stearic acid. Fatty acids of freshwater fish tend mainly to be composed of the somewhat shorter 16- and 18-carbon chains.

Certain species, notably the mullet, contain a large proportion of odd-numbered carbon fatty acids in both body and liver oils. Of these odd-numbered carbon fatty acids, the most predominant chain length is that with 17 carbon atoms. However, chains of 13-, 15-, 19-, and 21-carbon atoms occur in varying quantities. Similar degrees of unsaturation exist in these fatty acids as exist in the more common even-numbered chains. The greatest portion of fish oil is found as a triglyceride component; although free fatty acid, monoglyceride, and diglyceride fractions exist. Recently the existence of large quantities of wax esters of fatty acids has been confirmed in both the roe and flesh of certain species. Approximately 0.5–1% of the total weight of a fish is phospholipid in nature. In some cases, phospholipid material represents all or nearly all of the lipid material present in the flesh. Lecithins are generally found to be the most abundant type of phospholipid, with cephalins, sphingomyelins, and phosphoinositides following in descending order of abundance.

The unsaponifiable portion of the oil contains a number of different sterols, cholesterol and its esters predominating. The total amount of cholesterol, however, seldom exceeds 100 mg/100 g of body weight, the average quantity being somewhat closer to 70 mg. Squalene is found in unusually high amounts in the livers of several species, notably some species of the shark family. Saturated and unsaturated alcohols are found in considerable quantity in liver oils of some species. Also, certain fish-liver oils contain large amounts of the oil-soluble vitamins—A, D, E, and K.

Vitamin Composition. Vitamin A is found in the oil components of fish in the viscera, muscles, and membranes, but is usually most highly concentrated in the liver. Several types of vitamin A are found, vitamin A_1 (axerophthol) being present in almost all species. Neovitamin A_1 can be found in dogfish, cod, and halibut. Freshwater and anadromous fish contain vitamin A_2, whereas sharks contain subvitamin A. Retinene is also found in the roe and milt. Certain species contain over 20,000,000 I.U. of vitamin A/100 g of liver oil, whereas others contain virtually none. The flesh of lean fish is quite poor in vitamin A (less than 50 I.U./100 g,) whereas that of fatty fish may contain as much as 4500 I.U./100 g. The amount of vitamin D present in fish flesh varies from 500–3000 I.U./100 g, whereas that in the liver may reach as high as 76,000 I.U./100 g of oil. The distribution pattern of vitamin D closely resembles that of vitamin A. In the flesh of fish, vitamin E is largely found in the form of α-tocopherol. Vitamin K is also found in fish, fish meal being a somewhat rich source.

The water-soluble vitamins are found in rather large quantities in the flesh of most fish. The red meat of fish contains more thiamine than does the white meat. The values for this vitamin range from 0.01–0.25 mg/100 g. In certain species of fish, an antithiamine factor, thiaminase, is present. The presence of this substance, which is heat labile and therefore can be destroyed by heat, predominates in freshwater fish, but it occasionally occurs in marine fish. Riboflavin, vitamin B_{12}, and pantothenic acid are also found to a greater degree in the red flesh than in the white flesh. The amount of riboflavin varies from 0.001–0.330 mg/100 g; vitamin B_{12} 0.05–14 μg/100 g; and pantothenic acid 0.09–2.1 mg/100 g. Niacin is present in appreciable quantity (0.01–14.8 mg/100 g) as is pyridoxine (0.03–0.96 mg/100 g). Somewhat smaller amounts of biotin (5–8 μg/100 g) and folic acid (0.5–87 μg/100 g) are also found in the flesh of fish. Vitamin C (ascorbic acid) appears to be present in small quantities in the flesh (1–5 mg/100 g) of most fish, although the quantity can reach a greater amount in the liver (27 mg/100 g of cod liver) and in the roe (10–180 mg/100 g).

Pigment Composition. There is a tendency for fish to concentrate xanthophylls rather than carotenes. The pigments of fish are largely located in the skin, with only a few species, such as salmon, having a colored flesh. Several pigments are associated with the various colors represented in the skin—astaxanthin, zeaxanthin, and lutein being the most common. Other carotenoids that have been described as being present in fish are taraxanthin, violaxanthin, fucoxanthin, and tunaxanthin. The carotenes, β-carotenes and α-carotenes, are also sometimes present.

MARY H. THOMPSON

BIOLOGICAL CLOCKS IN MARINE ANIMALS

Biological clock is an expression which was coined to describe the unknown means by which living things time rhythmic variations in their diverse bodily processes—changes that normally harmonize them with their rhythmic physical environment. More specifically there are the solar daily "clocks" adapting activities of each species to the day–night rhythmic changes in such factors as light, temperature, and the availability of their natural foods. Other "clocks" adjust rhythmic activities of animals and plants to the ebb and flow of the tides of their home beaches, tides which themselves are predominantly timed to the 24 hour and 50 minute lunar day, though solar daily influences modulate the cycles to varying degrees. Comparably, there are monthly clocks of organisms and annual ones too, these longer-cycle ones most commonly regulating reproduction and breeding rhythms.

Among the most spectacular demonstrations of clock-timed phenomena are the reproductive rhythms of marine organisms that assure that reproduction will be an orderly regulated process in which the production of eggs and sperm will be synchronized in a population. Concomitant swarming rhythms assure that these reproductive elements will be present in abundance at the same time and place, thereby maximizing egg fertilization and species maintenance.

One well-known example of a breeding cycle in a marine species, timed by a solunar calendar clock system, is that of the "fireworms" of Bermuda. These worms swarm from their coral hideouts at a particular season of the year, at a specific phase of the moon and hour of the day. It is possible to predict times in the year to within a few minutes that one will be able to observe the females swarming at the water's surface, and luminescing a brilliant signal to the waiting males the moment they are about to discharge their eggs into the sea. The prediction depends only on a fore-knowledge of the specific rhythmic schedule of the species in a given area. The closely related palolo worm of the southwestern Pacific ocean is also an oft-quoted example.

Another very familiar case of a comparable, uncannily precise, solunar, clock-timed breeding behavior is that of the grunion of the California coast. At specific times of year and of evening high tide, so dependably regular that the times may be published in advance in local newspapers, these small, edible fish swarm in large numbers to sandy beaches where they ride the waves to their highest point just as the highest, high tides of the month are commencing to ebb. There the males and females deposit their gametes together in pits they drill in the sand with their tails. The young then develop in the warm, moist sand while awaiting the next equally high tide to permit them to escape to sea. Since the variations in tidal height are effected by simultaneous gravitational influence of sun and moon, a nice solunar clock system must be operating.

A Japanese sea lily has been reported to display an annual breeding rhythm correlated

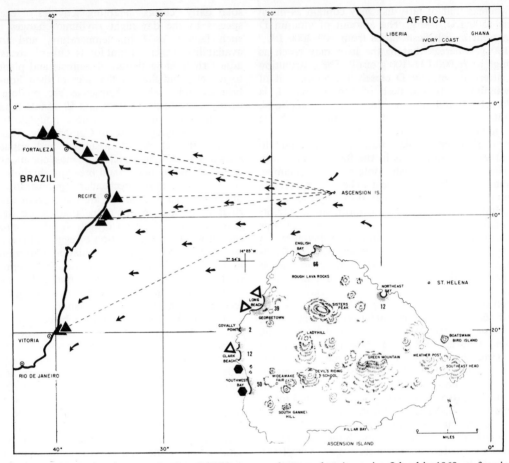

Fig. 1. Paths of migrating green turtles. Of 206 green turtles tagged at Ascension Island in 1960, so far nine were captured on the coast of Brazil, 14,000 miles away. In 1963, three of the turtles tagged in 1960 returned to the island and dug their nests in the same beaches they had used before. In 1964, two more tagged turtles appeared; one missed its home beach by a few hundred yards. All had presumably come from Brazil. (Courtesy of Dr. Archie Carr, University of Florida.)

in an extraordinary manner to a specific sequence of moon phases, time of year and of day. This echinoderm breeds progressively earlier over a monthly interval in the fall following an orderly pattern of specific moon phases for 18 years, at the end of which time it abruptly resets to its starting point. The "clock" which is involved thus appears to reflect the 18-year "beats" between the natural annual and synodic monthly periods.

Plants too, may depend upon calendar-clock-regulated reproductive cycles. The brown alga, *Dictyota*, breeds at particular times adjusted to time of year, phase of moon, and time of local tide.

Another kind of activity in which there is accumulating evidence that biological clocks normally participate is the diurnally rhythmic migrations of planktonic organisms. Still another type of role of the clocks is in permitting such organisms as fish, crustaceans, and birds to home or to navigate by using the sun, moon, or even the constellations as compass references. Such celestial orientation, termed astrotaxis, obviously requires some precise means of chronometry in order that the angle of the organisms relative to the reference points may be steadily corrected to compensate for the rotation of the earth and thus permit useful geographic orientation.

Living creatures clearly possess an extraordinary ability to tell time of day, tide, month, and year. Searching for the means by which organisms do this, biologists turned first to the possibility that these were just responses timed by the variations in sunlight and moonlight, or the solunar-regulated ocean tides. But early in the 18th century it was discovered by a French astronomer, M. DeMairan, that plants would continue the rhythmic daily "sleep-movements" of their leaves even when shielded from all light changes. Near the turn of the present century, two English biologists, F. Gamble and F. Keeble, and a French one, Georges Bohn, learned that the flatworm, *Convoluta*, which displays in nature a tidal rhythm of vertical migration to and from the surface of the mud, would continue to do so with remarkable tidal precision in a vessel on a laboratory bench away from the tides.

Bohn and a research associate also found similarly persisting tidal rhythms in several other marine creatures. A snail that inhabited the upper level of a seashore reached by the water only at the times of the semi-monthly spring tides exhibited a 15-day rhythm even when deprived of all obvious tidal information.

Emmanual Faure-Fremiet, another French biologist, found that even single-celled, littoral inhabitants—certain ciliates and diatoms—could continue their normal tidal activity rhythms while shielded from tides in the laboratory.

Fiddler crabs (*Uca*) that abound along seashores in many parts of the world offer an impressive illustration of a fundamental solunar biological clock system. These crabs have been intensively studied in recent years. In the field, the skin of these crabs changes color with time of day, darkening as the sun rises and paling as the sun sets. At the same time, the rhythmic habit of foraging for food is synchronized to the tides. The crabs wander actively over the beaches exposed by the ebbing tide and rest in their burrows over high tide. Shielded experimentally from both light and tidal changes, the crabs continue to change color employing their solar-day clocks and govern their running activity by their tidal clocks.

By periodically sampling from large populations of crabs kept away from all 24-hour light cues for periods up to 2 months, it has been shown that the daily rhythmic color changes of populations of crabs may remain accurately set to the outside times of dawn and dusk. This daily rhythm of color change is influenced by the tidal clock. The degree of daytime darkening was found to depend upon the tidal state on the animals' home beach. Greatest darkening occurred just as the tide was reaching its lowest. On a Cape Cod beach, with its semi-diurnal tidal pattern, low tide occurs at closely the same time of day once a fortnight; correspondingly, the crabs' solunar clock system provides them with an accurate semi-monthly rhythm in color change.

It is common knowledge that a biological system accelerates with rising temperature, and vice versa. Experiments were performed in an attempt to speed or slow the crabs' timing system by keeping crabs at different constant-temperature levels ranging from 6° to 26°C. The

Fig. 2. Fiddler crab. The male fiddler crab displays during periods of low tide by waving the large claw. The display is thought to be part of a ritualized courtship by which the male announces his presence to the female. The crabs are male specimens of *Uca rapax* photographed on the eastern coast of Florida. (Photo: Franklin H. Barnwell.)

periods of the clocks remained unchanged. Whatever comprised the solunar timing system, it was one that was most unconventional for a biological system with respect to such temperature independence. Indeed, the clock's very existence was contrary to the usual ecological concept of accounting for all in terms of a continuing interaction of organisms with such factors as temperature, light, foods, and the chemical character of their environment.

It is now widely conceded that there is no proof of the presence inside each organism of autonomous timers for the biological clocks. All these phenomena can be accounted for without need of postulating independent internal timers. In addition, the simultaneous existence of the many precise tidal, monthly, and annual rhythms, now well documented to persist under conditions the biologist has always presumed to be "constant," strains further the plausibility that these depend exclusively upon internal timing.

It has been impossible to date to get unambiguous evidence for the existence, or absence, of an independent internal timer complex for the persistent "calendar-clock" rhythms of organisms. All the evidence for basic internal timing could not exclude an interpretation of

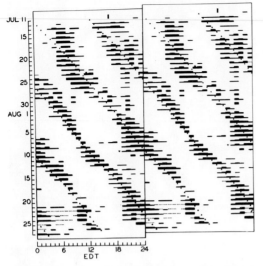

Fig. 3. Record of the walking activity in a fiddler crab, *Uca minax*, from Woods Hole, Massachusetts, automatically recorded for 48 consecutive days. The crab was confined to a small box in the laboratory in front of a window, where it was exposed to the natural daily changes in illumination. In the graph the height of the blocks indicates the amount of activity during each hour. The dotted lines signify periods when the recording system failed. The same triangles show the times of tide at the marsh where the crab was collected. The record is duplicated so that one can more readily visualize the movement of the periods of activity across the 24 hours of the day.

The record demonstrates a persistent tidal rhythm in the activity of the crab. The graph shows, however, that the level of activity of the crab is strongly influenced by the time of day, those times when the tide-related activity occurs at sunset being particularly favorable for high levels of activity. (From *Biological Bulletin*, **130**, 4 (1966).)

the same evidence in terms of basic external timing. In other words, the "clocks" remained interpretable in terms of either inherited, autonomous endogenous oscillators approximating all the natural terrestrial cycles or of the common employment by all living creatures of the public rhythmic, environmental information. As a result, a controversy continues to prevail as to whether the "endogenous" rhythm explanation lies in the existence in organisms of discrete physico-chemical systems of oscillations or instead, at the ecological level depending upon a continuing interaction of living things with their rhythmic physical environment.

Biological "clocks" have been demonstrated not only in the organism as a whole, but also in parts of organisms. The single-celled alga *Acetabularia* has "clocks" independently in each its nucleus and cytoplasm.

In view of the inability to exclude either internal or external timing, there is no proof that a biological clock has ever been experimentally speeded or slowed even though by numerous means one can suppress or reset the observed rhythms that are normally clock timed. Extensive, detailed searches for a chemical or physical system in the organism fulfilling the demands of the observed clock properties have been unsuccessful.

It is becoming increasingly probable that the clocks of life depend steadily upon responsiveness of living things to their subtle, rhythmic geophysical environment. Indeed, some characteristics of biological rhythms have been discovered that are either impossible to account for exclusively in terms of independent internal timing or seem improbably explained thus.

For example, oysters with maximum shell-opening occurring near times of local high tides in their specific habitat have been observed gradually to reset their times of maximum opening to times of upper and lower transits of the moon when kept in uniform conditions in the laboratory and thereafter to retain this relationship. The oysters seemed perhaps to substitute the atmospheric tides for the ocean ones.

Fiddler crabs collected on two beaches with times of low tide 4 hours apart, and kept separated in the laboratory, gradually became phase-synchronized in the particular lunar relationship of those crabs from one of the beaches where low tide had coincidentally occurred close to the times of upper and lower lunar transits.

In summary, possession of a remarkably precise calendar-clock system which continues to operate in the absence of all obvious cues is a very widespread, probably universal, attribute of terrestrial life. It is conceivable that living things possess fully autonomous, internal timing systems measuring relatively accurately the duration of all of these natural periods. However, the possibility that all the biological "clocks" depend upon external timing has not been excluded. A number of recently observed properties of persistent biological rhythms cannot be plausibly explained exclusively in terms of autonomous "endogenous" timers even should such timers be ultimately shown to exist. We are compelled to conclude that the solunar clocks of life are timed either by internal timers interacting with pervasive, subtle, external ones, or that they are fully dependent for their observed timing precision on steadily continuing response to the rhythmic, subtle, geophysical environment.

FRANK A. BROWN, JR.

References

1. Aschoff, J., Ed., "Circadian Clocks," 1964, Amsterdam, North-Holland Publishing Co.

2. Brown, F. A., Jr., "Biological Clocks and the Fiddler Crab," *Sci. Am.*, **190**, 34 (1954); "The rhythmic Nature of Plants and Animals," *Am. Sci.*, **47**, 147 (1959); "Biological Clocks," 1962, Boston, D. C. Heath Co.
3. Cloudsley Thompson, J., "Rhythmic Activity in Animal Physiology and Behavior," 1961, New York, Academic Press, 236 pp.
4. Sollberger, A., "Biological Rhythm Research," 1965, Amsterdam, Elsevier Publishing Co., 461 pp.

BIOTOXINS, MARINE

In a wide sense, marine biotoxins are specific substances produced by marine organisms that are capable of seriously impairing living processes and, in some cases, of destroying life. Usually the toxins exert their deleterious effects in species different from those in which they are produced. Indeed, many compounds which are highly toxic to some organisms are normal constituents of the chemical machinery of the bodies of others. In some organisms toxic compounds are produced which appear to serve no definite purpose and are possibly byproducts of metabolic processes peculiar to the organisms which elaborate them. However, numerous marine organisms produce toxins used specifically for offense or defense.

Among these organisms are the venomous marine animals, generally regarded as those in which a toxin or toxins are produced in association with an apparatus—the venom apparatus—capable of actively injecting the toxic material. Some venomous marine animals use their venom apparatus primarily to immobilize their prey, others solely as a defense mechanism. There are also numerous marine organisms which elaborate potent toxins but do not possess a venom apparatus. Some of these, particularly those which are sluggish, have sedentary habits or are attached to the substratum when adult, release toxic materials into their immediate surroundings. These may be termed toxin-releasing organisms. Some release toxic material when danger threatens in order to deter enemies from molesting them. Others exude toxic material continuously, thereby preventing the fouling larvae of various species from settling upon them and/or preventing harmful micro-organisms from invading their tissues and/or preventing other organisms from encroaching upon their living space. Because such organisms are poisonous when eaten, the species as a whole is afforded some degree of protection against predators.

Some species of poisonous marine organism do not themselves produce toxins but they acquire them from other organisms, either directly, or indirectly via a food chain. Accordingly, these species may be poisonous only at certain times and at certain locations.

Biotoxins have been isolated from representatives of most marine groups, but it would appear that a greater proportion of the fauna and flora of the shallow-water marine tropics possesses potent toxins than is the case with the fauna and flora from other marine environments. Of course, some species are immune to the toxins produced by others and this has probably led to the establishment of many specific predator–prey relationships and associations.

Protozoa. Mass mortality of fish and other marine animals is often associated with periodic "blooms" of marine flagellates but there is dispute as to whether the observed mortality is due to the liberation of toxic material by the flagellates or whether a physical phenomenon, such as gill clogging or oxygen depletion of the water due to the presence of enormous numbers of the flagellates, plays the major role. However, potent toxins have been isolated from some flagellates. Also, the flesh of bivalves such as mussels and clams, becomes toxic during the periodic maxima of certain species of flagellate presumably due to the ingestion of these flagellates by the bivalves. If the flesh of these bivalves is eaten by man, paralytic shellfish poisoning, characterized by muscular paralysis and often fatal, may ensue.

A toxin which has been isolated from the flesh of toxic mussels and clams and from the dinoflagellate *Gonyaulax catenella* has a molecular formula of $C_{10}H_{17}N_7O_4 \cdot 2HCl$ and a molecular weight of 372. This toxin which interferes with the production of action potentials in nerves and skeletal muscle is among the most lethal of biotoxins as far as mammals are concerned.

Another type of shellfish poisoning, called venerupin shellfish poisoning has stemmed from the ingestion of oysters (*Crassostrea gigas*) and tapestry cockles (*Tapes semidecussata*) occurring in Japanese waters. Over 100 fatalities stemming from this type of poisoning have been recorded. Little is known about the pharmacology or chemistry of the toxic material responsible but this material is believed to originate with a species of dinoflagellate belonging to the genus *Prorocentrum*.

The phytoflagellate *Prymnesium parvum* produces a number of toxins, and toxic material from this species has been shown to block transmission at the neuromuscular junction in vertebrates and crustaceans.

Porofera. Some species of sponge, such as *Tedania toxicalis* and *Haliclona viridis*, release materials into the surrounding water which are toxic to both vertebrates and invertebrates. Moreover, many species of sponge release

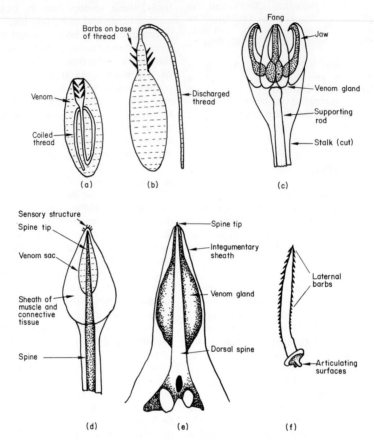

Fig. 1. Types of venom apparatus. (a) Diagram of an undischarged nematocyst of the injector type showing the coiled hollow thread lying in a venom-filled capsule. (b) Diagram of a discharging nematocyst with thread everted and with barbs at base of the thread erected. The mechanisms involved in discharge of nematocysts and expulsion of venom through the hollow threads are not well understood. The threads of the nematocysts of many species possess minute barbs which assist the penetration of the thread into the flesh of a victim. (c) A globiferous pedicellaria of *Toxopneustes pileolus* showing the head consisting of three jaws, each of which is associated with twin venom glands. A thin muscular sheath which covers each jaw and associated venom glands is not shown. The head is mounted on a stalk which is supported by a calcareous rod which articulates with the corona of the sea urchin. (d) A venomous aboral spine of a sea urchin belonging to the genus *Asthenosoma*. (e) A dorsal spine of the stonefish *Synanceja trachynis* showing twin venom glands in antero-lateral grooves of the spine surrounded by an integumentary sheath. (f) A serrated fin spine of a trachysurid catfish. In life, aggregations of venom-producing cells occur in the web of the fin in the vicinity of the spine and these cells rupture, releasing their contents when the spine penetrates the flesh of a victim.

terials with antimicrobial properties. Ectyonin from *Microciona prolifera* was the first of these antibiotics from sponges to be isolated and named and the isolation of others is in progress.

Cnidaria. The possession of minute stinging capsules called nematocysts (Fig. 1) is characteristic of this phylum. About 20 different types of nematocyst have been described. Each major group within the phylum possesses a characteristic type or assemblage of different types of nematocyst. Some types, the injectors, contain potent toxins which are introduced into the flesh of prospective prey via hollow threads when the injectors discharge. Although the majority of the Cnidaria are potentially venomous, the penetrating power of the threads of injectors possessed by different species varies and only a few species can be regarded as being venomous as far as man is concerned. These include hydroids (*Halecium beani, Aglaophenia cupressina, Lytocarpus philippinus*), siphonophores (*Physalia physalis, Rhizophora eysenhardti*), milleporines (*Millepora alcicornis*), anemones (*Segartia elegans, Anemonia sulcata, Actinodendron plumosum,*

Anthopleura xanthrogrammica), madrepores (*Acropora palmata*), and scyphozoans (*Chironex fleckeri, Chiropsalmus quadrigatus, Corukia barnesi, Linuche unguiculata, Rhizostoma pulmo, Cassiopea xamachana, Cyanea capillata, Sanderia malayensis*).

The symptoms exhibited by human victims of stings from cnidarians vary according to the species involved and the site of the sting. Contact with the tentacles of the jellyfish *Chironex fleckeri* can cause death within seconds, and this species is the most dangerous of all venomous animals Toxin from its nematocysts has a powerful action on the cardiovascular system of mammals, as has toxin from the nematocysts of the "Portuguese man-of-war," *Physalia physalis*.

Proteins, peptides, amino acids, 5-hydroxytryptamine, and quaternary bases including tetramine have been found in toxic material from cnidarians. However, some toxins are found in the tissues generally while others are confined to nematocysts. Moreover, evidence is accumulating to suggest that different toxins are found in nematocysts from different coelenterate groups.

Terpenoid substances possessing antibiotic activity have been isolated from extracts of gorgonians and alcyonarians but there is a strong possibility that these substances are produced by symbiotic dinoflagellates found in the tissues of gorgonians and alcyonarians.

Fatal poisonings have followed the ingestion of the anemone *Rhodactis howesii*.

Annelida. The annelid worm *Glycera dibranchiata* possesses venom glands associated with its fang-like jaws but little is known about the venom produced. A toxin possessing an anaesthetic effect on insects and an action on the nervous system of vertebrates is found in the tissues of the marine annelid *Lumbriconereis heteropoda*. The toxin, called nereistoxin, has the formula $C_5H_{11}NS_2$.

Nemertea. In some nemerteans the protrusible proboscis is armed at the tip with a small stylet and secretory cells which may produce a toxin are found near the base of the stylet.

Mollusca. Members of the large gastropod family Conidae elaborate potent venoms, used primarily to paralyze their prey, in a well-developed venom apparatus (Fig. 2). Most species are vermivorous, some molluscivorous, and a few are piscivorous. The paralytic actions of the venoms of the various species appear to be restricted to representatives of the phylum to which their prey belongs. Sometimes the venom apparatus is employed defensively and piscivorous species pose a threat to man who finds the shells of Conidae attractive. Thus *Conus geographus* has been involved in human fatalities and both *C. geographus* and *C. tulipa* have caused several near fatalities. The venoms of some vermivorous species contain proteolytic enzymes which cause local necrosis at the site of injection in vertebrates.

A direct action on vertebrate musculature is exhibited by the venoms of piscivorous Conidae. The venom of *C. geographus* elicits a flaccid paralysis of vertebrate skeletal musculature. That of *Conus magus*, however, elicits a sustained contracture of vertebrate skeletal and smooth musculature and modifies the behavior of cardiac musculature.

The chemistry of the active principles found in the venoms of Conidae has not been elucidated.

The salivary glands of gastropods belonging to the genus *Neptunea* contain tetramine. Many human poisonings have occurred in Japan stemming from consumption of *Neptunea arthritica*. Murexine, or urocanylcholine, is a toxin found in the hypobranchial glands of members of the gastropod family Muricidae. It possesses a potent neuromuscular blocking action. Another choline ester, acrylylcholine, has been found in the hypobranchial gland of a member of the family Buccinidae. However, the roles of these esters have not been elucidated.

Numerous human poisonings have been caused by the consumption of toxic bivalves but, as mentioned earlier, the toxic material present

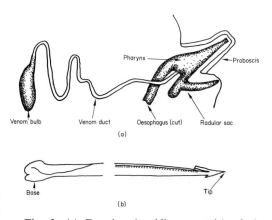

Fig. 2. (a) Drawing (semidiagrammatic) of the venom apparatus of *Conus geographus*. Contraction of the musculature in the walls of the venom bulb forces venom along the venom duct, through the lumen of the pharynx and that of the protrusible and mobile proboscis and into the tissues of prey via a hollow radular tooth held at the tip of the proboscis. The venom, which consists of numerous sausage-shaped bodies suspended in a viscous fluid, is produced principally in the venom duct. *C. geographus* preys on small fish.

(b) Drawing of a barbed radular tooth of *Conus geographus*. The tooth is hollow and possesses an opening near its base and another near its tip for the passage of venom. It functions in the manner of a hypodermic needle.

occasionally in bivalves originates, in most cases at least, from toxic flagellates filtered from the water in which the bivalves occur. A biologically active principle, called mercenene, is found in water extracts of tissues of the clam *Mercenaria mercenaria*. This principle has antitumor activity. It has an inhibitory effect on the growth of Sarcoma-180 and Krebs-2 ascites tumors in mice and on the human HeLa cell line but has a negligible toxic effect on the normal human amnione line. From the data available, it would appear that mercenene is probably toxic to many types of tumor and possibly non-toxic to normal cells. Elucidation of the biological role of mercenene is awaited with interest.

Octopuses possess a venom apparatus used essentially for inactivating prey seized by the tentacles. The prey is drawn to the mouth of an octopus and bitten by a beak-like structure. Venom, released in response to nervous stimulation, passes to the buccal cavity from well-developed salivary glands. The venom is introduced into prey via the puncture wound made by the beak. The venom is synthesized as spherical granules and carried in a mucous medium. Its paralyzing effect on crustacean prey is marked and the venom of some species of octopus is capable of paralyzing vertebrates. One Australian genus, *Hapalochlaena*, possesses two species which have been implicated in human fatalities. Paralysis of skeletal musculature was evident in all human victims of these species and respiratory failure seems to have been the prime cause of death.

There appears to be great variation in the pharmacological properties and chemical structures of the venoms of different species of octopus. Tyramine, histamine, octopamine, acetylcholine, taurine, 5-hydroxytryptamine and the polypeptide, eledoisin, have all been detected in the venoms of various species of octopus. However, there is evidence that proteins are the really toxic principles of octopus venom.

Crustacea. A toxic alkaloid is found in the tissues of the king or horseshoe crabs *Carcinoscorpius rotundicauda* and *Tachypleus gigas*. Ingestion of the flesh of these crabs causes a poisoning which may be fatal. There are reports which suggest that certain other species of crab may elaborate toxic materials but, on the whole, it is surprising that such a large and diverse group as the Crustacea contains so few representatives which produce biotoxins.

Echinodermata. The large multi-rayed sea star *Acanthaster planci* possesses prominent spines on the aboral surface. It is believed that certain secretory cells of the epidermis covering the spines produce toxic material. Contact with the spines results in immediate and severe pain, frequently followed by protracted vomiting.

Clinical evidence also indicates that the spines of many species of sea urchin are associated with toxic material and it has been shown that the tips of the primary oral spines of *Phormosoma bursarium* and the secondary aboral spines of species of *Asthenosoma Araeosoma*, and *Echinothrix* are each invested by a sac-like epithelium in which venom is elaborated.

The globiferous pedicellariae found in many species of sea urchin are venom organs. Each consists of a three-jawed structure (the head) mounted on a stalk which is articulated with a small tubercle on the corona of the echinoid. Each of the three jaws usually terminates in a hook-like process which is hollow and termed a fang. A single or double venom gland surrounded by muscular tissue is associated with each jaw. Ducts link the venom glands with the fangs and venom is injected when the fangs close on flesh. Globiferous pedicellariae are particularly well developed in the Toxopneustidae and there are reports of generalized paralysis, respiratory distress, and fatalities stemming from envenomations by *Toxopneustes pileolus*.

Little is known about the physiopharmacology and chemistry of the venoms of echinoderms. Noradrenaline has been tentatively identified in the toxic material extracted from the secondary spines of two species of *Echinothrix* and one of the active principles of the pedicellarial venom of the sea urchin *Tripneustes gratilla* is a protein which releases histamine from isolated tissues of the guinea pig.

When molested, species of the holothurian genus *Actinopyga* eject structures called Cuvierian tubules which are highly toxic to fish and other organisms. From these tubules a steroid saponin, holothurin, has been isolated. As well as being highly toxic to many animals, holothurin shows antitumor activity, causing regression of Sarcoma-180 and Krebs-2 ascites tumors in mice. Holothurin has haemolytic properties, blocks nerve conduction, and modifies developmental processes in animals. Other holothurians and asteroids such as *Asterina pectinifera*, *Pycnopodia helianthoides*, *Asterias forbesi*, and *Pisaster ochraceous* have been shown to possess saponins, similar to holothurin, in their body walls.

Chordata. Two sharks (*Heterodontus francisci* and *H. philippi*), the dogfish *Squalus acanthias*, the chimaeroids or ratfishes *Chimaera monstrosa* and *Hydrolagus colliei*, stingrays, and many different kinds of teleost fish possess venom glands associated with a traumatizing apparatus. This apparatus takes the form of dorsal spines in the sharks, the dogfish, and the chimaeroids, a few or many fin spines or opercular spines in the teleosts and one or more caudal spines in the stingrays. The glands are derivatives of the epidermis of the integumentary sheaths which surround the spines. In some cases (e.g., stingrays and many catfish) the glands are represented by aggregations of venom-producing cells in certain regions of the epidermis of the integumentary sheath investing each spine and the spines themselves are solid and serrated. In toadfish (Batrachoididae) a mass of venom-producing cells surrounds the base of each spine which is hollow and possesses openings both at the base and the tip. In most cases the venom glands associated with each spine are paired cylindrical bodies comprised chiefly of aggregations of venom-producing cells and each gland is lodged in one of a pair of grooves found in each spine.

If stingrays are seized, trodden on, or otherwise molested they usually lash vigorously with their tails and seek to drive their venomous caudal spines into the flesh of their tormentors. Zebrafish (*Pterois* spp.) tend to lunge with their bodies in order to jab their venomous dorsal spines into enemies but in most fish the venom apparatus is used passively. Envenomation occurs when a spine penetrates the flesh of a victim.

All fish venoms so far studied contain toxic proteins. In some scorpaenids (*Synanceja trachynis, Notesthes robusta*) discrete proteinaceous granules are present in the venom. A somewhat similar chemical picture is presented in human stingings involving different species of venomous fish. This is characterized by intense pain, swelling, discoloration, and necrosis of tissue at the site of the injury and systemic effects involving the cardiovascular system. An antivenene is available against the venom of the Australian stonefish *Synanceja trachynis*.

Groups of teleost fish known or suspected to possess venomous representatives are: Synancejidae (stonefish); Scorpaenidae (scorpion fish, *Pterois, Notesthes, Scorpaena*); Siluroidea (catfish); Teuthidae (rabbitfish); Trachinidae (weeverfish); Batrachoididae (toadfish); Scatophagidae (spadefish, butterfish); Uranoscopidae (stargazers); Chaetodontidae (butterflyfish); Holocentridae (squirrelfish); and Carangidae (jacks, pompanos).

Ostracion lentiginosus, a member of the boxfish family (Ostraciontidae), can secrete into its immediate surroundings a toxin capable of killing fish. This toxin, called pahutoxin, has been identified recently as the choline ester of 3-acetoxyhexadecenoic acid. Apart from its ability to haemolyse erythrocytes little is known about its biological activity. When disturbed, the soapfish *Rypticus saponaceus* releases a foamy secretion toxic to other fish. The results of preliminary studies indicate that a protein or polypeptide is responsible for the toxic effects.

Ciguatera is a type of fish poisoning, characterized by gastrointestinal disturbances and nervous malfunctions, which stems from the ingestion of certain tropical fish, particularly species associated with coral reefs. Species belonging to many families and some of them important food fish have been implicated in ciguatera. For years these fish may be eaten with impunity then, suddenly, their consumption results in severe poisoning and often fatalities occur. There is evidence that the toxic material present in the flesh and/or viscera of the fish is acquired directly, or indirectly via a food chain, from toxic algae and possibly toxic flagellates. Pharmacological studies with crude toxin isolated from the toxic fish indicate that it contains material which blocks transmission at the mammalian neuromuscular junction. The principal toxin is an anticholinesterase, but there is evidence that other toxins with different molecular structures and pharmacological activities are present.

Ingestion of certain species of goatfish (Mullidae) and mullet (Mugilidae) sometimes results in a type of poisoning referred to as hallucinatory fish poisoning. Muscular uncoordination, dizziness, and frequently hallucinations and nightmares are among the toxic manifestations.

It has long been known that fish called puffers (Tetraodontidae) contain a potent toxin which is concentrated in their livers, reproductive organs, intestines and, to a lesser extent, in their skin. Many fatalities have stemmed from consumption of these fish. The biotoxin tetrodotoxin has been isolated from puffers and also from porcupinefish (Diodontidae) and sunfish (Molidae). Tetrodotoxin suppresses action potentials in nerves and muscles of many species but there is dispute as to the exact mechanisms involved in this suppression. The formula of tetrodotoxin is probably $C_{11}H_{17}O_8N_3$.

Approximately 50 species of sea snake are known and all appear to be venomous. As in their terrestrial relatives, the venom apparatus consists of paired venom glands from which ducts leading to the fangs arise. Usually four fangs, each of which is grooved for the passage of venom, occur. Venoms from some species of sea snake have been shown to have marked effects on skeletal muscle and on neuromuscular junctions. Low molecular weight proteins have been found in some venoms.

At certain times, and at certain localities, ingestion of the flesh of marine turtles has resulted in many human poisonings. A large percentage (about 44%) of those poisoned die. The toxic material responsible for these poisonings has not yet been isolated.

Some marine biotoxins are incredibly potent and many possess highly specific biological activity. Consequently, some may find a use as tools in the investigation of biological processes. Indeed, a few (e.g., tetrodotoxin) are already being used in this way. Moreover, marine biotoxins provide a pool of new or poorly known molecular structures which warrant investigation by chemists interested in the synthesis of new compounds.

Among the known marine biotoxins are substances which can produce one or more of the following effects: block conduction in nerves, block transmission at nerve–muscle junctions, elicit a sustained contraction of muscles, relax muscles, lyse cells, and alter the permeabilities of biological membranes. There are others which affect the behavior of the cardiovascular system, some which possess marked antibiotic activity against micro-organisms, some which possess anaesthetic properties, some with psychopharmacological properties, and some which can cause regression of certain tumors in mice. Also, many new biotoxins will undoubtedly be discovered in marine organisms as a result of screening programs now being undertaken or planned. It is to be expected that some of these biotoxins or compounds derived from them will find a use in medicine.

Robert Endean

Cross references: *Mortality in the Sea; Toxicology of Marine Animals.*

TABLE 1. VOLUME AND DOLLAR VALUE OF U.S. BREADED FISHERY PRODUCTS, 1965

	Quantity (1000 lb)	Value (1000 dollars)
Fish sticks[a]	82,483	35,778
Steaks, portions, fillets[a]	139,505	55,808
	221,988	91,586
Fillets:		
cod	423	244
flounder	3,316	2,008
haddock	1,333	672
ocean perch	3,027	1,263
sea trout	84	35
yellow perch	606	391
	8,789	4,613
Shellfish:		
shrimp	98,144	77,091
clams	2,710	2,388
oysters	3,126	2,872
scallops	7,126	6,083
	111,106	88,434
Miscellaneous:		
smelt (whole)	91	28
marlin (cakes)	1,667	996
cakes[a]	3,481	1,474
dinners[a]	6,367	4,086
	11,606	6,584
Total	353,489	191,217

[a] Unclassified as to species.

BREADED FISHERY PRODUCTS

Frozen breaded fishery products consist of controlled amounts of fish or shellfish covered with a uniform coating of batter and breading material. These items are available to the consumer in the raw state, or they may be precooked to facilitate preparation by the consumer. Breaded fishery products were first introduced in the late 1940's. Since then, these products have had a marked effect on the fishing industries of the world. In general, breaded fishery products have met with a high degree of acceptance by the consumer public, primarily because of their convenience aspects. To the housewife, these items represent a varied menu with a minimum of preparation. To the restaurant and institutional buyer, they provide uniform servings of consistent quality with exact cost control. In addition, these items provide year-round availability, greater speed of serving, lower labor costs, and less storage space.

Many different types of breaded fishery products are on the market, and new products are being added constantly. Table 1 shows the 1965 volume and dollar value of some of the more common breaded fishery products produced in the U.S. The most important products being produced are breaded shrimp and breaded fish sticks and fish portions. In 1966, these items alone represented a total volume of 332 million pounds valued at 187 million dollars. The growth of these products has been increasing steadily. Figure 1 shows the increase in production volume of breaded shrimp, breaded fish sticks, and breaded fish portions from 1955 to 1966.

The rate of growth of these items has been so great that the processors are forced to rely on imported raw materials to meet their requirements. This is especially true in the fish stick–fish portion industry, where nearly the entire U.S. production is dependent upon imported fish blocks and slabs. Currently, frozen fish blocks are being exported to the U.S. from 17 different countries in Africa, Europe, and the Americas. Shrimp are being exported from more than 60 countries representing 6 continents and almost every ocean of the world. Table 2 shows the volume and dollar value of U.S. imports of shrimp and frozen fish blocks and slabs from 1955-1956.

In general, a breaded fishery product is processed in 5 separate steps: (1) preparing the raw material; (2) coating the fish with batter; (3) coating with breading material; (4) frying; (5) packaging and refreezing. These steps are essentially the same for all breaded fish and shellfish products. Manufacturers, however, do vary the process slightly, depending upon the item being processed and the amount of coating to be added to the fish.

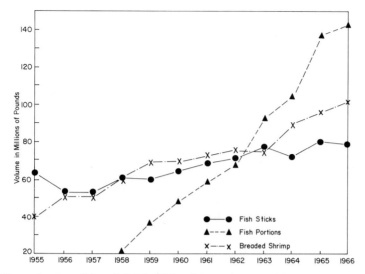

Fig. 1. U.S. production of breaded fish sticks, fish portions, and breaded shrimp, 1955–1966.

Preparing the Raw Material

Since all breaded fishery products are "portion controlled," the preparation of the raw material is an extremely important part of the production. In the case of shellfish and fillets, the raw material must be graded as to size and must be prepared in a manner as to ensure that all bones, shell, sand, and viscera are removed from the product. The raw material is carefully inspected prior to and during the breading process, and defective products are culled from the line.

Shrimp are peeled and deveined either by hand or by machine and are processed unfrozen. One method of preparation commonly used leaves the terminal segment of the shell (the fantail) still attached. The shrimp may be breaded either "round" or may be "butterflied," in which case the shrimp are split down the center and flattened out prior to being breaded. In the case of shrimp portions or sticks, the shrimp are first frozen into a shrimp-shaped mold and then cut with a band saw or guillotine cutter to the desired thickness.

Fish sticks and fish portions are prepared from frozen blocks or slabs of fillets. Since the sticks and portions are cut to predetermined dimensions to give a certain number of portions in a package of a given weight, the frozen blocks must have uniform densities and dimensions. To ensure this, the fillets are carefully hand-packed in a wax paperboard carton and uniformly arranged to avoid air spaces. The cartons are slightly overfilled so that when they are plate frozen, the pressure from the plates compresses the fillets into a uniform block, filling the corners and edges. For freezing, the filled cartons are placed in metal trays (2 cartons per tray) and then loaded into the plate freezer. The dimensions of the blocks vary according to the specifications of the buyer. Most blocks, however, are produced in the range 13.5–16.5 lb with thicknesses ranging 1.5–2.5 in. For shipment, four blocks are ordinarily packed in one master carton.

The sticks and portions are cut while the block is still frozen. It is common practice, however, to temper the blocks for a short time at some elevated temperature to minimize breakage during the cutting process. The fish blocks are ordinarily cut in three stages. Band saws or gang saws are used in the first two stages to produce slabs with the required portion thickness. In the last stage, the slabs are cut into the final size and shape by use of saws or of guillotine cutters. Guillotine cutters are used to reduce sawdust losses;

TABLE 2. U.S. IMPORTS OF SHRIMP AND FROZEN FISH BLOCKS, 1955–1965

Year	Shrimp[a]		Frozen Blocks or Slabs[b]	
	Pounds (1000)	Dollar Value (1000)	Pounds (1000)	Dollar Value (1000)
1955	53,772	24,532	48,236	9,466
1956	68,618	32,986	38,919	7,285
1957	69,676	35,415	50,233	9,605
1958	85,394	43,162	51,147	9,865
1959	106,555	52,306	85,290	17,039
1960	113,418	56,406	89,672	18,515
1961	126,268	68,538	118,609	24,606
1962	141,183	91,898	143,551	28,436
1963	150,138	101,911	153,270	31,387
1964	151,168	104,355	166,166	37,100
1965	160,287	111,277	214,807	52,506

[a] Fresh or frozen shrimp and prawn.
[b] Groundfish and ocean perch.

however, broken or chipped portions or sticks will sometimes result if the blocks are not tempered correctly. Normally, the internal temperature of the block should be about 15°F.

The cutting of the fish blocks is controlled carefully to ensure a minimum of waste. The amount of sawdust lost during a cutting operation using saws in all three stages ranges from 7–12%. A guillotine cutter used in the third stage may reduce this loss by as much as 60%. One of the major causes of waste during the cutting operation is due to defective portions resulting from the use of blocks of nonuniform dimensions and densities. Holes or air spaces in the blocks result in underweight and misshaped portions. Blocks of nonuniform dimensions or dimensions not meeting specifications produce either underweight or overweight portions. When the thickness of a block fails to meet specifications, as much as one-third of the product cut from this block may be defective. If the block is too thin, packages will not meet net weight requirements. If it is too thick, the required number of sticks or portions will not fit the package.

Coating with Batter

After the product has been processed to the correct size, shape, and weight, it is separated into evenly spaced rows on a conveyor. This operation can be carried out by hand or, in the fish-stick and portion industry, by a type of automated unscrambling device. The separated product is then conveyed to the automatic breading and battering equipment.

Many commercial types of batter mixes are available for use in the manufacture of breaded fishery products. All these contain such ingredients as corn flour, corn starch, corn meal, wheat flour, nonfat dry milk, eggs, and various seasonings. These ingredients are mixed in many possible combinations, depending upon the buyer's specifications. Since 1960, manufacturers of breaded fishery products have converted primarily to starch-based batter mixes. These batters adhere to the fish flesh much better than the flour-based batters, reducing the incidence of the coating peeling away from the flesh during the frying operation.

The battering portion of automatic breading equipment such as that shown in Fig. 2 typically consists of: (1) a mixing tank and reservoir; (2) a circulating pump; (3) a spill-over tray or manifold; and (4) a blower.

The product is passed along the conveyor belt under the manifold from which a curtain of batter flows down over the product. A small amount of excess batter collects in a depression under the belt and coats the undersides of the product as it

Fig. 2. Coating fish portions with batter.

passes through the curtain. The remainder of the excess drains into the reservoir and mixing tank from which it is recirculated into the manifold. Most manufacturers now employ a viscosity-control system that continuously measures the viscosity of the batter in the reservoir and adds dry batter mix or water as required. A typical batter material will be prepared by adding two parts of dry mix to three parts of water. The viscosity of this batter will measure between 70 and 120 centipoises.

Often, to aid the batter in adhering to the product, some processors dust the raw material with dry batter mix prior to battering. This method is commonly used in the production of breaded shellfish products.

After the product has been battered, it is conveyed under a blower, which removes excess batter. It is then conveyed to the breading portion of the equipment.

Coating with Breading Material

The type of breading material used for breaded fishery products depends primarily upon the type of product being produced. Breadings are formulated to attain particular intensities of cooked color; the final color of the product is almost entirely determined by the choice of breading. Breadings for precooked products are designed to attain a golden brown color quickly (20–45 sec at 400°F) and to retain good color upon subsequent heating by the consumer. Since raw breaded products are cooked entirely by the consumer, these products must brown more slowly; they, therefore, are designed not to reach full color until the product is completely cooked. Breadings may contain, either alone or in combination, wheat cereals, cracker meal, potato flour, soy flour, starch, and bread crumbs. The mixtures may be either granular or flaky, and the particle size can be varied, depending upon the preference of the manufacturer.

The breading operation is similar to the battering operation where breading material is applied to all sides of the product. Figure 3 shows a typical breading operation. The battered product is carried along the breading conveyor on a bed of breading material about $\frac{1}{2}$ in. thick. The product is passed under a hopper from which a curtain of breading material is continuously falling. After the product is completely covered, it is usually conveyed under a roller or a series of ascending and descending flexible rings, which press the breading to the product to give better adherence. Excess breading is removed from the product by means of a vibrating screen or conveyor. The excess breading is recirculated by conveyors in order to keep the hopper and the breading conveyor full. If the product is to be packed raw-breaded, it passes from this point directly to the packaging line; otherwise, it is conveyed to the cooker.

The amount of total coating applied to the product is determined by a number of interdependent variables, the most important being the surface area of the raw material, the temperature of the raw material, the viscosity of the batter, and the speed of the conveyor carrying the product from the batter curtain to the breading. All these variables can be changed to control the total coating pickup. Most manufacturers, however, try to hold everything constant, varying only the batter viscosity either to increase or to decrease the pickup of coating. In cases where a very small product is being breaded, some manufacturers, to increase the coating pickup, employ a "double dip" operation in which two batter and breading machines are placed in tandem. Often, instead of the product being conveyed through a second battering process, the product coming from the first breader will be sprayed with a fine mist of water before being passed to the second breader. This spray will wet the product just enough to enable more breading materials to be added.

Frying

Most breaded fishery products are fried in continuous cooking equipment similar to that shown in Fig. 4. The product travels on a conveyor belt through the cooker, which is about 30 ft long. The oil is held to 400°F, and the product is cooked 20–45 sec, depending upon the desired color and the size of the material. At these rates, about 2000 pounds of product can be processed per hour.

Fig. 3. A typical breading operation.

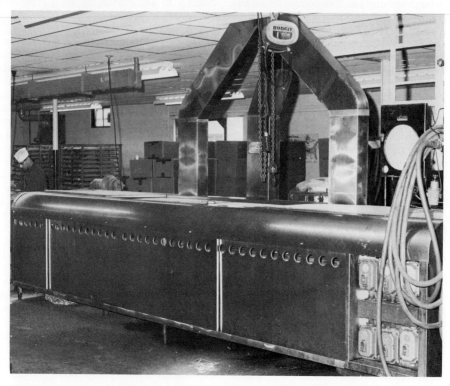

Fig. 4. A continuous fryer used in manufacturing precooked breaded fishery products.

In most installations, liquid vegetable oils rather than melted hydrogenated fats are used for frying. Although unsaturated fats often result in a shorter shelf life, the initial quality of the product is generally preferred over that of products fried in hydrogenated fats. Some consumers prefer the flavor resulting from the use of vegetable oil, and the vegetable oil does not solidify into unattractive irregular patches on the surface of the product as hydrogenated fats tend to do.

The oil is either heated directly in the cooker by means of a gas flame or by means of a heat exchanger. The volume of oil in the cooker is held to a minimum to ensure quick turnover. The oil absorbed by the product is sufficient to produce a complete change of oil in the cooker two or three times a day, thus preventing the buildup of free fatty acids in the oil. The amount of oil absorbed by the product depends largely upon the length of the cook and the physical characteristics of the breading material. During cooking, breading material is shaken loose from the product. These crumbs burn quickly and, in combination with the high temperature of the oil (400°F), promote a breakdown of the oil. For this reason, fryers are equipped with a continuous filtration system to eliminate breading from the oil.

Most manufacturers fry the product as fast as possible. By using a very quick browning breading and a very high temperature, they minimize losses in weight due to frying. The moisture lost from the product comes almost entirely from that moisture present in the coating which, during cooking, is replaced by oil. Longer fry times cause moisture to be lost from the flesh. This moisture loss is not replaced by oil, so the result is a net loss in weight. The major problem associated with quick frying is an increased breakdown of oil due to the high temperatures (400°F). Some manufacturers feel that a slightly longer fry at 375°F would be more economical in the long run.

Packaging and Freezing

Raw-breaded products are layer packed in wax-coated cartons. Usually, the layers are separated by inserts and overwrapped.

Precooked products are handled in two different ways. Some manufacturers refreeze the product in a continuous-blast freezer prior to packaging; others package the product in overwrapped cartons and then blast freeze or shelf freeze. As far as the quality of the product is concerned, freezing prior to packaging is preferable. This method reduces the incidence of loose breading, oil, and frost in the package, giving the product a much better over-all appearance. Products handled in this manner are often machine-packed into end-loading cartons without overwraps.

Just prior to the packaging process, the products are given a final inspection for defects, and all defective product is removed. Some manufacturers also employ a weighing machine that will detect all overweight and underweight cartons and automatically cull any cartons exceeding the allowed limits.

ROBERT J. LEARSON

References

1. Anon., "Fisheries statistics of the United States," U.S. Dept. of the Interior, Bureau of Commercial Fisheries, Washington, D.C., 1950–1964.
2. Lyles, C. H., "Fisheries of the United States—1966," C.F.S. 4400, U.S. Dept. of the Interior, Bureau of Commercial Fisheries, Washington, D.C.
3. Idyll, C. P., "The shrimp fishery," in "Industrial Fishery Technology," Stansby, M. E., and Dassow, J. A., Eds., 1963, New York, Reinhold Publishing Corp.
4. Peters, J. A., "The Bottom Fisheries," in "Industrial Fishery Technology," ibid.
5. Holston, J. A., "Weight changes during the cooking of fish sticks," Commercial Fisheries Rev., **17**, No. 4 (1955).
6. Cocca, F. J., "Some factors affecting sawdust losses during the cutting of fish sticks," Commercial Fisheries Rev., **19**, No. 2 (1957).

BREEDING IN MARINE ANIMALS

Although marine organisms display a bewildering variety of reproductive habit, certain features are common. They usually display some form of periodicity in breeding and possibly also a fairly well marked breeding season at a particular time of year. The aqueous environment permits retention of a primitive reproductive habit in a wide variety of groups, i.e., eggs and spermatozoa are spawned freely into the sea, where fertilization takes place apparently by chance. Active embryos tend to emerge at an early stage of development and a period of free-swimming larval life ensues which is particularly important for the distribution of sessile benthic and littoral organisms. These larvae contribute importantly to the plankton, particularly the temporary plankton of inshore waters.

It would be a mistake to overemphasize these common features. The most successful group of marine fish, the Teleosts, are mainly dioecious and discharge their eggs and spermatozoa into the water, but internal fertilization and copulatory devices do occur, as do hermaphroditism and self-fertilization plus all stages of oviparity, viviparity and ovoviviparity. According to Hoar[1] "the fishes ... provide examples of almost every means of reproduction known among animals as well as some curious situations which might be unique to this group."

Growth Versus Reproduction

In a sense, sexual development is initiated in the embryo with the appearance of the primordial germ cells and the subsequent development of sperm and egg mother cells in the differentiated gonad. However, gametogenesis and development of the gonad are slow during growth of the soma. Acceleration of sexual development usually occurs when growth is completed or during intervals between periods of growth. This, however, is not a general rule, Orton[2] cites many instances of coincident growth and reproduction and mentions species which may breed at an early stage of growth. Small specimens of the limpet *Patella* and tunicates such as *Clavelina* also may breed at an early stage of development. However, in the marine polychaet worms such as the Nereidae and the Nephtyidae, sexual maturation accompanies a sharp decline in segment proliferation. In the crustacean *Crangon vulgaris*, growth ceases at the molt at which egg laying occurs while in *Balanus balanoides*, the onset of maturation is delayed if environmental conditions are adverse and growth slow so that the onset of sexual maturity is more closely related to growth than to age. Male cephalopod mollusks continue to grow slowly after they reach maturity, but in females growth ceases with egg laying. In general, however, both males and females of vertebrate species complete their growth before reproduction.

Whatever the reasons for the apparent separation of growth and reproduction, it may be specifically ensured in a variety of marine organisms by the secretion of so-called juvenile hormones[3], or by inhibition of gonadotropin secretion. In the higher groups control appears to be exercised by the latter route. Thus in the octopus and possibly in other cephalopods, the central nervous system inhibits production of what appears to be a gonadotropic hormone by the optic gland in the juvenile organism.[4] While no work has been carried out on lower vertebrates, it seems probable that in mammals the hypothalamus exerts an inhibitory effect on pituitary gonadotropin secretion in juveniles.[5] What is clear is that in both cephalopods and vertebrates the juvenile gonad is capable of rapid maturation when inhibition is removed. There is evidence that in decapod crustaceans neurosecretory cells of the sinus gland-X-organ which are under the control of higher nervous centers produce a hormone inhibiting development of the gonad in juvenile stages. In recent years much experimental work has been done on the control of sexual development in marine polychaets. These organisms show a variety of somatic changes immediately prior to breeding which in the most extreme cases result in an epitokal transformation to form a specialized reproductive individual—the heteronereis. Nereids metamorphosing in this way display enlargement of the eyes and of the parapodia in the genital region of the body, with the formation of specialized chaetae and histolysis of the body wall musculature.

The Breeding Season

Some form of periodicity in reproduction is widespread among marine organisms but all sorts of variations occur. At one extreme are the epidemic spawners which may breed once annually over a period of only a few days. At the other extreme are organisms which breed all year round. This is perhaps more common under homothermic conditions in the tropics, but Thorson[6] lists a number of invertebrate species that breed throughout the year. Even these forms, however, tend to display a maximum of breeding activity at a particular time of the year. Within these extremes the breeding season may be variously prolonged and within the breeding season itself a lunar or semi-lunar periodicity of spawning is detectable.

Coincident spawning of a large number of individuals in a population is plainly of survival value because of the improved chances of fertilization of the eggs. Seasonal breeding may have arisen in part for this reason and in part by selection operating in favor of the progeny of individuals spawning at a time of year most favorable for survival of the larvae. This is probably an oversimplification, as environmental factors may operate at all stages of the reproductive cycle to limit breeding to a particular season. Thorson[7] reviewing the work of Orton[2] and others states "Marine invertebrates require very definite and normally much higher temperatures to ripen and spawn their sexual products than are necessary for them outside this season . . .; young cleavage stages need more precise and limited temperature intervals than older larvae and bottom stages."

Another potentially limiting factor is nutrition. Thorson remarks, "The food requirements of invertebrate larvae seem on the average to be 5–10 times higher than those of their parents at the bottom." A direct relationship between nutritional level and rate of egg laying has been demonstrated in *Calanus finmarchicus*, the dominant planktonic organism of northern seas,[8] while the number of broods per annum increases from one in the Barents Sea and east Greenland to three in the warmer waters of the west coast of Scotland and possibly more at Plymouth (Fig.1).

While many marine organisms begin breeding at a particular temperature or between characteristic temperature maxima and minima which Orton described as "physiological constants for the species," some species are capable of adapting their breeding temperature depending upon their location. The American oyster commences breeding at 25°, 20°, and 16.4°C in Delaware Bay, the Bideford River, and Long Island Sound, respectively.[9] Loosanoff and Davis[10] have suggested that the temperature prevailing during maturation may determine the temperature at which spawning first occurs. Presumably there are genetic limits upon the degree of adaptation which is possible in any species. Another principle enunciated by Orton is that high-latitude forms, towards the lower latitude fringe of their distribution, should breed at relatively low temperatures, while low-latitude forms in high latitudes should only breed in summer.

There are many exceptions to general rules of this kind, which are probably due to factors other than breeding capability that limit the distribution of species so that they never approach the extremes of lower or higher temperatures (prevailing all year round) which would effectively prevent breeding. We can conclude:

(1) That various conditions, but principally temperature and nutritional requirements, must be fulfilled at different stages of the life cycle for successful breeding. These may

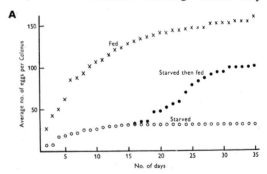

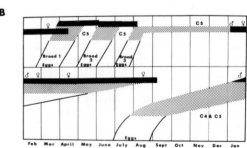

Fig. 1. The effect of nutrition and temperature upon breeding in the planktonic copepod, *Calanus finmarchicus*. (A) Number of eggs laid by starved and fed *Calanus*. (B) Number of broods per annum in Loch Striven, West of Scotland (upper diagram) as compared to East Greenland (below). (After Marshall and Orr, 1953, "Essays in Marine Biology," 1953, Oliver and Boyd.)

effectively limit breeding to a particular season, especially towards the extremes of distribution of a species.

(2) That at any particular station *within* the distribution of a species, many or all the phases of the life cycle may be conducted well within the limits of temperature and/or food supply necessary. Nevertheless, selection will favor breeding taking place under optimum conditions for successful completion of the life cycle, taking into account not only temperature and food supply, but other factors. This will result in a "matching" of the reproductive cycle to the most favorable sequence of changing environmental factors and breeding will still occur during a relatively restricted season and (perhaps only coincidentally) between characteristic temperature limits.

Regulation of Sexual Cycles

Breeding demands the integration of numerous processes within the organism. Starting with the sexually quiescent stage, gonadal development, gametogenesis, development of any secondary and accessory sexual characteristics and of breeding behavior must be completed before the organism is in breeding condition. In the higher vertebrates, internal integration of reproductive processes, and therefore control of the sexual cycle, is accomplished by secretions of the pituitary and the gonads which are released in a generally rhythmic fashion. These rhythms of secretion are inherent to a variable degree, i.e., they will continue for some time in the absence of external stimuli. The timing of the sexual cycle is thus independent of the day-to-day vagaries of the environment. The relationship between the environment and the secretory rhythm of the organism can be visualized as being similar to that between the astronomical and slave clocks of an observatory; the slave is corrected now and again by an impetus from the master instrument. Change of day length or lunar photoperiodicity seem to be the chief environmental factors influencing internal rhythms of secretion in the majority of vertebrate species examined.[11] In fish, it appears that both photoperiod and temperature are environmental regulators but their relative importance varies from species to species.[1] As these external stimuli impinge on all the members of a population, this insures that their sexual cycles are more or less in phase.

It is now generally accepted that external stimuli received by the sense organs induce a variable rate of secretion by neurosecretory centers in the hypothalamus. The axons of these secretory neurones are found chiefly in the hypothalamo-hypophyseal stalk, from whence at the median emminence (in amphibia and higher forms) the secretion may pass via a venous portal system into the adenohypophyseal region of the pituitary. In this way the brain (and less directly, external changes) can influence the secretory activities of pituitary gonadotrophs and subsequently the gonads (Fig. 2).

There is now a substantial body of evidence that "the same general processes appear to govern

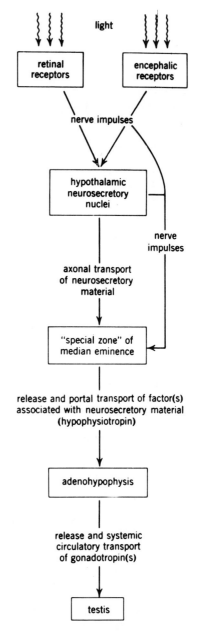

Fig. 2. "Route" whereby external stimuli, in this case light, may come to influence the gonads. (From Gorbman and Bern, "Textbook of Comparative Endocrinology," 1962, John Wiley and Sons.)

the reproductive physiology of fish as form the general corpus of vertebrate sexual endocrinology."[12] A similar pathway to that just described could exist in fish where the neurohypophysis, and at least parts of the adenohypoyphsis, are closely interdigitated and share a common blood supply. Information on the reproductive physiology and endocrinology of fish has been exhaustively reviewed in recent years by Ball,[12] Dodd et al.,[13] and Hoar.[1] Much of the information upon the Teleosts comes from fresh-water forms. However, observations on the shore-living form *Gobius paganellus* by Vivien[14] and more recently upon the Plaice, *Pleuronectes platessa*, by Barr[15,16] confirm that in marine Teleosts, as in the Elasmobranchs, hypophysectomy inhibits vitellogenesis and causes follicular atresia while spermatogenesis is halted at the spermatogonial stage. A persistent thread running through investigations of humoral influences upon gametogenesis, whether in vertebrates or in invertebrates, is that withdrawal of hormone appears to affect a particular stage or stages of gametogenesis.

In conformity with the rest of the vertebrates, the secondary sexual characteristics in fish are to some extent controlled by sex steroids secreted by the gonads under pituitary stimulation.[17]

In recent years research into the control of sexual cycles in vertebrates has concentrated upon the internal integrating mechanisms. In contrast, research upon the invertebrates, and in particular marine invertebrates, has tended to concentrate upon the environmental determinants. However, here too there is usually some cyclic development of the gametes, gonads, and gonoducts. This entails a degree of internal coordination related to the complexity of the reproductive apparatus, plus some means of matching the internal cycle to the changing environment.

In the Crustacea the androgenic gland provides the immediate stimulus for spermatogenesis and the development of the male secondary sexual characteristics. According to Charniaux-Cotton[18] "reproductive cycles in crustacean females depend on external conditions such as temperatures and light. Most probably these external factors act by way of the eye stalks." The ovary induces development of the structures necessary for incubation of the eggs.

The situation in the Crustacea is strongly reminiscent of that in the nereid polychaets. Here an inhibitory hormone secreted by the supraesophageal ganglion appears to control spermatogenesis, egg growth and, where it occurs, the epitokal transformation. In the total absence of the brain, although egg growth is accelerated, yolk accumulation is usually abnormal. Thus high titers of hormone are inhibitory but lower titers appear to be necessary for the successful completion of oogenesis. This would appear to provide a direct means whereby environmental change might serve to coordinate the sexual cycles of population, i.e., by causing simultaneous reduction in the rate of release of inhibitory hormone by neurosecretory cells situated in the central nervous system.

The control of the sexual cycle in organisms showing precise sexual periodicity is of special interest. Species displaying lunar or tidal periodicity may not be so frequent as was at one time supposed. Thorson[6] found little evidence for periodicity of this kind in his extensive investigations of reproduction in Danish marine invertebrates while much earlier Fox[19] found that popular belief in periodicity among a number of Mediterranean forms was unfounded.

There are, however, well-documented examples of species displaying lunar or tidal periodicity from a wide variety of groups:

Coelenterata —*Obelia geniculata*
Annelida —*Arenicola cristata* (Japan)
 (Polychaeta) *Platynereis dumerilii*
  (Mediterranean)
Mollusca —*Mytilus edulis* ⎫
 (Lammelibranchia) *Ostrea edulis* ⎬ (British
  Pecten ⎭ Isles)
  opercularis
Echinodermata—*Centrechinus setosus* (Red Sea)
Vertebrata (Actinopterygii)—*Leuresthes tenuis*
 (Pacific Coast, North America).

The mollusks and echinoderms tend to show rhythmic development of the gonad associated with the spawning periodicity while the Californian grunion or smelt, *Leuresthes tenuis*, which comes inshore to spawn on the three nights following each full moon during the breeding season from March to June, appears to have a distinct ovarian cycle with a fresh batch of eggs maturing on each occasion. Except in rather specialized instances, investigators are equally divided as to whether variation in tidal pressure or photoperiod during lunar periods provides the timing device for these reproductive rhythms. However, in the case of *Platynereis dumerilii*, Hauenschild[20] has convincingly demonstrated the influence of photoperiod. Metamorphosis takes about a week and the mature heteronereid swarms and spawns on the night following completion of the process.

Under natural light, whether in the sea or in the laboratory, the worms swarm mainly in between the last quarter of the moon and the first quarter. Spawning is at a minimum around full moon. Periodicity disappears under continuous illumination but can be maintained under a light regime of 24 days of 12 hours illumination and 6 days of continuous illumination. If the period of continuous light is shifted half a cycle, the swarming periodicity adjusts to the new regime after about 2 months (Fig. 3). The spawning

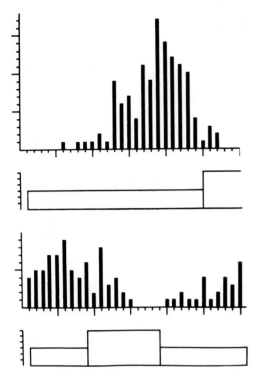

Fig. 3. A frequency distribution of swarming in *Platynereis dumerilii*, under experimental conditions. Upper diagram, 12 hours illumination each day with, every 30 days, 6 days of continuous illumination (25th to 30th days). Lower diagram, 2 months after shifting the continuous illumination to begin at the 9th day and altering the period of continuous illumination to last 10 days. (From Hauenschild, *Cold Spring Harbor Symposia Quant. Biol.*, **25**, 491 (1960).)

periodicity is maintained using levels of illumination corresponding to natural moonlight and daylight and Hauenschild's experiments have also indicated that the spawning maximum is determined by the end of the period of continuous illumination. Finally, he has shown that the reproductive rhythm is endogenous to the extent that periodicity can be maintained by exposing cultures kept normally under 12 hours of illumination and 12 hours darkness each day, to one period of two days of continuous illumination every 2–3 months. This would explain how, under natural conditions, such a rhythm can be maintained in the event of heavy cloud obscuring the moon during a full-moon period.

Breeding Migrations

Substantial population movements onto particular breeding grounds are a characteristic feature of the behavior of many marine species. In this way motile organisms are sometimes able to select the most favorable environments for various stages of the life cycle, e.g., for breeding, for larval development, and for growth. Hoar[21] points out that breeding is part of a complex behavior pattern involving apetitive behavior and consumatory acts. In the case of fish, migratory movements may be regarded as a component of the apetitive behavior prior to spawning. Marshall[22] quoting Heape states "it would appear that a breeding migration is almost universal among fish... and that the length of (some of) the breeding migrations is equalled by only a very few birds and is far beyond the capacity of any other animals except whales." The study of catches and the results of tagging experiments have shown that around the British Isles plaice migrate to the northern and southern extremities of the North Sea to spawn. Various spawning stocks of herring aggregate at specific times and localities along the eastern seaboard of Britain while hake move inshore to breed. Many migrations occur within a single geographical area; others involve prolonged directional movement between the growth area and the breeding grounds.

In the case of the cod (*Gadus callarias*) of the Barents Sea, a breeding migration of adults takes place between the feeding grounds of the Bear Island–Spitzbergen Bank to the Lofotin Islands, a distance of some 800 miles against the West Spitsbergen current.[23] Some of the most extensively studied migrations are those of the katadromous eel (*Anguila anguila*) and anadromous forms such as the various species of salmon. These migrations involve not only very prolonged journeys, in the case of the European eels of up to 3500 miles from the rivers in which growth occurs to the breeding grounds in the depths of the Sargasso Sea, but also the osmotic problems encountered in the movement from fresh to salt water or vice versa (see Refs. 21, 24, 25 for reviews).

A frequent but not universal feature of the physiology of the migrating fish is heightened thyroid activity and in the case of breeding migrations gonadial maturation is usually under way. Experimental evidence that gonadial and thyroid hormones increase locomotory activity in fish has been cited as evidence that apetitive behavior may be stimulated by these hormones, whether acting singly or together. Hoar[24] speculates that these hormones "may not only increase the activity, but also modify behavior with regard to variables such as temperature, light and salinity." In view of the possible influence of current movements in stimulating directional swimming, one might add to this list rheotactic stimuli. The thyroid may also be implicated in the physiological adjustments necessary during the spawning migrations of diadromous fish, but despite considerable research, the control process for readjustment of osmotic regulation is still uncertain.

Spawning migrations also occur among other nektonic and active benthic organisms, especially the crustaceans. Naturally, the latter are limited in terms of distance, but they are nonetheless well marked. Bainbridge[26] describes the movements of various species of decapods off the coast of Florida, in the Caribbean, and on the South African and Australian coasts. The general rule appears to be a movement into shallower water where mating occurs and egg laying may take place, followed by a migration into deeper water where incubation is completed and the larvae are eventually released. Inshore movements for spawning have been recorded among many benthic forms. The large nektonic euphausiids, *Meganyctiphanes* and *Thysanoessa*, normally oceanic in their distribution, also come inshore to breed. Antarctic species owe their wide distribution to a prolonged "passive migration" which incidentally results in a substantial non-breeding fringe to the population. Breeding takes place at the surface in low latitudes. The eggs descend and are carried into high latitudes by deep southward-moving currents where eventually the immatures come to the surface and move northward again to latitudes where breeding is possible. Spawning at the surface appears to be fairly common among nektonik and planktonic crustaceans.

The Consumatory Phase

In organisms with well-regulated cyclic development of the gametes, there is often a substantial time lag between the completion of gamete development and the onset of spawning as, for example, in the annelids *Nereis diversicolor* and *Arenicola marina*, in the mollusk *Pecten opercularis* and in various vertebrate species. Furthermore, some species contain mature gametes virtually all year round, although spawning may take place over a relatively restricted period. These observations suggest that spawning is triggered by quite specific external or internal stimuli which may, incidentally, be unconnected with stimuli required for other phases of the reproductive cycle. In the absence of a specific stimulus, mass or epidemic spawning would be difficult to explain, as would the many reports of organisms successfully completing reproductive development but nevertheless failing to spawn.

Spawning can usually be associated with some physical or biological change in the environment. However, in the absence of experimental tests, one should not assume that a particular environmental change is causative simply because it shows a good correlation with the onset of spawning. Most frequently spawning is closely associated with the attainment of a particular breeding temperature for the species or with a marked change in the annual temperature regime, e.g., a fall from the summer maximum or a rise from the winter minimum.[2]

In a careful experimental analysis of factors governing the spawning of the American oyster *O. virginica*, Galtsoff[27,28] found that mature females can be induced to spawn by exposure for a prolonged period to a constant temperature a few degrees higher than that of the normal environment, or, more effectively, to a very substantial rise in temperature, i.e., 20° to 34°C for 20 minutes. However, more than half of the oysters treated in this way failed to spawn, but almost all of them spawned successfully when a suspension of oyster sperm was added to the water.

Males respond to the presence of both oyster eggs and spermatozoa in the water and to a number of non-specific chemical agents. Interestingly, spawning in response to non-specific agents is almost immediate, whereas there is a latent period of 6–27 minutes between contact with the specific gamete suspension and the onset of spawning. It is suggested that the gamete suspensions release activating substances (gamones) which are actually absorbed either through the gills or gut before stimulating the spawning mechanism. Thorson[6] lists species in which it has been recorded that males spawn first, and that this stimulates shedding of eggs by females. He concludes that this is so common "in marine invertebrates . . . that it must be regarded as an ecological rule for species which shed their sexual products freely in the water."

The importance of this device in terms of improving the chances of fertilization of the eggs cannot be overemphasized, especially in view of the short period of survival of active sperm suspensions and fertilizable eggs. In many species the effect of gamones appears to be reciprocal between the sexes. *Nereis limbata* may display a rarer phenomenon among marine forms, namely spawning in response to pheromones. According to Townsend,[29] males will swarm and spawn in water "charged" by a swimming female.

It appears that a certain density of sperm suspension is required to induce spawning in female oysters, which implies that they must be situated close together for successful mass spawning. Among motile invertebrate species which release their gametes freely into the sea, there are records of males and females associating closely together during spawning, e.g., some nereids cohabit in the same burrow while in *Sagartia troglodytes* males and females will move close together before spawning. With the appearance of more highly developed central nervous systems, this process has been refined and various well-recognized and sterotyped behavior patterns serve to stimulated association of males an females. In many vertebrates courtship behavior results in the female becoming receptive to the male and in some cases may be responsible for the release of ovulation-inducing hormones. Reproductive behavior has been carefully described in many species of birds and mammals and also in fresh-water fish. Less is known about marine fish in this respect.

Hawkins, Chapman, and Symonds[30] have described courtship behavior and spawning in the Haddock (*Melanogrammus aeglefinus*) in captivity. This involved aggressive behavior among males with intense sound production and extension of all the fins, which appeared to stimulate the female to approach the male and resulted in leading behavior and a flaunting display by the male. Eventually a male and female came together in a sexual embrace, their ventral surfaces touching. At this stage eggs and milt were released simultaneously. At present there is little information on the internal control of ovulation and oviposition in fish although Barr[15] has obtained some evidence for pituitary stimulation of this process in the plaice.

It is perhaps safe to conclude that in the majority of marine organisms a particular temperature regime creates the necessary environment for spawning and in some individuals in a population may provide the external stimulus. However, in the majority of individuals, the external stimulus for spawning is most probably a biological one provided by the presence of gamones in the water or in some way through the association of one sex with the other.

There are, of course, a number of other stimuli said to induce spawning. The epitoke of the palolo worm, *Eunice fucata* becomes photopositive at particular light intensities which might be experienced at night during lunar quarters (>0.005 ft candelas) so that mass swarming to the surface occurs at this time. Physical disturbance or shock provided by rough weather and changes in tidal pressure may have some influence in inducing spawning in *Patella vulgata* and *Chaetopleura apiculata*, respectively. In certain environments salinity changes may be correlated with spawning. Finally, it has been noted that in some species in temperate and arctic climates the onset of spawning is more closely correlated with the spring phytolplankton increase than with the rise in temperature from the winter minimum which it precedes.

Close investigation suggests that in certain species there may be no immediate external stimulus determining the onset of spawning. The spawning of *Arenicola marina* in the British Isles coincides closely with the first sharp fall in air temperatures experienced during the autumn and the epidemic spawning of this species might suggest that a biological factor is also involved. However, males and females spawn simultaneously in the laboratory when kept under constant-temperature conditions and in individual containers (Fig. 4), although admittedly this spawning takes place a little later than on the shore.[31,32] Eggs in the body cavity are rejected by the nephromixia (gonoducts) during growth but are accepted automatically and shed whenever they display metaphase figures of the first maturation division. Similarly, free spermatozoa are accepted automatically by the nephromixia, whereas sperm morulae are retained in the body cavity.

These changes in the gametes, which result in spawning, take place suddenly in all the reproductive cells and are stimulated by a maturation

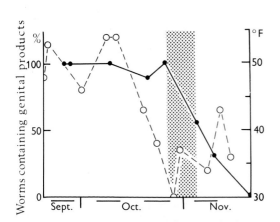 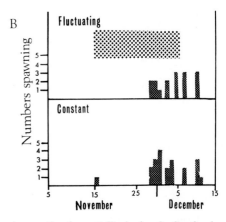

Fig. 4. Spawning of the lugworm, *Arenicola marina*. (A) Spawning on the shore at St. Andrew's, Scotland. Solid circles, percentage of worms containing genital products; open circles, air temperatures. (B) Frequency distribution of spawning among worms maintained in the laboratory at Dublin, Ireland, under either fluctuating or constant conditions. In both diagrams the light stipple indicates the period of main spawning on the shore.

hormone secreted by the supraesophageal ganglion.[33-35] It would appear that timing of release of the maturation hormone is dictated not by any immediate environmental signal but in some way by previous events. More widespread investigation of the spawning mechanism and the means whereby it is controlled internally is needed to yield a better understanding of the interaction of internal and external factors in determining the onset of spawning in invertebrate species.

D. I. D. HOWIE

References

1. Hoar, W. S., *Ann. Rev. Physiol.*, **27**, 51 (1965).
2. Orton, J. H., *J. Marine Biol. Assoc. U.K.*, **12**, 339 (1920).
3. Clark, R. B., in "Neurosecretion," Memoirs of the Society for Endocrinology, 1962, Heller, H., and Clark, R. B., Eds., London and New York, Academic Press, No. 12, pp. 323–327.
4. Wells, M. J., and Wells, J., *J. Exp. Biol.*, **36**, 1 (1959).
5. Donovan, B. T., and Harris, G. W., in "Marshall's Physiology of Reproduction," 1966, Parkes, A. S., Ed., London, Longmans Green, Vol. 3, pp. 301–378.
6. Thorson, G., *Medd. Komm. Havundersøg.*, Kbh. (Serie: Plankton), **4**, 523 (1946).
7. Thorson, G., *Biol. Rev.*, **25**, 1 (1950).
8. Marshall, S. M., and Orr, A. P., in *Essays on Marine Biology*, 1963, Marshall and Orr, Ed., Edinburgh, Oliver and Boyd, pp. 122–140.
9. Stauber, L. A., *Ecology*, **31**, 109 (1950).
10. Loosanoff, V. L., and Davis, H. C., *Science*, **111**, 521 (1950).
11. Amoroso, E. C., and Marshall, F. H. A., in "Marshall's Physiology of Reproduction," 1960, Parkes, A. S., Ed., London, Longmans Green, Vol. 1, Part 2, pp. 707–831.
12. Ball, J. N., *Symp. Zool. Soc. Lond.*, No. 1, 105 (1960).
13. Dodd, J. M., Evenett, P. J., and Goddard, C. K., *Symp. Zool. Soc. London*, No. 1, 77 (1960).
14. Vivien, J. H., *Biol. Bull. France Belg.*, **75**, 257 (1941).
15. Barr, W. A., *Gen. Comp. Endocrinol.*, **3**, 205 (1963).
16. Barr, W. A., *Gen. Comp. Endocrinol.*, **3**, 216 (1963).
17. Dodd, J. M., *Proc. Internat. Congr. Endocrinol., 2nd London*, Excerpta Medica International Congress Series No. 83, 124 (1965).
18. Charniaux-Cotton, H., in "The Physiology of Crustacea," Waterman, T. H., Ed., New York, Academic Press, Vol. 2, pp. 411–447.
19. Fox, H. M., *Proc. Roy. Soc.*, **B95**, 523 (1924).
20. Hauenschild, C., *Cold Spring Harbor Symposia Quant., Biol.*, **25**, 491 (1960).
21. Hoar, W. S. *Mem. Soc. Endocrinol.* No. 4, 5 (1955).
22. Marshall, F. H. A., in "Marshall's Physiology of Reproduction," 1960, Parkes, A. S., Ed., London, Longmans Green, Vol. 1, Part 1, pp. 1–42.
23. Woodhead, A. D., *J. Marine Biol. Assoc. U.K.*, **38**, 407 (1959).
24. Hoar, W. S., *Biol. Rev.*, **28**, 437 (1953).
25. Fontaine, M. M., *Biol. Rev.*, **29**, 390 (1954).
26. Bainbridge, R., in "The Physiology of Crustacea," 1961, Waterman, T. H., Ed., New York, Academic Press, Vol. 2, pp. 431–463.
27. Galtsoff, P. S., *Biol. Bull. Wood's Hole*, **75**, 286 (1938).
28. Galtsoff, P. S., *Biol. Bull. Wood's Hole*, **78**, 117 (1940).
29. Townsend, G., 1939. Chicago, Priv. Ed. University of Chicago Libraries, 58 pp.
30. Hawkins, A. D., Chapman, K. J., and Symonds, D. J., *Nature*, **215**, 923 (1967).
31. Howie, D. I. D., *J. Marine Biol. Assoc. U.K.*, **38**, 395 (1959).
32. Howie, D. I. D., *Gen. Comp. Endocrinol.*, **3**, 660 (1963).
33. Howie, D. I. D., *J. Marine Biol. Assoc. U.K.*, **41**, 771 (1961).
34. Howie, D. I. D., *Nature*, **192**, 1100 (1961).
35. Howie, D. I. D., *Gen. Comp. Endocrinol.*, **6**, 347 (1966).

BRITISH FISHERIES

This article discusses the fisheries of England, Wales, and Scotland as a whole, although the Scottish fisheries are administered separately from those of England and Wales. Scientific names for the principal British commercial species are given in Tables 2 and 3, but these are not repeated elsewhere. No attempt has been made to describe in detail the fishing methods mentioned, as these differ very little, if at all, from those of the same name used by other European countries. Fishing vessels are referred to by registered length, a dimension that is normally a little longer than the length between perpendiculars, but less than the over-all length. Weights of fish landed are expressed to the nearest 100 tons, and values to the nearest thousand pounds sterling.

Size of the Industry

In terms of total weight of fish landed, Britain ranks 11th among fishing nations of the world, and 5th in Europe. The USSR, Norway, Spain, and Iceland are the leading European fishing nations, followed by Britain, France and West Germany in that order. Table 1 gives the quantity and value of the sea fisheries catch landed by British vessels in 1965 and 1966.

Cod and haddock account for over half the weight and value of fish landings in Britain (1966) and, together with whiting, saithe, and plaice, constitute 2/3 of the British catch. The two most important pelagic species, the herring and the sprat, make up about 20% of the land-

TABLE 1. BRITISH LANDINGS OF FISH AND SHELLFISH

	1965		1966	
	Quantity tons	Value £	Quantity tons	Value £
Demersal fish	718,500	54,254,000	700,500	54,294,000
Pelagic fish	157,300	3,583,000	195,900	3,688,000
Shellfish and crustacea	26,200	2,859,000	32,000	3,485,000
Total	902,000	60,696,000	928,400	61,467,000

ings, and the Norway lobster, the lobster, and the crab are the most important crustaceans.

Forty-eight species of fish and shellfish are of sufficient commercial importance to appear in the landing statistics, and the 20 most valuable of these are listed in Table 2. In Table 3 the 20 species landed in the greatest quantity are given; all but three of these are among the top 20 of Table 2. The weight and value of the fish represented by the species listed in Tables 2 and 3 account for about 95% of all British landings.

The fish are caught in the north Atlantic and Arctic Oceans. Although commercial voyages have been made from time to time to the middle and south Atlantic, no permanent British fishery of any size has yet been established in these waters. The pelagic fishery is pursued mainly in waters fairly close to the British Isles, but demersal species are captured all over the north Atlantic from the Gulf of St Lawrence to the Barents Sea. Grounds remote from Britain are termed distant waters (anywhere north of latitude 63°N, west of longitude 17°W, or south of latitude 43°N).

The principal distant-water grounds are off the coasts of Iceland and northern Norway, Bear Island, Spitzbergen, west Greenland, Labrador, and Newfoundland. Nearer home, the principal demersal fisheries are in the North Sea, off the northwest coast of Britain and around the Faroe Islands; the fisheries in the English Channel and around the southern parts of Britain are much less prolific. Britain does not normally fish in the Baltic or its approaches. Table 4 shows the quantities of fish landed from the principal fishing areas.

It can be seen that almost half the demersal catch comes from distant waters, and that Iceland and the North Sea are by far the most important individual areas. The pelagic catch comes almost entirely from the North Sea and the west coast of Scotland.

Fishing Methods

The British fishery depends mainly on four methods of capture: the trawl, the Danish seine, the drift net, and the ring net. Line fishing produces something under 2% of the total catch

TABLE 2. THE 20 MOST VALUABLE BRITISH SPECIES

Name	Scientific name	Value 1965	Value 1966	% of sales (1966)
Cod	Gadus callarias	£25,024,000	£25,058,000	40.8
Haddock	Gadus aeglefinus	10,848,000	10,829,000	17.6
Plaice	Pleuronectes platessa	5,690,000	5,960,000	9.7
Herring	Clupea harengus	2,898,000	2,850,000	4.6
Whiting	Gadus merlangus	1,737,000	2,507,000	4.1
Saithe	Pollachius virens	1,656,000	1,576,000	2.6
Norway lobster	Nephrops norvegicus	954,000	1,347,000	
Hake	Merluccius merluccius	1,563,000	1,186,000	
Lemon sole	Microstomus kitt	1,162,000	1,130,000	
Skates and rays	Raja sp.	1,109,000	1,073,000	
Lobster	Homarus vulgaris	904,000	943,000	
Halibut	Hippoglossus hippoglossus	915,000	809,000	
Sprat	Sprattus sprattus	491,000	634,000	
Sole	Solea solea	664,000	630,000	
Dogfish	Squalus acanthias	387,000	499,000	
Turbot	Scophthalmus maximus	542,000	475,000	
Redfish	Sebastes sp.	525,000	410,000	
Crab	Cancer pagurus	355,000	395,000	
Catfish	Anarhichas sp.	368,000	328,000	
Angler	Lophius piscatorius	292,000	253,000	
		£58,084,000	£58,892,000	96%

BRITISH FISHERIES

TABLE 3. THE 20 SPECIES LANDED IN LARGEST QUANTITY

Name	Scientific name	Weight (tons) 1965	Weight (tons) 1966	% of total (1966)
Cod	*Gadus callarias*	311,200	319,900	34.5
Haddock	*Gadus aeglefinus*	165,900	149,700	16.1
Herring	*Clupea harengus*	96,400	112,700	12.1
Sprat	*Sprattus sprattus*	56,300	78,300	8.4
Whiting	*Gadus merlangus*	46,800	51,300	5.5
Saithe	*Pollachius virens*	46,800	44,400	4.8
Plaice	*Pleuronectes platessa*	39,100	39,600	4.3
Redfish	*Sebastes* sp.	15,900	13,000	
Skates and rays	*Raja* sp.	12,400	11,200	
Dogfish	*Squalus acanthias*	8,500	9,700	
Norway lobster	*Nephrops norvegicus*	5,500	7,300	
Catfish	*Anarhichas* sp.	7,900	6,500	
Lemon sole	*Microstomus kitt*	5,800	5,500	
Hake	*Merluccius merluccius*	6,500	4,900	
Crab	*Cancer pagurus*	4,300	4,700	
Ling	*Molva molva*	5,300	4,000	
Angler	*Lophius piscatorius*	4,400	3,400	
Mackerel	*Scomber scombrus*	3,100	3,200	
Halibut	*Hippoglossus hippoglossus*	3,200	2,500	
Dab	*Limanda limanda*	2,000	2,000	
		847,200 tons	873,800 tons	94%

and miscellaneous inshore nets and traps another 1%; purse seining, begun on a small scale in 1966, does not yet figure significantly in the landings (2600 tons of herring were caught by this method during the year).

In 1965 more than ⅔ of all British fish was caught by trawling, and three-quarters of the demersal catch were captured by this method. Danish seining accounted for nearly all the remainder of the demersal catch, the two methods between them producing over 97% of the catch of bottom-living fish.

The drift net and the ring net are used for capturing pelagic fish, and the two methods together accounted for about ⅔ of the pelagic catch in 1965. The remaining third, almost entirely sprats, were caught by pelagic trawl.

Bottom trawling is the only method of importance used by British ships in distant waters, long lining having become almost obsolete; Danish seining is practiced only on fishing grounds near to Britain. Table 5 shows the approximate amounts of fish captured by each method of fishing.

TABLE 4. QUANTITIES LANDED FROM THE PRINCIPAL FISHING GROUNDS (BASED ON 1965 LANDING FIGURES)

Fishing ground	Demersal catch tons	% demersal total	Pelagic catch tons	% pelagic total	All species tons	% total catch
Barents Sea	34,500	4.9	...	...	34,500	4.0
Norway Coast	30,600	4.4	...	...	30,600	3.6
Bear Is. and Spitzb'n.	36,000	5.2	...	...	36,000	4.2
Iceland	187,600	26.8	...	...	187,600	21.9
Greenland	13,000	1.9	...	...	13,000	1.5
Newfoundland and Labr.	35,300	5.0	...	...	35,300	4.1
All distant water	337,000	48.2	...	...	337,000	39.3
Faroe Islands	31,700	4.5	...	...	31,700	3.7
North Sea	233,500	33.3	94,000	59.7	327,500	38.2
Scotland W coast	80,300	11.5	54,300	34.6	134,600	15.7
England W coast	13,200	1.9	5,500	3.5	18,700	2.2
England S coast	4,200	0.6	3,500	2.2	7,700	0.9
All grounds	699,900[a]	100.0%	157,300	100.0%	857,200[a]	100.0%

[a] These totals are slightly lower than the figures given for 1965 in Table 1, since landings of fish livers are not included here.

TABLE 5. QUANTITIES LANDED BY METHOD OF FISHING (TONNAGES GIVEN ARE APPROXIMATIONS)

Fishing method	Demersal catch tons	% demersal catch	Pelagic catch tons	% pelagic catch	Total catch tons	% total catch
Trawl	527,600	75.5	52,400	33.3	580,000	67.7
Danish seine	154,800	22.1	1,200	0.8	156,000	18.2
Drift net			61,700	39.2	61,700	7.2
Ring net			37,000	23.5	37,000	4.3
Lines	8,000	1.1	3,000	1.9	11,000	1.3
Other	9,500	1.3	2,000	1.3	11,500	1.3
	699,900	100.0%	157,300	100.0%	857,200	100.0%

Fishing Vessels

There are about 2000 fishing vessels of more than 40 ft; a further 6000 craft under 40 ft are registered as fishing vessels, but many of these are used only part of the year for fishing, often by a variety of methods in inshore waters, and contribute only a very small part of the total landings.

What might be called the deep-sea fleet (vessels of 80 ft and over) is composed almost entirely of trawlers. At the end of 1965 there were just over 600 trawlers in this group, almost all steel-hulled; more than ¾ of them have diesel engines, the remainder being steam-driven. The deep-sea trawlers can be divided into three main size groups of roughly equal number, known as distant-water, middle-water, and near-water trawlers.

The distant-water trawlers, vessels of 140 ft and over, are based mainly at the ports of Hull and Grimsby, with a few at Fleetwood. They work almost entirely on the distant-water grounds and catch almost half of the demersal landings. Less than 10% were built before World War II, and these will disappear in the next few years. About ⅔ of this group are steam trawlers, but no new steam trawlers have been built since 1958, and whenever new diesel vessels are added to the fleet it is usual to withdraw at least as many of the older steam trawlers from service.

A typical motor trawler in this group is 190 ft long, 34 ft wide, and 17 ft deep from the main deck; with a gross registered tonnage of about 800 tons, a main propulsion engine of about 2000 hp and a speed of 15 knots, she can carry up to 300 tons of fish in crushed ice, and will make voyages of about 3 weeks with a crew of 20. The fishroom is insulated, lined with light alloy, and fitted with light alloy boards for constructing the portable compartments or pounds in which the fish are stowed in bulk with crushed ice. The most important piece of ancillary machinery is the 300-hp winch for hauling the trawl gear on board. The bridge is equipped with radar, radio transmitters and receivers, echo sounders and other fish-detection devices, gyro compass, and automatic pilot. The engine-room is aft, crew's accommodations are located midships and aft, and the fishroom is in the forepart of the ship. The cod end of the trawl is brought over the side of the trawler, the catch is discharged on the foredeck, gutted, washed, and put into the fishroom immediately below.

Because of the remoteness of the distant-water fishing grounds, and the limitations of storage of fish chilled in ice, British trawler owners have built a number of large distant-water trawlers that can freeze the catch on board and store it at low temperature until the vessel reaches port; this enables the ship to continue fishing until the hold is full without fear of spoilage, thus increasing the proportion of time spent fishing and also providing the processing industry with first-class raw material that can be stored on shore for long periods to maintain continuity of supply.

Britain's first trawler equipped for freezing on board entered service in 1954; this ship, a factory vessel of over 2600 registered tons, fillets the fish before freezing them, and carries a crew of over 70. Two sister factory ships have since been built. These ships were also the first British trawlers to fish commercially over the stern, and all Britain's freezer trawlers built since then have been stern trawlers also.

In 1961 the first British trawler for freezing whole fish went into service, and since then the number of freezer trawlers of this type has steadily increased; at the end of 1966 19 were in use, and several more were due to be completed in 1967. In addition, a smaller type of fillet-freezing trawler was built in 1965, followed by two sister ships. The total complement of distant-water trawlers fitted for freezing the catch at the end of 1966 was 25, made up of 3 factory trawlers, 3 smaller fillet freezers, and 19 whole-fish freezers, all fishing over the stern.

The 3 factory trawlers together catch about 12,000 tons of fish a year for conversion on board to blocks of frozen fillets, and the whole fish trawler freezers land about 2750 tons a year. In 1965 British freezer trawlers caught about

TABLE 6. THE BRITISH FISHING FLEET (1965)

Type	Method of propulsion	140 ft and over	Registered length 110–139.9 ft	80–109.0 ft	40–79.9 ft	Total	%
Trawler	steam	133	10			966	48
	motor	63	183	203	374		
Drifter/trawler	motor			19		19	1
Seiner	motor			1	651	652	32
Drifter	motor			1	82	83	4
Liner	motor			7	95	102	5
Ring netter	motor				75	75	4
Other	motor				111	111	6
		196	193	231	1388	2008	100%

31,000 tons of fish and in 1966 45,000 tons, about 90% of which was cod, accounting for 13% of all distant-water fish, or about 6½% of all demersal fish landed.

In the near-water and middle-water trawler fleets, steam propulsion has disappeared and the motor trawler reigns supreme. Almost all the 400 ships in these two groups are side trawlers stowing the catch in crushed ice, the near-water vessels working mainly in the North Sea and around the west coast of Britain, and the middle-water ships fishing as far away as the Faroe Islands and southeast Iceland. Stern trawling is not yet common practice in these fleets, and at the end of 1966 only 6 near- and middle-water trawlers fished over the stern.

Trawlers generally are single-purpose fishing vessels, but a very small number of the smallest size of deep-sea trawlers are sometimes converted to drift-net fishing for herring during part of the year.

The next most important group of British fishing vessels is the seiner type, of which there were over 700 at the end of 1966. These are usually wooden-hulled, diesel-driven, ranging mainly from 40 to 80 ft in length. This type is less rigidly tied to one method of fishing, and fairly frequent changes are made from seining to drifting or lining to suit a particular area or season. The seiner can also be readily adapted for towing a light trawl, so that the type is truly a multi-purpose one.

Apart from a handful of vessels of the near-water trawler type that still practice great lining, the remainder of the fleet is in the 40–80 ft, range and fitted mainly for drifting, ring netting, or lining; they are essentially seiner hulls equipped with the appropriate deck gear for the method.

Fishing Ports

The British deep-sea fleet is concentrated principally at eight fishing ports, all but two of which are on the east coast; all are primarily trawling ports. In addition, there are a number of ports, mainly in Scotland, at which considerable quantities of demersal fish are landed by seiners. Fish landed at the west-coast ports are often consigned direct by road to larger ports on

TABLE 7. THE 10 PORTS AT WHICH MOST FISH WERE LANDED 1965
(DEMERSAL AND PELAGIC, BUT EXCLUDING SHELLFISH)

Port	Demersal tons	% British landing demersal	Pelagic tons	% British landing pelagic	All species tons	% Total British landing	All species tons 1966
Hull	193,700	27.0	...	...	193,700	22.1	190,500
Grimsby	166,300	23.2	5,700	3.6	172,000	19.7	163,700
Aberdeen	102,800	14.3	7,900	5.0	110,700	12.7	107,400
Fleetwood	48,200	6.7	100	...	48,300	5.5	45,100
Fraserburgh	17,300	2.4	13,300	8.5	30,600	3.5	29,200
Lowestoft	19,500	2.7	1,800	1.1	21,300	2.4	23,100
Buckie	12,100	1.7	9,000	5.7	21,100	2.4	16,400
Oban	4,100	0.6	10,900	7.0	15,000	1.7	16,000
North Shields	12,300	1.7	2,000	1.3	14,300	1.6	20,800
Ayr	5,400	0.7	7,800	5.0	13,200	1.5	10,100
	581,700	81.0%	58,500	37.2%	640,200	73.1%	622,300

the east coast, around which the processing industry is concentrated.

Table 7 shows the 10 most important British ports in terms of quantity of the fish landed. Over 70% of the demersal fish, and 60% of all species, were landed at the first four ports listed in Table 7.

Cod Fishery

The cod is by far the most important species in British sea fisheries, both in quantity and value; more than ⅓ of the weight and the money is in cod. Nearly 90% of it is caught in the trawl,

The average price of haddock was about £65 a ton in 1965, and about £72 in 1966.

Plaice Fishery

British takings of plaice come principally from the North Sea, and ⅔ of the landings of this flatfish are made at the east-coast ports of Grimsby and Lowestoft. Nearly ¾ of all plaice are caught in the trawl, and almost all the remainder in the seine net. Although caught in far smaller quantities than the cod or the haddock, the popularity of the plaice makes it the third most valuable British species.

TABLE 8. WHERE COD ARE CAUGHT (1965)

Fishing ground	Catch tons	% of cod landed
Iceland and NE Atlantic	184,350	59
NW Atlantic	43,450	14
	227,800	73% from distant waters
North Sea	55,050	18
W coasts of Britain	17,450	6
Faroe Is.	10,950	3
	311,250	100% value £25,024,000

TABLE 9. WHERE HADDOCK ARE CAUGHT (1965)

Fishing ground	Catch tons	% of haddock landed
Iceland and NE Atlantic	45,900	27
NW Atlantic	900	1
North Sea	83,300	50
W coasts of Britain	27,500	17
Faroe Is.	8,300	5
	165,900 tons	100% value £10,848,000

and a further 9% comes from the seiners. The ports of Hull and Grimsby between them receive nearly ¾ of the cod catch, the bulk of it from distant waters.

Table 8 shows the principal grounds concerned. The average price of cod was about £80 a ton in 1965, £78 a ton in 1966.

Haddock Fishery

About ⅙ of Britain's catch is haddock, caught mainly by trawl and seine and landed at most British ports, but particularly in Scotland; it is the most important species on Aberdeen market, where ¼ of Britain's haddock catch is sold. About ⅔ of all haddock are captured in trawls, and almost ⅓ by seines (Table 9).

The average price of plaice was about £145 a ton in 1965, and £150 in 1966.

Whiting Fishery

The whiting is essentially a near-water fish, and occurs in large quantities around the northern coasts of Britain; it figures prominently in the landings at Scottish ports. Seining accounts for about ⅔ of the landings, and trawling for most of the remainder. Less esteemed than cod or haddock, it usually finds a less favorable market (Table 11).

The average price of whiting was about £37 a ton in 1965, and £49 in 1966; demand varies considerably, often depending on availability of haddock and other species.

Saithe Fishery

The saithe, or coalfish, although landed in considerable quantity, is not very popular with the consumer, and most of the landings are used for purposes other than direct sale to the public, for example as raw material for fish cakes and other products. Caught almost entirely by trawlers, and landed at the four big trawling ports, it comes from both near and distant waters and is often caught along with the cod when that fish is being hunted (Table 12). attempts are now being made to use the purse seine, following the pattern of Scandinavia. The catch is taken almost entirely in near waters, 38,900 tons from the North Sea (40%) and 57,500 tons (60%) from the west coast of Britain in 1965. The total of 96,400 tons realized £2,898,000 at first sale. Landings are seasonal as the fishery moves around the coast; most of the landings are transported immediately by road to processing centers such as Fraserburgh and Aberdeen.

TABLE 10. WHERE PLAICE ARE CAUGHT (1965)

Fishing ground	Catch tons	% plaice landed
Iceland and NE Atlantic	5,900	15
North Sea	28,300	72
Other near waters	4,900	31
	39,100 tons	100% value £5,690,000

TABLE 11. WHERE WHITING ARE CAUGHT (1965)

Fishing ground	Catch tons	% total landed
North Sea	35,400	75
West coasts of Britain	10,100	22
Other near waters	1,300	3
	46,800 tons	100% value £1,737,000

TABLE 12. WHERE SAITHE ARE CAUGHT (1965)

Fishing ground	Catch tons	% total landed
Iceland and NE Atlantic	19,700	42
NW Atlantic	600	1
Faroe Is.	7,700	17
North Sea and W Scotland	18,800	40
	46,800 tons	100% value £1,656,000

The average price of saithe was about £36 a ton in 1965, and about £35 in 1966.

Herring Fishery

Although of far less importance in British fisheries now than formerly, the herring is nevertheless still the most important pelagic species, and one that has been the subject of numerous attempts at improved catching methods. The drift net is still the most common fishing gear, accounting for about 55% of the 1965 landing, but on the west coast the ring net is also used to land a further 40%. A small amount, about 2%, is captured by means of the midwater trawl, and

Table 13 shows how the herring catch was disposed of in 1965; some of the traditional products like pickle cured and red herrings are now made only in small quantities, and the kipper is the most important herring product. Just over 20% of British herring products were exported in 1965, but at the same time 6800 tons of fresh and frozen herring for processing were imported from Scandinavia, about 7% of the total intake.

Processing of the Catch

There are no accurate figures available for the quantities of British-caught fish dealt with by the various methods of processing but, taking

TABLE 13. HOW THE HERRING CATCH IS PROCESSED (1965)

Method	% catch
Fresh and cold smoked	41
Quick frozen	17
Klondyked	4
Canned	7
Pickle cured	3
Redded	2
Marinated	2
Pet food	14
Fish meal	10
	100%

the demersal and pelagic landings of 1965 as a whole, the following very rough percentages apply: chilled 57%, frozen 20%, smoked 10%, canned 1%, dried, salted, marinated, and other miscellaneous methods 1%, pet foods 3%, fish meal 8%. Much of the fish is subjected to more than one of these processes; for example, fish may be frozen for subsequent thawing and smoking, or smoked and then frozen, so that there is sometimes overlap between the methods; in addition the amounts used for pet foods and fish-meal manufacture fluctuate from year to year; the foregoing figures are therefore no more than an indication of the pattern of fish processing.

The amount of frozen fish has been rising steadily for some years, and in 1965 about 73,000 tons were produced, mainly as fillets or fish fingers, from about 159,000 tons of demersal fish. Just over half of the production was consumer packs and the remainder was in the form of large blocks for catering and similar outlets. In addition a further 31,000 tons of frozen demersal fish were imported for distribution during that year, although this was counterbalanced to some extent by the export of about 16,000 tons of frozen demersal fish products.

An accurate total for smoked fish production is not available, but in Scotland, where fish landings amounted to about 42% of the British total in 1965, 11,000 tons of smoked fish were made from cod and allied species, roughly equal to about 25,000 tons of whole fish as landed, and 26,000 tons of herring were used for kippering. Thus about 14% of all Scottish fish was used for smoking, and it is estimated that about 10% of the British catch as a whole was used for this purpose.

Canning is confined principally to herring and sprat, although small quantities of demersal fish such as whiting are canned in bulk for use in other food preparations. About 9000 tons of fish were canned in 1965.

The white fish meal industry received about 64,000 tons of whole demersal fish in 1965, and in addition used about 324,000 tons of fish offal resulting from filleting and other processes to make 84,000 tons of white fish meal. A further 10,000 tons of herring were also used in the manufacture of meal and oil. In the same period Britain imported about 190,000 tons of white fish meal and about 170,000 tons of herring meal, mainly from South Africa, Peru, Norway and Iceland.

Administration and Research

Regulation and inspection of fisheries in England and Wales is the concern of the Fisheries Department of the Ministry of Agriculture, Fisheries and Food; in Scotland the Fisheries Division of the Department of Agriculture and Fisheries for Scotland serves the same purpose. Both organizations have research establishments and vessels for investigating the fisheries, the most important laboratories being at Lowestoft and Aberdeen. The Fisheries Inspectorate provides the link between the Government and the catching industry by having inspectors and fishery officers covering the whole of the British coastline.

Two statutory bodies, the Herring Industry Board and the White Fish Authority, financed partly from the Treasury and partly by levies on landings, are concerned with fisheries development; they give technical and commercial advice, have some powers to make regulations, and can give financial assistance towards approved schemes for new vessels, plants, and equipment. They also sponsor research and development work that will benefit British fisheries.

Problems concerning the handling, processing, and preservation of fish as food are investigated by a Ministry of Technology research station at Aberdeen.

The increasing European fishing effort in the north Atlantic is making it continually more difficult to pursue a profitable fishery; the declining yield in the northeast Atlantic has forced more and more distant-water trawlers to work off Greenland and North America, where the freezer trawlers have found their element. Greater interest is being shown in non-traditional grounds in the south Atlantic, already being fished by other European countries. Stocks of less popular species are being seriously studied, and ways and means of making use of these fish are being investigated, both for new food products and for direct reduction to animal feeding stuffs.

Britain imported 216,000 tons of fishery products in 1965, valued at £67,000,000, as opposed to 53,000 tons of exports valued at

£10,000,000, and in addition nearly £40,000,000 worth of marine meal and oil products were bought from overseas; any reduction of this imbalance of trade would be economically worthwhile.

The proportion of the British catch that is frozen and cold stored will probably continue to increase, and the recent marked expansion in the fishery for shellfish and crustacea should also continue. The advent of purse seining should help to augment supplies of pelagic fish, and improved design of trawlers will make manning problems simpler and shipboard preservation more efficient, but it seems unlikely at the present time that the rather small profit margin of the catching industry will markedly increase in the next decade or so.

J. J. WATERMAN

Cross references: *Fishing Vessels and Support Ships; Freezing Fish at Sea.*

TABLE 1. BROMINE CONCENTRATIONS IN SEAWATERS OF THE WORLD[3]

Numbers given are in ppm.	
Adriatic Sea	69
Antarctica	65
Arabian Sea	49
Arctic Ocean	46
Atlantic Ocean, eastern	69
northwest	71
southwest	68
Australian Coast	81
Azov Sea	12
Baltic Sea	24
Barents Sea	68
Bering Sea	65
Black Sea	29
British Isles	66
Caspian Sea	6[4]
English Channel	79
Gulf of Aden	48
Gulf of Alaska	62
Gulf of Mexico	66
Indian Ocean	67
Irish Sea	61
Japan, Sea of	68
Marmara	45
Mediterranean Sea	72
North Sea	44
Norwegian Sea	48
Pacific Ocean, northeast	58
Persian Gulf	110[5]
Red Sea	66
South China Sea	40
Suez Canal	92
White Sea	50
Average, all oceans	66

Note: The concentrations shown are averages of several analyses, some of which are many years old.

BROMINE FROM THE SEA

Of all the elements found in nature, bromine is relatively low in abundance. The average concentration of bromine in the earth's crust varies from 3 ppm in sedimentary rocks to 30 ppm in peat (the bromine content increases with increasing organic-matter content).[1] Bromine is also very dilute in seawater, averaging only 65 ppm. When considering the total volume of the oceans, however, and the fact that each cubic mile of seawater contains 607 million pounds of bromine, the world's bromine resources are inexhaustible. At present, less than one third of a cubic mile of seawater is processed each year for bromine recovery.

Bromine exists in the oceans as the bromide ion, fourth most abundant anion in seawater. Only chlorine, sulfate, and carbonate are more abundant.[2] The average bromine content of seawater is about 65 ppm, but concentrations vary around the world with ocean currents. Table 1 shows bromine concentrations in seawater at various locations around the world. Since the relative ratios of the various constituents in seawater are always the same, specific gravity or chloride content may be used as a measure of bromide concentration, except for locations near river outlets where dilution and contamination may be appreciable.[6]

Bromine Recovery

Many commercial bromine processes have been tried, but economical extraction of bromine from brines and seawater has always involved the oxidation of the bromide ion (Br^-) to free bromine (Br_2), removal of Br_2 from the brine, and recovery of liquid bromine or bromine combined in inorganic compounds.

A. J. Balard discovered bromine in 1825 by oxidizing a sample of seawater bittern with chlorine. Production processes using chlorine to oxidize the bromine were developed as early as 1877,[7-12] but difficulty in transporting this raw material prevented its use until later. Because of improved chlorine technology and materials of construction, however, nearly all bromine obtained today is oxidized by chlorine.

$$Cl_2 + 2Br^- \rightarrow Br_2 + 2Cl^-$$

Initial commercial bromine production was based on oxidation by manganese dioxide in the presence of sulfuric acid.

$$MnO_2 + 4H^+ + 2Br^- \rightarrow Mn^{++} + Br_2 + H_2O$$

Low yields and high consumption of raw materials and utilities, however, made this process uneconomical.[13] Two electrochemical oxidation processes were developed and used for a time,[14,15] but these processes employed either diaphrams or bi-polar carbon electrodes which faced problems of small cell outputs, clogging by $Mg(OH)_2$, and low current efficiencies.[13] Chlorates were also tried as oxidizing agents,[16] but the acid requirement was a major drawback because of the difficulty of transporting HCl and the problem of sulfate precipitation when H_2SO_4 was used.

$$6Br^- + 6H^+ + ClO_3^- \to 3Br_2 + Cl^- + 3H_2O$$

Only two important processes have been developed to remove bromine from its solution in brine or seawater once the bromide has been oxidized to free bromine. Steam stripping has been used since the beginning of commercial bromine production, and is used today in the Kubierschky process. Air stripping was developed by H. H. Dow in 1889 and is the basis of the "blowing-out" process.[17] Solvent extraction of bromine from chlorinated brines has been investigated but is not used for large-scale bromine production.

After free bromine has been removed from the brine or seawater, further treatment is necessary to produce a liquid bromine product. If the steam-stripping process is used, the bromine and water vapors may be condensed and separated in much the same way as in two-phase azeotropic distillation. Liquid bromine is recovered directly.

In the blowing-out process, however, large volumes of air accompanying the bromine vapors prohibit condensation unless extremely low temperatures are used. Several other means of removing the bromine from the air are employed, the most common of which involves reaction with a reducing agent.

In Dow's first blowing-out process, the air and bromine vapors were passed through a bed of iron filings, the resulting iron bromide being sold as a product.

$$2Fe + 3Br_2 \to 2FeBr_3$$

Ferrous bromide solutions were also used on occasion.

$$2FeBr_2 + Br_2 \to 2FeBr_3$$

Ammonia, sodium hydroxide, and sodium carbonate solutions were used for many years and are still used in some bromine plants. The result is a concentrated aqueous bromide solution.

$$8NH_3 + 3Br_2 \to 6NH_4Br + N_2$$

$$6NaOH + 3Br_2 \to 5NaBr + NaBrO_3 + 3H_2O$$

$$3Na_2CO_3 + 3Br_2 \to 5NaBr + NaBrO_3 + 3CO_2$$

If reduction is accomplished with an alkali hydroxide or carbonate, a liquid bromine product is obtained by treating the concentrated bromide-bromate solution with acid, followed by steam stripping.

$$5NaBr + NaBrO_3 + 3H_2SO_4 \to 3Br_2 \\ + 3Na_2SO_4 + 3H_2O$$

The reducing agent now most widely used is sulfur dioxide.[18]

$$Br_2 + SO_2 + H_2O \to 2HBr + H_2SO_4$$

Liquid bromine product is obtained from the resulting concentrated aqueous bromide solution by chlorine oxidation followed by steam stripping.

In the "cold process," as reported by Israeli research, bromine is absorbed from the airstream by a cold ($<0°C$) concentrated aqueous sodium bromide solution, from which the bromine is then liberated by heating in a subsequent steam-stripping column.[13]

The simplicity of the steam-stripping process makes it look most desirable for all bromine recovery plants. But since most of the steam used in the steam-stripping process is expended in heating the brine, this process is attractive for bromine removal only from more concentrated brines, i.e., from brines containing less water to be heated per pound of bromine. The blowing-out process, in one form or another, is the only economical process currently used in removing bromine from ocean water.*

The Seawater–Bromine Process

The world's largest bromine plant was built in Freeport, Texas during World War II, and currently produces almost ⅔ of the bromine that is recovered from ocean water. This plant uses Dow's blowing-out technology by which the bromide-containing brine is oxidized with chlorine and blown with air to strip the bromine from solution. This process functions according to the flow sheet shown in Fig. 1.

Seawater flows over a traveling screen and into a basin, from which it is pumped to the top of a blowing-out tower through a large pipe. This pipe also serves as a mixing chamber in which acid and chlorine are added. The acid is "recycle acid" supplemented by fresh dilute sulfuric acid, and is added before the chlorine in order to reduce the pH to about 3.5 and suppress undesirable side reactions of chlorine. Air, drawn up through the packing in the blowing-out tower

---

* Bromine has been removed from seawater and used directly in the production of tribromoaniline in a floating chemical factory on the U.S.S. *Ethyl*.[19]

$$3Br^- + 3Cl_2 + C_6H_5NH_2 \to \\ C_6H_2NH_2Br_3 + 3Cl^- + 3HCl$$

Low yields and high costs, however, shortened the commercial life of this process.

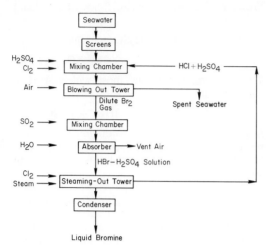

Fig. 1. Seawater blowing out process for bromine. (Courtesy The Ethyl-Dow Chemical Co., Freeport, Texas.)

by powerful fans, desorbs free bromine from the seawater which is returned to the sea. The moist bromine-laden air is mixed with sulfur dioxide, and passes into an adjacent absorption tower where the bromine is reduced and absorbed in aqueous solution.[18,20,21] The strong hydrobromic acid solution resulting from this absorption* is pumped to storage. This solution is then contacted with chlorine, and the oxidized bromine is stripped from solution in a steaming-out process, which is a modification of the Kubierschky process. The effluent liquid from the steaming-out process is recycle acid, $HCl + H_2SO_4$, and is used earlier in the blowing-out process to reduce the pH of the entering seawater.[22] Over-all bromine recovery is about 85%.

Acidification. At total bromine concentrations lower than 1000 ppm, the influence of pH becomes very important. Acidification of seawater is necessary because of the hydrolysis which, if not counteracted, may result in greatly reduced desorption efficiencies. Bromine is hydrolyzed as follows:

$$Br_2 + H_2O \rightarrow H^+ + Br^- + HOBr$$

Acid also reacts with ammonia and carbonates in the seawater thereby reducing the subsequent chlorine requirement. Two tons of concentrated acid per ton of bromine are generally used.[23]

Chlorination. About 0.44 lb of chlorine is theoretically required to oxidize each pound of bromine. In cases where brines contain reducing substances, usually organic matter, the chlorine requirement may increase by 50% or more. Compounds formed by the reaction of chlorine with

* This solution also contains H_2SO_4 and a little HCl.

bromide and bromine also result in increased chlorine demand.

$$Cl_2 + Br^- \rightarrow Cl^- + BrCl$$
$$Cl_2 + Br_2 \rightarrow 2BrCl$$

Seawater requires about 15% excess chlorine.[23]

Stripping. The second step in the blowing-out process is the stripping of the oxidized solution of its free bromine content. Bromine is transferred from the seawater to the air stream as a result of the vapor pressure exerted by the bromine. Because of the small concentrations of bromine in the seawater, a vast amount of air from high-capacity fans is required to do the stripping. Formation of addition compounds of bromine with chlorides and bromides (e.g., BrCl) is an obstacle to stripping.

Absorption. Since the low concentrations of bromine vapor in the air stream prevent separation by cooling, the bromine vapors are chemically reduced by the addition of sulfur dioxide vapor.[18] The reaction is caused to take place partly in the gaseous phase which gives rise to the formation of a spray of fine droplets of mixed acid. The mist is later trapped by a solution of HBr and H_2SO_4 circulating in an absorption tower. The contact area between the air–bromine mixture is made up of the surface of the droplets and of the liquid–gas interface in the absorption tower. Absorption is complete and the concentration of HBr in the absorbing solution reaches nearly 20%.

The Steaming Out Step. Since the absorbed bromine has been converted to non-volatile

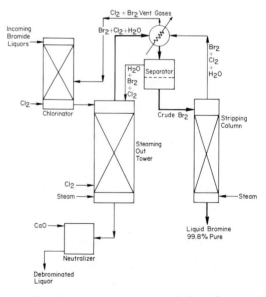

Fig. 2. Steaming out process for bromine.

bromide, it has to be liberated again before it can be recovered from the absorbing solution. Re-oxidizing with chlorine is done simultaneously with steaming out in what is essentially a Kubierschky process* (Fig. 2). The amount of steam required, however, is much less than that required when steaming out more dilute brines and bitterns. The acid bromine solution is heated by the steam inside packed towers to near boiling. The bromine set free is steam distilled, condensed, and separated.

Economic Aspects

The important variables to consider in locating a seawater bromine plant are bromine concentrations, seawater temperatures, and raw material and utility costs. Since about 15,000 lb of seawater must be processed to recover each pound of bromine, primary consideration should be given to brines with a consistently high bromine concentration, not diluted by various ocean currents, river outlets, or bromine-plant effluents. The seawater should also be free from organic contamination. Temperature is an important factor, since the vapor pressure of bromine in water is quite dependent on temperature. Almost twice as much bromine may be recovered from seawater at 25°C as opposed to seawater at 10°C.[30] Costs will vary with location and economic conditions. The principal operating expenses are for chlorine, acid, and electrical power. Investment capital is relatively high because of size and type of equipment needed.

World Production

More than ⅓ of the world's bromine production comes directly from the sea, the remainder coming from salt lakes, evaporates, and underground brines. Seawater is the source of over half the bromine produced outside the United States.

There are more than six plants in five countries that recover bromine directly from seawater.

France. A seawater bromine plant at Port DeBouc, near Marseille produces an estimated 20 million pounds of bromine per year using a blowing-out process.[24]

Great Britain. Two seawater bromine plants use the Dow blowing-out process. One is located at Hayle, Cornwall, and has a capacity of 9 million pounds per year. The other is located at Amlwch, Wales, and has an estimated capacity of 36 million pounds per year.[5,25]

Italy. An estimated 3 million pounds of bromine

* Although the original steaming-out process was developed in Germany for processing the Stassfurt deposits and with modifications is still used, the "hot" or "steaming-out" process as it is applied now, with certain modifications in many places, has been developed by Dr. Konrad Kubierschky.

TABLE 2.
WORLD BROMINE PRODUCTION [5, 13, 23–25, 26–29]

	Plant capacity, million lb/yr	
	Seawater plants	Other plants
France	20	5.4
Great Britain	45	
India		0.002
Israel		26
Italy	3	9
Japan	4	16
Spain		1
U.S.	115	250
U.S.S.R.		?
West Germany		5.8
Total	187	313

per year is produced from seawater at Priolo, near Syracusa, Sicily.[13]

Japan. There are several seawater bromine plants in Japan, the largest of which is at Nan Yo-cho, on the island of Honshou. These plants use the blowing-out process followed by alkaline absorption to produce more than 4 million pounds of bromine per year.[13,25]

U.S. The world's largest seawater bromine plant is located in Freeport, Texas, on the Gulf of Mexico, and produces 115 million pounds of bromine per year in a blowing-out process previously described.

Table 2 summarizes world bromine production.

M. P. MOYER

References

1. Rankama, K., and Sahama, Th. G., "Geochemistry," 1950, Chicago, Univ. Chicago Press, pp. 294–299, 766–269.
2. Horne, R. A., *Water Resources Res.*, **1**, No. 2, 263 (1965).
3. Goldberg, E. D., Associate Professor of Chemistry, University of California, 1958 (personal communication).
4. Lotze, F., "Steinsalz and Kalisalze," 1957, Berlin-Nikolasse, Gebruder Borntraeger, I, Teil, p. 37.
5. The Dow Chemical Co., 1966, Dow Estimates.
6. Osmun, R. H., The Dow Chemical Co., 1944 (personal communication).
7. Borsche, G., German Patent 9353.
8. Frank, German Patent 2241.
9. Kubierschky, K., German Patent 194,657.
10. Kubierschky, K., German Patent 269,995.
11. Kubierschky, K., German Patent 410,107.
12. Muller, R., Bockel, H., German Patent 7743.
13. Jolles, Z. E., Ed., "Bromine and its Compounds," 1966, London, Ernest Benn Ltd., pp. 1–42.
14. Kossuth, German Patent 103,644.
15. Wunsche, German Patent 140,274.
16. Phalen, W. C., *Bull. Bur. Mines*, **146**, 85 (1917).
17. Campbell, M., Hatton, H., "Herbert H. Dow—Pioneer in Creative Chemistry," 1951, New York, Appleton-Century, Crofts, Inc.

18. Heath, U.S. Patent No. 2,143,223 (1937).
19. Stine, C. M., *Ind. Eng. Chem.*, **21**, 434 (1929).
20. Hart, P., *Chem. Eng.*, **54**, No. 10, 102 (1947).
21. Hooker, U.S. Patent No. 2,043,224 (1929).
22. Shreve, R. N., "The Chemical Process Industries," 1956, New York, McGraw-Hill Book Co., Inc., pp. 425–427.
23. Stewart, L. C., *Ind. Eng. Chem.*, **26**, 361 (1934).
24. *European Chem. News*, **7**, No. 172, 4 (1965).
25. *Inform. Chim.*, p. 48 (December, 1966).
26. "Chemical Horizons," Overseas Report, p. 75 (May 13, 1965).
27. *Chem. Met. Eng.*, **38**, No. 11, 638 (1931).
28. Foxhall, H. B., The Dow Chemical Co., 1963 (personal communication).
29. "Plant Information Service," Temple Press Intelligence Unit.
30. Shigley, C. M., *J. Metals*, **1,** 3-7 (1951).

C

CANADIAN FISHERIES

General

By the year 2000, Canada should be producing as much as 2,000,000 metric tons of fish, double its present volume, if industrial development programs currently underway are carried forward at the present aggressive pace. Steady growth in harvesting and production capacity has marked the progress of the Canadian fishing industry during the 1960's, and despite temporary setbacks due to marketing problems such as occurred in 1967, industry representatives are confident that this forward trend will continue.

Preliminary estimates by the Federal Department of Fisheries show that 1967 landings of fish and shellfish amounted to 2.3 billion lb with an estimated value to fishermen of $147,000,000. This compares with the record landings in 1966 of 2.5 billion lb, worth $162,000,000. The Atlantic coast, which possesses the most important fish stocks in Canada, again showed an increase in landings for the 7th year in succession. Total landings for the area amounted to 1,971,000,000 lb with a total estimated landed value of $99,000,000.

The most spectacular single increase occurred in the herring catch, with landings in the Atlantic area amounting to 675,000,000 lb. Newfoundland alone landed 85,000,000 lb more than in 1966, and all the other Atlantic Provinces except Prince Edward Island showed appreciable increases.

The Atlantic catch of cod declined in all provinces except New Brunswick, which increased its catch to 23,000,000 lb from 22,000,000 lb in 1966.

Total catch of lobsters on the Atlantic coast amounted to 36,000,000 lb, worth an estimated $23,000,000, compared with a landed value of $22,000,000 in 1966. In Nova Scotia and New Brunswick, however, lobster catches were slightly below the previous year's totals while in Prince Edward Island there was a slight increase. In accordance with a general demand from fishermen, a program to limit the number of lobster traps per boat was being introduced in all lobster fishing districts in the Maritime Provinces in 1968. Further reduction in number of traps is planned for 1969. Consideration is also being given to means of restricting entry into the lobster fishery on an equitable basis.

On Canada's Pacific coast the estimated total landing for fish and shellfish was 318,000,000 lb with a value to fishermen of $47,000,000. This represented a considerable drop from the 574,000,000 lb valued at $61,000,000 landed in 1966.

Contributing to the over-all decline was the low catch of herring. Landings of other species fell also, including coho, pink and chum salmon, and halibut. However, the outlook in British Columbia was considerably brightened by healthy showing in sockeye salmon which went up to 36,000,000 lb in 1966, for a landed value of $14,000,000.

There is increasing interest in the development of fish-protein concentrate as a source of nutrition. A conference on fish-protein concentrate sponsored by the Federal-Provincial Atlantic Fisheries Committee was held in Ottawa and attracted over 200 scientists and other specialists from various departments of government both in Canada and the United States. Subjects discussed included the composition and nutritional value of the product, quality standards, food and drug regulations pertaining to the product, fish resources, and Canada's future role in meeting world protein needs. The availability of fish resources will not be a limiting factor in the development of a Fish Protein Concentrate industry in Canada.

The world catch of marine products has doubled during the last decade to a current annual yield of more than 50,000,000 metric tons, and even the most conservative biologists predict that landings can double again by the end of this century. An aggressive fisheries development program should enable Canada to double its present level of annual production (approximately 1,000,000 tons) within the same period.

The northwest Atlantic Ocean, with its huge continental shelf, offers the greatest opportunities for steadily increasing landings, and expansion is also anticipated in the northeast Pacific.

The potential is great for expanding production of small fish, particularly herring, capelin, and sandlaunce. Annual landings of Atlantic herring have sharply increased to more than 400,000 tons, and further expansion appears certain. Abundant also are such under-exploited species as silver hake, capelin, argentines, grenadiers, dogfish, squid, and lantern fish.

Through the Federal Department of Fisheries' industrial development program, the provinces and industry received wide-ranging technical and financial assistance in experimental fishing and production techniques. More than 100 projects of this type were in progress during the year, either being carried out independently by the Department or in cooperation with the provincial fisheries authorities and the industry.

An interesting development arose from an experimental herring fishing project. A 121,000 lb haul of pollock in one short tow was made by a typical Nova Scotia 100 ft scallop dragger converted to midwater trawling primarily for herring. Such a single haul of ground fish has never before been made on Canada's east coast by a vessel of this size. Midwater trawling for herring has been successfully carried out by large German vessels and the net successfully used by the Nova Scotia dragger is as big as those used by the Europeans, although the vessel is considerably smaller.

Attention was also focused on marine plants. The Marine Plants Experimental Station in Prince Edward Island, operated by the Federal Department of Fisheries, is currently concentrating its efforts on Irish moss from which carrageenin is extracted. (Carrageenin is used as a stabilizing agent in various food products, in sizing cloth, leather tanning, and in paper making.)

New markets for a number of marine plants are under investigation and, where warranted, programs will be set up to investigate the feasibility of marketing other plants. Since the future of the marine plants industry is dependent on the natural resources available, the station plans to include studies on conservation and cultivation in the future program.

East coast Canada is now entering the commercial tuna fishing business for the first time. With the recent establishment of a million-case-a-year canning plant at St. Andrews, New Brunswick, a fleet of five specially designed and equipped tuna clippers, each costing $2.8 million, is available. Owing to distances and the time involved for each trip, the crews of these vessels have been specially trained, and can, for example, mend the big 30-ton seine nets at sea, and even make new ones.

Since the best fishing grounds are off the coast of Peru and in the Gulf of Guinea, off the coast of West Africa, each round trip will cover 4500 miles or more and will take up to three months. The 1440 tons vessels will each have a capacity of 1000 tons of tuna with a potential landed value of up to $450,000.

CANADIAN DEPARTMENT OF FISHERIES

Cross references: *Labrador Fisheries; Newfoundland Fisheries.*

British Columbia

Growth of British Columbia fisheries since the late 1920's has paralleled the modest growth of other Canadian fisheries. As shown in Fig. 1 and in Table 1, commercial landings in Canada (excluding Newfoundland) increased from 1.14 billion lb annually in the 1926–1930 period to 1.65 billion lb annually in 1961–1964. In corresponding periods, average annual landings in British Columbia increased from 0.50 billion lb (44% of Canadian total) to 0.70 billion lb (43%

TABLE 1. AVERAGE ANNUAL COMMERCIAL LANDINGS AND VALUE OF LANDINGS TO FISHERMEN IN BRITISH COLUMBIA AND CANADA, 1926–1964[a]

Years	Average annual landings		Average annual landed value	
	British Columbia	Canada	British Columbia	Canada
	Millions of Pounds		Millions of Dollars	
1926–1930	501	1144	14.4	33.0
1931–1935	383	893	6.5	18.0
1936–1940	475	1102	8.2	22.7
1941–1945	521	1231	17.7	48.3
1946–1950	545	1357	28.0	69.9
1951–1955	539	1370	32.8	81.0
1956–1960	561	1443	36.1	90.7
1961–1964	702	1647	44.2	111.2

[a] All Newfoundland landings are excluded. During 1951–1964 the average annual landings were 548 million pounds valued at $15.0 million.

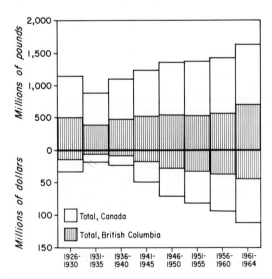

Fig. 1. Proportion of the total Canadian catch (excluding Newfoundland) and proportion of the total landed value of Canadian catch contributed by the Province of British Columbia, 1926–1964.

of total). The average annual value of landings to Canadian fishermen in 1926–1930 was $33,000,000; by 1961–1964 it had increased to $111.2 million. Average annual value of landings to British Columbia fishermen in 1926–1930 was $14.4 million (44% of Canadian total); in 1961–1964 it was $44.2 million (40% of total). When the fisheries of Newfoundland are included in the Canadian total, British Columbia's contribution in 1961–1964 is reduced to about 32% of the landings and about 34% of the landed value.

The landed values of major fisheries in British Columbia are contrasted in Table 2 for the 1936–1940 and 1961–1964 periods. Not included are values for the otter trawl fishery for bottomfish that have increased greatly in recent years. As shown in Table 2, salmon dominate the fishing industry in British Columbia, but the relative importance of the salmon fisheries has declined somewhat recently. In 1936–1940 salmon accounted for 69% of the landed value of all fisheries whereas in 1961–1964 salmon contributed 62% of the total value.

The second most valuable fishery in British Columbia is the Pacific halibut (*Hippoglossus stenolepis*) when landings by Canadian vessels in British Columbia and U.S. ports are included. Moreover, the relative value of the halibut fishery has increased considerably in recent years—from 11% of the landed value of all species in 1936–1940 to 18% in 1961–1964.

Other important commercial fisheries in British Columbia include those for herring (*Clupea harengus pallasi*) and for shellfish. Herring ranks as the Province's third most valuable fishery to the fishermen, and its share of the total landed value has increased from about 7% in 1936–1940 to 12% in 1961–1964. Shellfish—particularly oysters—have also shown strong gains in recent years. Absence of values for Pacific sardine (*Sardinops sagax*) in 1961–1964 reflects the virtual disappearance of this species from British Columbia waters after the 1940's.

Volume of landings in British Columbia by species and species groups is shown in Table 3. Salmon production averaged 156,000,000 lb annually in the period 1926–1965. The smaller production of salmon since 1956 is caused largely by a substantial drop in landings of chum salmon (*Oncorhynchus keta*). Herring, which have been fished in British Columbia since about 1877, have been the object of a steadily expanding fishery and in the period 1961–1965 accounted for 70% of all species landed. The fishery for Pacific sardines was relatively short-lived, but during its peak period, 1926–1945, it accounted for almost ¼ of total production in the Province. Reciprocal landing privileges for Pacific halibut enjoyed by Canadian and U.S. fishermen have contributed to the free flow of landings by nationals of each country into Canadian and U.S. ports. As shown in Table 3, landings of halibut by Canadian and U.S. fishermen in British Columbia have averaged about 26,000,000 lb annually in recent years. Shellfish constitute a small percentage of the total fisheries production but are quite valuable. Production of oysters in British Columbia is steadily increasing

TABLE 2. VALUE TO FISHERMEN OF CERTAIN MAJOR BRITISH COLUMBIA COMMERCIAL FISHERIES, 1936–1940 AND 1961–1964

Years	Salmon	Herring	Sardine	Halibut[a]	Shellfish	Total all species
			Thousands of Dollars			
1936	5155	383	332	802	136	7504
1937	5276	440	422	927	133	7838
1938	6331	316	453	855	146	8669
1939	5828	508	18	977	140	7891
1940	5504	1203	258	1103	137	9068
Average	5619	570	297	933	138	8194
% of total	68.6	7.0	3.6	11.4	1.7	
1961	26,152	4589	0	6035	1111	38,778
1962	30,742	4752	0	9890	1181	49,066
1963	22,790	6477	0	7431	1429	40,492
1964	30,244	6167	0	7603	1526	48,436
Average	27,482	5496	0	7740	1312	44,193
% of total	62.2	12.4		17.5	3.0	

[a] Includes value of halibut caught by British Columbia vessels and landed in United States ports.

TABLE 3. AVERAGE ANNUAL COMMERCIAL LANDINGS IN BRITISH COLUMBIA BY SPECIES AND BY SPECIES GROUPS, 1926–1965

	1926–1930	1931–1935	1936–1940	1941–1945	1946–1950	1951–1955	1956–1960	1961–1965
				Millions of Pounds				
Salmon—total	193.7	148.8	166.6	150.4	161.5	172.0	125.0	128.2
sockeye					26.1	31.9	29.0	19.6
chinook					15.2	13.8	12.9	10.9
coho					23.3	25.0	21.3	29.0
pink					35.8	52.5	34.4	52.8
chum					60.6	48.4	27.2	15.7
steelhead trout					0.5	0.4	0.2	0.2
Tuna	0	0	0	0.4	1.6	0.1	0.1	0.2
Sardine	143.5	106.8	71.4	123.3	1.8	0	0	0
Herring	142.0	107.8	208.8	205.8	325.6	304.0	364.7	482.7
Halibut[a]	27.6	17.6	20.5	22.2	23.0	26.1	28.8	33.6
"Sole" and flounders	1.2	0.6	0.9	3.1	9.6	8.9	7.5	6.3
Rockfish	0.4	0.2	0.3	1.9	1.4	0.6	1.1	2.2
Pacific cod, trawl-caught	[b]	0.8	1.5	1.2	3.1	4.3	6.1	9.2
Lingcod	4.7	4.8	5.1	6.1	5.9	3.9	4.5	4.0
Oysters	0.6	0.6	1.3	4.0	4.6	5.8	6.1	9.9
Crab	0.7	0.5	0.9	0.8	1.9	3.1	4.0	3.7
Shrimp	0.1	0.2	0.1		0.3	0.9	1.5	1.5
Total all species	519.6	393.2	483.2	529.9	548.6	540.4	560.7	692.6

[a] Includes landings by U.S. vessels in British Columbia ports.
[b] Not separable from lingcod in official statistics.

and with more intensive culture should continue to rise.

British Columbia's fishery for bottomfish (exclusive of halibut) involves at least 30 species, the most important of which are Pacific cod (*Gadus macrocephalus*), lingcod (*Ophiodon elongatus*), flounders or "sole" (mainly Pleuronectidae), and rockfish (Scorpaenidae). The bulk of the catches is taken by the otter-trawl fishery, which came into prominence during World War II; the remainder is taken by older methods such as long-lining and handlining. From 1946 to 1963 otter-trawl landings of foodfish generally ranged between 10,000,000 and 20,000,000 lb annually, with a slow increase over the latter part of this period. A very rapid increase then occurred when trawl landings of foodfish rose to about 28,000,000 lb in 1964, 40,000,000 lb in 1965, and 50,000,000 lb in 1966. Much of the increase has been from expanded trawling for Pacific cod—a close relative to the Atlantic cod (*Gadus morhua*).

As shown in Table 4, seine nets—mostly purse seines—account for about $\frac{3}{4}$ of the total harvest by British Columbia fishermen. Seines take about 99% of the herring catch and much of the salmon catch. Gillnets are primarily used for harvesting salmon. Lines account for almost the entire landed catch of halibut as well as for some of the salmon, sablefish (*Anoplopoma fimbria*), lingcod, and rockfish. Trawls account for most of the "cods," rockfish, and flounders. Traps and pots are mainly used for crab and shrimp. "Other gear" includes shovels, rakes, tongs, and dredges mainly used to harvest clams and oysters.

A. T. PRUTER

TABLE 4. COMMERCIAL LANDINGS IN BRITISH COLUMBIA IN 1964 BY TYPES OF FISHING GEAR EMPLOYED

Type of fishing gear	Millions of pounds landed
Seine nets	
Purse, haul, etc.	543.5
Trawls	
Otter, beam, etc.	30.4
Lines	
Set, troll, hand	59.3
Traps	
Pots, pound, weirs, floating, etc.	4.4
Gillnets	
Drift, set, etc.	57.9
Other	17.1
Total	712.6

CAPELIN (MALLOTUS VILLOSUS)

The capelin is essentially a cold-water pelagic fish occurring extensively in the north Atlantic and north Pacific and adjoining regions of the

Arctic. It is a soft-rayed fish and together with the smelts comprises the family Osmeridae. On the basis of a meristic character index the Pacific capelin and the Atlantic capelin were at first considered to be separate species, *Mallotus catervarius* and *Mallotus villosus*, respectively. However, examination of further material, including this time Arctic specimens, showed that the Arctic capelin were intermediate in meristic and morphometric composition between the Atlantic and Pacific capelin. Consequently, *Mallotus villosus* is now regarded as a monotypic species with the Atlantic and Pacific forms being consubspecific.

World Distribution

The capelin has a boreo-Arctic distribution in the northern regions of the Atlantic and the Pacific (Fig. 1). In the eastern Atlantic the capelin (*Mallotus villosus*) occurs abundantly from the Trondheim Fjord region of northwestern Norway northwest to Jan Mayen, Spitzbergen, and Novaya Zemyla at the eastern extremity of the Barents Sea. It also occurs sporadically in the White Sea and Kara Sea but the central part of its range in the eastern Atlantic is the Barents Sea. Iceland also has an abundance of capelin around its shores as does Greenland, where, in the last 40 years the center of the capelin distribution has moved north as far as Thule (76°N) on the west and Scoresby Sound (70°N) on the east. In the Canadian Arctic Archipelago capelin have been reported from the Melville Peninsula but not from Baffin Island. Individual occurrences of capelin have been reported from the Coronation Gulf, Bathurst Inlet, and Great Fish River of the Canadian Arctic. Capelin are reportedly very common in the southern half of Hudson Bay, rare in the northern part; they are not known in the western part of Hudson Strait. There is thus a gap in its distribution between eastern Hudson Strait and southern Hudson Bay.

From Saglek south along the Labrador coast, capelin occur in large quantities wherever suitable spawning beaches are found. The Newfoundland coast, the Grand Bank, St. Pierre Bank, and the Banks of the Labrador Shelf also possess large populations of capelin but they have not been reported from Flemish Cap where water temperatures are too warm. In the Gulf of St. Lawrence the capelin is most abundant on the northern shore although in colder years they also occur extensively around Gaspé. South of the Cabot Strait as far as Cape Cod the occurrences of capelin are rare and are related and restricted to the occasional influx of cold water into the Gulf of Maine.

In the Pacific the distribution of the capelin (*Mallotus villosus*) extends from Cape Barrow, Alaska around the Bering Sea south along the Pacific coast of Canada to Juan De Fuca Strait. On the Asiatic coast it extends from the Sea of Chukotsk south to Hokkaido Island, Japan, and Tumen River, Korea.

Anatomical Features

When capelin are in their breeding form the sexes differ so discretely that they are separable at a glance. The most obvious feature at this time is the appearance on the male of two pairs of spawning ridges, a dorsal pair and a ventral pair. The dorsal pair begins at the operculum (gill cover) and extends along the back immediately dorsal the lateral line to the caudal peduncle while the ventral pair extends from the pectoral fin back to the pelvic fin (Fig. 2). These spawning ridges begin to develop 4–5 weeks before the spawning season begins and arise from the elongation of scales which extend posteriorly like pyramidal projections to form villous plush-like ridges. These ridges are essential to the execution of the spawning act and disappear within a month after spawning has ended.

Sexual dimorphism is also exhibited in the structure of the anal fin whose base is strongly convex in the male but follows the contour of the body in the female. In addition the 5th–12th rays in the male are much thicker than in the female and the terminal branches of the rays along the mid-section of the anal fin fuse at their tips in the male to form a plate-like structure. In addition the pelvic and the pectoral fins are larger in the male and project out from the body whereas in the female they lie along the contour of the body. The male also tends to become pigmented during

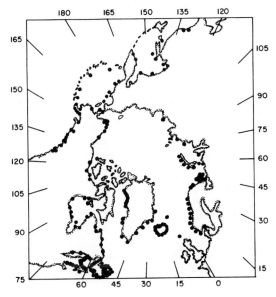

Fig. 1. World distribution of the capelin *Mallotus villosus*.

Fig. 2. Mature male (upper) and female capelin during the spawning season showing the spawning ridges, enlarged projecting fins, and greater size of the male.

Fig. 3. Spent male (upper) and female capelin several weeks after spawning has ended. The disappearance of the spawning ridges and the withdrawal of the fins to lie along the contour of the body in the male makes external sex separation very difficult at this time.

the spawning season with the operculum becoming pigmented first and then, as the spawning season progresses, the entire sides become blackened. Finally, the male tends to exceed the female in size, i.e., in length, depth, and weight.

With the exception of the size disparity these differences in form disappear after spawning is completed and at that time external separation of the sexes is difficult (Fig. 3). The transformation in appearance is so great that many fishermen refuse to recognize these fish as capelin.

Meristic Composition

There is a correlation between the latitudinal distribution of capelin and its vertebral number—the higher the latitude the greater the number of vertebrae (Table 1). The reason for such a correlation is not clear. Temperature, which generally affects the vertebral number to a marked degree, would seem the obvious explanation but this is not so. The spawning temperature of capelin in the Barents Sea is reported to approximate the $+2°C$ isotherm which is very near that required by capelin spawning on the Grand Bank of Newfoundland (2.8°–4.2°C); yet the capelin from these two areas have vertebral averages differing by about 3 vertebrae. But the British Columbia capelin are reported to spawn at temperatures between 10°–11°C, yet they have vertebral averages very close to those of the Grand Bank area. It is apparent either that temperature does not greatly influence the vertebral number of capelin or that some other factor has an overriding effect. Furthermore, the regional populations of capelin in the Atlantic each with its own spawning unit are so widely separated that genetic influences are likely to come into consideration.

Pacific capelin have vertebral averages very similar to those of the Newfoundland area which lies in the same general latitude. It may also be noted that there is more variation meristically within the Atlantic than between the western Atlantic and the Pacific.

In the Atlantic capelin the female vertebral averages exceed those of the male, whereas in the Pacific the opposite is true. The reason for this is not clear.

Age and Growth

Capelin are relatively small fish, the mature specimens being generally 13–20 cm in length although fish up to 24.5 cm have been recorded. Growth is greatest during the first 2 years of life after which it decreases until in the fifth year the size increment is negligible. During the first year both the male and female are the same size, but during the second year a differential growth rate sets in favoring the male, which is 1.0–2.5 cm larger than the female at sexual maturity. This may be reached in the second year of life, but it is not before the third year that mass maturation

TABLE 1. COMPARISON OF VERTEBRAL AVERAGES AND ANAL FIN COUNTS OF CAPELIN FROM VARIOUS PARTS OF ITS RANGE IN THE ATLANTIC AND PACIFIC

Area	Degrees North latitude	Number of Vertebrae		Anal fin (mm)
		M	F	
Newfoundland	43–55	66.51–66.72	66.53–66.97	23.48–23.66
British Columbia	48.55	66.32–66.76	66.05–66.24	...
Greenland	60–70	68.28	...	...
Iceland	63.5–66.5	69.40–69.48	69.45–69.53	...
Murman Coast	67–72	69.23–69.59	69.48–69.75	22.80–23.24

occurs; consequently, 3-year-old fish dominate the spawning schools. Because of the heavy mortalities incurred in spawning, few fish live beyond 5 years. Generally the older fish are predominantly females which reflects a heavier spawning mortality in the males. Since growth takes place mainly during the first 2 years of life and since the life cycle of the capelin is very short, an adverse year will markedly affect the abundance of capelin and this is often reflected in great fluctuations in catch.

Sex Ratio

Conclusions regarding the sex ratio of capelin are very contentious owing to the segregation of the mature capelin into schools of different sexes. European scientists report that, due to their earlier appearance on the spawning grounds, the males predominate at the beginning of the spawning run but towards the end of the spawning season the females become more abundant and tend to predominate. For beach-spawning capelin, as in the Newfoundland area, the situation is more complicated as the males tend to predominate near the beach whereas the females lie 30–40 m offshore and as they ripen squads of females leave the school, go into the beach, spawn, and retreat again. A more accurate estimate of the sex ratio would be obtained from the immature fish which presumably have no reason to segregate into schools of different sexes. Large-scale estimates of immature sex ratios, however, have not yet been made.

Spawning

Capelin are highly specific with regard to the conditions required for spawning. They may either be beach spawners or demersal spawners. The beach-spawning capelin as found along the beaches of the Newfoundland coast in June–July have a very narrow spawning temperature (5.5°–8.5°C); they must have the correct-size pebbles (0.5–2.5 cm); they require overcast conditions and diminished light intensity; the tides are important in that they determine the size of pebbles which are exposed.

The demersal spawners spawn on the bottom at depths between 50–100 m in the Barents Sea and 30–50 m on the Grand Bank. These spawning grounds are generally located in regions of strong horizontal currents (such as at the mouth of bays or offshore banks like the Grand Bank) which produce a substrate of fine gravel. The capelin on the southeast shoal of the Grand Bank were found to spawn on fine gravel from 0.5–2.2 m in diameter. As regards temperature, Grand Bank capelin spawn in temperatures 2.8°–4.2°C which are quite similar to spawning temperatures of Barents Sea capelin; capelin around the west and south coasts of Iceland are reported to spawn in temperatures between 6°–7°C; those around Greenland 1.9°–6.6°C; and finally, on the Pacific coast of Canada spawning temperatures of 10°–11°C are required.

The spawning act as exhibited in beach-spawning capelin is performed in the following manner. Generally one female is confined between two males in the groove formed by the ventral and dorsal spawning ridges and the enlarged, projecting fins of the males (Fig. 4).

Occasionally, for the lack of a suitable partner, only one male will participate and this forces the pair to take up a curved position (Fig. 5). Masses of capelin are deposited on the beach by the onrush of a wave (Fig. 6) and the recession of the wave causes the capelin to change direction and now face the beach swimming against the

Fig. 4. The spawning act as performed by two males and one female (in the middle).

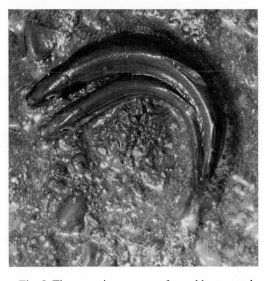

Fig. 5. The spawning act as performed by one male and one female (inside). Note the curved position of the pair.

receding water. The spawning contact is then made as the female is positioned into the hollow formed between the two males. The triplet or pair then carry out violent wriggling motions which scoop out a hollow into the sand into which are deposited the eggs of the female (up to 60,000) mantled by the milt of the male(s). After separating there is a few seconds respite before the next wave comes in, and if they are fortunate they will be carried back into the sea again. Many fail to regain the water and become stranded on the beach and die (Fig. 7). Owing to their constant attendance on the beach most of the stranded capelin are males which by the end

Fig. 6. Photo of a wave filled with capelin as it rolls in toward the beach.

Fig. 7. Photo of a beach strewn with dead capelin deposited there by waves as the capelin were spawning on the beaches.

of the spawning season are so exhausted and emaciated that many die anyway.

Such masses of dead capelin stranded on the beach or floating along the sea surface at the end of the spawning season have led to the impression that capelin are one-time spawners. This is not entirely true since spent fish in prime feeding condition have been caught at least a month after spawning had concluded. Some authors believe that the bulk of the first-time spawners survive to spawn a second time but because of the great predominance of a single age group (3-year-olds) in the spawning schools it must be assumed that the overwhelming majority of capelin die after spawning.

The eggs of the capelin are demersal and become attached to the sand pebbles on the beach or on the bottom. The incubation period varies with the temperature; hatching will occur in 55 days at 0°C, 30 days at 5°C, 15 days at 10°C, 9 days at 12°C. The larvae upon hatching are about 5 mm long and float to the surface and assume a pelagic life. The young of the year remain in the uppermost layers until the winter chill sets in at which time they descend to the bottom and ascend pelagically again the following spring.

Food and Feeding

There is a seasonal rhythm to the feeding activity of the capelin and correlated with this is a seasonal variation in the fat content. In the pre-spawning season in late winter and early spring the feeding intensity is high but begins to decline with the onset of the spawning migration and ceases altogether during the spawning season except for sand and capelin eggs apparently swallowed incidental to the breathing processes. The survivors of spawning resume feeding again several weeks after the end of the spawning period and feeding proceeds at a high intensity until early winter when feeding ceases. Fat contents as high as 23% have been reported in these post-spawning fish but are as low as 1% for spent fish.

Capelin are mainly filter-feeders thriving on planktonic organisms which are filtered out by the gill-rakers. Euphausiids compose the highest proportion by weight although copepods (*Calanus finmarchicus*) occur most frequently. In addition amphipods and a variety of other planktonic invertebrates are generally included in their diet.

Barents Sea Capelin

In the Barents Sea the capelin is an abundant fish and forms an important link for many food chains. This is especially true of the cod which migrates towards the coast in pursuit of the capelin during their spawning migration in early spring but also of the haddock and redfish.

The seasonal distribution and migration routes

of the Barents Sea capelin are well established. During the pre-spawning period the capelin are concentrated near the central plateau and the north-central and west-central area of the Barents Sea. At this time there are no concentrations of capelin in the coastal zones. In February the incipient migration towards the coast begins although the main bulk of the capelin continues to remain in the central area of the sea. During March, however, massive inshore migrations of capelin occur which remain inshore, where they spawn, until June when a movement away from the coast begins. The capelin then move northward to their summer feeding grounds near the Bear Island–Hope Island Bank region. There they remain until September when again the capelin begin their pre-spawning concentration in the central area of the Barents Sea.

Barents Sea capelin are demersal spawners which deposit their eggs at depths between 50–100 m. Spawning occurs in two phases with the first one taking place from March–April along the Finmarken and West Murman Coast at temperatures 1.9°–2.7°C and the second phase characterized by smaller, younger capelin taking place between May–June from the Kola Gulf to the White Sea at temperatures 2.2°–4.9°C. Spawning activity in the Barents Sea varies in position from year to year, these displacements being due to the water temperature of the Barents Sea. The position of $+2°C$ isotherm is the guiding factor during the inshore spawning migration of capelin and annual variation in the position of this isotherm causes tremendous fluctuations to occur in the Norwegian and Soviet capelin fisheries (see Table 2). In warm years (positive

TABLE 2. PRODUCTION OF CAPELIN BY VARIOUS NATIONS OF THE NORTH ATLANTIC IN METRIC TONS

Country	1960	1962	1964	1966
Norway	92,765	363	19,000	391,000
U.S.S.R.	31,000	2,600	...	[a]
Iceland	...	...	8,640	124,000
Canada (Nfld.)	770	490	535	510

[a] Statistics not available.

anomalies) the capelin leave the feeding grounds later, penetrate further to the east and approach the shores relatively late. Consequently, the Soviet catch tends to increase whereas the Norwegian catch declines. In cold years the reverse is true.

The capelin of the Barents Sea are exploited commercially only during the winter–spring spawning period. Both Norway and the U.S.S.R. catch the capelin although in recent years the Norwegian catch has increased tremendously until in 1966 it reached 390,000 metric tons.

The capelin vitally affect the location and productivity of the fisheries for demersal fish in the Barents Sea. It is an important food for the cod, particularly in February–May when the schools migrate inshore at which time the distribution, migration, and behavior of the cod depend to a considerable degree on the presence and migration of capelin. Years of good capelin catches are simultaneously years of an intense cod fishery in the coastal waters. Also the shifting of the capelin spawning to the western regions of the Barents Sea, as in years of negative anomalies, creates a general decrease in the yield of bottom fish in the Barents Sea owing to the rough bottom condition which hampers trawling in the western region.

Iceland Capelin

Capelin are extremely abundant in Icelandic waters but are found close to shore only for a brief period of the year during which they appear along the south and west coasts to spawn. Once used only as a bait for cod longlining or as a supplement to the diet of sheep and cattle, today it forms the basis of a significant commercial fishery (124,000 metric tons in 1966) for reduction into meal and oil.

The Icelandic capelin show an apparently quite simple migration pattern. The spawning run usually starts in March and ends in June along the south and west coasts where the required spawning temperatures of 6°–7°C are provided by the warming influence of the Atlantic drift. Spawning may also occur in individual years along the north coast. When spawning is completed the survivors migrate northward to the deeper water at or near the warm–cold boundary off the north and northeast coasts of Iceland (from approximately 67° to 68°N and 13° to 27°W) where they feed together with the maturing recruits. In late winter or early spring the capelin start migrating southwards again to the spawning grounds along the south and west coasts. The larvae when hatched are about 5 mm long and are pelagically distributed in the surface waters. They are then transported by the currents to the west and north of Iceland and by late summer arrive at their feeding and wintering grounds. These capelin larvae are very numerous around the coasts of Iceland and account for more than $\frac{1}{2}$ of all fish larvae caught in that area.

Greenland Capelin

The warmer water of the West Greenland current permits capelin to be distributed much farther north in West Greenland than in East Greenland and accounts for their greater abundance on the west coast. In the last 40 years the center and range of the capelin distribution has shifted to the north in West Greenland and associated with this was a northward shift in the cod population. Spawning occurs in May–June in the Godthaab Fjord region and in June–July farther north. This takes place in the coastal waters close to shore in temperatures ranging from 1.9°–6.6°C. The spawning capelin are pursued by great numbers of cod and, as in the Barents Sea, the productivity of the cod fishery is greatly affected by the availability and abundance of capelin. In addition capelin form an important part of the diet of Greenland halibut and redfish. As one proceeds north there is a trend towards an increase in mean size with increasing latitude. It appears that the capelin stocks in Greenland are rather stationary and that this increase in size from south to north is due to an increased growth rate.

Newfoundland Area

The capelin is of vital importance in the food spectrum of many commercially important fish species of the Newfoundland area. Not only do the capelin attract the cod into the shallow coastal water in June and July where they can be caught by stationary traps but the capelin also provide the entire sustenance of the cod at this time and fatten them to prime condition. In addition to this, capelin form an excellent source of bait by which the cod may be caught. Consequently the commercial success of the inshore cod fishery is often concomitant with the availability and abundance of capelin. Haddock, American plaice, halibut, and salmon also feed on capelin to a high degree.

In the Newfoundland area beach spawning begins around the first of June on the south coast and progressively later as one goes north along the east coast until in Labrador it is often mid-July before spawning begins. On the west coast of Newfoundland, where the cold Labrador current, which flows south along the east coast of Newfoundland to the tail of the Grand Bank, has less effect, water temperatures rise so fast that beach spawning occurs only sporadically and is usually replaced by deep-water spawning. Beach spawning lasts from 4–6 weeks at temperatures ranging from 5.5°–8.5°C although spawning has been reported in temperatures as high as 10°C. When beach spawning is completed spawning may continue in the deeper waters near shore for those fish which failed to spawn during the regular period.

Large concentrations of capelin form at the southeast shoal of the Grand Bank in June–August to spawn. At the same time beach spawning is being carried on inshore. Also, capelin have been trawled on the eastern slope of the Grand Bank and St. Pierre Bank at the same time. At least part of the Grand Bank capelin (those on western and Northwestern edge) cross the Avalon Channel region in dispersed schools in early June and migrate inshore where they aggregate into large spawning schools.

Exactly where the survivors of spawning retreat to feed and winter over is not known. Mass mortalities of capelin inshore in winter, vessel catches, and presence of capelin in stomachs of turbot, cod, and sea birds indicate that at least part if not all of the beach-spawning capelin spend their winters in the deeper waters of the bays. The presence of capelin on the Grand Bank and St. Pierre Bank at this time indicates that the inshore capelin are mainly discrete from those on these banks with limited intermingling occurring only at fringes where the banks are in close proximity to the inshore coastal waters (i.e., northwestern edge of the Grand Bank).

Pacific Region

Very little is known about the Pacific capelin. It extends from Juan De Fuca Strait on the American side to Tumen River on the Asiatic side. Although very common in these areas capelin are not fished in any commercial quantities yet. Having such a wide range it is to be expected that the species might have become differentiated into species or subspecies.

In the Strait of Georgia spawning occurs in late September and October on the sandy beaches at maximum high tide in temperatures between 10°–11°C. The adhesive eggs are buried in the sand. After spawning capelin are rarely seen until the next spawning run. Chinook salmon, coho salmon, as well as a variety of other piscivorous animals such as dogfish and water fowl, feed on capelin in great amounts.

Capelin Fisheries

The main producer of capelin is Norway (Table 2) which along with the U.S.S.R. alternates in importance as the chief exploiter of the Barents Sea capelin, depending on the position of the $+2°C$ isotherm during the spawning migrations. Iceland, which initiated the exploitation of its capelin resource in 1963, is rapidly rising in importance. All three nations depend mainly on purse seiners as the chief gear although pelagic and semi-pelagic trawls are increasing in importance in Norway. Nearly all of the Norwegian and Iceland catch as well as the greater bulk of the Soviet catch is reduced to meal and oil. In Newfoundland capelin have been traditionally used as a source of raw fertilizer and as bait but in recent years the amount of capelin used for these purposes has declined greatly. However, increasing amounts are being used for reduction purposes and this will likely increase in the future.

Although Greenland has no reported catch of capelin, large quantities are used as food for human consumption, bait, and as a supplement to the diet of sheep and domestic cattle.

GEORGE H. WINTERS

Cross reference: *Breeding in Marine Animals.*

CARRAGEEN—*See* IRISH MOSS INDUSTRY; PHYCOCOLLOIDS AND MARINE COLLOIDS; SEAWEEDS OF COMMERCE

CEPHALOPODA

The Cephalopoda represent the most complex of the classes of the phylum Mollusca. Sharing many molluscan characteristics in common with others of this non-segmented, schizocoelomate phylum, they are, however, of such diverse nature as to warrant individual consideration. As was stated by Meglitsch, "from the standpoint of complexity of structure and behavior, cephalopods stand at the apex of invertebrate evolutionary development."

Cephalopods are bilaterally symmetrical, with the dorso-ventral axis altered into a functionally anterior-posterior axis. The class Cephalopoda may be divided into two subclasses, namely, (1) Nautiloidea and (2) Coleoidea. The former is represented by only one modern genus, *Nautilus*, the chambered nautilus. All other nautiloids are extinct. The sub-class (i.e., *Nautilus*) is characterized by an internal many-chambered, siphunculate shell of both organic and inorganic fractions, numerous retractile tentacles without suckers, eyes possessing neither cornea or lens, a funnel or hyponome of two separate folds, and two pairs of nephridia and gills or ctenidia. This last characteristic accounts for the occasional use of the name Tetrabranchia for the subclass.

The subclass Coleoidea (Dibranchia) comprises three modern orders. These all possess a naked mantle that forms a sac enclosing the viscera. The hyponome is a closed tube. There are normally 8 arms, with or without a pair of longer retractile tentacles originating between the third and fourth pair of arms. Coleoids normally have an ink sac, eyes with a crystalline lens, and an open or closed cornea. There is one pair of nephridia and ctenidia, The three orders of the coleoids are:

ORDER OCTOPODA Mantle a muscular sac, without fins; shell (if present) internal and vestigial; 8 arms, no pair of retractile tentacles; non-pedunculate suckers. Octopi.

ORDER DECAPODA Mantle elongated, with fins; 8 arms augmented by pair of retractile tentacles; suckers armed and pedunculate; relatively well developed internal shell either organic or inorganic. Squids and cuttlefish.

ORDER VAMPYROMORPHA Long classed with the Octopoda, but differ primarily in arm pattern, in that the 8 arms are augmented by a pair of small retractile tendril-like arms. The vampire squids.

Cephalopods are all marine and there are approximately 600 living species. They are found in all the seas of the world, and at all depths, from the deep seas to the surface waters. They are extremely common and for this reason some species are of extreme importance for their commercial value. Despite their importance, very little is known about many vital aspects of their biology and life history.

The importance of Cephalopods as a marine resource cannot be denied. Reference to Table 1 and 2 will indicate the relative importance of Octopoda and Decapoda, with the leading national efforts being indicated as well. Japan leads the world in the fishing of squid, and the great majority of the catch is a single species. Canada, and especially the province of Newfoundland and Labrador, is next in importance to the Orient, again a single decapod species predominates, namely, *Illex illecebrosus illecebrosus*

TABLE 1. CATCHES OF CEPHALOPODS. THE WORLD,[a] EXCLUDING THE NORTH ATLANTIC (TAKEN FROM YEARBOOK OF FISHERY STATISTICS, FAO, 1960). AFTER CLARKE

Country	Type	Thousand metric tons 1955	1960
South Africa	Squid	...	0.1
U.S.A. (Pacific)	Squid	6.5	1.2
Argentina	Squid	0.2	0.5
Taiwan	Cuttlefish and Squid	3.7	6.8
Hong Kong	Cuttlefish	1.6	0.8
	and squid	0.4	0.5
Japan	Common squid		
	(*Todarodes pacificus*)	383.4	480.7
	Octopus	47.6	57.6
	Other squid	51.0	61.2
South Korea	Octopus	1.2	1.3
	Squid	18.3	42.1
Ryukyu Island	Octopus	0.2	0.2
Greece	All	2.0	...
Italy	Cuttlefish	...	6.6
	Octopus	...	5.4
	Squids	...	3.3
Yugoslavia	Cuttlefish	0.1	0.1
	Octopus	0.1	0.1
	Squid	0.1	0.2
Australia	Squid	0.1	0.2
TOTAL		516.5	668.9
Total including North Atlantic		540.0	704.9

[a] The U.S.S.R. has a developing fishery for Cephalopods; however, no figures were available.

TABLE 2. CATCHES OF CEPHALOPODS,
NORTH ATLANTIC (AFTER CLARKE)

Country	Type	Thousand metric tons	
		1955	1960
Morocco	Squids and sepiolids	0.1	...
	Cuttlefish	0.5	0.5
Canada (Atlantic)	Squid	7.1	5.1
U.S. (Atlantic)	Squid	1.9	...
Belgium	Squid	0.1	0.3
Norway	Squid	0.1	0.1
Portugal	Cuttlefish	1.5	3.1
	Octopus	0.9	0.6
	Squid	0.5	1.5
Spain[a]	Cuttlefish	1.7	5.8
	Sepiolids	0.1	2.1
	Octopus	4.2	e 8.1
	Squid	4.8	9.4
TOTAL (excluding the U.K.)		23.5	36.0
United Kingdom		0.4	0.8

[a] Includes catch from Mediterranean Sea.

(Lesueur) (Fig. 1). Spain exceeds Canada in total catch (all Cephalopods).

The catch of *I. illecebrosus* in Newfoundland averaged 7,700,000 lb over the period 1958–1962. Yearly fluctuations exist, and are little understood. For example, the total catch in Newfoundland was 4,995,761 lb in 1963. This was up to 24,000,000 lb in 1964. A cyclic pattern to the relative abundance is becoming evident. The catch in Newfoundland is, of course, dwarfed by the 650,000 tons of *Todarodes pacificus* (Steenstrup) in 1963 from Japan. According to Voss, a good year's catch for the U.S. would be approximately 10,000 tons and this is largely from California waters and is composed mostly of *Loligo opalescens* Berry.

The Cephalopods are used for a multitude of purposes, not the least important of which is as food for human consumption. It is used as such in a variety of forms or dishes, predominantly in the Orient and in Mediterranean countries, although there are other places where Cephalopods are considered as a human food stuff, either as a near staple or a delicacy. Cephalopods represent a good source of protein and can be prepared in a number of ways, and it is hoped that their acceptance as a food for the human population universally will be an early one.

Nutritionally, squid are superior to all other forms of mollusks that are prepared for human consumption. Squid exceed certain fin fish both in calorie and protein value, and are slightly inferior to beef and veal in these categories.

Cephalopods are used in great quantities for the production of animal food and fertilizer. The famous dietary supplement for budgies or canaries and other caged birds is really the cuttlebone or sepion (i.e., the internal shell) of the cuttlefish. In Newfoundland, the tremendous numbers of squid caught are, for the most part, frozen and used later as bait for the longline fishing that figures so largely in the island's economy. Squid has been used in this manner

Fig. 1. Living short-finned squid, *Illex illecebrosus* LeSueur, in captivity. (Photo: L. Moores, Memorial University of Newfoundland.)

Fig. 2. Squid-fishing boat equipped with six motor-driven squid jigging machines. (Photo: L. Moores, Memorial University of Newfoundland.)

throughout its 400 year history. The Madeiran fishery for espada (*Alepisaurus ferox*) also used squid exclusively as bait.

Several methods are employed in the catching of Cephalopods. These methods are varied, and in many cases are the result of local improvisations. Most of the world's commercial catch of squid is taken by a method known as "jigging," of the use of a "jigger." A jigger, whether it be in Newfoundland or in Japan, is basically the same design—a grapnel-type shaft of lead or plastic, 3–4 in. long. The jigger possesses one or more ciclets of unbarbed points or "hooks" at its free end. Traditionally, the jiggers—one to a line—were lowered into the water to a depth of several fathoms and upon a series of jerking vertical movements, brought back toward the boat. This is traditionally a hand-operated device, but the Japanese have mechanized it, as seen in Fig. 2 and have also mounted the jiggers "in series," as can be seen in Fig. 3. This latter figure shows two such jiggers "in series," one successfully having "jigged" an inking specimen of *Illex illecebrosus illecebrosus*. It should be pointed out that the hooks do not normally or necessarily pierce the body of the squid; rather, the arms and tentacles of the squid become entangled about the body of the jigger, the squid then being forcibly pulled from the water by the returning line to which the jigger is anchored. The use of the mechanized Japanese squid jigging machine is currently revolutionizing the squid fishery of eastern Canada, where one such rig can fish up to 4000 pounds of *Illex* in an hour.

The jiggers are highly colored. In Newfoundland they are usually painted bright red, as they are in Norway. The plastic Japanese models are all colors, ranging from white (the lead jigger) to shades of yellow-orange and red. It is not clear if color is a factor in fishing efficiency as was once believed. The modern jigger fishery techniques employ a light attraction system involving lights directed on the surface waters surrounding the boat.

Commercially successful methods of catching squid include a variety of seines, perhaps the one most successful being the lampara net used so profitably in the *Loligo opalescens* fishery of Monterey, California, baited traps, and nets. Some of the methods used involve waving "flags" of colored cloth as "bait" or a lure. Pound-stand-type traps are used to good purpose in Newfoundland, the traps being emptied during the dawn and dusk hours, into dories and the catch being brought to the bait freezing plants on the adjacent shore.

Fig. 3. Short-finned squid, *Illex illecebrosus* LeSueur, jigged on Japanese rig and expelling ink. (Photo: L. Moores, Memorial University of Newfoundland.)

Less successful are nets. Bottom trawls account for some of the catch of *Todarodes sagittatus* (Lamarck) in the North Sea and the waters around Iceland, the Faroes and northern Norway. Despite the statement by Clarke to the effect that midwater trawls fail to take larger squids, the author has demonstrated that *I. illecebrosus* can, in fact, be captured by midwater trawls at depths of 7–14 fathoms in the waters of eastern Canada. Finally, with respect to squid, it should be mentioned that some of the larger and more aggressive species have been taken by and afford good sport for the deep-sea angler using a game-fish rod and reel. The classic example of this is the Humboldt squid, *Dosidicus gigas* (d'Orbigny).

The octopod fishery is more specialized and less effective in many respects. The fishery for these organisms still involves hand operations—be it using the traditional *nassa* (basket), *mummarella* (earthenware pot), torch and spear, or tree-branch mazes or Scuba gear.

In general, octopods are docile and timid in manner and behavior. On the other hand, decapods are most often aggressive and extremely active. The source of locomotion is the expelling of water from the mantle cavity, via the funnel or hyponome. So efficient is this that squids are among the most rapid moving of all invertebrate organisms.

The expulsion of water is accomplished by the contraction of circularly disposed muscle fibers in the mass of the mantle. Longitudinal fibers contract during the "inhalant phase" when water enters the mantle cavity. That the jet of water is directed through the hyponome is accomplished by the close adherence of the open edge of the mantle against the base of the head. This is further enhanced by simple or ornate "studs" and articulating ridges and sockets that make for a tighter fit, hence a better seal. The fins of the tail configuration are basically for stabilizing and attitude, rather than propulsion. The rate of a squid's movement is the result of the force with which the jet of water is expelled from the hyponome, plus the hydrodynamic features of the squid anatomy.

The wall of the mantle is easily bruised, with necrotic areas forming and spreading rapidly. Some authors have speculated that the species coming from more shallow areas can more safely be handled than those taken from over greater depths. However, it is difficult to transport and successfully maintain most species of squid under laboratory conditions.

The complexity of this group of organisms can easily be seen by a consideration of reproductive and behavioral phenomena. Cephalopods are dioecious, with the gonads to be found near the posterior end of the body in the case of both males and females. The females possess a single, saccular ovary. The oviduct, of which there are two in many octopods and decapods, terminates in an oviducal gland that functions in secreting a membrane that encapsulates the eggs. Further protective layers are applied to the eggs by the nidimental gland. These secretions may be gelatinous and swell upon addition of water to a size several times larger than the original egg. Eggs may be deposited singly and be free-floating, as is the case with some oegopsid squid. Other squid, and octopods, produce attached clusters of eggs or egg masses. The clusters may be finger-like groups, as is the case in *Loligo*, or more grape-like in configuration as in the case of *Octopus* and the sepiods. The eggs are heavily laden with yolk and vary in number, from single individuals, as noted above, to clusters of 20, as in the case of *Sepiola atlantica* d'Orbigny, to over 2000 for *Loligo pealii* Lesueur. The tetra-branch *Nautilus* produces a single, attached, elaborately encapsulated egg.

The male Cephalopod possesses a single saccular spermary (testis), from which sperm migrate along a highly coiled vas deferens to the ciliated seminal vesicle. Here the sperm are formed into spermatophores, or packets of sperm. Accessory glands and organs of exact uncertain function are also present. The spermatophore is composed of (1) a closed outer sheath bearing a terminal filament, (2) an ejaculatory apparatus with a cement gland, and (3) an innermost sperm tube or spermatangium. Either by mechanical traction on the terminal filament or by the imbibition of water, the outer sheath ruptures, resulting in the explosion of the ejaculatory apparatus freeing the sperm tube and the enclosed sperm. The cement gland anchors the contents of the spermatophore to the body of the female during mating. The exact site of the anchoring differs in different Cephalopods, in that in octopods it is usually in the female reproductive tract, while in decapods it is on either the buccal membrane or elsewhere on the head. It is not known how sperm liberated from the spermatangium reach the eggs for fertilization. Mating is frequent, involves courtship behavior and is swift (10 sec in *Loligo pealii*). The actual transfer of the spermatophores from the male onto the body of the female is accomplished by aid of a modified arm known as the hectocotylus. This arm, the fourth right or fourth left arm in most decapods, the third right arm in octopods, is modified primarily by a reduction of suckers, resulting in a groove at the distal end, which confines the spermatophores. Hectocotylization occurs in varying degrees throughout the Cephalopoda.

The Cephalopods have well-developed nervous systems. The more complex systems are found in the octopods, where it is proper to refer to well-developed brains (really a circumesophageal ganglionic mass) composed of over 168 million neurons, in 30 distinct lobes, in the Octopus. Of this total a high percentage of the neurons are concerned with what may be described as benefiting from experience and fairly subtle visual and tactile stimuli. With the ability to employ analysis and recognition comes an enormous flexibility of response, resulting in rapid learning that figures largely in observed behavior.

Decapods are not characterized by the concentration of nervous elements into brain masses as noted for octopods. Rather, in decapods, there is a separation of ganglia, giving a diffuse pattern, with giant axons (in certain species) connecting several of the ganglia. There is a definite correlation between the degree of concentration of the nervous system and the mode of life of the different groups of Cephalopods. The complex behavior of octopods has been correlated with the extreme concentration into a brain mass, while the agility and intricate rapid movements of decapods have been seen as a result of the diffuse nature of the squid's chain of ganglia.

Cephalopods are also characterized by highly developed sense organs, particularly the eyes. These are image-forming eyes that can discriminate polarized light, small details of form, and possibly color. The information thus visually acquired is then used by the complex central nervous system. The eye is extremely close to the human eye in complexity of structure. The similarity is thought to be a classic example of convergent evolution.

Octopods may be blind, with no eyes, while others possess telescopic stalked eyes. Eyes are best developed in the squids, however, with some species possessing what must surely be the strangest eyes in the entire animal kingdom. The eye of *Opisthoteuthis depressa* Ijima and Ikeda is ⅓ the entire body length in size. *Histioteuthis* has a huge left eye 4 times the size of the smaller right eye. Several species, including the genera *Histioteuthis*, *Toxeuma*, and *Bathothauma* have eyes surrounded by light-producing organs which function in the illumination of the viewed field.

Near the eyes are small pits (in squid) or papillae (in octopods) that function in reception of taste or smell stimuli. There are also structures known as thermoscopic eyes scattered over the surface of some species, particularly certain squid. These are highly pigmented spots surrounded by transparent epidermal cells circling nerve endings which are believed to be sensitive to changes in water temperature.

Buried in the cartilaginous skeleton of the head is to be found the main sense organs for equilibrium, the paired statocysts. These are composed of structures known as maculae and cristae, which function in a manner similar to the semicircular

Fig. 4. Mandibles of giant squid, *Architeuthis dux* Steenstrup. (Photo: H. J. McCullough for Memorial University of Newfoundland.)

canals of the inner ear of humans. The maculae function in the detection of position and linear acceleration, while the function of the cristae is the detection of angular acceleration. These are best developed in decapod Cephalopods.

All Cephalopods are carnivores. The diet of octopods consists primarily of snails, crabs, and fish. Decapods, which are particularly ravenous predators, feed on crustacea and fish. In the case of decapods, prey is captured by the tentacles, drawn to the mouth by the arms and there held while the chitinous beaks or mandibles (Fig. 4) tear off triangular bites. In the buccal mass there is a rasping radula of doubtful function. The radula is lacking in deep-sea octopods.

Two pairs of salivary glands are present in all except the nautiloid, which have none. The two pairs in others may be fused. The first pair is known to secrete enzymes and mucus, and has been cited as the source of juices responsible for extracorporeal, digestion by octopods. The second pair of salivary glands is really a pair of poison glands, the toxin they produce being injected through the wound inflicted by the lower mandible. The toxin produced is a neurotoxin and that produced by octopods is particularly virulent. Gilpatric, the pioneer of free diving, correctly observed that squid and cuttlefish are more apt to bite than is an octopus. There are many instances of people being bitten by an octopus, however. The bite is reported to be painful, with considerable bleeding, but rarely is it fatal. The bite of the blue octopus, *Octopus rugosus* has been known to be fatal to humans. Anyone who has ever worked with live squid can most certainly describe the sharp, scissor-like bite of the decapod.

The food from the buccal cavity is carried by muscular peristalsis along the U-shaped digestive tract and is mixed with protease, peptidase, and lipase enzymes from the hepatopancreas. Absorption of nutrients takes place mainly in the large caecum, with fecal matter passed to the mantle cavity via the intestine. The terminal portion of the intestine, the rectum, bears a rectal gland modified into an ink gland which produces the ink, or sepia, which has as its chief component the pigment melanin.

Frank Lane wrote, "more marine animals prey upon squids than upon all other Cephalopods combined." The chief defense against this massive predation is the squid's great speed and agility, augmented by the release of ink. The ink may, as in some cases, coagulate, forming a "phantom squid" as described by several authors, which serves as a decoy permitting the flight of the true decapod.

Man is undoubtedly the greatest predator of Cephalopods. But he is not the only one. Octopods are prey for numerous species of fish, shark, the poisonous cone shells, certain sea mammals, and conger and moray eels. Certain echinoderms, particularly asteroids (sea stars), ophiuroids (brittle stars or serpent stars) and others, cause a paralysis of octopods by their very presence. This may be another example of an ectocrinal response, due to the effect of some externally liberated metabolite from the echinoderm. No less an authority than Dr. Waldo Schmitt of the U.S. National Museum reported that veteran octopus collectors in Alaska control their catches by placing a sea star in the collection boxes or pails.

Squid are the prey of a great list of marine species, such as dolphins, seals, and sea lions. Among fish, salmon, tuna, jewfish, and the broadbill swordfish are of considerable importance as predators. With respect to the lancet fish (*Alepisaurus ferox*), Voss has written, "stripping the stomachs of this fish is one of the approved methods of collecting deep-sea squid and octopods."

Apparently, most of the toothed whales feed on squid. It is certainly true of the bottlenose whale, *Hyperoodon*, the pilot whale, *Globicephala*, and the sperm whale, *Physeter*. The latter has been observed in combat with the largest of all Cephalopods, *Architeuthis dux* Steenstrup in the waters of Newfoundland as recently as 1966.

Squid are the chief, if not exclusive diet item of the elephant seal. The sea lions of the Falkland Islands are reported to have digestive tracts stained by the ink of consumed squid.

Surface-swimming species of squid are also devoured by sea birds, primarily the albatrosses and the giant petrels. The king penguin (*Aptenodytes patagonicus*) and the emperor penguin (*Aptenodytes fortesi*) both consume great numbers of squid. They consume so many, in fact, that the ground of a penguin rookery is littered with defecated beaks.

More than 300 years B.C. Aristotle reported his observations on the Cephalopods of the Mediterranean Sea. Today, teuthologists work in all the oceans of the world, but the number of specialist teuthologists is small. It is hard to understand why this should be so, for the Cephalopods are not only a fascinating natural group, but one that should assume ever-increasing importance as a marine resource.

FREDERICK A. ALDRICH

CHEMISTRY OF SEAWATER

Seawater is an electrolyte solution containing minor amounts of non-electrolytes and composed predominantly of dissolved chemical species of the 14 elements O, H, Cl, Na, Mg, S, Ca, K, Br, C, Sr, B, Si, and F (Table 1). The minor elements, those that occur in concentrations of less than 1 ppm by weight, although unimportant quantitatively in determining the physical properties of seawater, are reactive and are important in organic and biochemical reactions in the oceans.

The form in which chemical analyses of seawater are given records the history of chemical thinking about the nature of salt solutions. Early analytical data were reported in terms of individual salts (NaCl, $CaSO_4$, etc.). After development of the concept of complete dissociation of strong electrolytes, chemical analyses of seawater were given in terms of individual ions (Na^+, Ca^{++}, Cl^-, etc.), or in terms of *known* undissociated and partly dissociated ions, e.g., HCO_3^-. In recent years there has been an attempt to determine the thermodynamically stable dissolved ionic species in seawater and to evaluate the relative distribution of these at specified conditions. Table 1 lists the principal dissolved species in seawater deduced from a model of seawater that assumes the dissolved constituents are in homogeneous equilibrium, and (or) in equilibrium, or nearly so, with solid phases.

Both associated and non-associated electrolytes exist in seawater. The non-associated electrolytes, typified by the alkali metal ions Li^+, Na^+, K^+, Rb^+, and Cs^+, exist in seawater predominantly as solvated free cations. The major anions, Cl^- and Br^-, exist as free anions, whereas as much as 20% of the F in seawater may be associated as the ion pair MgF^+, and IO_3^- may be a more important species of I than I^-. Based on dissociation constants and individual ion activity coefficients, the distribution of the major cations in seawater as sulfate, bicarbonate, or carbonate ion pairs has been evaluated at specified conditions by Garrels and Thompson[2] (Table 2).

About 10% each of Mg and Ca is tied up as the sulfate ion pair. It is likely that the other alkaline earth metals Sr, Ba, and Ra also exist in seawater partly as undissociated sulfates. Notice from Table 2 that about 40% of the total $SO_4^=$ and HCO_3^- is complexed with cations, and $\frac{2}{3}$ of the $CO_3^=$ is present as the ion pair $MgCO_3^0$.

TABLE 1. ABUNDANCES OF THE ELEMENTS AND PRINCIPAL DISSOLVED CHEMICAL SPECIES OF SEAWATER, RESIDENCE TIMES OF THE ELEMENTS[a]

Element	Abundance (mg/l)	Principal species	Residence time (years)
O	857,000	H_2O; $O_2(g)$; SO_4^{2-} and other anions	
H	108,000	H_2O	
Cl	19,000	Cl^-	
Na	10,500	Na^+	2.6×10^8
Mg	1,350	Mg^{2+}; $MgSO_4$	4.5×10^7
S	885	SO_4^{2-}	
Ca	400	Ca^{2+}; $CaSO_4$	8.0×10^6
K	380	K^+	1.1×10^7
Br	65	Br^-	
C	28	HCO_3^-; H_2CO_3; CO_3^{2-}; organic compounds	
Sr	8	Sr^{2+}; $SrSO_4$	1.9×10^7
B	4.6	$B(OH)_3$; $B(OH)_2O^-$	
Si	3	$Si(OH)_4$; $Si(OH)_3O^-$	8.0×10^3
F	1.3	F^-; MgF^+	
A	0.6	$A(g)$	
N	0.5	NO_3^-; NO_2^-; NH_4^+; $N_2(g)$; organic compounds	
Li	0.17	Li^+	2.0×10^7
Rb	0.12	Rb^+	2.7×10^5
P	0.07	HPO_4^{2-}; $H_2PO_4^-$; PO_4^{3-}; H_3PO_4	
I	0.06	IO_3^-; I^-	
Ba	0.03	Ba^{2+}; $BaSO_4$	8.4×10^4
In	<0.02		
Al	0.01	$Al(OH)_4^-$	1.0×10^2
Fe	0.01	$Fe(OH)_3(s)$	1.4×10^2
Zn	0.01	Zn^{2+}; $ZnSO_4$	1.8×10^5
Mo	0.01	MoO_4^{2-}	5.0×10^5
Se	0.004	SeO_4^{2-}	
Cu	0.003	Cu^{2+}; $CuSO_4$	5.0×10^4
Sn	0.003	$(OH)?$	5.0×10^5
U	0.003	$UO_2(CO_3)_3^{4-}$	5.0×10^5
As	0.003	$HAsO_4^{2-}$; $H_2AsO_4^-$; H_3AsO_4; H_3AsO_3	
Ni	0.002	Ni^{2+}; $NiSO_4$	1.8×10^4
Mn	0.002	Mn^{2+}; $MnSO_4$	1.4×10^3
V	0.002	$VO_2(OH)_3^{2-}$	1.0×10^4
Ti	0.001	$Ti(OH)_4?$	1.6×10^2
Sb	0.0005	$Sb(OH)_6^-?$	3.5×10^5
Co	0.0005	Co^{2+}; $CoSO_4$	1.8×10^4
Cs	0.0005	Cs^+	4.0×10^4
Ce	0.0004	Ce^{3+}	6.1×10^3
Kr	0.0003	$Kr(g)$	
Y	0.0003	$(OH)?$	7.5×10^3
Ag	0.0003	$AgCl_2^-$; $AgCl_3^{2-}$	2.1×10^6
La	0.0003	La^{3+}; $La(OH)^{2+}?$	1.1×10^4
Cd	0.00011	Cd^{2+}; $CdSO_4$	5.0×10^5
Ne	0.0001	$Ne(g)$	
Xe	0.0001	$Xe(g)$	
W	0.0001	WO_4^{2-}	1.0×10^3
Ge	0.00007	$Ge(OH)_4$; $Ge(OH)_3O^-$	7.0×10^3
Cr	0.00005	$(OH)?$	3.5×10^2
Th	0.00005	$(OH)?$	3.5×10^2
Sc	0.00004	$(OH)?$	5.6×10^3
Ga	0.00003	$(OH)?$	1.4×10^3
Hg	0.00003	$HgCl_3^-$; $HgCl_4^{2-}$	4.2×10^4
Pb	0.00003	Pb^{2+}; $PbSO_4$	2.0×10^3
Bi	0.00002		4.5×10^5
Nb	0.00001		3.0×10^2
Tl	<0.00001	Tl^+	
He	0.000005	$He(g)$	
Au	0.000004	$AuCl_2^-$	5.6×10^5
Be	0.0000006	$(OH)?$	1.5×10^2
Pa	2.0×10^{-9}		
Ra	1.0×10^{-10}	Ra^{2+}; $RaSO_4$	
Rn	0.6×10^{-15}	$Rn(g)$	

[a] Adapted from Ref. 1.

TABLE 2. Distribution of dissolved species in seawater of chlorinity 19°/₀₀, pH 8.1 at 25°C, and 1 atm total pressure[a]

Ion	Total molality	% Free ion	% Me-SO₄ pair	% Me-HCO₃ pair	% Me-CO₃ pair
Ca^{2+}	0.0104	91	8	1	0.2
Mg^{2+}	0.0540	87	11	1	0.3
Na^+	0.4752	99	1.2	0.01	...
K^+	0.0100	99	1	...	...

Ion	Total molality	% Free ion	% Ca-anion pair	% Mg-anion pair	% Na-anion pair	% K-anion pair
SO_4^{2-}	0.0284	54	3	21.5	21	0.5
HCO_3^-	0.00238	69	4	19	8	...
CO_3^{2-}	0.000269	9	7	67	17	...

[a] Adapted from Ref. 2.

The activity coefficients of some major dissolved species in seawater at specified conditions are given in Table 3 with the activities of the free ions and ion pairs. The activities of Mg^{++} and Ca^{++} obtained from the model of seawater proposed by Garrels and Thompson have recently been confirmed by use of specific Ca^{++} and Mg^{++} ion electrodes, and for Mg^{++} by ultrasonic absorption studies of synthetic and natural seawater. The importance of activity coefficients and activities to the chemistry of seawater is amply demonstrated by consideration of $CaCO_3$ (calcite) in seawater. The total molality of Ca^{++} in surface seawater is about 10^{-2} and that of $CO_3^=$ is 2.7×10^{-4}; therefore the ion product is 2.7×10^{-6}. This value is nearly 600 times greater than the equilibrium ion activity product of $CaCO_3$ of 4.6×10^{-9} at 25°C and 1 atm total pressure. However, the activities of the free ions Ca^{++} and $CO_3^=$ in surface seawater are about 2.6×10^{-3} and 4.7×10^{-6} respectively; thus the ion activity product is 12.2×10^{-9}, which is only 2.6 times greater than the equilibrium ion activity product of calcite. Thus, by considering activities of seawater constituents rather than concentrations, we are better able to evaluate chemical equilibria in seawater; an obvious restatement of simple chemical theory but an often neglected concept in seawater chemistry.

Constancy and Equilibrium

The concept of constancy of the chemical composition of seawater, i.e., that the ratios of the major dissolved constituents of seawater do not vary geographically or vertically in the oceans except in regions of runoff from the land or in semi-enclosed basins, was first proposed indirectly in 1819 by Marcet and expanded later by Forchhammer and Dittmar.[3] The concept was established on a purely empirical basis, whereas in fact there is a theoretical basis for the concept.

Barth proposed the concept of residence (passage) time of an element in the oceanic environment and formalized this concept by the equation

$$\lambda = \frac{A}{dA/dt}$$

where λ is the residence time of the element, A is the total amount of the element in the oceans, and dA/dt is the amount of the element introduced or removed per unit time. Seawater is assumed to be a steady-state solution in which the number of moles of each element in any volume of seawater does not change; the net flow into the volume exactly balances the processes that remove the element from it. Complete mixing of the element in the ocean is assumed to take place in a time interval that is short compared to its residence time. Table 1 shows the residence times of the elements, and Table 4 compares the residence

TABLE 3. Activity coefficients and activities of dissolved species in seawater of ionic strength 0.70, chlorinity 19°/₀₀, pH 8.1 at 25°C, and 1 atm total pressure[a]

Dissolved species	Activity coefficient	Activity
Na^+	0.76	0.356
Mg^{2+}	0.36	0.0169
$MgSO_4°$	1.13	0.0069
K^+	0.64	0.0063
$NaSO_4^-$	0.68	0.00335
Ca^{2+}	0.28	0.00265
SO_4^{2-}	0.12	0.00179
HCO_3^-	0.68	0.000975
$CaSO_4°$	1.13	0.000964
$NaCO_3^-$	0.68	0.000311
$MgHCO_3^+$	0.68	0.000238
$MgCO_3°$	1.13	0.000199
$NaHCO_3°$	1.13	0.000195
KSO_4^-	0.68	0.000103
$CaHCO_3^+$	0.68	0.000047
$CaCO_3°$	1.13	0.000020
CO_3^{2-}	0.20	0.0000047

[a] Adapted from Ref. 2.

TABLE 4. THE RESIDENCE TIMES OF ELEMENTS IN SEAWATER CALCULATED BY RIVER INPUT AND SEDIMENTATION[a]

Element	Amount in ocean (in units of 10^{20} g)	Residence time in millions of years	
		River input	Sedimentation
Na	147.8	210	260
Mg	17.8	22	45
Ca	5.6	1	8
K	5.3	10	11
Sr	0.11	10	19
Si	0.052	0.035	0.01
Li	0.0023	12	19
Rb	0.00165	6.1	0.27
Ba	0.00041	0.05	0.084
Al	0.00014	0.0031	0.0001
Mo	0.00014	2.15	0.5
Cu	0.000041	0.043	0.05
Ni	0.000027	0.015	0.018
Ag	0.0000041	0.25	2.1
Pb	0.00000041	0.00056	0.002

[a] After Ref. 1, p. 173.

times of some elements on the basis of river input and removal by sedimentation.

For the major elements the results are strikingly similar and suggest that, at least as a first approximation, seawater is a steady-state solution with a composition fixed by reaction rates involving the removal of elements from the ocean approximately equalling rates of element inflow into the ocean. Thus, as a first approximation the steady-state oceanic model implies a fixed and constant seawater composition and provides a theoretical basis for the concept of the constancy of the chemical composition of seawater. However, it is possible that at any time t_0 (for example, the present) the ratios of the major dissolved constituents in the open ocean may be nearly invariant simply because the amounts of new materials introduced by streams and other agents to the ocean are small compared to the amounts in the ocean, and these new materials are mixed into the oceanic system relatively rapidly. But over time periods of 1000–2000 years or more, the major ionic ratios can remain constant only if the ocean is a steady-state solution whose composition is controlled by mechanism(s) other than simple mixing.

Further insight into the constancy concept can be gained by exploring possible mechanisms governing the steady-state composition of seawater. The steady-state solution could be simply a result of the rates of major element inflow into the oceans being equal to rates of outflow by biologic removal, flux through the atmosphere, adsorption on sediment particles, and removal in the interstitial waters of marine sediments. For example, Na carried to the oceans by streams is certainly removed, in part, in sea aerosol generated at the atmosphere–ocean interface and transported into the atmosphere, later to fall as rain or dry fallout on the continents. However, recent theoretical and experimental work suggests that seawater may be modeled as a steady-state solution in equilibrium with the solids that are in contact with it.

Sillén[4] has modeled the oceanic system as a near-equilibrium of many solid phases and seawater. Experimental work (Fig. 1) has shown that aluminosilicate minerals typical of those in the suspended load of streams and in marine sediments react rapidly with seawater containing an excess or deficiency of dissolved silica. Reactions involving these aluminosilicates may control the activities of H_4SiO_4 and other constituents in seawater. Thus it has begun to emerge that the composition of the oceans represents an approximation of dynamic equilibrium between the water and solids carried into it in suspension or precipitated from it by the continuous evaporation and renewal by streams.

Therefore, if seawater is a solution in equilibrium with solid phases, or even closely approaches such a system, then the *ion activity ratios* of the major dissolved species would be fixed and the chemical composition of the ocean would be "constant." Consequently, the activity of Na^+ in the ocean is not simply a result of removal processes involving sea aerosol, adsorption, etc., but is controlled by solid–solution equilibria. A model leading to nearly invariant ion activity ratios geographically and vertically at any time t_0 in the oceans based on mixing rates alone may be

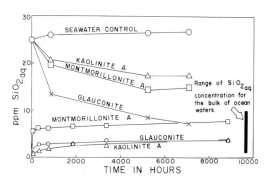

Fig. 1. Concentration of dissolved silica as a function of time for suspensions of silicate minerals in seawater. Curves are for 1 g (<62 μ) mineral samples in silica-deficient (SiO_2 in water was initially 0.03 ppm) and silica-enriched (SiO_2 was initially 25 ppm) seawater at room temperature. Notice that the minerals react rapidly and that the dissolved silica concentration for individual minerals becomes nearly constant at values within or close to the range of silica concentration in the oceans (from Ref. 5).

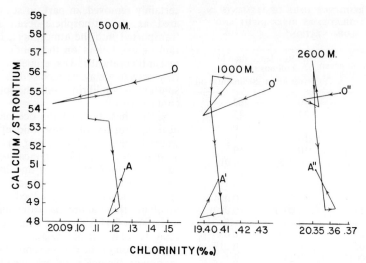

Fig. 2. Temporal variation of the Ca/Sr ratio at several depths as a function of chlorinity. The data were obtained from water samples collected at specified depths from June, 1966 through March, 1967 at an open ocean hydrographic station located 14 miles SE of the Bermuda Islands. 0, 0′, 0″, and A, A′, and A″ indicate the values of the Ca/Sr ratio and the chlorinity for the initial and final samples, respectively. The arrows illustrate the directions of change of these values as a function of time between June, 1966 and March, 1967 (from Ref. 6).

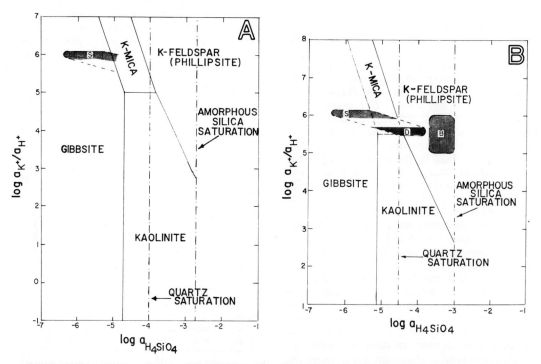

Fig. 3. Activity diagrams for a portion of the system K_2O-Al_2O_3-SiO_2-H_2O at 25°C (diagram A) and 0°C (diagram B), unit activity of water, and 1 atm total pressure. The areas labeled S, D, and B represent the composition of surface seawater, deep ocean water, and interstitial waters in marine sediments, respectively (from Ref. 8).

sufficient to explain the constancy of seawater composition, but is somewhat misleading and uninformative when considered in light of the recent advances in treating the oceans as an equilibrium system.

Some limitations of the equilibrium model of seawater do exist. Sillén[4] has pointed out that based on equilibrium calculations all the nitrogen in the ocean–atmosphere system should be present as NO_3^- in seawater; however, most of the nitrogen is present as N_2 gas in the atmosphere. Also, very recently the concentrations of the major alkaline earth elements Mg, Ca, and Sr in seawater have been monitored as a function of time and depth at a single vertical profile near Bermuda. The ratios of the concentrations of the alkaline earth metals varied temporally. The Ca/Sr ratios determined at several depths at this Bermuda station over approximately a 10 month interval of time are shown in Fig. 2 plotted against chlorinity. Notice that the ratios vary with time and that there are similar patterns of change in the ratios at the various depths as a function of chlorinity. Although the significance of these temporal variations and the processes causing them remain to be evaluated, the data show that *concentration ratios* of some of the major elements in seawater do vary at least on a temporal basis.

A further limitation to the equilibrium model is illustrated in Fig. 3, which shows phase diagrams for a portion of the system K_2O–Al_2O_3–SiO_2–H_2O at 25° and 0°C, and 1 atm total pressure as functions of the activities of K^+, H^+, and H_4SiO_4. The equilibrium model implies that the activity ratios of constituents in the oceanic system do not vary geographically or with depth. It can be seen from Fig. 3 that there is no single equilibrium composition of seawater for all the mineral assemblages that occur in marine sediments and (or) are transported to the oceans by streams. However, the phase diagrams do indicate that seawater is in partial equilibrium* with these mineral associations. For example, Fig. 3 shows that in some parts of the deep ocean, seawater contains concentrations of K^+, H^+, and H_4SiO_4 that are equilibrium concentrations with respect to the assemblages, K-mica (illite) + kaolinite + quartz, K-mica (illite) + K-feldspar (phillipsite) + quartz, or kaolinite + K-feldspar (phillipsite). However, in warm surface waters seawater contains concentrations of constituents that are equilibrium concentrations with respect to the association K-mica (illite) + gibbsite.

In the oceans the equilibrium seawater composition for any one of these mineral associations

---

* Partial equilibrium describes a state in which a system is in equilibrium with respect to at least one process or reaction, but out of equilibrium with respect to others.

differs from that of another association by only very small differences in the activities of H_4SiO_4 and (or) H^+. Stability diagrams as functions of dissolved constituent activities could be constructed for other systems (e.g., MgO–Al_2O_3–SiO_2–H_2O; Na_2O–Al_2O_3–SiO_2–H_2O), and seawater compositions plotted on these diagrams. Evaluation of such diagrams shows that seawater is not a homogeneous part of the oceanic system, but that there are equilibrium compositions of seawater or interstitial marine water corresponding to compositions predicted from thermodynamic data alone. The solids transported to the oceans, the suspended solids in the oceans, and marine sediments contain mineral associations for which there is an equilibrium composition of seawater or interstitial water. It appears that seawater may be viewed as a system in partial equilibrium with these mineral assemblages.

Buffering and Buffer Capacity of Seawater

The view has long been held that hydrogen ion buffering in the oceans is due to the CO_2–HCO_3^-–CO_3^- equilibrium. Within recent years this view

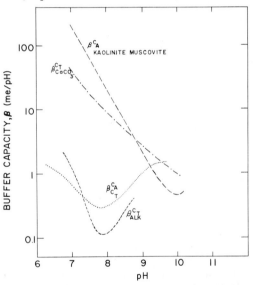

Fig. 4. Buffer capacity as a function of pH for some homogeneous and heterogeneous chemical systems. The buffer capacities are defined for $\beta^{C_A}_{\text{Kaolinite-muscovite}}$, addition of a strong acid (or base) to seawater in equilibrium with kaolinite and muscovite; $\beta^{C_T}_{\text{CaCO}_3}$, addition of total CO_2 in a seawater system of zero noncarbonate alkalinity in equilibrium with $CaCO_3$; $\beta^{C_A}_{C_T}$, addition of a strong acid (or base) to a seawater solution of constant total dissolved carbonate; and $\beta^{C_T}_{\text{Alk}}$, addition of total CO_2 to a seawater solution of constant alkalinity (data from Ref. 7).

has been challenged, and the importance of aluminosilicate equilibria in maintaining the pH of seawater emphasized. The buffer capacity of a system is of thermodynamic nature and is defined as

$$B_{C_j}^{C_i} = \frac{dC_i}{dp\text{H}},$$

where $B_{C_j}^{C_i}$ is the pH buffer capacity for incremental addition of C_i to a closed system of constant C_j at equilibrium. Homogeneous buffer capacities are defined for systems without solid phases, e.g., the addition of a strong acid to a carbonate solution, whereas heterogeneous buffer capacities are defined for systems with solid phases, e.g., the addition of a strong acid to a solution in equilibrium with calcite ($CaCO_3$) or with kaolinite and muscovite. Some homogeneous and heterogeneous buffer capacities for seawater systems as a function of pH are shown in Fig. 4. Notice that the homogeneous buffer capacities for the range of seawater and interstitial marine water pH values (7.2 to 8.3) are about 10- to 100-fold less than the heterogeneous capacities involving equilibrium between calcite and seawater or kaolinite, muscovite, and seawater. Both of these heterogeneous equilibria represent large capacities for resistance to seawater pH changes. Unfortunately, the kinetic aspects of these buffer systems have not been investigated quantitatively. However, it is apparent that aluminosilicate equilibria have buffer capacities equal to and perhaps greater than (the buffer capacities of most aluminosilicate equilibria in natural waters have only been qualitatively evaluated) the CO_2–$CaCO_{3(s)}$ equilibria in seawater. Small additions of acid or base to the oceans could be buffered by the homogeneous equilibrium CO_2–HCO_3^-–$CO_3^=$ shown in Fig. 4. However, large incremental additions of acid or base, or additions over a duration of time, would involve the heterogeneous carbonate and aluminosilicate equilibria; the relative importance of each would depend on the buffer capacities of the various equilibria and the relative rates of aluminosilicate and carbonate reactions.

FRED T. MACKENZIE

References

1. Goldberg, E. D., "Minor elements in sea water," in "Chemical Oceanography," 1965, Riley, J. P., and Skirrow, G., Eds., New York, Academic Press, pp. 163–196.
2. Garrels, R. M., and Thompson, M. E., "A chemical model for sea water at 25°C and one atmosphere total pressure," *Am. J. Sci.*, **260**, 57 (1962).
3. Dittmar, N., "Report on researches into the composition of ocean-water, collected by *H.M.S. Challenger* during the years 1873–1876," in "Report on the Scientific Results of the Voyage of *H.M.S. Challenger* during the years 1873–1876. Physics and Chemistry," 1884, Vol. 1, pp. 1–251.
4. Sillén, L. G., "The physical chemistry of sea water," in "Oceanography," 1961, M. Sears, Ed., Washington, D.C., A.A.A.S. Publication No. 67, pp. 549–581.
5. Mackenzie, F. T., Garrels, R. M., Bricker, O. P., and Bickley, F., "Silica in sea water: Control by silica minerals," *Science*, **155**, 1404 (1967).
6. Billings, G., Bricker, O. P., Mackenzie, F. T., and Brooks, A. L., "Temporal variation of alkaline-earth element/chlorinity ratios in the Sargasso Sea," (preprint).
7. Morgan, J. J., "Applications and limitations of chemical thermodynamics in natural water systems" (preprint).
8. Helgeson, H. C., Garrels, R. M., and Mackenzie, F. T., "Evaluation of irreversible reactions in geochemical processes involving minerals and aqueous solutions, Part 1, Thermodynamic Relations, Part 2, Applications," *Geochim. et Cosmochin. Acta* (in press).
9. Mackenzie, F. T., and Garrels, R. M., "Chemical mass balance between rivers and oceans," *Am. J. Sci.*, **264**, 507 (1966).

CLAMS

Soft-Shell Clams

Every coastline in the northern hemisphere, both sides of the Atlantic Ocean, and both sides of the Pacific Ocean, support great numbers of soft-shell clams—one of the exceptional food products man obtains from the sea (Fig. 1). Paradoxically, only on the Atlantic coast of North America has the soft-shell clam fishery attained any real economic importance, and only in New England is there any real market demand.

Fig. 1. The soft-shell clam; an important seafood of the Atlantic coast.

The soft-shell clam, *Mya arenaria*, is found from the coast of Labrador to the region of Cape Hatteras, North Carolina. On the European coast, it has been recorded from northern Norway to the Bay of Biscay. About 1879, young clams were accidentally introduced into San Francisco Bay, California, with shipments of seed oysters from the Atlantic coast. Since then they have spread from Monterey, California, to Alaska. The soft-shell clam is also found along the Western Pacific coast from the Kamchatka Peninsula to the southern regions of the Japanese islands. The southern range limits have always been well defined, but the northern limits have been confused by the presence there of the closely related *Mya truncata*, a northern form of which some populations exhibit similar external shell morphology.

Soft-shell clams, abundant on the coasts of Europe and Asia, have rarely been used for food, except for occasional local consumption. A brief and modest soft-clam fishery was developed in southern Japan in the post-war years, 1945–1947, but preference for other species, and possibly overfishing, limited the quantities marketed. European clam populations are plentiful, but have not been used, primarily because of lack of acceptance. Soft-shell clams in Europe apparently hold a similar position to the blue mussel, *Mytilus edulis*, in the U.S.; along the coasts, people find them plentiful and of excellent quality, but they remain unacceptable to commercial markets.

The soft-shell clam has played an important role in the history and economy of the eastern coast of the U.S. About 1850 the first real commercial fishery began with the demand for salted clams used as bait by cod fishermen on the Grand Bank. For the next 25 years this was the most important outlet for clams and was a source of wealth to some coastal communities. About 1875 a new demand for soft-shell clams arose as clam bakes and shore dinners increased in popularity. Clams became important as a consumer item for the first time, and this demand is still important to the northern fishery. Around 1900, canned clams became popular, and for the next 40 years most soft-shell clams went to the canneries. During this period clamming was prohibited in some areas during the summer by State law, thereby concentrating production in the

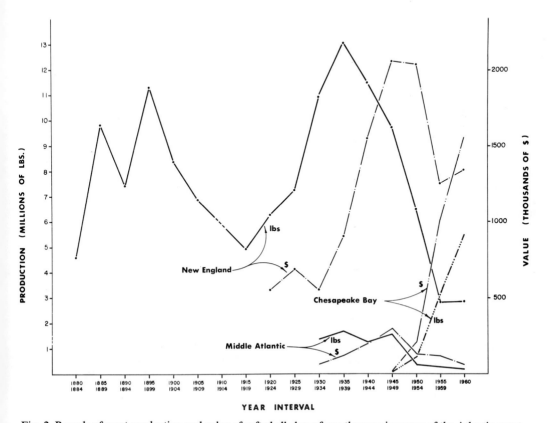

Fig. 2. Pounds of meat production and value of soft-shell clams from three major areas of the Atlantic coast.

winter when there was little competition from other industries for labor, and when clam prices were low.

Since the late 1930's most clams have been used in the fresh shucked trade, and recently cooked and uncooked frozen products have become popular. Clams are primarily used in the steamed-clam and fried-clam retail market. The increase in population, better preserving and marketing methods, and extensive publicity by large restaurant chains have brought a gradual increase in the demand for clams.

Early production records for soft-shell clams are not easily obtained, but the first figures given in the Federal statistics indicate that a fairly high production had been established by 1880 (Fig. 2). The New England fishery has been marked by consistently high, but variable, production until recent years. The value of the fishery reached the $2,000,000 mark during the 10 years from 1946 to 1955, even though this period saw a drastic decline in production of almost 6,000,000 lb.

Following World War II, particularly after 1950 when the "Maryland" escalator dredge was invented and introduced, the soft-shell clam fishery in Chesapeake Bay rose to dominate production. Production was almost double the New England output in 1960, although the total value of clams from the two areas was nearly identical. The southern fishery is now limited only by demand as great quantities of almost virgin stocks are available. Almost all southern clams are shipped to the New England area, for the market is still centered in the north. New England clams have a decidedly favored price differential and are valued at almost twice the price of their southern counterparts.

Natural History. The soft-shell clam is placed in the phylum Mollusca, further arranged, together with other bivalves into the class Pelecypoda and, because of certain characteristics of gill structure, placed in the order Eulamellibranchia. All of the structurally similar clams are then grouped in the family Myidae. The western Atlantic coast has only two species of clams in the genus *Mya*; they are *Mya arenaria*, our commercially important clam, and *Mya truncata*, a clam with a short, square-ended shell which lives in deeper northern waters and is of no commercial interest.

The external appearance of the soft-shell clam is an irregular ellipsoid with the two valves, or shells, of nearly equal size. The right shell is usually slightly larger than the left. When closed the two valves may touch along the ventral edge but gape widely in front, allowing passage of the foot, and in the rear, allowing extrusion of the siphons. The outer surface of the shell is covered with rough striations, some more pronounced than others. These annular marks or "rings" result from the change in growth rate during the year. Since cold winter temperatures restrict growth, these rings can indicate the clam age. The interior of the shell is relatively smooth but has certain areas scarred at the points where major muscles are attached. Along the dorsal edge of each shell are the prominent hinge teeth: one valve carries a large, projecting spoon-shaped structure, and the other valve has an inverted "bowl." The "bowl" lies directly over the "spoon." Between the two structures a tough, rubberlike ligament acts to keep the shells apart by opposing the closing action of the adductor muscles, much as a rubber eraser squeezed in a door hinge would force the door open.

If one valve is removed, the entire animal is seen to be enclosed in a thin membranous tissue, the mantle, which adheres very closely to the shell. This tissue secretes and shapes the shell. If the mantle tissue, the upper pair of gills, the upper palps, and the upper half of the siphons are removed, the clam appears as in Fig. 3. Some of the most obvious structures are the large central visceral mass (belly) with the small foot projecting from the anterior end, the open canals of the two siphons (each opening into a separate body chamber), and the large posterior and anterior adductor muscles whose contractions cause the shell to close. Lying under these structures are the two lower folds of the fragile gills. The juncture of all the gill folds provides a septum dividing the entire body (pallial) chamber within the mantle into two parts—a dorsal epibranchial chamber opening into the excurrent siphon and a ventral infrabranchial chamber opening into the incurrent siphon and the foot opening.

The interior of the visceral mass contains the large brown liver and the gonad (a tissue spread throughout the body). Embedded in this gonad and surrounding connective material lie the long, coiled intestine, the crystalline style within its sheath, and the stomach situated almost between the halves of the liver. The mouth opens on the anterior surface of the visceral mass just between the two pairs of palps, which gather and pass food to the mouth.

The intestine leaves the stomach, coils within gonad, then passes dorsally along the hinge line and through the heart. Finally, it passes around the base of the posterior adductor muscle and ends in an anal papilla suspended close to the junction of the excurrent siphon, where all fecal material can be discharged. Closely associated with the intestine in the mid-dorsal region are the two brown Keber glands, which perform excretory functions. The kidney mass, wrapped tightly around the ventricle of the heart, is just forward of the posterior adductor muscle.

Water is taken in through the incurrent siphon, by the action of ciliated cells lining the siphon and by definite respiratory movements of the clam. Water in the body chamber (pallial chamber) is circulated past the palps, the gills, and through the gill septum into the epibranchial chamber, where excretory products are collected. The water is finally passed out the excurrent siphon. Two alternate water paths may be observed. To aid in digging, the clam may force water out of the foot opening by closing the siphons and contracting the valves; to rid itself of unwanted materials gathered in feeding, the animal may reverse the entire flow of water by muscular contraction.

Unlike higher animals, the clam has blood circulating in an open system. The two-chambered heart pumps blood through a system of arteries which eventually empty into large, open lacunae. The blood is returned to the heart by a loose network of veins.

Life Cycle. Life, for a soft-shell clam, begins as sperm and egg join in the coastal waters. The fertilized egg develops into a swimming trochophore larva (Fig. 4) in about 12 hours in cold New England water, probably sooner in warmer, southern waters. The larva moves by hair-like projections (cilia) arranged in distinct bands around the body. The larva has a mouth and a minute shell gland which will give rise, within the next 24 to 36 hours, to the two calcified valves that envelop and protect the clam body throughout life. Shell formation produces a new larval stage (early veliger) that is typical of most bivalves. Early veliger larvae are characterized by a

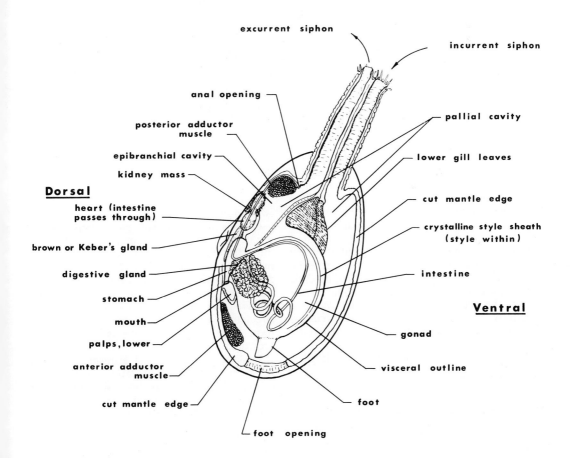

Fig. 3. Anatomy of the soft-shell clam.

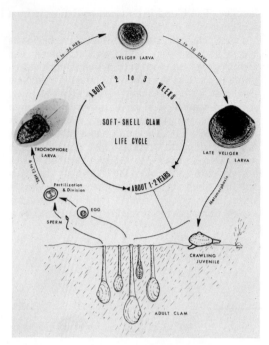

Fig. 4. Life cycle of the soft-shell clam. Larval stages are greatly enlarged, and not to scale.

straight hinge line and a swimming organ, the velum, formed by the modification of the trochophores' ciliated bands into a circular pad. This swimming organ keeps the animal suspended in the water where currents can carry the clam for great distances. Gradually the hinge line becomes stronger and humped as in the adult. The size and function of the velum are reduced and a muscular foot develops. The clam has, at this point, reached the late veliger or "setting stage" when metamorphosis is possible and the final adaption to sedentary bottom existence can begin. This period is extremely critical in the life cycle for, once removed from transportation by the water, the clam must establish itself within a relatively small area and its future success is influenced by the type of bottom material on which it settles.

Upon setting, the clam immediately attaches to sand grains, plants, or other materials, by a tough horny thread, the byssus. Byssal threads are formed from water-hardened secretions of the byssal gland, located on the underside of the large, muscular foot. The byssal "anchor" reduces further washing by wave action and currents, but can be cast off and re-formed elsewhere should the animal search for more desirable habitat. At the onset of cold weather the young clam often burrows and then becomes less active through the winter. In New England, the great spring tides stimulate juvenile clams to move, or may actively wash them from the sediment. Clams beginning their second growing season thus gain an additional exposure to new habitat and an opportunity to relocate. Movement decreases with size and age; mature clams are rarely found out of their burrows unless dislodged by storm or tide.

The soft-shell clam is found in bays, coves, estuaries, and other protected areas. It lives in a wide variety of sediments, ideally a mixture of mud and sand where it is less likely to be exposed to predation and climate. Clams live deeply buried in the bottom sediments, and the greatly extensible siphons may reach lengths of more than four times the length of the shell, drawing water and the food it contains from just above the bottom. Microscopic plants, animals, and other organic food materials are filtered from the water, possibly 12 gallons a day, drawn in through the siphons. Most soft-shell clam populations in New England are found in intertidal flats, exposed at low tide, whereas most clams in the Chesapeake Bay area are in subtidal regions and are seldom exposed. In productive areas, clams occur in vast but erratically distributed beds.

Sexual maturity may be reached at 1 year of age (at a length of $\frac{1}{2}$–$\frac{3}{4}$ in. in northern waters). The sexes are separate and can be determined only by microscopic examination. Each clam can produce millions of eggs or billions of sperm. At spawning sex cells are extruded from the gonad into the epibranchial chamber and out the excurrent siphon in successive puffs. Spawning is related to water temperature and occurs from early June to mid-August in northern areas. Clams north of Cape Cod have only one reproductive cycle each year, but clams south of Cape Cod may spawn twice in the same year.

Growth rates of soft-shell clams depend primarily upon temperature, although many other factors may be involved. Southern clams may reach a shell length of 2 in., an acceptable commercial size, within $1\frac{1}{2}$ or 2 years in Chesapeake Bay, but Maine clams may require 5 to 6 years to attain this length. Food abundance, water circulation, tidal exposure, crowding, and sediment type are some elements that influence growth.

Predators, Diseases, and Parasites. Probably the most serious clam predator in recent years, exclusive of man, has been the green crab, *Carcinus maenas*. Natural temperature cycles apparently control the northern range limits of this crab and several management control methods (pesticide barriers and subtidal fences) have been developed. Other soft-shell clam predators, not as destructive as green crabs but often of local importance, are moon snails, oyster drills, other crabs, ducks, gulls, horseshoe "crabs," starfish, and bottom-feeding fish such as flounders, skates, and rays.

Fig. 5. New England clam digger.

Mass mortalities of some soft-shell clam populations have been recorded, but practically nothing is known of the causative agents. The incidence of observed clam diseases is low and probably does not affect clam production, nor do any of these diseases affect man. A condition known as "water belly" is prevalent in some areas of New England; clams with the watery, thin meats typical of this condition are not acceptable for market. "Water belly" may be caused by nutritional deficiencies, but some biologists believe that other factors such as parasitic infection, enzyme deficiencies, and disease can be involved.

Incidence of parasites appears to be low and probably does not influence the abundance of clam stocks. Several trematodes inhabit soft-shell clams during some period of their life cycle, and a parasitic ciliate, *Trichodina myicola*, has been recorded.

Fishery—Methods and Management. Armed with clam hoe and hod the New England clam digger waits for the tide to expose his prey. As the water drains from the clam flats, he begins to dig, sometimes in long trenches or pits, when clams are abundant, or he may search for the individual depressions on the surface that indicate clams in the soil below (Fig. 5). He will have about 4 hours to gather his catch. Formerly, during the periods of great abundance, the fishermen could expect to dig a barrel or more of clams, but now 1 to 2 bushels per tide is considered an average catch. At present the digger can expect to receive about $6 for each bushel sold to the clam buyer, who, in turn, will receive a commission from the processing plant. At the processing plant the clams are cleaned, graded, and shipped directly to retail markets or may be shucked, washed, and canned.

Although it is believed that a great potential exists for shellfish farming and responsible management, only sporadic attempts have been made during the history of the soft-shell clam fishery. In New England, the failure to establish reasonable management techniques is the result of a continued demand that the fishery be free and publicly owned. The investment is modest—clam hoe and boots—for an individual to enter the fishery. Very few diggers work on a year-round basis; most are engaged in various seasonal work requiring unskilled labor, and take up clamming as a part-time source of income. Diggers rarely work for the maximum catch obtainable at each low tide, but rather aim for a daily income determined by the individual's need. Hand digging is inefficient and wastes the clam resource. Research workers estimate that 50% of the clams left in the flats will die from the effects of hand digging.

Recent management efforts in Maine and Massachusetts have been directed toward predator control, clam flat improvement, and protection of juvenile clams. In some areas, intensive management programs use frequent population surveys to establish digging periods, forecast production, and plan rotation of harvest areas. Unfortunately, these techniques are not in general practice and are not typical of the fishery.

About 1951 the escalator dredge (Fig. 6) was introduced into the Chesapeake Bay region to harvest subtidal soft-shell clams. The dredge is attached to a boat, which slowly pushes it through the bottom sediments. Clams, loosened from the sediments by the high-pressure spray of water, are washed or scooped onto the chain-mesh belt. This belt then carries the clams to the deck

Fig. 6. Typical Chesapeake Bay dredge-boat with escalator dredge.

where commercial clams are removed, and all debris and small clams fall back into the water. The method is efficient; it takes almost all commercial clams and does little damage to others, and does not appear to have detrimental effects on bottom sediments. Each boat takes about 30 bushels a day. Fishing is restricted to certain parts of a small number of open areas. The supply is greater than the present demand, and most dredges are restricted to a daily quota. Although an expanded market is needed, the Chesapeake fishery is thought to be in good condition at this time and no cultivation or extensive management of the fishery resource is being undertaken.

Special Problems of Paralytic Shellfish Poison and Pollution. Occasional outbreaks of paralytic shellfish poison are of serious concern to those responsible for the public welfare. The toxic agent is concentrated in clams and other bivalves when they feed on a small marine organism, *Gonyaulax* sp., that produces the poison and is noted for erratic population explosions. When *Gonyaulax* is in great abundance, the clams become potentially lethal to man. Fortunately, the species of *Gonyaulax* involved in paralytic shellfish poison are restricted to northern climates. U.S. and Canadian public health scientists continuously survey the toxicity of edible shellfish. Whenever shellfish from producing areas begin to show an increase in toxin towards scientifically established minimum levels, these areas are immediately closed to the clam fishery.

Vast supplies of soft-shell clams are unavailable to the fishery because of domestic and industrial pollution. Pollution problems are unfortunately increasing as our population grows, and each year additional shellfish areas are closed. Bacterial pollution is of primary concern to man, since clams tend to concentrate pathogenic organisms from domestic pollution and may transmit serious diseases. Chemical pollution, such as fuel and industrial effluents, may kill shellfish or may flavor the meats so that they are unacceptable to the market. Physical pollution, such as the dumping of sawdust, masonry, and other industrial wastes, may kill clam populations by smothering and may remove producing areas from further production by changing physical characteristics of the shore.

Potential as a Food Resource. The soft-shell clam has a great potential as a food resource from shallow marine waters. Near the bottom of the food chain, the clam converts microscopic plant and other organic materials directly into protein. In this respect the soft-shell clam is more efficient than many other shellfish (for every 100 lb of clams 18 lb of meat are obtained, whereas oysters produce only 6 lb). The clam is amenable to intensive management, when such steps are required, but at present exploitation of existing

TABLE 1. CATCH (THOUSANDS OF POUNDS OF MEATS) AND VALUE (THOUSANDS OF DOLLARS) OF SURF CLAMS IN THE UNITED STATES, 1943–1966

Year	New York		New Jersey		Maryland		Other states		Total	
	Pounds	Dollars	Pounds	Dollars	Pounds	Dollars	Pounds	Dollars	Pounds	Dollars
1943	475	50	170	11	...	...	250	64	895	125
1944	912	116	15	2	...	...	277	70	1,204	189
1945	3,982	500	527	47	...	...	616	83	4,780	630
1946	6,483	770	...	...	...	...	167	46	6,649	816
1947	3,314	346	169	21	...	...	214	58	3,698	425
1948	3,521	366	177	23	...	...	13	4	3,711	393
1949	4,904	470	407	39	...	...	18	4	5,329	514
1950	3,286	331	4,299	416	130	10	28	7	7,742	764
1951	4,046	422	6,419	622	...	...	22	6	10,487	1,050
1952	4,138	432	6,418	802	2,088	174	3	1	12,648	1,409
1953	3,345	418	6,878	790	2,454	204	...	...	12,677	1,412
1954	3,360	420	6,877	844	1,346	168	232	26	11,815	1,458
1955	2,026	253	8,278	967	1,695	141	180	22	12,179	1,383
1956	2,368	306	11,583	1,277	1,850	173	173	26	15,976	1,782
1957	1,599	220	15,224	1,867	934	134	196	19	17,953	2,240
1958	429	69	12,462	1,317	792	93	781	93	14,464	1,572
1959	514	61	20,164	1,622	850	70	1,752	171	23,280	1,924
1960	722	85	23,448	1,546	420	34	481	48	25,071	1,713
1961	722	65	26,697	1,693	71	6	12	2	27,502	1,766
1962	840	76	29,830	1,917	75	6	109	11	30,854	2,010
1963	974	91	37,548	2,580	64	5	...	...	38,586	2,676
1964	1,218	109	36,875	2,504	38	3	13	3	38,144	2,619
1965	1,505	127	42,307	3,048	275	22	...	...	44,087	3,196
1966	1,840	148	43,174	3,713	64	6	...	...	45,078	3,867

stocks in the U.S. and other lands, awaits market expansion. This expansion may result from new technological developments in packaging and transporting highly perishable marine products, or a greatly intensified requirement for marine protein resources.

ROBERT W. HANKS

Cross references: *Biotoxins, Marine; Pollution of Seawater.*

Surf Clam Fishery

The Atlantic surf clam, *Spisula solidissima* (Dillwyn), has inhabited the northwestern Atlantic coast for more than 10,000,000 years. The oldest known fossils are from deposits of upper Miocene age in North Carolina.[1] Surf clams are the largest bivalve mollusks in the region; some have a maximum length of 210 mm (over 8 in.). Their shells are a familiar sight along ocean beaches after storms.

The surf clam is a molluscan in the class Pelecypoda. It is a member of the order Eulamellibranchia, which is distinguished by large, platelike gills formed of a fine meshwork of filaments. Within the order is the family Mactridae which includes the surf clam and over 200 species with structural similarities. Many common names are in local use: bar clam in Canada; hen clam in Maine; sea clam in Massachusetts; and surf clam, beach clam, or skimmer clam in the middle Atlantic states.

The surf clam fishery is relatively new but has already reached a level of major commercial importance.[2] The industry has grown from a small bait fishery before World War II to a producer of 62% of all clam meats used in the U.S. in 1965. In 1966, 45,000,000 lb of surf-clam meats were processed (Table 1), attesting to the significant role this bivalve plays in the national shellfish economy.

Taxonomic and Anatomical Description. The equal-sized shell valves of surf clams have a somewhat triangular shape and gape slightly when closed. The clams have both an external and internal ligament. The latter, called the resilium, is large and is contained within a deep spoon-shaped pit, the chondrophore. The resilium forces the shells open and a pair of adductor muscles oppose the resilium when the shells are closed.

Several gross anatomical structures of the surf clam are seen after removing one valve: the mantle, gills, palps, and body wall covering the visceral mass (Fig. 7). Paired siphons protrude from the posterior end of the clam (the mouth is at the anterior end). Thus, like most clams, surf clams burrow headfirst into the bottom and extend their fused siphons into the water. One siphon (incurrent) draws in water with food and oxygen and the other (excurrent) ejects water carrying excretory products. The gonad, composed of a mass of root-like tubules, varies in size depending upon the degree of ripeness. In ripe clams, the gonad may completely surround the dark-brown liver mass. The tubules coalesce

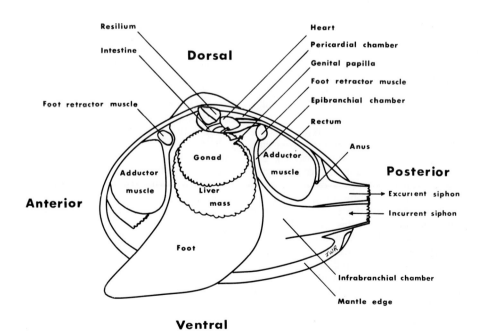

Fig. 7. Diagrammatic sketch of the anatomy of a surf clam.

into a pair of gonoducts which terminate at minute papillae on either side of the visceral mass. Sex products emitted from the gonoducts flow into the epibranchial chamber during spawning and are released from the clam through the excurrent siphon. From the stomach, an intestine coils throughout the liver mass, passes through the heart, and lies dorsal to the posterior adductor muscle. Excretory products, emitted from the anus, leave the clam through the epibranchial chamber and excurrent siphon.

Habitat. Surf clams inhabit bottoms of gravel, sand, or muddy sand. They are powerful, active animals and live buried just beneath the surface of the substrate. They burrow rapidly by successively probing the massive, triangular, wedge-shaped foot into the bottom and rhythmically contracting the valves. By closing the valves downward jets of water are forced out near the probing foot.[3] Each time sand is displaced by the force of the water, the clam extends its foot deeper into the sand and draws itself further into the burrow. Surf clams also propel themselves along the bottom and even jump off the bottom and glide short distances through the water. They are thus better adapted than most clams to live in the wave-disturbed surf zone near shore. Nevertheless, vast numbers are often washed ashore by winter storms. For example, Jacot[4] estimated 5,000,000 adult clams per linear mile were washed ashore at Rockaway Beach, New York, after "unusually severe storms." Countless clams were grounded on a New Jersey beach in 1951 after a storm (Fig. 8). More recently (1964) the Bureau of Commercial Fisheries investigated the fate of juvenile surf clams living in the oceanside beach at Wallops Island, Virginia. The number of clams averaged 20 per square foot near the low-tide level along a 2-mile stretch. Clams were dislodged and carried higher up the beach by incoming tides, eventually to perish and contribute to the windrows of empty shells high on the beach.

Gametogenesis. Sexes of surf clams are separate and hermaphroditism is a rare anomaly (one case was found in a sample of 2500 by Ropes[5]). The sex of a clam cannot be determined from external observations. Even a microscopic examination of gonad smears is not a foolproof technique for determining sex because the sex cells are not sufficiently differentiated during the early stages of the reproductive cycle. Preparation and study of histological sections of gonads are essential to determine sex and gonad condition. Surf clams mature rapidly. Many contain ripe gonads and spawn 1 year after setting, although full sexual maturity is reached during the second year.

Gonad conditions (Fig. 9) of more than 1500 clams collected from offshore New Jersey during 1962–1965 were studied to determine the annual

Fig. 8. Surf clams washed ashore on a New Jersey beach by a storm. (Photo: Wide World Photos.)

frequency and duration of spawning. This information is basic to an understanding of the reproductive cycle, larval production, and subsequent recruitment of juveniles. In 1962, 1963, and 1964, two annual spawnings occurred. The first (major) spawning was in mid-July to early August and the second (minor) spawning in mid-October to early November. In 1965, however, only a single spawning was observed—during mid-September to mid-October.

Bottom water temperatures (taken daily from the nearby Barnegat Lightship) seem to affect the spawning cycle. A gradual warming trend of bottom water during the first 6 months of 1962, 1963, and 1964 coincided with a gradual ripening of the clam gonads. In 1965, when only one reproductive cycle and spawning occurred, temperatures were generally lower than in previous years.

Larval Stages. Surf clams begin life when a sperm and egg unite after the gametes are released into the water. The fertilized eggs (about 53 μ in diameter) divide, forming trochophore larvae and later veliger larvae with minute shells. Early veligers can develop within about 28 hours after

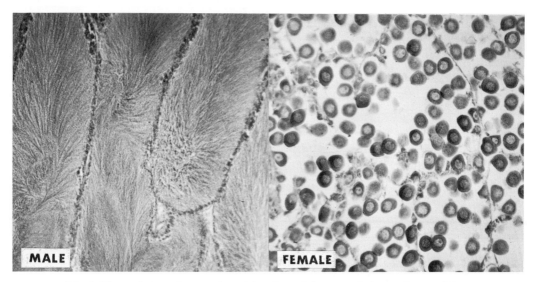

Fig. 9. Photomicrographs of male and female surf clam gonads in the ripe condition.

fertilization at a temperature of 22°C, but form more slowly at lower temperatures. Veligers have an organ fringed with cilia, called the velum, which protrudes beyond the shells. Active beating of the cilia propels the pelagic larvae through the water. Newly developed larvae (48 hours old) have shells with a straight hinge; later the larval shells develop prominent umbones and acquire a form resembling adult clams.

In the laboratory, surf clam larvae have been reared from eggs to a setting size of 230–250 μ in 19 days, at a temperature of 22°C.[6] During the transition from the planktonic to benthic existence, the larvae develop a very active ciliated foot, but still retain the velum (Fig. 10). At this stage, the larvae can either swim or crawl. In a completely metamorphosed clam, the velum is absorbed but the foot remains as an adaptation to a benthic existence. During their planktonic stages, surf clam larvae, like those of many bivalve mollusks, can be carried great distances by strong currents.

Relationships between shell length and age of surf clams have been studied (Fig. 11),[7] and a life span has been estimated as about 17 years. Verification of these data is needed, however, from extensive marking and tagging experiments. A marking experiment was carried out when the Bureau of Commercial Fisheries found large numbers of small clams about 6 months old and averaging 21 mm (ca. $\frac{7}{8}$ in.) long at Chincoteague Inlet, Virginia. After grinding notches in the shells of several thousand and returning them to the natural environment, growth was followed for 18 months, before the clams were lost. They grew at the same rate as unmarked clams from the original population. The ages and average lengths of unmarked clams observed were: 1 year, 45 mm (1$\frac{3}{4}$ in.); 2 years, 69 mm (2$\frac{3}{4}$ in.); 3 years, 91 mm (3$\frac{5}{8}$ in.). On the basis of these data, 4-year-old clams would be expected to average about 110 mm long (4$\frac{3}{8}$ in.). A few 4-year-olds may be taken commercially, since the fishery catches mostly clams 105–180 mm long (4$\frac{1}{8}$–7$\frac{1}{8}$ in.), but generally the clams recruited to the fishery stocks are 5 to 6 years old.

Predators. The moon snails, *Lunatia heros* and *Polinices duplicatus* (family Naticidae), bore countersunk holes in the shells of surf clams and

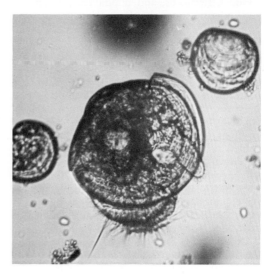

Fig. 10. A surf clam larva with both a ciliated foot and a velum.

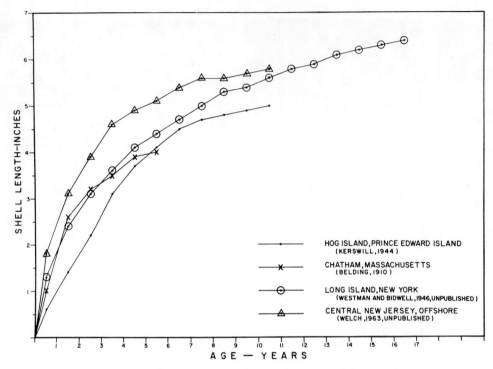

Fig. 11. Estimated ages and shell lengths of surf clams.

rasp out the meats with their radulae. The former species is usually found offshore in deepwater; the latter is a near-shore inhabitant. Predation by these snails on small clams may be extensive—50% of the shells examined in the windrows at Wallops Island in 1965 had been bored. Shells of surf clams have been found nested in cod and haddock stomachs, and undoubtedly other fish and crabs prey upon the clams. Seagulls and other birds at the water's edge consume very small surf clams.

Importance in U.S. Fisheries. The surf clam is the most recent and rapidly growing addition to our east coast shellfisheries; it has risen to a position of major importance during the past 20 years. In this period, total landings have increased almost 30-fold (Table 1). In 1966, total landings reached a peak of 45,000,000 lb of meats, valued at about $3.8 million.[8] The species ranked fifth in volume and tenth in value among 13 shellfish species (including Crustacea) landed in the U.S. Surf clams ranked twelfth in average ex-vessel price per pound of meat (8.5¢) in relation to other shellfish. The fishery, then, lands large quantities of meats of low initial value.

Description of the Fishery. In pre-colonial days the Indians gathered surf clams thrown up on beaches by storms and used them for food. The early settlers along the shore from Massachusetts to New Jersey learned of the clams from the Indians. The clams were a source of fertilizer and food for hogs and poultry, although some were eaten by the settlers. Cape Cod fishermen began the first organized fishery in the 1870's. The meats were salted down in barrels for bait and used in the handline fishery for cod and haddock. Catches were relatively small—3000 barrels valued at about $18,000 were produced in the peak year of 1877. Clams in the Cape Cod region were severely depleted after the turn of the century. An increase in the catch off Long Island, New York, followed the development of a scrape-type dredge in 1929. However, these dredges, used before late 1945, were inefficient and broke many clams. Consequently, less than 2,000,000 lb were landed in each year.

After World War II, certain circumstances stimulated rapid growth of the modern fishery. Among these were: depletion of the soft-shell clam *Mya arenaria,* in northern New England; an increased demand for clam products; and technological innovations that enabled the surf clam industry to meet this demand. Late in 1945, the industry on Long Island developed the jet dredge, an efficient tool for catching surf clams without breakage and at an unprecedented rate. The jet dredge has been continuously improved since then. A special problem in processing surf clams, the removal of sand, was overcome when the industry developed a thorough washing process for the chopped meats. Partly as a result of this processing innovation, 98% of all recent landings have been used for food and only 2% for bait.

Production from the Long Island offshore beds rapidly rose from 0.9 million lb of meats in 1944 to its peak, 6.5 million lb, in 1946 (Table 1). Nevertheless, this production was not equal to the demand. Far more extensive beds were discovered off New Jersey in 1950, and landings from these have greatly surpassed all previous records (Table 1). Although landings of surf clams have continued off Massachusetts, Rhode Island, New York, Delaware, and Maryland, the landings from New Jersey beds have increasingly predominated; in 1966, 96% of all U.S. landings were at New Jersey ports.

Today about 60 vessels dock at Point Pleasant, Barnegat Inlet, Wildwood, and Cape May, New Jersey. The fleet comprises 60–95-ft converted shrimp trawlers, small inshore draggers, oyster dredge vessels, and surplus military vesesls. The spacious decks of the vessels are used for holding a day's catch of clams and storing the hydraulic dredge and auxiliary equipment. The weight of the dredges on these vessels averages about 1 ton. Two power plants are employed—one to propel the vessel and operate the winch, and another to drive the pump for the dredge's hydraulic system.

Hydraulic surf clam dredges employ jets of water pumped through 5- or 6-in. diameter hoses at about 1500 to 3500 gallons per minute. The jetted water loosens the sand ahead of the digging blade and washes clams into the dredge. The dredge knife, digging into the bottom throughout a tow, breaks very few clams. The dredge knives are 30–84 in. wide, although most are 40 and 48 in. wide.

During the 1-day fishing trip, the vessels make repeated tows at a site and average four tows per hour. Catches per tow are often 10–15 bushels. After dumping the catch on deck and resetting the dredge, the 2-3-man deck crew selects the large clams from the pile. Clams are funneled into 1-bushel bags and the bags usually are stacked on the deck. Some vessels use heavy-gauge wire cages that hold 30 bushels of clams. Vessels return to port after 8 to 12 hours at sea carrying from 100 to 600 bushels of clams (average, 330 bushels).

Shuckers at the processing plant scoop out the clam meats and remove the digestive tract, reproductive organs, liver, heart, gills, and mantle. Thus only the muscle tissue of the foot and mantle edge, and the adductor muscles are retained for food. Removal of sand and bits of shell by washing and careful inspection insure a final product of stripped and minced meats suitable for many varied clam preparations. The surf clam meats are processed into canned chowder, canned minced meats for homemade chowders and party dips, or products for restaurants. The perishable surf clam is processed rapidly under strictly hygienic conditions. None are sold in-the-shell, as are oysters or hard-shell clams.

Geographical Variation of the Species. *Spisula solidissima* varies in size in different geographical areas. They commonly grow to lengths of more than 150 mm (6 in.) north of Cape Cod. In the sounds immediately south of Cape Cod and perhaps westward to near Long Island, they rarely reach 150 mm. From Long Island to Virginia, they grow as large as in the north, but tend to have a more triangular shell contour. Those found south of Cape Hatteras to the limit of their range off the northernmost coast of Mexico are rather diminutive, rarely more than 75 mm long (3 in.). Many books on marine animals list the southern populations of small individuals as a subspecies, *Spisula solidissima similis* (Say). Merrill and Webster[2] considered *Spisula solidissima raveneli* (Conrad) to be the appropriate name of southern surf clams. Jacobson and Old[9] believed that the southern populations are sufficiently distinct to warrant specific recognition as *Spisula raveneli* (Conrad).

The surf clam is found from the southern Gulf of St. Lawrence to the northern Gulf of Mexico. In the northern part of this range it is commonly abundant in the turbulent waters off outer beaches, just beyond the breaker zone, and is

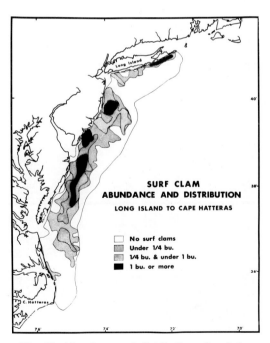

Fig. 12. Abundance and distribution of surf clams along the middle Atlantic coast in 1965.

occasionally found to depths of 420 ft on Georges Bank. Surf clams have been taken at depths of 200 ft on the middle Atlantic continental shelf, but commercially useful concentrations are in depths of 36–180 ft. South of Cape Hatteras, however, the clams are small and inhabit shallow water.

Data for mapping surf clam abundance and distribution from Long Island to Cape Hatteras were obtained during two survey cruises aboard a Bureau of Commercial Fisheries research vessel in 1965. Dredge samples were obtained at 591 stations in an approximately 5-mile grid pattern. For the purposes of this article, the quantities of surf clams caught at these stations are grouped in 4 classes: (1) 1 bushel or more, (2) ¼ bushel or more, but less than 1 bushel, (3) less than ¼ bushel, and (4) no surf clams. The distributions of these quantities of surf clams in the surveyed area are shown in Fig. 12.

The catches of 1 bushel or more during the survey should be of particular interest to fishermen, since these are equivalent to commercial catches of 5 bushels or more, if adjustments are made for the small size of the dredge used in the survey and the short duration of the tows. This dredge had a 30-in. wide knife (Fig. 7) and was towed 5 min; the usual commercial dredge has a 40-in. wide knife and is towed 20 min. Catches of at least ¼ bushel but less than 1 bushel indicate locations that might yield slightly less than the preferred commercial quantities. Catches of less than ¼ bushel are of scientific interest because they show the pattern of distribution of the clam.

The percentages of the survey stations that produced each of the 4 quantities of surf clams are shown for the following areas (Table 2): (1) Long Island, extending from Montauk Point, New York, south to the Hudson Channel off New York City; (2) New Jersey, extending from the Hudson Channel south to off the mouth of Delaware Bay; (3) the Delmarva Peninsula, extending from the mouth of Delaware Bay to the mouth of Chesapeake Bay; and (4) the Virginia-North Carolina coast, extending from the mouth of Chesapeake Bay to Cape Hatteras, North Carolina.

In the Long Island area, nearly all the clams were at depths of 60 ft or less, and none were taken from deeper than 85 ft. Catches of 1 bushel or more were made at only about 6% of the stations, and of ¼ bushel but less than 1 bushel at another 6% of the stations. No surf clams were caught at 66% of the stations, and less than ¼ bushel at 22%.

Surf clams were caught in the New Jersey area from depths of 25–192 ft but most were taken from 150 ft or less. The dredge caught clams at 68% of the stations; 1 bushel or more were caught at 12% of the stations, at least ¼ but less than 1 bushel at 18%, less than ¼ bushel at 38%, and none at 32%. The heaviest concentrations were off Point Pleasant, near Barnegat Lightship, and off Cape May in areas presently being exploited by the fishery.

In the Delmarva Peninsula area, 9% of the stations produced catches of 1 bushel or more and 18% produced ¼ bushel or more, but less than 1 bushel. The heaviest concentration of clams formed a north–southward band about 15–30 miles offshore. Several stations 40–50 miles offshore and in depths of nearly 200 ft produced catches but they were all less than ¼ bushel. The catch from the Delmarva Peninsula area was similar to that in the New Jersey area except that the concentrations of clams were somewhat farther from shore. Since fishermen land their catches daily, the greater distances to the fishing grounds would lengthen the fishing day.

The Virginia–North Carolina area was the least productive of clams. At no station did the catch equal 1 bushel, and only 2.6% of the stations produced ¼ bushel or more. At 23% of the stations ¼ bushel or less were taken, and at 74% no clams were caught.

In summary, it is apparent that the Virginia–

TABLE 2. CATCH OF SURF CLAMS WITH EXPERIMENTAL GEAR FROM FOUR AREAS OFF THE MID-ATLANTIC COAST, 1965

Area	Percentage of stations producing—			
	1 Bushel or more	¼ Bushel and under 1 bushel	Under ¼ bushel	No surf clams
Long Island	5.9	5.9	21.8	66.3
New Jersey	11.6	18.5	38.2	31.8
Delmarva Peninsula	9.4	18.3	37.2	35.0
Virginia–North Carolina	0.0	2.6	23.1	74.4

Fig. 13. An experimental jet dredge used to sample for surf clams along the middle Atlantic coast.

North Carolina area is of little commercial importance but our records have biological significance in outlining the general distribution of surf clams along the mid-Atlantic coast. An inshore population of clams in the Long Island area, although not extensive, can produce catches of commercial quantities. In comparison with the New Jersey and Delmarva Peninsula areas, however, the Long Island area is of minor commercial importance. New Jersey and Delmarva are by far the most productive areas.

JOHN W. ROPES
J. LOCKWOOD CHAMBERLIN
ARTHUR S. MERRILL

References

1. Gardner, J., "Mollusca from the Miocene and lower Pliocene of Virginia and North Carolina. Part 1. Pelecypoda," 1948, U.S. Geological Survey Professional Paper 199-A, pp. 1–168.
2. Merrill, A. S., and Webster, J. R., "Progress in surf clam biological research," in "The Bureau of Commercial Fisheries, Biological Laboratory, Oxford, Maryland, Programs and Perspectives." U.S. Fish and Wildlife Service, Circ. No. 200, pp. 38–47 (1964).
3. Ropes, J. W., and Merrill, A. S., "The burrowing activities of the surf clam," *Underwater Naturalist*, 3 (4), 11 (1966).
4. Jacot, A., "Notes on marine Mollusca about New York City," *Nautilus* 34 (2), 59 (1920).
5. Ropes, J. W., "Hermaphorditism in the surf clam, *Spisula solidissima*," 1966, Am. Malacological Union, Inc. Annual Report for 1966, p. 26.
6. Loosanoff, V. L., and Davis, H. C., "Rearing of bivalve mollusks," in "Advances in Marine Biology," 1963, London and New York, Academic Press, Vol. 1, pp. 1–136.
7. Yancey, R. M., and Welch, W. R., "The Atlantic coast surf clam—with a partial bibliography," U.S. Fish Wildlife Service, Fish. Circ. No. 288, pp. 1–14 (1968).
8. Groutage, T. M., and Barker, A. M., "The Atlantic surf clam fishery in 1966," *Comm. Fish Rev.* **29** (8–9), 64 (1967).
9. Jacobson, M. K., and Old, Jr., W. E., "On the identity of *Spísula similis* (Say)," Am. Malacological Union, Inc. Annual Report for 1966, pp. 30–31 (1966).

Ocean Quahog Fishery

For over a century, ocean quahogs have been known to exist off the east coast of the U.S. and Canada, and in recent decades these clams have supported a moderate to small fishery. The existence of this same species off the coast of Europe has been known much longer, but apparently it has never been harvested there for food. The natives of Iceland, however, are known to have eaten this clam in the 18th century, calling it *ku-skiael* and *krok-fishur*. This use of ocean quahogs by Icelanders is a basis for its scientific name *Arctica islandica* (Linne). In addition to *ocean quahog*, some other common names used in America are *black quahog* and *mahogany clam*.

Arctica islandica (also known as *Cyprina*), is placed by taxonomists in the phylum Mollusca, class Pelecypoda, order Eulamellibranchia, and family Arcticidae. Fossil records show that this species is a relict, the only living member of its family. The Arcticidae were abundant in earlier geological times; fossils of several genera and over one hundred species have been found, largely in the North Atlantic region. The oldest of these fossils classified in the genus *Arctica* are from the early Cretaceous (about 120,000,000 years ago). In Europe, fossils of *Arctica islandica*, itself, date back some millions of years to the Pliocene period. Pleistocene fossils of this species in Sicily show that it ranged farther south in glacial times than at present. Even more recent fossils in West Greenland have been used as evidence of a postglacial warm period. There is no evidence that the species lives in Greenland waters today.

Distribution and Habitat. *Arctica islandica* lives on the continental shelf of eastern North America from southern Newfoundland to just north of Cape Hatteras, North Carolina. Southward from Cape Cod, Massachusetts, its range is contiguous with that of the abundant surf clam, *Spisula solidissima*. The surf clams, however, are largely concentrated in depths of 20 fathoms (37 m) or less, whereas the ocean quahogs are in from 15 to 80 fathoms (27 to 146 m).

In Europe the known limits of *Arctica islandica* are from the vicinity of North Cape, Norway, southward to the Bay of Cadiz on the southwest coast of Spain (Nicol, 1951). It is abundant in the Faroes, the Shetlands, the British Isles, southern Norway, and northern France. It is also abundant in Iceland.

The locations where ocean quahogs have been taken in the western Atlantic by the Bureau of Commercial Fisheries and others are shown in Fig. 14. In the Gulf of Maine and the vicinity of Cape Cod, the species generally lives close to shore, but it also occurs on eastern and southern Georges Bank as much as 200 miles offshore. Along the outer coast of Long Island, ocean quahogs live both close to shore and offshore, but from there southward to Cape Hatteras, the species occurs only well offshore, usually in 20 to 80 fathoms. Extensive sampling of the ocean quahog in the Middle Atlantic has been done by the Bureau of Commercial Fisheries only in recent years. But in this work, no stations have been made deeper than 80 fathoms, and thus the deeper limits of bathymetric distribution of ocean quahogs in the Middle Atlantic are not known. In the 1880's, during explorations by the U.S. Fish

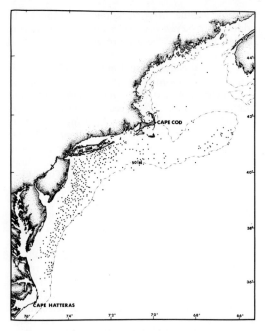

Fig. 14. Distribution of ocean quahogs based on various sources of data.

Commission along the Middle Atlantic Continental Slope, shells of *Arctica* were obtained from as deep as 482 m (264 fathoms). There is no proof, however, that these were of living individuals.

Biology, Life History, and Ecology. Attaining a maximum length of about 5 in. (127 mm), the ocean quahog is one of the larger bivalves of the northern waters of Europe and eastern North America. In special collections from recent stations on the Middle Atlantic shelf, by the Bureau of Commercial Fisheries, the specimens range in size from 30 to 120 mm (1.3 to 4.8 in.) long. They average 87 mm (3.4 in.) long and 71 mm (2.8 in.) wide.

The tightly-closing, rounded oval shells are thick and heavy, and have incurved, contiguous, prominent beaks. The hinge has four large interlocking cardinal teeth and a prominent external ligament.

The shells of ocean quahogs resemble those of the more familiar venerid hard-shell clam or quahog, *Mercenaria mercenaria*, which is, however, usually more pointed posteriorly. Adult ocean quahogs are easily distinguished from the latter species by the presence of a dark brown, amber, or black periostracum (the horny outside layer of the shell). The periostracum of hard-shell clams is pale amber and largely absent in the adults. The inner shell margins are smooth in the ocean quahog, but crenulated in the hard-shell clam. The ocean quahog has no pallial sinus, but does have posterior lateral teeth; the hard-shell clam has a well developed pallial sinus, but no lateral teeth.

Arctica islandica is dioecious, that is, has separate sexes; hermaphrodites occur, but very infrequently. Loosanoff (1953) studied the reproduction of the species from specimens off Rhode Island. The entire population did not reach ripeness at the same time. Spawning began in late June or early July when temperatures were about 13.5°C (56°F), reached a maximum in August, continued during September, when the water temperature was at a high of 15°C (59°F), and diminished early in October as the temperature slowly declined. The main period of gonad recovery was in the late fall and early winter. This recovery slowed during the late winter, but resumed at a rapid rate with the spring increase in temperature. Loosanoff did not succeed in artificially fertilizing the eggs of *Arctica*.

Jorgenson (1946) found the species spawning at Kristineberg, Denmark, in July. The eggs were around 75 μ in diameter. The prodissoconchs (completed larval shells), measured in juvenile individuals, varied between 270 and 300 μ. Jorgenson remarked that the small size of the egg and of the metamorphosed larva points to a planktonic stage in the larval development.

Nothing is known about the growth of *Arctica* after it starts to live on the bottom. Thus, it is not yet possible to estimate the age of the adults found in offshore beds.

Some information is available on environmental conditions influencing the distribution of *Arctica*. Turner (1949, 1953) believed that temperature may be one of the factors influencing the depth zonation of the species. Intolerance to high temperatures probably explains its absence from shallow water in the southern part of the range. Turner remarked that the difficulties encountered in keeping the species alive long enough to transport them to market during the summer months demonstrate that its thermal death point is low, probably about 55–60°F (13–16°C). He noted that specimens held in tanks at 70°F (21°C) or more died in a few days, and he mentioned learning that commercial operators had experienced similar mortalities when they bedded the clams in warm inshore waters for storage. On the other hand, Turner was able to hold them in tanks and inshore beds after the temperature had dropped below 50°F (10°C). Loosanoff (1953) reported that a steady salinity between 31.0 and 32.8 ppt prevailed in an area off Point Judith, Rhode Island, where a large population of ocean quahogs was established. He suggested that this may be an optimum salinity range for the existence and propagation of the species.

Additional information on the biology of the ocean quahog is available from anatomical and behavioral studies. Internally, there are anterior and posterior adductor muscles that close the shells. Clear scars on the inner shell surfaces show where these muscles attach. The mantle lining the inner surfaces of the shells is open ventrally from the anterior adductor muscle to near the base of the siphons. The siphons are short and fringed with small tentacles, of which those surrounding the incurrent opening are the longer. The foot is large and the labial palps are both large and wide. The intestine is long and convoluted—typical of filter-feeding bivalves that subsist largely on phytoplankton. Saleuddin (1964) gave many details of the anatomy and histology of this clam (Fig. 15).

The ocean quahog is necessarily a shallow burrower, because the siphons extend only a short distance out of the shell. Burrowing is accomplished by protrusion of the foot into the substratum and then pulling the shell down. Saleuddin suggested that water expelled from the mantle cavity assists this process by flushing away sand.

Some information has also been reported on the predators of *Arctica islandica* and on its commensal relationships. In European waters, the ocean quahog is apt to be infested by species of pinnotherid crabs and turbellarian flat worms; Brunberg (1964), dealing with the clam in European waters, reported that it is a host for the commensal nemertean, *Malacobdella grossa*. On the other hand, Ropes and Merrill (1967) have examined numerous specimens of the clam from the Middle Atlantic without finding such commensals. Empty shells of the species, taken in dredges, often contain the drill holes of naticid gastropods.

There is evidence that ocean quahog populations may experience heavy natural mortalities. On the Middle Atlantic shelf, we have taken full scallop dredge loads (about 32 bu) of empty shells still held together by their ligaments. In 1967, scientists aboard the Woods Hole Oceanographic Institution submarine *Alvin*, working off the mouth of Chesapeake Bay in depths of 45–55 m (25–27 fathoms), saw an ocean quahog "graveyard." The shells were aggregated in incredible numbers, and bottom currents were thought to

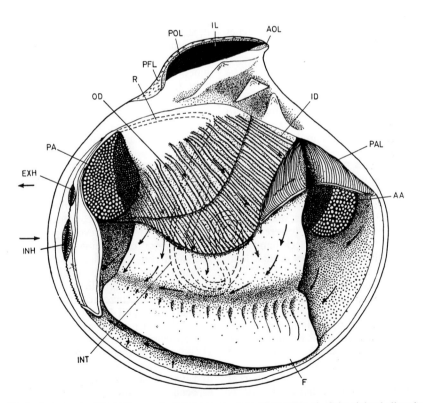

Fig. 15. General anatomical features seen in the ocean quahog after removal of the right shell and underlying mantle: AA, anterior adductor muscle; AOL, anterior outer layer; EXH, exhalant siphon; F, foot; ID, inner demibranch; IL, inner layer; INH, inhalant siphon; INT, intestine; OD, outer demibranch; PA, posterior adductor muscle; PAL, labial palp; PFL, posterior fusion layer; POL, posterior outer layer; R, rectum. Large arrows show inhalant and exhalant streams; small arrows show direction of ciliary currents. (After Saleuddin.)

be responsible (R. L. Edwards, personal communication).

History and Description. Ocean quahogs were not fished commercially until 1943 when the food production program of World War II stimulated investigation on the feasibility of utilizing lesser known species of fish and shellfish. The ocean quahog was then determined to be of value for processing into chowders and related products.

Arcisz and Neville (1945) found beds of commercial significance 3 to 12 miles off Rhode Island in muddy bottom at depths of 13 to 25 fathoms. The shells of the clams averaged 3.3 in. long (83 mm), and the shucked meat averaged 1.2 oz per clam (35 g). Turner (1949) surveyed Cape Cod Bay and Vineyard Sound and found ocean quahogs generally on sand-mud or mud bottom at depths of greater than 9 fathoms; the largest concentrations were in depths of 13 fathoms (24 m) or more. The shells of these clams averaged 3.4 in. (86 mm) in length and reached a maximum of 4.3 in. (108 mm). Turner stated that there appeared to be a sufficient supply to furnish fishing for several years if a market could be developed. MacPhail and Medcof (1959) explored for ocean quahogs in areas of southern Northumberland Strait, off Prince Edward Island, and made good commercial size catches at several stations in 15 to 18 fathoms (27 to 36 m). They felt little doubt that the species could be fished commercially if markets could be established.

The heavy concentrations of ocean quahogs off the coast of Rhode Island were the first to be exploited (Arcisz and Sandholzer, 1947). Nantucket-type dredges with 7-in. long (180 mm) teeth, spaced $1\frac{1}{2}$–2 in. (40–50 mm) apart were used to plow the clams out of the bottom. The fishing vessels were usually converted draggers or oyster boats operated by a skipper and a crew of two. The size of the dredge depended on the power of the boat. Two dredges, one on each side of the boat, were raised and lowered alternately at 15-min intervals. A deck-hand was usually able to cull and bag one dredge load before another was hauled in.

The size of the fleet fluctuated from 6 to 15 boats, and was limited by the market demand for the abundant harvest of this new fishery. A 40-ft (12 m) vessel could easily land 100 to 150 bushels in a 6-hr fishing day. Landings were principally in the vicinity of Point Judith and Newport, R.I., and the catches were trucked to nearby processing houses. Because the market was unable to absorb the potential catch, quotas were often applied to prevent overproduction.

At its peak in 1946, the Rhode Island fishery produced 1.5 million lb of meats with a dockside value of $126,000. Production declined in 1947 to less than $\frac{1}{3}$ of a million lb, and after further declines in succeeding years, levelled off at about 100,000 lb annually, with a value of about $10,000. The price has varied from a low of 6.2 cents per lb in 1948 to a high of 11.8 cents in 1965. In recent years the price has generally been stable at about 10 cents per lb.

Some other small ocean quahog fisheries have developed outside Rhode Island waters since World War II, but none has persisted. For example, a fishery started on Cape Cod in 1954 (Russell, 1955) lasted for five years. During that time over a million pounds of meats were landed and processed. The meat was frozen and sold in 5-lb bags, mostly to the Navy for use in chowder. The processor informed us that this fishery was discontinued because of the intense competition of the surf clam fishery.

Contamination and spoilage is more likely with ocean quahogs than with hard-shell clams. The latter will stay alive and remain tightly closed for prolonged periods out of the water, even in the summer time, but ocean quahogs may gape open soon after removal from the water. Prevention of contamination of ocean quahogs therefore requires not only icing and sanitary handling of the catch aboard ship, but also expeditious processing ashore. Arcisz and Sandholzer (1947) described in detail the hazards to sanitation of the fishery, as it existed during World War II, and recommended preventive measures. In conjunction with development of the booming surf clam fishery in subsequent years, modern shucking and processing plants have been built and environmental sanitation procedures put into practice both aboard ship and ashore. In consequence, the sanitary handling of ocean quahogs should now pose no unique or unusual problems to fishermen or processors.

The meat of this species sometimes has a fishy or "iodine" taste, which some find displeasing. Research is in progress at the Bureau of Commercial Fisheries Technological Laboratory, Gloucester, Mass., to develop processing methods to minimize or eliminate undesirable flavors (personal communication, F. J. King).

Importance in the U.S. Fisheries. The ocean quahog is an underutilized resource because (1) it has been less favorable for development than the surf clam and (2) the remarkable growth of the surf clam fishery has continued to meet the market demand for clam meats since World War II. At present there is only a small fishery for ocean quahogs, almost entirely in Rhode Island, and vast beds of the species are untouched, especially offshore in the Middle Atlantic. In comparison to the surf clam, the ocean quahog has the following unfavorable characteristics: (1) the species does not grow nearly as large, and thus yields only about one-third as much meat per clam; (2) it is not abundant as close to shore or in as shallow water; (3) it is more difficult to keep live and sanitary prior to processing; and (4) it does not consistently have as good flavor. For these reasons, the unutilized beds of ocean quahogs are a reserve that is not likely to be much harvested unless surf clams become scarce or the demand for clam meats increases.

It is conceivable, however, that an ocean quahog fishery might develop more or less in competition with the surf clam fishery. The modern hydraulic dredge, developed by the surf clam industry, is an excellent collector of ocean quahogs. Because the two species are not abundant in the same ranges of depth, there is little mixing of the catches. Even now, ocean quahogs might be utilized to a greater extent if handling and processing methods were developed that would yield products clearly acceptable to the consumer. Another immediate problem is that the great beds of this clam in the Middle Atlantic are so far offshore. Because of this it might not be economical to fish them and return on a daily basis, and therefore the development of methods of processing this perishable species at sea might be a necessity. To produce more accurate assessment of the long term resource potential of the ocean quahog, research is recommended on its biology: growth rate, longevity, mortality, and reproductive potential.

ARTHUR S. MERRILL
J. LOCKWOOD CHAMBERLIN
JOHN W. ROPES

References

Arcisz, W., and Neville, W. C., "Bacteriological and sanitary survey of the ocean quahog industry," in "The Ocean Quahog Fishery of Rhode Island," 16–25, R. I. Dep. Agric. Conserv., Div. Fish Game, (1945).

Arcisz, W., and Sandholzer, L. A., "A technological study of the ocean quahog fishery," *Commer. Fish. Rev.*, **9** (6), 1–21 (1947).

Brunberg, L., "On the nemertean fauna of Danish waters," *Ophelia*, **1** (1), 77–111 (1964).

Jörgenson, C. B., "Lamellibranchia," *Medd. Dan. Fish. Havunders*, 4 (Ser. Plankton), 277–311 (1946).

MacPhail, J. S., and Medcof, J. C., "Ocean quahog explorations," Dep. Fish. Can., *Trade News*, 3–6 (June 1959).

Nicol, D., "Recent species of the veneroid pelecypod *Arctica*," *J. Wash. Acad. Sci.*, **41** (3); 102–106 (1951).

Ropes, J. W., and Merrill, A. S., "*Malacobdella grossa* in *Pitar morrhuana* and *Mercenaria campechiensis*," *Nautilus*, **81** (2), 37–40 (1967).

Russell, H. D., "A new clam industry in New England," *Nautilus*, **69** (2), 53–56 (1955).

Saleuddin, A. S. M., "Observations on the habit and functional anatomy of *Cyprina islandica* (L.)," *Proc. Malacol. Soc. London*, **36** (3), 149–162 (1964).

Turner, H. J., Jr., "Report on Investigations of Methods of Improving the Shellfish Resources of Massachusetts," 1949, Mass. Div. Mar. Fish.

Turner, H. J., Jr., "A Review of the Biology of Some Commercial Molluscs of the East Coast of North America," 1953, Mass. Div. Mar. Fish., 6th Rep. Invest. Shellfish. Mass.

Cross references: *Biotoxins, Marine; Pollution of Seawater.*

COD FISHERY

Biology and Distribution

The common Atlantic cod, *Gasus morhua*, is well known on both sides of the north Atlantic. On the American coast it is found as far north as Greenland, Davis Strait, and Hudson Strait and south nearly to Cape Hatteras.[1] In Europe it is found from Novaja Zemlya, Spitsbergen, and Jan Mayen to the Bay of Biscay. It is also common near Iceland and the Faroe Islands. A close relative, the Pacific cod, *Gadus macrocephalus* occurs in the north Pacific.

The most noticeable external characteristics are its three dorsal and two anal fins, its protruding upper jaw, its almost square tail, and a pale line running along each side of the body from the head to the tail. There is a fleshy barbel under the lower jaw. In most fish the upper part of the body is thickly speckled with small, round spots somewhat darker than the body color which may range from reddish to brown, gray, or greenish.

Cod can grow to a very large size. Although fish from 25–50 lb are considered "large," several examples of cod weighing over 100 lb are known. Cod can be found from shallow water near shore down to 250 fathoms. Its usual habitat is within a few fathoms of the bottom but it also comes to the top of the water in pursuit of small fish or squid. It is most plentiful on the banks and in oceans of moderate depths. Cod live chiefly over rocky, pebbly ground, on sand or gravel, and seldom on soft mud. They go in schools but not in such dense bodies as mackerel and herring. The movements on and off shore and from bank to bank are due chiefly to temperature influence, the presence or absence of food, and the search for proper spawning conditions. Cod prefer temperatures of 32°–41°F (0°–5°C), but good catches can be made in waters up to 50°F (10°C).

Cod feed on almost all types of sea life. The most important food is fish, especially herring, capelin, and sand launce, but mussels, crabs, and other bottom animals are also eaten.

The majority spawn during the early spring months, each fish producing from 3,000,000 to 9,000,000 eggs depending on its size. The fertilized eggs rise from the bottom and drift with the

Fig. 1. Atlantic cod (*Gadus morhua*).

currents, hatching occurring in 10 to 50 days depending on the temperature of the water. The young live pelagically until they are 2–4 in. long when they settle near the bottom. Growth varies from area to area and from season to season depending largely on the food available and the temperature of water in which they live. Templeman[2] gives as an example the difference in size between Labrador and southern Grand Bank cod. At 5 years of age the difference is not too large, 1.3 lb for Labrador cod and 2.2 lb for the southern Grand Bank cod, but by 12 years of age the difference in weight is much greater, namely, 3 lb vs 13 lb. Sexual maturity is reached at age 2–4 in certain coastal populations and not before 7–12 years in cod in northern oceans.

Fishing Methods and Vessels

Several types of fishing gear and vessels are used to catch cod depending on the area being fished.

On the banks and fishing grounds some distance from shore, the otter trawl is by far the most commonly used fishing method today. It consists of a net being pulled behind a vessel along the bottom, the front part being held open with kite-like "otter doors" and the rest of the net gradually tapering off until the fish are driven into a bag called the cod end. Synthetic materials such as polypropylene have replaced the traditional natural fibers, i.e., cotton, manila and linen, in the manufacture of trawls, the result being greater strength and less maintenance. Midwater trawls which can catch fish at almost any distance from the bottom are also being used, although chiefly for other species of fish with more pelagic habits than the cod. The Spanish are using a single, large trawl pulled by two vessels of similar size (Pareja trawl).

The otter trawl can be used by vessels of almost any size as long as the bottom is fairly smooth and free of large rocks. In the Bay of Fundy, for instance, vessels as small as 40 ft are used as otter trawlers, and on the banks the numbers of large factory trawlers up to 250 ft are increasing. The most common size of otter trawler, landing fish preserved with ice, is between 110–150 ft with a capacity of 300,000–400,000 lb gutted fish.

Traditionally the trawl was hauled from one side of the vessel but the majority of new trawlers haul the net over a ramp in the stern. These stern trawlers can fish in rougher weather than the side trawlers and can usually haul and set the trawl faster.

The oldest fishing method and one which is still being used extensively is the hook and line. The hook can be on a single handline fished by one man, or when a large number of hooks are fastened to a line and set near the bottom, it is called a longline. The name trawl-line is still used for this type of gear, but should not be confused with the otter trawl. Longlines are very efficient, especially on rougher bottoms where the otter trawls may easily be torn. They also catch larger fish on the average than do trawls.

Longliners range in size from 50–100 ft in Canada, wheras most Norwegian or Icelandic vessels are from 100–160 ft in length. Up to quite recently a large fleet of Portuguese sailing vessels, fishing cod with longlines and handlines from dories, could be found on the Grand Banks and off Greenland and Labrador ("The White Fleet"). They have gradually been replaced by motor vessels and trawlers. The Canadian and United States cod fishery also was carried out with dories from schooners, the most famous of these being the "Bluenose" from Lunenberg, Nova Scotia.

The choice of bait is very important, especially when the cod are on their spawning migrations and eat very little. Squid, herring, and mackerel are most commonly used and are today supplied almost exclusively in the frozen state.

Another effective handlining gear is the jig. It has been in use for centuries and consists of a shiny metal (lead, zinc) imitation of a fish with two hooks being pulled up and down in the water. The "Norwegian Jigg," a triangularly shaped, stainless steel jig with triple hooks, has largely replaced the traditional jig in the last two or three decades.

Where the schools of cod come close to shore, as in the famous Newfoundland fishery, special cod traps are in use. These traps are square with walls and bottom made of netting of 35–85 fathoms circumference, and the fish follow a "leader" into a "chamber" from where it is difficult to escape.[3] The net can be lifted by the fishermen to remove the catch. In recent years cod 4–6 years of age comprise on the average from 80–95% of the trap catches.[2] Gillnets are also extensively used by inshore fishermen. These nets are now usually made of synthetic materials such as nylon, and the meshes are of such size that the cod are caught by the gills while trying to pass through the net.

Fig. 2. Lifting a Newfoundland cod trap.

Preservation and Utilization on the Vessel

The preservation of cod to produce a quality product must start on the vessels as soon as the fish are taken from the water. The fish are first gutted and thoroughly washed; on some vessels the livers are saved for steaming to cod-liver oil, but the rest of the viscera is usually discarded.

On the fresh-fish vessels the fish are then iced in the hold, which is divided into pens with removable boards; layers of fish and ice are alternated until the pen is filled. In some cases the cod are placed in single layers on ice, the so-called shelving. This is especially common on British trawlers. Icing in boxes of aluminum, wood, or plastic is also practiced in some areas. Well-iced fish should remain of good quality from 5–8 days. Additives such as antibiotics and sodium nitrite have been used in the ice, but have not found universal acceptance for many reasons.

The second important method of preservation on the vessels is freezing. Freezing at sea has become of greater importance in the last few years as fishing vessels seek the cod on increasingly distant grounds. A few, like the English "Fairtry" series of vessels, fillet the fish on board and land their catch as frozen fillets.

However, the large freezer-trawlers are becoming more and more popular. On these vessels the fish are gutted and washed and then frozen in vertical plate freezers into blocks weighing about 75 lb. Thawing facilities have to be provided ashore and can consist of warm air, dielectric, or water thawing. The only commercial installation in eastern North America at the present time is located in Catalina, Newfoundland, where the Fisheries Research Board of Canada has designed and installed a water thawer.[4] A dielectric thawer has also been used. Fillets cut from thawed cod, packed, and refrozen have been shown in these experiments to be of excellent quality.[5]

Many vessels still salt cod for later drying ashore. After gutting the head is removed, the fish is split, and part of the backbone removed.

Preservation and Utilization Ashore

Fresh and Frozen. Most of the cod is landed at filleting plants ashore. The filleting is carried out by machine or by skilled filleters, the skin is removed by a skinning machine, and the fillets

Fig. 3. Salted cod being dried on the beach.

further trimmed and inspected. Depending on the end usage they are:
   (1) Packed in boxes or other containers, chilled and shipped with ice in insulated trucks or railway cars for the fresh-fish market.
   (2) Packed in 1 lb, 2 lb, or bigger cartons and frozen in plate freezers for the retail frozen-fish trade.
   (3) Packed in cartons and frozen in plate freezers into blocks to be used for fish sticks and portions.

The advantage of the freezing of fillets and portions and the importance of low storage temperatures were studied as early as 1926.[6]

Precooked. A fairly new development in the marketing of cod has been breaded, frozen products such as fish sticks and portions. The latter, especially, has had an exceptionally rapid growth in the United States, the quantity going from practically zero to 146,600 lb in the 10-year period 1956–1966.

Fish sticks or fish fingers are prepared by sawing a frozen block of cod into sticks of approximately $1 \times 1 \times 5$ in., whereas the portions are flat pieces of fish approximately $3 \times 3 \times \frac{1}{2}$ in. These are dipped in batter, breaded, and cooked in oil until golden brown although still partly frozen in the middle. The sticks and portions are then cooled, packaged, and frozen; the housewife only has to heat them in the oven for 15–20 minutes to serve.

Salted and Dried. Salted cod fish is practically always dried before being marketed. The drying used to be carried out on rocks and cliffs; whenever rain or fog threatened they had to be gathered together, piled, and covered. Mechanical dryers have been in use in the last few decades so that drying can be carried out independently of the weather the year around.[7]

The product commonly available is quite heavily salted and has a white surface covered with salt. A typical example of a lightly salted product is the Gaspé cured cod with a hard, amber-colored surface.[8]

Another type of dried codfish with historical interest is the stockfish. The fish are split, the two halves being joined by the tail and hung on wooden poles to dry without salt. Since a dry cold climate is required for successful drying, only northern Norway, especially Lofoten, and Iceland produce this commodity. Due to the excellent keeping qualities of salted and dried cod and stockfish these products are very popular in tropical countries where refrigeration facilities are unknown.

Smoked. Smoked cod, especially fillets, are also a popular product in many countries. The fillets are first brined for a few hours and are then dried and lightly smoked. The short salting and smoking time does not preserve the fish materially, and the fillets are, therefore, often marketed frozen.

Nutritional Qualities

Cod is an excellent source of high-quality protein and since the fat content is low, so is the calorie value. Several essential vitamins and

TABLE 1. COMPOSITION OF PRODUCTS FROM COD[9]

	Chemical composition							
	g/100 g				mg/100 g			
	Water	Protein	Lipid (fat)	Ash	Ca Calcium	P Phosphorous	Fe Iron	I Iodine
Fillets	80.4	18.1	0.7	1.1	20	200	0.6	0.5
Roe	74.0	20.4	2.4	1.4	30	500	1.5	0.2
Liver	32.0	6.2	60.3	0.8	25	100	...	0.4
Salt cod	39.5	37.8	1.0	22.2	60	300	1.6	...
Dried cod (Stockfish)	14.8	78.5	1.4	5.9	160	950	2.5	1.2

	Vitamins					
	μ/100 g				mg/100	
	Thiamine B_1	Riboflavin B_2	Pantothenic acid	Vitamin B_{12}	Niacin	Calories per 100 g
Fresh fillets	50	110	180	0.8	2.0	70
Roe	750	700	3000	10.0	1.3	105
Liver	100	580	640	11.0	2.9	570
Salt cod	...	230	340	3.6	2.4	160
Dried cod (Stockfish)	...	240	1675	10.0	7.5	325

TABLE 2. COMPOSITION OF A COD SCRAP MEAL[10]

				Amino acids (g/100 g crude protein)	
Moisture	7.5%	Ca	7.3%	Isoleucine	4.64
Protein	68.8%	P	3.8%	Glycine	8.51
Fat	1.9%	K	1.1%	Cystine	0.63
Ash	21.3%	Na	0.97%	Tryptophane	0.83
		Mg	0.22%	Threonine	4.87
Thiamine	0.151 mg/100 g	Al	130 ppm	Arginine	6.79
Riboflavin	0.46 mg/100 g	Ba	10 ppm	Histidine	2.82
Pantothenic acid	0.47 mg/100 g	Fe	80 ppm	Phenylalamine	4.10
Niacin	3.81 mg/100 g	Sr	100 ppm	Methionine	2.73
Vitamin B_{12}	7.1 mcg/100 g	B	6.5 ppm	Leucine	7.01
Choline chloride	405 mg/100 g	Ca	8.0 ppm	Valine	4.73
		Zn	80 ppm	Tyrosine	3.33
		Mn	9.7 ppm	Lysine	9.61
		Cr	14 ppm		

minerals are also present in significant quantities as listed in Table 1.

By-products

Fish meal. The cuttings from the filleting lines, i.e., heads, backbones, skins, etc. are processed into fish meal. The offal is ground and cooked with direct or indirect steam in a continuous cooker. Due to the low fat content of cod muscle (0.5–0.7%) the pressing step is often omitted. It is then dried in a dryer, either heated by steam or direct flame.

The cod meal (white fish meal) is of a light color, has a low fat content (2–5%), and slightly lower protein content than meal made from whole fish (58–66% vs 72–75% for herring meal). It is an important ingredient in feed for poultry, pigs, and mink (Table 2).

Skins. Cod skins have been used for the manufacture of glue for centuries and still have a market in spite of the inroads of synthetic materials. The skins were traditionally salted, in which case a thorough soaking in fresh water was necessary. More recently freezing has been used to preserve the skins. The manufacturing process is briefly as follows: The freshened or thawed skins are cooked with dilute acid in steam-jacketed cookers in two stages for 5–10 hours each, with glue being withdrawn after each stage. The moisture content is reduced in triple-effect vacuum evaporators and preservatives added. When the product is made from top-quality skins under the best conditions, it can be used as photoengraving glue.

Isinglass. Isinglass is made from the swim bladder or "sounds" from the cod. They are first washed thoroughly and then air dried. Isinglass is the purest form of fish gelatin which can be prepared, and it is used as a clarifying agent, primarily in wine manufacture.

Cod-liver oil. The process of steam-rendering the livers to produce a pale-colored medicinal cod-liver oil is usually credited to the Norwegian Peter Möller (1853). However, there is evidence that Charles Fox of Scarborough, England, built a factory in St. John's in 1848 and rendered the oil from the livers by subjecting them to a steam-heated hot-water bath. Soon the improved oil prepared by these methods was in general use. High-pressure steam is now used to render the livers, although the production has been decreasing due to competition from synthetic vitamins.

Cod-liver oil was originally used as a lamp oil and for dressing leather, but in the 18th century

TABLE 3. PROPERTIES OF COD-LIVER OIL

Description: A pale, yellow liquid; odor and taste slightly fishy, but not rancid.
Solubility: Slightly soluble in alcohol, freely soluble in chloroform, ether, carbon disulfide, ethyl acetate, petroleum ether.
Refractive index: N_D^{40} 1.4705–1.4745.
Saponification value: 180–190 mg KOH/g oil.
Iodine value: 155–180.
Density at 20°: 0.917–0.927 g/ml.
Vitamin A: Not less than 600 International Units/g.
Vitamin D: Not less than 85 USP units/g.

Average Fatty Acid Composition[11]

Fatty acid	Wt. %	Fatty acid	Wt. %
14:0	3.3	20:1 ω 9	7.8
16:0	13.4	20:4 ω 6	1.4
16:1 ω 7	9.6	20:5 ω 3	11.5
18:0	2.7	22:1 ω 11	5.3
18:1 ω 9	23.4	22:5 ω 3	1.6
18:2 ω 6	1.4	22:6 ω 3	12.5
18:3 ω 3	0.6	Others	4.5
18:4 ω 3	1.0		

TABLE 4. ANALYSIS OF FISH PROTEIN CONCENTRATES[12]

Raw material	Protein (%) (dry basis)	Moisture (%)	Ash (%)	Fiber (%)	Fat (%) (ether extract)	Fat (%) (chloroform–methanol extraction)
Cod fillets	92.9	4.64	1.89	0.50	0.02	0.033
Whole cod	84.7	7.62	14.6	0.88	0.02	0.056
Headed eviscerated cod	90.26	5.25	8.37	0.81	0.02	0.02
Cod trimmings	87.2	3.54	11.42	0.34	0.039	0.04
Cod trimming[a] press cake	76.6	4.19	26.68	0.63	0.03	0.035
Cod trimming[b] press cake	70.7	5.66	23.78	0.60	0.01	0.032
Whole herring	89.7	8.24	7.13	0.94	0.09	0.18

[a] Cod trimmings cooked with indirect heat and pressed to 60% moisture.
[b] Cod trimmings cooked with live steam and pressed to 60% moisture.

TABLE 5. NUTRITIONAL VALUE OF FISH PROTEIN CONCENTRATES FED AT THE 10% PROTEIN LEVEL FOR 4 WEEKS TO GROUP OF 10 WEANLING RATS

	Source of raw material for fish protein concentrate				
	Cod fillets	Whole cod fish	Headed, eviscerated cod	Cod trimmings	Whole herring
Initial wt. (g)	49	49	49	47	50
1 Week wt. (g)	74	73	69	65	68
2 Week's wt. (g)	103	100	101	98	99
3 Week's wt. (g)	133	122	114	126	118
4 Week's wt. (g)	166	158	149	158	152
Weight gain (g)	117	110	100	111	103
Food (g) consumed	379	398	374	383	352
P.E.R.	3.09	2.75	2.68	2.91	3.21
Corrected[a] P.E.R.	2.97	2.64	2.58	2.57	2.74
Liver fat[b] (% dry basis)	7.86	10.2	7.33	7.84	9.05

[a] Protein efficiency ratio corrected to casein value of 2.50.
[b] Average liver fat for casein-fed rats—10.9% (dry basis).

TABLE 6. NOMINAL CATCH (LIVE WEIGHT). THOUSAND METRIC TONS[a]

Species, country	1938	1948	1958	1961	1962	1963	1964	1965
Cod (Atlantic) *Gadus morhua*	2085.0	2085.0	2560.0	2955.0	3010.0	2964.0	2676.0	2730.0
Canada	372.8	381.4	287.9	286.0	320.6	333.7	315.5	313.0
Greenland	4.7	15.8	26.7	35.2	36.3	24.2	23.1	25.2
United States	58.6	32.4	18.8	21.1	21.3	19.1	17.6	16.0
Denmark	19.3	49.8	58.1	65.5	62.9	69.0	68.3	79.3
Faroe Islands	57.7	91.3	76.0	87.2	116.3	106.1	103.2	95.1
France	199.9	131.3	172.5	169.0	172.6	157.9	164.7	163.5
Germany (Eastern)	...	...	...	...	...	41.6	45.1	...
Germany (Fed. Rep. of)	163.2	99.2	109.7	166.1	200.1	207.8	176.3	192.4
Iceland	95.3	181.9	296.1	248.7	223.4	240.4	280.7	244.0
Norway	328.2	272.4	373.4	346.5	296.5	277.1	224.6	262.6
Poland	1.6	33.6	38.6	40.7	47.2	57.4	53.5	66.2
Portugal	43.4	106.9	190.0	197.0	217.6	230.4	227.8	197.2
Spain	25.8	30.6	102.9	197.1	199.4	216.5	221.2	226.5
UK (England and Wales)	393.0	370.8	359.4	302.5	337.7	329.8	305.0	323.3
UK (Scotland)	33.4	40.9	51.2	47.4	47.6	55.7	55.6	56.6
USSR	243.9	191.3	319.2	638.1	608.5	536.5	340.5	343.6

[a] Yearbook of Fishery Statistics, 1965. F.A.O., Rome.

the medicinal qualities were recognized, However, only in the first decades of this century were the growth-promoting factor, Vitamin A, and the anti-rachitic agent, Vitamin D, discovered in cod liver oil. Some average values for these vitamins, fatty acid composition, and physical constants are given in Table 3. In spite of the inroads of synthetic vitamins, cod-liver oil is still recognized as an excellent source of vitamins and essential fatty acid both in human and animal nutrition.

Fish Protein Concentrate (FPC). A high-quality fish flour or fish protein concentrate can be prepared from whole cod, cod fillets, or cod trimmings. By using an isopropanol method developed at the Halifax Laboratory of the Fisheries Research Board of Canada, the water-soluble flavor components and fat are removed; the characteristics of the practically odorless and tasteless powder are given in Tables 4 and 5. It will be noted that FPC is comparable with egg protein and is better nutritionally than milk protein (casein). FPC can be used as a high-protein food supplement in human nutrition as well as in animal feeding, for instance, as a milk replacer in calf rations.

World Catch and the Future

In Table 6 the world catch of Atlantic Cod is listed by countries; it is evident that the quantity caught increased dramatically from 1948–1956, which was the peak year. Since then the quantity has fluctuated from 2,560 to 3,010 thousand metric tons per year in spite of greatly increased fishing effort. Several fishing areas are now producing only a fraction of the quantity caught only 10–15 years ago; it is estimated, for instance, that the cod population of the Barents Sea is only about 10% of what it used to be. The other chief fishing areas have as yet not reached this point, but increased fishing pressure on the stocks are showing in the smaller average size of the fish and the increased effort needed to catch the same amount of fish.

International commissions are active in the conservation of the cod stocks and implement regulations regarding trawl mesh size. ICNAF, International Commission for the Northwest Atlantic Fisheries and ICNEF, International Commission for the Northeast Atlantic Fisheries employ biologists to evaluate the statistics gathered from these areas. They then recommend measures to be taken to protect the stocks with a view toward producing a sustained yield.

With the continued rapid expansion of the fishing fleets of the countries involved in the cod fishery, it is likely that the pressure on the stocks will continue for some time. It is hoped that international cooperation will eventually stabilize catches at a point where all will be able to continue to harvest this valuable fish in future years.

D. R. IDLER
P. M. JANGAARD

Cross reference: *Breaded Fishery Products; Fish Glue.*

References

1. Leim, A. H., and Scott, W. B., "Fishes of the Atlantic Coast of Canada," *Fish Res. Bd. Canada, Bull. No.* 155, 485 pp. (1966).
2. Templeman, W., "Marine resources of Newfoundland," *Fish. Res. Bd. Canada, Bull. No.* 154 (1966).
3. Ronayne, M., "The Newfoundland cod traps," *Trade News* (Dept. Fisheries, Canada), **9**(4), 3(1956); **9**(6) (1956).
4. Anon. "Trade views thawing demonstration," *Trade News* (Dept. Fisheries, Canada), **19**(9), 7 (1967).
5. MacCallum, W. A., Laishley, E. J., Dyer, W. J., and Idler, D. R., "Taste panel assessment of cod fillets after single and double freezing," *J. Fish. Res. Bd. Canada*, **23**(7), 1063 (1966).
6. Huntsman, A. G., "The processing and handling of frozen fish, as exemplified by ice fillets," *Fish. Res. Bd. Canada, Bull. No.* 20 (1931).
7. Wood, A. L., "The artificial drying of salt fish," *Fish. Res. Bd. Canada. Prog. Rept. Atl. Coast Sta. No.* 37, 18 (1947).
8. Beatty, S. A., and Fougere, H., "The processing of dried salt fish," *Fish. Res. Bd. Canada, Bull. No.* 112, (1957).
9. Taarland, T., Mathiesen, E., Øvsthus, Ø., and Braekkan, O. R., "Nutritional values and vitamins of Norwegian fish and fish products," *Tidskr. Hermetikind*, **44**(11), 405 (1958).
10. March, B. E., Biely, J., Bligh, E. G., and Lantz, A. W., "Composition and nutritive value of meals from alewife, sheepshead, maria and tullibee," *J. Fish. Res. Bd. Canada*, **24**(6), 1291 (1967).
11. Jangaard, P. M., Ackman, R. G., and Sipos, J. C., "Seasonal changes in fatty acid composition of cod liver, flesh, roe and milt lipids," *J. Fish. Res. Bd. Canada*, **24**(3), 613 (1967).
12. Power, H. E., "Characteristics and nutritional value of various fish protein concentrates," *J. Fish. Res. Bd. Canada*, **21**(6), 1489 (1964).

COMMUNITIES—See ECOLOGY

CONCHS

The term *conch* is of Indo-European origin and refers to the shell of mollusks. In the molluscan literature of the 18th and 19th centuries, conch specifically applied to the class Bivalvia, including the clams, cockles, and mussels. Today, however, the term is predominantly used in reference to certain large prosobranch gastropods, particularly in the families Strombidae, Cassidae, and Galeodidae.

Fig. 1. The queen conch, *Strombus gigas* (length = 225 mm).

The most important species, from an economic standpoint, belong to the Strombidae. Distributed throughout the world in tropical and subtropical waters, this group is represented by nearly 40 species in the Indo-Pacific Region.[1] The most extensive fishery exists in the Caribbean area where some seven species of *Strombus* thrive.[2] The largest stromb, the rare and coveted *S. goliath*, which lives off the coast of Brazil and attains a length of over 350 mm has been purchased in native markets near Recife. The queen conch or pink-lipped conch, *S. gigas* (Fig. 1) is the most important commercial species in the family and is used as an important source of protein by the natives of Haiti and the Bahamas. Its biology and commercial value were recently discussed by Randall.[3-5]

Visitors to Florida and the Caribbean have seen the queen conch used to decorate the paths of gardens or to brighten flower boxes. As an attractive curio, *S. gigas* often finds its way back to northern climes where it is used as an ornamental doorstop or an aquarium piece. The export of this species—particularly from the Bahamas to the shell shops and hotels of Florida—is a business of no mean proportions. Many tons of shells are sent to Florida. The U.S. Commerce Department's statistics indicate that over $15,000 worth of shell and related objects are imported into Florida from the Bahamas each year. The real figure is larger since many tourists return from the Bahamas with these shells.

At one time the pink pearls formed by the mantle of *S. gigas* brought excellent prices on the world market. In the 19th century, specimens cost as much as $300 at the Great International Fisheries Exhibition in London and one fine individual fetched $5000 in Nassau. Since the bright coloration of the pearls fades and the substance loses its luster, their value has fallen to about $20 for a good specimen; no active fishery for the pearls of *Strombus* exists today.

The shells of *Strombus*, like those of the helmet

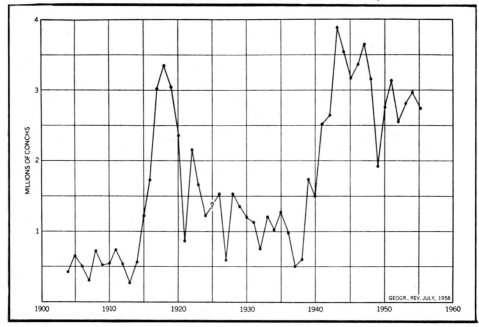

Fig. 2. Export of the queen conch from the Caicos Islands to Haiti from 1904–1955. (After Doran, 1958.)

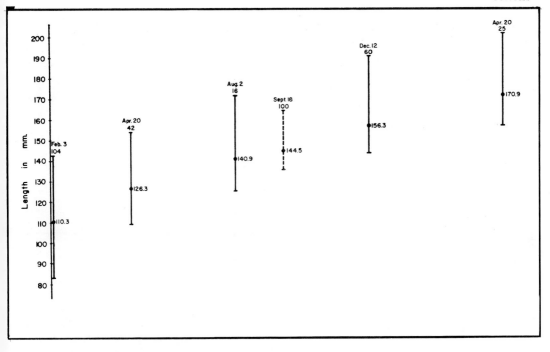

Fig. 3. Growth of tagged *Strombus gigas* in St. John, Virgin Islands. Initial tagged group of 104 (mean length, 110.3 mm) released on February 3, 1959. An additional sample of 100 (dotted on graph), with a mean length of 144.5 mm was released on September 19, 1959. (After Randall, 1964.)

conchs, *Cassis*, can be carved or finely etched into beautiful cameos, which, of course, vary in price according to their artistic excellence. In the last century, pulverized queen conch was shipped to England where it constituted an ingredient in porcelain. In the Bahamas, ground shells of strombs were used to make lime mortar for building. It is of interest to note that the relatively high copper content of *S. gigas* may contribute both to the low incidence of copper-deficiency anemia and poliomyelitis in the Bahamas. The shells are still employed as horns, tools, and ceremonial objects.

However, as an item of commerce their food value is of the utmost importance. Diced and cooked in broth with vegetables, *S. gigas* makes an excellent chowder which is said to rival that of New England's surf clam. Deep fried, the snails make a delectable fritter and in the French Antilles they are chopped, marinated raw in lime juice, and added to salads. The animal of *Strombus* is also used as fish bait.

In the Bahamas, the fishery for *S. gigas* is second only to that for the spiny lobster, *Panulirus argus*. In Nassau a fleet of over a dozen vessels ply the waters of the outer Bahama chain, fishing mainly in the vicinity of Eleuthera. It has been calculated that during 1959 over 150 tons of conchs were taken to Nassau by this fleet. Further, another 100–200 tons were estimated to have been consumed in the outer islands alone. With prices ranging from 14¢ for a small stromb to 21¢ for a large one, the fishery in Nassau was worth about $230,000 during that year.

Another aspect of the conch fishery is the export of dried conchs from the Turks and Caicos Islands in the southern Bahama Bank to Haiti. This active trading has been discussed by Doran.[6] The animals are taken from the sea by the islanders who use glass-bottom buckets and long poles with triple terminal prongs. They are removed from their shells, strung up to process overnight, tenderized by beating with a heavy club, and then dried in the hot sun for 3 days. The odiferous meats are then shipped to Haiti.

Figure 2 presents a summary of the Caicos–Haiti conch trade. The average annual export was 1.7 million conchs, the maximum 3.9 million, and the total for the half century 87.8 million. Although the conchs often sold for less than a penny each, more recently the selling price in Haiti has been 2 U.S. dollars per 100 or 2¢ each. At that rate, the business is roughly worth about $60,000 per annum. The importation of conchs into Haiti provides a valuable source of protein for the country, amounting to 0.16 lb of conch per capita—or nearly the total current figure for all meat imported into Haiti (0.2 lb per capita).

For a time, a trade in frozen conchs sprang up between the Caicos and Miami. Though once this business accounted for 30% of the total conch export, it has dwindled in recent years.

Biology of *Strombus gigas*

The usual habitat of adult *S. gigas* is in beds of sea or turtle grass (*Thalassia testudinum*) or manatee grass (*Cymodocea manatorum*) on sandy bottoms. The animals have been found in depths to 200 ft but normally occur from the low-tide line to 100 ft, largely because the plants are limited in their depth distribution. Juveniles are nocturnal and burrow beneath the substrate during the day. Almost exclusively herbivorous, *S. gigas* prefers a diet of algae and will move about to find better grazing areas—nearly 1000 ft in a relatively short time. Adults tend to migrate into deeper water except during the periods of reproduction.

Strombus gigas reaches maturity in about 2 years and it has been estimated that some individuals live as long as 10 or even 25 years. A dioecious species, the females attain a size slightly larger than the males. Normal populations have a balanced sex ratio of 1/1, but the natives, desirous of the largest specimens, tend to collect more females than males. Studies of their breeding habits confirm the observations of native fishermen. Copulation and reproduction take place

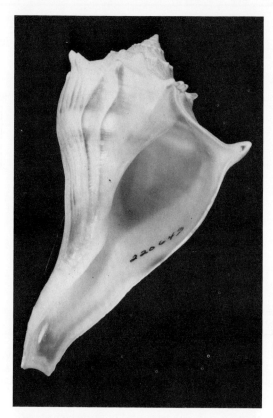

Fig. 5. The knobbed conch or whelk, *Busycon carica* (length = 115 mm).

during most of the year but a period of relative quiescence occurs from late November until spring or sometimes between later January and March, depending on the locality in the West Indies. At this time, the snails may burrow into the substrate. Egg cases are relatively massive and the species is exceedingly fecund. The female lays a continuous mucoid strand of eggs which is folded back upon itself and becomes invested with sand grains. This egg mass may attain a length of about 100 mm. The strand of eggs within the mass, variously estimated to include over 400,000 individual eggs, may be as much as 37 m long! In the ontogeny of the species, veligers are formed from the eggs within 7 days and settling follows shortly thereafter. The average rate of growth is 52 mm per year; the results of Randall's study in St. John is presented in Fig. 3. One of the largest queen conchs on record was over 250 mm in length and the heaviest, some 238 mm in length, weighed 7 lb, 5 oz. The length–weight relationship is demonstrated in Fig. 4.

Other than man, serious predators of the queen conch include carnivorous snails, cephalopods, and numerous fish, the most important of which

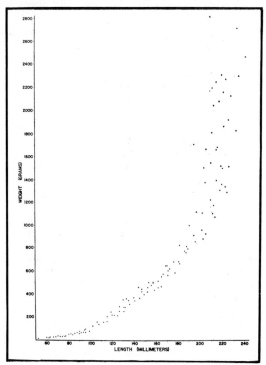

Fig. 4. Length–weight relationship of juvenile to young adult *Strombus gigas*. (After Randall, 1964.)

Fig. 6. The common periwinkle or conch, *Littorina littorea* (length = 20 mm).

TABLE 1. CATCH AND COMMERCIAL STATISTICS FOR *Busycon* ALONG THE EASTERN SEABOARD OF THE U.S. FROM 1940. STATES INCLUDED: NEW ENGLAND (MASSACHUSETTS, RHODE ISLAND AND CONNECTICUT); MIDDLE ATLANTIC (NEW YORK AND NEW JERSEY); CHESAPEAKE (MARYLAND AND VIRGINIA). DATA COURTESY BCF LABORATORY, OXFORD, MARYLAND

		New England	Middle Atlantic	Chesapeake
1965	lb	134,000	227,000	255,000
	$	31,000	58,000	28,000
1964	lb	147,000	232,000	293,000
	$	27,000	47,000	19,000
1963	lb	192,000	552,000	347,000
	$	34,000	118,000	30,000
1962	lb	167,000	184,000	412,000
	$	29,000	26,000	31,000
1961	lb	117,000	224,000	201,000
	$	20,000	30,000	16,000
1960	lb	156,000	340,000	179,000
	$	27,000	58,000	17,000
1955	lb	268,000	307,000	129,000
	$	38,000	45,000	6,000
1950	lb	109,000	129,000	171,000
	$	17,000	25,000	11,000
1945	lb	12,000	10,000	23,000
	$	1,000	1,000	8,000
1940	lb	425,000	270,000	15,000
	$	8,000	10,000	1,000

appears to be the spotted eagle ray, *Aetobatis narinari*. Hermit crabs, though never observed killing a *S. gigas*, may constitute a serious threat since the shell is a much-sought-after item. *Strombus* is equipped with a sharply pointed operculum which may act as a protective device.

Other Conchs

In Florida and along the Atlantic seaboard, *Busycon carica* and *B. canaliculatum* of the family Galeodidae are often referred to as "conchs" (Fig. 5). The fishery for these species is over 100 years old. In the middle of the last century, *Busycon* sold for $1 per 100 in New York City. Today complete specimens with their shells may be purchased in Italian markets in the larger cities on the eastern seaboard and in Italian restaurants one may eat *scungili*, a delicious dish prepared from the muscular foot of *Busycon*. In Connecticut, the so-called "winkle" chowder is made from *B. canaliculatum*. Catch and sales statistics for these species are included in Table 1.

In Maine, the common periwinkle *Littorina* (Fig. 6), locally called a conch, is easily collected intertidally along the rocky coast. This New England fishery is larger than one might expect and a résumé of the catch is presented in Table 2.

Other snails along the coasts of North America have an economic importance. The Louisiana conch, *Thais haemastoma*, which occurs from North Carolina to Texas, and the crown conchs, *Melongena melongena* and *M. corona*, which are common in the southern states and West Indies are predators on commercially valuable shellfish such as the oyster. The horse conch, *Pleuroploca gigantea* which lives from North Carolina to Florida is often eaten and the west Indian top shell or *burgao* as the islanders call it, *Cittarium pica*, is made into a chowder in Puerto Rico.

The sacred chank shell, *Turbinella pyrum* (Fig. 7) is considered a conch and is fished in Ceylon

TABLE 2. CATCH AND COMMERCIAL STATISTICS FOR LITTORINA IN THE STATE OF MAINE FOR THE PERIOD 1930 TO 1965. DATA COURTESY BCF LABORATORY, OXFORD, MARYLAND

	lb	value
1965	51,000	$12,000
1964	48,000	13,000
1963	34,000	11,000
1962	35,000	11,000
1961	34,000	10,000
1960	32,000	10,000
1955	9,000	3,000
1950	3,000	1,000
1945	166,000	35,000
1940	88,000	6,000
1935	159,000	7,000
1930	330,000	39,000

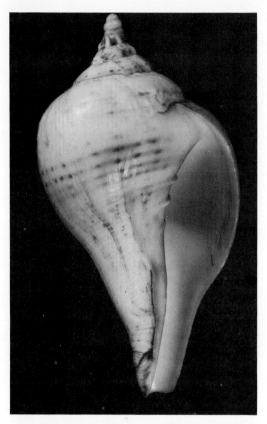

Fig. 7. The sacred chank shell or conch of India, *Turbinella pyrum* (length = 80 mm).

7. Hornell, J., "The sacred chank of India. A monograph of the Indian conch (*Turbinella pyrum*)," *Madras Fish. Bur., Bull.* 7, 181 pp. (1914).

COPEPODS

The copepoda are small primitive crustaceans belonging to Entomostraca, an artificial assembly of four distinct orders, the Branchiopoda, the Ostracoda, the Copepoda, and the Cirripedia. Each of these might be called a subclass as the Malacostraca, which form a true natural division. The copepods include marine and fresh-water and India. Hornell[7] wrote an excellent account of this fishery which has been going on for over 2000 years. For the last 150 years it is claimed that this fishery has closely paralleled the profit made in the renowned pearl fisheries of the Gulf of Mannar and is estimated to be worth around $700,000 a year.

KENNETH J. BOSS

References

1. Abbott, R. T., "The genus *Strombus* in the Indo-Pacific," *Indo-Pacific Mollusca*, **1**(2), 33 (1960).
2. Clench, W. J., and Abbott, R. T., "The genus *Strombus* in the western Atlantic," *Johnsonia*, **1**(1), 1 (1941).
3. Randall, J. E., "Monarch of the grass flats," *Sea Frontiers*, **9**, 160 (1963).
4. Randall, J. E., "The habits of the queen conch," *Sea Frontiers*, **10**, 230 (1964).
5. Randall, J. E., "Contributions to the biology of the queen conch, *Strombus gigas*," *Bull. Mar. Sci. Gulf Carib.*, **14**, 246 (1964).
6. Doran, E., "The Caicos conch trade," *Geog. Rev.*, **48**(3), 388 (1958).

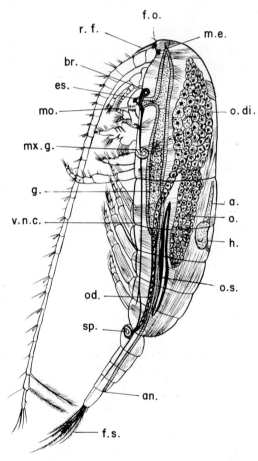

Fig. 1. Diagram of a female *Calanus* from the side: a., aorta; an., anus; br., brain; es., esophagus; f.o., frontal organ; f.s., furcal setae; g., gut; h., heart; m.e., median eye; mo., mouth; mx.g., maxillary gland; o., ovary; o.di., oviducal diverticula; od., oviduct; o.s., oil sac; r.f., rostral filament; sp., spermathecal sac; v.n.c., ventral nerve cord.

oida the first segment of the urosome is called the genital segment, as the genital duct opens in the segment both in the male and female. The last segment of the urosome is the anal segment, because it bears the termination of the digestive tract. The anal segment bears a pair of furcal rami, each of which is usually furnished with six furcal setae. In some species one of these setae is very long, and helps the animal as floating organ. The segments of the urosome are often reduced in number in the female; it is usually four-segmented, as the first and second segments are fused and form a genital segment.

Appendages. The first antennae (Fig. 1) which serve to maintain balance while swimming, are uniramous and long. They are usually composed of 25 segments, each furnished with several setae and sensory filaments. In most of Calanoida the male first antennae are alike on both sides, but there are the species in which one of the first antenna is modified into a geniculated antenna which serves as a clasping organ during copulation. The second antenna (Fig. 4a) are biramous, composed of two-jointed basal segments, seven-jointed exopod, and two-jointed endopod. The third pair of the head appendages are the mandibles (Fig. 4b). They are composed of the mandibular palp and cutting edge; the palp consists of two-jointed basal segment,

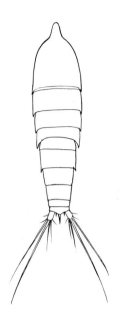

Fig. 2. *Longipedia coronata,* a harpacticoid copepod.

species and many parasitic species. They are very small, ranging from about 0.3–17 mm in length in adult stage. Over 6000 species of copepod exist, mostly in the sea, where some 800 species are planktonic; others are benthic or parasitic. The planktonic species occupy a most important place in the general economy of marine environment, namely, they are intermediary in food relationship between the primary producer (the phytoplankton) and larger consumer (fish and whales). The pelagic copepods are very useful as an aid to the study of water movements of the ocean. Finally, the copepods occupy an important systematic division among crustaceans.

External Anatomy

Metasome and Urosome. The body (Fig. 1) of a generalized form, such as *Calanus,* is divided into two divisions, metasome (cephalothorax) and urosome (abdomen). The metasome is elongate ovate in outline, consists of the head and first to fifth thoracic segments. The head carries two pairs of antennae, a pair of mandibles, two pairs of maxillae, and a pair of maxillipeds. The five segments of the thoracic region each bear a pair of swimming legs, but the last pair of legs may be much reduced or absent in the female. The posterior region is limbless and slender, articulated with the cephalothorax. It usually consists of five segments in the Calanoida. However, in Harpacticoidae (Fig. 2) and Cyclopoida (Fig. 3) the fifth thoracic segment with its pair of legs is included to the urosome. In the Calan-

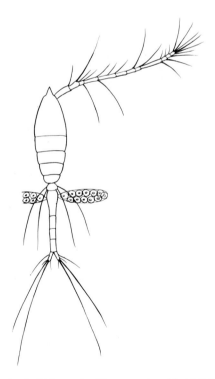

Fig. 3. *Oithora plumifera,* a cyclopoid copepod.

COPEPODS

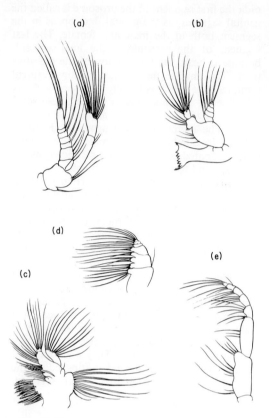

Fig. 4. The mouthparts of adult *Calanus finmarchicus*: a, antenna; b, mandible; c, first maxilla; d, second maxilla; e, maxilliped.

spermatophore to the female, and is usually asymmetrical in structure. In the Harpacticoida and Cyclopoida the fifth pair of legs of both male and female is always rudimentary.

Internal Anatomy

Muscular System. The muscle band of the metasome and urosome are clearly seen in the living animal. The main muscles are the pair of dorsal longitudinal and ventral longitudinal muscles. Besides these there are muscles which support the heart, those concerned in the expansion and contraction of the pericardium, and those which support the genital duct. Also there are conspicuous bands of muscles which run transversely from the appendages of the metasome fanning out on the dorsal exoskeleton.

Digestive System. A short esophagus connects the mouth with the main parts of the gut. The

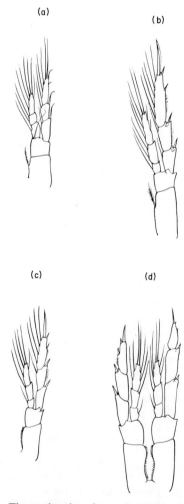

five-jointed exopod, and two-jointed endopod; the cutting edge is a chewing organ. Behind the mandible are the biramous first maxillae (Fig. 4c), the uniramous second maxillae (Fig. 4d), and maxillipeds (Fig. 4e). These five pairs of appendages, except the first antennae, act as a collecting basket in filter-feeding species, whereas, in the predatory species they are used in capturing prey. In the male the mouth appendages are usually reduced.

The thoracic segments typically bear five pairs of biramous swimming legs (Fig. 5a,b). In primitive form five pairs of legs are composed of two-jointed basal segments, three-jointed exopod, and endopod. In the majority of genera there is a reduction of the number of the segments. Such a reduction occurs frequently in the endopod of the first and second pairs of legs. The fifth pair of legs (Fig. 5c,d) of the female tends to be reduced in size and in the segmentation of both exopod and endopod; the endopod is often absent, resulting in a uniramous leg. Often the fifth pair of legs of the female is entirely absent. In the male the fifth pair of legs is used to transfer

Fig. 5. The swimming legs of adult *Calanus finmarchicus*: a, first leg; b, fourth leg; c, fifth leg; d, fifth pair of legs of male.

labrum immediately in front of the mouth forms part of its anterior wall and bears two rows of fine chitinous teeth as well as the eight openings of the labral glands. The esophagus opens into the mid gut which is wide in the head region and has a wide diverticulum stretching forward almost to the front end of the head. About the level of the second thoracic segment the gut suddenly narrows and continues backward to the hind gut which opens as the anus at the end of the urosome. In the middle portion of the mid gut the epithelial cells are highly vacuolated. The gut contents are cemented into fecal pellets by a mucous secretion through this vacuolated region.

Circulatory System. The extent of occurrence of a heart and actual circulatory system throughout the copepods has not been well established. It has been accepted that the structures are found only among the primitive calanoid species. However, the heart occurs widely not only among the primitive calanoids but in the advanced type of calanoids. The circulatory system involves a single heart, enclosed in a large pericardial space and an anteriorly directed aorta terminating in an anterodorsal aortic sinus. The latter communicates through three pairs of openings with the sinuses in the heart, which are in turn continuous with the perivisceral cavity, from which blood is returned to the pericardium. The heart has the form of a flask with an aortic valve at the tapered anterior end and a posterior ostium. The complicated musculatures of the heart have been described by Lowe (1935). It is enough to say here that when the muscles of the heart itself, both circular and longitudinal, contract, the aortic valve opens and the ostia closes. The muscles surrounding the aortic valve belong essentially to the pericardial floor and when these contract, the aortic opening is closed.

Reproductive System. In both sexes there is a single median gonad attached to the pericardial floor and bulging into the perivisceral cavity. The male has one and the female two ducts connecting the gonad to the single genital opening on the first segment of the urosome. In the male the genital organ (Fig. 6a) is composed of a single testis and a single duct which is divided into four differentiated sections, the vas deferens, the seminal vesicle, the spermatophore sac, and the ductus ejaculatoris. The female reproductive system (Fig. 6b) consists of a single ovary, two oviducts, each with several diverticula, leading to the paired openings into the vaginal cavity, a pair of spermathecae, and a pair of glands which open into the oviducts. In the mature female the oviducts are wide and sac-like, rise from the anterior end of the ovary and cross the body in posterior and ventral directions. They then run ventral to and parallel to the gut. As the ducts enter the abdomen they approach one another, joining as they reach the genital opening. A pair of small oval spermathecae are present at the sides of the genital opening and are connected to it by a pair of short canals. The oviducts open by a short median chitinized portion into a small cavity formed by the union of the spermathecal ducts. The genital opening is crescentic in shape with the horns extending forward as grooves in the chitin.

Nervous System. The calanoid nervous system was first described briefly by Claus (1863) for *Eucalanus*. The detailed anatomy of the calanoid nervous system was presented by Lowe (1935). The central nervous system composed of a large brain in front of the esophagus is connected by massive circum-esophageal connections to the ventral nerve cord which passes down the whole length of the body. From the front end of the brain nerves run to the eye and frontal organs.

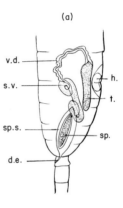

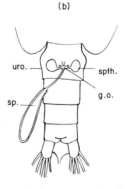

Fig. 6. (a), Genital system of male of *Calanus finmarchicus* from left side: d.e., ductus ejaculatorius; h., heart; sp., spermatophore; sp.s., spermatophore sac; s.v., seminal vesicle; t., testis; v.d., vas deferens. (b), Ventral aspect of first urosome segment of female of the same species showing spermatophore attached to genital aperture: uro., first urosome segment; spth., spermatheca; g.o., genital opening; sp., spermatophore.

Nerves are also given off to the appendages either from the brain or the ventral nerve cord.

Excretory System. The excretory system consists of a pair of maxillary glands. Each is composed of an end sac, a coelomic secretory tubule, and an ectodermal excretory duct. The end sac communicates with the tubule through a valvular opening. The antennal gland described by Grobben (1881) in the nauplius disappears in the adult.

Sense Organ. Most of the copepods have an eye which is referred to as a nauplius eye. It is a simple pigmented spot with a lense-like mass on either side. It occurs in the adult as a definite organ, as well as occurring in the developmental stages. However, there are copepods provided with highly developed dorsal eyes. Such are the species belonging to *Pontella, Labidocera, Pontellopsis, Corycaeus, Sapphirina, Copilia,* and *Clytemnestra*. The eyes of these species are furnished with large lenses of complicated structure. The lenses are much more developed in the male. The frontal organ is a pair of sensory filaments in the anterior end of the head, each of which is enervated by a frontal nerve. However, the function of the organ is unknown. The cutaneous gland and pores on the swimming legs are thought to be important for the systematic study of the copepods. In *Calanus finmarchicus*, minute pores are found on the anterior surface of the second to fifth pair of legs. The pores are very remarkable in the swimming legs of the genera *Metridia, Gaussia,* and *Lucicutia*.

Pigmentation

When alive, some copepods show a great wealth of color. *Temora longicornis*, which is very common in the southern North Sea, has a bright red eye and other red and orange pigments contrast with a distinctly blue hue of the rest of the body. *Calanus finmarchicus* is beautifully transparent except for a few spots of bright scarlet pigment. *Euchaeta norvegica* is an inhabitant of the deep water off the coast of Norway and west coast of Scotland. Its large size, about 8.0 mm, its deep rose-red pigment in the oral region, deep blue eggs, and the rainbow color of furcal setae make it a never-to-be-forgotten animal. In general blue and green are the colors of warm-water species, while red is the color of cold-water species.

Iridescence and Bioluminescence

In the genus *Sapphirina* the males are remarkable for their iridescence. The dorsal surface of the body display a metallic brilliancy, which surpasses that of any other copepods. This iridescence, after careful histological examination by Ambronn (1890), proved to be the interference color produced by a layer of closely set uniaxial anisotropic prisms just beneath the chitin integument.

In the sea most of major animal groups from the protozoans to the vertebrates are luminescent. The copepods are excellent light producers. Their light results from luminous slime secreted over parts of the body as in the case of the ostracod *Cyridina*. *Metridia lucens* is one of the most common copepod species in the boreal region of the Atlantic and Pacific Oceans. It has been reported the following genera are luminescent: *Chiridius, Euchaeta, Metridia, Pleuromamma, Lucicutia, Heterorhabdus,* and *Oncaea*. It is highly probable that many species not known to be luminescent will prove to be so as our information becomes more complete.

Parasites

The copepods are infected by various parasites, both internal and external. Marshall (1955) recorded the following parasites occurring in *Calanus*: Dinoflagellata, Blastdinium, Syndinium, Paradinium, Sporozoa, Ciliata, Nematoda, Trematoda, and Crustacea.

Feeding

The copepods feed on various organisms. Many copepods combine filter-feeding with predation. In examination of the gut content of *Calanus* over a year, Marshall (1924) showed that the food was very varied and consisted of both phyto-and-zooplankton. The other copepods examined have a consistently higher proportion of crustaceans and animal food. These copepods are carnivores as well as herbivores. There is close relationships between the structure of the mouth parts and feeding habits of the copepods. In herbivorous species, the second antennae, mandibular palp, first maxillae, and maxillipeds are well developed to produce a pair of "feeding swirls." The second maxillae are efficient filtering nets. The cutting edge of the mandible is provided with grinding teeth. In carnivorous species, such as *Temora, Centropages,* and *Tortanus*, the mouth parts have few setae and are on the whole much simpler in structure. The first and second maxillae, and mandibles are modified as prehensile appendages; the cutting edges of the mandibles have very sharp teeth. It has been estimated that, in a sea fairly rich in phytoplankton, *Calanus* would have to filter 72 ml of water a day.

Respiration

The interest in the respiration of copepods is focused on the effect of varying environmental factors, both animate and inanimate, on the metabolic activities of the animals. The first measurements recorded are those of Ostenfeld (1913) who estimated the oxygen used by adult *Calanus hyperboreus* and found it to be 0.68 μl/

Calanus/hour, a value which compares well with recent measurements on the smaller species. However, little is known about the manner of uptake of oxygen by copepods. According to Krogh (1941) copepods obtain the necessary oxygen by diffusion through the surface and exchange of gases probably takes place wherever the blood sinuses lie close to the cuticle. When a pelagic copepod makes darting movements the swimming legs are more vigorously moved and will furnish a good surface for oxygen exchange. It has been suggested that there may be an anal respiration. However, according to Fox (1952), the function of oral and anal uptake of water by crustacean is not respiration but is to extend the walls of the main part of gut for facilitating defecation.

Locomotion

Calanoid copepods have two distinct methods of swimming, namely, slow feeding movement and rapid escape movement. The former method of swimming is performed continuously as long as the animal is alive, and is brought by restless vibrations of the mouth appendages. During these movements the first atennae are held extended horizontally at right angles to the body, and at the same time the swimming legs are held directed forward. The rapid escape movement is a simple escape reflex resulting from a quick backward beating of all the swimming legs by contraction of the remoter muscles. In this reflex the animal can control movement toward any direction by contraction of the dorsal longitudinal trunk muscles and remoter muscles of the first atennae to result in a bending of the metasome and urosome, and at the same time by folding the first antennae against the body. Thus the urosome is very useful in controlling the speed of swimming and in keeping direction of swimming. For this reason the urosome is in the calanoid and is distinctly articulated from the metasome, while in harpacticoid the discrimination between the metasome and urosome is not distinct, as the animal is benthic, creeping on the surface of the bottom of the sea or on other substances.

Reproduction

The female, after copulation, usually carries a spermatophore attached to the genital segment. Occasionally, however, a female is found carrying more than one. In cases of abnormal deposition the neck of the spermatphore may be much longer than usual. A *Calanus* lays eggs which may number up to 20–60 in 24 hours. The total production of eggs varies considerably according to the feeding condition but is usually 200–300 eggs. Fertilization is reported to take place when the eggs pass through the vaginal cavity at the time they are laid, because the spermatozoa are non-motile and are only found in the spermathecae and the vaginal cavity. In *Calanus* there are no ovisacs; the eggs are released individually in spawning and they are already fertilized when laid. Some copepods carry their eggs in a single or a paired ovisac attached to the ventral surface of the genital segment. The number of eggs in ovisacs varies considerably. The female of *Valdiviella* carries a large egg in each of the ovisacs; deep species of *Euchaeta* has less than 20 eggs in an ovisac.

Development

The non-parasitic copepods go through no changes of form in their life histories that can be called a metamorphosis. The successive developmental stages are merely steps in growth from youth to maturity. The larva hatch out from the egg as typical nauplius (Fig. 7a) which is followed by second to sixth nauplius stages. The newly hatched nauplius is oval in shape, and is provided with three pairs of appendages, the first antennae, second antennae, and mandibles. In the final nauplius stage the animal increases in length and is furnished with, besides five pairs of the cephalic appendages, two pairs of swimming legs. At the next molt the larva begins to take on the form and structure of the adult, and is now termed a copepodite (copepodid). The first copepodite (Fig. 7b) acquires a third pair of swimming legs. With further growth it passes six copepodite stages until it becomes a sexually mature adult (copepodite Stage VI). Thus there is no abrupt change between the various stages of growth, but new segments are added and the appendages develop from simple rudiments to their definitive forms. On the other hand, the development of parasitic copepods is very different from that of the free-living species. The truly parasitic copepods include a large number of species, all of which undergo striking metamorphosic adaptations to the nature of the host or the part of the host attached. Some parasitic copepods undergo their metamorphosis during the larval development and become again free-living in the adult stage. On the other hand, others remain on the host and attain their highest degree of metamorphosis as adults. However, most of them hatch from eggs as typical nauplii, and in this stage or the following copepodite stage they must find their proper host.

Life Cycle

The life cycle of the copepods differs according to the nature of the species and conditions in which the animals live. Abundant food supply and rise of temperature will accelerate the development of the animals. In *Calanus finmarchicus*, by far the most thoroughly investigated of all copepods, the breeding occurs in spring and summer in boreal waters, and there are two or

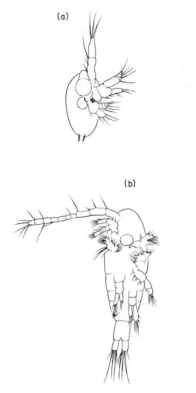

Fig. 7. Nauplius stage and copepodite: (a), nauplius Stage I, (b), copepodite Stage I.

three successive generations, each of which apparently may bear more than one brood. The generation arising from the first spring spawning appears to mature quickly in a period of about 28 days. The second generation grows up during June and spawns in July. The last generation produced in autumn is a relatively long-lived one, because it is this generation which carries the stock over the winter period. The winter stock is found in relatively deep water and is of uniform composition, consisting of copepodites in Stage IV or V. The stock is much reduced during winter, but with the return of spring the animals pass into Stage VI and spawn in the surface waters to produce the first generation of the season. The life cycle of other copepods has not been adequately investigated except those of several characteristic species.

Vertical Distribution

Copepods are distributed in every layer from the surface down to great depths. Each layer has its own characteristic species. However, it is difficult to seek a boundary in the water layer. Furthermore, the vertical distribution is often disturbed by diurnal and seasonal vertical movements of some species. Dahl (1894) first distinguished three layers in the vertical distribution of copepods in the Atlantic, namely, surface, intermediate, and deep layers. The surface layer, from 0–200 m is characterized by the wealth of different species. This is the layer into which relatively large quantities of light penetrate. Phytoplankton develops in this layer. Besides the surface-living species some deep-sea species rise at night to this layer. The next layers are the intermediate layer from 200–1000 m, and the deep layer from 1000 m to the bottom. Recently Soviet researchers divided the layers below 200 m in the following systems: the transition zone, 200–500 m, the upper deep-water subzone, 500–2000 m, lower subzone, 2000–6000 m, and zone of depression, 6000 m to the bottom. A greater part of the bathypelagic species are found in the upper deep-water subzone with the exception of some lower bathypelagic species.

Vertical Migration

A striking behavior characteristic of the pelagic copepods is the habit of changing vertical distribution. They migrate towards the surface at night and return to deeper water at or before the break of dawn. This rhythmic movement is known as diurnal migration. The significant factors affecting the vertical migration in one place is light. Some species are able to move vertically through several hundred meters of depth. For the diurnal migration of *Calanus finmarchicus* our knowledge has been summarized by Russell (1927). At present about 80 species of pelagic copepods are known to have the habit of vertical migration.

Seasonal Migration

We know that not only are there the nightly migrations, but that also the animal exhibits changes in vertical distribution throughout its life history according to its age. Some plankton seek the deeper levels in winter. A very good example of this phenomenon has been shown for the fifth copepodite stage of *Calanus finmarchicus* at the Plymouth area (Nicholls, 1933). The final brood produced in autumn has to pass from August until the following February when there may be a prolonged deficiency of their diatom food. It is obviously beneficial for them if they can reduce their metabolic activity to a minimum. A daily migration will be an unnecessary waste of energy when there is not a sufficient supply of diatoms at the surface. One of the most obvious ways for plankton to reduce the expenditure of energy is to seek the deeper layers and become irresponsive to external stimuli, especially to that of light.

Copepods as Food

It is well known that *Calanus finmarchicus* is a good food for herring. The stomach of a single herring has been found to contain more than 60,000 copepods. Hardy et al. (1936) investigated to see if there was any correlation between the number of herring caught and those of *Calanus*. The result shows that the total catch of herring in waters rich in *Calanus* was more than double that in the poorer waters. Cushing (1952) showed how, in a series of cruises, the herring were seen to aggregate in increasing numbers about patches of *Calanus*. It is also worth noting that the whales and the basking shark feed on planktonic copepods. Various biologists have thought it possible to collect plankton for use as food for livestock or for man. This question has been carefully investigated from the engineering point of view by Jackson (1955). According to his findings, the plankton may help to save a marooned sailor or airman on a raft from starvation, but it is unlikely to support a fishery except in those areas in which plankton is extremely abundant. On the other hand, Hardy (1964) says that one of the deep-sea species of copepod, *Euchaeta norvegica*, is a delicacy which one day might support a fishery for a luxury market.

OTOHIKO TANAKA

References

1. Bogorov, B. G., "Regularities of plankton distribution in the northwest Pacific," Proc. UNESCO Symposium on Physical Oceanography, Tokyo, 1955.
2. Brodsky, K. A., "Calanoida of the far eastern and polar seas of the U.S.S.R., Fauna of the U.S.S.R.," *Publ. Inst. Acad. Sci. Moscow*, **35** (1950).
3. Giesbrecht, W., und Schmeil, O., "Copepoda I. Gymnoplea," Das Tierreich 6(1898).
4. Hardy, A., "The open sea: its natural history, Part I: The world of plankton," 1956, London.
5. Marshall, S. M., and Orr, A. P., "The biology of a marine copepod," 1955, London.
6. Park, T. S., "The biology of a calanoid copepod *Epilabidocera amphitrites* McMurrich," *La Cellule*, **64**, 2 (1966).
7. Russell, F. S., "A review of some aspects of zooplankton research," *Rapp. Proc. Verb. Réunion*, **95** (1935).
8. Sars, G. O.,"An account of Crustacea of Norway", Vol. IV, "Copepoda Calanoida."
9. Snodgrass, R. E., "Crustacean metamorphoses," 1956, Smithsonian Misc. Collection, Vol. 131, No. 10.
10. Steuer, A., "Zur planmässigen Erforschung der geographischen Verbreitung des Haliplanktons besonders der Copepoden," *Zoogeographica*, **1**, 3 (1933).
11. Sverdrup, V., Johnson, M. W., and Fleming, R. H., "The Oceans," 1942.
12. Wilson, C. B., "The copepods of the Wood's Hole Region," U.S. Nat. Museum Bull., 158 (1932).

CORAL REEFS

A coral reef may be defined as a complex organogenic framework of calcium carbonate (primarily of corals) which forms a rocky eminence on the sea floor and customarily grows upward to the low-tide limit. It thus causes waves to break, and consequently, the internal spaces in this branching framework are packed with fragments of broken reef material, coralline algae, broken-up mollusks, echinoid debris, and foraminifera. The principal reef builders today are the Madreporarian (or Scleractinian) colonial corals of the class Zoantharia, phylum Coelenterata. Coral reef rock is often a very porous material with a large variety of components, with mainly inorganic cementing materials.

Both the outer reef periphery and the lagoon are marked by loose clastic debris that has largely been wave broken and washed off the reef. This detritus is known as bioclastic or coral sand when fresh, or as calcarenite when cemented. A special type of coral sand is found on beaches; after repeated concentration of the interstitial seawater under the hot tropical sun over many low-tide periods, the sand becomes cemented into a beachrock, beach sandstone, or littoral calcarenite.

Calcareous algae always contribute to coral reefs but also form independent *algal reefs* in some areas, particularly in places which are warm but for some reason are unfavorable for coral growth. Such spots include Shark Bay (Western Australia) which is mostly too saline for coral growth and parts of the mid-Queensland coast near Cairns where heavy rainfall leads to reduced salinity in the wet season and to excessive turbidity from sediment in suspension.

Reef Types

Fringing Reefs. These were regarded by Darwin as the basic reef type, forming a shallow veneer or shelf in shallow water at or near the shore of the mainland or around offshore islands. Heavy sedimentation and fresh water runoff along many mainland coasts tend to make these less attractive for fringing reefs than the offshore and oceanic islands. In heavy rainfall areas the nearshore surface of the reef (if present at all) is often so veneered by terrigenous sediments as to obscure the corals and inhibit growth in the pools and shallow lagoons.

Platform and Patch Reefs. These reefs are generally rounded or ovoid reefs, the large ones over a mile or so long being called platforms, while the smaller ones are generally referred to as patch reefs (also variously called shelf, bank, table, and hummock reefs). They are found growing up from water of moderate depth (generally 20–40 m) on the continental shelves. Sometimes they are dotted here and there in almost random manner, but more often in recognizable belts, suggesting an evolution from old (lower) shorelines. Probably they grew on headlands and capes during temporary steady phases of the post-glacial sea-level rise and then grew on upwards as the sea level continued to rise.

Barrier Reefs. According to Darwin the effect of crustal (tectonic) subsidence upon a fringing reef, if carried out slowly, would be to cause the corals to grow upward and as the land behind became gradually submerged, a lagoon would form between it and the up-growing barrier. According to Daly, it was the post-glacial rise of sea level that caused the submergence. In many parts of the world both factors are contributing.

In some places these reefs are long ribbon reefs, in others, discontinuous modified platform reefs. The passes or channels between them represent Pleistocene stream valleys that crossed the continental shelf before the post-glacial eustatic rise, and thus separated the initial fringing reefs. These passes have strong tidal currents and are thus subject to constant sediment scour and hence kept open.

Barrier reefs are always strongly asymmetric in plan and section, steep-to on the ocean side, often dropping abruptly away to 1000–5000 m, grading off gently to the interior with a sediment wedge, dotted by small reef patches, pinnacles, and coral heads. Depths in the lagoon may drop to 50–80 m.

The finest example of a barrier reef in the world is that of Queensland, which stretches 1200 miles from the Gulf of Papua to the Tropic of Capricorn. In eastern New Guinea (Papua) there is another fine barrier complex that also rings the Louisiade Group (the Tagula Barrier Reef). Others occur in New Caledonia, Fiji, Borneo, and in the Palau Islands. During the post-glacial rise of sea level, there were several important secondary oscillations of level, which led to coastal changes. With renewed submergence sometimes new fringing reefs become incorporated in the pattern and a double or looped barrier develops.

Off the Sahul Shelf (Timor Sea) and likewise off the Sunda Shelf (Borneo) there are examples of "drowned" barrier reefs. These, for some reason, did not maintain their regular upward growth during the post-glacial period and form submarine rims on the outer shelf edges.

Reports of barrier reefs offshore in the mid-latitudes (Morocco, southwest Africa, Western Australia) are incorrect; these are basically "sandstone reefs," though in places they have a veneer of coral.

Atolls. An atoll is a ring-shaped reef, morphologically like a ribbon reef bent into a circle, enclosing a lagoon. Darwin suggested that fringing reefs surrounding volcanic islands, during subsidence, passed gradually to become ring-shaped barrier reefs, and as subsidence continued, eventually to center-island-free ring reefs. This is certainly the origin of some atolls, especially as seen in the Society Islands, but there are several other types of atoll.

Shelf Atolls. On the Australian shelves there are shelf atolls that do not have volcanic foundations, but appear to have grown up as open platform reefs or from earlier platform reefs. Off northwestern Australia, there are several large atolls rising from depressed outer sectors of the shelf, from 400–500 m.

Compound Atolls. In several parts of the world large continental crustal segments have slowly subsided, so that barrier reefs and platform reefs have grown upwards in the same manner as in oceanic atolls, e.g., in the South China Sea, in the Tiger Islands of Indonesia, in the Maldives, and Laccadives, and in the Australian region— in the Coral Sea plateau. These reefs often have compound tops. Evidently they grew up during subsidence, but were exposed during the last glacial low sea level to form a differentially weathered crest. MacNeil has demonstrated that this crest would be higher on the outside and would predispose the old "stump" to the atoll form. Often many little rings form along the initial ring (called Faros in the Maldives).

On the Western Australian shelf a group of such compound atolls is found in the Houtman's Abrolhos Islands. Here parts of the old stump are exposed, deeply weathered, and penetrated by deep holes (former karst pipes). Here there has been no subsidence, so that they are classified as a "compound shelf atoll" group.

Oceanic Atolls. These are the so-called "mid-Pacific type," which rise from isolated volcanic cones in the deep ocean basins (seamounts or guyots), with up to 2000 m of accumulated reef growth, that may date back as far as the Cretaceous. It is interesting that oceanic atolls often have U-shaped gashes, attributed to landslides down volcanic slopes; shelf atolls are more often perfect and smoothly rounded.

Regional Distribution of Reefs

Reefs may be recognized in essentially four geotectonic groups:

(a) *Epicontinental reefs*, rather stable foundations, with eustatic features dominant, e.g., Queensland (Great Barrier Reef complex), Borneo, and some other islands in the East Indies.

(b) *Mobile Belt reefs*, with unstable or highly unstable foundations, eustatic factor often obscured, e.g., gently subsiding zone in the eastern Papua-Tagula-Louisiade Barrier Reef, in New Caledonia and the Banda Sea area of the East Indies; in contrast there are the strongly positive uplift zones as in Timor, Sumba and along the coast of northern New Guinea.

(c) *Quasicratonic reefs*, where there is *en bloc* subsidence of appreciable areas of former continental crust, e.g., Coral Sea Plateau, the South China Sea, the Tiger Islands, the Maldive-Laccadives, and the Bahamian block, east of Florida.

(d) *Oceanic Volcanic reefs*, where the foundations of atolls lie along the known volcanic trends and by implication are of similar nature (i.e., "mid-Pacific type"); the true atolls are subsiding examples. In some other examples, e.g., Christmas Island, Minami Daito, Nauru Island, Ocean Island, etc., subsidence has been followed by uplift, converting them into bird sanctuaries and thus phosphate islands.

Economic Aspects

The economic significance of reefs can be tremendous, but it can be either favorable or unfavorable. For example, vast numbers of ships have been wrecked on unmarked reefs. There are still broad areas in the Great Barrier Reefs of Australia, the Coral Sea, and the South China Sea marked on the hydrographic charts with the warning "dangerous ground, numerous uncharted reefs." Even in these modern days of air photography, radar, and other navigational aids, there are appreciable areas of the world where no ship's captain would dream of going except in the clearest weather and with extreme caution.

On the favorable side, the list is long and may be subdivided thus:

Materials. The raw materials available from coral reefs and their associated islands are numerous. The cemented beach sand that is ubiquitous around coral islands forms and hardens so quickly that it is a self-renewing resource, and in some areas builders take out a license to exploit a section of coast, one "crop" each year. It makes an excellent building stone that readily cuts out into rectangular blocky "free stone"; on exposure to the sun it rapidly hardens to an appreciable strength. Because of its somewhat porous nature, it makes a wonderfully cool house.

Coral rock itself is more difficult to cut up into building stones and is usually crushed for roadbuilding and airfield construction. Again, on exposure to the sun, the crushed coral rock tends to dry out and cement itself. Coral rock also provides a raw material for cement making, but usually requires some admixture of magnesia, etc. Some of the older coral limestones are already magnesium-rich (dolomitized) and hence of great value for cement making. Some algal reef limestones are also rich in magnesia.

Other materials found in the lagoon may include oyster shell, also useful for road construction or cement. On some reef islands there are loose sand dunes, useful for concrete mixing, but the carbonate sand is soft and less satisfactory than quartz sand.

Some of the older reefs that have long served as sea-bird sanctuaries have accumulations of bird droppings (guano), a valuable agricultural manure. Where such islands have been long elevated above sea level, the guano has been leached downward into the coral limestone, leading to phosphatization, and these rare "phosphate islands" (e.g. Nauru, Ocean Island, Christmas Island) are extremely valuable sources of industrial phosphate.

Unless the coral reefs surround the stumps of old volcanoes or outposts of continental land masses, the *soils* of coral islands are exceptionally unfertile, being predominantly lime and lack many vital minerals. Thus plants, animals, and man characteristically suffer from malnutrition and deficiency diseases unless effective soil enrichment or supplements are provided.

The coral-reef rims of volcanic islands, in contrast, frequently have exceptionally rich soils, receiving additions from either slope wash or airborne ash, favoring the "island paradise" types of vegetation, fruits, and crops. Some of the deeply leached laterite of the more mature volcanic islands are less satisfactory, but in places may contain minor iron or aluminum ore deposits.

Reservoir Potential. Coral-reef rock and coral sands of present-day coral islands always carry a thin sheet of fresh groundwater at shallow depths below the surface. Wells dug to this water table are usually adequate for domestic purposes. However, the porous nature of the reservoir permits the sea water to encroach *under* the freshwater layer, so that if strong pumps are used, only salt water is obtained.

Fossil coral reefs (and other organic reefs such as algal bioherms) often still retain a high degree of porosity and in ancient rock systems they may represent valuable oil reservoirs. The organic hydrocarbons that are derived from decaying organic matter in the dark muds of the adjacent marine basins, are gradually squeezed out of the compacting clays and migrate upward and laterally into potential reservoirs such as sandstones or reef limestones. The latter represent some of the world's greatest oil fields.

Fishing. Because of their ring or wall-like nature, reefs commonly form a protective barrier favoring rich submarine organic growth and an association of fish that may have useful economic value. Much spearfishing provides important dietary items in many of the Indo-Pacific reef areas. Most reef fish, however, are more colorful than palatable, and the open-ocean fisheries are generally better. The tourist value of the reef fauna, on the other hand, is tremendous. Shark usually do not enter the narrow lagoons but appear in the bigger ones; offshore shark fisheries are useful both for meat (good only when soaked in fresh water) and for shark liver oil. Other reef organisms of value include: octopus (best for eating when caught young), sea urchins (the sharp spines may cause painful wounds), sea slugs (or Holothuria, widely fished for drying as a Chinese delicacy, and known sometimes as trepang or beche-de-mer), trochus (a large coiled sea snail, formerly much used for making "pearl" buttons), rare shells in general (some of the gastropod types are extremely poisonous), oysters (edible types are most common on the roots of mangrove; large pearl oysters and pearl shell are rather uncommon in coral areas), turtle (fishing for turtle was formerly common, both for shell and meat, but owing to overkilling it is now largely banned).

The so-called "pink coral" of the Mediterranean that is used in some sorts of jewelry is not really coral at all. It is not found in the great reef regions of the world. However, it supports a small industry in Italy, chiefly of the traditional handcraft type.

Conclusions

Fossil reefs, which are of great economic significance (especially for the oil industry) in the Paleozoic and Mesozoic, are epicontinental in nature, platform reefs and shelf atolls, but *not* oceanic atolls.

Such non-oceanic reefs of today almost exclusively rise from "antecedent platforms" that were only recently established, i.e., during and since the late glacial stage. The antecedent streams and canyon development of the Sahul and Sunda shelves was probably of about the same age. Although these belong to long-stable continents, these marginal belts seem to have suffered very recent and vigorous reactivation. From this may be learned important lessons in the geomorphic history of continents.

In the history of the Earth's crust, it is often claimed that while continents may enlarge, ocean basins remain essentially constant. The disposition of shelf and compound atolls over such areas as the South China Sea and Coral Sea plateau strongly suggests that this is a section of what Stille calls "quasicratonic crust," that is, suffering regeneration, i.e., it is currently subsiding *en bloc*, to form a new oceanic depression, the site of "lost continents." Fundamental geophysical problems, continental stretching, global expansion, convection currents, subcrustal attenuation etc., are involved.

The dating of eustatic "highs" and "lows," are recorded by former reef terraces. The interrelations of alluviation and/or downcutting with the former reefs through thalassostatic reactions can be helpful in contributing to our ideas on the climatic history of the continent and indeed to fundamental problems of solar control and the principles of paleoclimatology.

RHODES W. FAIRBRIDGE

CRAB INDUSTRY

Pacific Coast

The crab industry of the Pacific Coast from California to Alaska involves two species of major importance, king (*Paralithodes camtschatica*) and Dungeness (*Cancer magister*) crabs, and two of minor importance, rock crab (*Cancer* species) and tanner crab (*Chionoecetes* species). In Hawaii, a small fishery exists for the Kona crab (*Ranina ranina*). King crab is harvested entirely in Alaskan waters (see King-Crab Industry, this article).

Of the minor species, the rock crab in southern California is the most important. In 1965, 329,000 lb of rock crab, which is similar to but somewhat smaller than Dungeness crab, were landed and marketed locally in the fresh-cooked form.

The tanner crab (Fig. 1) is a largely undeveloped resource in the deeper waters of the continental shelf of the north Pacific Ocean. In recent years, a small amount of tanner crab has been taken by otter trawl and landed in Alaska—for example,

Fig. 1. Auction of precious coral collected near the Ryukuan Islands near Miyako. Not true coral, this "pink coral" is used for decorative applications, such as costume jewelry.

Fig. 1. Adult tanner crab caught off the Oregon coast. The ruler is 6 in. long. (Photo: Bureau of Commercial Fisheries.)

11,000 lb in 1962. The substantial development of a domestic tanner crab fishery, however, depends on improved economics and techniques for catching and processing this deep-water crab in large volume.[1] The fresh-cooked meat of tanner crabs has a tender texture and a mild sweet flavor that should result in good public acceptance of the product when a fishery is developed. Both the Japanese and Russians catch tanner crabs in quantity from the north Pacific; for example, in 1963 the Japanese landed over 36,000,000 lb of tanner crab harvested mainly in the Japan Sea.

The Kona crab, a crab about 1 lb in weight and 5 in. across the carapace, is the most important species of Hawaiian crab; it is harvested by means of baited lift nets set in depths of 90–120 ft. In 1965, 28,000 lb of Kona crab were harvested commercially and utilized, primarily in restaurants, as a specialty crab meat.

In view of the relative importance and widespread distribution on the Pacific Coast of the Dungeness crab (Fig. 2), also called the market crab in California, it will be discussed here in detail.

The Dungeness crab gets its name from the area on the Straits of Juan de Fuca where an early crab fishery developed. The crab occurs from southern California along the coast north to Alaska and westward to the Aleutian Islands. The most productive grounds are off northern California, southern and central Oregon, the coastal district of Washington, and the central region of Alaska. The crab inhabits shallow waters inshore and the estuaries as well as offshore waters on sandy or mud bottoms up to 50 fathoms in depth.[2]

The annual catch of Dungeness crab in California, Oregon, Washington, and Alaska has ranged from 25,000,000 to over 40,000,000 lb during the period 1947–1966. In 1966, the catch of 37,000,000 lb was valued at over $7,000,000.

The abundance of Dungeness crab has fluctuated significantly in specific areas during the past 20 years, but biologists report that the resource is not endangered by the fishery as long as conservation regulations are adequate. The wide fluctuations in the abundance of crab are believed to result from natural causes and are not necessarily related to fishing pressure.

Fishery regulations are imposed by the various states and generally provide for retention of male crabs above a size at which they have spawned at least once. Owing to differing growth rates in several areas, this size limit is not uniform but ranges from $6\frac{1}{4}$–7 in. across the carapace. Regulations usually prohibit the retention of female, soft-shell, and undersized male crabs and provide for escape ports in the pots to enable the smaller females and undersized males to escape. A closed season is applied in some areas to protect the crabs further during the molt and soft-shell stages.

The annual distribution of Dungeness crabs is affected by their migration pattern from the deeper to more shallow waters during the spring and the reverse during fall and early winter. In regions having a fishery in protected waters, such as in Puget Sound in Washington and Prince William Sound in Alaska, the inshore crabs are generally distinct from those taken offshore and tend to be smaller in average size. During recent years, the exceptionally large, fine Dungeness crabs found in Shelikof Strait and waters adjacent to Kodiak Island, Alaska, have been marketed in the fresh and frozen cooked form. At present,

Fig. 2. A male Dungeness crab, dorsal view. The crab was 8 in. across the carapace (between the longest spines) and weighed 2 lb. (Photo: Bureau of Commercial Fisheries.)

the central region of Alaska, including the Kodiak area, offers the greatest potential for substantial expansion of the Dungeness crab fishery.

Harvesting Dungeness Crabs. Pots and ring nets are commonly used for harvesting Dungeness crabs.

Ring nets, sometimes called hoop nets, consist of two rings about 15 and 30 in. in diameter held together by cotton webbing to form a basket when suspended. These nets are used in shallow inshore waters and are a favorite gear for personal-use crab fishing.

A pot is an individual trap fished on the bottom and secured by means of a line to a colored buoy at the surface. Fishing depths range commonly from 5-20 fathoms, although crabs are taken occasionally in depths up to 40 fathoms. The crab fishing boats vary in size but, in the offshore fishery, are most commonly from 40-60 ft in length. Regulations in some areas limit the number of pots that can be fished; for example, in southeastern Alaska, not more than 150 pots may be aboard the boat or be fished by any one vessel.

Fig. 4. The live crabs are transferred to the plant. (Photo: Bureau of Commercial Fisheries.)

The pot consists of a circular iron frame 36-44 in. in diameter and 18 in. high and is heavy enough to hold in place on the bottom. The frame is wrapped with rubber strips to minimize electrolysis with the stainless-steel wire mesh covering the frame. Two entrance funnels are provided, a hinged lid is fixed on the top, and a bait box is hung in the middle. Such pots are baited with fish, squid, or clams, depending on area preference. Pots are hauled up every day or two depending on the rate of catch and the weather (Fig. 3).

The male crabs of legal size are placed in wells containing seawater on the boat and are delivered live to the plant (Fig. 4). If held too long in the pots or live wells, some of the crabs may become weakened and die. The crabs also may fight among themselves, causing physical injuries that weaken them. On occasion, crabs are weak and listless when landed but become strong and lively if transferred to a shore tank or live well. Overcrowding in the live wells may cause the oxygen in the water to become depleted, particularly if the temperature of the water is too high—for example, above 50°F. The increased temperature causes a higher rate of respiratory exchange in the crabs, a condition also noted in crabs caught shortly after they have been feeding. For these reasons, the condition of crabs in live wells should be checked frequently.

Species Characteristics. Dungeness crabs are members of the crustacean class of the order *Decapoda*. In common with other members of the order, the crab has a hardened outer shell or exoskeleton, five pairs of jointed legs, gills for respiration, and an open circulatory system in

Fig. 3. Aboard a crab fishing boat off the coast of Washington. The crabs are being transferred from the pot to the hold. (Photo: Bureau of Commercial Fisheries.)

which the blood or hemolymph circulates in open channels or sinuses before being returned to the single-chambered heart. The animal grows by shedding its exoskeleton periodically. This molt occurs 7 times or more during the first year of growth but, in a mature crab, occurs only about once a year, usually during the summer months. Following the molt and to provide for future growth, the soft-shell crab rapidly expands its body by absorbing water in the tissues. The new enlarged shell hardens quickly and becomes fairly rigid in the first few days. The Dungeness crab is not harvested commercially, however, until later in the season when the tissue growth is completed, the protein content is higher, and the yield of the meat is increased. On the average, prime crab will yield 24% of cooked meat; a recently molted crab may yield only 14–16%.

Mating occurs primarily in spring and early summer, shortly after the female crab molts. A male crab mates with as many as 5 females, according to observations in aquaria. Spawning occurs in the fall months. The female carries the eggs until they hatch, and the shrimp-like larvae are released in the water. Normally, this release of larvae occurs during winter or early spring, depending on the area. The crab grows to be $1\frac{1}{2}$ in. wide within a year and becomes sexually mature after the second year when it is about 4 in. in breadth. A male crab reaches the marketable size of 6–7 in. in breadth in about 4 years and will continue to grow until it is about 10 in. in breadth.

Only live healthy crabs should be utilized. One basis for this requirement is that, in a weak or dying crab, physiological changes adversely affect the quality and yield of the cooked meat. Meat from weak or dead crabs tends to have a chalky, friable texture and is difficult to remove from the shell after being cooked. The yield of meat thus drops, and the appearance and flavor are less desirable. Another basis for the requirement is that one cannot easily judge the condition of a dead crab even if it has been dead for only a few hours.

The hemolymph in crabs comprises up to 37% of the weight of the live animal as compared with about 4% blood by weight in fish. With such a high percentage of the weight of the body being in the circulatory system, crabs that are handled roughly and injured may not only die quickly but may lose weight rapidly after death. Injured crabs should be removed promptly from live wells and either processed immediately or discarded.

In crabs, the oxygen-carrying pigment in the hemolymph is a copper hemocyanin rather than an iron heme compound as in the blood of higher animals. The copper heme pigment in the crab is relatively colorless in the hemolymph but tends to develop an objectionable bluish color after the crab is processed. This bluish color may become especially noticeable in canned crab meat during storage. The formation of black sulfides, if the meat is in contact with iron, is also a problem. Because iron and copper tend to promote discoloration, crab meat should be handled and processed without exposure to metals or compounds containing iron or copper. This includes the metal ions present in water used for processing and any food additives. Interestingly enough, crabs from certain areas along the Pacific coast prove more troublesome with respect to bluing than do crabs from other areas. Generally, crabs taken in the more northern areas, such as southeastern Alaska, provide few problems of discoloration.

In crab canned in Oregon and northern California, the formation of the blue discoloration is a major problem that is minimized only by prompt, careful processing and by the use of a food additive such as citric acid, which inhibits the activity of the metal ions. Both the fresh and canned crabmeat are packed in cans having a C-enameled (Seafood Formula) lining to minimize iron sulfide discoloration.

Processing and Utilization. Dungeness crabs are marketed as cooked in the shell and as fresh, frozen, or canned. Depending on the product desired, the live crabs are cooked whole or are eviscerated and cooked. Usually, the crabs are then placed in baskets and batch cooked in a vat of boiling water for about 12–15 min for crab halves or sections and for about 15–25 min for whole crabs. Whole cooked crabs are chilled or frozen for shipment to market. The larger prime crabs with a clean shell free of discoloration are preferred for marketing as whole or eviscerated crabs.

In recent years, the production of chilled or frozen eviscerated crabs has increased because of the ready demand by restaurants, retail markets, and consumers for fresh-cooked crabs. This demand, coupled with reduced air-freight rates introduced in 1965 for transcontinental shipment, has opened markets in the Midwest and Atlantic coast for fresh-cooked Dungeness crab. Another recent trend has been a developing market for live Dungeness crabs in areas where live Maine lobsters are difficult to obtain or are too expensive. Air shipment of live Dungeness crabs to Hawaii was pioneered by Washington firms but was not successful until studies in 1967 by the Bureau of Commercial Fisheries Technological Laboratory, Seattle, demonstrated improved procedures. Proper handling and temperature conditioning of the crabs were found to be important, along with the use of a light, insulated container to keep the live crabs cool.

The cooked sections are used primarily to produce crab meat for sale as chilled or frozen meat and for canning. The sections are cooled and the

Fig. 5. Dungeness crab meat is removed by breaking the sections and shaking the meat into stainless steel pans. (Photo: Point Chehalis Packers, Inc.)

meat is removed by breaking the shell and shaking the meat into stainless steel pans (Fig. 5). The leg meat is about half the total meat recovered, which averages 24% of the weight of the live crab. Leg meats are the more desirable parts and are kept separate from the body meat for later packing. Usually, the meat is washed in a strong brine solution in which the meat floats and the pieces of shell sink. The meat is then rinsed with cold fresh water and inspected carefully to remove any remaining bits of shell, curd, and other undesirable material. The meat is packed into No. 10 cans holding 5 lb, hermetically sealed, and chilled for shipment as fresh meat to wholesalers, retail stores, and restaurants. Retail stores repack the meat into trays or plastic containers holding 4–8 oz.

Fresh chilled meat is perishable and should be used within 7–10 days. Alternately, the sealed cans of meat may be frozen and stored at from $0°$ to $-10°F$ for up to 6 months. If the meat is to be held more than a month or two, it should be stored at $-10°F$ or lower to minimize adverse changes in crab flavor and texture.

In 1965, the production of chilled and frozen Dungeness crab meat was 5,100,000 lb, which had a wholesale value of $4,500,000. Chilled and frozen meats are the main products of the industry.

For canning, the washed, inspected meat is packed into $\frac{1}{2}$ lb cans with the addition of salt for flavor and citric acid to keep the meat slightly acid (pH 6.4–6.8) and to minimize the development of metallic discolorations. The cans are vacuum sealed and heat processed under steam pressure. Current process recommendations of the National Canners Association are 55 min at $240°F$ for $\frac{1}{2}$ lb cans of crab meat. The distinctive and desirable flavor and texture of canned crab deteriorate gradually in storage, especially at higher warehouse temperatures. Therefore canned crab should be stored at cool temperatures and marketed within a year for best results.

The production of canned Dungeness crab has fluctuated greatly during the period from 1946–1965. In 1955, production declined to a low of about 19,000 standard cases (48 $\frac{1}{2}$ lb cans) from the high of about 170,000 cases in 1948. Both the supply of the crabs and the market demand for the canned crab are important factors in these production fluctuations. In former years, California and Oregon firms canned a substantial volume of crab, but in recent years, production has shifted to the fresh and frozen meat sections. Currently, canned king crab dominates the domestic canned crab market; however, an upsurge in the production of canned Dungeness crab may be expected in Alaska as the production of king crab levels off.

Crab meat is classified as a speciality food that adds variety to the diet; nevertheless, its nutritional value is important, as in the low calorie diet. Dungeness crab meat is high in protein, from 18–20%, is very low in fat, from 0.7–1.1%,[3] and has about 90 calories food energy per 100 g. The natural sodium level of the fresh-cooked meat (not brined) is fairly high, ranging from 153 to 339 mg per 100 g in comparison with an average level of 68 mg per 100 g of sodium in marine fish, such as Pacific halibut and salmon. Commercially prepared crab meat has a much higher sodium content (from 547–1244 mg per 100 g) because of the use of brine solutions in its preparation.

JOHN A. DASSOW

References

1. Pereyra, Walter T., "Tanner Crab—An Untapped Pacific Resource," *Natl. Fisherman*, p. 16A, July 1967.
2. Schmitt, Waldo L., "The Marine Decapod Crustacea of California," University of California Publications in Zoology, Vol. 23, 470 pp. (May 21, 1921).
3. Nelson, Richard W., and Thurston, Claude E., "Proximate Composition, Sodium, and Potassium of Dungeness Crab," *J. Am. Dietetic Assoc.*, **45**, 41 (July 1964).
4. Benarde, Melvin A., "Technology of crabmeat production—a bibliography," *J. Milk Food Tech.*, **24**, 7, 211–217 (July 1961).
5. Dassow, John A., "The crab and lobster fisheries," Chapter 14 of "Industrial Fishery Technology," 1963, M. E. Stansby, Ed., New York, Reinhold Publishing Corporation.

6. Hipkins, Fred H., "The Dungeness crab industry," Fishery Leaflet 439, 12 pp., Fish and Wildlife Service, Washington, D.C. (May 1957).
7. Lyles, Charles H., "Fishery Statistics of the United States, 1965," Statistical Digest 59, Bureau of Commercial Fisheries, Washington, D.C. (1967).
8. Rees, George H., "Edible crabs of the United States," Fishery Leaflet 550, Bureau of Commercial Fisheries, Washington, D.C. (December 1963).
9. Waterman, Talbot H., "The Physiology of Crustacea," Vol. 1, "Metabolism and Growth," 1960, New York, Academic Press.

King-Crab Industry

The king crab, *Paralithodes camtschatica* (Tilesius) is an eight-legged, spider-like arthropod covered with spiny projections. Two degenerate legs specially modified for breeding are tucked under the rear carapace margin. The carapace and upper surfaces of the legs are red to purplish in hue; the abdomen and undersides of the legs are white. An adult male king crab measures up to $9\frac{1}{2}$ in. across the carapace; females are considerably smaller. The meat is white with reddish covering and firm in texture. The subtle flavor of king-crab meat makes it popular in salads or as a main course.

Meat content represents about 29–30% body weight with most meat being in the legs and shoulders. 25% recovery of meat is considered excellent by king-crab processors.

King crabs are found in ocean waters from southeastern Alaska to Siberia. Commercial concentrations of king crab have been located in all areas of Alaska with fisheries conducted inside the continental shelf from southeastern Alaska westward to the Aleutian Islands, in the Bering Sea, and Cook Inlet. Asian fisheries are centered primarily in the Okhotsk Sea and western Kamchatka.

Individual migrations of king crab have been recorded at a maximum of about 100 miles. The norm, however, is dictated according to the habits of the creature, which find it moving into shallow areas to spawn in late March and April and returning to deep-water trenches during the summer and early fall. Winter migrations are shoreward or toward offshore shallows where molting and breeding take place in the spring, thus completing the cycle.

Biology and Life History. Biology of the king crab has been studied most extensively in the Kodiak area of Alaska. Tagging projects have been incorporated with skin diving and other research studies to determine growth, behavior, breeding requirements, movements, and, in general, life history information vital to proper management of the king-crab stocks.

The female king crab carries 150,000–400,000 eggs under her abdominal flap for about one year. In late March or April the eggs hatch, the female sheds her exoskeleton and breeds. Male crabs normally molt earlier than females and preceed them to the spawning grounds.

A larval king crab is nearly microscopic, physically resembling a shrimp and free swimming. Within 12 weeks, characterized by heavy mortalities, the larva metamorphoses into a bottom-dwelling, eight-legged miniature king crab less

Fig. 6. A deck-load of king crab.

than $\frac{1}{10}$ in. long. Young crab band together in pods consisting of thousands of individuals, apparently for protection against enemies. Podding, as a function of survival, continues until about the third or fourth year when the carapaces of both sexes are about $3\frac{1}{2}$ in. long. In the fifth year of life, king crabs attain sexual maturity and ecdysis, or molting, begins to occur once a year. Most creatures having an external skeleton molt, shedding their skeleton and replacing it with another, to accomplish growth. Young king crabs molt an average of seven or eight times the first year, four or five times the second, two or three times the third, and one or two times the fourth year.

From the fifth year to about the ninth year, molting occurs annually. Biennial or triennial molting begins about the ninth or tenth year and continues throughout life, which may be 13–14 years. At maturity, a male king crab may weigh as much as 25 lb, have an over-all leg span of 6 ft, and contain 4 lb of edible meat.

The king crab industry dates back to 1892 when the Japanese established a shore-based cannery at Hokkaido, Japan. The product was exported in part to England as early as 1893 where a market still exists. In 1920, America entered the fishery. A small shore plant at Seldovia canned a few cases of king crab. Over the next 30 years only a few hundred cases at a time were put up at Seldovia, Hoonah, and Kodiak. In 1966, over 160,000,000 lb of raw king crab were processed in Alaska by a total of 40 processors.

Processing. Early processing efforts were dependent wholly upon canning. In recent years, technology has advanced so that king crab is now processed in a variety of ways. Frozen king-crab legs in the shell, frozen meat in glazed ice blocks, cooked whole king crab sold fresh, and canned king crab are now available to the consumer. Each method of processing varies according to the desired product; however, basically, processing consists of butchering, pre-cooking or blanching, final cooking, shaking or squeezing meat from the legs and shoulders, canning or freezing. If the product is to be canned, final cooking is accomplished in retorts after the cans are sealed.

Gear Development. Methods of obtaining the raw product have varied throughout the years. American fishermen initially employed tangle nets in 1938 when the floater, *Tondeleyo*, worked the Kodiak, Shumagin Islands, and Bering Sea areas. The tangle net is essentially a large-mesh gill net. The lead line holds the net to the bottom; the corks hold the net upright, thereby creating a fence to migrating or moving crabs. Crabs become entangled in the mesh and often die due to damages sustained from cutting by the thread of the webbing.

Another type of gear to evolve in the king-crab fishery was the otter trawl. In 1943, the trawl was first used but did not gain popularity until about 1953 when fishermen discovered a trawl could be employed quite effectively during March and April when king crab congregate in the shallow-water bays for breeding and molting.

As early as 1938, crab pots, or traps, were being used by Americans in attempts to establish a crab industry. The *Tondeleyo* used some pots, and the small plants operating at Kodiak Island from 1941 to 1944 depended entirely on pots.

Early king-crab pots were round, resembling shellfish pots used for Dungeness or blue crab with larger tunnels. The metal frames were covered with chicken wire, stainless steel, or seine webbing. Some pots were even constructed with wooden frames. Hemp rope was used as buoy lines with strings of seine corks being employed as buoys.

The standard king-crab pot in use today is square or rectangular and is any combination of sizes from $6 \times 6 \times 2\frac{1}{2}$ ft to $8 \times 8 \times 3$ ft in depth. The pot has two ramps or tunnels. Instead of stainless steel wire mesh or seine webbing, pots are covered with a mesh of heavy nylon thread. The mesh measures $9\frac{1}{2}$ in. and enables small male and female crabs to escape. Most pots are constructed of heavy gauge iron and weigh up to 850 lb; $\frac{5}{8}$ or $\frac{3}{4}$ in. nylon, leaded and unleaded polypropolene, or other synthetic fiber lines are

Fig. 7. King crab being unloaded from a fishing vessel into a processing vessel's live tanks. (Photo: Alaska Dept. of Fish and Game.)

Fig. 8. Typical king crabber *Shellfish,* owned by George Johnson of Alaska. (Photo: Ray Krantz.)

attached to two or three inflated plastic floats for buoy lines. One or two 1 qt perforated plastic containers are filled with chopped herring or bottom fish and placed inside the pot for bait.

Vessel Development. Much of the gear development can be attributed to the increase in average fishing-vessel size. Fishermen with boats large enough to venture to the more profitable offshore banks found that heavier, more substantial pots were needed.

Until the late 1950's, Alaska fishermen harvested salmon in the summer and king crab during the winter. The fleet was comprised of salmon purse and beach seiners having a maximum length of 58 ft. These vessels were capable of handling tangle nets, trawls, and most of them were equipped with seine winches which could pull pots. Crabs were loaded into the vessel hold or carried on deck. Since king crabs cannot normally survive for more than 24 hours out of water, boats were forced to deliver daily or hold crabs in live pens near the fishing grounds.

By 1960, many of the purse seiners were fishing king crab through the summer. Finding that crabs migrated offshore into deep-water trenches during the summer, fishermen were forced to move farther offshore to effect profitable harvests. As a result, larger fishing boats proved to be more suitable than limit seiners. In 1961, several converted halibut boats and herring seiners of the 70 ft class arrived in Kodiak and Cook Inlet to engage in the summer king-crab fishery. These vessels had tanks with circulating systems for maintaining a constant flow of seawater, allowing crabs to be held alive for several days.

By July of 1962, power scows made their appearance as king-crab vessels. With twin engines and a large working space on deck, they rapidly became popular. About 400 king-crab vessels are currently fishing Alaskan waters. The most recent king-crab vessels are 90 ft in length employing many innovations—hydraulic power blocks to pull pots, long-range radar, Loran, depth finders, circulating tanks, hydraulic lifting booms for moving pots, and pot launchers for dumping crabs onto the deck and sliding pots into the ocean.

Catch History. King-crab production has risen from 141,000 lb of raw product in 1947 to 160,267,127 lb in 1966 (Table 1). Increased processing capacity and improved gear and fishing vessels have contributed primarily to the growth of the king-crab fishery. Kodiak and the areas west have produced over 92% of the entire Alaska king-crab pack.

Initially king-crab production was primarily from trawling efforts in the Bering Sea and bays of the south side of the Alaskan Peninsula. A small fishery was devoloping in the bays of Kodiak Island, but did not gain importance as the major king-crab fishery until 1959 when 14,300,000 lb of king crab were delivered, as compared with only 6,100,000 lb for the entire area west of Kodiak, including the Bering Sea.

Kodiak's importance as the foremost producer of king crab has never been relinquished since 1959, except for 1964 when tidal waves from the Good Friday earthquake destroyed many of the Kodiak processing plants and the fishing vessels based there.

TABLE 1. ALASKA KING CRAB PRODUCTION (LB)

Year	Bering Sea Peninsula and Aleutians	Kodiak	Cook Inlet and Prince Wm. Sound	Southeastern	Totals
1947	141,000				141,000
1948	3,362,855				3,362,855
1949	2,476,250				2,476,250
1950	2,123,880				2,123,880
1951	3,460,525	335,337			3,795,862
1952	2,229,715	579,707			2,809,433
1953	2,628,665	2,531,120	1,359,854		6,519,639
1954	2,830,873	2,491,536	1,275,852		6,598,288
1955	3,852,488	3,717,145	1,915,821		9,485,454
1956	6,117,823	7,015,988	2,129,035		15,262,846
1957	7,245,526	5,070,638	620,858		12,937,022
1958	7,245,947	7,137,529	752,990		15,136,466
1959	6,116,974	14,348,110	2,191,437		22,656,521
1960	5,668,330	16,797,428	4,466,741	3,424	26,935,923
1961	8,238,480	29,257,677	4,251,896	429,600	42,177,653
1962	9,009,442	33,870,725	6,305,902	1,289,550	50,475,619
1963	24,191,510	40,699,978	8,249,197	1,112,220	74,252,885
1964	48,226,741	30,641,638	6,679,971	820,530	86,368,880
1965	50,495,946	76,641,933	2,780,065	579,300	130,497,244
1966	65,688,285	90,527,348	3,945,595	105,899	160,267,127
Totals	261,351,255	361,663,837	46,970,646	4,340,503	674,326,241

Regulations. The history of king-crab regulations shows at the outset restrictions in the minimum legal size and to the taking of males only. Prior to 1949, the minimum legal size for male king crab was $5\frac{1}{2}$ in. in greatest width of carapace. In 1949, the minimum carapace width was increased to $6\frac{1}{2}$ in. The only reference literature supporting this size is Marukawa who notes that the Japanese were prohibited from taking crab under 165 mm (or $6\frac{1}{4}$ in.). This width represents the approximate maximum size of female crab on the west side of the Pacific and is a size at which males have been sexually mature for two or more seasons.

As growth data were collected in the Kodiak area, it became apparent that king crab in this region experienced a greater increment per molt than in the Sea of Okhotsk and Bering Sea. In 1962, the minimum size limit was raised to 7 in. in carapace width in the Kodiak and Cook Inlet areas. Factors influencing the increase in size were that an increase in weight could be realized from each crab and that at least one additional year for breeding could also be realized from each male crab.

Gear regulations were first effected in 1955 when tangle nets were outlawed. The major reason for prohibiting the use of tangle nets was their non-selectivity and destructive nature to smaller and female crabs. King crabs are extremely difficult to remove from tangle nets and normally sustain damages. As a result, many non-commercial crabs were being damaged or killed.

The otter trawl was the next type of gear to be banned from the king crab fishery. In similarity to tangle nets, otter trawls fish non-selectively. Otter trawls were especially effective during the spring months when king crab congregate in bays for molting and breeding, and their use resulted in considerable mortality on newly molted crabs. Otter trawls were outlawed in the king-crab fishery in 1960.

When the question arose as to the legality of applying state laws outside the 3 mile limit, a landing law was enacted in 1960 which requires that crabs transported into territorial waters or delivered at Alaskan ports must be taken in conformance with state laws.

Two other regulations, a 30 pot limit per boat in the Kodiak area and area registration were imposed in 1960. The 30 pot limit was a measure designed to preserve the economics of the Kodiak king-crab fishery. The pot limit served its purpose and was subsequently rescinded in 1964. Area registration is still in effect. King-crab fishermen must register in one of two registration areas prior to fishing.

Another important type of regulation which is a valuable tool in king-crab management is the emergency closure. This regulation allows the closing of isolated fisheries or the entire fishery if, in the opinion of the management biologist, the stocks are being adversely affected. In recent years, this form of regulation has been used successfully in the Kodiak and Cook Inlet areas during the breeding and molting season in the spring of the year.

In 1965, a new regulation closing the Kodiak area to the taking or possessing of newshell king crab from April 1 to July 1 was enacted. In 1966, an additional regulation prohibited fishing in Kodiak and Cook Inlet during the months of May and June. This two-month closure is deemed necessary to prevent the extensive damage which occurs when newly molted king crab are handled. The carapace and leg shell continues to harden and strengthen after molting, returning to complete hardness by about October 1.

International Negotiations. As early as 1930, the Japanese and Russians were fishing the eastern Bering Sea for king crab. It became apparent that, in order to preserve Alaska's king-crab stocks, negotiations with these countries were necessary. The 1958 Geneva Conference labeled king crab a creature of the continental shelf.

In 1964, Soviet trawlers moved into the Kodiak Island area and began fishing king crab. American fishing gear (pots) were being destroyed by Russian drag gear. Some destruction was unquestionably deliberate. Economic losses in terms of thousands of dollars were being sustained by American king-crab fishermen. Then, in February of 1964, American and Russian fisheries representatives met in Moscow to resolve fishing-gear conflicts. A subsequent meeting in Juneau, Alaska, the following June resulted in six areas around Kodiak Island being set aside for pot fishing only.

International negotiations regarding king-crab stocks have mostly applied to the eastern Bering Sea. Japan, in 1964, agreed to a quota of 185,000 cases of king crab and agreed to set aside an area in the eastern Bering Sea for pot fishing only. Also, minimum tangle-net mesh size was set at 50 cm, stretched measure.

The Russians agreed to a quota of 118,600 cases of king crab for 1965 and 1966.

In 1967 talks resumed with the Russians and Japanese. Both nations agreed to stay out of the six areas near Kodiak Island and accepted reduced quotas in the eastern Bering Sea. Russia's 1967 and 1968 quota was reduced to 100,000 cases of king crab and Japan's quota was reduced to 163,000 cases each year.

Future of the Fishery. The expansion of the king-crab industry has been as rapid as any fishery in the world. Most of the available fishing grounds in the Kodiak area have been harvested to a degree approaching the sustained yield. With the possible exception of a few small areas, expansion of the Alaska king crab fishery is deemed unlikely. Catch statistics show that 1967 fishing success in the Kodiak area has declined almost 50% from the two previous years. The average weight per king crab in Kodiak's commercial catches has dropped from over 10 lb in 1960 to about 8 lb in September of 1967. With the minimum size restrictions and the taking of male king crab only, breeding stocks are insured for future fisheries; however, exploitation will continue to reduce stocks to a sustained yield of minimum legal-size crabs.

Fishing and processing potential in the Alaska king-crab fishery is greater than any possible yield of the resource. Perhaps the best possibility for expansion of the fishery is in the Bering Sea. Should American fishing efforts expand extensively into the Bering Sea, perhaps Russia and Japan will relinquish their present fisheries there.

MELVAN E. MORRIS, JR.

Red Crab Fishery

The deep-sea red crab (*Geryon Quinquedens*) occurs along the edge of the continental shelf from Nova Scotia (Canada) to Cuba, and samples have been found in the Gulf of Mexico and off the coast of Brazil. The crab is generally found where the water temperature is between 38°–41°F. South of New England, it is rarely found in waters less than 170 fathoms deep and the larger concentration seems to be at a depth between 250–300 fathoms. Further south along the coast, the crab will stay deeper and samples found in the Gulf of Mexico and off Brazil have been caught at 700 fathoms.

The red crab is about twice the size of the blue crab and grows to a size of 2¼ lb or more. Females are more slender and seem only to grow to 1¼ lb. The body of the crab is squarish and the walking legs are long and slender. On each side of the front edge of the carapace are five short spines or teeth to which the scientific name, quinquedens, refers. This is not a swimming crab. The color is red or deep orange. (Some albinos have been

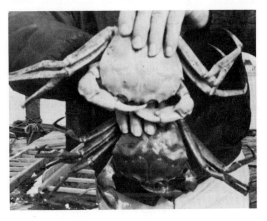

Fig. 9. Male and female red crabs.

found.) The small ones are always bright red and the color of the crab might be related to time period since last shedding. There are indications that the red crab might generally shed in late fall or early winter.

The red crab was first described by Smith in 1879.* Until recently, little research was concentrated specifically on the species, and most available data resulted from general explorations. In the 1950's, large concentrations of offshore lobster were found south of New England which started a commercial fishery, and whenever the trawler went deeper than 200 fathoms, significant quantities of red crabs were taken. Because of the lack of a market, the crabs were, and still are, thrown overboard.

The Bureau of Commercial Fisheries conducted exploratory cruises for red crabs in the north Atlantic region in 1959–1960. Their catches were small, but the best were taken at depths between 200 and 300 fathoms and the heaviest concentration was found east-southeast of Ocean City, Maryland. More intensive research on the red crab was not undertaken before 1966, when the University of Rhode Island started a project to determine the commercial potential of this species. This project is still in progress, but over the last year much more has been learned about this species.

The red crab, despite being widespread, seems to congregate in pockets, and segregates both by sex and by size. In one area, one might find only large males, in another only medium-sized males, and in a third only females. It is found both on muddy and hard bottoms. The crab is more hardy than previously indicated. It will live in shallow aquariums; will remain for weeks in temperatures up to 70°; and will survive out of water for days if kept cool and moist. The crabs held in shore tanks lose their red color after a couple of weeks and become light orange. While a fresh crab does not normally change color in cooking, one which has been in shallow tanks for some time turns partially black.

The crab is caught with otter trawls and the average catch has been 30–60 bushels per 3-hour tow. A bushel of crabs weighs 36–38 lb. This fishing method is hard on the crabs; when the catch is hauled onboard, a large percentage of the crabs have lost a number of legs. To store these crabs in seawater (tanks) onboard the vessel therefore does not seem to be a feasible solution. The death losses experienced have been up to 30% after 5 days. Crabs placed in boxes or in burlap bags (coarse weave) and put on ice, without letting the crabs come in direct contact with the

* For morphology of the crab, see Mary J. Rathbun, *Oxystomatons and Allied Crabs of America*, Bulletin 166, United States National Museum.

Fig. 10. Picking meat from red crabs.

ice, seems like a better method. Despite the fact that the crabs handled this way can die in a day or two also, the commodity is still in good condition for some additional days. No difference could be detected by laymen between crabs cooked while alive or cooked 5 days after they were killed and put on ice.

Under laboratory conditions, a batch of crabs weighing an average of about $1\frac{1}{2}$ lb in live weight, gave a meat yield of 24%. Large crabs give better yield than small ones, and a big crab carefully picked can yield $\frac{1}{2}$ lb of meat. Under commercial picking conditions in Crisfield, Maryland, by labor used to picking blue crabs, but entirely unfamiliar with the red crab, the average meat yield from $\frac{1}{2}$ ton of crab was 11.4% of live weight. It is felt that this yield can be improved as the pickers gain experience or if machines can be designed to pick the whole or part of the crab. The best cooking time has been found to be 15 min at 250°F (15 lb pressure). After cooking, the crabs should be cooled to room temperature before being placed in a cooler to reduce the temperature to 40°F. Crabs are more easily picked when cool. The crabs decrease 26–28% in weight in cooking.

On the average, 24% of the meat is in the claws, 36% in the legs, and 40% in the body. The crab meat has the same color as lobster meat (white and pink) but it is difficult under handpicking to remove all the shell. Small shell fragments often occur in the finished product.

A taste panel at the Bureau of Commercial Fisheries Technological Laboratory in Gloucester, consisting of 13 judges, rated the characteristics of fresh red crab meat and fresh regular blue crab meat. The test took place in February, 1967. The results are given in Table 2.

TABLE 2. QUALITY EVALUATION OF CRAB MEAT STORED AT 40°F (9 POINT RANKING SCALE)

Product	Post-cooking storage time	Appearance	Odor	Flavor	Texture	Over-all
Fresh blue (regular)	8 days	5.9	5.8	5.2	5.3	5.5
Fresh red (mixed)	8 days	6.5	6.8	6.6	6.4	6.6

The crabs caught had been processed on the same day in the same plant. The blue crabs had been caught by dredging and were cooked the day after capture. The red crabs were caught south of New England, had been stored in burlap bags on ice and had been 6 days out of the water. Most of the red crabs were dead at the time of cooking.

Red-crab meat lends itself to pasteurization, and taste-panel evaluations have been carried out for pasteurized red-crab meat both right after pasteurization and after 2 months storage. Red-crab meat always scored higher than regular blue crab meat on every characteristic evaluated.

The water temperature during pasteurization is 190°F. The temperature of the meat in the center of the cans is raised to 186°F. This requires 40 min for a 5 oz can and 110 min for a 1 lb can. If a paper liner is used, pasteurization time for a 1 lb can is about 150 min. After pasteurization, the cans are put into ice water. The cooling takes about 30 min.

The average output of 12 pickers was 3.3 lb of crab meat per hour (range 2.8–4.1). Under present Federal minimum wage rates, this output must be improved to make processing red crab meat commercially feasible. Whereas the blue crab contains three entirely different grades of meat, the meat in the red crab is uniform. The red crab should therefore better lend itself to machine picking. Even a crude tool such as the wringer of a washing machine easily squeezes the meat from the legs, and a machine has just been designed which will remove the carapace, clean the crab, and cut off the legs and claws. Another machine designed to remove the body meat has been tested and works satisfactorily. The Division of Conservation of the State of Rhode Island is as of this writing engaged in research to determine whether or not pot fishing for the crab is superior to trawling.

The red crab may not be able to support a fishery of its own because it has no lump meat (back fin) which is the high-priced meat in the blue crab, but it could be utilized in combination with deep-sea lobster. At this writing, the deep-sea red crab is still an untapped resource of unknown value and extent.

ANDREAS HOLMSEN

D

DESALINATION OF SEAWATER

Evaporation

"Salt water, when it turns into vapor, becomes sweet; and the vapor does not form salt water again when it condenses." Although Aristotle wrote this thousands of years ago, it is only some hundreds of years since the first simple evaporators used the principle to make drinking water for sailors at sea. Today, evaporation processes produce most desalinated water, and evaporation will be the important method for the predictable future.

Solar Evaporation. Evaporation by solar energy first desalinated salt water much more concentrated than that of the sea, at a mine in Chile. A large, open basin had a glass roof, on the underside of which, cooled by the mountain breezes, some 6000 gallons per day condensed. Commercial units from seawater have been built in Greece by Delyannis. He had the advantage of modern plastics to give a black, heat-absorbing, waterproof floor. Again, the fresh water condenses on the underside of a glass cover—transparent to the sun's rays, but cooled by the breeze. Condensate runs down to a collection trough, and the residual brine is returned to the sea. More of such units will use larger areas and a fraction of the available large amount of solar energy. On sunny days, a maximum of 1 gallon for each 15 ft^2 is condensed—on rainy days, the rain falling on the top of the glass runs into the same troughs for collection. Because of their great size and the amount of materials required, solar stills are much more expensive than evaporators using other forms of heat.

Multiple-Effect Evaporation. Steam from boilers supplies the heating tubes of the usual still or single-effect evaporator at a very high cost for energy. Such submerged surface evaporation depends on the steam condensing on one side of a metal tube. On the other side, the heat of this condensation evaporates the seawater, to give steam at a lower pressure, which is then condensed on other tubes. Several such stills may be operated at successively lower pressures, with vapors from one evaporator body or "effect" passing their heat to condenser tubes in the next effect, and thus evaporating about the same amount in each successive effect. Much of the evaporation of the chemical industry, the sugar industry, the commercial salt industries, and others, uses such multiple-effect evaporators, with from 4 to 6 effects as the usual maximum. One large desalination unit of 1,000,000 gallons per day, uses as many as 12 effects (probably the practical maximum) to give a net evaporation or "gain ratio" of about 10 lb water/lb of boiler steam used. Large tonnages of metals are required to make up the bodies of the effects, the shells of the heaters, the tube sheets, and the heat-transfer surfaces—tubing, etc. The amount of hardware goes up almost proportionately to the gain ratio. No major improvements are expected in multiple-effect evaporators for large desalination use.

Vapor Compression. Instead of passing to a second effect, the vapors from the boiling seawater at one pressure may be passed to any one of several devices for mechanically compressing them. Meanwhile, their temperature also increases, so that they may be passed back to the other side of the tubes to condense thereon, and thus continue the evaporation. Condensate is the fresh water. Mechanical or electrical power drives the compressor. Some heat also may be used, but such an evaporator may require no make-up of thermal energy. If electrical power is available, it is used; or, where power is generated by boiler steam, this may pass through a turbine driving the compressor and the exhaust steam utilized for another evaporating process.

Vapor compression has advantages for small production or at inland installations using brackish water, since no condenser cooling is necessary. Internal combustion engines are used in isolated locations, and thousands of such units were used by the U.S. military in landings on South Pacific Islands during World War II. The heat from the engine block and from the exhaust gases, added to the mechanical energy driving the compressor, gives a combined thermal–mechanical compression unit. Newly developed improvements for large units may make possible the production of 1000 gallons of fresh water by only 25 KWH of power.

In order to minimize the compression ratio (from the pressure of the vapors formed, to the steam on the other side of the tubes) and thus minimize power consumption, the temperature drop from the steam to the boiling water should be as low as practical. This usually is minimized by increasing the velocity of the seawater through the tubes by forced mechanical circulation to increase the coefficient of heat transfer. The temperature difference which is required and the corresponding compression ratio is thereby reduced.

Scale Formation on Evaporator Surfaces. Most of the resistance to heat transfer when steam condenses on a tube surface to evaporate water, is on the side of the tube where the liquid is evaporating, and this is greatly increased by the formation of scale. Scale formation is of great importance in the design and operation of all types of evaporators having heat-transfer surfaces adjacent to the boiling liquid.

Scale is formed on heat-transfer surfaces in boilers and evaporators by the heating due to: (a) the concentration of slightly soluble materials beyond their solubility to cause their precipitation; (b) the decrease of solubility in water at the higher temperature of some few dissolved materials, such as calcium sulfate; (c) the decomposition of relatively soluble bicarbonates, which become insoluble carbonates, by chemical action. Usually, the seawater is treated with various chemicals to prevent scale formation; e.g., sulfuric acid decomposes carbonates and bicarbonates. This produces more calcium sulfate. The evaporation therefore is conducted at a temperature below that at which scale will form. The tendency for precipitation thus is minimized for those materials such as calcium sulfate with "inverted solubility curves," and other materials which form scale at higher temperatures by chemical interaction. This lower operating temperature may reduce the gain ratio.

A related consideration is the removal of air dissolved in water, by heating in a partial vacuum. Air which leaks into equipment operating below atmospheric pressure also must be removed, by venting from the condensing zones to a vacuum. Air in condensing vapors to the extent of even $\frac{1}{2}\%$ may reduce the rate of their condensing by 50%.

Multi-Stage Flash Evaporation with Metallic Condensing Surfaces. To eliminate the effect of scale formation caused by the liquid evaporating at the heating-tube surface, multi-stage flash (MSF) evaporation was developed in the chemical process industries about the turn of this century.

MSF evaporation is diagrammed in Fig. 1. Seawater is heated to its highest temperature and pressure in a prime heater (or brine heater) with boiler steam or other source of prime heat. The seawater then passes through a series of flashing vessels 1, 2, 3, etc., at successively lower pressures and temperatures on the evaporating–cooling side. Arrows pointing downward on the left side

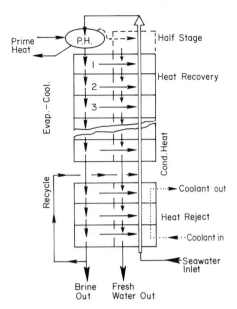

Fig. 1. Multi-stage flash evaporation. MSF evaporating and cooling of hot-feed brine (vertical arrows down) on left side at successively lower pressures after heating to highest temperature in prime heater at top; vapors (horizontal arrows) from MSF, passing to preheat seawater and recycle brine by condensation–heating tubes on right side; fresh water condensate passing stagewise from top to discharge at bottom; additional seawater coolant (dotted line) rejecting heat in lower stages; withdrawing of vapors from prime heater to be condensed in "half-stage" (dashed lines) for increasing the production of fresh water.

indicate the liquid passing to an open "flash" zone of the stage, where vapors form due to the change of sensible heat into latent heat. Because of the violence of the flash evaporation—almost an explosion—considerable liquid mist is entrained with the vapors. This usually is removed by passing the vapors through screen mesh "demisters" before reaching the condensing surfaces.

Finally, the brine, concentrated 2–1, is discharged from the bottom stage, much of it to recycle and join the seawater feed at an appropriate condenser–heater where the temperature is about the same; some is discarded as waste back to the sea.

The vapors, formed in each MSF, pass as shown by the horizontal arrows to individual condensers in a counter current series on the condensing–heating side, through tubes in which the incoming seawater is being passed countercurrently at high velocity. These condenser tubes are indicated by the double line on the right side; in modern practice, tubes 70 ft or more in length may pass through many stages in series.

The system is substantially a counter-current heat exchanger, with the liquid from the prime heater being cooled by the MSF evaporations, to form vapors which condense to preheat the seawater before it is passed to the prime heater.

An additional coolant stream of seawater is used in additional condenser tubes in the lower stages of Fig. 1, and this seawater, after being partially preheated, is passed back to the sea to waste or reject the heat available in these lower stages. Some of this heat rejected may be recovered as indicated below under vapor reheat MSF. Alternately, a single stream of seawater is passed through these heat rejection stages, part is discarded and the balance acidified to decompose carbonates, and the carbon dioxide and air are removed. The treated seawater then enters the lowest one of the heat recovery stages.

For each pound of fresh water produced, there may be recycled as much as 10 lb of seawater-brine being reheated. Of this amount, usually only 2 lb come from the make-up seawater (1 for product, and 1 for concentrated blowdown at about twice the brine strength of seawater.) The balance recycles from the last flash stage back to the raw seawater stream entering the lowest of the heat recovery stages.

Figure 1 diagrams a vertical or tower arrangement. Most large MSF plants built to date are spread out with a horizontal flow sheet over much more area than the vertical arrangement requires.

Controlled-Flash Evaporation. In the very large MSF plants which have been built, there is considerable waste of energy as well as entrainment of brine during the flash evaporations done simply in huge open chambers. The incomplete approach to equilibrium between vapors and liquid here causes important inefficiencies.

More recently, a controlled-flash evaporation (CFE) method has been developed wherein the flashing stages, by correct hydraulic design, allow vapors to form relatively quietly, so that they are always very near equilibrium with the liquid seawater. Various designs may be utilized to give this important result. One has long, thin, rectangular passageways with an expanding cross section for water to flow downward from the higher pressure above, to the lower pressure of the stage below. These are called "chutes," wherein the water flowing downward vaporizes smoothly throughout a controlled distance of flow and pressure drop. The vapors form in the middle, with the water on the sides evaporating slowly compared to the near explosion in the usual flash chamber.

Such controlled-flash evaporation minimizes the energy losses due to turbulence, and maximizes the approach to equilibria between seawater and vapors. Thus, the seawater is preheated more nearly to the temperature of the prime heater, and less prime heat is used, so that the gain ratio is increased. Also, a much smaller volume is required for the flash units. Of more importance, the low velocity of disengagement of vapor from liquid prevents entrainment. Thus, a higher quality of fresh water is produced even without demisters.

By the improvement in the design which is possible for the condenser–heaters attached to each stage of the CFE, a large increase in the heat-transfer coefficient is obtained, and a much smaller amount of condenser tubing is required. Large conventional MSF evaporators now being engineered require as much as 15,000 miles of copper-nickel alloy tubing in a single unit. Fractional saving of condenser surface thus becomes very important, since condenser tube cost may amount to 40–50% of the over-all cost of the evaporator.

A still further advantage of CFE is the fact that it lends itself well to vertical design of the units, with a large reduction of the ground area and piping required for conventional plants which are horizontally disposed. Only a small fraction of their large land requirements are needed for such vertical installations. Since desalination plants usually are required in areas of high population density, adjacent to large cities, plant sites are expensive. This saving in real estate costs may be important. As an example, for one large desalination plant being built in a densely populated area, no adequate site was procurable; a man-made island had to be built offshore to accommodate the large MSF installation with the horizontal layout.

Evaporation in Prime Heat Vapors to Half Stage. The conventional prime heater brings the seawater to its highest temperature. Besides this simple addition of sensible heat, the prime heater also may evaporate a small amount of seawater to give vapors which are at a higher temperature and pressure than those from the highest of the MSF. This increases over-all capacity without increasing the amount of steam used. These vapors are condensed in what is indicated as the "half-stage" in dashed lines at the top of the condensing–heating side. These vapors from the prime heater, on condensation, preheat the seawater further in this additional condenser–heater. This is called a "half-stage," since it does not have the other half (a flashing zone). The total heat supplied to the prime heater is the same, and these vapors, passed to the half-stage, merely add an equivalent part of the sensible heat otherwise added directly in the prime heater. But their condensation gives an increase in production—the amount of the condensate formed. A higher gain ratio of fresh water produced is achieved by what amounts to a "double-effect" heating. This improvement in heat economy may vary from 4–20%, depending on the number of stages.

In the largest MSF unit yet designed, and after a very thorough optimization of conventional practice because of its major cost, it has been shown that 6,000,000 gallons of water daily would be produced additionally without cost by this use of the prime heater also as an evaporator to pass vapors to the half-stage.

MSF Evaporation Without Metal Heat-Transfer Surfaces—Vapor Reheat. The standard evaporator, whether single-effect, multiple-effect, or vapor compression, has heat-transfer surfaces on which vapors are condensed to supply heat to boil the liquid on the other side (usually, but not always, the inside). The MSF evaporator removes the heat-transfer surface from the evaporating liquid but it is still required for the condensation.

The vapor-reheat modification of MSF eliminates metallic heat-transfer surfaces of both evaporation and condensation. Because of the greatly reduced plant cost, optimization of design using vapor-reheat MSF to minimize the sum of capital and operating costs will show an increased number of stages possible and a substantial reduction of heat costs. The optimum may be reached between 50 and 60 stages, depending on the relative costs of energy and of the plant itself.

Vapor-reheat MSF, diagrammed in Fig. 2, is much the same on the left, the evaporation-cooling side, as the conventional MSF of Fig. 1. Again the arrows downward on the left side indicate flash evaporation of liquid which is superheated compared to the pressure and temperature conditions of each successive flash zone which it enters. The vapors formed are again indicated by the horizontal arrows from each flash zone. In vapor-reheat, they are condensed on the condensation-heating side by a recycling stream of colder, fresh water. Vapors condense directly on the surfaces of this fresh water, in sprays or films, to heat it as it flows countercurrently to the evaporating seawater, from a stage of lower temperature to one of next higher temperature. This is indicated by the arrows pointing upward on the right side, to show an open spray, for example, in the condensing zone of each stage, which may be simply an open chamber.

The fresh-water stream leaving the top stage may be heated by direct contact with vapors coming from evaporation in the prime heater to the half-stage, as shown in Fig. 2, or there may be no half-stage, and only sensible heat may be added by the prime heater to the seawater. The heated fresh water then would be cooled in the heat exchanger to transfer its heat to the cold seawater feed, which then enters the top stage immediately for flashing.

The hot fresh-water stream is cooled by cycling in a special dual heat exchanger by contact with a water-soluble oil, e.g., a naphtha fraction, or a chlorinated heat-transfer oil. This stream of oil is consequently heated, and in turn, it is then countercurrently contacted with the seawater feed, where it is cooled and recycled to the hot water. This liquid–liquid–liquid heat exchanger (LLLEX) has intimate contacts of the two insoluble liquids in the form of droplets or otherwise, and it is very inexpensive compared to the alloy heat-transfer surface which it replaces (i.e., per cubic foot of relatively inexpensive *volume*, the LLLEX may transfer per hour per °F, 10 times the heat as that transferred by one square foot of a copper–nickel tube). The product water

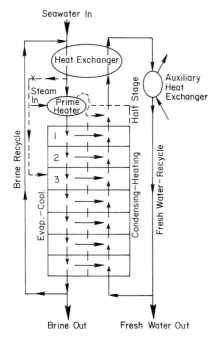

Fig. 2. Vapor reheat MSF evaporation, without metallic heat-transfer surface. MSF evaporating and cooling of hot-feed brine on left as in Fig. 1; vapors (horizontal arrows) from MSF condensing by direct contact with and heating of cooler fresh-water stream on right in open sprays or films and countercurrent flow; heated fresh water being cooled in heat exchanger to preheat seawater feed and brine recycle; withdrawing fresh-water product before recycle on right; increasing production of fresh water by withdrawal of vapors from prime heater (dashed lines) to be condensed in half-stage; additional seawater coolant in auxiliary heat exchanger (dashed lines) cooling the hot fresh-water stream for rejecting heat; or bypassing some part of preheated brine around prime heater (by dotted lines) to upper stage of MSF as preferred method of heat reject.

is equal to the condensate formed in this cycling stream of fresh water, and it is removed continuously.

An alternate method of supplying prime heat to that of Fig. 2 is possible because of the low cost and close temperature of approach of this LLLEX method of heat exchange. Prime steam may be added directly to the half-stage, and the water there reaches the highest temperature in the system. This fresh-water condensate is passed directly to the LLLEX for preheating the seawater, and no separate prime heater is used. Thus, there is no metallic heat-transfer surface for heating, evaporating, or condensing.

The vapor reheat process, among its other advantages, cannot have scale formation on heat-transfer surfaces, because it has no metallic heat-transfer surfaces. Any precipitation of materials which normally give scale comes as a sludge in the seawater or brine being heated in the LLLEX. It can be readily removed by filtering or settling, rather than as the hard scale so difficult to remove from tubes.

In Fig. 2, the auxiliary heat exchanger may be cooled by seawater to reject heat in a manner comparable to the condensation of steam in the lower or heat rejection stages of Fig. 1. This has been shown not to be essential for the most economic operation of vapor reheat MSF.

The heat supplied in vapors coming from the left side must be removed by the condensing action on the right side. Heat may be rejected to bring the condensation–heating side of the stages into balance with the heat supplied in the vapors of the MSF from a higher temperature level, and thus allow the recovery of much of this heat otherwise rejected. One way is to divert a part of the stream of preheated seawater around the prime heater, as shown by the dashed lines of Fig. 2. The preheated stream enters a flash chamber corresponding in temperature; thus little of its sensible heat is lost.

Calculations for the optimization of heat utilization show a substantial saving for vapor reheat, and a somewhat smaller one for conventional MSF, when such a system of saving heat otherwise rejected is used. Of the total heat added in the prime heater, a larger amount must be removed in the "brine-out."

Materials of Construction. Seawater is quite corrosive to metals at the high temperatures of evaporation; thus, corrosion-resistant materials of construction are necessary. Relatively inexpensive metals such as steel and cast iron may be considered for shells of the evaporators, particularly for low-temperature service, and for those parts that can be lined with cement or plastics. Pre-stressed concrete and fiberglass-reinforced plastic shells, vessels, etc. have also been proposed. Currently, some alloy of copper is generally used for tubes. Nickel, aluminum, zinc, or tin are the alloying metals.

Stainless steel is not recommended because of the chloride concentration of seawater. Titanium appears to have great merit. However, a sufficient number of plants have not been built with titanium. Titanium is very expensive, but tubes may be made quite thin because of the excellent mechanical properties. Metallurgical developments appear to be possible in the near future, which will lower titanium costs greatly; until then titanium will probably not be widely specified. Since large deposits of titanium ores are available, this metal may well be the material of the future for desalination evaporators.

The large amounts of fresh water which will have to be produced in the balance of this century will require, if copper-alloy tubes must be used, more copper than will be available. Thus, some system must be developed such as vapor-reheat MSF or other, where no metallic heat-transfer surface is required.

Freezing

Seawater freezes to give ice, which is pure water. To utilize this phenomenon, various processes have been evaluated to study (1) the rate at which ice crystals can be frozen without occlusion of substantial amounts of brine; and (2) economical methods by which the ice crystals may be removed, drained, and washed to remove the surface film of the brine.

Some basic advantages appear in using freezing instead of evaporation as a method involving a phase separation of water from brine: (a) the latent heat of freezing is only about $\frac{1}{7}$ that for evaporation; (b) the temperature of freezing is closer to the ambient than that of evaporation, and so there is less heat loss or gain; (c) the scale-forming materials are so much less apt to precipitate cold than hot, that there is no danger of scale formation from the usual materials; (d) the corrosion of metals at the freezing point is very much less than at the boiling point, and thus mild steel, cement, and many plastics may be used satisfactorily, and, on the whole, much cheaper materials of construction may be specified; (e) the LLLEX allows the development of freezing processes which require no metallic heat-transfer surfaces.

The basic disadvantages of freezing as a process are (a) the time required for freezing ice may be tens or hundreds of thousands of times greater than that for steam formation; (b) the handling and processing of the solid ice crystals is more difficult than the handling of the fluids in evaporation processes; (c) a clean separation of ice from brine cannot be obtained to give the very low content of salt in the final product water as obtained in evaporating steam from brine; (d) the thermal energy for evaporation as usually practiced is cheaper than the mechanical energy usually required for refrigeration (however, absorption of refrigerants also may be used, with subsequent distilling of refrigerant from the absorbing liquid so that thermal rather than mechanical energy may be used for freezing); (e) no system seems possible for using multiple freezing steps to use the thermal energy over and over, as in multiple-effect evaporation; nor for using a multi-stage process. This is because the freezing point does not change appreciably with changes of pressure in a practical range.

Three general types of processes for ice separating and melting have been developed for desalination:

(1) The seawater may be cooled and frozen by transfer of its heat through a metallic surface to the colder refrigerant on the other side. This requires a relatively large temperature difference, with a corresponding higher pressure difference and compression ratio on the refrigerant gas; thus, a high power cost.

(2) The water vapor itself may be the refrigerant and by its low-pressure evaporation from the seawater, ice is frozen therefrom. Since the specific volume of water vapor at the freezing point is about 150 times that at the boiling point, a mechanical compressor of very large volumetric capacity must be used. The so-called Zarchin process is the best developed example of this process, using water vapor as refrigerant. The very simple, inexpensive, and effective Zarchin compressor gives advantages for relatively small plants, and large plants would require a multiplicity of such compressors.

(3) Water may be used as refrigerant, and instead of the mechanical compressor, an absorbing hydrophilic liquid may be used, such as lithium chloride, to absorb water vapor by the usual absorption system of refrigeration. Thermal energy reconcentrates the lithium chloride solution to remove a small amount of fresh water, in addition to that formed by melting the ice formed in the freezer.

(4) Direct-contact refrigerants are used, such as butane, which boil at about the freezing point of the seawater, while bubbling through it. These freeze part of the seawater in the evaporator-freezer; the vapors are compressed to a higher temperature and pressure, thereby contacting the ice crystals which have previously formed and separated from the brine; the refrigerant is condensed for recycle as the ice is melted for product water.

In each of these systems, for which many mechanical devices have been tried, the heat removed from seawater to cause the ice to form may be added back to the ice previously made after this ice has been separated from the seawater by settling, centrifuging, or other methods. Thus, the ice previously made is melted. This may be compared to the vapor-compression system of evaporation

Some new systems have been developed which control and optimize the rate of freezing and hence of production of ice crystals, thus minimizing the time in the freezer, and the required volume of the freezer. Other developments project cyclic systems, which will minimize the separation, handling, or transport of the solid ice crystals from freezer to melter—a major problem of all freezing processes. So far, none of the freezing processes has been able to compete successfully in large-scale desalination of seawater.

In the so-called "Hydrate" freezing processes, the effective freezing point is raised by the presence of molecules of the refrigerant (hydrocarbon or similar materials) in the solid crystal structure, which form with the water a loose molecular structure called a hydrate. The hydrate may freeze considerably higher than ice, as high as normal seawater temperatures. Hydrate crystals freeze even more slowly than ice, and thus the time of freezing and the volume of the freezer is greatly increased. Also, the separation of the crystals from the film of concentrated seawater always present, is more difficult. These disadvantages of the hydrate processes have prevented any commercial use to date.

Extraction and Hyper-Filtration (Reverse Osmosis)

Water is separated from seawater in the gas phase as steam, by evaporation; and in the solid phase as ice, by freezing. Two processes have been attempted to separate water as a liquid: extraction and hyper-filtration, sometimes called reverse osmosis. Here again, the phase separation is of the fresh water from seawater; not the salt from the seawater, since there is no solvent and no filter media which will separate out salt. The chemical methods of salt removal are much too expensive for commercial use.

Most non-aqueous liquids have some solvent action for water out of brine, and this effect is more noticeable with a more concentrated brine. However, thermodynamic considerations show that extraction processes are fundamentally less efficient than evaporation or freezing.

Liquid solubilities vary with temperature, and the relative solubilities for water of solvents at different temperatures are used for the separation. Water is extracted by the solvent at the temperature of maximum solubility; then the temperature is changed to that of minimum solubility in order to precipitate out the water. The solvent used is one which has practically no solubility for salt.

The extraction processes developed require

various distillation and solvent recovery systems, heat exchangers, etc., and there are usually relatively extensive and expensive processing steps. Furthermore, non-common liquids are used as solvents. The solvent liquids on which most work has been done are the substituted amines, which are expensive, and have other objectional features. Common solvents such as hydrocarbons and higher alcohols also may be considered, with advantage being taken of their difference of solubility for water, hot and cold.

So far the inefficiencies associated with the various necessary solvent-recovery operations have made questionable the possibility of large-scale use of extraction to separate liquid phase water from seawater.

Filtration at extremely high pressures—depending on what may be regarded as molecular size separation—is called hyper-filtration or reverse osmosis, and plastic membranes with extremely fine pores are used. Some of the water molecules may be filtered in effect from dilute brines since they are smaller than salt molecules or ions. The fresh water discharges through pores in the plastic film which is supported by wire mesh or perforated plates, and the concentrated brine continually flows past the membrane and out of the high-pressure side.

Cellulose acetate membranes prepared by specially developed techniques are presently favored but membranes of other polymeric materials may be developed. Very large sums of money currently are being spent to develop membranes, operable processes, and staging of processes, since it is impossible to make a salt-free water from a concentrated brine in a single hyper-filtration step.

The membranes are supported on screens or perforated plates to withstand the pressures, 1000 lb/in.^{2} and higher. The capital cost of course is high for equipment to withstand the pressure and the maintenance cost is high because of the frequency with which the membranes must be replaced. From 10–50 ft^2 may be used for each 1000 gallons fresh water per day; this increases rapidly if sediments are present to plug the pores.

The mechanical-energy cost required now may be in the range of 25–35 KWH per 1000 gallons, and this is no lower than for the best evaporating processes or freezing processes which may be devised. The hope is that the hyper-filtration will be reduced to the range of 15–20 KWH per 1000 gallons, from brackish waters.

Ionic Processes

In the ionic processes, salt, as brine, is separated from water. Acids, bases, and salts are ionized in aqueous solution. The ionic processes remove, either chemically or electrically, the ions of the salt from the brackish water, and replace them in another stream which may be wasted. An electrochemical equivalent of the ionic change, or a chemical equivalent or more of the salt must be used. Thus, ionic processes may be much too expensive to use with seawater, but may be considered for brackish water, estuaries of rivers, or inland waters. Improvements in processing may use less-expensive chemicals than now, and the simplicity of the several processes is attractive. In each case, the ionic processes depend on specially developed polymeric resins in special forms.

Ion Exchanger. The sodium ion of salt may be removed when the saline water passes through an ion-exchanger bed of a synthetic resin. By its nature, this resin replaces the sodium ion with a hydrogen ion, to let the chloride go through in an acid form. After the resin is entirely charged with the sodium ion, it is washed with an acid solution, the hydrogen ions of which replace the sodium ions to "regenerate" it for a new cycle. Dilute hydrochloric acid is made in the first bed from the sodium chloride; and this passes through another bed where the chloride ion is interchanged with the hydroxyl ion on the resin, to give the deionized or fresh water. Here again, when the second bed is "saturated" and allows acid to go through, it must be taken off stream, and washed with an alkali solution, to replace the chloride ions. The cycle is repeated. Stoichiometric amounts of acid and of alkali are required for the chemical usage. In general, the chemical cost may be several times the theoretical, but for small installations of brackish water, this high chemical cost may be tolerable. Ionic exchange, as now developed, has no possibility of economically desalinating seawater.

Electrodialysis. This is the other system of removing salt as ions from water. Under the influence of electric currents, a membrane which is permeable to sodium ions forms the wall on one side of a channel of flowing saline water, and a membrane permeable to chlorine ions forms the wall on the other. The water being deionized flows between the two membranes. The other aqueous streams on the other side of the respective membranes may countercurrently flow out in the other direction. A large number of such passages are in parallel, somewhat like a plate-and-frame filter press.

Since alternate membranes are used, the water streams lose sodium ions to one side, and chloride ions to the other. The alternate channels between the ones where the water being purified passes, receive from their respective two sides the opposing ions, which unite in these streams, so that these channels discharge a more concentrated brine. Here again, it may be impossible to have a pure water separating from a concentrated brine in one step, and staging may be needed. The electrical energy required is proportional to the concentration, and the process is economic only with brackish waters with lower amounts of salt than seawater.

The process is governed by the distance of the flow in the compartment, the density of electric current flow, the permeability and other specifications of the membranes, their distance apart, and some other considerations.

In ion-exchange, electrodialysis, and also in hyper-filtration, sedimentary and other impurities in the water can spoil the surfaces and greatly reduce the capacity. Careful pretreatment of the water to remove undesirable materials is usually necessary.

Summary

Regardless of the type of process used for desalination of seawater, the principal costs, aside from labor, which is relatively unimportant for very large plants, are the capital costs for equipment, the energy costs, and the maintenance, replacement, or chemical costs.

At the present time, evaporation is the best developed and best understood system, with numerous variations. The conventional multistage flash is currently the optimum, and usually in optimizing performance, the desired gain ratio is reached at about 10 lb water/lb steam used. This may be increased by controlled flash evaporation, where, because of the lower equipment costs, and the lower energy cost, a substantially higher gain ratio, and a lower over-all cost, may be reached at the optimum point. Modifications of the MSF flow sheets and heat balances allow the prediction of lower heat costs in the future.

The vapor reheat MSF process has not been demonstrated on a large scale, but it may more than double this gain ratio and increase it still further with various accessory systems. All of the various MSF systems, including vapor reheat, may be placed in multi-effect with each other, to increase considerably the steam economy, while adding greatly to the capital costs.

With the large amount of fresh water required for the increasing population of the world, and particularly the rising requirements of water, especially in the undeveloped countries, there is not enough copper or nickel in the world to provide all the copper–nickel tubes of evaporators presently specified. Titanium may be used. At present, it is very expensive, but improved metallurgy may lower titanium costs. Alternately, it appears that developments will make it possible to heat-exchange liquids with other liquids very efficiently in relatively inexpensive equipment—consisting substantially of simple vessels—with very few internal parts, and without metallic heat-transfer surfaces. Developments of this type appear essential if future generations in this century are to have water, more and more of which will come from the sea.

If thermal energy is used, evaporators use it without reference to whether it comes from fossil or nuclear fuels. Various plants using nuclear energy are being built, but until there is demonstration of the long-term availability of nuclear fuels at energy costs comparable to the costs of fossil fuels, the situation is not resolved, and the vast fuel supply necessary for future water demands is not demonstrated.

With many exotic new processes being offered, there still seems to be the standard one of evaporation, which, to date, has little serious rivalry for large plants.

DONALD F. OTHMER

DIATOMS—*See* **PHYTOPLANKTON**

DINOFLAGELLATES

Dinoflagellates are single-celled microscopic organisms, mainly planktonic, found in marine, brackish, and fresh waters throughout the world. The largest are visible to the naked eye as dots; the smallest are visible only in a high-powered microscope. Size varies from 2–3 μ to 2 mm.

Individual organisms may be oval, spherical, or elongated with various kinds of appendages. The cells are nearly always asymmetrical. A typical cell of a dinoflagellate (Fig. 1) consists of the

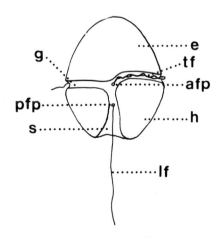

Fig. 1. Typical naked dinoflagellates; only external characteristics shown (after Ref. 1): e, epicone; h, hypocone; g, girdle; s, sulcus; tf, transverse flagellum; lf, longitudinal flagellum; afp, anterior (ventral) flagellar pore; pfp, posterior (transverse) flagellar pore. (Courtesy U.S. Dept. of the Interior.)

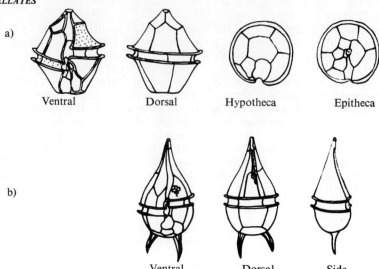

Fig. 2. Two armored dinoflagellates (after Ref. 1): (a) *Gonyaulax polyedra* Stein, 42 μ long; and (b) *Gonyaulax longispina* Lebour, 56 μ long. (Courtesy U.S. Dept. of the Interior.)

epicone (upper part) and the hypocone (lower part). The epicone is separated from the hypocone by a transverse groove, known as a girdle. The tip of the epicone is the apex, and the tip of the hypocone the antapex. A longitudinal groove, called the sulcus, lies almost perpendicular to the transverse groove, and divides the cell into right and left halves. Each cell has two flagella—the longitudinal flagellum, which emerges from the ventral pore in the sulcus, and the transverse flagellum, which originates from the transverse pore in the girdle. The transverse flagellum usually has a thin membrane. The girdle may encircle the cell; its ends may either meet or be displaced so that one end lies below the other. If the right end is below the left, it is a descending spiral (left-handed); if it is above the left, it is an ascending spiral (right-handed), as in Fig. 1.

Dinoflagellates are divided into two forms—naked and armored (thecated). All naked and most armored forms have an ascending spiral. The cell of the naked form is covered by a thin, cellulosic pellicle. In the armored form, the cell is covered by an epitheca (covering the epicone), a

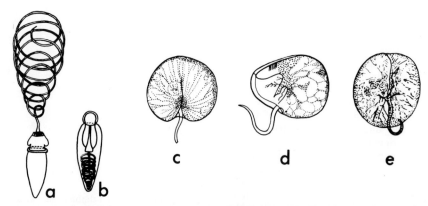

Fig. 3. Nematocysts of *Polykrikos schwarzii* Butschli (a, b) and various views of *Noctiluca scintillans* Macartney (c–e): a, Nematocyst capsule with ejected thread; b, Nematocyst containing coiled thread and fluid; c, *Noctiluca*, dorsal view, 350 μ diameter; d, lateral view; e, postero-lateral view (a and b after Ref. 2; c after Ref. 1, drawing by Miss G. E. Webb; d from Ref. 2, after Allman; e from Ref. 2, modified after Robin). (Courtesy U.S. Dept. of the Interior.)

girdle plate (covering the girdle), and a hypotheca (covering the hypocone), as shown in Fig. 2. The epitheca and hypotheca consist of many plates that may be arranged in different patterns. Classification of armored dinoflagellates is based on the number and arrangement of these plates.

The internal structure of a dinoflagellate consists mainly of protoplasm, nucleus, chromatophores, vacuoles, and (in some) a stigma and nematocysts. Nematocysts are minute, rounded capsules filled with fluid and containing a coiled, threadlike tube that may be everted to aid in either the capture of prey or in locomotion (Fig. 3). Locomotion is accomplished by the vibration of the flagella. Forward movement in a spiral course results from whiplike motions of the longitudinal flagellum and from continuous beating of the transverse flagellum. Speed and type of movement vary according to the shape of the cell, and are often species-specific.

Many dinoflagellates can produce light, and often cause the frequently observed phosphorescence of the ocean. The luminescence produced by the organism consists of a bright flash that lasts about $\frac{1}{10}$ sec, and is produced after mechanical, chemical, or other stimulation. Fishermen often locate schools of sardines by the luminescence in the dinoflagellate-filled water around them. The path of a torpedo or the location of an underwater mine may be visible because of swarms of dinoflagellates in the water.

The most common mode of reproduction in dinoflagellates is asexual, by longitudinal or oblique fission; one of the daughter cells retains the original flagellum and the other grows a new one. Division may take place either while the cells are in motion (sometimes resulting in the formation of chains of individuals; see Fig. 4), or while they are encysted (static). Reproduction also may be accomplished by formation of spores. Sexual reproduction is rare.

Cysts are a physiological state in the life cycle of dinoflagellates. Formation of cysts has been observed in only a few species. The appearance of the encysted organism is entirely different from that of the parent cell; it does not have flagella, sulcus, or girdle, and has a relatively thick outer cell wall. Three types of cysts have been observed[4]: (1) resting cysts, formed during periods of adverse ecological conditions; (2) reproductive cysts, in which the formation of a cyst is followed by the formation of new daughter cells within the cyst; (3) digestive cysts, formed to permit undisturbed digestion of food (after the organism has consumed hard-to-digest food).

Most dinoflagellates are free-living and possess almost every mode of nutrition encountered among protists (see Fig. 5): (a) They synthesize their food like the plants (autotrophy); (b) they ingest prey organisms (phagotrophy); (c) they supplement their nutrition with, or require, organic nutrients (mixotrophy, heterotrophy). Dinoflagellates and diatoms serve as food for some of the phagotrophic forms. Some dinoflagellates are

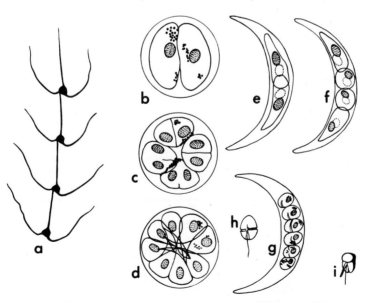

Fig. 4. Various forms of reproduction in dinoflagellates: a, *Ceratium contrarium* (Gourret) Pavillard, four-celled chain; b–d, *Gymnodinium lunula* Schütt, division in round cysts, 80 μ across; e–g, semilunar cysts, 130 μ from tip to tip, showing division; h, free-swimming form, 22 μ long; i, *Noctiluca scintillans* Macartney, zoospore (a after Ref. 3; b–h after Ref. 1; i, after Ref. 2). (Courtesy U.S. Dept. of the Interior.)

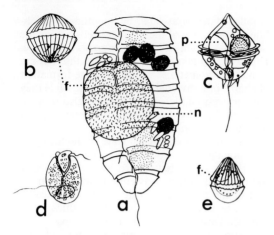

Fig. 5. The food of various dinoflagellates (after Ref. 1): a, *Polykrikos schwarzii* containing *Peridiniopsis* (f); b, *Gymnodinium* sp. containing *Cochlodinium* (f); c, *Peridinium pellucidum*, probably saprophytic, showing a fluid reservoir (p); d, typical holophitic form, *Amphidinium klebsii*, with yellow chromatophores; e, *Gymnodinium rhomboides* containing a diatom (f); f = food; n = nematocyst; p = pursule, or fluid reservoir. (Courtesy U.S. Dept. of the Interior.)

parasites of copepods and copepod eggs, salps, and other planktonic organisms.

Dinoflagellates are claimed by both botanists and zoologists. Because plant characteristics predominate over animal characteristics in most species, dinoflagellates are commonly listed as phytoplankton. Presence of photosynthetic pigments makes most of the forms plantlike. The colorless forms that lack photosynthetic pigments are phagotrophic-like animals. Some animal-like dinoflagellates have specially modified appendages, like the characteristic tentacle in *Noctiluca*, and nematocysts in certain Polykridae (Fig. 3).

Marine dinoflagellates live in coastal and nearshore waters (neritic species), or in the open sea (oceanic species). Neritic species include both naked and thecated forms and show great adaptability to environmental changes in temperature, salinity, and other ecological factors. Most neritic dinoflagellates are simply formed and lack conspicuous appendages. Oceanic forms are especially sensitive to environmental changes, evidently because of the basic stability of their open-sea ambience. The shapes of oceanic forms are more variable than those of neritic forms. Some species can live in both neritic and oceanic zones.

Some groups of dinoflagellates are found only in specific water masses, in different areas of the world.[4] These organisms are useful in ecological studies, as biological indicators of seawater movement. Wood[4], listed species of dinoflagellates that are specific for the Antarctic, for low temperate and subarctic zones, for the east Australian current, for the Coral Sea, and for the east and south coasts of Australia.

Other dinoflagellates, strictly arenaciphilous (sand-loving), live exclusively on sandy beaches periodically covered by tidewater. These organisms are either laterally or dorsoventrally compressed, an adaptation that enables them to move rapidly between particles of sand. Some forms are colorless, but most are pigmented. Discoloration of beaches by the pigmented forms has been reported from Port Erin (Isle of Man), Cullercoats (Northumberland), Woods Hole (Massachusetts), and California. Beaches have been discolored green to brownish on the west coast of Florida by the presence of dinoflagellates of the genus *Amphidinium*.

The abundance and seasonal distribution of dinoflagellates depend on temperature, light, salinity, currents, and the availability of nutrients. Ecto- and endo-metabolites and vitamins may play an important part in the physiology of the species.

Growth of photoautotrophic dinoflagellates depends on the quantity and quality of the light that penetrates the surface and the upper strata of the sea. Most oceanic dinoflagellates are at depths of less than 50 m; greatest densities usually are between 5–40 m. Light requirements vary among species. Estuarine and neritic forms need less than the oceanic forms, and vertical distribution of colorless forms is not limited by light. Dinoflagellates have been found in depths to 500 m in the Mediterranean and Black Seas.

In early classical studies of marine dinoflagellates, temperature was considered the chief ecological factor governing geographical distribution. The importance of temperature in the ecology of dinoflagellates has been verified by numerous experimental observations. Thus, the dinoflagellates generally are divided into arctic, boreal, temperate, and tropical forms. Some can withstand wide temperature changes, however, and these are probably best adapted to the colonization of estuaries and coastal regions.

Salinity tolerance largely determines whether the distribution of dinoflagellates is estuarine, neritic, or oceanic. In neritic regions—particularly in estuaries—salinity varies greatly. Because the nutrient level is usually much lower in oceanic waters than in estuarine and neritic waters, any addition of nutrients to these waters by land drainage may be beneficial to euryhaline dinoflagellates (those able to withstand wide variation in salinity).

Laboratory and field studies of *Gymnodinium breve*, the causative organism of the red-tide phenomenon in the Gulf of Mexico, indicate that geographic distribution of the species may be influenced by salinity. Growth of G. breve was best, under controlled conditions, at salinities between 27–37 parts per thousand; field observations indicated that growth was good between 31–37 parts per thousand, and that the upper salinity limit varied with temperature. Salinity and temperature were also the controlling factors in the mass occurence of three species of *Ceratium* near the coast of Norway in the North Sea.

The fluctuations in abundance of dinoflagellates in the sea reflect changes in environmental conditions and, to some extent, the rate at which zooplankton consume dinoflagellates. In the northern latitudes, combinations of ecological factors governing the growth of dinoflagellates are most favorable during the warmer part of the year. Usually, blooms of dinoflagellates follow blooms of diatoms (another form of phytoplankton). Abnormally high concentration of phytoplankters per unit volume of water is called a bloom. In the tropics, which lack pronounced seasons, production may persist throughout the year. Under exceptionally favorable ecological conditions, blooms of certain species are so dense that they discolor the seawater, changing it to yellow, red, green, or a combination of these colors.

One manifestation of bloom is the "red tide." Various theories have been evolved to explain red tides, but research has shown that they are caused by the sudden appearance of enormous quantities of micro-organisms. According to Kofoid[5] and Hayes and Austin,[6] the oldest references to the "red water" or red tide go back to the time of Moses (about 1500 B.C.).

Dinoflagellates are the organisms that most frequently are responsible for red tides. In different parts of the world, one or more species may cause red tides; the discoloration is sometimes caused by non-toxic dinoflagellates and sometimes by toxic ones. The mass mortality of marine organisms concomitant with dinoflagellate blooms may be attributed directly to production of toxic material or indirectly to depletion of dissolved oxygen. The most reliably documented reports of red tides and mass mortalities of marine organisms come from the Malabar coast of India, the coast of southwest Africa, Japan, southern California, and the Gulf of Mexico.[7]

Dense concentrations of the naked dinoflagellate, *G. breve*, sometimes discolor the bays and coastal waters of the Gulf of Mexico, changing the color from light green to deep amber, rust, or distinct red. Blooms of *G. breve* red tide are most common along the west coast of Florida, appearing less frequently along the southern coast of Texas and off the coast of Mexico. In the following discussion of the *G. breve* red tide, consideration is confined to the waters off the west coast of Florida.

Outbreaks of red tide usually occur during or soon after periods of heavy rainfall. The discolored water appears as slick patches or streaks, in diameters of several square yards to hundreds of square miles. The viscosity of the discolored water increases until it resembles a thin syrup. Fish entrapped in these patches die quickly, and the result can be mortality on a grand scale. The number of fish killed by *G. breve* red tide depends on the density of the *G. breve* cells in the areas affected. Concentrations during normal (no red tide) periods are about 1000 cells per liter; concentrations of 250,000 cells per liter or more are lethal to fish. In 1957, concentrations as great as 100,000,000 cells per liter occurred in Tampa Bay, Florida.

The degree of discoloration of seawater depends on cell density, the physiological state of the dinoflagellates, light conditions (cloudy or sunny), and suspended particles in the water. The color of seawater is unchanged at a density of 250,000 cells per liter, but the water becomes visibly discolored at concentrations of a few to several million cells per liter.

Red-tide outbreaks are generally classified as minor and major. Minor outbreaks affect restricted localities such as a bay, or an area limited to

Fig. 6. A representative sample of dead fish deposited on the beach in the lower part of Tampa Bay, Florida during the 1963 red tide.

a few square miles, for a short time—from a few days to a week. Major outbreaks that encompass large areas (a few to several hundred square miles) and last for several weeks, may cause mortalities devasting to marine populations (Fig. 6). Bottom fish are usually the first to die, then the pelagic species. Grunts, porcupine fish, snake eels, pinfish, catfish, batfish, and mullet are commonly observed victims of *G. breve* red tide. Fewer large fish—tarpon, groupers, cobia—are killed. Dead mackerel, bonito, and little tuna are sometimes found. When the wind is onshore, enormous quantities of dead fish are washed onto beaches —as many as 175 per front foot of beach.

The full impact of fish mortality caused by red tide in Florida has not been studied adequately. The destruction of fish by red tide has not been nearly as severe as the layman imagines, however, when he learns that millions of fish have been killed; Rounsefell and Nelson[7] stated, "The kill of fishes by the 1946–1947 red tide was estimated . . . as 500,000,000 fish. This figure may sound large, but actually it is not. If we realize that these fish will probably run no less than 10 to the pound, the total is only 50 million pounds. Every year about 1 billion pounds of menhaden are caught along a 300-mile stretch of the northern Gulf, but this enormous catch of a single species does not appear to be harming the supply."

G. breve red tide is a menace not only to fish, but to barnacles, coquinas, oysters, shrimp, crabs, porpoises, and turtles. Sea gulls and pelicans have also died after eating fish killed by red tide. Even man can be affected indirectly by red tide. Hay-fever-like sneezing, spasmic coughing, irritation of mucous membranes of the eyes and nose, burning sensation in the throat and nostrils, and breathing difficulty are some of the symptoms caused by red-tide "gas," which is produced by the decaying fish that litter the beaches, but as far as is known, no humans have been killed.

Because outbreaks of red tide (which generally coincide with the Florida tourist season) have a detrimental effect on the tourist business, commercial and sport fisheries, and public health, several federal state agencies are studying the problem. Among them are the U.S. Bureau of Commercial Fisheries, the U.S. Public Health Service, and the Florida State Board of Conservation. Several private institutions have also engaged in the research.

Culturing of both unialgal and bacteria-free cultures of *G. breve* has yielded considerable information on the physiology of the organism.[19] Seasonal occurrences of red tide, the life forms of *G. breve*, and the relation between oceanographic conditions and abundance of the dinoflagellate have pinpointed certain ecological aspects of the

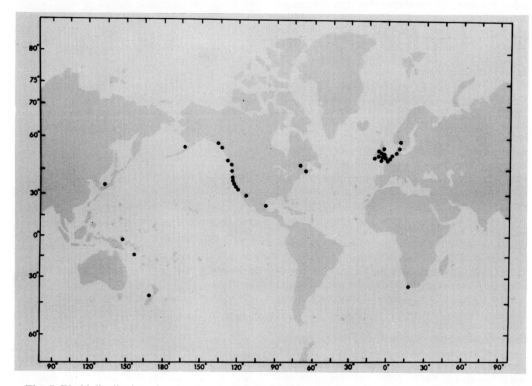

Fig. 7. World distribution of outbreaks of paralytic shellfish poisoning (after Ref. 12). (Courtesy U.S. Dept. of the Interior.)

red tide (Dragovich[8]; Dragovich and Kelly[9]; and Rounsefell and Dragovich[10]). Marvin and Proctor[11] screened more than 4000 compounds in an attempt to discover a chemical or chemicals that might be toxic to G. breve at levels sufficiently low to kill the organism but leave unharmed other marine biota, under natural conditions. Fifty-five of these compounds were toxic at concentrations as low as 0.01 ppm; several seemed promising, but none has been tested in the natural environment. Meanwhile, effective measures have been devised to clear dead fish from beaches—bulldozers and other mechanized equipment collect and remove or bury the debris.

Species of *Gonyaulax*, an armored dinoflagellate, are responsible for the red-water phenomenon in North America and elsewhere in the world, and for paralytic shellfish poisoning (Fig. 7). Several toxin-producing dinoflagellates are shown in Fig. 8.

Gonyaulax polyedra is associated with red tide off the coasts of California and Lower California,[20] and in the Black Sea. In California, concentrations of about 1–40,000,000 cells of *G. polyedra* per liter sometimes discolor the seawater and cause red tide. Fish mortalities are not as great as those caused by the *G. breve* red tide, but losses can be serious. Oxygen depletion rather than toxicity is considered the cause of death. Species most affected—primarily in California harbors—are northern anchovies, mussels, and barnacles. Benthic plants are also harmed by the anaerobic conditions that accompany red-tide outbreaks. Damage is considerable to inshore fisheries, the bait industry, shoreside recreational facilities, and tourist enterprises.

Since toxin has been found in *G. polyedra*, it is highly probable that this dinoflagellate causes paralytic shellfish poisoning in coastal waters of southern California. Carlisle[20] maintained, however, that *G. polyedra* is not toxic and does not cause shellfish poisoning in California waters.

Gonyaulax catenella is a primary source of shellfish poisoning on the Pacific coast; *G. acatenella* was observed in red-water blooms in the Strait of Georgia (British Columbia), and was a source of toxin in shellfish; *G. tamarensis* caused paralytic shellfish poisoning on the Atlantic coast; *G. monilata*, seen in red-water blooms in the Gulf of Mexico, was judged toxic to fish but toxicity to shellfish has not been established.[13]

Probably the first account of paralytic poisoning in a human, caused by eating shellfish, appeared in *Ephémérides des curieux de la nature* (1689), and cited in 1851. North American Indians suffered from such poisonings on the Pacific coast centuries before white settlers arrived.[12] A single serving of infected shellfish containing only a few milligrams of toxin may be lethal to man.[14] Ordinary cooking heat does not destroy this very stable poison; it is soluble in water and alcohol, and it has no known antidote. For some time, studies of the illnesses and fatalities following the consumption of contaminated shellfish by humans have occupied the attention of health and fishery authorities the world over. Many laboratories are attempting to develop effective controls over the disease. The ultimate goal of these investigations is to help public health authorities and shellfish-canning industries to eradicate the causes of paralytic shellfish poisoning. The annual shellfish production of the United States and Canada combined amounts to about 900,000,000 pounds; the total world catch of shellfish is just under 9 billion pounds.[15]

Despite tremendous advances in modern methods of culturing micro-organisms, knowledge of the physiology of dinoflagellates is still rudimentary. To date, only a few dinoflagellates are known to be definitely toxic to marine life and to man; the list of marine dinoflagellate species suspected of being toxic to man is much longer.[12]

As stated earlier, mass mortality of marine organisms is not caused entirely by toxic dinoflagellates; blooms of non-toxic dinoflagellates

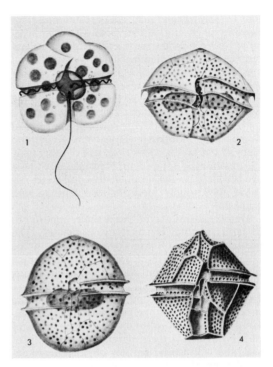

Fig. 8. Several toxin-producing dinoflagellates (after Ref. 12): 1, *Gymnodinium breve* Davis × 1,800; 2, *Gonyaulax catenella* Whedon and Kofaid × 1,800; 3, *Gonyaulax tamarensis* Lebour × 1,500; 4, *Gonyaulax polyedra* Stein × 2,000. (Courtesy U.S. Dept. of the Interior.)

like *G. polygramma* may also create conditions detrimental to marine life. Contrasting with destructive red tides are harmless red tides, generated by non-toxic dinoflagellates. A number of species of *Ceratium, Gymnodinium, Prorocentrum, Peridinium, Gyrodinium* discolor the sea but cause no damage to marine biota. Discolored blood-red water caused by blooms of *Ceratium* was observed in the upper portion of Hillsborough Bay (part of Tampa Bay, Florida) by Dragovich, Kelly, and Kelly,[16] but no toxic effects on fish or other marine organisms were evident. Red tides like these are beneficial to the productivity of the sea because they supply food for zooplankton and shellfish.

Dinoflagellates are of inestimable importance in terms of total productivity in the sea. The greatest amount of primary production in the oceans is in the form of unicellular microscopic plants—phytoplankton. Almost all sea-life forms depend on energy produced by phytoplankton. Dinoflagellates are second only to diatoms in the production of new organic matter in the sea. Until man has learned to duplicate the process of photosynthesis in the laboratory, his supply of sea food will depend entirely on the energy derived from phytoplankton—the primary link in the oceanic food chain.

Though it is not yet possible to reap phytoplankton, herbivores (plant-eaters) e.g., anchovies, menhaden, shad, etc. can be gathered, thus drawing off energy from the sea in the form of food. The average efficiency of transfer of energy from phytoplankton to herbivores is estimated to be 10–20%.[17] "Transfer of energy" means the ratio between food consumed by an organism and its conversion into animal tissue, or energy. For instance, estimates set the conversion rate among plankton as seven plankton plants for each plankton animal; about 10 lb of plankton are required for each 1 lb of small fish. It may be possible, through perfection of methods for mass culturing of some non-toxic dinoflagellates, to increase the rate of energy transfer from primary to secondary producers. In other words, herbivores with heightened nutritional content may eventually be reared in marine cultures, or "sea farms."

The more than 1,100 species of marine dinoflagellates have long been known as producers of some of the most biologically active substances known to science[12]. Studies have pointed out that some of these dinoflagellates also constitute an important source of new tools for pharmacological studies on the cell and tissue level.[18]

ALEXANDER DRAGOVICH

Cross reference: *Toxicology of Marine Animals*.

References

1. Lebour, M. V., "The dinoflagellates of northern seas," Marine Biological Laboratory, Plymouth, England, 172 pp., 1925.
2. Kofoid, C. A., and Swezy, O., "The free-living unarmored Dinoflagellata," Memoirs of the University of California, Vol. 5, 562 pp., 1921.
3. Graham, H. W., and Bronikovsky, N., "The genus Ceratium in the Pacific and North Atlantic Oceans." Carnegie Institution of Washington Publications 565, 209 pp., 1944.
4. Wood, E. J. F., "Dinoflagellates in the Australian Region," Australian J. Marine Freshwater Res., **5**, No. 2, 171 (1954).
5. Kofoid, C. A., "Dinoflagellata of the San Diego Region. IV. The genus Gonyaulax, with notes on its skeletal morphology and a discussion of its generic and specific characters," Univ. Calif. Publ. Zool. **8**, No. 4, 187 (1911).
6. Hayes, H. L., and Austin, T. S., "The distribution of discolored seawater," Texas J. Sci., **3**, No. 4, 530 (1951).
7. Rounsefell, G. A., and Nelson, W. R., "Red tide research summarized to 1964 including an annotated bibliography," U.S. Fish and Wildlife Service, Special Scientific Report—Fisheries No. 535 (1966).
8. Dragovich, A., "Hydrology and plankton of coastal waters at Naples, Florida," Quart. J. Florida Acad. Sci., **26**, No. 1, 47 pp. (1963).
9. Dragovich, A., and Kelly, J. A., Jr., "Distribution and occurrence of *Gymnodinium breve* on the west coast of Florida, 1964–65," U.S. Fish and Wildlife Service, Special Scientic Report—Fisheries No. 541 (1966).
10. Rounsefell, G. A., and Dragovich, A., "Correlation between oceanographic factors and abundance of the Florida red-tide (*Gymnodinium breve* Davis), 1954–61," Bull. Marine Sci., **16**, No. 3, 404 (1966).
11. Marvin, K. T., and Proctor, R. R., Jr., "Preliminary results of the systematic screening of 4306 compounds as "red-tide" toxicants," U.S. Fish and Wildlife Service, Data Report, 2, 3 microfiches (85 pp.), 1964.
12. Halstead, B. W., "Poisonous and venomous marine animals of the world," Vol. 1, "Invertebrates," 1965, Washington, D.C., U.S. Government Printing Office.
13. Gates, J., "Toxicity of *Gonyaulax monilata* to fish," in "Annual Report of the Gulf Fishery Investigations for the year ending June 30, 1958," U.S. Fish and Wildlife Service, pp. 90–93, 1958.
14. McFarren, E. F., Schafer, M. L., Campbell, J. E., Lewis, K. H., Jensen, E. T., and Schantz, E. J., "Public health significance of paralytic shellfish poison," Advan. Food Res., **10**, 135 (1960).
15. Lyles, C. H., "Fisheries of the United States, 1965." U.S. Fish and Wildlife Service, Bureau of Commercial Fisheries, C.F.S. No. 4100, 65 pp., 1966.
16. Dragovich, A., Kelly, J. A., Jr., and Kelly, R. D., "Red water bloom of a dinoflagellate in Hillsborough Bay, Florida," Nature, **207**, No. 5002, 1209 (1965).

17. Chapman, W. M., "The ocean and human food needs" (an address prepared for delivery at Gordon Research Conference 1966, Colby Junior College, New London, New Hampshire, 22 pp.), 1966.
18. Schantz, E. J., "Biochemical studies on paralytic shellfish poisons," *Ann. N. Y. Acad. Sci.*, **90**, 843 (1960).
19. Ray, S. M., and Wilson, W. B., "Effects of unialgal and bacteria-free cultures of *Gymnodinium brevis* on fish, and notes on related studies with bacteria." U.S. Fish and Wildlife Service, Fishery Bulletin 57, 1957; also U.S. Fish and Wildlife Service, Special Scientific Report-Fisheries No. 211.
20. Carlisle, J. G., Jr., "California red tide," in James Sykes, Bureau of Commercial Fisheries Symposium on Red Tide, U.S. Fish and Wildlife Service, Special Scientific Report—Fisheries No. 521 (1965).

E

ECOLOGY

Fish

Marine fish live in an environment which encompasses approximately 71% of the earth's surface and circumscribes the globe in two planes, north–south and east–west. The greatest depth, 10,915 m, exceeds the highest continental mountain while the mean depth at 3795 m represents a significant volume of the ocean. Shallow areas occur at the junction with the land masses, whereas general bottom topography rises and falls to form plains, trenches, plateaus, canyons, seamounts, gayouts, terraces, and continental shelves.

Physical and chemical characteristics of marine waters range in magnitude as though to keep pace with the ocean's immense size and bulk. Temperatures near the poles register $-2.0°C$ and in some surface waters of the equatorial regions $30.0°C$; approximately 76% of all the ocean is comparatively cold at an estimated $3.9°C$. Salinity, as an example of chemical variation, ranges from 0.5% or less at river mouths to over 4% in hot equatorial regions; the mean is estimated at 3.47%. The other physical and chemical parameters, such as pressure, currents, or dissolved oxygen, all tend to emphasize the dynamic complexity of this habitat.

Living marine sharks and teleosts, estimated at 20,000 species, have occupied every habitat type in the ocean save one, the hot ($56°C$), highly saline (27%) holes of the Red Sea. From the floating gardens of the Sargasso Sea in the tropical western Atlantic Ocean, to the frigid arctic waters, and to the cold, high-pressure, inky blackness of the deepest trenches, at least one species of fish has made the adjustment to existing conditions and survived. Some species have changed little through millenniums of time; for example, the Coelacanth off the east coast of South Africa. Other fish are so plastic we can measure their response to relatively small environmental changes. The vertebrae number in the California sardine (*Sardinops sagax*), for example, is influenced by temperature during larval development; higher temperatures result in fewer vertebrae.[1]

Fish, together with their abiotic and biotic environment, form integral systems. Whereas these systems are usually self-sustaining, occasionally they are subject to modifications by external factors. For example, the nutrient-rich coastal waters off Peru sustain an almost inestimable planktonic growth which supports millions of tons of anchovies and in turn are fed upon by thousands upon thousands of birds (cormorants, gannets, terns, and gulls). The key to this classic pyramid of life is the prevailing southerly winds which are responsible for upwelling of cool nutrient-rich bottom waters. The north-flowing Humboldt Current also exerts some influence. Certain atmospheric changes from time to time disrupt the system resulting in catastrophic mortalities throughout the chain. The condition, known as "El Nino," is characterized by winds from the north (equator) and is also responsible for radical meteorological and physical changes on land.

To study so broad and massive a system as marine fish and their environment is a difficult, time-consuming, and costly task. The most prevalent approach has been to examine the parts (autecology) followed by a synthesis into a whole (synecology). Relatively few species of fish, mostly those of commercial importance, have been studied in sufficient detail to permit the construction of the broader picture. Fewer still have been the broad all-encompassing studies of the marine ecosystem, again usually in connection with economically important fish.

Abiotic Environmental Factors. *Temperature.* Marine fish are poikelothermos (cold-blooded); thus they are dependent upon and strongly influenced by the temperature of the surrounding waters. The body temperature of most marine teleosts differs from that of their environment in the neighborhood of 0.5 to $1.0°C$, except such groups as the tunas (*Scombridae*) whose internal temperatures have been recorded at $10°C$ or more above their surroundings.

Temperature influences feeding and metabolic rates, rate of egg and larval development, the rate of growth, and may act as a stimulus to migration and spawning. Each of these physiological responses is influenced by temperature limitations. The total of these tolerances forms patterns characteristic of the species. Some fish are stenothermal (narrow range) while others are eurothermal (broad range). In general, the physiological process increases with rising temperatures.

Temperature is not only associated with the primary distribution of each species but apparently causes relatively large shifts in their geographical occurrence. The latter stages of the "El Nino" condition off Peru is marked by the appearance of equatorial species such as hammerhead sharks (Sphrynidae), manta rays (Mobulidae), flyingfish (Exocoetidae), and dolphinfish (Coryphaenidae). Similarly, shifts in fish populations have been correlated with warmer or cooler than normal waters in other oceans. For example the Atlantic mackerel (Scomber scomber) in the western North Atlantic and barracuda, (Sphyraena argentea) in the eastern North Pacific Ocean (see second part of this article).

Density. Fish density is reported to range from 1.01 to 1.11 while the seawater averages about 1.026; thus the over-all tendency is for the fish to sink. To maintain their preferred position in the water column with a minimum of energy, fish depend on swimming, on fatty deposits, on bone modification, and primarily on an air or gas bladder. The gas bladder forms 4–6% of body volume in those marine fish possessing one.

Hagfish (Myxiniformes), lampreys (Petromyxontiformes), sharks (Squaliformes), rays (Rajiformes), and various teleosts, such as mature flatfish (Pleuronectiformes) do not have a gas bladder. These fish tend to live at or near the bottom. Sharks and rays that inhabit the surface waters are presumed to regulate body buoyancy by adjusting metabolic water (i.e., by swallowing or eliminating water).

In general, the free-swimming fish that occupy the epipelagic zone of the nertic and oceanic provinces have gas bladders. Mesopelagic forms that make diurnal vertical migrations either possess a gas bladder or are neutral buoyant. Lanternfish (Myctophidae) are unique examples of this latter group. Juvenile lanternfish have an active gas bladder which aids in adjusting density and thus vertical movement. In the adult lanternfish, the bladder is proportionally reduced, filled with a "cottony" substance and is surrounded (invested) by fatty tissue. Thus, mature fish are nearly neutral buoyant and must change position by active swimming.

The gas within the gas bladder may be composed of oxygen, nitrogen, carbon dioxide or various combinations of all three. In some species of fish this organ is used in respiration. The pressure within the bladder may be considerable in some species; Nicol[2] reported a pressure of 138 atm in an eel (Synaphobranchus pinnatus), taken at 1380 m.

The position of the gas bladder in relation to the fish's center of gravity is important in swimming and in its orientation in the aquatic environment. The upside-down swimming of some species is attributed to the position of the bladder. Creation of sound and reception of sound and pressure, notably by the grunts (Pomadasyidae), is another use of this organ.

Salinity and Osmotic Regulation. Ocean salinity is higher than that of the body fluids of fish inhabiting it; thus water is drawn out and salts tend to diffuse inward. The process by which marine fish maintain their metabolic balance of fluids and salts is as varied as it is complex; the stomach, kidneys, gills, skin, and ingested food all play a part in the process. The hagfishes are nearly iso-osmotic since their blood has approximately the same osmotic pressure as that of sea water. Body fluids are maintained from ingested food. The bony fish eliminate surplus salt, which comes from swallowed seawater and ingested food, through their gills and gut, only traces leave through urine. Large quantities of water are swallowed deliberately to aid in maintaining body fluids. Marine elasmobranchs regulate water intake through their gills. Salts are excreted in the feces and in the urine. Body fluids are high in urea content which tends to equalize the osmotic pressures. The osmo-regulatory organs of catadromous (eels) and anadromous (salmon) fish are modified to achieve a rapid change in osmotic pressures necessitated by their movement into and out of waters that change abruptly from high to low salinity and vice versa. Fish that withstand great changes in salinity are called euryhaline, while those that cannot are considered stenohaline. Eels (Anguilla rostrata), salmon (Salmonidae), American shad (Alosa sapidissima), and striped bass (Morone saxatilis) are examples of euryhaline fishes. Deep-sea fishes are stenohaline.

Respiration. Metabolism of marine fish, as in all animals, is dependent upon the intake of oxygen and the elimination of carbon dioxide. The gill, in one form or another, is the principal organ where the exchange of gases takes place. To accomplish this most effectively the gills must be constantly exposed to water. The normal openings are protected by large flattened opercular bones and associated with the mouth to provide a channel for constant circulation. The gill cavities of fast-swimming fish, tunas and mackerels, are usually smaller than the sedentary fishes. Rays are illustrative of gill openings modified to a bottom habitat; water is inhaled from the dorsal-oriented spiracle to avoid intake of sediments and exhaled ventrally against the substrate on which they rest. Trunkfish (*Ostraciidae*) respire at a high rate to overcome the restriction imposed by a rigid bony skeleton. Hagfish have sac-like openings not connected to their mouths because it is modified for clinging (and feeding) to their parasitized host. Some fish can respire through their skin as does the mudskipper of Asia, Africa, and Australia. A fair number of fish utilize their gas bladder, intestine, or have special sacks as auxiliary avenues for the intake of oxygen.

Light. Below 1500 m the waters of the open sea are essentially black because most sunlight is absorbed or scattered by the water column above. Water transparency varies from sea to sea and from the shallows to the depths. Again the morphological and physiological adaptations of teleosts and elasmobranchs are as varied as the physical parameters of their habitat.

Illumination has a common bearing on all species of fish (even though some live in its absence) because phytoplankton, the organisms at the base of the food chain, depend upon it. Light may also be a stimulus for certain metabolic processes, maturation of gonads, migration, etc. The ability to see aids fish in feeding, orientation to surroundings, and in the detection of predators. Experiments in the laboratory have demonstrated that many species of fish can distinguish relatively large sections of the light spectrum, including shades of gray. Some deep-sea fish have visual gold (chrysopin) in their rods which is more sensitive to light than visual purple (rhodopsin).

Most fish have eyes: a few primitive groups (hagfish) also have light-sensitive spots on the body. Many mesopelagic and deep-sea fish have their own light-producing organs (photophors) or are involved in a symbiotic relationship with certain light-emitting bacteria.

The eyes of pelagic fish in the surface waters of the ocean are usually semi-spherical organs more or less recessed within the bony portion of the skull and oriented so as to gather light stimuli from two directions roughly 180° apart. As we proceed down the water column fish eyes usually become proportionately larger to accommodate for decreasing light and in many instances protrude and point upward in the direction of the light source. Some fish of the abyssal and hadal zones have greatly reduced eyes or merely light-sensitive areas.

Sound and Pressure. Fish are able to detect or sense mechanical and sonic vibrations by means of the lateral line system, the lower section of the auditory labyrinth (the sacculus and lagena), or the gas bladder. Each system is sensitive to various vibration frequencies: the lateral line, 5–25 Hz and the labyrinth 16–13,000 Hz. The response to sound by the gas bladder has, apparently, not been measured. However, it is known that the bladder functions as a resonator and in some species there is a positive connection with the labyrinth. Experiments have also shown that when the gas bladder does connect with the semi-circular canal system the auditory threshold is lowered and the range of response extended.

Many species of fish produce sounds for a variety of reasons, the most obvious being communication during spawning, feeding, or in time of peril. It has been suggested that sound emission may be used as an echo-sounding system to aid in orientation to the abiotic as well as the biotic environment. Fish sounds vary in character and intensity: drum beats, croaking, snorting, whistles, gnawing, and rattles. The sounds may be produced by the gas bladder (*Sciaenidae*); by the fins (*Batridae*); bones of the pectoral girdle (*Siluroidae*); or by the pharyngeal and buccal teeth (*Tetrondidae*).

The adaptive significance of sound production and reception has not been fully explored except as noted above. Some clues are provided by the unusually prominent lateral line organs in some of the bathypelagic fishes (brotulids and macrourids). Since this system is sensitive to low-frequency pressure waves it may be used to detect the presence of moving prey or predators.

The gas bladder, which is responsive to changes in pressure, apparently also acts as a barosensitive organ. For example, as the water pressure increases some fish will become more active and swim upward and conversely when it decreases they become quiescent and sink. Change in pressure also produces changes in the rate of gas secretion or resorption within the gas bladder. The biological value of these responses is not entirely clear.

Locomotion. Body form in fish is the result of the interacting forces of skeleton, muscles, behavior patterns, and the fluid medium in which they live. The most generalized, and perhaps the most primitive, body shape is fusiform. Its leading edge reduces resistance and the tapering posterior portion minimizes drag. Departure from this shape represents adaptations to specialized modes of life. Many epipelagic fish, usually swift and powerful swimmers, exhibit the fusiform shape; the tunas (*Scombridae*) and members of the shark group (*Carcharhinidae*) are excellent examples.

Deviation from the ideal fusiform shape can be categorized into general groups; the most prominent being lateral compression, dorso-ventral depression, elongation, and spherical.

Lateral compression, as exemplified by the butterfly fish (*Chaelodonidae*), Pompanos (*Carangidae*), or dolphinfish (*Coryphaenidae*), finds its utility in strong quick turns. Typical habitats include the breaker zone of sandy beaches, coral reefs, kelp beds, and the open sea, where the sunfish (*Molidae*) wanders.

Dorso-ventrally depressed fish characteristically live on or near the bottom. The skates and rays (*Rajiformes*) all exhibit this form, including adaptations for feeding off the bottom and for protection by burrowing into or mimicking the substrate. In teleosts, the stargazers (*Uranoscopidae*) and goosefish (*Lophiidae*) also live on the bottom but look upward for their food.

Elongation is exemplified by barracudas (*Sphyraenidae*), lancetfish (*Alepisouridae*), cusk eels (*Ophidiidae*), and morays (*Muraenidae*). No specific habitat is associated with elongated fish for they are found in the epipelagic zone, in reefs, and on the bottom.

Spherical fish are weak swimmers, depending on fin movement for mobility; pufferfish (*Tetraodontidae*) and trunkfish (*Ostraciidae*) are typical examples of this group. They are usually found in nearshore waters of warm seas.

There are many exceptions to the above groups usually representing extreme cases of specialized morphological adaptations to specialized habitats or behavior patterns; the sargassum angler (*Histrio histrio*) or the extremely poisonous blob-shaped stonefish, (*Synanceja verrucosa*), are striking examples.

Non-swimming locomotion, burrowing, crawling, soaring, jet propulsion, and hitch-hiking, are common among marine fish and tend to illustrate the wide variety of habitats they have occupied. Cusk eels (*Ophiidae*) burrow into the substrate. Batfish (*Ogcoecphalidae*) crawl on the bottom on pectoral fins. Flying fish (*Exocoetidae*) soar in the air on enlarged pectoral fins. Remoras (*Echeneidae*), attach themselves to other fish, usually sharks, by means of a sucking disk which is a modified dorsal fin. Lampreys inadvertently move about while feeding on their hosts. Jet propulsion is used by some fish while most of them must counteract its effect as a result of the expulsion of water from the gill chamber during respiration.

Coloration. An obvious manifestation of habitat occupancy is fish coloration. Out of context the colors and patterns are brilliant, startling, and beautiful to the human eye. Within its own niche, however, the fish blends with the background providing essential protection in a harsh survival of the fittest environment. Aside from the obviously apparent protective use of color, other functions have not been precisely delineated despite frequent and numerous anthropomorphic assignments. Cott[3] groups color function as follows: concealment, disguise, and advertisement.

Despite the wide range of colors and patterns exhibited by fish, some generalizations can be drawn apropos their habits and habitats. Epipelagic fish primarily develop whitish bellies, silvery sides, and blue-green to black backs. Many of these fish have vertical or diagonal bars or stripes, such as Pacific bonito (*Sarda chiliensis*) and California barracuda (*Sphyraena argentea*). In the open sea as one proceeds down the water column, fish become red, black, or silvery, changing to violet and black in the abyssal and hadal zones.

Bottom dwellers of the continental shelf and shoreward are usually white underneath and black to brown above. However, they possess the ability to change their dorsal marking to suit their surroundings; halibut, flounders, and soles (*Pleuronectiformes*) are typical examples. Reef-dwelling species of tropical areas are the most brilliant and even bizarrely marked of all fish. In the cold arctic areas the fish are less conspicuously colored or marked, though of no less intricate design.

Biotic Environmental Factors. Biotic interrelationships among fish are not only variable but extremely complex. They are inexplicably interwoven with the abiotic factors of temperature, light, salinity, currents, pressure, etc. Although some marine biologists have attempted to construct models of the biotic environment, the complexities of the system are far from being understood—let alone its dynamics.

Some of the more basic biotic interrelationships are those of feeding, survival, and reproduction. The predator–prey relationship is perhaps one of the most dominant features of the biotic world of fish. Food is required periodically, if not daily, for maintenance, growth, maturation, and reproduction. Survival is part and parcel of the predator–prey picture, as are the abiotic factors which influence individual as well as group reactions. Reproduction is, of course, the driving force which perpetuates the species and is at the apex of the life process.

Intraspecific Relationships. Aggregations of individual fish at the intraspecific level are necessary for reproduction. The act of reproduction ranges from free extrusion of eggs and sperm for chance union to the pairing off between males and females for internal insemination and protection of the developing eggs and larvae. Any number of intermediate reproductive processes can be found. Male sharks and rays possess claspers for the transfer of sperm to the female. Sardines group together for free extrusion of eggs and sperm into the surface waters. Salmon pair off over specially prepared nests or redds. Some deep-sea fish, the angler fish, (*Linophtynidae*), have a unique method of insuring fertilization. The dwarf-like male becomes permanently attached and parasitic on the larger female; thus he is always on hand to fertilize the eggs.

The question of how the male initially finds the female has yet to be elucidated.

Some species of fish form schools or shoals to facilitate food gathering. Typically these fish are planktonic feeders; however, those that prey on larger organisms may also travel and hunt together, as in the case of tuna. The size of a shoal is frequently related to the pattern and concentration of the available prey.

Aggregation of species into schools also serves as a protective mechanism for the group. Predation appears to be minimized when a school begins to mill about in a circle. There is some evidence that groups of fish will survive under adverse abiotic conditions where individuals would succumb. The precise mechanism of this latter phenomenom is not known.

Schooling has distinct advantages, aside from protection, in migrating to or from specific feeding or spawning areas. The factors here seem to be mutual stimulation, and the ability of the group to function better than individuals in picking routes, etc.

Interspecific Relationships. Interspecific relationships of fish are characterized by predation, parasitism, commensalism, symbiosis, competition for food and space, and a number of other associations not clearly understood.

In the predator–prey type of relationship evolutionary adaptations lead to increased efficiency of the predator with corresponding protective modifications by the prey. Neither process is 100% effective. Predator adaptations are characterized by greater agility or speed in swimming; special adaptations of teeth, modifications of body shape, and lures to attract. Prey species evolve such protective adaptations as increased fecundity, coloration, thorns, spikes, toxicity, and sound.

In a general way, predator–prey relationships form patterns associated with geographical areas. In the warm seas of the lower latitudes there is a tendency for a greater diversity of species; in body form, in care of young, in armor, and in color patterns. In the cooler waters of the higher latitudes the number of species decreases but the number of individuals per species increases; protective coloration is more uniform, and body shape less bizarre. The relationships in the temperate regions are intermediate between the two extremes, typical of a transitional zone.

Commensalism, a form of relationship between animals in which one derives benefits while the other is unaffected, is exemplified by remoras and sharks. Remoras (*Echeneidae*) attach themselves to sharks by means of their dorsal sucking disk, travel along with them, and feed on the remnants of the shark's food. The sharks do not appear to prey on the remoras, probably because the remoras are too fast.

Symbiosis, where two organisms derive mutual benefit from their association, is not a pronounced relationship among fish; however, there are several notable examples. The clown anemone fish (*Pomacentridae*) of tropical seas associates with various species of anemones. Grooming or cleaning activities by certain fish, prominently wrasses (*Labridae*), angelfish (*Chaetodontidae*), and shrimp, is wide spread throughout the tropical and temperate waters of the world. Recent evidence indicates that cleaning symbiosis may also be important in the higher latitudes. In this relationship the cleaners maintain stations to which different fish come and submit to examination and removal of external parasites. Bacterially infected areas may also be cleaned. The degree of health enjoyed by fish within an area can be traced to cleaning symbiosis.

The light-emitting organs of some deep-sea fish (*Anomalopidae* and others) are in reality patches of luminous bacteria. In this symbiotic relationship of the deep ocean waters it is presumed that the bacteria derive nourishment from the fish and in turn function as a photophore.

Competition for Food. The greatest competition for food is intraspecific and probably between distinct age groups. Competition between larvae, juveniles, and adults is generally minimized though not eliminated by morphological and behavioral differences. Sometimes three to four species may feed together; Pacific sardine (*Sardinops sagax*), northern anchovy (*Engraulis mordax*), Pacific mackerel (*Scomber japonicus*), and jack mackerel (*Trachurus symmetricus*) make up one such assemblage. The question of whether or not there is conflict or differential cropping has yet to be resolved.

Marine Ecosystem. The marine ecosystem is by earthly standards an immense, dynamic, and in places continuous system of land, saline water, and living things. The application of autoecological techniques is a convenient method of depicting the whole. To this end various oceanographic disciplines have found it of mutual benefit to use a classification system that embodies the water and the bottom.[4]

In this two-dimensional classification the water mass is the pelagic zone and the land mass beneath it the benthic zone. Subdivisions of the two are integrated into recognizably distinct units although boundaries are not precisely defined. A third dimension to this model of the oceans can be achieved by superimposing either the geographical concepts of arctic, temperate, and tropical areas (i.e., essentially latitude) or the zoogeographical regional concept based on animal assemblages. The ultimate subdivision is the individual and the niche he occupies.

Pelagic Environment. The pelagic zone is divided horizontally into a neritic province and an oceanic province. The neritic province is composed of all the waters above the continental shelf extending offshore to a depth of 200 m. The oceanic province consists of all the waters with depths greater than 200 m. Vertical stratification or divisions, with approximate depth ranges, are: epipelagic, 0–200 m; meso-pelagic, 200–1000 m; bathypelagic, 1000–4000 m; and abyssopelagic 4000 m.

Neritic Province. The neritic province, being contiguous with the continental and insular land masses, is a well lighted but turbulent section of the ocean. The basic pattern is that of warm equatorial waters that gradually become icy cold in the polar regions. Seasonal variations, coupled with the physical influence of the moon and Coriolis force, create a dynamic situation characterized by strong wave action, marked currents, and broad swings in temperature, salinity, and nutrients. Dissolved-oxygen levels are high as a consequence of this turbulent activity. Detritus in suspension plus extensive micro- and macro-biological production imparts a characteristic green, grey, brown, or red color to these nearshore waters. Representative habitats include waters of the shoreline, kelp beds, coral or rock reefs, and estuaries. The number of potential habitats is thus large and varied. Representative fish groups include: herring (*Clupeidae*), mackerels and tunas (*Scombridae*); barracudas (*Sphyraenidae*); sharks (*Squaliformes*); croakers and drums (*Sciaenidae*); wrasses (*Labridae*); surgeonfish (*Acanthuridae*); triggerfish (*Balistidae*); and goatfish (*Mullidae*).

Oceanic Province. The oceanic province extends from pole to pole, from one continental shelf to another, and from the surface to the greatest depths. Seasonal fluctuations influence only limited portions of this section, mainly the upper layers. The remainder of this province is fairly stable, uniform, and with relatively few habitat types. Productivity is markedly less than in the neritic province.

Epipelagic Zone (0–200 m). The epipelagic zone of the oceanic province is characteristically blue in color with good light penetration, a factor associated with the low level of detritus and biotic activity. Exceptions are found in areas of convergence between two water masses or currents; for example, the polar seas and the central water mass. The cool nutrient-rich waters provide the basis for exceptional planktonic growth which in turn supports substantial populations of larger animals. Seasonal variations, most pronounced in the higher latitudes, are manifest by changes in temperature, salinity, current, and plant and animal population. The fish groups in this zone include such types as: sauries (*Scombersocidae*); tunas (*Scomberidae*); salmons (*Salmonidae*); flyingfish (*Exocoetidae*); sharks (*Squaliformes*); sailfish (*Istiophoidae*); bluefish (*Pomatomidae*); lanternfish (*Myctophidae*); opahs (*Lamprididae*); and viperfish (*Stomiatidae*).

Mesopelagic Zone (200–1000 m). The mesopelagic zone is characterized by decreasing light intensity, predominantly in the blues; decreasing temperatures and dissolved oxygen; and steadily increasing water pressure. In general, conditions are relatively stable. Fish are black, red, or silvery and dark-adapted (i.e., large eyes, light organs, etc.). Fish in this area feed on the rain of detritus from the epipelagic zone or migrate into it at night to forage for food. Fish groups are represented by: lanternfish (*Myctophidae*); hatchet fish (*Sternoptychidae*); stomiatoids (*Stomiatidae*); bristlemouths (*Gonostomidae*); and deep-sea swallowers (*Chiasmodontidae*).

Bathypelagic Zone (1000–4000 m). The bathypelagic zone is characterized by darkness, cold, high pressure, and low biological activity. Seasonal variations are essentially nil. Knowledge of fish of this zone is considerably less than for the one above it; however, representative groups include: deepsea anglers (*Lophiiformes*); gulpers (*Eurypharynigdae*); and swallowers (*Chiasmodontidae*).

Abyssopelagic Zone (4000+ m.) The abyssal area of the ocean extends over half of the earth's outer perimeter. It is characterized by darkness, high pressure (200–1000 atm), cold (less than 4°C), and low levels of dissolved oxygen. Physical change or variation appears slight if any. We know very little of this hard-to-observe area of the ocean. Fish life appears to be at a low level in number as well as in species. It is well to note, however, that current investigations are yielding new evidence that is rapidly changing our thinking and concepts of life in these regions.

The known fish in this area have greatly modified eyes (some are essentially blind by epipelagic standards) and enlarged lateral lines that take on the functions of sight. Fish groups in this area are represented by deep-water eels (*Halosauridae*) and the deep-sea anglers (*Melanocetidae*, and *Linophrynidae*).

Benthic Environment. The benthic or land mass forming the bottom of the sea ranges from high tide to the maximum depths of 10,000+ m. Convenient subdivisions are related to the units of the pelagic environment.

Supralittoral (*high tide to the spray zone of the beach*). A region of extreme variation—dry to wet to moist in a matter of moments. Few marine plants or animals occupy this habitat. Gobies (*Gobiidae*) are about the only fish found in this habitat.

Littoral Zone (high to low tide). This is essentially the intertidal area and is characterized by extreme changes in water depths (0 to several meters) and turbulence. The substrate ranges from mud to sand to rock and is subjected to hourly, daily, monthly, and seasonal changes which impose harsh conditions for marine fish. Fish utilizing this zone include: grunion (*Atherinidae*); stingray (*Dasyatidae*); flatfish (*Bothidae* and *Pleuronectidae*); sculpins (*Cottidae*); gobies (*Gobiidae*); pipefishes (*Syngnathidae*); and some sharks and rays (*Squaliformes* and *Rajiformes*).

Sublittoral Zone (0–200 m). The sublittoral zone is the underlying or bottom portion of the neritic province and as such is subjected to or influenced by the seasonal variations and conditions that impinge upon the latter. The substrate consists of sand, mud, biological oozes, rocky outcroppings, coral reefs, canyons, cliffs, etc. These varying physical conditions act to create a great number of biotopes, habitats, and niches. Correspondingly the area is rich in plant and animal life, both in number and species.

The fish of the shallow sublittoral zone, low tide to a depth of 50 m, are exemplified by: sharks (*Squalidae, Carcharhinidae, Sphyrnidae*); skates (*Rajidae*); stingrays (*Dasyatidae*); surfperch (*Embiotocidae*); flatfish (*Bothidae, Pleuronectidae,* and *Soleidae*); bonefish (*Albulidae*); morays (*Muraenidae*); seahorses and pipefish (*Syngnathidae*); croakers and kingfish (*Sciaenidae*); wrasses (*Labridae*); rockfish (*Scorpaenidae*); butterflyfish (*Chaetodontidae*); parrotfish (*Scaridae*); surgeonfish (*Acanthuridae*); triggerfish (*Balistidae*); trunkfish (*Ostraciidae*); puffers (*Tetraodontidae*); and midshipmen (*Batrachoididae*).

The deeper sublittoral zone (50–200 m), where light intensity diminishes to that of the mesopelagic zone, is characterized by a mixed fauna composed of mesopelagic types as well as shallow sublittoral fauna. Typical fish types in this area include: cods and hakes (*Gadidae*); halibuts (*Pleuronectidae*); chimeras (*Chimeridae*); hagfish (*Myxinidae*); various sharks (*Squaliformes*); and rockfish (*Scorpaenidae*).

Bathyal Zone (200–4000 m). The bathyal zone of the benthic environment corresponds to the mesopelagic and bathypelagic areas of the oceanic province. The substrate is composed of large areas of biological ooze and detritus (including terrestrial vegetation). Darkness, cold, high pressure, and low dissolved-oxygen levels are the dominant features of this region of the bottom. The fish fauna is composed of groups that occur in the deeper sublittoral zone (halibuts, cods, and chimeras) and those from deeper regions (rat-tails, brotulids and deep-water eels).

Abyssal Zone (4000–6000 m); Hadal Zone (deeper than 6000 m). The abyssal and hadal zones of the benthic environment form the bottom of the abyssopelagic portion of the oceanic province. Little is known of these regions except that there is life of various types, including fish. The substrate consists of biological ooze and detritus while the waters are black and cold with high water pressures, and contain low levels of dissolved oxygens. The few fish captured or observed in these zones are represented by rat-tails (*Macrouridae*), brotulids (*Brotulidae*), rayfins (*Sudidae*), and an unidentified flatfish. The latter was observed by Jacques Piccard and Lieutenant Donald Walsh (of the U.S. Navy) during their historic descent to the Marianas Trench in 1960 aboard the bathyscaphe *Trieste*.

LEO PINKAS

References

1. Clark, F. N., "Analysis of populations of the Pacific sardine on the basis of vertebral counts," *Calif. Div. Fish Game, Fish Bull.* No. 65 (1947).
2. Nicol, J. A. C., "The Biology of Marine Animals," 1960, London, Sir Isaac Pitman and Sons, Ltd.
3. Cott, H. B., "Adaptive Coloration in Animals," 1940, New York, Oxford University Press.
4. Hedgpeth, J. W., Ed., "Treatise on marine ecology and paleoecology," in "Ecology," Vol. 1, Geol. Soc. Am., Mem. 67, (1) 1296 pp., 1957.
5. Fairbridge, R. W., Ed., "The Encyclopedia of Oceanography," 1966, New York, Reinhold Publ. Corp.
6. Herald, E. S., "Living Fishes of the World," 1961, New York, Doubleday and Co.
7. Idyll, C. P., "Abyss: the Deep Sea and the Creatures that Live in It," 1964, New York, Thomas Y. Crowell Co.
8. Lagler, K. F., Bardach, J. E., and Miller, R. R., "Ichthyology," 1962, New York, John Wiley & Sons, Inc.
9. Marshall, N. B., "Aspects of Deep Sea Biology," 1954, London, Hutchinson's Sci. Tech. Publ.
10. Moore, H. B., "Marine Ecology," 1958, New York, John Wiley & Sons, Inc.
11. Nikolsky, G. V., "The Ecology of Fishes," 1963, Trans. from Russian by L. Birkett, New York, Academic Press.

Cross references: *Abyssal Zone; Benthonic Domain.*

Marine

Marine ecology began as a branch of fisheries biology in the 1870's with the studies of the Baltic Sea and the oyster beds of the North Sea by Moebius, and of Vinyard Sound by Verrill. It was Moebius who established the concept of the biocoenosis and coined those terms whose wide current use indicates the major interest of many marine ecologists and physiologists: eurythermal, stenothermal, euryhaline, and stenohaline. The aim of marine ecology is not, however, to categorize the phenomena of the marine (and estuarine) environment with complicated terms, but to understand the processes related to the abundance of organisms, their occurrence in various parts of the seas, and the transfer of energy in the natural systems of the sea in terms of rate and turnover and accumulation of mass. From the viewpoint of the fisheries biologist, marine ecology is the discipline which should provide the basic information to enable man to exploit the renewable resources of the sea to their maximum extent consistent with sustained yields, although this is only one of many phases of marine ecology.

The ramifications of marine ecology as related to fisheries problems may best be understood by recalling that Petersen set out to ascertain the natural food base for bottom-feeding fish by studying the composition of the bottom communities in the Danish seas, not only in terms of the species but also of their numbers and mass (Fig. 1). This study has led to investigations of the characteristics of the seawater and the composition of the sediment as well as of the animals, and, in recent years, to studies of reproductive cycles and behavioral interrelationships of organisms. Along more classic lines investigations rising from the decline of the California sardine fishery in the fourth decade of the century have become a major part of the activity of one of the largest oceanographic research institutions, Scripps Institution of Oceanography.

The oceans comprise the largest of all biospheres or regions of life on our planet, being some 300 times greater by volume than the living space on and over the land. The range of the various environmental conditions is much narrower than on land; temperature, for example, ranges from about $-1.6°C$ in Antarctic and deep waters to about $30°C$ at the surface of the Red Sea (as opposed to recorded air extremes of $-68°$ and $+58°C$), and salinity or salt content ranges from about $3.2-4.0\%$ in various parts of the open sea (ranges are greater of course in estuaries and enclosed lagoons). A highly saline, hot brine source has recently been discovered at the bottom of the Red Sea, but on the whole the marine environment is remarkably constant over large areas, so much so that the major characteristics of the world ocean may be measured in terms of relatively minor differences in specific gravity or density. Often these minor differences may be associated with the distribution of characteristic aggregations of plants and animals, the so-called indicator species. Some marine species, such as *Euphausia superba*, the Antarctic krill (restricted to south of the Antarctic Convergence), and the northern hemisphere species of pteropods of the genus *Limacina*, are among the most abundant species on this planet, although it is possible that the number of some of these species may soon be exceeded by the total population of man. While there is a greater variety of major types of organisms in the sea, since some phyla such as Echinoderma and Chaetognatha are exclusively marine and several others (Porifera, Coelenterata, Bryozoa among animals and Phaeophyta and Rhodophyta among plants) have only a few freshwater representatives there is evidently a smaller number of species in the sea than in terrestrial and fresh-water habitats combined. Nevertheless, insects have not colonized the sea, and most of the arachnid groups are also absent, although at least two lines of terrestrial mammals have invaded the marine environment. In Mesozoic times there was a similar evolutionary invasion of the sea by reptilian stocks. Man is currently making progress towards invasion of the sea with prosthetic devices, and physiological modifications of his internal systems, perhaps by surgical intervention, may not be beyond future technology.

Environment of the Seas. The environment of the seas is divided ecologically according to depth, distance from land and degree of light penetration and, in general, characteristic types of organisms occur in these various environmental divisions (Fig. 2).

The primary division is overlying water (pelagic) and bottom (benthic). The pelagic realm is divided into neritic, or nearshore (usually this implies that these waters have heavier phytoplankton crops, and hence are greenish, and that the nutrients from the bottom are more directly circulated), and oceanic (the blue water of the

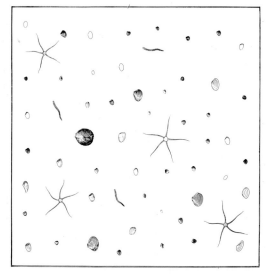

Fig. 1. An example of early bottom community analysis: the Abra community with Echinocardium; animals occurring on $0.25m^2$. (From C. G. J. Petersen, "Valuation of the Sea".)

ECOLOGY

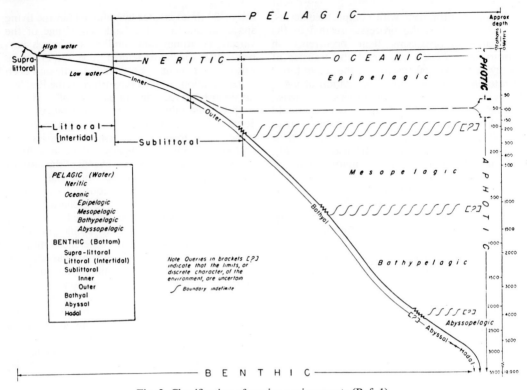

Fig. 2. Classification of marine environments (Ref. 1).

open ocean). Related to this in part is the division according to light penetration. In tropical seas removed from the influence of land, the photic zone may be as deep as 150 m, whereas it may be only a few meters deep near shore in temperate regions near river influences. The subdivision of the benthic zone is related also to light penetration, since the division between the two parts of the sublittoral are related to the attachment of macroscopic algae or active coral reef growth (the infralittoral or inner sublittoral) while beyond this, but still within continental influences, we recognize the circalittoral or outer sublittoral region. The classic littoral zone, however, is that between the tides (intertidal), which in tideless areas may be a very narrow zone. In such places as the Gulf of Mexico the seasonal differences in sea level exceed the tidal range. In areas of pronounced tidal range the actual levels may vary from predicted because of winds, barometric pressure, or other perturbations.

The pelagic regions are inhabited by floating (planktonic) or swimming (nektonic) organisms, and the benthic regions by sedentary, crawling or creeping organisms; among the animals, the benthos is further subdivided into those that live within the sediment (infauna) and those that are on the surface or attached to rocks, etc. (epifauna) Many organisms in or on the bottom, however, have larval stages which are planktonic, especially in the nearshore or neritic regions of the sea. Furthermore, many benthic organisms feed on planktonic life or detritus derived from overlying waters, and fish whose young stages live near the surface feed as adults on bottom organisms. In the deep parts of the sea (bathyal to hadal), however, there is little interchange between the upper levels, and the benthos at least is ultimately dependent upon what may descend from the surface layers, although there may be some production by autotrophic organisms in the deeps.

Life in the Seas. Life along the shores, in the shallow seas, and the surface layers of the oceans conforms in a general way to the temperature conditions of the seas; thus we refer to arctic, temperate, or tropical floras and faunas, characteristically associated with certain temperature ranges (Fig. 3). The actual extent of these various biogeographic regions or provinces varies with season or from year to year, but the general pattern is that indicated in Fig. 4. In the deeper parts of the ocean, temperatures are uniformly cold and many abyssal and bathypelagic animals are worldwide in distribution. Animals of the trenches (hadal), however, appear to have a higher degree of endemism. With the exception of organisms near shores, in estuaries, and in landlocked seas of reduced salt content, salinity is less significant than temperature in influencing the distribution of animals in the sea. Many planktonic organisms, however, are associated with specific types of water (water "masses"), with rather narrow differences of temperature and salinity, which in turn reflect the major current systems of the pelagic regions.

Temperature exerts its greatest effect on distribution during the reproductive stages of many marine organisms, which often have more precise temperature requirements for spawning and development than the adults. Accordingly, comparatively minor shifts in current patterns, bringing about changes in temperature regimes, may cause great changes in the success or survival of such mass species as sardines and anchovies. Within the tidal regions where daily extremes of air and water temperatures occur, and in shallow seas where turnover may be comparatively rapid, actual temperatures governing reproduction or survival may vary considerably from long-term temperature means inferred from incomplete records.

Communities. The primary ecological unit is the community, a grouping of organisms characteristically occurring together, possibly dependent upon each other or at least upon common environmental requirements. It would follow from the environmental control hypothesis that these natural groups should be recurrent given sufficiently similar conditions; if control of community composition depends on interspecific interactions, stability would depend on species composition as well as on environmental conditions. Communities may be discrete and easily circumscribed, as the classic example of the oyster community, or encompassing vast areas of the ocean such as the recurrent plankton groups,[2] or such complexes as the Antarctic community based on the krill. Figure 5 is an illustration of such communities.

In terrestrial ecology, communities are considered to succeed each other in time, resulting in the climax, the stable community associated with, but to some extent modifying, the prevailing environmental conditions over large areas. This concept, developed from the study of vegetational complexes in the northern hemisphere, appears to be less applicable to the marine environment with the possible exception of the coral reefs. Mangroves have frequently been cited, but again this is a vegetational situation at the land–sea interface. There is a succession (or procession) of different types of organisms on newly exposed surfaces in the sea, but it is not clear that this entails a succession of dependencies or is simply stabilization by the hydrographic regime.

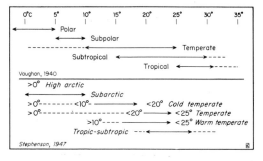

Fig. 3. Approximate temperature limits of major biogeographic regions (Ref. 1).

Communities are composed of various trophic or feeding levels beginning with the producers (phytoplankton and larger algae) which fix the organic nutrient matter, and proceeding through various levels of herbivores and first and secondary carnivores to be returned via the scavengers, detritus feeders, and bacteria. The feeding levels of the herring are indicated in Fig. 6. The original estimate by Petersen that about 10% of the organic material produced at one level is transferred to the next still remains unverified although it is frequently quoted. There is growing evidence that some organisms at intermediate trophic levels may be capable, under certain conditions, of utilizing dissolved organic substances directly, as proposed long ago by Pütter. In any event, efficiencies of better than 10% obviously exist since some marine food chains are tremendously productive, yielding millions of tons of fish or bottom invertebrates annually to the fisheries. Even the most massive fishery stocks may be

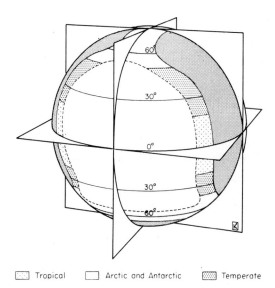

Fig. 4. Idealized symmetry of the marine realm (Ref. 1).

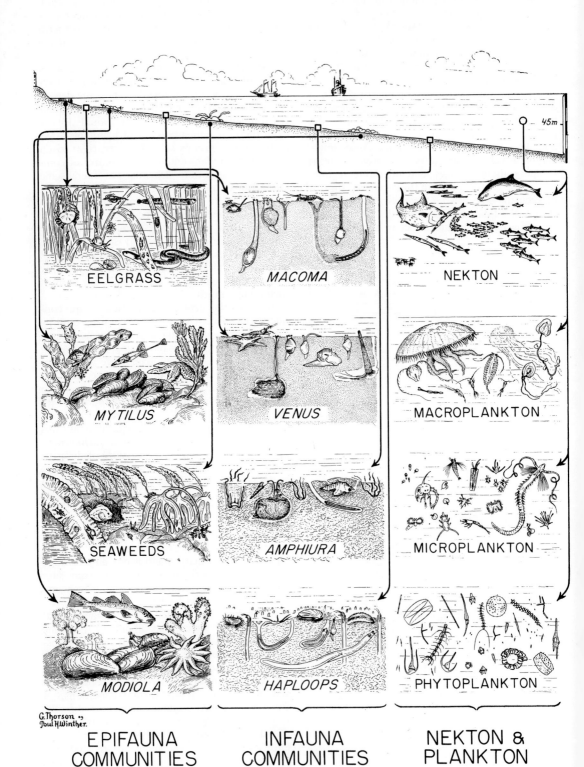

Fig. 5. The communities of the Øresund. (Courtesy Gunnar Thorson.)

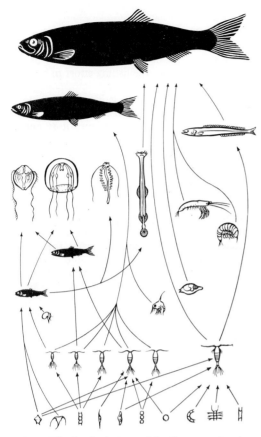

Fig. 6. The food relations of the European herring. (From R. and M. Buchsbaum, "Basic Ecology," Boxwood Press, Pittsburgh, 1957; reprinted by permission.)

drastically influenced by man, who has perhaps irrevocably tampered with the populations of the California sardine and reduced the blue whale, the largest animal on earth, to the verge of extinction.

The California sardine fishery has become the classic example of an ecologically complex community modified by an intensive fishery. A simple matter of overfishing might, on theoretical grounds, be overcome by abstention from fishing for an appropriate period, but apparently an ecologically related and evidently competitive species, the anchovy, has occupied the gap left by the exploitation of the sardines (the gap was evidently increased by harvesting during a period of conditions unfavorable for reproductive success). In the absence of a similar market for anchovies, a reduced technology for processing sardines, and legislative restrictions on harvesting anchovies, the situation has reached the stage where a sardine fishery of almost any magnitude further decreases the stock. This imbalance could possibly be redressed by an unpredictable altera- tion in natural conditions in favor of the sardine, but this does not appear likely. The moral of the sardine fishery, as pointed out by Murphy[4] is "that judicious utilization of all ecologically similar species within a trophic level offers the only hope for sustained yields."

Ecosystem of the Sea. The ecosystem is a complex interaction of community and environmental processes (Fig. 7). The prime mover of the ecosystem is of course the sun, but of the radiant energy that reaches the surface of the sea, only about 0.2% is converted. This process, the fixation of carbon by photosynthetic or autotrophic organisms, is dependent primarily upon microscopic organisms in the sea, especially diatoms and dinoflagellates. The contribution of macroscopic algae and marine spermatophytes, even nearshore, is very small and may be disregarded for the sea as a whole, although it is significant in intertidal regions and in shallow bays (see Table 1).

At the present rate of knowledge, all such estimates are admittedly approximations, and the debate as to whether the sea can produce vastly greater amounts than the land or is approximately equal in total production of organic material proceeds without adequate data. A better idea of the approximate magnitude of production of some parts of the marine ecosystem is to be obtained from statistics for the heavily fished and more completely studied European and Asiatic seas (see Table 2).

Man's expectations of greater harvests from the sea cannot be realized without closer attention to ecological circumstances, as the history of the sardine fishery has abundantly demonstrated. One must view with reservation, for example, the suggestion that the blue whale fishery may be replaced by an intensive utilization of the krill on which it fed. The krill is not only the food resource of the decimated whale stocks, but of almost every major Antarctic consumer, especially

TABLE 1. ESTIMATES OF PRODUCTION OF ORGANIC MATERIAL FOR 30,000 SQUARE MILES OF OCEAN OFF SOUTHERN CALIFORNIA (IN MILLIONS OF TONS DRY WEIGHT, ANNUAL BUDGET)[5]

		Regeneration of nutrients	
Phytoplankton	42	From the bottom	1.0
Attached plants	1.7	From the trophic levels	42.3
	43.7		43.3
Zooplankton	3.4	Organic matter in sediment	0.4
Fish	0.1		
Bathypelagic organisms	2.02	Organic matter lost	0.27
Benthos	1.5		

TABLE 2. APPROXIMATE MAGNITUDE OF PRODUCTION OF ORGANIC MATERIAL IN VARIOUS SEAS[6]

	Average benthic biomass (g/m^2)	Fish catch (kg/ha)
Sea of Azov	321	80
Sea of Japan	175	28.8
North Sea	346	24.5
Baltic Sea	33	6
Barents Sea	100	4.5
White Sea	20	1.2
Mediterranean Sea	10	1.5

the penguins. There are indications that the penguins, especially the Adélies, may be increasing in response to the reduction of whale stocks, and it is obvious that an intensive Antarctic krill fishery might have far reaching effects on the entire Antarctic ecosystem.

The recent development of "pulse" fishing, in which a massive short-term fishery is directed against certain stocks poses further ecological problems. The implications of this type of fishery in which a year's (or several years') stock may be harvested within a few days, have yet to be examined. The long-term effect could be a sort of unplanned selection for stocks or races that do not school so densely, and the effect of this would be to increase the fishing effort in space as well as time.

Pollution Ecology. There is increasing evidence that the sea cannot be considered a convenient stabilizer for human indiscretions; it is not, in other words, a self-renewing septic tank and its use as such cannot be considered economically beneficial. The complete overturn of the sea may be in the order of a single mellennium instead of five as formerly suspected, while mixing may in some cases be even more rapid because of biological processes. The appearance of DDT in Antarctic fishes and penguins, the detection of radioactivity in organisms far from the site of radioactive waste disposal, and the accidental spillage of crude oil, accompanied, as in the Torrey Canyon disaster by an overuse of detergents, are all circumstances indicating the critical need for thorough knowledge of ecological processes in the sea, and for eventual control of additives potentially disruptive of marine ecosystems.

JOEL W. HEDGPETH

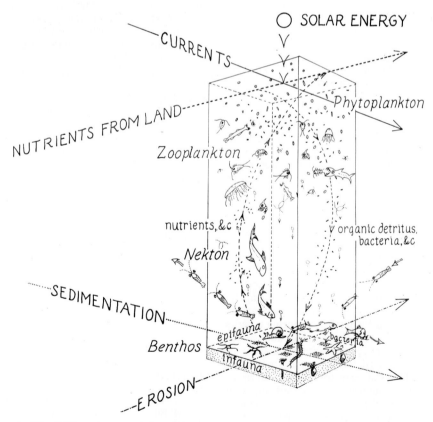

Fig. 7. The processes of the ecosystem. (Courtesy Stanford University Press.)

References

1. Hedgpeth, J. W., Ed., "Treatise on marine ecology and paleoecology, Vol. 1, Ecology," 1957, Geol. Soc. Am. Mem., 67.
2. Fager, E. W., and McGowan, J. A., "Zooplankton species groups in the North Pacific," *Science*, **140**, 453 (1963).
3. Hardy, A., "The Open Sea: Fish and Fisheries," 1959, Boston, Houghton, Mifflin Co.
4. Murphy, G. I., "Population biology of the Pacific Sardine (*Sardinops caerulea*)," *Proc. Calif. Acad. Sci., Ser.* 4, **34**, No. 1 (1966).
5. Emery, K. O., "The Sea off Southern California," 1960, New York, John Wiley & Sons, pp. 139–179.
6. Pérès, J. M., "Océanograph Biologique et Biologie Marine. Tome 1. La Vie Benthique," 1961, France, Presses Universitaires de France.
7. Hardy, A., "The Open Sea: The World of Plankton," 1956, Boston, Houghton, Mifflin Co.
8. Longhurst, A. R., "A review of the present situation in benthic synecology," *Bull. Inst. Oceanog.*, **63**, No. 1317 (1964).
9. Marr, J. W. S., 1962, "The natural history and geography of the Antarctic krill (*Euphausia superba Dana*)," *Discovery Reports*, **32**, 33 (1962).
10. Olson, T. A., and Burgess, F. A., "Pollution and Marine Ecology," 1967, New York, Interscience Publishers Inc.
11. Pérès, J. M., and Deveze, L. "Océanographie Biologique et Biologie Marine. Tome II. La Vie Pélagique," 1963, France, Presses Universitaires de France.
12. Zenkevitch, L., "Biology of the Seas of the U.S.S.R.," 1963, New York, Interscience Publishers Inc.

Cross references: *Benthonic Domain; Estuarine Environment; Pollution of Seawater.*

EDUCATION FOR THE COMMERCIAL FISHERIES

Educational and training facilities for fishermen and shore-based supporting groups vary widely in type, direction, and level, among the 40 or more countries in which they exist. In the vast majority of cases, training at a vocational level only is available.

In countries having a fishing industry of long standing, as the United Kingdom, short formal pre-sea courses may be available for young men who wish to enter the industry; further study may then be arranged between periods spent at sea with the fleet. The periods ashore and afloat usually coincide with sea-time requirements for the various levels of certification issued, and the training is either along a "deck" or "engineering" line. Often the industry will have its own scheme similar to an apprenticeship arrangement, or there will be a recognized manner of progressing within the fleet. The depth and content of subject matter is usually of a vocational nature and includes only certification requirements, covering perhaps seamanship, navigation, and engineering, it being left to the individual to pick up his trade.

Alternatives to this type of arrangement provide for more lengthy pre-sea training, or shore training for certification after several years experience at sea. In countries with developing fisheries, there is often a more fisheries-oriented training program, but again due to the general level of education, it is at a vocational level. Educational arrangements which continue beyond the knowledge required for certificates of competency or beyond the vocational level are available in only a very few cases.

In areas of the world faced with rapid expansion and technological developments in their commercial fisheries, particular cases being Japan at the turn of the century and U.S.S.R. within the past 20 years, a much higher standard of formal education has been introduced to offset the natural growth and experience possessed by fishermen in countries where fishing is traditional. These offerings cover all branches of the industry from catching through the supporting industries, and with the worldwide deployment of the modern and technologically advanced fishing fleets of these two countries and their success on an international scale, it is doubtful whether the argument for higher education in the fisheries can be faulted. The arrangement in these two countries are notable in that while U.S.S.R. follows essentially a practical technological approach, Japan has favored, at the higher levels, an academic attack. In both countries, however, recognition is given to the graduates of the fishery colleges and universities, and the Captains of many vessels are expected to possess degrees.

In the case of Japan, Dr. E. B. Slack of New Zealand has listed 315 of the annual 1400 graduates from the fisheries universities as entering the catching sector of the industry, 110 as engineers, while 275 go into fish processing ashore; other than the 150 entering fish farming, the remainder are absorbed by the supporting groups ashore, by the government, in research laboratories, in experimental stations, and in teaching.

Slack also compared the influence of the universities on fisheries development in Japan, and agriculture in New Zealand, which developed in the same period along similar lines; in both cases the university responded to the needs for an intensely practical approach to training methods while at the same time retaining research schools dedicated to the pursuit of new knowledge in the traditional academic manner: "The results of the combination of practical and intellectual endeavor have been quite remarkable. The rapidity of growth and development of New Zealand's agriculture and Japan's fishing industry set an example to the rest of the world. Surely, here is a lesson from which any country which wishes to expand and improve either its agriculture or fisheries has much to learn."

University-level education has been available in fisheries biology and oceanography for a number of years in the U.S. and Canada, and technologists for research and design in commercial fisheries have come either from such graduates or from men trained in other disciplines, such as naval architecture, engineering, and food technology, but who have the interest to enter the commercial fisheries field. Commercial fishermen, self-educated and with practical ability, have also filled posts in the area.

A similar situation has existed in the industry—in fishing operations and in the supporting industries serving the fleet and handling the products, both at the technical and managerial levels.

The lack of suitably educated and trained personnel for the industry and associated endeavors has been recognized during the past 4 years by both Canada and the U.S., although western European countries generally have as yet taken no definite steps in the direction of fisheries higher education.

In 1964 the College of Fisheries, Navigation, Marine Engineering, and Electronics was established in St. John's, Newfoundland, jointly by the Governments of Canada and Newfoundland, to provide education on an even wider base than that suggested by its title.

Diplomas in Technology are offered in Nautical Science (Fisheries or Merchant Marine), Gear Technology, Food Technology, Naval Architecture and Shipbuilding, Marine Electrical and Electronics Technology, Marine Engineering, and Plant Engineering Technology. All are heavily fisheries and marine-oriented, of 3–4 years duration, and based on a sound framework of written and oral communication, mathematics and science common to all options.

Depending on the discipline, programs are operated either on a "Sandwich System" in which the student spends alternate 6-month periods at college and in undertaking supervised industrial experience, or on an academic year basis.

The Diploma-level programs are open to students across Canada and from abroad, and the college is recognized as the center for higher fisheries training in the Atlantic Provinces. Despite its youth, the establishment has gained an international reputation, and welcomed a large number of students from overseas, many financed by their own government or under the Canadian Government Aid schemes; a number, however, are paying their own way.

In addition to the technological education, training for potential commercial fishermen, tradesmen, and industrial operatives is provided at the vocational level for the Province of Newfoundland in each discipline, together with extension and residential programs enabling working fishermen, skippers, engineers, and merchant marine personnel to upgrade themselves.

Students entering the Diploma courses will have completed high school or passed through preparatory courses at the college, the latter designed especially to upgrade men having less than high school education. The complete system of upgrading, Vocational and Diploma courses is so arranged that a student can proceed to the level of general education of which he is capable, and then by passing through one of the terminal-vocational courses prepare to enter industry.

In the U.S., the University of Rhode Island established a Department of Fisheries and Marine Technology in early 1967, the first task of which was to establish and operate a 2-year Associate Degree Program in Commercial Fisheries financed in the main by a grant from the U.S Government.

The first students, from all parts of the U.S., entered the program in September 1967 and strong interest from overseas indicates the likelihood of students from abroad entering in future years.

There was good reason to choose the University of Rhode Island as the home of this first venture in higher fisheries education in the U.S. During the last decade, this Institution has become marine-oriented over a wide base—it has a Graduate School of Oceanography, a University-wide Marine Resources Program, a Department of Ocean Engineering, and a Law of the Sea Institute. The Faculty works in very close cooperation with the local fishing industry whose representatives are members of committees at the University.

With the great emphasis on education, research, and extension within the whole marine area, which is destined to follow implementation of the Sea Grant Bill of 1967, the commercial fisheries 2-year associate degree program may well herald the way to similar centers in other regions of the U.S., lead within the foreseeable future to a full undergraduate program, and eventually to graduate study. Future captains of long-range trawlers and other U.S. fishing vessels may well hold Bachelor's Degrees, and so join their equally qualified opposite numbers from U.S.S.R. and Japan.

Organization of a Fisheries College or Technical Institute

In common with other endeavors which rely heavily on a combination of theory and practice embracing knowledge from a number of disciplines, a fisheries technical institute or college, if it is to play its full part in industrial development, must have available a wide range of facilities and faculty representing specialized and general areas of education.

This might be achieved by a completely separate institution appropriately financed and organized, by integrating the fisheries education with existing universities, higher education establishments having competence in a number of the areas, or by close association between a separate college and existing universities, research establishments, and other institutions. Perhaps an even better home, remembering the essentially technological nature of the fisheries, would be in association with a College of Technology.

Usually existing institutions will be either a navigation or nautical college, a university with marine-oriented programs, government research and experimental laboratories, or technical colleges. Whether to integrate, associate, or establish a separate college will depend on geographical and financial conditions, on local educational arrangements, and the manner in which all areas of endeavor can be pursued most efficiently. Local conditions and arrangements may dictate the areas of endeavor which an institute should pursue.

The establishment of a practical program such as that required for the fisheries is time consuming for all the faculty concerned, involving as it does the design and construction of specialized teaching laboratories and equipment. Unlike many other fields of education, shelf items of laboratory equipment are, for the most part, nonexistent, and have in many cases to be built by the instructors themselves. Display material has to be cajoled from manufacturers, and there are always the ever-present budgetary worries, because the expense of commercial fisheries education, with its combination of academicism and practical training, is rarely recognized by governments or institutions responsible for the initial effort.

After several years, however, perhaps an even better case can be made for a broadening of objectives and activities; participation of faculty members in basic and applied research assists in keeping both them and the instruction offered in line with, and at the head of the rapid developments in fisheries science and technology. Practical activities, including teaching, experimental work, and advisory services maintain that contact with firms and individual fishermen which is so vital in an industry of so practical a nature while so technological in background.

The wide range of specialty faculty members needed could lead in a number of cases to marginal utilization of their talents if only teaching were considered; in many cases faculty appointments might include several of the areas discussed and so maintain the cross connections between academics, teaching, and the field. This has the additional merit of providing a "built-in" system allowing teachers continually to update themselves technologically and practically.

In many countries, government laboratories exist which are expected to undertake the research, and extension workers may be responsible for the fieldwork, so that the College is expected to occupy itself entirely with education. If this is the case, the vital contact with industry and the updating of teachers may be maintained by the faculty being encouraged to undertake consulting duties either privately or under college control.

The scope of the scheme of studies must be expected to vary widely to conform with local requirements and conditions. Generally, observing the present tendency, it might be expected that a technician program would occupy 2 years while a full technological program might require 4 years.

Experience in other areas of technology and technical education indicates that the technicians and technologists best prepared to meet present and future demands and developments, possess a thorough grounding in the basic education and science of their discipline. It is reasonable to hope, therefore, that the scheme of studies would allow considerable effort to be placed in this general education area at all levels, while not neglecting the training concept.

As the whole area of commercial fisheries is one of a close alliance between the theoretical and the practical, the so-called "Sandwich Scheme" would appear to have direct application to this field of education. Under such an arrangement, students spend part of each year's study in college, and the remainder in supervised industrial experience. This establishes the program on a calendar year rather than academic year basis, and ensures that students, who normally expect to work during the long vacation in any case, do so in a manner which assists their training. Experience indicates little difficulty in placing students so that they receive a reasonable financial return for their labor during these periods.

If a Fisheries College or Institute is associated with a University, the academic portion of the studies will usually consist of some 28–32 weeks each year, thus providing periods of 16–18 weeks for students to undertake the industrial "filling" of the "sandwich" while allowing for vacation time.

If the college is a separate institution, considerably more freedom in arranging programs is available, and it may be preferable to equalize the periods of academic and industrial training. If it is necessary that the greatest number of

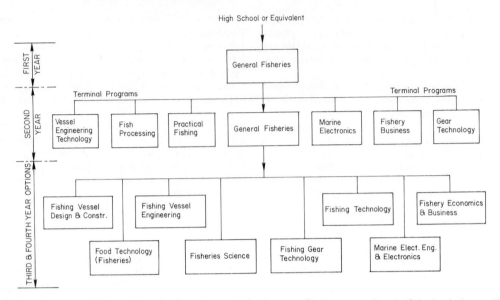

Fig. 1. Integrated scheme of studies for a 2-year and a 4-year college program for the fisheries industry. The college year is two semesters, each having 14 weeks. Practical training between each college year is 18 weeks: between first and second years, practical fishing at sea; between second and third years, seatime to suit option; between third and fourth years, practical experience ashore or afloat to suit option.

students possible be passed through, this arrangement allows two groups to be entered each year, and permits greater flexibility and efficiency in the use of facilities in terms of student numbers, class sizes, utilization of buildings, equipment, and vessels. On the debit side, it could be expected to lead, however, to an increase in the number of faculty required, and usually to teaching being the only concern of most faculty members.

Figure 1 illustrates an integrated scheme of studies for a 2-year and 4-year program including a wide range of options covering the operations and supporting areas of the fisheries industry. The scheme for a particular college might be expected to include all or several of these options depending on local requirements and the presence of existing programs which have ready application. Examples of such existing programs might be Food Technology, Naval Architecture, Marine Engineering, Electrical Engineering, and Business, where students from existing universities or technological colleges could be provided with a fisheries application program of 1 year or more either at the technician or technological level in the Fisheries College after completing their professional courses.

A general commercial fisheries program may be preferred, in order to provide a background in all areas, and in this case the Fishing Technology path might be the most appropriate, being that which interested captains and operations people would follow.

Using a scheme similar to Fig. 1, all students would enter into a common first year of studies consisting principally of basic education: written and verbal communication, physics, mathematics, economics, and business, while introducing fisheries and practical work in a general manner.

Students may then be divided into those who would benefit more from technical courses at a technician level, or those of greater technological ability who would proceed along the full 4-year path.

Between the first and second year's study, an ideal opportunity is available for students to experience commercial fishing operations at sea; no matter whether a student wishes to enter the seagoing or shore-supporting section of the industry or related fields, understanding of the work, life, problems, and general atmosphere of commercial fishing will prove invaluable. Experience indicates that this practical training should be on board a vessel which is fishing commercially and of several months duration if students are to gain full benefit.

Courses in the second year at the technician level should include studies in associated areas of fisheries, a continuation of basic education, and both basic and applied work appropriate to the option. At this stage a balance between the general fisheries and the basic and applied areas of the option is essential.

Students who indicate by their aptitude during the first year's study that they can profit from the more academic and technological nature of the longer program, may enter upon the second year of this stream. Work during this year might profitably consist of further basic education, particularly in mathematics and engineering subjects (e.g., mechanics, strength of materials, hydraulics, hydrostatics, thermodynamics), or of fisheries-related subjects (e.g., fishing gear, business, applied engineering, navigation) and of applied practical work in the fisheries field.

Following a further period of seagoing experience extending over several months and arranged to suit individual interests, students could then be allowed the choice of several or all of the options indicated in Fig. 1 when they continue their studies during the third year.

Subjects for this third year's study could be a combination of further basic education, general fisheries courses, and elective fisheries subjects, both basic or applied, depending on the option.

On terminating their third year's study, and following an industrial training period spent at sea or ashore in work appropriate to the option, students enter the fourth college year, concentrating studies on the areas particular to their options; at this stage about half the time might profitably be spent on individual problem study by each student resulting in the production of theses.

Extension. An extension service provides many very real advantages to industry, college, and government.

(a) Providing fishermen and the supporting industries with information on developments, methods, and techniques, and assistance in their practical application.

(b) Assistance in the solutions of everyday problems, both extensive and peripheral in nature.

(c) Maintaining contact and mutual confidence between the college and industry.

(d) Feedback of information from the industry.

(e) Preparing the industry for development through courses of both a basic and applied nature.

The way in which these duties are organized and practiced will vary greatly in both natural and local conditions; some may already be available through government sponsorship.

Placement of the extension service in the hands of a college, provided that close government contact is maintained, allows the participation of both academic and practical faculty members to assist in the work, acting as a team in the solution of problems and the application of developments.

Fieldworkers prove invaluable, and in many cases can assist with the day-to-day problems of individual fishermen and industry, but are able to refer questions beyond their scope or which need further study to pertinent persons at the college. In addition they are able to collect data needed for teaching and research at the institute, and can promote the use of new equipment and methods through their day-to-day contact.

Information dissemination must be considered one of the most important aspects of extension work; all too often discoveries and developments are lost in the transfer from the research area to those involved in their practical application by being written up only as formal or semi-formal papers, which in most cases are incomprehensible to the possible user. Clear, simple, well-written writeups containing the information needed by the users has been proved in other industries to bridge this gap.

Where the user requires an upgrading of his general education in order to keep abreast with modern developments, this might be made available either by traveling schools, or by residential courses, depending on the level, the applied nature of the instruction, and the equipment required. Type of instruction would normally be expected to vary from the making and repair of nets or engine maintenance, through basic physics or engineering, to such topics as the mathematics of sonar, depending on the requirements.

Extension courses are also of considerable use in preparing fishermen for statutory government examinations to gain certificates of competency, if these exist.

Faculty. Perhaps the greatest problem in effecting a higher education establishment for commercial fisheries is the procurement of qualified faculty and top administrative personnel.

The top administrative officer, responsible for the complete operation, must have a knowledge of world and local fisheries, technology, and some experience at a reasonable rank in education; formal qualifications required may well depend on the regulations of, for example, a university with which the establishment may be affiliated.

Because of the present lack of higher education establishments in fisheries, very few men fulfilling these requirements are available.

A similar situation exists regarding faculty appointments; academically qualified men rarely possess the technological and practical aptitudes necessary for faculty in a school of this nature, while practically-oriented personnel often are unacceptable academically.

As the growth of education in the area gains momentum, the void is likely to be filled gradually, but in the meantime a compromise which has proved successful in several colleges is to recruit from the Navy and Merchant Marine, the fishing industry, engineering, and similar fields, in addition to academically qualified men. Technicians from the various areas may also be useful, bearing in mind the blend of theory, technology, and practice that builds a graduate in fisheries.

The majority of these men will lack one or more parts of the education and training desirable, often academic qualifications, but have considerable experience in their own field. Although the approach is perhaps contrary to usual policy in higher education establishments, the small amount of experience available suggests that careful selection will bring in faculty whose interest and ability will overcome most of their deficiencies. A real problem exists with the practicing fisherman whose knowledge and experience is so essential to a college, but who will in many cases have passed the time when he can make up his lack of formal education; practically oriented men are essential, however, in the practical side of the programs, in extension duties, and in maintaining contact with the industry.

Laboratories. The combination of the practical with the theoretical as part of each discipline when applied to fisheries leads to a requirement for extensive teaching laboratories both of the normally accepted engineering, physics, and hydrodynamics type, and of the practical workshop type. As an example, facilities are as necessary for the teaching of basic practical seamanship, net building and repair, as for the hydrodynamics and structures aspect of fishing gear.

This combination of accepted academic education and training leads to the necessity of installing many expensive teaching aids, such as a radar simulator to prepare future captains for their task. One of the latest developments is a simulator for training captains in fish finding and capture; developed in Norway, this is estimated to cost up to $200,000.

Training Vessels. Common agreement exists among schools and international bodies who have studied the question that seagoing training plays an extremely important part in fisheries training at all levels. The validity of this in the field of higher education is illustrated by the large number of vessels employed for the purpose in both U.S.S.R. and Japan. The type and arrangement of training to be given by training vessels at any college must be integrated carefully into the over-all program.

A training vessel, operated effectively as a "floating laboratory" is invaluable for translating classroom study to seagoing practice in all subjects of a program. While such a vessel should be outfitted to allow laboratory-type application, it must have the practical equipment for demonstration and practice of all principal fishing methods. Vessels between 50–80 ft appear ideal for such use; they require only one or two permanent crew, allow rapid evolutions using scaled-down fishing gear, and permit students to work cooperatively in their handling and management. Thus the expense is reduced to a minimum, while the range of usefulness can be equivalent to or greater than a much larger vessel.

Such a vessel would be used during regular laboratory sessions on a daily basis and would not undertake long cruises, thus obviating the necessity for large accommodation spaces, and permitting the equipment to be of a size suitable for laboratory-type work of both a teaching and experimental nature.

JOHN C. SAINSBURY

EELS

A family of fresh water eels (*Anguillidae*) and several families of salt water eel-like fish belong to the order *Anguilliformes*, formerly known as Apodes. The latter term signifies fish without limbs, since they lack a pelvic girdle and pelvic (ventral) fins. Representatives of some families (*Muraenidae*) also do not have the pectoral fins. The electric eel (*Electrophorous electricus*) of the family Electrophoridae from South America is not a true eel and belongs to a different order of bony fish, *Cypriniformes*.

Skulls of all the members of Anguilliformes have many peculiarities. Several bones of the ethmoid region lose their individual identity and become fused; hence the praemaxillaries, vomer, and one or several ethmoid bones form a single dentigerous bone known as the "praemaxillo-ethmo-vomer."

The gill arches and the pectoral girdle are displaced posteriorly and have lost the attachment to the cranium, due to the absence of the first pharyngobranchial arch and the post-temporal bones. The bones of the opercular cover are reduced in size and modified. The gill openings are small, usually located on the sides of the head, and in some, on the ventral surface. The smallness of the gill openings enables the fish to retain sufficient moisture in their gill cavity to survive a considerable period out of water. The body is elongated, cylindrical, or in some compressed, and flexible, thus allowing the fish to move in a serpentine fashion. Representatives of three families (*Anguillidae, Simenchelyidae*, and *Synaphobranchidae*) have small cycloid scales while others are naked. The number of vertebrae (or myomeres in larvae) is very high, varying in different species from 100–260. The smallest number of vertebrae, 68–78, is found only in

Cyema atrum Günther of the family *Cyemidae*. The highest number of vertebrae, 400–700, characterizes the family *Nemichthyidae*.

Fins of eels are supported only by soft rays. The dorsal and anal fins are usually very long and confluent with the caudal fin. Only snake eels (*Ophichthyidae*) lack the caudal fin and the tip of their tail is a hard sharp point.

Somber gray or green hues without distinct spots prevail among eels. However, many morays living on coral reefs sport round spots or vertical bands of green, black, yellow, and other colors giving them a striking and beautiful appearance. Some of them are mimics of the sea snakes (*Hydrophiidae*).

All eels breed in the sea in areas with warm and highly saline water, and undertake extensive migration for this purpose. Maturing individuals display several changes in appearance, particularly an increase in the eye diameter. They produce pelagic eggs, whose size varies from 5.5 mm in *Muraena helena* Linnaeus to 1.4 mm in *Anguilla anguilla*. They are usually prolific fish. The number of eggs laid by a European *Conger conger* (*Linnaeus*) was estimated to be from 3,000,000–7,000,000 per female, while for the Atlantic eels (*Anguilla*) the number was 10,000,000 per female.

All eels pass through the larval stage known as *leptocephalus*. These larvae are completely transparent and are provided with very long, sharp larval teeth. The shape, pattern of pigmentation, number of myomeres (corresponding to the number of vertebrae in adults), position of anus, and size help to distinguish and identify them with their respective adult fish. The number of eel families which belong to the order Anguilliformes is estimated to about 25. The number of genera is about 220 and the number of species totals several hundreds. Eels are predominantly warm-water fish since they are found in all tropical seas. Although the majority of eels are saltwater dwellers, some are found in brackish and fresh waters.

Eels vary greatly in size. The smallest is probably *Cyema atrum*, the adults of which may attain a length of only 126 mm. The largest individuals are found among morays (*Muraenidae*) and conger eels (*Congridae*). Capture in Australian waters of probably the largest eel in the world, the long-tailed moray *Evenchelys macrurus* (Bleeker) measuring 12 ft 11 in. has been reported. It is preserved in the Queensland Museum. This dangerous eel is found in other parts of the tropical Pacific, near Taiwan, and in the Indian Ocean and Red Sea. Among Congridae, the largest species is *Conger conger* (Linnaeus) living in European waters. The capture off the English coast in 1904 of a specimen 9 ft in length and weighing 160 lb was reported in 1925. Smith states that "large congers are much dreaded as they are difficult to kill and most vicious, and many a fisherman has been mutilated." Such large eels and the occasional captures of leptocephali of considerable length give rise to the belief in the existence in tropical seas of "giant sea serpents." All eels are carnivorous, feeding indiscriminately on different groups of invertebrates and fishes. By their habits, eel families can be arranged in the following groups.

Deep-Water Eels

About 10 families of cosmopolitan eels can be included in this category. Although not one species is of economic importance, the representatives of the following three families, consisting of a single species, are most interesting. *Cyema atrum* (*Cyemidae*) is characterized by unusually deep-bodied larvae, few vertebrae, and small-sized adults with forked tails, a unique feature among Anguilliformes. The snipe eel *Nemichthys scolopaceus* Richardson (*Nemichthyidae*), has an extremely slender body (the fish may be 75 times as long as deep) with its tail tapering to a thread. Its jaws, as in *Cyema* are bill-like. The snipe eel has the highest number of vertebrae among all the eels. It reaches a maximum size of 1500 mm. The snub-nosed eel, *Simenchelys parasiticus* Gill (*Simenchelyidae*), is parasitic in its habits and reaches a length of about 2 ft. Like the hagfish (*Myxine*), its skin is very rich in slime glands, and it uses its sharp-cutting teeth to chew away the body wall of various large fish, penetrating into the body cavity and feeding on the viscera.

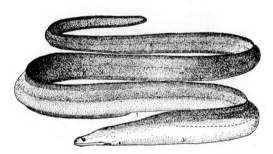

Fig. 1. Long-tailed moray, *Evenchelys macrurus* Bleeker, the longest eel species. (After Chen and Weng, 1967.)

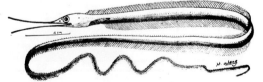

Fig. 2. Snipe eel, *Nemichthys scolopaceus* Richardson, a type of deep-water eel with a very high number of vertebrae. (After Cheng and Weng, 1967.)

Sand-Dwelling Eels

Eels which normally do not swim freely by day are chiefly sand-dwellers. They favor muddy areas and are found in estuaries and fresh water. None of these eels are of economic importance.

There are a few little-known species of so-called garden eels (*Heterocongridae*) found in the tropical Pacific and Atlantic. These small eels live in colonies, but each is solitary, anchored almost all of its lifetime to a tube bored in the sand from which they stretch to catch food or into which they withdraw from predators.

In contrast to rare garden eels are the spotted or striped snake eels, *Ophichthyidae,* whose 200 species abound in tropical waters around the world. Their hard, pointed tails, without the caudal fin rays, help them burrow tail-first in the sand or mud. Most are small, but some species grow to 5 ft. They have no economic importance.

Three of the important eel families can be included in this category. The moray eels (*Muraenidae*) are represented by about 30 genera with numerous species. They can easily be distinguished by smooth, scaleless skin, absence of pectoral fins and, in the majority, lateral line pores on the body, narrow gill openings, and rather small eyes. They have large mouths with sharp, powerful canine teeth, mostly raptorial.

Lurking in crevices among coral rocks, they prey on almost any marine animal. They are among the most dangerous of marine creatures: aggressive, fierce, and powerful. Morays must be treated with respect; there are known cases of them attacking divers. It is believed that the bite of the moray is venomous, but there is no sign of any venom apparatus. The unpleasant effects are due mainly to virulent infection causing sepsis rather than to any specific venom.

Although moray eels have rather tasty flesh, they are not generally sought for food. In Japan, the thick skin of the moray *Gymnothorax kidako* is tanned for leather. In the Indian Ocean, several morays of the genus *Lycodontis* McClelland attain a size of 5–6 ft and are considered very dangerous. In the western Atlantic, the largest species is the green moray, *Gymnothorax funebris* Ranzani, its name derived from the color of its slime. In 1952, a specimen measuring 5 ft 8 in. and weighing 33 lb was taken in Nova Scotia.

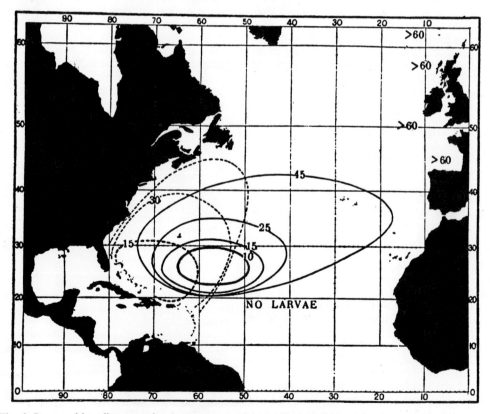

Fig. 3. Presumed breeding areas for the European eel (*Anguilla anguilla*) and the American eel (*A. rostrata*). Solid curves refer to the European eel, and dotted ones to the American eel. The curves indicate the limit of occurrence of leptocephali of particular length; those less than 10 mm have been found only inside the 10 mm curve, those less than 15 mm, only inside the 15 mm curve, etc. (After Schmidt, 1922.)

The European species, *Muraena helena*, was well known to and much esteemed by the ancient Romans, and there are numerous references to it in the classics. There is a fine picture of it in a Pompeian mosaic now in the Naples Museum.

The members of two other eel families, *Congridae* and *Muraenesocidae*, also have scaleless skin, but are easily distinguished from Muraenidae by their uniform color, large pectoral fins, distinct lateral line, large gill openings, and rather large eyes. Whereas Congridae lack canine teeth, Muraenesocidae have them and also are provided with much more attenuated snouts. There are about 8 genera among Muraenesocidae compared to 45 in Congridae. Members of both families are valued as food. Although some genera, such as *Pseudoxenomystax* Breder, are found in deep water, the conger eels are essentially bottom dwellers in shallow inshore waters. According to Okada, in Japan, where a very great variety of congers are found, *Astroconger myriaster* (Brevoort) is most highly esteemed for culinary purposes.

Among pike eels, Muraenesocidae, the largest is *Muraenesox arabicus* Bloch and Schneider, silvery gray in color and attaining a length of at least 5 ft. It is valued as food, but is difficult to handle. It is found in the sea and also in brackish and freshwater streams. Its range is from the Red Sea to India, China, Japan, and Australia.

Fresh-Water Eels

The family of *Anguillidae* (fresh-water eels) consists of a single genus *Anguilla* Shaw with 16 species. Although spending the greater part of their lives in fresh water, all *Anguilla* species descend to the sea to breed, the salt water being necessary for the stimulating of the generative organs and the ripening of the resultant ova and spermatozoa. Having remained in fresh water or along the seashore, feeding and growing, the eels attain maturity at different ages for males and females. They then commence the long journey down to the sea for reproduction. The adults never return to fresh water after spawning.

The life-history of all *Anguilla* species is practically the same. Hence the details given for the best studied, the European eel, *A. anguilla*, can apply with some restrictions to other species. The scientific world owes a debt of gratitude to the late Johannes Schmidt of Copenhagen for his valuable discoveries and contributions on the breeding habits of the freshwater eels of the Atlantic and Pacific Oceans.

Economic Importance

Among eels of different families, the Anguillidae are the most valuable commercially. They are caught by different methods (baited hooks, eel pots, eel weirs, etc.) and marketed in great numbers, their sale realizing substantial revenue. The white, rich, succulent flesh is in great demand all over the world with the exception of North America. Fat mature eels (silver or bronze) make the best smoked product, a choice morsel for the connoisseur. The half-grown (yellow) eels are prepared in different ways, especially jellied. Glass eels and elvers of earlier runs, still rich in "baby fat," are caught in large quantities in fresh-water streams in southern Europe and in certain parts of Great Britain. They are fried in oil.

In recent years annual catches of the American eel in Canada amounted to about 1,500,000 lb and in the U.S., a little less than 1,000,000 lb. It is evident that the North American eel fishery has not been developed to its full potential. The eel fishery in Europe nets 16 times the number taken in North America at the present time. In 1963 66,000,000 lb of eels were taken around the world (FAO statistics).

VADIM D. VLADYKOV

Cross reference: *Sea Lamprey*.

ELECTRONICS IN DEVELOPING MARINE RESOURCES

The role of electronics in developing marine resources is substantial and increases daily. In the short span of 30 years, electronics have revolutionized the fishing industry and provided the means for exploiting the principal natural harvest of the sea to an extent undreamed of earlier this century. Its applications to the study of the sea bed, tides, currents, and temperatures, are bringing a new and much needed knowledge in the broad field of oceanography. Electronics plays a leading role in finding rich oil and gas deposits beneath the sea bed and, less spectacularly, in improving the efficiency of dredgers in gathering sea-bottom materials such as sand or gravel for the service of man.

Because of the immediate and fundamental importance of the provision of food (plus a number of important byproducts) for man, this survey deals largely with electronics in fishing; but many aspects of electronics, particularly navigation using electronic aids and radio communications, are basic to most other activities at sea.

Electronics in fishing is directly concerned with maximizing the yield of fish in terms of quality, quantity, and productivity (i.e., the yields rated against the cost of maintaining the ships and crew at sea), with the safe navigation of the ship and the safety of its crew, and protection of the fishing gear against damage.

The importance of electronics in fishing is strikingly illustrated by the amount of money spent on equipping the latest vessels. Distant-water freezer trawlers based in the U.K. are currently being fitted with an electronics installation costing over $57,000 (Fig. 1). The radio equipment accounts for nearly $11,000, navigation equipment $23,000, fishing equipment $15,400 and intercom and other smaller items nearly $8000. A modern passenger liner has electronic equipment costing about $80,000—only 30% more on a ship probably costing 25 times more to build. Costs quoted are typical at 1965 price levels and are based on actual, not hypothetical installations. Almost certainly both types of ship would also be fitted with Decca Navigator at an annual rental of from $1260–$1870 depending on facilities required.

The provision of such expensive equipment has been fully justified. The Silver Cod Championship Trophy awarded annually in Britain for side trawlers was won, in the last year for which figures are available, by the 789 ton *Somerset Maugham*. Her aggregate catch was 2462 tons for 343 days at sea. Both figures are significant. The electronic aids helped both in finding the fish and keeping the trawler at sea fishing in all weathers, day and night, in winds up to force 9.

Good electronic navigation equipment is now regarded as a primary requirement in fishing. Duplicated radar installations (Fig. 2) are becoming more usual on larger craft.

In addition to one or two radars the navigational aids will generally include Loran and Decca. The radar is frequently used not only in the classical anti-collision role but also for determining position in relation to the shore when fishing close to the national limits of foreign governments. Loran provides good fixes of position at long range and Decca gives sufficient accuracy to tow back over areas where fish have been located. It is also an effective aid in navigating clear of underwater obstacles which have been recorded from past experience. Conversely, the high accuracy of the Decca system enables seine netters to fish with confidence within yards of obstructions which frequently form an attraction for large shoals of fish.

In general the navigational aids assist the safety and efficiency of the vessel and crew but are of special value in certain practical aspects of the fishing operation itself.

Fish Detection

Experience acquired over hundreds of years of fishing has defined regions where a good harvest may be anticipated. Electronic fish detection has greatly reduced the element of chance. Today's fisherman still operates in traditionally good fishing areas but, with electronic aids, he is now an efficient hunter of fish rather than a gambler shooting his trawl and relying largely on intuition to provide a good yield. Electronics gives him facts upon which he can act.

All fish detectors are based on echo-sounding techniques in which a pulse of energy is trans-

Fig. 1. A modern trawler wheelhouse equipped with modern fish-finding and other electronic aids.

ELECTRONICS IN DEVELOPING MARINE RESOURCES

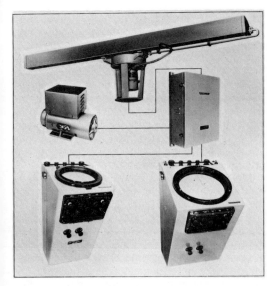

Fig. 2. A modern radar for fishing vessels. Extensive use of radar enables fishing to continue in bad visibility and provides fixes from the shore when fishing near national limits.

This introduces questions of how broad the ultrasonic beam should be if all echoes are to be satisfactorily received back at the transducer. This point will be taken up later. The third is the variability of the sea as a transmission medium. Such factors as temperature and salinity have an effect on transmission speed, and accuracy can also be affected by aeration of the water due to the ship's own movement. Fourth, there is the problem of reliability. The environment for electronic gear on a hard-worked trawler has never been favorable. The development of solid-state electronics has improved reliability considerably in the past few years.

The basic elements of an echo sounder are the transmitter, the transmitting transducer, the receiving transducer, the receiver, and the indicator. The transducers, which convert electrical energy into ultrasonic sound and vice-versa, are either inserted in the hull in contact with the water or fitted outside the hull in limpets (Fig. 3). The transmitter and receiver are electronic units concerned with generating power and amplifying the faint return echoes to usable proportions. The indicator is normally an electro-mechanical recorder but cathode-ray tube displays are becoming increasingly popular as supplementary indicators with special features.

mitted through the water and received back from any reflecting object, the length of time between the original pulse and the received echo being proportional to the distance from the target. The principle is akin to radar, the main differences being the use of ultrasonic (typically 30–50 kHz) instead of radio frequencies and the slow rate of transmission (typically 4800 ft/sec) of the signal in a water medium compared with that of radar pulses in free space.

Electronic echo sounders were developed in the 1930's as a navigational aid. By 1934 at least one observer had detected fish using the equipment and a paper by Sund entitled "Echo sounding in fishery research" published in the British scientific journal "Nature" in 1935 included a recording of cod made on board a Norwegian research vessel at Lofoten. Since then there has been intensive development of the technique in a fish-finding role as well as great advances in underwater ranging in general.

Before discussing the more recent developments, it is important to understand the natural difficulties and limitations of fish detection by electronic instruments. The first major difficulty is that a fish is a very poor reflector of ultrasonic sound compared with the sea bed. Typical echoes from the sea bed are received back at up to 1000 times the intensity of an echo from a fish, bringing problems of discrimination. The second is that a fishing vessel is generally small and even in only moderate sea states is liable to have substantial roll and pitch.

The technology of electronic fish detection has advanced to the stage where it is possible to detect a single fish only 1 ft from the bottom down to 300 fathoms. The layman's concept of great shoals of fish being scooped up by a trawl is, unhappily, all too rare. An average catching rate, for example, in Arctic fishing is one fish for every 30 to 50 yards traversed by the trawl. The fish detector is used not only in the initial hunting role to discover fish but also to indicate to the skipper the rate at which his trawl is filling. For the last-named use it is equally important to know that the catch is proceeding at a steady but slow rate as at a fast rate or, for that matter, that nothing is being caught at all.

The diversity of fishing methods has provided a great challenge to the equipment designer. One of the finest commercial dual-purpose instruments is the CERES (Combined Echo Ranging Echo Sounding) Fish Finder specially developed for the smaller general-purpose vessel. It has all the features required for both horizontal and vertical detection of fish and can detect worthwhile shoals at ranges of almost 1000 yards.

Skillful design of the transducers has reduced the amplitudes of sidelobes, above and below the main beams, from a typical 20% to less than 5%. This advance alone gives greatly improved target-to-background ratio by substantially reducing unwanted echoes from both the sea bed and surface when transmitting horizontally.

First echoes of a distant shoal are heard on a

(a)

(b)

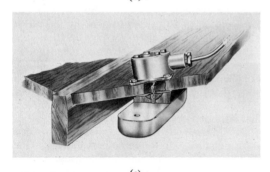

(c)

Fig. 3. Transducers. Typical limpet installations: a, wooden vessels; b, steel vessels; c, small craft.

loudspeaker and an indication of range is provided by a neon lamp, which flashes with each echo, on the pen arm of the recorder. After the first detection and indication of range it is a simple matter to adjust the recorder range to bring the echoes on to the recording paper and build up the trace patterns in the usual way.

CERES has proved to be among the most popular of fish-finders because of its great versatility and reasonable cost.

A more powerful equipment is the Fisherman's Asdic (Fig. 4) developed mainly for midwater fishing (i.e., horizontal or forward search) but also including echo sounding and bottom-fish detection. The beam can be used for automatic horizontal search over any arc between 15° and 165° in 5° steps, and range and depth are, respectively, 0–2000 yards and 0–100 fathoms.

Most recent work, however, has been concentrated on improving detection capability for the all-important bottom fishing of demersal species (cod, haddock, and all flat fish) which live and feed either on the bottom or very near it. Because demersal species are commonly caught at great depth the equipment needs to be powerful in range (up to, say, 500 fathoms) and highly discriminating in that the faint echoes from fish within a few feet of the bottom are not lost among the powerful returns from the sea bed itself, or among stong echoes from mid-water targets in which the bottom fisher is not interested.

Among the most highly developed equipment for bottom-trawling is the Humber Fish Detection System. A full Humber system includes three displays for the skipper and two important technical developments, the Sea-bed Lock and the White Line System (Kelvin Hughes patented) which, between them, provide the most advanced equipment commercially available at the time of writing. It is capable of detecting a codling of 12 in. length at depths of 250 fathoms in any weather which allows fishing operations.

A high-power (8kW peak) transmitter is used with either one of two transducers. One transducer is located in a limpet attached to the hull and is used for general search. The second is mounted on an extensible and retractable arm which can be lowered below the hull and is found more effective when the trawl is being towed as it is below the layer of aerated water. The transmitter and receiver can be switched to either transducer at will.

The indicators comprise a normal-depth-presentation recorder, a sea-bed-locked expanded cathode-ray-tube presentation, and a sea-bed-locked expanded-presentation recorder. The last two items show in great detail only the 4-fathom range immediately above the sea bed (i.e., the operational depth in which the catch lies). If required the CRT display can also be used to look at any 5 or 10 fathom layer of water between the surface and sea bed.

The sea-bed lock is a significant advance over previous systems. It will be appreciated that the inspection of a layer of water only a few fathoms

ELECTRONICS IN DEVELOPING MARINE RESOURCES

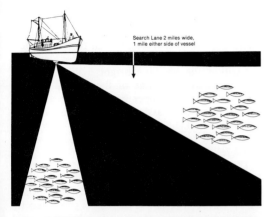

Fig. 4. Diagrammatic representation of the operation of an Asdic type search sonar commonly used for both mid-water and bottom fishing.

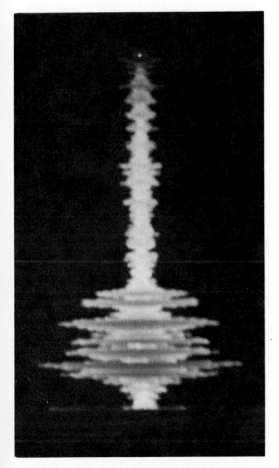

Fig. 5. A typical expanded CRT display. Picture shows echoes from a small fish shoal and the sea bed at a depth of 40 fathoms.

deep at the sea bed would need constant adjustment of the equipment to keep the layer in view if any combination of vertical movement of the trawler and irregularities of the sea bed were greater than the depth layer being observed. The displayed range is, therefore, locked electronically to the sea-bed echo by storing these echoes on a magnetic recording drum and using these to trigger the displays from a uniform point.

The advantage of the CRT display (Fig. 5) is that any part of the range may be readily expanded, electronically, to fill the screen. Its disadvantages are that it needs to be continually observed during trawling and mental notes made of the fish echoes detected, but this is overcome by the expanded presentation recorder which operates in parallel with the CRT and provides a record of all echoes against a time scale.

The White Line System (Fig. 6) electronically introduces a white line on the recorded trace giving improved discrimination of fish echoes from sea-bed echoes even if the fish are hard on the bottom.

The Humber Fish Detection System was tested experimentally at sea before being put in production and is now virtually standard for British "stern freezer" trawlers. It is found to give a greatly increased and far more reliable indication in both the search and actual trawling phases of the quantity of catchable fish and the time for which the trawl should remain down. Experience showed that larger catches occur when all fish echoes are within one fathom of the sea bed.

The latest development is the Steered Narrow Beam Echo Sounder which is the subject of research by the British White Fish Authority, the technical development of the system being undertaken by Kelvin Hughes. Three systems are being operationally evaluated at sea at the time of writing.

Basically the Steered Narrow Beam System is a standard Humber equipment but instead of an athwartship beam of some 17° (at half-power points) the beam is narrowed to about 3°. The narrow beam is also gyrostabilized so that even should the trawler roll by as much as 15° from the vertical the beam will remain with its axis vertical within 1°.

The advantages of the Steered Narrow Beam are dramatically illustrated by the diagrams (Figs. 7, 8, 9). In Fig. 7, a bottom-fishing trawler fitted with normal wide-beam equipment adjusted for a trawl headline height of a few feet above the sea bed, is rolling heavily. Three fish are detected which could appear on the display as if they were about to be caught when, in fact, they are outside the headline height of the trawl.

In Fig. 8 the beam is concentrated but there are still strong possibilities that fish indicated are

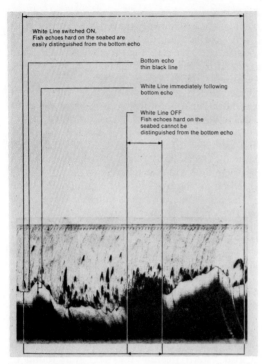

Fig. 6. The White Line system. Diagram shows the marked advantage of the white line presentation.

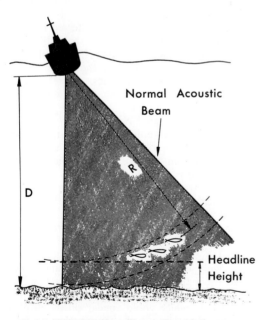

Fig. 7. With the ship rolling and a wide-beam sonar adjusted to the range R, the fish would be detected even though they are above the headline height of the trawl (see Figs. 8 and 9).

not entering the trawl and that echoes from the sea bed along the line D could mask any genuine fish echoes. If, however, the beam is maintained vertically, irrespective of any roll of the trawler, as in Fig. 9, then all fish indicated, provided they are within the headline height, will be netted. The principal advantage of the steered narrow beam is that the skipper has far more accurate data from which to assess his catch, thus avoiding the pitfalls of hauling in too soon, giving a part-wasted tow; or hauling in too late which could result in a torn trawl belly.

The special transducer used in Steered Narrow Beam is divided into 18 segments. The firing pulse to each segment is delayed by an amount which varies in response to signals generated by a vertical-keeping gyro, the effect being to produce a sonar beam which remains vertical though the trawler may be rolling badly. The form of electronic stabilization employed is considered superior to using a conventional transducer stabilized by electro-hydraulic or electro-mechanical means.

Research is also being conducted on towed-body methods in which stabilization is achieved by fitting the transducers in a towed body of a hydrodynamic shape (Fig. 10) which keeps below the surface and is comparatively stable. The method has some advantage as the transducers are separated from the hull with a consequent reduction in hull-borne noise and freedom from aeration effects.

Longer-term projects being studied include horizontal searching for demersal species of fish but the technical difficulties are considerable. Searching ahead involves a slanted beam and any fish detected would always be at the same range as part of the sea bed, leading to problems of discrimination between wanted and unwanted echoes. Very narrow beams and some really sophisticated electronics would be required. If the technical problems could be solved, the practical advantages of a forward search system for demersal species would be great, as the skipper could steer his ship and trawl over the more profitable concentrations of fish. He would have the same advantages as with Asdic in searching for herring and whales.

Auxiliary Aids to Fishing

A comparatively simple device, the Trawl Warp Tensionmeter, has come into general use on stern trawlers. Installation is easy since all that is required is to reroute the towing warp from the winch through a pulley instead of directly to the rear gantry. The warp makes an angle of 168° over the pulley and the pulley reaction is measured by an electrical resistance strain gauge (load cell) and transmitted to the wheelhouse where an amplifier and direct-reading scale give an indication of warp load in tons. Accuracy is within 3%. The value of the Trawl Warp Tensionmeter is in warning of the trawl becoming fast, in which case the skipper may be able to ease back before severe damage results. It is most important that at all times the warp tension is known.

An interesting new development under investigation is a telemetry system from the trawl which transmits to the trawler such information as headline height of the trawl, mouth spread of the trawl, water temperature, and the strain on a main structural member of the trawl. Other data such as salinity or light intensity at the sea bed could also be transmitted.

Because of the commercial impracticability of having cable connections for over a mile between the ship and trawl, a small battery-operated underwater communications link is used operating at 40 kHz and with the information transmitted as pulses which vary in repetition frequency over a range of 10–40 Hz in accordance with the data from the sensors. The pulses are received at

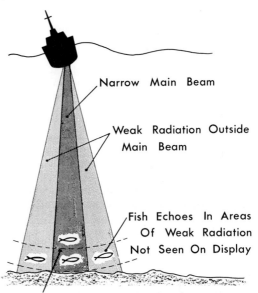

Fish Echoes In Area Insonified By Main Beam Are Displayed In Correct Positions Relative To The Sea Bed

Fig. 9. The Steered Narrow Beam system. The sonar beam is now stabilized and gives an accurate indication of fish within the headline height of the trawl and in their correct positions relative to the sea bed (compare with Figs. 7 and 8).

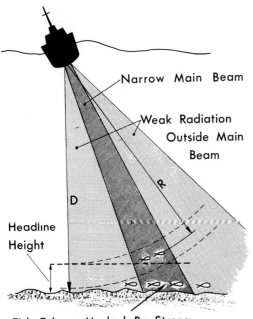

Fish Echoes Masked By Strong Bottom Echo At Range 'D'

Fig. 8. The sonar beam has now been narrowed but echoes from fish near the bottom could be masked by reflections of echoes from the sea bed via the shorter route D. These, although outside the main beam, are still probably stronger than the fish echoes. It will also be seen that fish can again be detected above the headline height of the trawl, giving misleading indications (see Figs. 7 and 9).

Fig. 10. Transducers. Typical fibrous glass, towed-body transducer of the self-stabilizing type.

the ship and converted into dial readings of the quantities being measured.

Whether such telemetry systems will come into general use will, of course, depend on the results of the trials. Although unquestionably valuable in fisheries research, telemetry systems may not prove sufficiently beneficial in day-to-day fishing to warrent their expense.

The Future

Without question the application of electronics in solving all manner of marine problems will continue at an accelerated rate. As the technology advances, so each step forward becomes more costly. But there are heartening indications in the new awareness among nations that the oceans are still largely unknown and, therefore, untapped, and that their exploration and exploitation are in the common interest.

There is more scientific research being undertaken now and more money invested in it than at any other time in history. Meantime, established techniques are being constantly improved technically, and operationally substantial advances have been made which are proving of immense practical value not only in further exploitation of marine resources but also in our understanding of them.

R. G. HAINES

ESTUARINE ENVIRONMENT

An estuary is the meeting place of sea and river water and may be defined as the area in which seawater and fresh water have mutual influence. The movement, level, salinity, and temperature of the water are four of the more obvious parameters, as they all influence life within the estuary. Equally important are the materials forming the bed of the estuary and in suspension. Pollution of an estuary by discharges of sewage, sewage effluents, and industrial waste waters can cause fundamental changes in the environment. The interplay of the factors mentioned may vary within wide limits; thus each estuary has its own characteristics which are not exactly repeated elsewhere.

Where the difference between the level of the sea at high and low water—the tidal amplitude—is large, say over 25 ft, there is usually good mixing of salt and fresh water within the estuary and only a slight vertical salinity gradient. During the run of the tide this may amount to a difference of perhaps 0.1–0.2 g of salt per 1000 g of estuary water between the salinity at the surface and that near the bottom. At low water the tide may start to flood possibly 20 min earlier on the bed than on the surface, and differences of about 2 parts per thousand between bed and surface salinities may occur.

If the tidal amplitude is moderate there may be an appreciable vertical salinity gradient at all states of the tide. Where there is a very small variation in the level between high and low water, an extensive salt wedge may develop. An extreme example of this is in the Mississippi[1] where seawater penetrates more than 120 miles inland in the lower layers. The surface layers of water are fresh and move continuously seaward. Sandwiched between the fresh and the salt water is a layer in which some mixing takes place, so that a proportion of the inflowing seawater is transferred to the lower part of the outflowing fresh water.

The distribution of salinity is also influenced by the shape of the estuary. Thus, in an estuary having sharp bends, pockets of water of relatively high salinity may remain in eddies at the seaward side of sharp points during the ebb tide. On the turn of the tide this water is swept into the main stream. On the inland side of the point, an eddy containing water of low salinity may develop during the flood tide. This is also swept into the main stream after high water. Thus there may be slight discontinuities in the relation of salinity to distance along the estuary. At the seaward end of an estuary there are flora and fauna conditioned to a seawater habitat, and at the other end are fresh-water plants and fish. In between there is life which is conditioned to living in water which is neither fresh nor fully saline or which may adapt itself to living in either.

The smelt (*Osmerus Eperlanus*) is a euryhaline fish which seems to move endlessly to and fro in some estuaries so that it is always in water having a salinity of about 20–25 parts per thousand. Migratory fish such as the salmon pass through the estuary once as smolts on their way to the sea and again as mature salmon on their way back to the fresh-water spawning grounds. Different problems of adaptation are posted by different patterns of salinity distribution.

Other types of fish may be able to adjust themselves to varying salinity to a limited extent so that some sea fish will penetrate into the seaward end of the estuary and some fresh-water fish to the water containing a proportion of seawater.

Shell fish, which for much of their lives remain in one place, are subjected to a limited variation in salinity. In parts of the estuary near to the open sea the variation in salinity during a single tide is not as great as it is higher up the estuary. Where there are oyster beds, the salinity rarely drops below about 80% of that of full seawater.

In parts of the estuary where the proportions of salt and fresh water are about equal, the salinity may vary by about 8 or 9 parts per thousand during a single tide. The variation is greater during spring tides than during neap tides. Over longer periods there are much greater changes of salinity. Toward the end of a long, dry summer the penetration of seawater will have reached its maximum. But long periods of heavy rainfall in winter and early spring in the catchment areas of the rivers discharging fresh water to the estuary will push back the seawater, so that places where the salinity may be about half that of seawater in the early autumn may be wholly fresh in the spring.

In the absence of discharges of heated water the temperature of the water in an estuary will be intermediate between that of the sea and that of the rivers. In the summer river water is several degrees warmer than the sea and during the winter it is colder.

Power stations are often built on the banks of estuaries and use estuary water in large volumes for cooling purposes. For a given consumption of fuel at a power station the amount of heat transferred to the cooling water will be some inverse function of the efficiency of the station. The rise in the temperature of the cooling water in the power station will be directly proportional to the fuel consumption and almost inversely proportional to the discharge of cooling water. As a result of the discharge of heated effluents there may be some parts of an estuary where the temperature is over 5°C higher than that of the sea or of the river.

Temperature is distributed in an estuary in the same way as any dissolved constituent, except that there are some losses and gains due to conduction, convection, and radiation. A heated discharge is distributed into water of higher salinity, usually to seaward, in the same way that fresh water is distributed seaward, and into water of lower salinity (usually in the inland direction) in the same way that salt water is distributed in the same direction.

In discussing salinity the movement of water has been mentioned. The movement of particles in suspension and changes in the bed of the estuary are also important. The distribution of dissolved substances and temperature in an estuary are governed by the salinity distribution. In a different way the movement and deposition of silt is influenced by salinity, but the movement of silt is not directly related to the salt concentration of the water.

In an estuary where there is only a small vertical salinity gradient the results of a large number of velocity measurements at a series of depths have indicated that near the surface there is a net movement of water in the seaward direction[2] at all parts of an estuary, except perhaps near sharp bends.

The size of the net movement decreases with depth and in the seaward half a depth is reached at which for flood and ebb tides of equal amplitude there is no net movement of water. In this part of the estuary the water near the bed has a net movement toward the head of the estuary. There is one position where, at the bed, there is no net movement of water either seaward or landward, and on either side of which the net movement of water near the bottom is toward this point. This is a position at which the formation of deposits is to be expected.

The movement of silt in suspension is complicated by several factors which include tidal amplitude, the distribution with depth of velocity during flood and ebb tides, the distribution of the concentration of suspended silt with depth during flood and ebb tides, and the history of the deposits. Thus, in a well-mixed estuary the distribution of velocity with depth during the ebb tide is similar to that in a unidirectional river, the greatest velocity occurring at or near the surface and the least near the bottom. The concentration of suspended solids increases fairly uniformly with depth. During the flood tide, the maximum velocity may be found at about mid depth. With this kind of velocity distribution most of the silt in suspension during the flood tide is in the lower layers. To determine the movement of silt at any point in a cross section, the concentration of silt must be multiplied by the corresponding velocity and the product integrated over the whole of each half tide. The difference between the values for the flood and ebb tides gives the net movement of silt per unit area of cross section at the chosen point. If the same exercise is carried out at several points in the depth at the same vertical, integration of the product of velocity and concentration over the depth gives the net transport of solids per unit width of cross section. Finally, the results from many verticals may be integrated across the width of the estuary to give the net transport of silt at the chosen cross section.

Since the concentration of silt and the velocity near the surface are greater during the ebb than during the flood tide, there is a net movement of silt seawards in the upper layers. Near the bed the net movement of silt may be toward the head of the estuary.[3]

The position of the cross section where there is no net movement of silt changes with alterations in the flow of fresh water into the estuary. It is to be found further to seaward in the winter and early spring than it is in the summer. In fact, if the rivers feeding the estuary are in high flood for several days in succession, deposits in the middle reaches of the estuary may be scoured away and carried downstream.

Oyster farming and silt deposition do not go well together. Oysters beds are usually situated near the mouths of clean estuaries or, if the estuary is one in which appreciable silt is transported, an area is chosen where there is little or no silting. Adult oysters can live where there is a small amount of silt but breeding is inhibited by the presence of silt. A change in the regime in an estuary resulting from the erection of a barrage or other obstruction may alter the salinity distribution and the positions where silt is deposited.

These changes may make it necessary to change the position of oyster farms.

What are the sources of oxygen dissolved in water in an estuary?[4] Some enters in seawater, some in fresh-water streams and effluents which, however, have an oxygen demand, some in rain, some by photosynthesis when sunlight acts on algae near the surface particularly in the spring, and some by the interchange of oxygen between water and air at the water surface. There is also a store of oxygen in compounds such as nitrates and sulfates and this combined oxygen under particular circumstances can be made available.

When clean water is saturated with dissolved oxygen at a given temperature and atmospheric pressure, equal amounts of oxygen are entering and leaving the water surface. If the water is polluted and is no longer saturated with dissolved oxygen, oxygen will be entering the water at a greater rate than that at which it is leaving the water. The net rate at which oxygen will be transferred from the atmosphere to unit surface area of the water will be proportional to the oxygen deficiency in the water, the depth of the water, the partial pressure of oxygen in the atmosphere and will depend on the temperature. For a given surface area in plan the actual surface area of water will depend on the roughness of the surface, which in turn is a function of wind strength and fetch or other surface-disturbing factors.

Now, the degree of pollution suffered by an estuary may be judged by the concentration of dissolved oxygen in the water. If the rate at which pollutants are removing dissolved oxygen from the water is greater than that at which fresh oxygen is being supplied, the concentration of dissolved oxygen will decrease.

This has the effect of increasing the rate at which oxygen enters the water through its surface and it is possible that at a given decreased level of oxygen concentration equilibrium will be attained. With higher charges of pollution or with lower flows of fresh water the water may become completely deoxygenated over considerable lengths of the estuary. Oxygen will be passing through the surface of the water at a maximum rate but cannot be detected in the body of the water. Just as the last trace of dissolved oxygen is disappearing the reserves of combined oxygen in the form of nitrates are attacked by denitrifying bacteria and for the most part nitrogen gas is produced. There is some evidence that a small proportion of the nitrate is reduced to ammonia. Where there is no dissolved oxygen in an estuary there is no nitrate except at the fringes of the deoxygenated area. When all the dissolved oxygen has gone the concentration of dissolved oxygen can no longer be used as a criterion of the condition of the water. We can either consider the notion of a negative concentration of dissolved oxygen, i.e., the concentration of oxygen necessary to bring the water to the threshold of containing a trace of oxygen, or can consider another parameter such as the redox potential. A high positive redox potential is indicative of conditions suitable for oxidation reactions to take place, whereas a low positive or a negative potential indicates conditions suitable for reduction reactions.

Measurements of redox potential may be made not only in the water but in the deposits on the bottom. When silt falls through water containing a proportion of seawater, the deposit holds interstitial water containing sea salts. The silt, particularly in a polluted estuary, contains organic substances which rapidly use up the oxygen dissolved in the occluded water. The redox potential falls and reaches a value at which sulfate-reducing bacteria start to operate. The sulfate in the water is reduced to sulfide and a reaction between the sulfide and iron sesquioxides in the deposit yields iron sulfide, which is a black substance. This is the cause of the black coloration of highly polluted estuaries. When this reaction is complete any further sulfide produced dissolves in the water as a solution of hydrogen sulfide and causes an objectionable smell. Simultaneously, carbon compounds in the deposit are reduced anaerobically to methane. These reactions are in progress whatever the state of the tide but it is only toward low water when the hydrostatic pressure on the deposits is reaching its lowest point that the minute bubbles of gas can expand, coalesce, and escape from the mud.[5] In parts of some estuaries at the time of spring tides as much as 50 liters of gas can be collected in 24 hours by a hood covering 1 m^2 of surface.

If the water of an estuary contains some dissolved oxygen, then the surface of the deposits will be sandwiched between the main body of anaerobic mud and the aerobic water. The surface of the deposit is sand-colored to a depth of about $\frac{1}{4}$ in. and below that it is black. The division is clear-cut. The light and dark parts of the deposit contain particles having the same size distribution and the samples have the same chemical composition except for the presence of sulfide in the black part of the deposit.

The gravity of the pollution in an estuary depends on the pollution load, the temperature, the size and shape of the estuary, the discharge of fresh-water streams into the estuary, and the position of the effluent outfall.

If an outfall is at a position such that part of the effluent can escape from the estuary during a single ebb tide, then the polluting effects of the given effluent will be less than those caused by discharging the same effluent at a position nearer to the head of the estuary. In some estuaries it can be shown that no effluent can attain the open sea for several weeks after its discharge. During the intervening period the oxidation of the sewage effluent is completed. This is indicated by the quality of the water and the dissolved oxygen content at the seaward limit.

Under reasonably steady conditions of freshwater flow, effluent discharge, and composition, and in the absence of gales, a condition approaching that of an equilibrium is set up in an estuary. Thus, if for the moment we disregard any decomposition and consider an effluent discharged at a position where at half tide the salinity is 10 parts per thousand, then, as far as dissolved constituents are concerned, the concentration of a constituent will vary toward the head of the estuary in a way exactly parallel to the variation of salinity. Thus the concentration of the constituent will be twice as great where the salinity is 2 parts per thousand than it is where the salinity is 1 part per thousand and no dissolved constituent can reach that part of the estuary containing only fresh water. The concentration of a dissolved constituent will vary with the proportion of fresh water on the seaward side of the outfall so that if the salinity of the sea outside the estuary is 35 parts per thousand, the concentration of a constituent is three times as great where the salinity is 20 parts per thousand as it is where it is 30 parts per thousand, i.e., $(35 - 20)/(35 - 30) = 3$. In fact, if we know the discharge of fresh water into the estuary, the discharge of the effluent, and the average salinity of the water near the outfall, the distribution of the effluent in water of different salinities can be calculated.[6] This is only possible for steady conditions in a relatively long estuary and if the effluent has a long period of retention in the estuary.

The position where the pollution is greatest moves up and down the estuary with the flood and ebb of the tide. At mean tide it will not be far from the effluent outfall. The greater the pollutional load the greater will be the depression of the concentration of dissolved oxygen in the water near the outfall and if it is great enough the water will become deoxygenated.

The volume of deoxygenated water will expand with increasing load or with decreasing supplies of oxygen caused either by low discharges of fresh water or lack of wind to ruffle the water surface. The rate at which oxygen enters through the water surface is influenced by the presence of synthetic detergents. As little as 1 part of detergent per million of water has a marked effect on the rate at which oxygen enters. Investigations are being made to discover new detergents which will undergo decomposition in sewage works and which should therefore not add to the difficulties caused by the discharge of effluents into estuaries.

The surface area of deoxygenated water is greatest at low water when a given volume of polluted water is present in a smaller depth. The width of the water surface is also less at low water so that the length of estuary where anaerobic conditions prevail expands and contracts with the ebb and flood of the tide. At the end of a dry summer in a seriously polluted estuary the length of water devoid of dissolved oxygen can be as much as 30 miles.

All forms of flora and fauna require oxygen. The effects of organic pollution on life in an estuary are to a great extent the effects of a reduction in the availability of oxygen. Different species of fish require different minimum percentage saturation levels of dissolved oxygen for survival. Fish of the salmon family will not live long in water containing less than 30% of the saturation concentration of dissolved oxygen, nor do they like warm water. Other fish, such as eels, and coarse fish such as tench, can live in water containing little dissolved oxygen. When the water becomes devoid of dissolved oxygen, the only organisms which can exist are those capable of obtaining their oxygen from compounds in solution. These are the denitrifying and sulfate-reducing bacteria.

Migratory fish will not be able to pass through an estuary which at particular times of year is devoid of dissolved oxygen, nor will fish which are frequently found in estuaries stay there. Shrimp can only be found much further seaward than is the case when there is no pollution. Shellfish may be contaminated by bacteria. This particular difficulty can be overcome by keeping them alive in clean seawater for a few days before they are eaten.

J. GRINDLEY

References

1. "Evaluation of present state of knowledge of factors affecting tidal hydraulics and related phenomena," Corps of Engineers, U.S. Army, Committee on Tidal Hydraulics, Report No. 1, February 1950.
2. Inglis, C., and Allen, F. H., "The regimen of the Thames estuary as affected by currents, salinities, and river flow," 1957, Institution of Civil Engineers, England, pp. 827–878.
3. Allen, F. H., and Grindley, J., "Radioactive tracers in the Thames estuary," *Dock & Harbour Authority*, January 1957.
4. "Effects of Polluting Discharges on the Thames Estuary," "Technical Paper No. 11 of the Water Pollution Research Laboratory," H.M.S.O., 1964.
5. Grindley, J., "The evolution of gas from the bottom deposits of the estuary," *Dock & Harbour Authority*, July 1955.
6. Grindley, J., "The Determination of the Salinity of Water in Estuaries," Publication No. 51 of the I.A.S.H. Commission of Surface Waters, pp. 379–386

Cross reference: *Pollution of Seawater.*

EXPLOITATION OF MARINE RESOURCES

Mankind in the future will be engaged in two crucial races: the race between population growth and natural resources availability, and the race to reverse trends widening the gap between living standards in the poor and rich nations. The recent and current attention at industrial, state, national, and international levels being devoted to increasing the exploitation and development of marine resources is largely concerned with these two problems.

What will be the most important uses of the ocean, and therefore the demands for the sea's resources, in the future? Will it be for animal protein—either in a natural state or as a protein concentrate? Will it be for energy resources—petroleum, gas, or possibly the utilization of the tides or currents? Or might it be for non-energy mineral resources—copper, nickel, cobalt, manganese, phosphorites, etc? We also know that there are, and undoubtedly will be, many other uses of the ocean—transportation, military, recreation, waste disposal, weather modification and control, etc. One can easily perceive many possible conflicts between potential uses and potential users. But the answer to the question as to which uses will become most significant, and therefore dominant, will largely depend upon the state of existing technology and the rate of technological development in the various uses. In terms of the methodology of economics, the future uses of the ocean will be largely determined by the *supply* of marine resources relative to the *demand* for these resources. Furthermore, the interrelationships between demand and supply, or the potential value of the respective marine resources, will have an important bearing on the establishment of a particular regime for utilizing the ocean resources.

The basic distinction for economic analysis between "ocean" resources and what are generally classified as "land" resources is due to the former's inappropriateness—in a legal sense—or what has been characterized as "common property" resources. Technically, the distinction is related to mobility versus immobility of the resources. That is, land resources are essentially immobile in the sense that they have been "nationalized." The marine resources, however, are mobile in the sense that they are freely available to—and therefore transferable between —the nation-states who might desire to utilize them. This, then, makes marine resources conceptually the same as what are generally classified by economists as "labor" and "capital" resources. They are free to move across national-state boundaries in seeking higher returns encompassing economic, political, and social forces. The question then is: Why is it not in the national interest (for marine resources within the national domain) or in the international interest (for marine resources within the international domain) —from an economic welfare point of view—to create an oceanic regime which would allocate the sea's resources according to those uses which yield highest returns, giving proper consideration to the aforementioned forces? The problem of exploiting marine resources then becomes essentially one of allocation rather than of distribution.

Demand Function for Marine Resources

In terms of the compact language of mathematics, we can state the general demand relationship for a marine resource as follows:

$$D_1 = f(p_1, p_2, \ldots p_n; \bar{Y}; P; T)$$

This says that the commodity demand for a marine resource D_1 is a function of its own price p_1, the prices of other substitute resources, or products $p_2, \ldots p_n$, the average community income level $\bar{Y}$, the community population level P, and the specific tastes or preferences for the marine product itself T. The law of demand is an hypothesis concerning the functional relationship between D_1 and p_1, assuming $p_2 \ldots p_n$, $\bar{Y}$, P, and T remain constant. Specifically, it says that $D_1 = D(p_1)$ and $\partial D_1/\partial p_1 < 0$ for a small change in p_1.*

Within this condition, the demand curve for a marine resource may have a wide range of characteristics and may be classified by the concept of price elasticity for demand. This is defined as:

$$\eta_d = (\triangle D_1/\triangle p_1)(p_1/D_1) \quad (-\infty < \eta_d < 0)\dagger$$

If $-\infty < \eta_d < -1$, the demand for the marine resource is said to be elastic, while, if $-1 < \eta_d < 0$, the demand for the marine resource is said to be inelastic.

The relationship between elasticity of demand for the marine resource η_d and total receipts of the producers R_1 (or total expenditures of buyers) can be formulated as follows: We have $R_1 = p_1 D_1$. Then the rate of change in R_1 is $\triangle R_1/\triangle p_1 = D_1 + p_1(\triangle D_1/\triangle p_1)$. But, since $\triangle D_1/\triangle p_1 = \eta_d (D_1/p_1)$, therefore $\triangle R_1/\triangle p_1 = D_1 + p_1(\eta_d D_1/p_1) = D_1(1 + \eta_d)$, which shows how the rate of change in total receipts of producers depends upon η_d.

---

* The partial derivative, $\partial D_1/\partial p_1$, is the rate at which D_1 changes with respect to incremental changes in p_1 while all other variables are held constant.

† Since a change in the quantity demanded for a product almost invariably moves inversely with a change in its price, the elasticity coefficient of demand is normally negative.

Supply Function for Marine Resources

We can state the general supply relationship for a marine resource as follows:

$$S_1 = g(p_1, p_2 \ldots p_n; w_1, w_2, \ldots w_n; N)$$

This says that the total quantity of a marine resource which will be supplied is a function of its own price p_1, the prices of alternative resources, or products, which might be supplied $p_2 \ldots p_n$, the prices of factors, e.g., labor, capital, etc., used in its production $w_1, w_2 \ldots w_n$, and the state of technology N. The law of supply is a hypothesis concerning the functional relationship between S_1 and p_1, assuming $p_2 \ldots p_n, w_1, w_2 \ldots w_n$, and N remain constant. Specifically, it says that $S_1 = S(p_1)$, and $\partial S_1 / \partial p_1 > 0$ for a small change in p_1.

In measuring the responsiveness of the potential supply of a marine resource to a change in its price, the concept of elasticity may be used in the same manner as it was used in connection with demand. The elasticity of supply, then, is a measure of the ease or difficulty of expanding output of a marine resource or product, in response to an increase of its price, or of reducing output in response to a decrease of its price. For the supply curve of a marine resource, the coefficient of elasticity (η_s) is defined in exactly the same way as it was for demand:

$$\eta_s = (\triangle S_1 / \triangle p_1)(p_1 / S_1) \quad (0 < \eta_s < \infty)*$$

If: $1 < \eta_s < \infty$, the supply of the marine resource is said to be elastic; while, if: $0 < \eta_s < 1$, the supply of the marine resource is said to be inelastic.

Equilibrium of Supply and Demand

Given a competitive market, we can say that the price of the marine resource will tend to a level which equates the total quantity demanded with the total quantity supplied. This is the *equilibrium* price which will clear the market. The stable equilibrium price, therefore, is where: $S_1 = D_1 = S(p_1) = D(p_1)$. The elasticity of demand, in the relevant range, is clearly of more than passing interest to an industry potentially affected by changes in technology, e.g., the ocean industry.

That the elasticity of demand is an important consideration in evaluating alternative methods and means to stimulate, or promote, marine resources development can be illustrated as in Fig. 1. The same downward shift in the supply curve of a marine resource, from S to S' (e.g., a decrease in costs resulting from application of a technological improvement) may induce a different response in output $q - q'$, depending upon the

* Since supply curves almost invariably rise with an increase in price, the elasticity coefficient of supply is normally positive.

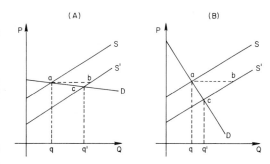

Fig. 1. A permanent decline in the cost of producing a marine product.

elasticity of demand for the marine resource, or the responsiveness of demand to price changes in the product. The question is: How much price change does it take to choke off the excess supply of the marine resource, shown by the distance $a-b$, brought about by the downward shift in the supply schedule? Or, asking the same question in a different way: How much increase in quantity will be induced by the cost reduction?

As can be seen in (A) and (B), the answer to this question largely depends upon the responsiveness of demand to the ensuing price and quantity adjustments. In (A), the elasticity of demand being *elastic*, demand is quite responsive to price adjustments and most of the excess is absorbed by the market as a result of a relatively small decline in price ($a-c$). In (B), on the other hand, the elasticity of demand being *inelastic*, the excess supply is choked off by a rather sharp decline in price ($a-c$) resulting in only a small increase in quantity potentially supplied at this price.

The effect of a direct governmental subsidy, tax relief, or an extension program (e.g., the sea grant college program) would be to reduce the operating costs of the industry, thus shifting the supply schedule downward with much the same effect as any technological improvement in operating techniques. As can be discerned from the previous analysis, the success or failure of governmental policy in this area of economic activity would depend upon the particular shape, or elasticity, of the demand curve and whether the policy is for a substantial price reduction or a substantial quantity increase in the production of marine resources. In the agricultural sector, for example, much of the pressure for price supports arose because of declines in farm income. These declines could have been caused by a rapid rate of technological improvement in agriculture if the demand for agricultural products had been relatively inelastic.

Externalities in Marine Resources Exploitation

Economic activity, whether public or private, as related to the development and exploitation of marine resources can involve what economists call external economies or diseconomies of production. These essentially involve shifts in the supply, or cost curves of enterprises—downward in the case of the former and upward in the case of the latter. Some examples of external economies to firms in the ocean industry might result from: (a) a form of tax relief or other governmental subsidy; (b) direct public expenditures on oceangraphic research and development; or (c) indirectly through public ocean-oriented expenditures (e.g., naval operations) which induces transfer of technology to private industry. It has even been pointed out that fisheries management, by increasing the catch per unit of fishing effort, acts as an external economy to the fishing industry. Some examples of external diseconomies to ocean enterprises might be: (a) polluting the oceans through sewage flow and other wastes; (b) creating hazardous operating conditions in the ocean environment through radioactive waste disposal; or (c) thermal pollution of estuaries resulting from nuclear power plant discharge. Again state and local legislation, or various management practices which tend to impede efficiency in the exploitation of marine resources —especially in the fisheries—acts as an external diseconomy.

It can be shown as in Fig. 2 how an improved marine technology might be *induced* through an over-all increase in the demand for marine resources—for either private or public uses. As a result of the increase in demand from D to D', the consequent improved techniques of exploitation are reflected by a permanent downward (irreversible) shift in the supply curve from S to S'. This means that even if demand should fall back to its original level, the equilibrium price of the marine product would not return to E, but instead would fall to E'', hence reflecting the new level of technology at which the industry is now operating.

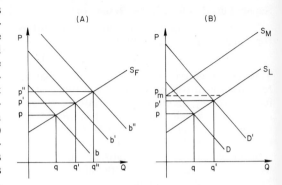

Fig. 3. Supply and demand relationships for living and non-living marine resources.

Economic Potential and Problems of Exploitation

A distinction can be made between exploiting living resources (e.g., fisheries) and other non-living resources of the ocean. As far as economics is concerned these relate to the final adjustments to changes in demand and/or supply of the marine resources vis-a-vis alternative land resources. As can be seen from the (A) portion of Fig. 3, an increase in the demand for fish, with a given supply schedule, will raise price and thus induce a larger output for this particular marine resource which is a unique product of the oceans and hence not available on land. For other non-living resources of the seas (B), because the costs of exploiting these marine resources are so much higher than for comparable land resources an increase in demand may not be enough to induce exploitation. In fact, only when the price increases beyond p_m will it pay to begin exploiting non-living marine resources, and then relatively insignificantly. The only other possibility for utilizing the oceans for, say, energy and non-energy resources, is if technological advances permit exploitation costs to fall toward land costs. The narrowing price discrepancies will then encourage the movement toward looking to the oceans for man's natural resources.

Exploitation of non-living marine resources simulates the extensive margin of cultivation in Ricardo's model of economic development which was the original formulation of the theory of economic rent. Since exploration and recovery costs are higher on inferior land tracts, they should, therefore, earn no economic rent. In fact, as shown in Fig. 3, economic rent earned by non-living marine resources is actually negative as long as the prevailing price is below minimum costs of recovery. This is true, for most potential ocean resources at present, except petroleum and gas and relatively small amounts of other mineral resources which are being exploited in offshore areas on the continental shelves. This is not true,

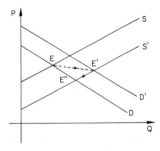

Fig. 2. An improved marine technology induced by an increase in demand.

however, for the living marine resources, as shown in (A). There, it is seen that a price such as P' returns economic rent to the inframarginal factors of production whose costs are below P'; the higher the price the greater the economic rent the resources are capable of earning.

Economic rent is a surplus over opportunity costs because the supply of "land"* is fixed, or completely inelastic. Consequently, economic rent, which is determined entirely by the strength of demand for the resources themselves, serves no incentive function in the sense of determining the extent of utilization of the resources. In other words, economic rent could be eliminated without affecting the productive efficiency of the enterprises engaged in exploitation or the productive efficiency of the economy in general; it is not a necessary payment to ensure that the natural resources will be available to the economy. The fact that some types of "land" resources, e.g., non-living marine resources, are "free" to anyone who might want to exploit them means that the demand for these resources is low relative to the supply (or the costs of production) and therefore economic rent is zero, or even negative, in these circumstances. Although rent is not necessary as an incentive to utilize the resources, it *is* necessary to provide a rational means for allocating the "land" resources among alternative users when demand becomes significantly intense and these resources therefore begin to become scarce. This is the rationale for economic rents paid to scarce factors of production.

The economic problems of exploiting marine resources relate to the controversy as to whether they should remain common property or come under the jurisdiction of private or public ownership, whether national or international. It has only been since economists have become concerned with the peculiar problems of marine resources that the real costs and benefits of these alternative regimes have been laid open to serious discussion. The fact that there is no price, or rental charge, for exploiting a marine resource indicates that there is no way through the market mechanism to reflect its true scarcity value. When there is relatively little pressure upon marine resources, or demand is small relative to its supply, this factor may be of little consequence. But, with the rapidly increasing interest in developing and exploiting the resources of the oceans, not only to feed the world's hungry, but also to sustain the industrial machine of nations, increasing attention is being given to developing some mechanism for limiting utilization to prevent overexploitation and external diseconomies* among various users. The Law of the Sea conferences in 1958 and 1960 and the resulting conventions are a recognition of this concern.

Living Marine Resources

The nature of the marine fisheries is such that they are "common property" resources, that is, they are open to exploitation to all who want to fish, subject only to limitations imposed upon foreign nationals fishing within the territorial waters of a nation and those imposed upon domestic nationals relating to allowable gear and equipment. The inevitable consequence of freely permissible utilization of common property resources is the inducement to excessive entry of productive inputs (or users) that such a regime generates. In the case of the fisheries, this results in higher than necessary costs of operation to take any given level of catch. Due to diminishing returns† resulting from excessive capital and labor inputs relative to the fishery resource, the eventual outcome is low labor productivity and low returns to capital. There is, thus, inherent in such an industry a lack of willingness to invest in long-term undertakings and new technology but, rather, in quick, short-term investments with a high turnover. Another consequence is that fishing industries have become traditionally protectionistic in reacting to new techniques and means of competition. The general tendency has been to seek governmental protection against more efficient gear and techniques, or from foreign competition.

These consequences have led to a questioning of the proper objectives of fisheries management which inevitably must cope with the economic questions involved. The traditional technique of management is biological in nature; i.e., the variables are population density and catch per unit of gear fished with little or no consideration given to the costs and returns from the fishing operation. The objective is primarily to maximize

---

* The "land" factor in economics denotes the original endowment of all natural resources, including the oceans.

* That is, the increase in costs imposed upon the operations of one producer as a consequence of the operations of another producer.

† The hypothesis of diminishing returns states that if more and more of a variable factor is used in combination with a fixed factor, the amounts added to total production by successive increases in the variable factor will eventually begin to decline. The reason for this is that as the firm uses more and more of a variable factor with a given amount of a fixed factor, it increasingly departs from the optimal combination of factors. This means that equal additions of the variable factor bring smaller and smaller increments of output or, what is the same thing, the cost of producing each additional unit of output must be rising.

the sustained yield from the fish stock. Economists have questioned the rationality of this objective in terms of economic welfare to the fishing industry and to the economy in general. That is, unless the marginal cost* of fishing is zero or negative, taking the maximum sustained yield would imply that the marginal value product of fishing is less than the marginal cost of fishing effort—clearly a misallocation of economic resources.

The controversy between marine biologists and/or fisheries administrators and economists relates to "rational" objectives of fisheries management. The economist's solution to the persistently high costs and low returns in the fisheries is to cut back on operating capacity by requiring operators to bid for licenses to fish. The number of licenses issued by the management authorities would attempt to equate the catch capacity with some "optimum" sustainable yield. The rationale is that by making licenses transferable and/or periodically putting up licenses for bidding, the efficient operators would eventually buy out the inefficient ones thus approaching a position which maximises the *net economic yield* of the fishery. Under this regime the state authority would capture the economic rent which was previously lost in excess fishing capacity. This whole approach, and the awakened recognition of the basic economic problems involved, has recently raised the controversy over the particular regime which is most rational for exploiting international fishery resources, and, going one step further, international marine resources in general.

Non-Living Marine Resources

Petroleum resources have been extensively developed in offshore waters and along coastal areas. Based upon supply and demand factors, future developments of offshore oil and gas resources appear promising. However, except for the relatively few minerals extracted from seawater, other marine mineral resources have not been extensively developed. The prospects for future development will depend largely upon supply—discovery and costs of recovery—and demand conditions.

Deuterium is a radioactive material which could be used in controlled thermonuclear reactors to produce 1000 times more energy for the next one million years than is generated in the world today. Unfortunately, this type of reaction produces temperatures too high for any known materials to withstand. As deuterium is found in abundance in the sea, if the technology

* Marginal (or incremental) cost is the increase in total cost resulting from raising the rate of production by one unit.

can be worked out, there will undoubtedly be a demand.

The oceans can furnish energy in a mechanical form through movements of the tides. By damming tidal basins and installing turbine generators in the sluices, it is possible to generate electricity with the rise and fall of tides. This has already been accomplished at the River Rance and on an experimental basis at St. Malo, both in France. It has been proposed as a United States-Canadian venture for the Bay of Fundy. There are factors, however, that would seem to minimize the possibility of this method ever being a significant source of energy. In the next four decades, atomic power is expected to eventually supply 50% of the electricity and 15% of all the energy requirements of the United States. Atomic power will probably have a lower over-all cost and is not geographically confined as is tidal electricity.

Since minerals are one of the bases of an industrialized society, as more and more parts of the world become industrialized, the rate of depletion of land-based reserves will grow steadily. There is the hope that the seas will provide new reserves to supply the needs of society. Research and development will play a vital role in this search by locating deposits and developing the unique technology required for effective exploitation. There are relatively few minerals which can be economically exploited currently or in the near future. Of course, seawater contains in a dissolved ionic form every element known to man, 40 of them in measurable quantities. Besides the vast amounts of sodium and chloride, others, e.g., calcium, potassium, sulfur, magnesium, and bromine are in sufficient amounts to permit extraction. In most cases, however, the cost of pumping the seawater to be processed exceeds the value of the minerals extracted. It is conceivable that as more water is pumped for desalination, minerals will be extracted as a by-product. On the whole, there seems to be very little prospect in the next four decades except for sodium, chloride, magnesium, and bromine. The so-called "flotation process" shows promise as a low-cost method of extracting aluminum but is not yet well established in comparison with the cost of bauxite extraction.

There has recently been increasing attention and interest directed toward the mining of minerals from the sea bottom of continental-shelf regions and the ocean floor. Most abundant as a commercial ore are the manganese nodules formed from the discarded wastes of a bacteria type. Unfortunately, these resources are found only at great depths at which technology is presently not equipped to work. An active research effort is being undertaken to devise effective suction dredges for their retrieval. It is estimated that one dredge could supply the United States with half of its manganese demand, all of which is imported at the present time. Because of the high cost of extraction from the complex of metallic minerals contained in the nodules, production of the manganese alone will not make the mining profitable. However, the value of the byproducts—copper, cobalt, and nickel—can make the investment worthwhile. Presently, the main advantage to be derived from exploitation is self-sufficiency. It is estimated that the world supply of manganese should be enough for the next 40 years.

Phosphorite nodules seem to be a more likely prospect as a market commodity in the near future. Located in shallower water and closer to shore, there are good deposits off southern California and Florida. This is an advantage especially for the large West Coast fertilizer market which presently must pay for phosphates transported from the mid-continent. An estimated $10–20 million per year industry can be developed. Other deposits off Australia and India could supply the Far East fertilizer market. Other known mineral deposits such as Globigerina Ooze (cement rock), red clay, barium sulfate, and magnetic spherules are not expected to be economically feasible or necessary for a very long time.

The continental shelves of the world show great promise as mineral-producing areas. More easily accessible for current technology, they will probably be exploited as soon as demand and/or costs will allow. Sulfur is already taken from wells in the Gulf of Mexico. Diamonds are being dredged off southwest Africa of gemstone and commercial value. Tin, an increasingly difficult metal to find, is currently being dredged from water less than 200 ft off the coasts of Malaysia, Thailand, and Indonesia. Tin-bearing sands have also been discovered off Cornwall in Great Britain.

Private enterprise is potentially the biggest customer in ocean exploitation and development, whereas in space, with the exception of Comsat, the Federal government is the only customer. It is basically the opportunities for investment of private capital in search of maximum profit potential which bear out the proposition that a rational oceanic regime is properly based upon private initiative in allocating economic resources among alternative uses. Projections indicate that basic research will prevail for the next 5 years, followed by 5 years of applied engineering, and fulfilled by the actual exploitation of many economic commodities from the seas. This places much emphasis upon the market for hardware and manpower in the immediate future. With the exception of petroleum, fish, and a few minerals, substantial commodity markets appear at least a decade in the future. These future commodities from the sea are most likely to be algae products and increasingly scarce land metals. It is generally agreed that the future exploitation of the seas will depend heavily upon the development of a unique ocean science, engineering, and technology to work and live under the sea. However, to utilize these capabilities for the benefit of man will require solving the political, economic, and legal impediments—some national and some international in scope—which have heretofore hindered a more rapid development.

SALVATORE COMITINI

F

FACTORY SHIPS—*See* **FISHING VESSELS AND SUPPORT SHIPS**

FATS AND OILS

Oils in Fish. Oils occur in the flesh of fish as essential components of cells where they perform important metabolic functions. In such instances the oils make up at least several tenths of a percent of the flesh and do not exceed about 0.7% by weight. With some species of fish such a form of oil represents most if not all of the oil present. Such fish are referred to as lean or non-oily species. They usually have white or very light-colored flesh; cod and haddock are typical of such species which usually are more or less of a sedentary type and usually feed at the bottom of the ocean floor.

Most species of fish have a larger or smaller portion of their oil in the form of depot fats. These depots lie just beneath the skin at well-defined layers, often along the lateral line, sometimes along the dorsal region. The flesh around the belly wall also often contains a layer of fat. The amount of oil contained in the depot fats of fish varies considerably with the season at which the fish is caught. As fish feed, fat is laid down building up the oil content. Then during other times of year, often during the winter, fish often feed very little so that their depot fats may serve as a supplementary energy source. If, as may often be the case, the fish make a spawning migration also in the winter or spring, they may not feed at all for long periods of time causing rapid depletion of fat reserves. In such extreme cases, the oil content may vary from a high value of 30% or more during peak feeding times to a minimum of 1–2%. The oil content of the flesh of a few species (e.g., some lake trout) may reach as high as 75% or more. Ordinarily, however, 30% is considered very high.

In the flesh of fish, the oil content is highest at cross-sectional slices or steaks near the head and lowest near the tail; a difference of two to three times may exist. The oil content of dark flesh (which usually occurs beneath the lateral line) is ordinarily several times as high as that of the light-colored flesh.

Oil also occurs in the nonedible portions of fish. The head usually contains relatively high oil content. Of the visceral contents, the liver may contain much oil. With species of fish having lean flesh, usually large livers (making up as much as 10% of the weight of the fish) occur with high oil content in the range of 35–80%. Species having oily flesh, on the other hand, generally have small livers with only a few percent oil.

Sources of Supply. Commercial fish oils are generally manufactured from certain species from which the whole fish is rendered into oil and fish meal. In some cases, however, trimmings from fish canneries or from preparation of fillets or steaks comprise the raw material. Species such as the herring, anchovies, sardines, pilchards, or menhaden make up by far the largest source of raw material. At one time whales were the principal supply from which fish oils were obtained. In recent years, however, the supply of whales has dwindled, and at the same time species like anchovies have been utilized in greater and greater proportions. Today whales are still a significant source of oils.

Liver oils have been used for hundreds of years for their medicinal properties long before it was known that these properties resulted from the high content of vitamins A and D. With the successful and relatively cheap chemical synthesis of vitamin A, the use of fish-liver oils for their medicinal properties is no longer economically feasible in the U.S. In many countries elsewhere in the world where labor costs are somewhat less, e.g., Japan, Norway, and Great Britain, such oils are still manufactured in quantity, although at reduced production levels.

South America, especially Peru and Chile, is today the largest producing area for fish oils. The exploitation of anchovy for manufacture of fish meal and oil in South America got under way only in the past decade or so, and this has changed the pattern of fish-oil production. In North America, the menhaden fishery conducted along the Atlantic coast from New York southward around Florida and into the Gulf of Mexico furnishes the fish, accounting for over 80% of the fish-oil production. The whole menhaden are rendered into meal and oil and are not used as a food fish. The remainder of U.S. production stems principally from three minor sources: (1) trimmings from tuna canneries, mostly in California, (2) from herring from Alaska and Maine, and (3) from "trash fish" (unwanted species mostly of a non-oily character).

Principal European producers of fish oils are Norway, Iceland, Denmark, and West Germany. Herring is the most important single species of fish used as raw material. Mixed species of trash fish, however, form a major group for reduction to meal and oil, and, in fact, such fish represent 95% of the raw material used for oil manufacture in Denmark and a substantial portion in West Germany. Liver oil production continues especially in the United Kingdom, Iceland, Norway, and West Germany; fish livers represent the third most important source of European fish oils, but the production is slowly declining. South Africa and southwest Africa produce substantial quantities of fish oil with a major portion derived from pilchard. Other important species utilized include maasbanker (or horse mackerel), mackerel, anchovy, hake, and shark.

Japan, the world's second largest fish-producing country, also ranks high in fish-oil production, and first in production of fish liver oils. Species of fish used for oil production varies. Recently saury has been the leading source, but several others, notably true cod livers and pollack livers, are of considerable importance. Sardines, which prior to 1940 was by far the leading species used for oil production in Japan, declined in availability to a point where this species is today one of the minor sources.

USSR and China, although third and fourth in world fish production (exceeded only by Peru and Japan), produce mainly food fish. No statistics are available on their fish-oil production, but it is believed to be relatively minor in proportion to their huge fish landings. USSR is a leading producer of whale oil.

Manufacture of Fish Oil. Fish oils are prepared as a byproduct of fish-meal manufacture. The method most commonly employed, known as the wet rendering process, consists of cooking the fish in either direct (used mostly in the U.S.) or indirect steam cookers; pressing to remove moisture and oil; separating oil, water, and solids by centrifuging; and generally giving the crude oil a final polishing by high-speed centrifuging. The cooking, which coagulates the protein and releases the oil, is the most critical stage of the operation. It is carried out by manual empirical methods and requires an experienced operator. Direct steam cookers, usually about 30 in. in diameter by 30 ft in length, are loaded through a hopper connected to an internal screw conveyor. The loading rate is controlled by a variable-speed motor which operates the conveyor screw. Small pipes at the bottom of the cooker provide inlets for the steam. A direct steam cooker may cook as much as 60 tons of fish in an hour. Indirect cookers, although having smaller fish-cooking capacities, can provide important economies in steam requirements. They may provide indirect heating both for the cooker jacket and for the hollow screw conveyor. Usually steam can also be injected directly into the cooker when needed for faster cooking.

Oil is separated from moisture and protein in a continuous screw press which may be of the single- or multiple-stage type or of the more recently introduced twin-screw variety. The pressed liquor, in addition to oil and water, also contains considerable coagulated protein, bone fragments, and scales. Three-phase centrifuges or sludgers are therefore used for preliminary separation of oil from the other components. These may be of the continuous or periodic self-desludging types. These machines operate on the same general principle as any liquid separator of the disk-bowl type. Ordinarily the oil phase from the sludger

TABLE 1. WORLD FISH OIL PRODUCTION IN 1964, PRODUCTION IN MILLIONS OF METRIC TONS

Country	Fish body oils	Fish liver oils	Whale oils including whale liver oils	Total	Principal species
Peru	212.1	...	...	212.1	Anchovy
Japan	19.0	9.7	155.0	183.7	Cod
Norway	83.6	10.8	44.0	128.4	Cod
USSR	a	a	115.6	115.6	Whale
Iceland	80.7	10.3	...	81.0	Cod
USA	80.3	...	...	80.3	Menhaden
South and southwest Africa	71.1	...	...	71.1	Pilchard
Denmark	34.8	...	...	34.8	Trash fish
Canada	26.6	...	...	26.6	Herring
Germany	18.5	4.0	...	22.5	
United Kingdom	0.01	19.7	...	19.8	Cod

a Data not available.

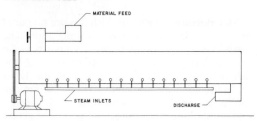

Fig. 1. Direct steam cooker used in the preparation of fish meal and oil.

passes through a "polishing" centrifuge after first being stripped with steam or hot water. This process removes small quantities of finely divided protein and traces of moisture.

The dry-rendering method, much in use for drying meat scraps, is used for fish in cases where the raw material is low in oil and where also the supply of fish is irregular or small. The fish are cooked and dried in one operation usually in a batch-type vacuum steam-jacketed cooker–drier using a rotating paddle-type agitator. Five hours or so of operation will render a 5-ton batch in equipment about 5 ft in diameter and 16 ft long.

Fish livers of the oily type are processed by cooking to release oil followed by centrifuging. Low oil type livers or even more oily types with very high vitamin content are usually digested with alkali using 1–2% by weight sodium hydroxide or 2–5% sodium carbonate, with equal weights of liver and water. Heating at 82–88°C for 1 hour liquefies the livers and the oil is separated in a three-phase sludge-type centrifuge. With very high potency livers, a wash oil process may be used, often in a countercurrent manner. Low vitamin A oils are built up in vitamin potency by use as wash oils in several stages.

Refining. Fish oils are often "winterized" by chilling them to a point where some of the more saturated fatty acids (stearine) solidify. The oil pressed from such a treatment then remains clear when held at reduced temperatures. Fatty acids are removed by alkali refining, either by a batch or continuous process. Sodium hydroxide solution in the concentration range of 1–6 N (depending upon fatty acid content) is agitated with the oil, followed by heating to about 55°C. The clear oil is removed after settling for several hours. The oil is then washed several times, first with very dilute alkali, then with several batches of water. By this process soaps are removed. The refined oil should then have not more than a few hundredths percent free fatty acid. Alkali refining also improves the color of the oil by removing some pigments. Further bleaching can be accomplished by agitating at about 95°C, sometimes under vacuum with activated clay using up to 2% of clay. The clay is removed by filtering through a plate-and-frame-type press.

Hydrogenation, which may be considered a form of refining, will remove all odors and produce a bland-tasting solidified fat suitable for incorporation in margarine or shortening. It also has some industrial uses. Nickel formate or other commercial nickel compounds are used as the catalyst; it is added initially at a concentration of only about 0.1%. As hydrogenation is continued, because of poisoning of the catalyst, more and more must be added up to several percent before discarding and starting with fresh materials. The fish oil must have no more than 0.1% free fatty acid content. Hydrogenation is carried out at temperatures usually below 200°C. Sometimes a two-step processing, first at 150°C then at 200°C, is used. Hydrogenation is stopped when the melting point reaches the desired point, determined by the ultimate use of the product.

Composition. Fish body oils consist for the most part of triglycerides. A relatively small phospholipid content is always present. Other substances, ordinarily totaling under 1% and including cholesterol pigments and vitamins are also present. Fish liver oils may contain quite different classes of lipids. Hydrocarbons, especially squalene and to a lesser extent pristane and zamene, are commonly present. In the liver oils of a few species of shark, these hydrocarbons may exceed 50% of the oil. Another common component of some liver oils such as that of the common dogfish is alkoxydiglycerides. Another class of lipids found in some fish oils are the wax esters which, with a few minor species, make up a sizable proportion of the body or liver oils.

Fatty acids of fish oils are more numerous, covering a wider range of types than occur in animal and vegetable fats and oils. Approximately 35–40% of the fatty acids of most fish oils are of chain lengths exceeding 18 carbon atoms (C_{20}, C_{22}, and sometimes small amounts of C_{24}). Usually 20–35% of the fatty acids of fish oils contain more than three double bonds per molecule, five and six double-bonded fatty acids being the most common. On the other hand, 20–50% of the fatty acids are usually saturated. Sizable quantities of monoenes also occur. There are always some odd carbon chain length fatty acids present, predominately C_{15} and C_{17} acids. They usually occur up to a total content of $1\frac{1}{2}$ to $2\frac{1}{2}$%. In mullet body oil, however, usually 10–15% of the fatty acids have odd-numbered carbon chain lengths. Table 2 shows the fatty acid composition of a few typical marine oils.

Chemical Reactivity. Fish oil fatty acids undergo reactions at the carboxyl group typical of fatty acids in general, and reactions occur also at the double bonds. Reactions for fish oil fatty acids often differ somewhat from those for vegetable or animal oil fatty acids due to such considerations as the somewhat higher molecular weight

TABLE 2. FATTY ACID COMPOSITION[a]

Fatty acid[b]	Cod body oil	Cod liver oil	Halibut body oil	Herring oil	Mullet body oil	Salmon (pink) body oil	Trout (rainbow) body oil	Whale finback (outer blubber)
14:0	1.8	2.8	2.8	6	3.9	3.4	2.1	5.7
15:0	0.5	0.4	0.3	+	7.4	1.0	0.8	0.2
16:0	33.4	10.7	15.1	13	29.6	10.2	11.9	7.1
16:1	2.4	6.9	8.9	6	6.3	5.0	8.2	8.4
17:0	0.9	1.2	0.7	+	5.5	1.6	1.5	0.4
18:0	4.0	3.7	3.4	1	4.8	4.4	4.1	1.4
18:1	11.8	23.9	25.7	22	5.8	17.6	19.8	28.9
18:2	1.2	1.5	0.9	1	2.6	1.6	4.6	2.0
18:3	0.8	0.9	0.3	+	1.5	1.1	5.2	0.8
18:4	1.2	2.6	3.6	2	NK	2.9	1.5	0.5
20:1	1.6	8.8	8.0	15	NK	4.0	3.0	19.6
20:4	3.2	1.0	2.5	+	3.6	0.7	2.2	0.6
20:5	12.4	8.0	10.1	6	4.7	13.5	5.0	0.9
22:1	0.7	5.3	5.1	16	NK	3.5	1.3	17.9
22:5	0.6	1.3	1.6	1	3.1	3.1	2.6	0.7
22:6	21.9	14.3	7.9	6	10.3	18.9	19.0	1.1

[a] Only most commonly occurring fatty acids are listed here. At least 50 others occur in very small or trace amounts.
[b] First figure is chain length; second is number of double bonds. Thus 20:5 indicates a fatty acid with 20 carbon atoms and 5 double bonds. NK = unknown; not reported by analyst. + Indicates presence of a trace amount too small in quantity to estimate.

and greater degree of unsaturation. Thus, when a reaction is directed toward the carboxyl group, in some instances side reactions at the double bonds may occur.

Typical carboxyl reaction derivatives of fish oil fatty acids include: salts prepared by neutralization; alcohols prepared by reduction with lithium aluminum hydride or with sodium metal and alcohol; esters from reactions with alcohols and catalyzed by mineral acids, by zinc dust, by boron trifluoride, or by other substances; and preparations of many derivatives using fish oil fatty acid alcohols as intermediates. These include such compounds as amines, amides, quaternary ammonium salts, xanthates, and nitrates.

While most of the above-mentioned reactions are carried out under controlled conditions, oxidation takes place spontaneously as a form of deterioration both for extracted fish oils and with the oil while still present in the fish tissue. Under the latter condition, fish oils oxidize at quite different rates depending upon the presence of various anti-oxidants and especially certain pro-oxidants as well as the degree of unsaturation of the particular marine oil. Certain areas of the fish contain blood pigments such as hematin compounds. The rate of oxidation in the dark muscle of fish, such as just beneath the lateral line where such pro-oxidants are located, may be hundreds of times as great as in the light muscle. Fish oils in fish flesh oxidize relatively more rapidly when stored in the frozen or irradiation-pasteurized form than when iced and stored. There is some evidence linking the slower oxidation rate in the latter case with bacterial activity such that, whenever bacteria are multiplying, the rate of oxidation is diminished.

Anti-oxidants slow down the rate of oxidation of fish oils but to a much less extent than is the case with other oils and fats. Fish oils oxidize according to a typical induction-type curve, but often the break in the curve is not clearly defined, with a small increase in oxidation going on during the induction period followed by a gradual increase in rate toward the end of this period. The best way to diminish oxidation of fish oils is to reduce storage temperature and especially to prevent access of air.

Nutritional Properties. The polyunsaturated fatty acids of fish oils with three or more double bonds are predominantly of the linolenic acid family; i.e., the first double bond occurs at the 3,4 position counting from the terminal methyl group end rather than in the 6,7 position, as with the linoleic acid family of fatty acids. Only members of the latter family possess full essential fatty acid (EFA) activity; specifically they are the ones curing dermal symptoms. Fish oils are not completely deficient in EFA, however; when used in large enough quantity the small proportion of linoleic and arachidonic acid present will cure dermal symptoms. The fish-oil type fatty acids do, however, support growth, and they transport serum cholesterol to a greater extent than do the linoleic acid family of fatty acids. Whole fish oils, because of their high content of these polyunsaturates and despite the presence of considerable quantities of saturated fatty acids, will transport serum cholesterol when the fish oil in the diet is at much lower (as little as 1/10) levels than that brought about by linoleic acid.

Uses. Industrial fish oils find widespread use as a source of raw material for hydrogenation for use in shortening and margarine. More than half the fish oil produced in the U.S. is shipped to Europe for this purpose. The next most important application is for use in protective coatings. The fish oils may be used directly or modified by forming a derivative having better characteristics for some particular application. For example, a copolymer produced by reaction of fish oil with cyclopentadiene at about 275°C under pressure has improved drying and bodying characteristics; water and chemical resistance, as well as resistance to yellowing are improved when this polymer is substituted for fish oil, but flexibility and durability are impaired. Another derivative from fish oil for use in paints is a maleated oil prepared by reacting fish oil with maleic anhydride followed by esterification with glycerine. The use of fish oils in alkyd resins is the most important application where a modified derivative is first prepared. For example, a fish oil may be first reacted with glycerine to prepare a monoglyceride which, in turn, is reacted with phthalic anhydride, with isophthalic acid, or with another dibasic acid or anhydride to produce an alkyd resin. Such fish-oil resins find uses in aluminum paints, flat, gloss, and semi-gloss paints, enamels, house paints, rust inhibitive primers, and numerous other applications.

Fish oils and fish oil fatty acids find use in many other applications. Some of these are: lubricants, as caulks and sealants, for animal feeds, for use in treatment of leather, in printing inks, and as plasticizers and surfactants. In most of these applications the fish oil is modified to make some chemical derivative which possesses the properties desired for the particular application. The use of fish oils as the starting point rather than some other oil may be dictated as a result of the higher proportion of long-chain fatty acids which modifies physical characteristics, by its content of very highly unsaturated fatty acids, or from a price standpoint.

MAURICE E. STANSBY

FERTILITY OF THE SEA

The phrase "fertility of the sea," is broadly equivalent in meaning to "the biological productivity of the world ocean." It suggests such questions as: what is the distribution of the standing crops of marine organisms; what are their rates of production; what environmental factors mediate or limit this production; and, ultimately, what amount is harvestable and thereby available to man? The concept is essentially an evaluation of the synecology and autoecology of the living marine resources of the world.

Evaluation of Living Marine Resources

While many different approaches have been used in evaluating these resources, each based on different forms of data and producing different estimates, they can be categorized as follows.

(a) *Quantitative Sampling:* The attempt to evaluate directly the standing crop or standing stock by sampling, which is usually severely limited by unknown inefficiencies of sampling gear and by the difficulties inherent in adequately and simultaneously sampling large areas.

(b) *Physiodynamics:* The quantification of elements in energy-flow models of ecosystems, with an attempt to account for the anabolism and catabolism of organic material from the initial photosynthate through utilization by several animal groups to ultimate breakdown by bacteria and/or burial in sediments. The approach is based on an estimation of the rates of initial synthesis as controlled by environmental parameters, and is most useful in providing a broad-scale evaluation of biological production.

(c) *Trophodynamics:* The quantification of all steps in a food chain, with a critical evaluation of the number of trophic levels supporting a harvestable standing stock. It is based on estimations of standing stocks (often at the herbivore level), predator–prey relationships, and feeding coefficients.

(d) *Population Dynamics:* An approach based on an initial estimate of the harvestable stock, modified by data on changes in the age and size composition of the stock, recruitment, mortality, and growth rates.

These approaches are not mutually exclusive and marine resource engineers may combine various elements from each when studying a particular region and its living resources.

The synthesis and conversion of organic materials is usually compared on the basis of the common element, carbon, and from several different viewpoints.

(a) *Gross Primary Production:* The total amount of organic material synthesized by the activities of photosynthetic organisms in unit time per unit volume of water or beneath unit area of sea surface. It is expressed as: mgC/m^3 hr, mgC/m^3 day, mgC/m^3 day, or gC/m^2 day, etc., and is most useful in relating primary production to environmental parameters.

(b) *Net Primary Production:* The amount of organic material synthesized by the activities of photosynthetic organisms, *in excess of that respired or excreted.* It is usually expressed as $mgC/m^3/hr$, $mgC/m^3/day_{12}$, $mgC/m^2/day_{24}$, gC/m²/yr, etc., and represents the organic material in particulate form available to support dependent food chains.

(c) *Production:* Represents the rate at which organic material is converted at a designated trophic level, under the influence of all environmental factors, including all losses from death or predation. In addition to the units used for primary production, it is often expressed in tons/yr.

(d) *Productivity:* Represents the capacity to produce, and is used to evaluate the relative fertility of different regions, it is often based on primary production measurements made under "standard" conditions that differ from the actual natural environment.

It is important to distinguish between production, the standing crop of marine plants, and the standing stock of marine animals. Production is determined on the basis of at least two observations, from which the amount of organic material synthesized or converted in the time interval between observations is determined. The standing crop or stock is the amount of organisms per unit volume or area of the ocean, and is usually expressed by weight, e.g., grams, wet weight, per cubic meter. It is not a dynamic measurement, and since it is determined from a set of single measurements can only represent the quantity of organisms at a given time. It provides no information on the *rate* of synthesis or conversion of organic material.

Primary Production

The plants of the sea—the primary producers—alone are able to convert the light energy in sunlight into energy-containing plant materials that can form the base of other food chains. Near shore, the bulk of the primary production is effected by the attached plants, which are dependent upon the depth of light penetration. Light in sufficient quantities for photosynthesis penetrates to an average depth of 30–50 m, which limits the growth of these benthic plants to an extremely small area fringing the shores—an area representing less than 2% of the ocean's area. Therefore, on a world-wide basis marine primary production is predominantly accomplished by the phytoplankton.

During the last decade, partially as a consequence of the increased interest in the sea, but primarily as the result of new techniques (the use of radioactive carbon isotopes) for estimating the rate of primary production, sufficient data have been gathered to allow an evaluation of the variation in primary production throughout the world ocean (see Fig. 1).

It should be noted that the ocean varies as much, if not more, by area in its production as does the land; as with land, there are ocean "deserts" as well as extremely fertile ocean regions.

Although gross primary production estimates may serve as an index to the production of harvestable marine resources, the production of the different steps in a food chain *is not* directly proportional to the primary production. This arises because variations occur in the density of the phytoplankton standing crop and therefore its availability to herbivores, because variations occur in the herbivorous standing stock independent of phytoplankton densities, and because

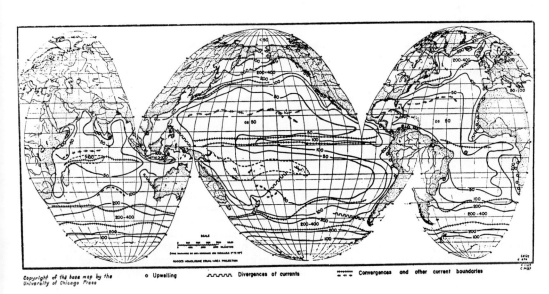

Fig. 1. Estimated amounts of basic organic production in the oceans as grams of carbon per square meter of sea surface (g C/m²/yr). (Courtesy Fishing News (Books) Ltd.)

the smaller carnivores tend to occur in schools which move from one area to another thereby cropping the herbivores unevenly.

The average gross primary production of the world ocean is usually estimated at about 150 mgC/m² day, ranging from about 50 mgC/m² day in the Sargasso Sea to about 5500 mgC/m² day in the Peru and Benguela current regions. This represents a total production of about 1.9×10^{10} metric tons per year for the world ocean (some oceanographers have produced estimates up to 15×10^{10} metric tons per year). Primary production on the land is usually estimated at about 2.5×10^{10} metric tons per year. Thus the marine plants are producing slightly less than or up to six times more than their terrestrial counterparts.

Ecological Factors Affecting Primary Production

Two major factors establish the basic pattern of primary production: the availability of light energy, and the availability of plant nutrients. Light energy is more important in determining the seasonal patterns of primary production especially in the higher latitudes, and the availability of nutrients establishes the large-scale features of primary production when it is considered on an annual basis.

In most open ocean, the low concentrations of available nutrients impose a brake on the rate of primary production. Consequently, especially in tropical areas, the rate at which nutrients are supplied through regeneration and the physical processes of mixing tends to determine the rate of primary production. When the rate of nutrient supply is low, the population is usually limited by a deficiency of some essential element. More important, under these circumstances the net primary production may be a small fraction of the gross primary production, most of which is required to balance the respiratory needs of the plants themselves, leaving only a small amount to enter other food chains.

A simplified example of the control and interaction that radiation and nutrients impose on primary production is apparent in the cycle of primary production in temperate waters. The spring phytoplankton bloom in higher latitudes tends to coincide with the spring increase in radiation. It is usually considered to be triggered by that increase. This rapid outburst of phytoplankton growth tends to deplete the nutrients that have been introduced into the euphotic zone by winter vertical mixing, which greatly retards further population growth. During the summer, regeneration within the euphotic zone may again raise the level of nutrients to the point where a secondary phytoplankton bloom can occur, a bloom which subsequently is slowed and stopped by the autumn decrease in radiation.

However, specific parameters will combine to control the rate of primary production in a given area. These factors, and their *day-by-day* interaction, can be categorized as follows.

(a) The size of the initial standing crop of viable phytoplankton, and

(b) the initial concentration of nutrients within the euphotic zone

allow the continuation of plant growth.

(c) The size and feeding efficiency of the standing stock of zooplankton cropping the phytoplankton, and

(d) the rate of loss of phytoplankton cells due to sinking below the euphotic zone

determine the size of the standing crop of phytoplankton.

(e) The mean depth of the euphotic zone throughout 24 hours, which is a function of the quantity of light falling on the sea surface and the transparency of the water,

(f) the rate of regeneration of nutrients, from organic compounds, within the euphotic zone,

(g) the rate at which nutrients enter the euphotic zone from below by vertical mixing,

(h) the effect of temperature on the respiration rate of the phytoplankton, which is partially offset by its effect on the rate of regeneration of nutrients and upon the rate of photosynthesis,

(i) the qualitative photosynthetic characteristics of the phytoplankton population present in the euphotic zone, and,

(j) the growth rate of the standing stock of herbivorous zooplankton

determine the growth rate of the phytoplankton population.

Secondary Production

In one of the most fertile marine areas, the Peru Current region, the daily net primary production is about 5000 mgC/m² day$_{24}$; which represents about 12.5 g of plant material if the carbon is fixed as a carbohydrate. If it were possible to harvest this organic material directly (which is highly unlikely, since such tremendous quantities of water would have to be filtered that the energy expended would be greater than the energy gained) it would only represent a daily yield of about ½ oz. Consequently, the harvesting of this organic material must take place at some higher step in the food chain where other organisms have already concentrated the material.

Unfortunately, only rarely do harvestable marine resources feed directly upon the phytoplankton; there are usually one to several links in the food chain between these and the primary producers (Fig. 2). As the organic material moves from one trophic level to the next, less is available

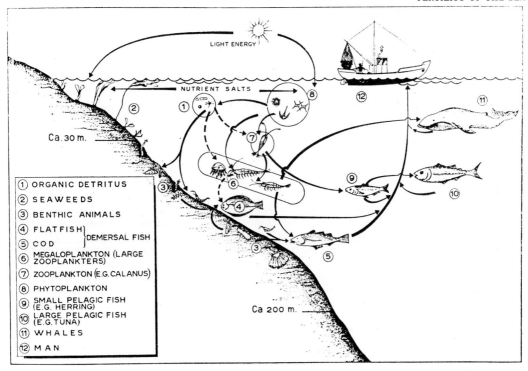

Fig. 2. Links in the food chains between harvestable marine resources and the primary producers. (Courtesy Fishing News (Books) Ltd.)

at each new trophic level. This loss of organic material is due not only to metabolic activity and waste products (i.e., respiratory carbon dioxide, urea excretion products, moulted exoskeletens, etc.), but also—especially with the herbivorous zooplankton—to unassimilated ingested material. As a result the efficiency of conversion at each trophic level is usually estimated in the 1–10% range. However, this conversion factor may be entirely too conservative. Recent studies have demonstrated that shrimp can convert their food to body weight with an efficiency of about 25%, and one important group of zooplankton, the *Euphausia* have food-to-body-weight efficiencies ranging from 11–44%. With a food chain of several steps, only some 1/1000 to 1/10,000 of the original photosynthate may be harvested by man. This is illustrated in Table 1, which presents an organic material budget for the Plymouth (England) sea area, a high-latitude coastal shelf regime.

On land, a certain amount of "natural fertilization" takes place. Dead organisms decay *in situ* and the nutrients they contain are released directly to serve subsequent plant generations. However, in the sea moribund plants and animals tend to sink and the organic material they contain is constantly being removed from the euphotic zone. While this material ultimately decomposes, and plant nutrients are regenerated, the process takes place primarily below the euphotic zone. Thus the remineralization process, which on land often tends to enrich the very strata in which plants grow, does the opposite in the sea, and tends to deplete the strata in which the phytoplankton exist. This downward flux of nutrients is

TABLE 1. MEAN QUANTITY THROUGHOUT THE YEAR OF PLANTS AND ANIMALS BELOW UNIT AREA, THE AVERAGE DEPTH BEING 70 M

Wet weight of tissue containing 80% of water per acre (lb)	Type of organism	Dry weight of organic matter, g/m^2	
		Quantity below a square meter (g)	Daily production per square meter (g)
180	Phytoplankton	0.4	0.4–0.5
70	Zooplankton	0.5	0.15
80	Pelagic fish	1.8	0.0016
...	Bacteria (water column)	0.04	...
50	Demersal fish	1–1.25	0.001[a]
800	Epi- and in-fauna	17	0.03[b]
...	Bacteria (sediments)	0.1	...

[a] Based on a natural mortality of 30% per annum, due to being eaten.
[b] Based on a natural mortality of 60% per annum, due to being eaten.

accelerated by the diurnal vertical migration of zoo-plankton which tend to graze in the euphotic zone during the night and defecate in the deeper underlying water during the day.

Since the subeuphotic waters are always relatively rich in nutrients, euphotic-zone fertilization takes place by oceanographic processes which lead to the mixing of underlying with surface waters.

Processes Affecting Production

Several basic types of oceanographic processes bring about vertical mixing. In some regions, usually mid-latitude, wind drives the surface waters away from a coast or an internal boundary, and nutrient-rich waters are upwelled from mid-depths. In higher latitudes, surface waters are cooled in the winter, become more dense, and sink and mix with the underlying water. Throughout the oceans, but especially in the high latitudes, the density gradient—which often isolates the euphotic zone from underlying waters—may be broken down by wind-induced mixing during severe storms, resulting in an upward nutrient flux. Lastly, vertical mixing often takes place along current boundaries, or where oceanographic features such as domes outcrop the surface.

However, even with these processes there are substantial nutrient reserves which remain unutilized and effectively lost to the primary production process. These *might* possibly be brought into use in certain areas by artificial upwelling, and oceanographers have speculated on the effects to be produced by a slowly reacting atomic pile placed at depths where the resulting temperature gradient would effect upwelling.

The distribution of the major oceanographic processes tend to concentrate production of harvestable protein in the temperate waters of the northern hemisphere, mostly within a few hundred miles off the coast. The continental-shelf areas of the mid-latitudes are characterized by moderate levels of production with the exception of some well-known areas where upwelling occurs, as for example, off the west coast of Africa and South America. The low latitudes are generally unproductive, as are the mid-ocean areas of all oceans.

In the mid-ocean regions, primary production may be entirely dependent upon and limited by the processes of regeneration and vertical diffusion of essential nutrients. Even in these regions of low-nutrient concentrations, many essential elements are in organic combination, and it is not known to what extent these organic compounds can be used directly by the phytoplankton or how rapidly they are decomposed.

Harvesting Marine Production

Lastly, other nonbiological factors combine to determine how much of the biological production of the oceans is harvested. Since the major fishing nations are usually the major industrial nations,

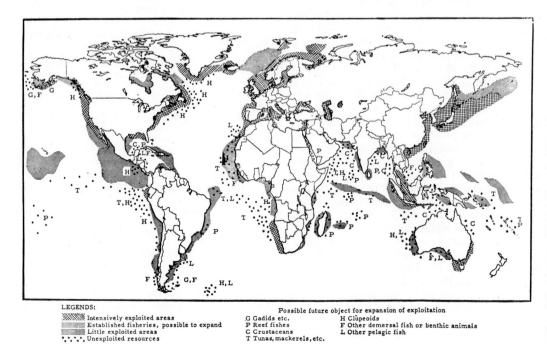

Fig. 3. Marine fisheries of the world. (Courtesy Fishing News (Books) Ltd.)

the major sea fisheries (Fig. 3) tend to be located in the relatively shallow coastal-shelf regions of the northern hemisphere, even though the southern oceans contain about 57% of the world's sea area. The biological production of the sea is made available to man only where oceanographic parameters allow high primary production; where this high primary production is concentrated by herbivores and/or carnivores which congregate in such a manner that they can be cheaply harvested; and where this secondary production can be processed and transported to population centers in a form and at a price acceptable to man.

These sociological limitations combined with the biological limitations of oceanic production indicate that approximately half of the protein being produced in the oceans is being harvested. Under present conditions and utilizing current fishing techniques, the annual harvest of marine resources can probably be no more than doubled.

MALVERN GILMARTIN

FISHERY ENGINEERING

Probably the most dramatic changes in high-seas fishing operations, until post-World War II years, started in the 19th century. These changes occurred from about 1880 when the conversion from sail to steam engines aboard fishing vessels commenced. In additon to allowing fishermen to conduct their activities without being dependent on the vagaries of the weather for vessel movement, the use of engines made available reliable and controllable power sources which led to mechanization of deck and fishing-gear handling equipment. Tradition dies hard in the fishing industry, and it was not until about 1938 that the sail had largely disappeared from offshore vessels. During the past 20 years, steam has been largely replaced by internal combustion engines (gasoline and diesel).

With regard to fishing methods, since about 1900 four basic types of gear have accounted for the greatest share of the fish harvested from the world's oceans. These include: seines, gill nets, hook and lines, and trawls. Many of the present individual units, however, have little similarity to their predecessors. The modern herring or tuna purse seine is as comparable in efficiency and size to previous versions as a factory stern-ramp trawler is to an old-time New England sailing schooner.

Although growth of world fisheries was halted during World War II when large portions of the fleets of various nations were destroyed, a rapid rebuilding took place. This was given impetus by the realization in many countries that the sea is a most logical source for increasing the supply of animal protein. The trend in the post-war fishing fleet was towards larger and more mechanized vessels equipped with electronic devices and nets of synthetic fibers.

Electronics

A wide variety of war-surplus electronic devices became available to the general public in 1946 and were quickly adapted by the fishing industry. Of these devices, those involving communication, navigation, and fish detection have had the major impact on world fisheries. Present electronic equipment has little resemblance to the rather cumbersome original surplus equipment because of the advances in the electronics industry during the last 20 years. These have resulted in the development and construction of compact radiotelephones, echo sounders, radars, and Lorans.

Practically all the world's high-seas fishing vessels have communication systems that provide ship-to-shore and ship-to-ship communications. Radiotelephones are most commonly used, although the larger vessels usually carry CW equipment. Radiotelephones are used to conduct the business of fishery operations at sea in much the same manner as telephones are used to conduct normal business operations ashore.

Loran and Decca navigation systems, radar and echo sounders, have increased the navigational capabilities of the world's fishing fleets tremendously. Loran was quickly adopted by the U.S. fishing fleet since it provides a simple and low-cost method of offshore navigation. Many sectors of the U.S. fishing industry have become so dependent on Loran that they carry standby sets in case one breaks down and repairs cannot be effected at sea. Radar provides the artificial eyes necessary to see through the fog, thus enabling the fisherman to continue fishing during periods of reduced visibility. Since water depths appear to have a direct relationship in the occurrence of many species of fish, the use of echo sounders to provide accurate water-depth measurements has increased the efficiency of fishing operations considerably.

Schools or concentrations of desirable species of fish are detected primarily by visual search, thermometric techniques, and underwater sound. From the standpoint of electronics, underwater sound is probably the most universal technique currently employed to detect fish below the surface of the ocean. Its use in fish detection during commercial operations is almost entirely restricted to active "pinging" rather than passive techniques. Sound pulses are generated in the water and echoes from objects which the sound strikes are amplified, detected, and portrayed on various recording devices. Echo sounders, which now serve the dual purpose of "depth finders" and "fish finders" aboard fishing vessels, include the one-directional vertical type or multidirection echo-ranging type which we call sonar. Vertical sounders are universally used aboard trawlers in many countries to search for fish concentrations near the ocean floor. They employ frequencies ranging from about 20–50 kHz with a detection capability reaching a maximum of about 3000 ft. Horizontal sounders are commonly used by vessels of many major fishing nations engaged in fishing species such as herring, sardines, and anchovies, which frequently occur in compact schools. The sounding principles are identical to those of the vertical sounder except that the sonar transducer is mounted so that it can rotate in the horizontal plane. Most units are fixed to search from the beam of the ship to the bow on either side of the vessel, thus providing a 180° search pattern. Sound frequencies used are within the general range of 20–50 kHz and the scanning range may vary from about 1200–9000 ft depending on the model.

Further development of acoustical devices for fish detection is continuing. Efforts are underway to refine present equipment to identify accurately the species located and to determine the size of individual fish as well as the number of fish comprising the targets. Fishery engineering could also play an important role in determining the possibilities of using passive sonic (listening) methods of locating subsurface schools of fish. A passive system might consist of an array of listening devices set across areas where fish are known to migrate. Such a system would be operated from shore stations and would serve to alert the fishermen to the presence of fish.

Since the distribution of many commercially important fish is influenced by changing oceanographic conditions, the use of thermometric means to locate potential areas of fish concentrations is receiving considerable attention. This has received impetus through the development of the portable radiofacsimile recorder. This device, which receives and records synoptic environmental data in chart form from shore station broadcasts is currently being used in some Japanese fishing operations. In the U.S. it has been used aboard the Bureau of Commercial Fisheries exploratory fishing vessel *Delaware* during tuna explorations in the western North Atlantic and ashore between tuna trips to assemble data for planning future explorations. Information received of potential use to fishermen includes:

1. Sea surface temperature analyses and forecasts.
2. Sonic layer depth analyses and forecasts.
3. Selected bathythermograms.
4. Surface weather analyses and forecasts.
5. Sea wave analyses and forecasts.

Sea condition and weather information pertain primarily to vessel operations; sea surface temperatures (SST), layer depth (LD), and selected bathythermograms (BT) can be useful in selecting areas that have environments favorable for fish. Such an environment, for example, is the warm convergence that shows as a characteristic gradient pattern on the synoptic SST chart, and a characteristic thermocline structure on the LD and BT charts. This information is the basis for advance selection of the fishing area. Succeeding verification of the structure by vessel transect across the area is followed by test or commercial fishing.

Net Materials

New synthetic net materials, such as polyamide (nylon), polyvinyl alcohol, and polyethylene which originated in the textile and plastics industry, have had a tremendous impact on the construction of fishing nets throughout the world. The use of synthetic fibers in the fishing industry had a rapid growth, beginning in about 1948. In a few years synthetic fibers replaced natural fibers in the construction of many kinds of fishing nets in numerous countries. Synthetic fiber netting was quickly adopted because it has the following advantages over natural fiber netting—no drying is needed, it does not rot, has greater tensile strength per weight unit, more durability, does not require preservatives, and has greater fishing power or catchability.

Although new synthetic materials are continually appearing on the market, their efficiency can no longer be judged by comparing their properties with those of conventional natural fiber twines, such as linen and cotton. They now have to compete with well-proven synthetics and acceptance is considerably more difficult than in the early 1950's when their efficiency was judged by comparing their properties with those of natural fiber twines. For these reasons, indications are that the introduction of new net materials into the fishing industry in the future will not have as much impact as in the past. An ideal synthetic fiber, suitable for all types of fishing is not available, but efforts to develop such a fiber will continue.

Mechanization

Improvements in deck machinery have provided for mechanization aboard nearly every type of fishing vessel. The main purpose of mechanization is to increase the productive capacity of the individual fisherman. This means that the same amount of fish, or in some cases, more fish, will be produced with fewer fishermen. In many nations the most serious problem in operating fishing vessels is the lack of trained fishermen so that the social impact of mechanization is minimal. The results of mechanization are apparent in many of the world's fisheries, particularly in the California tuna fishery. This fishery is able to compete profitably with less expensive foreign-caught tuna through the conversion of pole, hook and line fishing to mechanized purse seining. The application of mechanical equipment has increased the efficiency of menhaden vessel operations so that the crew now averages 14–16 as compared to 27–29 a few years ago. The first large shift to purse seining in the eastern Atlantic was made by the Norwegians and Icelanders. Some Norwegian vessels commenced purse seining off northern Denmark and in 1964 caught 250,000 tons of herring. This led to additional conversions and the demand for purse seiners in Norway still exceeds the supply.

The increased use of hydraulic systems that operate whole arrays of deck machinery and gear-handling apparatus has had a major influence in mechanizing vessels. Early hydraulic drives worked poorly, but after improvements, hydraulically driven deck machinery was readily accepted. The advantages of hydraulic equipment are its compact size, light weight, reliability, stepless speed control, and the ability of hydraulic relief valves to absorb sudden shocks.

Impact on Fish Production

In 1938 (Table 1) the world fish catch totaled 21,000,000 metric tons. By 1948 it was recovering its loss in production caused by World War II and the total was 19.6 million tons. By 1957 it had reached 31.5 million tons, and most of this increase could be attributed to the developments we have described; electronics, synthetic net materials, and mechanization. From 1957–1964, however, as indicated in Table 1, much of the increased catch could be attributed to the geographic expansion of fishing activities. Production from South America, only slightly over 1,000,000 metric tons in 1957, increased to over 11,000,000 metric tons in 1964. Soviet production during this period increased from 2.5 to approximately 4.5 million metric tons. Much of the increased Soviet production could be attributed to the development of large-scale fisheries in the previously lightly exploited eastern Bering Sea and Gulf of Alaska. It is also interesting to note that the world catch in 1965 increased only 400,000 metric tons over 1964, the smallest yearly increase since at least 1953.

The forecasted potential of living resources that can be harvested from the world's oceans varies greatly, depending on the individual making the forecast. A potential ranging from 80,000,000 to 1,000,000,000 metric tons annually has been suggested. Regardless of the estimates, it is apparent that fishery production can be increased considerably before even the minimum potential is reached.

Fishery engineering will play a leading role in future increases in the world's fish catch just as it did in the period from 1946–1957. Some of the increase will come from species inhabiting fishing grounds on the continental shelves but whose harvest depends on the development of economically efficient harvesting methods. An even greater source is the open ocean. The problems in developing gear and techniques for locating and harvesting the pelagic species inhabiting the open oceans, where fish occur in low concentration levels, are much greater than on the shelves. Forms not presently utilized, such as the small shrimp-like organism called euphasiids, offer great potential but their efficient harvest will require development of highly efficient extractive technology.

Fishery Engineering Efforts in Various Countries

Most of the fishery engineering efforts in various countries after World War II and until recently were considered to be in the gear research phase since application of a total systems concept for fishery resource utilization, from locating fish through landing the catch at

TABLE 1. WORLD FISH CATCH BY REGIONS

Continent	1938	1948	1955	1957	1959	1962	1963	1964	1965
					(million metric tons)				
World total	21.000	19.600	28.900	31.500	36.400[a]	46.400	47.600	52.000	52.400[b]
Africa	0.630	0.950	1.820	2.070	2.250	2.630	2.750	3.020	3.060
America, North	3.160	3.590	3.950	3.890	4.260	4.490	4.380	4.300	4.430
America, South	0.250	0.480	0.830	1.170	2.950	8.290	8.420	11.010	8.980
Asia	9.700	6.850	11.900	13.720	15.870	18.630	18.970	19.290	19.950
Europe	5.680	6.150	7.840	7.880	8.170	8.640	8.980	9.740	10.810
Oceania	0.800	0.900	0.100	0.110	0.120	0.140	0.140	0.150	0.150
USSR	1.523	1.485	2.495	2.531	2.756	3.617	3.977	4.476	4.980

[a] Includes fresh water catch of 5.98 million metric tons.
[b] Includes fresh water catch of 7.15 million metric tons.
Source: Yearbook of Fishery Statistics, 1965, Vol. 20, Food and Agriculture Organization of the United Nations.

dockside, was approached in individual steps, each with a unique problem to be solved, rather than looking at the entire process as a system. There was a tendency to place emphasis on the physical engineering properties of the fishing gear and on the behavior of the gear itself with little attention to the importance of considering fish behavior, which is now a must for successful gear development. A review of recent efforts in the field of fishery engineering as related to the harvest of fishery resources follows.

Electricity. The use of electricity to harvest fish has been experimented with by many nations but only indirect application has been adopted by the fishing industry. Various applications of electric fishing currently undergoing engineering research and development include: electropumping from nets, light in conjunction with electropumping devices, and electrotrawling. Fluorescent and mercury lamps utilizing ac and dc power, however, are used by the Japanese and Russians in the western North Pacific to aggregate saury (a close relative of the flying fish) for capture with stick-held dip nets. As a result of experiments by Russian scientists and engineers, a system of catching anchovy using lights in combination with pumps was devised. This was introduced into the Russian anchovy fishery of the Caspian Sea, and by 1961, catches by the "pump light" method had reached almost 93,000 tons of a total anchovy catch of about 200,000 tons. Recently, the Soviets have introduced a new system of "pump light" fishing for anchovies in the Caspian Sea, which, in addition, utilizes electrical current. This system combines light attraction with an electrical anode reaction system whereby the fish are attracted and held in a sphere within a sphere. After being attracted and concentrated in the inner sphere, the fish are removed by means of a pumping system. This device is placed in the midst of fish concentrations that have been detected with echo sounders. Catches up to 70 tons a day have been reported using this system. The electrical field is developed with a 20–25 kW current. The direct application of electricity to catching fish has long intrigued engineers and fishermen. Although attempts at direct application were made as long ago as 1895, success was not attained in a commercial fishery until the 1920's and even now the success is limited to fresh water.

Electrical fishing depends upon the change in behavior patterns of fish caused by passing the proper form of electricity through the water. When proper conditions are established, fish will turn in the direction of the positive electrode (the anode) and swim toward it. They will continue swimming until they arrive within a certain distance of the anode where, depending upon the species of fish, the size of the individual, and the intensity of electrical current, they become immobilized, turn on their backs or sides, and remain in this stunned state until the electric power is turned off.

The major obstacle to widespread adoption of this technique is that electrical fishing in salt water requires approximately 500 times more current than fishing in fresh water. Therefore, if simple ac or dc is used under existing circumstances, fishing would be uneconomical. For example, an electrical fishing range in salt water of 33 ft—spherical radius from an electrode—would require 10,000 kW. Search for a solution to these astronomical electrical power requirements led to the general use of one or another form of condenser discharge pulses or bursts of electrical power applied at intervals. The current is made to increase very rapidly and then decline more slowly to get the proper pulse shape.

Sporadic efforts have been made in the U.S. to electrically guide bottom-dwelling fish toward the opening of an otter trawl being towed across the sea bottom. Using a total power input of 125 kW, with 87 kW the effective power of the net electrodes, the fish are shocked sufficiently to become immobilized and swept easily to the back of the trawl. This method has not been adopted commercially because the equipment is too complex and costly, and sustained increases in catches have not been made.

Electricity is also being used with trawls to shock shrimp from the bottom where they normally burrow during daytime. The system uses ac generated primary power transmitted by electrical cables to an electronic pulse generator on a trawl door. It is there converted to dc and released at a fixed pulse rate through an electrode in front of the trawl net. This low-voltage field which requires a 3 kW generator causes an involuntary kick-in response which forces the shrimp out of the bottom and into the path of the trawl. At the present time, trawling for pink shrimp and brown shrimp in the Gulf of Mexico is regulated by the shrimp's daily activity cycle. These shrimp burrow in the bottom during daylight. Commercial fishing, therefore, is restricted to night trawling when the shrimp are available, effectively reducing by almost 50% the fleet's fishing activity at sea. The use of electricity was investigated in an effort to make the shrimp come out of the bottom involuntarily where they could be captured. The system offers considerable potential for increasing the efficiency of shrimp trawling and its adoption by commercial vessels is expected soon.

Mid-water Trawling. Considerable effort has been devoted to development of mid-water trawls and trawling systems in various nations during the past 20 years. Since many schooling fish inhabit the water column outside the reach of conventional bottom or surface gears, mid-water trawls appeared to be the most logical gear for fishing this zone. Development of a commercial fishery based on the use of this gear is an excellent example of the need for the systems approach in fishery engineering efforts. Such an approach was used in developing a mid-water trawling system for Pacific hake, which generally are too far above the ocean floor for bottom trawls and too deep for purse seines, on the west coast of the U.S. This involved analysis of the behavior of the fish; developing sonar search modes for locating fish concentrations; designing the net and specialized aluminium trawl doors; developing telemetry gear to determine the depth of the net; and determining the fishing tactics. This system of fishing, utilizing a drum trawler, has provided for effective harvest of Pacific hake from Puget Sound, Washington, as well as waters off the Washington coast. For example, between January and June 1966, in Puget Sound, a 53-foot vessel equipped for drum trawling made 126 drags and caught 1,750,000 lb of hake with an average hourly catch of 19,000 lb. The superiority of this type of gear over similar drum trawlers fishing conventional bottom trawls in the same vicinity was emphatically demonstrated since their catches averaged several times less than the mid-water trawl even though the drags were longer. One of the vital keys to the success for any mid-water trawling system is to know the depth of the net so that the fish schools can be intercepted. This is done with a telemetry system which continuously transmits the net depth to the wheelhouse through electrical conductors in the trawl cable from a sensing unit contained in a stainless steel housing. The unit is attached to the end of the trawl cable. In addition to this hard-wire type telemetry system, remote acoustical links have been developed which provide similar types of information. Mid-water trawls have also been developed and successfully used in the herring fisheries in various areas of the world.

Conclusions

These are but a few of the examples of the potential fishery engineering holds for efficient harvest of fishery resources. Fishery engineering programs related to fish harvesting currently underway in various nations, in addition to those previously mentioned, include such aspects as: Fish behavior in relation to gear; sytems for harvesting euphasiids; methods of artificial aggregation; fish pumping systems for unloading nets, and computer-operated vessels. Remote-sensing instrumentation technology has offered a new approach and dimension to fisheries resource development. Feasibility studies now being conducted in the U.S. for NASA by the Navy's Oceanographic offices' Spacecraft Oceanography project, are investigating the usefulness of a host of instrumentation techniques for gathering oceanographic and marine resource data from satellites. Techniques are being developed for direct detection and assessment of fish stocks. Although these programs are currently classed as "feasibility studies," it is not improbable that useful data on fishery resources gathered by spacecraft will be routine within a decade.

The future rate of advance resulting from applying fishery engineering to fisheries will depend on the extent to which the talents of naval architects, engineers, fishermen, economists, equipment manufacturers, and fishery scientists are combined into coordinated efforts aimed at improvement of the total harvesting process. If the extent of applying this system approach is great, the rate will be fast.

EDWARD A. SCHAEFERS

Cross reference: *Fishing Gear and Methods.*

FISH GLUE

Fish glue has been manufactured since 1870, and it may be possible to trace its history before that time. The production of fish glue (and resultant sales) has been subject to the problems of raw materials, changing source of supply, and technological developments.

Manufacture of fish glue in the U.S. began in the area of Gloucester and Boston, as this was the center of the salt fish industry. Here the fresh fish was cleaned and split, and put through a salting process to cure and preserve the fish for marketing. After salting, the fish was skinned and trimmed before being packed for shipment. The skins and trimmings were all parts that could be used in the manufacture of fish glue. Thus the glue manufacturer situated his plant near the source of the raw material.

Other references have mentioned the use of cuttings, heads, and bones in the manufacture of fish glue. These parts could be used, but generally the yields were poor and the grade of glue was difficult to control. The production from this waste was then a marginal operation which became uneconomical as time progressed and new adhesives were developed.

Although there were other fishing centers around the U.S., a large supply of fish skins was not available at these centers. Neither canned nor fresh fish was skinned. The best glue source was the skin of the cod, and these were obtained in volume as salted skins in the New England area.

In recent years (since 1960) there has been a change in the location of ground fish (such as cod, haddock, and cusk) and also in the processing of this fish. New England waters do not have the volume of cod that were caught in the past. With the advent of frozen fish, as fish sticks, fish steaks, or fillets, much of the fishing industry has moved to the Maritime Provinces of Canada. The fishing areas off the Maritimes are abundant with ground fish. With the encouragement of the Canadian government, new frozen fish processing plants have been established around this long coastline. Here, the fish are filleted and skinned, the skin being collected for fish glue. It is also possible to freeze the skins in blocks, to preserve them until ready to use. Since salting cod requires considerable hand labor, this has likewise moved to the Maritime Provinces near the source of the fish. Because of this change in the fishing industry, fish glue plants that once existed in Gloucester as recently as 1958 have closed down. Presently there are no fish glue plants in the U.S. Some fish glue is obtained from England and France, but the bulk of it is produced in the Maritimes.

Another factor which affected fish glue was the changing technology in adhesives. Fish glue has always been more expensive than animal glue, but since fish glue is liquid, it became a good household item which was easy to use. More recently newer adhesives such as polyvinylacetate emulsions, or "white" glue, have been developed and heavily promoted by chemical manufacturers. These are quicker-drying, practically colorless in wood glueing, and worked well in household applications. As a result the bulk of the consumer market for household use now has shifted over to white glues. Fish glue does have special properties (described later) that make it unique as an adhesive. Where fish glue is made with special attention to these properties and with proper quality control, it meets a need which has not been met by synthetics or other natural products.

Chemistry

All gelatin or glue is derived from collagen, which is found mostly in the skin and bones. Collagen itself is not soluble in water, but can be broken down with heat in the presence of water and other chemicals to produce a water soluble product. Whether the end product is gelatin or glue will depend upon how it is processed and how it will be used. Since glue is used mostly as an adhesive, its qualities will be looked upon differently than a gelatin, which is used for edible or photographic purposes.

Collagen is one of the most prevalent and important proteins in the animal kingdom. It is a long-chain protein made up of varying amounts of 20 amino acids. The proportion of these amino acids will differ slightly depending upon the source, whether fish or animal, or a particular species of fish or animal. Fish skin collagen breaks down more easily than animal skin collagen, both by heat and by enzyme activity. The latter accounts for the odor found in an improperly run fish glue plant. Fresh fish skins (or salted skins that have been washed) will spoil very quickly in warm weather if steps are not taken to prevent bacterial growth.

One of the major differences between fish and animal glue is that a water solution of fish glue is a liquid at room temperature, whereas a water solution of animal glue is a gel. Conversely, a dried fish glue will dissolve in water, whereas a dried animal glue will swell and does not dissolve until the water is heated.

Glue and gelatin are amphoteric materials—they will behave either as a base or an acid. This is because there are both carboxylic acid and amino end groups in the molecule. Fish glue is a long-chain molecule with an estimated molecular weight between 30,000 and 60,000. The isoelectric point, which might be termed the pH of neutrality, will be in the range of 5.0–6.5, depending upon processing conditions. The chemical structure of animal gelatin differs from fish gelatin as follows. The amino acid content of animal gelatin is higher in hydroxyproline and proline and lower in methionine, serine, and threonine. The reason for the gelling of animal gelatin is usually attributable to more hydrogen bonding which holds the water solution in a more rigid structure.

Production Processing

Washing. The first step in the manufacture of fish glue is the preparation of the skins prior to cooking. They must be washed thoroughly to remove salt and any soluble material. Salted skins will take a longer time to freshen, and the wash water drawn off can be tested either by chloride content or conductivity of salt to determine the progress of the washing cycle. The washer is common to both animal and fish glue manufacturers. It consists of a circular wooden tub, about 15 ft in diameter and 4 ft high, into which a 1 ft depth of skins is charged. A cone-shaped wooden roller is dragged around the tank from a pivot in the center as water is added and continually withdrawn. The wooden roller moves the skins about the washer, and tends to squeeze them as it rides over the bulk of them. Considerable cold water is required (preferably under 50°F) and a 12–24 hour cycle is not unusual.

Cooking. Following the washing cycle, the skins are pitched into rectangular cooking vats. These vats may be 4 × 12 × 4 ft high, with a false bottom. A limited amount of water is added to the skins and then heated with live steam from steam spargers located in the false bottom. Acids such as acetic or hydrochloric may be added to assist in the breakdown of the skins. The skins are cooked for a given time and at a given temperature, and the resultant broth or fish glue liquor pumped to the next step. The broth of a well-run fish glue plant will have a mild and very pleasant odor. The concentration of the glue liquor at this point will vary between 3–8%, probably averaging around 5%. After the first run is drawn off, a second cook is made under similar conditions. The residue is then dried and produces a fish meal used in animal feed supplements.

Intermediate Processing

Depending on the end use of the finished glue, the glue liquor can be put through further processing where a clear gelatin grade is required. Special chemical steps are taken (proprietary in nature) which will remove the impurities and insoluble products from the glue liquor. Where special high sensitivity is required for photographic purposes, the liquor can be treated to insure a controlled end product.

Evaporation

Finish evaporation consists of removing the excess water and concentrating the glue liquor to the 45–50% solids of a commercial product. Either a coil evaporator or a vacuum evaporator can be used, the latter having the advantage of not exposing the glue to a high temperature for a long period of time.

Finish Blending

After evaporation the glue is tested for viscosity and solids. In use as an adhesive, the glue is adjusted for viscosity; if for use as a photographic base, the glue is adjusted for solids. The glue cannot be adjusted for both viscosity and solids. The cooking process will determine the breakdown of the glue, which controls the relationship between viscosity and solids.

Other additives such as preservatives and reoderants are also added at this time. Since fish glue is a protein, it must be protected against bacterial attack. Cleanliness is essential throughout the plant, to avoid a sudden buildup of bacteria which are difficult to erradicate and will cause undue breakdown of the glue as well as an undesirable odor. Most of the preservatives are phenol types, and these are added in ranges of $\frac{1}{8}$–2%. The reoderants help mask any odor that might be present, but are not necessary if the glue or gelatin is processed under special care.

Adhesive Properties and Uses

As an adhesive, fish glue has the advantage of high initial tackiness. Once coated and allowed to dry, it has excellent remoistening properties. This allows for easy reactivation of the adhesive by water at a later time for bonding. Although fish glue has good solvent and heat resistance, it has poor water resistance. Other properties of a typical fish glue as an adhesive are as follows:

Color	Light caramel
Odor	Mild, usually that of a reodorant that is added to the glue
Viscosity	4000–7000 centipoises
Solids	43–50%
Temperature range	30°F to +500°F
Shear strength	3200 psi with 50% wood failure (ASTM D 905)
Open time	$1\frac{1}{2}$–2 hours
Time to tack	1 min
Weight/gallon	$9\frac{3}{4}$ lb

There are several advantages which make fish glue unique for specific applications. It has good adhesion to many surfaces, such as glass, metal, wood, cork, paper, and leather. It has high initial tack when first coated or when remoistened, but it has slow setting for wood-bonding applications when several joints are to be positioned before clamping. Fish glue is resistant to solvents and can be used in applications where solvents are used to clean off inks. The dried film does not soften with heat. From a production standpoint, a notable advantage is that fish glue is easily thinned and cleaned up with water. Typical applications are as follow:

(1) As an additive to adhesive formulations in the manufacture of remoistenable gummed paper packaging tapes.
(2) Wood glueing when long open times are needed for assembly operations.
(3) Paper bonding of heavy-grade box board in packaging.
(4) Bonding of manila paper for identification tag manufacturing.
(5) Any application where it is desirable to supply an adhesive coated surface which is to be reactivated much later by simple water remoistening.
(6) In die rooms as an adhesive when corking steel rule cutting dies and when used for make-ready on cutting presses.

Fish Glue in Photography

Photoengraving glue can be obtained in various degrees of purity, the gelatin grade being crystal clear. This grade is easy to use as a coating. The gelatin can be made sensitive to light by the addition of varying proportions of chromic acid or salts such as ammonium bichromate, potassium bichromate, or ammonium chromate. Exposure to light (especially actinic rays from an arc lamp or other light source high in ultraviolet) will make the gelatin insoluble. Fish gelatin can also be used as a base for silver halide coatings. The gelatin has a protective colloid effect, in that silver chloride can be precipitated and will remain in suspension in the gelatin.

Because photoengraving glue is liquid and can be made light-sensitive, it has several unique applications. One interesting application is in wash-off emulsions for light-sensitive papers used in engineering. A silver halide emulsion with a light sensitizer is coated on a film backing. The resultant sheet can be exposed, washed out with water to remove the soluble area, and then developed with a photographic developer to darken the exposed areas.

Photoengraving glue has been used for many years in the processing of printing plates. An image can be printed photographically on a metal plate which is then baked to make the image insoluble. The open areas of the metal plate can then be etched away with acid to produce a raised image which is the printing surface.

The bichromated coating is known in the trade as a photo resist, and it has led to a new industry known as chemical blanking or chemical machining. Thin metal parts can be made by photographically exposing identical images on the front and back of the metal sheet. The open areas are then etched out, leaving a metal part which corresponds to the photo resist image.

The process has many advantages. Very complicated thin metal parts can be produced and controlled to close tolerances. There are no burrs left on the edges, as would be in stampings. The process does not require any complicated dies.

The following is a list of parts that can be produced by chemical machining or blanking: printed circuits, electric shaver heads, fine-mesh screening, razor blades, name plates, instrument panels, and special electronic parts.

Photoengraving glue has the following advantages as a photo resist base over many synthetic photo resists:

(1) Since water is the solvent, it does not require special equipment for volatile solvent.
(2) A photo resist made from photoengraving glue can be removed easily with caustic after the acid etching is completed.
(3) Cost of a photo resist prepared from photoengraving glue is very reasonable.

ROBERT E. NORLAND

Cross reference: *Cod Fishery.*

FISH HARVESTING

The vessels of the western world engaged in the North Atlantic are fishing on a strictly commercial basis in which profit and loss accounts take a governing place. With so many nations involved, it is most unlikely that any measure of voluntary sacrifice or restraint would be acceptable to all. However, with the present escalation in catching power it is quite definite that, before long, the unrestrained "hunting" will have to be curtailed by international agreement.

It may be appropriate to indicate the types of fishing vessels which Great Britain is using in the waters of the northern hemisphere. Most of the types mentioned have their counterparts in the fleets of other maritime nations.

Britain has approximately 1,000 seine net vessels, ranging from 45–80 ft in length. Most are built of hardwood and powered by engines ranging from 150–350 hp. A large proportion are operated from Scotland and are mainly skipper or family owned. Many of them are not operated in the same way as the Danish seine netters; usually they tow their gear after the trawl and

Fig. 1. Near-water trawler, 103 ft over-all. Five of this class were built for Boston Deep Sea Fisheries Limited in 1965–1966.

Fig. 2. Middle-water trawler 154 ft over-all. Two of this class were built for Boston Deep Sea Fisheries Limited in 1965–1966.

seine net ropes have been shot. This is a type of fishing that has expanded very greatly since World War II and is being successfully operated. These, fishing vessels, mainly inshore, are now deriving considerable benefit from the new limit lines which came into operation in 1964, but further benefits are expected.

Some small development with purse seining in Britain has taken place during the last 2 years. Following the Norwegian successes, a large development in this new type of fishing is not likely to take place here in the near future because British fish meal factories cannot, at present, deal with large quantities of industrial oily fish.

Four types of vessel comprise the main trawling fleet. First, there are the near-water trawlers (Fig. 1). These lie within the range of about 90–120 ft over-all; 22–26 ft beam; 10 ft 6 in.–12 ft 6 in. molded depth; gross tonnages ranging from 150–300 and horsepowers from 300–800 giving speeds of 9–$11\frac{1}{2}$ knots. These vessels operate in the North Sea, the Irish Sea, and other waters fairly close to the coasts of Great Britain and the other maritime countries of western Europe.

Second, there are middle-water trawlers (Fig. 2) which in length range from 120–150 ft over-all; 25–29 ft beam; 12 ft–15 ft 6 in. molded depth; gross tonnages range from 300–500 and horsepowers from 700–1200 giving speeds of 11–15 knots. These vessels mainly fish in the deep waters of the continental shelf on the west and north of Scotland and on the fishing grounds off the

Fig. 3. Distant-water trawler of the Boston (England) fleet; named after D. B. Finn, former director, Fisheries Division, F.A.O.

Fig. 4. Freezer stern-trawler, 240 ft over-all, the flagship of the Boston (England) Fleet. Two of this class were built in 1966.

Faroe Islands and around Iceland which are still open to British trawlers.

Third, there are the distant-water wet-fish trawlers (Fig. 3). from about 150 ft to a little over 200 ft over-all, with a beam of 28–33 ft; tonnages range from 500–900 and speeds vary from 12–15 knots. These vessels spend most of their time fishing off Iceland, the Norway Coast, in the Barents Sea, Bear Island, Spitzbergen, Greenland, and, occasionally, on the Newfoundland Banks and at Labrador. Very few distant-water wet-fish trawlers have been built during the last 3 or 4 years as owners have been considering the economics of stern-fishing freezer trawlers which constitute the fourth main type (Fig. 4).

Britain now has about 36 such freezer stern-trawlers operating or on order. 20% of this total fillet and freeze their entire catches while the remainder are designed to freeze all their catches as gutted, whole or headless fish. However, the design of the whole-fish freezers lends itself to the installation of a filleting line or small fish-meal plant should future circumstances call for this equipment.

The dimensions of these ships vary from 210–250 ft over-all; 36–42 ft beam; 16–24 ft molded depth; with powers ranging from 1500–2500 hp, giving speeds of $13\frac{1}{2}$–$15\frac{1}{2}$ knots. While a few are diesel–electric-powered, the majority are direct driven by a single motor. Most are capable of freezing from 24–40 tons of whole fish daily.

The great problem concerning this class of vessel is whether there will be enough fish in the North Atlantic to enable them to work profitably. It is estimated that a ship must catch a minimum of 12 tons per day while on the fishing grounds to break even. Some of the ships of other European countries differ from the British, partly because they are fishing for a public with different tastes and hence are fishing for different stocks. Nevertheless, the freezer stern-trawler described here is not peculiar to Britain and may be found in France, Belgium, Holland, Spain, and West Germany, save that most of the freezers used by West Germany fillet their fish at sea rather then freeze it whole or headless.

It seems likely that, for some years to come, British freezer stern-trawlers will follow present trends and that future freezer vessels will either freeze whole fish or freeze fillets processed on board, with only the odd vessel attempting a combination of the two end products.

It is still an open question whether a really large factory trawler should operate with a number of satellite catchers. In Britain, research has been carried out into the most appropriate methods of transferring catches; maintaining quality is an important part of this operation and much has been learned from the commercial trials that have already taken place. But much remains to be done. Consideration has also been given to a rotating mother-ship system whereby, say, six trawlers transfer all their catches to a seventh which would soon depart for home with a hold full of high-quality (newly caught) fish, stowed in ice in the traditional manner. One of the remaining trawlers would then assume the role of mother ship and its place as a catcher taken by a newly arrived trawler and so on. Some feasibility studies were made on this question but, even assuming satisfactory answers to the technical problems involved, conclusions were not always optimistic. However, the techniques of operational research are now being brought to bear in further analysis of the issue.

The word "harvesting" is very expressive. It may well be that "hunting" is a more appropriate word to use at present, particularly when applied to North Atlantic fishing. On all the known fishing grounds in this region it is becoming increasingly hard to find a commercial quantity of sizeable and edible fish of the type in popular demand. Many grounds around the coasts of Europe which were prolific 30 years ago are now commercially barren. It is well known that the main reasons for this are overfishing and the failure to take the necessary precautions to allow small fish to pass through the nets.

Good husbandry simply means economic management. In the North Atlantic management is virtually absent, but fishery scientists have repeatedly sounded the warnings. Britons sincerely hope that both the North-East Atlantic Fisheries Commission and the International Commission for the Northwest Atlantic will heed these very clear and serious warnings. There is an urgent necessity that the growing intensity of exploitation be curbed and effectively regulated to maintain fishing yields at commercial levels.

In the past, all international conventions concerning conservation in the North Atlantic have suffered from, first, a restricted number of signatory countries; second, the meagreness of the measures themselves and, third, the lack of effective enforcement in some signatory countries. These inadequacies must be reviewed in the not too far distant future if there is to be any commercial deep-sea fishing available on the high seas.

To offset the increasing difficulties of catching, reasearch and experiment into new methods, new techniques, and new aids have been greatly increased. Most European trawlers are now carrying three sounding devices. At least two of these are usually graph-type sounders, which give not only the depth of water but also a record of both demersal and palagic fish. Many of these devices are able to magnify a particular range of depths to enable skippers to see how near the fish are to the sea bed or how dense the shoal is.

Many trawlers are fitted with a speedometer. If, after towing at 4 knots, the vessel suddenly steadies down to $3\frac{3}{4}$ knots, the skipper, on reading this, will realize that he has either a lot of fish or mud in his net, or that he is in trouble for some other reason. He will then decide to haul, lest he loses his gear together with all the fish in it. Without the speedometer he might continue towing until serious damage was done to the gear.

British distant-water trawlers carry very powerful radio sets, easily capable of world-wide transmission and reception. Radio telephone sets are also fitted. Owners often speak to their ships in the Arctic, and across the Atlantic to Greenland, the Newfoundland Banks, and beyond. A smaller set with a range of possibly 500 miles is normally fitted in addition for ship-to-ship operation. Most fleets also have short-range VHF sets and, in this connection, private frequencies are allotted. Thus all the vessels within a company's fleet can talk to one another without the ships of other companies listening in.

Most of the larger trawlers now carry two radar sets, both capable of utilizing a range of from $\frac{1}{4}$–60 miles. A reserve against breakdown is a highly prudent precaution. It is often difficult to know, in the waters adjacent to many countries, where the limit line is situated without a really good radar. Straight base lines are often drawn, not merely from headland to headland, but from one rock to another in the ocean. As a result, a 12-mile fishery limit may be 50 miles from the nearest land. For similar reasons British trawlers usually carry two direction finders.

There is no doubt whatever that owners have benefited greatly by making trips to sea on their trawlers. But there is also a great deal to be learned from research and development scientists, particularly those in the mechanical and electrical fields as distinct from marine biology. Nowadays, such scientists are to be found on ships as a result of the new emphasis Britons have given to research.

Considerable sums of money have been spent on attempts to develop more efficient trawls. In the process a great deal has been learned about the hydrodynamics of towing a trawl through the water. Indeed, a trawl of much wider mouth area has been devised. But, despite all the expenditure, all the pure and applied research, the new trawl caught little more fish then the old type.

Many different types and shapes of otter boards have been tried but no great change has taken place yet. As of old, wooden bobbins are still used; though most vessels now use mainly steel or rubber bobbins. There has been, however, a distinct improvement in net materials. Most nets are made of synthetic fibers instead of manila or sisal. Several vessels have actually made a synthetic trawl last 12 months. Fishing is carried out on such rough ground that quite a lot of good gear is lost. It is not unusual for distant-water trawlers to lose up to $6000 worth of gear in one trip and freezer trawlers up to $30,000.

British fishermen adhere strictly to the regulation size mesh in the nets of their trawls; this reduces the quantity of fish caught in comparison with countries which are not governed by or do not adhere to the regulation mesh sizes.

The decline in the rate of catch has caused voyages to lengthen. This has raised problems of maintaining standards of quality. Boxing the catch at sea in up-to-date containers has been tried out on a pilot scale from several trawling ports in the past year or so. Much valuable information has been gained about the suitability of box design and methods or handling, stowage, and landing of the boxes.

A possible alternative method to straightforward stowage of fish in ice, which stops short of freezing at sea, has been the subject of recent intensive study. Known as superchilling, the technique is already in use aboard Portuguese trawlers and the possibility of applying the principle on British trawlers is being examined. It has long been known that bacterial spoilage, which is slowed down by keeping fish at ice temperature, can be reduced still further by lowering the temperature of stowage a few degrees below 32°F. It is now possible to devise fish rooms that can maintain the iced fish at an accurately controlled temperature of about 28°F, with the result that the useful storage life of the fish can be extended by about a week.

Recently, commercial tests of this new method of storage have been carried out on a British trawler with some measure of success so far as the quality of fish is concerned. One disadvantage of the method is that the fish is partially frozen at this low temperature and needs to be allowed to warm a little, either just before or just after landing, to make it soft enough for filleting and processing. Another drawback is that there is some loss of textural quality, with slightly increased drip loss on thawing. During the British tests some difficulty was experienced in determining the exact weight of fish in the boxes on board ship, but the evolution of a new type of weighing machine will overcome this shortcoming. Moreover, the costs involved are quite substantial.

A number of distant-water trawlers now carry antibiotic ice for fish stowage as a means of extending the shelf-life of the earliest-caught part of the catch.

Research into all aspects of freezing at sea is continuing apace. Experimental work is now being concentrated on improving the design of a freezing plant for use on trawlers and on further development of a number of types of thawing plant for use on shore to handle the frozen catches coming in.

As a result of an extensive series of measurements taken during normal operating conditions, a much more comprehensive picture is now available of the exact requirements of the trawl winch and main propulsion systems of stern-fishing freezer trawlers. These investigations also led to suggested changes in operating practice that should bring useful savings in time, fuel, and wear and tear of fishing gear. They also showed the need for improved winch brakes and controls as well as the desirability of instruments to indicate to the skipper the tension on the trawl warps. Warp meters have been installed in trawlers already and the skippers have been enthusiastic in their reception of this new fishing aid.

The British fishing industry has recognized that the taking of measurements of the kind just described in commercial fishing conditions can form a sound basis for the development of improved ships and equipment. For this reason several owners have provided permanent facilities for research workers in their vessels. The study of the speed of these ships in a seaway as compared with measured mile performance and the recording of stresses, motion, fuel consumption, and other factors affecting cost and performance have begun using automatic recorders and experimental data loggers as well as human observers. Moreover, work is now in hand to develop a system by which the skipper will be continuously informed of net spread, headline height, and other matters concerning the trawl and its behavior when being towed along the sea bed.

The British are far from being alone in giving this added emphasis to matters of research and development. The pace of technological change is quickening throughout the fishing industries of the world and there is now a much greater interest in finding out what the other man is doing. Continued improvements can be expected in fishing gear and methods of fishing; these improvements are already increasing rapidly with the growth of the Japanese and Communist fishing fleets. While nothing must be done to arrest or to slow down the rate of economic progress, this cannot be allowed to lead to ruin. The international control of the fishing industry together with effective measures of conservation and international means of enforcement are imperative if commercial fishing industries are to continue to make their contribution to the satisfaction of growing world demand for protein.

What is now happening in the North Atlantic may happen one day in the South Atlantic where ships of many nations are making good hauls. Britain has not entered this latest field so far but experimental voyages are being made during the winter of 1967–1968.

Undoubtedly, there are many areas of the sea which are virtually virgin. The east coast of South America readily comes to mind, but there are difficulties in establishing shore bases and factories in this area. Again, there are many prolific grounds in the Arabian Sea and Arabian Gulf but the varieties found there are not those to which most consumers of fish are accustomed. It may be necessary, and even essential, however, to re-educated the public. In any event, much more needs to be done to explore and to assess the value of fishing grounds the world over. While the intensive fish farming of inland lakes may provide a long-term answer to increasing supplies of certain varieties on a commercial basis, it must be the world's oceans that continue to provide the bulk of the world's fish supplies.

BASIL A. PARKES

Cross reference: *Freezing Fish at Sea.*

FISHING GEAR AND METHODS

There have been many attempts to classify and categorize fishing gears and methods. It can be done in very general terms, but individual variations and the mixtures of various components of the gears do not provide the basis for sound and rational organizational patterns.

There is extensive literature on the so-called primitive fisheries. There are hundreds of descriptions of gear such as jigs, tangles, small traps and weirs, spears, and other hand tools, the use of animals such as cormorants in the Far East, etc. It will be found in reading these that a very confusing terminology exists in referring to fishing gear and that the same names have different

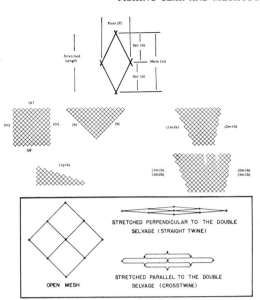

Fig. 2. Netting tapers and mesh nomenclature. (From "North Atlantic Travel Nets," U.S. Fish. Wildl. Service, Sep. 600.)

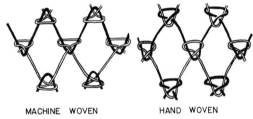

Fig. 3. Fish netting or webbing is a sequence of *loops* or *half meshes* interwoven by knots to form a series of meshes.

Fig. 1. Modern netting machines installed in Japanese fish net factory. The Japanese lead the world in production of synthetic netting. (Photo: "Modern Fishing Gear of the World," Vol. I, FAO.)

meanings in different geographical areas. As of now there is no adequate glossary to solve these problems for the casual reader.

The invention of netting marked a great epoch in human advance. By what stages it came or how the knowledge spread, no one knows; but it is intriguing to know that somehow primitive peoples scattered all over the world contrived fish netting which, compared to modern netting, was certainly crude in material and design but still basically the same general type, and knit with the same netting knot in common use today.

Net crafting as we know it, however, dates from the introduction of netting machines (Fig. 1). There is considerable uncertainty regarding the origin of the first machine, but records show that in the last half of the 18th century and the first of the 19th, individuals in different countries were striving to perfect a netting machine. These earlier machines were not successful because of slippage in the knots they tied, which were of square or figure-eight types. About 1840 the first machine was perfected that tied the knot generally used by hand-knitters and in universal machine netting use today. This knot has several names, the most common being "Trawler's Knot," "Sheetbend," "Fisherman's Knot," "Weaver's Knot."

There can be no doubt that we are in the midst of a major evolutionary change in the materials used to construct fish netting. The use of nylon, polypropylene, and other man-made synthetics has spread throughout the world and is rapidly replacing manila, cotton, and other natural fibers in the construction of netting, lines, and cordage. Synthetics are also being used for floatation devices, replacing cork, wood and glass, and have contributed to the confusion in fishing-gear terminology such as lead line, cork line, etc.,

since in many cases the name was originated on the basis of material used rather than function.

Although synthetics are generally stronger, more resistant to deterioration and abrasion, and have greater resistance to chemical attack and fouling than natural fibers, their use must be planned judiciously, for certain synthetics, while possessing outstanding qualities for one type or section of gear, may be totally inadequate for another.

Monofilament nylon, popular with sports fishermen for spinning tackle since its introduction, has proven to be an extremely effective netting material for gillnets and is now in wide use where not prohibited by local statute. Conservationists, however, have shown some concern in the use of monofilament gill nets since they deteriorate very slowly and lost gear continues to fish with deadly effectiveness. Large quantities

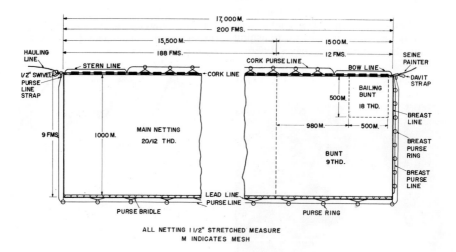

Fig. 4. Detail of purse seine construction and operation.

of such gear lost in a bay or lake could have serious longrange impact on the stocks of fishes susceptible to capture.

With few exceptions such as marquisette weaves, all modern netting is characterized by a 4-side mesh opening configuration, each mesh of equal length. In the net chandler's language, each knot is referred to as a *point* and each length of twine between two points is a *bar*. In this way mathematical formulas may be devised to shape pieces of netting for desired configurations (Fig. 2). Straight edges may be formed by cutting netting with an all-point edge or an all-bar edge. Various tapers can be made by cutting different combinations of points and bar edges. Mesh sizes are most frequently referred to as stretch-mesh length. Sometimes the sizes are described as the stretched opening size or just the bar length.

While net manufacturing machinery has been in use for many years, the hand weaving of netting is still widely practiced, particularly for the very large stretch-mesh netting constructed of heavy twines. The basic difference between machine and hand-made netting is in the alignment of knots (Fig. 3).

The selvage or edge of most netting is reinforced with either heavier or double twine, referred to as *double selvage* and is primarily for strengthening in way of the hanging lines since netting is almost always attached to framing and supporting lines. The framing gives the net its proper shape, while the supporting lines help absorb the main strains acting on the net. The attachment of the netting to the framing together with any attachment, spacing, etc., of fittings such as corks, leads, rings, and chain is called *hanging* and the amount of netting hung to a given length of line influences the mesh configuration.

Netting used for different purposes and in different localities may be hung in with different degrees of fullness and is expressed as a percentage. For example, netting hung in $33\frac{1}{3}\%$ means that 3 ft of stretched netting measures 2 ft hung on the rope.

The stress reaction in netting is such that stretch at one point will cause contraction at right angles to the direction of stretch at another point and is the primary reason that netting must be hung to framing lines; otherwise the stresses encountered would deform or break the net with a resultant decrease or loss of catching efficiency. Hanging the netting to the framing and supporting lines is accomplished in various ways. Sometimes the selvage meshes are hung directly to the line (stapling) and sometimes with intermediate lines (norcelling). The important thing is that the hanging does not shift, thus creating an imbalance in netting configuration. The hang-in coefficient greatly influences the catching efficiency of the net and is usually determined by past experience and by experimentation.

The most important fishing gears and techniques employed today, in terms of directional trends in the expanding harvest of aquatic resources, fall within the following categories: encircling or encompassing, entrapment, entangling, and lines and hooks.

Encircling or Encompassing Gear

Seines. A seine is a portable roped and fitted net generally used to drag in or to encircle fish. The top of the seine is kept on the surface of the water by a floatline (corkline) strung with corks and is held in vertical position by a weighted, leaded, or chained footrope (leadline). They are principally divided into two groups: drag seines and purse seines (Fig. 4).

Drag seines are also called sweep seines, haul seines, beach seines, or shore seines, and are usually used in shallow waters. Drag seines are used, just as the term denotes, to be dragged around a school of fish, thus sweeping in the fish in their path. They may be swept into shore or operated from a boat or boats. The larger seines are hauled by seine winches and tractors located on the shore. They range in length from 15 to several thousand feet with the larger seines usually fitted with a bunt or bag of smaller mesh and heavier twine to hold the catch. When haul

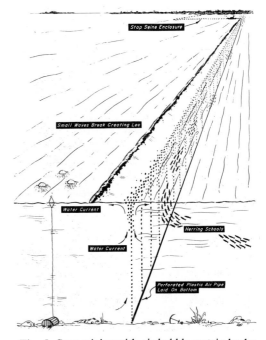

Fig. 5. Stop seining with air-bubble curtain leader. (From: "Air Bubble Curtains," Bur. Comm. Fisheries, Fishery Leaflet 614.)

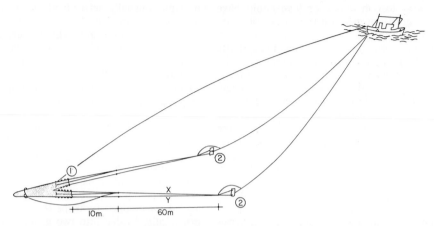

Fig. 6. One-boat midwater trawl. The third wire terminating at the headline is the conductor cable for the headline echo sounding transducer, which records (in the wheelhouse) the vertical opening of the trawl, the position of the trawl in relation to the bottom or surface, and fish as they enter the trawl mouth. When towed by two vessels, the otter doors are omitted. (From: "FAO Catalogue of Fishing Gear Designs.")

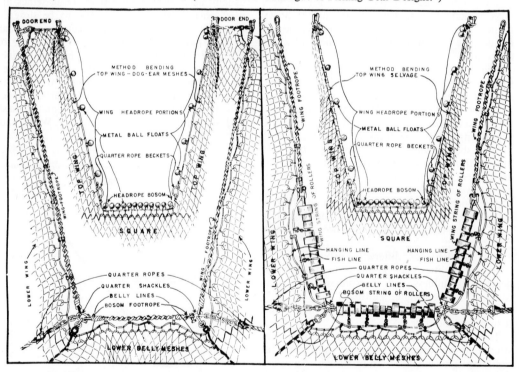

Fig. 7. Diagram of roller attachment to footrope. On extremely rough bottom the rollers are extended to the wing tip.

seines are used to close off the mouth of a bay during high tide to permit harvesting of the entrapped fish at low tide, they are called *stop seines* or *stop nets*. A recent innovation has been the substitution of an air-bubble curtain for netting (Fig. 5).

The lampara net is a type of drag seine consisting of a bag with two large mesh wings. They are considerably more intricate in design than the usual drag seine. In use the bag with the two wings is laid out in a circle, somewhat after the method of setting drag seines. The wings are then hauled into the boat which distorts the circle causing the net to operate as a huge scoop. The wings closing in frighten the fish toward the bag. The net is hung with the lead line in advance of

FISHING GEAR AND METHODS

the cork line which further aids in the scooping process as the wings are closed and the fish scooped into the bunt.

Purse seines are constructed with rings on the leadline so they can be gathered or pursed at the bottom like an old-fashioned money bag, thereby preventing downward escapement. In use, the purse seine is set in a circle around a school of fish, thus fencing in the fish from lateral escapement. Purse seines are used in a wide range of lengths, mesh sizes, and twine sizes, but are all similar in the principle of operation.

Fish schools are located on the surface by means of lookouts or spotter planes and beneath the surface by sonar. The seine is then set around the school and the bottom of the seine pursed. The wings of the seine are brought aboard the vessel either manually or by power block and the fish crowded into the bunt. The fish are removed by brailing with a large dip net or pumped directly from the water into the fish hold.

Purse seines are usually identified by the species of fish they are designed to capture. Thus we have tuna seines, menhaden seines, herring seines, salmon seines, mackerel seines, etc. There is even a record of a lobster seine. Purse seines are perhaps the most efficient gear from the standpoint of fishing power—40–50 tons or more per set is not uncommon in some fisheries.

A compromise between the lampara seine and the purse seine is the ring net, sometimes called the half-purse seine, which, like the purse seine, has purse rings along the foot but whereas the purse seine is practically uniform throughout its length and practically square at the ends, the ring net is made in three parts—a central bag of fine webbing and two wings of coarse mesh and heavy twine closely gathered at each end.

Trawl Gear

An otter trawl or trawl net is a conically shaped bag of netting fitted with various devices

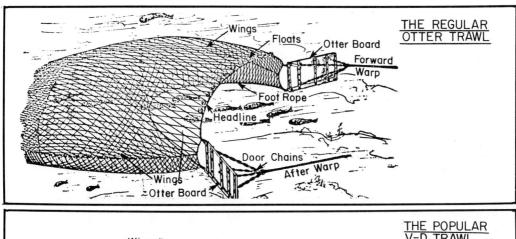

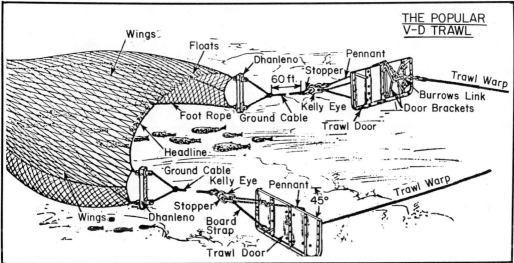

Fig. 8. Evolution of the otter trawl. (After Symonds, New England Trawler Equipment Co.)

FISHING GEAR AND METHODS

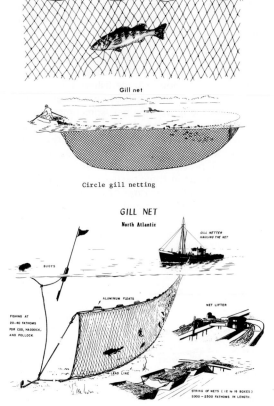

Fig. 9. Gill net operation. (From "Commercial Fishing Gear of the United States," U.S. Fish Wildl. Service, Circ. 109.)

In order of their development came paranzella nets (a link in development between the drag seine and a primitive trawl net), beam trawls, and otter trawls. The paranzella net, like the present-day Spanish "parega" and the Japanese "bull" trawl was towed by two vessels pursuing parallel courses and did not require otter boards to maintain horizontal trawl spread.

The name "beam trawls" refers back to the days of sailing vessels when the trawl net was spread by a heavy oak beam and towed before the wind. The beam was about 12 in. in diameter and 50–60 ft long with high sled runners at either end, keeping the top of the net 4–5 ft off the bottom. The beam formed what is now the headline on a modern trawl with the "square" fastened directly to it. The footrope attached to the runners was about 100–120 ft long and assumed the form of a catenary when towed, thus providing "overhang" or "set back" (Fig. 8).

The beam trawl is no longer used in ocean fishing, its use being restricted to shrimp fishing in certain areas. Nevertheless, in some areas, principally New England, large otter trawlers

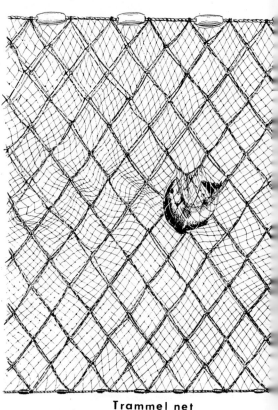

Fig. 10. Trammel net. (From "Commercial Fishing Gear of the United States," U.S. Fish Wildl. Service, Circ. 109.)

to maintain its vertical and horizontal opening. It is towed on or over the bottom of the sea by a trawler or pair of trawlers. It is mostly used for the capture of demersal fish species such as cod, flounder, redfish, etc. Trawl design and rigging varies greatly between different countries and for the capture of different species. Nevertheless, the same general shape prevails regardless of innovations in design and rigging.

Variations in "otter trawl" design and use include mid-water trawls (Fig. 6), dougle-rigged shrimp trawls whereby two small trawls are towed side-by-side on bridles instead of one large trawl, wing trawls, roller-rigged trawls (Fig. 7), herring trawls which employ kites to raise the headline and lead the fish down to the trawl, Danish and Scottish seines (actually a submarine drag seine), and a host of others which are designed to capture a specific species or to work on a particular type of bottom. Regardless of how they are rigged or their purpose, they still assume the shape of a huge funnel while being towed in quest of fish.

FISHING GEAR AND METHODS

are sometimes erroneously referred to as beam trawlers. This is because prior to the introduction of otter trawling the term "trawler" was used to describe a vessel engaged in bottom "longline fishing." To avoid confusion, fishermen put the prefix "beam" to describe this new type trawler, a name which has carried down through the years in certain fishing communities.

The otter trawl evolved from the bean trawl by substituting a headline and a pair of otter boards for the beam. The word "otter board" and the idea back of its use comes from a device used by poachers on English estates. The boards are actually underwater kites and a pair hung right and left serve to spread the trawl wings and keep the mouth open. The trawl in turn, serves

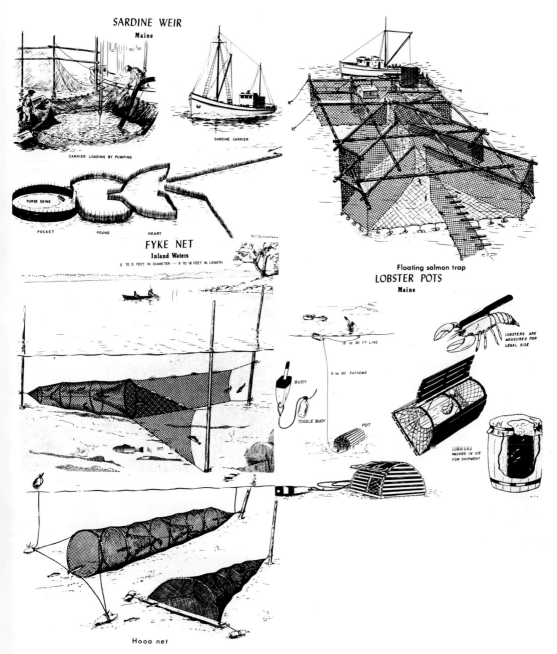

Fig. 11. Common types of entrapment gear. (From "Commercial Fishing Gear of the United States," U.S. Fish Wildl. Service, Circ. 109.)

FISHING GEAR AND METHODS

the same function as the tail of a kite by stabilizing the boards. Originally the trawl was shackled directly to the boards but this method had many shortcomings since the reaction of the headline and footrope on the door varied with their respective tensions.

About 1925 the Vigneron-Dahl system, a vastly improved method of connecting the trawl net to the otter boards, was introduced. The V-D or "dandyline" rig as it is commonly known in many areas, or variations thereon, greatly improved the catching efficiency of the trawl and is in universal use today aboard both small and large trawlers.

Entanglement Gear

Gill Nets. It has been speculated that gill nets were the first nets made by man. They are in widespread use throughout the world and since they are among the most portable of nets and can be used where trawls, seines, or traps are not practical, they account for a large part of the fish catch in many areas. The advent of synthetic materials, especially nylon, has revitalized gill-net fishing in many countries, even in those with highly capitalized fisheries. Gill nets (Fig. 9) may be described as vertical walls of netting in which the actual meshes of the net capture the fish. Any fish striking this wall of netting will become caught in the mesh unless they are small enough to pass through or too large to get their heads in beyond return.

The fishing efficiency of gill nets depends to a large degree on the construction and color of the nets as well as on the size and body shape of the fish to be caught. Therefore special attention must be given to the choice of mesh size to insure optimal size for a given species. It is generally considered that the success of a gill net is inversely proportional to its twine size, so that the finest size twine that will hold the fish is almost always the most productive. One of the great advantages of monofilament nylon netting is the fine twine size possible due to nylon's added strength. The color of the netting has a marked effect upon the fishing efficiency of gill nets. Nets that are properly dyed to afford a fair degree of camouflage generally catch more fish than do nets of unsuitable color.

In recent years the gill-netting operation has become highly mechanized by the development of powered reels and gurdies and net shaking

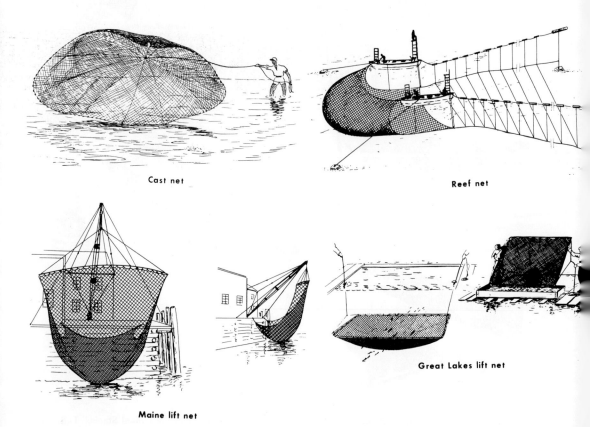

Fig. 12. Cast nets, lift nets, and reef nets.

FISHING GEAR AND METHODS

machines to speed the time-consuming task of removing the fish from the net.

Gill nets may be called set nets or drift nets depending on whether they are anchored or staked into a set position or allowed to drift with tide or current or from boat or buoy. They may be fished at or near the surface, on or near the bottom, or at intermediate levels. They may be fished singly or in small groups or they may be fished in strings or shackles of considerable length. It is not unusual for vessels to fish 100–150 nets joined end to end for a length of 5–10 miles. In fishing they may be lifted to be reset at once after the fish are removed or to be replaced by other nets or groups of nets, or they may be "underrun," thus removing the fish without lifting the net from the water.

A technique called "circle gill-netting" is employed in many areas. In this method the net is used to encircle a school of fish. When the fish are completely surrounded they are frightened into gilling themselves by splashing the water with an oar or some such similar contrivance.

Trammel Nets. A trammel net (Fig. 10) is an entangling net constructed in two or three layers or panels, the inner or middle panel hung slack and of small mesh with the outer panels hung stretched and of large mesh. Fish swim into the net striking the small mesh. In trying to extricate themselves, they push through the large mesh making a pocket of small mesh netting from which they cannot escape. Trammel nets may be used as drift nets or set nets.

Entrapment Gear

Entrapment gear includes traps or weirs, hoop nets, fykes, and pots (Fig. 11). They are all enclosures of netting or other material. They are so shaped as to permit fish to enter with comparative ease but not to escape. In this way they are similar in principle to many animal traps. They vary in size from quite small to very large and are either floating traps or submarine traps, which are anchored, or stake traps which are fixed in position by being fastened to stakes or poles driven into the bottom.

The use of enclosures for trapping fish is very ancient. Evidences of fish weirs or traps among primitive peoples is common. It is recorded that one of the early governors of the Virginia Colony sent details to England of the fish traps in use by the Indians. In Boston, while excavating for the foundation of the public library in the Back Bay section (which is made up of fill), the workmen uncovered vertical wooden poles in a definite arrangement. Investigation by archaeologists revealed that they were the relics of an Indian fish weir. Primitive weirs were generally made of brush or stone, or a combination of both. Large brush weirs are still common in some areas of the world.

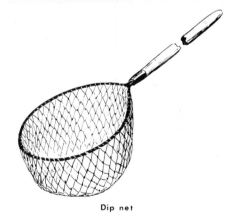

Dip net

Fig. 13. Hand and power-operated dip nets.

Netting traps are a very marked advance over brush or stone traps, since they are portable and can be better located with respect to fish runs. Traps vary considerably in shape, size, and complexity, not only for the different fish they are to catch but also because of local conditions such as bottom topography and consistency, depth of water, tidal action, etc., to say nothing of the individual ideas of the fishermen.

Despite the large variety of trap net designs in use, they are all characterized by three basic

Whale harpoon gun

Spear

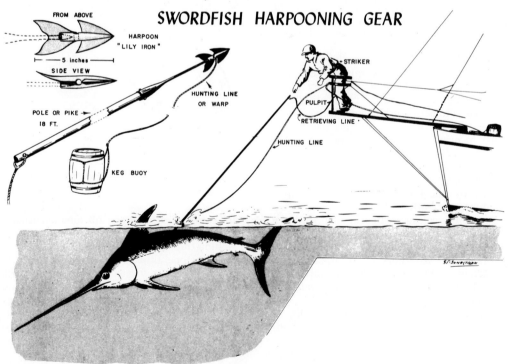

Fig. 14. Harpoons and spears.

sections: (1) The "leader" which acts as a barrier and guides the fish into the trap from distances as great as half a mile from the trap; (2) the "heart," which derives its name from its shape, functions as a nonreturn valve to keep the fish from escaping once they pass beyond the leader; (3) the bowl, pound, or crib in which the fish are held until removed from the trap.

Most traps derive their name from the species of fish they are designed to catch or the area in which they are fished. Thus we have tuna traps, salmon traps, mackerel traps, cod traps, eel traps, crab traps, New Jersey traps, Cape Cod traps, Great Lakes deepwater traps, etc.

Hook and Line Gear

Hook and lines are fished in a variety of ways. Originally used to capture a single fish, they have been developed to the point where many fish may be taken on a single line. Hook and line fishing may be classified as: handline, fished by one man, with or without mechanical reels; baited longline or set line; troll lines; trot lines; and jig line or snag line. Lines can be anchored or allowed to drift and can be set on the bottom or surface or at any intermediate depth. They can be towed behind the vessel as in trolling or attached to poles as in tuna live-bait fishing.

Although not as important a fishing method as before, having been supplemented to a large degree by otter trawling and purse seining in some areas, hook and line fishing is still pursued extensively throughout the world.

Miscellaneous Fishing Gear

In addition to the fishing gear and methods already described, there is a multitude of other harvesting gears used throughout the world. These include scooping gear (dip nets, lift nets, reef nets, cast nets) (Fig. 12), impaling gear (harpoons, spears), and shellfish gear (rakes, dredges, tongs).

Dip nets (Fig. 13) are mesh bags constructed from fiber netting which are hung on a circular, oval, or rectangular frame and usually attached to a handle. Small dip nets are hand operated. The larger ones, sometimes called "kill devils" in some areas, are power operated. Dip nets are also called "brail," "scoop," or "bully" nets. Their main use is to remove fish that have been impounded in other types of gear. When used to capture fish or shellfish they are bridled and submerged and then retrieved rapidly so as to capture any fish or crustaceans which are over the net. Bait is sometimes used to attract the prey over the net.

Lift nets are constructed of netting hung to metal or wood frames. They may be round or rectangular in shape and are power or manually operated. They are suspended below the surface of the water by a haul line and bridles and are sometimes baited. They can be fished on the bottom or at intermediate depths and entrap fish and shellfish when lifted rapidly to the surface.

Reef nets are a type of fishing gear used mainly by certain tribes of American Indians for capturing salmon. The reef net is a type of trap designed to guide the fish by means of rope leaders into a bunt. They are constantly monitored and when the fish are observed to have passed over the square netting in front of the bunt the square is raised to the surface, impounding the fish.

A cast net is essentially a circular piece of netting with lead weights secured around the lower perimeter, which is tucked under to form a pocket. A retrieving line is attached to the center. Cast nets are usually hand woven but may be constructed from machine netting.

Two different styles of cast nets are in common use—the Spanish type and the English type. The main difference lies in the length of the tucks or brails, the Spanish net having tucks of approximately 6 in. length—the English net having

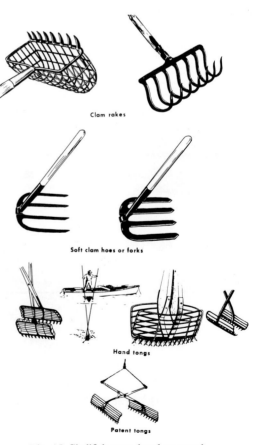

Fig. 15. Shellfish gear, hand operated.

FISHING GEAR AND METHODS

tucks several inches longer than the radius of the circle.

In practice the cast net is thrown over the fish—the outer edge being weighted sinks rapidly to the bottom or to the extent of the retrieving line if the cast net is used in deep water. The leaded edge is then drawn together by pulling on the retrieving line which is attached to the tucks, closing the net and entrapping the fish.

Harpoons (Fig. 14) are used to capture large fish such as tuna and swordfish and mammals such as whales. A harpoon usually consists of a barbed head or "lilly" to which a retrieving line is attached. The harpoon head is mounted on the harpoon shank which is in turn attached to a pole or shaft and detaches itself when embedded in the fish. Harpoons are thrown by hand or propelled from a gun.

Fish spears are implements provided with a number of barbed or barbless prongs. Unlike the harpoon, the prongs are rarely detachable from the shaft. They are hand thrown and have a retrieving line fastened to the shaft.

Shellfish gear consists of rakes, tongs, hoes, grabs, and dredges. They are used in shallow water and on tidal flats to rake in clams and oysters (Fig. 15). Their fishing scope is limited to the length of the handles to which the rake head is fastened.

Dredges, on the other hand, are towed from a boat and are almost unlimited in scope of operation. They are all characterized by a metal frame to which is attached a chain-link or webbing bag. Clam and oyster dredges normally have teeth attached to the cutting or raking bar, and are named for the species they are designed to capture (Fig. 16). Scallop dredges utilizing a cutting bar with no teeth are called "bar dredges"; when used with a combination of cutting bar and chain sweep they are called "chain dredges" (Fig. 17). A variation of the "bar dredge" is the tumbler dredge which is fished singly or in gangs.

Hydraulic or jet clam dredges are employed in some areas for capturing clams. With this type equipment the clams are washed out of the bottom by the action of high-pressure water from a series of nozzles that are attached in front of the tooth bars. The clams are retained in a metal or netting cage. The complete rig is hoisted aboard the vessel to empty the catch. A

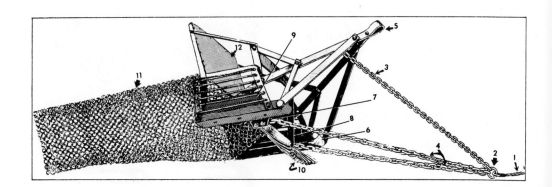

Oyster dredge

Fig. 16. Towed dredges. Top, "Fall River" clam dredge with the chain towing bridle (accumulator not shown): (1) tow warp; (2) tow ring; (3) control chain (for changing the angle of attack of the dredge); (4) chain bridles; (5) bail; (6) runner (shoe); (7) chain bridle connection point (same on both sides); (8) steel bar, vertical stiffener; (9) lead weights (approximately 200 lb distributed on both sides); (10) teeth (7–9 in. in length); (11) chain bag constructed of $\frac{1}{4} \times 2$ in. rings and $\frac{5}{16}$ in. connections; (12) pressure plate. Bottom, towed oyster dredge.

FISHING GEAR AND METHODS

variation of this dredge, used in shallow water, employs a conveyor to transfer the clams to the vessel.

A special type of dredge called a suction dredge is used in the oyster industry to remove the oysters from the bottom to the vessel. Its action is similar to that of a household vacuum cleaner.

Summary

Volumes have been written describing fishing gear and methods in detail; therefore, no attempt has been made to describe any of these gears in great detail. Rather an attempt has been made to provide the reader with a basic knowledge and understanding of the various fishing gears and methods used by man in his pursuit of food from

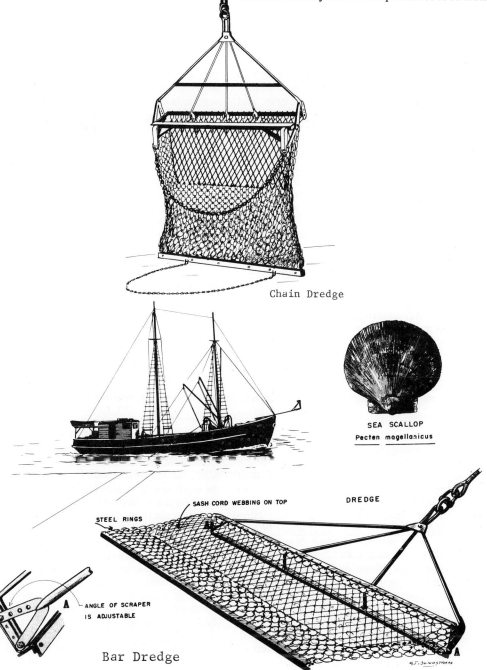

Fig. 17. Scallop dredges. (From "Commercial Fishing Vessels and Gear," U.S. Fish. Wildl. Service, Circ. 48.)

the sea. We have further attempted through description and illustration, to outline the evolution of fishing gear from the primitive to the modern.

The massive array of gears and techniques that are deployed today in marine harvest well testifies that fishing gear is still in active evolution. The application of 20th century technology is just starting to make inroads into the harvesting of fishery products.

Fishing gear and methods which developed slowly over the years solely through empirical methods have advanced rapidly since the turn of the century. This is especially true in regard to the mechanization of gear handling methods and has done much to increase efficiency and relieve the back-breaking labor associated with fishing. The dory schooners and the whalers have passed into tradition.

However, despite the considerable scientific research being carried out today, the design of fishing gear is still for the most part arrived at by conjecture and proven by trial and error methods, since the behavior of the fish is beyond numerical interpretation, making it extremely difficult at this stage to express the catching efficiency of gear purely by calculation. Nevertheless, new fishing methods are on the horizon which are based on the reaction of fish to light, sound, electric currents, and other outside stimuli. Only continuing research utilizing a combination of biological and engineering knowledge will provide the breakthrough necessary to bring fishery harvesting gear and methods in line with the present-day technological advances that have already been incorporated into both the heavy industrial and agricultural areas.

Harvey R. Bullis
Francis J. Captiva

FISHING VESSELS AND SUPPORT SHIPS

About 1.5 million fishing vessels and their supporting ships make up the fishing fleets that harvest the fish and sea animal resources of the world. These vessels range in size and complexity from the dugout to a 715 ft whale factory ship. The wide variety of small boats fishing rivers, lakes, and beach areas will not be discussed. The larger fishing vessels, which take the bulk of the oceans' catch, will be discussed first under five categories according to the type of fishing gear used. Support ships, which do no fishing but are necessary to keep the fishing fleet functioning properly, will be discussed last. Throughout this treatment one must keep in mind that many of these ships are designed for more than one job.

Several terms are used extensively for comparing sizes of vessels and should be explained. The length of a vessel is given as the length overall. Tonnage is used to express the vessel's weight as well as its internal space. Naval architects calculate the weight by determining the volume of water displaced by the hull. The volume is calculated from the plans of the vessel and is divided by 35 cubic feet, the approximate displacement of a long ton of seawater, to determine the displacement tonnage, a term which usually refers to the weight of a fully loaded vessel, including water, fuel, cargo, and stores. This term is also known as loaded displacement tonnage, whereas light displacement tonnage is the weight of an empty ship. The difference between these two weights is deadweight tonnage. Government surveyors use the registered ton, which equals 100 ft^3 of space to determine the internal capacity of a ship for registration. The total inside space is calculated by stringent government rules which delete certain areas, such as the wheelhouse, heads and galley. The space remaining is converted to registered tons and is known as gross tonnage. Net tonnage is gross tonnage minus the internal space of non-earning compartments such as crew's quarters, engine room, ballast tanks, stores, and compartments for steering gear and navigational installations.

Fishing Vessels

The fishing vessels include trawlers, seiners, gill netters, hook and liners, and miscellaneous fishing vessels.

Trawlers. Virtually all fishing nations use trawlers. Although there are more trawlers than any other type of fishing vessel, their total landings rank second to the seiners. The name trawler comes from the baglike net, which is held open by trawl doors and towed by cables from the vessel. Of the several types of trawlers, the side and stern trawlers depicted in Fig. 1 are the most commonly used throughout the world.

The side trawler gets its name because the net is set, hauled, and towed from the side, usually the starboard. Side trawlers have the house aft, a raised forecastle deck, a fore and aft gallows frame for lifting the trawl doors, and a large trawl winch partially covered by the bridge. The engine room is in the aft part of the hull below the house and the fish hold is forward. The crew's quarters are in the superstructure of the house since the forecastle is usually for storage of fishing gear (Fig. 2). The largest side trawlers are the salt cod trawlers, which were built in the late 40's and early 50's by the French and Spanish. Their function is to take white fish, such as cod, on the Grand Banks and salt the catch. The vessels range from 230–275 ft long, have loaded displacement from 2450–3000 tons, deadweight from 1100–1850 tons, and gross tonnage from 1400–1900 tons.

FISHING VESSELS AND SUPPORT SHIPS

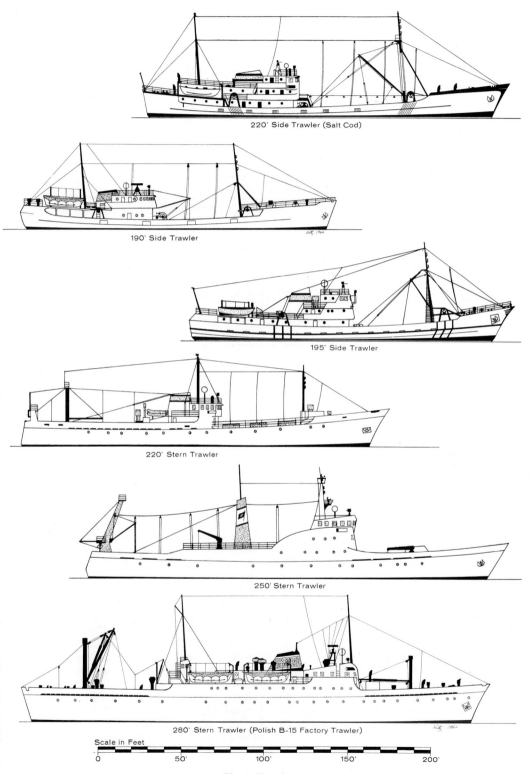

Fig. 1. Trawlers.

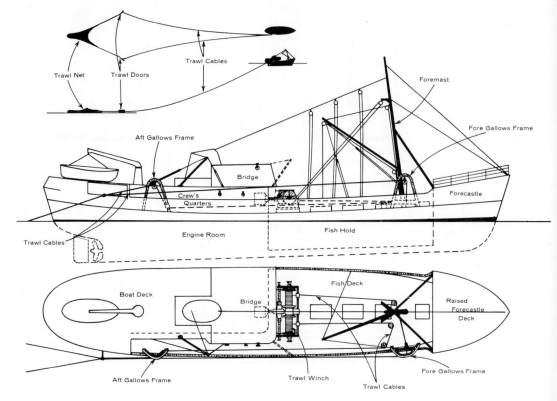

Fig. 2. Arrangement of a side trawler.

In many nations the revolutionary stern trawler is replacing the side trawler. The stern trawler is usually a vessel with a stern chute or ramp where the net slides when the gear is set and is hauled back after fishing. The design first was used in the mid-1950's and the term factory trawler was commonly applied, since the catch was processed on board.

Large factory trawlers are about 280 ft long and have a displacement of 3700 tons. Deadweight is about 1300 tons and gross tonnage about 3200 tons. They utilize their catch to the maximum by freezing the edible parts and processing the wastes into fishmeal and oil. These vessels (for example, the Polish B-15 trawler) have a large superstructure which gives them the appearance of cruise ships (Fig. 1).

Although superstructure arrangements vary greatly, factory trawlers have certain characteristics in common (Fig. 3). They have a stern chute, a stern bridge over the chute, large posts on the afterdeck, aft gallows blocks, and a trawling bridge and cargo booms on the foredeck. The house shields the trawl winch. The afterdeck is for the fishing operation. Below deck in the stern section is the fish processing room, where fish are sorted, cleaned, and headed or filleted, often by machines. The waste material goes into a fishmeal plant which converts it into fishmeal and oil. The dressed fish or fillets are packaged and sent forward on conveyors to a freezer where they are quickly frozen and then stored in the freezer holds.

Small stern trawlers, under 200 ft, usually lack a ramp but have a low and wide aft deck. The arrangement of the deck equipment varies from vessel to vessel, but basically winches are aft of the house and the towing cables lead through the gallows blocks attached either to a gallows frame or to a movable gantry at the stern. A mast and boom or a movable gantry is necessary to handle the net with a large catch. A variety of other accessories, such as rollers on the stern rail, net reels for rolling up the net and movable stern rails, are often seen aboard these boats.

Shrimp trawlers use the double-rigged trawling method that permits two small nets to be simultaneously towed, one from each side. The towing cable is led from a drum of the winch through a block attached to the end of an outrigger boom which holds the nets away from the boat and each other. Small trawl doors are used to spread the net and a short bridle joins the doors to the single cable.

Another way of fishing the trawl is by using two vessels to spread the trawl without otter doors. These craft, called pair trawlers, have been used largely by Spanish and Japanese. Their appearance is similar to the side trawler except that they lack gallows frames. They are generally small, seldom exceeding 130 ft in over-all length.

Other fishing boats, which drag different types of catching devices along the bottom, are grouped with the trawlers. The beam trawler tows a bag of netting held open by a wood or metal beam. The scallop dragger tows a dredge to catch scallops.

Seiners. Seiners (Fig. 4) use two distinctly different nets, the Danish seine and the purse seine. The latter is a large net set around a school of fish to form a circle of webbing. The top of the net floats on the surface and the bottom is drawn together to form a purse around the fish. The Danish seine is a baglike net with long wings. The net, along with its hauling cables, is laid on the bottom in a circle. The seiner tows the net slowly forward by both cables until they come together, closing the net around the fish. Danish seiners are about 60 ft long and are similar to a small side trawler. The Japanese and Russians use Danish seines aboard their side trawlers when fishing for some species of fish. The Pacific coast seiner uses the purse seine. The largest of these are tuna seiners which are about 130 ft long and have a gross tonnage of about 400 tons. Many of these are converted tuna clippers and some of the larger ones are as long as 190 ft.

Purse seiners are characterized (Fig. 5) by a wide flat fishing deck aft, house forward, a seine winch amidships, a pursing davit, and a mast with a boom and power block attached. Below decks the engine is forward under the house and the hold is in the center and aft. The crew's quarters are in either the forecastle or deckhouse. When the vessel is rigged for seining, the net is stacked on the stern and a seine skiff is carried on top of the net.

Some of these 40–100 ft purse seiners are called Pacific coast combination vessels, because they are easily converted by slight alterations of their deck equipment to fish in the most lucrative fishery at the time. They are converted to stern trawlers by attaching a pair of gallows frames to the stern quarter and using the combination seine–trawl winch for storage of the trawling cable. They are also easily made into gill netters, long liners, trollers, bait boats, crab boats, and tenders.

A number of nations use the seiner to capture fish. Off the west coast of South America a large anchovy fishery has developed in recent years and many seiners have been built to harvest this species. The South American seiners are similar to the Pacific coast combination vessel in most

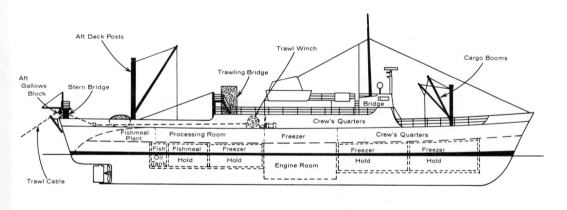

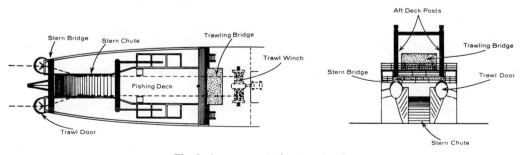

Fig. 3. Arrangement of a stern trawler.

FISHING VESSELS AND SUPPORT SHIPS

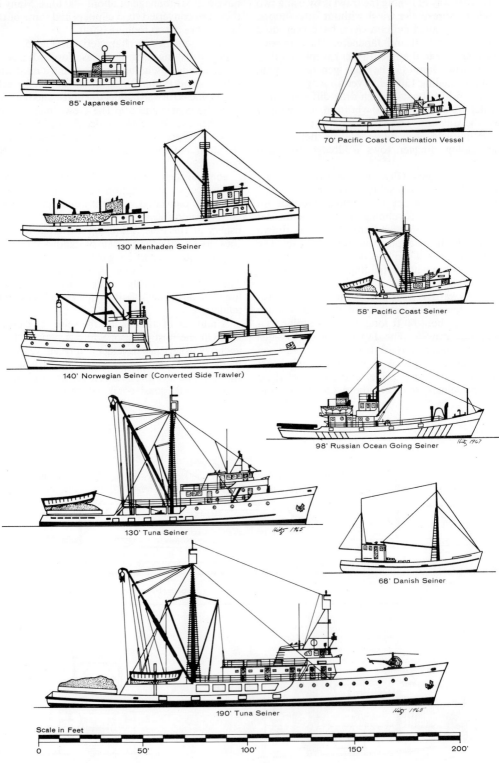

Fig. 4. Seiners.

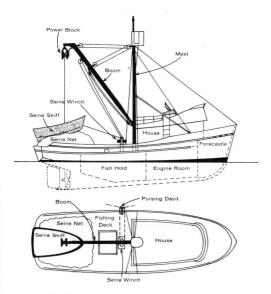

Fig. 5. Arrangement of a purse seiner.

Gill Netters. The third group of vessels also uses nets but these nets are designed to capture fish by entanglement. The vessels are known as gill netters and rarely exceed 135 ft in length.

Gill nets are set in a straight line to form a wall of webbing. The fish swim into the net and become entangled. This method of fishing is popular in the coastal areas but is declining in popularity on the high seas. The best known gill-net fisheries on the high seas are the European herring fishery and the Japanese Pacific salmon fishery.

The European gill-net vessels fish for herring and are known as drifters because they drift with nets, letting the nets act as a sea anchor. Vessel length does not exceed 130 ft since larger vessels may break the attachments to the nets while drifting. The boats look like side trawlers but differ in several ways (Fig. 6). The foredeck is flush without the characteristic raised deck. Instead of a large trawl winch there is a powered capstan to pull in the gill-net warp while a powered roller on the rail pulls in the net. A fair lead and a small roller built into the starboard rail are used for setting the gear. The engine room is aft below the house superstructure. The crew, which is unusually large for the length of the vessel, is lodged in the house and forecastle. The hold is divided into three sections; one for the nets, one for the warp (cable for attaching the net), and the last for the catch, which is salted in barrels.

Since the herring fishery is seasonal and the boats are not used during the off season, a combination vessel evolved, the drifter–trawler. Some of these are longer than 130 ft. Examples are the Russian 167 ft *Okean* and 178 ft *Mayak* class trawlers which were modified for fishing with drift nets and trawls. Both are characteristic side trawlers.

The Japanese use a salmon drifter in the Pacific. It ranges from 75–90 ft in over-all length and has a gross tonnage of 65–96 tons. The salmon drifter differs from the European drifter by the raised forecastle deck and the large net bin and net roller on the stern.

Hook and Liners. The fourth group uses the hook and line instead of nets. This gear is probably used by more people to capture fish than any other type of fishing gear. Vessels utilizing this method are known as hook and liners (Fig. 7), and include primarily longliners, pole and line boats, and trollers.

The long line is constructed of a main line with a number of branches to which baited hooks are attached. The gear is set on the bottom or in mid-water. Before 1940, this method was widely used in European waters but now is rarely used because of the increase of trawler-caught fish of good quality. In the Japanese tuna fisheries and north Pacific halibut fishery, however, long lining is an important method of capturing fish.

respects except that they have a greater freeboard and carrying capacity for the length, which varies from 40–60 ft. Russia has a number of ocean-going seiners about 100 ft in total length. Some of these vessels are similar to the Pacific coast seiners, whereas others have the house aft. All of them have a low flat stern to carry the seine and can be converted to trawlers. Norwegians also use purse seiners on the high seas and in recent years have converted side trawlers to seiners. These are in the 145 ft class and have a pursing davit instead of a forward gallows frame. On the boat deck where life boats were formerly carried, there is a bin for the purse seine and power blocks are arranged to bring the net aboard. The Japanese have a different type of seiner, about 85 ft long with a gross tonnage of 80 tons. It has a high bow, a central house with the engine room below, a hold forward, and a crew's space aft under the flat stern deck, which is used for carrying the seine. They also use a vessel similar to the Pacific coast seiner.

A two-boat seining system in which each boat carries half of the seine has been used in several fisheries. With this method boats run alongside each other until they are ready to set, at which time they spread apart to form a circle of net. The Japanese store the catch in the holds of both vessels, which are about 60 ft long. In the menhaden fishery off the east coast of the U.S., the seine boats are only 32–36 ft long. Their catch is placed aboard a third and larger vessel, the menhaden seiner which ranges from 85–200 ft in length. In the menhaden seiner the house is forward, the hold in the center, and the engine room aft. The seine boats are carried in davits.

FISHING VESSELS AND SUPPORT SHIPS

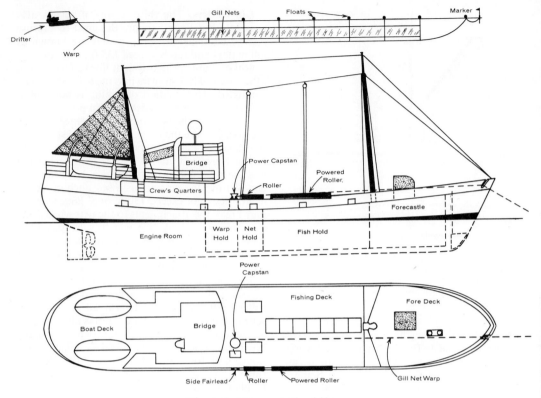

Fig. 6. Arrangement of a drifter.

The Japanese have a very large fleet of long liners, the largest being tuna long-line mother ships that do no fishing but carry 4–6 catcher boats. Catcher boats are 50–55 ft long and weigh from 31–38 tons. The gross tonnage ranges from 12–22 tons. A smaller version of the mother ship is capable of long lining and carries 1–2 catcher boats. Large tuna long liners are 127–245 ft long, have a displacement of 2100–3020 tons, and a deadweight of 1240–1630 tons. The gross tonnage ranges from 1000–1500 tons. Long liners less than 215 ft have a light foremast instead of the large and heavy booms used on the long-line mother ships for lifting catcher boats.

The Japanese long liners are similar to the side trawlers. Long lines are stored on the boat deck over the crew's quarters. A long-line hauler and a conveyor for moving the long lines aft are on the fish deck between the house and forecastle. A roller is on the starboard rail near an opening in the bulwarks to ease the landing of fish. The freezer room is below the bridge.

In the north Pacific halibut fishery the Canadian and U.S. long liners are mostly Pacific coast combination vessels. A baiting table and a small stern chute for setting the gear are on the port side of the stern deck. A skiff or dory on the starboard side of the stern is used for storage of gear as well as a life boat. A roller is mounted on the guardrail amidships and a gurdy is attached to the winch drive for hauling the gear. The boom is usually lowered and a steadying sail placed on the mast.

Pole and line boats usually fish for tuna and mackerel. The basic gear consists of a short bamboo pole attached to a fishing line with a hook. This gear is used by the U.S., French, and Japanese fishermen.

Boats used to catch tuna off the Pacific coast by the U.S. fishermen and off the western coast of Africa by the French fishermen are known as bait boats. Tanks off the afterdeck carry live bait which are used to chum the tuna to the stern of the boat where they are caught by feathered lures. These vessels have the house and engine room forward and the fish holds aft and amidships. The afterdeck is close to the water and the fishing is carried out here or from portable racks attached outside the rail. French pole and line vessels are about 100–113 ft long and have a gross tonnage of 260 tons. The larger American vessels of 90–170 ft long are known as tuna clippers.

Japanese use a similar method on their skipjack tuna pole and line boats. These have a distinct appearance because of their flush decks and extended platforms on the bulwarks which run forward to form a bowsprit. The rest of the vessel is arranged similarly to the Japanese tuna long liner, and most of these boats long line for other tuna after the skipjack season. Fishing for skipjack is carried out on the platforms and the bait is taken from one of the holds flooded with seawater and used as a bait tank. These vessels are about 70–130 ft long and have a gross tonnage of 40–240 tons. Mackerel pole and line vessels are similar except that they are smaller and fish at night.

The troller, another line boat, is one of the most common fishing vessels off the Pacific coast of the U.S. and Canada. Trollers tow moving baits

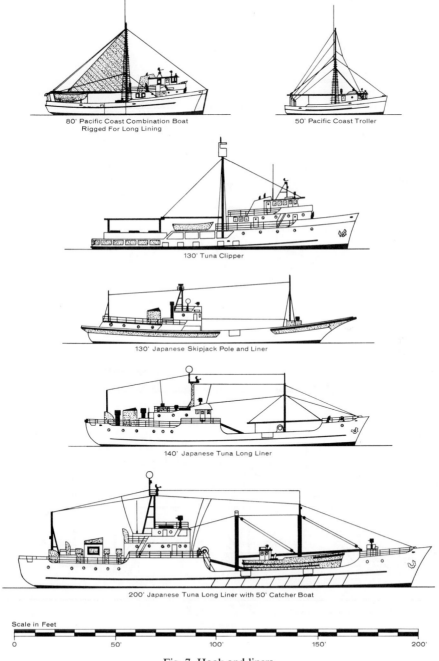

Fig. 7. Hook and liners.

FISHING VESSELS AND SUPPORT SHIPS

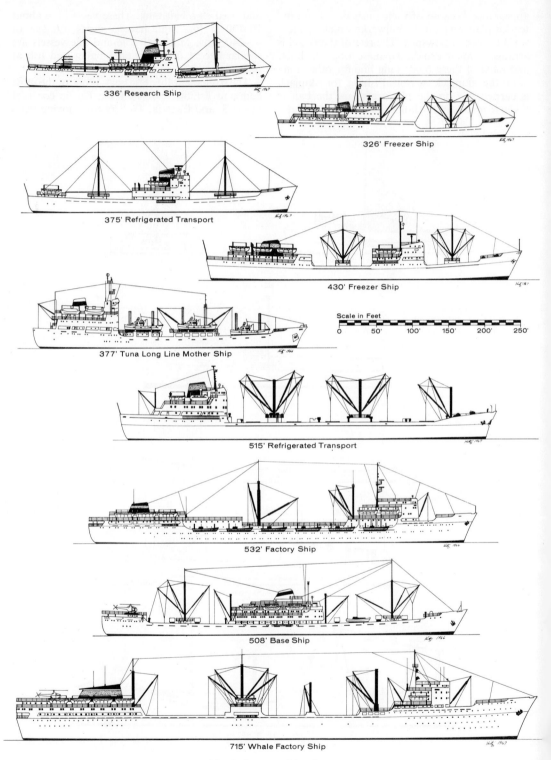

Fig. 8. Support ships, large.

FISHING VESSELS AND SUPPORT SHIPS

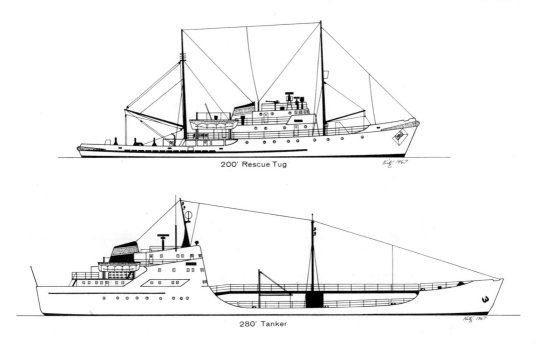

200' Rescue Tug

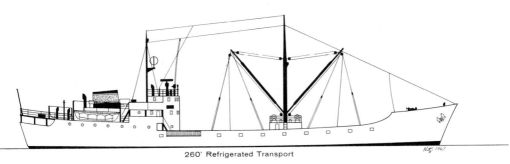

280' Tanker

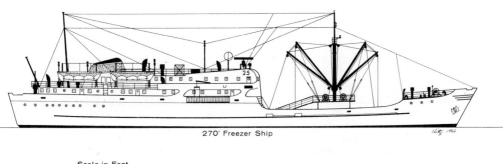

260' Refrigerated Transport

270' Freezer Ship

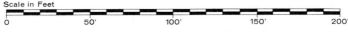

Fig. 9. Support ships, small.

or lures to catch salmon and albacore tuna. Length of these vessels varies from 25–60 feet. The house is forward and one or two pairs of outrigger poles are lowered when fishing. Two or three trolling lines are towed from each pole.

Miscellaneous Fishing Vessels. The only major additional types of fishing gear used on vessels are traps and harpoons. These devices are generally used to catch animals other than fish and are grouped in a fifth category.

Traps, which are called pots, are used to capture lobsters, crabs, and shrimp. They are baited and set on the bottom to attract the catch into the pot. Boats fishing this gear are usually vessels converted from some other fishery or combination vessels; however, a few specific boats have been built for these fisheries. One is the king crab boat of the Pacific northwest and another is the lobster boat of the east coast of the U.S.

Whales, the largest animals taken from the sea, are captured with harpoons. Whale catchers are built for speed, maneuverability, and seaworthiness. Boats range from 150–200 ft and have displacement tonnage of 1100–1200 tons. Some catchers can reach speeds of 18 knots during a whale chase. All have a raised bow for the harpoon gun and most have a catwalk from bow to bridge.

Support Ships

A number of boats in some fishing fleets do no fishing (Figs. 8 and 9) but are necessary for the fleet's operation at great distances from its home port. These are grouped under support ships.

The term mother ship is often used for the vessels accompanying a number of fishing vessels or catcher boats to fishing grounds where they stay for months. Supplies necessary for sustained operations of the catchers are contained within the mother ship and distributed during the season. Besides having a supply function, mother ships have hospital and recreation facilities and often act as command ship for the fleet.

An example of a true mother ship is the tuna long-line mother ship which carries her catchers on the deck. These vessels are 320–377 ft long, have a displacement tonnage of 4200–5100 tons, a deadweight of 1140–2800 tons, and a gross tonnage of 2800–5300 tons. The largest is now being built in Russia. She will be over 700 ft long, will have a gross tonnage of 40,000 tons, and will carry 14 catchers, which are 56 ft combination boats.

Whale factory ships are mother ships which work with catchers large enough to be independent except for the supplies needed for continuous operation. The largest vessels in the fishing fleets of the world are whale factory ships ranging up to lengths of 715 ft and tonnages of 43,800 displacement tons, 26,700 deadweight tons, and 32,000 gross tons. Recently several whale factory ships have been converted into fish-meal factory ships for operation off southwest Africa.

A fish factory ship, whether a conversion or a new construction, manufactures fish meal, oil, canned products, and frozen products and functions as a mother ship. They are large, ranging from 9000–19,000 gross tons and 460–600 ft in length. Some of these ships have stern ramps to ease the hauling of bags of fish left floating by catcher boats.

Unlike the factory ship, the base ship does not usually manufacture fish products, but functions solely as a mother ship to support 20–40 catchers. On the catchers the catch, generally herring, is salted in barrels and the preserved product delivered to the mother ship for storage.

They are usually in the 500 ft class with a gross tonnage of 11,000 to 13,000 tons although a few have been built in the 350 ft, 5000-gross-ton class.

Some of the large distant-water fishing fleets freeze their catch but do not use the large mother ship for necessary services. They use, instead, a number of merchant vessels, such as tankers and transports, for specific duties. The tankers supply fresh water, a valuable commodity on the high seas, and fuel and lube oil. Transports bring new crews to relieve tired ones, allowing the fishing vessels to stay on the grounds an entire season. Refrigerated transports shuttle between the home port and the fishing grounds carrying supplies to the fleet and frozen cargoes to the home port. Some of these transports are modified to freeze fresh fish on the grounds and are known as freezer ships.

The fleets have other support ships, such as rescue tugs and hospital ships. Rescue tugs, which carry high-powered radio communication equipment, divers, salvage equipment, and repair materials, do salvage and repair work. If the fleet lacks hospital services in the mother ship, a hospital ship usually is present.

Many of the leading fishing nations have scientific research vessels as well as training vessels. These vessels study the ocean's environment, scout for new fish resources, experiment with new techniques of fishing, and train seamen and scientists to man new ships.

C. R. Hitz

References

Traung, Jan-Olof, "Fishing Boats of the World," Vol. 1, 1955, Vol. 2, 1960, Vol. 3, 1967, Fishing News (Books) Ltd., London.

Cross reference: *Fish Harvesting.*

FISH MEAL

The early 1950's saw the advent of efficient high-energy diets and later the use of computer techniques for diet formulations by the manufacturer of mixed feeds. Research expanded until more was known about the nutritional requirements of the chicken than of any other animal. As a result, chickens grew bigger in a shorter time and at less cost. Competition increased, and the margin of profit decreased. The poultry industry developed into its present highly efficient, automated operation. As a necessity, the feed formulator today carefully checks and balances his sources of feed ingredients to arrive at the most economical formulation.

The fish-meal industry, because of competitive necessity, has had to become more efficient and to put out a better product over the years. Today, fish is caught with less labor at lower costs, and product quality is much improved and more consistent. Its nutrient quality as compared with similar nutrients in other feed sources is well known. Properly priced fish meal, as a supplemental source of high-quality protein, minerals, vitamins, and fats (energy) is used in animal-feed formulations only because it is economically advantageous to do so to supply critically needed nutrients. If the price of fish meal goes too high in relation to other feed supplements, a familiar pattern develops. The formulator stops buying fish meal, and meal prices drop—often below actual value. Then, because fish meal is a good buy, sales and prices gradually increase.

In this fluctuating economic picture, the manufacturer of fish meal is faced with two difficult problems. The first of these concerns the raw material. He cannot use expensive fish, such as tuna or salmon. The price he can pay is critical. And even an inexpensive fish or a fish waste produced as a byproduct has to be available in large quantity and at a low unit of effort per catch or per handling. The second problem of the manufacturer is how to manufacture the fish meal. Thus, in a competitive market, he has to produce a quality product—but at low cost.

Raw Materials

In the U.S. the principal species of fish used as raw material in reduction are menhaden, tuna, groundfish (also called bottomfish or white fish), herring (also called sardine), industrial fish, and sardine (also called pilchard).*

* Alewives, salmon, mackerel, and anchovies are also used. Processing wastes from crab, seal, shrimp, and whale contribute to the total production as do whole fish and the scrap discarded in canning and filleting. However, these sources of raw material contribute little to the total.

Menhaden. Menhaden is by far the most important species, since more reduction products are derived from it than from all the other species combined. Menhaden are small, oily fish caught almost solely for reduction. They are taken on the Atlantic coast from New England to northern Florida and on the Gulf coast from west Florida to east Texas. Reduction centers are located at Amagansett, New York; Port Monmouth, Tuckerton, and Wildwood, New Jersey; Lewes, Delaware; Reedville and White Stone, Virginia; Beaufort, Morehead City, and Southport, North Carolina; Megget, South Carolina; Fernandina and Apalachicola, Florida; Moss Point and Gautier, Mississippi; Empire, Morgan City and Cameron, Louisiana; and Sabine Pass and Port Arthur, Texas.

The catch may vary in a given area during a single year and from year to year. However, poor catches in some areas usually are accompanied by above-normal catches in others. The 5-year average catch (1960–1964) was 2 billion lb per year. During 1966 and 1967, this volume was considerably less.

Much of the catch is made near the plants (within 25 miles) and is landed late on the day that the fish are caught.

Tuna. Although whole tuna are not caught for reduction, such enormous numbers of them are canned that the waste from the operation results in a large quantity of raw materials for reduction. The tuna industry is largely concentrated on the Pacific coast, most landings being in California. San Pedro and San Diego are the principal ports. Since some of the fishing vessels travel as far as the coast of South America, landings do not necessarily represent a local catch.

Groundfish. Most of the groundfish caught are marketed for human consumption, but some species are used mainly for reduction. Those marketed for human consumption are sold whole or, as is more common, are filleted. The waste produced in the filleting operation is reduced to meal. The yield of oil from fillet waste is relatively low. (The trimmings and scrap from alewives, salmon, and anchovies, which are primarily canned are also used for reduction.) The principal species of groundfish are haddock, ocean perch, whiting, cod, and pollock. Massachusetts generally leads in the landings of all these species. A considerable quantity of Atlantic ocean perch is landed in Maine. The groundfish are taken primarily in waters off New England, but quantities are caught off Nova Scotia and some are caught in the Gulf of St. Lawrence and on the Grand Banks. Large landings of groundfish are made at Boston, Gloucester, and New Bedford, Massachusetts, and at Portland and Rockland, Maine.

Herring. Herring are landed primarily in Maine and Alaska. The catch tends to be about the same order of magnitude in both areas. Most of the Maine herring are canned, whereas most of the Alaska herring are reduced to meal, condensed solubles, and oil. Recently, offshore resources of herring from the northwest Atlantic are being considered for reduction purposes.

Industrial Fish. An industrial fish is any fish from which industrial products are prepared. They are edible but are usually considered as being less desirable food species—for example, monkfish, sculpin, sea robins, squirrel hake, sharks, and rays. Some are from an area where certain edible species do not grow to commercial size, such as in the shallow waters off the coast of Alabama, Mississippi, and Louisiana, where enormous populations of croakers, spot, butterfish, and similar species occur. In other areas, where the fish are larger, many of these species are considered to be commercially important food fish. Most of the industrial fish, however, are used either for canned pet foods or are sold raw (frozen) as feed for fur animals, such as mink.

Sardine. The great volume of sardines that were caught in past years in Pacific waters has now decreased to a very small size. Today, by far the largest catch of sardine is from Maine. Surplus alewives or river herring—principally from Massachusetts, Virginia, and North Carolina, which are not used for the relatively small cannery market—are sold for reduction or for use as bait. Huge resources of alewives appear to exist in the Great Lakes.

Methods of Capture. The way fish are caught depends largely on their habits. For fish that school near the surface—such as menhaden, herring, and sardine—the most efficient method of harvest is by use of encircling nets, such as purse seines. For fish that live near the floor of the ocean, such as haddock and ocean perch, the usual method of capture is by otter trawl. Longlines, troll lines, gill nets, pound nets, traps and pots, dredges, haul seines, and stop seines are also used in the capture of various fish.

The kind of fishing gear used determines, at least in part, the type of fishing vessel employed. In general, both purse seiners and otter trawlers are large vessels, but size and design are often dependent on whether the fisheries are close to or far from shore.

The method used to preserve the quality of the fish after harvest is important. When a purse seine is used, a great number of fish usually are caught at one time. Preserving this large quantity of fish is difficult. Accordingly, an attempt is made to catch the fish close to the processing plant and to bring them in quickly after capture. Some menhaden vessels are equipped with refrigeration, which permits fishing at considerable distances at sea with assurance that menhaden of acceptable quality will be landed.

If fish cannot be returned on the day they are caught, they usually are iced for preservation.

Methods of Manufacture

In the U.S., two principal methods of reduction are used: wet rendering and dry rendering. In wet rendering, the oil is removed before the fish material is dried. In dry rendering, the oil is removed afterward.

Wet rendering is most commonly employed for, and is particularly well adapted to, the rapid production of meal and oil from oily fish. In addition to the meal and oil produced, condensed solubles may also result from this method of manufacture.

Dry rendering is well adapted to production on a small scale from fishery materials of low oil content, such as fillet waste from haddock and cod. Continuous dry reduction is used with shrimp and crab scrap and with some industrial fish products.

Wet Rendering. In the menhaden industry, the fish are generally removed from the vessels by partially flooding the hold with fresh water and removing the fish by centrifugal pumps or positive displacement. These pumps are large and can empty a hold of 500,000 fish in less than an hour. The pump water used is held either for reclamation at a later time of the soluble protein dissolved in it or for suitable disposal. The fish are separated from the pump water by means of a large rotary sieve. From here, the fish discharge to a "quarter box," which measures the quantity of fish by volume. Sometimes the fish are drained free of water, pumped directly onto conveyor belts, and passed over an automatic scale (or weighing conveyor) to determine the weight landed.

From the quarter box or automatic scale, the fish are either carried directly to a cooker or to temporary storage in a "raw box." The raw boxes usually are large enough to hold up to several million fish and are designed so that the floors slant toward the middle of the box, where a screw or flight conveyor can pick up the fish and transport them, as needed, to the cooker.

If scrap or fillet waste is used, it must be gathered in quantity from one or several locations for conveying in some manner to the processing location. This material then is measured in various ways and held, if necessary, prior to being processed.

For the purposes of this article, wet rendering as practiced in the menhaden industry will be discussed. The various stages of the process may be divided into cooking, pressing, centrifuging, drying, deodorizing, and curing.

Cooking. The fish are cooked so that the oil and water in the fish can be separated from the solid protein easily and economically in the subsequent pressing operation. Cookers in the menhaden industry vary in size (16–30 in. in diameter) and in length (30–40 ft). A continuous steam-cooker screw press of 10-tons-per-hour capacity consists of a steel shell 24 in. in diameter and some 30 ft in length.

Fish are introduced through a hopper at one end and are slowly moved through the cooker by a revolving screw. Steam is introduced through a series of jets in the wall of the cooker from two manifolds paralleling the cooker. As the screw pushes the mass of fish toward the discharge end of the cooker, any oil and water freed in the process is permitted to escape through a 4–6 ft section of screen on the bottom half of the cooker shell. At the discharge end, the cooked fish fall through a hopper to a press directly below.

The conditions of cooking can be altered to suit the varying raw materials by the use of a suitable steam pressure in the cooker (usually 5–10 lb) and by changing the speed of the screw. The cooking process breaks down the oil cells and coagulates the protein. Overcooking or undercooking the fish results in an unsatisfactory product from the pressing operation.

Pressing. Pressing is an economical way to squeeze out sufficient oil and water from the fish so that the resulting material has a low oil content and is economical to dry. One type of press used most consists of a cast-steel screw, surrounded by a cylindrical screen, suitably reinforced at intervals along the area of pressure. The screw is designed to fit tightly against the screen, and either the pitch or the flights progressively decreases or the screw is tapered, resulting in compression and development of pressure on the solids as they progress towards the discharge end of the press. The discharge throat can be varied in size. The cylindrical screen making up the press may have perforations of about $\frac{3}{64}$ in. diameter at the inlet end and of about $\frac{1}{32}$ in. diameter at the discharge end. The press liquor (liquid from pressed fish) is screened or centrifuged to remove fine solids and then sent to the settling tanks or centrifuges for the recovery of the oil. Other types of presses are sometimes used, but the principles of their operation are more or less the same as those for the screw-type press.

The solid pressed fish, called presscake, that discharges from the screw press is transported to the driers for further treatment, which will be discussed in a subsequent section after consideration has first been given to the removal of oil from the liquid separated from the fish by the screw press.

Centrifuging. Centrifugal separators have generally replaced settling tanks for recovery of the oil from the liquid portion, called the press liquor. Several companies have designed and manufactured centrifuges. Essentially, these centrifuges are of only two different types. One is a three-phase machine called a sludger, which is designed to handle liquor containing oil, water, and some suspended solids. The solids, and part of the water, are continuously eliminated through the ports in the wall of the separator bowl. Usually the machine is adjusted so that all the oil and a minor amount of water present are discharged at the oil spout. This adjustment ensures that no oil will be lost, but does produce a wet oil, or emulsion. This emulsion then is fed, with added fresh hot water, to the second type of centrifuge, the oil purifier, in which the last traces of solids and water are removed. The sludger machine can be adjusted so that with careful operation, an oil may be obtained suitable for direct marketing. However, oil may be lost into the water discharge in the separation process. Furthermore, the processing capacity of the centrifuge is less if this technique is used. For these reasons, processors usually use the two types of centrifuges sequentially.

If gravity settling is used instead of centrifuging, the liquor passes to a series of settling tanks. The operator may have to use heat, acids, bacteria, or enzymes to break the emulsion of oil and water that often comes from the presses. In some operations, the liquid portion coming from the sludger machine, called stickwater, is transferred to tanks, where acid is added in sufficient amount to stop bacterial and enzymatic action. Precipitated solids and entrapped oil are removed by centrifugation, and the stickwater is dehydrated to a content of 50% solids. The resulting product is called condensed solubles. In the menhaden industry, however, the stickwater is usually first concentrated. Acid is then added to the finished solubles in an amount sufficient to reduce the pH to about 4.5. The acid acts primarily as a preservative.

The solubles are stored in large steel tanks and are shipped in tank cars, tank trucks, or, in some areas, by tank barges.

If the solubles are added back to the presscake, the resultant product, after being dried, is called full meal in contrast to regular meal, which has no added solubles.

Drying. Drying the product is necessary to bring the moisture content of the meal down to about 9% to prevent spoilage and to make the product easier to handle, ship, and use. Several different types of dryers are used.

In a direct-heat dryer, the dryer unit consists of a dutch-oven type of furnace, as a source of heat, and a horizontal dryer shell. The dryer shell is mounted on trunions and is slowly rotated by means of a motor, the speed of which has been reduced through a jack-shaft arrangement. From the jack shaft, the power is applied through a gear to a bull ring attached to the shell. Wet meal is introduced through an overhead hopper or an under-feed screw conveyor. The inlet end of the dryer shell is mounted to turn in a collar in the rear of the furnace. Combustion products, and the air drawn through the dutch-oven heating area, pass through the dryer parallel to the path of the fish meal. Steel plates or lifters attached to the inside wall of the dryer shell cause the meal to be lifted up the sides and then to be spilled continuously into the path of the hot gases. A fan draws the spent gases and the suspended fine meal through a cyclone separator. Here the meal is recovered, and the gases are discharged. The major portion of the dried meal is discharged from the end of the dryer and is then sent to the grinder.

In a steam-tube dryer, heat is applied through steam tubes mounted in circles concentric and adjacent to the inside dryer wall. As the dryer rotates, the meal falls through the spaces between the heated pipes and a slight inclination of the dryer toward the discharge end results in the meal moving in that direction. A fan and a cyclone separator may be used to collect fine particles of meal and to remove the moisture-laden air from the dryer.

In an air-lift dryer, wet presscake is introduced into the bottom of a funnel-shaped tower. At this point, upward air velocities are sufficient to entrain and bring all materials into suspension. As the material dries, it is carried through a blower to a cyclone. Here the dried product is separated.

Deodorizing. The moisture-laden gases resulting from the removal of the fine meal particles contain undesirable odors. Generally, the odor components of the gases must be removed before they may be discharged to the outside air. The condensable portion of the gases may be removed in scrubbers; the non-condensable portion may then be burned.

Curing. The dried product, called scrap or unground fish meal, is conveyed to storage sheds, where it accumulates in large piles. Some residual oil remains in the scrap, which is a factor influencing the method of handling the scrap. In general, fish oils are highly reactive, being characterized by a high degree of unsaturation, that is, the oil molecules do not contain as many atoms of hydrogen or other elements as they are capable of reacting with. This unsaturation permits easy combination of the oil with oxygen in the air, releasing considerable heat, which may result in charring or even in fire. To retard the rate of oxidation and thereby lower the rate of heating, many operators add antioxidants, usually to the dried scrap but occasionally to the presscake before it is dried. Thus, depending on the rapidity of accumulation of the scrap, the amount of residual oil present, and whether or not anti- oxidant had been added, the piles of scrap are turned and aerated, in some cases several times a day, so that they cool sufficiently to prevent spontaneous combustion. The turning of the scrap pile, which presents a logistics problem when large quantities of scrap accumulate rapidly, is accomplished by use of tractors and overhead conveyors. The scrap, after being turned and cooled, is left unground or is ground into meal and is sold either in bulk or in burlap or paper bags usually of 100 lb size.

Dry Rendering. Although many variations in this method of meal production are possible, the fishery material usually is loaded into a large, steam-jacketed, cylindrical dryer. Inside the dryer is a rotating scraper, which brings all the material into quick contact with the hot inside wall, yet prevents the material from sticking. The drying is done either under vacuum or at atmospheric pressure. The oil is separated from the dried scrap by batch pressing in hydraulic presses. No product other than oil is produced from this pressing operation. After the oil has been pressed out, the remaining solid material is ground into meal (called whole meal) or is left unground as cake. The product is sold either in bulk or, as is more common, in burlap or paper bags.

When material containing very little oil, such as shrimp and crab scrap, is to be dried, it often is charged directly into a rotary-tube hot-air dryer such as is used for presscake in the wet-rendering method. This process is not usually practical for raw fish because the dry material tends to stick on the dryer walls.*

General Comment

A comprehensive account of the fish reduction industry would consider matters in addition to the catching and processing of fish as was done in this chapter. For instance, a detailed discussion on just the nutritive quality and factors affecting the nutritive quality of fish meal would alone fill a text. Nevertheless, a general overview has been provided on two critical elements of the industry.

For those interested in obtaining more information, see the publication by F. Bruce Sanford and Charles F. Lee published in 1960. Aside from providing much of the material used in the present report, it contains an excellent bibliog-

---

* Homogenized condensed fish is another product produced in small quantities, by patented processes. The processes consist of treating ground fish with combinations of acids or alkalis, enzymes, heat, and pressure to form a slurry that is then evaporated to a 50% solids. In another method, a slurry formed from fish digested by stomach enzymes is mixed with a cereal carrier, such as soybean meal, and dried. This kind of product is made from fish of low oil content.

raphy up to 1960 covering such items as analyses and composition, feed for animals, fishing vessels and gear, manufacturing methods and equipment, nutrition, odor control, oil, solubles, and other topics. Other older publications by Charles Butler and C. F. Lee are useful also.

Factors affecting the quality of fish meal and discussions on the use of fishery reduction products are covered rather well in Ref. 3. Later information can be obtained from well-known journals, such as *Poultry Science*, and from trade publications, such as *Feedstuffs*. All publications concerning the fish meal industry and related research are abstracted in Ref. 2. The Fishery Technological Laboratory of the Bureau of Commercial Fisheries at College Park, Md., is the Government body responsible for fish meal research in the U.S. Inquiries on specific questions should be addressed to the Director. Information on statistics and exploratory fishing should be addressed to the Chief, Branch of Fishery Statistics, and Chief, Branch of Exploratory Fishing, Bureau of Commercial Fisheries, U.S.D.I., Washington, D.C. 20240. Questions related to the industrial side can be directed to the Director, National Fish Meal and Oil Association, 1614—20th Street, N.W., Washington, D.C. 20009.

Internationally, two laboratories conduct significant research on fishery reduction products: The Torry Research Station, 135 Abbey Road, Aberdeen, Scotland, and the Fishery Industry Research Institute, University of Cape Town, Rondebosch, South Africa. The Food and Agriculture Organization of the United Nations correlates international matters concerning fish meal. Inquiries should be addressed to the Department of Fisheries, Division of Technology, FAO, Rome. The Food and Agriculture Organization also published an abstract of current literature in the field of fisheries, including reduction products.

DONALD G. SNYDER

References

1. Butler, C., "Fish reduction processes," U.S. Department of the Interior, Fish and Wildlife Service, Fishery Leaflet 126, 1945.
2. Commercial Fisheries Abstracts. Bureau of Commercial Fisheries, Branch of Reports, Building 67, U.S. Naval Station, Seattle, Washington 98115.
3. "Fish in Nutrition," (1962), Fishing News (Books) Ltd., Ludgate House, 110 Fleet Street, London, E.C.4, England.
4. Lee, C. F., "Menhaden industry—past and present," U.S. Department of the Interior, Fish and Wildlife Service, Fishery Leaflet 412, 1953.
5. "World Fisheries Abstracts," Food and Agriculture Organization, Rome, Italy.

Cross reference: *Animal Feeds.*

FISH PROTEIN CONCENTRATE

Fish Protein Concentrate (FPC) is defined as an inexpensive, stable, wholesome product of high nutritive qualities, prepared for human consumption from whole fish by sanitary food processing methods. FPC is more concentrated in proteins and certain other components of nutritional importance than the raw material from which it is prepared. FPC is intended for use as a high-protein food supplement; it includes products of various characteristics, ranging from tasteless, odorless, light-colored and flourlike materials, through coarse meals having fish taste and odor, to highly flavored dark-colored pastes or powders resembling meat extracts.

The significance of this product, intended to be used as a food additive, becomes clearly apparent when viewed against the deepening world food crisis. The present situation is due to many interrelated factors: A universal population explosion, more acute and more difficult to combat in developing countries; migration of populations to urban centers, depleting agricultural manpower, reducing agricultural productivity, and exacerbating already inadequate food distribution problems; a general shortage of purchasing power in developing countries, making it increasingly difficult for them to buy surplus foods from abroad; resistance of most population groups to changes in their food habits; a general absence of refrigeration or other food preservation facilities in most food-deficient countries, coupled to inadequate internal transportation systems and aggravated by a continuous and vast destruction of food due to insects and rodents; consumption of diets overwhelmingly based on single types of cereals and consequently a general lack of foods containing balanced proteins of high nutritive value.

Against the background of these conditions, the significance of FPC is clearly outlined. Establishment of industrial FPC endeavors and incorporation of FPC into diets of developing countries can contribute directly to improve the over-all economic situation in general and the food situation in particular by: (1) Supporting and furthering the expansion of national fishing industries, thus increasing their purchasing power; (2) creating a new and diverse processing industry, new jobs, and income; (3) utilizing a plentiful natural and hitherto incompletely exploited resource fish; (5) producing an inexpensive food additive of high nutritive value which, because of its amino acid composition can, by the addition of small amounts to cereal diets, upgrade their nutritive value without changing their organoleptic characteristics; (6) utilizing the whole fish as raw material, thus reducing processing costs and waste; (7) affecting the nutritional status of large numbers of people, irrespective of their geographic location, since FPC does not require either refrigeration or special handling or packaging and can be stored almost indefinitely if protected against excessive moisture.

The keeping qualities of foods in general, and of fish in particular, are directly related to the water content of the product; thus salting and sun-drying techniques were long ago developed for the conservation of fish and other foods. Such dehydration procedures, although not very efficient, have, over the centuries, had considerable success; the wide-spread usefulness of these methods is based upon the fact that the removal of water from plant and animal tissues causes certain chemical changes to occur and reduces enzyme action. Concomitantly, by sufficiently lowering the amount of free water in the tissues, micro-organisms that are ubiquitously present and responsible to a large extent for the processes that we recognize as decomposition and decay are prevented from growing and multiplying. The removal of water, then, just like the application of very low or very high temperatures for different reasons, is important for the establishment of bacteriostatic conditions. Another reason why drying has always been of importance is the reduction in weight and volume of the raw material, thus reducing transportation costs of the dried products. Thus, dehydration constitutes a means of preservation, and is the first and perhaps the most important process in the preservation of fish.

Apart from water, which makes up about 80% by weight of most common species of raw fish, reference must be made to fish oils, fats, and related compounds which are present to the extent of 2–20% by weight.

These so-called "lipids" are of great importance because of their highly reactive nature, especially as far as their tendency to react with oxygen is concerned. Under suitable conditions, fish lipids become rapidly rancid and can lead to the formation of compounds that are considered, by most, to be extremely unpleasant, and frequently reduce the nutritive value of the finished product.

In the simplest terms, then, an effective preservation of fish is based upon the removal of water and lipids from the raw material, or conversely, the isolation of a water- and lipid-free protein fraction. In practice, however, the picture becomes more complicated because of the fact that the dehydration and defatting process must be so designed as to lead to endproducts that have nutritional properties closely similar to those of the raw material. The processed material, to be useful must, of course, also be suitable for incorporation into different diets. A closer look thus reveals a number of unique and closely interrelated engineering and biochemical problems requiring intense and well-coordinated scientific research: certain proteins, for example, can be easily damaged by a variety of causes and must be preserved and separated from complex lipids and other components to which they are firmly linked; at the same time, a multiplicity of almost mutually exclusive processing requirements is needed to separate these components. There are also the elusive problems of taste preferences and acceptability, novel storage and transportation parameters and delicate questions of digestibility of the products by infants and expectant mothers; there are, finally, amino acids balances, physical characteristics and a host of new economic and sociological problems to be considered.

Processing Methods

The foremost objective of an FPC manufacturing endeavor is the production of a food supplement of the highest nutritive qualities. Within the present context of dietary supplementation, the fish proteins are the most important. To obtain a product in which the proportion of protein is significantly greater than in the raw material, the proteins themselves can be extracted from the original material, or constituents such as water, lipids, etc., can be separated from the protein-rich part of the raw fish, and a protein concentrate so obtained.

To achieve extraction of proteins *per se*, chemical means including those that induce drastic pH changes can be employed to render the proteins water-soluble and extract them in solution; biological agents such as enzymes, micro-organisms, etc., can also be applied to break down the proteins into smaller water-soluble units which are then separated from the raw material.

Methods achieving protein solubilization by the addition of acids or bases frequently cause serious damage to the nutritive value of the proteins and result in products of very limited nutritional usefulness. These methods will not be further considered here. Biological procedures that employ micro-organisms or enzymes, whether native to the fish or isolated from another source, have been used with some success in the large-scale manufacture of fish sauces and pastes in southeast Asia.

Biological Methods. Over the years, numerous methods have been developed to render the protein of fish water-soluble and remove them in this form from the raw material. Solubilization is achieved by the activity of proteolytic enzymes, bacteria, or yeasts. In many cases, enzymes and micro-organisms present in the fish are the responsible agents; in other instances, either specific enzymes isolated from other sources, or bacteria and yeasts, are introduced into a fish slurry where they break down the fish proteins into smaller units such as peptides, amino acids, etc. These components are usually separated as aqueous solutions from the undigested residue, and the solutions concentrated by the removal of water, yielding pastes or crystalline powders. Almost all the products manufactured in this way have strong characteristic odors and flavors, varying over a very wide range: according to the processes employed, products have been obtained that resemble meat protein concentrates, or cheese products; others may remind one of the fish from which they were produced.

By far the greatest volume of protein breakdown products commonly used as food additives or condiments are the so-called fish sauces and pastes manufactured in the Far East where they are known by different names: "nuoc-mam" in Vietnam; "nam-pla" in Thailand; "bagoong" in the Philippines, etc. The economic importance of the wide-spread industry manufacturing these products is extremely great: according to FAO, at least 80,000,000 liters of "nuoc-mam" was recently the estimated yearly production in Vietnam.

Unfortunately, fish sauces and pastes have, until now, been rather expensive in southeast Asia. They are, furthermore, produced under primitive conditions and are of rather low nutritive value. Also, large quantities of salt are added to the product as a bacteriostat to prevent putrefaction and this makes the food unsuitable for consumption by small children and expectant mothers. The best products have a rather strong cheese-like odor and salty taste. Although "nuoc-mam" and like products have traditionally been prepared from small sea fish, the use of fresh-water fish has continuously increased since World War II.

According to the most primitive method of fish-sauce manufacture, the fish are first kneaded and pressed by hand. They are then salted and placed in earthenware pots that are tightly sealed, buried in the ground and left there for several months. At the end of this time, the pots are dug up and the supernatant liquid that has formed is carefully decanted. On the average, 1 part of fish gives from 2–6 parts of "nuoc-mam" with the following approximate composition: one liter of the liquid contains 15.9 gm total nitrogen, 270 gm sodium chloride, and 0.5 gm calcium oxide.

Fish pastes differ from fish sauces in their method of manufacture, consistency, and by the fact that carbohydrates such as roasted cereals, rice, flour, and bran may be added to the raw fish before it is allowed to "ferment." According to van Veen, these products can make important contributions to the protein and calcium content of the local diet.[1] Salt is also added here as a preservative, usually in the proportion of 1 part of salt to 3 parts of fish. The mixture is allowed to "ferment" for at least 3 months before it is consumed in the form of a paste. A good review of various manufacturing methods for the production of fish hydrolysates, pastes, and sauces is given in a paper by Lahiry and Sen.[2]

A number of biological processing methods have recently been investigated that promise to be of considerable interest. As an outstanding example, the work of Bertullo of the University of Uruguay in Montevideo should be mentioned. The patented process[3] utilizes a proteolytic marine yeast that has the capacity, in the presence of a carbohydrate source, to break down the proteins of comminuted fish to an almost uniform liquid containing proteins and protein break-down products, only small bone and scale residues remaining undigested. Experiments with this material have shown that the liquid can be centrifuged and sludge and oil separated from a solution consisting of amino acids, peptones, etc. After evaporation of excess water, a paste or crystalline powder is obtained that is instantly water soluble and can be used as a valuable food supplement. Specific isolated enzymes have also been used to carry out protein hydrolysis, resulting in high-quality end products. Thus papain, trypsin, and several proprietary enzymes were investigated by workers in Canada, India, and other countries.[4]

Sen, Sripathy, and others have investigated the rate of hydrolysis, a method for the standardization of digestion conditions and the effect of degree of hydrolysis on the nutritional value of hydrolysis products prepared from a fresh-water fish with papain. A useful summary is contained in a later paper by Sripathy.[5] It has been suggested in India to use buttermilk for the fermentation of comminuted fish in order to mask some of the fish odors and flavors that may be considered unpalatable.[6]

Chemical Methods (Solvent Extraction). Many ethnic groups prefer foods that have strong, characteristic flavor profiles, other groups like bland-tasting foods; by and large, however, all groups have in common that they are highly sensitive to anything that brings about a change in the flavor of the foods they are used to. It is, therefore, frequently advantageous to produce a food additive that has no distinctive organoleptic properties and that can be added to various diets without risking flavor and texture alterations.

In order to obtain a bland-tasting, stable, and yet wholesome FPC, it is important to extract from the raw material water and certain usually water-soluble odor-bearing compounds such as ammonia and amines, as well as the large and varied class of lipids.

Removing water with a solvent (or by other conventional procedures) does not appear to present a major problem. But in removing lipids from the raw material, many new difficulties arise.

These are due to the almost contradictory requirements of exhaustive lipid removal and simultaneous preservation of the nutritive value of the proteins to which certain lipids are firmly attached.*

In general, the number of different lipid solvents utilized in the food processing industry is small. Also, in almost all presently used commercial oil and fat extraction processes used in the food industry, single solvents are employed because of the desire to reduce solvent recovery problems to a minimum. However, as a result of the general inefficiency of almost all single solvents (with, at present, the possible exception of isopropyl and ethyl alcohols) to effect exhaustive extraction of lipids in general, FPC production methods frequently suggest the use of solvent mixtures or solvent sequences.

Solvents used in extraction procedures should, ideally, have a variety of important characteristics, among which the following are of particular importance: they should be non-toxic; available in pure form; easy and safe to handle; low in price; constitute a low fire or explosion hazard; be efficient in the removal of water, triglycerides, other lipids, and odor-bearing compounds, without dissolving or reacting with proteinaceous components; they should have a low boiling point to allow for inexpensive desolventization of the concentrate and avoid the necessity to heat the extracted material to excessive temperatures; solvent recovery and deodorization should be easily feasible; and, finally, the solvent should be inert towards the usual materials of construction.

A partial list of solvents that are being used or that have been proposed for use is contained in Table 1.

In general, the selection of the fish to be extracted must follow the dictates of common sense. More than 20,000 species of fish inhabit the oceans; until much more is known about the properties of most of them, only those should initially be chosen for the manufacture of FPC that are well known as suitable for human consumption. Even then, caution must be exercised: A number of fish (over 200 species), many living in tropical regions, are or can suddenly become toxic. The toxins, most of which remain as yet unidentified, can be immensely dangerous. It is possible that extraction with one of the solvents used in the manufacture of FPC destroys or eliminates some of these toxic elements, but until and unless this is proved, such fish must be avoided. For the same reason, it is advisable to select a schooling type of fish such as the hakes or herrings as raw material for the production of FPC: such schools of fish are usually composed of many millions of individuals of the same species, with almost no foreign interlopers among them, thus assuring a homogeneous raw material.

Another reason why single species of fish, rather than mixtures, should be selected is due to the fact that the composition of fish differs significantly between species, some, for instance, containing more lipid material than others. The lipids themselves may also vary greatly as far as their fatty acid and general composition is concerned. These variations are likely to affect the processing conditions that must be chosen to insure a satisfactory FPC product.

To insure optimum quality of the finished product, the fish must be as fresh as possible and indeed be of food-grade quality. This is particularly important in order to meet the most exacting bacteriological standards. In this connection, it is also essential to utilize only fish caught in deep waters, as far removed as possible from the sea shore where bacteriological contamination is most likely to occur.

Although the use of fresh, iced, or frozen fish will give products with the best nutritional qualities, some FPC manufacturing processes use fish meal as the raw material, produced by conventional cooking, pressing, and hot-air drying techniques. Some of the types of raw material used by different processes are shown in Table 1.

In order to gain an idea of the variety of approaches that have been made to develop methods for the production of FPC by solvent extraction methods, a few selected processes are briefly described.

Wiking Eiweiss.[7] According to this process, developed in Germany during World War II, macerated whole fish is heated with stirring to 70°–80°F for 1 hour in an 0.5% acetic acid solution. After reducing the water content by pressing, the mass is extracted with alcohol and hydrolyzed with alkali; the protein solution is finally neutralized with acetic acid and then spray dried. A pure white water-soluble powder is obtained, and can be used as an egg-white substitute. During World War II, the Germans manufactured and used several thousand tons of this material.

Dabsch, V.[8] Dabsch developed a process, utilized in a modified version in the UNICEF-sponsored pilot plant in Quintero, Chile, according to which fish meal or comminuted fresh fish is extracted first with hexane and then with ethanol.

---

* The extraction of lipids from fish, and indeed from any other animal tissue, is further complicated by many facts, such as: lipid-protein and lipid-carbohydrate complexes are frequently insoluble in fat solvents; certain lipids are only soluble to a limited extent in the usual fat solvents; some conventional fat solvents are efficient solvents also for certain non-lipid constituents of the tissue; the original wet tissue cannot be efficiently extracted by many otherwise good solvents and must first be dehydrated; many solvents react with proteins or protein constituents, others have undesirable physical properties, etc.

FISH PROTEIN CONCENTRATE

TABLE 1. LIPID SOLVENTS NOW IN USE OR PROPOSED (PARTIAL LIST)

Process	Raw Material	Solvent	Other Processing
Dabsch/UNICEF	Fish meal or fresh fish (hake)	(1) Hexane; (2) Ethyl alcohol	Dehydration by heat
Fisheries Board of Canada	Fresh fish (cod, herring) Fresh trimmings Fish filets	Isopropyl alcohol	Acidification of the initial aq. slurry
Vogel	Fish meal Fresh fish	Ethyl alcohol Ethyl alcohol	pH change of initial aq. slurry
Morocco	Fish meal (sardines)	Mixt. of ethyl alcohol and hexane	
C. Verrando Bruera Astra, A.B.	Fish meal (anchovy) Fish meal Fresh fish (herring)	Hexane vapor N-butyl alcohol sec.-butyl alcohol and/or iso-butyl alcohol	
Lever Brothers	Fresh fish (cod)	Ethyl alcohol	Addition of sodium sulfite to the aq. slurry of comminuted fish; pH change to 10 by addition of NaOH. Roller drying of gelatinous mass before extraction.
VioBin	Fresh fish (hake)	(1) Ethylene dichloride (2) Isopropyl alcohol	Addition of sodium sulfite and anti-oxidant to initial slurry, simultaneous dehydration and extraction by 1st solvent
General Foods Corp.	Fresh fish	Tert.-butyl alcohol or isopropyl alcohol	pH change to 4.0–5.5 by addition of sulfuric acid to aq. slurry of comminuted fish; addition of anti-oxidant
Bureau of Commercial Fisheries	Fresh fish (hake)	Isopropyl alcohol	Dehydration of raw fish with solvent at room temperature followed by extraction at temp. close to boiling point of solvent

Fisheries Research Board of Canada.[9] According to the Canadian process, the comminuted fish is acidified with phosphoric acid and the mixture heated to 150°F. After centrifugation and washing, the material is twice extracted with isopropyl alcohol, heated to 180°F, and filtered. The final product is dried at 100°F. The method was originally developed by Guttman and Vandenheuvel in 1957. It is upon this method that a great deal of subsequent FPC processing developments have been based.

Vogel, A. G.[10] The Vogel method incorporates pH changes of the macerated whole raw fish and uses ethanol as the solvent for the extraction step.

Fishing Industry Research Institute.[11] The South African contribution to the development of FPC manufacturing processes is mainly based upon the careful extraction of fish meal with ethanol.

Morocco.[12] The Moroccan process utilizes fish meal as the raw material and a solvent mixture of isopropanol and hexane for extraction.

Bruera, C. B.[13] This process also utilizes fish meal as the raw material and effects solvent extraction by contacting the meal with hexane vapors.

Astra, Sweden.[14] This process presently utilizes fish meal as the raw material but claims that raw fish can also be utilized. The principal feature of the process consists of the use of a polar organic solvent that is partly soluble in water and less dense than water, such as secondary butyl alcohol, normal butyl alcohol and/or isobutyl alcohol. The extraction is carried out at a temperature somewhat lower than that of the azeotropic mixture.

Lever Brothers.[15] The Lever Brothers Company utilizes fresh comminuted fish which, after its pH has been changed, is drum dried and then extracted with ethanol.

VioBin Corporation.[16] The pioneering work that has been done by the VioBin Corporation of Illinois utilizes a process according to which whole raw fish is simultaneously dehydrated and defatted with ethylene dichloride. After solvent removal, a second solvent such as methyl or isopropyl alcohol is used for a second extraction stage. This process was approved by the United States Food and Drug Administration (FDA) for manufacture and sale of FPC in the U.S.

General Foods Corporation.[17] In this process the comminuted raw fish is slurried in water and the pH of the mixture lowered by the addition of sulphuric acid. An anti-oxidant is added to the slurry. After 15 min, the slurry is screened and pressed. Solvent extraction is carried out with tertiary butyl alcohol or other similar alcohol.

Bureau of Commercial Fisheries.[18] The process developed by the United States Government at the Bureau of Commercial Fisheries, Technological Laboratory in College Park, Maryland, consists of dehydrating whole comminuted fish (hake) by contacting it with isopropyl alcohol at room temperature, separating the miscella by suitable means such as centrifugation, and extraction of the dehydrated material with isopropyl alcohol at a temperature close to the boiling point of the solvent. This process was approved by the FDA for use in the U.S.

Properties of FPC Prepared by Solvent Extraction. FPC manufactured by a suitable solvent extraction is a non-hygroscopic, free-flowing, greyish-white or tan product which, depending on the method employed for final milling, may be granular or flour-like in consistency. It is only slightly water-soluble, but can be easily dispersed and suspended in water. FPC, manufactured according to the process approved by the FDA, is almost tasteless and odorless, contains at least 80% crude protein, no more than 7% moisture, a maximum of 100 ppm of fluoride, and a residual level of isopropyl alcohol that does not exceed 150 ppm. The total lipid content of the FPC must not be higher than 0.5%. The nutritive value of FPC must be equivalent to or higher than that of casein. FPC shall not contain any coliform organisms, no Salmonella, and a total bacteria count of no more than 10,000 micro-organisms per gm. The most important property of FPC is, of course, its nutritive value. A number of investigations have been made in this country and overseas to evaluate the nutritive value of this product by itself, and as a food additive.

Use and Acceptability of FPC

The degree of acceptability will more profoundly determine the extent to which FPC will be introduced into the world's food stream than almost all the other technical and technological processing problems put together. FPC is a food supplement that will exert its greatest nutritive value when added in the right proportions and under the correct conditions to other foods; it is not a complete food by itself; it has a low caloric value since it consists, after all, mostly of protein, some mineral matter, almost no lipids or carbohydrates, only traces of vitamins, etc.

Also, FPC as produced by solvent extraction is not as yet an attractive product with salesmarket appeal and image; it will for some time occupy the attention of the socio-anthropologist, biochemist, public relations man, food technologist, engineer, and physiologist before the combined efforts of these scientists will have given FPC the image and desirability that it needs, and its place on the world food counter.

Since FPC at this stage is a bland, free-flowing powder with no specific food texture, and is likely to retain these characteristics, most of the work that has been done to investigate its use has centered around (a) supplementation of staple foods, such as wheat, corn, rice, etc., and (b) the addition of FPC to common processed food items such as bread, noodles, soup, etc.

Sidwell[19] added FPC in various proportions to the wheat flour used in the manufacture of Vienna bread. In her experiments, Sidwell added 5, 10, 15, 20, and 25% FPC at the expense of wheat flour to the standard bread formula. She observed significant changes in the crumb texture and loaf volume of the baked bread; the former became less fine and regular, the latter decreased with increasing proportions of FPC. Sidwell has also shown that a 40 gm slice of the 10% FPC bread —almost indistinguishable in taste from the control—can contribute a quarter of the daily protein needs of a child. In the same paper successful incorporation of FPC into cookies is described. Pasta products such as noodles are consumed in large quantities in many parts of the world and can be manufactured in simple equipment. The addition of FPC to noodles, described in the same publication, tends to result in a product that is somewhat darker and firmer than the control containing no FPC, but taste tests have pointed to a high degree of acceptability of these products. Very attractive FPC-supplemented soups and milk-like drinks have also been prepared and described. Many attempts, too numerous to describe here, have been made to supplement various staple diets with FPC.

Although the use of fish as food is probably as old as man himself, a considerable body of resistance exists today in many parts of the world to the more extensive consumption of this commodity. This specific barrier, together with the even more ubiquitous resistance to change in food habits, has to be overcome before it can be hoped that FPC, even with improved organoleptic properties and image, will be generally accepted.

The problem of acceptance is a fundamental and formidable one, even if only the diets of the most food- and protein-sensitive groups of developing populations—babies between the ages of 1 and 5—are to be reached; success here can only be achieved indirectly by convincing the mothers that they themselves will accept and like FPC-supplemented formulas and foods. This, then, is the most difficult task in the whole FPC picture, and indeed in all large-scale attempts at food supplementation. The target groups that are most directly in need of assistance are usually not only the most conservative as far as their taste patterns are concerned, but also least is known presently about those very patterns that require change.

The Potential of FPC

It is appropriate now to attempt to translate various estimated figures relating to an over-all and specific fisheries resource into terms of the weight of FPC that could be manufactured from such a resource by solvent extraction. It seems, furthermore, appropriate to translate the figures so obtained into terms of protein weight and then into the number of daily protein rations that could be manufactured from a unit weight of raw fish. It is proposed to modify this figure further in order to indicate how much FPC, how much protein, and therefore how much fresh fish are needed to supply a certain number of people with a certain level of protein. In order to come up with such a rough estimate, which should only be taken as a point of reference and comparison, a number of assumptions and general simplifications must be admitted:

It will be assumed, (1) that the whole fisheries resource is composed of edible fish, i.e., that the number of toxic fish in existence is either insignificantly small or that the toxicity in certain fish can be eliminated by suitable means; (2) that, unless otherwise specified, all fish including marine animals that belong to other phyla and classes, such as whales, porpoises, sharks, squids, mollusks, etc., contain 20% by weight of protein; (3) the gear; vessels, and manpower can be created to catch all the fish in this estimate; (4) that all fish can be processed successfully into FPC with a protein content of 80% by weight and a yield of finished FPC of 15% by weight (very low figures chosen to include processing losses; this means, for the purpose of this calculation, that 100 metric tons of fish will yield 15 metric tons of FPC equivalent to 12 tons of 100% protein), (5) that a reasonable average daily consumption of 20 g of protein as supplied by 25 g of FPC represents a sufficient daily supplement of high-class protein; (6) that a yearly ration, per person, of this supplement is represented by 365 20 g protein supplements.

For the sake of simplicity in operating with the large figures involved, we will define a mega ration (MR) as that amount of FPC (80% protein) that will provide 1 million persons with 365 20 g portions of pure protein.

One MR is therefore, equivalent to:
$365 \times 20 \times 10^6$ g = 7.3×10^9 g, or
7.3×10^3 metric tons of pure protein,
9.1×10^3 metric tons of FPC, or
60.52×10^3 metric tons of fresh fish.

Bearing the above assumptions in mind and converting various educated estimates of total specific fisheries resources into FPC, protein quantities, or "mega rations," certain equivalents can be obtained and these are summarized

TABLE 2. TOTAL SUSTAINABLE ANNUAL FISHERIES HARVESTS AND THEIR "MEGA-RATION" EQUIVALENTS, AS ESTIMATED BY VARIOUS AUTHORS

Author	Estimated total sustainable annual harvest in thousand metric tons	Mega-ration equivalents MR[a]
Edwards and Graham[20]	115×10^3	1.9×10^3
Pike and Spilhaus[21]	190×10^3	3.1×10^3
Schaefer[22]	200×10^3	3.3×10^3
Larkin[23]	500×10^3	8.3×10^3
Kesteven and Holt[24]	500×10^3	8.3×10^3
Chapman[25]	2000×10^3	33.0×10^3
Schmitt[26]	2000×10^3	33.0×10^3

TABLE 3. ESTIMATED UNUTILIZED AND UNDER-UTILIZED FISH STOCKS IN U.S. COASTAL WATER AND THEIR "MEGA-RATION" EQUIVALENTS

Region	Estimated total sustainable annual harvest		Mega-ration equivalents MR[a]
	million lb	thousand metric tons	
Atlantic coast	420	190	3.1
Gulf of Mexico	5300	2400	40.0
Pacific coast and Alaska	6458	2900	47.8

[a] Mega-ration: One "MR" or mega-ration is that amount of FPC (80% protein) that provides every day for 365 days one million persons with 20 g of pure protein. 1 MR = 7.3×10^9 g FPC = 9.1×10^3 metric tons of FPC = 60.52×10^3 metric tons fresh fish.

in Table 2. From these figures it will be seen that, irrespective of whether we base ourselves on the somewhat theoretical predictions of global fisheries resources, or the more concrete availability estimates of particular fish species closer to home, the volume of FPC that can be produced and the number of diets supplemented (even after taking conservative material losses into account) is impressive.

E. R. Pariser

References

1. van Veen, A. G., *Advances in Food Research*, **IV**, 209–229 (1953).
2. Lahiry, N. L., and Sen, D. P., *Reviews in Food Technology*, **2**, 14–28 (1960).
3. Bertullo, V. H., U.S. Patent 3,000,789, Sept. 19, 1961.
4. Sen, D. P., et al., *Food Technology*, **XVI**: 5, 138–141 (1962); Sripathy, N. V., et al. *Food Technology*, **XVI**: 5, 141–142 (1962).
5. Sripathy, N. V., et al., *Food Science*, **28**: 3, 365–369 (1963).
6. Krishnaswamy, M. A., *J. of Food Science and Technology*, **1**: 1, 1–3 (1964).
7. Myles, H. H., et al., British Intelligence Objectives Sub-Committee Final Report No. 4R3, Item No. 22 Target No. C22/328.
8. Pariser, E. R., and Odland, E., *Commercial Fisheries Review*, **25**: 10, 6–13 (1963).
9. Guttman, A., and Vandenheuvel, F. A., *Prog. Reports Atlantic Coast Stations*, Fish. Res. Board, Canada, **67**, 29 (1957).
10. Vogel, R. and Mohler, K., U.S. Patent 2,875,061, Feb. 24, 1959.
11. Dreosti, G. M., and van der Merwe, R. P., Progress Report No. 64, *The South African Shipping News and Fishing Industry Review*, August, 1961.
12. FAO Unicef, "Report on the Processing of Fish Flour From Sardines," Sept., 1960.
13. Bruera, C. V., U.S. Patent 3,064,018, Nov. 13, 1962.
14. Astra Nutrition, AB., Dutch Patent Appl., 6,609,272 (cl. A.23 j), January 6, 1967.
15. Galliver, G. B., and Holmes, A. W., U.S. Patent 2, 813,027, November 12, 1957.
16. Levin, E., and Finn, R. K., *Chemical Engin. Progress*, **51**: 5, 223–226 (1951).
17. General Foods Corporation, U.S. Patents 3,099,562 July 30, 1963; 3,252,962 May 24, 1966; 3,164,471 Jan. 5, 1965.
18. U.S. Department of Interior, Bur. of Comm. Fisheries "Marine Protein Concentrate," Fishery Leaflet No. 584, April, 1966.
19. Sidwell, V. D., *Activities Report*, **19**: 1, 118–124 (1967).
20. Edwards, R. L., and Graham, H. W., "Fish in Nutrition," 1962, Fishing News (Books) Ltd., London.
21. Pike, S. T., and Spilhaus, A., NAS/NRC, Publ. 100–E, pp. 1–8, 1962.
22. Schaeffer, M. G., *Trans-Amer. Fish. Soc.*, **94**: 2, 123–128 (1965).
23. Larkin, P. A., *The Fish Boat*, **10**: 7, 21–23 (1965).
24. Kesteven, G., and Holt, S. J., "A Note on the Fisheries Resources of The Northwest Atlantic," FAO, No. 7.
25. Chapman, W. M., paper given at the Marine Technological Society, Washington, D.C., 1966.
26. Schmitt, W. R., *Annals of the New York Academy of Sciences*, **118**, 645–718 (1965).

FOOD CHAIN—See BENTHONIC DOMAIN

FORAMINIFERA

Foraminifera (Order Foraminiferida) are dominantly marine shelled protozoans. Widely distributed, numerous and diversified, they represent about 2.5% of all known animals, both living and fossil. About 34,000 species have been described, of which some 4000 are living. Although these organisms are unicellular, they are complexly organized, and show a broad range in size, composition and morphology of the shell, nature of their life cycles, and ecologic distribution. Like other members of the protozoan Subphylum Sarcodina, Class Reticularea, the animal itself continually changes in shape and form, but the firm, mineralized skeleton (test) of the foraminifer is characteristically formed, adult specimens ranging in size from 20 μ–12 cm or more.

The protoplast of the organism is differentiated in structure, an inner region of darker-colored endoplasm being surrounded by a thin outer layer of more transparent ectoplasm. The one or more nuclei commonly increase in size with growth. All nuclei may be alike in form and function (homokaryotic), or in some species, only the generative nuclei take part in reproduction, the others being merely somatic, or vegetative (heterokaryotic). Sexual and asexual generations alternate, the asexually produced forms being uninucleate, and sexually produced ones typically multinucleate, although in some species of primitive structure both generations are uninucleate.

The pseudopodia, temporary or semipermanent cytoplasmic projections from the surface, serve for locomotion or attachment, for capture and digestion of food and expelling of debris, in the construction of the shell, and as temporary protective cysts. Foraminiferids have granuloreticulose pseudopods, greatly elongated, readily bifurcating and anastomosing. A relatively solid axis (stereoplasm) is surrounded by the more fluid layer (rheoplasm) in which granules of plasmatic origin stream continually. Protruding from the shell aperture of some monothalamous forms is a distinct pseudopodial trunk from which arise the finer pseudopodia.

Free-living species move slowly about over the bottom, or along algal surfaces, by means of their elongate pseudopodia. On gravelly or sandy

Fig. 1. *Elphidium macellum* (Fichtel and Moll). Superfamily Rotaliacea. (a) Side view of living specimen, showing elongate and anastomosing granuloreticulose pseudopodia; (b) edge view of shell. × 72. Recent, Italy. (After Schultze, 1854.)

a shell wall of foreign particles (suborder Textulariina), of quartz and other minerals, or of agglutinated shell fragments, sponge spicules, or the shells of smaller foraminiferans, all held in a secreted organic cement containing organically bound ferric iron, calcium, and perhaps silica. The resultant shell may be firm and strongly cemented, or soft, delicate, and easily destroyed, depending on the proportions of the various components. The agglutinated shell may surround an inner organic shell, 1–10 μ thick, that resembles the tectinous shell of the Allogromiina, but not all agglutinated foraminiferans have an organic lining. Some agglutinated species are highly selective, utilizing only certain components from the surrounding sediment, and orienting them within the shell. Others are selective as to particle size, but not composition, or the material used in shell construction may range widely both in nature and coarseness of fragment size, according to its availability in the local substrate.

The morphologic form of the agglutinated shell ranges from a singular globular, irregular, or tubular chamber with one or more openings (superfamily Ammodiscacea), to the highly complex multilocular forms in which succeeding chambers are of increased size and complexity

bottoms they may be anchored to the substratum by the pseudopodial network, which may also act as a binding agent in the soft sediment. Some species are temporarily or permanently attached to seaweeds, coral, or mollusks. Others live within the empty shells of other foraminifers, or in polychaete worm tubes, and a few are even parasitic. A small group of genera are planktonic (superfamily Globigerinacea); others may have short pelagic stages in their ontogeny (e.g., *Iridia*, *Rosalina*).

The shell or test may be simple or extremely complex in composition and morphology. It may consist of a single chamber, or many interconnected ones, through which the cytoplasm streams. Although constructed by the pseudopodia, which can largely be retracted within it, the foraminiferan shell is an internal skeleton. An extramural layer of cytoplasm flows from the opening to cover the outer surface.

The supposedly most primitive shell is membranous or tectinous (suborder Allogromiina), the latter consisting of protein and an acid mucopolysaccharide. The shell wall may be thick and relatively firm, or thin, delicate, and easily deformed. Other foraminiferans construct

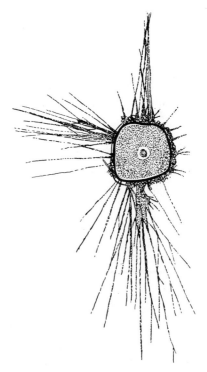

Fig. 2. *Myxotheca arenilega* Schaudinn. Suborder Allogromiina. Living specimen with gelatinous shell externally covered by cytoplasm and pseudopodial network. × 300. Recent, Adriatic Sea. (After Schaudinn, 1893.)

(Lituolacea). The latter may have a rectilinear, planispiral, trochospiral, or concentric arrangement, and be triserial, biserial, or uniserial. All combinations of these are also found. The chambers may be simple, or various modifications of the wall and cavity may result in a reticular meshwork, labyrinthic interior, partial partitions, or distinct chamberlets. Some complexly constructed forms attain a size of 6–12 cm.

The greatest variety in the Foraminiferida occurs in those with secreted calcareous shells. Most commonly these are of calcite, although some magnesium may replace part of the calcium, and a few are constructed of aragonite. The range in mineralogical composition, wall microstructure, and gross morphology provides a basis for the recognition of three suborders.

The Fusulinina are characterized by a wall of microgranular calcite. Important zonal indices for the Late Paleozoic, all are extinct.

In a second major group of calcareous foraminiferans (suborder Miliolina), the minute calcite crystals of the wall are so arranged as to present a porcelaneous appearance in reflected light. In transmitted light, the shell appears brown and homogeneous, the color perhaps resulting from the presence of the organic, acid mucopolysaccharide, components of the wall or the lining of its inner surface. A few porcelaneous shells may have foreign matter agglutinated to the exterior. The wall of post-embryonic stages is always imperforate. Commonly the globular first chamber (proloculus) of the shell is followed by an enrolled tubular chamber. The latter may be subdivided into separate chambers of a half coil in length, and the coiling may be planispiral, or lie in various specific planes. Some Miliolina also have complex organization, labyrinthic interiors, partitions, and chamberlets, and produce large flabelliform, discoidal, or fusiform shells.

The last major group of foraminiferids is that with perforate hyaline lamellar calcareous walls (Rotaliina). The distribution and size of the fine wall perforations, through which pseudopodia may extend, are highly characteristic. Some pores may contain pore plugs, or sieve plates, at the level of the successively formed wall laminae. Even the perforate foraminiferans may have some imperforate areas, most commonly as apertural modifications, lips, toothplates, and flanges, or peripheral keels, surface ribs, plugs, and spines. Last of the major wall types to appear in the geologic record, the hyaline perforate groups, also include a great variety of differing microstructures. Typically they are many chambered; the few unilocular representatives are regarded as later derivatives of multilocular forms, because of various aspects of those studied, such as their complex ornamentation, internal apertural modifications, reduced life cycles, and even a parasitic habit.

On the basis of various subordinate wall characters, and other features of shell morphology the foraminiferal suborders and superfamilies have been divided into about 100 families and some 1400 genera. Between 50 and 75 new generic taxa are proposed annually, together with a proportionate number of additional species, partly as a result of the stratigraphic and oceanographic interest in these abundant and useful organisms. Although foraminiferal taxonomy, like that of other shelled protozoans, is based largely on skeletal features, these differences are paralleled by others in the nature of the nuclei, cytoplasm, gametes, and other features of the reproductive cycle. To date, too little information is available for an evaluation of these aspects of the living organism, scarcely two dozen of the more than 4000 living species having been thus examined in any detail.

Morphologic dimorphism of the shell had been long known in foraminiferids, one shell form (microspheric) possessing a relatively small first chamber (proloculus), but a large adult shell, whereas the other (megalospheric) has a larger proloculus, and smaller adult shell. In most foraminifers studied to date, the microspheric form represents the diploid asexual generation, also termed agamont, or schizont, since it reproduces by schizogony, or multiple fission. The megalospheric form represents the haploid sexual generation (gamont), which produces the usually flagellated gametes. Meiosis occurs immediately prior to schizogony. This intermediate meiosis, between distinct diploid and haploid generations, is unique in the animal kingdom. This correspondence between shell form and generation is not complete, however, since in some species that produce relatively large amoeboid gametes (*Spirillina*, *Patellina*) instead of the smaller flagellate ones, the larger resultant zygote gives rise to an agamont generation with larger proloculus than that of the gamont. Thus the terms microspheric and megalospheric, based on shell morphology, cannot be used interchangeably for the agamont and gamont generations. The basic alternation of gamont and schizont is also modified in some species. Under less than favorable conditions, such as temperature or salinity, repeated generations may be asexually produced, and the sexual reproductive phase omitted. Some species apparently have completely lost the ability to reproduce sexually, and continue only by schizogony. Gamont and schizont generations may differ morphologically even to a greater extent than that of proloculus or adult shell size. They may differ in chamber arrangement, or even in habitat. The gamont of the benthic *Rosalina* develops a large float chamber just prior to gamete formation, perhaps as an aid to dispersal; this species occurs widely on coasts of nearly every continent. The distinctive appearance of the gamont had led to its description as a separate species and genus (*Tretomphalus*) before studies were made of the life cycle. Undoubtedly additional examples of morphologic dimorphism will become known following a greater emphasis on life-cycle studies.

Gametes also differ in character within the Foraminiferida. Most are biflagellate and those of the Miliolacea possess a distinctive axostyle; triflagellate gametes occur in *Glabratella*, and those of the Spirillinacea, and of *Allogromia*, are amoeboid. The fusion of gametes may occur within the parent shell (*Allogromia, Rotaliella*), or within a space formed by solution of the umbilical surfaces of the closely adhering parents (*Glabratella*), or within a reproductive cyst constructed by two or more parent individuals in association (*Patellina, Spirillina*). The numerous tiny biflagellate gametes produced by many foraminiferans are expelled into the surrounding water, where they may persist for days before fusion occurs (*Iridia, Elphidium*). Some adult individuals are reported to show sexual differentiation of the gamont (*Glabratella*).

Normally, both the nuclei and cytoplasm of the parent individuals are completely utilized in the production of young gamonts or gametes. The process of reproduction thus terminates the existence of the parent individual as such, although no death occurs. The empty tests are contributed to the bottom sediment, and the cytoplasmic components of the parent continue as the new generation. An apparent exception to this lies in the reproduction of a few large agglutinated foraminiferids by asexual budding (*Halyphysema*), or fragmentation (*Astrorhiza, Bathysiphon*). In laboratory culture, *Spiroloculina* may asexually produce some young, but the adult resumes vegetative existence for some days or weeks thereafter, before continuing to produce another brood of young individuals. Whether or not the production of multiple broods occurs outside the laboratory is as yet unknown.

The entire life cycle, with the alternation of two generations, may require as little as two weeks for completion, but more commonly represents an annual cycle, but may even require 2 years or longer for completion.

Although characteristically marine, some species are relatively tolerant of brackish conditions, and a few species (of Lagynacea) occur in fresh water. The occurrence of such ecologically restricted species may be used in interpretation of past changes in water depth and salinity.

Most foraminiferans are benthonic, different species being characteristic of different water depths, from the tide pools to the deepest levels sampled of the sea floor. The known depth distribution of modern species is utilized in paleoecologic interpretation of ancient sediments. Submarine transport of bottom sediments by slumping or turbidity currents is indicated by an assemblage that contains individuals of differing depth habitat, or displaced faunas. Long-range changes in sea level can also be suggested from a study of changing foraminiferal assemblages.

Foraminiferans occur at all latitudes, although some genera and species are restricted to the tropics and others to polar regions. Distribution of the typically eupelagic planktonic foraminifers is largely controlled by temperature; hence they are zoned latitudinally as well as more diversified in the lower latitudes.

Because planktonic species do not occur in coastal areas except where these are immediately adjacent to the open ocean, the ratio of planktonic to benthonic individuals in sediment samples has been used as an indication of the direction and distance from shore. Planktonic species occur at different levels in the water column, the maximum number being concentrated between 6–30 m. Vertical migration may be diurnal or related to ontogenetic development. Various depths may be sought by a species in different regions, in adjusting to the local water temperatures and density.

Because of their wide distribution and abundance, foraminiferans undoubtedly play an important role in the food web structure of the oceans, for example, in the concentration of very fine particles. Food utilized by foraminiferans consists dominantly of diatoms when these are available, but may include other organisms, both plant (bacteria, phytoflagellates, fragments of larger algae) and animal (ciliates, radiolarians, small crustaceans, nematode worms, and even echinoderms). Food material may be carried into the interior of the shell by the protoplasmic streaming, particularly in those species with a relatively large aperture (Miliolidae). Some large species (e.g., *Elphidium*) with small apertures digest the food externally by means of the pseudopodia, and only the dissolved nutrients are carried into the shell interior. Waste may remain in the cytoplasm as condensed pellets (stercomata) until just prior to reproduction (*Peneroplis*), or be extruded through the aperture or through openings of the canal system (*Elphidium*).

Little information is available as to selective predators upon the foraminiferans, although many are undoubtedly consumed by detritus feeders on the sea floor. Some gastropods are known to utilize foraminifers as food, and foraminiferal shells that have been perforated by some predator are common. A few parasites (protozoans, nematodes) have been reported to affect foraminifers, and planktonic species appear particularly susceptible to bacteria. Some reportedly contain symbionts, a cryptomonad in *Peneroplis*, and zooxanthellae in *Orbitolites*, *Buccella*, *Elphidium*, *Buliminella*, *Nonion*, and the planktonic *Globigerina* and *Orbulina*.

In addition to their role in the organic carbon cycle of the sea's economy, foraminiferans also markedly influence the presence of dissolved carbonate. Because of their great numbers, vast areas of the ocean floor are covered with the calcium carbonate deposits resulting largely from the accumulation of foraminiferal shells and debris. These are commonly termed *Globigerina* ooze or *Orbulina* ooze, although they may contain remains of many additional taxa as well. Lithified planktonic oozes of similar origin are abundant in later Mesozoic and Cenozoic rocks. The rise of planktonic foraminferans has been suggested to have changed the major site of carbonate deposition from the near-shore region and epicontinental seas of the Paleozoic, to the present occurrence in the open ocean.

Most studies of modern foraminiferal distribution have been based on sediment samples, and little data are available as to the rate of production. Based on the rate of reproduction of the species in one assemblage, approximately 650,000 foraminiferal skeletons were added annually to the sediment at depths between 20–50 fathoms, and about 100,000 annually at depths of 300–400 fathoms. Periodic sampling of limited areas, to determine the total number of living individuals, their rate of growth, longevity, and reproductive cycle, will be necessary but should provide useful data for evaluating sea-floor fertility at various depths, as well as the rate of sediment deposition.

Because of their wide distribution, distinctive evolutionary changes, abundance, and the ease of studying the well-preserved small fossils, foraminiferans are probably the most widely used of all fossils for geologic age dating. Countless numbers of individuals may be obtained from a small oceanic sediment core, or piece of subsurface rock obtained in drilling for petroleum, or from an outcropping stratum. The resultant concentration of studies of these organisms has allowed the accumulation of a wealth of data as to their geologic and ecologic ranges. They serve as useful indicators of water depth and temperature, paleocurrents, water masses, proximity of ancient shore lines, and rate of sedimentation.

Helen Tappan
Alfred R. Loeblich, Jr.

FREEZING FISH AT SEA

Significant increases in the world catch of fish and shellfish over the past 20 years would not have come about without successful development of a technology for freezing and storing fish on the vessel. Freezing has provided a viable means of extending the keeping quality of fish so that fishing vessels can engage in distant-water fisheries and remain at sea until fully loaded.

As far back as the early 1900's freezing was used as a means of increasing the productivity of fishing vessels. One of the earliest records of freezing at sea in North America dates back to 1929 when Zarotschenzeff[1] observed salmon, halibut, haddock, and cod being frozen successfully aboard the *S.S. Blue Peter*, a mothership serving a small fleet of fishing vessels operating off the coast of Labrador. The frozen fish were exported to Great Britain by the Hudson Bay Company.

The next major development occurred in the early 1930's when the *American Beauty*, the first tuna clipper to be outfitted with a brine immersion freezing system, was used to freeze tuna at sea on a commercial basis off California. In the following years freezing at sea became common in the U.S. tuna fleet because of the need to extend fishing to distant waters. The brine refrigeration system employed today on modern tuna seiners is basically the same as that developed over 30 years ago.

Soon after World War II a number of vessels in the Pacific Northwest were converted for freezing whole salmon and halibut at sea. During this period a large refrigerated mothership named the *Pacific Explorer* was used for freezing tuna in the south Pacific and for processing and freezing king crab and bottom fish in the Bering Sea. Although this venture did not culminate in a viable commercial enterprise, it did nonetheless contribute materially to development of the present U.S. king-crab fishery.

In the post-war period it became necessary for many countries to extend the scope of their fishing operations to fulfill the world demand for seafood. Two systems were advanced for freezing groundfish on trawlers—factory processing at sea and freezing in bulk for processing ashore. The former provides for filleting, packaging, and freezing on the vessel whereas the latter is limited to freezing whole or eviscerated fish on the vessel for later thawing and processing on shoreside factories.

The English vessel *Fairtry I*, constructed in 1953, was the forerunner of today's modern factory trawlers which are usually equipped with sorting and filleting machines, low-temperature freezing and cold-storage facilities, and fish-reduction equipment. Construction of factory trawlers has risen sharply during the past 25 years and it is estimated that 300–400 freezer vessels of this type are now in operation. The Soviet Union, Poland, West Germany, Japan, Spain, and other countries operate factory trawlers which range from 250 to over 300 ft in length, have a gross tonnage of 1500 to over 3000 tons, and can return to port with over 500 tons of frozen fish fillets.

An example of a large freezer factory vessel is the recently constructed Soviet vessel *Vostok* which is capable of producing in a single trip 20 million cans of fish in oil, 600 tons of frozen fish, 180 tons of fish meal, and 12 tons of industrial oil.

The freezing of whole or eviscerated fish aboard trawlers for processing ashore has been set forth as a less costly alternative to the more complex factory vessel operations. According to Ranken[2] the scarcity and high price of fish in the Mediterranean in the early 1950's led Italy and Greece to convert and later build freezer trawlers for fishing off the coast of Africa. The *Evanglistria*, one of the earlier converted vessels operating out of Greece, was equipped with four low-temperature blast freezers and had a capacity of 250 tons of frozen whole fish. After being unloaded from the vessel the frozen fish were usually air thawed and marketed without further processing.

The technology of freezing whole or eviscerated fish on trawlers was further advanced as a result of pilot studies conducted on the British trawler *Northern Wave* and the U.S. research vessel *Delaware* which employed vertical plate freezers and a brine-immersion freezing system, respectively.[3] These early experiments and those later by British, American, Canadian, and other workers[4,5] demonstrated rather conclusively that whole fish such as haddock and cod could be frozen, pre- or post-rigor, in brine or refrigerated plates, thawed with circulating air or water or electronically, and refrozen without significant loss of quality. In fact, frozen fillets prepared from sea-frozen fish compared favorably with the quality of frozen fish fillets produced from iced fish that were handled properly.

Pilot studies on the *Northern Wave* and subsequently research conducted at the Torry Research Station (Aberdeen, Scotland) led to construction of the first commercial British part-freezer stern trawler *Lord Nelson* in 1961. This vessel was followed a year later by the *Junella* and since then a number of freezer trawlers have been built throughout the world.

The Spanish vessel *Mar Di Vigo* commissioned in February 1966 is West Europe's largest freezer trawler to date. She is 344 ft (105 m) long and is equipped with low-temperature blast freezers capable of freezing 45 tons of headless fish blocks per day. The fish room has a capacity of 1850 tons of frozen fish.

Major developments in freezing fish at sea have been made by the United Kingdom, Spain, USSR, Poland, and Japan. Interest in North America has been primarily limited to tuna, king crab, and salmon. Two groundfish freezer trawlers are operating out of Canada, and it is anticipated that this technology may find wider application in that country. In the U.S. two factory freezer trawlers are being constructed under the Federal Government subsidy program.

Interest has been renewed in the freezing of shrimp at sea aboard relatively small trawlers operating out of ports in the Gulf of Mexico and in Central America. The immersion freezing systems used today represent a significant improvement over those employed over a decade ago on shrimp vessels but discarded because of operational and maintenance problems.

Freezing Methods

A number of variables are involved in the freezing of fish at sea. The species of fish, the form in which they are submitted to the freezer, and the particular freezing method employed may differ widely. In designing a freezing system one must also take into consideration the nature of the fishery in which the vessel is to operate, the rate at which the fish come aboard the vessel, the labor available for handling at sea, the storage capacity needed to insure a profitable operation, and the end use of the frozen product.

In view of the many different considerations involved it is not possible to suggest a freezing method universally applicable to all fisheries. Each particular method of freezing therefore must be appraised in relation to the special nature of fishing and scope of operations being considered. It is generally agreed that whole or dressed fish should be frozen prior to onset of rigor mortis at a speed of 0.25 in. (0.6 cm) or more per hour, except in the case of large fish such as tuna or halibut, and stored at a temperature of at least 0°F (-18°C). If the fish are to be further processed and refrozen ashore, particular care must be taken to make sure they are properly frozen on the vessel; otherwise serious loss in quality can occur in the refrozen product.

In freezing fish fillets at sea more attention has to be given to the state of rigor than when freezing whole or eviscerated fish. Since filleting machines experience difficulty in handling firm and twisted fish still in rigor, the fish should be filleted and frozen prior to or after rigor mortis for best results. If fish are filleted prior to rigor, however, the fillet, in the absence of the skeleton frame, may contract before freezing or upon thawing. A review by Jones[6] indicates that fillets cut from cod, if frozen prior to rigor, tend to discolor and lack gloss and translucency. Fillets frozen in rigor or post rigor present relatively little trouble, however, provided they are handled properly. In freezing fillets particular care must be taken to package the product adequately after freezing or guard against moisture loss during frozen storage.

The following is a discussion of immersion, contact plate, and air blast freezing systems used on fishing vessels, with examples of commercial applications.

Immersion Freezing. Immersion systems have received special attention in freezing unpackaged fish on the vessel because of low initial cost, compactness of equipment, fast rate of freezing, and high efficiency of operation. The first consideration is that the freezing medium, which is the heart of the immersion freezing system, must be of food grade and must not in any way adulterate the fish. Although work has been conducted on a number of freezing solutions involving alcohols, glycols, glycerin, glucose, and salt mixtures, sodium chloride brine is still by far the most widely used immersion freezing solution.

The *Delaware* experiment indicated that groundfish such as cod and haddock could be frozen whole in a 22% solution of sodium chloride brine provided the freezing time was controlled to prevent excessive salt penetration. Small haddock frozen in 1 hour then subsequently water thawed were of low salt content and the thawed refrozen fillets were quite acceptable after 9 months of frozen storage at 0°F (−18°C). However, fish that had been inadvertently left in the freezer for three or four times the required freezing time contained excessive salt, and were not acceptable. Further, the presence of salt on whole or eviscerated fish accelerated quality loss during frozen storage. For the above reasons, brine freezing is not used commercially for freezing groundfish at sea. Immersion freezing has, however, proved quite satisfactory in freezing whole tuna aboard the vessel for later thawing and processing into canned fish ashore. Over 150,000 tons of tuna are frozen annually aboard U.S. vessels. The American tuna fishery presents a number of rather special requirements for freezing fish at sea. For example, the weight of the fish may vary from 5–130 lb depending on the species, a single catch may range from less than 10 tons to as high as 100 tons, and the water temperature may be 85°–90°F (29–32°C) in the area where the fish are caught.

A modern U.S. tuna seiner is 150–170 ft long, carries 500–900 tons of frozen fish in 8–12 wells used for both brine freezing and dry storage of the frozen fish. The insulated wells are cooled by direct expansion of ammonia in coils lining the innersides, bottom, and ceiling of each well. Prior to freezing the fish are loaded into wells of prechilled seawater to a density of about 50 lb of fish per ft^3 of space. After the fish have been precooled to about 30°F (−1°C) salt is added gradually to make a 17% saturated brine. The brine is then cooled to 15°F (−9°C) by contact with the refrigerated pipe coils.

Although brine coolers are needed they are not usually used to assist in cooling the brine. After the fish are frozen the brine is pumped overboard and the frozen fish are stored in the dry well maintained at 10°–15°F (−12–9°C) by the evaporated pipe coils lining the inside surfaces. Prior to being unloaded, the fish are partially thawed by adding seawater and salt to the well. Fish unloaded frozen or partially thawed are thawed at the canneries in circulating water.

Immersion systems for freezing tuna have been appraised by Slavin and Finch[7] who concluded that heat transfer could be greatly improved with the use of external heat exchangers and by redesigning the wells so they are reduced to 3–4 ft (0.91–1.2 m) in width. It was also suggested that the design of the storage area be modified to permit containerization of the catch and thereby overcome present problems of excessive time in unloading fish and the need for thawing prior to unloading.

Immersion freezing is also used for the freezing of shrimp on the vessel. Preliminary work done over a decade ago indicated that shrimp could be satisfactorily frozen in a salt brine or in a glucose–salt solution. The glucose–salt mixture was preferred because of its lower freezing temperature and the lack of salt penetration in the product.

In recent years immersion-freezing systems capable of handling 500 lb of shrimp per hour have been installed on vessels operating out of the U.S. and Central American ports. On one commercial system, freezing is accomplished in a deck-mounted cylindrical unit containing a special brine solution which is circulated over plates cooled by expansion of refrigerant 502. The shrimp are frozen in round mesh baskets each with a capacity of 38–50 lb. Approximately 6 min are required to accomplish the necessary freezing. After the shrimp are frozen they are packaged in bulk and stored at −10°F (−12°C) in a hold refrigerated with circulating air.

Contact Plate Freezing. Fish are frozen at sea in horizontal or vertical contact plate freezers. Horizontal freezers are employed to freeze packaged fillets on factory ships and are very similar to equipment used throughout the food industry for freezing packaged products. Vertical plate freezers have, however, been designed especially for the freezing of whole or eviscerated fish in bulk at sea.

The original purpose of developing the vertical plate freezer was to provide a means of freezing groundfish with as little change as possible in methods of handling on board British distant-water trawlers. This freezer was developed by the Torry Research Station. It was designed so that the fish could be loaded in between the vertical plates from the top; after freezing, the plates opened and the blocks of headless eviscerated fish ejected into the cold-storage hold located below the freezer deck.

The prototype freezer tested on the *Northern Wave* contained a hot gas system for defrosting the plates in order to release the frozen blocks of fish at the end of the freezing cycle. On the *Lunella* defrosting was said to be eliminated by coating the plates with a plastic substance. According to Ranken[5] commercial installations of vertical plate freezers on freezer trawlers consist of 6, 8, or 10 freezers each with 12 stations. The freezing cycle depends on the product freezing time and the number of freezers. For example, if 8 freezers are used, a freezing cycle of about 3 hours and 20 min is necessary to permit emptying of one freezer and reloading of another one every 25 min. Freezer-station sizes vary from 48 lb blocks for herring to 116 lb blocks for headed cod or haddock. The blocks of fish may be 3–4 in. (7.6–10.2 cm) thick, 30–42 in. (76–107 cm) long, and 15–21 in. (38–53 cm) wide. Although vertical-plate freezers are designed for loading from above, unloading can be arranged in any direction to suit the layout of the vessel. The blocks of headless fish are usually dropped by gravity into the hold or discharged horizontally onto a conveyor or chute to the storage area. The frozen blocks of fish are stacked by hand in the fish hold which may be cooled with circulating cold air or with pipe coils installed along the walls and ceilings of the storage space.

Freezing in Air. Since circulating air was one of the first methods used to freeze fish ashore, it is natural to expect this method to be used at sea. Systems employed on fishing vessels vary from natural or forced circulation of cold air in the fish hold to modern blast freezers in which the product is loaded and unloaded automatically.

According to Ranken[5] the majority of freezer vessels use air-blast freezing tunnels of the batch or continuous type. Air temperatures may vary from $-20°$ to $-40°F$ (-29 to $-40°C$) and velocities may range from 500–1000 ft/min (153–305 m/min). Freezing cycles for fish blocks approximately 4 in. (7.6 cm) thick vary from 7–10 hours, depending on the system employed.

Considerable innovation is used in design of air freezers. In one of the earlier, more inefficient systems used on a European vessel, the whole fish were loaded by hand into wire baskets and frozen in a vertical blast freezer. After freezing the fish were removed from the baskets with great difficulty and stacked loose in the fish hold. This system was capable of freezing 14–15 tons of fish per day in four freezers.

A more modern but somewhat cumbersome air freezer is employed on large USSR factory vessels. One system used consists of a conveyorized freezer containing bucket-shaped freezer trays, secured to the conveyor chain. In operation the bucket trays are moved under a hopper outlet, the fish are loaded, the tray is covered, and the fish moved into the freezing tunnel. When the freezing cycle is completed the trays pass under an electric heater which releases the lids, the trays still on the conveyor chain, are turned over and pass under another set of heaters to release the blocks and the frozen fish are chuted to a glazing or packaging station. The blocks of frozen fish are then chuted or conveyed to the fish hold where they are stacked by manual labor.

Blast freezers are used principally to freeze whole or eviscerated fish. They are not suitable for freezing fish fillet blocks because of lack of pressure which is needed to produce a uniformly sized rectangular block. Other disadvantages of blast freezers include high freezing cost, high space requirements, and long freezing time, unless relatively thin products are used.

Fluidized bed freezers have not been used on shipboard, but they may offer advantages in freezing shellfish or fillets individually. Single freezing with ultrahigh-velocity air would permit handling and packaging of items of widely different sizes in a manner so they could be marketed frozen from the vessel or thawed ashore for further processing. The time required to freeze individual items in low-temperature circulating air is considerably less than that achieved when freezing in vertical contact plates. Problems of salt penetration found in immersion freezing are non-existent in air freezing; for this reason circulating air is used for king crab and other products which are suited to this method.

Thawing Methods

Sea-frozen whole or eviscerated fish can be thawed by immersion in circulating water, water spray, circulating warm air, heated plates, dielectric, or electric resistance. In his recent review of this subject Merritt[8] pointed out that all the methods previously mentioned can produce satisfactory results in thawing blocks of whole or headed fish with no appreciable difference in eating quality. Electric methods, however, represent substantial equipment costs when employed on a continuous basis for large blocks of sea-frozen fish.

Operating costs compare favorably with continuous air thawing and are about $12–$18 per ton. The continuous water thawer which stems from work done over a decade ago in the U.S. and more recent studies by Canadian and English workers shows considerable promise in offering substantial reduction in costs of thawing equipment and operation.

The technology of freezing at sea has developed rapidly in the past 25 years. It is possible to freeze successfully fish in bulk on the vessel for processing and re-freezing ashore or to process when on the vessel with retention of quality. Differences of opinion exist regarding the merits of factory trawlers as compared to freezer trawlers. The freezer trawler will be more favorably received until factory processing techniques on vessels are sufficiently refined to deliver a product which can be marketed directly to the consumer. Availability and cost of labor will also greatly influence decisions in determining which operation is the most economical. Development of improved fishing and processing methods and the need to travel farther to catch large quantities of fish will tend to increase the number of freezer vessels. Freezing at sea will surely play a more dominant role in the U.S. in the future.

JOSEPH W. SLAVIN

References

1. Zarotschenzeff, M. T., "Between Two Oceans, the Cold Storage and Produce Review," Empire Havre St. Martins, le Grand, London, 1929.
2. Ranken, M. B. F., "Progress in the adoption and application of freezing at sea," unpublished, 1965.
3. Slavin, J. W., "Icing vs. Freezing," "Fishing Boats of the World," Vol. 2, p. 227, London, Fishing News, 1960.
4. Slavin, J. W., and MacCallum, W. A., "North American experience in chilling and freezing on board vessels," International Congress of Refrigeration, Proceedings XIth Congress, Munich, Germany, 1963.
5. Ranken, M. B. F., "Evaluation of modern techniques and equipment for freezing whole fish at sea," paper presented at FAO Technical Conference on the Freezing and Irradiation of Fish, Madrid, Spain, 1967.
6. Jones, W. R., "Fish as a raw material for freezers; factors influencing the quality of products frozen at sea," paper presented at FAO Technical Conference on Freezing and Irradiation of Fish, Madrid, Spain, 1967.
7. Slavin, J. W., and Finch, R., "The design of systems for freezing tuna at sea," paper presented at the FAO Technical Conference on the Freezing and Irradiation of Fish, Madrid, Spain, 1967.
8. Merritt, J. H., "Evaluation of techniques and equipment for thawing frozen fish," paper presented at FAO Technical Conference on Freezing and Irradiation of Fish, Madrid, Spain, 1967.

FUNGI, MARINE*

Three classes of fungi, the Phycomycetes, Ascomycetes, and Deuteromycetes are widely distributed in inshore and offshore marine areas. Comparatively little is known about the marine

*Contribution No. 993 from The Institute of Marine Sciences, University of Miami, Florida.

occurrence of the fourth fungal class, the Basidiomycetes. Ecologically, three types of fungi may be found: (1) those from terrestrial sources that are unable or have limited abilities to inhabit marine areas; (2) fungi that successfully inhabit both terrestrial and marine regions; and (3) fungi indigenous to the sea whose habitat may or may not be highly specific.

Many of the marine fungi are economically important because of their destructive activities. They have been implicated in wasting disease of eel grass; mass mortalities of oysters, fish, commercial sponges, and the blue crab; destruction of wood and cordage; and enhancing the successful colonization of fouling and boring organisms. Many of the mycological studies therefore, have centered on these parasitic, pathogenic, and wood-inhabiting fungi. By contrast less is known about the distributions, concentrations, and roles of other marine fungi.

The following discussion is a synopsis of the occurrence of the major fungal groups in the sea. Detailed information may be found in the reviews by Johnson and Sparrow,[1] Wilson,[2] Kohlmeyer,[3] van Uden and Fell,[4] and Meyers.[5] Alexopolous[6] should be referred to for a basic taxonomic treatment of fungi.

Phycomycetes are considered the most primitive fungi. The principal characteristics of most species are formation of an aseptate mycelium and/or motile cells. Phycomycetes occur both as parasites and saprophytes. The parasites are widely distributed; however, individual species often have a range of habitation that is host dependent. For example, the saprolegnid *Eurychasma dicksonii* was found in a Swedish fjord at depths of 10–30 m in the same habitat range as the host alga *Striaria*; in contrast, occurrence of the saprolegnid *Ectrogella perforans* with a eurybathyl diatom was independent of depth. While parasites usually have a specific host, exceptions occur, such as the entomophthoralid *Ichthysporidium hoferi* which invades several species of fish and the copepod *Calanus finmarchicus*. The latter probably acts as the intermediate host. A partial list of Phycomycetes and their host species is presented in Table 1.

The occurrence of saprophytic forms depends on concentration and availability of organic materials. Inshore intertidal sediments therefore have more dense fungal populations than do offshore sediments. These Phycomycetes, along with other fungi, probably have a significant role in degrading organic materials. Ten orders of Phycomycetes are found in the ocean:

(1) The Chytridiales are widely distributed and mainly saprophytic. Most species are cosmopolitan, inhabiting terrestrial, freshwater, and marine regions. There are a few marine endemics, most of which are algal parasites.

(2) The Hyphochytridales parasitize brown algae (Phaeophyceae) and may be pathogenic. *Anisolpidium ectocarpii* produces a thallus in the filaments of *Ectocarpus*, apparently invading only healthy intact cells. Following infection the host declines in vigor.

(3) The Plasmodophorales are obligate endoparasites of marine vascular plants, invading sea grasses, e.g., *Diplanthera*, *Halophila*, and *Zostera*. The fungi usually cause a localized hypertrophy, often at the internodes, with subsequent dwarfing of the plant. The host nucleus degenerates after being surrounded by a multinucleate plasmodium.

(4) The Saprolegniales are important marine parasites and pathogens. Several species attack larval invertebrates. *Haliphthoros milfordensis* is found in ova and embryos of the oyster drill *Urosalpinx cinerea*. Following ingestion of *Leptolegnia baltica* hypha and planospores by the copepod *Eurytemora hirundoides*, an infection spreads rapidly through the viscera and destroys and distorts all the internal organs. Due to a large production of oil and gas, the copepods rise to the water surface. This epidemic infection of copepods severely affects the herring fisheries in the North Baltic. Algae also are infected by saprolegnids. *Eurychasma dicksonii* was considered responsible for depletion of populations of *Ectocarpus* near Plymouth, England and *Striaria* in a large Swedish fjord. *Ectogella perforans* initiates an infection which ultimately destroys populations of the diatoms *Licmophora* and *Striaella*. Saprolegnids are not particularly successful saprophytes, occurring infrequently in brackish water and rarely in seawater. An exception is *Thraustochtyrium* which is prevalent at depths of 40–600 m in North Sea bottom sediments.

(5) The Legenidiales are usually parasites and pathogens and are often destructive to eggs and larvae of invertebrates. *Sirolpidium zoophthorum* destroys larvae of the northern quahog *Venus*

TABLE 1. PARTIAL LIST OF PHYCOMYCETES ASSOCIATED WITH PLANTS AND ANIMALS

Fungus	Host	Remarks
Chytrids		
Chytridium polysiponiae	variety of algae, e.g., *Polysiponiae*, *Ceramium* spp.	widespread occurrence
Rozella marina	chytrid *C. polysiponiae*	occurrence possibly concomitant with host
Coenomyces consuens	blue-green algae: *Celothrix crustacea*, *Rivularia alta*	in algal gelatinous sheaths
Hyphochytrids	brown algae	invades healthy, intact algal cells with decline in vigor of host
Anisolpidium ectocarpi	*Ectocarpus siliculosus* E. *mitchella*	
A. sphacellarum	*Chaetopteris*, *Cladosterphus Sphacelaria*	
A. rosenvingii	*Pylaiella*	
Plasmodiophorales	vascular plants	induces hypertrophy and hyperplasia; envelops and disintegrates host nucleus
Tetramyxa parasitica	various species of *Zannichellia*, *Ruppia*, *Potamogeton*	produces greenish-brown galls on stems, peduncles, and leaf margin
Plasmodiophora diplantherae	*Diplanthera wrightii*	hypertrophy of inner cortical cells
P. halophilae	*Halophila ovalis*	hypertrophy of petioles
P. bicaudata	*Zostera nana*	arrests internode elongation causing marked dwarfing of host
P. maritima	*Triglochin maritima*	dwarfing of host
Saprolegniales		
Eurychasma dicksonii	Algae *Ectocarpus*, *Striaria*	causes hypertrophy
Haliphthoros milfordensis	oyster drill—*Urosalpinx cinerea*	infects early development stages of ova
Atkinsiella dubia	crustacean *Pinnothers pisum*	infects ova, zoeae, and pre-zoeae
Ectrogella perforans	diatoms—*Licmophora*, *Striatella*	
Eurychasmidium tumefaciens	alga *Ceramium*	hypertrophy, hyperplasia, and destruction of nodal cells
Leptolegnia baltica	copepod *Eurytemora hirundoides*	destroys or distorts all internal body organs

TABLE 1.—Continued

Fungus	Host	Remarks
Lagenidiales		
Olpidiopsis feldmanni	algae *Falkengergia, Trailliella*	host distribution may be modified by pathogen
Sirolpidium zoophthorium	Northern Quahog *Mercenaira mercenaria* Oyster—*Crassostrea* Clam—*Venus mortoni*	destroys motile larval stage (veliger)
Pontisma	alga *Ceramium*	facultative parasite
Lagenidium callinectes	blue crab, *Callinectes sapidus*, barnacle *Chelonibia patula* oyster crab *Pinnotheres ostreum*, mud crab *Neopanope texana*	infects and destroys ova
L. chthamolophilium	barnacle *Chthamalus fragilis*	attacks ova and embryo
Peronosporales		
Pythium thalassium	crustaceans *Pinnothers piscum, Leander serratus, Portunus depurator*	contributes to destruction
Eccrinales		
Enterobryus halophilus	mole crab *Emerita talpoida*	commensal—attaches in hind- and mid-gut
Arundinula capitata	hermit crab *Pagurus cuanensis*	occurs in stomach and intestine
A. incurvata	hermit crab *P. prideauxi*	stomach and intestine
Toeniella longa	*P. excavatus*	
Asellacia ligiae	isopod *Ligia mediterranea*	intestine
Entomophtorales		
Ichythosporidium hoferi	herring *Clupea harengus* flounder *Pseudopleuronectes americanus* alewife *Pomobolus pseudoharengus* plaice *Pleuronectes platesia* sea trout *Trutta trutta* mackerel *Scomber scomber*	virulent pathogen—destroys body and heart muscles, central nervous system, gonadal function, and interferes with blood circulation.

mercenaria and the American oyster *Crassostrea virginica*. Larvae of all ages are susceptible to infection. The fungus often develops a thallus which completely fills the larval shell. This fungal disease is a serious problem in the commercial development of clams and oysters in hatcheries. *Lagenidium callinectes* attacks ova of the blue crab *Callinectes sapidus*, the oyster crab *Pinnotheres ostreum*, the mud crab *Neopanope texiana*, and the barnacle *Chelonibia patula*. A germ tube penetrates the ovum membrane, and hyphae rapidly develop within the egg. Although the fungus is highly pathogenic and destroys the egg, the incidence of infection is low (14%) and infects only 25% of an egg mass. The infection therefore is considered a minor factor affecting crustacean populations. In contrast *L. chthamalophilum* may be responsible for reduction of populations of the barnacle *Chthamalus fragilis*. This fungus can destroy any embryonic stage from late gastrulation to the nauplius. The fungus penetrates the ovum and becomes established as a hyphal rudiment within 25 min after inoculation and infects an entire egg mass within 2 days. Algae also may be infected by the Lagenidiales (see Table 1).

(6) The distribution of saprophytic Peronosporales has been investigated in the North Sea where the relative abundance of sexual forms was highest in brackish water and decreased toward the open ocean. In offshore regions, asexual organisms, probably introduced from brackish waters, were found on nets, baskets, and ropes. Parasitic peronosporalids also occur, e.g., *Pythium thalassium*, which penetrates ova of various invertebrates. Destruction of ova is usually in conjunction with other organisms, often bacteria.

(7) The Mucorales have been reported from littoral sediments but almost nothing is known about their origin and ecology.

(8) The Eccrinales are commensals that attach to the walls of the alimentary canal, stomach, and intestine of various arthropods. The thallus

is an unbranched hypha that secures to the host by a holdfast.

(9) *Ichthyosporidium hoferi*, whose taxonomic position is uncertain, is a virulent pathogen of several species of fish. The infection reached epidemic proportions in 1930 and 1931 when herring populations in the Gulf of Maine were severely depleted. This pathogen probably enters the host through the alimentary canal. The sites of infection and subsequent reactions may vary in different fish. In general the effects are destruction of body and heart muscles, interference with blood circulation and degeneration of the central nervous system, loss of gonadal function, and necrosis of body organs.

(10) The Labyrinthulales appear to significantly affect plant populations. *Labyrinthula macrocystis* was associated with the disappearance of the eel grass, *Zostera marina*, from the coasts of America and Europe during the 1930's. However, it has not been determined whether the destruction was solely by the fungus or combined with environmental factors. Other species of *Labyrinthula* also occur, e.g., *L. roscoffensis* with *Ectocarpus* and *Taonia*; *L. minuta* with *Ulva*, *Bryopsis*, and *Zostera*; *L. vitellina* with *Zostera* and the phyto- and zooplankton *Chaetoceras*, *Rhizosolena*, *Nitzchia*, and *Coscinodiscus*.

Ascomycetes are characterized by the presence of a sac-like structure (ascus) containing sexually produced spores (ascospores). Opinions differ on the systematics of this class; however, for this discussion the Ascomycetes will be presented under two general subclasses: the Euascomycetes (asci produced within an ascocarp); and the Hemiascomycetes (ascocarp not formed).

Euascomycetes are saprophytes, parasites, and commensals of marine plants. Most species are cosmopolitan but there is evidence that the habitat of certain species is limited by environmental conditions. Temperature and salinity are two factors affecting distribution limits. Fungi with high temperature optima for growth and reproduction, such as are found in inshore temperate and tropical regions, apparently are unable to compete in the open ocean with the more psychrophilic fungi. Salinity probably has only a minor effect on fungal distributions since most fungi can tolerate and reproduce at, and above, salt concentrations of 3.5% and few, if any, have an obligate requirement for NaCl. Compound effects of temperature and salinity also must be considered. *Ceriosporopsis halima*, for example, develops in salinities to 3.5–3.7% at 25–29°C, but does not occur above 1.5% at 7°C.

Because of the economic interest involved in wood decay of pilings, boats, etc., considerable research has been devoted to lignicolous fungi. An extensive list of these fungi has been recorded and discussed in detail by Johnson and Sparrow.[1] Some species, e.g., *Ceriosporopsis* and *Lulworthia*, cause a "soft rot" in wood. The hyphae penetrate the walls of the wood elements and permeate the cellulose layer of the secondary wall. Other species soften the surface layers exposing the wood to erosion. Some lignicolous fungi, e.g., *Ceriosporopsis cambrensis* and *Zalerian* spp. have been associated with isopod gribbles and teredine borers. It has not been determined conclusively if the infestation of wood is dependent on plant–animal interactions or if the association is merely fortuitous.

The genus *Lulworthia* is perhaps the largest reported genus of Euascomycetes. The species are widely distributed, occurring on such cellulosic substrates as wood and intertidal spermatophytes.

Members of the genera *Guignardia* and *Haloguignardia* occur with marine algae. *H. decidua* and *H. longispora* are parasites on *Sargassum*, inducing simple and compound clavate galls. The relationship of *Guignardia* with algae is possibly symbiotic rather than parasitic and has been considered similar to lichen associations. In the composites, *G. ulvae* with *Ulva californica* and *G. alaskana* with *Prasiola borealis*, the algal cells become surrounded but not penetrated by mycelial networks.

Sporogenous yeasts are the most abundant Hemiascomycetes, occurring widely, with numbers generally low, averaging less than 5 cells/liter. The genus *Debaryomyces* (commonly *D. hansenii*) is probably the most prevalent genus. Isolates have been recorded from inshore and offshore waters of most oceans. Species of *Saccharomyces* are rare, although *S. cerevisiae* has been recorded from several oceanic regions. The genus *Metschnikowia* may be highly adapted to the marine habitat. Initially the genus was observed as parasitic on the fresh-water crustacean *Daphnia magna*. The asci are ingested by healthy animals and the ascus wall is dissolved, freeing needle-shaped spores that perforate the gut wall and penetrate the body cavity. The spores then germinate and initiate an infection. Two species have been obtained along the California coast. *M. zobellii* from seawater (2–58 cells/100 ml), fish intestinal material (25–5730 cells/ml), and giant kelp (520–39,200 cells/gm) and *M. krissii* from seawater (1–57 cells/100 ml). A third species, *M. bicuspidata*, has been isolated from Antarctic Ocean waters near the South Shetland Islands.

Deuteromycetes (or Fungi Imperfecti) usually are the conidial stages of Ascomycetes (or less often Basidiomycetes) whose sexual stages are no longer formed or have not been discovered. Two orders are predominant in the ocean: the Sphaeropsidales and the Moniliales. The Sphaeropsidales reproduce by means of conidia in a globose or flask-like fruiting body (pycnidium). The Moniliales produce conidia directly from vegetative hyphae or on a conidiophore without a pycnidium.

The Deuteromycetes have a distribution similar to that of the Ascomycetes; they are found in sediments, waters, and associated with plants and animals. These fungi may have a commercial importance through destruction of cordage and wood products. Most of the reported Fungi Imperfecti, as well as Ascomycetes, have been isolated inshore, particularly in tidal regions where they apparently are abundant because of high levels of organic materials and the influx of terrestrial fungal species. The lack of offshore fungi may be artificial as there has been a concentration of inshore fungal research.

Sediments contain a diversity of species, e.g., *Pencillium, Aspergillus, Cephalosporium, Trichoderma, Alternaria*, and *Cladosporium*. Fungal concentrations in sediments range from 2–1600 cells/gm. Most of the species are common terrestrial soil saprobes and it has not been determined if these organisms are significant in the degradation of sedimentary organic material in the ocean.

Some fungi occur in a constant association with algae, e.g., *Blodgettiomyces borneti* with the green alga *Cladophora fuliginosa*. The fungus produces parallel strands of anastomosing hyphae with large conidia between the outer pectin-like wall and inner true wall of the alga. *B. borneti* does not penetrate into the algal cell lumen. Similarly a *Blodgettiomyces*-like organism inhabits the green alga *Siphoncladus rigidus*, but does not produce conidia. These associations have been compared to lichens; however, little is known about the relationships.

Sea foam appears to be a dispersal agent for various fungal spores. The species composition in foam varies with habitat; for example, foam from open ocean beaches contains spores of *Corollospora maritima, C. trifurcata*, and *Varicosporina ramulosa*, while the spores of species that develop on substrates in salt marshes such as *Alternaria* sp., *Lignicola laevis, Pleospora pelagica*, and *Leptospaeria discors* were present in foam in estuaries.

The pelagic occurrence of Fungi Imperfecti is poorly understood, partially due to the lack of sterile collecting techniques. Deep-sea collections usually have been taken with Nansen or van Dorn bottles which are not sterilized and remain open until the water is sampled. Both types of bottles can contain residual fungal contaminants and give false representation of apparent fungal populations. The recent development of aseptic devices, such as the Niskin biosampler, capable of collecting large quantities (2 liters) of water, have allowed more accurate observations of oceanic fungi.

One of the groups that has been most extensively studied is the yeasts and yeast-like organisms. Quantitative distributions appear to fluctuate with the organic content of the water. Inshore polluted areas, such as estuaries, often contain as many as 5000–10,000 cells/liter of seawater. The numbers of cells diminish offshore and generally are in the range of tens of cells/liter, occasionally hundreds and rarely to thousands of cells/liter. Alterations of population densities are dependent on such factors as quantities of plant and animal standing crops, presence of decaying materials, and proximity to land. Qualitative distributions are often specific. Regions with terrestrial influxes contain terrestrial animal and soil associated organisms such as *Candida tropicalis, C. parapsilosis*, and *Torulopsis glabrata*. Intermixed are species of *Cryptococcus, Rhodotorula*, and *Candida* spp. that become dominant in near-shore nonpolluted regions. Water temperature appears to be one of the major factors controlling species distributions, e.g., warm inshore waters, such as found in the tropics, harbor species capable of growing at up to 37°–44°C, whereas offshore and other cooler waters have species with maximum growth temperatures of 26°–30°C. The other extreme is the abundance of psychrophilic yeasts in Antarctic waters with maximum growth temperature of 15°C.

Distributions of yeasts in the open ocean are similar to those found for other marine organisms; some species are cosmopolitan while others are restricted to particular oceanic regions. This type of distribution has been found in the Indian Ocean where there were three ecological groups in a study area that consisted of a transect along 60° E from 12° N to 40° S: (1) cosmopolitan species (*Rhodotorula rubra* and *Candida diddensii*) that were present throughout the study area; (2) regionally distributed organisms found either in northern Indian Ocean waters (*Sporobolomyces hispanicus* and *Sp. odorus*) or southern waters (*Rh. crocea*); and (3) widely distributed species (*C. polymorpha* and *Rh. glutinis*) that were excluded from a particular region (viz. Red Sea water that extends into the northern Indian Ocean). While such studies indicate species distributions, little is known about the specific role of these fungi in the marine ecosystem.

Basidiomycetes have been considered "rare" in the ocean, in fact only one "marine species" has been described. This is the smut (Ustilaginales) *Melanotaenium ruppiae* that invaded the rhizoides of the salt water monocot *Ruppia maritima* along the coast of France; however, recent evidence indicates that a group of Ustilaginales are abundant in the ocean. Certain of the yeasts which would be classified in the genera *Rhodotorula* and *Candida* have been found to be haploid and diploid organisms with life cycles typical of the smuts. These life cycles have been observed in artificial culture; however, the *in vivo* parasitic and/or saprophytic activities are not known.

J. W. FELL

References

1. Johnson, T. W., Jr., and Sparrow, F. K., Jr., "Fungi in Oceans and Estuaries," 1961, Hafner Publishing Co., New York.
2. Wilson, I. M., "Marine fungi, a review of the present position," *Proc. Linnean Soc. London*, Session 171, 1958–59, pp. 53–70 (1960).
3. Kohlmeyer, J., "The importance of fungi in the sea," in Oppenheimer, C. H., Ed., Symposium on Marine Microbiology, 1963, pp. 300–314.
4. Van Uden, N., and Fell, J. W., "Marine yeasts," in Droop, M., and Wood, E. J. F., "Advances in the Microbiology of the Sea," 1968, Academic Press, New York.
5. Meyers, S. P., "Observations on the physiological ecology of marine fungi," *Bull. Misaki Mar. Biol. Inst., Kyoto Univ. Proceedings U.S.–Japan Conference on Marine Microbiology*, Tokyo Japan, August 1966 (1968).
6. Alexopoulos, C. J., "Introductory Mycology," 2nd Edition, 1962, John Wiley & Sons, New York.

Cross reference: *Marine Borers.*

G

GOVERNMENT (U.S.) DEVELOPMENT OF MARINE RESOURCES

Before World War II

In this period governmental involvement in marine natural resources was limited primarily to coastal navigation and fisheries investigations. Congress enacted several navigation regulatory functions. The Rivers and Harbors Acts of 1890 and 1899, respectively, prohibited deposition of refuse in navigable waters and required anyone wishing to dredge, fill, bulkhead, or build in navigable waters to obtain permission from the Corps of Engineers. While authorities under these Acts have been used primarily to protect navigability, their provisions have increasingly been used to protect other public interests. The Oil Pollution Act of 1924 prohibited discharge of oil in U.S. territorial waters, the Corps of Engineers having enforcement authority (transferred in 1966 to the Department of the Interior).

After the Civil War, disputes over the apparent decline in New England fisheries led to studies at Woods Hole, Massachusetts, by Dr. S. F. Baird of the Smithsonian Institution. His work led to establishment of the U.S. Fish Commission in 1871. The Commission and its successor agencies undertook increasingly detailed and comprehensive studies of fishery resources in the years that followed.

In 1911, the North Pacific Sealing Convention was signed by Japan, Russia, Great Britain, and the U.S. This action came as a result of depletion and threatened extinction of the fur seals by unrestrained hunting on the high seas for their excellent pelts. Under management by the Bureau of Commercial Fisheries, the Pribilof herd has grown from about 200,000 to nearly 2,000,000.[1] This was the first of several conventions establishing international commissions to conserve fishery and related marine resources. Such treaties are a recognized instrument for management of fishery stocks in the open ocean. They provide a means for regulating the harvest of "common property" resources, resources that are available to all until reduced to individual ownership by capture.[2] Other commissions are International Pacific Halibut (1923), International Whaling (1935), International Pacific Salmon (1937), Inter-American Tropical Tuna (1950), Northwest Atlantic Fisheries (1950), International North Pacific Fisheries (1953), and Great Lakes Fishery (1955).

The advance of basic oceanography prior to World War II was marked largely by a few government-financed exploratory oceanographic cruises, which steadily expanded basic understanding of the oceans, and by establishment of Scripps and Woods Hole Oceanographic Institutions from private sources. These institutions provided a vital nucleus for the vast expansion of oceanographic work related to war problems. The few trained personnel available when war broke out were pressed into service to aid in many aspects of naval warfare, such as weapons systems design, beach topography, and sea state prediction.

Post-War Period: 1946–1966

This period, in contrast to that before World War II, was marked by rapid expansion in the number of Federal programs related to marine natural resources. Some of these are discussed below.

Fig. 1. Seal rookery in Alaska. (Photo: U.S. Dept. of Interior.)

Shore Protection. Although the Beach Erosion Board of the Army Corps of Engineers was established by Congress in 1930, its functions were limited to making studies. In 1946, Federal cost-sharing was authorized for constructing beach erosion projects for public shores. In response to a realization of the importance of public shores and beaches to public recreation, this authority was broadened in 1962.

Following a comprehensive study of hurricane storms begun in 1955, the Federal role in shoreline protection was expanded when the first of a series of projects was authorized in 1958 for construction by the Corps of Engineers to protect coastal areas from hurricane and storm-surge damage. Several of these projects have been combined with beach erosion control projects.

Nuclear Development. Because of the Atomic Energy Commission's responsibilities to protect public health and safety, AEC mounted a substantial oceanographic program to investigate the fate of radioactive elements in all environments, including the ocean. From research on the effects of low-level wastes, weapons tests in the South Pacific, and other studies using radioisotopes, a great fund of knowledge has been obtained regarding the uptake, accumulation, and transport of various radioactive elements in the marine environment. This and other work has provided striking insights into the transfer of nutrients and energy in living communities and their environment.

Water Pollution. The first general Federal legislation on water pollution control was enacted in 1948. The authority has steadily expanded since then. Specifically, under the 1965 Water Quality Act amendment, the states must establish, in inter-state and coastal waters, water quality standards that meet the approval of the Secretary of the Interior. Violations of these standards are subject to enforcement procedure by the Federal Water Pollution Control Administration.

Offshore Minerals. Although some oil had been obtained from strata under the ocean as early as the 1890's in California, only modest efforts were made to tap undersea pools until after World War II. Progress was hampered, too, by the uncertainty engendered by the jurisdictional dispute that followed the Supreme Court decision in 1948 in the California tidelands case. Much of the uncertainty was dispelled by passage of the Submerged Land Act in 1953, setting the seaward boundaries of the coastal states. This was followed, also in 1953, by passage of the Outer Continental Shelf Act, providing conditions for exploration, development, and leasing of continental shelf areas beyond state boundaries. The Act is administered by the Department of the Interior, through the Bureau of Land Management and Geological Survey.

Beginning in 1954, 13 competitive oil and gas lease sales have been held. Over 1000 leases comprising 4.6 million acres have been issued, with bonus-bid income of $2 billion. Extensive drilling has taken place and production has steadily increased in both State and Federal areas. Federal royalty income has exceeded $700,000,000 a year. Practically all the production has been off Louisiana and California.

Activity under the Outer Continental Shelf Act has been almost entirely for petroleum, except for sulfur production in the Gulf of Mexico. Leases for phosphorite off California have been issued, but no production has resulted. Interest has been shown in leasing areas on which manganese nodules are found, particularly on the Blake Plateau off Georgia. No formal requests have been made, however.

Desalination. While a small number of successful seawater distillation plants have been put into operation to convert seawater to fresh water, these have been in special situations and the water produced has been costly. In 1952, the Department of the Interior began a systematic program through the Office of Saline Water to develop the technology of desalination to permit water from the sea and other saline sources to be converted economically into fresh water. A primary aim has been to advance technology sufficiently to permit the unlimited resource of seawater to be a feasible alternative source for water-short coastal areas. Under this program

Fig. 2. Offshore oil-drilling rig. (Photo: U.S. Dept. of Interior.)

of research, several processes have been explored and industry has been encouraged to develop them. Successful demonstration plants using seawater have been constructed in Texas and California; the latter is now providing a secure water supply to Guantanamo Naval Base in Cuba. In 1967, contracts were signed with the Metropolitan Water District of Los Angeles for a full-scale desalting plant that will eventually produce 150,000,000 gallons per day. This plant will be powered by a nuclear reactor which is also used to generate 1.8 million kW of electric energy.

Electric Power from Tides. Harnessing the power of the tides has been an age-old dream. One project is already in successful operation in France. For several years, engineers of the U.S.-Canadian International Joint Commission studied the possibilities of utilizing the tidal energy of the Bay of Fundy to generate electric power. Further studies carried out by the Department of the Interior have indicated that a combination of conventional hydroelectric power with tidal power would be economically attractive. At the present time, however, the tidal project could not be judged economically feasible in comparison with other sources of peaking power, although technical developments give hope of eventual success.

Fish Protein Concentrate. In 1961 the Bureau of Commercial Fisheries of the Department of the Interior began to develop methods of producing a stable, inexpensive, high-protein product from whole fish. Such a product is commonly known as fish protein concentrate (FPC). Originally regarded mainly as a means of finding profitable use for under-utilized stocks of fish immediately offshore to improve the position of domestic fishing industry, FPC has taken on added significance. As a highly concentrated (80–85%) source of balanced animal protein, FPC is an attractive possibility as a low-cost supplement for protein-short, cereal-based diets typical of poor people throughout the world.

A method has been developed for manufacturing FPC from lean fish using chemical extraction. The product received approval by the Food and Drug Administration in early 1967, and pilot demonstration plants have been authorized by the Congress for market feasibility studies in the U.S. and selected countries abroad. Research is being carried out to develop processes suitable for fatty fish and to explore various biological and physical processes.

Mineral Mining. In contrast to petroleum exploration and production on the continental shelf, technology related to solid mineral recovery has remained relatively primitive. Significant exploitation of near-shore deposits of sand, gravel, and other construction materials already occurs, and highly promising deposits at greater depths are known. Extensive layers of manganese and phosphorite nodules have been known for a long time. Promising submerged placer sands have been found offshore of Alaska and several other locations. There is no reason to believe that the submerged continental shelf is any less rich in mineral resources than adjacent continental areas above sea level.

In 1963 the Department of the Interior, through the Geological Survey and Bureau of Mines, began a systematic program of reconnaissance, with emphasis on the Atlantic and Alaskan Continental Shelf areas.

Estuaries. These areas, where the waters of rivers and the ocean mingle, are important natural resource systems. Their vital role as breeding and nursery grounds for sport and commercial marine fish has been increasingly documented by careful biological studies. Estuarine waters and marshes are highly valuable for recreational boating, wildlife, and scenic attractiveness. On the other hand, estuaries have high value for commercial, industrial, and residential development, and—with channel deepening—protected waterways and harbor sites.

Only recently has there been a wide awakening to the extent of the conflict in uses of estuaries and the importance of establishing a balanced policy at Federal, State, and local levels. The need for fundamental knowledge on which to base rational management is imperative. Information to assess future economic pressures is scanty. While a number of studies of various aspects of estuaries have been made, few have been of sufficient scope to meet policy–development needs. In 1966, the Corps of Engineers was authorized to carry out a comprehensive study of Chesapeake Bay, the largest estuary in the U.S. Also in 1966, the Department of the Interior, through the Federal Water Pollution Control Administration, was authorized to make a three-year study of pollution and other aspects of the nation's estuaries. These studies will undoubtedly need to be supplemented.

Present Period

The present period is one of active transition for marine affairs in general as well as for marine natural resources in particular. Congress held extensive hearings on Federal oceanographic activities in 1965. Repeated dissatisfaction was expressed over the lack of clear national goals, the absence of unified top-level consideration of the various agency ocean programs, and inadequate prominence given to marine affairs.

Enactment of the Marine Resources and Engineering Development Act in 1966 for the first time set forth, as national policy, the development of a coordinated, comprehensive, long-range national program in marine sciences applicable to the oceans and Great Lakes.

The Act established two temporary organizations:

(1) The National Council on Marine Resources and Engineering Development, which is composed of five cabinet officers and three agency heads having significant marine-related activities and is headed by the Vice President. It assists the President by identifying issues, evaluating policies, and formulating a balanced set of marine science programs and priorities.

(2) The Commission on Marine Science, Engineering and Resources, which is composed of 19 Federal and non-Federal members. It is charged with making a full study of all aspects of marine science in order to recommend an overall oceanographic plan and governmental organization structure to meet present and future needs.

The Act stimulated Federal agencies to review their marine programs. The Department of the Interior, principal Federal agency responsible for conservation and development of natural resources, established a Marine Resources Development Program to draw together a unified program from its various ocean-related activities. In this program, the Department is planning on the basis of a long-range view of national needs for marine natural resources. By such planning, the groundwork can be laid now to meet increasing—even critical—needs for some marine resources in the near future and to obtain the knowledge and technical capability necessary to exploit and manage all marine resources when this becomes necessary or desirable.

Also enacted in 1966 was the Sea Grant College and Programs Act, administered by the National Science Foundation. The program is designed to encourage academic participation in the application of the various marine sciences to practical problems.

Almost simultaneously with passage of the Marine Resources and Engineering Development Act, two comprehensive reports on ocean science appeared: one by the Panel on Oceanography of the President's Science Advisory Committee;[3] the other by the National Academy of Science.[4] Both placed strong emphasis on the need to develop ocean resources and pointed to the need for a national ocean program and organizational structure capable of carrying out a unified program.

The first report of the Marine Science Council was issued in February 1967.[5] It reflects the transition of marine science from concentration on research to development of broader and more productive technology. Nine priority programs were identified as part of the President's budget request for ocean programs. Included in these were three important marine natural resources efforts:

(1) Launching a pilot program to aid protein-deficient countries to increase their capacity for using the fishery resources of the seas.

(2) Studying Chesapeake Bay as an example of the many problems of estuaries.

(3) Exploring offshore solid mineral deposits to identify potential new mineral sources and develop pilot mining procedures.

While it is not possible to predict future developments accurately, it appears that the development and management of marine natural resources will receive increasing emphasis. It also seems certain that the role of the Federal Government in marine natural resources will include greater emphasis on cooperation with academic institutions and on creating opportunities for industry. There will undoubtedly be more systematic efforts to aid the flow of defense-related marine technology into civilian use. Lastly, the use of marine resources as an instrument of public policy, particularly to promote peace and security in the world community, will become a recognized challenge for the U.S. and other developed nations of the world.

STANLEY A. CAIN

References

1. Baker, R. C., Wilke, F., and Baltzo, C. H., "The northern fur seal," *Fish and Wildlife Circ.* No. 169 (1963).
2. Christy, F. T., and Scott, A., "The Common Wealth in Ocean Fisheries," 1966, Johns Hopkins Press, Baltimore.
3. President's Science Advisory Committee, "Effective Use of the Sea," 1966, U.S. Government Printing Office, Washington, D.C.
4. National Academy of Science, "Oceanography 1966—Achievements and Opportunities," 1967, Washington, D.C.
5. National Council on Marine Resources and Engineering Development, "Marine Science Affairs—A Year of Transition," 1967, U.S. Government Printing Office, Washington, D.C.

Cross references: *Desalination of Seawater; Estuarine Environment; Fish Protein Concentrate; Heat and Power from the Ocean; Minerals; Pollution of Seawater; Radioactivity in the Ocean.*

GULF FISHERIES

When the Spaniards first went into Cuba they utilized some of the fishery resources of the Gulf of Mexico. Somewhat later they went to Mexico itself, but advanced mostly to the central highlands, and for a long time there was little utilization of marine fisheries. Still later, the Spaniards killed and harassed the West Indian seal almost to the point of extermination for its oil and blubber, which was the only common source of marine oil in the West Indian islands in the colonial days. In any case, the early history of the Gulf of Mexico fisheries is lost in the mists of time, but there is no reason to assume that the resources were much different a few hundred years ago from what they are now.

Even so, when the French settled the Mississippi coast, as a prelude to their expansion into the vast territory called Louisiana, the city-bred colonists suffered from starvation when supplies did not arrive on schedule. They were able only to "wade" oysters and pick them up from the shallow bottoms, and many colonists became ill with pneumonia during the winter because of exposure. One hundred and sixty years later, knowledge of fishery resources had changed considerably, and the coastal residents stated that they could not be starved out by the Union forces because "Biloxi bacon," the striped mullet, was always available.

Use of the oyster dredge slowly spread around the south Atlantic and Gulf coasts from Virginia; this apparatus was first pulled by sailboats, which went out of use last of all in Mississippi in 1933. Along the Gulf coast, the only fish used commercially in the early days were those of very fine keeping qualities such as the redfish, *Sciaenops ocellata*. These fish were surrounded and penned by seines and taken out in smaller quantities by smaller nets as they were needed for sale. Later, when ice became available, following the invention of an ice-making machine by Dr. John Gorrie of Apalachicola, and the more or less concomitant extension of railroads, other species of fish such as the speckled trout, *Cynoscion nebulosus*, and the southern flounder, *Paralichthys lethostigma*, were utilized in the commercial fishery. In Florida the mullet, *Mugil cephalus*, became an important food fish, and it was sold smoked or salted over most of the southeastern United States. Strangely enough, this excellent food fish was never accepted in the other Gulf states except Alabama and Mississippi, and even in these it has declined in popularity in recent years.

During the last century, the redfish, *S. ocellata*, the black drum, *Pogonias cromis*, the speckled trout, *C. nebulosus*, and the flounder, *P. lethostigma*, were the chief commercial fish produced within the bays and estuaries of the Gulf coast. For a number of reasons, some of them unreasonable whims of the consumer public, production of food fish has never been great on the Gulf coast, even though sometimes inferior products are shipped from elsewhere and consumed in fairly large numbers. At the present time there is little doubt that sports fishery produces several times as much poundage of food fish as commercial fishery.

Gunter and Guest[1] noted that a catch of a few thousand pounds would glut the Mississippi local market for speckled trout for several days. Sports fishermen's catches of croakers, flat croakers, sheepshead, and the more marine mackerels and mackerel-like fish add to the fish consumption of all states of the U.S. Gulf coast and quite probably exceed commercial fish production. The same influences are much less important in Cuba and Mexico, but in these countries there is considerable local fishing for personal and family consumption.

In past years the sheepshead, *Archosargus probatocephalus*, was quite abundant and was utilized to considerable extent in commercial fisheries, but the availability of this fish declined markedly between 1912 and 1915, for reasons unknown, and the population has never recovered.

About the same time fishery for marine turtles, chiefly the ridley and the green turtle, began to decline, and by the late 1920's it was gone. Formerly turtles were caught and penned on the Texas coast until gathered and carried alive on decks of steamers to New York City. Fishery for the diamond-back terrapin ceased at about the same time. This species has made a comeback in numbers, but it is no longer in demand.

All around the Gulf of Mexico, including its northern shores and the Gulf of Campeche, are large numbers of various red snappers of the genus *Lutjanus*, which are received with favor by the fish-consuming public. They are caught by hook and line outside the territorial waters of the nearby countries mostly. They are still imported into the U.S. by boats of American registry. In recent years Cuban boats have ceased fishing off U.S. shores, but U.S. boats still fish off the Mexican coast. This fishing is all done by hand lines, but today automatic and hand reels are also used.

A net fishery for the Spanish mackerel, *Scomberomorus maculatus*, has been of some importance on the lower Florida Gulf coast for a number of years, and a desultory fishery for the clam, *Mercernaria campechiensis*, has been of some interest in the same region.

The two most important fisheries for fish, as differentiated from mollusks and crustaceans, have been for menhaden, and lately for the so-called industrial fish.

The menhaden fishery in the Gulf of Mexico began as an abortive development in World War I, and was not revived until 1948. It grew until the Gulf production was over 1 billion lb in 1964 and again in 1966, equaling the Atlantic production from Maine to Florida. Menhaden production in pounds for the whole U.S. was over 2 billion, the only time any genus of fish has been produced in such great quantity in U.S. fisheries. The species *Brevoortia tyrannus* on the Atlantic, and *B. patronus* and *B. gunteri* on the Gulf were involved. This fishery has small production from west Florida, Albama, and east Texas, and the major production has been confined to Mississippi and Louisiana. On the Gulf coast the fishery is almost wholly estuarine, and average salinity where catches were made was 1.95%.

The Gulf menhaden season extends from April to October. Spotter airplanes aid the large fishing vessels (125–200 ft long). The fish are approached from opposite sides by two smaller boats, and are caught in large purse seines that go to the bottom and are then drawn around the schools, virtually encasing them from below. The large boat pulls alongside and the fish are pumped into the hold. The fish are refrigerated and taken to the factory as soon as possible, since the value and quality of the oil and other products are directly correlated with the state of freshness of the raw fish. Menhaden are the only fish sought because of their oil.

Although unwanted, a few predators on the edge of the schools are sometimes taken. Menhaden school tightly and the fishery is one of the cleanest in the world with regard to absence of other species in the catch. Nevertheless, it is under continuous attack by so-called sports fishermen who cannot believe that the large menhaden nets do not surround significant numbers of other fish as well.

Menhaden belong to the Clupeidae, herring family, and, like all members of this group, are plankton feeders. On the Gulf coast their heavy production is centered around the nutrient-rich waters at the mouth of the Mississippi River. Menhaden were originally processed for oil, but in later years the ground, cooked, and dried flesh and other residues became desirable commercial products. Menhaden oil is used in paints, waxes, cleansers, lipstick, margarine, certain lubricants, and leather dressings. The meal and protein derivatives are used in animal feeds, especially for poultry.

The idea concerning the use of the so-called industrial fish was some 30 years or more in developing, and came about as an offshoot of the shrimp fishery. Otter trawls were first introduced into the central Gulf in 1917 and they came into general use by 1927. These took considerable numbers of fish with the shrimp, especially when shrimping was not too good, and the trawlers were "scratching" for a small shrimp catch. Many proposals were advanced for the use of the "trash fish" to be collected by buy boats and brought to a central processing plant. Moreover, there was no market and the idea was sterile. However, in the early 1950's, various pet food companies decided that they could utilize the total trawl catch of fish of the bays, sounds, and shallow offshore Gulf. Crustaceans and the fish with hard spiny rays and hard bony parts are removed. The remainder is ground whole, cooked, and then canned by the usual methods.

The industrial fishermen trawl for fish in large amounts; they do not search for shrimp, which would cause a problem of separation if they were taken in quantities. This same method of fishing with otter trawls is used by large Russian fishery boats fishing for human food all over the world. In their case the catch is separated and better processed before it is canned or otherwise preserved.

In spite of the fact that as many as 100 or more species of fish may be taken in one industrial fish catch, the bulk of it is made up of the common croaker, *Micropogon undulatus*, and the flat croaker, *Leiostomus xanthurus*, these two species comprising about half the weight of the catch. A small part of this catch is shipped to the northern mink farms in a frozen state, and the remainder is canned and processed as stated above. The consumption is mostly by pet cats, but most dogs will eat this material as well and sometimes it is recommended by veterinarians.

It has been estimated that there are approximately 1 billion lb of industrial fish on the continental shelf to about ten miles offshore between southern Texas and Apalachicola, Florida. Some fisheries men hold that this is a low estimate and that a better figure would be 4 billion lb. Nevertheless, the only industrial fish plants are in the state of Mississippi. The future potentialities for this resource probably lie in the manufacture of fish protein concentrate. At present 100,000,000 lb of industrial fish are caught and processed on the Mississippi and Louisiana coasts each year, and the industry is expanding.

A fishery for the blue crab, *Callinectes sapidus*, has been carried on on the Gulf coast since prehistoric times. The blue crab is an estuarine crustacean and the greatest production is in Chesapeake Bay and off the Louisiana coast. On the Gulf coast, these animals were taken originally by a piece of meat on a line and, in some small numbers, in seines for fishes. After 1917 they were sometimes taken in considerable numbers in trawls that were fishing primarily for shrimp. Then as the fishing became more extensive, long trotlines were used with pieces of bait tied to the end of the line. Fishermen ran along this line and used a small paddle to bat the crabs into the boat. Later a rig was devised that dragged the crab up over a small inclined plane and shuttled him off into the hold as he released the bait. Trotline fishing for crabs is still practiced in the low-salinity brackish waters of Louisiana.

However, for the most part, the crab fishing has been taken over by the so-called Virginia crab pot, which is essentially a cube of chicken wire with funnels going in or with a grooved inset running along the top. This apparatus is baited and the crab can find its way in but seldom, if ever, find its way out. These pots are thrown down every 25–50 yards in a line with a small buoy extending to the surface. Approximately 90% of the Gulf coast commercial crabs are caught with this rig. The crab dredge from Chesapeake Bay has not been put into use on the Gulf coast. A few crabs on the Gulf coast are sold whole and alive to consumers or restaurants who wish to boil and serve them whole. However, most crabs are boiled and the meat is picked out in fish houses and then refrigerated in small 1 lb or $\frac{1}{2}$ lb containers. A few crabs are held in a pen and fed until they burst their shells and then are sold as soft-shell crabs. Nevertheless, these soft-shell crabs seem to be in relatively small demand on the Gulf coast and the percentage of production is small.

The stone crab, *Menippe mercenaria*, a member of the family Xanthidae, is produced in the amount of about 2,000,000 lb per year on the Gulf coast, most of this being in Florida with some little production for local consumption in Texas. This crab cannot swim and it lives mostly on shallow flats in high salinity areas near coon oyster reefs. It has terrific strength and can crack the shell of a large oyster. Nevertheless, it has a mild disposition and is often taken by hand, the catcher extending his arm to almost full length within the crab's hole. Sometimes only the claws are broken off and the crabs are left to grow other claws. Thus the total weight of the stone crab production as recorded is not comparable to the blue crab or other species.

A small amount of the spiny lobster, *Panulirus argus*, amounting to about 15,000 lb a year is taken on the Florida west coast. Similarly, approximately 20,000 lb of commercial sponges are hooked on the west Florida coast today. The total value of these two species is about $100,000 a year. The great decline of the Florida sponge fisheries because of disease is a sad story in the history of American marine products. Space limitations preclude recounting it here.

In 1967 the U.S. shrimp fishery exceeded 100,000,000 dollars in dockside value, the first time any American fishery has attained such value. This fishery depends almost solely upon the family Penaeidae, which extends from Cape Hatteras, North Carolina to Brownsville, Texas with the addition of catches from offshore waters in Mexico. Fisheries for the Pandalidae in Maine, Oregon, Washington, and Alaska contribute a small percentage of the total shrimp production of the U.S.

The white shrimp, *Penaeus fluviatilis* Say, has been caught in Louisiana since prehistoric times. When the white man came he introduced large seines that lasted until the 1930's and were at that time up to a thousand fathoms long, requiring a crew of 30 or more to drag them. Fishing was done primarily in the bays, and after 1900 a great many shrimp were sun dried on platforms built up over the marshes. The sale of this product was chiefly in China, and it declined precipitously with the onset of the Japanese invasion of China and the subsequent Communist takeover. Today a few dried shrimp are sold in the various Chinatowns of the U.S. and for local consumption in bars and saloons. Shrimp canning also began long ago; the centennial of shrimp canning was celebrated in Louisiana in 1967. Nevertheless, many troubles bedeviled this industry and it did not really get on its feet until the early 1900's when cans with an enameled interior came into general use. Spoilage was fast and refrigeration was poor in the early days, and shrimp were never shipped inland in large quantities with a mere addition of ice as oysters were. With the increase in the size and power of boats and refrigeration, which included the flash freezing of shrimp, this product pervaded the whole U.S.

The shrimp fishery of the Gulf and South Atlantic was taken over by otter trawls from about 1927 on, and Louisiana was always the prime producer with regard to *Penaeus fluviatilis*. Many studies have shown that this is a low-salinity shrimp and that the vast area around the mouth of the Mississippi River is a primary consideration in its production.

The white shrimp production could not support the demand for shrimp in perpetuity, and, following overfishing plus an extensive drought beginning in 1948, a large decline of the production of this shrimp came about. This was followed by a rise in production of the grooved shrimp, *Penaeus aztecus* and *Penaeus duorarum*. The pink shrimp, *P. duorarum*, extends from North Carolina around the coast to Campeche and Yucatan as a high-salinity shrimp of continuous distribution. The highest production is around the Dry Tortugas and Campeche Bay in southern Mexico. The brown shrimp, *P. aztecus*, is found on the south Atlantic coast and from western Florida to south Texas and in southern Mexico. The white shrimp is found on the south Atlantic coast of the U.S. around the Mississippi River to southern Texas and locally in the bay waters of the Gulf of Campeche. The South American white shrimp, *Penaeus setiferus*

Linnaeus, is found on the southern shores of Cuba, but not in the Gulf of Mexico.

Brown shrimp production is greatest in the state of Texas, and in some years production of this shrimp exceeds all others. In general however, the North American white shrimp, *Penaeus fluviatilis*, is the leading producer. Partly as a result of the American shrimp industry, the consumption of frozen and canned shrimp has increased all over the world, and now a great many countries are shrimp producers and consumers and exporters to the U.S. and Europe.

Present Fisheries Production in the Gulf

Three nations fish the Gulf of Mexico; Cuba, the U.S., and the Republic of Mexico. Cuba's current production from the Gulf is not available. Cuban boats still fish red snappers and shrimp from Mexican waters, but these figures are also unavailable. The greatest production of fishery products is along the U.S. coast. Some of the U.S. landings come from Mexico, but most of them are from the U.S. With regard to the U.S., the Gulf of Mexico now produces year in and year out 30% of the fishery products of the nation. The percentage increase has been 700 in the 27 years from 1936–1962. In fact, the Louisiana and Mississippi coast, which has been called the Fertile Fisheries Crescent, produces $\frac{1}{5}$ of all the fishery products of the U.S. in an area less than 350 miles in extent. This area stands in the same relative position with regard to the value of its fishery production. The high-poundage production lies in menhaden and the industrial fishes, while the high-value production lies in shrimp, oysters, and crabs. The state of Louisiana is the leader in fishery production of all the 50 states. This is because of the paramount importance of the Mississippi River and the nutrients it brings into the Gulf of Mexico. Furthermore, as has been shown before, the greatest fisheries on the North American continent are those which must be classified as estuarine, and 97.5% of the Gulf coast fisheries are estuarine.[2]

Table 1 gives the known production for the U.S. during 1965. The Mexican fisheries are not fully exploited, but, due to the absence of large estuarine areas, they can never be expected to be as great as those lying off the more extensive land areas of North America.

GORDON GUNTER

TABLE 1. FISHERY PRODUCTION OF THE GULF OF MEXICO FOR 1965 (THOUSANDS OF POUNDS)[a]

	West Florida	Alabama	Mississippi	Louisiana	Texas	Total
Blue runner	1,210					1,210
Groupers	7,662					7,662
King mackerel	1,314					1,314
Mullet	35,541	1,072				36,613
Spotted trout	2,799					2,799
Red snapper	6,532	2,393	1,849		2,250	13,024
Spanish mackerel	3,880					3,880
Black drum					1,409	1,409
Red drum				312	447	759
Menhaden			237,833	599,538	66,686	904,057
Flounder	204	162	57	190	305	918
Spiny lobsters	2,845					2,845
Stone crab	751					751
Blue crab	14,081	1,762	1,288	5,892	2,484	25,507
Pink shrimp						
Brown shrimp	39,966	7,215	6,416	59,382	66,053	179,032
White shrimp						
Clams	72					72
Oysters	2,793	1,005	4,829	11,401	3,357	23,385
Scallops	18					18
Squid	15	4		4	24	47
Turtles	31			20		51
Terrapin				6		6
Sponges	44					44
					Total	1,205,403

[a] Only those species taken in amounts greater than 100,000 lb are listed, except for a few mollusks, turtles, and sponges. Figures for North American production are taken from Lyles.[3]

References

1. Guest, W. C., and Gunter, G., "The sea trout or weakfishes (*genus Cynoscion*) of the Gulf of Mexico," Gulf States Marine Fisheries Commission Technical Summary No. 1, pp. 1–40, 1958.
2. Gunter, G., "Some relationships of estuaries to the fisheries of the Gulf of Mexico," part IX, *Fisheries*, pp. 621–638, "Estuaries," Lauff, G. H., Ed., Publ. No. 83, American Association for the Advancement of Science, Washington, D.C., 1967.
3. Lyles, C. H., "Fishery Statistics of the United States, 1965," Sec. 6, "Gulf Fisheries," pp. 321–379, Statistical Digest No. 59, 756 pp., Bureau of Commercial Fisheries, Washington, D.C., 1967.

Cross references: *Menhaden Fishery; Sea Turtles; Shrimp Fisheries; Sport Fishing.*

H

HAWAIIAN FISHERIES

Hawaii lies in the path of the northeast trade winds, which profoundly affect its climate and oceanography.[1] Although average temperatures range between 72.0° F in winter and 78.5° F in summer, the weather is only rarely uncomfortably warm, for through most of the summer the trade winds blow at an average speed of 14 mph. In winter, the center of the trade winds moves to the south of Hawaii and the area is visited by occasional storms from the north. Sea surface temperatures range between 73° F in winter and 80° F in summer. All the islands are partially fringed by coral reefs that extend to a distance of a few hundred yards offshore. Beyond the reefs, the ocean floor slants steeply to the abyss.

Although it is the 47th of the states in land area, being about ⅓ larger than Connecticut, the state of Hawaii ranks fourth in the length of its sea coast. No point in the state is more than 30 miles from the shore. Hawaii lacks the shallow rich banks that characterize most of the world's great fishing areas, however (Fig. 1). Off the principal islands only 1560 square nautical miles of water are shallower than 100 fathoms.

The living marine resources of Hawaii, then, are found on the reefs and the inshore area—both relatively small—and the all but illimitable open sea beyond the 100-fathom line. The people of Hawaii draw upon all three habitats for their food, but unequally, since it is the open sea that provides by far the greatest portion of the locally caught marine products.

Modern Hawaii is a blend of several ethnic strains. Numerically predominant are the Caucasians and Japanese, who account for more than 2/3 of the people in the state. Since the state is surrounded by the sea and since in its population it has many citizens whose forebears came from cultures in which fish constituted a major portion of the ordinary diet, Hawaiians consume more seafood than the citizens of many of the other states. The per capita consumption of commercial marine food is about 40 lb a year, as compared with a national average of about 10 lb. Half or more of this, however, is imported. In 1965, the Hawaiian fish catch per capita was 19 lb. In the same year, the state imported about 20 lb per capita of marine foods, mostly from foreign nations. Some of these imports were specialty items, such as rock lobster and shrimp, that are not harvested in quantity from local waters. Yet, lying at the center of a region that provides the Japanese longline catch with an estimated 100,000,000 lb of tuna a year within 1500 miles of Honolulu,[2] Hawaii has notably failed to take full advantage of its marine resources.

There are several reasons for this situation. One that might be stressed, because it is not generally recognized, is that Hawaii is a wealthy state. Per capita income in 1965 was $2906, bringing Hawaii to 12th rank within the United States. Fishing remains by and large a primitive and uncertain business, and this is particularly so in Hawaii, where the fleet is small and old and better-paying jobs are immediately available to the fishermen. As a result, production is low and prices are so high that fresh fish flown from the mainland is more than competitive with local

Fig. 1. Cross section of the island of Hawaii, showing the peak of Mauna Loa (13,746 ft above sea level) and the steep descent to the abyss offshore. Water a mile deep lies within only a few miles of the shore, affording only a restricted habitat to shallow-water animals.

fish in the retail market; it is often very much cheaper. The few Hawaiian fishermen, despite, or perhaps because of, the paucity of their catch, are among the best-rewarded in the U.S. and probably the world, but their annual income remains far lower than that of men in other trades. Thus it remains as true today as in 1900 that "the most noticeable feature in this market is the excessively high prices charged for fishery products. As compared with other retail markets in the U.S., and possibly of the world, Honolulu ranks first as regards high prices."[3] Fish is a luxury food that prosperous Hawaii can afford—not a daily necessity, as it was to the Polynesians. Until recently it has not been looked upon as a potential means of adding further substantial wealth to the state.

As in other Polynesian communities, fish formed a staple in the diet of the early Hawaiians. By tradition, the ordinary Hawaiian lived on fish and poi (the root of the taro plant, baked and pounded). A fascinating body of lore grew up around fish and fishing. The Hawaiians had names not only for the hundreds of species of fish they knew, but also for the same fish at different stages of their lives. These names survive. The Hawaii Division of Fish and Game carries them in its statistical reports. One of the unusual features of the early Hawaiian fisheries was the existence of many fishponds along the shores of several of the islands, notably Molokai. Privately owned, these fishponds were used for the nurture of fish captured at sea or for those that entered the ponds as fry and could not escape to the sea when they grew larger. One or two of the ponds are still used, but most have become silted up or their stone walls have been breached by the sea.

With the decay of the Hawaiian culture and the successive waves of immigration to the islands in the 19th century, the customs of the land rapidly changed. Small-scale commercial fishing developed relatively early. The first survey of the fisheries, conducted shortly after Annexation,[3] showed that although Hawaiians, i.e., Polynesians, still predominated in the fisheries, persons of other ethnic backgrounds were playing an increasingly important role. Most of the offshore fishing was in the competent hands of the Japanese, who brought traditional and efficient methods of harvest from their homeland. These methods are still used.

At that time (1900) 138 kinds of fish and invertebrates were sold in Hawaiian markets. The total catch for the year was 6.2 million pounds, which brought $1.1 million. In 1966, 13.0 million pounds of 85 kinds were taken, with a value to the fishermen of $3.1 million. The increase, over 66 years, it will be noted, is slight. Indeed, if the purchasing power of the dollar is taken into account, there has been essentially no gain in the value of the catch. Even then, Hawaii was supplementing its own supply of sea food with stocks imported from abroad. Especially important was salted salmon from the Pacific Northwest. The Hawaiians had immediately taken to this novelty when it was first imported in the 19th century and had made it the basis of one of their dishes, lomi lomi salmon, a favorite then as it is today.

The marine fauna of the Hawaiian Islands belongs to the great Indo-Pacific region; indeed, constitutes its northeasternmost extension. Gosline and Brock[4] listed 584 species of native Hawaiian fishes. As is true elsewhere, only a few of these are used, or probably could be used, as food for humans.

Of all the fishery resources of Hawaii, the one which is most heavily harvested and which is commonly agreed to be capable of sustaining the greatest increase, is the skipjack tuna (*Katsuwonus pelamis*). Throughout the world, tunas command a premium price. Although skipjack tuna is not as valuable per pound in the markets as some of the other tunas, it is widespread and exists in immense quantities. About 36%, by weight, of the tunas taken in the Pacific Ocean in 1965 were skipjack tuna. There is a very large fishery off Japan and another off North America. Compared with these, the Hawaiian fishery is minuscule. In 1965, its peak year, it took 16,000,000 lb, as against almost 300,000,000 lb in Japan and 170,000,000 in the eastern Pacific Ocean. Nevertheless, it is by far the largest fishery in Hawaii and experts believe that the central Pacific Ocean may eventually contribute a catch that could equal or exceed those of the eastern and western shores combined. The evidence has been summarized by Manar.[5] Hawaii is strategically

Fig. 2. The present Hawaii skipjack tuna fleet catches the fish by pole and line. Schools are often located by sighting bird flocks that compete with the fish for prey. The schools are attracted to the ship by throwing live bait overboard.

located to harvest this largely under-utilized resource. The market is, for practical purposes, almost unlimited. The potential is great, being estimated as hundreds of thousands of tons.

The present fishery for the skipjack tuna, or "aku" as it is called in Hawaii, is carried out close to the islands by pole-and-line boats that attract fish by throwing live bait in the water (Fig. 2). Fish schools are usually sighted directly or are located by watching for the bird flocks that are often associated with feeding tuna schools. The number of boats in the fleet decreased from 32 in 1948 to 17 in 1967. The annual catch is about 10,000,000 lb. Improvements in the present fishery are being initiated, but if the larger body of the resource lies far from the islands, and particularly if it is not available to present techniques of fishing, new vessels and gear may be required to take advantage of it.

The second largest fishery in Hawaii is that conducted by the longline fleet, which uses a scaled-down version of the Japanese longline method of fishing. The principal catch consists of yellowfin tuna (*Thunnus albacares*) and bigeye tuna (*T. obesus*)—both much larger in size than the aku—and some of the billfishes. The number of boats decreased from 59 in 1948 to 30 in 1967. The catch has declined from about 3,000,000 to 2,000,000 lb.

Both the pole and line and the longline fleets fish close to the islands, usually within 20 miles (Fig. 3). Only occasionally does one of the longliners venture out of sight of the chain. These expeditions have been attended with considerable success; catches were better several hundred miles away from the islands than near them.

Of the inshore fish, the akule, or bigeye scad (*Trachurops crumenophthalmus*), is the most important in the Hawaiian commercial catch. In 1965, 449,000 lb were taken. The opelu, or mackerel scad (*Decapterus pinnulatus*), which supports the next most important fishery, yielded 190,000 lb in 1965.

The bottom fisheries consist mainly of several snappers. These fish are taken by handline in fisheries that appear to be highly localized. Catch rates are often very good over a small area of bottom, very poor beyond.

As for other fisheries, they amount to little in weight and value at present. A few spiny lobsters are taken (8100 lb in 1965, a figure which represents a substantial decline from the yield of 131,000 lb in 1900). The only shrimp used as food is the fresh-water opae, much valued in the Polynesian cuisine, of which 34 lb were reported sold in 1965.

None of these fisheries, except for that for the aku, has been sufficiently studied to permit estimates of the true abundance of the resource and

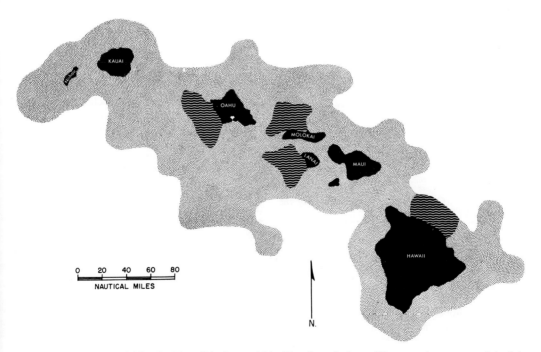

Fig. 3. Most of the fishing in Hawaii is done within 20 miles of shore. Here are shown areas fished in 1964. Intensity of shading denotes the relative success of the catch. The most productive areas lie within a few miles of the principal port, Honolulu.

its maximum yield. The Bureau of Commercial Fisheries Biological Laboratory, Honolulu, has located a small shrimp resource in the islands. The Hawaii Division of Fish and Game is conducting research on some of the local inshore fishes. It has also actively studied the possibility of oyster culture in Hawaii. The heavily polluted lochs of Pearl Harbor shelter beds of eastern oysters (*Crassostrea virginica*), worth perhaps a million dollars. Some work has been done on the possibility of transplanting some of this valuable stock into cleaner water.[6] The Hawaii Division of Fish and Game is also investigating the possibility of establishing the giant prawn (*Macrobrachium rosenbergi*) from Southeast Asia. A large local market exists for shrimps and prawns.

The fishery resources of Hawaii are: the tunas of the high seas, which might produce one of the great fisheries of the world; the unknown fisheries of the shallower depths beyond the reef, which might become of considerable local importance; and the reef fisheries, which are probably more important from the recreational standpoint than as commercial fisheries.

THOMAS A. MANAR

References

1. Blumenstock, D. I., "Climate of the States—Hawaii," *U.S. Dept. Commer. Weather Bur., Climatogr. U.S.* No. 60-51, 24 pp., 1961.
2. Brock, V. E., "A proposed program for Hawaiian fisheries," *Univ. Hawaii, Hawaii Mar. Lab., Tech. Rep.* 6, 18 pp., 1965.
3. Cobb, J. N., "Commercial Fisheries of the Hawaiian Islands," *U.S. Comm. Fish and Fish., Rep. Comm.*, 1901, pt. 27, pp. 381–499, 1902.
4. Gosline, W. A., and Brock, V. E., "Handbook of Hawaiian fishes," Univ. Hawaii Press, Honolulu, Hawaii, 372 pp., 1965.
5. Manar, T. A., Ed., "Proceedings of the Governor's Conference on Central Pacific Fishery Resources," State of Hawaii, Honolulu, 266 pp., 1966.
6. Sparks, A. K., "Survey of the oyster potential of Hawaii," *State of Hawaii, Dept. Land Natur. Res., Div. Fish Game*, 44 pp., 1963.

Cross references: *Coral Reefs; Trust Territory of the Pacific Islands; Tuna Fishery.*

HEAT AND POWER FROM SEAWATER

The ocean covers 71% of the earth's surface (140,000,000 square miles) with a volume of 330,000,000 cubic miles. This water varies considerably in temperature at different latitudes and at different depths in the same latitude. Many related factors influence ocean temperatures, including particularly absorption of solar energy near the surface, thermal-syphon action caused by the change of density with temperature; winds and currents which carry large amounts of water thousands of miles due to unequal distribution of solar radiation, the earth's rotation, and other factors. Thus, the flow of water in the Gulf Stream off the east coast of the U.S. has been estimated as over 540 cubic miles per day off the coast of Florida; as about 1700 cubic miles per day off Chesapeake Bay; and about 800 cubic miles per day off the Grand Banks.

Solar Heat

The solar constant is the radiant heat from the sun which would fall on a surface normal to the sun's rays at the earth's distance from the sun without losses due to the atmosphere. This is about 10,000 BTU per day per square foot. On the spherical earth, the area is 4 times the area normal to the sun's rays, so these figures, as an *average* value over the whole earth, must be divided by 4 to give about 2500 BTU per square foot per day, or somewhat over 100 BTU per square foot per hour. This includes the facts that no radiant heat is received at night, and quite large variations occur with the seasons.

Deductions are necessary for losses due to the atmosphere, clouds, etc. *Average* figures have little interest; data are needed for particular areas at different seasons. Thus, from measurements made in many stations, charts give contour lines of different constant values of solar energy actually received during each month of the year. Since few fixed stations are in the middle of the oceans, data on radiation received by the oceans are not nearly as complete as data for land areas.

The solar energy received by the great mass of the ocean's water of high specific heat varies from day to night and summer to winter for any one locality, and averages the climate to within the range of the extremes tolerable by man on land. It is a major cause of the winds and the ocean currents which warm one coast while cooling another. Besides these tempering, or averaging, effects on the temperatures of the land masses, this heat *input* into the oceans must be balanced by a heat *output* from the oceans; otherwise the temperature would increase. Neglecting many other factors, the average temperature of the ocean's surface rises to that value where the vapor pressure at this temperature causes sufficient water to evaporate so as to maintain a dynamic equilibrium. This water vapor is condensed by the cold upper atmosphere and comes back in the fresh-water cycle to the sea.

In the ceaseless daily currents of many cubic miles of warm water in different directions, and many other cubic miles of cold water in other directions, often there may be proximity of a current of cold and a current of warm water, having considerably different temperatures. Often there may be a temperature substantially lower at a considerable depth from that at the surface, with two steady currents of quite different temperatures, possibly at cross flow, or even in countercurrent to each other.

Most of the heat from the sun is absorbed in the surface water which, on heating, expands to a lower density and stays above the colder, heavier water below. In the tropic seas, there is a dramatic thermal-syphon effect. The cold water from the direction of the Poles flows near the ocean floor, or in other currents. It becomes heated in the tropics and rises, due to its lower density. The water so warmed becomes another current back toward the Poles. In the polar areas, the absorption of solar energy is low because of the low angles of the sun's rays, and the much greater amount of the earth's atmosphere to be penetrated. The water is chilled and sinks. Either as cold water or as a melting iceberg, it starts its return trip toward the tropic seas to repeat the cycle.

Direct Use of Heat from Warm Seawater

For many years, seawater in shallow lagoons —often man-made—has been evaporated by solar energy, to give dry salt. A desirable location will have a steady, dry wind from inland over the shallow water to carry off the vapors. More or less additional processing of the concentrated brine gives pure salt. The effective evaporation rate is the *net* evaporation after accounting for rain. This may vary from about 40 in. of seawater during that part of the year when there is no important rain in the latitude and under the conditions of San Francisco Bay—where much solar salt is produced—to several times this much per year in lower latitudes and dryer climates. Much of the world's salt is produced by harvesting the crystal salt after most or all of the water has been evaporated by the sun. A small amount of blue dye in the water increases the absorption of solar energy.

Survival kits for airmen down at sea include a simple plastic device to utilize the sun's heat to evaporate seawater and condense it for drinking. A black, shallow pan of seawater receives the sun's rays through a sloping, transparent glass or plastic cover. Sometimes a flexible plastic cover is inflated by very slight air pressure. The cover is cooled by the air above, the water vapors condense on the underside, and run down to a trough on the lower edge.

Such plants, acres in extent, have been built and are proposed for fresh-water supply, particularly in underdeveloped countries. Seawater is circulated, with approximately $\frac{1}{2}$ evaporated for condensation, and the doubly concentrated brine is returned to the sea. Production rates from about 0.2–1.5 lb of fresh water per day per square foot of surface may be expected, depending upon the latitude and the time of the year, both of which determine the angle of incidence (hence intensity) of the sun's rays. Also important are the cloud factor, the design of the still, and other considerations. Salt and fresh water are not produced simultaneously, since the rate of evaporation is greater from the less concentrated brines used for desalination than from the supersaturated brines used to produce salt.

All such absorption of solar energy is in seawater after it has been taken from the body of the sea. It is conceivable that such a solar distiller could be constructed by inflating a plastic bubble, or half bubble, over an area of seawater, particularly in a lagoon, where most of the heat absorbed would not be carried away by currents of water; but it is more practical to use a black surface on land such as the bottom of a pan. However, this transmits a small part of the solar heat into the ground.

Shallow seawater lagoons, with only narrow inlets, obtain a relatively higher temperature, and are used for increasing the rate of growth of some forms of sea life, e.g., oysters in Japan for pearl culture.

Stored Thermal Energy

Much more vast amounts of thermal energy that are absorbed in ocean water "on the spot" give the warm currents in the tropic seas and flowing therefrom. Such energy is at a relatively low temperature, and seawater is seldom warmer than the air or the tropic lands it washes. However, the solar heat absorbed in the tropic seas, by being carried in ocean currents, may warm lands thousands of miles away. The amount of heat stored in warm seawater is tremendous. If released as a source of energy for electric power or for desalinating water, major problems of the future may be solved.

Thus, the Caribbean Current washes the north shores of South America, then flows northwesterly through the Florida Straits between Yucatan and Cuba, with a clockwise turn around the Gulf of Mexico, to escape back toward the Atlantic Ocean between Cuba and Florida, only to be turned northward by the Great Bahama Bank. Then, joined by the Antilles Current flowing westerly on the north side of the Bahamas, it becomes the Gulf Stream, which continues north and east as the North Atlantic Current. This branches into several currents, one circling Iceland, with a westerly branch that gives what green there is in Greenland; another which passes Ireland, then Norway, well up into the Arctic Ocean, to keep ports there from ever being closed by ice; still another branch turns clockwise around the Bay of Biscay to join another which has diverted more directly to wash in a southerly flow the coast of Portugal.

These tremendous volumes of comparatively warm water are well known to mariners and mapmakers. However, the return trip of these waters from the Arctic Ocean to their source of heat—the rays of the tropic sun—is less well charted, since much of it is at much lower velocities, generally southerly and westerly in the great deeps. One is the Cold Wall or Coastal Countercurrent flowing southwesterly from the Labrador Current against the Atlantic Coast of the U.S.

Energy From Temperature Differences

Heating and cooling involves a change in temperature referred to a first temperature. So also, the availability of energy from heat has been known for almost 150 years to depend on the difference between the high temperature from which the heat is supplied and the low temperature to which the heat is rejected.

Therefore, in utilizing heat, a cold body must be available to receive heat, as well as a hot body to make it available. Physical proximity is also necessary to allow heat transfer. The ice of Antarctica would be a good absorber to pass tropic solar energy in generating power, but it is not available. There are, however, many places in and near the tropics where two ocean currents, having as large a temperature difference as 35°–45°F, and of great (and for this purpose almost infinite) magnitude, are very close together; sometimes within 2000–3000 ft between the surface water and a deep supply or current of cold water.

To heat one cubic mile of the colder water per day to a temperature 45°F higher would take an amount of thermal energy 25 or 30 times that of all the electrical energy produced in the U.S. The reverse is more staggering; and it has been estimated that the heat in the warm water of the Gulf Stream could generate over 75 times the entire power production of all the U.S.

Theoretically, mechanical energy (and from it electrical energy) may be developed from heat at any temperature level being used thermodynamically to heat any material which may receive it, because of its lower temperature. Practically, however, the production of such energy becomes more difficult and less efficient the lower the temperature difference of the source of the heat, compared to the low-temperature reservoir. Carnot expressed the maximum possible efficiency as $(T_1-T_2)/T_1$ where T_1 is the temperature at which heat is available, and T_2 is the temperature at which it can all be discharged. Absolute temperatures, i.e., above absolute zero, are used here.

Heat engines usually work at a temperature difference between the average temperatures of heat input and of heat output of many hundreds, or even thousands, of Fahrenheit degrees. The closer the input temperature approaches the output temperature, the less the maximum possible thermodynamic efficiency. In this case, if 10°F of the temperature of the warm water coming in at 80°F is utilized, the availability of the energy thus may be at 80°–10°, or 70°F. Similarly, if the cold water at 40°F is heated 10°F to 50°F, this temperature may be regarded as the low temperature at which all heat may be discharged.

Thus, if the heat supply is at 70°F or 460 + 70 = 530° above absolute zero and the corresponding value for the heat rejection is 50°F or 50 + 460 = 510°, then the maximum thermodynamic efficiency is (530–510)/530, or about 3.8%. Practically, because of many energy requirements in related machinery and many energy losses, the efficiency obtainable could not be more than about 2.5%. Of equal importance, the amount of equipment required to obtain the energy when available at such a low temperature difference usually *increases* greatly with a decrease of the temperature difference. Thus the trillions of BTU's in a cubic mile of warm seawater which may be passed to colder seawater may develop mechanical energy during the process to make electrical power. However, the heat available at this low temperature can be converted to power only with a large, costly plant, at a very low efficiency, and by the handling of extremely large amounts of the warm seawater to give up its heat and of extremely large amounts of the cold seawater to absorb the heat. It has been pointed out, however, that the total amount of water to be handled may be less than that required to produce the same amount of power in a hydroelectric plant. Dams, penstocks, and machinery of a hydroelectric plant are also expensive.

There are many places known in the Caribbean area, within a few miles of land, where seawater has a surface temperature of 80°–85°F, while at 2000–3000 ft below the surface it may be only 40°–45°F. Elsewhere, the ocean floor drops off very steeply indeed to an ocean deep within a few hundred feet of land. The ideal location for utilizing this temperature difference so as not to change the temperature of the source of supply might be in a land-based plant at which warm surface water would be available, say on one side of a peninsula, with a great sea depth close to shore on the other side of the peninsula. The contour of the bottom should be favorable to the installation of a large suction pipe to supply cold water. The great depth from which it is necessary to pump the cold water does not represent suction head on the pump. Only the friction head must be considered, plus the small static head caused by the difference in density of the cold water and the average density of the water from the surface to the bottom of the pipe.

Low-Pressure Steam Cycle

Various power cycles have been proposed to generate power from the thermal energy of warm seawater available by removing heat with cold seawater. Of these cycles, two are cited. The first converts some part of the sensible heat of warm seawater in cooling it by flash evaporation to give steam, as proposed by D'Arsonval (1882) and Claude (1926), and more recently improved by the Andersons. Steam will be at the low pressure corresponding to the saturation temperature of the warm seawater after it has been cooled.

The steam may then do work by passing through a turbine, and it must be condensed at an even lower temperature and pressure by the cold water, the temperature of which is raised in absorbing heat. Thus, the effective (or average) temperature difference available for the generation of power between the cold and warm streams initially would hardly be more than one-half of the over-all temperature drop.

A very rough example may indicate this direct steam cycle. The warm water may be cooled from 80 to 70°F in flash evaporating to give steam at an absolute pressure of 70°, neglecting the boiling-point elevation due to the 3.5% salt which is present. From this 10°F drop of the warm water approximately 10 BTU per pound is available to evaporate water in this adiabatic balance between sensible heat in cooling water and latent heat in evaporating a small part of it. Latent heat of water is approximately 1000 BTU's per pound, so that, for every pound of warm water, about 0.01 lb of steam is formed at 70°F and 0.36 lb per in.^{2} absolute (psia). This steam may be expanded through a special steam turbine to a lower pressure and temperature to give mechanical energy which will drive an electrical generator.

The cold water removes the heat of the steam by condensation. It is available at 40°F and, for simplicity, an amount equal to that of the hot water may be assumed. The temperature rise of the cold water approximates that of the temperature drop of the warm water. Thus, the cold water is heated to 50°F.

If the steam is assumed to condense with a temperature of approach to this 50°F of only 2°F, the vapors after the expansion in the turbine will be at 52°F and 0.19 psia. The pressures on the two sides of the turbine thus are 0.36 and 0.19 psia, a working pressure drop of only 0.17 psia. The specific volume of the steam on the high side is 869, and on the low side 1588 ft^3 per lb. This compares with 27 ft^3 per lb at the atmospheric boiling point, or 0.67 ft^3 per lb at 500°F and 681 psia—more reasonable conditions for a steam power plant. Thus, the volumetric capacity of the turbine must be extremely large. About 100 lb of steam will have to be formed using about 10,000 lb each of warm and cold water, to make 1 kWh of electric power. (The important consideration is that there are practically unlimited supplies of both warm and cold water for the taking.)

Of this gross amount of power, about $\frac{1}{3}$ will be required for pumps and other equipment, to give as salable power only about $\frac{2}{3}$ of that generated. The boiler may be an open vessel in which occurs a single flash of the warm water; the condenser may be a simple open vessel with sprays or films of water in open flow to give a large amount of surface for contacting and condensing the steam. The vapor volumes and connections are extremely large, as is the size of the turbine, because of the very high specific volume of the steam.

An important accessory of any plant operating below atmospheric pressure is an air removal system. This is particularly important here because of the very large amounts of both cold and warm seawater to be handled. Both contain dissolved air. In the process, the warm water is partially evaporated, which removes air; and the cold water is being warmed, which makes air less soluble therein. Hence, an air removal system must follow the condenser.

There is a large turbulence of the tremendous volume of warm water which must be flash evaporated to convert the 10°F of sensible heat to latent heat and produce about 1% of the water weight as steam at the very high specific volume. Thus, for every cubic foot of warm seawater cooled in the flash evaporation, there is approximately $62 \times 1\% \times 869$ ft^3 per lb, or 540 ft^3 of steam formed. A substantial amount of the energy may be lost in the turbulence of this evaporation because of the non-equilibrium conditions. Also, the hydrostatic head of most of the water below the surface in any usual flash chamber might be much larger than the available steam pressure; and much of the water would never have a chance to come to equilibrium with steam at the surface.

To minimize these losses, the system known as the Controlled Flash Evaporation (CFE) has been developed. Here the warm seawater at the higher temperature goes down a multitude of parallel vertical channels. Each is a long, narrow "chute" with an expanding cross-section leading to an open space of lower pressure beneath. As the hydrostatic pressure decreases on the water going down, evaporation is caused, starting in the middle of the stream in the chute, with the water flowing in contact with the tapering sides.

This expansion of the cross-section accommodates, to some small extent, the large increase in the volume of water plus vapor. Thus, from the calculation above, the water volume is changed about 1% and is still about 1 ft^3 to which is added 540 ft^3 of vapor. The velocity at the bottom of the chute is therefore very large for the vapors in the center and much slower for the water on the sides. The vapors separate from the water in the designed discharge space below the chute almost completely with practically no mist or carryover.

Instead of taking all this flash evaporation in one stage, it may be divided economically with the Controlled Flash Evaporation system in 2 or 3 stages of successively lower pressures, the final one giving steam at the same 70°F. In such a multi-stage CFE, the vapors available at temperatures above 70°F have higher pressures. By using these different vapor streams to operate separate turbines, or passing in to different stages of blades on the same turbine shaft, a higher net power is achieved.

Another advantage of the CFE, evidenced from its use in flash evaporators for desalination, is that the vacuum production and deaeration can be taken care of automatically in the evaporation and condensing without added vacuum pumps and the expense of their operation.

The advantages of the CFE system have been demonstrated in the flash evaporation of seawater while cooling it from 10° to 20°F and producing vapors very close to equilibrium with the seawater. The turbulence and mechanical losses are very low, as is seen by the extremely low entrainment—with no demisters—of the condensate of the steam which has a maximum of 1 ppm of solids.

The Low Pressure Steam Cycle gives desalinated or fresh water condensate if a surface condenser is used. Similarly, and without the expense of the surface condenser, the Vapor-Reheat system may be used, whereby a fresh water stream, chilled to as close an approach temperature to that of the cold seawater as is possible, is then cycled in open flow of films or sprays to condense vapors. Such recycle of cold fresh water gives added fresh water as condensate by heat transfer surfaces, as is known in desalination processes by evaporation for desalination.

Indirect Vapor Cycle

Power may be generated using a turbine supplied with vapors other than water, having a more nearly conventional density and pressure for power generation and operating in a closed cycle. Barjot (1926) proposed such a system, and the Andersons recently have suggested that warmer water be passed through a standard tube and shell or other surface heat exchanger (boiler) where its sensible heat is used to vaporize a low-boiling liquid, e.g., propane, although ammonia or other liquids may be used. Vapors of propane at a pressure of 150 psia or more are passed through the turbine and exhausted to a surface condenser fed with the cold seawater and condensing liquid propane at a pressure of about 90 psia.

Thus, heat is transferred (a) from warm seawater, (b) to and through the tube wall, (c) to boiling propane; and after expansion with accompanying temperature drop, the latent heat of the vapors must be transferred (d) to the condenser tube wall as they condense (e) through the tube wall and (f) thence to the cold water cycling on the other side of the condenser tube.

Here are six temperature drops, plus the temperature drop of the expansion across the turbine; and the sum of all seven can only be the available difference of temperature between the cold and the warm water. Since the temperature drop in the expansion should be as large as possible to maximize energy conversion, it follows that each of the others must be as small as possible. This means huge amounts of boiler and condenser surfaces.

Standard tube and shell designs becomes intolerably expensive, and the only possibility would be the use of plate heat exchangers. To withstand the pressures on the propane sides of these plates, the Andersons have applied for patents on a most ingenious design, wherein a plate exchanger acting as the boiler is to be lowered to a depth in the sea of 290 ft and the plate condensers are to be lowered to a depth of 150 ft. Presumably any time the thermodynamic liquid is not in use, both sides of both units will be opened to the sea at the respective depths, to maintain a pressure balance. During start-up, the water will be blown out of all parts as the unit goes on-stream.

Other thin-plate heat exchangers may compensate or balance for the large pressure difference on the two sides, so that only the cold-water suction line would be submerged. Thus, the plant and its operation would not need to be inside a steel submarine submerged several hundred feet below the surface of the sea, as in the Anderson proposal.

Advantages noted for the closed propane cycle are: (a) turbines can be of reasonable size and (b) vacuum and deaeration problems are eliminated, as they are also with the CFE. However, there must be large areas of heat exchange surface, and no fresh water condensate can be produced.

Donald F. Othmer

Cross references: *Desalination of Seawater; Salt; Solar.*

I

ICELAND FISHERIES

About 150 fish species have been found in Icelandic waters within the depth of 400 m, a tentative boundary for the continental shelf of Iceland. Only 30 species can be considered commercial fish, but they are far from being equally important. Besides, some of the species are important as food for various commercial fish.

Table 1 shows the Icelandic annual catch of the 6 most important fish species since 1947. As indicated in the table, there are considerable fluctuations in the catches of the various species, not only caused by natural fluctuations in the stock size, overfishing, the protection of specific grounds or from natural causes, but also due to the introduction of better vessels, and especially due to new fishing techniques. A short presentation of the biology of the most important commercial fish and the condition of the stocks is given below.

Herring (Clupea harengus). Three stocks of herring are taken in Icelandic waters. Two of them are of Icelandic origin and the third of Norwegian origin. The Icelandic spring spawners mostly spawn in March off the south coast, but the other Icelandic stock, the summer spawners, mostly spawn in July, off the south coast and to some extent off the west coast. Parts of these stocks, especially the spring spawners, migrate to waters north and east of Iceland after spawning, where they are caught along with the Norwegian stock in summer and fall.

Figure 1 shows Iceland's annual catch of these three stocks since 1950. It is evident that catches of Icelandic herring stocks have greatly deteriorated in recent years. From recaptures of tagged herring, it is estimated that the size of the Icelandic stocks have decreased from 931,000 tons in 1962 to 457,000 tons in 1964. In 1962 more than $\frac{1}{2}$ of the herring taken off north and east Iceland was of Icelandic origin but in 1966 the catch of Icelandic herring in the same waters was only 3% of the total catch. The annual mortality rate of the Icelandic stocks due to fishing is thought to be 33% per year, and further fishing effort is not expected to increase the catch.

TABLE 1. THE ICELANDIC ANNUAL CATCH OF THE MOST IMPORTANT COMMERCIAL SPECIES DURING 1947–1966 IN THOUSANDS OF TONS, LIVE WEIGHT

Year	Herring	Cod	Haddock	Saithe	Redfish	Capelin	Total
1947	216.9	189.3	16.3	32.6	10.0	...	483.6
1948	150.1	181.9	21.1	67.2	25.1	...	478.1
1949	71.4	205.4	23.5	40.5	32.7	...	407.7
1950	60.6	189.8	19.9	15.6	71.4	...	373.3
1951	84.6	187.3	16.4	16.3	95.6	...	417.8
1952	32.0	269.1	12.3	30.9	36.7	...	401.6
1953	69.5	262.3	10.0	27.9	36.0	...	424.7
1954	48.5	300.0	15.6	16.6	59.5	...	455.4
1955	53.6	311.1	14.9	12.1	72.1	...	480.3
1956	100.5	295.0	20.3	23.6	58.6	...	517.3
1957	117.5	251.5	25.1	18.0	61.6	...	502.7
1958	107.3	296.1	23.5	14.9	116.3	...	580.4
1959	182.9	290.1	23.4	15.0	99.3	...	640.8
1960	136.4	304.3	42.1	12.9	55.9	...	592.8
1961	325.9	248.7	51.4	14.8	28.5	...	710.0
1962	478.1	223.4	54.3	13.5	22.3	...	832.6
1963	396.5	240.4	51.9	14.8	35.4	1.1	784.5
1964	544.4	280.7	56.9	21.8	27.7	8.6	972.7
1965	762.9	244.0	53.7	24.9	29.9	49.7	1,199.0
1966	769.2	231.4	36.0	21.0	23.1	124.9	1,238.4

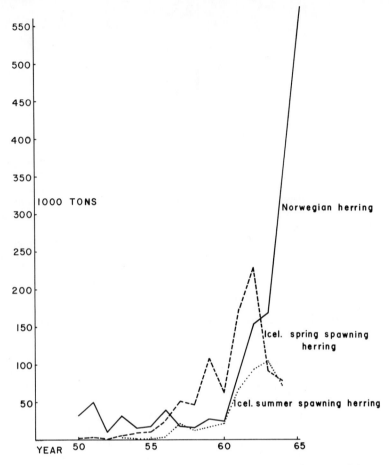

Fig. 1. The Icelandic herring catches in 1950–1965 according to stock.[1]

Cod (Gadus Morrhua). Cod are by far the most important demersal fish species in Icelandic waters. The cod catch fluctuates less than the catches of some other commercial fish.

The cod spawn in the warm waters near Iceland's south and west coasts, from the beginning of March into May. A relatively small part of the stock spawns off the north and east coasts. The eggs and larvae move with currents until the fry begin to seek bottom when they are about 3 months old. Immature fish are stationary in shallow waters until they have matured at 4–14 years, but mostly, however, when they are 6–8 years old. The cod that grow up in the cold waters off the north and east coasts have a much slower growth rate and later maturity than fish off the south and west coasts.

Cod feed on various animals. Besides bottom animals they also take pelagic food, such as crustaceans, squids, sea butterflies, and various small fish, especially sand lance and capelin.

Icelanders catch about 60% of cod taken in Icelandic waters, but foreign vessels, mostly trawlers, take about 40%. From 1954–1964 the total catch decreased by almost 22% while the fishing effort nearly doubled. It is therefore evident that the mortality rate of the stock has risen considerably, the mean mortality rate of mature cod in the years 1960–1964 having reached almost 70%—17% due to natural causes and the rest due to fishing. A higher total mortality rate than 65% is considered undesirable. The annual mortality rate of immature fish is about 60%, $\frac{2}{3}$ of which is supposed to be caused by fishing. Thus it is clear that more is taken from the cod stock than it seems to be able to bear.

Haddock (Melanogrammus aeglefinus). The haddock only spawn near the south and west coasts, usually about 2–3 weeks later than the cod. The rest of the year they are rather localized in the shallow waters near the coast. The haddock are more demersal than cod and feed mainly on bottom animals. In the years before World War II, the haddock stock had become very small due to overfishing, but during the war the stock increased considerably since fishing effort was

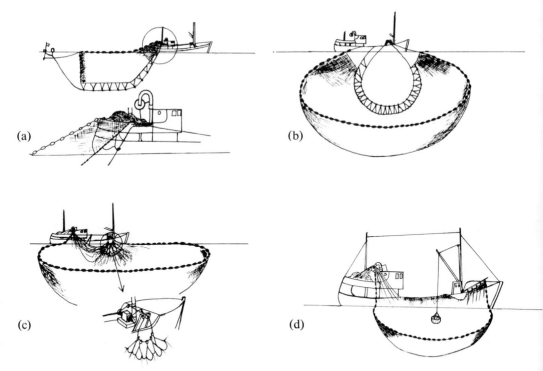

Fig. 2. The four phases of a purse seine shot: (a) shooting, (b) pursing, (c) hauling with the power block, and (d) brailing.

small. After the war the catch increased at first but after 1950 there was a decline in the catch once more because of too intense fishing effort. In 1952 important haddock nursery grounds were protected from bottom trawling, and since then catches have increased greatly. Increased mesh size in bottom trawls has also had a favorable effect on the catch. At present the haddock stock does not seem to be endangered by increased effort.[1]

Saithe (Pollachius Virens). Saithe spawn at similar localities as the haddock. They spawn earlier than other Icelandic gadoid fish, mostly in February–March. Saithe are more pelagic than cod and feed mainly on krill and small fish, mostly in the upper waters. Saithe require higher temperatures than cod and therefore stay mainly in the warm water near the south and west coasts. Saithe catches are characterized by great fluctuations which are partly thought to be caused by migrations from Iceland to the Faroes and Norway. Icelanders take about $\frac{1}{3}$ of the saithe catch in Icelandic waters. The stock is not endangered by overfishing.[1]

Redfish (Sebastes Marinus). Redfish are distributed all over the Icelandic continental shelf but are most common, however, at 200–500 m depth off the west coast. Redfish differ from other Icelandic commercial teleostian fish in bearing living young, which are usually released in May off southwest Iceland about 200–500 m below the surface in areas of great depth. Various details in the biology of redfish are still uncertain, such as their growth and age.[2] During the last 10 years the Icelandic redfish catch near Iceland amounted to 15–20,000 tons.

Capelin (Mallotus Villosus). Capelin are among the most common fish in Icelandic waters. They are found all around the island but are most common, however, near the north and north-east coasts through most of the year. The majority of the capelin spawn near the south coast in late winter and in spring, the fish being 2–3 years old at that time. Capelin are said to die in great quantities after the first spawning, but there are indications that part of the stock spawns twice. The maximum age of capelin is therefore only 4 years.[3]

Up to 1964 capelin were caught for use as bait only, mostly for cod and haddock, but in the winter of 1965 they were for the first time caught on a large scale in purse seine nets for reduction.

Other Commercial Fish. Catfish (*Anarhichas lupus*) are common near the west coast especially and of considerable economical importance. During the preceding 10 years the annual catch has been 6600–17,300 tons.

In the past 4 years the catch of flatfish has increased from 6000 to 9000 tons. However, only the catch of the most important species, plaice (*Pleuronectes platessa*), has increased (from 54% to 81%). Flatfish are mostly caught in summer, in Danish seine, usually in the shallow waters at the south and west coasts.

Of gadoid fish not mentioned specifically above, these are of greatest importance: ling (*Molva molva*), blue ling (*Molva byrkelange*), and tusk (*Brosmius brosma*). These fish are mostly caught with line or trawl, the first two at considerable depths.

Lumpfish (*Cyclopterus lumpus*) are caught in gillnets in spring, mostly at the west and north coasts. The roes are the most sought-after part of the fish.

Three species of Elasmobranchii are of importance: skate (*Raya batis*), starry ray (*Raya radiata*), and Greenland shark (*Somniosus microcephalus*). The first two are salted and consumed locally. The shark liver is often reduced to oil, but the meat is consumed after a rather long fermentation.

Freshwater fish are of small importance. Besides the European eel (*Anguilla anguilla*), three salmonoids are found in Icelandic freshwater, and they are mostly caught by sportsmen. Farming of freshwater fish is increasing, however, so they might grow in importance in coming years.

Crustaceans

Only two species of crustaceans are caught in Icelandic waters. The deep-sea shrimp (*Pandalus borealis*) are mostly caught at the northwest and north coasts, and the Norway lobster (*Nephrops norwegicus*) mostly at the south and southwest coasts. Both species are caught in special types of trawl, and the fishing season is limited to keep the stock balanced.

The Fisheries

The beginning of the 20th century saw the first real changes in the development of the fisheries when Icelanders began to purchase steam trawlers, as used by other nations. Besides this, the building of motorboats began, and now the fleet of fishing vessels grew rapidly. In 1925 Iceland possessed 47 trawlers and close to 600 motorboats (with $\frac{1}{3}$ over 12 BRT) and 27 longliners. Together these vessels made nearly 30,000 BRT. During the last decades there has not been such a rapid increase in the number of fishing vessels as in their size. As of December, 1966, the Icelandic fishing fleet consisted of the following:

756 fishing vessels	59,571 BRT
32 trawlers	22,876 BRT
7 whale catchers	2,858 BRT
	85,305 BRT

The winter season lasts from January to early May. This fishery is mainly based on the cod stock located near the south and west coasts seeking food and later spawning in these areas. Other important species are mostly haddock and saithe, but ling, tusk, and catfish are caught to some degree. During the last few years capelin have been taken on a large scale.

Various types of gear are used, depending on the size of the boats, the size and behavior of the fish in the various areas, and finally on the owner's financial status.

Longline is the most effective fishing gear in the beginning of the season, while the fish are still scattered. Besides cod great quantities of haddock—which are more valuable—are taken. In some areas considerable quantities of ling and tusk are also taken. Fishing with longline is often resumed in spring at the time when the fish begin to scatter after spawning. Catfish catches are sometimes especially good near the northwest coast when the fish move shoreward after having spawned at great depths off northwest Iceland.

The most common and practical length of a line boat is 20–30 m. The crew consists of 6 men which daily lay a line of more than 20 km, with about 16,000 hooks. Five men bait the line ashore, and the boats return to port every night to land

TABLE 2. DISTRIBUTION OF THE WINTER SEASON CATCH IN 1963–1966 (EXCLUDING HERRING, CAPELIN, AND CRUSTACEANS) THE CATCH IS GIVEN IN THOUSANDS OF TONS, LIVE WEIGHT

Year	longline 10^3 tons	%	gillnet 10^3 tons	%	purse seine 10^3 tons	%	bottom trawl 10^3 tons	%	handline 10^3 tons	%	other[a] 10^3 tons	%	total 10^3 tons
1963	54.4	27.1	131.4	65.4	5.4	2.7	3.4	1.7	6.3	3.1			200.9
1964	39.7	14.8	175.4	65.4	40.8	15.2	7.9	3.0	4.3	1.6			268.1
1965	22.0	9.6	157.9	69.0	35.7	15.6	8.7	3.8	4.6	2.0			228.8
1966	43.0	15.4	153.2	54.9	16.1	5.8	24.1	8.6	21.6	7.7	21.0	7.5	278.9

[a] Danish seine and crustaceans' trawls.

the catch and pick up baited line. The line is constructed like ordinary longlines. It is mostly made of sisal of about 7 mm in diameter, but the use of synthetics is growing. The snoods are usually of monofilament nylon and 0.5 m long. The longline is used at 30–350 m depth over all kinds of bottom except a hard lava floor. Fishing is possible up to winds of 8 Bft if the boat is seaworthy.

Cod fishery with gillnets usually begins about the middle of February. At that time line boats ordinarily take up this fishing method. The gillnets, constructed as other gillnets, are laid near the bottom—seldom in the upper layers—and 15 nets are combined to form the so-called fleet. Each boat, with a crew of 10–12, can lay 7 fleets at a time, and hauls them and lands the catch daily. Ashore the catch is immediately gutted and then goes into processing. The net fleets are left anchored on the grounds while the boat heads for port.

The gillnets are made of greyish, continuous polyamide of 210 den/12. Nets of 0.50 mm diameter monofilament have been tried and given good results on a hard bottom (up to sevenfold catches), but have not proved good on a muddy floor. They are also found to be awkward to handle. The mesh opening is about 180–200 mm. The nets are 600 meshes in length (50 m), 32 meshes deep, normally with a hanging rate of 50%. The lines are made of polypropylene, the leadline 16 mm in diameter and the floatline 10 mm. Since the nets' catchability mostly depend on their invisibility, they are made of fine, greyish twine, almost invisible in the water. Ordinarily, the nets last only about 2 weeks, so the catches have to be good to cover the great cost. It sometimes occurs also, in the peak season, i.e., April, that one single boat takes 50 tons or even more in one trip.

Since the nets are left in the sea while the boat returns to port, it often happens that they remain anchored for 2 or more days when unfavorable weather prevents sea-going. Since the gills are clamped down in the nets, the fish die, and if the nets cannot be hauled within 2–3 days the catch is classified as inferior raw material. It also frequently occurs that complete fleets are lost in bad weather. It is debatable whether the so-called ghostnets continue to catch as they drift across the sea floor, but in any case considerable amounts of fish are lost when these nets with gilled cod are not recovered.

The gillnet season usually ends in late April or early May when most of the boats discontinue this fishing and begin to prepare for the herring fishery. The smaller boats, however, continue with longline for some time or soon begin to prepare for fishing with trawl, Danish seine, crustacean trawl, or handline.

During the winter season in 1965 the new purse seine fishery* was a turning point in the history of Icelandic fishing technique. Then boats began fishing capelin on a large scale for reduction purposes, using small-meshed purse seines. Later in the season the boats obtained good catches of cod in purse seine. The catch usually consisted of very big cod above the selection range of the gillnets, i.e., they were too big to entangle their heads in the meshes. During the last two winter seasons this fishery has not been as successful, as the shoals of cod were not as concentrated as previously.

This cod fishery is remarkable for the fact that the purse seine is lowered almost to the bottom or on the bottom when it is smooth. For this purpose some cork is removed from the floatline. The sinking depth of the purse seine can be controlled by attaching floats with suitable loops to the floatline. Thus the purse seine sinks until the loops have given way. When the pursing is over the net eases and the floatline rises.

Another very important tactic is not to shoot the net in a complete circle as usually done, but only half a circle and then tow the latter end on to the line first shot. In this way it is possible to fish a larger area on one shot. Besides cod, considerable amounts of haddock are also taken in this manner.

Bottom trawling from boats during winter has increased considerably in recent years. A great many boats use handline but this kind of fishing is more common at other times of the year when the weather is more calm.

The Herring Fishery. It is most difficult to describe the Icelandic herring fishery in detail because the stock size of the herring and their migrations are subject to great variations, which to a great extent affect the course of the fishing. Generally speaking, Icelanders catch herring anywhere near Iceland and at any time, so long as they gather into shoals dense enough to make fishing profitable. Herring are caught in purse seines only. Driftnets are not used any more, and experiments with midwater trawls have been unsuccessful so far.

The herring fishery can be divided into two seasons, the fishing of Icelandic herring off the south and southwest coasts and the fishing off northeast Iceland where mainly Norwegian herring are taken.

The fishing of Icelandic herring usually begins after the winter season, most often in late May and continues as long as there are herring to

---

* The purse seines and the relevant fishing method described in the section on the herring fishery apply roughly to the method used in the capelin fishery, although the herring purse seines are somewhat larger as a rule.

catch, but in recent years the season has hardly lasted longer than through September. Immature, as well as mature herring, are taken. The Icelandic herring stocks are very small today so this fishery is expected to become less important.

The herring fishery off the east coast usually starts in May, but many boats, especially the smaller ones, do not start until June or July. In the beginning of the season the best catches are often taken near Jan Mayen or even northeast of the island. Although the time at which the herring begin to move southward varies greatly, they have usually moved relatively close to Iceland by the middle of August. About the middle of October they have generally reached their place of overwintering, which is about 60–100 miles east of Iceland. There they remain up to Christmas but then start for Norway. At that time most of the boats stop fishing, but sometimes the biggest boats follow the herring in January, and they often take some herring near the Faroes.

Over 200 boats take part in the herring fisheries off the south and east coasts of Iceland. The duration of the vessels' activity varies considerably. Many of the bigger vessels are engaged in fishing for more than 200 days, but most of the smaller ones have shorter periods of activity since they are not suited for fishing far offshore as required early in the season. And in fall, when the herring are closer to land, fishing is often prevented by foul weather.

The development in the herring fishery technique has been very rapid. Up to 1944 the purse seiners used 2 dories, each setting $\frac{1}{2}$ of the net. As a result of an abrupt decline in catches in 1945, the vessels began to use one dory only during the next few years, since this meant a smaller crew and less cost on the whole. The net was always hauled manually. The purse seine was only set around schools of fish visible with bare eyes. Aerial scouting as an aid for the herring fleet began in 1928.[4]

Since ordinary depth recorders were of relatively small use in this fishing, it was not until the horizontal echo ranging equipment (sonar) had been developed (first used by an Icelandic vessel in 1954) that modern fishing techniques were introduced into the herring fishery. In 1958 almost all vessels were equipped with this important fishing aid, and now the boats could shoot at submerged, and thus invisible schools. It takes considerable dexterity and practice to shoot by sonar, but most Icelandic skippers soon learned to do so. Generally, shooting starts with the wind straight ahead, the school located off starboard, and the net finally shot around the school. The most difficult part is to decide how far from the school shooting should start and from what angle. This depends on in what direction and how swift the school moves. Figure 3 shows some shooting methods at various interactions of wind, current, and the movement of the school.[4]

The nets have grown enormously in size. While only surfacing herring were taken before the introduction of sonar, the nets were most often 150 fathoms long and only 30 fathoms deep. Now they are usually about 300 fathoms long and more than 100 fathoms deep. It is of vital importance that the nets sink as quickly as possible so pursing can begin before the herring submerge. For this purpose the nets are weighted with lead up to 15 kgm per fathom. Polyamide twine of different strength is used in the nets, and they vary in construction.

Trawling. Today trawlers fish mostly in Icelandic waters and on the nearer fishing grounds at East Greenland. The Newfoundland and West Greenland banks have proved unsatisfactory in recent years. All attempts made by trawler owners to be granted permission to fish in parts of the territorial waters at certain seasons have failed.

Icelandic trawlers use only polyethylene nets. This material is lighter than water (sp.gr. 0.95), and trawls made of this material have proved better on a hard bottom than manila trawls previously used. Polyamide, a fiber in great use by various nations, has never been of widespread use in Icelandic trawls, probably because of their tendency to get caught and tear on a hard bottom, due to high density (1.14). Otherwise Icelandic trawlers use similar trawls as used by foreign trawlers in the North Atlantic.

The catch is iced and must be landed about 2 weeks after fishing has begun. Only one trawler freezes the catch on board. The Icelandic trawlers are all side trawlers. Should new vessels be added to the fleet before too long, they are expected to be stern trawlers.

Whaling.[5] Whaling in Icelandic waters was introduced by Norwegians in 1891 and continued by them until 1915. Up to 30 catchers operated simultaneously, and close to 1300 whales were caught annually at the peak of this period, mostly blue whales and fin whales. This was too great an effort for the stocks, and whaling became unprofitable and was finally prohibited.

Icelanders did not begin whaling until 1935 and then with 2–3 catchers. This whaling was, however, discontinued in 1940 and was not resumed until 1948. Since then whaling has been permitted with 4 catchers with a season of 4 months, i.e., from late May through September. These measures have kept the stock in balance. The humpback and blue whale stocks could not stand up to the fishing effort, so the former were protected in 1955 and the latter in 1960. In 1948–1964 a total of 6429 whales were caught:

4011 fin whales, 1347 sperm whales, 902 sei whales, 163 blue whales, and 6 humpback whales. Of the three species now permitted to be caught, the fin whale is the most sought-after, since it is of greatest value, but sei whales are gunned only when there is no other choice. During 1948–1954 there was an average catch of 300 whales, but since then catches have increased to an average catch of 433 per year in 1955–1964. This increased catch is probably due to the renewal of the whale boats, but added experience is undoubtedly also an important factor.

The catchers are about 450 BRT each, with 1800–2000 hp engines and equipped with every modern gear. Catching is carried out off the west coast of the country, between 64° and 66°N. Prior to 1960, on the other hand, the best grounds were found between 63° and 65°N.

Processing and Value

The distribution of the catch according to method of processing during 1958–1965 is shown in Table 3.

Generally, the cod catches are mostly frozen, since this processing is most profitable. Fish meant for freezing must be very well preserved. The freezing plants are scattered all around the coast, although mainly concentrated on the southwest coast. Most part of the year, there is insufficient raw material available to keep the

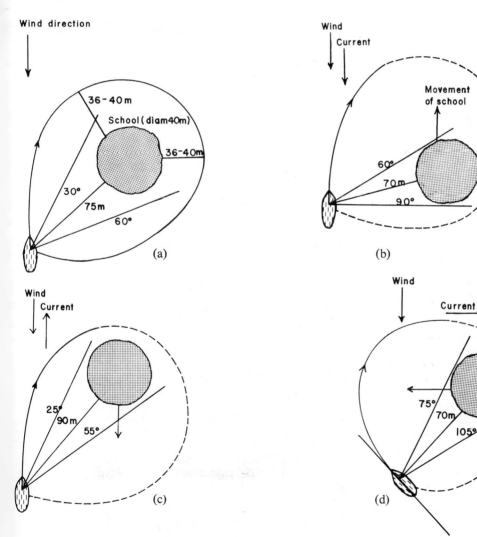

Fig. 3. Starting positions of shooting herring purse seine at varying conditions of current and relative movements of the schools. In all cases the school is expected to be in the position shown in (a) at the completion of the circle.[4]

TABLE 3. THE FISH CATCH IN 1958 AND 1965 IN THOUSANDS OF METRIC TONS, LIVE WEIGHT

	freezing	drying	salting	canning	reduction	fish on ice exported	domest. consump.	Total
1958								
White fish	301.1	51.5	96.7	0.075	5.8	11.8	6.5	473.5
Herring and capelin	16.0	...	53.5	...	37.9	...	...	107.4
Crustaceans	...	...	...	...	...	...	...	...
1965								
White fish	185.4	54.4	88.8	0.032	3.2	35.4	14.6	381.8
Herring and capelin	33.0	...	61.1	0.962	714.7	2.9	...	812.7
Crustaceans	4.4	...	...	0.190	...	...	...	4.6

freezing plants going at full capacity, but in peak fishery during the winter season the fish landed cannot all be frozen so the more expensive species are chosen for freezing.

The salting of demersal fish is the second most important processing method, but has become relatively less important in past decades. Thus salted fish constituted 70% of Iceland's export in 1920, and average of 26.5% during 1951–1955, but only 10% during 1961–1965. The total production of salted fish during 1951–1955 is of course somewhat less than 10 years ago, but the annual mean export quantity of the last 10 years is nevertheless about twice the quantity of the years 1920–1940. The decrease in the percentage export value can be explained by the fact that salted fish is now mostly exported without having been dried and the fish is of worse quality because of the salting of inferior gillnet-caught fish. Furthermore, relatively smaller proportions of the catches have been salted.[6]

Drying is the third most important processing method. This method is the least profitable one, but since processing is rapid and undemanding in manpower, it is often used when great quantities of fish are landed during the winter season. The head is cut off the gutted fish and it is cleaned thoroughly, a pair of fish tied together by the caudal fins and hanged across special beams where the fish are allowed to dry.

As much of the herring catches as possible is salted. In the beginning of the season the fat content of herring is usually too low for salting, since it must be 20% in order to be salted. It also must be caught close to land so that it is fresh when salted. For these reasons a relatively small part of the catch is fit for salting. Herring are also frozen, but by far the greatest part of the catch is processed for production of herring oil and meal.

With increasing catches the export value of fish products has risen and amounted to 5600 million Icelandic Kronur in 1966. The export value of fish products has increased about 40% from 1958–1965, compared with the fixed rate of exchange in 1962. The prospects for still increasing export values are far from good today, since there has been a drastic reduction in the price of fish meal, fish oil, and indeed frozen fish as well. Possibilities for favorable sales of other fish products are somewhat better, but a number of important markets have deteriorated because of competition from other fishing nations, or even been closed temporarily because of disturbances in various parts of the world. This past year the export of fish products has amounted to more than 90% of the total export.

Even though the inhabitants of Iceland number only about 200,000 its annual fish catch in the past 3 years has been 1,000,000 metric tons. No recent figures are available on the number of crew manning the fishing fleet, but as of April 1961 there were 6082 fishermen. At the same time 6961 persons (2628 of them women) were engaged in the fish processing. The number of fishermen has probably increased by about 10–20%, but it is clear that the production value contributed by each fisherman is nevertheless incredibly high.

Future Prospects

Iceland is more dependent on fisheries than most other nations. It is therefore natural that attempts are made to promote other occupations, especially industrial production, since it must be considered hazardous to base almost the entire national economy on only one branch of industry. The waterfalls and geothermal heat provide great possibilities for power production for industry. It can, however, be taken for granted that Iceland's economy will in the near future be based mostly on the fishing industry. It is therefore of vital importance to continue unceasing research on the fish stocks, aiming at keeping the catches as constant as nature permits. Simultaneously, research in fish processing must continue in order to increase the value of the catch. Finally, it is to be expected that gear research will become increasingly important in the future. The work of the scientists engaged in these branches of research has met with increased understanding in recent years, which is the prerequisite for good results.

G. THORSTEINSSON

References

1. Jónsson, J., "Helztu fiskstofnar á Íslandsmiðum og áhrif veiðanna á þá," *Tímarit Verkfræðingafélags Íslands*, **52** (1967).
2. Magnússon, J., "Um lifnaðarhætti karfans," *Náttúrufræðingurinn*, **26**, 3 (1956).
3. Vilhjálmsson, H., "A Contribution to the Knowledge of the Icelandic Capelin (Mallotus villosus O. F. Müller)," *International Council for the Exploration of the Sea*, Symposium on "The Ecology of Pelagic Fish Species in Arctic Waters," Paper No. 20, 1966.
4. Jakobsson, J., "Recent developments in Icelandic herring purse-seining," *Modern Fishing Gear of the World*, **2**, 294 (1964).
5. Jónsson, J., "Whales and Whaling in Icelandic Waters," *Norsk Hvalfangst Tidende*, **54**, 245 (1965).
6. Loftsson, L., "Saltfiskiðnaður Íslendinga," *Tímarit Verkfræðingafélags Íslands*, **52** (1967).
7. Verkfræðingafélags Íslands, "Töflur 1–14 frá efnahagsstofnuninni um framleiðslu, fjármunamyndun og fjármunaeign í fiskveiðum of vinnslu sjávarafurða," *Ráðstefna um vinnslu sjávarafurða*, 1967.

ICHTHYOLOGY

By 1900 over 12,000 species of fish had been described. Present-day estimates vary, but 25,000 seems a reasonable estimate—the same order of magnitude as all other vertebrates (Amphibia, Reptilia, Aves, and Mammalia) put together.

Classification. This subject has long been and will continue to be controversial at all levels from species to class. The vagaries of taxonomists are, however, by no means unique to those engaged in ichthyological pursuits. In this century the range of systems is considerable. Nonetheless, there is general but not complete agreement, no matter the category or scientific name, about the fact that there are several major groups: (1) Lampreys and hagfish (Cyclostomata); (2) Sharks, rays, and chimaeras (Chondrichthyes —cartilaginous skeleton); and (3) Bony fishes (Osteichthyes). The last commands particular attention on the basis of numbers and diversity. In this conglomeration, the lungfish (Dipnoi) and lobe-finned fish (Crossopterygii) are probably within ordinal relationship; the "Crossops" are of special interest because of the capture in 1938 of a 127 lb Coelacanth and subsequent infrequent catches off South Africa, a "living fossil" from a group thought to be extinct over 50,000,000 years ago. There follow, by way of convenience, the primitive paddle-fish and sturgeons (Chondrostei), the intermediate bowfin and garpikes (Holostei), and the enormous assemblage of modern bony fish (Teleostei). Since 1940 the classification by Leo S. Berg received wide acceptance, but it, like its predecessors, derived the teleosts monophyletically from the holosts.

In 1966 Greenwood, *et al.* produced a new "... conception of the evolutionary relationships of the principal groups of teleostean fishes." They proposed a polyphyletic derivation from the fossil pholidophoroid holosteans of three divisions of soft-rayed (malacopterygian) fish, two of which lead to the spiny-rayed, acanthopterygian, level. The major groups in terms of numbers and commercial importance are the Clupeomorpha (Div. I, including anchovies, herrings, menhaden, sardines, shads), and from Division III the Salmonoids, Gadiformes (cods, haddock, hakes), Perciformes and their derivatives (perch-like fish of all forms from striped bass to mackerels and their relatives), and the Pleuronectiformes (the flat fish such as flounders, halibut, plaice, soles, turbot).

Adaptations. As might be expected in such a large and diverse group as the fish, adaptations are manifold.

It can only be by physiological means that polar fish survive in temperatures below those of the freezing points of their blood. Similarly, the physiology and endocrinology of anadromous and catadromous fish are built-in adaptations that allow the transfer between fresh and salt water. Surely too, it must be a physiological adaptation that allows *Nomeus* to live without harm surrounded by the poisonous tentacles of the jelly-fish Portuguese Man-of-War.

Morphological adaptations run the gamut; however, we will confine the subject to fins. The sucker of the Remora, which enables it to remain attached to sharks and other fish, is a modification of the dorsal fin. So is the luminescent lure of the angler-fish an adaptation of the (first ray) dorsal. And the neurotoxin of the stonefish is injected via the dorsal fin hypodermic rays. Pectoral fin rays are adapted to "walking" in gurnards. Flying fish are enabled to glide by their wing-like pectorals—also the modification of the caudal and in some species the pelvics. Lumpsuckers and snail fish have "plumber's helpers" which are adaptations of pelvics and pectorals that enable them to survive in turbulent waters. An extreme adaptation of fins (pelvics) is their various modifications to serve as intromittent organs, occurring both in cartilaginous and bony fish.

Structural adaptations of the mouth and dentition in fish are of comparable complexity and variety.

Just as we may discuss physiological and anatomical adaptations in fish, so may we consider social habits and behavior as adaptations: courtship, methods of reproduction (see below), schooling, the cleaning of larger fish by smaller fish (perhaps a truly symbiotic relationship), and other complicated relationships.

Distribution. The fish of today live in all parts of the ocean: surface and bottom, shoreline and midocean, as well as the polar seas.

The Antarctic Zone, bounded on the north by the 6°C isotherm, is dominated by perch-like species (Nototheniids); others are eel-pouts (Zoarcidae) and sea snails (Liparidae); all told there are about 100 known species.

The Arctic ichthyofauna is poor by comparison: sculpins (Cottidae) are most common; there are some cods (Gadidae), flatfish (Pleuronectidae) and one Zoarcid genus (completely distinct from its Antarctic counterpart, but an elegant example of convergent evolution).

The Temperate Zones are delimited by the 20°C isotherms, a pair of wavy lines pretty much between 30°–15°S and 30°–20°N. The broad belt in between is the Tropical Zone. Further subdivision of oceanic zoogeographic zones is a *reductio ad absurdum* with fish; there are a multitude of transgressors from one zone to another. However, the distribution of many pelagic and coastal fish does correspond to temperature and salinity. Eurythermal types have broad temperature tolerances (e.g., winter flounder, *Pseudopleuronectes americanus*, with a resident habitat from the Straits of Belle Isle to the Carolinas—a range of over 21°C [70°F]). Stenothermal types have narrow temperature tolerances with generally circumscribed geographical ranges (e.g., tilefish). Similarly, stenohaline fish have narrow salinity tolerances (the open-sea, pelagic forms such as mackerels and flying-fishes: $\pm 35‰$). Euryhaline fishes have the ability to survive broad ranges of salinity (*Fundulus* spp., sticklebacks, winter flounder, and those that make the total change from fresh to salt water or vice versa).

In zonal distribution, the Tropical has by far the greatest diversity of genera and species—open-sea, inshore, and coral-reef types. But in terms of sheer numbers the Temperates dominate —herrings, cods, flatfish, etc.

Longitudinal generalizations require division between coastal and oceanic fishes. The majority of the littorals are herrings, cods, percomorphs, and flounders. The oceanics include such forms as the tunas, swordfish, etc., the lesser-known lanternfish (Myctophidae) and associated types in the upper layers (± 150 m).

Vertical oceanic zonation, apart from the surface waters, can be divided roughly into: (1) *bathypelagic* forms such as the scaly dragonfish (Stomiatidae) and anglers (Ceratiidae), and (2) the *abyssals*, for example, certain ratfish or grenadiers (Macruridae).

Generalizations about the distribution of marine fish are fraught with difficulty in the present state of knowledge.

Reproduction, Survival, and Commercial Abundance. Bisexuality is the rule in fishes; hermaphroditism is the exception, most frequently documented in the perch-like forms, with both proterandry (more common) and proterogyny being represented. Parthenogenesis (gynogenesis) occurs in a viviparous Amazonian "molly." There is an extraordinary range in the types of courtship. So also with modes of reproduction: oviparity (by far the most common), ovoviviparity, to true viviparity.

There is generally an inverse relationship between the number of eggs a single species lays and the amount of parental care given to the eggs and developing young: the greater the number of eggs the less the care, and vice versa. An arbitrary division of fish into three categories with regard to number of eggs and parental care follows.

(1) There are the fish which produce few eggs per individual—50 more or less; the developing young receive a high degree of parental care. Thus, the marine catfish (South Atlantic and Gulf States) lay eggs as big as marbles which are picked up by the males after fertilization and carried in their mouths for several months—a type of oral incubation protecting the young against predation throughout the most vulnerable part of their life history.

(2) There is a large group of fish with intermediate numbers of eggs per individual—several thousands more or less; these species provide intermediate care. Thus the salmon bury their eggs several inches below the surface of the gravel bottom very shortly after fertilization; while these eggs do not have the degree of care provided by marine "cats," they are protected to a certain degree during their early stages.

(3) There are the species which produce huge numbers of eggs per individual—50,000 to several or more million. Examples are herring and shad (at the lower end of the range and with demersal and semi-demersal eggs), cod, haddock, mackerel (in the middle and upper part of the range and with pelagic eggs), and the ocean sunfish with a reported 50,000,000 eggs per individual. These forms scatter their eggs broadside, with no parental care whatsoever. In this category there are often wide fluctuations in abundance of individual species from year to year.

These kinds show the occasional phenomenon of dominant-year classes—the production and survival of so many individuals in a single spawning season that this age-group dominates the whole population for a series of years. The classic example of a dominant-year class is the Norwegian herring, where the young born in 1904 were so numerous that they were a dominant element in the catch from 1907–1921 and virtually supported the fishery for 15 years. It would appear, then, that in fish with large numbers of eggs the fluctuations in abundance are due more to the particular environment in which the eggs and larvae develop than to the size of the adult stock.

As an example of poor survival, mackerel (*Scomber scombrus*) mortality from the fertilized egg to a 2 in. stage in 1932 was 99.9996%, i.e., the survival was roughly 1–10 fish per million newly spawned eggs. Hence, a fluctuation in survival of several thousandths of 1% may be the difference between a dominant- or a weak-year class. In the last 50 years fishery biologists have sought, with meager results, to discover what environmental factor or combination of factors (temperature, salinity, turbulence, drift, predation, etc.) was determinant in a variety of commercial species.

The ability of a small stock of prolific adults to produce large numbers of successful offspring under proper conditions is well known. There were no striped bass on the West Coast until 1879 and 1881, when 435 small stripers were seined in New Jersey, transported across the continent by train, and planted in San Francisco Bay; 20 years later (1899) the commercial net catch was 1,234,000 lb. Another example is the stenothermal tilefish, which lives along the continental shelf of the East Coast where the temperature is normally 8.4°–10.0°C (47°–50°F). In 1882 a marine catastrophe overtook the population; dead and dying fish were sighted from Nantucket to Delaware Bay. At least a billion-and-a-half dead tilefish were seen. The cause of this mass mortality was probably the invasion of a low-temperature current. No tilefish were seen from 1882 to 1892 despite extensive searches; then the species began to reappear in limited numbers. With food shortages looming large 50 years ago (World War I), the commercial fishery took 11½ million pounds of tilefish in one year.

The prolificity of our most abundant species, as well as their ability to recover from severe decimation, either man-made or from natural causes, indicates that we have by no means reached the limit of the exploitation of marine protein resources on a world-wide basis. This is not to say that careful observation and, where possible, management of the commercial harvest of each species is not essential, or that over-fishing or the failure of certain fisheries (notably the California sardine) have not occurred.

In recent years Peru had become a leading nation in fish landings, due in large measure to the take of anchovy-like fish; judging by the depth and other characteristics of the guano caps on islands off the Peruvian coast, the huge depredations by cormorants (guanay) of anchovies probably date back to about 500 B.C. For those who are concerned with the problems of overfishing by man, here is an example of a fish which has had the reproductive potential, in the proper environment, to maintain its population over centuries despite a fiercely intensive avian fishery. Now man has put additional strain on the anchovy population, and it remains to be seen whether or not, or for how long, these fish can hold their level. Fortunately there is an economic check; man's effort diminishes as the catch becomes less lucrative, and the resilience of such prolific fish asserts itself granting that the environment remains unchanged and that a competitive species has not usurped the ecological niche.

Since the late 1930's and the immediate post-World War II period the world production of fish has essentially tripled (40 to 115 billion lb). In the same 30-year span the U.S. contribution has dropped from 11 to 4%, remaining roughly constant at 4–5 billion pounds. Yet within the past 10 years the consumption (in round weight) of fishery products in U.S. has doubled (over 12 billion lb in 1966); imports fill the gap. It is of interest in this connection that it has been estimated that the yield from the U.S. "domestic fishing banks" could reach an annual sustained level of 20 billion lb. The presence of numerous far-away foreign fishing vessels off our coasts is a further indication that we are not impoverished in our easily available fisheries resources.

Migrations. This subject alone can be a lifetime study for the ichthyologist. The vertical, horizontal, and seasonal movements of fish, both fresh- and salt-water are generally, but not necessarily, related to feeding or spawning. Certain repetitive patterns must have been known to man since the earliest days of fishing, e.g., salmonids (whose artistic depictions are easily identifiable in the palaeolithic caves of Dordogne), members of the herring tribe, striped bass, certain flatfish, and tunas (whose seasonal Mediterranean appearances have been known since Roman times). But even today our precise knowledge of the open-sea movements on a year-round basis of such valuable species as the various bill-fish is much less than adequate—especially at a time when the proper harvesting of our protein marine resources is so important and despite technical advances such as acoustic tagging.

In the present century two particular types of migrations have been the object of intensive study. Into one sort fall the anadromous fishes, those that spawn in fresh-water streams or rivers, but spend most of their lives and make by far the most of their growth in salt water. In this extraordinary life history, involving two very different physiological environments, certain Clupeids (shad and alewives), both Atlantic and Pacific salmons, and striped bass are prime examples.

The other sort are catadromous fishes. The common North American and European eels are classic types. They spawn in the Sargasso Sea; they accomplish migration first at the whim of prevailing currents as pelagic eggs, later as mobile leaf-shaped larvae to the east and west in one or several years. They metamorphose into small but adult form when they approach their respective coasts and as they enter the fresh-water phase of their lives. Now they are called elvers, delectable morsels particularly appreciated in Europe. To Johannes Schmidt, eminent Danish ichthyologist, goes the credit for discovering the general area of reproduction. He managed this geographical feat by plotting the diminishing sizes of leptocephali as captured in stremen nets towed from a variety of vessels; the smaller the larvae, the closer to the source of spawning. As to the marine migrations of sexually mature anguillas well en route back to their mating and final destination, we know little.

On the other hand, a lot has been learned about the marine phases of the life histories of the anadromous salmons in the last 15 years. There are two genera. *Salmo* (one species, *salar*) spawns in rivers on both sides of the Atlantic. In the Pacific there are six species of the genus *Oncorhynchus*, distinguishable from their Atlantic cousins anatomically by their greater number of anal fin rays, and physiologically by the fact that they all die after their one and only upstream migration to reproduce their own kind precisely where they were born.

Until recently we knew little or nothing about the migrations of either Atlantic or Pacific salmons at sea. Tagged fish on both sides of the Atlantic had been recovered more than 1,500 miles from the point of release. But it was anybody's guess as to whether or not *Salmo salar* of North American or European origin ever intermingled. The present decade has provided some preliminary answers. The source of information, as is to often the case, comes from the commercial catch. The Danes and Greenlanders developed an intensive gill-net, close-to-shore fishery on the southeast end of Davis Strait. The 1964 take was of the order of a million fish, at least 25 times as many as in 1960. There appears to be just one Greenland river (the Kapisigdlit, near Godthaab) which produces its natives; this single place of reproduction could not possibly account for such a catch. Then where did the salmon come from and why did the fishery make such a dramatic increase?

We can answer the first question with some assurance from the results of international tagging programs. A small percentage came from Maine, a much larger quantity from Canada, and substantial amounts also from Ireland, Scotland, England, and the Scandinavian Peninsula. "It is safe to say that in recent years, during late summer and autumn at least, feeding salmon from Europe and North America occur together in considerable numbers in west Greenland waters."

The question as to why the fishery took such a fantastic leap in so short a time has no easy answer. It is said that Danish fishermen came on this agglomeration of salmon while setting their nets in time-honored places for cod (*Gadus morhua*, Linnaeus 1758), a year-round, cold-water species. But then there is that Irminger Current with its vagaries of warmer water influxes: less cod, more salmon? Who is to say until one knows much more about the physical oceanography of the situation as well as the predaceous habits of these two distantly related fish—or far better, the life histories of the animals on which they feed and their accommodation to changing temperatures? There is no doubt that the north Atlantic waters, south to north, over recent decades have warmed; the case is well documented by the northern extension of the previous limits of all kinds of marine animals from crabs to fish. Yet it is interesting to note that the Greenland catch of salmon in 1965 appears to have been roughly half that in 1964.

Commercial catch and export statistics, though useful and necessary for an understanding of oceanic harvesting, are not alone reliable measures of abundance. Quite aside from their inaccuracy, there are a multitude of other factors to be considered: the introduction of new gear such as braided nylon gill nets; the fishing intensity or, more accurately, the catch-per-unit-effort, i.e., how many men, vessels, days fished, and in this case how many thousands of yards of netting; and most certainly the variations in demand and price. All these and similar concerns are vital in measuring a population. They are quite aside from the problems posed by environmental changes, long-term, short-term, man-made, or just in the course of nature on this planet.

Even more recent than the reports on the oceanic movements of salmon from both sides of the Atlantic is the release of precise, though still incomplete, information about the long-distance migratory movements of its close relatives, the various *Oncorhynchus* species. The tagging of more than 200,000 Pacific salmon under a 1953 treaty involving Canadian, Japanese, and U.S. fishery biologists has provided us with hitherto unknown insights. The north Pacific surface waters have a series of common patterns in their eddies or gyres; all are counterclockwise, as, coincidentally, is the Irminger Current in the north Atlantic. Gulf of Alaska, Bering Sea, Kamchatka, subArctic, et al., all have at least superficially similar circular flows. All five species of North American Pacific salmon, as well as steelheads (rainbow trout gone to sea), swim "downstream" or with one or another of these counterclockwise currents mostly in the upper 35 ft, and when homing to their parent stream, at rates up to 40 miles a day for hundreds of miles on end. The different species of salmon have different migratory patterns, but it is clear that Asian and North American stocks intermingle freely in the open ocean.

Bristol Bay, on the north side of the Alaskan Peninsula, produces huge runs of sockeye salmon. This species typically spends its first 1–2 years in fresh water, then 2–3 in salt water; in the marine phase of their life histories the sockeyes make from 4–6 huge counterclockwise swings around the north Pacific Ocean covering anywhere from 7,000–11,000 miles on their travels.

Both genera of salmon and the anadromous Clupeids, perhaps to a lesser extent, are prime examples of the parent-stream instinct. We know most about the Pacific salmon, simply because of their abundance and commercial importance. Here we can say with confidence that of the very small percentage of potential progeny that ever make their way to sea, those that survive the oceanic periods of their lives (a quarter more or less) return to spawn not only in the same river in which they were born but to the same branch of the same tributary of the same river where their parents spawned and died. How do they do it? All manner of theories have been proposed. It now seems clear that olfactory and visual senses are important once the fish are within the sphere of influence of their respective rivers. Far out at sea celestial navigation has been suggested; why not if the fish are in the surface layers and if this method of finding one's way around has been so highly developed in such diverse groups as bees and birds? On the other hand, University of Washington scientists have suggested ". . . that their findings on salmon migration appear to support the electromagnetic theory." And since we know that other fish can detect very small voltages, it is not implausible to entertain the thought that salmon may actually make use of them as directional cues.

Size, Growth, and Age. Fish, unlike mammals, grow throughout their lives, though generally at a slower rate after sexual maturity. Some breed when an inch or less long; the classic example is a pygmy goby from the Philippines (maximum length $\frac{1}{2}$ in.). Among the modern fish-like vertebrates, the whale shark is the biggest: up to 60 ft and probably over 50,000 lb; the basking shark reaches 45 ft, and the manta ray weighs over 3,000 lb. Ironically, all three of these elasmobranchs are essentially planktonic feeders, just as is the largest of mammals, the Blue Whale.

Among the bony fish the sea-sturgeon, the groupers (jewfish), tunas, swordfish, marlins, halibut, and ocean sunfish must be reckoned as some of the largest, ranging from several hundred to several thousand pounds.

The rates of growth in fish range over such a wide spectrum that generalization is next to impossible. Temperature is a dominant external factor in the egg and larval stages, just as the pituitary is regnant internally. Males tend to become sexually mature at younger ages, but females quite commonly reach the larger sizes (e.g., male striped bass mature at 2–5 years and seldom live beyond 5 years when they are approximately 20 in. long and weigh 4 lb; females mature at 4–6 years and live to over 30 years when they are over 50 in. and 65 lb). Sexual maturity is reached in spans ranging in different species from months to more than 12 years.

Age is determined by reading annuli on scales, otoliths, opercula, and cross-sections of fin-spines and vertebrae, as well as by confirmatory evidence from length-frequencies and the recapture of tagged individuals over long intervals, etc. Generally speaking, the older the fish the more difficult it is to determine the age with accuracy. Nor are the respective sizes of different species any indication of age. Thus Mediterranean tuna are reportedly over 50 lb at 4 years (compare with striped bass above) and 650 lb at 14, while the anadromous sea-sturgeon caught in the Hudson River takes 18 or more years to reach 225 lb. To the best of our knowledge few fish live more than 30 years, the majority of species quite certainly less than 10.

DANIEL MERRIMAN

INDIAN FISHERIES

India has a V-shaped coastline of about 3000 miles, running north–south. The peninsula projects into the northern part of the Indian Ocean to separate the Arabian Sea from the Bay of Bengal. The west coast forms the eastern boundary of the Arabian Sea, while the east coast forms the western and northwestern border of the Bay of Bengal. The continental shelf is estimated at about 100,000 square miles with an average width of about 35 miles. The shelf off the west coast, particularly at the northern end where it exceeds 180 miles in a few regions, is wider than off the east coast. The west coast with pronounced upwelling of nutrient-rich deep water during the southwest monsoon accounts for over $\frac{3}{4}$ of the annual catch of about 900,000 metric tons of marine fish. Among the causes for the relatively poor fisheries of the Bay of Bengal are: (1) the large amount of silt due to the enormous drainage from the major rivers of India, East Pakistan, and Burma which all open into the Bay, reduces transparency so that the rate and quantum of photosynthesis per unit area is relatively low; and (2) absence of large-scale upwelling or other mixing processes so that the inorganic phosphate content of surface water is relatively poor.

As in other tropical regions of the Indo-West Pacific, the marine fisheries of India are characterized by the occurrence of a great number of species, a feature which distinguishes them from those of the temperate and cold-water regions of the northern hemisphere. Thus there are no predominant fisheries comparable to those of the herring, cod, plaice, etc., of the north Atlantic or to the anchovy of the Peruvian coast. However, species of the two suborders Clupeoidei, Scombroidei and shrimps are important in terms of quantity and value.

Species

Among the clupeoids which account for $\frac{1}{3}$ of the total catch, sardines (*Sardinella* spp.), and engraulids (*Stolephorus* and *Thryssa*) are abundant. The oil sardine *Sardinella longiceps* Val. is, along with the mackerel (see below), the most important species. Restricted to the west coast of India, in some years over 150,000 tons of this fish are caught between August and March, mostly along the southern half of the west coast. Like many other pelagic plankton-feeding fish in other parts of the world, it shows periodic natural fluctuation in abundance. The indiscriminate capture of juveniles and maturing adults as a factor contributing to the fluctuation remains to be assessed. In normal years, other common clupeoids (over 20 species) together account for about $\frac{1}{2}$ the catch of oil sardine. The oil sardine is captured by boat seine, drift net, and shore seine operated from unpowered craft within a 10-mile wide stretch of coastal water.

Among scombroids, the mackerel *Rastrelliger kanagurta* (Cuv.) is practically as important as the oil sardine. Although not restricted to the west coast like the latter, it forms a major fishery only along that coast, where the fishing grounds and seasons of the two species overlap. However, the main fishing grounds for mackerel (north Kerala, Mysore, south Maharashtra) are a little north of those for oil sardine (Kerala and south Mysore). The mackerel which also is a pelagic, plankton-feeding fish moves in dense shoals into coastal waters toward the end of the southwest monsoon and is captured in shore seine (*rampani*), drift net, boat seine, and purse seine. When the catch of mackerel in the rampani is very large, the fish are impounded for up to a week by staking the net in a semicircle from the shore, and the fish bailed out as per demand. There is ample scope for definitive work on the spawning season, grounds, and behavior of this fish.

The spanish mackerel or seer fish *Scomberomorus commersoni* (Lac.), *S. guttatus* (Bl. and Schn.) and *Acanthocybium solandri* (Cuvier), are widely distributed along both coasts and captured in good quantity by hand line and in gillnet; not infrequently they are captured in boat seine. They are excellent table fish and ideal for filleting. On the fishing grounds off both coasts they are caught practically throughout the year, though in different areas there are well-defined periods when they are more abundant.

The landing of tunas like the mackerel tuna *Euthynnus affinis* (Cantor), the frigate mackerel *Auxis thazard* (Lac.), the blue-fin *Kishinoella tonggol* (Blkr.), the skipjack *Katsuwonus pelamis* (Linn.), the albacore *Thunnus alalunga*, and the yellow-fin *Neothunnus macropterus* (Temm. and Schl.) is limited, although there is a good tuna fishery by hand line using live bait around Minicoy Island. At present the larger tunas and other large scombroids like *Xiphias gladius* Linn., *Istiophorus gladius* (Brouss.), *Makaira indica* (Cuvier) are taken only in an area within 10 miles from the coast by hand line and accidentally in drift net and boat seine where they do considerable damage to the cotton and hemp nets. There are good prospects for exploitation of the tuna resources in the two seas.

In terms of value and development, the sea-based shrimp fisheries of India occupy a significant position. At present around 80,000 tons of shrimp are captured along the continental shelf. The shrimp fisheries are particularly well developed on the west coast. In addition to traditional fishing by boat seine, dip net, and cast net, shrimp trawling is quite extensive off Kerala coast. Along the northern end of the west coast *Metapenaeus monoceros* (Fabr.) and *M. brevicornis* (M.Edw.) are the dominant species, whereas in the southern half *Metapenaeus dobsoni* (Miers) is the dominant species followed by *M. affinis* (M.Edw.), *Parapenaeopsis stylifera* M.Edw. *M. monoceros*, and the large-sized *Penaeus indicus* M.Edw. Cochin, with its relatively good harbor facilities for fishing vessels, is the center of a thriving export industry for frozen and canned shrimp. The spiny lobster *Panulirus homarus* (Linn.) captured in small quantities along the southwest coast is frozen for export.

A variety of sharks and rays are captured

along both coasts throughout the year (over 50,000 tons) by hand line and boat seine operated in inshore waters from unpowered craft as well as in trawls and bottom-set gillnets operated from mechanized craft. They often do considerable damage to drift nets. Species common in the landings are *Chiloscyllium indicum* (Gmelin), *Galeocerda cuvieri* (Le Seuer), *Scoliodon* spp., *Eulamia* spp., *Sphyrna* spp., *Pristis* spp., *Dasyatis* spp., *Rhinoptera* spp., and electric rays. Catering especially to the lower middle classes and poorer sections of the coastal populations, elasmobranchs represent a rich field for the persevering systematist. There are shark-liver oil factories at Bombay and Calicut on the west coast, but since the capture of sharks is widespread and the fishing villages are often in remote regions, the livers rich in vitamins A and D are not utilized to the extent that there is demand for them in a country where malnutrition and deficiency diseases are common.

The Bombay duck *Harpodon nehereus* (Ham.) constitutes an important fishery off the coasts of the states of Bombay and Gujarat. In recent years the landings of this fish have been estimated at around 100,000 tons. The main gear for this fish is the stake net. A greatly relished soft-bodied fish, it is consumed both fresh and after sun-drying. This fish begs definitive study on biology and behavior.

In addition to the above significant fisheries, mention must be made of numerous species of marine catfish (Ariidae), croakers (Sciaenidae), horse mackerel (Carangidae), ribbon fish (Trichiuridae), ponyfish (Leiognathidae), pomfrets (Stromateidae) and flatfish (particularly *Cynoglossus* spp.) which are extensively distributed and captured in fair quantities in coastal waters along both coasts, within range of the unpowered craft.

Mechanized Fishing

The last two decades have witnessed an appreciable mechanization of craft and gear with the help of FAO experts and the Indo–Norwegian Project. Mechanization of craft has involved the fitting of engines in improved versions of local craft as well as design of new small and medium craft (below 50 ft over-all length) to suit the requirements of the Indian coasts and harbors and the skill of Indian fishermen. Technological know-how available in India apart from the structure of the fish populations exploited do not yet warrant the introduction of large trawlers and factory ships. Even with powered craft the 50 fathom line is the common range and the edge of the continental shelf is the maximum limit of present fishing operations. The resources of the adjacent seas beyond this limit are as yet untapped by Indian fishermen.

Engines, winches, and navigation, communication, and fish-finding equipment are imported and, as can be judged from reports of FAO experts under the Expanded Program of Technical Assistance, there are few ancillary industries for maintenance, overhaul, and repair of boats and above-mentioned equipment. This results not only in enormous loss of time on repairs and replacement of parts and consequent loss of fishing hours, but also in an uneconomic increase in the cost of production.

In spite of these impediments, mechanized fishing has caught on and the mostly illiterate fishermen have taken enthusiastically to it. Trawling (bull- as well as otter-trawling), bottom-set gillnet and drift-net fishing have proved successful whereas purse seining has limited application because of the absence of large dense schools.

Among the more valuable fish captured along the northern half of the west coast in trawls and bottom-set gillnets are the thread-fins *Polydactylus indicus* (Shaw) and *Eleutheronema tetradactylum* (Shaw), *Pseudosciaena diacanthus* (Lac.), *Otolithoides brunneus* (Day) and other sciaenids, the grunter *Pomadasys hasta* (Bl.,), the eel *Muraenosox telabonoides* (Gronov) Gray, pomfrets, sharks, and rays. Shrimp are captured by trawls along the southern half of the west coast. A considerable proportion of the catches consists of low-value and trash fish.

On the east coast, the fish taken are generally smaller (except sharks) than comparable species on the west coast. Catfish (Ariidae), sciaenids, threadfins the thread-fin bream *Synagris japonicus* (Bloch), goatfish (Mullidae), flatfish, sharks and rays, pomfrets (particularly from October to February), shrimps, and a great variety of trash fish constitute the catches in trawls. Sharks constitute more than 80% of the captures by bottom-set gillnet. Tunas, larger carangids, and scombroids are captured by gillnet (drift and bottom-set). There is no shrimp trawling as along the southwest coast, though there are prospects for this fishery beyond the mouth of the rivers Mahanadi and Godavari and at the head of the Bay.

Marketing

Except in Bombay, Madras, Calcutta, and a few other larger cities which have established fish markets, marketing of catches is done through middlemen (who exploit the socio-economically backward communities of fisher folk) or by fisherwomen. Most of the fishing is done from seaside villages and small towns which have limited or no facilities for cold storage, hygienic preservation and processing, and roadrail transport; the picture, however, is rapidly changing and governmental agencies as well as fishermen's cooperatives are providing the above facilities in increasing measure. A peculiar feature of traditional fish marketing is that the catches are not weighed either at the landing centers or in the local markets and are not sold by weight except in cities. This makes estimation of landings and value difficult. There are wide fluctuations in price determined by supply, demand, and buying capacity of the customer. Fishing is negligible or at a standstill during the southwest monsoon and in squally weather. There is another reason for the wide fluctuation in availability and price.

Export

India exports over 20 items of marine products to about 50 countries. During the fiscal year April 1966 to March 1967, exports were just over 20,000 tons valued at a little over $22,000,000. Of this, frozen products accounted for nearly 10,000 tons valued at a little over $15,000,000 (slightly over 68% of total value of exports). The chief countries to which frozen and canned products (mainly shrimp) are exported are the U.S., Japan, and Australia, followed by France, the U.K., and West Germany. Ceylon is the main importer of pickled and dried products, followed by Burma, Hong Kong and Singapore.

Processing

It is estimated that about 50% of the landings are processed by a variety of methods, since edible marine products deteriorate very rapidly in high atmospheric temperatures (28°–40°C daytime temperatures) prevalent in India. The majority of fishing vessels being unpowered, they are too small to have fish holds or even to carry ice. Ice and cold storages are available only at a minority of the fish landing centers, although better facilities are available along the west coast. Most processing is by rough-and-ready methods.

The following methods of processing are widely employed:

Sun Drying. (a) *Without salt* Small and lean fish like white-bait (*Stolephorus* spp.), sardines other than oil sardine, ribbon fish, bombay duck, malabar sole (*Cynoglossus semifasciatus* Day) are dried without salting or gutting. While bombay duck and in some regions ribbon fish are skewered through bamboo poles or coir rope fixed between stout bamboo frames for effective drying action of sun and air, white-bait, sole, etc., are just spread on coir or palm-leaf matting or even right on the sandy beach so that the latter product is of very poor quality. Fluvial and estuarine shrimps are dried for consumption in the interior as well as for export to East Asian, African, West Asian and other countries.

(b) *After salting* Scombroids (except mackerel), carangids, elasmobranchs, and other larger fish with low fat content are dried in the sun after gutting and application of crude salt, by being spread out on mats or directly on the beach; the product is of poor quality.

(c) *Cooked-dried shrimp* Shrimps in the shell are boiled in seawater and dried in the sun. The dry shrimps are shelled by threshing them in jute sacs. After winnowing, the shrimp is exported to Hong Kong, Malaya, Singapore, etc.

Pickling. (a) In south India, oil sardine and mackerel are pickled in salt after gutting. They are stacked either in mat-lined pits in sand (pit-curing), in bamboo baskets, on the floor, or in shallow cement tanks, in different regions along the southeast and west coasts.

(b) Along the coast of north Kerala and south Mysore, sardine, mackerel and seer fish are pickled by the Colombo method for export to Ceylon. After gutting, a small quantity of the fleshy pods of the malabar tamarind *Garcinia cambogea* is inserted into the bony cavity; the fish are then stacked in wooden barrels between layers of salt and malabar tamarind. Research on the common methods of processing has determined that tamarind (*Tamarindus indica*) improves the quality of the product. The valuable results of investigations aimed at improving the quality of processed fish have not been adequately disseminated among fishermen and the trade.

Semidried Shrimp. Shrimps are blanched for 2–3 minutes in 4–6% brine, shelled manually, soaked in cold brine (25° Be) for 15 minutes and then dried in the sun. Semidried shrimp have excellent consumer appeal because of pleasing color, appearance, texture, and palatability.

Canning. In recent years factories for canning sardine, mackerel, and shrimp have been started at Malwan, Malpe, Cochin, and a few other cities on the west coast. The price of the canned product is beyond the reach of the average Indian, but canned shrimp particularly have a good export market.

Freezing. The port city of Cochin is the main center of a quick-freeze industry for shrimp and lobster tails which are exported mainly to the U.S. and Japan.

Other Products. (a) Fish maws and shark fins are dried for export to Singapore and Hong Kong. (b) Swim bladders of eels, sciaenids, thread-fins, and grunters from Bombay coast are dried for export to Europe. (c) Sardine oil, turtle meat, *beche-de-mer*, and seaweed are processed for export.

Problems of the Industry

The fishing villages are widely dispersed and not always within easy reach of road and rail transport facilities so that fresh fish, particularly during the peak period of the various fisheries, do not reach the inland areas of consumption, except in a few regions.

The fishing communities are socio-economically backward and illiterate and slow in appreciation of modern standards of hygiene and sanitation. These factors affect the quality of the products.

The indebtedness of fishermen to money lenders and their exploitation by middlemen have stood in the way of improvement of their economic status and opportunity to improve craft and gear.

Considering the length of the coast, there are comparatively few natural harbors and no fishing ports *sensu stricto*.

Basic requirements like clean fresh water and good quality salt are inadequate to meet the demand of the processing industry.

The paucity of harbors and ancillary industries has hindered the growth of fishing fleets and modernization of the industry. Mechanized fishing particularly is handicapped by the limited facilities for maintenance, overhaul, and repair of craft and gear and because of the dependence on import of sosphisticated navigation, communication, and fish-finding equipment.

The seafood export industry faces the problem of rising costs of raw material, labor, tinplate, and industrial facilities in a highly competitive market.

While good work has been done on improved methods of processing, studies on population dynamics have lagged behind. Nothing is known of the spawning grounds and the route and range of migrations of the oil sardine, mackerel, sole, etc., at the end of the fishing season in coastal waters.

Considering the resources and the market, relatively few financiers and entrepreneurs have taken to this industry except in the states of Kerala and Bombay.

The conservative tastes and food habits of the fish-eating section of the population make the introduction of unfamiliar but economically feasible methods of processing like marinating and smoke-curing difficult.

Future Potential

Although the problems facing the industry are formidable, the progress during the last two decades gives hope for further development of the industry and for exploitation of the outer half of the continental shelf and the open sea beyond.

There are good prospects for intensive exploitation of scombroids (particularly tuna) and petch (*sensu lato*) resources of the Arabian Sea and Bay of Bengal by longlining and trolling. The Arabian Sea also appears to support large schools of flying fish. There are indications of lobster grounds in some regions off both coasts; at present there is no lobster fishery as such. Trawling, particularly along the east coast, brings in a high percentage of low-value and trash fish. At present considerable quantities of trash fish are thrown overboard because they do not fetch an economic price and at the same time the cost of production is so high that it is also uneconomical to convert them. Efforts must be directed at better utilization of these fish by converting them at low cost into fish flour, fish meal, and guano, and for extracting body oils for which there is a demand. Small-scale conversion plants spread along the coast would serve the purpose better than large centralized industries, because in each region the quantity of raw material is limited. These plants should be so designed that they could also produce byproducts like animal charcoal from scales, pearl essence from guanine, glue, etc.

There is scope for developing fleets of carrier launches with insulated holds for transport of raw material like trash fish and shark liver from the fishing grounds and landing centers to processing factories. These launches could also supply ice to fishing vessels and landing centers and distribute fish during a period of glut over a wider area than at present or transport them to railheads for transport to the interior.

S. DUTT

INTERSTITIAL FAUNA

Considerable areas of the sea coasts of the world consist of sandy beaches, composed to a large extent of terrigenous sediment. In the sublittoral and at greater depths there is sand too. Similar sediments may be formed of crushed skeletons of corals, bryozoans and echinoderms, and of shells of snails and mussels, so-called shell sand. Marine sand provides an interesting biotope for marine organisms, and as such it was studied during the 1920's by Remane, who was the first to publish clear ecological definitions of the sand fauna.

The fauna of marine sand was divided by Remane into the *epipsammon*, the *endopsammon* and the *mesopsammon*. The epipsammon comprises the organisms living on the surface of sandy bottoms (e.g., *Crangon*, gobies); the endopsammon organisms that dig actively or live in tunnels in the sand (*Arenicola*, many mussels, Enteropneusta), while the mesopsammon is the fauna living in the interstitial water. In the English literature this fauna is called the *interstitial fauna*, a term introduced by Nichols. There is also an interesting interstitial fauna in coastal subsoil water, i.e., in the narrow zone where fresh subsoil water meets salt interstitial water in a sandy beach. The coastal subsoil fauna was studied first by Remane and Schulz. Among the interesting components of this specific fauna are the Mystacocarida, a new order of animals discovered by Pennak and Zinn. In addition to such steneocious forms, the coastal subsoil water also contains euryoecious marine interstitial species, including representatives of limnobiotic organisms.

The study of the marine interstitial fauna has, ever since Remane's pioneering work in the 1920's and 30's, enriched systematic zoology more than the study of any other marine biotopes. Marine sand research, which is now the concern of more than 100 biologists, has led to the discovery of many new animals, several of which represent hitherto unknown types of organization. Many of these discoveries are of great taxonomic value; families, orders and classes.

Interstitial Environment

The dimensions of the interstitial space are of great importance for the organisms living there, and are a decisive ecological factor, determining as they do the upper limit for the size of the bodies of the organisms that can live there.

Some idea of the dimensions in the interstitial space may be obtained by measuring the volume of the interstitial water per volume unit of sediment. Biologists, however, prefer to use granulometric methods, by which the proportions of different grain sizes in fractions of sand from samples sifted in a series of sieves with different mesh sizes are determined. The granulometric measurements are then illustrated in cumulative diagrams.

Interesting studies on the preference of interstitial species for certain grain sizes have been published by Boaden on the archiannelid *Trilobodrilus heideri*, and Gray on the archiannelid *Protodrilus symbioticus*. In experimental situations, in which the animals tested are able to choose among different sediments representing different grains sizes, these scientists observed that the organisms increase in number in definite types of sand with respect to grain size. The experimental results are in conformity with experience of many psammologists, that the horizontal distribution of the interstitial fauna, on a beach for example, is irregular, and that the frequency of the organisms varies greatly, which is most likely due to grain size.

The light factor is interesting, and it seems highly probable that in most types of sand, light cannot penetrate for more than a few centimeters into the sediment. Very few interstitial species of animals have well-developed organs of sight, which must be associated with this circumstance. Temperature, salinity, and content of oxygen are of great importance for the interstitial fauna.

The temperature of the interstitial water is measured with ordinary thermometers inserted into the sediment or with the aid of thermistors. Especially great variations of temperature occur in sand in the intertidal zone, where the interstitial water in the surface layer is quickly affected by the temperature of the air. In the surface layer of the sand, therefore, the temperature of the interstitial water varies with the rhythm of the tides, but also with the seasons of the year. On the west coast of Europe the amplitude of the variations in the temperature of the interstitial water is greater in summer than during the cold months, owing to the difference between the temperatures of the air and of the seawater.

The vertical migration in the sediment made by many members of the interstitial fauna has been related to the temperature gradient in the sediment. Renaud-Debyser studied, among other things, the vertical migration of tardigrades in sandy beaches at Arcachon (Gironde, France), and found that they occur at relatively great depths in the sediment (50–70 cm) during winter, but that in the warm periods of the year their maximum frequency is at lower depths. These vertical migrations may be seen as an adaptation which makes the hibernation of the fauna possible.

The interstitial fauna in the tidal zone is definitely eurythermal. The significance of salinity for the interstitial fauna is revealed in the geographical distribution. In distinctly brackish water, some groups of interstitial animals, at least, are less richly represented than in the marine littoral. The great variations in salinity in many intertidal zones with a rich meiofauna show that the interstitial fauna is euryhaline.

Morphological Adaptation

The dimensions of the interstitial environment make it impossible for any but very small organisms to live there. The upper limit for body size is given usually as 2–3 mm. This means that, as a rule, it is in the interstitial sand fauna that the smallest invertebrates are found; for example, the cnidarian *Psammohydra*, 0.5 mm, the archiannelid *Nerillidium simplex*, 0.6 mm. and the opisthobranch *Hedylopsis loricata*, 0.8 mm. In the interstitial sand fauna the Protozoa and Metazoa have bodies about the same size.

The small size of some interstitial organisms has been interpreted by some authors as the result of regressive evolution, in which neotony is said to be an active mechanism. This is supposed to explain, among other things, the larval morphological characters found in some species, such as, for example, provision of transversal cilia rings in some archiannelids (*Trilobodrilus, Diurodrilus*), and extensive epidermal ciliation (the Cnidaria: *Halammohydra* and *Otohydra*).

The neotenous features, the smaller number of cells and the loss of organs observed in certain regressive developmental series in the interstitial sand fauna suggest that we are here concerned with the smallest possible body sizes for certain groups of invertebrates.

In the body shapes of interstitial organisms we find a distinct adaptation. Elongated forms are very suitable in an interstitial environment, and are found in a strikingly high proportion. Remane compared the length–width index (maximum width × 100: body length) of tubellarians, nematodes, ostracodes, and copepods in different biotopes, and found that the sand microfauna had much stronger vermiform tendencies than, for example, the phytal fauna. Many astonishing examples of vermiformity are found in the sand microfauna; for instance, in animal groups in which elongated bodies are unusual, such as the cnidarian *Halammohydra vermiformis*, the opistobranch *Pseudovermis* and the copepod *Cylindropsyllis laevis*. Another type of body shape is that of wide and flat organisms (many ciliates, turbellarians, and tardigrades).

Different types of skin reinforcement are common, and useful, on account of the mobility of the sand and the unceasing restratification, and provide mechanical protection and good adaptation in these otherwise fragile species. Epidermis reinforcements may consist of cuticular differentiations in the form of scales or spicules (gastrotrichs, Solengastres) or epidermal calcareous spicules (the turbellarian *Acanthomacrostomum spiculiferum*, the order Acochlidiaceae among the Opistobranchia).

Adhesive organs are very common in various groups of interstitial animals. The importance of such organs may also be related to the dynamic environment. The adhesive organs have been formed in different ways in various groups of animals.

Some turbellarians and all gastrotrichs have unicellular mucus glands with orifice tubes (so-called adhesive tubes), the secretion of which has adhesive properties. In other groups of animals, particularly among the Crustacea, claws and spines are used to attach the animals to the grains of sand. The tardigrade *Batillipes mirus* has adhesive disks on its extremities.

In many cases the adhesive organs are of importance for locomotion. Side by side with the normal cilia-gliding method of locomotion, the gastrotrichs can move with the help of cephalic and caudal groups of adhesive tubes in the same way as a leech, when these adhesive groups work alternately and the body is simultaneously contracted or estended.

As in cavernicolous organisms, well-developed organs of sight are lacking in the majority of the interstitial animals. Statocysts and similar organs of equilibrium occur in many interstitial animals, which can be related to the dynamic features of the environment. In the interstitial fauna the occurrence of such organs is the rule in cnidarians, turbellarians, nemertines, and mollusks, but they also occur in some annelids.

Biology of Interstitial Fauna

Locomotion. The great majority of interstitial organisms are free-swimming, and only a smaller group consists of sedentary forms (foraminifera, the brachiopod *Gwynia capsula*). In addition several semi-sessile forms (e.g., the polychete *Psammodrilus balanoglossoides*) are known. It is interesting to find that in the interstitial fauna there are free-swimming or semi-sessile species which represent groups of animals that are normally sedentary (e.g., the bryozoan *Monobryozoon ambulans* and the interstitial ascidians).

In the free-living forms, cilia-gliding is a common means of locomotion (the ciliate *Halammohydra*, gastrotrichs, turbellarians, archiannelids, mollusks). The cilia-gliding animals move with the help of cilia only and with the ventral side in contact with the substratum.

Some elongated animals (nematodes, *Polygordius*, the harpacticoid *Cylindropsyllis laevis*) move by writhing, and locomotion is effected with the help of the muscles. The writhing species are dependent on the mechanical support provided by grains of sand for locomotion. If a *Cylindropsyllis* is transferred to seawater without the presence of sand, the animal performs the writhing movements, but it does not move forward.

No sedentary or semi-sessile animals are known in very mobile sand, where the waves stir up the sediment. All the organisms in such regions are free-swimming and many of them (e.g., several turbellarians, the gastrotrich genus *Xenotrichula*) move very rapidly in this environment.

Nutritional Biology. The principal nutritional biological types are (a) predators, (b) diatom feeders, (c) detritus eaters, and (d) suspension feeders. The predators include the cnidarians *Halammohydra* and *Psammohydra* and a large number of turbellarians and nematodes. The diatom feeders from a large group which includes many gastrotrichs (especially the Dactylopodalidae and Thaumastodermatidae), archiannelids, etc. Detritus eaters are found among gastrotrichs, nematodes, and archiannelids. Suspension feeders are almost exclusively sedentary and semi-sessile forms (*Monobryozoon*, the brachiopod *Gwynia capsula*, and interstitial ascidians).

Reproduction. The biology of reproduction is relatively unknown. But the life-cycles of some representatives of various groups of animals have been studied and found to reveal certain characteristic features. The number of eggs and spermatozoa produced by these small organisms is usually small. Many species produce only one or two eggs at a time (*Psammohydra*, some gastrotrichs, the archiannelids *Nerillidium* and *Diurodrilus*, the *opisthobranch* Hedylopsis loricata). A very rough estimate gives about 10 eggs for most species, and only exceptionally are more than 100 eggs produced on each reproduction occasion. The number of spermatozoa is also often small; in many species, e.g., gastrotrichs belonging to the family Thaumastodermatidae, the spermatozoa are large in relation to the size of the animal.

A certain degree of adaptation is often required of these low-producing animals to ensure the survival of the population. Among these adaptations are:

(a) Prolonged reproduction periods; in many species reproduction goes on all the year round, with only a pause or a minimum during the cold season.

(b) Spermatophores. Such formations must be regarded as a means of ensuring fertilization, and occur in, among others, the gastrotrich *Dactylopodalia*, some members of the archiannelid genus *Protodrilus*, and the opisthobranchs *Hedylopsis brambelli* and *Microhedyle lactea*.

(c) Various types of brood protection occur. The cnidarian *Otohydra vagans* incubates one or two eggs, which develop in the female to a very advanced stage. The larvae of the branchiopod *Gwynia capsula*, 2–4 in number, develop in the mother animal's coelom up to the stage when three segments have been formed. The archiannelids *Mesonerilla* and *Nerillidium* spawn 1–6 eggs which are attached to the female's body by a mucous band and develop there until four setigerous segments and the digestive gland are completely differentiated, after which they become detached.

(d) Spawning modalities. The eggs of many species are spawned in gluey envelopes or cocoons (several turbellarians, *Protodrilus symbioticus*, the mollusks *Pseudovermis*, *Philinoglossa*, Acochlidiacea). The egg's cocoons adhere to grains of sand and are thereby kept within the range of the population area until they hatch.

(e) Larval development. Larval development with a pelagic, planktotrophic phase, is found only in the archiannelids *Polygordius* and some species in the genus *Protodrilus*, which produce more than 200 eggs. As a rule, larval development is benthic, with lecithotrophic larvae. In the interstitial mollusks belonging to the order Acochlidiacea, and the genera *Philinoglossa* and *Pseudovermis*, the veliger larvae, when hatched, are without photic reaction, and it may therefore be assumed that their free-swimming phases are spent in an interstitial environment. This is probably necessary since each animal produces only about 50 eggs; a pelagic phase in the open water would here imply far too great dispersion to retain the population within its boundaries.

Composition of Interstitial Fauna

Since the small sizes of the interstitial environment restrict the fauna to a microfauna, it is understandable that the fauna is recruited from groups of animals in which small bodies are normal. Among these are the Protozoa, and among the Metazoa are, among others, gastrotrichs, nematodes, tubellarians, tardigrades, ostracodes, and copepods. In addition there are, in the interstitial fauna, numerous reduced representatives of other groups of invertebrates. Another category must be included, too, benthic larvae and juvenile stages of animal species which, as adults, live in another environment.

Protozoa. In addition to a few genera of Foraminifera, it is above all the ciliates that dominate. Important works on interstitial ciliates have been published by Kahl, Dragesco, and others; in Dragesco's monograph, 301 species, belonging to 85 genera are dealt with. Important genera are *Centrophorella*, *Condylostoma*, *Geleia*, *Loxophyllum*, *Remanella*, *Trachelocerca*, and *Tracheloraphis*.

Cnidaria. In the interstitial fauna the cnidarians are represented by a few aberrant forms. *Sphenotrochus wrightii* (Madreporaria) is a solitary, bipolar Caryophyllidae, 1–2 mm in size, found in sediment of shell debris in the Mediterranean and the English Channel at depths of 20–60 m. Most studied of the interstitial cnidarias are the order Actinulida with the genera *Halammohydra* (Fig. 1) and *Otohydra*, which are Hydrozoa 1–2 mm long. The Actinulida are free-living and solitary, and retain as adults the bipolar organization of an actinula larva. They have direct development. All the known species have lithostyles.

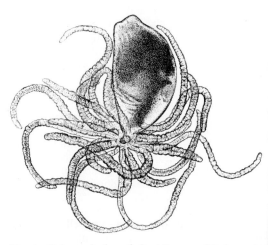

Fig. 1. *Halammohydra schultzei* Remane (Hydrozoa).

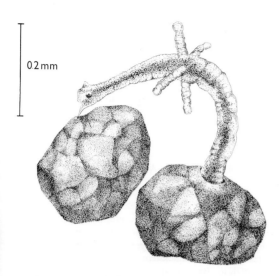

Fig. 2. *Psammohydra nanna* Schulz, an interstitial corymorphid polyp. (Swedmark.)

Annelida. Among the polychetes, several genera belonging to the families Syllidae, Pisionidae, and Hesionidae are interstitial. These species are about 2 mm long. The Psamodrilidae family comprises two genera, each of one species, *Psammodrilus* and *Psammodriolides* (Fig. 3), which are strongly aberrant polychetes. They are semi-sessile and in many respects adapted to the interstitial environment. Most of the archiannelids (*Polygordius, Protodrilus, Nerilla, Nerillidium, Mesonerilla, Nerillidopsis, Thalassochaetus* and *Meganerilla*) belong to the marine sand microfauna.

Tardigrada. Almost 20 species of this otherwise limnobiotic group live in the interstitial environment (*Batillipes, Actinarctus, Stygarctus, Tanarctus, Orzeliscus*).

Psammohydra nanna Schulz (Fig. 2) is a polyp, barely 1 mm long. It can move in sand by releasing the basal adhesive disk at the same time as it adheres with the front end by means of sticky oral nematocysts. The polyp contracts, and the movement is continued by the basal disk adhering to a new spot.

Turbellaria. The turbellarians are numerous in the sand microfauna. Some groups, e.g., the Kalyptorhynchia, show several morphological adaptations to the interstitial environment, which have been studied especially by meixner.

An aberrant form is *Acanthomacrostomum spiculiferum*, which has an epidermal skeleton of dense spicules. The family Otoplanidae is characteristic of certain interstitial environments, primarily the breaker zone.

One group, regarded by some authors as an order in the class Turbellaria, and by others as an independent class, is the Gnathostomulida. Although this group was discovered only recently, new studies have shown that it has a wide geographic distribution. That these animals were not discovered earlier probably should be attributed to the small size, about 1 mm.

Gastrotricha. Both the orders Chaetonotoidea and Macrodasyoidea occur in the marine interstitial fauna, the latter exclusively there. More than 100 species of the macrodasyoid gastrotrichs are known and about 30 of the chaetonotoid in marine sand. The literature on marine gastrotrichs is relatively extensive. One important work is Remane's, and a survey of more recent literature is given in Swedmark.

Nematoda. This group, which has been studied by Gerlach, Wieser, and others, is richly represented in interstitial environments.

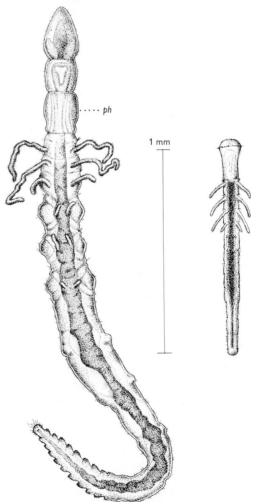

Fig. 3. *Psammodrilus balanoglossoides* Swedmark (left) and *Psammodriloides fauveli* Swedmark (right). Interstitial polychaets. *ph*: pharyngeal apparatus. (Swedmark.)

Crustacea. Only one group of crustaceans, the Mystacocarida, is completely restricted to the interstitial environment. Otherwise Ostracoda and harpacticoid copepods are very common. The families Microparasellidae and Microcerberidae have some marine interstitial representatives which are definitely thigmotactic and also in other ways adapted to the interstitial environment.

Mollusks. Several groups of mollusks are represented in the sand microfauna. The genus *Caecum* (Prosobranchia) is, with its cylindrical shell, well adapted to life in the interstitial environment. Most of the interstitial mollusks belong to the Opisthobranchia. Among them are the genus *Rhodope*, a regressive form without radula, and *Philinoglossa*. Further, the order Acochlidiacea with 11 species belonging to the genera *Hedylopsis*, *Microhedyle*, *Parhedyle*, and *Unela*. Another hitherto unnoticed group of mollusks in the interstitial fauna is the Solenogastres. This group of animals is represented here by a number of still undescribed species with bodies 1–3 mm long.

Bryozoa. The solitary bryozoan from Helgoland, *Monobryozoon ambulans*, described by Remane, can move very slowly by contracting the muscles of the stolons. It is the only interstitial bryozoan known at present.

Echinodermata. The synaptides *Leptosynapta minuta* and *Rhabdomolgus ruber* are known in the microfauna of so-called Amphioxus sand on the coast of Europe.

Ascidiacea. Weinstein discovered and described interstial species in Amphioxus sand at Banyuls. *Psammostyela delamarei* is able to move in the sediment by muscle contraction.

Technique of Collection

Most psammologists, especially those working mainly on morphological and taxonomical problems, collect their material in a very simple way. Experience has shown that the greater part of the microfauna is in the upper layer of the sand, the top 3–4 mm. Sand is collected during the ebb period by skimming off the surface of the sand with, for instance, a rectangular plastic tray, and transferring it to a cylindrical vessel holding approximately a liter. The sample of sand is allowed to stand in the laboratory for a day or two, during which time the organisms move towards the surface of the sample. A few deciliters of the surface layer are then transferred to a separate vessel, a liter or so of seawater is added, and the mixture is stirred vigorously. The supernatant liquid is then decanted through a funnel over which a nylon net, mesh size 200–300 μ, has been stretched as a filter. The organisms caught in the net are transferred to a petridish with seawater, and can later be isolated with the help of a binocular microscope and a pipette. For quantitative studies of the microfauna, coresamplers are often used for field sampling. Suitable core-samplers have been described by Renaud-Debyser and Muus. Attempts have also been made to construct apparatus to separate the microfauna from samples of sand.

BERTIL SWEDMARK

INVERTEBRATES AS FOOD

Recent reviewers place the number of phyla in the animal kingdom between 25 and 30. All but two of these are represented in the sea, and several, such as the echinoderms, are found only in the sea. With 25 or more phyla to choose from, it is perhaps surprising that only three, the Chordata (vertebrates), the Mollusca, and the Arthropoda are of any real significance as a source of food for man. Similarly, with more than 70 classes of animals represented in the sea, it is also surprising that only seven contribute significantly to man's diet: Osteichthyes, Chondrichthyes, and Mammalia in the Chordata; Pelecypoda, Gastropoda, and Cephalopoda in the Mollusca; and Crustacea in the Anthropoda. In view of the surprisingly few animal groups utilized by man, the question arises, can we look forward to other marine animals as significant supplements to our diet? A brief review of the animal phyla may indicate some of the potentials that exist and at the same time may dispel false hopes. It may also point to areas that need increased research if new animal groups are to be tapped.

No protozoans are used for food by man. Although some marine Foraminifera reach a diameter greater than 1 in., the minute size of the vast majority of protozoans makes it very unlikely that man will find any significant source of food in this very diverse and widespread phylum.

No sponges (phylum Porifera) are eaten by man although they have long had an economic value. With all-pervading, indigestible mesh-like skeletons and their common development of noxious chemicals to reduce predation, their use other than as "sponges" does not seem promising.

In the phylum Coelenterata three species of jellyfish (class Scyphozoa) are eaten by the Japanese. The most common, *Rhopilema esculenta* or Bizen kurage, is collected from a small area in the Inland Sea in early summer. The manubrium is pickled in alum. Kurage (jellyfish), as it is called, is rinsed in fresh water, cut into thin strips about $\frac{1}{8}$ in. wide by 3 in. long and with a bit of vinegar is served as a small side dish at rather elaborate meals. The texture is crunchy or crisp; the taste essentially that of alum and vinegar. Its nutritional contribution to the meal is very minor if indeed there is any. The utilization of other genera and species does not seem promising.

Sea anemones (class Anthozoa in the Coelenterata) are occasionally eaten in small numbers in southern Japan, Samoa, and France. Apparently they are cooked in oil, although there is little information available about the edibility and nutritional contribution of this large and widespread group.

As far as can be determined, the Ctenophora, Platyhelminthes, Aschelminthes, Nemertea, Echiurida, Sipunculida, Ectoprocta, Entoprocta, Brachiopoda, Chaetognatha, Pogonophora, Hemichordata, and a variety of parasitic phyla are never used as food by man. There remain the five large phyla, Annelida, Mollusca, Arthropoda, Echinoderma, and Chordata as the most likely sources. The West consumes a rather limited fare from these phyla. It eats no annelids whatsoever (at least intentionally), but everyone knows of the annual two-day feasts of the Samoans and other south sea islanders on the palolo worm, a nereid polychaete (*Eunice viridis*) related to the familiar clam worm of New England. This type of polychaete reproductive behavior occurs in the Pacific, Atlantic, and Indian Oceans and a variety of species are eaten raw or cooked or used for bait in such widespread places as Fiji, the Gilbert Islands, Malaya and Japan. Despite their great abundance and concentration while spawning, they do not look promising as a future source of food. Even if the western loathing of worms as food could be overcome, questions of extent of the resource, efficient methods of harvesting, methods of preservation and marketing are completely unknown.

Western man eats a greater variety of Mollusca but only five groups, oysters, clams, mussels, squids, and abalone are eaten in any significant quantity (see articles in this encyclopedia). A few daring epicureans eat octopus, conch, and other mollusks when they are in seaside communities where these are regularly offered. Because they have a greater variety available and because the motivation is there, the Japanese, without doubt, eat more kinds of Mollusca than any other people in the world. There is no need to list here the great variety of gastropods, pelecypods, and cephalopods they utilize. There must be more than 100 species.

Among the Arthropoda, a great variety of decapod crustaceans have long been used for food. Lobsters, shrimps, and crabs are highly esteemed, and almost any shrimp, lobster, or crab can be eaten. However, catching them in sustained quantities is a major problem. For example, the Japanese eat five species of spiny lobster, but *Panulirus japonicus* is most common.

They eat six species of scyllarids or sanda lobsters, but only two are commonly seen, *Parribacus antarcticus* or Zori ebi and *Scyllarides squamosus* or Semi ebi.

Some minor crustacean groups are eaten in local areas. Gooseneck barnacles have long been eaten by the Basques of northern Spain and the forays of jai-alai players in Tijuana against *Pellicipes polymerous* (formerly *Mitella*) are well known to southern California marine biologists few of whom, however, have ever tried them. The Chileans are reported to eat a large sessile barnacle, *Balanus psittacus*. The Japanese also eat a gooseneck barnacle, *Mitella mitella* or Kamenote. According to some Japanese it is eaten "only by fishermen starving on islands." However, where abundant, large clusters of them can be torn from the rocks and boiled for a few minutes in seawater. Broken open while still hot, the pink meat tastes like crab with an acid component.

Mantis shrimps of the crustacean group Stomatopoda are eaten in Hawaii, Japan, and other parts of the orient. Because of their rather solitary habits and deep burrowing they are difficult to catch and thus an expensive delicacy. The Japanese catch "shako" incidentally in gillnets set out at night for fish.

Of the several orders of Echinoderma still living in the sea, only the eggs of sea urchins are eaten in the West and these primarily around the Mediterranean. Although the eggs can be pickled, they are much tastier when fresh. In the orient and in parts of the south Pacific several species of sea cucumbers (order Holothuroida) are dried and made into soup. The Japanese eat one species, *Stichopus japonicus* (Manamako) raw. After gentle rubbing with salt during which the animal may eviscerate, the outer portion is cut into $\frac{1}{8}$ in.-thick circular slices and eaten with soy sauce, vinegar, or mustard. It is strongly acid.

Of the three subphyla in the phylum Chordata, the Vertebrata provide man with his greatest source of food from the sea. The two invertebrate subphyla, the Cephalochordata and Tunicata (Urochordata) are unknown to most laymen. However, zoologists are familiar with the cephalochordate, amphioxus, though it is generally considered a rare or inconspicuous animal. It occurs abundantly enough in some parts of the world to be economically harvestable. Light reports that 1 ton is caught daily from the months of August to April off the Island of Amoy in south China. They are eaten fresh or dried. Gibbons has eaten specimens he caught off southern California.

The much larger and more widespread subphylum Tunicata with hundreds of species furnishes surprisingly few edible forms. Walford reports that species of the family Cynthiidae are eaten in Japan, China, Siberia, and southern France. In Japan two species, *Halocynthia roretzi* (Maboya) and *H. aurantium* (Akaboya) are eaten, although the first is by far the more common. Almost all are taken by diving and during the summer are common in the seaside markets of northern Honshu. Some are pickled and sold in the south. The thick tunic is cut open and the orange-colored soft parts removed. The tunic is discarded. The soft parts are cut into small pieces and eaten raw after dipping in vinegar. The flavor is extremely acidic.

A vast number of invertebrate species are edible and are used by man for food. However, only a few major groups, namely the Crustacea, the Cephalopoda, the Pelecypoda, and the Gastropoda, already well utilized by man, hold any promise as sizable food sources. Several species have been under successful mariculture for many years (mussels in France and Spain, oysters in France and Japan). Few westerners realize, however, that all of these are essentially dietary luxuries or delicacies. In Japan only squids, among all the invertebrates eaten, are a basic element in the diet of the common man. Despite more than a century of experimentation in molluscan mariculture, the Japanese still find the harvesting (hunting) of wild stocks (squids) their most significant source of marine invertebrate protein. More tons of squid are harvested than oysters, clams, and crustaceans combined. There is a fundamental problem illustrated here. Despite extensive experience in several countries, no cultured marine invertebrate is yet more than a dietary luxury.

ROBERT BIERI

IODINE—See **ALGAE, MARINE; NUTRITIVE VALUE OF SEAFOODS; MINERALS OF THE OCEAN**

IRISH MOSS INDUSTRY

Irish moss, a seaweed harvested on the east coast of the U.S. and Canada, as well as elsewhere, is important primarily because it is the source of carrageenin, a colloidal material that is used as an emulsifier and thickener, chiefly in foods. The following discussion deals with two main aspects of the Irish moss industry: (1) the raw material used in the production of carrageenin; and (2) the carrageenin itself.

Raw Material

Irish moss, *Chondrus crispus*, is one of a class of red alga. This species is a small seaweed growing attached to rocks from just above low-water level down to a depth of about 20 ft. The plants are short, sturdy perennials, repeatedly branched near the tips, the multiple branching giving them a crisp tufted appearance somewhat resembling that of parsley. They range from 2 to 10 in. in height; those exposed to wave action in the intertidal zone tend to be the shorter and the more densely branched. Their color depends on the season, the chemical constitution of the water, and the amount of light they get. In summer, plants that grow in shallow water and are exposed to sunlight are yellowish green. In the autumn, their color changes to brown. Plants growing in deep water or in shade are a reddish purple and are frequently irridescent.

Irish moss can be found along the shores of the North Atlantic from Newfoundland to Long Island and from Norway to the coast of North Africa. Large quantities have also been found off the coast of Chile. The major supply of Irish moss for the United States' commercial industry is harvested along the shores of Massachusetts, Maine, Nova Scotia, and Prince Edward Island; the European industry obtains its raw material primarily from Norway, France, and Ireland. Secondary supplies for both the European and the American industries come from Spain, Portugal, Indonesia, and the Philippines.

Harvesting and preparation for sale. Making Irish moss available to the manufacturer of carrageenin involves removing it from the sea and then drying and baling it.

In the U.S., Irish Moss is gathered from May until about September 1. Because it is usually obtained from the sea by costly hand-gathering methods, harvesting is one of the biggest single cost factors in the price of the extracted colloidal material.

Yet harvesting by hand, though slow and difficult, is presently unavoidable. Only about one-half of the upright fronds can be removed—the holdfast must not be removed or injured. If the harvester tears away whole plants or damages the holdfast, as mechanical devices are apt to do, the supply will be depleted for many years. Harvesting, therefore, must be done with care.

One of the most common hand-gathering methods used in North American waters is that in which long-handled rakes are manipulated from a skiff or a dory. Use of a suitable rake not only enables the harvester to obtain a supply of moss but, where the substratum is hard, also preserves the beds. Such a rake will remove only the large branches, leaving the others to increase in size for a later harvest. More than one crop can be harvested a year if proper techniques are used.

Ordinarily the rake is made of tapered steel tines welded to a crossbar in such a way that their bases are close and their interspaces wedge-shaped. The angle of the rake, which has a 12- to 20-ft wooden handle, is such that the harvester can draw the tines horizontally across the ocean floor. As bunches of moss are wedged between the tines, they are torn loose from the holdfast. An experienced mosser can gather from 400 to 500 pounds, fresh weight, of moss during the 3- or 4-hr period of one low tide.

Recent efforts to develop a mechanical harvesting device have been sparked by the increasing demand for raw material. The first commercial rig for collecting Irish moss is expected to appear soon.

After Irish moss is harvested, it is washed and then dried as rapidly as possible, lest it heat up and spoil. Usually it is spread out thin and dried by means of sunlight and wind. The washed moss may be spread on portable racks to facilitate drying. In good commercial practice, it is never spread on sand or grass. In certain places along the coast, extensive areas of clean flat rock are used as a platform for drying. Otherwise, wooden platforms are made from laths nailed to two side rails.

Drying is difficult in moist, foggy weather, for moss that is partly dry can absorb much moisture; the consequence is a danger of spoilage. In dry, windy weather, one or two days is sufficient to reduce the moisture content to the 18–20% level that commercial processors seek. Moss that is sun dried to this extent will feel crisp and springy; it will not crumble when pressed in the hand.

Mechanical drying stations are being used increasingly. The wet moss is taken to a centralized drying station, where it is unloaded and run through a large flame-heated drum dryer. Drum dryers turn out a more rapidly dried, more uniform raw material than a natural drying method can. The result is a reduction in potential processing problems during extraction.

Before large-scale refining processes were developed, all harvested Irish moss was sun-bleached during the drying step. Now, bleaching is usually avoided. The unbleached material is called "black moss." If bleaching is required, the moss is spread in thin layers as for drying, but it is kept moist with salt water, turned frequently, and covered at night or when rain is falling. From one to three days is usually needed for bleaching in midsummer; a longer time is needed in autumn.

Moss destined for the processing plants where carrageenin is extracted is usually packaged in 100- to 200-lb burlap-covered bales. However, shipping space can be saved by grinding the moss and packaging it in heavy paper bags or cartons. If it is shipped in this form, the buyer may require that it be given a special washing before it is ground.

The baled, or otherwise packaged, material must be accumulated in quantities large enough to warrant its being purchased and shipped. At some point, then, an individual or agent must take responsibility for accumulation and sale. He may hire one group of people to do the gathering, paying them by the hour or by the pound, and he may then hire other groups to do the drying and baling, paying them, too, by the hour or the pound. He may purchase wet seaweed or the dried Irish moss from many free agents—rarely does one individual agent do both the gathering and the baling.

Carrageenin

Extraction and refinement. The procedure for extracting carrageenin from Irish moss varies slightly from one processor to another. The following is the one generally used. After the moss is given a preliminary cleaning by the use of cold water and mechanical devices, the carrageenin is extracted by cooking the moss in hot water and chemicals. The time, temperature, pH, and amount of agitation used vary with the product to be produced and the judgment of the processor. Usually, the carrageenin is extracted with dilute alkaline solutions at temperatures close to the boiling point and with slight agitation for from one to several hours. Because of carrageenin's high degree of hydration and viscosity, the ratio of dry plant material to extraction water is about 1 : 50; the resulting extract solution contains about 1% of soluble solids by weight.

When a batch of Irish moss has been cooked, the extract solution is clarified by the use of centrifuges, mud screens, and filters. The use of these ordinarily simple chemical engineering devices is complicated by the high viscosity of the extract solution and its low content of solids. The clarified extract may be subjected to either of two procedures: it may be concentrated, before the carrageenin is recovered, by being dried on hot rolls or drums; or it may be dehydrated with isopropyl alcohol or some other hydrophilic organic solvent that will precipitate the carrageenin. This latter procedure leaves a stringy precipitate from which residual water and solvent must be removed in conventional vacuum pans or rotary dryers. Grinding and blending complete the process. A yield of from 60 to 80% is obtained from clean, thoroughly dried Irish moss.

Because of the quantity and quality of the native seaweed hydrocolloid varies with such factors as location, season, and weather, the maintenance of a uniform end product depends upon the selection of the raw material, the control of the process, and the final blending of various production lots to specifications. Quality-control laboratory tests are essential for the production of uniform carrageenin products.

Many refinements have been added to the processing techniques, some of which are trade secrets, to enable the processor to obtain higher yield, lighter color, and better controlled viscosity and gel strength.

Characteristics and uses. The uses of carrageenin depend, of course, upon its characteristics. Accordingly, in the following paragraphs its structure and properties are described before its principal applications are discussed.

Carrageenin is generally considered to be composed of two distinct types of galactosan sulfates, kappa and lambda. Both kappa and lambda carrageenins consist of sulfated D-galactose units, which apparently have a distinct and definite structural relation with one another, and cannot occur in large separate aggregates.

Kappa carrageenin can be precipitated with potassium ions; thus it can easily be separated from lambda carrageenin, which is not precipitated by potassium.

Researchers recently have demonstrated that kappa and lambda fractions contain subfractions. For example, lambda contains D-galactose-2-sulfate and D-galactose-2, 6-disulfate; kappa contains D-galactose-4-sulfate and 3, 6-anhydro-D-galactose.

The presence or absence of anhydro groups or ester sulfate groups, or both, along with the positions occupied by the sulfate groups, and the sequential arrangement of the units, exerts a marked influence on the stereo-chemical configuration of the molecules, and results in products of widely divergent properties.

Products such as kappa carrageenin, which contain anhydrogalactose groups, have the ability to form thermally reversible gels in aqueous media. Those that are characterized by the presence of ester sulfate, as are both types of carrageenin, can be differentiated from other hydrocolloids, such as agar-agar and algin, by the amount of sulfate present. Both kappa and lambda carrageenin contain a minimum of 20% of sulfate; in contrast, agar-agar averages about 3%.

Carrageenin is used primarily in foods, but it also has other valuable uses. During the past ten years, the quantity used in the U.S. has doubled; previously it was used primarily in ice cream, chocolate milk, and toothpaste. Although these applications are still important, they have been supplemented with a wide variety of others. The following are a few of the newer applications.

Two carrageenin-based water-dessert systems have been recently developed, each capable of producing a finished, nonrefrigerated, clear-water gel. Such materials are especially useful in the preparation of such ready-to-eat products as gelled fruit juices and fruit salads. Desserts prepared with these systems can be stored at temperatures of up to 125°F without losing their excellent eating qualities or exhibiting syneresis (contraction of a gel on standing, with or without exudation of liquid).

One system is based on the use of a *Eucheuma spinosum* extract, which yields elastic gels free of syneresis. This system, which provides strong but brittle gels, is ideally suited for the preparation of dry mixes. It forms a rapidly setting gel with the ability to suspend fruits during preparation.

Another gel system recently developed is that of carrageenin and clarified locust-bean gum. This system provides a rapid gelling medium that contributes to clear products of excellent eating qualities.

Either of the above systems may be used for making various types of sweetened dessert gels or for preparing dietetic and low-calorie gels. Only since the development of high temperature-short time processing systems has it been possible to use carrageenin in finished ready-to-eat gel preparations. Conventional long-time cooking or heating of carrageenin in acid products, such as fruit juices, breaks down the carrageenin molecule and decreases its ability to form satisfactory gel structures.

The following is a brief list of some of the products in which carrageenin is used:

Bakery items

Bread doughs	fruit and creme fillings
cake batters	sugar glazes
fruit cakes	icings
yeast-raised doughnuts	meringues

Packaged desserts

milk puddings (cooked and instant)	pie fillings
	dietetic products
water gels	

Dairy products

chocolate and other flavored milks (hot and cold process)	frozen ices
	variegating syrups
	egg nog mixes
ice creams	cottage cheeses
sherberts	cheese spreads

Fountain confections

fruit toppings	milk shakes
syrups	whipped creams
flavoring emulsions	

Meat, poultry, fish products

jellied packs	antioxidant coatings
jellied coatings	sausages
batter mixes	

Condiments

relishes	mustards
spaghetti sauces	salad dressings

Miscellaneous items

beers and other alcoholic drinks	infant feeding formulas
	bulk laxatives

soups
fruit juices
soft drinks
candies
tooth pastes
tooth powders
medicinal and
  lubricating jellies

tablet binders
emulsions and
  suspensions, both
  medicinal and
  cosmetic
blood anticoagulants
culture media

Industrial applications

leather finishings
ceramics
water-based paints
wax emulsions

graphite suspensions
electroplatings
hectograph
  compositions
rubber latexes

Research is continuously being undertaken by seaweed processors, research institutes, and universities to find additional uses for carrageenin and other hydrocolloids. As new types and combinations of these products are discovered, and as their unique properties are increasingly used in food and industrial applications, development of additional convenience foods and industrial products will necessarily follow.

NORMAN W. DURRANT
F. BRUCE SANFORD

Cross references: *Agar; Phycocolloids; Seaweeds.*

J

JAPANESE FISHERIES

Organization and Economics

For 50 years Japan has been one of the world's leading fishing nations. Although records are sparse for the prewar years, Japan ranked first in 1938 and during all of the 1930's and part of the 1920's. Later, during World War II, the Japanese fishing industry suffered disaster (the catch fell to a minimum of 1.8 million tons in 1945), but after the war, Japan rapidly rebuilt its fisheries and, by 1948, had regained its former lead, which it maintained for the next 15 years (Table 1).

Japan is comprised of a group of islands extending in a north-south direction from 25° to 46°N for 1200 miles or more. In many ways, the length of coast line, the oceanographic features, and the variety of fish are similar to the eastern coast of the U.S., but the waters are much richer in food and more productive in fish and shellfish. (In 1965 the catch of fish from the east coast of the U.S. amounted to 910,000 metric tons; the fisheries along the Pacific coast of Japan (excluding the distant water fisheries) totaled 2,254,000 metric tons.)

Oceanographic conditions that provide an abundance of fish in the waters around Japan are a most valuable natural resource. The northern branch of the warm Kuroshio passes northward along the shores of Japan. The major portion flows along the Pacific coast to intercept the southerly moving Oyashio off Hokkaido and northern Honshu, while the remainder flows westward through the East China and Yellow Seas to mix with waters of the cold Liman current in the Tsushima Straits and the Japan Sea. Remnants of the Kuroshio pass from the Japan Sea through the La Perouse and Tarter Straits to disperse in the cold waters of the Okhotsk Sea. The numerous eddies formed along the boundaries between these warm and cold currents produce an abundance of food, and provide some of the best fishing waters in the world.

TABLE 1. THE WORLD CATCH OF FISH

Year	Japan	Peru	Comm. China	Soviet Union	United States	Norway	Other	Total
1938	3.7			1.5	2.3	1.1	12.4	21.0
1948	2.5	0.1		1.5	2.4	1.4	11.7	19.6
1949	2.8	0.1	0.4	1.8	2.5	1.3	11.2	20.1
1950	3.4	0.1	0.9	1.6	2.6	1.5	11.0	21.1
1951	3.9	0.1	1.3	2.0	2.4	1.8	12.1	23.6
1952	4.8	0.2	1.7	1.9	2.4	1.8	12.4	25.2
1953	4.6	0.2	1.9	2.0	2.7	1.6	12.9	25.9
1954	4.5	0.2	2.3	2.6	2.8	2.1	13.1	27.6
1955	4.9	0.2	2.5	2.5	2.8	1.8	14.2	28.9
1956	4.8	0.3	2.6	2.6	3.0	2.2	15.0	30.5
1957	5.4	0.5	3.1	2.5	2.8	1.7	15.5	31.5
1958	5.5	1.0	4.1	2.6	2.7	1.4	15.5	32.8
1959	5.9	2.2	5.0	2.8	2.9	1.6	16.0	36.4
1960	6.2	3.6	5.8	3.0	2.8	1.5	16.6	39.5
1961	6.7	5.3	5.8[a]	3.2	2.9	1.5	17.6	43.0
1962	6.9	7.0	5.8[a]	3.6	3.0	1.3	18.8	46.4
1963	6.7	6.9	5.8[a]	4.0	2.8	1.4	20.0	47.6
1964	6.4	9.1	5.8[a]	4.5	2.6	1.6	22.0	52.0
1965	6.9	7.5	5.8[a]	5.1	2.7	2.3	22.1	52.4
1966	7.1	8.8	5.8[a]	5.3	2.5	2.8	24.5	56.8

[a] Current statistics for Communist China not available, 1961–1966.

The seas off Japan are made even more productive for fish and shellfish by the numerous inlets, islands, and channels through which the waters flow and mix. The length of the Japan coast line (including the Ryuku Islands) is 5090 miles or almost equal to the 5016 miles of coast for the United States (excluding Alaska).

Contributing to the intensive development of the fisheries by Japan is the limited amount of land available for agriculture. Japan consists of four major islands and probably 10,000 or more small islands of various sizes. The total area is 142,700 square miles—a little less than that of Montana. The islands are volcanic in origin, mountainous, and of relatively poor soil. About 20% of Japan's population is concentrated in the plain areas around Tokyo, Nagoya, and Osaka-Kobe, occupying land that was formerly available for crops; as the population grows, the amount of available land for raising food slowly decreases (Table 2). In 1965 only 16% of the area of Japan was considered arable. Under such pressure, Japan has been forced to turn to the sea for food, and she depends heavily upon the catches from her vessels now fishing in every ocean of the world.

The importance of fish to the Japanese people as a predominant source of food is dramatically emphasized by the per capita consumption given in Table 3. The average Japanese eats more than six times as much fish per year as a resident of the U.S.

Expansion of the Fisheries. Japan's entry into world fisheries began with the Meiji period in 1868. Throwing off the restrictions of isolation, the Japanese government authorized construction of high seas fishing vessels, established schools to train fishermen, built hatcheries for rearing salmon, and adopted modern methods to fish and to process the catch. For all practical purposes, this "industrial revolution" lasted about 40 years.

Fishing rights of the Japanese in the Okhotsk Sea were recognized in Article 11 (2) of the Treaty of St. Petersburg (1875). These rights were strengthened by the more precise terms of the Treaty of Portsmouth (1905). In 1907 the first fishery convention was signed between Russia and Japan. This continued in effect until the beginning of World War II in 1941, and opened the way for Japan to develop its fisheries in the Japan, Okhotsk, and Bering Seas.

Other treaties were signed during this period; the most important were the Fur Seal Protective Treaty (1911) and the Whaling Treaty (1937). These were conservation-type treaties aimed at obtaining a maximum yield from the resources.

The War and Post-War Developments. World War II basically destroyed Japan's distant water fisheries. The number of powered fishing boats in 1945 dropped to less than 24% of the total operating in 1940 and to less than 40% of the tonnage; the 1945 catch was only half that for 1940. By the end of the war in 1945, Japan's fisheries had declined to the level of the 1918 catch, and was confined almost entirely to the already heavily exploited coastal waters.

During the postwar occupation, Japan's high seas fisheries were rapidly rebuilt. Japan sent a small whaling fleet into the Antarctic in 1946, resumed her fishery for tuna in the tropical western Pacific in 1950, and began to fish salmon and king crab again in the North Pacific and Bering Sea in 1952. By 1952, the catch had exceeded the prewar peak of 4.3 million tons (1936), and by 1966 had reached a total of over 7,000,000 tons—almost double its prewar high (Table 4).

In the most recent ten years (1956–1965), landings of fish and shellfish from coastal waters have increased 16% while the catches from the offshore (land-based) fisheries have increased about 50% and catches from distant waters have almost doubled. Although still small in volume, production from marine farms (oysters, seaweed, yellowtail, etc.) has also doubled during the same period of time.

TABLE 2. AMOUNT OF LAND UNDER CULTIVATION IN JAPAN

Year	Land Area (square miles)
1960	23,400
1961	23,500
1962	23,500
1963	23,400
1964	23,300
1965	23,200

TABLE 3. PER CAPITA UTILIZATION OF FISH AND MEAT PRODUCTS (KILOGRAMS/YEAR)[a]

| Year | United States | | | Japan | | |
	Fish	Total	Percent	Fish	Total	Percent
1960	4.9	93.2	5.3	27.6	30.9	89.3
1961	4.8	94.6	5.1	31.1	35.3	88.1
1962	4.8	95.6	5.0	30.5	35.9	84.9
1963	4.8	98.7	4.9	30.5	36.3	84.0
1964	4.7			27.4	34.0	80.6
1965	4.9			30.5	37.4	81.5
1966				31.1		

[a] Excludes eggs and dairy products.

TABLE 4. ANNUAL CATCH OF FISH AND SHELLFISH BY JAPAN (1000 METRIC TONS)

Year	Fishery	Marine Culture	Total	Fishery	Fresh-Water Culture	Total[a]	Total[a]
1915	1999	28	2027	[b]	4	[b]	2032
1920	2433	42	2475	[b]	5	[b]	2482
1925	2794	39	2833	[b]	8	[b]	2843
1930	3135	37	3172	[b]	13	[b]	3186
1935	3863	93	3956	[b]	19	[b]	3977
1940	3427	78	3505	[b]	19	[b]	3526
1945	1750	61	1811	[b]	12	[b]	1824
1946	2075	24	2099	[b]	6	[b]	2107
1947	2257	23	2280	[b]	4	[b]	2285
1948	2477	36	2513	[b]	4	[b]	2518
1949	2666	53	2719	37	3	40	2761
1950	3255	48	3303	63	5	68	3373
1951	3774	88	3862	66	6	66	3930
1952	4646	113	4759	53	9	62	4823
1953	4387	144	4531	57	8	65	4598
1954	4304	145	4449	82	9	91	4541
1955	4658	154	4812	82	11	93	4907
1956	4488	180	4668	90	13	103	4772
1957	5067	244	5311	81	14	95	5407
1958	5198	214	5412	78	15	93	5506
1959	5568	225	5793	75	15	90	5884
1960	5817	284	6102	74	15	90	6192
1961	6287	322	6609	81	18	100	6710
1962	6397	362	6760	84	20	104	6864
1963	6200	389	6590	84	23	118	6350
1964	5868	362	6231	89	29	118	6350
1965	6382	380	6761	113	33	146	6908

[a] Some discrepancy in totals due to rounding to the nearest 1000.
[b] Catch from fresh-water fisheries included with the marine fisheries.

Stable production of the coastal fisheries is perhaps of greatest interest; these stocks of fish and shellfish have been intensively exploited for at least 300 years, and, although some species (herring and sardines) have fallen to a very low level of yield (even disappeared) from some areas, the total catch has been maintained.

Any real increase in Japan's catch of fish comes from the more distant waters where competition with fishing fleets from other countries and higher costs are critical.

The characteristics of the fleet are changing too. Before World War II, it was mostly of wooden construction, of short life, and hazardous during storms. By 1966, 69% of the fishing boats of 50 tons or larger were built of steel, and were usually equipped with the most modern navigation equipment. From 1957–1965, the number of powered fishing vessels and the average tonnage increased by about 50%. Although the Japanese have traditionally fished with a variety of gear, about $\frac{1}{3}$ of the marine catch in 1965 was taken by trawl and a total of about 60% of the catch by four types of gear: trawl (30%), purse seine (20%), tuna long line (7%) and squid angling (6%).

The types of manufactured fishery products being produced in Japan—mainly for the domestic market—are also showing some change. Briefly, the production of frozen products is increasing while other products seem to be remaining about the same. This change is due to the general economic well being of the Japanese people at the present time and the growing consumer preference towards western foods and living. There is some exception. The fish paste products (kamabuko, chickawa, fish ham, sausage etc.) have developed rapidly within the past five years—due mainly to the use of Alaska pollock and other fish for fish minced meat suitable as a base for manufacture of these types of products.

Table 5 summarizes the trends in supply and demand for marine products in Japan from 1960 to 1965. During the six years, the population of Japan increased 6%, production of fishery products 11.4%, exports 14.0%, and imports 617.6%; thus the supply for domestic consumption has increased by 11.8%. Most significant is the sixfold increase in imports during this period compared with the very modest increase in exports.

TABLE 5. SUPPLY AND DEMAND OF MARINE PRODUCTS (1000 METRIC TONS)

Year	Total Production	Exports	Remainder	Imports	Total Supply	Animal Food	Total Supply Available for Domestic Consumption	
							Gross	Net
1960	6034	556	5478	108	5586	983	4603	2590
1961	6545	526	6019	147	6166	1196	4970	2810
1962	6689	662	6027	227	6254	1091	5163	2908
1963	6551	608	5943	460	6403	1150	5253	2946
1964	5257	758	5499	606	6105	1400	4705	2657
1965	6801	715	6086	686	6772	1429	5343	2999
1966	6954	778	6167	667	6834	1361	5473	3066

Economics. A number of moves are being made to offset the economic difficulties of the fishing industry. Most obvious is the attempt to reduce manpower by automating many of the traditional ways of fishing. However, to date automation has not resulted in any savings in size of crew—only easier work, which is some help in recruiting. A great deal of attention is being given to the development of more effective ways to fish. For example, eight or ten years ago, the gill net fleets began to use nylon nets, and their rate of catch increased by perhaps 50%. Similarly, the Japanese industry is sytematically developing methods for purse seining tuna and skipjack; they have had some success in the waters off west Africa, where tuna are concentrated close to the surface, and also in the tropical waters of the western Pacific, where they are using a new "fast setting" purse seine. Japan took 66% of the world catch of tuna from 1958–63. But Japan still finds it increasingly difficult to match the cheaper labor costs of other rapidly developing countries—such as Taiwan or the Republic of Korea.

However, all this is overshadowed by a basic cost that is difficult if not impossible to eliminate —management, transportation, crew replacement, and other costs related to operating great distances from a home port. Even now, the costs limiting distant operations are apparent. Japanese trawlers can operate economically in the Bering Sea or in the western part of the Gulf of Alaska, but find it impossible to fish economically off the coasts of Oregon, Washington, or British Columbia. This same problem of cost makes operations off Africa marginal, and governs whether tuna boats fish in the Indian Ocean or off southern Australia.

Culture. Japan is turning to culture to satisfy the domestic demand for fish. At the present time, Japanese scientists and industry are culturing many types of fresh water and marine fish and shellfish, but few have actually reached the stage of a true commercial venture. Historically, the pearl oyster has probably attracted the greatest attention. The Japanese were first to learn how to insert small seeds of mother-of-pearl into the tissue of the oyster; these seeds ultimately grow into a "cultured" pearl. Shrimp culture is also attracting the attention of many people. After about 30 years of work, a method has been developed for rearing the prized kuruma ebi (*Penaeus japonicus*) from the egg to commercial size. Rearing techniques are simple, and the farms either distribute the young shrimp (about 1 in. in length) to Fishery Cooperatives for release on the fishing grounds, or rear them to market size to sell at $4–5 a pound, heads on. The same techniques have been applied with some success to other varieties of penaeid shrimp especially the Taisho ebi (*Penaeus orientalis*) now being reared in the Republic of Korea.

For marine fish, the culture of yellowtail (*Seriola quinqueradiate*) is perhaps of greatest importance. The very young of this species are captured in May by purse seine along the east coast of Shikoku Island (Japan) and placed in large screened areas (perhaps an acre or more in size) for rearing. Fed on ground scrap fish, the young yellowtail grow rapidly, reaching a marketable size in the fall or early winter. Because the cultured fish are both fresh and of uniform size, the growers obtain a high price for them.

These are but three examples of commercial culture in Japan. There are, of course, many other kinds of marine organisms being cultured— oysters, clams, abalone, lobsters, octupus, eels, salmon, trout, and seaweed are perhaps the most common. With the high demand for fish products in Japan, the depleted stocks of fish in coastal waters, and a shrinking availability of fish in distant waters, even more effort will be placed in the future on the development of new culture techniques for rearing other kinds of fish and shellfish.

Treaties. International treaties have imposed added restrictions on Japan's high seas fisheries. In Article 9 of the Treaty of Peace with Japan, signed at San Francisco in 1951, Japan agreed to enter promptly into bilateral or multilateral agreement with the Allied Powers for regulation and conservation of fisheries on the high seas. Almost immediately thereafter, Canada, Japan, and the U.S. negotiated a Tripartite Treaty, ratified 1952, in which Japan agreed to abstain from fishing salmon (east of 175°W), North American halibut, and herring; king crab was made a subject of study and joint regulations. Although the treaty protects most North American salmon and halibut but quite adequately, there are two areas of major disagreement between the U.S. and Japan: (1) The valuable Bristol Bay red salmon migrate into the western Pacific and are taken in significant numbers each year by the Japanese mothership fishery; (2) Halibut between Canada and the U.S. are taken in some quantity by the Japanese long line vessels and trawlers operating in the Gulf of Alaska and the Bering Sea. Although the Treaty has been in effect for more than 14 years, no completely satisfactory solution has been found to these problems.

The Japan-Republic of Korea Fisheries Agreement was finally concluded after ten years of tedious negotiation, and came into force on December 18, 1965. This treaty is a conservation-type agreement, which establishes a Joint Fishery Commission with broad powers to study and regulate the fisheries, to insure safe operations for vessels operating in the conservation area, and to determine penalties for violations of the agreement. Although frequent disputes still arise, the Commission has gone far to stabilize fishing relations between the two countries.

Although not a signator to the Peace Treaty, the Soviet Union in 1956 unilaterally declared the Okhotsk Sea and a large part of the western Bering Sea and North Pacific Ocean a conservation zone, and banned high seas fishing for salmon within the area. This was the area historically developed by the Japanese under treaties beginning in 1875 and intensively exploited by a high seas salmon fisheries since about 1930. Although Japan immediately disputed the right of any nation to unilaterally close such areas to fishing, a Japanese delegation had little choice but to go to Moscow within a month to negotiate a Japan-Soviet Fisheries Convention for the Northwest Pacific Ocean. Although the treaty calls for decisions based on scientific fact, agreement has rarely been reached among the scientists, and decisions on quota and other regulatory controls are the result of annual political negotiations. Each year Japan's catch quota has been reduced and the fishing area further restricted; it is apparent that the Soviet Union, in fact, is slowly eliminating Japanese fishing from these northern waters.

Other international treaties also affect Japanese fisheries. In 1962 the Convention on the Continental Shelf became effective. Shortly thereafter, the U.S. took steps to declare king crab a creature of the continental shelf, and, through bilateral negotiations, the catch of king crab taken by Japan and by the Soviet Union in the eastern Bering Sea has been restricted. In 1968 the Soviet Union took similar steps to limit Japanese king crab fishing off Sakhalin and east Kamchatka.

Restrictions on Antarctic whaling have posed very difficult economic problems for Japan. In order to save the whale from extinction, the International Whaling Commission prohibited the killing of blue and humpback whales, and severely restricted the annual quota for other whales (Table 6). This reduction in kill has meant

TABLE 6. ANNUAL QUOTA FOR ANTARCTIC WHALES

Year	Quota (Blue Whale Units)	
	Japan's Share	Total
1953/59	6,574	15,500
1962/63	6,149	12,000
1963/64	4,600	10,000
1964/65	4,125	8,000
1965/66	2,340	4,500
1966/67	1,633	3,500
1967/68	1,493	3,200

the elimination of one fleet from Japan's Antarctic whaling operations—a mothership and killer boats not readily adaptable to other use.

An increasing number of nations are declaring a 12-mile exclusive fishing zone in order to protect their own coastal fishing industries. Although Japan does not recognize such unilateral action, negotiations have been carried out with the various countries for recognition of Japan's historic fishing rights and an agreement that will allow Japanese boats to continue to fish within the exclusive fishing zone. Negotiations were first held with the U.S. (1967), then New Zealand (1967), Mexico (1968), and Australia (1968); Spain, Mauritania, and France are among the countries with whom Japan hopes to negotiate similar agreements in the future.

The Japanese fishing industry is also feeling the effect of foreign boats competing with their own fishermen in waters traditionally fished by them and close to their home ports. Since 1963 Russian boats have intensively fished saury off northern Honshu and Hokkaido and, more recently, have begun to fish mackerel and squid. Consequently, there is growing pressure among Japanese fishermen, the industry, and the prefectures for Japan to adopt a 12-mile exclusive fishing zone—perhaps even broader, or to enter into some kind of an agreement with the Soviet Union that will protect the Japanese fishermen and industry.

A problem related to exclusive fishing zones is the "headland to headland" method of defining base lines, which determines the extent of territorial waters; most critical to Japan is the Indonesian claim encompassing the Banda, Flores, Makkasser, and other seas, and restricting Japanese tuna boats from working in these favored areas. Agreement will be reached between Japan and Indonesia, but quite different from the other treaties. The treaty will be signed by officials of the Fishery Cooperatives and the Indonesian Government and although not termed as such, each Japanese boat will pay a "license" fee to fish within the area reserved for Japan.

Some precedence for the Indonesian agreement is found in the agreements between a Hokkaido Fishery Cooperative and the Soviet Union for the harvesting of seaweed on certain of the Kurile Islands, or between the Japan-Communist China Fisheries Council and Communist China to permit fishing within certain areas off mainland China. Although it may assist or even negotiate the agreement, the Japanese government cannot sign such agreements because of serious conflict with the established government position on other fishery matters.

Japan is an active participant in several other International Fishery Commissions and Organizations—a member of the Atlantic Tuna Commission, the Marine Resource Advisory Committee (FAO), and the Southeast Asia Fishery Training Center. Although not a member, Japan voluntarily honors regulations on yellowfin tuna set each year by the Inter-American Tropical Tuna Commission.

Joint Ventures and Agreements. In order to solve the economic problems of distance and international restrictions, and to satisfy Japan's demand for fish, Japanese interests have worked out a variety of agreements and contracts with foreign companies. For example, Japanese fishing companies have established foreign bases that provide supplies for the fleet, storage and transshipping facilities for the catch, and rest for the crew. The largest base of this kind is at Las Palmos; it serves most of the Japanese boats fishing in the Atlantic.

Japanese companies annually negotiate contracts with foreign interests to purchase fish or fish products. One of the most important contracts of this type is with U.S. and Canadian canners for the purchase of salmon roe, processing them, and shipping them to Japan for the domestic market; 2500 to 3500 tons are exported to Japan each year. Contracts with the Soviet Union for the purchase of Alaska pollock and seaweed, or with the Ross Group (Great Britain) for shrimp from the Persian Gulf are other examples.

Fishery Organization. Organization of the fisheries in Japan falls into three categories. Perhaps most basic is the authority given the Japan Fishery Agency for regulation and control of all phases of fishing activity. The Agency is divided into four major divisions: Administration (planning, fishery cooperatives, insurance, promotion, etc.); Production (management of the fishery and design, construction and licensing of the fleet); Fishing Ports (planning and construction of port facilities); and Research. Somewhat unique in organization is the control that the Fishery Agency exercises through its licensing system of Japanese fishing activities all over the world.

In 1962 Japan established the Fishery Cooperative Act, which has proven especially effective in maintaining a modern and stable fishing effort. Although the Cooperatives are mainly concerned with loans and subsidies for constructing new boats and purchasing gear, marketing fish, and procuring fuel and other operating supplies, they have also proven to be effective media for testing and adopting new findings from research. There are about 4000 fishermen's cooperative associations in Japan with a total membership of over 700,000.

Japanese fishing companies are among the largest businesses in Japan, and are politically powerful. Their interests and investments are diversified in a variety of ventures both domestic and international in scope. To reduce competition for foreign markets, and to protect their basic supply of fish, groups of companies have formed a number of active industry associations of which the All-Japan Fisheries Association is most prominent; the larger industry associations are usually subdivided into a number of committees, each working on a specific phase (or problem) of a fishery. This effort is supplemented by a myriad of independent, highly specialized associations made up of perhaps six or eight companies organized for very specific purposes. (Fish Culture Feed Association, Overseas Trawl Fisheries Association, Council for Safe Tuna Fishing in the Indonesian and Philippine Waters, etc.) Finally, in order to minimize competition, several fishing companies will frequently join together to form a single joint sales company (e.g., The Canned Salmon and Crab Joint Sales Company) to handle sale of their products; such joint sales companies are usually in a very advantageous position since they basically have a monopoly of the available product and can hold sales until demand assumes the highest price.

CLINTON E. ATKINSON

Management System

The legal regulatory system for Japanese fisheries is composed of two primary parts: the Fishery Right System and the Licensing System. The former applies to coastal fishing and the latter to offshore and high seas fishing. Under the law, fishing rights in coastal fisheries are reserved for the coastal fishermen. The larger, more efficient vessels are prohibited from intruding in coastal fishing areas. Offshore fisheries are controlled by a licensing system that prevents intensified fishing operations in excess of the productivity of the fishing area. The number and tonnage of fishing vessels, operating areas, and operational frequency are subject to the approval of the Ministry of Agriculture and Forestry or prefectural governors. Number, tonnage, and operational frequency of fishing vessels operating on the high seas are brought under the direct control of the Ministry of Agriculture and Forestry.

The traditional administrative structure of fisheries regulation and control has undergone little change during the postwar political transformation of Japan.[3,7] The Fisheries Agency is responsible for centralized fisheries adminstration and management.[4] Although given a certain amount of indepence in the conduct of day-to-day affairs, the Director-General of the Agency is responsible to the Minister and must obtain ministerial approval on important questions of policy and proposed legislation. Major policy decisions, therefore, are not unaffected by political considerations. The Agency is divided into three divisions and 13 sections. In addition, each of the 46 prefectures acts as an administrative agency in implementing and enforcing the law on the local level.

The divisions of the Fisheries Agency most directly concerned with fisheries management are the Administration Division and the Production Division. Within the former, the Planning Branch is responsible for planning and coordinating fisheries policies and their administration, while the Fisheries Adjustment Branch is responsible for granting fishery rights and licenses, supervising and coordinating coastal, offshore, and inland fisheries, and adjustment of the utilization of interprefectural or other fishing grounds. Within the Production Division, the First Ocean Branch is responsible for licensing and supervising the following pelagic fisheries: whaling, salmon, and crab fishing, and mothership and other type fishery operations in the northern Pacific. It is also charged with administering international fishery agreements with respect to Japanese operations on the high seas and in foreign territorial fishing waters. The Second Ocean Branch is responsible for licensing and supervising the trawl and tuna fisheries and pearling in the Arafura Sea.

The Fisheries Management System. All types of fishing are regulated by the Fisheries Law and the Marine Resources Conservation Law.[5,9] Under these laws, regulations may be issued for purposes of aggregate planning of fisheries operations, for controlling and preventing wasteful competition between fishing operators, and for conservation and maintenance of the marine resources. Although the Fisheries Law is defined as the basic legal system concerning fisheries production, it is applicable only to the collection and cultivation of marine resources as carried out in those seas, lakes, and rivers that are publicly owned. Operations carried on in waters located on privately owned land, such as those involving goldfish, eel and carp cultivation, are not subject to the Fisheries Law, but to the law concerning land ownership. These latter, however, comprise less than 1% of all fishery operations.

From a regulatory standpoint, fishing operations are classified into three categories: (a) fishing based on fishing rights; (b) fishing for which licenses are required; and (c) fishing for which neither fishing rights nor fishing licenses are required.[12] Ordinances issued under (a) and (b) are promulgated for purposes of adjustment of fishing operations as well as for conservation of resources. In recent years, fishery right fishing has accounted for approximately 25% of total fisheries production, licensed fishing for slightly over 60%, and "free" fishing for a little less than 15%. This may be compared with the prewar proportions when fishery right fishing is estimated to have comprised approximately 65% of fisheries production. This sharp reversal reflects the rapid expansion of offshore and high seas operations relative to coastal and nearshore fishing.

Before presenting a detailed description of both the fishery right system and the permission (license) system, it may be well to observe some of their distinguishing characteristics. Any *operator* desiring to create (legalize) a fishery right is required to secure a grant from the proper prefectural government. This gives the holder of the fishery right an exclusive right to utilize a given area of water. Although, legally, the fishery right is deemed to contain the same provisions as a land right, it does not entail the right to "own" the aquatic resources within the "right" waters. Instead, limitations are particularized as to the type of fishing permitted, location and area of the fishing ground, the time of fishing, and similar matters. With regard to the permission system, each fishing *vessel* is required to be licensed by the competent authority—the Prefectural Governor or the Ministry—to operate in a specific fishery.[6] Although in each system there is a common requirement to be issued a right, or a license, to engage in fishing, the former is a right to exclusive use of a given area, whereas the latter is, technically, a release from a general prohibition to operate in a given area, whether in nearshore water or on the high seas. Unlike fishery right fishing, permission fishing entails no exclusive right; therefore, the distinction between the two is primarily in the manner in which a fishing area can be utilized rather than on any distinction between property rights. In actuality, the "real property" provisions of a fishery right are quite limited since leasing or mortgaging the right is prohibited and its transfer is strictly limited. On the other hand, given the nature of the permission system, real property aspects tend to accrue to the licensed fishing operation. Since these fisheries are regulated as to number of vessels allowed to operate, fishing range, capacity, and so forth, within any specific fishery, licensed operators own a real property interest, which can be bought and sold merely through purchase and sale of the licensed vessel.

Free fishing includes all fishing not specifically defined under either fishery right or permission fishing. These fisheries, however, are not "free" in the sense of their being nonregulated. In the free saury fisheries, for example, vessels are not permitted to operate beyond a certain number of days. Other restrictions could be imposed, since operations within these fisheries must conform to the provisions of the Marine Resources Conservation Law.

The Fishery Right System. Fishing rights are applicable to those fishing operations that, by their nature, occupy a certain coastal area exclusively.[13] These include: (1) set net fishing; (2) aquicultural operations in demarcated areas; and (3) shellfish and seaweed collecting, and other small-scale fishing carried on mainly in certain coastal areas by fishermen living within that area. A fishery right is a right to carry out a particular type of fishing operation on a particular body of water in an exclusive manner. It is, in effect, a real property right and is thus treated in accordance with the land law. However, the private ownership character of the fishery right is considerably limited since a right cannot be mortgaged, leased, or transferred by the holder. If the holder becomes ineligible to hold the right, it is cancelled by the Prefectural Governor. The law separates fishing rights into two distinct classes. One, applicable to (1) and (2), pertains to a method of fishing that requires exclusive, private use of a water area. The other, applicable to (3), represents a right to monopolistic use of a fishing area by a specific group of fishermen, and is known as a "common fishery right."

(1) The Set Net Fishery Right. The set net fishery right applies to fishing with fixed equipment and, as a rule, in which the deepest part of the main net is below 27 m from the water surface. Because set net fishing requires an exclusive right to utilize an area along the coast, set net operators must be licensed by the Prefectural Governor. Since the Law is based on effective utilization of a fishing area, the regulation of set net operations is applied mainly to maintain an aggregate balance to other operations rather than to protect existing rights. Eligibility for a set net fishery right is determined through voting by the commissioners of the Sea-Area Fisheries Adjustment Commission of each fishing district. In actuality, seniority is the most common criterion of eligibility. Due to the necessity to operate with relatively large amounts of capital and labor, the set net fisheries are larger in scale than other coastal fisheries and thus comprise a significant proportion of the coastal catch.

(2) The Demarcated Fishery Right. The demarcated right is a right to engage in aquiculture, defined as an activity involved in keeping a marine resource in a given water locale for purposes of increasing its numbers or its weight by artificial treatment. While the common and set net rights involve the catching of a marine resource, demarcated rights involve the cultivation of a marine resource. Since these operations, like set net fishing, require exclusive use of a part of the coastal area, they are licensed by the Prefectural Governor, and must be balanced against other fishing rights from the standpoint of aggregate utilization of a given fishing area. Eligibility for a demarcated fishery right differs according to the type of cultivation. For shellfish culture, a fishermen's association has first priority, but private individuals are allowed second priority. As regards pearl culture, priority is given only to private individuals, and is determined according to experience, ability, residence, and the relationship to other local fishing operations from the standpoint of aggregate utilization of the marine resources of the area.

(3) The Common Fishery Right. The common fishery right is a right to carry out, in common, particular types of fishing in a particular body of water. The common right is primarily held by a local fishermen's association. However, instead of the association itself engaging in fishing operations, the members fish independently, subject to regulations embodied in the constitution of the association. Although they do not represent a substantial proportion of the total catch, almost all small-scale operations are dependent upon this type of fishery right. The specifications of a common fishery right include the kind of fishing permitted, the defined area, and the fishing season. Unless specified, no exclusive right on an individual basis can be claimed.

The term of existence of a set net right and a demarcated right is five years; the holders of these rights must renew them to retain eligibility. The term of a common right extends for a period of ten years. In special circumstances, the Prefectural Governor may designate shorter terms for purposes of fisheries adjustment. The historical development of set net and aquicultural operations within the institutional structure of the nearshore fisheries meant that they became ingrained within the basic management regime concerned with the coastal fisheries.[2] Since they both require exclusive rights to an area of coastal waters overlapping the common rights, inevitable conflicts develop among these diverse rights.

The Licensed Fisheries Regulation System. Basically, the objectives of licensed fisheries regulation are to promote conservation of the fishery resources, discourage overinvestment of capital in fishing operations, and allocate fishing enterprises between different areas and fishery resources as a means of preventing excessive competition. The administrative method of balancing the existing catch capacity to the existing marine resources is through licensing of fishing enterprises. Licenses, or permits, for operating in certain types of fisheries are granted by either the Ministry or by the Prefectural Governor, depending upon the nature of the fishery. The respective authorities can thus issue administrative codes relating to control over the fishing area, and regulation of such matters as fishing effort and fishing seasons, in accord with the aforementioned objectives. Moreover, under the Marine Resources Conservation Law, the authorities can issue regulations almost at will on the grounds of conservation. For example, any application of explosive chemicals as a means of catching or collecting resources is absolutely prohibited. Also, to protect reproduction of the species, regulations can be issued relating to the allowable size and catch of fish, duration of fishing seasons, and establishment of protected areas.

(1) The Ministry Permission Fisheries. Those fishing operations requiring licensing by the Ministry are the "designated high sea fisheries" and others that may be prescribed for purposes of conservation of marine resources, and fisheries management or adjustment of fishing effort. The Cabinet Ordinance, issued in 1963, designating fisheries in accordance with Article 52^5 lists 17 fisheries in this category. The designated high sea fisheries are defined as large scale whaling, otter trawling operating west of 130°E longitude, medium dragnetting (over 50 gross tons) operating west of 130°E longitude, high seas salmon and crab fishing, high seas tuna and bonito operations over 100 gross tons, pearl shell operations in the Arafura Sea, and certain purse seine operations considered main fisheries.

The aggregate capacity of the designated high sea fisheries is regulated in two ways. An individual or enterprise establishing a fishing operation is required to be licensed by the authority. Eligibility is determined by capability to carry on the enterprise within the legally defined regulations, the adequacy of the fishing vessel, and sufficient financial capacity. In addition, authorization to commence fishing with each vessel must be obtained before actual fishing can begin. Under the Marine Resources Conservation Law, the Ministry can determine the maximum number of boats that can operate in a particular area and the types of fishing.

The term of a license for the designated high sea fisheries is for five years, although the Minister may limit the term for purposes of fisheries adjustment. Whereas other types of fishing are not clearly defined regarding extension or renewal of a license, for the designated high sea fisheries it is clearly specified. If there is no change in the contents of a specific permission, the authority is obligated to renew the license if the holder re-applies. This is applicable even when the claimant applies with a new boat or with a boat already having an issued license. Thus, extension of permission to operate is dependent not on the judgement of the administrative authority, but on whether the applicant previously held a license. It follows, therefore, that transferring a license is possible through sale or purchase of a licensed vessel, in spite of the fact that transferring any fishing right is prohibited by law.

In addition to the designated high seas fisheries there are nine other fishery operations that are earmarked for special regulation under Article 65^5 and thus require licensing by the Ministry. Generally speaking, the trawling and drag-net fisheries are regulated by demarcating a regional fishing area into different subareas and permitting vessels to operate only in an assigned area. On the other hand, the pelagic fisheries, e.g., tuna-bonito, are regulated by demarcating each type of fishery into different tonnage classes and issuing regulations for each class. Medium trawling is defined as an operation with a vessel more than 15 gross tons. Medium trawlers over 50 gross tons must operate west of 130°E which is a designated high sea fishery. Those between 15–50 gross tons, therefore, are permitted to operate only east of 130°E. Small trawlers are defined as those less than 15 gross tons; they come under the authority of the Prefectural Governor as to permitted fishing area. With regard to tuna-bonito fishing, the regulatory classification is not geographical but, instead, according to the capacity of the vessel, e.g., less than 40^t, $40–100^t$, and above 100^t.

Generally, ministerial ordinances and regulations tend to prohibit medium dragnet fishing in near- and offshore waters where small dragnetters are concentrated for reasons of both conservation policy and fisheries adjustment. Each of the medium dragnet fishing operations is restricted to a particular area as a condition for obtaining a license. However, the specification of each case is very complex since these operations were formerly regulated by the Prefectural Governors who issued licenses on the basis of local interests without considering the national interest. Even after a recent reclassification, about 40 different operational areas still exist throughout the country. In addition to the area restriction, medium dragnetters are regulated as to capacity in respect to aggregate tonnage and number of boats allowed to operate within each area. To meet the general objectives of the national fisheries policy, the plan has been to halt the issuance of new licenses and to encourage existing vessels to shift to other fisheries. With respect to tuna-bonito fishing, the problem in recent years has been one of deciding how to rationally expand fishing capacity due to the growth of these fisheries.

(2) The Prefectural Government Permission Fisheries. Fishing operations requiring permission from the Prefectural Governors fall within two general categories: (a) those operations defined by the Fisheries Law, e.g., small and medium trawling, and all dragnetters operating in the Inland Sea; and (b) those operations defined by the Marine Resources Conservation Law and related prefectural ordinances. The former are regulated by the Prefectural Governor with respect to the number of boats and the maximum tonnage capacity. Legally, the small dragnetters are regulated with respect to area and fishing equipment through ordinances issued by the Ministry from the standpoint of national policy. The Prefectural Governors then act in the capacity of national officers in issuing regulations under those ordinances.

When a Prefectural Governor intends to issue a code for an adjustment within the authorised area, he is required to obtain permission from the Ministry. These prefectural codes have been interpreted by the Supreme Court as applicable to any Japanese national in any fishing area—even in the Atlantic Ocean. In case of a controversy, the Ministry decides the jurisdictional boundary between and among the different prefectures. An inevitable consequence of the prefectural permission system, however, is a tendency toward administrative conflicts between related prefectures. The vague wording of Article 65[5] tends to delegate an indefinite amount of responsibility to the prefectural governors according to tradition and convenience, which is not specified in the law. Thus, there is no necessary predisposition toward uniformity in prefectural licensing procedures. Some types of fishing require licensing in some prefectures but not in others. In other cases, boats fishing off the shores of several prefectures must obtain separate licenses in each one. The solution to conflicts that threaten to get out of hand has been to leave the final authority for adjustment within the competence of the Ministry, which presumably renders a decision from the standpoint of the national interest as against any particular local interest.

The Technique and Strategy of Fisheries Management. Japan had a large coastal fishery population long before the drive toward industrialization during the latter part of the 19th century. Coming from an agricultural background, the coastal fishermen were infused with a cultural tradition in feudalistic institutions characterized by hereditary obligation and deference to higher authority. Following the revolution of 1867–1868, the new leaders of the Meiji regime decided against a reform of the feudalistic structure in the coastal fisheries based upon the institution of fishing rights. The strategy of this unparalleled move was primarily to avoid friction and social unrest in the fishing communities. Productivity and efficiency were thus sacrificed for allegiance to the new regime.

The legal control system was designed primarily to protect the income and economic welfare of the coastal fishing communities and the local fisherman's cooperative organizations. This policy significantly affected the government's attitude towards the introduction of new techniques and larger-scale operations into the traditional nearshore fisheries. Basically, more efficient fishing techniques were restricted from operating in coastal waters so as to prevent conflicts with the traditional operations. The fishery right system was an instrumental factor in preventing disputes among nearshore operators and in stabilizing the yield of fisheries dependent upon fixed gear and, more importantly, in protecting the inshore operators against incursions from offshore operators. However, the system tended to preserve small-scale operations based on family fishing, and thus impeded and discouraged the introduction and adoption of new fishing methods in the coastal fisheries.

While, under the fishery right system, an exclusive right is given to a fishing operator to exploit a certain fishing ground, under the license system, permission is granted to a fishing vessel to engage in a specific type of fishery. Licensing thus plays the major role in allocating fishing effort among the various offshore and high seas grounds, with the express purpose of limiting entry in order to stabilize conditions within each operating sector or tonnage category. The development of regulation of offshore fisheries followed a pattern similar to that of the coastal fisheries. When these grounds came under heavy exploitation, licensing was used for roughly the same purposes as for the coastal fisheries and with somewhat the same results. That is, exclusive licensing tended to prevent disputes among offshore operators, to stabilize the yield from these fisheries, and also to protect offshore operators against incursions from the larger-scale high seas operators. The expediency of adopting the licensing method to control the offshore fisheries was essentially due to the relatively high mobility of these fishing operations, and, therefore, in the ability of the authorities to reallocate them and reduce their number if necessary in order to minimize conflicts between themselves and with the nearshore fisheries.

Since locally entrenched operators had a vested interest in maintaining the status quo, the larger-scale, more efficient operators were legally prevented from exploiting the rich offshore resources by being steadily "pushed out" into the high seas grounds, e.g., the exclusion of otter trawlers and, later, medium trawlers over 50 gross tons from grounds east of 130°E longitude. Thus, the development of the high seas fisheries has been separate and distinct from the devolpment of the other domestic fisheries. There are basically two types of licensing regulation applied to the high seas fisheries. One type is applicable to large-scale fisheries requiring heavy capital investments, which can have a substantial impact on the sustainable yield of the fishery resources. Thus, control was necessary to prevent overfishing, eventual depletion, and financial disaster to the fishing enterprises. Subject to control of this type are the whaling, otter trawling, and salmon and crab operations. The method of control consists primarily of limiting the number of vessels and vessel tonnage with the objective of ensuring profitability of fishing operations. The other type applies to large-scale vessels fishing over extensive areas, where the effects of fishing effort on the fishery resources are not clearly discernible. Subject to this type of control are the tuna–skipjack fisheries, which are migratory for the most part and dependent for abundance as much on environmental changes as on fishing effort. Control is aimed chiefly at stabilizing economic conditions in the fishery, and thus the regulations generally provide for limiting aggregate operating capacity.

In the Japanese management system there is no *direct* relationship between regulation and the maintenance of an equilibrium between fishing effort and a maximum physical yield from the fishery resources. Rather, important weight is placed upon economic conditions in the various fisheries as a general guide to the imposition of restrictions.[1] However, the strategy of regulation aims not only to increase the rate of return on capital, but also to protect the livelihood of the coastal village population, reconcile conflicts between inshore and offshore fishing operations, promote conservation, and prevent excessive competition.[4,7] Because management is substantially geared toward repressing competition between different regions and class interests, the control system has become exceedingly complex and has had the effect of hindering the technological improvement of the domestic fisheries. This can be easily perceived from the fact that the tremendous increase in modern vessels and gear since the war has not been accompanied by a corresponding decrease in historical methods of fishing. This situation results from the manifold objectives of the management system. As new techniques are introduced into the fisheries, competition is raised to a new dimension and new crises develop. These are resolved through the licensing method by reallocating the more efficient vessels to new grounds. Since technological advances tend to outrun the regulations, new ordinances and regulations are constantly needed, thus complicating the whole structure of fisheries management.

The provisions of the Fisheries Law emphasize regulations concerned with coastal and nearshore fishing operations, and are relatively less specific regarding management of the licensed fisheries. This has led to serious conflicts in the regulation of certain fisheries, whereby the Ministry's control could not nullify the licensing practices and procedures of the prefectural authorities and, consequently, overexpansion of fishing operations could not be forestalled.[5,6]

In awarding licenses, the law gives priority to applicants who have fishing experience and sufficient capital to engage in fishing operations. In order to stabilize fishing effort on the traditional grounds, licenses for new vessels currently require scrapping or converting an equivalent tonnage. This requirement may be waived in cases where a vessel agrees to operate either wholly or partly in newly opened distant grounds. For example, new licenses are granted only for trawler operations in the Bering Sea and other deep sea areas on the condition that otter trawling in the East China Sea be given up.[6]

The rationale for restricting the legal transferability of licenses has been defended on grounds of securing the survival of small, inefficient operators. This general policy, however, tends to induce instability into the management system; i.e., as demand and supply shifts occur between different fisheries, there is no mechanism within the system that automatically readjusts the allocation of economic resources. This is because fishing operators are forbidden to buy and sell licenses, and, therefore, operating units, *per se*. If operators were allowed to transfer their fishing licenses, the exchange price would reflect the profitability of the particular operation and would, therefore, differ for different fisheries. Economic resources, in the form of capital and labor, would then automatically shift to the most profitable operations through the action of the market price mechanism. The government, however, does not consider this element of resource allocation in determining the number of licenses to issue. Generally, when a particular fishery is profitable, the Fisheries Agency is subjected to strong political pressure to issue more licenses and, therefore, capacity tends to be built up rapidly. The basic policy objectives of maximizing fisheries production, capital investment in the fisheries, and the aggregate income of fishermen thus inevitably leads to conflicts with other objectives that might be sought. Given the diverse objectives sought, the control system lacks a general decision standard on the proper number of licenses to issue to check excessive entry. Without an explicit criterion of management—whether biological or economic—there is no specific target, or equilibrium, toward which the control system tends to move.

If one reviewed the accomplishments of the Japanese fisheries he would have to admit that this unique management system has been eminently successful in leading Japan to preeminence among the world's leading fishing countries.[2] However, just as the old prewar management system required a renovation to take account of the postwar realities, the steadily increasing importance of the high seas fisheries will eventually necessitate a major reform of the management system to take account of contemporaneous international forces. Until recently, the Japanese fisheries have been controlled exclusively by domestic laws, except for a few fisheries regulated by international agreements. This has made it easier to establish and effectively enforce regulations more in tune with traditional methods of operation. In the past, Japan considered the high seas a "safety valve," which was endlessly capable of absorbing the excess capacity generated by a combination of economic growth, technological progress, and the basic objectives of the management system. In the postwar world, however, Japanese high seas fisheries expansion has been increasingly limited due to the existence of international fisheries conventions and the trend toward expanded territorial fishing limits by coastal nations. In view of the potential international conflicts, the desirability of having a centralized policy in the management of the Japanese fisheries and one that stresses "rational" control through licensing will undoubtedly gain serious consideration.

SALVATORE COMITINI

References

1. Comitini, S., "Economic and legal aspects of Japanese fisheries regulation and control," *Washington Law Review,* North Pacific Fisheries Symposium Issue, 1967.
2. Comitini, S., "Marine resources exploitation and management in the economic development of Japan," *Economic Development and Cultural Change,* 1966.
3. Croker, R. S., "*Japanese Fisheries Administration,*" GHQ, SCAP, 1951, Natural Resources Section, Tokyo.
4. "Fisheries Agency Establishment Law" (Law No. 78 of 1948).
5. "Fisheries Law" (Law No. 267 of 1949).
6. "Fishing Vessel Law" (Law No. 178 of 1950).
7. Hasegawa, Y. "Public administration in fisheries," *Japanese Fisheries,* Asia Kyokai, Tokyo, 1960.
8. Kasahara, Hiroshi. "Japanese Fisheries and Fishery Regulation," *California and the World Ocean,* California Museum of Science and Industry, Los Angeles, 1964.
9. "Marine Resources Conservation Law" (Law No. 313 of 1951).
10. Oka, N., Watanabe, H., and Hasegawa, A., "The Economic Effects of the Regulation of the Trawl Fisheries of Japan," *Economic Effects of Fishery Regulation,* R. Hamlisch Ed., FAO, Rome, 1962.
11. *Organization of Fisheries Administration in Selected Countries,* Economics Branch, Fisheries Division, FAO, Rome, January, 1964.
12. "Present Condition of the Basic Regime of Fisheries" (translated), 1958, Fisheries Agency, MAF, Tokyo.
13. *Some Aspects of the Fishery Right System in Selected Japanese Fishing Communities,* GHQ, SCAP, 1948, Civil Information and Education Section, Tokyo.

Cross references: *Tuna Fisheries.*

KELP—See ALGINATES FROM KELP

KING CRAB—See CRAB INDUSTRY

L

LABRADOR FISHERIES

With the exception of waters over the deep ocean, the continental shelf off Labrador may be regarded as the last frontier of commercial fisheries exploitation in the northwest Atlantic, at least in a geographical sense. This is not to say that fishing is either very recent or very light. A shore-oriented fishery for cod (*Gadus morhua*) has existed for several hundred years. Offshore fishing for cod and other species began only in the 1950's, but has quickly attained major proportions. Nevertheless, the fishing banks off the northern half of the coast are only roughly charted and seasonal distribution and abundance of most species are incompletely known.

The Labrador coast presents a barren, rugged, and much indented coastline to the north Atlantic, extending from 52 to about 60°N (Fig. 1). For most of its length, the coastline is dotted by hundreds of small islands. The 200-fathom contour varies from about 60 to 120 miles offshore,

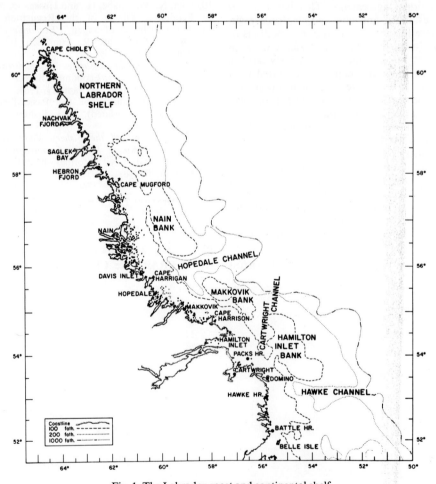

Fig. 1. The Labrador coast and continental shelf.

but the continental shelf is interrupted by several deep channels, forming a number of more or less well-defined fishing banks (Fig. 1). The shelf descends more quickly beyond 200 fathoms and the 1000-fathom contour is less than 150 miles from the coast.

The dominant hydrographic feature is the cold, southward-flowing Labrador Current. The shoreward portion of this current consists of very cold water flowing southward from the east coast of Baffin Island, joined by similarly cold water flowing southward from Hudson Strait. This inner portion of the Labrador Current bathes the whole coastline and extends over the continental shelf to depths of at least 100 fathoms. Early spring temperatures are usually less than 0°C in depths to 100 fathoms over the shelf, and often are less than $-1°C$. The surface is ice-covered from January–February to May. Off Baffin Island and northern Labrador, this cold current is joined to seaward by warmer water of west Greenland origin, which flows southward along the deeper parts of the continental shelf. Even here, temperatures seldom reach 4.5°C in depths to 400 fathoms. Solar warming in summer produces a shallow surface layer over the shelf with temperatures above 0°C extending to 30 fathoms, and surface temperatures up to about 10°C. The southward flow is greatest in spring and early summer, when the surface velocity approximates 12 miles per day.

The commercial offshore fisheries are almost entirely for cod and redfish (*Sebastes* sp.). Cod are especially important, accounting for about 90% of the landings in 1965–1966. Other commercially valuable species are taken incidentally in relatively small amounts, and include American plaice (*Hippoglossoides platessoides*), witch flounder (*Glyptocephalus cynoglossus*), Atlantic halibut (*Hippoglossus hippoglossus*) Greenland halibut (*Reinhardtius hippoglossoides*) and wolffishes (*Anarhichas lupus* and *A. minor*) The inshore fishery is almost entirely for cod, with smaller fisheries for Atlantic salmon (*Salmo salar*) and Arctic char (*Salvelinus alpinus*).

Cod

With the possible exception of Ungava Bay, cod do not occur in commercial quantities north of the Labrador coast. The northern limit in North America is at latitude 63°N, though they extend to 73°N off west Greenland. A relict population is present in Ogac Lake, Frobisher Bay. Trans-Labrador Sea and trans-Atlantic movements of cod are rare, though a few are recorded. Cod are distributed in greatest quantity between the surface and about 200 fathoms, depending on area and season, though the adults are usually demersal.

Northwest Atlantic cod are separated into a number of more or less well-defined stocks, i.e., groups distributed within a restricted geographical range. One of the largest and least well-defined of these is the so-called "Labrador-Newfoundland" stock, occupying a vast area from northern Labrador to eastern Newfoundland. Tagging and growth studies have shown that fish tend to separate out into smaller units within this large area, but the boundaries of such units are ill-defined.

Cod from Labrador, as elsewhere, undertake pronounced seasonal migrations in response to reproductive, feeding, and hydrographic influences. Spawning occurs in April–May in relatively warm (2.5–3.1°C) and deep (125–180 fathoms) water on the continental shelf. Spawning concentrations are difficult to locate, as much of the area is ice-covered at this time, but have been reported off Cape Chidley and on the southeastern edge of Hamilton Inlet Bank. Onshore movement begins in late May to early June, apparently a migration in the surface layers following the onshore spawning migration of capelin (*Mallotus villosus*). Most fish undertake this migration, though a few remain on the shallower parts of the offshore banks. Time of appearance inshore is progressively later from south to north, extending from mid-June to early August. On arrival the cod are distributed in the surface layers and may sometimes be seen in pursuit of capelin at the very surface. Movement is progressively deeper over the summer, culminating in eventual return to the deep offshore grounds in winter.

Egg and larval distribution, following spawning, is poorly known. There must be a significant transfer of eggs and larvae to more southern areas, but evidence exists to indicate that this may be offset by a return northward migration of small fish. It is possible that recruitment may also occur through drift of larvae from west Greenland spawning areas.

The food spectrum is extremely wide, and changes from smaller to larger animals as the cod increase in size. Twenty-two species of fish and over 70 invertebrate species have been listed from stomachs collected offshore in summer. Growth is relatively slow compared to that of cod from other areas. Maturity occurs mainly over ages 5–6 in males and 6–7 in females. Most males are mature at 50 cm in length; most females at 55 cm. Egg production varies from a few hundred thousand to over 10,000,000 per mature female, depending on size. Moderate annual variation occurs in survival of young, though an occasional year-class may experience extremely poor or extremely good survival.

Redfish

Two species of redfish occur in the Labrador fisheries; *Sebastes mentella* and *Sebastes marinus*. These are not separated by the fishing industry.

S. marinus grows faster and attains a larger size, but *S. mentella* accounts for most of the commercial catch. *S. marinus* is distributed in shallower water than *S. mentella*, and differs most obviously in the following characters: blunt chin, smaller eye, and more orange-red than rose-red color. The following discussion applies particularly to *S. mentella*.

Redfish, preferring deep and fairly warm water, occur off Labrador in commercial size and quantity in 150–250 fathoms in temperatures of 2.5 to 4°C. They occur in smaller quantities as deep as 400 fathoms. They are bottom-oriented, but apparently feed some distance off bottom, especially at night. While redfish occur all along the Labrador continental shelf, commercial quantities are not present north of Hamilton Inlet Bank. The presence of large numbers of larvae over very deep water in the Labrador Sea indicates that a large pelagic redfish population may be present in this area.

Redfish usually reach the surface in a condition unsuitable for tagging and subsequent release; thus it is difficult to obtain direct evidence for their movements. It is believed, however, that redfish are relatively sedentary, and do not undertake extensive migrations. They may often be segrated by sex and size in closely adjacent areas and depths. They do move deeper in winter and may sometimes be taken in quantity at 300–350 fathoms at that time.

Fertilization is internal in redfish, and the females bear live young. In Labrador, larval extrusion occurs between March and July, each female releasing from 25,000 to 100,000 larvae about ¼ in. in length. These inhabit the surface layers for a time, but gradually descend as they increase in size. There are apparently no special spawning areas. Considerable variation in annual survival of larvae has been found in some areas, but little is known about larval survival in Labrador. Redfish grow extremely slowly; few reach a length greater than 50 cm, and many of the fish in the commercial catch are older than age 20. Maturity does not occur in females until about age 10.

The combination of slow growth, late maturity, and localized distribution makes redfish particularly susceptible to overfishing. Little is known of the population size and age structure and mortality of Labrador redfish. There is nothing to suggest that these are overexploited at the moment, but neither are estimates of maximum sustained yield yet available.

International Fisheries

After World War II, the development of fishing fleets by many European countries gave rise for the first time to an offshore Labrador fishery. Almost all these countries belong to the International Commission for the Northwest Atlantic Fisheries (ICNAF), and have reported statistics of catch and fishing effort in the area since 1949. The Commission meets annually to review the status of the fisheries and fish stocks, and is responsible for regulatory measures for purposes of conservation. The countries now adhering to the Commission are Canada, Denmark, France, Federal Republic of Germany, Iceland, Italy, Norway, Poland, Portugal, Romania, Spain, U.K., U.S.A. and U.S.S.R. Of these, all except Denmark, Italy, Romania and U.S.A. fished off Labrador in 1966. Virtually all offshore fishing is done by bottom-fishing otter trawls. The Canadian fishery operates inshore only, using a variety of fixed or hand-operated gears.

The international fishery off Labrador began in the early 1950's with the appearance of French and Portuguese vessels fishing for cod. By 1960 these countries had expanded their operations in the area, and were joined by ships from Spain, West Germany and U.S.S.R. (Fig. 2). Landings increased almost tenfold between 1957 and 1960, declined from a peak of 297,000 tons in 1961 to 223,000 tons in 1963, but with the addition of fleets from Poland and other countries increased beyond 300,000 tons in 1965 and 1966.

Cod and redfish account for virtually all the landings, with cod by far the more important

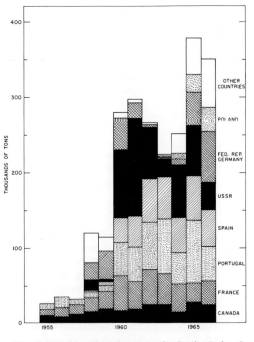

Fig. 2. Total landings by countries in the Labrador fisheries, 1955–1966. Figures are in thousands of metric tons round fresh weight (1 metric ton = 1,000 kg).

Bank, the spring fishery being based largely on spawning and postspawning concentrations on the southeastern edge of the bank. By June, these concentrations break up and most of the fish move inshore where they are fished in July and August. The autumn offshore fishery builds up as fish move away from shore toward the spawning area, being interrupted in winter because of bad weather and ice. Areas to the north of Hamilton Inlet Bank produced good cod catches for the first time in 1966, but little is known of seasonal distribution offshore in these areas.

The traditional northwest Atlantic cod fishing nations (France, Portugal, and Spain) salt most of their catch at sea. Countries beginning fishing more recently process and freeze the catch on larger vessels. These sometimes may be factory vessels only, being supplied by smaller trawlers fishing nearby. A total of 806 European otter trawlers totalling over 800,000 gross tons fished the northwest Atlantic in 1965. Vessel sizes ranged from less than 500 to about 5000 gross tons. Vessels fishing off Labrador would typically be larger than 500 gross tons. The offshore cod fishery is particularly suited to large fleets of large trawlers, capable of remaining at sea several months at a time, and continually scouting the area while maintaining radio contact. The newer fleets fit this pattern.

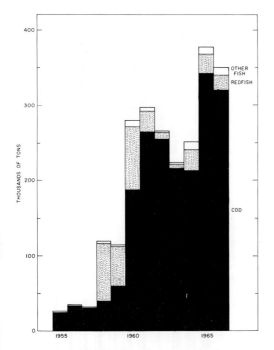

Fig. 3. Landings of principal species from the Labrador fisheries, 1955–1966. Units as in Fig. 2.

(Fig. 3). Redfish landings have actually decreased from a high of 83,000 tons in 1960 to 25,000 tons in 1965. Cod landings declined from 265,000 tons in 1961 to 213,000 tons in 1964, but exceeded 300,000 tons in 1965 and 1966.

Significant quantities of redfish are taken only by Germany, Poland, and U.S.S.R. These nations operate large factory trawlers, which process and freeze the catch at sea. The redfish fishery began (1959–1961) essentially as a winter–spring fishery, most of the catch being taken in deep water east of the surface ice cover during the months of January to June. The reduced catches of 1964–1966 were more evenly spread out through the year, but were mainly in the March–October period. Also, the countries that account for most of the redfish have recently taken greater catches of cod than redfish from Labrador. It is apparent that the countries fishing for redfish divert considerable effort to cod fishing when redfish yields are unsatisfactory.

All countries fishing off Labrador fish most heavily for cod. The offshore cod fishery originally developed (1954–1958) as a relatively small autumn fishery. Beginning in 1959, and with the appearance of a large Russian fleet in 1960, fishing was extended throughout the year. The autumn fishery increased, but the new spring (April–June) fishery accounted for most of the increased landings from 1960 onward. Most of the fish are taken from the area of Hamilton Inlet

Canadian Inshore Fisheries

The inshore Atlantic salmon fishery is relatively small in comparison with the cod fishery, but significant in the Canadian commercial salmon fishery as a whole. The catch from Labrador in 1965 and 1966 exceeded 300 metric tons each year; about 15% of the whole Canadian catch. Fishing is done with surface gillnets, usually on a part-time basis by fishermen fishing primarily for cod. The salmon are usually held in ice and collected frequently for shipment southward.

In northern Labrador a gillnet fishery for Arctic char is carried out by Eskimo fishermen in July and August. Catches in 1965 and 1966 averaged 110 tons. The catch is salted in barrels and sold at the end of the season.

The inshore cod fishery takes place mainly in July and August. This fishery had its beginnings at least as early as the early 1700's when French vessels fished on the coast. These were replaced by English vessels during the late 1700's and the latter by vessels based in Newfoundland in the early 1800's. By this time a shore-based fishery was also established. The fishery grew rapidly until the early 1900's, and three categories of fishermen were recognized: livyers, resident fishermen; stationers, those who came from Newfoundland each summer and operated from premises on shore; and floaters, Newfoundland fishermen who fished along the coast from schooners. The catch is salted, and disposed of at the end of the season.

Fishing is best around the headlands and islands, and many small, isolated communities have been established in these barren and rugged areas. Most of these are deserted in winter when the livyers move into more sheltered and permanent settlements and the stationers return southward to the island of Newfoundland. The fishermen operate from small open boats, the schooners being used only for storage of the catch and accomodation. Four main gears are used, the most important being the cod trap, a rectangular box of netting open at the surface, closed at the bottom, and attached to the shore by a "leader" of netting directed toward the trap opening. Second in importance are nylon gillnets (a recent innovation), followed by jiggers (lead weights in fish form containing large hooks) and longlines (locally called linetrawls or simply trawls). Traps are set in shallow water (8–15 fathoms) early in the season with the other gears being used later as the fish move progressively deeper over the summer.

No catch statistics are available for the early period, but on the basis of numbers of ships engaged in the floater fishery, catches must have been substantial, even by recent standards. Thus in 1908 about 1400 schooners fished along the shore. Catch per ship in the 1940's was about 200 tons annually. On this basis the catch from the floater fishery alone may have approached 300,000 tons annually.

For economic reasons the inshore fisheries experienced a rapid decline to the mid 1950's. Many of the summer communities were abandoned, and by 1955 the floater fishery had completely disappeared. There has since been a relatively minor resurgence. Livyers and stationers (hereafter combined as stationers for convenience) increased from about 350 to 1500 men from 1956 to 1965, while floaters reached 500 men in 72 ships by 1966 (Fig. 4). In fact, the number of men fishing has levelled off from 1963–1966 at about 2000. In the past, fishing was carried on along the entire coast, but with reduced floater fishing and the resettlement southward of some Eskimo communities, the northern third of the coast is no longer fished. The most northerly fishing community at the moment is the Eskimo settlement of Nain.

The recent increase in inshore fishing occured over the same period as the rapid development of fisheries offshore. Both the inshore and offshore cod fisheries operate on the same stock of fish. However, the offshore fishery now takes about 15 times the amount caught inshore, as opposed to nearly equal amounts in both regions in 1957–1958.

This greatly increased competition appears to have adversely affected inshore production. While

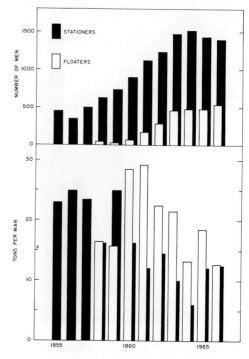

Fig. 4. Effort and landings per unit effort in the inshore cod fishery, 1955–1966.

the number of men fishing inshore has quadrupled since 1957, annual catch per man has declined to little more than half the level at that time (Fig. 4). This has occured in spite of the addition of a new and productive gear inshore (nylon gillnets). The inshore fishery is at a competitive disadvantage because of the short season and use of fixed gears, and if offshore fishing is further increased, or even maintained at present levels, further inshore, expansion is unlikely to produce proportional increases in catch.

A. W. MAY

References

1. Black, W. A., "The Labrador floater codfishery," *Annals Assoc. Amer. Geographers*, **50**(3), 267–293 (1960).
2. May, A. W., "Effect of offshore fishing on the inshore Labrador cod fishery," *Int. Comm. NW Atl. Fish., Res. Bull.*, No. 4 (1967).
3. Templeman, W., "Redfish distribution in the north Atlantic," *Bull. Fish. Res. Bd. Canada*, No. 120, 173 pp. (1959).
4. Templeman, W., "Marine resources of Newfoundland," *Bull. Fish. Res. Bd. Canada*, No. 154, 170 pp. (1966).

Cross references: *Cod Fishery; Newfoundland Fisheries; Redfish.*

LAMPREY—*See* **SEA LAMPREY**

LAW OF OCEAN SPACE

Nature and Sources

The law of ocean space is composed of national and international law. The *customs* of seafaring nations and mariners are the source of much of the traditional national and international law of the sea. Custom, in the legal sense, is habitual practice adhered to from a feeling that violation of the custom is legally wrong.

International agreements relating to ocean space are a source of law for the parties to such agreements, which supersede, supplement, or derogate from the otherwise applicable law. International agreements may also be evidence of customary law when they are made for the purpose of declaring the parties' understanding (codification) of the law, or they may become the source of customary law.

A United Nations conference on the Law of the Sea at Geneva in 1958 produced four international agreements, which are partly codification of customary international law and partly new international legislation:

(1) The Convention on the Territorial Sea and the Contiguous Zone
(2) The Convention on the Continental Shelf
(3) The Convention on the High Seas
(4) The Convention on Fishing and Conservation of the Living Resources of the High Seas.

These four conventions, and their *travaux preparatoire* are a most important source of law because even nations that do not become parties to them follow many of their most important general provisions. Other sources of the law of ocean space include *national legislation*, decisions of national and international *courts*, the *writings of jurists*, and *general principles* of law common to the major legal systems.

Ocean Space Zones, Boundaries, and Jurisdiction

Two basic principles of the law of ocean space are freedom of the high seas and national jurisdiction in a coastal belt. These principles give rise to problems regarding the nature and seaward extent of the coastal state's jurisdiction. The law of ocean space has evolved, and is still evolving, a pragmatic solution to these problems. Ocean space is divided into zones in which the coastal state's authority becomes less absolute seaward until it becomes merged into the common right of all to the freedom of the high seas.

The zones of world ocean space are: (1) internal waters; (2) territorial sea; (3) contiguous zones of high seas; (4) high seas; (5) continental shelf; and (6) deep sea bottom.

Internal Waters and the Territorial Sea. The territory of a state extends beyond its land territory to certain ocean areas that form part of its domain as either internal waters or territorial sea. Law relating to these zones has been evolving for the past three centuries. Many controversial points were finally resolved in the Convention on the Territorial Sea and the Contiguous Zone.

There are certain areas, such as harbors, bays, estuaries, and canals, over which the coastal state exercises sovereignty in every respect identical with its rights over inland lakes and rivers. Such areas are called *internal waters*, as distinct from the *territorial sea*.

Article 1 of the Convention on the Territorial Sea and the Contiguous Zone says that "The sovereignty of a State extends, beyond its land territory and its internal waters, to a belt of sea adjacent to its coast, described as the territorial sea." Article 2 says that "The sovereignty of a coastal State extends to the air space over the territorial sea as well as to its bed and subsoil."

Under customary international law, the sovereignty of a coastal state extends to the bed and subsoil of its internal waters. Thus a coastal state exercises in its internal waters and territorial sea the same legislative, executive and judicial authority as in its land territory. Hence the coastal state can exclude aliens and foreign ships except for the right of foreign ships to innocent passage in the territorial sea.

The line between internal waters and the territorial sea is the low-water line following the sinuosities of the coast, except for bays and coasts fringed with islands. This line is called the normal baseline. Rules are formulated for the drawing of straight baselines in the case of bays and coasts fringed with islands (see below).

The width of the territorial sea is one of the most controversial questions in contemporary international law. The rule of three nautical miles, which for more than a century and a half has been followed by the great sea powers, has never been universally accepted. In view of the large number of states now claiming a territorial sea of from four to twelve nautical miles, and in view of the position of the overwhelming majority of the 73 coastal states represented at the United Nations Conferences on the Law of the Sea in 1958 and 1960 that no state is entitled to extend its territorial sea beyond twelve nautical miles, it would be difficult to defend the proposition that a state violates international law by extending its territorial sea to a width no greater than twelve nautical miles, at least as against other states claiming a similar width for themselves.

High Seas and Contiguous Zones. The high seas are those water areas that do not fall within internal waters or territorial sea. The principle of freedom of the high seas means that the high seas are in law common to all (*res communis*), they are not capable of being subjected to sovereignty, and every state and its nationals has equal rights and

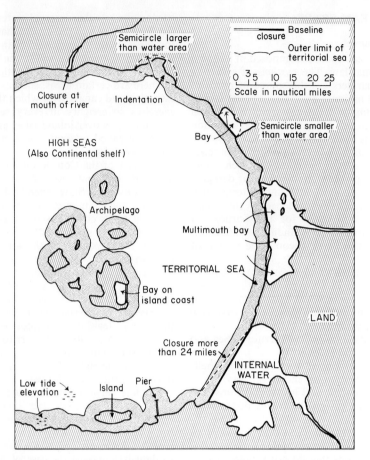

Fig. 1. The baseline from which the territorial sea is measured. (From U.S. Dept. of State, "Sovereignty of the sea," Geographic Bulletin No. 3, April 1965.)

duties in connection with them. Freedom of the high seas in regard to use means that each state has an equal right to use and exploit the high seas, subject to the rights of use of others. Freedom of the high seas in regard to jurisdiction does not mean the high seas are subject to no jurisdiction. Every state has an equal right to exercise jurisdiction on the high seas, but, except for piracy, this right in general extends only to a state's own ships and nationals.

Contiguous zones are zones of high seas contiguous to the territorial sea, within which a coastal state may exercise over others' uses a limited control for such special purposes as security and to prevent and punish violation of its customs, fiscal, immigration, sanitary and fishing laws. The exercise of control in contiguous zones of the high seas is necessary in view of the inadequacy under modern conditions of any reasonable width of territorial sea. Whatever the width of territorial sea accepted under international law, there are occasions and purposes for which control must be exercised farther out from shore. This differs from an attempt to declare such areas to be territorial sea subject to the sovereign jurisdiction of the coastal state.

The Convention on the Territorial Sea and the Contiguous zone permits zones 12 nautical miles wide from the baseline of the territorial sea for customs, fiscal, immigration, and sanitary purposes. A similar 12-mile exclusive fishing zone has received general recognition in customary international law in recent years.

Despite disagreements as to details, military security contiguous zones are also accorded general recognition in customary international law.

Coastal states' exercise of control for special purposes in contiguous zones has sometimes been criticized as an exercise of jurisdiction violative of freedom of the high seas and hence unlawful. The sounder view is that since freedom of the high seas in regard to use can be abused at the expense of the coastal state at some geographical point in the *res communis* before reaching the territorial sea where the coastal state may assert superiority, contiguous zones are a necessary limitation guarding coastal states against abuse of the right to put the *res communis* to a particular use.

Continental Shelf and Deep Sea Bottom. For years international jurists have discussed whether the seabed and subsoil of high seas is *res communis*, like the water above it, or whether it is *res nullius* (territory of no one) over which rights can be acquired. In fact, state practice has long recognized that rights to land beneath high seas may be acquired by occupation and contiguity provided that freedom of navigation is not thereby impaired. For many years, in some cases for centuries, claims have been asserted to the exclusive exploitation of such sedentary fisheries as oysters, pearls, chanks, and sponges on the bed of high seas off Ceylon, India, Bahrein, Ireland, Tunis, Australia, and elsewhere. These claims appear to have become established by acquiescence and to be acknowledged by other states.

The development of technological capability for intensive exploitation of natural resources of geological continental shelves began shortly after World War II. The geological continental shelf is the submerged land mass that declines moderately from most continental coasts before descending steeply into the deep ocean. In some places it is a gently sloping plain; in others it is moderately mountainous, comparatively deep close to shore and shallow beyond. Off some coasts there is practically no geological shelf; off others it may extend seaward over 200 miles. By 1958 some 30 states had claimed exclusive rights to exploitation of the seabed and subsoil under the high seas off their coasts. Such claims, provided they do not interfere with navigation and fishing, were well on the way to becoming accepted in customary international law when they received qualified recognition in the Convention on the Continental Shelf. The rationale of the freedom of the high seas is that they are an international highway securing freedom of communications and commerce between distant lands; there is no reason for extending this concept to the seabed or subsoil.

The *continental shelf*, as a legal zone, is defined in Article 1 of the Convention as "the seabed and subsoil of the submarine areas adjacent to the coast" of mainland and islands "but outside the area of the territorial sea, to a depth of 200 m or, beyond that limit, to where the depth of the superjacent waters admits of the exploitation of the natural resources of the said areas."

The foregoing definition is not precise as to the outer limit of the legal continental shelf. However, the basis in legal theory for recognizing that the coastal state has exclusive rights in the continental shelf zone is contiguity and propinquity, not effective or theoretical occupation or proclamation, just as in the case of internal waters and the territorial sea. The legislative history of the Convention indicates that, regardless of the presence or absence of a geological shelf, the outer limit of the shelf zone extends to at least the 200-m isobath regardless of its distance from the coast; beyond that point it may extend to whatever distance and depth the coastal states' exclusive rights will be acquiesced in by other states. The still developing practice of states suggests that each coastal state will extend its shelf jurisdiction seaward for specific existent or immediately potential uses, and that conflicts of interest that develop will be solved by the drawing of a political outer limit boundary only when there is no other preferred solution.

Another problem not wholly settled by the Convention on the Continental Shelf is the content of the coastal states' rights. Article 2 of the Convention recognizes that "The coastal state exercises over the continental shelf sovereign rights for the purpose of exploring it and exploiting its natural resources" and that "if the coastal State does not explore the continental shelf or exploit its natural resources, no one may undertake these activities, or make a claim to the continental shelf, without the express consent of the coastal State." The intent of this Article is that property rights in natural resources are vested exclusively in the coastal state, that ownership of them cannot pass to others except by express grant, and that the coastal state has all rights reasonably necessary for exploration and exploitation. It seems to create a new and unrealistic legal concept, viz., that a state can have sovereign rights for certain purposes in a certain area without having sovereignty over that area.

However, the Convention does not prohibit broader claims to rights over the shelf unrelated to natural resources. Before as well as since the 1958 Geneva Conference, the practice of states seems to be moving in the direction of a new rule of customary law: that the coastal state may exercise over its adjacent sea bottom plenary jurisdiction for all purposes tantamount to sovereignty.

The *deep sea bottom* may be defined as the seabed and subsoil seaward of the continental shelf zone. The legal regime of this area is practically untouched by present written law and therefore lies almost wholly in general principles of customary law and their progressive development. Pragmatic inherent political realities have made contiguity and propinquity the accepted bases for sovereignty and similar rights in the zones of internal waters, territorial sea, and continental shelves. By the same reasoning, contiguity and propinquity are not apposite for acquisition of rights in the deep sea bottom. It must be anticipated that state practice will regard the deep sea bottom as *res nullius* in which rights can be acquired by occupation. It must also be anticipated that no mere paper claim of occupation will be respected by others who will have perfected the nascent technology of human requirements and capabilities under water.

States and individuals may acquire rights in the deep sea bottom by effective occupation, which means only the minimum activity of actual occupation that the nature of the deep sea bottom calls for.

Contiguity and propinquity as the basis for allocating rights in the continental shelf have the virtue of preserving shelf resources for the benefit of a technological "have-not" coastal state until it can exploit them for itself or by lease to others. Effective occupation as the basis for allocating rights in the deep sea bottom will mean a race between the technological "have" states with the technological "have-nots" as spectators. The undesirability of such a race is self-evident.

Baselines. The division of ocean space into zones makes necessary some definite base reference from which zonal boundaries are delimited. In general, this function is served by the coastline itself, i.e., the general area of land-water contact. However, the configuration of coastlines may vary from a smoothly curving shoreline to a highly complex interface of land and water made up of tortuous embayments and myriads of islands, reefs, shoals, and rocks awash. Under these circumstances, the coastline as commonly understood is not susceptible of the precise definition that is essential for a base reference for legal boundaries. Therefore it is necessary to choose an arbitrary base, which is known as the *baseline*. Moreover, the single term baseline is too inexact for precise analysis of particular situations. It must be determined from the context whether the reference is to a normal baseline, a straight baseline, or a system of straight baselines.

The *normal baseline* is the low water line as marked on official charts of the coastal state. It is not necessarily a continuous line along a coast, because for many purposes it may be sufficient to identify only selected points on the low water line, e.g., a headland or offshore rock. The normal baseline is drawn as a continuous line only where it is necessary for a particular purpose to draw a boundary between internal waters and the territorial sea.

A *straight baseline* (or *closing line*) is a straight line drawn between salient normal baseline points. It is not a normal baseline that is straight, but a concept for simulating the coastline seaward of the normal baseline.

A *system of straight baselines* is a series of related straight baselines along a land-water interface of myriad indentations or fringing islands in its immediate vicinity. Under the Convention on the Territorial Sea and the Contiguous Zone, the normal baseline must be used except where particular coastal configurations justify the use of straight baselines. Where a coastline is broken, as by a bay, mouth of a river, or other indentations, it becomes necessary to construct a straight baseline across the opening by arbitrary means. Solutions may be extremely simple, or may involve intricate computations, depending upon the complexity of the shoreline. Shoreline segments with indentations may fall into types, but in no instances are two ever alike. In all cases a closing line must be drawn across the seaward opening of the indentation to simulate the coast. Otherwise the baseline would extend into land bodies along the banks of rivers and around the shores of bays and inlets.

Rivers seldom present problems of any magnitude. A line drawn directly across the mouth of a river where it empties into the sea usually provides a suitable baseline across a break in the shoreline. Where a river flows first into an estuary or other embayment before emptying into the sea, it is treated as a bay.

Along any irregular coastline, indentations other than the mouths of rivers are either "bays" or "curvatures of the coast." (See Fig. 1.) The normal baseline is used in the latter case, but in the former case closing lines are drawn. A definite technique exists by which a coastal indentation may be identified as a bay. Article 7 of the Convention on the Territorial Sea and the Contiguous Zone stipulates the requirements by which a bay is determined:

". . . a bay is a well-marked indentation whose penetration is in such proportion to the width of its mouth as to contain landlocked waters and constitute more than a mere curvature of the coast. An indentation shall not, however, be regarded as a bay unless its area is as large as, or larger than, that of the semi-circle whose diameter is a line drawn across the mouth of the indentation."

Regardless of its configuration, however, the mouth of any bay may not exceed 24 miles in width. As set forth in the same Convention, where the natural entrances of a bay are more than this distance, ". . . a straight baseline of twenty-four miles shall be drawn within the bay in such a manner as to enclose the maximum area of water that is possible with a line of that length."

Where a closing line across the mouth of a bay transects an island or islands, the accumulated water distances alone may not exceed 24 miles. Individual segments of the closure must be straight lines but not necessarily aligned one parallel with another. Finally, the water of bays within bays may be included as water surface of the outer bay in determining the dimensions of any coastal indentation.

The foregoing rules regarding bays do not apply to "historic bays," which are coastal indentations over which a state has traditionally asserted dominion as a bay with the acquiesence of other states.

The coasts of islands have baselines, just as has a mainland coast and generally according to the same rules. As the only exception of note, low tide elevations (reefs, shoals, drying rocks) in certain instances require modification in baseline construction. The low water line of these features may or may not serve as a baseline from which to measure offshore claims.

Examples of coastlines where a system of straight baseline might be applied with validity are found in relatively few areas in the world, such as along the highly irregular and fragmented coasts of Yugoslavia, Norway, and southern Chile. In these instances the margins of the ocean are not well-defined in the sense that the economic regime of a state encompasses nearby offshore islands and the water passages that separate them from the mainland and from each other.

In contrast, most coastlines do not lend themselves to the construction of straight baselines. Even though a number of states have made unilateral claims for additional segments of territorial water by the use of systems of straight baselines, this technique can be supported neither by logic nor in accordance with the Convention.

A system of straight baselines has also been unilaterally claimed by certain island states, though necessarily of a somewhat different geometric design than those constructed along a continental mainland. This use of straight baselines, known as the "archipelago concept," adopts the idea of a perimeter around an island or a group of islands, usually midocean. Such a line around an island would touch on capes, peninsulas, offshore isles, or other prominent points along the coast. Such a line around a group of islands, or archipelago, would "box in" the ensemble, the straight baseline normally touching at the more prominent geographic features of the outermost islands.

This type of straight baseline is no more justified than a corresponding line along the mainland. Again, each island has its own normal baseline and where islands are close together their territorial seas tend to coalesce and form a continuous zone of territorial water. Otherwise the situation is that sufficient water distances exist between or among the islands to justify their status as high seas.

Boundaries in Ocean Space. Delimitation of seaward jurisdictional or sovereign limits depends upon specific distances from the baseline. The standard method for plotting the territorial sea is by compass on hydrographic charts. As an example, to plot a territorial sea with a breadth of 3 miles, the compass is set at a scale to indicate that distance on the chart, and arcs of circles swung seaward from salient points along the baseline. The envelope formed by these arcs of circles makes up the outer limits of a state's territorial waters, and hence the limit of its sovereignty in the water. The result is a geometrically precise line that can be plotted regardless of any complexities of the baseline.

A highly irregular coastline or one fringed with islands will have a territorial sea, the outer limits of which are correspondingly, but to a lesser degree, irregular except in those places where straight baselines are used. Geometrically the outer limits of the territorial sea under any conditions will not be as irregular as the baseline from which it is measured, for the method of construction smoothes out such a line—the greater the radii of the arcs swung, the smoother the contour of the envelope of arcs.

Another set of boundaries also comes into play, namely those that separate the offshore territories of two coastal states:

(a) boundaries separating the territorial seas of adjacent coastal states;
(b) boundaries separating the territorial seas of opposite states;
(c) boundaries separating the continental shelves of adjacent coastal states; and
(d) boundaries separating the continental shelves of opposite states.

Any two countries with contiguous offshore waters may agree on a common line of demarcation between them. Frequently median lines are the means of expressing boundaries between adjacent states, starting at the baseline and extending seaward, first between territorial seas and then between continental shelves of the two states concerned. They also serve to separate the waters of opposite states that have merging territorial seas and/or continental shelves. A median line is defined as a line, or boundary, every point of which is equidistant from the nearest points on the baselines from which it is measured. Oddly enough, the technique upon which the construction of such lines depends is purely trial and error, that is, establishment of points contingent upon being so placed that they be no farther from one than from the other fixed points representing the two sovereignties. The median line concept does not preclude other offshore boundary agreements between states.

Irregular and undemarcated land boundaries as well as complicated coastal configurations produce situations that create many problems apart from straight geometrical computation of median lines. Of particular note, sovereign exclaves and enclaves along a coast may bring about problems extremely difficult to resolve. Impasses may easily arise in cases of disputed territory, whereby basic premises for constructing median lines then become unacceptable to one or more of the states involved.

The spirit of the median line principle is to provide a means whereby boundary agreements between states may be facilitated. Since median-line boundaries are objective they can frequently be used at least as a point of departure in the reaching of agreement. Site of known or potential resources, location of a navigation channel, or traditional offshore practices of a state are among special circumstances that may give rise to modifying or even disregarding completely a median line in affixing a boundary. For example, a boundary in the territorial sea may only roughly approximate a median line, compensating for loss of an area in one place by gain in another.

WILLIAM L. GRIFFIN

References

McDougal, Myres, S., and Burke, William T., "The Public Order of the Oceans," 1962, Yale University Press, New Haven.

Shalowitz, Aaron L., "Shore and Sea Boundaries," Vol. I, 1963, Vol. II, 1964, U.S. Government Printing Office, Washington, D.C.

Griffin, William L., "The Law of Ocean Space—Text, Treaties, Statutes, Cases and Other Materials," 1968, Washington, D.C.

LOBSTERS

General

Lobsters are crustaceans that belong to the order Decapoda, which includes the true lobsters, spiny lobsters, crayfish, shrimps, and crabs. The three species of true lobsters are the American lobster (*Homarus americanus*), European lobster (*Homarus gammarus*), and the Norwegian lobster (*Nephrops norvegicus*), which is also called Dublin Bay prawn and scampi. These three species are found in the temperate waters of the North Atlantic and support large fisheries. The two species in Europe are fished from Norway south to North Africa and in parts of the Mediterranean. The American species ranges from Labrador to North Carolina.

In both poundage landed and value of the catch, the American lobster fishery is much more important than the fisheries for the other two species. Each year about ten times more American lobsters (60,000,000 lb in 1967) are landed than European lobsters. The catches of American lobsters and Norwegian lobsters are about equal, but the value of American lobsters (about $45,000,000 to fishermen in 1967) is several times greater.

The principal fisheries for the American lobster are between Cape Cod, Mass. and the Gulf of St. Lawrence, Canada. Maine is the most important U.S. producer—about two-thirds of the catch of 25,000,000 lb were taken by Maine fishermen in 1967. Nova Scotia and New Brunswick are the principal lobster provinces of Canada, which produced 36,000,000 lb in 1967. The catches of American lobsters have been declining in recent years—since 1964 in the U.S. and since 1960 in Canada. These declines have occurred despite the greater efforts of the fishermen, who are not only more numerous but are fishing more traps over larger areas.

Decreases in the lobster catches have occurred despite a wide assortment of regulations in both Canada and the U.S. Various laws, which may include closed seasons, closed areas, minimum and maximum size limits, prohibitions of certain gears, and protection of egg-bearing females, have been in effect for decades. Recently, limitations on the number of traps fished are being placed on Canadian fishermen. It is not clear how much the decrease in the catches may be due to a changing environment and how much to excessive fishing, yet legal efforts to check the decline continue to be made.

Most of the two species of *Homarus* are caught in much shallower water than the Norwegian lobster, which are taken in depths over 100 fathoms. Otter trawls are used to take these deep-water lobsters; the traps used for *Homarus* would be difficult and expensive to operate at such depths.

The wooden trap is the principal gear used for the American and European lobster. This boxlike or semicylindrical gear has remained basically the same for many decades. In recent years otter trawls have taken increasing quantities of lobsters in offshore waters as deep as 200 fathoms, but the traps, which are used in much shallower waters, still take over 90% of the catches. Any improvements in fishing methods and gear in state or provincial waters face the opposition of the fishermen who use the traditional traps. Much less opposition to change would occur in the Norwegian lobster fishery, which has a much shorter history, and, therefore, less tradition.

Trap fishing has changed somewhat during the years, but essentially it still involves setting traps that are baited with dead fish. The lobster is attracted by the bait and enters the trap from which it cannot escape easily. The traps may be set on the bottom as single traps or in strings that may have two to ten or more traps. The traps are lifted every day or so, the lobsters are removed, and the traps are rebaited and dropped to the bottom.

Improvements in fishing have resulted principally from the use of larger, more seaworthy, and better equipped boats that can operate over larger areas and be less influenced by bad weather than the smaller boats they have replaced.

Because most European and American lobsters are marketed alive, there is an elaborate, yet often somewhat disorganized, system of buyers, storage facilities, transportation, wholesalers, and retailers. Some lobster meat is marketed, but essentially the meat represents a salvage operation—lobsters that are weak, dying, or clawless are cooked and their meat is extracted. The usual trade channels desire two-clawed lobsters that are strong enough to stand shipping or holding. After the live lobsters are purchased from the fishermen, they are held in floating live cars, pounds, or indoor tanks with circulating sea water. From buyer to wholesale dealer the lobsters are transported in vessels with live wells, or, more often, in crates in insulated trucks. From wholesaler to retailer, the lobsters are shipped in barrels or in second-hand wooden baskets and crates that are used commonly for fruit and vegetables. Special efforts are made to keep the lobsters cool and damp; otherwise they die readily.

At the wholesale and retail levels, American lobsters are sold on the basis of their sizes and condition. The 1-lb lobster is most common because the minimum size laws make it illegal to catch a lobster less than about 1-lb (with two normal claws). Most of the lobsters landed have just moulted to the legal size so the 1-lb lobsters dominate the market and are sold more cheaply than the larger $1\frac{1}{2}$–2-lb lobsters that are much scarcer. Over 2-lb, the lobster declines in value, because the consumer does not like so large a lobster. American lobsters grow to be 50 lb, but these are rarely caught, for the fishery is so intensive that few reach so large a size. Offshore lobsters, which are subject to a less intensive fishery, are, on the average, much larger than the inshore lobsters.

The greatest recent advance in marketing lobsters in North America has been the improvement and widespread use of recirculating seawater systems for keeping lobsters alive. Live lobsters are now readily available any time throughout the country, and demand for these crustaceans has soared. The Norwegian lobsters are not sold alive, for they are less hardy than the European and American lobsters. The Norwegian lobsters are seldom sold whole; instead, the tail is marketed. In contrast, most *Homarus* are sold alive and whole. Much of the *Nephrops* catch is sold as frozen tail meats; few *Homarus* are sold frozen.

Many problems remain to be solved before the lobster resource can be handled most efficiently. More must be known about how to manage the fishery for the most economic yield. A great deal can be done to reduce the lobster mortalities that occur between the time the lobster is caught by the fisherman and the time it is cooked and eaten. Of special interest is the need for more information on how to store live lobsters for long periods of time in tanks and pounds, for regulation of the supply is an important factor in the economics of the industry.

LESLIE W. SCATTERGOOD

Giant Lobsters

Along the New England coast, 30 lb of lobster usually means 20 to 30 lobsters. In contrast, lobsters taken in the offshore fishery on the edge of the Continental Shelf may weigh 10, 20, or even 30 lb each. These giants are brought to market alive, as are their inshore cousins, but seldom reach the table intact, having previously been cooked for use in stews or salads. Large lobsters are not as desirable in the live market as smaller ones, but as the giants are most certainly filled with succulent meat, more and more are being captured to meet the growing demand for this popular crustacean.

Lobsters in the offshore waters are caught with large nets—called *bottom trawls*—that are dragged across the sea floor for periods up to 3 hrs. Vessels participating in this fishery are usually 80 ft or more in length. Though catches are made year-round, the greatest effort is expended in late fall and early spring. Summer fishing is avoided by some of the fishermen because the lobsters are shedding during this period and because, during warm weather, the lobsters do not survive as well aboard ship. The average fishing trip lasts 12 days, and the lobsters are held in live tanks until the vessels return to port. Catches average close to 10,000 lb during a trip, but may range from a few thousand to 20,000 lb. Lobsters less than 3 lb each are termed *selects* and command a premium price, as much as $1.00 per lb during periods of short supply. The jumbos—over 3 lb—usually market for $.20 less per lb. On an annual basis, the price in recent years has been about $.80 per lb for selects and $.60 per lb for the jumbos.

The history of offshore lobster catches is fragmentary, but has been pieced together by Firth (1940) and Schroeder (1959). The earliest records date back to the 1900's when the beam and otter-trawls were introduced in American waters. These catches of lobsters were incidental to the various species of finfish that were being sought. Though an active lobster fishery exists today, the incidental catches in other fisheries continue and are not regularly recorded in the statistics of the landings. Perhaps the earliest otter-trawl catch of lobsters that is recorded was 8,000 lb landed in New York in 1921—the exact location of the catch is unknown. Even through 1960, little information was gathered on the exact location of offshore catches. Statistics of the U.S. Bureau of Commercial Fisheries do list the catches of lobsters by otter-trawls from the New England region to the Chesapeake, and these date back to the late 1920's. Though crude, these data can be assumed to document the increase in offshore lobster landings over the years. Through 1946 these catches remained less than 100,000 lb; from 1947, the landings increased steadily and surpassed 1,000,000 lb for the first time in 1955. This growth continued, and the peak landings—$5\frac{1}{2}$ million lb—occurred in 1965.

A greater fishing effort has been responsible for this increase in catch, and it has been coupled with the exploration of new areas and fishing in deeper waters. The vessel *Caryn* of the Woods Hole Oceanographic Institution recorded catches to 250 fathoms in 1948, and the *Delaware* of the U.S. Bureau of Commercial Fisheries recorded catches to 400 fathoms in 1956. Today, the commercial operations are concentrated between 100 and 250 fathoms, and extend from Corsair Canyon along the edge of the Continental Shelf to Hudson Canyon, and occasionally off the coast of Delaware. There are between 20 and 30 full-time lobster trawlers, and the major ports of landing are New Bedford, Sandwich, and Plymouth in Massachusetts, and Newport and Point Judith in Rhode Island. Although precise measures of catch per unit of effort are lacking, there are indications that it has declined in some areas. Fishermen report that the average length of the trips has increased and that the duration of the tows is somewhat longer. Present research by the U.S. Bureau of Commercial Fisheries is designed to obtain more exacting data on these aspects of the fishery so that changes in fishing effort can be documented and evaluated.

Because the lobsters are taken in international waters, 50 to 100 miles beyond the 12-mile national limit, there are no restrictions on the method of capture or the size or condition of the lobsters. But state laws can and do regulate the *landings,* so, in essence, the offshore fishermen are subject to the same laws that pertain to the inshore fishermen. Since the state laws are not uniform, the landings from offshore catches are concentrated in Massachusetts and Rhode Island where the laws permit the landing of large lobsters. These states have a minimum size regulation and do not permit landings of female lobsters that are bearing eggs—commonly called *berried lobsters,* but there are no restrictions on

Fig. 1.

maximum size. Only Maine has a maximum size limitation and thereby prohibits landing the offshore giants. The size measurements are made from the eye socket to the end of the back or carapace, and the minimum size ranges from 3 to $3\frac{1}{2}$ in. Maine's maximum is 5 in.

The size of lobsters occurring in an area is, in part, a function of the fishing pressure that has been applied to the stock. The offshore fishery is relatively recent and its lobsters have not been as heavily fished as those inshore. Historically, large lobsters were found in the coastal waters, and records of lobsters weighing 20–30 lb were not infrequent at the turn of the century. Even today, an occasional giant of 10–15 lb can be found very close to shore. These individuals somehow have escaped the intensive pot fishery that usually captures the lobsters as soon as they attain legal size. Offshore, there is no concrete evidence that the average size has changed drastically, though in certain seasons and on certain fishing grounds, only smaller lobsters are taken. As the fishing effort increases offshore, fewer giant lobsters will remain and the average size will eventually approach that of the coastal fisheries. The lowered abundance of the large lobsters probably will not be detrimental to the population, rather, the production is likely to increase. With fewer giants, competition for food and space will be reduced; and a higher survival and recruitment of small lobsters is probable.

In the present commercial fishery, the average weight of the offshore lobsters approximates 5 lb. Many lobsters less than 1 lb are taken, but are returned to the sea because of state restrictions. Though 30-lb lobsters probably are not as abundant as in former years, they still occur frequently in the commercial catches offshore. The largest lobster authentically recorded is on display at the Museum of Science in Boston, Massachusetts. Its total length, from claw tip to tail fan, is 38 in., and its weight is reported to have been 45 lb. The reports of larger lobsters—up to 60 lb—have never been substantiated.

The age of these giant lobsters remains a mystery, despite the specific ages listed under specimens displayed in restaurants and other public establishments. Unlike fish, whose ages can be estimated by enumerating the annual growth rings on their scales or on other hard parts, there are no such patterns that will divulge the exact age of lobsters. Biologists can estimate age by the size of the lobsters, but this method is not reliable beyond seven years. During their first years, lobsters moult several times a year. After they reach maturity they moult only once a year. After five to seven years of age, lobsters moult only once every two years. Whether lobsters continue to moult at this rate or once every three or four or more years is not known, so that estimates of the ages of the giants are purely speculative. At best, we can "guesstimate" that a lobster of 30 lb is between 25 and 35 years of age. There is certainly no evidence to support the figures of 100–200 years that are often proposed for lobsters of this size.

Research cruises by the Biological Laboratory of the U.S. Bureau of Commercial Fisheries in Boothbay Harbor, Maine, and the Bureau's sampling of the commercial landings provide information on all sizes of lobsters and their distribution by depth and area of catch in the offshore waters. The examination and measurement of 15,000 lobsters has revealed several important facts regarding size and the sexes. The sex ratio of small lobsters (less than 3-in. carapace length) was usually 1:1, but the ratio of females increased with size and in some samples of larger lobsters (5 to 7 in.) accounted for 70% of the total. Above 7 in., the dominance of females decreased, and the males dominated above 9 in. In all, the females accounted for 60% of the lobsters sampled, and their dominance was greatest in the winter. In samples from July to November, females usually accounted for less than 50% of the catch. This seasonal variation in sex ratio and the dominance of females probably reflects changes in behavioral patterns (including migration) that influence catchability, sexual differences in moulting rate after maturity, and the effects of regulations requiring the release of egg-bearing females.

The smallest egg-bearing female that has been observed from the offshore catches was 3 in. (carapace measure). Between 4 and 7 in., 50% of the females were egg-bearing. Relatively few lobsters over 8 in. were egg-bearers. One means of estimating female maturity is the comparison of the second abdominal segment with the length of the carapace. The width of this segment in small lobsters (2–3 in.) has been compared with the size and development of internal eggs, and validates the method as a means of estimating minimum size at maturity. Relative to carapace length, the segment has a greater increase in width prior to the first extrusion of eggs. In the specimens examined, the carapaces ranged from 2 to 8 in. The width of the abdominal segments ranged from 1 to 5 in.; in the smallest egg-bearing female it was $2\frac{1}{4}$ in. All lobsters with a carapace length less than 3 in. had abdominal widths less than $2\frac{1}{4}$ in. and only 2% of the lobsters with a carapace measure longer than 3 in. had abdominal widths less than $2\frac{1}{4}$ in. Using these guidelines, it is possible to determine whether a lobster is approaching maturity or has previously spawned.

The size distribution of lobsters varies with the canyon areas that are fished. For example, the mean carapace length of the catches from Veatch Canyon average between 4 and $4\frac{1}{2}$ in.; whereas, in Oceanographer Canyon the range is between 5 and 6 in. Over 50% of the lobsters from Veatch have carapace measures of less than 4 in., and at Oceanographer less than 10% of the catch is this size. These results pertain

Fig. 2.

only to the catches of legal-sized lobsters, but hold for the data collected from the commercial landings as well as those from the research cruises. Research cruises did, of course, capture prerecruits not landed by the commercial vessels. The causes of these differences between areas are not understood. The differences could reflect varying fishing effort or natural causes such as migration. A planned tagging study and additional catch data to supplement the three years' records now available should help explain these differences.

A long-term tag—one that is retained through moulting—has been developed by the U.S. Bureau of Commercial Fisheries, and was first used on offshore lobsters in 1968. The tag, which is inserted in the musculature between the carapace and the tail, has an external piece of plastic that is readily visible and can be coded so that individual lobsters can be identified. Previously, tags were attached directly to the carapace and were lost when the lobster moulted. The new method adds considerable dimension to a given tagging experiment. The tag was tested near Monhegan Island which lies 10 miles off the Maine coast. Of some 2000 lobsters tagged, over 60% moulted and 90% of these retained the tag. As far as can be determined, they showed normal growth patterns. The planned offshore tagging will provide information about the individual fishing areas and may give direct evidence as to whether the inshore and offshore stocks intermingle.

The relationship of the inshore and offshore stocks has been a point of interest to fishermen and scientists alike, and has become a matter of particular consequence as the offshore fishery grows in intensity. Proper management of the lobster fisheries is dependent on the knowledge of whether these populations are discrete. In 1964, the U.S. Bureau of Commercial Fisheries was asked by the Atlantic States Marine Fisheries Commission to study this relationship, and funds for this purpose were provided by the Congress. Bureau scientists at Boothbay Harbor, Maine have approached the problem in several different ways to provide a mosaic of evidence. The methods are completely independent so that the results will give a decisive answer as to the degree of intermingling, if any, between the stocks.

In addition to the planned tagging studies, the other approaches include a biochemical study, an examination of lobster parasites, and larval and morphometric studies. If the stocks are different and genetically isolated, they should reflect this in their genetic characters. In the biochemical study, starch gel electrophoresis is being used to compare serum proteins from inshore and offshore lobsters. The structure of proteins has been shown to be under genetic control and should serve as useful markers of genetic differences. Quantitative differences in serum proteins have been found among inshore, offshore, male, female, and recently moulted lobsters. Work with other crustaceans has shown that these proteins vary considerably among the different stages of the moult–intermoult cycle, so that the results on lobsters must be interpreted with caution until it is determined whether these differences are actually genetic rather than physiological.

The distribution of parasites is often restricted by environmental conditions that do not influence the distribution of the host animal. Under such circumstances, the parasites assume the status of "natural tags," and provide evidence of the geographical origin and migratory history of the host. To determine whether natural marks exist in the normal array of lobster parasites, lobsters have been examined throughout their geographic range—both inshore and offshore. Special attention has been paid to those parasites that exhibit geographical discontinuity. Two such faunal associates appear to characterize the stocks. *Ascarophis* sp. (Nematoda) is generally restricted to the offshore stocks, and *Corynosoma* sp. (Acanthocephala) typically occurs in the inshore stocks. These parasites are mutually exclusive north of Cape Cod, but their distributions have been found to overlap in Rhode Island and Block Island Sounds—areas that, according to fishermen's reports, have seasonal intermingling of coastal and offshore stocks.

The morphometric studies, which include the measurements of various anatomical features and their relative proportions to one another, have also shown that differences do occur between stocks. Interpretation of these results also requires caution, for the difference may well be associated with the environment.

Lobster larvae have also been studied in conjunction with the inshore–offshore problem. During one of the Bureau's cruises, early August 1966, the cruise track zigzagged across the Gulf of Maine and along the edge of the Continental Shelf. Less than one larva per plankton tow was captured in the Gulf, whereas 6 to 12 larvae per tow were taken in the vicinity of the canyons. Most of these were taken in surface tows; oblique tows, made simultaneously with those at the surface, were less productive. Larvae in developmental stages I and II were most numerous in 16°C to 18°C waters; stages III and IV were most frequent at temperatures of 21°C to 22°C. Catches of larvae, with the same gear, along the Maine coast have been smaller than those offshore, but catches are not really large in either area. There is no information, at present, to indicate that ocean currents redistribute larvae from offshore to coastal areas or vice versa.

The research program conducted by the Bureau of Commercial Fisheries was stimulated by the belief of many inshore fishermen that the large offshore lobsters serve as the brood stock for the coastal lobsters. Though the several methods have indicated that there are differences between inshore and offshore populations, demonstration that these differences are due to genetic isolation remains to be established and the possibility of intermingling must be more closely examined. The present lines of research are scheduled to continue so that a definitive answer to this question can be reached. Careful management will be necessary if the stocks are not discrete, to prevent over-exploitation. On the other hand, if the indications of discreteness that now exist in the Gulf of Maine hold true, the development of the offshore fishery has the potential of adding another dimension to the industry, which even now has difficulty meeting the demand for lobsters. With continued sampling of the commercial landings and returns from the tagging experiment, which can provide estimates of population size, biologists will be able to assess the potential of these stocks along the Continental Shelf. Further development and added fishing pressure will undoubtedly spell the end of the giant lobster era, although that is inevitable in any productive fishery—and is the way of all giants.

BERNARD EINAR SKUD

References

Firth, Frank E., "Giant lobsters," *New England Naturalist*, No. 9, 11–14 (1940).
Herrick, Francis H., "The American lobster; a study of its habits and development," *Bull. U.S. Fish Comm.* 1895, **15**, 1–252 (1896).
Herrick, Francis H., "Natural history of the American lobster," *Bull. Bur. Fish. Wash.* 1909, **29**, 149–408 (1911).
McRae, E. D., Jr., "Lobster explorations on the continental shelf and slope off northeast coast of the United States," *Comm. Fish. Rev.*, **22** (9), 1–7 (1960).
Perkins, Herbert C., and Skud, Bernard E., "Body proportions and maturity of female lobsters," *American Zoologist*, **6** (4), 615 (1966).
Scattergood, Leslie W., "The American lobster, *Homarus americanus*," Fishery Leaflet No. 74, 1958, U.S. Fish Wildl. Serv., 9 p.
Schroeder, William C., "The lobster, *Homarus americanus*, and the red crab, *Geryon quinquedens*, in the offshore waters of the western North Atlantic," *Deep Sea Research*, Vol. 5, 1959, pp. 266–282.
Skud, Bernard E., "Size composition, sex ratios, and maturity of offshore lobsters," *American Zoologist* **6** (3), 362–363 (1966).

Offshore Lobsters

Small quantities of lobsters have been caught in the Northwest Atlantic each year since the advent of the otter trawl, which was introduced shortly after the turn of the century in American fishing. However, deliberate fishing for lobsters with otter trawls did not begin until shortly after World War II. Since this time, the fishery has grown from a few boats operating seasonally to a substantial fishery currently producing more than 10% of the total U.S. lobster landings per year. During the present time more than 3,000,000 lb of offshore lobsters are landed annually in the U.S.

The economic importance of the entire fishery for the northern lobster is steadily increasing in the New England states and the Canadian maritime provinces. In both Canada and the U.S., the inshore pot fishery has been reasonably stable in terms of annual landings. The inshore stocks are heavily exploited, in contrast to the offshore (trawl) lobsters, which have shown a consistent increase in landings since the advent of deep water trawling. The developing offshore lobster fishery and controversies concerning it have prompted a large scale regional investigation of various aspects of lobster life history and management regulations. Much of the present work in the U.S. is being coordinated by the U.S. Fish and Wildlife Service, Bureau of Commercial Fisheries, Biological Laboratory, West Boothbay Harbor, Maine. The Canadian work in similarly being coordinated by the Fisheries Research Board of Canada, Atlantic Biological Station, St. Andrews, New Brunswick, Canada. In spite of the fact that several of the research projects in the regional research program are still in progress, it seems desirable at this time to attempt to synthesize available information and point out significant gaps in our knowledge concerning the offshore lobster.

Distribution and Abundance. The so-called deep sea or offshore lobster fishery consists primarily of an otter trawl fishery in depths to approximately 1500 ft near the edge and on the slope of the Atlantic continental shelf. The presently defined limits of the distribution of offshore lobsters are between the eastern part of Georges Bank and the offing of Delaware Bay. The depth distribution of the resource is believed to occur roughly between 60–250 fathoms. The length of the area is on the order of 400 miles.

There is reason to believe that the above-mentioned limits of distribution have been significantly affected by limitations in the sampling gear as well as the frequency and seasonal distribution of sampling effort. Available information from recent limited tagging studies as well as length–frequency analyses of offshore catches clearly suggest a low rate of exploitation and a greater depth distribution than that indicated above. However, it remains to further delimit geographic distribution and relative abundance by means of better sampling gear or direct observations.

Recently (September, 1967) the author had an opportunity to utilize a research submersible (Deepstar 4000) for limited direct observations of lobster and red crab habitat in the vicinity of Corsair Canyon (41° 21.7′ N latitude 66° 09.2′ W longitude) in water depths approachnig 3000 ft. Although the results of this work are preliminary, direct observations along the slopes of the canyon suggested a habitat preference similar to that observed in shallow waters. That is, burrows of large crustacea were observed in depths exceeding 2700 ft along the steep slopes of the canyon. These observations suggest that an adequate sampling and observational program utilizing research submersibles with sufficient operational depth and endurance can do much to qualitatively delimit the distribution of large offshore crustacea.

The results of exploratory fishing and commercial trawling suggest that the relative abundance of deep sea lobsters along the continental slope is far from uniform. Certain areas, such as the vicinity of Veatch, Lydonia, Oceanographer and Corsair Canyons have been consistently more productive (in terms of catch per unit of effort) than others. Furthermore, the success of trawling is related to water temperature in the trawled area. The limited direct observations made by the author in Deepstar 4000 suggest that the distribution and abundance of offshore lobsters may be habitat limited. That is, in areas with precipitous slopes and suitable burrowing habitat, relative abundance may be expected to be higher than in other less diverse areas.

It seems reasonable to conclude that the limited amount of sampling effort and the inherent bias in exploratory fishing with otter trawls do not provide adequate information to clearly delimit offshore lobster distribution in terms of depth or geographical area. Furthermore, only limited inferences on absolute abundance are possible at present. However, the evidence obtained by limited tagging studies, trends in landings, and length-frequency analyses suggest that the fishery will be capable of producing sustained yields considerably in excess of those now taken. Finally, direct observations with submersibles, together with properly executed tagging and exploratory fishing programs, should permit more precise estimates of stock magnitude in the near future.

Stock Identification. From almost the inception of the deep sea lobster fishery, some segments of the inshore lobster fishermen have contended that the offshore stocks contribute significantly to the recruitment on inshore grounds. The relations between inshore and offshore stocks have been carefully studied, and a definitive answer to this question is now at hand. The following experiments and observations are related to the stock identification problem.

Large scale marking experiments, extending over several years, with inshore lobsters indicate that, with few exceptions, the marked individuals do not make extensive migrations. Indeed, the average home range of inshore lobsters is circumscribed by a circle with a radius of approximately 7 miles. Available evidence to date also indicates that the inshore lobster fishery consists of several discrete populations with very little mixing between areas.

A recent tagging study involving the displacement of offshore berried female lobsters from the vicinity of Veatch Canyon to Narragansett Bay, Rhode Island, has been completed. Tagged, displaced female lobsters tended to remain in shoal waters in suitable spawning habitat until they had shed their eggs, molted, or both. After this there was a pronounced directional tendency in the movements of the displaced animals. The average vector of directional movement was very nearly in the theoretical home direction. Furthermore, within a period of one year, four of the displaced females were recaptured in the vicinity of the home area, approximately 135 miles from the point of release. The results of this study clearly support the concept of discrete inshore and offshore stocks with well defined home ranges as well as a pronounced homing tendency.

Limited tagging studies of offshore lobsters that have been tagged and released at the point of capture all indicate restricted movements by the offshore stocks. That is, the results are similar to inshore marking and tagging studies. However, there is some suggested evidence that the size of the home range may be somewhat larger for the offshore stocks. The offshore stocks also seem to consist of several discrete populations with relatively little overlap among them.

Sampling of coastal and offshore lobster stocks for parasitological studies has produced interesting results. This work has been conducted by the U.S. Fish and Wildlife Service, Biological Laboratory at West Boothbay Harbor, Maine. Two tracer parasites *Ascarophis* sp. (Nematoda) and *Corynosoma* sp. (Acanthocephala) have been found, and they seem to be almost mutually exclusive. *Ascarophis* seems to characterize the deep water stocks, and *Corynosoma* the inshore stocks. Very little overlapping has been found except for waters in the vicinity of Rhode Island Sound and Long Island Sound. It is possible that some intermingling of inshore and offshore stocks occur in these areas on the basis of these data.

In addition to the above studies on stock identification, a multivariate morphometric technique of assigning unknown individuals to specific groups has been developed by the author. The results of the morphometric study (based on simultaneous comparisons of the separate body parts) are in good agreement with the parasitological and behavioral studies. A correct classification probability of more than 0.8 has been achieved in separating inshore offshore samples.

In summary, the following inferences appear reasonable on the basis of existing evidence:

(a) Both inshore and offshore lobster stocks are composed of several discrete populations with relatively little overlap.
(b) The inshore and offshore lobsters are recognizable as discrete and separate populations. Vital statistics should be gathered for both groups for management purposes.

Other Aspects of Biology and Management. At this time it is not possible to provide reasonable estimates of the numerical magnitude of the offshore population. However, the apparent low rate of exploitation indicated by tagging and length frequency analyses suggests that the offshore lobster fishery is still in a state of development.

The large size of the offshore lobsters compared to coastal lobsters is not believed to be related to peculiar genetic or physiological properties of these stocks. Instead the larger sizes of offshore lobsters merely reflect a fishery still in the early phases of exploitation. Old colonial records indicate that large (10–20 lb) lobsters were found near shore prior to any significant exploitation by man.

One of the practical problems encountered in offshore lobstering is that of keeping the lobsters alive at sea. First, there is a certain amount of mechanical damage resulting from the trawling operation, and secondly, the partial occlusion of the respiratory apparatus (gills) by mud and silting is common. Finally, the temperature difference of surface water and that of the water at 150 or more fathoms is fairly great during summer months. At present, lobsters are being returned to port in tanks through which sea water taken near the surface is pumped. The lack of refrigerated circulating sea water systems on lobster boats during the warm months undoubtedly is a factor in high summer mortality in transit. High summer mortality after capture actually prohibits or restricts summer offshore lobstering at present.

Some efforts are in progress at present to extend the scope and size of stationary lobster gear (pots and net trawls) to permit fishing deeper areas. Limited work in this area has met with success, and it seems reasonable to speculate that the future offshore lobster fishery will consist of stationary gear somewhat analogous to the present Alaskan king crab gear. It is believed that stationary gear development will not only minimize mechanical damage to lobsters, but will also increase the efficiency of exploitation.

A theoretical study has been completed by the author to assess regulations pertaining to the protection of large lobsters (the so-called double guage regulation) as well as regulations pertaining to the protection of egg-bearing females. The results of this study indicate that there is no biological justification for the double guage regulation. That is, protection of large animals beyond a prescribed maximum length is neither in the interests of good conservation nor good economics. On the other hand, there seems to be some reason, on the basis of available information, to protect the female lobster to which the developing eggs are attached externally. The reasons for this include: (a) the probability of significantly altering sex ratios of mature animals in favor of males seems high without protecting berried females; and (b) the probability of a higher total fecundity and consequent higher stock strength, at least in those areas with favorable environmental conditions, is increased by protecting berried females.

There are a number of problems remaining to be resolved in order to effect a rational management program for the offshore lobster fishery. Some of these problems include:

(a) Better estimates of the offshore population size, and geographic and seasonal distribution.
(b) Improvements and innovations in both gear and vessels.
(c) Precise estimates of vital statistics of offshore lobsters including, growth, natural mortality, rates of exploitation, and movements.
(d) Design of a stock and yield assessment model to establish optimum size limits for maximum sustained yields, and to assess effects (both short term and long term) of various restrictive regulations.

SAUL B. SAILA

Spiny Lobsters

"Spiny lobsters" has been recently adopted as the common name for the marine family Palinuridae to remove former confusion resulting from the use of names such as crawfish, lobster, and crayfish, which in Europe and America usually apply to the family Astacidae (Sims 1965).

The marine spiny lobsters (Palinuridae) can easily be distinguished from the true lobster because they have a pair of horns above the eyes, and the first four legs do not have claws. The true lobsters, which belong to the family

Astacidae, have a single central horn between the eyes, and claws on all the front legs. Useful references to facilitate the identification of particular species of spiny lobster are given in George and Main (1967) who have attempted to interpret the past evolution of the world's spiny lobsters from an examination of the present distribution and the morphological affinities of the 50 species of spiny lobsters known today.

Size. Spiny lobsters are, as a group, one of the largest in size of all the crustacea; *J. verreauxi* of New Zealand attains weights of 30 lb and lengths of 3 ft. Most species average about 1–2 lb in total weight, from which 30–40% may be recovered as tail meat when processed.

Spiny lobster contributes a substantial proportion of the total marine crustacean catch (Table 1), being greater than lobster or Norway lobster, but much less than the top products such as prawns, shrimps, and king crab. However its high market price enhances its value considerably. The spiny lobster fisheries are based mainly in temperate and subtropical countries (Table 2) where the species of the genera *Jasus* and *Panulirus* are the basis for the fisheries. Except for Cuba, the tropical countries produce little toward the world's catch.

In view of the continued high value of the properly processed product, earlier methods of catching by spearing are now discouraged, unless processing facilities are very close to the fishery. Divers in some regions (east coast of Africa, Galapagos, West Indies) take spiny lobsters by hand, either because of available labor force or because the local species cannot be caught economically by other means; some species have not been enticed into traps, so capture by hand is the only known method. In other regions spiny lobsters are captured at night when they leave their daytime retreats to forage over the reefs. Men with lanterns walk across the reefs picking them up with gloves or scooping them with a bully net from a boat. The above methods usually provide a very low level of catch on the world market, and the large regular national fisheries of spiny lobster employ nets or traps of various designs, which requires a higher capital outlay to operate and maintain the gear.

Basically the traps are baited wire, stick-, or wooden-covered cages with a cone-shaped entrance funnel that prevents escape. Some are round, beehive-shaped with an opening at the top; others are square or rectangular with top or side openings or both. Hoop nets are also successful. In South Africa for instance, each baited hoop has a loose netting bat suspended from it; these bags are periodically raised to surround the spiny lobsters attracted to the bait. Tangle nets consist of runs of loosely hung coarse-mesh nets set on the bottom; the head line is lightly buoyed to facilitate tangling of the spiny lobsters.

Boats vary from dugout canoes to ocean-going catching and processing vessels, which may serve as mother ship to a fleet of smaller boats. These are fitted with all the modern navigational equipment, as well as line haulers or winches and tipping and lifting aids to recover the traps.

Processing methods. There are three forms in which spiny lobsters are prepared for marketing:
(1) whole uncooked, alive, or frozen;
(2) whole cooked;
(3) tails uncooked, frozen.

There is a regular trade in airfreighting live lobster from Africa to the European market. The lobsters are held in tanks awaiting dispatch.

TABLE 1. WORLD MARINE CRUSTACEAN PRODUCTION, THOUSAND METRIC TONS BASED ON FAO YEARBOOK OF FISHERY STATISTICS FOR 1966, VOL. 22

Prawns and shrimps	595.0
King crab	149.0
Various marine crustaceans	106.0[a]
Other marine crabs	85.0[a]
Blue crab	75.0
Common shrimp	60.0
Spiny lobster	57.0[a]
Norway lobster	32.0
Northern lobster	30.0
Deepwater prawn	21.0
Dungeness crab	19.0
Edible crab	8.0
European lobster	4.0[a]

TABLE 2. 1966 WORLD SPINY LOBSTER PRODUCTION THOUSAND METRIC TONS BASED ON FAO YEARBOOK OF FISHERY STATISTICS FOR 1966, VOL. 22

JASUS	Total	31.2
South Africa		8.0
Southwest Africa		8.0
New Zealand		7.2
Australia		5.0
St. Pauls		1.0[a]
Tristan da Cunha		1.0[a]
Juan Fernandez		1.0[a]
PANULIRUS	Total	27.9
Cuba		9.0
Australia		8.6
Brazil		3.3
U.S. Atlantic		2.4
Japan		1.6
Mexico		1.4
France		1.4
U.S. Pacific		0.2
PALINURUS	Total	1.5
Spain		0.5
Italy		0.5
Ireland		0.2
France		0.2[a]
Portugal		0.1[a]

They are then removed and bound lightly in wrapping or tied to avoid movement. This reduces their oxygen consumption and increases the period they can live out of water. The wrapped lobsters are packed in cartons filled with dry sawdust to cushion any movement.

To supplement the live lobster market, countries that cannot provide suitable air transport freeze the live lobsters. In this case the lobsters are drowned in fresh water first and the individuals are frozen in about 2 hrs. This rapid freezing is necessary to avoid spoilage.

For whole cooked lobster, they are drowned in fresh water and boiled in brine for 15 min if 1 lb to 1½ lb in size. Large lobsters require additional cooking—approximately 5 min extra per ½ lb. After cooking they are scrubbed clean in running water, drained, and cooled before quick freezing.

The process of tailing lobsters is perhaps the most popular method of preparation. The live lobster is pushed on to a fixed knife on the processing table, and the tail is cut away from the carapace by rotating the carapace. The tail is then washed and the digestive canal or intestine is removed by instruments that force it out by water pressure or suck it out through a vacuum-assisted circular knife. The tail is trimmed and rewashed before grading and draining. It is then wrapped in cellophane or similar material, graded for size, boxed, and frozen. The freezing is mainly by use of low temperature air blast. It is important that the product is frozen within 6 hrs of tailing.

Quality varies according to the specifications set down by the Government Inspection Department in each country. In Australia the regulations stipulate that only live lobsters can be processed for export and that the product must be under refrigeration within 2 hrs of being processed.

In many other countries the regulations are not as stringent, and lobsters are either tailed at sea and packed in ice or partially slow frozen or else kept alive in circulating wells on the decks for eventual processing 3 to 4 days after catching. The variation of handling methods and the resultant variation in quality has produced big differences in the selling prices of lobsters exported from different countries.

Packaging procedure is fairly common to all the producing countries. Whole cooked lobsters are packed in polyethylene bags, individually frozen, and stacked into wooden cases to protect them from damage. Lobster tails are packed in 25-lb wooden cases and 10- or 20-lb cartons, depending on market requirements.

All lobsters are frozen to 0°F in the processor's freezers before transfer to a holding room held at 0° to 50°F. This temperature is maintained whether shipped by refrigerated road transport or by ship. Properly prepared lobster tails have been stored for up to 24 months with little or no deterioration.

Lobster products are transported from most of the world's producing areas by conventional refrigerated ships. The consignments are transported from cold stores by insulated or refrigerated trucks as it is most important that there is no temperature variation after the product has been reduced to 0°F. Any temperature variation can affect the storage life and quality of the lobster.

Markets and marketing. The U.S. is by far the largest user of spiny lobsters in the world. In addition to its own production, the U.S. absorbs almost the total production of South Africa, New Zealand, Australia, Brazil, and Mexico. Most of the production is in the tailed form except for the Mexican product, which is cooked and transported packed in ice into the southwestern U.S. cities. The lobster tails imported into the U.S. are marketed in two divisions;

(1) cold water tails—South Africa, New Zealand, and Australia;
(2) warm water tails—Brazil, Mexico, and other equatorial countries.

The cold water tails are preferred, and generally sell at a much higher level. South African and Australian lobsters are recognized as the best imported into the U.S., mainly because of the very stringent processing regulations enforced by their Government Food Departments.

Spiny lobsters are also marketed in Europe, France being the main consumer. This country imports most of the live lobsters airfreighted from North and South Africa, and generally prefers to handle the lobsters in their whole form. Many other countries produce spiny lobsters, but do not export them because the production is very small or else it can be absorbed at world prices within the country of origin. In some cases governments enforce restrictions on imports to insure that the local product is consumed.

Species distribution. There are eight genera and about 48 species recognized living in the temperate, subtropic, and tropic waters of all the major oceans of the world. Species of two genera, *Panulirus* and *Jasus*, support the major fisheries on spiny lobsters. Spiny lobsters occur from the shallow waters of the coastal reefs down to about 700 fathoms, but most of the present commercial species are fished in depths less than 100 fathoms. Not a great deal is known about the species living at the greater depths, but the recorded species are usually of moderate size (¼ to 1 lb) and have been trawled occasionally in some hundreds at a time.

Panulirus. The species of the genus *Panulirus* are the only spiny lobsters of any quantity in the subtropic and tropic shallow waters of the coastal reefs and corals. The deepest species inhabit depths to about 90 fathoms. The most important species occur in relatively restricted regions of the subtropics whereas the less important species (either because of their low abundance or the difficulty of catching them) occur in the tropic regions. The number of species occurring near a particular locality in the tropics is much greater than in the subtropics, and each tropic species usually has a particular and reasonably narrow environmental preference compared with the more tolerant subtropic species. The six widespread tropic species of the Indo-West Pacific region can be arranged in a series from those preferring clear water, through moderately dirty water, to those that prefer very turbid water. As a result of these ecological preferences, each species tends to remain physically separate from the others. The distance separating the different species may be by yards or miles, but ecological differences of turbidity or type of shelter or type of bottom can usually be detected fairly readily. None of the *Panulirus* species apparently tolerates lowered salinities such as occur immediately outside large river mouths.

Of the four species in the west Atlantic region, *P. argus* has the widest distribution from North Carolina and Bermuda to Rio de Janeiro, and throughout this range is generally the most abundant and the most important commercial species. Two other species, *P. guttatus* and *P. laevicauda*, occur in the central part of the *P. argus* range; (they are rare in the Florida-Cuba region and do not appear to extend to Brazil). *P. echinatus* extends eastwards from its only mainland occurrence in Brazil to the islands of Cape Verde, St. Helena, and St. Peter and St. Paul Rocks (Chace 1966). In this respect *P. echinatus* achieves a remarkable oceanic distribution. Only one species of *Panulirus* occurs on the east Atlantic coasts, and French fishermen visiting Mauritania and Cape Verde Archipelago use tangle nets for its capture. There are no *Panulirus* fisheries of any importance on the Mediterranean or the Red Sea.

In the vast Indo-West Pacific region, *Panulirus* species are widespread, and six are restricted to small or moderate sized areas. The widespread species are:

(1) *P. penicillatus*—prefers clear, turbid-free water;
(2) *P. longpipes*—prefers clear waters;
(3) *P. versicolor*—prefers clear to moderately turbid water; juveniles prefer very clear water;
(4) *P. ornatus*—prefers moderately turbid water;
(5) *P. homarus*—juveniles prefer very turbid water; adults moderate turbidity and slightly cooler water;
(6) *P. polyphagus*—adults prefer very turbid waters.

All species of *Panulirus* with restricted distributions within the Indo-West Pacific regions, except *P. stimpsoni* of Hongkong, which is tropic, live in subtropic temperatures among reef or coral in clear to moderately clear conditions, and are commercially important. They are *P. japonicus*, Japan; *P. marginatus*, Hawaii; *P. pascuensis*, Easter Island; and *P. cygnus*, west coast of Australia.

In the east Pacific, the three species of *Panulirus* have adjacent distributions with little overlap; *P. interruptus* inhabits the subtropic water temperature zone of the California-Baja California region, and is less frequently found in the Gulf of Carpentaria, the region that is dominated by *P. inflatus*. *P. gracilis* dominates the Central American coast from the Gulf of California to Peru. It also occurs at the Galapagos Islands, where it is displaced from the very shallow water part of its range by *P. penicillatus* (Holthuis and Loesch 1967).

Jasus. The six species of *Jasus* completely dominate the temperate region of the Southern Ocean. They live among rocks and kelp and, unlike *Panulirus*, move about during the day as well as the night. Each species is of considerable commercial importance. All except one (*J. verreauxi*) are very closely related, and form the very distinct "lalandii" group, recently reviewed by Holthuis and Sivertsen (1967), who also recognized two sub-groups: a "lalandii" subgroup of three species—*J. lalandii* on the lower west coast of Africa—*J. novaehollandiae* along the southern Australian coasts, and *J. edwardsi* of New Zealand—and a "frontalis" subgroup *J. frontalis* from Juan Fernandez and Desventurados Islands, *J. tristani* of the Tristan Group, Gough I., and Vema seamount, and *J. paulensis*, occurring around St. Paul and New Amsterdam Is. These six species probably represent now isolated parts of a once continuous, circumpolar species.

The commercial importance of this group of spiny lobsters is very significant, and surprising catches have been made (and in some places maintained) from remarkably small geographic areas. Islands such as St. Paul, New Amsterdam, Tristan da Cunha, Chatham, Juan Fernandez, and even seamounts (Vema) are examples of successful fisheries.

Palunirus. Three of the four species of *Palunirus* are found in the Atlantic Ocean, the best known being *Palinurus elephas* of the European and Mediterranean coasts. *P. mauritanicus* occurs in the offshore waters of west Africa, and *P. charlestoni* has recently been discovered in the Cape Verde Archipelago. In the Indian Ocean, *P. gilchristi* lives at depths of 60–520 m along the southern and southeastern African coasts. The species of *Palinurus* are exploited by tangle net (French fishermen off West Africa), trawl (South African fishermen off Durban), and lobster pot (the fishermen of the European and Mediterranean coasts).

The remaining genera in the family Palinuridae are not of commercial importance at the present time mainly because they occur at greater depths than the existing commercial species and are usually much smaller. The genera are *Projasus*, with one species living off the southeast African coast at 360-m depth; *Linuparus*, with several species in the Indian and west Pacific Oceans at 100–250-m depth; *Justitia* at depths of 25–75 m in Indo-West Pacific; *Palinustus* with three or four species in West Indies and Indo-West Pacific at 100–250 m, and a much deeper group of species of the genus *Puerulus* at 250–1400 m in East Indies and in the Indian Ocean.

Life history and larval recruitment. The young larvae that hatch from the eggs carried by the female under the tail are very distinctive in shape. They are circular, flattened, and transparent, have large eyes on stalks, and feathery paddles on the legs to swim up or down.

Just how long the larvae of each separate species of spiny lobster take to complete their many larval moults is still unknown for most species, but for *Panulirus interruptus* off California (Johnson 1960) and *P. cygnus* off Western Australia it is about 9–11 months, passing through approximately 30 moults. During this time the larvae of these two species are able to maintain an offshore distribution roughly approximating the latitudinal extent of the coastal adult stock. How this is achieved is not clearly understood but "riding on the local ocean currents by up and down movements" has been suggested as the answer for the successful return of these floating larvae to the coastal populations. Just what happens to the larvae of the *Jasus* species in the isolated island situations is difficult to imagine—perhaps they have a quite different larval behavior and just hang on the bottom for most of the larval life.

R. W. GEORGE
T. G. KAILIS

References

Chace, F. A., "Decapod crustaceans from St. Helena Island, South Atlantic," *Proc. U.S. natn. Mus.*, **118**, 622–662 (1966).

George, R. W., and Main, A.R., "The evolution of spiny lobsters (Palinuridae): A study of evolution in the marine environment," *Evolution*, **21**, 803–820 (1967).

Holthuis, L. B., and Loesch, H., "The lobsters of the Galapagos Islands (Decapoda, Palinuridea)" *Crustaceana*, **12**, 214–222 (1967).

Holthuis, L. B., and Sivertsen, E., "The Crustacea Decapoda, Mysidacea and Cirripedia of the Tristan da Cunha Archipelago with a revision of the "frontalis" subgroup of the genus *Jasus*, *Results Norw. scient. Exped.*, *Tristan da Cunha* No. 52, 1–50 (1967).

Johnson, M. W., "Production and distribution of larvae of the spiny lobster, *Panulirus interruptus* (Randall) with records on *P. gracilis* Streets," *Bull. Scripps Instn. Oceanogr.*, **7**, 413–462 (1960).

Sims, H. W., "Let's call the spiny lobster "Spiny Lobster," *Crustaceana*, **8** (1), 109–110 (1965).

LONGLINE FISHING FOR SWORDFISH

The objective of the U.S. Fish and Wildlife Service in introducing Japanese longline fishing gear and methods to the U.S. was to test the fishing gear for possible application in the development of a commercial fishery for bluefin tuna (*Thunnus thynnus*) in the North Atlantic area. Starting with the original project in 1952, and continuing in 1953, longline fishing for tuna was employed by the Exploratory Fishing and Gear Development Section of the Bureau of Commercial Fisheries in the waters off the New England and Middle Atlantic States.

Fig.1. Setting basket of longline gear for tuna. (Photo: Bureau of Commercial Fisheries, R/V *Delaware*.)

Bluefin tuna catches on the longline gear were extremely low in both 1952 and 1953, ruling out the establishment of a commercial tuna longline fishery in the region. In 1952 a total of 118 sets with 8453 hooks produced 311 tuna and 683 sharks. In the following year the catch was markedly lower with only 2000 lb of tuna caught by the longlines. (Murray 1952, 1953). Additional tuna exploratory projects from 1954 through 1956 employed purse seine and pole-and-line fishing methods. Longline fishing operations were not resumed until 1957, when explorations for tuna were carried out in the northwestern Atlantic beyond the 100-fathom curve of the Continental Shelf in the comparatively warm waters of the Gulf Stream. While valuable information on the distribution and composition of the tuna stocks evolved from this project, the results from a commercial viewpoint did not justify adoption of tuna longline fishing by the commercial fishing fleet.

In 1960 and 1961, under the aegis of the National Science Foundation, the Woods Hole Oceanographic Institution conducted exploratory longline fishing for bluefin tuna and other large fish in the waters between the 100-fathom curve and the Gulf Stream from Cape Hatteras to the Nova Scotia Banks. (Mather and Bartlett).

Employing a former Coast Guard cutter, the R/V *Crawford*, for the project, fishing operations were conducted solely during the daylight hours. Moderately heavy catches of large bluefin tuna and small catches of albacore and yellowfin tuna were recorded. During a 1961 cruise off the Virginia Capes, after setting the longline gear, the vessel was forced to return to port without completing the set. The vessel returned after 72 hrs and retrieved the gear with a surprising catch of 16 swordfish (*Xiphias gladius*), presumably related to the fact that the gear had fished nocturnally. Subsequent fishing was confined entirely to night operations, and swordfish were caught in ever increasing quantities.

The commercial fishing fleet soon learned of this serendipitous event and hastened to enter the fishery. The following year, 1962, a catch of 400,000 lb of swordfish was reported by ten Canadian longline vessels. By 1964, a total of 30 U.S. vessels had converted from otter trawling to swordfish longline fishing. In 1963 and 1964, a total of over 2,000,000 lb of fish was landed by the fleet.

Longline gear operated by the vessels was basically the same as the Japanese style longlines introduced into New England in 1952. Some variations and modifications of the number, length, and spacing of the gangions, and employment of various types of line haulers were noted in the U.S. fleet. Typical of the vessels entering the fishery were the *Chilmark Voyager* and *Chilmark Sword*, steel vessels designed primarily for longline fishing.

The discovery that longline fishing could produce commercially rewarding quantities of swordfish revived interest of U.S. fishermen in this fishing method, accounted for a rapid changeover from otter trawling to longlining of scores of vessels, and provided a new and lucrative fishery to the northwestern Atlantic area.

Initial entry of approximately 30 vessels to the newly opened fishery during 1962 and 1963 was reduced to a fleet of 14 in 1964 and in 1967 four. The center of the fishery in 1967 was at Portland, Maine; New Bedford, Massachusetts; Hampton, Virginia; and Wanchese, North Carolina.

Based on production figures and market potential, the prospects for longline swordfishing appear to be extremely promising. An ominous note in the fishery is manifested by the small average size of the fish captured. Whereas the harpoon-prosecuted fishery produced large fish in the 200–300 lb range, longlining produces appreciably smaller fish averaging not over 150 lb, especially during the winter months when operations are carried out in the Gulf Stream waters.

Undoubtedly longline fishing would not have survived except for its ability to catch swordfish. Judging by the success of the fishery it appears reasonably certain that the long line method has supplanted the harpoon as the principal fishing method for capturing swordfish and that longlining will continue as an integral though subordinate part of the fisheries of New England and the Middle Atlantic areas.

FRANK E. FIRTH

Cross reference: *Tuna Fisheries.*

M

MACKEREL FISHERY, ATLANTIC

Of all the numerous salt water food fishes, the Atlantic mackerel, *Scomber scombrus*, has presented marine researchers with one of the most fascinating and elusive puzzles. Its curious periodic cycle from dearth to superabundance was known to American colonists of the 1600's, who considered mackerel an important staple commodity. This valuable source of food at their very doorstep was free for the taking, and return of the schools each spring was eagerly anticipated. Unpredictable fluctuations in the catches occurred from year to year, however, and caused apprehension among the colonists. As early as 1670, the American settlers had enacted laws to prevent over-fishing by regulating the season and method of capture.

In many parts of the world changes in the mackerel's migration habits bring these same periodic fluctuations, and affect the economy of ports to which this fishery is adjacent. A transitory prosperity, frequently having the appearance of permanence, will encourage a local industry until the spring arrival of the schools fails to occur.

Among the many marine fishes some species bear a striking resemblance to others. The mackerel, however, is readily distinguished by its streamlined conformation and coloration. Iridescent greenish-blue covers most of the upper body, turning to blue-black on the head, while the belly is silver white. The body is barred with from 23 to 33 bands running down from the dorsal in a wavy course to the lateral line region. Brilliant as the colors are in life, they fade quickly after capture. Adult fish are between 14 and 18 in. long and weigh $1\frac{1}{4}$ to $2\frac{1}{2}$ lb. Occasional individuals measuring 22 in. may weigh as much as 4 lb and, in 1925, a $7\frac{1}{2}$-lb fish was caught.

The mackerel is considered a fish of the open sea, but seldom ventures beyond the limits of the continental shelf. Occasionally some of the smaller fish will enter estuaries and harbors in search of food, but they avoid fresh water entirely. Its range apparently is between northern North Carolina to the northwestern shores of the Gulf of St. Lawrence. In 1871, after mackerel had been completely absent for 40 years, Bonnes Esperance and Meccatina, on the Labrador coast, packed a total of between 400–500 barrels, but that region is considered well beyond its general range.

During an average spawning life of 4 years, the mackerel may produce up to 4,000,000 eggs. A moderately prolific fish, one female may spawn 500,000 ova in a single season, although not more than 50,000 are extruded at one time. There is no definite evidence that these fish have a preferred breeding ground. The eggs, apparently, are shed wherever the fish happens to be when the ova ripen. Some investigators contend otherwise, believing that spawning is not so haphazard, but that there is a definite pattern, and once the reproductive products have been discharged the movement of the school only then becomes more leisurely.

Spawning occurs off the American coast from spring through early summer. Starting in mid-April in the latitude of Cape Hatteras, it progresses northward. By May it takes place off New Jersey and in June off southern Massachusetts. In late June and early July, spawning will occur in Nova Scotian waters and the lower Gulf of St. Lawrence. Almost the entire breadth of the continental shelf is covered, the most productive area being between the Chesapeake Bay and southern New England.

Research vessels, growing in importance as the heavily populated earth turns more than ever to the sea, are an indispensable tool in the investigation of oceanic resources. Marine biologists aboard these vessels, collecting plankton in special tows, study the concentration of mackerel larvae and fry. Length frequency graphs of mackerel taken by seine indicate the preponderance, or relative absence of certain length groups, or year classes. By further studies of the relationship between ova in the plankton sampling and mackerel larvae present these biologists have a means of determining how many fish must survive per million eggs.

Out of every million eggs one fish must survive to keep a population constant. To be considered a successful year class, survival of more than four fish at the 2-in. size per million eggs is required. In the egg and larval stages, many factors influence mortality. Adverse winds may push the waters in an unfavorable direction during the time when the fry lack sufficient motility and prevent them from reaching suitable nursery grounds. A lack of desirable zooplankton when needed reduces survival. Both conditions occurring together in one season are disastrous, and the resulting havoc is felt throughout the industry for many seasons. Depleted year classes would have insufficient spawners for a swift restoration of the population. A spawning stock can also grow too large and, by the very weight of numbers, become detrimental to the development of the recruitment body.

Besides contending with the caprices of nature, mackerel are preyed upon by many forms of sea life such as whales, porpoises, sharks, tuna, bonito, bluefish, and striped bass. Cod, squid, and other fish destroy great numbers of young fish less than 4 to 5 in. long, and sea birds devour multitudes of the smaller ones.

The earliest hatched fish will grow to between 1¼ and 2½ in. in length by June, 2½ to 5 in. by August, and will reach the "blink" length of 6½ in. by the end of August. Young fish caught in October, known on the American northeast coast as "tacks" or "spikes," measure about 7 to 8 in. Most fish of the year will be from 8 to 9 in. by late autumn. When the second summer has passed, the average fish of this year class, known as "tinkers," will be from 12 to 14 in. Growth from then on is gradual, and maturity is reached in the third year when reproduction begins.

Virtually all marine animals neither too large to be swallowed, nor too small to be seen, make up the diet of the mackerel. At times they are caught packed full of Calanus, the "red feed" or "cayenne" so named by fishermen. Any small fish larvae and minute pelagic crustacea, such as crab larvae and copepods, are eaten. Euphausiid shrimps rank high on its menu and squid, launce, and annelid worms have also been found in the stomachs. Feeding is by two methods, depending on the food available. Smaller pelagic organisms are filtered out of the water by the rows of gill rakers. Visual selection is made of larger individual organisms.

Instead of being one vast homogeneous stock as was once supposed, the mackerel native to the American coast are actually contained in two populations, a southern and a northern. These have been distinguished by analyses of size composition records. The southern vanguard appears from offshore in early April, advancing towards Virginia, Maryland, and New Jersey, and moving slowly northward to spawn off New Jersey and Long Island. In late May, the northern contingent enters southern New England waters for a short period, mingling with the other contingent, but soon moving on again, and in June or early July, spawns in the vicinity of

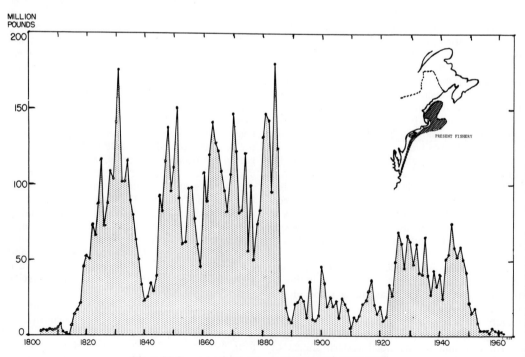

Fig. 1. U. S. catch of Atlantic coast mackerel, 1804–1962.

Nova Scotia and the Gulf of St. Lawrence. Such spring movements appear to be spawning migrations, and are probably triggered by water temperature, about 46°F being critical. For many years Gloucester fishermen took advantage of the early appearance of the southern population, sailed their boats down the coast to fish off New Jersey, and remained with the schools during the northward migration.

As autumn approaches, fish that summer along the Maine coast, mostly of the southern contingent, work back southward toward Cape Cod and disappear, after October, off Block Island. The northern contingent returns through the Gulf of Maine in November or early December, finally vanishing off Cape Cod. Infrequent catches have been made throughout the winter by otter trawlers and New Jersey pound nets, but, in general, by Christmas all mackerel have disappeared, probably into deep water, where they winter in a more acceptable temperature.

Fresh mackerel is considered by many to be one of the choicest of food fishes. In the 1800's mackerel were caught close inshore, dressed, and placed in tubs of salt water, which was changed frequently to keep the fish cooled. The object was to catch the mackerel and get them to market before daylight where they were sold in the cool of the morning.

Today, the freshly caught mackerel are immediately iced and packed in boxes for shipment. Improved handling and transportation methods have considerably increased the radius of fresh fish sales.

The first commercial voyages were strictly for salting purposes, and about 1818 the fleet began to grow. More than 900 vessels operating from New England brought in as much as 100,000,000 lb during 1838–39. In years of local scarcity some seiners even sailed to Europe for their fares.

Earliest records (1626) indicate that the principal commercial method of capturing mackerel was by haul seine. This continued as the major fishing gear until 1800, when "drailing" was favored. For this method of fishing a vessel was fitted with a number of outriggers (poles) to which lines were attached. On each line was tied a sinker and a hook, which was generally baited with pork rind. The vessel had to be underway before the mackerel would bite. About 1812, the sinker was attached to the shank of the hook, forming a unit called a "jig", which had the advantage of being more durable and effective.

The purse seine came into general use about 1850, and by 1870 the hook and line had largely been displaced as commercial gear. Mackerel school differently from other gregarious pelagic fishes, such as the herring and menhaden. Swimming schools disturb myriad tiny organisms, creating a "firing" or luminescence, which indicates the presence of a school to an observer on the seiner's masthead. The towed seine boat is manned and the school is encircled. Not only did the purse seine affect productivity in the mackerel fishery, the coming of power to replace sail gave impetus to the growing efficiency of the fishermen. This gear brought a freedom from bait supplies, and permitted fishing earlier in the season when prices were usually higher. To the present, the purse seine has continued to be the prime mackerel gear when this species is abundant.

During recent years, mackerel have been exceedingly scarce, and seines have not been able to operate profitably. In this period of scarcity the fishing gears that have produced most mackerel are the pound net and the floating trap. There is also a small gill net fishery, which concentrates on the spring and fall runs of fish, although in recent years this gear also has not been very successful.

A market demand for fresh mackerel in preference to the salted product occurred during the changeover from sail to power. The expanding fresh mackerel market required more rapid landings, even though the vessel's capacity had not been reached. By 1920 most fishermen were in power-driven vessels. Such quick trips resulting from power frequently flooded the market and depressed prices most when fish were most abundant. The canneries, however, depended upon these low-priced fish and produced a very palatable product that found wide acceptance.

The mackerel catch phenomenon in the Gulf of Maine may have a variation of 100 times as much more in a good year than in a poor one. When mackerel were plentiful in the 1940's, they were among the four most valuable Gulf of Maine fishes, surpassed in dollar value only by lobsters, haddock, cod, and the ocean perch (redfish). Massachusetts was by far the heaviest producer of mackerel, but recent years of extreme scarcity have forced fishermen and processors away from dependence on this species. In 1960, Massachusetts purse seiners sold only 300 lb valued at $38. Pound nets and floating traps, however, did account for almost 2,000,000 lb, which brought $258,000.

The astonishing changes in abundance are due to survival of the young. Sustained good years depend entirely on successful year classes year after year. A break in this chain for any reason will be followed by a sharp decline in the catches. A year when most of the fish caught are large, with very few small fish among them, indicates that returns in the next few years will be discouraging, whereas a year in which great numbers of small fish are observed will be followed by several years of heavy landings.

There is no record that large amounts were landed in Massachusetts before 1815. In 1810, Massachusetts had passed an inspection law requiring that barrels containing pickled fish be branded as to content: salmon, mackerel, shad, etc. Records kept by the inspection department point to the period 1825–1835 as one of great abundance of mackerel, averaging over 65,000,000 lb (round weight) a year. During the following eight years, 1835-1845, the average annual production was only 24,000,000 lb in Massachusetts. The catch then soared to 100,000,000 lb, to be followed by a disastrous scarcity that almost ruined a burgeoning industry.

Canadian landings data are available from 1876 and, in general, undergo the same fluctuations apparent in the United States catch, but to a more moderate degree. During the period from 1876–1949, the U.S. landings were the greatest in 57 of the 74 years. The Canadian landings have been greater in each of the years since 1949.

In 1910, the entire eastern United States coast produced only 6,000,000 lb, almost none coming from Massachusetts and the Maine coast. The catch started climbing in 1911, and by 1917 had reached 37,000,000 lb, still far short of the tremendous landings of the 1880's. Another decline set in after 1917 and the catch fell to 10,000,000 lb in 1921. In keeping with a see-saw pattern of unpredictability, mackerel returned to the Gulf of Maine in 1925, and fishermen harvested 49,000,000 lb. Fluctuations like these continued with 1932 a high year (59,000,000 lb) while 1937 fell to 20,000,000 lb. The 13 years 1933–1946 brought to the markets landings averaging 37,000,000 lb per year, with 1944 a high year of 74,000,000 lb, after which production dropped steadily to a low 2,100,000 lb in 1962. There are presently no signs that the spectacular recuperative powers of the mackerel are operating to restore the species to its former abundance.

George M. Clarke
Dwight L. Hoy

References

1. Dominion Bureau of Statistics, Canadian fishery statistics.
2. Gloucester Master Mariner's Association, "The Mackerel Fishery, Its Yearbook," 1947. pp. 17-27.
3. Goode, G. B., and Collins, J. W., "The Mackerel fishery of the United States," in G. Browne Goode, "The Fisheries and Fishing Industries of the United States, section V, History and methods of the fisheries, 1887, vol. 1, pp. 247-313, U.S. Commission of Fish and Fisheries, Washington, D. C.
4. Sette, O. E., "Biology of the Atlantic mackerel of North America, Part 1," U.S. Fish and Wildlife Service, Fishery Bulletin 38, 1943, vol. 50, ii pp. 149-237.
5. Sette, O. E., and Needler, A. W. H., "Statistics of the mackerel fishery off the east coast of North America, 1804 to 1930," Bureau of Fisheries, Investigational Reports No. 19, 1934, vol. 1, 48 pp.

MAGNESIUM FROM THE SEA

Magnesium is the third most plentiful chemical element in seawater. Only chlorine and sodium occur in greater amounts. The development of a practical and economical process for extracting metallic magnesium from the waters of the sea has guaranteed the world an inexhaustible supply of this very versatile metal, which has a multitude of both structural and nonstructural uses.

Magnesium is silvery white in color and has a density of 1.74 gm per cm^3 or 108.6 lb per ft^3. Magnesium is in group IIA of the periodic table of the elements. The atomic number is 12, and its atomic weight is 24.32. The low density of magnesium gives it the distinction of being the lightest of the metals used for structural purposes. Aluminum is 1½ times heavier. Zinc and iron weigh about four times more, and copper is five times heavier.

In the pure state, magnesium has many chemical and metallurgical uses throughout industry. For structural usage, magnesium, like other pure metals, must be alloyed with other metallic elements. The alloying metals in common use include aluminum, lithium, manganese, rare earth metals, silver, thorium, zinc, and zirconium. The addition of these metals in various combinations provides magnesium with mechanical

Fig. 1. Each cubic mile of seawater contains 6,000,000 tons of magnesium. With the earth's surface containing an estimated 331,000,000 cubic miles of seawater, magnesium is truly inexhaustible.

Fig. 2. Close-up view of three of the huge settling tanks from which the magnesium hydroxide slurry is drawn off and converted to magnesium chloride, which is electrolyzed to produce metallic magnesium and chlorine gas.

strength as well as other properties and characteristics that are needed in a variety of applications.

Man has always depended on the sea as a source of food, but scientists have also found indications that primitive man utilized the salts of the sea. Some attempts were made over the centuries to obtain various other compounds from the sea, but available records do not indicate any substantial progress until 1923, when magnesium chloride and calcium sulphate were produced from the bitterns obtained by solar evaporation of water from San Francisco Bay. From that time on, progress was rapid, and the sea has since been a source for chemical elements as well as compounds. Bromine was produced from the San Francisco Bay bitterns in 1926. Similarly, potassium and bromine were recovered from the Dead Sea in the early 1930's. It has been said that the annual value of substances extracted from seawater throughout the world totals about $136,000,000, with the United States producing 82% or $112,000,000.

By 1933, the ion exchange process had replaced solar evaporation as a means of concentrating the desired sea salts, and this led to the use of lime to precipitate magnesium hydroxide. The Dow Chemical Co. of Midland, Michigan has been a producer of metallic magnesium since 1916, using initially what might be termed pre-historic seas existing as naturally-occurring brines underlying Midland, Michigan. The experimental work that was done to develop Dow's successful process for the extraction of bromine from seawater disclosed that magnesium was present in an amount 20 times greater than bromine. Foreseeing the need for a greater production capacity for magnesium, the company began an extensive research program that resulted in the development of a workable process. Successful on a laboratory scale, it went to pilot stage and finally a production plant was constructed at Freeport, Texas. On January 21, 1941, the plant went into operation. On this date, Dow poured the first ingot of any metal ever to be produced from seawater. Dow remained the sole producer of metallic magnesium from seawater until 1951, when Norsk Hydro Elektrisk at Heroya, Norway began the operation of its process.

Production

The combined production of two world producers of magnesium from seawater, The Dow Chemical Co. in the U.S. and Norsk Hydro in Norway, accounts for well over half of the total world production of magnesium today. Dow's plant is still located in the Gulf of Mexico at Freeport. At this plant the first step in obtaining magnesium is to treat the seawater with lime obtained by roasting oyster shells, another product of the sea. This converts the magnesium in the seawater to magnesium hydroxide (milk of magnesia), which is insoluble. The magnesium hydroxide precipitate is collected in huge settling tanks from which it is drawn off as a slurry. In the next step, filters are used to obtain a magnesium hydroxide cake, which is mixed with a magnesium chloride solution so that it can be pumped to the neutralizers where hydrochloric acid prepared from chlorine and natural gases converts the magnesium hydroxide to a solution of magnesium chloride. The next step is the evaporators, where most of the water is removed, leaving magnesium chloride, which then goes to the driers prior to being fed to the electrolytic cells. Electrolysis of the magnesium chloride yields magnesium metal, which is cast into ingots, and chlorine gas, which is returned to the process to make hydrochloric acid.

The production of magnesium from seawater as practiced by Norsk Hydro Elektrisk is a modification of the I.G. Farbenindustrie process introduced into Norway at the time of the German occupation during World War II. Norsk Hydro uses dolomite, a mineral comprised of

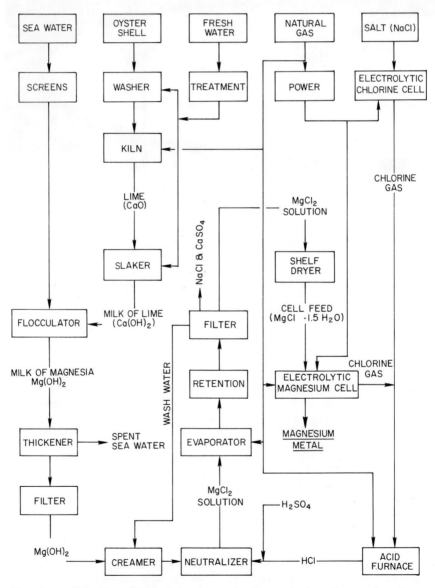

Fig. 3. Flow sheet of The Dow Chemical Co. process for the production of magnesium from seawater at Freeport, Texas.

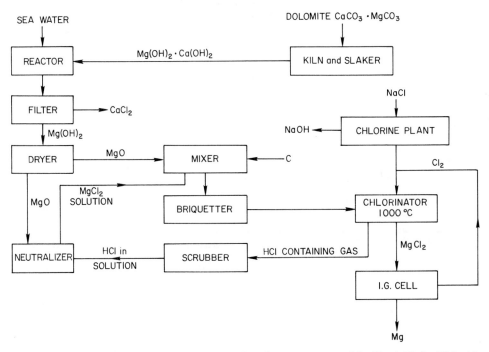

Fig. 4. Flow sheet of the process for producing magnesium from seawater used by Norsk Hydro Elektrisk at Heroya, Norway. (After "Magnesium and Its Alloys," by C. S. Roberts, p. 210, John Wiley and Sons, New York.)

calcium and magnesium carbonates, instead of oyster shells to precipitate the magnesium hydroxide out of the seawater. The magnesium hydroxide goes to a drier where it is converted to magnesium oxide. The oxide is then mixed in proper proportion with carbon, in the form of powdered coal or coke, and is then wetted with magnesium chloride brine. This mixture is then fed to briquetting machines, and the resulting briquets then go to the chlorinators, which operate at 1850°F. The result is fused or anhydrous magnesium chloride, which is fed to electrolytic cells to be electrolyzed into its component parts of magnesium metal and chlorine gas.

It is significant that magnesium is available to all countries having a sea coast. The essential ingredients needed to establish a plant for the production of magnesium from seawater include:
(1) a source of seawater undiluted by nearby rivers;
(2) availability of limestone, dolomite, or sea shells;
(3) salt beds, brines, or other sources of chlorine;
(4) low cost power, natural gas or coal;
(5) a plentiful source of fresh water.

The availability of magnesium from the waters of the sea gives it a big advantage over other metals. For these other metals, we must depend upon deposits of ore bodies, which are widely scattered throughout the world. Eventually, ore deposits will be depleted, and, in fact, many countries today have only a few years' supply of critical metals.

Abundance

As already noted, magnesium, with seawater as a source of supply, can be considered virtually inexhaustible. Scientists have stated that the sea probably contains every known element. The magnesium content, exceeded only by chlorine and sodium, is 1,272 ppm. This is equivalent to 12,000,000,000 lb or 6,000,000 tons per cubic mile. Estimates for the number of cubic miles of seawater on the face of the earth range as high as 331,000,000, making the total magnesium content the astronomical amount of approximately two quadrillion tons. In addition to the amount of magnesium already in the sea, it has been estimated that the content is being replaced by the soluble magnesium salts being carried into the sea by rivers at the rate of 137,000,000 tons per year. From this, it is apparant that the current world usage of magnesium at the rate of 189,000 tons per year is exceeded by the amount of magnesium washing into the sea each year. These statistics confirm the fact that magnesium, due to its presence in seawater, is forever available to the world.

Uses

A versatile metal, magnesium is used for many nonstructural applications, as well as for those where structural strength is required. In nonstructural applications, magnesium is used in its pure state. Several grades of primary magnesium are produced; it is supplied in the form of pigs, ingots, sticks, and turnings to best suit the intended use.

Metallurgical applications account for the greatest tonnage for nonstructural uses, and the greatest volume in this field is its use in aluminum alloys. Other metallurgical applications include its utilization as an alloying constituent in copper, lead, nickel, and zinc alloys. Appreciable magnesium also is used in the production of nodular iron, also commonly known as ductile cast iron. Magnesium has been found to be the ideal metal for use in the thermal reduction process to produce such other metals as beryllium, hafnium, titanium, uranium, and zirconium.

For structural uses, magnesium is alloyed with other metals to produce a series of alloys with high strength-to-weight ratios. The structural uses of magnesium alloys include many items in a variety of industries such as aerospace, business machines, electronics, consumer goods, graphic arts, industrial machinery, materials handling, portable hand tools, ground transportation, and the tooling industry. The lightness of magnesium in its various forms has always made it attractive to the aircraft market where it has been used in many secondary structures. These include engine parts, flooring, control surfaces, skin and landing wheels. In recent years, missiles and space vehicles have utilized magnesium extensively. As a result, it has been estimated that there is more magnesium than any other metal now floating out in space.

WILLIAM H. GROSS

References

Beck, A., "The Technology of Magnesium and its Alloys," 2nd ed., 1941, transl. of "Magnesium und seine Legierungen," F. A. Hughes & Co. Limited, London.
Ryder, David W., "Magnesium Salts from the Sea," 1947, Marine Magnesium Products Corporation, San Francisco.
Gross, W. H., "The Story of Magnesium," 1st ed., 1949, American Society for Metals, Metals Park, Ohio.
Schambra, W. P., and Ballard, D. A., "Can our critical minerals be produced from sea water?", *Yale Scientific Magazine*, pp. 7–14 (January, 1959).
Roberts, C. S., "Magnesium and Its Alloys," 1960, John Wiley & Sons, New York.
Comstock, H., "Magnesium and Magnesium Compounds," 1963, U.S. Dept. of the Interior, Bureau of Mines, Washington, D.C.
Tangen, Thorstein, "Magnesium Production in Norway—Present State and Future Growth," Paper #63-6, *Proceedings of the Magnesium Association*, 1963.
"Magnesium—metal on the move," *Precision Metal Molding*, 23, No. 7, 35–61 (1965).
Church, F. L., "Magnesium: starting to make out in mass markets," *Modern Metals*, 22, No. 6, 57–81 (1966).
Emley, E. F., "Principles of Magnesium Technology," 1966, Pergamon Press Ltd., London.

Cross references: *Bromine From the Sea; Chemistry of Seawater; Minerals of the Ocean.*

MAINE MARINE WORM FISHERY

Probably the most valuable, regularly harvested, marine animal in the world per unit of weight is the annelid *Glycera dibranchiata* Ehlers, the bloodworm. Prices paid diggers for this salt water, sport fishing bait are as high as $5.25 a lb, or roughly seven times the maximum annual average price paid fishermen for lobster, which brings the highest value per lb of any major North American fishery.

For approximately 40 years bloodworms have been dug from the tidal flats of Maine, and during most of those years the average lb contained 44 worms, which were sold to buyers and dealers on an individual basis. The increase in the market for bait worms and the limited available supply have recently combined to decrease the average length of the individual worms that the market will accept by 50% or more. Sampling of bloodworms for acceptable sizes indicates that a pound may contain as many as 150 worms. Since harvesting is by custom limited to the intertidal zone during ebbtide, the number of worms that can be dug fluctuates from day to day with variations in tidal amplitude. This means that larger worms will be found only during spring tides when those more seaward areas, not regularly available to the digger, will be exposed. The number of worms that can be dug will also vary with tidal amplitude. Fifteen years ago marine biologists in Maine observed that a + 1 ft low tide reduced the take of marine worms on the average about 30% in comparison with a 0 low tide.

Although worms are known to inhabit the ocean bottom well beyond the drainage area of maximum spring tides, men regularly employed in harvesting bait are not interested in using hydraulic or mechanical equipment, nor will they tolerate others using such methods. It is obvious that subtidal populations of both bloodworms and of *Neanthes (Nereis) virens*

Sars, the sandworm, will have to be harvested if bait demand is to be met.

In addition to the bloodworm, another polychaete, *Neanthes virens*, the sandworm or clamworm, is also used as salt water angling bait. Currently sandworms bring the digger $1.90 per 100 or from $.75 to $1.50 per lb, depending upon the weight of the individual worm.

Both species of marine annelids are sold for bait, the principal market area being from Long Island Sound to Chesapeake Bay. Commercial diggers specialize in, and almost invariably will harvest, only one species, even though in some growing areas both species will be found in commercial concentrations.

Both worm fisheries are largely seasonal. By reason of reduced winter markets, unfavorable weather conditions, frozen and ice-covered flats, the fisheries are limited to the period of the year from March through November. Table 1 shows the average monthly catch of marine worms by species for two five-year periods, 1949–1953 and 1962–1966.

Sandworms had for many years averaged 40 per lb, but in recent years the market has tolerated a decrease in size to approximately 60 or, even at times, to as many as 75 per lb. But always sandworms have had to be larger than bloodworms in order to be commercially acceptable. With the wide range of tolerance, conversion factors have for purposes of convenience remained the same, 40 for sandworms and 44 for bloodworms; while in actual practice the market has accepted up to more than 150 bloodworms and nearly 80 for sandworms.

Both species appear to have a wide salinity tolerance, from relatively brackish to full seawater. Yet exposure to fresh water by diggers who utilize this method of washing to increase the apparent size of their worms, will usually result in a high mortality rate during storage and shipment to market.

Both species are dug by means of short-handled spading forks whose tines have been bent approximately 90° to form a hoe. Two tines are added by welding at either end to increase the width of the implement. Digging is done by holding the hoe in one hand and turning over the sediments of the exposed tidal flats with a quickly executed chopping motion. As worms are exposed they are picked up with the free hand and placed in a container. Bloodworms may be collected in galvanized pails, but a wooden bucket or comparable container must be used for sandworms, since this species is very sensitive to toxic metals.

As both species are efficient burrowers, this ability accounts for the high rate of escape from diggers each time an area is dug. Some flats are dug as many as 20 times a year. The productivity of some areas is indicated by the record of one of the most intensively harvested flats in Maine. During each of the past 20 years, the average value to the diggers of the worms from this less than 20-acre area has been $75,000, or $3,750 per acre per year (Fig. 1).

Sporadically the bait worm industry in Maine, like other common property commercial fisheries, has supported restrictive legislation under the guise of conservation. Until 1955 more and more coastal municipalities requested state regulations prohibiting municipal nonresidents from digging marine worms. Although marine biologists had repeatedly observed that marine worms were fully capable of crossing political boundaries, many coastal residents held to the underlying assumption that such restrictions were good conservation.

Following industry support of the repeal of these regulations, many have persuaded themselves that minimum size limitations should be established. Only anesthetized worms can be measured accurately, and available anesthesias have a generally permanent effect on the behavior

TABLE 1.

Month	1949–1953 Number of Bloodworm	Sandworm	1962–1966 Number of Bloodworm	Sandworm
January	14,124	840	91,203	52,192
February	36,080	600	114,294	70,456
March	478,896	41,200	1,575,050	959,320
April	1,498,904	260,440	3,987,553	2,390,064
May	1,869,956	584,760	4,848,474	5,133,840
June	1,364,660	1,047,720	4,255,909	5,714,992
July	1,503,656	880,080	4,800,004	5,240,720
August	1,524,512	811,720	4,477,070	4,851,048
September	1,696,024	551,920	3,107,192	2,789,184
October	1,044,868	559,720	2,216,183	2,263,384
November	865,172	254,040	1,146,191	687,792
December	365,112	113,240	247,597	147,752

Fig. 1. Bloodworm (top) and sandworm. (Photo: Ivan Flye.)

of both species. In a report on marine worm conservation and management in 1955, Dow and Wallace concluded: "On the basis of the information so far accumulated it does not appear desirable to establish any more restrictions and the restrictions presently in force do not appear to have any management or conservation value whatsoever." With the biological research done on marine worms since 1955, all evidence supports the validity of this assumption.

In some coastal communities restrictions on worm digging have been supported by commercial shellfishermen. The basis for this support appears to be the belief that digging marine worms will destroy large quantities of shellfish as, in some instances, it has. It is also alleged that if a choice is to be made between shellfish and worms, it should go to the shellfish, even though many of these animals may also be used for angling bait. Hydraulic harvesting, as has been demonstrated, would obviate the necessity of a choice.

Interspecies harvesting competition is then one of the basic components of conservation regulation. In general, commercial concentrations of shellfish and marine worms do not occur in identical geographic areas.

Restrictive legislation increased rapidly between 1937, in the early years of the fishery, and 1955, when all local regulations were repealed. Much of the legislative argument presented for repeal by the industry was based on the presumption that curtailment of free-roving activity of the diggers had reduced the catch of worms. After repeal of these laws, there was no appreciable increase in production of worms for three years, suggesting that this had not been an entirely valid argument and that, in all probability, other factors were influencing abundance and production.

With consideration of market restrictions on minimum sizes, the life cycle of the two species, the coastwide dispersion of an adult population, the migratory capability of marine worms, and the apparently good biological condition of the resource there appears to be, for the predictable future, no defensible grounds for establishing any restrictions on the use of the resource.

Reasons for fluctuations in production suggested by the industry, as well as by marine biologists, have ranged about as widely as has production itself. Cyclic changes in environment (Dow 1951, Dow and Wallace 1955), gradual changes in soil composition (Klawe and Dickie 1957), expansion of area dug (Dow and Wallace 1955), and changes in tidal exposure because of bridge and highway construction (Ganaros 1951) are some factors that have influenced production.

Deviations from high production levels between $46.6°$ and $48.8°F$, which occurred in 1949 and 1950—although still higher than any other year outside the optimum range—can be accounted for by a bridge and causeway construction project in those two years, which drastically reduced tidal exposure in one major producing area. Estimates made independently by both the industry and the Maine Department of Sea and Shore Fisheries of annual production losses resulting from this construction ranged from 25 to 30%.

Dow and Wallace (1955) concluded that year-to-year fluctuations in production were indicative of short-term natural fluctuations in abundance, and Dow (1964) demonstrated a consistent

Table 2. Annual Seawater Temperature and Worm Production[a]

Year	Temp. in Declining Order, °F	No. Bloodworms in Millions	Year of Production	No. Sandworms in Millions
1953	52.0	7.5	1956	11.3
1951	51.4	10.6	1954	11.4
1954	50.3	10.5	1957	11.6
1952	50.2	8.9	1955	7.2
1955	50.0	13.6	1958	10.8
1950	49.3	11.2	1953	9.7
1957	48.8	24.2	1960	24.5
1947	48.6	13.7	1950	—
1956	48.6	18.8	1959	21.5
1960	47.9	32.2	1963	32.5
1963	47.9	31.5	1966	31.8
1958	47.4	26.1	1961	25.7
1946	47.3	17.7	1949	—
1961	47.3	33.4	1964	30.9
1945	47.1	25.0	1948	—
1959	47.0	25.7	1962	27.1
1962	46.6	33.9	1965	29.5
1944	46.4	7.2	1947	—
1943	45.3	2.6	1946	—

[a] Annual temperature data in 1948 and 1949 are incomplete.

correlation between seawater temperature during the year of spawning and supply three years later. Klawe and Dickie (1957) concluded that bloodworm catches in Nova Scotia, the other important source of supply, consist largely of three-year olds.

In recent years the average size of worms in the catch has declined, indicating both that diggers are taking younger worms and that the market is accepting smaller sizes. It is assumed that no worms less than a year old are being harvested, but it is very likely that an increasingly greater number of two-year old worms is entering the fishery.

Since 1953, seawater temperature, as measured by the United States Fish and Wildlife Service at Boothbay Harbor, has steadily declined—from an annual average of 52.0°F in 1953 to a mean of 45.7° F in 1966. Previous studies (Dow 1964) have suggested that (1) abundance of worms for the fishery is related to favorable temperatures during the year of spawning and (2) the optimum temperature range during the year of spawning is from approximately 46.6 to 48.8°F (Table 2). On the basis of this evident relationship, it was predicted in 1965 that the abundance and production of bloodworms would probably begin to decrease in 1968.

The less intensively exploited sandworm is also influenced in its abundance by favorable temperatures during the year of spawning, but there is not the same evidence of as high a fishing mortality rate for this species as there is for the bloodworm. It can be anticipated that the actual abundance of sandworms will decline at the same time and for the same reason, but, with a potentially greater available supply, this decline may not become evident as early.

Marine worms reach minimum market size some time after their first year. The larger worms taken in the fishery are probably three years or older. According to Gustafson, "Neither the time required to reach sexual maturity nor the life span of Nereis (Neanthes) virens are known

Fig. 2. Packing worms for shipment. (Photo: Ivan Flye.)

TABLE 3.

Year	Bloodworm Worms in Millions	Bloodworm Value	Sandworm Worms in Millions	Sandworm Value	Total Number	Total Value
1946	2.6	$ 57,125	2.3	$ 47,188	4.9	$ 104,313
1947	7.2	144,530	2.0	37,086	9.2	181,616
1948	25.0	305,044	3.1	57,307	28.1	362,351
1949	17.7	297,021	1.4	18,910	19.1	315,931
1950	13.7	242,081	2.3	37,158	16.0	279,239
1951	9.5	157,966	5.9	88,412	15.4	246,378
1952	9.2	178,312	6.3	91,109	15.5	269,321
1953	11.2	217,966	9.7	148,499	20.9	366,465
1954	10.6	200,518	11.4	167,196	22.0	367,714
1955	8.9	167,004	7.2	110,283	16.1	277,287
1956	7.5	150,748	11.3	177,672	18.8	328,420
1957	10.5	246,436	11.6	214,344	22.1	460,780
1958	13.6	309,678	10.8	193,853	24.4	503,531
1959	18.8	371,832	21.5	334,285	40.3	706,117
1960	24.2	482,100	24.5	365,850	48.7	847,950
1961	26.1	515,979	25.7	387,066	51.8	903,045
1962	25.7	516,362	27.1	421,267	52.8	937,629
1963	32.2	696,887	32.5	506,578	64.7	1,203,465
1964	33.4	745,315	30.9	450,544	64.3	1,195,859
1965	33.9	759,582	29.5	447,341	63.4	1,206,923
1966	31.5	731,335	31.8	509,018	63.3	1,240,353

with certainty. Copeland kept isolated specimens living under laboratory conditions for at least three years."

It is likely that nearly all members of the two species spawn only once—toward the end of their third year, and that, after spawning, all, or nearly all, of the spent spawners die. For a short period prior to spawning, mature females are not acceptable for market because of their poor survival and the mortalities they are blamed for causing to other worms in the same container. Spawning of bloodworms was first photographed by research personnel in Maine in 1967. Males spawned first, discharging their semen into the water near the surface. The females then appeared and swam madly about discharging their eggs in the water that had been exposed to male spawning.

The natural mortality rate of either species is not known. It is obvious that year class populations will decline with time, and that the supply of two-year-olds will be greater than that of the same year class as three-year-olds.

The acceptance of smaller size worms by the market will increase the total supply available and, even with a decrease in abundance because of less favorable seawater temperature, the increased use of younger and smaller worms may initially serve to obscure this decline.

Marine worm production data were not differentiated by species until 1946 (Table 3). Since then, temperature data are the only data that can be used consistently to account for fluctuations in the abundance of bloodworms as indicated by commercial production. Fluctuations in the supply of sandworms, because of a lower level of exploitation, did not become evident until the mid-1950's. Since that time, the temperature-supply relation appears to be approximately the same for sandworms as it has been for a longer period of time for the bloodworm population.

From 449 in 1948, when worm licenses were first issued, the number of diggers has increased to well over 1000. Increased production of marine worms between 1959 and 1967 reflects not only more favorable spawning year temperatures, 46.6 to 48.8°F, but also further exploitation of growing areas, which had not been dug at all or only sporadically. The marked increase in catch in 1959 resulted from further expansion of the sandworm fishery.

ROBERT L. DOW

References

Dow, Robert L., and Wallace, Dana E., "Marine worm management and conservation," Fisheries Circular No. 16, 1955, Maine Department of Sea and Shore Fisheries.

Dow, Robert L., "Changes in abundance of the marine worm *Glycera dibranchiata,* associated with seawater temperature fluctuations," *Commercial Fisheries Review* 1964, Fish and Wildlife Service, Sep. No. 708.

Ganaros, Anthony, *Commercial Worm Digging,* 1951, Maine Department of Sea and Shore Fisheries.

Gustafson, A. H., "Some Observations on the Dispersion of the Marine Worms *Nereis* and *Glycera,*" Fisheries Circular #12, 1953, Maine Department of Sea and Shore Fisheries.

MARINE BORERS

Marine borers have been a menace ever since man has had anything to do with the sea. Ancient civilizations were well aware of the "shipworm" and its ravages. Man has continually been striving to find a successful way to combat this damage. Borers have been found in treated and untreated wood, concrete, rocks, cordage, and plastics; their distribution is worldwide. Borer attack is most prevalent in waterfront structures and wooden vessels, but they have been recorded from depths of more than 5600 ft.

Description

The marine borers that cause the greatest amount of damage are found in two main categories—bivalve mollusks and crustaceans, commonly referred to as shipworms and gribbles, respectively. The molluscan borers may be separated into two families—the Teredinidae, or the wood-boring shipworms, and the Pholadidae, or rock borers.

Teredinidae. The shipworm is not really a worm but a near relative of the clam. Stripped of its shell and protective tube, the adult *Teredo* or *Bankia* has a wormlike appearance (Fig. 1). The small paired shells cover only the foremost part of the body. Muscular contractions move the shells in a rasping motion in such a way that the rows of minute teeth on the outer shell surface excavate the tunnel in which the animal lives. The posterior end of the body has two siphons that extend into the water. The incurrent siphon draws in water containing microscopic organisms for food, and dissolved oxygen for respiration. The excurrent siphon expels the waste products. When the siphons are withdrawn, the openings to the water are plugged with small calcareous pallets. These pallets also serve as the most reliable means of identifying the separate species. As it advances into the wood, the animal secretes a protective calcareous lining for the burrow from the entrance hole to just behind the shells. As the Teredinidae grow in length, they also increase in diameter so that they are confined within their own tunnel, and the destruction that takes place in the wood often goes unnoticed until it is too late (Fig. 2).

Pholadidae. The Pholadidae, or rock borer, will bore into wood, clay, soft rock, and other shells, and have also been found in plastic and poorer grades of concrete. In appearance the pholad looks very much like a small clam (Fig. 3). Unlike the Teredinidae, which are tube-forming, the pear-shaped pholad is complete within its bivalve shells, but like the teredinid, it, too, is confined within its burrow.

The crustacean family, which includes the well-known lobster, contains three major groups of boring organisms.

Limnoria. Of the crustacean borers, the gribble, or *Limnoria*, causes the greatest damage to waterfront structures, and is fairly well known. This small animal, approximately $\frac{1}{8}$ in. long, has a small head with very strong mandibles, a segmented body ending in a broad tail plate or telson, and somewhat resembles the woodlouse. It is free-swimming and can leave its burrow at will.

Fig. 1. *Teredo* specimen removed from its tube, showing shells, siphons, and pallets and *Limnoria* specimens. Enlarged approximately 4X.

Fig. 2. *Teredo* attack in untreated piling, showing outer surface with minute entrance holes and inner area filled with tubes. This attack occured in 90 days in a moderate temperature zone.

Fig. 3. *Pholad* in wood. Enlarged approximately 2½X.

Sphaeroma. The lesser known *Sphaeroma*, a relative of the pill bug, resembles *Limnoria* in appearance, but is longer and broader, averaging about ⅜ in. in length (Fig. 4). Only one species, *Sphaeroma destructor*, causes any extensive damage, and its occurrence is limited.

Chelura. A third crustacean borer, *Chelura* (Fig. 5), is approximately the same size as *Limnoria*, but it is more cylindrical and turns pink when removed from the water. It usually is found in association with *Limnoria*, and major damage is caused by its enlargement of tunnels made by *Limnoria*.

Damage caused by marine borers can be separated into two types: surface and internal. The pholads and crustacean borers cause surface damage; the teredinids, internal damage. During initial attack by pholads, the entrance hole is small as the animal first bores into the "host" material. As the animal grows, the hole increases in size until the average burrow reaches 2–2½ in. in length, and the surface entrance hole ¼ in. in diameter. Because holes of this size may be easily recognized, protective measures can be taken before excessive destruction occurs. Attack by *Limnoria* is somewhat different (Fig. 6). When the animals bore into the surface wood, they make an extensive network of tunnels, which is readily eroded by wave action. However, because *Limnoria* are free-swimming and can come and go from their tunnels at will, they can continue to attack the existing surface area. A heavy infestation of *Limnoria* can completely destroy wood, from the surface to a depth of more than ½ in. in one year. In a round timber pile, this is the equivalent of 1 in. a year. The method of destruction of wood by *Chelura* and *Sphaeroma* is similar.

The entrance holes of the Teredinidae remain quite small, usually $\frac{1}{16}$–⅛in. in diameter, and the intensive damage within the wood may easily go unnoticed. The length of the animal depends upon several factors, the most important of which are optimum growing conditions and the number of organisms present. The size of an adult *Teredo* does not affect its potency, as a stunted 2 in. specimen is capable of producing just as many young as a 2 ft specimen. Such specimens are known as stenomorphs. The teredinids usually bore with the grain of the wood, but will turn to avoid other tunnels or unfavorable sections of the wood, such as knots, without reappearing on the surface. Consequently, the inner portions of a timber may become completely riddled with these borers without its being detected on the surface.

Distribution

Marine borers have been observed throughout the world from Spitzbergen, above the Arctic Circle, to Tierra del Fuego, in the southern

MARINE BORERS

Fig. 4. *Sphaeroma*, dorsal view. Enlarged approximately 25X.

hemisphere, and from the tide line to depths of more than a mile.

The numbers of species and specimens are more prevalent in the warmer waters, and, in general, borer activity increases as the water temperature rises. The presence of borers at a given location depends upon the physical condition of the water more than the geographical location. Each species has its own normal as well as optimum occurrence range. Most of the borers require a salinity of 9 to 35 parts per thousand, but a few species such as *Teredo healdi*, *Teredo senegalensis*, and *Sphaeroma destructor* thrive in potable water.

Borers may be introduced into new areas by natural or mechanical means. Tides and water currents may carry the free swimming forms long distances from their normal habitat into areas of susceptible environments. Floating wooden debris, such as logs, driftwood, ship hulls, or other vessels, which has borer attack, can become the spawning area for attack at new locations.

Economic Importance

It has been estimated that marine borers cause over $500,000,000 damage to waterfront structures annually in the U.S. It is impossible to estimate the dollar value of destruction occurring annually in other countries. Some of these losses are spectacular; for example, the destruction of 50% of a 700 ft pier in less than six years, or the $1,000,000 loss to one firm in South America. The amount of damage to small private docks and boats is seldom reported, but is quite extensive.

Damage by marine borers is widespread throughout most of the inhabited areas of the world, although the extent of damage varies with the immediate environment. Where marine structures are concerned, one should consider that heavy attack is to be expected in subtropical or tropical areas. In colder environments, the attack is more seasonal and, consequently, less intense; the danger here lies in the incorrect assumption that due to longer periods of borer inactivity the structures do not require careful surveillance or proper protection. The cumulative destruction caused by the reappearance of borers in these areas can result in extensive damage to structures with insufficient protection.

Marine borers do not come under the usual classification of edible shellfish except in certain areas in the south seas where the natives harvest the Teredinidae for food.

Prevention

Finding ways to control marine borers is an ever-present challenge to man. The ancient Phoenicians coated the hulls of their ships with pitch and, later, copper sheathing. An oil mixture of arsenic and sulfur was in use by the 5th century B.C., and lead sheathing was used by the Greeks as early as the 3rd century B.C. Some of these same substances, in refined form, are still in use as contemporary techniques for borer control.

Knowledge of conditions in the locations where materials are to be used, whether on the shore or in deep water, is of prime importance, since all these destructive agents react differently. When the conditions of an area have been ascertained, the most satisfactory kind of preservative, coating, or type of material to be

Fig. 5. *Chelura terebrans*, dorsal view. Enlarged approximately 20X.

Fig. 6. Typical *Limnoria* attack. Protruding knots indicate original size of piling.

used may be determined. It is well known that the conditions of temperature, salinity, oxygen, food supply, and pollution affect the control of the borer population. The maintenance of exposure panels at a large number of varied sites provides much pertinent information concerning the differences in borer activity. Such panels also serve as a valuable tool for monitoring the presence and kinds of borers at a specific location.

For protection against borer attack, there are numerous methods in use today, all of which conform to rigid specifications of production and quality control. Wood to be exposed in borer-infested waters should always be given the type of protection determined to be best for that location. The use of chemical preservatives in treating wood for borer protection over the years has been shown to be the most satisfactory method, providing the proper preservative and type of treatment are used. Since marine borers react differently to different preservatives, considerable research is currently being conducted concerning the use of combination treatments as a single combatant for more than one type of borer. Protective coatings of various compositions containing toxic material are used to prevent the accumulation of surface marine growth. These coatings, when applied to wood, are also used to prevent the entrance of marine borers and, when applied to metals, to prevent corrosion. A third protective method, which has not been in extensive use, is the impregnation of wood with synthetic resins, resulting in a material of high density and structural strength. For on-shore installations, piling may also be given protection by sheathing. This method is often used in repairing damaged structures. A natural form of protection may be obtained for mobile structures, such as small wooden vessels, that can be removed from the water periodically for considerable periods of time, or are used in fresh water areas for extensive intervals.

Research and development throughout the years have led to great progress in improving borer control methods, but there still are no simple, foolproof methods to completely insure marine structures against these destructive organisms.

BEATRICE R. RICHARDS

References

DePalma, J. R., "Marine Fouling and Boring Organisms in the Tongue of the Ocean, Bahamas-Exposure 11," 1962, Marine Sciences Dept., U.S. Naval Oceanographic Office, Washington, D.C.

Muraoka, J. S., "Deep-Ocean Biodeterioration of Materials—Part 1, Four Months at 5,640 feet," 1964, U.S. Naval Civil Eng. Lab., Port Hueneme, California.

Richards, B. R., "A Study of the Marine Borers and Fouling Organisms at Wilmington, N.C., 1943. William F. Clapp Laboratories, Inc., Bull. No. 12.

Snoke, L., and Richards, A. P., "Marine borer attack on lead cable sheath," *Science*, **124** (1956).

Clapp, William F., Laboratories, Inc., "Sixth Progress Report on Marine Borer Activity in Test Boards," 1952.

"Seventh Progress Report on Marine Borer Activity in Test Boards," 1953.

"Eighth Progress Report on Marine Borer Activity in Test Boards," 1954.

"Thirteenth Progress Report on Marine Borer Activity in Test Boards," 1959.

U.S. Department of the Navy, Bureau of Yards and Docks, "Marine Biology Operational Handbook," 1965, prepared by A. P. Richards, William F. Clapp Laboratories, Inc.

Woods Hole Oceanographic Institute, "Marine Fouling and its Prevention," 1952, prepared for Bureau of Ships, Navy Dept. U.S. Naval Institute, Annapolis.

MARINE FISHERIES COMMISSIONS

The U.S. Congress has granted its consent and approval to three interstate compacts establishing the Atlantic States Marine Fisheries Commission, the Pacific Marine Fisheries Commission, and the Gulf States Marine Fisheries Commission. The purpose of each of these organizations, as defined by the statutes, is to promote the better utilization of marine, shell, and anadromous fishery resources, to promote the protection of such fisheries, and to prevent their physical waste from any cause.

To accomplish these objectives, each of the Commissions has the responsibility of making inquiry into methods and practices for preventing resource depletion and promoting conservation. They have authority to draft and recommend to the Governors and Legislatures of the signatory states legislation dealing with resource conservation. They also may recommend to the administrative agencies of the states such regulations as are deemed advisable. Recommendations may be made for stocking the waters of the states with marine, shell, or anadromous fish and eggs, and, where joint stocking is undertaken, the Commissions may act as coordinating agencies. The Commissions are specifically denied authority to limit production of fish or fish products for the purpose of establishing or fixing prices, or creating and perpetuating monopoly.

The compacts and the authorizing state legislation provide for essentially the same type of organization for all three Commissions. Except in the Pacific Marine Fisheries Commission, each member state has three members, one of whom must be the executive of the administrative agency charged with conservation of fishery resources, one a member of the legislature, usually named by the Governor, and the third a citizen having knowledge of, and interest in, fisheries, named by the Governor. On the Pacific Marine Fisheries Commission, each of the member states, except Oregon, has three members, one of whom is the administrative officer of the fish or fish and game agency. In Oregon the implementing legislation requires participation by its three Fish Commissioners and its five Game Commissioners. In all the Commissions, however, each member state is entitled to only one vote, and action can be taken only by the affirmative vote of a majority of the compacting states present at any meeting.

Provision is made for each Commission to elect from its members a chairman and vice chairman, to appoint and remove employees and define their duties and compensation, and to adopt rules and regulations for the conduct of business. At least one meeting a year is mandatory for all of the Commissions. The Fish and Wildlife Service of the U.S. Department of the Interior is designated as the primary research agency for both the Atlantic States and Gulf States Commissions, but for the Pacific Commission, the fishery research agencies of the signatory states act in collaboration as the official research agency. All Commissions are authorized to establish advisory committees representative of the commercial fishermen, commercial fishing industry, and such other interests as may be deemed advisable, but only the Pacific Commission has done so.

Atlantic States Marine Fisheries Commission

This Commission was the first to be established when, on May 4, 1942, P.L. 77–539 gave consent and approval to the compact that at that time had been approved by the states of Maine, New Hampshire, Massachusetts, Rhode Island, New York, New Jersey, Delaware, Maryland, and Virginia. South Carolina joined the compact in 1942, Georgia and Pennsylvania in 1943, Connecticut and Florida in 1945, and North Carolina in 1949. The original enabling act was amended on August 19, 1950, (P.L. 81–721) to authorize any two or more member states to designate the Commission as a joint regulatory agency for the control of their citizens and vessels fishing in waters in which they have a common interest. This additional authority has not been exercised.

Meetings of the Commission are held annually, and are usually preceded by meetings of the standing committees—executive, legal, and biological. The first day of the meeting is devoted to subjects of general interest and importance; the second to meetings of the four sections into which the Commission is divided—North Atlantic, Middle Atlantic, Chesapeake Bay, and South Atlantic—for the handling of more local matters; and the last is an executive session for the Commission to hear reports from the sections, the standing committees, and to attend to its financial and other business affairs.

The Commission publishes an annual report covering its work, financial status, minutes of the section meetings, committee reports, and resolutions considered and adopted. A supplement to the 16th Annual Report was compiled by the Biology Committee in 1958; it described 31 important fisheries on the Atlantic seaboard and summarized existing knowledge of the resources supporting them. The Estuarine Subcommittee of the Biology Committee in 1966 published, with Commission approval, a statement of policy and guidelines for effective action in developing and managing estuaries. In addition to these publications, the Commission, through its Biology Committee, initiated a Marine Resources Leaflet Series in 1965 and has issued eight leaflets describing the Atlantic menhaden, soft-shell clam, Southern shrimp, American lobster, summer flounder, American shad, and striped bass. These are available from the Commission's office.

The resolutions of this Commission have been influential in directing the course of Atlantic Coast marine fisheries research and legislation. Perhaps most significant was the part it had in the preparations leading up to the International Treaty for the Northwestern Atlantic, in the negotiation of the convention itself, and in the preparation of the implementing legislation. It had an active part also in securing the passage and funding of the Commercial Fisheries Research and Development Act, and has sponsored cooperative projects among its member states under this legislation. The office of the Atlantic States Marine Fisheries Commission is at 3965 West Pensacola Street, (P.O. Box 2784) Tallahassee, Florida 32304.

Pacific Marine Fisheries Commission

On July 24, 1947, this Commission came into being when P.L. 87–766 was signed into law. The original member states were California, Oregon, and Washington. Idaho joined on July 1, 1963, and there is provision for Alaska, Hawaii, and any other state having rivers or streams tributary to the Pacific Ocean, to be admitted to the compact for the development of joint conservation programs of mutual concern.

Meetings of the Commission are held annually, usually in November, and are rotated between the four member states. These meetings, which are open to the public, are preceded by meetings of the Advisory and Executive Committees and the Research Staff. Additional committee meetings are held as necessary, with independent annual spring meetings by the Research Staff and Executive Committee being the general practice. At the annual meeting, a report is made on the Commission's activities, financial status, and developments in the fishery field during the year. Status reports are made on the albacore, crab, groundfish, chinook and coho salmon, and shrimp fisheries, and special reports are made and discussion panels held on timely subjects. Proposals for cooperative research are discussed, and proposals for adoption as resolutions are considered.

The Commission publishes an annual report, and intermittently publishes bulletins containing research results of investigations by the member states. A data series was inaugurated in 1964 to make available raw data for fishery workers having need for them. The Commission issues annually a listing of fin marks being used on salmon and steelhead, and sponsors an annual meeting for the purpose of agreement on marks to be used. It is also the clearing house for catch statistics for Pacific salmon and steelhead, by area, for the entire Pacific Coast, beginning with the 1965 data. This statistical report is sponsored by the Pacific Salmon Inter-Agency Council, composed of Federal and State agencies.

The Commission, by sponsoring cooperative research, has had an important part in the acquisition of life-history and management information leading to (1) similar, if not identical, seasons and size limits for the chinook and coho troll fisheries off the Pacific Coast States and Canada; (2) more uniform groundfish or trawl regulations for the Pacific Coast States and Canada; (3) protection of spawning concentrations of petrale sole off Oregon, Washington, and British Columbia; (4) general uniformity of fishing seasons and minimum size limits for Dungeness crabs; and (5) compatible fishing regulations by California and Oregon for a stock of shrimp fished by vessels from both states.

This Commission has its headquarters at 741 State Office Building, 1400 S.W. Fifth Avenue, Portland, Oregon 97201.

Gulf States Marine Fisheries Commission

The Gulf States Marine Fisheries Compact was ratified as P.L. 81–66 on May 19, 1949. Alabama, Florida, Louisiana, and Texas became parties to the compact at the first scheduled meeting of the Commission on July 16, 1949. Mississippi entered the compact in January 1950 with the passage of enabling legislation.

The Commission meets twice a year, usually in March and October, rotating among the five member states. Special meetings have been held to handle emergency situations. Four standing committees have been established; Committee to Correlate Fishery Laws, Committee to Correlate Research and Exploratory Data, Shellfish Committee, and Estuarine Technical Coordinating Committee. These committees meet as required and periodically report their findings and recommendations at the regular meetings of the Commission. Progress reports are made to the Commission also by both Federal and State agencies on their activities affecting the fishery resources and the fishing industry of the Gulf of Mexico. These reports form the background for most of the resolutions adopted by the Commission.

The Commission publishes an annual report summarizing the year's activities, the actions taken and recommended, and its fiscal status. In addition to the annual report, the Commission publishes an Information Series containing current knowledge that might be useful in formulating protective measures, and a Technical Summary Series reporting the results of research of particular interest.

The Commission, through resolutions and coordinated action by the congressional delegations of the member states, has been instrumental in having the Bureau of Commercial Fisheries of the Department of the Interior undertake comprehensive programs in biological research, exploratory fishing, technological research, and statistics in Gulf waters. It actively supported legislation that authorizes federal aid on a matching fund basis, for fisheries research and development, and, as a consequence, has substantially strengthened the state fishery agencies. Particular attention has been directed to the problem of protecting the estuarine environment, and the Commission has served as coordinator of a Gulf-wide Federal-State estuarine inventory project, including preparation of a film demonstrating the importance of these inshore waters to the fishery resources and industry dependent upon them. The address of the Gulf States Marine Fisheries Commission is 400 Royal Street, Room 225, New Orleans, Louisiana 70130.

SETON H. THOMPSON

MARINE GEODESY

To explore and economically exploit the ocean resources, man must be able to map the oceans accurately as a first step toward mastering the hostile environment. One of the prerequisites for accurate mapping is the establishment of precise geodetic control. The solutions of hydrographic, bathymetric, and marine geophysical mapping problems fall in the domain of marine geodesy, one of the two main branches of the science of geodesy. Marine geodesy is relatively new as a special branch of study; therefore, a short background of general geodesy is presented here before the concept of marine geodesy and its role in expanding ocean resources is discussed.

Geodesy is an applied science dealing with the precise determination of the size and shape of the earth and its gravity field. The practical task of geodesy is the establishment of networks of control points for precise mapping of the earth's surface. A more descriptive definition is that geodesy deals with accurate observation or measurement of certain quantities and adjustment computations based on these measurements. The development and application of appropriate techniques for land measurements is called land geodesy; for sea measurements, marine geodesy. The basic geodetic quantities measured are of two types:

(1) Geometric quantities—involving measurement of distances, angles, or both for the establishment of geodetic control. The exact coordinates for horizontal control are determined through four basic techniques: (1) astronomic; (2) triangulation; (3) trilateration; and (4) traverses. Celestial and satellite methods are two of the newest methods available. For vertical control determination, leveling techniques are available by which heights of points are determined.

(2) Physical quantities—involving determination of the gravity field of the earth. Using airborne, land, shipborne, and underwater instruments, gravity measurements are obtained over the earth.

Geodesists deal with three basic surfaces: (1) the topographic surface of the earth, on which most of the measurements are made; (2) the geoid (mean sea level), an equipotential surface to which measurements are referred; and (3) the ellipsoid, on the basis of which computations are made. The ellipsoid is a mathematically defined surface approximating the geoid (see Fig. 1).

Despite the great efforts geodesists make to obtain accurate measurements, errors are practically unavoidable. Therefore, an important task in geodesy is adjustment computation.

Land geodesy has made significant achievements in the art of world surveying and mapping.

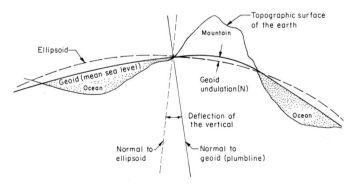

Fig. 1. Relationships between reference surfaces.

Many of the techniques that have been developed may also be useful for marine geodesy, although practical experience in using them at sea is very limited. Until a few years ago, the basic tools of geodesy were such instruments as measurement tapes, levels, transits, and logarithmic tables. At present, in employing electronic techniques for distance and angle measurements, land geodesy has taken a giant step forward. The camel-velocity techniques have been replaced by techniques based on the velocity of light or electromagnetic wave propagation. The wooden rods have been replaced by Invar wires and, more recently, by light-interference and microwave electromagnetic systems. The logarithmic tables have been replaced by electronic computers, which quickly perform the adjustment calculations that were formerly so tedious. The triangulation methods have been almost completely replaced by trilateration methods (measuring all sides of triangles by electromagnetic or light-wave techniques). Today, using aircraft, distances of the order of hundreds of kilometers can be measured; by means of satellites, precise measurement of thousands of kilometers is possible.

Since geodesy is concerned with accurate measurements and establishment of controls on a world-wide basis, it does not seem reasonable to neglect the ocean areas, which represent more than 70% of the earth's surface. In the past, the difficulties associated with establishment of control points at sea, and the lack of precise methods of measurements on water, have precluded the construction of accurate ocean maps. Furthermore, the need existed only for general-purpose maps to be used in navigation and shipping. However, as man moves into the seas for exploitation of their resources, the development of accurate and detailed maps will be a major requirement—already demonstrated in a recent continental shelf study[5]—to be satisfied by marine geodesy. Hence, man must now solve the more difficult problems associated with the complicated ocean environment.

Objectives of Marine Geodesy

Specifically, the main technical objectives of marine geodesy are: (1) the development of precise measurement technology; (2) the establishment of networks of accurate marine control points connected to the existing land geodetic system; (3) the utilization of this control network for gravity measurement, mapping, precise positioning and navigation, etc.; and (4) the accurate measurement of gravity all over the oceans.

Establishment of Marine Control Points and Networks. In marine geodesy, control points will be points of *fixed coordinates located on the ocean floor*. Their exact position will be determined in terms of geographic latitude, longitude, and depth from sea level. Marine control points could be of various types such as surface platforms, moored devices, and bottom-mounted instruments, either active or passive. The best approach would be to use instruments fixed on the ocean bottom. For example, control points might be identified by acoustic signals transmitted to a ship's receiving equipment by a set of three transponders permanently placed at suitable locations on the ocean floor as shown in Fig. 2. However, the placement on the ocean floor of a set of acoustic transponders would not in itself provide a control point. A control point would be established only after the exact coordinates had been determined according to acceptable geodetic procedures. In 1968 an experimental marine geodetic control point was established for the first time in the Pacific Ocean about 130 miles west of Los Angeles in 6,000 ft of water.

Whatever the hardware involved, a control point must also have the capability of acquisition and identification of signals. Signal transmission would be influenced by such factors as water depth, condition of water, beam width of transducer, frequency, and power. Transponders have been used on the ocean floor with spacings varying from hundreds of feet to two miles, depending on the purpose and on the depth of the water. Several transponder designs, some using lead-acid batteries, are available at present with reported lives ranging from a short time to three years. Radioisotope power sources are also available with 10 to 20 years life time.[12] Active acoustic devices in the marine environment require power sources with power outputs of 1 mW to 100 W. The desired characteristics of these power sources are long life per fuel charge, no maintenance, and reliable unattended operation. A survey of underwater power sources for use in marine geodesy was described in 1966 by Kortier, et al.[7]

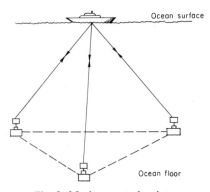

Fig. 2. Marine control point.

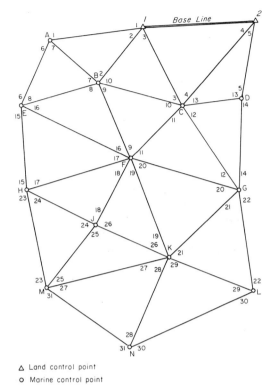

△ Land control point
○ Marine control point

Fig. 3. Marine geodetic network.

A series of control points established geodetically in a given area would form a marine control network (see Fig. 3). Although a form of a triangulation network has been established in the Gulf of Mexico, where the U.S. Coast and Geodetic Survey tied several of the existing offshore oil platforms to the U.S. land geodetic system, marine geodetic networks do not exist at present. Marine geodetic networks, which would provide a rigid framework for surveying and mapping, could be of several orders similar to land networks—1st, 2nd, and 3rd order, etc.—depending on accuracy of measurement, density of control, and other factors. It will be necessary, however, to redefine these orders for the purposes of marine geodesy. To provide a basis for redefinition, a marine geodetic range could be established for use as a standard for calibration and evaluation of instruments and for checking measurements.[11] The establishment of such a range or of a marine control point network would involve network design, network measurements, and adjustment computations.

Network Design. Before any other work is done, it will be necessary to make a network design that will meet the requirements of those who will use the network. The more carefully the design is constructed, the more efficiently and economically the actual network can be established and the more useful it will be. The basic network design should incorporate provisions for growth, for the integration of new systems or technological advances, for expansion into 2nd and 3rd order networks, and for eventual connection to a general geodetic reference system covering the land as well as the ocean. The physical parameters, e.g., geodetic, acoustic, or oceanographic, as well as the equipment to be used in establishing and using the network must be considered. Analysis of a model of a possible network for the Gulf of Mexico indicated the importance of the design in improving accuracy and reducing cost.

Network Measurements. In the establishment of marine control networks, distances between control points must be measured. Land control points, being on the surface, are visible and physically accessible, and distances between them can be measured directly by placing appropriate electronic equipment over them. However, it will not be possible to see marine control points located on the ocean floor, and they will not be physically accessible. Therefore, it will not be easy to measure distances between them.

If the methods developed for land geodetic distance measuring are to be used, additional techniques appropriate to the oceans must be devised. For example, an accurate means must be found for relating the measurements made on the ocean floor to those made on the surface. This can be accomplished by positioning a ship carrying appropriate equipment with respect to the bottom point. Also, the distance between the terminal points (ships) of a surface network line must be measured simultaneously with measurement of the relative location of these terminals to the bottom control points. Therefore, timing circuits connecting electromagnetic distance measurements (above water) with underwater acoustic measurements should be used.

The lengths of the lines forming the network or the geographic coordinates of the control points can be measured using surface-based, airborne, and satellite systems. However, accurate instrument-ship positioning is fundamental to the application of all these systems.

Ship Positioning. Positioning a ship, or an instrument vehicle, at the surface relative to a bottom control point is done by several methods. One method is similar in principle to the solution of a three-dimensional intersection problem in geodesy. To obtain a unique solution for such an intersection problem, three transponders would be required for each control point. The coordinates of the three bottom transponders would be predetermined as x_i, y_i, z_i ($i=1, 2, 3$), if three transponders are used. The distances R_i ($i=1, 2, 3$) from the unknown ship position,

$P(X, Y, Z)$, to the transponders would then be measured. These distances would be measured using the product of time (spherical) or time differences (hyperbolic) and the velocity of sound propagation in water. The unknown ship position, X, Y, Z, would be determined by a unique solution of the intersection of three spherical surfaces in terms of x_1, y_1, z_1, and R_1, where $P(X, Y, Z) = f(x_1, y_1, z_1, R_1)$. For redundancy and increased reliability, four or more transponders could be used.

Two basic transponder interrogation systems have been used[17]. In one system the transponders will respond to a call made from the ship at one of three frequencies (10.0, 10.5, or 11.0 kcps); the reply will be at 12 kcps. In the other system[3,4,12] the ship will interrogate with one frequency of 16 kcps, and the transponders will reply at different frequencies varying from 9.5–12 kcps. Relative accuracies achieved in positioning the ship over bottom transponders in the deep ocean have been reported as ± 3 m to ± 10 m.[3,12,17]

Other methods that could be applied involve acoustic line-crossing techniques or variations of these techniques to determine minimum distances. Both the acoustic line-crossing technique and the spherical intersection technique were used in the Pacific Ocean experiment that has been mentioned; at the 6,000-ft. depth, standard deviation of ± 3 m was achieved.[12] The use of a single source unit is also possible, as was done in the Mohole Program, if interferometry equipment can be provided on the ship.[1] The selection and use of any one of the methods available would be determined by requirements for accuracy and by the budget.

Surface-Based Systems. A number of surface-based electronic systems have been used to provide distance measurements to position hydrographic and other marine surveys with respect to the land. However, their use for the establishment of marine control points or their use with control points has only recently been considered. These systems are capable of measuring short ranges within 100 miles from shore. Examples are Raydist, HiFix, autotape, Lorac, etc. At ranges of the order of 1,000 miles or more, only two navigation systems are available, Loran-C and Omega. Although they are used to provide marine survey control, their accuracy is not satisfactory for establishing geodetic control.

Airborne Systems. In the airborne category, for example, two systems are available: Shiran/Hiran and Lorac systems. Both can employ line-crossing techniques for distance measurements. Shiran/Hiran has been used extensively in world-wide mapping, and has the capability of measuring great distances of the order of 500 miles on land. These systems can be used to measure distances between control points.[10] An airborne Lorac system was used in the establishment of the first marine geodetic control point in the Pacific Ocean experiment. Six lines (distances) connecting three known land points and the unknown marine control point were measured satisfactorily with this system.[12]

Satellite Systems. Satellite methods could offer a distinct advantage over other methods in the establishment of marine control networks in that they have the range capability and singularity of reference datum required for mapping. Distances of the order of thousands of miles can be spanned using satellite geodetic techniques. It is also recognized that establishment of large marine networks, although desirable, may not be economically practical, at least immediately. Instead, mapping of the oceans will probably be conducted on a priority basis. Therefore, a single measuring system to provide control for present and future mapping referenced to the same datum is highly desirable. Satellite methods could be ideal. A recent study on satellite applications to marine geodesy indicated that satellites have great potential. Five GEOS II satellite methods that have been used on land were evaluated: Doppler, SECOR, radar, photographic, and laser methods. It was found that at present the Doppler and radar methods bore the greatest potential.[13] Only the Doppler and photographic methods have been experimented with at sea. The Doppler investigations at sea have been mainly for navigation purposes.[14,19] The feasibility of using the Doppler method for marine-control-point establishment is being evaluated with actual data obtained during the Pacific Ocean marine geodesy experiment in 1968.[12] The photographic method involves the use of photogrammetric ocean survey equipment (POSE)[8] which consists of a gyrostabilized stellar-orientated camera and associated timing equipment mounted aboard ship. In an experiment with the POSE the ship was used as an unknown station, but its relative position to the bottom transponders was known. The possible application of the SECOR method to marine geodsey has been discussed by Rinner.[16]

Adjustment Computations. The science of geodesy is not complete without adjustment computations to determine errors of measurement and the best means to minimize and represent them. Adjustment computation is particularly significant in marine geodesy. Even with the best and most sophisticated techniques and equipment, errors in measurement are unavoidable, as evidenced by discrepancies in repeated careful observations. Adjustment computation is designed to give the most probable value of differing redundant measurements, to fit the small alterations (discrepancies) into a mathematical structure, and to determine the size and distribution of errors.

Determination of Gravity at Sea

Gravity measurements—both on land and at sea—are important for many reasons. For example, in exploration geophysics, gravity techniques (along with other geophysical methods) are used for detailed geological structure mapping and in the search for oil and minerals. Weapon system effectiveness and satellite orbit determination are greatly dependent on accurate gravity measurements.

From measurement of gravity on the surface of the earth, the gravity anomalies Δg's (differences between properly reduced observed and theoretical values) are determined and the quantities N (geoid undulations), ξ and η (deflection of the vertical components) are computed; see Fig. 1. These quantities are important in the accurate determination of the radius of curvature and shape of the earth and in the establishment of an absolute coordinate system.[6] On land, these quantities have also been determined by the astrogeodetic method (relative N, ξ, and η). The application of the astrogeodetic technique at sea has been made possible through the use of an inertio-optical system to sense deflection of the vertical by direct measurement of the tilt of the sea surface.[20] The accuracy of this method as reported, ± 10 arc seconds, while useful in navigation, is not yet satisfactory for geodetic applications. The determination of these quantities, even for land values, requires gravity measurements all over the earth's surface, including the ocean areas.

The measurement of gravity at sea is made through the use of surface-ship, submarine, underwater, and airborne gravity instruments. Most of these instruments are gravimeters used with stabilized platforms or gimbal suspension. Pendulum instruments have been used in submarines by most investigators and aboard surface ships only by the Soviets.[2] The accuracies of the gravity measurements at sea have been reported to be on the order of 0.1–1.0 mgal for bottom measurements, 1–5 mgal for submarine instruments, 2–10 mgal for surface-ship instruments, and 5–10 mgal for airborne instruments. Most of the errors in gravity measurements at sea on surface ships are attributed to inaccuracy in navigation. Several marine gravity ranges have been established on the continental shelf near shores for evaluation of shipboard gravitymeter systems.[15]

Gravity base stations and ranges in the open oceans would be of great importance for the control, evaluation, and improvement of accuracies of marine gravity observations.[9, 10] For effective measurements in land geodesy and geophysics, many base stations are needed. It can easily be seen that for gravity measurements at sea, many accurate marine control points and gravity base stations will be needed.

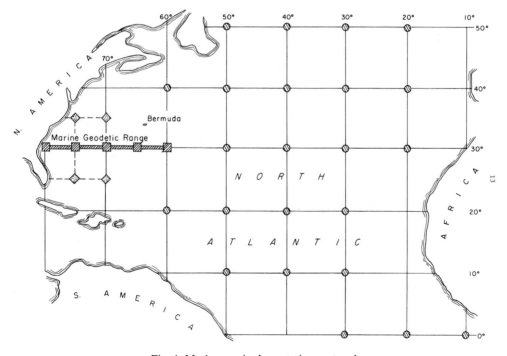

Fig. 4. Marine gravity base stations network.

For gravimetric surveys, marine control points would provide accurate base stations in the open ocean to control airborne and surface-ship-based surveys. Figure 4 shows the number and distribution of gravity base stations that would be needed to cover most of the northern Atlantic Ocean. Such stations would probably be spaced at 10° intervals (approximately one station every 600 miles). At each base station, surface-ship or submarine instruments could be employed to measure in detail the gravity over an area of perhaps 1° by 1° or 60 miles by 60 miles. At the control point station, for a more precise base value, an underwater gravity instrument could be used. The end result would be a systematic network of detailed gravity surveys over the oceans that could also be used to control airborne measurements.

Advantages and Use of Marine Control Networks

The advantages marine geodetic control networks would provide for surveying and mapping are twofold; (1) control of the surveys by providing the exact coordinates (latitude, longitude, and depth) of certain points in the surveyed area, with assurance that these points were precisely determined with respect to the national geodetic system of the country involved; and (2) capability for precise positioning on the basis of these points.

Marine control points or networks would be helpful in many practical operations. For example, they could be used in boundary determination for mineral deposits or oil leases. At present the determination of these boundaries on the continental shelf is the responsibility of the user, who is usually granted a number representing his lease on a map. Unfortunately, there are no actual markers in place on the continental shelf to identify or correspond with those on the map as is the case on land. As a result, the user, to be safe, allows several hundred or thousand feet on all sides of the lease without development and exploitation. Sometimes this could make the difference between a producing and a nonproducing deposit.

In hydrographic and bathymetric surveys, the accuracy of the survey depends on the accuracy and adequacy of the horizontal and vertical control available. Furthermore, the accuracy of a nautical chart is highly dependent on the accuracy of the hydrographic survey from which it is compiled. At present, horizontal control is extended through the use of electronic surveying systems from land control without the use of bottom markers. This method is acceptable but not desirable. Extreme care must be exercised in the use of electronic surveying systems. The resulting charts are not geodetically or geographically connected because the coordinates obtained are electronic coordinates, although they may be transferred mathematically into geographic coordinates. The variations in electromagnetic wave propagations cause errors often undetected in the final results unless the coordinates of predetermined check points (control points) are available. As distances from shore increase, the uncertainties also increase, and fewer systems can be used for extension of control. In fact, for distances over 200 miles, only navigation systems are now available for this purpose.

Positional fixes made with present navigation systems at sea are often in error by several miles, depending on the area and the system used.[11] If a marine standard range were established at sea (Fig. 4), navigation systems and, more important, surveying systems could be calibrated, tested, and evaluated.[11]

The establishment of a marine control network would not only increase the accuracy of charts, but would provide the necessary information to relate various charts to the same system. Hence, the propagation of errors could be controlled. Moreover, the use of the control points in conjunction with present surface-based electronic systems would limit the propagation of errors to the areas located between adjacent control points. Most of these errors, therefore, could be eliminated through simple adjustment computation procedures.

The control points could also be used with acoustic, inertial, and satellite systems. An acoustic technique employing side-looking sonar could be used with even passive acoustic reflectors to map the ocean floor.[18] The function of the side-looking sonar would be to map the given area, with the reflectors, as well as the bottom topographic features, being identified on the map.

Summary. The most important task of marine geodesy is the establishment of marine control points and networks tied to the land geodetic system. Such marine control points and networks would assist man in locating and exploiting ocean resources in that they would:

Aid in establishing and identifying lease boundaries, locating drill sites, and delineating mineral-deposit structures.

Serve as exact calibration points at great distances from shore in actual areas of operations, and permit frequent updating or position fixing of surrounding surveys.

Furnish independent accurate control for bottom mapping and other oceanographic missions.

Provide gravity base stations in the oceans for shipborne and airborne gravity measurements.

Make possible the establishment of a national marine geodetic range as a standard for evaluating and calibrating precise navigation and surveying systems, including inertial and electronic systems.

Provide a test range for improving SOFAR (long range sound transmission and propagation).

Provide precise positioning capability off of these points, independent of time and environmental conditions.

A. G. MOURAD

References

1. Anonymous, "Vessel positioning control systems," Staff Report, *Undersea Technology,* April, 1967.
2. Aleksandrov, S. Ye., Sukhodol'skiy, V. V., and Izmaylov, Yu. P., "New pendulum instrument for the determination of the force of gravity at sea," *Trudy Institut Fiziki Zemli,* No. 8 (175), 3–77 (1959).
3. Campbell, A. C., "Geodetic methods applied to acoustic positioning," Proceedings of the First Marine Geodesy Symposium," 1966, sponsored by Battelle Memorial Institute and USC&GS; Government Printing Office, Washington, D.C., 1967.
4. Cline, J. B., "Acoustic navigation surface and subsurface," *Ibid.*
5. Frazier, N. A., and Buttner, F. H., "Activities and developments in the U.S. continental shelf regions—motivation for marine geodesy," *Ibid.*
6. Heiskanen, W. A., "Geodesy," Encyclopedia Britannica," 1959.
7. Kortier, W., Pobereskin, M., McCallum, J., and Gates, J., "Underwater power sources for marine geodesy," Proceedings of the First Marine Geodesy Symposium," Battelle Memorial Institute, Columbus, Ohio.
8. Jury, H., "Photogrammetric ocean survey equipment," *Ibid.*
9. Kivioja, L. A., "Significance of open ocean gravity base stations and calibration lines," *Ibid.*
10. Mourad, A. G., "The concept of marine geodesy," *Ibid.*
11. Mourad, A. G., and Frazier, N. A., "Improving navigational systems through establishment of a marine geodetic range," *The Journal of the Institute of Navigation,* 14 (2) (Summer 1967).
12. Mourad, A. G., Frazier, N. A., Holdahl, J. H., Someroski, F. W., and Hopper, A. T., "Satellite Applications to Marine Geodesy," National Aeronautics and Space Administration, NASA CR-1253, January 1969.
13. Mourad, A. G., Frazier, N. A., and Holdahl, J, H., "The Battelle/Industry Marine Geodesy Experiment in the Pacific Ocean - - Feasibility and Geodetic Application," *Transactions of American Geophysical Union,* 50, 1 (1969).
14. Newton, R. R., "The Navy navigation satellite system," Proceedings of the First Marine Geodesy Symposium, Supra.
15. Orlin, H., "Marine gravity surveying instruments and practice," *Ibid.*
16. Rinner, K., "SECOR satellite ranging system and its application to marine geodesy," *Ibid.*
17. Spiess, F. N., "Navigation using underwater acoustics," *Ibid.*
18. Stephan, J. G., "Mapping the ocean floor," *Battelle Technical Review,* July-August, 1966.
19. Talwani, M., Dorman, J., Worzel, J. L., and Bryan, G. M. "Results on satellite navigation observed at sea," Proceedings of the First Marine Geodesy Symposium, Supra.
20. Von Arx, W. S., "Relationship of marine physical geodesy to physical oceanographic measurements, *Ibid.*

Cross references: *Navigation.*

MARKING OF FISH

The primary reason for marking fishes and other organisms is the need for distinguishing individuals or a small group of individuals from the remainder of the population so that biologists may arrive at certain inferences concerning them when they are recaptured.

Occasionally a group of fish bears a natural mark by which they can be distinguished from others of the same stock. Thus, the very abundant 1904 year class of Norwegian herring was distinguished by an extra large initial growth band on its scales. Similarly, salmon sometimes have scales with characteristic growth bands that distinguish fish of the same species, even from different tributaries of the same river. The difference in the species of their parasites, or in the degree they are parasitized, is also used to distinguish fish of the same species but of different origin.

Occurrence of natural marks is too uncertain, and the marks usually too difficult of interpretation, however, to serve in more than a few special cases, and intentional marking or tagging must be resorted to.

Several facts or attributes can be inferred from the recapture of marked fish.

(1) Species distinction. Perhaps the earliest use of tagging was the marking of immature salmon and sea trout in Britain to discover whether the young, which were difficult to distinguish, always returned in the adult form as the same species identified in the immature stage.

(2) Frequency of spawning. The first really successful tagging was of Atlantic salmon in the Penobscot River in 1873 by Charles G. Atkins. He discovered, by marking kelts, (spawned out adults) that the majority spawned only every second year.

(3) Race distinction. Marking is employed to determine whether groups of fish marked in different localities intermingle to any degree or whether subpopulations or races exist in different areas.

(4) Geographical distribution. Marking and recapture of fish reveals the geographical distribution of any group of fish (provided fishing provides adequate opportunity for recapture over the whole area). When young stages can be marked and recaptures continue over a considerable period of time, one may also determine the areas occupied by fish of successively greater size or age. This is often important in determining which areas suitable for young fish are contributing to specific offshore fisheries.

(5) Age and growth. Recapture of marked fish over a long time may show the increased size at successive ages. For some species, such as shrimp, which periodically shed their exoskeletons, or fish in tropical waters, which may not form definite annuli on their scales, this is an extremely valuable method of age assessment. Where sufficiently large numbers of carefully measured fish are liberated, growth rates may be determined. Even for species with decipherable annuli on their scales, the corroborative evidence from recaptured fish may be very useful in scale interpretation.

(6) Spawning migrations. Many fishes make long spawning migrations. Thus, tagging experiments have shown that many of the king salmon taken as far north as southeastern Alaska enter and ascend the Columbia River to spawn. Similarly, mature halibut from as far away as the Aleutian Islands migrate to the Gulf of Alaska off Yakutat to spawn.

(7) Migration routes. Marking is especially valuable here. For example, in southeastern Alaska, pink salmon can enter the myriad waterways of the Alexander Archipelago through several straits on their way to their natal streams to spawn. To assure adequate seeding of each stream, intelligent conservation requires knowledge of which routes are used by salmon spawning in different streams. Continuous marking at the entrances of these straits during the migration period has shown the routes used. In some cases, the early and late salmon runs to the same stream use different routes.

(8) Speed of migration. The speed of migration is important in certain circumstances. Thus, in determining the effect of barriers in delaying the upstream migration of salmon, it becomes vital to know the rate of progress, for the salmon must reach its spawning beds before exhausting its supply of stored fuel.

(9) Mortality rates. The most useful and necessary information concerning any population is knowledge of the rates of mortality. For humans, these mortality rates are very accurately calculated by insurance actuaries. Without the birth and death certificates, and occupational and geographical data available to the actuary, the fishery biologist does the next best thing; he marks individual members of the population. Circumstances may warrant the making of certain assumptions, such as the representativeness of the sample marked, its even distribution through the population (or, at least, all of the population being equally vulnerable to the gear fished), and the tagged individuals acting and being acted upon as wholly normal individuals. It is then often possible to calculate from the rate of recapture of the tagged individuals relative to the untagged portion of the population at the time of tagging, the rates of fishing and natural mortality.

(10) Rearing methods. Marking has been widely used to check the success of various methods of rearing fish, especially the salmonids. Thus, fast growth of young fish in the hatchery up to the time of release may appear desirable, but growth alone is not always a sufficient criterion of future survival. By marking and releasing numbers of fish reared under different conditions of feeding, handling, and temporal and spatial methods of release, it is possible, through comparison of the recaptures, to evaluate methods to improve hatchery efficiency in terms of adult fish.

Types of Marks

The choice of marks depends on such factors as size of the organism to be marked, speed of marking, degree of permanency desired, ease of handling the fish, manner of recovery, etc. The chief types are mutilation, vital stains, and tags.

Mutilation. This method is used chiefly for marking large numbers of very small fish, especially when recovery may be a long time hence, when the fish are much larger. Consequently, a mark with great permanency is desired. This method has been used on several species but chiefly on salmonids. Here, because the young are very small in comparison to the adult, few tags are suitable for marking the young. Thus, pink salmon fry $1\frac{1}{2}$ in. long were successfully marked by excising fins, and the adults were recaptured a year and a half later when they returned from the sea to spawn.

Because of the natural occurrence of fish with one fin missing, it is usual to excise two fins. The fins excised can include the dorsal, adipose, anal, left and right ventral (pelvic), and left and right pectoral. It has been shown that fewer salmon smolts survive to return as adults with a pectoral fin excised than with other fins removed; therefore, excision of the pectoral is not recommended, this leaving but ten two-fin combinations. Some biologists have attempted to increase the number of combinations by use of a half-dorsal or a half-anal mark, clipping off half the fin at the base, but fins not cut off at the very base tend to regenerate. Experience would suggest that these latter two marks are fraught with uncertainty, since only extremely careful and slow marking of fairly large fingerlings can guarantee against portions of fins regenerating. These partially regenerated fins cause little difficulty in recognition in the normal two-fin combination. However, the distinction between a partially regenerated dorsal fin and one supposedly half excised can be tenuous.

In addition to fin removal there have been attempts to mark some of the more bony fishes by clipping notches in the edges of the opercle or maxillary.

Fish are occasionally marked by punching holes in the fin membrane or cutting the tip off a fin, but such marks are usually not distinguishable for more than a few weeks. Lobsters also are sometimes marked temporarily by punching or notching the telson or uropods.

Clams and other hard-shelled mollusks are sometimes marked by notching or etching the shell with a file or drill, or by use of waterproof paint.

Another form of mutilation is branding. Both hot and cold branding has been tried, but the marks tended to become illegible as the fish grew. Sea herring were branded in Maine by burning through scales and skin with several resistance wires heated with electricity from a 12-volt battery. The marks of the individual wires were not distinguishable for more than 2–3 days, and are not useful for more than very short-term experiments.

Vital Stains. Use of vital stains to mark fish and shellfish has only recently been employed on a large scale. Early experiments involved staining starfish by immersion in a weak solution of stain. Others tried staining salmonids and invertebrates by mixing stains in their feed. More recently, shrimp have been marked in the Gulf of Mexico by hypodermic injection of small amounts of dye dissolved in distilled water.

The dye colors the entire shrimp first, but, in about 24 hours, it is all concentrated in the gills so that the head (actually the thorax) is brightly colored, and the shrimp can be readily separated from its nondyed companions. This successful technique can be used for other invertebrates. The chief advantages over earlier methods are that the mark is not affected by molting and that it can be used on very small individuals.

For experiments covering a very short period of time, immersion staining still has its uses. For instance, pink salmon fry are marked by immersion and returned to a stream. Within a few hours or days some of them are recaptured as they migrate downstream to the sea, and the proportions of colored fry to the noncolored fry in the samples captured permits an estimate of the total number leaving the stream. The use of stain by immersion is particularly suitable for temporary marking of larval forms both because they are too delicate to permit handling and because it permits the staining of the vast number that must be marked to obtain sufficient recaptures.

Recently biologists have been experimenting with fluorescent or phosphorescent dyes for marking. An advantage of this technique is the ability to discriminate between experimentally dyed specimens and organisms merely discolored or naturally possessing color that may cause confusion. Very minute quantities of different fluorescent dyes can be separated with a fluorometer by their difference in wave length.

Tattooing has also been used. It consists of forcing small quantities of inert material beneath the skin by means of needles. A single area may be marked, using different colors for different experiments, a letter or number may be inscribed or a combination of different colors may be used. Halibut have been marked on the white side, by India ink injected with a hypodermic syringe, but the marks become unidentifiable after three months. More recently, colored latex injections at the base of the dorsal fin have been successful over a period of a few months. Tattooing by electric needles is currently employed, because of its rapidity, to mark large numbers of small salmonids when a temporary mark suffices.

Tags. The most prevalent marking method is to attach to the exterior or a place inside an organism some readily identifiable foreign object spoken of as a tag. Tags can be variously classified according to several criteria, including material used, method of attachment, place where attached, and method of recovery. A general system of classification by Rounsefell and Kask (1945) lists 18 types; Rounsefell and Everhart (1953) list 21. Some of these types were never used extensively or have become obsolete as new and more efficient tags have been developed; the obsolete types will not be mentioned here.

With the development of new materials, especially plastics, many of the materials formerly used (see Rounsefell and Everhart, 1953) have been abandoned. Materials presently in vogue include various plastics, nickel, monel metal, silver, aluminum, stainless steel, titanium and tantalum wire, magnetic steel, and nylon. The choice of material depends upon several factors.

(1) Time before recovery. For short-term experiments, one may use material such as aluminum, that, although tough and having the great advantage of light weight, is subject to corrosion, especially if not of high purity. Aluminum, therefore, cannot be recommended

for experiments in which recoveries are expected over many months or even years.

Both Vinylite and cellulose acetate disks for Petersen tags are inferior to cellulose nitrate, because the first two tend to become brittle and crack. Cellulose nitrate disks, however, also become brittle if held a few years in storage. Nickel, monel metal, and silver used for pins or wire are inferior to stainless steel or tantalum.

(2) Place of attachment. For external attachment noncorrosive material is imperative for any long-term experiments. For body cavity tags, loose within the body cavity, nonstainless steel can be used.

(3) Method of recovery. The material used is somewhat dependent upon the method of recovery. For recovery of tags by electromagnets from fish meal, obviously only metals with magnetic properties are usable.

Recovery methods. Because divergent types of tags have been developed to suit the manner in which they can best be recovered by the fishery, the following recovery methods are given:

(1) By sight. The use of tags that are visible is most common. Detecting tags on live fish is restricted to special experiments, such as identifying bass on their nests or observing live salmon on their shallow spawning beds. These external tags must be large and conspicuously colored. Oversize Petersen disc tags with a sharply contrasting colored spot in the center have been used successfully.

The recovery of tags from fish in catches requires tags that can be easily spotted by the fisherman while sorting or handling his catch. It is essential to take into account the exact manner in which the fishermen in a particular locality handle their catch in order to insure placing a tag where it is not easily overlooked. For instance, it may be vital to successful recovery that an opercle tag be placed on the left or right cheek of a fish according to how the fish are held in cleaning. If fish of a certain species are customarily cleaned individually on the fishing vessel soon after catching, an internal (body cavity) tag may be used.

(2) By transmittal of underwater sound. Tags have been developed that emit low-frequency sound from a small battery-powered transducer for several hours. These permit following the individual fish from a boat. Such tags have been most useful in determining the movements of anadromous fish in finding and passing through fishways and the quiet forebays above large dams.

(3) By electromagnet or electronic detector. This method of detection finds greatest use in small species taken in enormous quantities and processed without individual handling. Tags with magnetic properties are used, usually in the body cavity. During the processing an electromagnet separates the tags from the fish meal. The use of the electronic detector, placed in a conveying line prior to processing, is a superior method for determining the exact locality of capture, and has the added advantage of separating out the whole fish for examination.

(4) By radioactivity. Tags have been developed with a very low level of radioactivity, much like the radium dial of a wrist watch. These tags can be detected with an instrument that measures radioactivity near the conveying line on which the fish are moved from the vessel into the processing plant.

Selection of Marks for Specific Experiments

A marking experiment should always be designed to answer some specific questions. Once the minimum evidence required has been determined, the method of marking can be chosen. There are advantages and disadvantages of the several general methods of marking. Mutilation by fin clipping, for instance, can be used on very small fish, and, when properly done, is quite permanent. However the individuality of the mark is very low, since only a few marks are available. Recovery of marks is also rather difficult, requiring either intensive canvassing of the fishery, or carefully scheduled representative sampling of the catches.

For some experiments the marking of very large numbers of small individuals is of such overriding importance that tattooing, for instance, may be in order.

In experiments in which the individual fish *must* be identifiable, there is no good substitute for tagging, as each tag can be given an individual number. A great variety of tags have been devised, and from experimental evidence they are being continually improved.

Even after selection of the specific type of tag, certain details bear watching. The color of an externally visible tag may be very important. Although there is some conflicting evidence, a red-colored external tag appears to be the most attractive to predators, according to most investigators. For the herring-type Atkins tag, it was found that with yellow plastic tubing the returns were five times better than with green, and no red tags were recovered.

For attaching Petersen tags, stainless steel pins or wire have been shown to be superior to those made of nickel or silver. The same is true of wire bridles for Atkins tags.

Experiments have shown that the hydrostatic Atkins tag fastened through the dorsal muscles yields better returns when attached by a bridle than by a curved loop. Others have found that a single nylon filament used to attach Atkins tags to the muscles is inferior to the heavier braided nylon.

For tagging the skipjack (striped tuna), the dart tag yielded several times higher recoveries over those obtained using the best spaghetti tag (tubing has a monofilament core of nylon, and the ends are fastened with a clamp in place of a knot). This is attributed to the ability to hook, tag, and release a skipjack in 4–7 sec using the dart tag, against 20 sec with the spaghetti tag.

The manner in which fish are recaptured may influence the proportion and the sizes recaptured. Thus, salmon marked with Petersen tags are more easily held by gill nets. As a result, smaller salmon that would otherwise pass through the nets are retained. Although this fact may tend to yield a larger proportion of recoveries, it may be undesirable from the standpoint of interpretation of the data; since in analyzing the data from a marking experiment, it is important that one be able to assume that marked individuals do not differ from the remainder of the population to any significant degree, including their chances of being captured.

Capture and Handling of Live Fish

The success of a marking experiment, *especially* one for the purpose of determining mortality rates, often depends on how the fish are captured and handled. Thus, 22% recoveries were obtained of pink shrimp captured in 15- to 20-min hauls with a small otter trawl (try net) of 15- to 20-ft spread, while recoveries from shrimp captured in 1-hour hauls with a large commercial otter trawl were only 14%.

Fish caught by gill net can sometimes be tagged successfully, but on the high seas tagging recoveries are consistently poor from salmon taken in gill nets. This is largely because of the need to use long nets to take the scattered fish and the impracticability of hauling the nets from a small boat. Baited longlines gave higher returns than gill net; however, much the highest proportion were returned when salmon were taken by purse seines.

As important as the method of capture is the subsequent handling. Large agile fish easily injure themselves against hard surfaces, and should usually be held in some type of padded cradle. For exceptionally large fast-swimming fish, the total elapsed time between capture and release may be the most important factor. As mentioned above, very much better recoveries were obtained by marking skipjack tuna with a dart tag that required 7 sec, than with a spaghetti tag that required 20 sec for the whole operation.

Many investigators have used various narcotizing solutions to quiet fish prior to marking. Others prefer not to use such means, and opinion remains divided.

When marking shrimp at sea, abnormal predation is avoided by releasing batches of shrimp in plastic containers locked with a lump of salt; the salt dissolves slowly enough for the container to sink to the natural habitat on the bottom, where the shrimp quickly bury themselves, before opening.

GEORGE A. ROUNSEFELL

References

Rounsefell, George A., "Marking fish and invertebrates," U.S. Fish and Wildlife Service, Fishery Leaflet 549 1963.

Rounsefell, George A., and Everhart, W. Harry, "Fishery Science: its Methods and Applications," John Wiley and Sons, New York, 1953.

Rounsefell, George A., and Kask, John L., "How to mark fish," *Transactions of the American Fisheries Society*, **73**, 320–363 (1945).

MENHADEN FISHERIES

Menhaden (genus *Brevoortia*) along the Atlantic and Gulf of Mexico coasts of the U.S. support the largest commercial fishery in North America. Since 1946, more pounds of menhaden have been landed annually by U.S. fishermen than any other fishery resource. In 1962, the total menhaden catch of 2.3 billion lb amounted to 44% by weight of the total U.S. commercial catch of all species; it was worth $26,000,000 to the fishermen. Since 1962, the catch has declined significantly, particularly in the Atlantic Coast fishery, and in 1966, the total catch was only 1.3 billion lb. (Fig. 1).

Of the four species of menhaden occurring along the coasts of the U.S., two contribute over 99% of the commercial catch. The Atlantic coast catch consists principally of the Atlantic menhaden (*B. tyrannus*), which ranges from Nova Scotia, Canada, to northern Florida. The Gulf menhaden (*B. patronus*), which ranges from the west coast of Florida to Mexico, contributes most of the catch from that area. Two other species, the yellowfin menhaden (*B. smithi*), which occurs mainly along the east and west coasts of Florida, and the finescale menhaden (*B. gunteri*) from the western Gulf of Mexico, are of minor importance. In addition, two other species of menhaden (*B. aurea* and *B. pectinata*) have been reported from the east coast of South America, but little is known about them.

Menhaden are small, oily, herringlike fishes closely related to shad, alewife, herring, and sardine. The largest authenticated specimen was about 19 in. long, but most of those caught are less than 12 in. long and weigh less than 1 lb. The average size of Gulf menhaden at any particular age is considerably less than that of the Atlantic species. In 1966 over 90% of the Atlantic menhaden caught were 1 and 2 year old

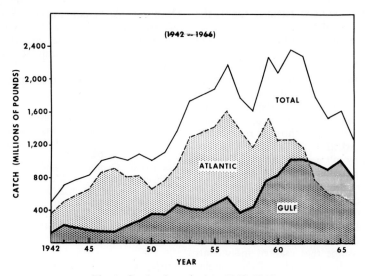

Fig. 1. Catch of menhaden, 1942–1966.

fish; the older fish were taken in the more northern waters. Atlantic menhaden live to an older age than Gulf menhaden—10 year old fish have been recorded. Most Gulf menhaden are 1 or 2 years old, and fish older than 4 years are scarce.

Menhaden are prolific spawners; the number of eggs produced by a single female varies depending on the age and size of the fish. Individual estimates of the number of eggs spawned by Atlantic menhaden range from 38,000 for a fish $8\frac{1}{2}$ in. long to 631,000 for a 13-in. fish. For Gulf menhaden, the average number of eggs produced by a 1 year old fish (average length about 6 in.) is reported at 21,960 and by a 3 year old fish (average length about 8 in.) 122,062.

Adult menhaden feed on microscopic plants and animals rather than on other fish. As the fish swim through the water the planktonic particles are effectively strained from the water by the sievelike gills.

The abundance of menhaden fluctuates greatly. Atlantic menhaden were reported to be abundant 300 years ago, but were reported to be scarce in the Gulf of Maine as far back as 1847. Although these fluctuations in abundance in certain local areas are undoubtedly controlled to a large extent by such environmental factors as water temperature and food, variations in abundance of the entire resource appear to be largely caused by varying survival of individual yearly spawnings (year classes). The highly successful 1958 spawning of Atlantic menhaden yielded about seven times as many fish to the fishery as the 1964 spawning, the poorest during the ten years following 1954. Since 1958, year classes have been poor constantly. The decreased catches in the Atlantic fishery since 1962 have been a coastwise phenomenon reflecting a definite reduction in the number of fish available.

Migrations of menhaden are extensive; those of the Gulf menhaden are not as fully understood as those of the Atlantic species. Atlantic menhaden—at least the older fish—appear to migrate seasonally along the coast northward in the spring and summer, going farther northward each year as they become older, and southward in the fall. They appear in the waters north of Chesapeake Bay only after the ocean water begins to warm in the spring and leave when the water begins to cool. Thus, fish are later in arriving in the more northern waters and leave these waters sooner. Along the New Jersey coast, menhaden generally appear in April or May and leave in October or November, whereas it is May or June before they reach Maine, and they usually depart from the waters off Maine before the middle of October. The migration in the fall is composed of maturing fish moving south to spawn off North Carolina and other South Atlantic states.

Throughout the summer, the fish are generally in water less than 20 fathoms deep, and are in greatest concentrations in localities with extensive estuarine drainage systems, such as Chesapeake Bay and the Mississippi River Delta area. The menhaden in Chesapeake Bay are reported to inhabit the bay throughout the year while they are immature 1 and 2 year old fish. When they reach maturity, at an age of about three years, they leave the bay and enter into the migrations and fisheries along the coast.

The migration of menhaden caught off North Carolina is rather complex. The summer fishery depends principally on immature 1 and 2 year old fish migrating from the south. The fall fishery, beginning about the middle of October, depends principally on maturing fish migrating from the north. In addition, about December, large catches are often made of very small fish whose average size is the smallest for fish of the same age caught anywhere along the coast. The source and ultimate destination of these small fish are still not known. The mature fish disappear from these waters about January and are not seen again until the following spring, when most of them have returned to the more northerly areas.

A commercial fishery for Atlantic menhaden has existed since colonial times when the catch came from gill nets, haul seines, and weirs, but it was not until about 1850, when the purse seine was introduced, that the fishery as it exists today really started. The Atlantic menhaden fishery was not fully developed until after World War II, and production reached a peak in 1956 of 1.6 billion lb. In 1966, 20 plants were processing menhaden along the Atlantic coast: 1 in New York, 2 in New Jersey, 1 in Delaware, 7 in Virginia, 7 in North Carolina, and 2 in Florida.

The fishery for Gulf menhaden started much later than that for the Atlantic species; the annual catch in the early 1940's was less than 200,000,000 lb. The fishery has grown steadily since that time, and in 1963, for the first time in history, the Gulf of Mexico menhaden catch of 980,000,000 lb exceeded that from the Atlantic fishery. The peak catch in the Gulf fishery occurred in 1962, and amounted to 1 billion lb. In 1966, 13 menhaden plants were operating along the Gulf of Mexico coast of the U.S.: 1 in Florida, 3 in Mississippi, 8 in Louisiana, and 1 in Texas. Menhaden fishing along most of the Texas and all of the Alabama coasts normally is prohibited by law, although catches made in Louisiana waters may be landed and processed in Texas.

Fishing usually takes place during the daytime in relatively shallow inshore waters. In 1966, however, over half the catch in the North Carolina fall fishery came from 15 to 40 miles offshore. On the Atlantic coast the fishery is a singleday operation; the boats leave early in the morning and return to the plants the same day. In the Gulf of Mexico, however, the vessels may be at sea more than one day even though fishing is done only in the daytime; for this reason many of the Gulf menhaden vessels are refrigerated.

The fishing season varies in different fishing areas, depending on when the fish arrive. Most of the fishing is during May to October on the Atlantic coast and during April to September in the Gulf. The fishing season in the four major Atlantic fishing areas is about as follows: the South Atlantic area from Florida to Cape Hatteras, North Carolina—April to October; the Chesapeake Bay area—May to November; the Middle Atlantic area from Chesapeake Bay to Long Island, New York—May to October; and the North Atlantic area from New York to Maine—June to October. In addition to the summer fishery in these four areas, the fall fishery off the coast of North Carolina extends from November to January. In 1966, 53% of the catch came from the Chesapeake Bay area, 33% from the North Carolina fall fishery, 11% from the South Atlantic, 2% from the Middle Atlantic, and 1% from the North Atlantic. This situation contrasts sharply with conditions before 1963, when almost half the total Atlantic catch each year came from the Middle Atlantic area. The fishing in the Gulf is in relatively shallow inshore waters off the Mississippi River Delta and the coast of Louisiana. In 1966, 70% of the menhaden catch from the Gulf of Mexico was landed in Louisiana, 24% in Mississippi, 5% in Texas, and 1% in Florida. The percentage landed in Louisiana has been continually increasing in recent years.

The principal fishing gear used to catch menhaden is the purse seine, which accounts for about 98% of the catch (Fig. 2). Pound nets contribute most of the remaining catch. Because menhaden move in dense schools, which normally are homogeneous as to size of fish, purse seines have proved to be the most effective method of

Fig. 2. Menhaden purse seine net being washed down on drying reel at the plant. (Photo: J. W. Reintjes, Bureau of Commercial Fisheries.)

Fig. 3. Menhaden carrier vessels at the dock. Note characteristic crow's nest, formerly used extensively for locating fish schools. Small spotter airplanes are now used almost exclusively for locating the schools. (Photo: J. W. Reintjes, Bureau of Commercial Fisheries.)

catching them. A menhaden purse seine is about 1200 ft long, 60–90 ft deep, and usually is made from synthetic webbing. Corks or buoys along the upper edge of the net keep it afloat, and metal ring weights along the bottom of the net cause it to form a vertical wall in the water. A line that runs through rings on the lower edge of the net acts as a drawstring enabling the bottom of the net to be closed, or pursed. The purse seine is placed aboard two small purse boats, normally all steel or aluminum, about 34–36 ft long; half the net is in each boat. These purse boats with the nets on them are hauled to and from the fishing grounds aboard a larger vessel. Most of the carrier vessels are about 100–200 ft long, and can hold from 125 to over 500 tons of fish (Fig. 3). A tall mast in the forward part of the vessel with a small platform near the top of the

Fig. 4. Menhaden purse seine set. The purse boats have met and are hauling in the net with the power blocks. Carrier vessels are in the background. (Photo: C. M. Roithmayr, Bureau of Commercial Fisheries.)

mast, usually surrounded by a rail, forms the crow's nest from which men can watch for schools of fish. With few exceptions, these boats are built especially for menhaden purse seining. In recent years small airplanes have been used extensively to search for and spot menhaden schools.

When a school is sighted, the carrier vessel approaches it and lowers the purse boats. The purse boats maneuver into the best position ahead of the moving school on direction by radio from the spotter plane pilot. The boats circle the school in opposite directions, letting the net out to surround the fish (Fig. 4). After the purse boats meet behind the school, joining the two ends of the net, the two purse lines are passed through pulleys on a heavy lead weight called a tom, which is then thrown overboard. As the tom, which weighs about 550 lb, falls to the bottom, it pulls the lower line through the rings and purses the net. Pursing is completed by bringing the tom weight, rings, and bottom of the net back into the boats. Most boats now have an automatic purse reel to reel in the lower line to purse the net. This device is particularly helpful in shallow water where the tom may not fall far enough to close the net completely. After the net is pursed, the cork line and much of the net is hauled aboard the boats by power blocks to concentrate the fish. The carrier vessel then comes alongside to form a triangle of the three boats with the concentration of fish in the center. The fish are further concentrated by pulling more net into the purse boats; this procedure is known as "hardening the catch." On most boats hydraulic fish pumps empty the fish from the net into the hold of the carrier vessel. In this operation a reinforced rubber hose of about 1 ft diameter is lowered into the concentration of fish in the net. Water and fish are pumped into the carrier vessel where the fish travel by a chute into the hold and the water is dumped back into the ocean (Fig. 5). The mouths of some of the hoses have a positive electrical field to attract and stun the fish.

The catch for each setting of the net may vary from only a few thousand fish for a poor set to several hundred thousand for a good set. The average catch per set the past few years has been between 60,000 and 75,000 fish.

When the carrier vessel returns to the fish plant, water is pumped into the hold, and the fish are pumped ashore through huge hoses, usually going into a "quarter box" or into an automatic scale, where they are weighed in units of 1000 standard fish. This method of weighing is unique with the menhaden industry, and is usually rather confusing to those outside the industry. Each quarter box holds a volume of

Fig. 5. Loading menhaden into carrier vessel. The water pumped aboard is discharged overboard through chute to the right, while fish are dumped into the hold. (Photo: J. W. Reintjes, Bureau of Commercial Fisheries.)

22,000 cu in., which constitutes an arbitrary measure of 1000 standard fish. One standard menhaden occupies a volume of 22 cu in. Of course, the actual number of fish per unit of measure varies with the size of the fish; the quarter box holds more small fish than large ones. Inasmuch as the fish are packed very tight, however, each measure of 22,000 cu in. will weigh very close to 660 lb.

The efficiency of the fishing operations has been greatly increased in recent years through the use of improved fishing gear and methods. The newer carrier vessels are larger and faster, permitting them to fish greater distances from the plant. The vessels fishing in the Gulf are refrigerated. The use of airplanes has increased the area that can be searched for menhaden schools. The power blocks for hauling in the net not only made this job easier and faster but also reduced the number of men previously required on the crew when the net was hauled in by hand. The lighter synthetic webbing permitted larger nets to be used. The use of the suction hose in place of a large dip net for unloading the net is another fairly recent improvement in the fishing operation. All of these innovations have not only increased the efficiency of the operation, but also have reduced the time required to make a purse seine set, thereby making more sets per day possible.

The increased fishing effort in Chesapeake Bay, coupled with the lesser abundance of recent year classes of menhaden, has caused some radical changes in the Atlantic menhaden fishery. The percentage of the total Atlantic menhaden catch coming from Chesapeake Bay has increased from about 14% in the early 1940's to 53% in 1966, whereas the percentage from the Middle Atlantic area decreased from 42% to only 2% during the same period. The increased catches in Chesapeake Bay, when the fish are 1 and 2 years old, reduce the number of fish that will be available in the other fishing areas as older fish. The disappearance of the older fish off New England has eliminated the fishery there, and all fishing effort north of Chesapeake Bay was considerably reduced in 1965 and 1966.

Most menhaden go into the manufacture of oil and fish meal, although some other uses have been attempted. In colonial days menhaden were used principally as fertilizer; the whole fish was placed in the soil. Although excellent crops resulted for a few seasons, the oil from the fish ultimately parched the soil, making it unfit for several years. Menhaden also have been used to a limited extent as food fish, but they never have been very popular. During the middle of the 19th century, considerable quantities were salted for export to the West Indies and for home consumption, and some small menhaden were canned as "American sardines," but the product was never generally accepted. During World War II a small canning plant was established near Morehead City, N.C.; nearly the entire production was used for food in the Lend-Lease program. Most people found the fish so unpalatable however, that, as soon as other foods were available, this market for menhaden collapsed. The eggs, or roe, of menhaden have had a limited acceptance as food.

The menhaden processing industry is reported to have started about 1811. At that time, after the oil was removed, the remaining fish and water were used as fertilizer. This procedure is in sharp contrast with conditions in 1967, when the catch is processed into oil and fish meal. Upon arrival at the menhaden plant, the fish are unloaded by suction hoses and conveyed into continuous cookers where they are processed with live steam. The cooked fish then pass into huge screw presses, which drive out the oil and water. "Press cake," the solid portion of the fish, is put into large rotary driers to remove the remaining moisture, and the dried "fish scrap" is stored in large warehouses to cool. Grinding reduces this dried material to fish meal. An antioxidant material is normally added to the fish meal before it is put into bags. The liquids from the screw presses are centrifuged to separate the oil. The remaining liquid, called "stickwater," is condensed by large evaporators into menhaden solubles, a syrupy product that contains large quantities of dissolved protein, vitamins, and other essential nutrients. Rich in protein, menhaden meal and solubles make excellent food supplements for poultry, hogs, mink and other animals. The oil is used in a variety of products including paints, soaps, cosmetics, resins, putties, caulking compounds, and lubricants, and for tanning leather. Substantial quantities of oil are exported to Europe principally for making margarine.

The menhaden resource is of great economic value to the U.S. The total catch of menhaden since the formation of the U.S. has exceeded 65 billion lb—more than the production of any other species. In 1962, about 3300 fishermen, 180 vessels, 360 purse boats, and 70 other boats were engaged in the fishery. Furthermore, menhaden supply about 75% of all fish meal, 80% of the marine oils, and nearly 80% of all fish solubles produced in the U.S. In a typical year, this resource yields over 445,000,000 lb of fish meal, 20,000,000 gals of oil, and 20,000,000 lb of solubles.

KENNETH A. HENRY

MICROBIOLOGY, MARINE

Traditionally the field of marine microbiology includes the bacteria, fungi, yeasts, and viruses that occur in the marine environment. The study of the minute animals and plants in the sea is normally treated as a separate discipline. Until recently, most microbiological investigations have been concerned with the bacteria in the oceans and, consequently, more is known about these organisms than yeasts, fungi, and viruses. The first studies of marine microorganisms were made toward the end of the 19th century, and there have been numerous but sporadic studies since then. Because marine biological research is expensive and technically difficult, most investigations of marine microorganisms have been limited in scope, time, and achievement. In view of the vastness of the oceans and the many different environments they contain, the available knowledge on marine bacteria is meager indeed; moreover, the information on yeasts, molds, and viruses in the sea is fragmentary at best.

Bacteria occur quite generally in the sea since they have been isolated from every type of marine environment sampled. They are most abundant in surface waters, particularly near land, in the sediment, even at great depths, and on the skin, and gills and in the intestines of marine animals. Yeasts have been found in somewhat smaller numbers in the water, in mud, and associated with animals. Fungi are probably most abundant in inshore waters, and are usually associated with wood or wooden structures, decaying plant materials, and, sometimes, with fish or other marine animals, which they may parasitize. However, there seems little doubt that fungi occur in much lower numbers than bacteria or (probably) yeasts in the oceans. All types of microorganisms are found in greatest numbers in those marine environments close to land, particularly in estuaries and areas subject to terrestrial drainage or fresh water intrusion. This is probably due to the increased level of bacterial nutrients in these areas. The number of bacteria in the water column of inland sea areas, such as Puget Sound, may reach several thousand per milliliter at certain times, while in the open ocean, numbers are generally much lower ranging from a few hundred per milliliter to one or two per cubic meter of water. The bacteria in the upper layer of sediment may be in millions per gram in estuarine areas, which show wide fluctuations in number, but are generally less abundant in offshore sediments where populations of 10,000 to 100,000 per gram are common. Populations of bacteria resident in the surface slime of fish skin and gills vary between several hundred and several thousand per square centimeter but the intestinal contents of fish may contain up to 10,000,000 bacteria per gram. There seems to be a regular pattern of occurrence of bacteria in the water column and in sediments that is related to nutrient supply and environmental conditions. In the water, there are moderately high bacterial numbers in the surface layers, decreasing numbers down the water column, and relatively high numbers near the sediment surface. In sediments, there are high numbers of bacteria in the upper layers and a progressive reduction in numbers below the surface. Some typical figures reported by investigators for bacterial numbers in various marine environments are shown in Table 1. Virtually all the quantitative data available refer to heterotrophic bacteria, and nothing is known with certainty about the quantitative distribution of autotrophic bacteria.

Representatives of all the major groups of bacteria and many of the yeast and fungal families have been isolated from the sea at one time or another. However, this is probably indicative more of the ubiquity of these microorganisms than of great diversity in marine populations, since many types isolated are clearly unable to grow or compete effectively in the marine environment. One puzzling observation is the consistent isolation of large numbers of sporing mesophilic bacteria from sediments the temperature of which is consistently below the minimum growth temperature for this type of organism.

Autotrophic and photosynthetic bacteria belonging to the families Athiorodaceae, Thiorodaceae, Chlorobacteriaceae, Nitrobacteriaceae, Methanomonadaceae, and Thiobacteriaceae have been isolated from seawater and marine sediments. There is no reliable information on their abundance or the actual importance of their chemosynthetic capabilities in situ.

An interesting group of nitrifying bacteria, which seems to be of widespread occurrence in seawater, has been isolated recently. Some of these nitrifying bacteria seem to be uniquely

TABLE 1.

Area	Sample	Bacterial Count per Gram Wet Sample	
		min.	max.
Atlantic			
Woods Hole	deep sediment	5.0×10^3	10^5
Scotland	deep sediment	2.0×10^4	2.0×10^4
Nova Scotia	deep sediment	10^4	10^6
Caribbean			
West Indies	deep sediment	1.6×10^7	1.8×10^8
Pacific			
Australia estuary	intertidal sediment	8.0×10^4	2.0×10^7
Puget Sound	intertidal sediment	10^6	10^8
Australia	deep sediment	10^5	10^6
Southern California	deep sediment	10^5	10^7
North Central Pacific	deep sediment	10^3	10^5
North East Pacific	deep sediment	10^3	10^5
Pacific trenches	deep sediment	10^4	10^6
Puget Sound	deep sediment	10^4	10^6
Gulf of Mexico	deep sediment		10^6
Indian Ocean	deep sediment	10^3	10^4
Antarctic Ocean	deep sediment	10^3	10^4

marine (e.g. *Nitrosocystis oceanis*), and electron microscope photographs of thin sections indicate complex multiple membrane structures that may be related to the nitrification process. Purple sulfur bacteria that oxidize hydrogen sulfide are abundant in the Black Sea and may occur in other areas where reduced sulfur is present.

Strictly anaerobic organisms of the genus *Desulfovibrio* are apparently widely distributed in marine sediments. They reduce oxidized sulfur compounds to hydrogen sulfide (H_2S), and rapidly produce anaerobic conditions in their immediate environment. They are probably a major cause of the permanent anaerobiosis of the lower layers of the Black Sea, and may cause temporary anaerobiosis in semi-enclosed areas of sea water, such as sea lochs or fiords, in which water remains stagnant during summer months. This condition usually results in fish kills. Numerous *Clostridia* have been isolated from sediments and intestinal contents of animals. However, most of these organisms have proved to be incapable of growth at the ambient temperatures in most of the sea (less than 4°C) when tested in the laboratory. The question of their actual activity in situ remains moot.

Gram-negative rods, belonging to the genera *Pseudomonas*, *Achromobacter*, *Vibrio*, and *Flavobacterium*, are the most prevalent type of bacteria isolated from all marine environments. Low numbers of *Cytophaga* also appear to be consistently present in many marine environments; however, they are often mistakenly identified as *Flavobacteria* by inexperienced investigators. Gram-positive cocci, *Bacillus* and *Corynebacteria*, are frequently present in marine samples cultured at environmental temperatures, and will sometimes appear to dominate the flora if count plates are incubated at temperatures above 30°C. In the latter case, many of the organisms are actually unable to grow at environmental temperatures. Occasionally *Spirilla* or some of the higher bacteria are isolated from marine environments and, indeed, in suitable circumstances they may be very abundant (e.g., the association of *Spirilla* with decomposing organic material). These organisms frequently will not grow well on the normal types of media used for bacterial counts, and this may explain the relatively few reports of their occurrence in the oceans.

Terrestrial bacteria, particularly coliform bacteria and enterococci, are not normally present in seawater taken off shore or in areas free from terrestrial contamination. Such bacteria seem to die rapidly in unpolluted seawater and this has been shown to be due to a substance or substances in seawater that may be inactivated by heat. Antibiotics, bacteriophage, halide, and heavy metal toxicity have all been proposed as the "active substance." The most recent evidence indicates that heavy metals are probably involved in the killing process.

Currently, there are no substantiated explanations for the dominance of gram-negative rod-shaped bacteria in the sea. There is probably a complexity of interacting environmental factors, including salt composition of the water (complex), nutrient supply (poor but varied), temperature (generally low), and pH (usually high), which selectively favors organisms of the type found in largest number. Certainly there does not seem to be any single factor that prevents gram-positive bacteria from occurring in greater numbers.

The general physiological and biochemical characteristics of marine bacteria are positively related to the environment in most cases. The vast majority of the bacteria (and yeasts) are psychrophilic or psychrotrophic, grow best in media containing seawater and are nonexacting nutritionally: they utilize a wide range of organic chemicals as growth substrates. The seawater requirement of bacteria isolated from the marine environment is not usually absolute, and the majority of them will grow on normal laboratory media containing 0.5% sodium

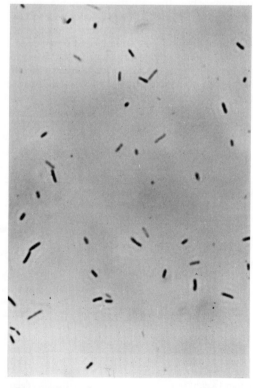

Fig. 1. Light micrograph of terrestrial *Escherichia coli* grown at 15 psi. (Photo: E. S. Boatman.)

MICROBIOLOGY, MARINE

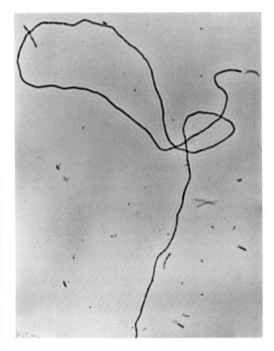

Fig. 2. Light micrograph of *Escherichia coli* grown at 6,000 psi. Note filament of 260 μ in length. (Photo: E.S. Boatman.)

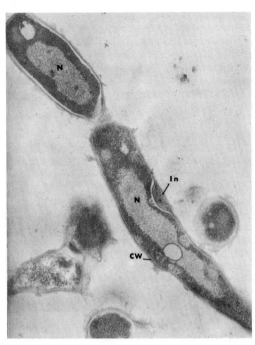

Fig. 4. Electron micrograph of an ultrathin section of terrestrial *Corynebacterium sp.* grown at 4,000 psi. Note cell wall inclusion, In. (Photo: E. S. Boatman.)

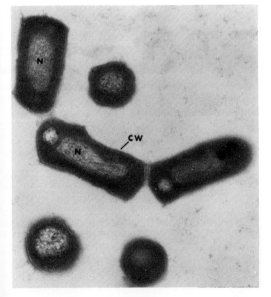

Fig. 3. Electron micrograph of an ultrathin section of a marine barosensitive *Corynebacterium sp.* grown at 15 psi. N = nucleus; CW = cell wall. (Photo: E. S. Boatman.)

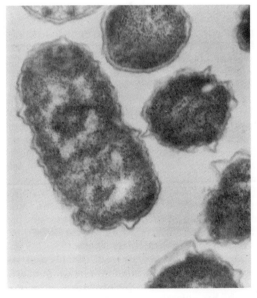

Fig. 5. Electron micrograph of an ultrathin section of a marine *Aerobacter sp.* grown at 15 psi. (Photo: E. S. Boatman.)

chloride. In many cases where a sea water requirement is present on initial isolation, it is lost during subculture in the laboratory. Recent studies on seawater-requiring marine bacteria have indicated that they possess a sodium requirement and, in some cases, a need for magnesium also.

Mineralization is the reconversion of complex organic materials to simple mineral salts and carbon dioxide. This process, which supplies essential nutrients for the growth of photosynthetic algae, is one of the vital activities of bacteria in the sea. The breakdown of protein and lipid material and the dissimilation of the numerous small molecule compounds of which plant and animal cell remains are composed, is carried out by oxidative, hydrolytic, and other mechanisms similar to those of common terrestrial bacteria. Pseudomonas types are probably most often involved in this process. Starch hydrolysis seems to be due in many cases to marine *Vibrios*. The hydrolysis of the complex polysaccharides, such as chitin and "agar-agar" is effected by *Vibrios*, *Pseudomonas*, and *Cytophaga* species. Agar digesters are most common near land, but chitin-digesting species are widely distributed (though they occur in greatest number in inshore areas). Agar- and chitin-digesting bacteria frequently hydrolyze starch also. *Cytophaga* species seem to be the principal bacterial types that hydrolyze cellulose, though marine fungi are also active against this polysaccharide. Again, cellulolytic activity is most common among inshore isolates. One striking biochemical characteristic shown by certain marine bacteria is the ability to produce light. These "luminescent" or "phosphorescent" bacteria are mainly *Vibrios* or *Aeromonas* species, but are often referred to as *Photobacterium*. Luminescence results from the oxidation of a dodecyl aldehyde in a system involving flavin mononucleotide, and is analogous to, but different from, that of the firefly. Phosphorescent bacteria produce light continuously when oxygen and a proper nutrient supply are available. Free-living phosphorescent bacteria seem to be associated mainly with crustacea, but also occur on fish and in sea water. They may cause dead fish to luminesce. Luminescent bacteria are present as symbionts in the light organs of certain deep sea fish. These organisms are very difficult to culture in the laboratory, and little is known of their properties. Some may be obligately symbiotic.

Many of the deeper regions of the sea and sea floor are under considerable hydrostatic pressure, and bacteria from these environments are commonly barotolerant in nature. Electron microscope photographs of sectioned barotolerant and normal terrestrial species of bacteria cultured at elevated pressures (up to 8000 psi) have shown normal structures in the former and impairment of the cell replicating mechanisms and membrane abnormalities in the latter. Some deep sea bacteria seem to be obligately barophilic, growing only under conditions of elevated pressure. These bacteria may have unusual enzyme systems (perhaps unusual protein structures) that enable them to grow and metabolize at low temperature and high pressure.

At present, only two species of indigenous marine bacteria are known to be dangerous to man. These are *Clostridium botulinum type E* and *Vibrio parahaemolyticus*. However, there is some question whether either or both are truly marine bacteria. Both have been implicated in food poisoning outbreaks due to consumption of seafoods. Other microorganisms occurring as contaminants in the sea can cause sickness in man. Examples of these are *Erysipelothrix insidiosa*, which causes localized infections in fish handlers and *Salmonella spp.*, which may contaminate shellfish and cause food poisoning. Recently evidence has accumulated that shellfish may act as vehicles of infection for enteric viruses such as infectious hepatitis. Viruses apparently will survive in the sea for short periods of time.

Various rather poorly defined species of bacteria, mainly *Vibrios*, *Aeromonas*, *Pseudomonas*, and *Mycobacteria* and some fungi (or fungi-like organisms) are known to cause disease in marine fish. Virus infections of marine fish have also been reported.

Only a few bacteriophages causing lysis of marine bacteria have been isolated and studied. These viruses have shown certain "marine" characteristics, most notably a requirement for sodium chloride and low temperatures for maximum lytic effect, and extreme sensitivity to heat. The microorganisms in the sea are thus numerous and varied in type and in activity. It is probable that most of the known functional activities of terrestrial bacteria, fungi, and yeast are duplicated in the sea, though probably by a smaller number of species.

The principal difficulty in investigating marine microorganisms is in the collection of samples, since this requires a ship and specialized equipment—except in the case of littoral samples. Material for microbiological analysis must be collected aseptically if possible or at least with the minimum of contamination by adventitious microorganisms. The majority of marine sampling operations must be done by remote control from the deck of a research vessel using a cable and winch, since the environments themselves are usually more or less inaccessible to man.

Sediments are most difficult to sample, and so far no satisfactory equipment is available for completely aseptic sampling of the sea bed.

Normally, cylindral "geological" covers are used. They usually contain a plastic sleeve that is sterilized prior to lowering into the water. The core is allowed to fall freely through the water column and drive into the sediment. During its return to the surface, a nonreturn valve closes off the upper end, and the lower part is plugged by the column of sediment trapped in it. Various devices may be used to prevent loss of sandy sediments through the open end of the corer during recovery. The sample obtained in this way is not completely free from contamination since the container is exposed to the water bacteria on the way to the bottom. However, the populations of bacteria on the sea bed are usually so much greater than those in the water column that there is little quantitative problem. Moreover, sterilization, using alcohol, limits contamination in the upper layers of water in which bacteria are most abundant. Samples of sediments have also been taken using grabs or dredges. In these cases there is considerable mixing of sediment layers and exposure of the sediment sample to contamination.

Water samples may be taken aseptically at different depths by means of sterile water samplers of various designs, usually activated by a series of messengers, heavy brass cylinders that slide down the cable, tripping a release mechanism by force of impact. The most widely used samplers are modifications of the ZoBell water sampler; they consist essentially of a partially evacuated bottle or rubber bulb sealed by a drawn out capillary tube. The sterile sealed apparatus is lowered to the desired depth, and the capillary is broken by impact of the messenger, causing the sampling tube to spring away from cable and bottle, and enabling a water sample to be taken (by pressure difference) without contamination. The Niskin sampler, which is now being used extensively for aseptic water sampling, operates by a positive pressure principle. A presterilized plastic bag with a sealed rubber outlet tube is held closed on a metal rack similar to a large hinge. The hinge is held closed under tension, which is released by impact of the messenger. The outlet tube opens, springs away from the structure, and the bag is forced open by the hinge movement, filling with water. When the cable begins to reel in, the outlet tube is automatically clamped shut, and the water sample is sealed within the bag, which is then brought to the surface. This positive action sampler has two advantages over the various types of ZoBell samplers—it provides a large volume of water (several liters), and it operates at depths at which evacuated bottles tend to implode and rubber bulbs may fail to open because of the high ambient hydrostatic pressure.

Samples of subsurface waters at a few meters depth may be obtained by using sterile, thick-walled bottles with stoppers that can be pulled off by jerking a string tied to them. On occasion, metal- or glass-lined or plastic water samplers of the Nansen Bottle or Frautchy bottle type have been used to obtain water samples for microbiological analysis. These samplers are open when lowered, and are closed, at the appropriate depth, by a messenger. Thus, they are not sterile samplers, and they tend to build up on the inside surfaces contamination that is difficult to eliminate.

Samples of the surface film of water may be obtained by use of a fine mesh stainless steel screen, which is lifted through the upper layer of water.

The microbiological populations on marine plants and animals must, of necessity, be sampled after removal of the living creature from its environment. The methods used in catching or harvesting will obviously affect the purity of the sample obtained. Generally, line-caught fish are subject to the least contamination during capture. However, most sampling is carried out on fish caught by nets. The actual sampling methods most common involve swabbing (with cotton or alginate swabs), aseptic excision of skin or gill material, aseptic dissection of the alimentary canal, or intracloacal swabbing, and, occasionally, direct heart puncture to obtain blood samples. Mollusca may be sampled by cleaning the surface of the closed shell with disinfectant and aseptically sampling the contents. Qualitative sampling of waters and (shallow) sediments has been attempted by suspending glass slides in the water column or burying them in the sediment. Microorganisms that grow on the surfaces provided may be directly examined microscopically. This method is difficult to use in the open sea, however.

Samples of water, sediment, or animal material are generally analyzed as promptly as possible after recovery, since considerable qualitative change in the microbial content can take place within a few hours, particularly if the holding temperature is high (above 10°C). Except where a sanitary survey is being made, it is best to conduct the analysis at low temperatures close to the in situ level by using prechilled media, dilution blanks, etc. It is customary to use sterilized seawater as the diluent material to make primary inoculations on seawater based media.

The highest counts of bacteria and the greatest diversity of species is obtained by incubation at the ambient in situ temperature. For most marine environments this is below 5°C. However, satisfactory results can be achieved more quickly by using a slightly higher incubation temperature of 10–15°C; of course, in the case of surface waters from tropical areas or estuarine waters in summer, a higher temperature may be desirable. It is rare, however, that the incubation temperatures commonly used by clinical microbiologists can be safely used with marine bacteria, many of which are quite sensitive to prolonged exposure to temperatures above 30°C (above 20°C in the case of true psychrophils). Because of the heat sensitivity of marine bacteria, it is best to use surface inoculation of prechilled plates or direct inoculation of liquid media for their primary isolation and enumeration.

The methods subsequently used to grow and test microorganisms from the sea depend more on the nature of the organisms themselves than on their marine origin. Particular attention must be paid, however, to cation requirements and thermosensitivity.

From the foregoing, it is apparent that despite a scientific history extending over 80 years, marine microbiology is an infant science in which even the techniques of sampling and isolation are poorly developed. Yet, it is equally clear from the small amount of information available, that the microorganisms in the sea play a role at least as important as that of microorganisms on land. They are essential components of the cycle of synthesis and decay on which life in general and human life in particular depends. Moreover, they have enormous economic significance ranging from the synthesis of petroleum to the decay of seafoods. The application of modern techniques of microscopy, biochemistry, and genetics to marine microorganisms is beginning to yield tantalizing glimpses of the occasionally unique characteristics of these creatures. The increasing technological significance of the oceans to man as a source of food and raw materials and as an environment that must be understood and controlled will undoubtedly necessitate a much more vigorous and, hopefully, more productive program of investigation in marine microbiology in the future.

JOHN LISTON

Cross references: *Antimicrobial Substances; Bacteria in the Sea; Fungi, Marine.*

MINERALS OF THE OCEAN

Potential Resources

Annually, the rivers of the world dump hundreds of millions of tons of material into the sea. Most of the clastic sediments rapidly settle to the seafloor in near-shore areas, in some cases forming valuable placer mineral deposits. The dissolved load of the rivers, however, mixes with seawater, and is gradually dispersed over the total oceanic envelop of the earth. While the kinetics of chemical reactions in the sea is generally almost immeasurably slow, because of its immense size, the ocean is acting on a significant scale to separate, recombine, and eventually concentrate on the seafloor many of those

TABLE 1. MINERAL DEPOSITS OF THE SEA

Region	Minerals of Interest
Marine beaches	Placer deposits of gold, platinum, diamonds, magnetite, ilmenite, zircon, rutile, columbite, chromite, cassiterite, scheelite, wolframite, monazite, quartz, calcium carbonate, sand and gravels.
Continental shelves	Calcareous shell deposits, phosphorite, glauconite, barium sulphate nodules, sand and gravels, placer deposits in drowned river valleys of tin, platinum, gold, and other heavy minerals, metal rich sediments such as those underlying the hot brines in the Red Sea containing iron, manganese, copper, zinc, silver, lead, etc.
Seawater	Common salt, magnesium metal, magnesium compounds, bromine, potash, soda, gypsum, and, potentially, sulfur, strontium, borax. Most other elements can be found in seawater and, given recently developed extraction techniques, seawater is a potential source of uranium, molybdenum, gold, etc. Bodies of water within the sea that have abnormally high concentrations of various elements such as gold, iron, copper, zinc, lead, etc.
Subseafloor rocks	Oil, gas, sulfur, salt, coal, iron ore, and possibly other mineral deposits in veins and other types of deposits as in continental rocks.
Deep-seafloor	Clays—for clay mineral uses, possibly also as a source of alumina, copper, cobalt, nickel, and other metals. Calcareous oozes—for limestone applications. Siliceous oozes—for silica and in diatomaceous earth applications. Animal remains—as a possible source of phosphates and metals such as tin, lead, silver, and nickel. Zeolites—as a source of potash. Manganese nodules—as a source of manganese, iron, cobalt, nickel, copper, molybdenum, vanadium, rare earths, and possibly many other metals.

minerals necessary in an industrial society. As a result of this concentrating process, the ocean floor holds many mineral deposits that, if found on the continents, would be considered high-grade ores. In addition, seawater itself is a potential source of many industrially important minerals.

Because of the nature of the minerals contained therein, it is convenient to consider the deposits of the sea as occurring in several environments: (1) marine beaches; (2) seawater; (3) continental shelves; (4) subseafloor consolidated rocks; and (5) the deep-seafloor. Minerals are now mined from all these regions except the deep-seafloor, which has only recently been recognized as a repository for mineral deposits of unbelievable extent and truly significant economic value. The types of minerals found in the offshore areas are indicated in Table 1.

Marine Beach Deposits. Because of the crushing, grinding, and concentrating action of the ocean surf, much of the minerals processing in the beaches has been done by nature. What mining and processing is left to do is, in general, uncomplicated and inexpensive. From marine beaches are mined such minerals as diamonds, gold, magnetite, columbite, ilmenite, zircon, scheelite, monazite, platinum, and silica, to mention the most important commercial minerals. Beaches have been mined for many years in some areas of the world, with periodic storms, in some cases, replenishing the mined out areas. In general, however, present beach deposits tend to be limited, although, in the offshore areas, one can find a series of beaches that contains minerals of the same type as the onshore beaches. During the Ice Ages, sea level was appreciably lowered as the ocean water was transferred to the continental glaciers. Because of the cyclic nature of the Ice Ages and the intervening warm periods, a series of beaches was formed in areas offshore present beach deposits. Thus, the offshore potential is generally much greater than that of the onshore for mineral production. With recently developed sonic devices, it is not difficult to locate and delineate these submerged beaches. An interesting aspect of beach deposits is that they tend to contain mineral paystreaks that are abnormally rich and are, thus, very profitable to mine.

Seawater. Seawater is generally considered to have all the natural elements dissolved in it. Covering an area of about 140 million square miles at a mean depth of about 2.5 miles, the sea holds about 330,000,000 cubic miles of water. Seawater contains an average of about 3.5% of elements in solution; thus, each cubic mile of seawater, weighing about 4.7 billion tons, holds about 166,000,000 tons of solids. Nine of the most abundant elements constitute over 99% of the total dissolved solids in seawater. Only four of these elements are commercially extracted at the present time to a notable extent: sodium and chlorine in the form of common salt; magnesium and some of its compounds; and bromine. Several calcium and potassium compounds are produced as byproducts in salt or magnesium extraction processes. Table 2 lists the elements whose concentration in seawater has been measured.

The commercial extraction of any element or compound from the sea immediately makes reserves of that material unlimited, as measured against present world consumptions. Bromine is almost purely a marine element; over 99% of the bromine in the earth's crust is in the ocean. There is about $0.02 worth of bromine per ton of seawater; thus, when being processed for this element, seawater is probably the lowest grade ore known. More significantly, the $0.02 per ton of contained element value is probably the lower limit of ore grade when considering seawater as a source of any element extracted by conventional techniques; that is, by taking the water from the ocean, pumping it through a plant, adding some reagent to it, performing a chemical or physical separation technique, and exhausting the barren seawater in a manner that will not contaminate the incoming seawater.

Ion exchange and chelating techniques, recently developed, may permit the recovery from seawater of many elements with a concentration much less that of bromine. The National Physical Laboratory in Great Britain recently announced the development of one such technique that permits the recovery of uranium from seawater at a cost competitive with that of winning uranium from low-grade continental ores. There is about $0.0001 worth of uranium per ton of seawater; however, considering the nature of this process, we cannot assume that this concentration is at all a lower limit to the value of an element that can be extracted from seawater using these techniques. Table 3 lists several of the elements that may be extracted by this process. Extension of this technique to any other of the elements in seawater immediately makes reserves of that element, as far as world consumption is concerned, unlimited.

Another concept that should be evaluated when considering seawater as a source of minerals is that of orebodies within seawater itself. As concerns the major elements dissolved in seawater, seawater is fairly uniform in composition. Where the trace elements are concerned, seawater is not of a uniform composition. Values of gold in seawater as high as 60 mg per ton have been noted. Such a concentration works out to be about $0.06 worth of gold per ton. Even using conventional extraction techniques, a body of

TABLE 2. CONCENTRATION AND AMOUNTS OF SIXTY OF THE ELEMENTS IN SEAWATER[a]

Element	Concentration (mg/liter)	Amount of Element in Seawater (tons/mile^3)	Total Amount in the Oceans (tons)
Chlorine	19,000.0	89.5×10^6	29.3×10^{15}
Sodium	10,500.0	49.5×10^6	16.3×10^{15}
Magnesium	1,350.0	6.4×10^6	2.1×10^{15}
Sulfur	885.0	4.2×10^6	1.4×10^{15}
Calcium	400.0	1.9×10^6	0.6×10^{15}
Potassium	380.0	1.8×10^6	0.6×10^{15}
Bromine	65.0	306,000	0.1×10^{15}
Carbon	28.0	132,000	0.04×10^{15}
Strontium	8.0	38,000	$12,000 \times 10^9$
Boron	4.6	23,000	$7,100 \times 10^9$
Silicon	3.0	14,000	$4,700 \times 10^9$
Fluorine	1.3	6,100	$2,000 \times 10^9$
Argon	0.6	2,800	930×10^9
Nitrogen	0.5	2,400	780×10^9
Lithium	0.17	800	260×10^9
Rubidium	0.12	570	190×10^9
Phosphorus	0.07	330	110×10^9
Iodine	0.06	280	93×10^9
Barium	0.03	140	47×10^9
Indium	0.02	94	31×10^9
Zinc	0.01	47	16×10^9
Iron	0.01	47	16×10^9
Aluminum	0.01	47	16×10^9
Molybdenum	0.01	47	16×10^9
Selenium	0.004	19	6×10^9
Tin	0.003	14	5×10^9
Copper	0.003	14	5×10^9
Arsenic	0.003	14	5×10^9
Uranium	0.003	14	5×10^9
Nickel	0.002	9	3×10^9
Vanadium	0.002	9	3×10^9
Manganese	0.002	9	3×10^9
Titanium	0.001	5	1.5×10^9
Antimony	0.0005	2	0.8×10^9
Cobalt	0.0005	2	0.8×10^9
Cesium	0.0005	2	0.8×10^9
Cerium	0.0004	2	0.6×10^9
Yttrium	0.0003	1	5×10^8
Silver	0.0003	1	5×10^8
Lanthanum	0.0003	1	5×10^8
Krypton	0.0003	1	5×10^8
Neon	0.0001	0.5	150×10^6
Cadmium	0.0001	0.5	150×10^6
Tungsten	0.0001	0.5	150×10^6
Xenon	0.0001	0.5	150×10^6
Germanium	0.00007	0.3	110×10^6
Chromium	0.00005	0.2	78×10^6
Thorium	0.00005	0.2	78×10^6
Scandium	0.00004	0.2	62×10^6
Lead	0.00003	0.1	46×10^6
Mercury	0.00003	0.1	46×10^6
Gallium	0.00003	0.1	46×10^6
Bismuth	0.00002	0.1	31×10^6
Niobium	0.00001	0.05	15×10^6
Thallium	0.00001	0.05	15×10^6
Helium	0.000005	0.03	8×10^6
Gold	0.000004	0.02	6×10^6
Protactinium	2×10^{-9}	1×10^{-5}	3000
Radium	1×10^{-12}	5×10^{-7}	150
Radon	0.6×10^{-15}	3×10^{-12}	1×10^{-3}

[a] After Mero, Ref. 4.

TABLE 3. STATUS OF ECONOMIC EXTRACTION TECHNOLOGIES FOR VARIOUS MINERALS FROM SEAWATER

Material	Concentration in Seawater (mg/liter)	Total Amount of Material in Seawater (Tons)	Value of Material in a Ton of Average Seawater ($)	Reserves of Material at Present World Consumption Rates (Millions of Years)	Status of Extraction Technology[a]
NaCl	29,500	45.6×10^{15}	0.31	+1,000	E
Magnesium	1,350	2.1×10^{15}	1.00	+1,000	E
Sulfur	885	1.4×10^{15}	0.03	200	E
KCl	760	1.2×10^{15}	0.024	300	E
Bromine	65	100×10^{12}	0.02	1,000	E
Borax	50	70×10^{12}	0.003	700	NE
MoO_3	0.1	25×10^9	0.0003	1	PIE
U_3O_8	0.006	10×10^9	0.0001	0.5	EIE
Silver	0.0003	5×10^8	10×10^{-6}	0.1	EIE
Gold	0.000004	6×10^6	4×10^{-6}	—	PIE
Tin	0.003	5×10^9	6×10^{-6}	0.1	PIE
Nickel	0.002	3×10^9	4×10^{-6}	0.01	PIE
Copper	0.003	5×10^9	6×10^{-6}	0.001	PIE
Cobalt	0.0005	1×10^9	1×10^{-6}	0.1	PIE

[a] E = economic with present extraction techniques; NE = not economic with present conventional extraction techniques; PIE = possibly economic with ion exchange extraction techniques; EIE = economic at present with ion exchange extraction techniques.

seawater containing 60 mg of gold per ton could be considered a high-grade gold mine if that body of water were of sufficient size. Thus far, very little work has been done to assess the size of these "high-grade" bodies of water within the ocean. In his general reconnaissance of the ocean, Haber (Ref. 3) found that the average content of gold in seawater was only about 0.04 mg per ton, or about 4×10^{-6} worth of gold per ton.

Recently, a body of water was found in the Red Sea that contains from 1000 to 50,000 times as much of such elements as iron, copper, manganese, and lead as does normal seawater (Miller, et al., Ref. 5). Thus, it is possible to consider the seawater in the same context as the land, as far as mineral deposits are concerned; that is, the sea is not uniform in its composition any more than the solid crust of the earth is uniform in its composition. Deposits, or bodies of water, of unusually high concentrations of elements exist in the sea as in ore deposits on land.

Continental Shelves. The continental shelves of the world cover an area of about 10,000,000 square miles or about 20% of the above water continental area of the earth. As the rocks of the continental shelves are basically similar to those of the adjacent continental areas, it can be expected that the mineral producing potential of the shelf rocks should be similar to that of the onshore rocks. In the case of petroleum, at least in the Gulf of Mexico and off California, such is proving to be the case. Because of their cover of water, however, the continental shelves hold other types of mineral deposits, which are economic to exploit with present day techniques.

Calcareous Deposits. Calcareous shells are mined from offshore deposits in a number of locations, notably in the Gulf of Mexico and off Iceland. The shell deposit off Iceland is located about five miles off the southwest coast in Faxa Bay. A shoal of rock lying west of the deposit contains a large population of various kinds of shelled animals. The winter waves break the shells free, crushing and grinding them. Tidal currents sweep the particles into Faxa Bay, where they are mined from a depth of about 120 ft at a rate of about 5000 tons per day. As the influx of the shell material far exceeds the rate of extraction, this deposit is a replenishing mineral deposit as opposed to land mineral deposits, which are generally considered as depleting resources. This renewing feature is a common, and highly significant, characteristic of mineral deposits in the sea. Calcareous deposits are also mined off the coast of Hawaii and the coast of certain of the islands in the Bahamas.

Placer Deposits on the Shelf. In Malaysia, Thailand, and Indonesia, tin is mined from river placers onshore to the present sealevel. These deposits, in many cases, extend into the offshore areas. Such offshore placer deposits, like offshore beach deposits, are beginning to show up with increasing frequency, largely due to the development of seismic techniques of locating the placer traps and offshore drilling techniques to delineate the ore locations. Drowned river valleys may also hold substantial placer deposits of gold and platinum off Alaska, of diamonds off the mouth of the Orange River in Southwest Africa, and of tin and titanium minerals off the coasts of Australia. In fact, any area of the world in which placer deposits were mined in river channels near the present coast, probably also contains placer deposits of like minerals in the offshore area.

Sand and Gravel. Sand and gravel are probably the most prosaic of mineral commodities. However, from a tonnage standpoint, they are easily the most important mineral commodities mined in the world. All industrial societies need copious quantities of these materials in the manufacture of concrete, for fill, and for myriad other purposes. Until recently, such material was generally extracted from pits near the market. Zoning regulations, however, are making the opening of new pits near cities impractical.

Offshore areas are generally well supplied with sand and gravel deposits, and, in a few areas, notably off England, exploitation of these deposits has been initiated. Because much of the world's population is concentrating in the seacoast areas, the offshore gravel deposits will become increasingly important sources of this material.

Glauconite. Glauconite is an interesting authigenic mineral found in considerable quantity in various offshore locations. Frequently, it is found in deposits relatively uncontaminated with other materials. Being fine-grained and having no overburden would allow relatively inexpensive mining costs for this material. Containing from 4 to 9% of K_2O, the sea floor glauconite might be a source of potassium for agricultural fertilizers.

Phosphorite. Phosphorite deposits are found on the shelves off the coasts of many nations of the world. Thus far, phosphorite deposits have been found off Peru, Chile, Mexico, the west and east coasts of the U.S., off Argentina, South Africa, Japan, and on the submerged parts of several islands around the Indian Ocean.

Off California, the seafloor phosphorite occurs as nodules, which vary in shape from flat slabs, several feet across, to spheres of oolitic size (Fig. 1). The nodules are commonly found on the seafloor as a monolayer at the surface of coarse-grained sediments. The composition of the phosphorite from the California offshore area is surprisingly uniform. As indicated in Table 4, these offshore phosphorite deposits contain economically attractive amounts of phosphorous. While it is estimated that there are about 1 billion tons of phosphorite in the California offshore area (Emery, Ref. 1), probably not more than 100,000,000 tons of this rock would be economic to mine at the present time. Around the world, the economic reserves of offshore phosphorite probably can be counted in the several billions of tons.

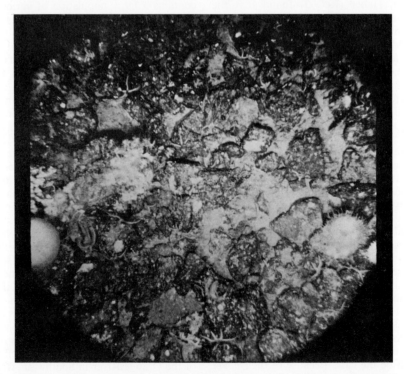

Fig. 1. Phosphorite nodules on the seafloor about 40 miles west of San Diego, California. The nodules, in 648 ft of water, and covering an area 4 by 4 ft, are lying on a coarse-grained calcareous sand. (Photo: U.S. Navy Electronic Laboratory, San Diego.)

TABLE 4. CHEMICAL ANALYSES OF SEAFLOOR PHOSPHORITE OFF CALIFORNIA

Element	Southern California Offshore Area	Northern California Offshore Area	Florida Phosphate Rock, Typical
P_2O_5	31.9	32.2	31.3
CO_2	4.8	4.8	3.67
Fe	1.9	0.88	1.0
Mg	0.56	0.44	0.22
SiO_2	4.25	7.43	9.55
Al	3.46	1.74	0.67
F	3.51	3.34	3.70
Cl	0.028	0.027	0.020
Ca	33.08	32.0	32.8
Na	1.41	1.45	0.16
K	0.36	0.26	0.13
Moisture	1.54[a]	1.26[a]	—

[a] Samples were crushed, pulverized, and dried at 100°C to constant weight prior to analysis.

Another type of seafloor phosphorite deposit has recently been discovered off the west coast of Mexico. This fine-grained, unconsolidated deposit of marine apatite lies in about 70 m of water. Although its total extent has yet to be assessed, its known dimensions are at least 60 by 160 km in lateral extent. The grade of this deposit is not overly spectacular, averaging possibly about 15% apatite; one section of the deposit, however, covering an area of about 20 by 30 km, assays as high as 40% apatite. Considering only the high-grade section of this deposit, there should be about two billion tons of recoverable phosphate rock in this single deposit.

Red Sea Deposits. Underlying the hot brines in certain deep holes in the center of the Red Sea are gel-like sediment deposits, which contain extraordinarily high percentages of such elements as iron, manganese, copper, zinc, and silver. As high as 5% of zinc and 1% of copper have been found in dried samples of these sediments. Silver has been found in several hundredths of a percent in these sediments. Not including the value of the iron and manganese, it has been calculated that the value of the metals in the sediments of only one of the locations, the Atlantis Deep, is in excess of $1.5 billion (Grice, Ref. 2). This calculation includes only those sediments above the 10-m depth, and there is some indication from seismic records that these sediments may be as much as 300 ft in total thickness. These sediments will probably constitute a high-grade mine in the near future, and there is a possibility of finding other such deposits elsewhere in the world's oceans.

Subseafloor Consolidated Rocks. In areas offshore England, Japan, Newfoundland, and Finland, coal and iron ore are mined from subseafloor rocks. These deposits are generally exploited by sinking shafts in the onshore rocks and driving drifts out under the sea and into the orebodies. Offshore methods, are used, however, to discover and delineate these orebodies.

Sulfur deposits are also found in the offshore areas, notably in the caps of salt domes. One such deposit was located while drilling for oil about seven miles off Grand Isle, Louisiana. This deposit is now mined by the Frasch process from a steel island constructed over the deposit in 60 ft of water. Much of the sulfur produced in the world is taken from the caps of salt domes in the Gulf Coast area. Indications are that the concentration of salt domes offshore in this area is about the same as that onshore, and it can be expected that the proportion of those domes containing mineable deposits of sulfur is about the same also.

The rocks of the continental shelf are basically similar in their geologic character to the rocks of the adjacent land area, and it can be expected that the mineral producing potential is also the same. While vein and bedded mineral deposits can be expected to be found on the shelf, unless they are located relatively near shore or in shallow water and under a watertight and competent cover of other rock, it is unlikely that such deposits can be exploited on an economic basis with the present level of mining technology.

The Deep-seafloor. It is the pelagic areas of the ocean, however, where nature is working on a truly grand scale to separate and concentrate many of the elements that enter seawater. The minerals formed in the deep sea are frequently found in high concentrations, as in these areas of the ocean there is relatively little clastic material deposited to dilute the chemical precipitates.

Red Clay. Red clay covers about 102,000,000 km^2 of the ocean floor. At an average depth of about 200 m, there would be some 10^{16} tons of red clay on the ocean floors. At an average rate of formation of 5 mm per 1000 years, the annual rate of accumulation of the red clays is about 5×10^8 tons. Table 5 lists some statistics concerning the amount of elements and the rate at which they are annually accumulating in these sediments. In addition to having some possible value in clay mineral applications, red clay may, in the future, serve as a source of various metals. While the average assay of red clay for alumina is 15%, individual samples of the clay have assayed in excess of 25% alumina. Copper contents as high as 0.20% have been found in some red clays.

Calcareous Oozes. Calcareous oozes cover some 128,000,000 km² of the ocean floor or about 36% of its total area. The average thickness of the calcareous ooze layers has been estimated to be about 400 m. Thus, there should be at least 10^{16} tons of calcareous oozes in the ocean. Limestone, for which these oozes could be substituted, is now mined at an annual rate of about 0.4 billion tons. If only 10% of the ocean-floor deposits proved mineable, the reserves would be about five million years at our present rate of consumption.

Siliceous Oozes. Siliceous oozes cover about 38,000,000 km² of the ocean floor. At an assumed thickness of about 200 m, there would be some 10^{15} tons of these oozes on the ocean floor. Normally, these oozes could serve in most of the applications for which diatomaceous earth is used, e.g., for fire and sound insulation, in lightweight concretes, as filters, and as soil conditioners.

Manganese Nodules. Probably the most interesting of the oceanic sediments, especially from an economic standpoint, are the manganese nodules (Fig. 2). These small, black to brown, friable concretions were discovered to be widely distributed throughout the three major oceans of the world almost 100 years ago by the famous *Challenger* and *Albatross* expeditions. It is estimated that there are some 1.5 trillion tons of manganese nodules on the Pacific Ocean floor alone, and that they are forming in this ocean at an annual rate of about 10,000,000 tons. Averaging about 4 cm in diameter, and lying loose at the surface of the seafloor sediments in concentrations as high as 100,000 tons per square mile, the manganese nodules, from all calculations thus far made, are indicated to be highly economic to mine at the present time. Grading as high as 2.5% copper, 2.0% nickel, 0.3% cobalt, and 36% manganese, all in the same deposit, or as high as 2.5% cobalt or 50% manganese in other individual deposits, the seafloor manganese nodules would be considered as high-grade ores if found on the continents. Another interesting aspect of these nodules is that their composition varies markedly over large lateral distances. Thus, the mine site can be shifted into those deposits with a mix of metals most amenable to market conditions. Flexibility such as this in choosing the grade of material to be mined is a great advantage and one that the mining industry does not normally have in land mines.

Table 6 lists statistics concerning the amounts of various elements in the nodules and the land deposits. Even if only 10% of the nodule deposits prove economic to mine, it can be seen that many elements are accumulating in the manganese nodules now forming on the Pacific Ocean floor faster than they are presently being consumed—in fact, three times as fast in the case of manganese, twice as fast in the case of cobalt, as fast in the case of nickel, and so on. As is the case with many mineral deposits of the sea, the manganese nodules would be a renewable resource. The fact that many deposits of the

TABLE 5. STATISTICS ON AMOUNT OF VARIOUS ELEMENTS AND THEIR RATE OF ACCUMULATION IN RED CLAY[a]

Element	Abundance in Red Clay (Weight Percent)	Amount in Red Clay (Trillions of Tons)	Rate of Accumulation in Red Clay (Millions of Tons/Year)	World Rate of Consumption (Millions of Tons/Year)	Ratio (Amount in Red Clay) (Annual Consumption)	Ratio (Rate of Accumulation) (Rate of Consumption)
Al	9.2	920.0	46.0	4.72	200.0	10.0
Mn	1.25	125.0	6.3	6.7	19.0	1.0
Ti	0.73	73.0	3.7	1.3	56.0	3.0
V	0.045	4.5	0.23	0.008	550.0	28.0
Fe	6.5	650.0	32.5	262.5	2.5	0.1
Co	0.016	1.6	0.08	0.015	110.0	5.0
Ni	0.032	3.2	0.16	0.36	8.9	0.5
Cu	0.074	7.4	0.37	4.6	1.6	0.1
Zr	0.018	1.8	0.09	0.002	900.0	45.0
Pb	0.015	1.5	0.08	2.4	0.6	0.03
Mo	0.0045	0.45	0.023	0.040	11.0	0.6

[a] After Mero, Ref. 4.

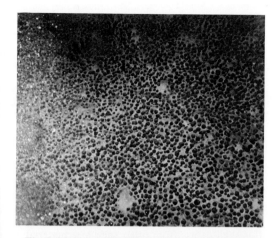

Fig. 2. Manganese nodules, about 4 cm in diameter, on the seafloor 250 miles north of the island of Tahiti. Taken in about 12,000 ft of water, the photo shows a concentration of approximately 30,000 tons of nodules per square mile. (Photo: S. Calvert, Scripps Institution of Oceanography, San Diego.)

sea are renewable resources is, of course, of academic interest only, for the reserves of the minerals contained in currently mineable deposits are generally measured in terms of hundreds of thousands or millions of years.

JOHN L. MERO

Mineral Exploration Technology

Every element that occurs naturally has been identified in the sea or in the sediments and rocks of the seafloor. Most of these elements are so widely dispersed that it is unreasonable to hope to exploit them commercially, but a few are sufficiently abundant or valuable to be economically important; the more important are listed in Table 2 (above). The purpose of marine mineral exploration is to identify such deposits, determine their value and extent, and suggest methods for their extraction.

Today mineral exploration at sea is essentially at the "prospecting" stage. We know that potentially valuable deposits of many minerals exist in the oceans, but we do not know, with any precision, the distribution or concentration of these minerals. In addition, international law has not yet made clear the national ownership of minerals recovered from the sea; exploration and exploitation have been hampered by this uncertainty.

There are four major classes of marine mineral concentrations that are, or may become, valuable resources:

(1) soluble salts that accumulate in solution in seawater;
(2) placer minerals that are physically concentrated by current and wave action in shallow water;
(3) authigenic minerals that are precipitated chemically or biochemically from seawater;
(4) continental shelf mineral deposits that are similar to terrestrial occurrences.

Since each class of marine deposits is controlled by different physical and chemical processes, a wide variety of prospecting techniques must be used.

Soluble Minerals. Undoubtedly the first marine minerals to be exploited were the salts. In

TABLE 6. RESERVES OF METALS IN MANGANESE NODULES OF THE PACIFIC OCEAN[a]

Element	Amount of Element in Nodules (Billions of Tons)	Reserves in Nodules at Consumption Rate of 1964 (Years)	Approximate World Land Reserves of Element (Years)	Ratio of (Reserves in Nodules) / (Reserves on Land)	Rate of U.S. Consumption of Element in 1964 (Millions of Tons/Year)	Rate of Accumulation of Element in Nodules (Millions of Tons/Year)	Ratio of (Rate of Accumulation) / (Rate of U.S. Consumption)
Mg	25.0	600,000	L[b]	—	0.04	0.18	4.5
Al	43.0	20,000	100	200	2.0	0.30	0.15
Ti	9.9	2,000,000	L	—	0.30	0.069	0.23
V	0.8	400,000	L	—	0.002	0.0056	2.8
Mn	358.0	400,000	100	4,000	0.8	2.5	3.0
Fe	207.0	2,000	500	4	100.0	1.4	0.01
Co	5.2	200,000	40	5,000	0.008	0.036	4.5
Ni	14.7	150,000	100	1,500	0.11	0.102	1.0
Cu	7.9	6,000	40	150	1.2	0.055	0.05
Zn	0.7	1,000	100	10	0.9	0.0048	0.005
Ga	0.015	150,000	—	—	0.0001	0.0001	1.0
Zr	0.93	100,000	100	1,000	0.0013	0.0065	5.0
Mo	0.77	30,000	500	60	0.025	0.0054	0.2
Ag	0.001	100	100	1	0.006	0.00003	0.005
Pb	1.3	1,000	40	50	1.0	0.009	0.009

[a] After Mero, Ref. 4.
[b] Present reserves so large as to be essentially unlimited at present rates of consumption.

prehistoric times, shore-dwelling people learned to extract salt from the water by simple solar evaporation; evaporation, even today, remains a major source of the world's salt. During the 18th century, it was discovered that many minerals, besides common salt, existed in seawater, and since then varied techniques have evolved for extracting the more useful salts. Sodium, potassium, bromine, and magnesium are currently being produced from seawater. It appears that very soon some of the less common salts may be extracted commercially by the use of specific ion exchange columns. As concentrated brines from salt water conversion plants and cheap energy from nuclear reactors become available, more extensive exploitation of dissolved minerals from seawater seems likely. Since the production of minerals from seawater is primarily a problem of chemical technology, not of exploration, this discussion will concentrate upon the solid resources on and beneath the seafloor.

Placer Minerals. Waves and currents can sort sediment into placer deposits the way miners pan for gold. With agitation and a flow of water to separate the mineral grains, high concentrations of the heavier or more abrasion-resistant minerals can build up. Many of the richest gold, platinum, and tin deposits mined on land were formed in this manner by streams. Similar concentrations can develop on sea beaches where waves attack the sand. Since worldwide sea level has been rising as a result of the melting of glaciers, many ancient beaches have been preserved under the shallow waters of the continental shelves. Also, ancient stream beds, which may contain alluvial minerals, have also been drowned and preserved by the rising sea level. At the present time, submarine placers are being developed off the Malayan and English coasts for tin, off Alaska for gold, and off South Africa for diamonds. There are many beach prospects known now that may become economic to exploit in the near future, particularly if related placer deposits can be located on the nearby continental shelf.

Authigenic Minerals. A diverse group of minerals is classified as authigenic. These are all materials that are chemically or biochemically precipated from seawater. On the continental shelf, in areas where there is slow sedimentation and high organic productivity, phosphorite may accumulate. This is a complex calcium phosphate mineral often including fish and marine mammal bones. At the present time phosphorite is produced from deposits that have been raised above sea level, but, as extensive submarine deposits off Florida and California are known, they may soon be mined for phosphate fertilizer. Glauconite, an iron and potassium-rich clay mineral, accumulates on the continental slopes, usually at water depths greater than 300 m. Glauconite is not now worked, but deposits on land were mined for potassium when other supplies were cut off during World War I. Ultimately it may become of economic importance. On the deep-seafloor in many parts of the world, one finds a rather enigmatic accumulation of manganese and iron oxides. These are the "manganese nodules," which actually occur either as discrete lumps of ore or as crusts on boulders. (See above).

On the continental shelves, quite pure deposits of calcium carbonate sand may occur. These are usually composed of the broken shells of such creatures as clams, oysters, barnacles, and bryozoans. Where the physical conditions segregate this shell debris from other sand size material, it may be pure enough to use for cement manufacture. A small cement industry at Faxa Bay in Iceland is based entirely upon a deposit of this sort. There the conditions are such that the calcium carbonate deposit more than renews itself each year as storm waves carry in new shell fragments.

Continental Type Deposits. By far the most important mineral resource now produced from the seas is petroleum. Oil and natural gas occur in ancient marine sedimentary rocks, which are often found under the continental shelves. Oil drilling at sea started as a simple extension of pools on land into the relatively calm waters off Venezuela and Texas; technology has now developed to the stage that drilling can be done from fixed platforms in water depths as great as 60 m, in almost any climate. There is some possibility of petroleum production from many parts of the world's continental shelves. The most productive areas today are the Persian Gulf, the Gulf of Mexico, the North Sea, the southern California coast, and Cook Inlet in Alaska.

Some normal vein-type mining has extended beneath the sea. In Nova Scotia, on the east coast of Canada, coal seams have been mined for several miles under the continental shelf. Submarine tin mines off the coast of Cornwall, England, were mined to exhaustion in the early years of the 20th century, despite many accidents. As a general rule, unfortunately, mineralization of the sort required to produce ore vein deposits does not occur on the continental shelf, for the subaerial weathering that enriches many terrestrial deposits is impossible under the sea.

Marine Survey Techniques. Ships equipped and staffed for survey work are extremely expensive. In order to guarantee productivity, surveys must be planned with great care. Before planning any work, one must determine which valuable minerals are likely in the study area. To some degree, these can be predicted from a knowledge of the geology of adjacent lands, from a review of any available bottom samples, and by analogy to other similar submarine prospects. Because bottom sampling is a particularly time consuming operation, it is usually wise to first make a reconnaissance with more rapid tools before attempting extensive sampling. Ultimately, of course, it is necessary to actually sample the bottom to permit positive evaluation of a prospect.

The first step in a survey is to prepare a reliable base map showing the form of the submarine topography (bathymetry). This, in itself, can become a major project because of the ship-time required and the normal uncertainty of navigation. Most modern survey vessels are equipped with a sonar continuous depth recorder. These produce short bursts of sound and record the time required for the bursts to echo from the seafloor as a profile. If navigation is well controlled, excellent detailed maps of the bottom can be constructed by plotting closely spaced profiles. Unfortunately, the normal quality of navigation is such that the ship's position is usually known only within a circle several nautical miles in diameter. This uncertainty often makes it impossible to correlate submarine features with confidence from one profile to the next. At the present time, there is much experimentation with satellite navigation systems; ultimately, we should be able routinely to determine geographic location within a few tens of meters.

A potentially very useful adaptation of standard sonar sounding has recently been developed. This is the side-scan sonar system. In this system a pair of transducers (combined transmitters and receivers) are mounted on a "fish," which is towed near the bottom. Sound waves spread out obliquely across the seafloor, and rebound from objects that stand above the general level, producing a record similar to low-angle aerial photography. Side-scan sonar thus permits a much broader section of seafloor to be studied. Isolated, pronounced features such as rock outcroppings, fault scarps, and wrecks can readily be distinguished for study by other techniques.

Submarine photography is becoming a more and more valuable survey tool, and stereo and television cameras have been successfully adapted for use at all depths in the ocean. Visual information can be most useful in distinguishing the small scale seafloor features that may be of economic interest. If, for example, direct sampling has proved that a deposit of manganese nodules or phosphorite is found in an area, the most rapid way to determine surface concentration of the minerals is by a photographic survey. Television is now being used for inspection of seafloor oil drilling sites; it may soon be adapted to function like the side-scan sonar, to permit the immediate evaluation of a seafloor prospect during an underway survey.

Various techniques have been devised to produce information about the sediments and rock beneath the seafloor. Because no actual samples of the sub-bottom material are available normally, much subjective interpretation must be made, but often useful information about the thickness of deposits, their composition, and the geologic structure of the bottom can be derived from indirect techniques.

One source of useful sub-bottom information is a refinement of the normal shipboard continuous depth recorder. Depth recorders produce a long, loud, high-frequency burst of sound in order to define the seafloor with precision

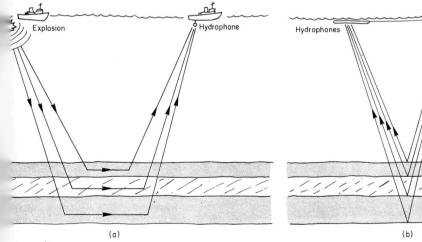

Fig. 3. Techniques of sub-bottom survey by refraction (a) and reflection (b).

without requiring a constant monitor. If a man can be assigned to adjust the depth recorder constantly, the duration and intensity of the sonar signal can be varied, and often the thickness and stratification in the first 1–2 m of the superficial sediments can be determined. Surveys of marine placers, for example, can be conducted with great precision by this technique.

Sonar uses high-frequency sound waves because they give maximum resolution, but, unfortunately, high-frequency sound is rapidly attenuated in the water column and the rock beneath. To penetrate more deeply into the sea floor, it is necessary to use a higher-energy, lower-frequency sound source (Fig. 3). Explosives are used in some of these systems, particularly in very deep water, but other sound sources are preferred for reasons of convenience and safety, where very high energy is not essential. The most popular sub-bottom profilers used today create the original pulse of sound with an electrical arc, or a series of simultaneous arcs. Compressed air, propane explosions, and a variety of magnetostriction sound sources are also in use. Such gear is particularly used for continental shelf petroleum exploration because, under good conditions, these profilers print out a cross-section of sub-bottom geologic structures to depths as great as 2000–3000 m below the seafloor. Unfortunately, the energy required to penetrate into the sub-bottom rocks is so great that the first echo from the seafloor tends to obliterate reflection from the superficial layers. This reduces the utility of acoustic profilers in studies of placer deposits on the shelf, although submerged and filled stream valleys can often be identified.

Acoustic profiling techniques depend upon the direct reflection of shock waves by the strata of the bottom and sub-bottom, but it is also possible to get subsurface information by the refraction of sound. In this system two boats, rather widely separated, are required. One vessel carries the recording gear, the other is used to drop explosives. Shock waves from the explosions can travel horizontally through the bottom strata and then up through the water to the recording ship. Since the velocity of sound in any medium is a function of the physical characteristics of the medium, rock types beneath the sea floor can often be identified from refraction studies.

Some survey techniques measure variations in the earth's gravity and magnetic fields. Both gravity and magnetism are influenced by the density or magnetic susceptibility of the local rock types, so large-scale, deep geologic structures can be surveyed by these techniques. Because gravity measurements require very complex apparatus, they are rarely used in prospecting. Marine magnetometers are not as delicate or complicated, and they promise to be of considerable use in mapping heavy mineral placers, which often contain a large percentage of magnetic grains.

Bottom Sampling. The only positive proof of the value of a prospect is samples, so eventually a complete suite of samples of the sediment and rock must be collected. Inevitably, bottom sampling in waters of any depths is a blind, rather haphazard process. From bathymetric and other information, a general idea of what sort of material to expect may be gained, but the recovery of useful samples is always somewhat a matter of luck. The time factor is often critical at this stage of a survey, for sampling requires that gear be lowered to the bottom. Typical light sampling winches operate at 200 m/min, while heavy duty winches are often slower than 50 m/min. This means that the round trip of a piece of heavy sampling apparatus in a water depth of 2000 m will take almost an hour and a half. Clearly it is essential that the right gear be selected for each sample and that the sampler function correctly each time it is lowered to the bottom.

Where the sediments are unconsolidated and interest is in the surface material, one of the various models of grab should be used (Fig. 4). There are many designs, but the basic system is a simple twin cup arrangement that is lowered to the bottom open. A trigger mechanism is tripped by contact with the sediment, the grab closes and it is withdrawn. A recent variation has a camera mounted within the jaws of the grab, arranged to photograph the bottom just before the sample is collected.

Many grabs do not function well in coarser sediments (sands and gravels) because they cannot dig in or because particles wedge their jaws open. To eliminate this problem some

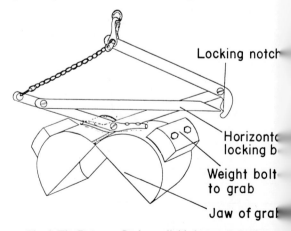

Fig. 4. The Petersen Grab, a reliable bottom sampler for unconsolidated silts and mud. (From Barnes, Ref. 6.)

any thin strata penetrated; a recent development is the box corer, which takes a square core perhaps 25 × 25 cm. Most box corers cannot penetrate more than 1–2 m into the sediment, but they provide useful samples of thin surface strata.

Sands are particularly difficult to sample, because they are hard and noncohesive. Normal coring tubes either will not penetrate or lose the sample when the tube is withdrawn. To sample sands, such as the important continental shelf placers, it may be necessary to use a powered coring device, a variety of which have been developed. The most successful are the vibratory corers; these are similar to piston corers, but are mounted on a frame that rests on the sea floor during coring and have an electric or hydraulic motor that vibrates the core barrel, permitting it to sink into the hard sediments. Cores as long as 10 m have been collected in sands with this apparatus.

When solid rock or very coarse sediment must be sampled, the only available tool is a dredge. Dredges are simple iron boxes or chain nets that can be dragged along the bottom to scoop up or break off a sample (Fig. 7). They are designed to be heavy, rugged, and cheap, for they must be pulled blindly through the bottom material. Dredges are attached to the towing cable with a bridle system equipped with

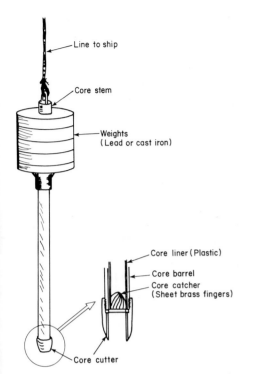

Fig. 5. A typical gravity corer that can take cores up to 2–3 m long in soft bottom sediments.

grabs are equipped with spring or weight-driven devices to insure a positive closure. One of the best modern samplers, the Shipek Grab, takes its samples in a half-cylindrical bucket, which is rotated through the sediment by a powerful spring. Grabs have also been designed for use while the ship is underway, but only small samples can be collected with these devices.

To obtain information about the thickness of layers of unconsolidated sediment, it is necessary to use some sort of a coring device. The simplest and most reliable of these is the gravity corer (Fig. 5). This apparatus consists of a heavily weighted pipe that is lowered close to the seafloor and permitted to drop into the sediment. The pipe is often lined with a plastic tube and has a corecatcher of flexible plastic or brass fingers, which fold down at the bottom of the tube to retain the core. Boundary friction in the tube causes the samples to be compressed. To reduce the deformation of cores, a piston can be installed, arranged to be withdrawn in the core tube as the sediment is penetrated (Fig. 6). This creates a slight vacuum and permits much longer cores to be collected. Gravity cores are rarely longer than 4 m, but piston cores as long as 25 m have been collected. Most cores collected with conventional gear have a diameter less than 10 cm, so they yield a very small sample of

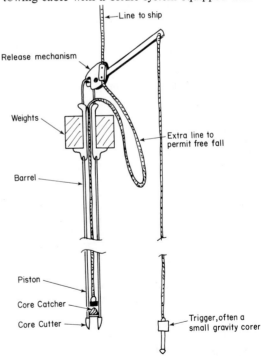

Fig. 6. Modern version of the Kullenberg piston corer, which uses a piston to overcome wall friction and take long cores of bottom sediments.

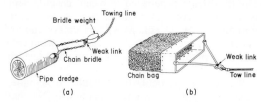

Fig. 7. Dredges for sampling boulders or hard rock exposures on the seafloor: (a) pipe dredge; (b) chain dredge.

a weak link and a safety line, to permit retrieval if the dredge gets wedged or buried in the bottom. Despite all precautions, dredges are regularly lost during sampling attempts.

Divers and Survey Submarines. Where conditions and economic factors permit, the ideal prospecting tool is the trained observer. In the shallow waters of the continental shelf divers equipped with scuba gear can actually prospect as they might on land. There are, of course, special problems in such underwater work, especially the depth and time limitations of the human body. Although divers can work in waters as deep as 65 m, their use at depths greater than 20 or 30 m is often impractical because of the time lost in decompression after each dive. Furthermore, in many interesting placer areas, the utility of divers is limited by poor visibility in turbid waters. Experiments are being conducted to determine the practicability of maintaining small working parties for weeks at considerable depth, eliminating the time consuming decompression necessary after each deep dive from the surface.

Small submarines equipped to sample and map the seafloor are now being developed. Most such submersibles are as yet too delicate and expensive to operate for routine surveys, but their technology is advancing rapidly, and, within a few years, most marine mineral prospecting will probably be conducted from such vehicles.

Conclusion. At the present state of technology and economic pressure, most mineral exploration at sea is frankly speculative. Petroleum production and the extraction of dissolved salts are practical, but the potential of most of the other mineral resources in the sea is largely unknown. We do know, however, that our supplies on land are finite, and that production of many basic elements from marine reserves is inevitable. The technology, information, and experience in marine operations that are now accumulating in a rather haphazard fashion will someday be immensely valuable. Uneconomic as present surveys are, they represent a very wise investment in the future.

F. F. WRIGHT

References

1. Emery, K. O., "The Sea Off Southern California," 1960, John Wiley and Sons, New York.
2. Grice, C. F., "The Red Sea's hot brines and heavy metals," *Ocean Industry*, **3**, no. 3 (1968).
3. Haber, F., "Das gold in meerwassers," *Z. Angew. Chem.* **40**, 303–317 (1927).
4. Mero, J. L., "The Mineral Resources of the Sea," 1965, Elsevier Publishing Co., New York.
5. Miller, A. R., Densmore, C. D., Degens, E. T., Hathaway, J. C., Manheim, F. T., McFarlin, P. F., Pocklington, R., and Jokela, A., "Hot brines and recent iron deposits in deeps of the Red Sea," "Woods Hole Oceanographic Institution Publication No. 65–38," 1965, Woods Hole, Massachusetts.
6. Barnes, H., "Oceanography and Marine Biology," Allen and Unwin, London.
7. Byrne, J. V., "The oceans: a neglected mining frontier," *The Ore Bin*, **26**, 4, 57–69 (1964).
8. Hopkins, R. L., "A survey of marine bottom samplers," in M. Sears (Ed.), "Progress in Oceanography," Vol. 2, 1964, Pergamon Press, London.
9. Steere, R. C., "Exploiting the Oceans," "Transactions of Marine Technology Society Meeting, 27–29 June 1966, Washington, D.C., 1966.
10. Terry, R. D., "Ocean Engineering," Vol. IV, Part 1, Mineral Exploitation, Western Periodicals, North Hollywood, California, 1966.

Cross references: *Bromine From the Sea; Salt, Solar.*

MORTALITY IN THE SEA

Mortality by natural causes includes loss of life by both abiotic and biotic factors, human agency excepted. Biotic factors are lack of food, predation, red tide, disease, spawning runs, and senescence; the principal abiotic factors are changes in the hydrographic conditions. Most fish have enormous fecundities; a female cod in her lifetime may produce millions of eggs, but an average of only two will survive to maturity. The bulk of the mortality occurs during the first few months of life by the combined effect of biotic and abiotic factors.

The existence of a critical period immediately after hatching was postulated long ago by Hjort (1914). Not finding food of a suitable size after yolk absorption, the young larvae of fishes would die of starvation. Evidence of a mortality of this kind is reported by Bainbridge (1965) for *Sebastes* in the Irminger Sea, which in the first weeks after hatching has a remarkably uniform diet, consisting nearly exclusively of *Calanus* eggs; survival of the larvae is largely dependant on the availability of this monotonous diet. A critical phase may also occur in later developmental stages, e.g., at the larva-postlarva transformation point (Strasbourg, 1959, Lillelund, 1965).

The life of most animals is terminated by their being captured and devoured by other animals, death by senescence being rare. Mortality by predation is highest among young animals. At a more advanced age predator attack may be enhanced by debility of the prey-animals as a result of disease or of extreme temperatures rendering the victim insensitive to the stimuli that normally provoke escape reactions (Dickie & Medcof, 1963).

Predatory invertebrates may cause serious losses of commercially important mollusks, such as oysters and scallops. The deadliest predators of oysters are some gastropods. The common oyster drill, *Urosalpinx cinera*, is particularly destructive to young oysters; in many localities in Long Island and on the shores of Virginia, drills kill 60–70% of the seed oysters and sometimes annihilate the entire crop (Galtsoff, 1964). Starfish are another highly destructive predator of oysters and scallops; in the state of Connecticut, the loss caused by starfish in 1887 was estimated at 634,246 bushels of oysters, i.e., nearly half the total harvest of the year.

Concerning predatory vertebrates, some statistics are available about the damage done by seals and birds. The numbers of prey-animals swallowed have been calculated from an examination of the stomach contents of the predators. In many regions seals and fish-eating birds are direct competitors of fishermen. A seal consumes about 5 kg of fish per day. In the waters around Scotland, seals eat about 100,000 tons of fish per year, valuable fish like salmon and cod being one of the main items in their diet (Rae, 1966). The oystercatcher (*Haematopus ostralegus*) is a specialized predator on mollusks; in its adult state it eats 350 to 500 cockles per day. In Great Britain there are some 150,000 oystercatchers in the winter season; in the Burry inlet in Wales, where there is an important cockle fishery, the average number of the overwintering birds is 8000, eating 500,000,000 to 700,000,000 cockles between October and March (Mason, 1965). On the Murmansk and Novaya Zemlya coasts, 760,000 tons of fish are eaten by guillemots (*Uria lomvia* and *U. aalge*) and kittiwakes (*Rissa tridactyla*) during four months of their living in nesting colonies (Golovkin, 1963). In the coastal waters of Southwest Africa, the Cape gannet (*Sula capensis*) and the Cape cormorant (*Phalacrocorax capensis*) take 405 and 375 lb of fish per year respectively, their main diet consisting of pilchards (*Sardinops ocellata*). The estimated 70,500 gannets and cormorants living in the vicinity of Walvis Bay swallow 27,000 tons of fish per year (Matthews, 1961). These figures for different regions are not directly comparable, but they may show that the role of seals and fish-eating birds, as consumers of fish and invertebrates, may be immense.

Red tide and lack of oxygen

Under certain circumstances one or a few species of plankton multiply to such a degree as to discolor the water, often to the virtual exclusion of all other organisms. Such monospecific discolorations, often called waterbloom, may be of any shade between red, brown, yellow, and green. The blooms may cause great destructions among invertebrates and fish. In fresh water and brackish water, mortality has been recorded during blooms of blue-green algae and of the toxic chrysomonadine *Prymnesium parvum*; blooming of dinoflagellates is rare in fresh water. In the sea the destructions are nearly always caused by a bloom of dinoflagellates, the so-called red tide. Some minor mortalities in the sea are due to blooms of blue-green algae. A chloromonadine flagellate, *Hornellia marina*, discoloring the water green, causes fish mortality on the west coast of India (Subrahmanyan, 1960).

Many red tides, however, are harmless; the reason why is not wholly clear. It may be that only certain species of dinoflagellates are injurious, that they are noxious only during part of their life, or that the concentrations are not in fish-killing densities. The density during destructive tides is in the order of several millions of dinoflagellates per liter. During the 1946–47 Florida outbreak, the numbers in the surface water varied from 13,000,000 to 56,000,000 per liter. Steidinger (1964) mentions a count of 75,000,000 of cells per liter in a Florida tide. In the Southwest African coastal waters, where mortalities occur repeatedly, counts ranging up to 84,000,000 and in Walvis Bay even to 100,000,000 of cells per liter have been recorded Kollmer, 1962, 1963).

Species of dinoflagellates made responsible for destructive red tides are: *Gonyaulax polyedra* (Pacific coast of North America, Portugal), *Gonyaulax monilata* (Offatts Bayou, Texas), *Gymnodinium breve* (Gulf of Mexico, Gulf of Paria), *Gymnodinium galatheanum* (Southwest Africa; probably also other species of *Gymnodinium* are involved, cf. Kollmer, 1962, 1963), *Gymnodinium polygramma* (South Africa near Cape Town, Japan), *Exuviella baltica* (Angola), *Glenodinium trochoideum* (Brazil), *Noctiluca miliaris* (India), and a few others.

In addition to destructions among invertebrates and fish, dinoflagellates have been blamed for another plague: they may render shellfish toxic to man (paralytic shellfish poisoning, usually called mussel poisoning). The filter-feeding shellfish derive their toxin from feeding on dinoflagellates; they store and concentrate the toxin with no apparent ill-effects to themselves. However, consumption of the poisonous mollusks by man may be fatal. So far only few species of dinoflagellates have been associated with mussel poisoning—*Gonyaulax catenella* and *G. acatenella* on the Pacific coasts of North America, *G. tamarensis* on the Atlantic coasts (Prakash & Taylor, 1966). *Prorocentrum micans* caused toxicity of *Cardium* on the coast of Portugal (Pinto & Silva, 1956).

There has been much discussion about the nature of the death of marine animals during red tides. There are two schools of thought. One is that the tide is only indirectly noxious, the destruction being brought about by depletion of dissolved oxygen. The high fertility of the surface water in many red tide areas may lead to a marked depletion of oxygen in the waters below. This argues for the opinion that part of the mortalities near Walvis Bay (SW Africa) are caused by lack of oxygen. The other view is that fish and invertebrates are killed by virulent poisons produced by the living plankton and set free during lifetime or after the cells are ruptured. At first it was suggested that the marine fauna is killed by toxins of the mussel poisoning type, mussel poisoning being caused by a concentration of dinoflagellates sufficient to render mollusks toxic to man, but insufficient to kill the mollusks and other cold-blooded animals.

Grindley & Taylor (1964) consider that the destructions of the marine fauna are caused by species of dinoflagellates different from those causing mussel poisoning in man; it is, however, uncertain whether this always holds. *Gonyaulax polyedra*, the causative organism of marine mortality in Southern California waters, possibly may be associated with mussel poisoning (Prakash & Taylor, 1966). In Florida, illnesses in 1962 were thought to have been caused by toxic shellfish; a relation with the red tide organism was suggested, but so far there has been no definite tracing to *Gymnodinium breve* (Anon., 1964). It has become known that dinoflagellates produce different toxins with distinct pharmacological actions. Abbott & Ballantine (1957) showed experimentally that *Gymnodinium veneficum* liberates a powerful toxin rapidly killing fish and some invertebrates; these authors were unable to render shellfish poisonous with it. The production of a toxin that kills fish has also been demonstrated for *Gymnodinium breve* (Rae & Wilson, 1957, Starr, 1958) and for *Gonyaulax monilata* (Gates & Wilson, 1960).

Blooming of dinoflagellates has been recorded from warm and temperate regions, but in the latter the phenomenon is rare, and appears only in the warmer time of the year. The majority of the destructive tides occur in the subtropics. They have been recorded from nearly all regions of upwelling water, and from other eutrophic parts of the sea. The causes leading to the development of red tides are not fully understood. In regions of upwelling, nutrient-rich water is brought to the surface; however, the inorganic nutrients do not appear to be the direct cause of the red tide, as the tide occurs after the cessation of the upwelling. The high nutrient content induces a mass development of diatoms; possibly compounds produced by these diatoms or by bacteria developing during the decay of organic matter are the triggering mechanisms of the red tide. *Gymnodinium breve* requires vitamin B_{12} for growth; extensive investigations on the sources of B_{12} in Florida waters are in progress (Stewart et al., 1967). Iron might be another stimulating growth factor (Donelly et al., 1966).

In many regions the appearance of the tide is seasonal. In the coastal waters of Southwest Africa during the greater part of the year, there is strong upwelling causing low temperatures and high nutrient contents. Red tide appears only in the southern summer, when upwelling is minimal; this implies high temperatures, a strong thermocline, and a low supply of inorganic nutrients. The bloom appears immediately after a heavy bloom of diatom plankton (Kollmer, 1962).

On the west coast of India the period of maximum phytoplankton (diatoms dominating by far) production is during the wet southwest monsoon, May–September, with production peak in July–August, but most red tides have been observed during September–November (Prakash & Viswanatha Saroa, 1964).

The hydrographic conditions under which the Florida red tide occurs, are less clear. Here the occurrence is more or less seasonal; the outbreaks commence between August and October, and usually end between November and February. A possible relationship has been suggested between red tide outbreaks and precipitation or river-runoff, but the evidence is not conclusive. Convergences in the water are considered important by concentrating the dinoflagellates, or by concentrating compounds that might favor their development (bibliography in Rounsefell & Nelson, 1966).

Disease

Marine animals may suffer from both non-contagious and infectious diseases. The first are due to organ malfunction or to deficiencies in the environment, such as lack of food or adverse hydrographic conditions. The infectious diseases are caused by pathogens and parasites. Widespread mortality is often caused by an adverse factor in the environment combined with infection.

A few microorganisms infecting oysters have been identified as pathogens. The fungus *Dermi-*

cystidium marinum is the cause of periodic oyster mortalities in North America from Delaware to Mexico (Mackin, 1962, Galtsoff, 1964). Shell disease, an infection of the oyster shell by a fungus of unknown identity has been reported to cause high mortality among oysters in the Oosterschelde, the Netherlands. This disease has been recorded from Dutch waters since 1902; it spread rapidly in the years following 1930. This was probably due to great quantities of cockle shells spread in former years to serve as spat collectors (Korringa, 1951a). Parasitic copepods of the genus *Mytilicola* have often been found in the gut of mussels, oysters, and other mollusks. *M. intestinalis* is known to cause widespread mortality in commercial mussel beds in the Netherlands (Korringa, 1951b). On the Pacific coast of North America oysters may show heavy infection by *Mytilicola orientalis*, but there is no apparent relationship between this infection and mortality (Chew et al., 1965).

Few pelagic animals have been adequately examined for diseases. *Ichthyosporidium hoferi*, a fungus responsible for "pepper-shot" disease of herring has been epidemic among herring of the North Atlantic, resulting in widespread mortalities (Sindermann, 1965).

Spawning runs

A regular dying after the first spawning is a rare phenomenon among vertebrates. It has been recorded for the Pacific salmons (genus *Oncorhynchus*), the lampreys (*Petromyzontidae*) and the eulachon (*Thaleichtys pacificus*), anadromous fishes developing to maturity in the sea and entering rivers in order to spawn, and for the katadromous eels (genus *Anguilla*). Death of the adult salmon occurs in nearly all instances within one or two weeks after spawning. The immediate cause of the post-spawning death is unknown. Physical exhaustion from the long migrations appears not to be the principal cause of death, since salmon running up short streams may come to the spawning place in good condition, and undergo degeneration which rapidly leads to death after shedding their sex products (Robertson, 1957).

High mortality after spawning has been recorded for the capelin (*Mallotus villosus*) as a result of stranding or wounding during the act of spawning. The pelagic-living capelin enters the tidal zone in order to spawn. The fishes come to the beach in front of the crest of an advancing wave, the spawning act is completed, and they go back with the returning wave. When coming short of the reach of the returning wave the capelins are stranded. Many others are injured by the vigorous motions during the act of spawning (Templeman, 1948). In this way incredible numbers are destroyed annually in arctic regions.

Temperature

In the oceans the temperature ranges from $-2-+30°C$; in landlocked waters the temperature may rise to higher values. The range of tolerance of marine animals varies between different species; further, it is largely dependent on the thermal prehistory of the specimens. If the animals are gradually acclimatized to extreme temperatures they have a better chance to survive than in case of a sudden chilling or heating.

In arctic and subarctic regions, cold winters with temperatures below $0°C$ and the formation of ice are normal, and the fauna is adapted to the prevailing conditions. In the littoral area many species live one summer only, and are annually recruited from deeper waters. The perennial species evade the low temperatures by spending the colder time of the year in deeper layers, unless they are very resistant to cold. The freezing point of seawater is about one degree lower than that of the teleost blood. Several teleost fishes living at subzero temperatures have their body fluids supercooled. They can tolerate supercooling if seeding with ice crystals is avoided; by contact with ice crystals their body fluids change from a supercooled to a frozen condition and the fish are killed. Low temperatures induce a rise in the blood salt content, which is considered to be an adaptation to winter cooling.

In spite of these adaptations, widespread mortality is often recorded. On the eastern coast of Newfoundland, down to 170 m, temperatures sometimes fall to $-1.7°C$. The larger cod (*Gadus morrhua*) evade the cold coastal water by passing deeper, but they may be trapped and killed by cold in some deep holes having no passage seaward. The resistance of haddock (*Melanogrammus aeglefinus*) to low temperatures is less than that of cod; moreover the haddock has a greater tendency to re-enter shallow water early in the year, where large numbers are destroyed by chilling and ice-seeding. The capelins (*Mallotus villosus*) are very resistant to cold; their ability to live in the cold upper layer becomes fatal if they enter surface water containing abundant ice crystals. Even if some life remains, the capelins cannot re-enter the deep water, as the ice that covers the fish keeps them floating (Templeman, 1965; Templeman & Fleming, 1965).

In boreal areas, winters are normally mild and permit the existence of an abundant and perennial littoral fauna; in these regions widespread mortality occurs during a cold and prolonged winter. In the severe winters of 1928-29, 1946-47, and 1962-63, unusually cold water extended from the coasts of the North Sea causing destruction among fish and invertebrates. Animals reaching the northern limit of their distribution suffered the highest mortalities, soles (*Solea vulgaris*) in the southern part of the North Sea (Woodhead,

1964), the lancelet (*Branchiostoma lancedatum*) in the German Bight near Heligoland (Courtney & Webb, 1964).

A sharp drop in temperature occurs in various parts of the Gulf of Mexico during heavy northers, in some years causing widespread mortality among the shallow-water fauna of the Texas coast and the west coast of Florida. The rapid onset of the northers combined with the great extent of the shallow bay waters along these coasts favors the frequent occurrence of catastrophes. In shallow waters the temperature drops more rapidly than in the deeper parts of the Gulf; in bays along the Texas coast surface temperatures may drop from 20 to 4°C within a few days. The destructiveness of the cold waves depends more upon the rapidity of the temperature drop than upon the minimum attained. In some years destruction reaches catastrophic proportions. In the winter of 1845–46 Army troops camping near Corpus Christi were fed on the fish and turtles thrown up on the beach after a cold wave. The weight of the fish killed along the Texas coast in 1947 was estimated at 16,000,000 lb. Widespread mortality occurs here at short intervals; on an average once in every six years (bibliography in Brongersma, 1957).

Animals living in the cold water below the thermocline may be killed if this layer is suddenly depressed. In the Southwestern Gulf of St. Lawrence, the warm surface layers are separated in summer by a strongly developed thermocline from the deep layers that continue to be cold. Summer winds may produce giant oscillations of the thermocline. The temperatures of the deep layers rise from 4 to 20°C or even higher when the thermocline is depressed, causing widespread mortality among scallops (*Placopecten magellanicus*) (Dickie & Medcof, 1963).

Currents

Eggs and larvae of many species of fish and invertebrates are planktonic. Currents often cause a slow drift of these planktonic stages to other marine areas, where they develop to maturity or perish. Hjort (1914) suggested that larval drift away from the littoral region might be a major cause of death. Stocks spawning on narrow shelves or on isolated banks surrounded by deep water seem to be most vulnerable to this cause of mortality. In the North Atlantic there are several banks that maintain a self-contained stock of ground fish and they have their own plankton population, which is quite different from the typical oceanic plankton of the surrounding waters. This is the result of an anticyclonic eddy system retaining the plankton on the shallow banks. In some years, however, when the wind blows from one quarter for a fairly long time, the plankton overlying the banks is pushed over the oceanic depths (Saville, 1965); being unable to return to the plateau and to settle subsequently the majority will die.

Sharp changes in hydrographic conditions occur on the boundaries of cold and warm currents. Shifting of the currents may cause catastrophic mortalities. The disaster of the tilefish (*Lopholatilus chamaeleonticeps*) near the edge of the Gulf Stream off the Massachusetts coast in 1882 was presumably caused by an unusual invasion of cold water into the warm zone. Destruction by an invasion of warm water into the cold coastal water has also been observed. In May 1956, tongues of warm Gulf Stream water extended farther than usual over Georges Bank causing high mortality among fish larvae. Only boreal forms were affected, indicating that the larvae were killed by temperatures above their limit of tolerance (Colton, 1959).

In the strong Agulhas Current, the isotherms slope upward toward the coast of South Africa. This is not usually perceptible at the surface, but occasionally a narrow band of very cold water appears in the inshore area causing mass mortality of fish. A destruction of fish and cephalopods observed in 1964 in an inshore area on the Somali coast has been attributed by Foxton (1965) to a similar cause. The greatest concentration of the fish coincided with an area of exceptionally cold surface water in the strong Somali Current.

Salinity

In the oceans the salinity varies from about 3.3–3.7%. In semi-enclosed bays the range is much greater, from fresh water to over-salinity. In areas where changes in salinity are of regular occurrence most animals are adapted, but if in certain years the change is greater or continues much longer than usual, widespread mortality occurs. In Chesapeake Bay, heavy mortality among oysters occurs when the Susquehanna river discharges unusually large volumes of fresh water. In the Laguna Madre, a marine lagoon on the Texas coast, fish die in very dry and hot summers when the water becomes oversaline (bibliography in Pearse & Gunter, 1957; Brongersma, 1957). Mass mortality of oysters in shallow bays in central Texas is attributed to high temperatures (above 37°C) and hypersaline (over 40%) conditions (Copeland & Hoese, 1966).

Currents may carry animals into water of lethal salinity. Freshwater plankton is killed when carried by rivers into the sea. On the east coast of the Caspian Sea, water of medium salinity flows through a narrow channel into the highly saline Gulf of Kara Bugaz; accumulations of dead fish are found in the Gulf near the entrance of the channel. Probably the fish enter in a layer of Caspian water floating on the highly saline water of the Gulf, and they perish when this layer evaporates or is mixed with underlying water.

MARGARETHA BRONGERSMA-SANDERS

References

Anonymous, "Adverse chemicals and toxins in the marine environment and shellfish," U.S. Publ. Health Serv., Shellf. Sanit. Res. Plan. Conf., 1–9, 1964.

Brongersma-Sanders, M., "Mass mortality in the sea," in Hedgpeth, J. W., Ed. "Treatise on marine ecology and paleoecology," *Grol. Soc. Am. Mem.*, **67**, 1, 941–1010 (1957).

———, "Mass mortality in the sea," in Fairbridge, R. W., Ed. "The Encyclopedia of Oceanography, Earth Sci. Ser., Vol. 1, 1966, Reinhold Publishing Corp., New York.

Colton, J. B., "A field observation of mortality of marine fish larvae due to warming," *Limnol. Oceanogr.*, **6**, 280–291 (1959).

Fonds, M., and Eisma, D., "Upwelling water as a possible cause of red plankton bloom along the Dutch Coast," *Neth. J. Sea Res.*, **3**, 458–463 (1967).

Foxton, P., "A mass mortality on the Somali coast," *Deep-Sea Res.*, **12**, 17–19 (1965).

Grindley, J. R., and Taylor, F. J. R., "Red water and marine fauna mortality near Cape Town," *Trans. R. Soc. Afr.*, **37**, 111–130 (1964).

Gunter, G., Christmas, J. Y., and Killebrew, R., "Some relations of salinity to population distributions of motile estuarine organisms, with special reference to penaeid shrimp," *Ecology*, **45**, 181–185 (1964).

Lillelund, K., "Effect of abiotic factors in young stages of marine fish," Icnaf Environmental Symp. Rome 1964, Spec. Publ., 6, 673–686, Dartmouth, Canada.

Medcof, J. C., "Shelfish poisoning—another North American ghost," *Canad. Med. Ass. J.*, **82**, 87–90 (1960).

Pearse, A. S., and Gunter, G., "Salinity," in Hedgpeth, J. W., Ed., "Treatise on marine ecology and paleoecology," *Geol. Soc. Am. Mem.*, **67**, 1, 129–157 (1957).

Rae, B. B., "Seal damage to white fish fisheries," *Scott. Fish. Bull.*, **25**, 25 (1966).

Rae, B. B., Johnston, R., and Adams, J. A., "The incidence of dead and dying fish in the Moray Firth, September 1963," *J. Mar. Biol. Ass. U.K.*, **45**, 29–47 (1965).

Rounsefell, G. A., and Nelson, W. R., "Red-Tide Research summarized to 1964 including an annotated Bibliography," U.S. Fish. Wildl. Serv., Spec. Sci. Rep. Fish., 1966, 535, 1–85.

Strasbourg, D. W., "An instance of natural mass mortality of larval frigate Mackerel in the Hawaiian Islands," *J. Cons. Int. Expl. Mer*, **24**, 255–263 (1959).

Templeman, W., "The life history of the capelin (*Mallotus villosus* Muller) in Newfoundland waters," *Bull. Newfoundl. Gov. Lab.*, **17**, 1–151 (1948).

———, "Mass mortalities of marine fishes in the Newfoundland area presumably due to low temperature," Icnaf Environmental Symp. Rome 1964, Spec. Publ., 6, 137–147.

Templeman, W., and Fleming, A. M., "Cod and low temperatures in St. Mary's Bay, Newfoundland," Icnaf Environmental Symp. Rome 1964, Spec. Publ., 6, 131–135.

Cross references: *Biotoxins, Marine; Dinoflagellates.*

MUSSELS AS A WORLD FOOD RESOURCE

Mytilidae or mussels have great possibilities as a world food resource, particularly in protein-starved countries. In Europe, mussels are an important natural resource, whose fisheries yield well over a quarter of a million tons annually; they are known as "the poor man's oyster." Most of this output is due to mussel culture, which is necessary when natural beds are too small to support an expanding fishery. It is a two-part process: (1) the quality of natural stocks is improved by moving them to good growing areas; (2) when supplies of seed mussels run short, new supplies can sometimes be obtained by bringing about the settlement of planktonic mussel spat, by various methods still being developed. As planktonic spat is extremely abundant in European waters, the vast size of this resource, plus the fact that mussels live on phytoplankton, gives mussel culture an outstanding biological potential.

Of all the fish eaten by man, edible bivalves are especially suited for cultivation, for two reasons in particular: (1) they are sessile, and need no caging to prevent escape; (2) they live on phytoplankton and possibly detritus, which gives them great potential productivity, as is reflected in the abundance of other phytoplankton feeders.

In addition, mussels have three special features that, in Europe, give *Mytilus edulis* a biological advantage and a fishery potential far greater than that of other bivalves: (1) their overwhelming abundance as planktonic larvae and plantigrade spat, available for seed supplies in culture; (2) their high productivity in culture; (3) their byssus, which facilitates their hanging culture in depth, the advantages of which are escape from benthic predators and better phytoplankton exploitation. The byssus also enables mussels to clump together, which makes them the easiest of shellfish to dredge. These cultural advantages, combined with a world-wide distribution, suggest that mussels may be an underexploited food resource of major importance.

European Mussel Fisheries

Mytilus edulis occurs intertidally and sublittorally in coastal waters, on rocky or stony ground, sometimes on sandy ground, and often on deep mud. In Britain, where mussels are widespread, the stocks amount to 50,000–100,000 tons of intertidal beds; however, most of these are largely unmarketable because of poor growth due to tidal exposure. Until recent years the fishery was restricted to the hand picking and raking of the small parts of the beds lying low enough to produce good mussels. Because of the scarcity of good stocks, a native system of mussel culture has grown up in Britain, as in France. The simplest kind of culture is the relaying of unmarketable intertidal mussels at lower levels, where longer immersion speeds growth and improves flesh quality. In Norfolk, England, this relaying is done by hand-net, rake, fork, and small boat. Plots as small as $\frac{1}{8}$ of an acre can yield $8\frac{1}{2}$ tons of mussels a year, and annual output per man may reach 50–60 tons. In Wales and Holland, mechanized culture with 60 ft powered boats and two to four dredges can yield over 150 tons of mussels per man in the same period. Production on this scale can create a crucial problem in mussel seed supplies. In Holland the mussel growers rely on the large regular natural settlements of young mussels in their offshore waters, which they relay to better growing grounds inshore. The control of seed production has been successfully achieved in French and Spanish mussel culture systems. In France, mussel spat settles heavily on wooden posts planted on intertidal mud flats and is spread in nets over brushwood fences, called bouchôts, to grow to marketable size.

In Spain, a big mussel fishery has been developed in recent years, based on the production of mussel settlements on ropes hanging from floats. These clusters of mussels are thinned out for better growth by spreading them with netting on to other ropes, to which they cling. In both hanging and bottom culture systems, the basic essential conditions for success are (1) shelter from waves and tidal scour; (2) good growth, dependent on tidal immersion, water exchange, temperature, salinity, and plankton; (3) abundance of seed mussels.

Shelter

In Britain the area of sheltered waters suitable for mussel culture may be restricted by low temperatures or by tidal exposure and scour. The stability and shelter of a ground can be judged from the bottom sediments; coarse clean sand is usually associated with strong tides and surf, whereas silt and mud generally indicate quieter waters. To test culture grounds, trial plots of mussels can be relaid at or below low-water mark of spring tides; when growth conditions have been found to be satisfactory in these small trials, large-scale trials are required for productivity tests. For example, in trials of culture grounds in North Wales, small plots of less than $\frac{1}{4}$ ton showed 50–100% losses, compared with 10–20% losses in an 80-ton plot over a similar period. In large-scale mussel culture on the bottom, the grounds should be smooth enough for dredging, with up to 10 m maximum depth at low water. Deeper waters, and those with foul bottoms, may be suitable for floating or hanging culture. Shelter may also be a factor in growth, as well as in culture ground stability.

Growth

Tidal immersion and exchange. It has been well established by Baird (1966) that tidal exposure to air reduces mussel growth rates, by stopping feeding. In North Wales, mussels kept at low-water mark of ordinary spring tides and exposed briefly for 12 to 14 tides a month showed only about 80% of the volumetric growth of mussels continuously under water, and took longer to reach marketable size. At half-tide level, above which mussels cannot survive, growth rate is only about 7% of the sublittoral rate. In large-scale culture, growth may be affected by the tidal exchange rate, through plankton depletion.

Temperature. Temperature seems to be an important factor in the growth rate of mussels. In Wales, where the mean sea temperature is about 10°C, sublittoral mussels take about $2\frac{1}{2}$ years to reach the marketable size of 60–65 mm. In northwest Spain, with a mean sea temperature of 14–16°C, mussels grow to this size in less than a year. The part played by temperature in causing this difference in growth rates can be found by plotting volumetric growth against "day degrees," or number of days multiplied by the mean monthly temperature in degrees Centigrade. In each case there emerges, over the 25–75 mm length range, a nearly linear relationship which is possibly part of a sigmoid curve. This suggests that doubling the number of degrees above freezing point doubles the growth rate, within the ranges observed. However, on comparing growth over the same day-degree period, that of the Spanish mussels is nearly twice that of the Welsh mussels. Since Spanish sea temperatures are not double those in Wales, some other growth factor seems to be involved, such as that of available plankton. This growth-temperature relationship is presumably valid only up to a maximum temperature for optimum growth.

Plankton. The relation of plankton quality to growth in mussels has been studied by Bøje (1965) in the Kiel Canal area, Germany; he found that good growth was associated with an abundance of nanoplankton and detritus. It is a question of how far the plankton factor is likely to limit the development of mussel fisheries in new areas. The easiest and most direct way of testing plankton suitability anywhere might be to try a few mussels to see how they grow. In Britain, for example, good quality mussels of 60–65 mm marketable size should yield 35–40% by weight of raw flesh, or 20–25% of cooked flesh after 2–3 min boiling. At this size a mussel should yield 2–2½ g dried flesh, or 10–12 ml wet flesh occupying 60–70% of the shell cavity. These rather arbitrary standards of quality need not necessarily apply to other kinds of mussel in other countries, where the shell/flesh ratio may vary, and where tastes may differ as to what size of mussel is worth eating. In France, for instance, 45–50 mm mussels are acceptable.

Salinity. The effect of salinity on mussel growth is not clear. There is a common belief that the best quality mussels occur where marine and fresh waters mix. This has not been established, nor, if it were, could it necessarily be attributed to the effects of salinity rather than to those of plankton, detritus, shelter, and perhaps temperature. Mussels of the highest quality are certainly produced in full-salinity inshore seawater of 32–34‰. Motwani (1955) has shown that *M. edulis* can thrive in salinities as low as 20‰, and exist at 10‰ for short periods. This lower salinity tolerance may be important in increasing survival rates, in that mussels can live in waters too brackish for predators, such as starfish (*Asterias rubens*).

Seed Supplies

Besides shelter and good plankton productivity, large-scale mussel culture demands an abundance of seed, either as natural settlements or as planktonic spat. The reproduction of *M. edulis* has been well covered in the literature (e.g., Bayne 1964); in brief, spawning is followed by about three weeks' planktonic larval development to plantigrade spat. According to Bayne, at 250–300 μ length these settle on filamentous algae, from which they may take off and settle more than once before final settlement at a length of about 1 mm. Conditions affecting the first and final settlements are not fully known, and need further study in relation to spatfall production on artificial materials.

Mussel settlements may appear on intertidal beds, or in deep water offshore, where their presence is only realized when they come up on fishing gear. These offshore beds are very vulnerable to starfish attack and do not usually last long, unless the mussels are relaid by dredging to inshore culture grounds. If no visible beds of young mussels exist in otherwise favorable culture areas, seed mussels can either be relaid there from distant beds, or produced from spat settlements on special collectors, as in the French and Spanish systems previously described. Around the coast of Britain mussel spat is widespread, and shows itself as isolated settlements on rocks and navigation buoys, miles from any big mussel bed. Its abundance is such as to cause an industrial problem when power station cooling pipes become clogged by mussel settlements.

For reasons not fully understood, mussel beds may never appear on some good growing grounds, despite an abundance of planktonic spat. Part of the reason may be lack of suitable settlement surfaces, and also predation by crabs (*Carcinus maenas*) and starfish (*Asterias rubens*).

Culture Problems

Predation. Given the right conditions of shelter, growth, and seed supplies, predation by bottom-living animals may be the outstanding problem in mussel culture, especially in seed mussel production. The fact that many suitable culture grounds are naturally bare of mussels suggests this possibility, which is confirmed in practice. The effect of predation by crabs (*Carcinus maenas*) on mussel settlement survival has been described above. In Norway concentrations of small starfish (*Asterias rubens*) of up to 3,000/m^2 on 0–10 mm mussel settlements have been observed by Bøhle (1965), and concentrations of 400–800/m^2 have been observed on beds in the Wash in England, on larger mussels. Heavy losses of marketable mussels due to starfish predation have been found on sublittoral culture grounds, especially in Wales and England. Offshore mussel settlements are usually destroyed by starfish in a few months. Hancock (1965) has shown that large starfish of 240 mm diameter can eat 20–50 mm long mussels at the rate of 30 per month. The starfish wraps itself around the mussel, pulls on the valves until they open, turns its stomach inside out, then slips it inside the mussel, and digests the flesh.

Control of starfish is possible either by destroying them, as in bottom culture, or by avoiding them, as in hanging culture. In Holland the Dutch control starfish infestation by roller dredging; fishing trials in Wales showed this to be 30–50% efficient, and it was found that starfish could almost be fished out of infested mussel grounds. But the timing and intensity of roller dredging may be tricky, to avoid excessive labor costs and also damage to mussels by the heavy dredge. In America, damage to mussels by *Asterias forbesii* can be reduced by the use of special mops made of long bunches of cotton waste; when dragged over infested ground these entangle starfish, which drop off when the mops are dipped in strong brine or hot water (Moore 1897). In other parts of the U.S. it is claimed by Mackenzie (personal communication) that starfish can be effectively controlled by spreading quick lime over sublittoral oyster beds; this method sounds promising, and seems a good deal safer than the use of organic chloride and other persistent predator poisons.

The occurrence of mussel settlements on buoys and boats floating over bare ground suggests a high predation rate on the sea bottom. The main drawback in floating culture is the cost of floats, settlement materials, and handling. If these costs could be drastically reduced, then the prospects of mussel seed production and of culture expansion would improve; investigations into these possibilities are already under way in Venezuela and in Norway, but there is a great dearth of adequate data on the economics of mussel culture systems.

The control of predation by crabs (*Carcinus maenas*) needs further development. In Holland, according to Korringa (personal communication), crabs migrate offshore in winter, and the relaying of mussel seed to shallow inshore grounds is delayed to take advantage of this. Mussels are most vulnerable to crab attack when small, but are fairly resistant after about a year's growth.

Young mussels are also eaten by flatfish (*Pleuronectes* spp.) such as plaice and flounders, as is well known by fishermen, who fish for these on offshore mussel beds. It is not known how much damage these fish do to mussel settlements, compared with that caused by starfish. On offshore beds, according to Olney (personal communication) diving ducks take their toll of mussels, while oystercatchers (*Haematopus ostralegus*) and other birds prey on intertidal mussels (Drinnan 1957, Dare personal communication). As far as is known, birds and flatfish are not as serious a problem in mussel culture as are starfish and crabs, but this view may change.

Parasitism. The parasitism of *M. edulis* by the copepod *Mytilicola intestinalis* has received much attention in biological literature (e.g., Hepper 1955). Heavy parasitic infestation is often, but not always, associated with poor flesh condition, and sometimes with mass mortalities. The parasites are spread by tidal streams or by the movement of infested mussels, for example on ships' bottoms, or during relaying, and possibly when used in the line fisheries as bait. According to Andreu (1965), infestation can be reduced by complete harvesting of the host mussels each year, to break the parasite's life cycle. Restocking culture units with clean spat, and keeping culture well offshore away from muddy waters, also reduce the infestation risk. These measures are particularly applicable to floating culture systems in deep water which are restocked with planktonic spat, and in which harvesting is annual and complete.

Apart from *Mytilicola* there may be other parasites such as trematodes and other organisms causing mortalities in mussels. In mussel samples kept in the sea for growth observations, 35–45% mortalities can occur in a year, for unknown causes other than predation and *Mytilicola* infestation. This suggests that parasitism in mussels may be a more serious problem than was thought, unless some other explanation is found.

Sanitation. If mussels are grown in sewage-polluted waters it will be necessary either to purify them or to heat-sterilize them before they are eaten. This is no problem when mussels are cooked for marketing, or when culture grounds are far from sewers. However, dinoflagellate toxicity is more difficult to avoid, in that dinoflagellate blooms are apt to occur in quiet waters such as fjords, which may also be suitable for mussel culture.

Allergy to mussels is worth further investigation in medical research, to find out whether this is commoner than allergies to other foods commonly eaten. Fear of allergy could well be a psychological barrier to mussel fisheries development on the scale suggested in this paper; the problem needs to be faced squarely, to find out whether mussel allergy is any commoner than other allergies in the population.

Future Development

Having investigated the methods, conditions, problems, and comparative productivity of mussel culture in Europe, one can now see in what other parts of the world it might succeed, and how far it might help to solve world food shortages.

Edible mussels, of the family Mytilidae, occur all over the world. The distribution of *Mytilus edulis* or *M. galloprovincialis* ranges from the Pacific and Atlantic coasts of North America to most of Europe and the African Mediterranean coast. *Mytilus perna* (or *Perna perna*) is found in Brazil and Venezuela, while on the Pacific coast of the Americas there are *Mytilus californianus* in the north, and *Choromytilus choros* and *Aulocomya* species in Chile. Along Asiatic coasts are found *Mytilus crassitesta* in Japan, and *Mytilus smaragdinus* in the Philippines as well as in India, where *Mytilus viridis* and the Brown Mussel also occur. *Mytilus viridis* has been reported in Malaya, and *Mytilus canaliculus* in New Zealand. In many of these places mussel fisheries occur, and the prospects of developing these are worth examination.

In considering the possibilities of mussel culture in Asia and South America, in these regions there are three special factors favoring its development: (1) the likelihood of higher sea temperatures and higher growth rates than those in Europe; (2) lower labor costs than in Europe; (3) the widespread food shortage and big potential demand. Information on growth rates is limited. Ronquillo (personal communication) reports that in Manila Bay 4- to 6-months-old *Mytilus smaragdinus* is harvested at lengths of 2–4 in. In India Murti (personal communication) mentions specimens of *Mytilus viridis* 150–200 mm in length. Iversen (1966) reports that in Venezuela 10–20 mm mussels, *Perna perna*, grew to 100–150 mm in five months, and yielded 50% of flesh by weight. By comparison, *Mytilus edulis* takes a year to grow from 10 to 75 mm in Spain, and well over three years to reach this size in Wales. These figures suggest that mussel culture in Asia and South America may be much more productive than in Europe, provided that other conditions are suitable.

High growth rate is important in that it brings quick returns from the high productivity of culture grounds or materials, and also because it shortens the mussels' exposure to predation and parasitism. Were it not for mortality losses, low growth rate would not matter much in bottom culture in areas where ground was cheap and plentiful. In hanging culture on costly artificial materials, high growth rate is vital for economy in capital outlay, maintenance and depreciation costs. This is one reason why floating culture has not been adopted in northern European countries, labor costs apart. Therefore, hanging culture methods may be more applicable in areas of high growth rate.

Trials of floating mussel culture have, in fact, recently been conducted in Venezuela (Griffiths 1967, Iversen 1966), where low labor costs mean that the laboriousness of floating culture and hand cultivation need not hinder fisheries development as much as they do in European countries. Capital costs need not be high in starting small-scale mussel culture. In Norfolk, England, one man needs only a rowboat with oars, a hand net, a hand rake, and a shovel to stock and harvest mussel lays of about ⅛ acre in area, yielding about 10 tons a year from 6 months' work. Although such primitive efforts may seem ludicrous in view of the vast scale of production needed to change the food situation appreciably, in India for example, the potentialities of this approach are to be measured in the area of culture ground, and the number of people available to work it. Small beginnings can pay off in experience, by getting the feel of the problem, and in developing new ideas from fishermen and scientists on adapting culture to local conditions, just as the Norfolk men did.

The most important reason why mussel culture might arouse interest in Asia and South America is that of food shortages. Although the potential mussel productivity is so much higher than that of other bivalves and of livestock, the quantities needed to make any important difference to food resources are daunting. In this one must reckon on survival ration rates, such as 1 g of dry protein per day per kilogram of body weight, say 50 g per head. *Mytilus edulis* yields about 35–40% by weight of raw flesh, containing 17.65% of dry matter, or 11% of protein. Hence, a ton of mussels should yield about 40 kg of dry protein at 4% conversion rate, a day's ration for 800 people, or a day's ration for 16 people once a week throughout the year. The question is, how many people could be fed in this way?

The culture potential of mussels, developed to the full, would be limited by the area of suitable sheltered waters available. At general estimate, based on a 1 : 2,000,000 scale map, the sheltered waters in northwest Spain producing 100,000 tons of mussels a year cover about 100,000 acres. On this basis a 1 : 4,000,000 scale map shows that the coasts of India and Pakistan, for example have about a million and a quarter acres, roughly 2000 square miles, of creeks and gulfs between Karachi, Ceylon, and the Ganges. At a yield of a ton per acre, this area, if cultivable, would produce one and a quarter million tons of mussels, enough to give 20 million people a day's protein ration once a week throughout the year. Despite the dangers of such speculation, it is as well to get some measure of culture ground potentialities in any kind of fish culture, to judge how far these are worth investigating in the face of such serious food shortages. It is very likely that the calculations do not do justice to the high productivity of the Spanish culture areas, and also that they may underestimate the growth potential of Indian mussels. Even so, the only apparent way of increasing the cultivable area would be by the development of open sea culture. Too little is known of this possibility to make it worth pursuing in this article.

The potentialities of mussel culture are seen in better perspective by comparing its productivity with those of other cultivated marine animals. According to Schuster (1952), Javanese tambak lagoons then covered 143,000 acres and yielded 30,000 tons a year of prawns, shrimps, *Chanos*, and other fish. Average annual yields were about 200 lb per acre, with maximum rates of 1000 lb per acre. When it is realized that Welsh and Dutch mussel cultures produce 10,000–11,000 lb of raw mussel meat per acre per year, or 10–50 times more than the Javanese tambaks, the high potentialities of mussel culture take on a new significance. If marine fish culture is to make any real difference to world food resources, then mussel culture seems the likeliest to do so, at this stage of our knowledge. It may well happen that those marine animals that are most productive and most easily cultivated may eventually take pride of place in fisheries production.

The problem of fish production cannot be studied in isolation, without considering also the problems of processing, preservation, transport, costing, and preparation as food. Boiled mussels can be made into a variety of simple dishes, curried with rice, mixed with spaghetti, or made into a fish pie with potato.

For large-scale distribution, shelling is necessary to economize on transport. The costs of labour, cooking, and preservation in acetic acid raise the price of mussel meat by over four times. The development of mechanical shelling might, in Europe, lower the cost, but this would have less effect where labour costs were lower. The technology and economics of fish preservation and processing would need careful study to produce a palatable, easily transported form of mussel flesh cheap enough for mass consumption.

With the development of adequate transport for marketing, and of increasing agricultural and other production to pay for the mussels, it is clear that the possibility of such fishery developments would depend upon the economics of the countries concerned. It may be that the development of mussel culture would start as a semi-luxury industry for export, as with prawns in India, which are presumably too expensive for mass home consumption. Even so, this experience might eventually lead to such lowering of production costs as to enable most people in the producing countries to eat the mussels themselves.

G. DAVIES

NATIONAL SEA GRANT PROGRAM

The National Sea Grant Program, administered by the National Science Foundation, was initiated by the U.S. Congress in 1966 for the purpose of aiding development of marine resources through institutions of higher education, institutes, laboratories, and public and private agencies.

The Sea Grant concept originated in 1963 when Dr. Athelstan Spilhaus, Dean of the Institute of Technology at the University of Minnesota and Chairman of the National Academy of Sciences Committee on Oceanography, proposed to the National Fisheries Society that the pattern of the Land Grant Colleges be followed in developing the resources of the sea. The concept was refined and elaborated at a conference in Newport, Rhode Island, on October 28, 1965. Educators and scientists throughout the country attended this meeting to voice their opinions on what should constitute a "Sea Grant College."

Legislation was introduced in 1965 by Senator Claiborne Pell of Rhode Island, and passed in October, 1966, as "The National Sea Grant College and Program Act of 1966," Public Law 89-688. While it was not possible to follow the Land Grant pattern to any great extent, the Act provided for stimulating and supporting scholars and academic institutions to tackle practical problems of marine resource development.

The National Science Foundation created an Office of Sea Grant Programs early in 1967, and established two program elements: (1) Sea Grant Institutional support is to be awarded to major institutions of higher education with existing broad based programs in marine science and education. Such support will aid the institutions to expand their activities into applied research, training and education of marine technicians and ocean engineers, and advisory, demonstration, and extension services for sea industries; (2) Sea Grant Project support is designed to support specific projects in any activities related to marine resource development.

Sea Grant programs and projects may cover legal, economic, marketing, business administration, and other activities essential to development of marine resources in addition to research in science and technology and the education of manpower to work in the marine resources field. Such resources include fish, shellfish, marine plants, minerals, recreation facilities, energy, and transportation. Included in the "marine environment" under the Act are the Great Lakes as well as the ocean waters, the continental shelves, and the bottom and sub-bottom of oceans adjacent to U.S. territory.

While the Sea Grant concept is similar in many ways to existing ocean-related programs, it differs in emphasis and scope. Traditional programs have focused on ocean science, evolving from the basic disciplines as physical oceanography, marine chemistry, submarine geology, marine biology, and air/sea interaction. Sea Grant carries these a step further in, first, applying their technologies to practical application, i.e., exploitation of the ocean's resources, and second, developing the concept of ocean engineering.

Further, the Sea Grant Program embraces a much greater number of disciplines, including law, economics, sociology, and administration—addressed to institutional as well as scientific and technological problems. The strength of the Sea Grant Program is expected to derive from the collaborative blending of all of these disciplines and skills.

The National Sea Grant Program provides a means through which institutions, industry, business, and State and Federal agencies may cooperate in solving practical problems of marine resource development. Eventually, some institutions receiving Sea Grant support may be designated "Sea Grant Colleges," based on the quality and scope of performance in achieving the objectives of the program. This is a cost-sharing program, wherein the government furnishes a maximum of two-thirds of the total amount planned for any project. The federal funds may be utilized for almost any purpose connected with the program except for purchase, rent, or maintenance of ships and facilities.

While the National Science Foundation has been given responsibility for the program's execution, the National Council on Marine Resources and Engineering Development, under the chairmanship of the Vice President, gives policy guidance.

The National Science Foundation envisions several ideal features for a Sea Grant institution, including: defined study curriculum in ocean engineering and other applied fields; commitment of top management to the Sea Grant concept; a full time program coordinator; location in, and sensitivity to, regional problems; emphasis on applications of research and engineering; related information and extension activities; and large scale collaboration among departments, among institutions, between schools and industry, and between the school and state government.

The National Science Foundation has published a brochure offering guidelines to institutions desiring to participate in this program. Among the criteria by which proposals will be judged are: existing resources; capacity for development; commitment to program goals; and regional factors.

The scope of Sea Grant activities in existing institutional programs ranges from location of such new resources as manganese in Green Bay, Lake Superior, to improvement of shrimp processing in Washington. The universities attempt solutions to marine problems peculiar to their areas, whether these are economic, technological, legal, social or scientific. Production of ocean engineers and marine technicians is directed both to national and regional manpower needs. In many cases, industries and state agencies cooperate directly with the Sea Grant Institution, often exchanging personnel and sharing facilities.

Sea Grant Projects are even more diverse. They include aquaculture of commercially valuable species, restoration of damaged marine resources including seaweed beds, improvement of engineering criteria for offshore platforms and boats, methods of utilizing marshes, "factory" operations for production of shellfish from seed to finished product, improvement of seafood handling methods, exploration for sand and gravel deposits on drowned beaches, marine technician training, curriculum development for ocean engineers, isolation of useful drugs in marine organisms, and management of shrimp stocks.

Any college, university, public or private research institute or laboratory, or state agency may be eligible for Sea Grant Project Support for a project of high quality that will contribute to the use of marine resources in any of the possible fields from recreation to production of living and mineral products. Soundly based innovations are particularly welcome.

ROBERT B. ABEL
HAROLD LELAND GOODWIN

NAVIGATION

Navigation is the process of directing the movement of a craft from one point to another. The scientific talent of the centuries has combined to provide various devices to assist the measuring, recording, and displaying information useful to the navigator, but the interpretation and utilization of the information is still basically a human function. Only the anticipated can be programmed into an automatic system, and in many instances the requirements do not justify a great amount of sophistication. The principles discussed here apply whether the process is a manual or automated one.

Marine navigation is concerned primarily with finding answers to the questions, "Where am I?" and "How do I get to my desired destination?". It involves, then, position, direction, and distance, with speed and time being elements often useful in determination of the basic information.

Navigation is essential to the completion of any mission involving travel, not only in going to and from the area of operations, but also in conducting many operations once the vessel is on site.

Elements of Navigation

Position. Basic to virtually any marine operation is the determination of present position of the craft. For most purposes position is stated in geographic coordinates, latitude and longitude, to a precision of a tenth of a minute of arc, although the position of a vessel at sea is seldom known to an accuracy of better than several *miles*.

Where two or more vessels are in company, position relative to one another may be of greater interest than geographic coordinates. Similarly, position relative to land or offlying dangers may be of prime importance when the safety of the vessel is involved. In certain operations the prime consideration is the ability to return to a point previously occupied. This requirement is termed *repeatability*.

Two distinct philosophies are employed in the determination of position. One, called *dead reckoning*, is the extrapolation forward, or integration of motion, from a previously determined position. Accuracy of the method is affected by error in the initial position, errors in measurement of direction and speed or distance of travel, disturbing influences such as current and wind, if measurements are made relative to the environment, and inaccuracies in the method of applying the change of position to the starting point. Although the error of dead reckoning tends to be cumulative, the method is always continuously available in some form and is widely considered the primary method of position determination.

The other method, called *fixing*, involves determination of position relative to external aids such as lighthouses, bottom topography, electronic transmitters, satellites, or natural celestial bodies. A fix obtained by adjusting data to a common time may be called a *running fix* (except where observations at sea are made within an elapsed time of 30 min), one of doubtful accuracy or based upon incomplete information an *estimated position* (an expression also applied to a position determined by estimating the effect of current and wind on a dead reckoning position). A position established by the application of judgment to conflicting information may be termed a *most probable position*. An *assumed position* may be used for plotting a celestial line of position.

Using both dead reckoning and fixing concurrently provides a complementary redundancy that is considered desirable in avoiding gross errors. It also provides means of establishing the "current" (in this case the total effect of all disturbing elements) acting on the vessel.

Direction. In navigation, direction is involved primarily with respect to the motion of the vessel and in determining position relative to an aid to navigation or celestial body. Direction is customarily expressed to the nearest integral degree (tenth of a degree when expressing the direction of a celestial body or checking the accuracy of a compass). Adjectives may be applied to indicate the reference direction used: *true* if the geographical meridian, *magnetic* if the magnetic meridian, *compass* if the axis of a magnetic compass card, *gyro* if the axis of a north-seeking gyroscopic compass, *grid* if the fictitious meridians sometimes used in high latitudes, and *relative* if the axis of the vessel.

The forward direction of the axis of a vessel is called its *heading*, which, because of current or leeway, may not be the same as the intended direction of motion, called *course*, or the actual direction of motion between two points, called *course made good*. The actual path of the vessel, and sometimes its direction, is called *track*. The direction of a terrestrial point from the vessel is its *bearing*, while that of a celestial point its *azimuth*.

Distance. Navigators commonly use *nautical miles* of 1852 meters (m) (about 6076.1 ft) in expressing distance. This unit is particularly suitable because it is approximately the value of one minute of arc of a great circle of the earth. Near an aid to navigation or a coast line, yards, feet, or meters may be used for distances. Heights are expressed in feet or meters; depths in feet, meters, or *fathoms* (of 6 ft).

Speed. Navigators invariably express speed in units of *knots*, one knot being one nautical mile per hour, except for certain special purposes when feet, yards, or meters per second may be more appropriate.

Time. Navigators use several kinds of time. That based upon the *average* rate of motion of the sun along the celestial equator (the *mean sun*) is called *mean time*, while that based upon the actual, but variable, motion of the actual sun along the ecliptic is called *apparent time*. In either case the adjective *local* is applied when the observer's meridian is the reference, and *Greenwich* when the prime (0°) meridian. *Zone time* is the mean time of the meridians at 15° increments from Greenwich, usually the nearest one to the observer. Zone time is customarily used for indicating the position of a vessel and noting its progress. Almanacs of ephemeridal information of celestial bodies use Greenwich mean time except for rising, setting, and twilight information, which is given as local mean time. *High noon* occurs when the apparent sun is over the observer's meridian, at 1200 local apparent time.

Navigators customarily express time in four figures on a 24-hr basis without punctuation when the nearest minute is adequate (0001 is 1 min after midnight and 1524 is 24 min after 3:00 p.m.). When greater precision is needed, as in celestial navigation, hours, minutes, and seconds are customarily used.

Time based upon apparent motion of stars is called *sidereal time*, either local or Greenwich. It is useful for locating positions of stars and in referring to star charts. It, too, is expressed on a 24-hr basis, but starts when the reference point, the *vernal equinox* (the point in the sky occupied by the center of the sun when it crosses the celestial equator in March), is over the *upper branch* (sidereal "noon") of the reference meridian. There is one more sidereal than solar day per year. There is no sidereal date.

Dead Reckoning

Plotting. Except in sophisticated automated systems, the common method of performing dead reckoning is by hand plotting on a chart. Positions are shown as dots enclosed by small circles (some navigators prefer semicircles for dead reckoning positions) and labeled with the time, usually in four figures, and the type of position (e.g., 0516 DR). Course lines are drawn on the chart and appropriately labeled. By this means the dead reckoning can be run ahead as an aid in anticipating events such as times for changes of course, sighting aids to navigation, and making landfall.

At sea, out of sight of land and beyond shoal water, plotting is generally done on a blank *plotting sheet*, which shows the *graticule* of latitude and longitude lines and *compass roses* for

NAVIGATION

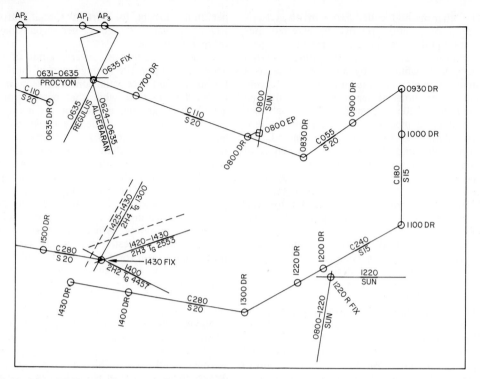

Fig. 1. A typical plot at sea showing dead reckoning, a celestial fix at 0635, a morning sun line and estimated position at 0800, a running fix at 1220, and a loran fix at 1430, with changes of course at 0830, 0930, 1100, and 1300, and changes of speed at 0930 and 1300.

measuring directions, but no other details. Lines of position used in determination of fixes or estimated positions, and the positions themselves, are generally drawn on the same plot with the dead reckoning. Figure 1 shows a typical plot of the progress of a vessel at sea.

Computation. On small vessels where space is a problem and motion of the vessel may make plotting difficult, as well as in special applications, dead reckoning may be performed by computation. A number of different methods, called *sailings*, have been devised for the purpose, depending upon the circumstances. These are named for their distinguishing characteristic: *parallel, plane, traverse, middle-latitude, Mercator, great-circle,* and *composite*. Tables and devices to facilitate hand computation have been developed, and the sailings serve as the basis for mechanical dead reckoning devices and electronic computers.

Automation. A mechanical device may take inputs from the direction and distance (or direction, speed, and time) measuring equipment and provide an automatic plot of the dead reckoning. Such a device may be called a *dead reckoning tracer* or *dead reckoning equipment*. The latter generally has, in addition to the plot, counters showing the present dead reckoning position.

Two types of dead reckoning systems providing measurement of progress with respect to the bottom are in limited use. One of these is *inertial* equipment, which uses accelerometers and an accurate directional reference to measure accelerations of the vessel in mutually perpendicular directions. The integral of acceleration is speed, and the second integral is distance. When coupled with time and direction, the components can be combined to produce an indication of change of position, or position relative to a reference position.

The other type directs one or more beams of acoustic energy downward at an angle from the vessel. The return signal reflected from the bottom has a slightly different frequency because of the *Doppler effect* due to motion of the craft relative to the bottom. The extent of the Doppler effect is directly related to the component of speed in the direction of the beam. This effect, with a good directional reference and time, provides a means of performing dead reckoning relative to the bottom. Because of power requirements and the vagaries of acoustic propagation in seawater, this device is practically limited to use over the continental or insular shelf, unless reduced accuracy by return of acoustic energy from the water mass is acceptable.

Tools for Dead Reckoning. Two categories of equipment are needed for dead reckoning: the sensing devices and the data reduction and display equipment.

Direction measurement is made by some form of compass, which provides a reference direction. The *magnetic compass* utilizes the directional properties of the earth's magnetic field. Some form of adjustment or compensation is needed to neutralize the local magnetic effects caused by magnetic materials and electric currents in the vessel itself. The residual effect is called *deviation*, which is applied as a correction to readings of the compass. A second effect needing correction is the angle between the geographic and magnetic meridians. This is called *variation* or *magnetic declination*. A north-seeking *gyro compass* is not subject to these errors, but may have a small residual error and requires a constant source of electrical power and a period for settling down after being turned on. *Repeater compasses*, generally associated with gyro compasses, may be located at convenient places throughout the ship. A *pelorus*, a compass-like device without a sensing element, may be used for sighting on objects to obtain bearings.

Distance measurement is made by some form of *log*. An older device, little used today, consisted of a rotor towed through the water, with a counter attached to it or secured to the taffrail. Modern logs, in a variety of forms, are attached to the hull of the vessel. Depth of water is measured by means of a *hand lead* or *sonic depth finder*.

Speed measurement may be combined in the same instrument that measures distance, or it may be determined by calibration of the average speed of the propeller shafts of the vessel. Speed may also be determined by computation, using distance and time. All of these methods provide a determination of speed relative to the water. Inertial and Doppler equipment determine speed over the bottom.

Time measurement is made by one or more accurate *marine chronometers*. The error and rate of these instruments are determined by comparison with radio time signals transmitted at stated times from a number of locations throughout the world. Hand watches (*hacks*) or stop watches, checked against the chronometers, are generally used for timing celestial observations, to avoid disturbing the chronometers and thus possibly changing their rates.

The data reduction and display equipment depends upon the method used. For hand plotting, charts or plotting sheets are needed, together with pencils, erasers, and a plotter, parallel rulers, or drafting instrument for determining direction and drawing straight lines. A pair of dividers is generally used for measuring distance, and a pair of compasses for drawing circles. Tables and work forms are needed if dead reckoning is performed by computation. Special equipment is needed if dead reckoning is performed by mechanical device or by a sophisticated inertial or Doppler system.

A *nautical chart* is a map intended primarily for navigation, showing latitude and longitude lines and scales, depths of water, heights of land, aids to navigation, and other information of use to navigation. The map projection most commonly used for nautical charts is the Mercator, although other projections may be used for special purposes.

Position Fixing

Piloting. The term piloting refers to navigation relative to reference points. Before electronics extended the range to virtually any point on the surface of the earth, piloting techniques were limited to a narrow strip near the coast, where visible landmarks, lights, and manmade aids to navigation were available, or where the water was shoal enough to permit soundings by hand lead. Although these limitations no longer apply, the techniques are still in common use.

Positions are generally determined by establishing two or more lines of position, on each of which the vessel is located, within the accuracy of measurement. A line of position may be a great circle established by measuring the direction or bearing of an identifiable, charted object; a small circle determined by measuring the distance to such an object; a hyperbola representing the difference in time of reception of radio signals from two synchronized transmitters; or some other identifiable line.

Other techniques are sometimes used when conditions are favorable. In shoal water where bottom topography is distinctive, a series of *soundings* may be compared with the chart to establish position. Certain combinations of bearings of the same object as a vessel steams past it provide positional information without plotting if one maintains a constant course and speed. A position may be determined by passing close aboard a buoy. A pair of horizontal angles between visible objects can be used to locate the craft. A limiting *danger bearing* or horizontal or vertical *danger angle*, may permit safe passage without determination of position. A vessel may be kept safely within the limits of a dredged channel by means of *range lights* placed one behind the other in line with the center of the channel, or one may steer for the centers of successive pairs of buoys marking the outer safe limits of the channel.

Charts, light lists, coast pilots, tide and current tables, and other useful navigational publications are available from the U.S. Naval Oceanographic

Fig. 2. A widely used marine sextant. (Photo: Weems & Plath Inc.)

Office and the Coast and Geodetic Survey of the Environmental Science Services Administration, Department of Commerce.

Celestial Navigation. Celestial bodies are used in a somewhat different way than terrestrial landmarks. Because of their great distances from the earth, with lines of sight to any one body being virtually parallel from all points on the earth at which the body is visible, position is determined relative to its *geographical position*. This is the point on the earth at which the body is in the zenith, vertically overhead. The altitude (the elevation angle above the horizontal) is measured by means of a sextant (Fig. 2) and compared with the computed value for an assumed position in the vicinity of the actual position. Each altitude represents a small circle of position with the geographical position as the common center of both. By drawing a radial line from the assumed position in the direction of the body, or the reciprocal, and measuring off the altitude difference in minutes of arc (one minute of arc being very close to one nautical mile on the surface of the earth), the navigator establishes one point on the circular line of position. A line through this point and perpendicular to the azimuth line is considered a part of the circle of position through the observer. Measurement is not made directly from the geographical position because at altitudes commonly used, this point is thousands of miles away. Azimuth is not used as for a landmark within visible range because neither the accuracy of measurement nor the accuracy of a reference direction is adequate for practical use in this manner. A celestial line of position (sometimes called a *Sumner line* after Thomas H. Sumner, who discovered it) can be used with any other line of position to obtain a fix.

Because the geographical position is in constant motion due to rotation of the earth, accurate timing of each observation is essential. For a body on the *prime vertical*, due east or due west, an error of one second of time introduces an error of a quarter of a minute of longitude (0.25 mile on the equator). Nautical and air almanacs are published by the U.S. Naval Observatory to provide information on positions of celestial bodies at any time. Because of the relationship of time to the positions of celestial bodies, a practical method of finding longitude at sea, within the capacity of the mariner, awaited the availability of accurate time, which was provided in the 18th century by the invention of the marine chronometer by John Harrison.

Special situations are sometimes utilized. A body on the prime vertical can be used to determine longitude. One on the celestial meridian, due north or south, indicates latitude. In this situation, time is relatively unimportant. A body perpendicular to the shore line will provide information on the distance of the vessel offshore. One dead ahead or dead astern provides a check on vessel speed, while one abeam provides a check on leeway.

Although the computations involved in establishing a celestial line of position can be made by electronic computer or mathematical calculation using natural or trigonometric functions, graphically, or mechanically, the most common method is to use *sight reduction tables* providing tabulated values of altitude and azimuth, published by the U.S. Naval Oceanographic Office. Also useful is some form of star finder, a device to aid in the identification of celestial bodies or provide a simple means of anticipating where a given body will appear.

Electronic Aids. A great many techniques for utilizing electronics in navigation have been proposed. A few of these have resulted in operational systems. Special equipment, both transmitting and receiving, is generally needed. A radial or azimuthal type provides lines of position similar to bearings of a visible landmark. The *radio direction finder, consol,* and *consolan* are examples of this type of aid. *Radar* can be used near suitable targets to provide distance measurements, as well as bearings. Long distance systems most frequently establish hyperbolic line of position by synchronized transmitters operating in pairs.

In *loran* the actual difference in time of arrival of signals is measured. In *Decca* and the VLF *Omega* system, the difference in phase of arriving signals is measured. Because of repetition of readings in narrow adjacent *lanes,* some method of ambiguity resolution is needed with phase measurement systems.

Navigation Satellites. The U.S. Navy has developed a navigation system involving special purpose satellites in 600-mile polar orbits (Fig. 3). Measurement of the Doppler effect is made during a 6–16 min period as the satellite moves across the sky. This information and orbital data transmitted periodically by each satellite are fed into a special purpose computer for determination of position of the vessel in terms of distance along a line through the subsatellite point at the nearest point of approach of the satellite and normal to the path of the subsatellite point. This method is not suitable for general navigation because of (1) the high cost of user equipment, (2) the need for an accurate velocity vector of the vessel for accurate results, (3) the relatively long integration time, and (4) the interval of approximately 108 minutes (occasionally double this) between available fixes.

A number of other techniques have been proposed to provide a system more suitable for non-military use, but none is in operational use. In general, these methods reduce drastically the sophistication and cost of user equipment by relocating the complexity of the system at ground

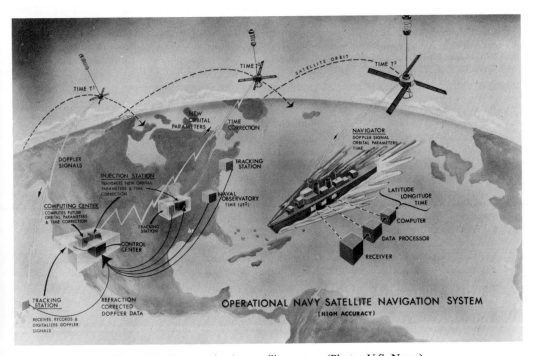

Fig. 3. The Navy navigation satellite system. (Photo: U.S. Navy.)

stations or in the satellites, or both. Ranging or angle-measuring techniques provide frequent fixing capability and add versatility to the systems. Participation of shore-based facilities open the possibility of additional uses relating to search and rescue, weather and oceanographic data collection, weather routing, and surveillance.

Hybrid Systems. As computer technology matures, the combining of outputs of several aids into a sophisticated overall shipboard navigation system becomes a practical but expensive reality. Among the purposes to be achieved are relieving the navigator of some of his duties, thus freeing him for other tasks; providing complementary redundancy for additional reliability; providing real-time output of present position and other useful information; and improving accuracy by filtering out errors of individual aids. The cost of hybrid systems is such that it can be justified for marine use only for special applications with requirements that cannot be met by other means.

Conclusion

The principles of navigation are applied in a variety of ways to fit individual requirements. Too often the navigation of a vessel is given less attention than it deserves. The large number of groundings (nearly three ships of 500 gross tons per day) and collisions (four ships per day) attest to the inadequacy of present navigation. Eleven percent of all ships of 500 gross tons and larger file an insurance claim each year for one of these causes. A large majority of these could be avoided by alert crews utilizing the means available to them. No figures are available to indicate the number of near misses or the amount of data of limited value collected by ships whose position was in doubt. On the other hand, navigation is not an end in itself. It is true, however, that "the price of good navigation is constant vigilance," to quote the Court of Inquiry investigating the Point Honda disaster when seven Navy destroyers were lost due to stranding on the rocky California coast north of Point Conception in 1923. The tools and knowledge are available to all who will use them, but intelligent use is essential to adequate navigation. Intelligent, alert, knowledgable man, properly equipped, is the key to successful navigation.

ALTON B MOODY

References

Bowditch, N., "American Practical Navigator," 1966, U.S. Naval Oceanographic Office, Washington, D.C.
Hill, J. C., II, Utegaard, T. F., and Riordan, G. "Dutton's Navigation and Piloting," 1969, U.S. Naval Institute, Annapolis, Md.
Mixter, G. W., "Primer of Navigation," 1967, D. Van Nostrand Co., Princeton, N.J.
H. O. Pub. No. 220, "Navigation Dictionary," 1956, U.S. Naval Oceanographic Office, Washington, D.C.
Weems, P. V. H., and Lee, C. V., "Marine Navigation," 1958, McGraw-Hill Book Co., New York.

Cross references: *Marine Geodesy.*

NEKTON—See PELAGIC DISTRIBUTION

NEWFOUNDLAND FISHERIES

The Newfoundland fishing industry had its beginning with cod, which have continued to provide the greatest share of the landings. However, at present, some 25 different marine species are fished commercially in the waters around Newfoundland and Labrador and the offshore fishing banks. The quantities caught, the methods of catching them, and their commercial values are quite different. Trends in the landings and values of the main commercial species groups from 1940–1965 (Fig. 1) and the listing of the main landings and values for 1965 (Table 1) give an indication of their relative importance in the industry.

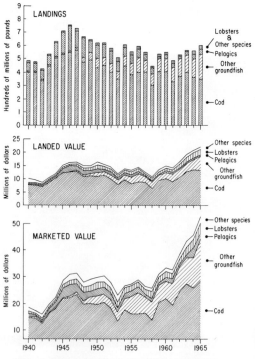

Fig. 1. Total landings (weights in form usually landed), landed and marketed values of the main species and species groups in Newfoundland, 1940–1965.

TABLE 1. LANDED WEIGHT (millions of pounds), landed and marketed product values (millions of dollars) of the main species groups in the Newfoundland commercial fishery, 1965

Species Group	Landed Weight	Landed Value	Marketed Value
Total	609.6	23.2	52.3
Cod	345.2	13.5	27.8
Other groundfish	189.7	5.4	16.4
Herring	28.9	0.3	0.9
Salmon	2.6	1.4	2.0
Lobsters	3.7	2.2	4.1

The Newfoundland fishing industry employed about 22,000 fishermen during 1965, contributed 35% of the Canadian Atlantic landings, and 25% of their landed value.

Inshore Fisheries

Cod. The inshore fisheries are carried out from shore premises in numerous coastal settlements. Over 95% of Newfoundland's fishermen operate in the inshore area, using small, generally open boats up to 35 ft in length, powered by gasoline engines; relatively few boats are larger than this, up to 50 ft in length, generally powered by diesel engines. These usually operate within 10 miles of shore and customarily return to port with their catch each day. Processing plants in various ports purchase the catch landed within their collecting range for production of fresh, frozen, or salted products. Where plants are not available, fishermen salt their catches at their own or community premises for later sale to commercial plants or exporters. In 1965 there were 46 plants for processing fresh and frozen fish and 34 for producing salted fish in Newfoundland. In addition, bait depots operated by the Canadian government are located in strategic points around the coast to ensure supplies of frozen bait being available at all times.

The principal species being caught in the inshore fishery is cod (*Gadus morhua*), the landed weight in 1965 amounting to about 70% of the total weight of all species landed from the inshore area. From 80 to 90% of Newfoundland's cod catch is from the inshore area.

Various gears are used in the inshore cod fishery. In the spring when the cod arrive in the inshore waters, they are in pursuit of food, usually capelin in most areas, or herring in others. Gears using bait most often procure small catches at this time so that handlines with jiggers (baitless hooks) in the shallow water and gillnets set on bottom in the deepest part of the shallow water are used. By late June the cod are close to shore in schools and are available to codtraps, large box-shaped nets with a door at one side from which a long net (the leader) leads to a shoal area or to the shore, so that fish swimming by it are directed through the trap doors. Traps are set on bottom, usually in depths of 10 to 20 fathoms. This is one of the most successful fixed gears used in the cod fishery and although in operation from June to August, a maximum of 7–8 weeks, traps account for about 50% of the cod catch annually (Fig. 2). Handlines (single lines with baited hooks, pulled by hand), longlines (long lengths of line set along the bottom, with baited hooks attached at intervals by short lines, and pulled either mechanically or manually), and

Fig. 2. Hauling a codtrap near St. John's, Newfoundland.

gillnets are used extensively from summer to autumn in various depths depending upon seasonal movements of cod.

In the western part of the south coast of Newfoundland there is a winter fishery. Cod that spend the summer migrating northward along the coastal areas of western Newfoundland, where they are available to traps, nets, and line gears, migrate southward out of the Gulf of St. Lawrence in the fall to the coastal waters of the western half of the Newfoundland south coast. There they spend the winter and are the basis for a fishery using line gears before they return to the Gulf in the following spring.

In the inshore cod fishery of the shallow water areas there are smaller amounts of other species taken. In the deeper areas the longline gears generally catch small amounts of other species: American plaice (*Hippoglossoides platessoides*), wolffish (*Anarhichas* sp.), and Greenland halibut (*Reinhardtius hippoglossoides*), amounting to about 10% of the total longline catch.

Herring. The present Newfoundland herring fishery is entirely coastal, particularly concentrated in the west and south coast regions. The fish are caught during winter and spring with gillnets and, since 1965, an increasing number of purse seines. With the introduction of purse seiners to the fishery, supplying plants for processing the fish into meal and oil, the catch has increased from below 20 million lb annually to nearly 30 million in 1965, over 60 million in 1966, and will continue to increase as additional purse seiners are added to the fleet each year, providing greater effort and mobility in the catching of herring. The purse seine fishery is concentrated on the west and south coasts of Newfoundland from Bonne Bay to Fortune Bay.

Lobsters. Lobsters (*Homarus americanus*) are fished in the Newfoundland inshore area, the season, extending from April to early summer, being controlled by the Department of Fisheries of Canada. Lobsters are coastal animals in Newfoundland waters and are most abundant on the west and south coasts and in Notre Dame and Bonavista bays on the east coast. The catch is taken by traditional wooden-lath traps baited with herring, flatfishes or other fish bait, set in shallow depths mainly to 5 fathoms and seldom greater than 10 fathoms. Fishing begins in the deeper part of the range at the beginning of the season and moves to the shallow depths toward the end. By this time the catch has decreased due to a reduction of the population by the fishery and to the approach of moulting, when the lobsters eat little. Since 1949 landings have usually been between 3 and 5 million lb. Practically the entire catch is sold live, mainly in the U.S. Lobster fishing in Newfoundland is already very intensive and all areas of lobster abundance are being fished. Any increase in this intensity would certainly be unprofitable and would be unlikely to increase the long-term landings.

Salmon. Atlantic salmon (*Salmo salar*) are fished in areas around the entire coast of Newfoundland and in Labrador from May through the summer, with landings being highest in southern Labrador and the northeast coast of Newfoundland.

The most common gear used is the gillnet, generally fixed at the surface, but there has been a surface drift-net fishery at the southwestern tip of Newfoundland for a number of years. The fishery for Atlantic salmon is a very old one, with the catches being salted up till the 1920's, when an increasing amount was exported either fresh (in ice) or frozen. Landings reached their highest level in 1930, close to 7 million lb, but have since declined to about 3 million lb annually, practically all of which are exported fresh or frozen to the U.S. and the United Kingdom. Although some of the salmon caught in the Newfoundland sea fishery are produced as smolts in local rivers, many of the large salmon originate in New Brunswick and Quebec rivers. No great increase in landings of salmon from the commercial fishery seems probable, because of increasing deleterious effects on river habitats resulting from industrial and domestic expansion in regions of many rivers used by salmon. In addition, an increasing fishery for Atlantic salmon off west Greenland catches many fish, some of which apparently originate in Canadian rivers. This could help reduce Newfoundland landings.

Capelin. Capelin (*Mallotus villosus*) are important food fishes for cod, and when cod come to the coast in spring they are often in pursuit of the capelin schools, which are migrating shoreward where they spawn in late June to early July on the beaches and in shallow water near shore. Capelin are found in most areas of Newfoundland and in southern Labrador. The present fishery for capelin is right at the beaches where they are caught with small cast-nets thrown by hand or by seines. An estimated 10 million lb are taken annually, mainly for use as cod bait, but to a limited extent for human consumption. Indications are that many times the present catch could be taken, both in the inshore area and by extending the fishery over a longer period to the offshore areas, principally the Grand Bank where capelin are known to be in quantity, particularly on the Southeast Shoal area.

Squid. The short-finned squid (*Illex illecebrosus*) is fished commercially in Newfoundland, the entire catch being taken close to shore particularly in Conception, Trinity, and Bonavista bays. The fishery occurs in the late summer to autumn period, but mainly in August in most areas of the coast. Squid are taken on barbless hooks (squid jiggers) either pulled singly by hand, or as multiple jigs hauled mechanically. In recent years the squid have been used for bait in the local line fishery for cod, and exported to Portuguese, Norwegian, and Faroese ships longlining in the Northwest Atlantic for cod and sharks. Large quantities of dried squid were once exported to China for human consumption.

The catch of short-finned squid is subject to extensive fluctuations, in recent years between 2 and 20 million lb. The squid appear on the offshore banks in May or June, presumably having come up from the deep slopes. They appear inshore in July and August, but in varying abundance annually. They retreat from the coast in November, presumably en route southward to spawn. Inshore, in years of abundance, landings of squid can be increased greatly, but, for a large expansion, the fishery will have to be extended beyond the coastal regions.

Greenland Halibut. The Greenland halibut (*Reinhardtuis hippoglossoides*) is caught in gillnet and longline fisheries in deep water off many Newfoundland bays, mainly Trinity, Bonavista, and Notre Dame bays, although in past years White Bay and Fortune Bay had fisheries as well. The fishing period is determined partly by the seasonal weather, the landings being principally between May and December. In recent years the amounts of the Newfoundland landings have been largely determined by available markets and have generally fluctuated between 1–2 million lb. The fish were exported salted in barrels, typically, but, since 1964, the chief product has been frozen fillets marketed principally in the United Kingdom and U.S., resulting in greatly increased landings reaching 15 million lb in 1965, and nearly 30 million lb in 1966.

Whales. A limited fishery for small whales is conducted in Trinity and Bonavista bays fishing the pilot (*Globicephala melaena*) and the minke (*Balaenoptera acutorostrata*). The fishery began in 1947, with small numbers being harpooned, and continued until 1950, after which the practice was begun of surrounding and driving herds of whales ashore to be slaughtered. The number of pilot whales captured has fluctuated widely between 700 and 9800 annually with the capture of minkes averaging about 30. The whales are processed in a small shore factory where the fat is used for oil and the meat is frozen for mink food.

Other Species. In addition to the inshore fisheries described, there are minor commercial fisheries that are generally of only local importance to parts of Newfoundland. These would include fisheries for: Atlantic mackerel (*Scomber scombrus*), using nets, mainly in Notre Dame, Trinity, and Placentia bays; giant scallops (*Placopecten magellanicus*), using bottom drags in Port au Port Bay; and eels (*Anguilla rostrata*) and smelts (*Osmerus eperlanus mordax*), in various estuaries, using eel traps and nets. Harp seals (*Pagophilus groenlandicus*) are also taken in varying numbers according to their abundance in the winter to spring period in the inshore areas of northeastern and northwestern Newfoundland. They are shot or killed by clubs when on the ice, or caught by net when in the water.

Offshore Fisheries

In the Newfoundland offshore area the principal fishery is for various species of groundfish. In contrast to the inshore fisheries where daily fishing trips are generally made, in the offshore fisheries the trips are of longer duration. The offshore fleet up to the conclusion of World War II consisted of large schooners up to 150 tons (commonly called "bankers") equipped with dories from which the crews fished for cod with longlines, using the schooners as floating bases. The catches were split and salted on board the schooner and were later brought to port for drying. The advent of the diesel-operated fishing vessels equipped with otter trawls in the late 1940's heralded the demise of the schooner fishery, as fishermen were happy to give up the hazardous long hours of fishing from a tossing dory for shorter trips on the firm decks of the more comfortable otter trawler. From the early beginnings in the 1940's, the otter-trawler fleet operating from Newfoundland ports has increased to 48 ships in 1966, of which 7 were modern stern trawlers (Fig. 3). These ships range in size between 200 and 700 gross tons, and are equipped with modern electronic navigational and fish-finding apparatus. They operate throughout the year from processing plants in 14 or 15 ports on the south coast and southern part of the east coast of Newfoundland. They generally fish on the Grand Bank, St. Pierre Bank, and in the Gulf of St. Lawrence, but can range as far as the fishing banks on the Nova Scotian Shelf in the south and to areas of Labrador in the north.

Cod. Although cod is the principal groundfish species caught in the Northwest Atlantic, it is not predominantly important in the catches of the present Newfoundland offshore fleet. In 1965 cod accounted for 24% of the offshore groundfish landings. Only a small portion of the offshore fishing effort is directed toward catching cod, and this species is generally caught along with larger quantities of other species. Present catches of cod by offshore ships are mainly from off Nova Scotia in the Cabot Strait area and from St. Pierre Bank in the spring, and from the north and east parts of the Grand Bank throughout the year.

Fig. 3. A modern Newfoundland stern trawler. (Photo: Bonavista Cold Storage Co., Ltd.)

The Newfoundland otter-trawler fleet has not engaged in the lucrative winter and spring cod fishery off the Labrador coast because the ships are generally neither large enough for the long trips nor constructed to withstand the rigorous icing conditions which are likely to be encountered in the northern fishery.

Haddock. The Newfoundland offshore fishery for other groundfish species began with the otter-trawler fishery in the 1940's. The species most sought in the beginning was the haddock (*Melanogrammus aeglefinus*). This was mainly on the Grand Bank and in some years on St. Pierre Bank where in winter to spring the fish are concentrated in warm water on the southwestern slopes, spreading shallower in the spring. By summer the fish have spread over the shallow plateau, generally in concentrations too small to be fished commercially. The catch rose until 1949 when over 20 million lb were landed by the Newfoundland offshore fishery, following which there was a decline until 1951 as the virgin stock was reduced. After this, because of the retention of small haddock formerly discarded by the fishery, and an abundant new year class of fish entering the fishery, catches rose steadily until 1957. However, because of a heavy fishery by European nations, particularly Spain and, from 1959, Russia, even the small haddock present were severely reduced in numbers so that the fishery has virtually ceased. The Newfoundland offshore catch in 1965 was about 6,000,000 lb, the lowest since the early 1940's.

In Newfoundland, haddock are at the northern limit of their range in the northwest Atlantic. The population is at a very low level, and small landings are to be expected over the next few years. Even if a new, abundant year class should enter the fishery, the fishing effort directed toward the catching of haddock would be so intense that good catches could not be sustained more than 2–3 years.

Flounders. The species for which there is the greatest effort, and as a result the highest landings, is the American plaice (*Hippoglossoides platessoides*). The annual Newfoundland otter-trawler catch up to 1958 was from 10,000,000 to 15,000,000 lb. Following this, the landings increased annually due to greater effort being directed toward catching plaice as haddock became scarcer. The 1965 landings amounted to 84,000,000 lb, 40% of the total offshore landings. The main otter-trawler fishing grounds are on the northwestern, northern, and eastern slopes of the Grand Bank. There is also a small fishery on St. Pierre Bank.

In addition to American plaice two other species of flounder are caught by otter trawlers, the witch (*Glyptocephalus cynoglossus*), caught on the deep-water slopes, and the yellowtail (*Limanda ferruginea*), caught in the shallow areas. Most of the witch landings, generally below 5,000,000 lb annually, have been caught during the winter and spring fishery for haddock on the southwestern slope of the Grand Bank. The present fishery by otter trawl is on the southern tip of the Grand Bank during summer and a limited one on the western side of the Laurentian Channel in spring.

Yellowtail flounders have only recently been

caught in quantity (3,000,000 lb in 1965) by otter trawlers, generally on the Southeast Shoal and other shallow areas of the Grand Bank.

Redfish. After American plaice, the most important species landed in the offshore fishery is the redfish (*Sebastes mentella*). Landings in the beginning were at a low level and have fluctuated between 15–65,000,000 lb. In 1965 the 63,000,000 lb landed amounted to 30% of the total offshore groundfish landing.

Redfish live in the deep warm water usually below 125 fathoms and down as deep as 400 fathoms in practically all areas around Newfoundland, but are most abundant on the outer slopes of the banks. The Newfoundland otter-trawler fishery up to recent years was limited to the shallower part of this range, generally 160–180 fathoms, in the Gulf of St. Lawrence and southwestern and eastern slopes of the Grand Bank. However, after the beginning, in 1956, of a European and especially Russian fishery for redfish, which extended off the eastern Newfoundland and Labrador area into deeper water, some of the Newfoundland catches were taken in deeper water, especially southwest of St. Pierre Bank in the Laurentian Channel.

Redfish are very slow-growing fish and, apart from the possibility of pelagic populations existing in the open ocean, it seems unlikely that any great new populations of redfish remain to be discovered in the northwest Atlantic. Therefore, it is unlikely that the maximum annual landings can continue to increase much longer.

Other Groundfish. Other groundfish species are caught by the otter-trawler fleet, although they are generally not the direct object of the fishery, but are caught incidentally while fishing for the main commercial species, and constitute only about 1% of the landing.

Atlantic halibut (*Hippoglossus hippoglossus*) are caught in small quantities during the deep-water otter-trawl fishery for haddock and redfish on the slopes of the Grand and St. Pierre banks. A longline halibut fishery occurs from time to time in the Gulf of St. Lawrence and in the St. Pierre Bank area. The Atlantic wolffish (*Anarhichas lupus*), and the spotted wolffish (*A. minor*), commonly called catfish when marketed, are part of the commercial catch, small amounts being taken in most fisheries. Pollock (*Pollachius virens*) and white hake (*Urophycis tenius*) are caught in otter-trawl fisheries for haddock and redfish on St. Pierre Bank and Grand Bank. They are landed only in small quantities because of low demand by the processing plants. It seems unlikely that any of these incidentally-caught species will ever increase in importance, although if the demand by the local processors should increase undoubtedly much larger quantities could be landed.

Seals. There is an annual fishery for harp seals (*Pagophilus groenlandicus*) and hooded seals (*Cystophora cristata*) off the east coast of Labrador and Newfoundland and in the Gulf of St. Lawrence. Harp seals migrate to Newfoundland waters in winter and spring from their summer habitat on the coast of West Greenland and in the eastern Canadian Arctic. By January they have reached the east coast of Newfoundland and the northern Gulf of St. Lawrence, and continue moving southward, feeding until February when they turn north. When they meet the sea ice drifting south, they mount suitable ice and give birth to their young during the first half of March. One group of seals produce their young off the northeast coast of Newfoundland, or off southern Labrador—the Front herd—whereas those in the Gulf of St. Lawrence whelp in the western area of the Gulf. The female harps attend their helpless pups (whitecoats) on the ice for about three weeks, then leave them. The herds have now drifted southward along the east coast, and eastward in the Gulf of St. Lawrence. As the ice recedes the herds move again farther and farther northward in the water and on the ice, and by June they are usually north of the Strait of Belle Isle. Hooded seals occur only in small numbers with the main harp seal herds so that they contribute only about 5% of the numbers in the catch.

The fishery begins in March, and the greatest part of the catch consists of whitecoats, up to three weeks old, with pelt (skin and blubber) weights averaging around 50 lb. The fat is used for production of oil and the skins in the fur trade.

Seals were taken by Newfoundlanders from the time of the earliest settlers, operating first from the shore and later from ships that sailed to hunt the seals herds on the ice floes. The fishery developed rapidly in the 19th century, the ship fishery, before the 1860's, being conducted by sailing schooners with 400 ships carrying about 13,000 men being involved in 1857. In 1863, the first steam vessels were used and these rapidly replaced the sailing ships. The number of steamers was greatest in 1881, 27 vessels carrying 5815 men, but, following this, a gradual decline in numbers of steamers occurred. At the beginning of World War I, the Newfoundland sealing fleet consisted of nine steel and 10–12 wooden ships, but by the war's end most of these had been sold or were sunk.

Airplanes were first used for finding the herds on the ice in 1921. Since 1962, in addition to searching, they have been used in the Gulf of St. Lawrence to transport men daily to the killing areas on the ice, and to bring the seal pelts back to land (generally the Magdalen Islands).

There have been large catch fluctuations in the seal fishery, ranging from as high as 500,000–600,000 pelts to extreme lows during both world wars and the 1931–1932 depression. The Newfoundland vessel landing between 1954 and 1964 ranged from 20,000–50,000 pelts.

The seal fishery in the Newfoundland area at present is carried out by ships from Norway, Nova Scotia, and Newfoundland. Unless there is some prospect that conservation measures will be adopted to sustain and increase the herds, which have been reduced since 1950, and more efficient vessels are used, which can profitably be employed before and after the seal fishery, the present competitive fishery leaves the future in doubt.

Whales. A fishery for large whales has occurred periodically in Newfoundland. The earliest whaling was begun by the Basques, apparently prior to the mid-16th century. From 1796 to 1807, American ships engaged in whaling on the south coast of Newfoundland, and between 1840 and 1888, there were occasional periods when whaling was carried out.

Modern whaling using a harpoon-shooting cannon began in Newfoundland in 1898. The number of land stations and whale catchers increased up to 1905 following which, because of a decrease in landings, the fishery diminished and eventually ceased during World War I. In 1904 the catch numbered 1275 whales. In periods after 1918, landings were only close to or greater than 500 whales on three occasions. Since 1951 the fishery for large whales in Newfoundland has been limited to a small effort from 1956–1959, taking 20–60 whales annually. Whales caught principally have been fin (*Balaenoptera physalus*), humpback (*Megaptera novaengliae*) in the early years, with blue (*Balaenoptera musculus*), sei (*Balaenoptera borealis*), and sperm (*Physeter catodon*) being caught in small numbers throughout most of the periods.

Although catches between 1945 and 1950 suggested that the stocks of large whales occurring in the area may withstand an annual kill of 400, the periods of continuous operation have been too short to confirm this. The revival of a Newfoundland fishery will depend upon higher prices for oil and meal, and the use of the meat for animal and human food.

A. M. FLEMING

References

Colman, J. S., "The present status of the Newfoundland seal fishery," *J. Anim. Ecol.*, **6**, 145–149 (1937).
Innis, H. A., "The Cod Fisheries; the History of an International Economy, 1954, University of Toronto Press, Toronto.
Royal Commission on Canada's Economic Prospects, "The Commercial Fisheries of Canada," 1956, prepared by the Department of Fisheries of Canada and the Fisheries Research Board, Queen's Printer, Ottawa.
Templeman, W., "Marine resources of Newfoundland," *Bull. Fish. Res. Bd. Canada*, No. 154 (1966).

Cross references: *Labrador Fisheries.*

NEW ZEALAND FISHERIES

Geographically, New Zealand is isolated from other land masses and comprises a group of islands extending from 34 to 47° south latitude, i.e., from the subtropics to the subantarctic. Extending from the coast is a narrow continental shelf. Trawling, Danish seining, and fishing by set nets and by lines, are the principal methods of taking the demersal fish that form the basis of New Zealand's fishing industry. Crawfish are mainly caught in baskets or pots, and oysters, mussels, and scallops are taken in dredges. As the fishery is still based on demersal species, the industry has been confined until recently to the continental shelf. Attention is now being directed to the harvesting of pelagic species principally by means of purse seining. The landed value of

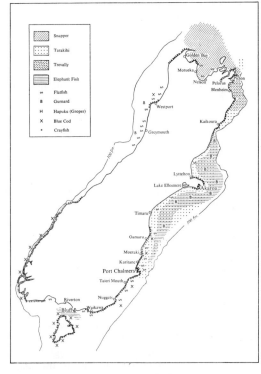

Fig. 1. Principal species and fishing grounds, South Island.

NEW ZEALAND FISHERIES

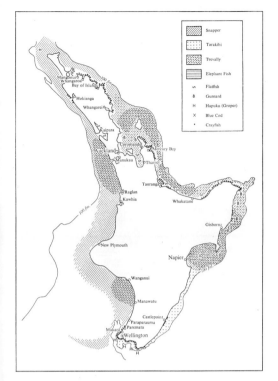

Fig. 2. Principal species and fishing grounds, North Island.

all fisheries for the calendar year 1966 was nearly $10,000,000.

The estimated total quantities and values of the principal classes of fishery products marketed in the calendar year 1966 are shown in Table 1.

Until December 1963, a restrictive licensing system operated. Following an extensive inquiry by a Parliamentary Select Committee in 1962, the licensing procedure was changed to a non-restrictive registration and permit system for vessels owned and operated by New Zealand nationals or New Zealand domiciled companies.

Legislation Affecting the Fishing Industry

Management and conservation of fish stocks generally is controlled by the Fisheries Act 1908, and pursuant to this Act, separate regulations are enforced covering a wide range of fisheries activities. Management and conservation of fisheries stocks, where required, are achieved by way of regulatory control, and the provisions of the regulations are administered by the New Zealand Marine Department. New Zealand's fishing limits were extended to 12 miles in 1965, which meant that from April 1, 1966, vessels not registered in New Zealand could not fish in this 12-mile limit. As Japanese vessels had been long-lining between 3 and 12 miles from the coast, in 1967 the New Zealand and Japanese Governments agreed to a phasing out of Japanese fishing within this zone by 1970.

To assist in the expansion of New Zealand's fishing industry, a Fishing Industry Board was established in April 1964 to promote the expansion and the efficiency of the fishing industry and to promote the export of quality fish and fish products.

Fishing Industry

The fishing industry in New Zealand is based mainly on demersal species and is confined to the continental shelf, which in many places is comparatively narrow. The most common class of vessel used in the New Zealand fishing industry is the motor trawler, from 40–60 ft in length, ranging many hundreds of miles in the course of their fishing. In 1966 the Marine Department commenced purse seining operations using a 92 ft stern ramp trawler, and late 1966

TABLE 1.

	Quantity	Value
Wet fish	666,274 cwt	2,194,058
Whitebait—West Coast, South Island only	941 cwt	68,544
Oysters		
Dredged—Bluff	160,382 sacks	561,337
Nelson	526 sacks	1,641
Rock	2,592 sacks	13,575
Mussels	31,550 sacks	24,260
Other shellfish		
Cockles, pipis, paua, etc.	1,053 sacks	2,882
Scallops, squid, sea eggs, etc.	5,762 cwt	17,944
Crawfish	128,981 cwt	1,922,154
Fish livers and fish meal	218,972 lb	7,521
Roes	85 cwt	465
	Total value	£4,814,381

and early 1967 saw the advent of large ocean going trawlers, which, in addition to undertaking deep sea trawling, are also equipped with fishmeal plants. The introduction of these vessels was a radical departure from the traditional fishing vessels, and their impact is being felt on both the domestic and export markets. Technical advances, such as the radiotelephone, refrigeration, more powerful or more reliable engines, and electronic fishing aids, are responsible for the increased range of fishing activities in recent years. In some ports the vessels still operate daily but usually where trawling is concerned and in all vessels fitted with freezers, the trips extend to a week at a time, while some refrigerated crawfish vessels in the southern part of New Zealand, make trips of approximately 1–2 months duration.

In 1966 there were 1912 licensed fishing vessels and 3288 fishermen employed in the industry. Of these vessels, 339 were motor trawlers, 14 were Danish seiners and 1410 were line and net boats; 29 boats were licensed to dredge oysters, and 865 vessels were licensed for crawfishing. Apart from the Marine Department's purse seining vessel, the remaining licensed fishing vessels were engaged in scallop and mussel dredging and other industries including pauas. Vessels normally are licensed for more than one method of fishing. Trawlers landed 80% and Danish seiners 4% of the total wet fish, while 9% was caught by lines and 6% by set and drag nets.

Principal Commercial Species

Since 1945 the total landed catch has doubled, 666,000 cwt being landed in 1966. The principal species have been snapper and tarakihi, which

Fig. 3. The New Zealand fishing fleet is growing in strength and acquiring modern vessels, including stern ramp trawlers built locally. In 1966, the number of boats increased by over 10%. The number of men employed in fishing has almost doubled in recent years.

Fig. 4. A mixed catch of kahawai, mackerel, and tuna being brought alongside the *W. J. Scott* ready for brailing. These fish, along with trevally, abound in New Zealand waters.

together comprise approximately 44% of the total quantity landed. The approximate percentages of the most abundant species are as follows:

Snapper – 28%	Tarakihi – 16%
Flounder – 10%	Hapuku – 8%
Sole – 7%	Gurnard – 6%
Elephant Fish – 6%	Blue Cod – 4%

Apart from blue cod and hapuku, which are mainly line-caught, the other species are taken principally by trawling. Snapper occur mainly from Tasman Bay northward on the west coast to North Cape, and southward on the east coast from North Cape to Napier, in the water out to about 40 fathoms. Tarakihi occur principally in the waters beyond 40 fathoms and are taken mainly off the east coast from the Bay of Plenty to Dunedin. The sea is shallow on the west coast, where fishing takes place, and this could be the reason why catches of tarakihi have not been so great. Gurnard, too, has a general distribution in the shallower waters, but show particularly in the commercial catch at Napier and Timaru. Elephant fish, which occur principally in the South Island east coast waters, form the basis of an important trawl fishery in the Canterbury area, both to the north and to the south of Banks Peninsula. Blue cod and hapuku, which are mainly associated with rocky bottoms, are caught principally by line methods. Cook Strait is the principal center of the hapuku fisheries; the Bluff-Stewart Island area and the Chatham Islands are the principal line fishing areas for blue cod.

TABLE 2.

Ports	Total Landings (cwt)	Hundredweights of Principal Species[a]
Auckland	132,850	Sn 92,287 Tv 18,229 G 8,048 Ta 6,798
Gisborne	75,467	Ta 41,588 Sn 7,816 G 5,868
Timaru	54,073	Ta 13,929 G 11,081 EF 10,579 Rc 5,065
Manukau	51,506	Sn 23,101 Tv 12,062 G 10,766
Napier	46,810	Ta 10,692 G 9,175 Sn 6,315
Nelson	42,835	Sn 21,282 G 5,421
Wellington	41,789	Ta 17,117 H 5,692
Tauranga	40,287	Sn 15,873 Ta 9,948 Tv 9,249
Lyttelton	31,576	Ta 12,431 G 6,265 EF 5,187
Port Chalmers	14,840	So 7,108
Thames	14,805	Sn 7,665
Bluff and Stewart Island	11,542	Bc 5,627
Akaroa	10,824	G 2,495 EF 2,326

[a] Sn = Snapper; Tv = Trevally; Ta = Tarakihi; H = Hapuku; EF = Elephant Fish; G = Turnard; Bc = Blue Cod; So = Sole; Rc = Red Cod.

Landings. Table 2 shows ports where landings were in excess of 10,000 cwt of wet fish and the principal species landed, i.e., in excess of 5000 cwt (1966 production figures).

Crawfish. This is the most valuable single species taken in New Zealand at present, the 1966 catch having been valued at $3,800,000. As New Zealand's major source of fishery export revenue, export receipts for a single recent year totalled $4,570,000. Where conditions are suitable, crawfish occur right around the New Zealand coast; recently, the industry received fresh impetus from the exploitation of prolific crawfish beds situated at the Chatham Islands, some 400 miles off the east coast of the South Island. With the exploitation of these beds the total catch of crawfish increased from 98,000 cwt in 1965 to 129,000 cwt in 1966, a record year. To insure a high quality export product in this rapidly expanding fishery, quality controls are currently being introduced; these will require all export crawfish to undergo examination by government inspectors.

Shellfish. In Foveaux Strait the dredge oysters, which are the most abundant and most important shellfish species, are taken mainly from depths of 10–18 fathoms. Production in 1966 was a record with a total of 160,000 sacks being produced. A season limit catch is determined annually from pre and postseason dredging and diving surveys. Work is currently in progress on the biology of the dredge oyster, and may provide information that could lead to cultivation as a means of expanding production. The rock oyster industry received impetus in 1964 when an Act was passed permitting rock oyster cultivation and farming. This infant industry is already showing signs of developing into a valuable fishery. Rock oysters, in their natural state, are picked and sold under Marine Department supervision, last year's harvest realizing nearly 2600 sacks. Toheroa are found on west coast beaches and are prone to seasonal fluctuations. The Marine Department is at present examining proposals for long term research into this shellfish.

Byproducts and Fish Processing

With the recent interest in the processing of pelagic species, the canning industry has made considerable use of the less popular kinds of fish, and this aspect of the industry is now growing in importance. The production of fish meal as a livestock food is now established in both ship- and land-based plants, and it is confidently predicted that fish meal production for animal consumption will feature in the future expansion of the industry. Fish sausage was developed under Government grants and is being marketed successfully both in New Zealand and overseas.

Frozen and prepared fish and frozen crawfish tails comprise the most import exports of fishery products in New Zeland. In 1966 the total value of fishery products exported was nearly $6,000,000, while crawfish receipts accounted for 75% of this total. The establishment of industrial plant purely for export of wet fish, regular supplies, attractive packaging, and the recent introduction of Government quality standards and control is resulting in flourishing export markets being established. Imports of fish products into New Zealand comprise almost entirely canned and prepared fish, mainly salmon, herrings, pilchards, and sardines.

With emphasis shifting to the small pelagic species and development work by the Marine Department's technological vessel to perfect and demonstrate techniques in purse seine fishing and

Fig. 5. Vacationers with a 16 lb crawfish. Demersal fishing and crawfishing have long been the mainstays of the New Zealand fishing industry, although the main developmental emphasis is now on the pelagic species. The crawfish catch increased to 128,981 cwt in 1966 from 97,933 in 1965. The New Zealand "packhorse" crawfish is the largest in the world.

Fig. 6. The filleting line at the factory of New Zealand Sea Products Export Ltd. at Nelson. All filleters wear white protective clothing and waterproof aprons and boots. Each filleting station is supplied with water from overhead tubes.

other modern techniques, the industry has realized that the field of greatest expansion lies in pelagic fishing. Private companies have already commenced purse seine fishing, and it is confidently expected that once adequate shore establishments for processing catches are provided, this aspect of the industry will dramatically expand.

Prompted by the intrusion of foreign fishing vessels into New Zealand's traditional fishing grounds, and consequent upon the lead given by the Select Committee report in 1962, the fishing industry is currently going through a period of rapid expansion. A serious handicap within the industry has been the shortage of capital, but with money being made available through the Development Finance Corporation and by public subscription, the value of New Zealand's fishing fleet and investment within the industry doubled between 1964–1967.

Research

A program is being followed that will provide the necessary information for development and management of certain aspects of the fishing industry. Within the framework of basic research, work is being done on snapper and tarakihi to determine their habits, life history, growth rate, and natural mortality, as well as to assess the effect of fishing on various populations. Similar work has been done on flatfish populations in Lake Ellesmere, Tasman Bay, and on the east coast of the South Island. A comprehensive study of crawfish, to elucidate their habits, life history, growth rates, mortality, and effects of fishing is underway, while detailed studies into the life history of the Bluff oyster is also undergoing study.

Management investigations include a study of the elephant fish population to determine the effect of fishing pressure on the population and potential of the species. At present, an extensive biological survey of the Foveaux Strait oysters is underway. With respect to whales, work has been confined to tagging whales on the northward and southward migration to study the age composition of the catches of the whaling stations although this research has now ceased. Population studies of toheroa and mussels is continuing, as well as short term biological programs on various fish species. The cultivation of rock oysters is being thoroughly investigated and experimental farms started with the view of establishing a major industry. The occurrence of tuna in New Zealand waters is also receiving constant attention aimed at determining whether the fishing and canning of tuna can be developed in New Zealand.

While the traditional methods of fishing will continue to play their part, purse seining, ocean going trawlers, factory ships, and larger fishing units, are expected to revolutionize the industry. That fish stocks to support this expanding industry frequent New Zealand's waters is undoubted. The industry is now indicating that it has the potential to realize these resources and recast their mould as a major contributor to New Zealand's economy.

J. G. WATKINSON

NORWEGIAN FISHERIES

Hammerfest, Norway, the world's northernmost town, on the island of Kvaloy, depends upon fishing for its economic way of life. Hammerfest, 300 miles above the Arctic Circle, is the cod liver oil capital of the world. In recent years more than 30,000 tons of cod, halibut, and haddock have been caught annually, yielding this town of 7000 people one of the highest per capita incomes in Norway. Hammerfest is the home port of over 300 registered motor fishing vessels and contains the most modern fish processing plants in the country.

Hammerfest's activities are representative of Norway's fishing industry, which, in 1967, ranked fifth in world fish production, representing a total of over 3 million metric tons of fish landed.

Landings by Norwegian fishermen averaged about 1.3 million tons annually between 1960 and 1964. These landings made Norway the foremost fishing nation in Europe (excluding the U.S.S.R.), a lead maintained today. Three features characterize the Norwegian fisheries: domestic consumption of fishery products is among the highest in the world, estimated to be about 45 kg (nearly 100 lb) per person per year on a live-weight basis; the Norwegian population is comparatively small—about 3,750,000 in 1967 —and domestic consumption absorbs only 12–15% of total landings; the remainder of the catch is prepared or processed for export as edible fishery products, and many species, principally herring, are reduced to meal and oil, which are also exported.

The substantial increase in 1967 was accounted for by the North Sea mackerel catches, which

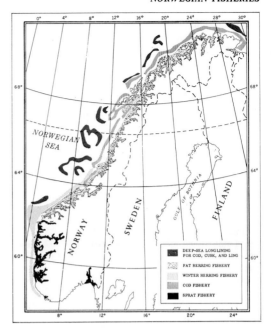

Fig. 1. Principal fishing grounds along the Norwegian coast.

increased from 455,000 metric tons in 1966 to 867,000 metric tons in 1967. This was largely attributed to the maturation of the purse-seine fleet and the widespread use of power blocks affixed to the relatively small (15 m) seiners. Power block equipped purse seiners fishing for mackerel, herring, and some cod, reached a total of slightly over 500 vessels in 1967. The resultant increase in fishing efficiency, requiring less manpower, was reflected in the record catches.

TABLE 1. NORWAY'S LANDINGS, 1966–67; USE OF LANDINGS, 1967

Species	Landings 1967	Landings 1966	Fresh	Frozen	Dried	Salted	Canned	Reduction	Bait
				(1,000 Metric Tons)					
Capelin	402.8	379.6	—	—	—	—	—	402.8	—
Herring:									
Winter	371.6	460.9	17.1	32.3	—	20.2	8.1	292.4	1.5
Fat	346.0	148.1	1.2	—	—	2.2	0.2	337.6	4.7
Small	106.4	78.5	0.4	—	—	0.1	10.9	94.8	0.2
Fjord	1.2	1.3	1.0	—	—	0.2	—	—	—
North Sea	335.8	454.9	5.0	2.0	—	0.9	0.3	327.5	—
Icelandic	52.1	42.2	—	0.3	—	7.7	—	44.1	—
Total[a]	1,213.1	1,185.9	24.7	34.5	—	31.3	19.5	1,096.4	6.4
Mackerel	866.6	484.0	5.2	12.0	—	3.0	1.8	841.3	3.2
Cod	196.9	197.0	18.5	49.2	71.8	54.0	2.4	0.8	—
Saithe	119.8	142.6	7.6	44.7	31.4	33.7	0.8	1.5	—
Haddock	40.0	62.5	7.8	27.1	2.9	—	1.0	1.1	—
Other	172.5	204.1	27.5	35.8	9.2	24.4	16.0	58.7	1.7
Total[a]	3,011.7	2,655.7	91.3	203.3	115.3	146.4	41.5	2,402.6	11.3

[a] Totals may not add due to rounding.
Source: "Fiskets Gang," published by the Norwegian Fishery Directorate, March 7, 1968, No. 10.

Foreign Trade

In 1967, income from exports of fish products rose 13% to $244,000,000. This was 14% of Norway's exports, slightly higher than 1966 (Table 2).

Table 2. Exports of Selected Fishery Products

	1967	1966
	Metric Tons	
Frozen Fillets:		
Haddock	10,966	14,602
Cod	25,583	26,056
Coalfish	19,565	17,828
Herring	6,689	8,435
Other	6,289	5,875
Total frozen fillets	69,101	72,796
Frozen herring	13,167	16,691
Canned fishery products:		
Brisling	5,963	7,539
Small sild sardines	13,463	12,637
Kippers	3,348	3,386
Shellfish	523	787
Other	4,133	4,539
Total canned fish	27,430	28,888
Fish meal	494,785	257,289
Herring oil, crude	165,721	80,841

Source: "Fiskets Gang," Jan. 20, 1968, and Jan. 26, 1967.

As in 1966, fish meal was the primary fish export in 1967 in volume and value. Such exports nearly doubled to a record 495,000 tons. Export income for fish meal was $75,000,000, up 55%. This implies an average price reduction from $184 in 1966 to $150 per ton in 1967. Exports of fish oil rose 105% in volume and 45% in value—165,700 tons and $20,000,000.

In 1967, canned fish exports increased 5.4% in volume to 38,000 tons.

Exports to U.S.

Exports of fish products to the U.S. rose over 50% to a record $32,000,000 in 1967. Fish meal accounted for it: exports rose from 22,700 tons in 1966 to 100,800 tons in 1967. Shipments of canned fish products, the principal fish product in value, remained at 1966's $10,000,000. In frozen fish fillet exports to the U.S., fierce competition, and sharply lower prices, reduced volume 21%, to 7700 tons, and value 25% to $3.9 million.

Norwegian Imports

In 1967, imports of fish and fish products were 23,200 tons and $9,000,000, compared to 41,200 tons and $11.6 million in 1966. The most important fish products imported were salted cod for klipfish industry, and salted herring and canned fish delicacies for domestic consumption.

Fig. 2. This 15 m long Norwegian fishing boat can operate as a purse seiner through the use of her hydraulic power block. Note the simple method of stowing the net.

As in 1965–1966, imports of U.S. fish products were negligible.

Aid to Fishermen

In 1967, the average price received by fishermen per ton of winter herring fell 20%, paid by the fish reduction industry. The smaller portion of the 1967 catch of winter herring was marketed fresh or frozen and so was eligible for price support.

Fishing Craft

Because most of the grounds fished by the Norwegians are in coastal and inshore waters close to numerous fishing ports, a large part of the fishing fleet consists of small and medium craft. During the last decade, however, the number of larger craft has increased as purse seining, longlining, and trawling have been extended into distant offshore waters.

In 1962 the Norwegian fishing fleet of 39,705

registered craft consisted mainly of 28,429 small open motorboats and 8799 decked, wooden motor craft less than 50 ft. Fourteen decked steel craft made up the rest of the motor-powered fleet under 50 ft. The remaining motor craft (2440 of 50 ft and over) were mostly in the 50–80 ft class. There were 23 steam craft.

Open undecked motorboats and small decked craft are used a great deal by fishermen who do not practice fishing full time. These small boats, usually operated near home ports, may also travel great distances along the coast to participate in the different seasonal fisheries, such as those for winter herring or Lofoten cod. The smaller craft are not as highly specialized as the larger vessels, and may be used for purse seining, gill netting, longlining, or trawling with slight changes in deck gear and winches.

A growing fleet of steel vessels over 80 ft in length includes large purse seiners, high-seas longliners, and trawlers. Since 1960 the trend toward building vessels of larger size has intensified. In 1965 Norway had 473 trawlers, of which 34 were over 300 gross tons. Some of the newer vessels, over 150 ft long, are stern trawlers equipped with a ramp for hauling nets on board to be unloaded. Since 1961 about 20 stern trawlers have been added to the Norwegian fishing fleet; three of these vessels were registered at 1000 gross tons each, three at 800 tons, two at 600 tons, two at 400 tons, and six between 200 and 300 tons.

Fishermen

Because of the abundance of fish in nearby waters, people early in Norway's history settled along the coast. The first settlements were made largely in the fjord districts, but soon people moved to the outermost islands within easy reach of the coastal fishing banks. Today, although the fishing population is scattered along the entire coast, the principal fishing centers are in western and northern Norway. Furthermore, almost half the Norwegian fishermen are located in the northern third of the country.

The fishing population has been declining steadily. According to the most recent census, the number of fishermen decreased from about 86,000 in 1948 to 56,890 in 1962. The latter

Fig. 3. Hydraulic power block. The whole of the net is drawn through the block until the bag is sufficiently dried up to enable it to be hauled aboard with a gilson. The three men simply stow the net ready for shooting again.

number was less than 2% of the total population, and included 21,475 who fished as their sole occupation, 19,756 as their main occupation, and 15,659 as their secondary occupation. The large number of part time fishermen in relation to full time fishermen is characteristic of highly seasonal coastal fisheries. Many part time fishermen have small farms or work in fish processing or other industries.

Fishermen's Organizations

Most fishermen are organized into a trade union, known as the Norwegian Fishermen's Association (Norgas Fiskarlag). Members are drawn mainly from fishermen engaged in coastal and nearby offshore bank fisheries. A distinct feature of the Norwegian fisheries is the fish marketing organizations, which represent fishermen in their dealings with buyers. The fish marketing organizations work in close association with the Fishermen's Association. Fishermen on vessels operating in distant waters generally are members of the Seamen's Trade Union.

Marketing

Marketing in the Norwegian fishing industry is conducted under laws that regulate and control firsthand sales of fish. The basic legislation authorizing this is the Law on the Marketing of Raw Fish (Rafiskloven) of 1951, which stipulates that fishermen's marketing organizations have exclusive rights to sell almost all fish landed in Norway. The basic reasoning behind this legislation is that fish are landed at a great many ports in relatively small quantities, and fishermen are at a disadvantage in dealing with buyers, who had previously exercised virtually complete control over prices.

Fishermen have now established 15 marketing organizations with firsthand marketing rights either for a specified area or for certain species of fish. For example, Norges Rafisklag, the most important in terms of membership and fish handled, markets practically all fish (except herring) landed at ports from Nord-More to Finnmark, which is an area corresponding to the northern half of Norway. In this area, the organization sells almost all cod and similar species used to prepare dried fish, $\frac{2}{3}$ of the fish that is frozen, and most of the fish that is salted and dried. Norges Sildesalslag, the second most important sales organization, sells the entire catch of winter herring regardless of where the fish are caught and landed.

The fishermen's marketing organizations stipulate ex-vessel prices through negotiations with groups of fish buyers. Buyers must obtain a license from the marketing organization, and are required to grant guarantees for payment. The prices established and the marketing policies of each organization vary widely. Norges Rafisklag confines its operations to setting fixed or minimum prices for the different species it handles and ensuring compliance with the rules of the organization. On the other hand, Norges Sildesalslag, which handles large amounts of winter herring within a short time, has a more elaborate price and marketing system. Prices are determined beforehand, depending on the final use of the herring, and the catch is distributed according to a quota system.

Whaling

Norway has three types of whaling operations. Pelagic factoryship operations for baleen and sperm whales in the Antarctic is the most important. Coastal whaling from Norwegian shore stations is also conducted for baleen and sperm whales. These two types of operations are under regulations adopted by the International Whaling Commission. The third type, a small whale fishery off the Norwegian coast, is regulated by the Norwegian Government.

Production of oil from Antarctic factory ship operations has been declining. At one time Norway was the leading producer, but this position has been taken over by Japan. During the Antarctic whaling season of 1954–1955, Norway operated nine factoryships and 101 catcher boats in the Antarctic. That fleet produced about 62,400 tons of whale (baleen) and sperm oil, or nearly 40% of total Antarctic production in 1954–1955. In the 1964–1965 season, Norway operated four factoryships and 32 catcher boats; production was 17,926 tons of whale and sperm oil, or 20% of a lower Antarctic total.

Higher operating costs and competition from the Japanese (who, in addition to oil, utilize whale meat and other whale products more fully than the Norwegians) have been given as reasons for Norway's curtailment of Antarctic operations. Also, the Norwegians have found it more difficult, as have other Antarctic whaling countries, to fulfill their quotas for baleen whale oil because of serious depletion of the Antarctic whale stocks. Beginning with the 1962–1963 season, an international quota system for taking baleen whales was adopted by the International Whaling Commission, and Norway was given a 28% share of the Antarctic quota of 15,000 blue-whale units. For the 1963–1964 season, the baleen catch limit was reduced by the Commission to 10,000 blue-whale units. Norway was unable to reach its quota of 2800 units, producing only 1485 blue-whale units. For the 1964–1965 season, the limit was further reduced to 8000 blue-whale units by the countries participating in Antarctic whaling. Again Norway was unable to reach its quota, producing only 1273 of the 2240 blue-whale units allotted to it.

Coastal whaling from Norwegian shore processing stations is conducted on a small scale. In 1962 the catch of eight catcher boats, operating from three shore stations was 91 sperm whales and 149 baleen whales. These yielded 690 tons of sperm oil, 800 tons each of whale oil, whale meat, and animal feed, and 500 tons of whale meal. The International Whaling Convention prohibits the taking of blue and humpback whales along the Norwegian coast, but other baleen whales can be taken between May and October. The Convention permits the taking of sperm whales during any consecutive eight months, and the Norwegian Government has set the season for March to October.

The small whale fishery is conducted along the Norwegian coast and in waters as far distant as those off western Spitsbergen, Jan Mayen Island, East Greenland, and Iceland. With several exceptions, the season established by the Norwegian Government extends from Match 15 to September 14, with a three week closed period in July. The season ends July 1 north of lat. 70°N. and east of long. 0°, including the area off the northeastern coast of Norway south of lat. 70°N. Between 3700 and 4800 small whales were taken in recent years. In 1964, however, the number declined to 3170; these whales yielded about 4500 tons of meat, 2000 tons of blubber, and 28 tons of bone, valued at 9.8 million kroner (nearly $1.4 million). The principal species taken was the minke (or little piked) whale; other species were the bottlenose, killer whale, and pilot whale. The 165 vessels engaged in small whaling in 1964 were mostly typical fishing boats, 40–80 ft in length, with a small harpoon gun mounted on the bow.

Sealing

Norway has the largest sealing operations in the North Atlantic. The take of seals during the past decade has ranged between 200,000 and 300,000 pelts annually. In 1962, 58 vessels and 1116 men took 238,830 pelts and 3750 tons of blubber, for a total value of about 16.7 million kroner ($2.3 million). The vessels were mostly large, up to 180 ft long and averaging about 200 gross tons. The principal sealing grounds have been off Newfoundland and in the Jan Mayen or "West Ice" area of the Norwegian Sea; a few seals have been taken north and northeast of Norway.

Four species of seals—members of the family Phocidae—are taken by the Norwegians. The Greenland seal is by far the most important, providing 60–80% of the total pelts taken. Most of the reaming pelts are those of the hooded seal; only small numbers of bearded and ringed seals are taken. Although these seals are not fur seals, Greenland seals retain their baby fur for a few weeks and young hooded seals for about a year. The hides of the young are classed as furskins and used for clothing and trimming. Hides of mature seals are used for leather.

F. E. FIRTH

NUTRITIVE VALUE OF SEAFOODS

Human Nutrition

Archaeological digs at the settlements of early man often reveal evidence of his consumption of fish—both finfish and shellfish—but especially shellfish, owing to the durability of their shells. Piles of mussel, clam, and oyster shells, for example, frequently are found near riverside and seaside campsites of early American Indians. Table 1 indicates the wide acceptance of fish as a protein food by modern man.

Early use of marine animals for food was perhaps dictated by both the necessity for eating any protein food available and early man's appreciation of seafoods. Today, with the increasing studies of human nutrition and its relation to several cardiovascular diseases, seafoods are coming to the front as a highly desirable and tasty source of protein. Fresh oysters, shrimp, and certain fish that were once restricted to a few gourmets are now in general distribution all over the world.

With the public's demand for more knowledge of the nutritive value of its foods, government, university, and private research laboratories have made extensive studies of the nutritive values of many species of fish. These studies have revealed fish as being an ideal protein food.

Calories. An important point for those on a low-calorie diet to know is that proteins contain only about 4 calories per gram, whereas fats and oils contain about 9 calories per gram, i.e., about twice as much as proteins contain. For this reason, the relative amounts of fat or oil and of protein in foods is significant to the dieter.

Stansby (1962) has pointed out that, although meats from farm animals are considered to be relatively lean, they contain from 5 to 15% oil.

TABLE 1. ANNUAL PER CAPITA CONSUMPTION OF FISHERY PRODUCTS IN SEVERAL COUNTRIES

Country	Year	Pounds in Edible Weight
Japan	1963	61
Norway	1964	45
Portugal	1963	44
Denmark	1963	35
Taiwan	1963	30
Finland	1963	20
Greece	1962	20
United States	1964	11

Almost no meats are so lean as to be placed in the low-oil content category of less than 5% in which most species of fish are found.

Though most species of fish have a low oil content—less than 5%—some species do contain greater amounts. Table 2 shows how the oil content varies among certain species, and supplies data on protein and calorie contents as well. The values given in Table 2 are for the raw or canned fish. Fish cooked in a rich sauce, e.g., one containing butter or margarine, will, of course, have a higher calorie content.

The ideal methods of preparation for high-protein, low-calorie seafoods are by broiling, baking, or steaming. Generally speaking, in free-choice feeding, people do not consume excess amounts of calories in broiled or baked seafoods. Seafoods have the quality of satisfying hunger, since they supply needed bulk and essential amino acids. Fish therefore should be an important part of any weight-reduction diet.

It is of interest to note that Eskimos and certain northern Indians stay healthy and vigorous on a ration composed mainly of fish. We could add much needed amino acids to our diet, and at the same time cut down on those harmful excess calories, by eating more seafoods. The 1800 to 2500 calories per day suggested for sedentary persons is easily achieved by including considerable broiled, baked, or smoked fish in the ration.

Amino Acids. At present, eight amino acids are accepted by most nutrition researchers as essential to the continued health of the human being: tryptophan, phenylalanine, lysine, threonine, valine, methioine, leucine, and isoleucine. Other amino acids may be supplied in the diet periodically in very small amounts. The continued absence of any of the eight essential amino acids will result in malnutrition.

Seafoods contain favorable amounts of each of the eight essential amino acids, and are especially good sources for lysine, threonine, valine, leucine, and isoleucine. Seafoods are valuable in providing most of the other recognized 22 amino acids that go to make up animal protein foods. These 22 amino acids are cystine, histidine, arginine, taurine, hydroxyproline, aspartic acid, serine, glutamine, proline, glutamic acid, glycine, tyrosine, hydroxylysine, alanine, and the eight essential amino acids named above. It is supposed that animals can manufacture these 14 nonessential amino acids, since they are found in animal tissues. Seafoods in general are especially valuable in supplying a balanced amino acid portion of the diet.

Protein or amino acid availability or digestibility is now known to vary with certain protein foods. Connective tissues may tie up the muscle fibers so as to prevent the easy digestion of the protein or muscle. In general, seafood proteins and muscle are bound up much less with connective tissue and are, therefore, more easily digested. The connective tissues are easily broken down by the digestive enzymes, thus making the amino acids readily available for absorption into the blood stream. The availability of protein or its component amino acids may be determined by controlled animal-feeding tests. It is of interest to note that protein "scorched" by excess cooking is much less readily digested than protein lightly broiled is. The digestibility of fish protein is high. Researchers have found that 90 to 96% of fish protein is digested as compared with 87 to 90% of beef.

Fish Oils. The lipids in seafoods are usually referred to as oils rather than fats, since they are liquid at room temperature (68°F). Their being in the liquid state is due to their high content of polyunsaturated fats. In fact, fish oil contains more polyunsaturated oils than do natural liquid vegetable oils. However, polyunsaturated oils from fish are different in molecular structure from those in vegetable oils. The polyunsaturated phase in fats is the result of the presence of several double bonds in the chemical structure of the molecule.

These polyunsaturated oils from fish aid in the avoidance of arteriosclerosis. Several studies, including those sponsored by the American Heart Association, have shown that a diet high in fish results in a lower rate of coronary attacks among middle-aged males.

Another study, which was made by the New York City Department of Health, placed 814

TABLE 2. COMPOSITION AND CALORIE CONTENT OF FISH OF VARIOUS SPECIES (RAW EDIBLE FLESH UNLESS OTHERWISE STATED)

Fish species	Calories Per 100 g.	Protein, %	Oil, %
Sardine, Atlantic (canned with added oil)	338	21.1	27.0
Tuna (canned with added oil)	290	23.8	20.9
Salmon, king (canned)	203	19.7	13.2
Herring	191	18.3	12.5
Eel	162	18.6	9.1
Halibut	126	18.6	5.2
Swordfish	118	19.2	4.0
Mackerel	102	11.0	6.2
Haddock	79	18.2	0.1
Sole	67	13.0	1.3
Cod	60	14.6	0.6
Smelt	60	11.3	1.6
Whiting, Atlantic	43	11.3	0.1
Turbot	43	8.8	1.3

TABLE 3. OIL AND CHLORESTEROL CONTENT OF SHELLFISH

Name of Shellfish		Oil Content, %	Cholesterol Content, mg/100 g
Popular	Scientific		
Blue crab, eastern	Callinectes sapidus	1.1	98
Blue, crab, southern	Callinectes sapidus	1.2	76
Clam	Mercenaria mercenaria	2.0	82
Dungeness crab, body	Cancer magister	1.2	63
Dungeness crab, claw	Cancer magister	1.0	52
Oyster, Chesapeake Bay	Crassostea virginica	2.0	58
Oyster, Biloxi Bay	Crassostea virginica	2.4	37
Scallop, sea	Aquapectid grandus	1.6	60
Shrimp, brown	Penaeus aztecus	1.1	156
Shrimp, white	Penaeus setiferus	1.2	157

middle-aged males on a diet low in saturated and high in polyunsaturated fats. There were only one-third as many coronary attacks among the men following this so-called "Prudent Diet" as there were among the control group of 463 men who did not follow the diet. The Prudent Diet called for a minimum of four fish meals per week. Significant reductions in obesity and high blood pressure were noted among the study group.

Feeding test animals on diets high in hard fats results in a high level of blood-serum cholesterol. Excess blood-serum cholesterol is deposited in the walls of arteries, where it becomes calcified into plaques, thus occluding the lumen of the arteries. This condition, known as arteriosclerosis, causes high blood pressure and coronary attacks. Feeding of fish oils or other polyunsaturated oils has been reported to lower the blood-serum cholesterol greatly without a change in diet. Fish and shellfish are much lower in cholesterol and cholesterol-forming lipids (Table 3) than are red meats and poultry.

Minerals. Seafoods have been favored in the diet because of their high content of certain essential minerals. The ash content of fish flesh varies from 0.9 to 2.6% (Ref. 4). The greater part of the minerals in the ash consist of sodium, potassium, calcium, phosphorus, zinc, sulfur, and chlorine (Ref. 5). The lesser portion consists of iron, manganese, magnesium, iodine, copper, molybdenum, cobalt, selenium, fluorine, arsenic, and lead, plus others said to occur as trace minerals.

Among the trace minerals regarded as essential to the continued health and well being of the individual are iodine and fluorine.

Iodine is recognized as a component of the thyroid glandular secretion. A lack of iodine in the diet produces a compensatory hypertrophy of the thyroid gland known as goiter. Oysters are recognized as a good source of available iodine, as are other shellfish.

Fish are an excellent source of fluorine. One of the effects of this essential mineral is in the prevention of tooth decay by strengthening the enamel. Chemical analysis reveals that there may be up to 0.5 mg per 100 g of fluorine in salt water fish.

Many diets for individuals with a coronary condition prescribe a low sodium intake. Foods having more than 100 mg of sodium per 100 g of food are usually excluded from the diet. All fish, even those from marine environments, contain less than the limiting 100 mg. It is the custom in some areas, however, to wash fish fillets in a salt water brine during processing. Persons on a low sodium diet should be warned to read the label, since U.S. Food and Drug Administration regulations make it mandatory to so label brined fillets.

Several other trace minerals are regarded as essential in human nutrition. Their exact minimum levels, however, have not been agreed upon. The continued absence of copper, cobalt, manganese, molybdenum, selenium, and zinc from the diet results in certain clinical manifestations, especially in a failure of children to grow properly. Salt water fish contain most of the minerals found in sea water and will improve the diet of inland peoples living in areas where minerals have largely been leached from the soil.

Vitamins. Vitamins in the edible portion of seafoods are those of the water-soluble group—thiamine, riboflavin, niacin, pyrodoxine, biotin, pantothenic acid, folic acid and vitamin B_{12}. In general, fish livers are no longer an important source of the oil-soluble vitamins A and D, at least in the U.S., since they are now synthesized. In this country, shark liver, once a primary source of vitamin A, can no longer compete economically with synthetic vitamins A and D.

The small amount of assay work done on seafoods indicates that they are as good a source of the water-soluble vitamins as any other protein food. Researchers have only lately turned their attention to the bioassay of the vitamins in fish flesh. Our best indication of the presence of these essential components of the diet is that peoples obtaining the bulk of their protein from fish are healthy and energetic.

Total Nutritive Qualities. Seafoods, then, are an all purpose protein food. They contain the 22 amino acids commonly found in animal protein food and a good proportion of the eight regarded as essential to human health. Since seafoods contain little or no carbohydrates and little fat, they fit into a low-calorie diet. Seafoods are especially valuable in avoidance of arteriosclerosis, since the fat is polyunsaturated and low in cholesterol. Fish and shellfish are high in essential minerals, and can supplement a diet low in these components. They supply water-soluble vitamins in amounts at least equal to those in other meats. And, finally, the public will eat seafood readily when it is available in a fresh condition and properly cooked. The American Heart Association found that large test groups of middle-aged males from inland cities consumed up to ten seafood meals a week for six months during a test period.

For many years, menhaden, or "pogies" as they sometimes are called, have been harvested for their valuable oil. In the 19th century, the oil was used primarily for lamps and other fuel, but in the early 20th century, uses in paints and waterproofing compounds became paramount. Considerable fish oil is used for making oleomargarine throughout the world. Fish oil was used for this purpose in the U.S. until the early 1950's. Unfortunately, the manufacture of oleomargarine involves hydrogenation, which destroys most of the double bonds that make the oil polyunsaturated. This does not affect the flavor, odor, keeping quality, or edibility of fish oil margarines.

Animal Nutrition

During the early years of oil manufacture, the protein residue, or fish cake, was used for fertilizer. Fish meal is reported to have been used as animal feed in Europe as early as the 19th century. In the early 20th century, fish meal was tried as a protein supplement in poultry and swine rations in the U.S. The response of the animals was good. Today, fish meal is used all over the world as a protein supplement in animal feeds. The water from cooking menhaden, which contains soluble protein, is condensed into "solubles" having 50% solids, for additional use in animal feeding. Researchers have shown that there is an unknown growth factor in fish meal and solubles that causes faster growth in poultry than does vegetable protein alone.

Since the early 1950's, there has been a steady growth in the use of fish for canned pet food and for mink feeding. Millions of cans of cat food containing 65% whole fish are canned each year in the U.S. Fish oil and vitamin and mineral supplement make canned cat food one of the most nutritious animal foods available. This huge industry uses a so-called mixed trash fish that, at present, is less economically suitable for any other use. However, since the perfection of a method for preparation of FPC (fish protein concentrate) by U.S. Bureau of Commercial Fisheries researchers at College Park, Maryland, these small fish may be diverted, in the future, to human food. Trash fish and offal from fish processing plants are used in the preparation of pelletized fish foods for the feed of trout and catfish in fish farming. Most marine animals may be converted to FPC to feed the world as a population crisis grows more acute.

TRAVIS D. LOVE

References

1. U.S. Department of the Interior, "Fisheries of the United States—1965," C.F.S. No. 4100, 1965, U.S. Government Printing Office, Washington, D.C.
2. Stansby, M. E., "Proximate composition of fish," in "Fish in Nutrition," Heen and Kreuzer, editors, 1962, Fishing News (Books), Ltd., London.
3. Borgstrom, G., "Fish as Food," Vol. II, 1962, Academic Press, New York.
4. Thompson, M. H., "Cholesterol content of various species of shellfish. 1. Method of analysis and preliminary survey of variables," U.S. Fish Wildl. Serv., *Fish. Ind. Res.*, **2**, 11–15 (1964).
5. Thompson, M. H., "Proximate composition of Gulf of Mexico industrial fish," U.S. Fish Wildl. Serv., *Fish. Ind. Res.* **3**, 29–67 (1966).

Cross references: *Fish Meal; Fish Protein Concentrate; Minerals of the Ocean.*

OCEAN

Potential Resources

For nearly 100 years research ships from all the maritime nations have been at sea making regional surveys and collecting facts and figures with instruments that have become increasingly sophisticated. Thus, from the scientific point of view oceanography is now in quite good shape. The chief trouble at present is the legal, social, and economic problems that nobody knows quite how to solve, especially at the international level. There is also a lack of good engineering experience.

The potential resources of the sea are vast. After all, it is a huge chemical sump that contains nearly all the materials that have washed off the land since the beginning of time. Most of these renew themselves naturally and faster than man could hope to extract them. However, traditionally there is no clear title to deep-sea resources. The old idea of the freedom of the sea, which developed before there was any comprehension of the wealth of the oceans, is diametrically opposed to their efficient exploitation.

Lately much has been written about the potential mineral and oil or natural gas resources of the ocean. In relatively shallow areas a considerable number of oil wells have already been drilled, but few of these are profitable yet. Petroleum and associated products are still available on land in considerable quantities, and the cost of production there is appreciably less. Except in a few special circumstances, several hundred years may pass before marine sources become economically competitive. Their full extent has not yet been adequately explored. The drilling technology is such that the limit of test wells is at about 2000 m. Lesser depths than this cover only a small fraction of the total bottom of the ocean. There is also limited activity in offshore mining of sulfur off the coast of Louisiana.

At present, the land is the source of nearly all the fresh water used by man. However, the demand is increasing very rapidly as industry expands and agriculture is being carried on at an ever-increasing scale. Since the sea is the source of nearly all our fresh water and is in fact the reservoir on which everyone has to draw, the question arises as to how the near-surface water could be modified to increase evaporation, and as to how this moisture might be distributed more evenly and in perhaps greater quantities on the land.

Some useful measure of climate control and weather modification is no longer beyond our technical capabilities. Leaving aside for the moment the difficult international agreements that would have to be resolved, how would one go about promoting a more effective rainfall?

A most important factor as to whether an active cloud produces rainfall is the number of so-called giant salt nuclei present within the cloud and in the surrounding clear air. These are in fact very small particles of sea salt that have been picked up by the atmosphere as it passes over the ocean. They originate as small bubbles of air that have been entrapped by white caps and then break through the sea surface. As the air rises within a cloud, these hygroscopic particles form slightly larger droplets of water vapor than do other kinds of condensation nuclei. Thus they are heavier and fall relative to the neighboring droplets and consequently grow by collision until they are large enough to reach the land surface without evaporating. Measurable amounts of sea salt particles are detectable in clouds over Chicago, for example. Thus, one possible method is to help white caps at sea to produce more salt particles. This could be done by artificially aerating the near-surface water or possibly by lowering sea surface tension over considerable areas. Another possible method is to promote convection over the land, as by black-topping areas of hopelessly unproductive soil.

In the case of weather modification and control, there are a number of important near-shore areas where one can hope to tap potential energy already available in large quantities within the ocean. The Florida Straits is a good example. Here exists about 2000 times the energy of the Mississippi River in flood stage, all nicely concentrated to become a huge air conditioning plant for the coastal cities of Florida in summer. Who is going to do such things? Certainly not science. The U.S. Government is much more concerned with inland problems involving conservation and pollution of both soil and atmosphere.

Experiments should start on a small scale and should be designed to be "self-healing", that is, able to be quickly discontinued. This is especially true of experiments that could indicate ways of breaking up potentially destructive storms, such as hurricanes, before they become organized.

Science has the knowledge and, on the whole, adequate instrumentation; what is needed is the authority to go ahead. Can this authority be a single government or should it be an international body? In neither case is adequate technical leadership available.

We come next to useful predictions both of "weather" within the ocean and weather over the land. Until recently, the calculations involved in transforming theory to practice seemed hopeless, but now electronic computers of great speed and capacity are available. If only one of these is employed, nine-tenths of the problem is solved.

The computer will need up-to-date input data, especially from oceanic areas. Thus, an internationally maintained data-gathering network will have to be organized, just as was done many years ago for ordinary short term meteorological forecasts. This will probably be a multiple sensing system. Anchored or drifting radio telemetering buoys can report stored information, either directly to land stations or to monitoring aircraft. Scientific satellites can also be employed in a number of ways. Only one computer will be needed to provide for world-wide service.

Such a data-gathering network could be organized rather quickly, certainly in fewer than ten years. Nearly everyone could make good use of both accurate short term weather predictions and of, perhaps at first less reliable, long range forecasts. It is interesting to note that so vital is the weather in man's activities that each year considerably more is being spent on telephone calls to local weather stations than is being appropriated by the U.S. Congress for maintaining the Weather Bureau, let alone for improving it.

Finally, there are the potential biological resources. Although during the past ten years or so the world fisheries have been increasing rapidly and now amount to about 50,000,000 tons per year, this is still far less than that of the land. It is not that fish are particularly difficult to harvest. They move around and vary greatly in quantity because of ecological factors. They are also a crop that spoils quickly and is expensive to preserve through salting, canning, or freezing, for example. The lack of ownership of the crop and the lack of skilled management to the fishermen have kept them, with few exceptions, relatively poor men. Properly designed fishing vessels are beyond their means for the most part.

Traditionally, fisheries experts have first been trained as conservationists. Their preoccupation is to try to understand the goings and comings of fish, rather than to develop the marine equivalent of agriculture.

A large percentage of the fish now being landed is being processed crudely as a supplement to cattle and chickenfeed. Fishermen have become highly skillful, especially in the more advanced nations, in going at their problem in reverse. They carefully seek out the "flowers" and leave the "weeds." Thus, in effect they are doing the reverse of weeding the garden. The so-called trash fish, which are too bony or too perishable for human consumption, eat just about the same foods as the commercially desirable species. Naturally the fishermen concentrate on the more valuable species.

Recently there has been much loose talk and uninformed writing about the development of marine aquaculture. Before this happens, a host of problems will have to be solved. Inevitably, marine aquaculture will start in shallow waters near the land. Here, along all well populated coastlines, there are active and opposed competing interests that demand political protection. Those who want to use the inshore waters as an inexpensive sewer have so far been the most successful, because taxes are thereby much reduced. Others want to use coastal waters for recreation: swimming, boating, and sports fishing. Conservationists feel that coastal waters and estuaries should be reserved as nursery grounds for the young of potentially valuable future fish. Commercial fishermen would like to harvest whatever they can still yield. It is interesting to note that when New York City was first settled it looked out on the largest and most productive shellfish beds of our entire coastline. Shellfish cannot only be overfished, for they like company, but they can also become polluted or killed by lack of oxygen. They are a crop that needs some elementary management.

So far fishermen need the equivalent of cheap fencing. Where would farmers be without barbed wire? A number of methods have been suggested, but it seems to be nobody's job either to do the engineering or to carry out convincing practical trials.

Inshore waters are also the victims of many artificial regulations and local statutes that have been devised over the years by people having no contact with science. In general, each town claims jurisdiction in the waters inside a straight line connecting the local headlands. To say the least this makes for great confusion along a highly irregular coastline. Outside this narrow but sometimes quite extensive coastal zone is a strip of water claimed by the state. Next comes an equally narrow strip in which the Federal Government claims to exert jurisdiction. It is a wonder that anyone dares to set out a lobster pot. The fact that commercial fishermen can operate in coastal areas almost at will is an indication of how poorly they are policed. In only very few areas is it necessary for them to fish at night.

So far, of course, we have been discussing only the special problems of the waters within a few miles of the coast. Although these are potentially considerably more productive than oceanic waters, they constitute only a small percentage of the total area of the earth covered by salt water where almost no restrictions are generally agreed to.

As on land, there are vast desert areas, and the crop that does exist is much more three dimensional in nature. In other words, the oceanic plants and animals are diluted by a great deal of seawater. Plants have insufficient sunlight to grow and reproduce below a depth of about 100 m. Vast reserves of nutrient chemicals are stored over long periods below some such level. In other words, to make the open ocean reproduce more vigorously, we should assist it in overturning. This is not quite a hopeless job. The energy requirements are low. It is much easier to turn over a few hundred feet of water than a foot or so of soil. Furthermore, in a good many cases, we could tap potential energy stored in the upper part of the water column.

What is likely to be the future of the profitable exploitation of the ocean? It will certainly be slow in coming into being. Even for transportation, the seas are being gradually bypassed. In deep water their active exploration is mainly being financed as part of national defense. Although the deep ocean is quite permeable to low frequency sound, fully effective surveillance systems are very expensive to design, build, and operate. The immense present efforts of the military to exploit these potentials, especially in the case of defense, are making the deep ocean more than ever a no man's land. To devise ways of exploiting its peacetime resources is the only assurance that our huge salt water envelope will not become a military preserve and proving ground for advanced weapon systems. In short, prospects for future uses of the deep ocean are far from being bright. It will not be easy to put them into use for the benefit of mankind.

C. O'D. Iselin

Ocean Engineering

Ocean engineering is that branch of engineering and technology that encompasses the design, construction, maintenance, and operation of structures, equipment, tools, devices, and systems to explore, utilize, and develop the resources of the ocean. It has also been described as being "Concerned with the development of new equipment concepts and the methodical improvement of techniques which enable man to successfully operate beneath the surface of the ocean."* Ocean engineering combines the engineering technologies that enable man to do useful work in the ocean and develop its resources for the benefit of mankind.

In a general sense, ocean engineering may be viewed as the seaward extension of land engineering. It encompasses engineering performed in the coastal boundaries and the deep ocean. On land, the design and construction of bridges, roads, structures, machinery, and power and communications systems are classed as elements of civil, mechanical, and electrical engineering. Moving to the boundaries of the sea, the contruction of seawalls, piers, harbors, breakwaters, groins, ramps, etc., are classed as coastal engineering; in the deep ocean, is the realm of ocean engineering where the structures, equipment, vehicles, and devices are no longer directly connected to dry land.

While ocean engineering employs many conventional engineering techniques and scientific principles, it is subject to many influences, restraints, and unusual obstacles The effects of great pressure, highly corrosive environment, biological attack, currents, temperature, and meteorological variations, etc., present special problems of great complexity to ocean engineers.

Ocean engineering includes the extraction of minerals, chemicals, food, and power from the sea; the design and building of structures, equipment, and vehicles with which man can live and work safely in the deep ocean; communications, storage, and transportation systems; and systems and equipment for military and naval use.

In the future there will be underwater laboratories, working platforms, observation and recreation centers, aquaculture establishments, fish herding areas, and mobile food processing plants designed by ocean engineers and operated by underwater technicians and ocean workers. Already there are many important ocean engineering operations and projects in existence including:

(a) Offshore oil drilling operations from both fixed (attached to bottom) and floating platforms for drilling to depths beyond 600 ft (Fig. 1).

(b) Underwater exploration and mining operations for gold and diamond placer deposits; tin, coal, and phosphates on the ocean floor.

(c) Extraction of chemicals and minerals such as bromine, magnesium, fluorine, and fresh water from seawater.

* Lee Hunt and Donald Groves, "A Glossary of Ocean Terms," 1965, Compass Publications, Inc.

Fig. 1. Southeastern Drilling Co. semisubmersible drilling rig (SEDCO 135) partially submerged. (Photo: Jesse T. Grice and Southeastern Drilling Co.)

(d) Underwater habitats for divers from which they can work at deep depths (400–600 ft) for underwater construction, maintenance, repair work, scientific observation, and, at lesser depths, for recreation (Fig. 2).

(e) Submersible work vehicles (non military submarines) that enable man to work at shallow and deep depths where the cold waters and great pressures of the deep ocean will not permit him to work efficiently (Fig. 3). Slave arms or prehensile devices enable men in the submersibles to do work outside the vehicle and pick up materials or scientific samples. Submersibles with lockout chambers have recently been built for transporting divers safely to and from their working areas in deep depths.

(f) Remote-controlled devices with television or camera viewers enable operators on the surface to see and do underwater work by remote control from surface ships in areas where such deep submersibles or scuba divers cannot work safely or efficiently.

(g) New deep sea fishing equipment, systems, and devices to lure fish to nets and lines; new methods of rapid processing and packaging of fish at sea permit increasingly rapid and efficient distribution of fish products for human consumption.

(h) New underwater detection, tracking, and communications equipment and devices are being developed for antisubmarine warfare applications. Research, test, and evaluation ranges for developing and testing newer, more effective systems against enemy submarines, weapons, and underwater swimmers are important ocean engineering projects today. The U.S. Navy's AUTEC (acronym for Atlantic Undersea Test and Evaluation Center) Range in the Tongue Of The Ocean, Bahamas Islands, is an example of such an installation.

(i) Improved instrumentation, photographic devices, communications, and research equipment to advance underwater exploration and development of marine resources.

(j) New construction materials, processes, and techniques to meet the demands for increased strength and engineering reliability during emplantment and operation in the ocean environment where the materials are subject to strong corrosive forces, marine growth, and attack by large fish and marine animals.

(k) New equipment and techniques for deep ocean salvage and recovery in which the U.S. Navy's Deep Submergence Group already is deeply involved. This work includes design of Deep Submergence Rescue Vehicles (DSRV) for rescue of personnel trapped in sunken submarines, recovery of sunken ships or large objects from deep depth, and the development of new salvage techniques for very deep depths.

(l) Design of special scientific equipment for deep ocean work such as the Floating Deep Ocean Drilling rigs used in Project Mohole where scientists made deep corings into the earth's crust in order to determine its composition and structure. Since the known thinnest portions of the earth's mantle can be found in deep depths of the ocean, future scientific experiments of this nature will require major ocean engineering operations from massive floating platforms.

(m) Long-life floating and underwater sensors, buoys, recorders, telemetering, communications links, and computer systems to read, store, transmit, and analyze atmospheric and oceanic data for rapid, accurate weather and ocean predictions used by mariners, offshore underwater operators, and naval commanders for their sea operations.

(n) Large underwater storage facilities for concentrating bulky or strategic materials for economic or defense purposes.

Mentioned above were areas of *present* ocean engineering capabilities or developments already underway. Looking into the future, one sees many new ocean engineering prospects that one can predict will become important areas of future development. These are:

(a) Construction of extensive underwater working and recreational areas of great size and complexity.

(b) Manned deep ocean underwater laboratories and observation stations at depths of 12,000 ft or more. Such installations would require long endurance, high power sources (probably nuclear), long endurance self-contained life support systems, and deep depth submarine docking facilities for personnel transfer and supply.

(c) Deep manned stations probably will be needed at strategic points on the ocean floor for economic development, defense, and occupation "recognition," under future international laws for deep ocean territories.

(d) Extensive systems to reduce or eliminate dangerous pollution of ocean waters from industrial, human, and nuclear waste discharged into the sea. Although some work has already started in this area, the control of pollution will become more urgent as time goes on, particularly in the disposal of nuclear waste in ocean areas.

(e) Development of safe, fast, and efficient underwater transportation systems including accurate navigation systems, beacons,

Fig. 2. General Electric "Tektite" underwater habitat. (Photo: General Electric Missile and Space Division.)

Fig.3. Deep submersible work vehicle *Alvin*. (Photo: Woods Hole Oceanographic Institution.)

markers, depth and obstruction warning devices for these fast moving submarine vehicles to permit them to operate efficiently and economically at depths not affected by surface storms and turbulent seas.

(f) The development of international laws for the deep ocean to establish legal bases for occupation, ownership, control, defense, and use of underwater areas and their resources. International agreements made in 1958 and 1960 grant sovereign rights to countries bordering the ocean to the limits of the adjacent continental shelves (to depths of 200 m) and to those areas further seaward *that can be exploited by that country*. This loose interpretation will require early clarification by international lawyers to permit engineers and industry to work confidently in the deep oceans.

(g) Development of new marketing techniques and procedures for products of the sea also will be required to encourage ocean industries to reach their full potential in future world markets.

At present, there are few formally trained ocean engineers in the U.S. or the world. Most engineers working in the ocean today were educated and trained in other engineering disciplines or the physical sciences, and they obtained their oceanographic knowledge empirically at sea or through special study at ocean institutes or schools. In other cases, trained oceanographers studied or learned engineering techniques by experience at sea that now permit them to do engineering work in the oceans.

In 1965, a report of the Interagency Committee on Oceanography of the Federal Council of Science and Technology, showed that there were only 167 ocean engineers in the U.S. out of a total of more than 633,000 engineers in all other categories. The rapid development of ocean industries and the need for ocean engineers in government laboratories and agencies clearly show the demand for increased training of ocean engineers. In 1967, several universities including Hawaii, Miami, Rhode Island, New Hampshire, California, are offering graduate programs and courses in ocean engineering; Florida Atlantic University at Boca Raton, Florida, offers an undergraduate program in ocean engineering leading to a degree of Bachelor of Science.

In general, ocean engineering curricula include various combinations of mechanical, electrical and civil engineering courses, plus oceanography, acoustics and supporting mathematics and science courses to provide basic technical engineering knowledge and a thorough understanding of the oceanographic environment in which the engineer will work. Interest is increasing in the education of ocean engineers in many universities that already have scientific oceanographic programs and the American Society for Engineering Education has established an ocean engineering committee to guide and strengthen the university programs in this new engineering discipline.

Several technical and engineering societies and associations are directly associated with ocean science and engineering. Among these are the Marine Technology Society, the National Oceanographic Society of America, the American Society of Limnology and Oceanography, and the International Oceanographic Foundation of Miami, Florida. In addition, many large engineering societies, such as the Institute of Electrical and Electronic Engineers, the American Society for Chemical Engineers, and the American Society of Petroleum Engineers, now have ocean engineering sections.

Within the Federal Government, ocean engineering programs and interest have increased in recent years. The ocean engineering portion of the national oceanographic budget increased from $62,000,000 in 1965 to $66,000,000 in 1967. The Marine Resources and Engineering Act of 1966, passed by Congress in June 1966, established a National Council on Marine Resources and Engineering which includes the Vice President of the U.S. as chairman, together with the secretaries of all the major departments of the government as members. It also provides for a Commission on Marine Science, Engineering and Resources composed of 15 representatives from government, industry, and university or research institutions. The Council and Commission have the tasks of developing long range national programs in marine science, engineering, and technology for increased protection of national health and property; enhancement of commerce and transportation; national security; rehabilitation of commercial fisheries and increased utilization of ocean resources. This Council's first report entitled "Marine Science Affairs—A Year of Transition" (February 1967), includes recommendations to "strengthen oceanographic engineering and improve technology and engineering for work at great ocean depths." All the major departments in the Federal Government including the Departments of Defense, Commerce, Interior, the National Science Foundation, and the Atomic Energy Commission, conduct large oceanographic programs and maintain closely associated ocean engineering programs. The Navy's Deep Submergence project, Office of Naval Research Civil Engineering Laboratory, National Oceanographic Data Center, and the Environmental Science Service Administration are all deeply involved in important national ocean engineering programs and development.

Many major industrial companies such as the Westinghouse Corporation, Underseas Division; General Electric Company; General Motors Corporation; General Dynamics Corporation; Lockheed Aircraft; North American Aviation Company; Ocean Systems, Inc., of Union Carbide; Corning Glass Company; Reynolds Aluminum; Grumman Aircraft; Ocean Science and Engineering Company; Perry Submarine Builders; Global Marine Company; Alpine Geophysical Corporation; the offshore oil and drilling companies; plus many manufacturers of oceanographic instruments; marine service companies —all are becoming heavily involved in ocean engineering projects and operations.

Many men of vision have provided great ideas, innovations, and inspiration in ocean engineering developments in the past and recent years. Examples include such leaders as Dr. Piccard whose Bathysphere and deep submersible *Trieste* have explored the deepest known depths of the ocean, 35,000 ft in the Mariannas Trench in 1960; Captain Cousteau, whose invention of scuba equipment and early Man-in-the-Sea Habitation work has done much to accelerate man's personal conquest of the ocean. Edward Link, inventor of the famous Link Trainer that became a vital part of every aviator's training and requalification program for years, now has put his great talents toward increasing man's capability to live and work safely and efficiently at deep depths in the ocean. His project for man's longest deep dive (two days at 440 ft) is still a record. Dr. George Bond, leader of the Navy's Sea Lab project, has made great contributions. Dr. John Craven, Scientific Director of the Navy's "Polaris" program now heads that service's Deep Submergence project, which, as a result of the tragic loss of the *U.S.S. Thresher* in 1963, is now developing ocean engineering systems that will greatly enhance man's capabilities, safety, and efficiency in deep ocean work. The combined efforts of the many unnamed engineers in the offshore oil drilling and pumping industry have contributed greatly to this new discipline. Ocean explorers and developers include names such as Willard Bascom, whose pioneering efforts are leading the way in ocean mining and exploration; Allyn Vine, whose *Alvin* vehicle is the forerunner of successful underwater work vehicles; the Perry-Link Deep Diver vehicle designed by Perry and Link, which offers the first successful "Lock-out" submersible for ocean engineering work. The list grows daily and will continue to do so as the ocean engineering discipline expands.

CHARLES R. STEPHAN

Cross references: *Heat and Power From Seawater; Oceanography.*

OCEAN FARMING—*See* SEA FARMING

OCEANOGRAPHY

Exploration and Exploitation

The oceans have excited man since he first viewed these gigantic bodies of water. Man has been slow in attempting to understand their composition and nature, and only in the last few hundred years has he considered the oceans a wealth of resources. Today scientists are investigating the oceans and seas through the science of oceanography. It is the composite of all the sciences involved in the marine environment. It can be divided into five principal disciplines: (1) physical oceanography; (2) geological oceanography; (3) chemical oceanography; (4) biological oceanography; and (5) geophysical oceanography. Each discipline is dependent upon and related to the others. Studies on currents, upwelling, sea level, underwater sound, waves, hydrography, and air-sea interaction describe physical oceanography. Geological oceanography can be abstracted by such topics as sedimentation, sedimentary processes, foraminiferal ecology, and topography. Chemical nutrients, chemical exchange between atmosphere and sea surface, and radioactive material in the marine environment are the concern of the chemical oceanographers. Biological oceanography is devoted to the study of organisms such as phytoplankton, zooplankton, nekton, and benthos. Scientists in geophysical oceanography try to understand better the structure and composition of the earth through investigations in seismology, magnetics, and gravity.

Men have been known to list and describe catches of marine life since the days of Aristotle. With the beginning of the 19th century, scientific thought turned more toward understanding life in general and the food chain within the oceans. Researchers developed better techniques for capturing sea life, and tows were made at different levels in the oceans. The effects of currents and temperatures on fish populations were beginning to be recognized. The area of scientific inquiry and the science of oceanography brought about the founding of coastal laboratories and undertaking of purely scientific expeditions.

In 1831, the *Beagle*, a small warship from the British Fleet, embarked on a survey cruise around the world. The ship's naturalist throughout the expedition, which lasted nearly five years, was a young man named Charles R. Darwin. The two main scientific contributions from the voyage of the *Beagle* were Darwin's theory of the formation of coral atolls and reefs and his theory of evolution.

In 1828, J. Vaughan Thompson of the British Army Surgery Corps, a naturalist, attached a jar to the end of a fine mesh net with the hopes of collecting specimens as he towed the device through the water. The microscopic life Thompson collected in his net was not thoroughly examined until several years later. In 1844, German naturalist Johannes Müller used the tow net and was able to impart his enthusiasm for the net to his colleagues. Although scientists were aware of the existence of plankton from Thompson's catches, they did not realize the important role plankton plays in the food chain.

British naturalist Edward Forbes dredged for sea life around the British Isles and the Aegean Sea in the 1830's and 40's. Even though sea life had been recovered from 1000 fathoms in 1818, Forbes' theory that no life existed below 300 fathoms was held as common belief from 1843 until the *Challenger* Expedition (1872–1876).

Meanwhile, in 1845, J. M. Brooke, U.S. Navy, developed a coring device to penetrate the sediments on the ocean bottom. Mud and ooze collected by the new core sampler contained microscopic shells. From these findings questions arose as to the origin of the shells.

Advances in marine techniques and knowledge of marine life allowed the mid-19th century scientist to tackle basic questions of the food chain in the ocean. It was soon apparent that plankton drifted with ocean currents, and that currents could be plotted by studies of plankton. With investigations of plankton, the importance of changes in temperature and salinity was discovered. The effect of this knowledge led to the field of fisheries research; the marine biologist, who, before 1845, had to be satisfied with describing and classifying the inhabitants of the sea, could now map areas of distributional patterns.

Matthew Fontaine Maury, the first American oceanographer, shared Edward Forbes' belief of a unit ocean, but from a physical and mechanical point of view. Lieutenant Maury, U.S. Navy, was assigned to the Depot of Charts and Instruments, where he compiled and evaluated all available data from ships' logs. At his request, Navy vessels made deep-sea soundings and took weather observations and water samples. From these data, Maury was able to save navigators a tremendous amount of time by plotting new sailing routes. At his suggestion, a world conference was held in Brussels in 1853 to discuss research in the seas. This international meeting was important to the fields of navigation, meteorology, and oceanography, and it led to the organization of the International Hydrographic Bureau and the World Meteorological Organization. In 1855 Maury published *The Physical Geography of the Sea*, the first book on oceanography as a science.

Scientific cruises were becoming the order of the day. Englishman C. Wyville Thomson participated in deep-sea dredging and sounding cruises aboard the *Lightning* in 1868 and the *Porcupine* from 1869 to 1870. In 1873, Thomson published the results from these cruises in a book entitled *The Depths of the Sea*. Even after these cruises, the controversy of life below 300 fathoms still existed. Thomson, with the support of the Council of the Royal Society, was able to persuade the British admiralty to sanction a four-year, round-the-world expedition aboard the wooden corvette HMS *Challenger* to determine if life existed below 300 fathoms. From 1872 to 1876, the *Challenger* expedition occupied 360 stations over a 140,000,000 square mile area, while sampling the ocean's depth and breadth.

This pioneer voyage propelled man into the new era of oceanography. Scientists aboard the *Challenger*, for the first time in man's history, were interested in the ocean from biological, chemical, and physical points of view. Naturalists collected 4417 living organisms from all depths of the oceans and discovered 715 new genera. Investigators recovered deep-sea manganese nodules rich in copper, cobalt, and nickel from the floor of the Atlantic, Pacific, and Indian Oceans. Official reports of the expedition number 50 volumes, and are still a basic work in use today. Summary volumes of the voyage were written by the noted scientist Sir John Murray.

In 1893, Norwegian explorer Fridtjof Nansen set out on one of the boldest and most remarkable adventures of polar exploration. To prove his theory that a current flowed across the Arctic Ocean without interruption by land, Nansen allowed his specially built ship, *Fram*, to be frozen in the polar ice to drift with the current from Siberia toward Greenland. Finally, in August 1896, the *Fram* sailed free of ice after three years of being a captive of the frozen north. Not only had Nansen proved his theory, but he had also collected valuable physical data on the cold Arctic waters.

Important contributions toward the advancement of the biological science in the oceans from the time of the *Challenger* to 1900 included work by the U.S. Coast Survey *Blake*, in the Gulf of Mexico and the Caribbean; the French ships *Travailleur* and *Talisman*, in the Mediterranean and the eastern Atlantic; the U.S. Fish Commission's *Albatross*, in the tropical Pacific, the yacht *Hirondelle* of Prince Albert of Monaco, in the Mediterranean and the North Atlantic; the German ships *National*, in the Atlantic, and *Valdivia*, in the Atlantic and Indian Oceans; the Danish ship *Ingolf*, off Iceland; and the Dutch ship *Siboga*, off the Dutch East Indies.

From 1900–1910, the Norwegian expeditions of the *Michael Sars* made many contributions to biological oceanography. The German expedition *Deutschland*, in the Antarctic during 1911–1912, boosted studies of the physics and the chemistry of the sea. The Carnegie Institute of Washington sponsored the *Galilee* (1905–1909) and the *Carnegie* (1909–1929), which performed extensive oceanographic studies in the tropical Pacific. After these voyages most oceanographic surveys and cruises were on a smaller basis and usually for specific purposes.

Coastal Laboratories. Museums of zoology existed in several countries, but they did not allow observation of living specimens. British interest in marine studies was aroused near the mid-1800's by naturalist Philip Henry Gosse, whose efforts led to construction of public aquariums. Swiss-born American Louis Agassiz, founder of the museum of Comparative Zoology at Harvard, had participated in several cruises aboard U.S. Coast Survey ships during 1847. His enthusiasm for the marine environment caused him to seek financial aid for a seaside laboratory. Finally, Agassiz received financial support from a wealthy merchant, and established a summer shore station at Penikese Island, Massuchusetts in 1873. At the same time, German naturalist Anton Dohrn appealed to his government for funds for a year-round marine station at Naples, Italy. His request was denied but with private contributions and funds from his own personal estate, the first building of the Zoological Station of Naples was completed in 1874. These two sea laboratories played an important role in the development of oceanography and initiated a chain of laboratories around the world. Today, scientists at 500 such marine stations are working on problems in the marine environment.

Organizations. Organizations have been established for the advancement of science, many interested in oceanography. The International Council of Scientific Unions (ICSU) was organized in its present form in 1931. The ICSU has sponsored such operations as the International Geophysical Year (IGY) and has been affiliated with such organizations as the Special Committee for Oceanic Research (SCOR), established to promote marine science activities such as studies on the Indian Ocean; International Union of Biological Sciences (IUBS), to stimulate interest in biology, including biological oceanography; International Union of Geodesy and Geophysics (IUGG), to promote geographical and geophysical studies; and special Committee for Antarctic Research (SCAR), to encourage scientific inquiry in the Antarctic.

Other organizations concerned with oceanographic programs are the International Atomic Energy Agency (IAEA), interested in handling and disposing of radioactive wastes; the World Meteorological Organization (WMO), which collects, interprets, and exchanges data; International Hydrographic Bureau (IHB), which standardizes nautical charts and exchanges hydrographic data; International Council for the Exploration of the Sea (ICES), primarily interested in fisheries research and synoptic studies of the sea during all seasons; and the International Ice Patrol (IIP), organized because of the menace of icebergs, which has added vast knowledge on currents and physical properties in the North Atlantic Ocean.

By the late 1800's, the refinement of fishing techniques, the knowledge of breeding grounds of mammals and fish, and the growth in human population had put a serious strain on fishery resources. Among those hunted and threatened with extinction were the whales. With a growing concern for depletion of whales in the Antarctic waters, the Government of the Falkland Island Dependencies formed a Discovery Committee in 1923. By 1925 the British had commissioned a ship, *Discovery*, to survey marine life, especially the whale, in the southern oceans. In 1926 and 1929, the *Scoresby* and the *Discovery II* respectively, were added to support the *Discovery* in her biological and oceanographic investigations.

Another method man used to protect marine life was to formulate and respect agreements established by international agencies. In 1948, the International Whaling Convention established the International Whaling Commission for the

Fig. 1. FLIP (Floating Instrument Platform) developed by Scripps Institution of Oceanography. The 355-ft vessel is towed in a horizontal position to the research site, where, by flooding ballast tanks, she is flipped to a vertical position; only the stable 55-ft. platform from which studies are conducted remains above the surface.

conservation of the whale, and, in 1949, the International Commission for the Northwest Atlantic Fisheries resulted from a convention to conserve fisheries. Other organizations that followed were Inter-American Tropical Tuna Commission (IATTC), International North Pacific Fisheries Commission (INPFC), International Pacific Halibut Commission (IPHC), International Pacific Salmon Fisheries Commission, and North Pacific Fur Seal Commission.

Sea Invasion. Although man was gaining knowledge and developing new techniques to sample the ocean's depths, he still needed to invade the scene of his investigations. In 1930, Otis Barton, an American, developed the first deep-sea submersible, the bathysphere. After many preliminary dives in the bathysphere, which connected with a cable to a parent ship, Barton and oceanographer William Beebe descended to a record depth of 3028 ft in August 1934.

Professor Auguste Piccard, with the financial support of the Belgian government, set out to build a submersible free of connecting cables. He designed the first deep-diving vehicle and made a record dive, in 1954, of 13,284 ft. Auguste and Jacques Piccard improved the original device and, in 1960, Jacques Piccard and Lieutenant Don Walsh, U.S. Navy, descended 35,800 ft into the Mariana Trench off Guam in the bathyscaphe *Trieste*. More recent developments include a two-man diving saucer, *Denise*, and the two-man research submarines, *Alvin* and *Aluminaut*.

A unique research vessel, FLIP (Floating Instrument Platform), launched in 1962 under the direction of Scripps Institution of Oceanography, fills the gap between the surface vessel and the submersible. FLIP is towed horizontally to location, then by flooding ballast tanks, it is stationed in a vertical position (Fig. 1). This allows scientists to collect wave data as well as chemical data from the sea. However, FLIP has no propulsive power of its own.

Deep-sea cameras have been used to photograph the ocean floor since 1893. Early exposures failed because of insensitive film and poor illumination. In 1940, scientists from Woods Hole Oceanographic Institution perfected the automatic camera for deep-sea use. During the International Geophysical Year, 1957–1958, investigators from Scripps Institution of Oceanography used the camera extensively in their surveys of the ocean floor. From photographs and dredges, they were able to locate high concentrations of manganese nodules in the eastern Pacific. Although these nodules have been known since the days of the *Challenger*, scientists have not been able to determine a way of mining them profitably.

The number of institutions in the world engaged in oceanographic work increased from some 245 in 1937 to more than 850 in 1964. During 1964 the U.S. accounted for the most oceanographic institutions with some 122, while Japan was second with some 66.

In 1882 the 234-ft vessel *Albatross* was built for deep-sea exploration. The first research ship built by the U.S. for oceanographic work was the *Atlantis*. Built in 1931 in Copenhagen, the *Atlantis* was operated by Woods Hole Oceanographic Institution for 35 years before being sold to the government of Argentina. The vessel berthed 19 crew members and scientists, and sailed mainly in the waters of the Gulf of Mexico, the Caribbean, the western half of the North Atlantic, the Mediterranean, and in the coastal waters off Africa and South America. Some 342 vessels were classified as performing oceanographic work during 1963. Relatively few of these vessels were built especially to do oceanographic work; most research vessels of today are converted warships or fishing vessels.

Sea Exploitation. Little effort has been given to the exploitation of mineral resources in the sea. American John L. Mero, in 1965, estimated that the oceans store some 5×10^{16} tons of mineral matter.

Salt was the first resource extracted from seawater, by the Chinese before 2200 B.C. By 1965, approximately 6,000,000 tons of solar salt was being produced annually on a world-wide basis. In 1825, Frenchman A. J. Balard discovered bromine in the byproduct of salt production. The first extraction of bromine from seawater occurred in 1926 by treating the bitterns from a salt operation in the U.S. Magnesium was discovered in seawater in England in 1808; the Dow Chemical Company of the U.S. constructed the first large scale plant for extracting magnesium from seawater in 1941. Seaweed has been used throughout history as a source of iodine, alkali, potash, and food.

In 1967, bromine, calcium, magnesium, and sodium were extracted commercially from seawater by some 145 operations on a world-wide basis. The unconsolidated deposits of diamonds, gold, sand and gravel, tin sands, lime shells, iron sands, and heavy mineral sands were mined from the sea by some 71 mining operations in 12 countries; and consolidated deposits of iron ore, coal, and sulfur were mined by eight countries in some 60 operations.

On a voyage into the South Seas in 1593, Sir Richard Hawkins converted seawater to fresh water. Although some mariners were aware of the saline water conversion, neither the process nor the knowledge of converting salt water to fresh water was available to all mariners. In 1791, statesman Thomas Jefferson suggested "that knowledge about sea water distillation be made available to all mariners in the first official scientific document by the United States Government." However, the recommendation was not accepted. Techniques for saline water conversion were finally recognized during World War II, when it was necessary for armed forces to be self-sufficient both at sea and on the beaches.

The increase in human population each year has put a serious strain on the world food supply. To increase the fish harvest, Asian countries have used ponds for fish culturing since before the 1700's. The Dutch have also used fish culturing methods in transplanting mussels from waters off the Friesian Island to the richer waters southwest of Holland. Also, in 1932, the Dutch started reclaiming land from the sea. Thousands of acres of cultivated land exists in areas that were under water a few years ago. The Dutch hope to gain more than a half million acres from the sea before their plan is completed.

In 1872, E. Sonstadt discovered gold in seawater. Later C. A. Münster measured 5 to 6 mg of gold and 19 to 20 mg of silver per ton of water from Kristiania Fjord, Norway, while analysis of Australian waters showed 32.4 to 64.8 mg of gold per ton of seawater. From 1925–1929 the German ship *Meteor* traversed the South Atlantic intending to extract gold from seawater. It

Fig. 2. Research vessels. HMS *Challenger* (above) made her historic expedition nearly a century ago. The first research ship built by the U.S., the *Atlantis* (facing page, top), engaged in oceanographic exploration for 35 years. USC&GSS *Oceanographer* (facing page, bottom), recently built, is a completely automated research ship; 303 ft long and 52 ft at the beam, she draws 18 ft and displaces 3800 tons. She has a cruising range of 13,000 miles, and can remain at sea for 150 days. (*Atlantis* photo: Woods Hole Oceanographic Institution. *Oceanographer* photo: Environmental Science Services Administration.)

proved not economically feasible, but the expedition returned with invaluable information on phytoplankton, sea floor bathymetry, and distribution of water masses.

Latest experiments in undersea technology include human dwelling on the sea floor. For example, five men lived below the surface of the Red Sea for four weeks in 1963. In the waters off California, aquanauts have maintained sealabs on the ocean's floor. In August 1965, divers lived 205 ft below the surface for 15 days while experiments were made on their body conditions as well as on underwater specimens. Similar experiments of undersea colonization are preludes to man's total use of the sea.

Wayne V. Burt
S. A. Kulm

References

Cywin, A., and Finch L., "Federal research and development program for saline-water conversion," *J. Am. Water Works Assn.* **52**(8), 983–992 (1960).

Daughtery, C. M., "Searchers of the Sea: Pioneers in Oceanography, 1961. The Viking Press, New York.

Deacon, G. E. R., (Ed.), "Seas, Maps, and Men," 1962. Doubleday and Company, Inc., Garden City, New York.

Oceanography and Marine Fisheries

Fisheries oceanography is concerned with all aspects of the ocean, its boundaries, and its contents that affect the abundance, location, and behavior of the harvestable living resources. Thus, it comprehends the topography of the ocean basins; the upwellings, horizontal currents, and other motions of the oceans; the effects of atmospheric processes exerted on the ocean at the air-sea boundary; the distributions of temperature, salinity, and other physical and chemical properties; and the abundance, rates of production, behavior, and interrelationships of the populations of living elements. The fishery oceanographer is particularly concerned with: the fixation of organic material by photosynthesis and its transfer through the food web, which determines the magnitude of the available harvest at various trophic levels; ocean conditions that bring about economically catchable fish concentrations; how the locations and sizes of fish populations vary with changing conditions in the sea, and with the intensity of fishing; and those aspects of behavior of marine organisms that can be exploited to reduce the costs of catching them.

Fishery oceanography especially helps to increase the harvest of the sea in four ways: (1) location of new, highly productive fishing areas; (2) identification and location of promising unutilized fishery resources; (3) providing to fishermen information that they can use to improve their tactical scouting and catching operations; (4) forecasting of space and time variations in the abundance and catchability of fish populations.

Location of New Fishing Areas. In the sea, as on the land, all the organisms in the food web depend upon the organic matter produced by plants through the subtle chemistry of photosynthesis. Thus, the richest fisheries occur where there are large blooms of phytoplankton, or at locations down-current therefrom, where large crops of forage organisms feed on the phytoplankton. These forage organisms, in the open sea, are zooplankton or small herbivorous fishes upon which the larger carnivores, which are often the objective of commercial fishery, depend. Consequently, it has been found that the rich fisheries occur where large quantities of organic matter are produced by the phytoplankton, due to fertilization of the sunlit upper layer of the sea by upwelling, mixing along current boundaries, winter overturn, stirring of nutrients up from shallow bottoms, and other physical processes.

Fig. 2. Continued.

With increased capability for studying the oceanic circulation, as well as for measuring directly phytoplankton productivity, and for directly assessing the abundance of zooplankton and other larger organisms, and with increased understanding of the processes whereby the upper layers of the ocean are fertilized, oceanographers are becoming increasingly capable of predicting promising new fishing areas.

For example, the northwest coast of Africa has been known for some time to be a region of strong coastal upwelling. Measurements of basic productivity and standing crops of phytoplankton in this region indicated that there should be abundant fish populations that might be harvestable by commercial fishing methods. This led to exploratory expeditions by many nations, including the U.S. and the U.S.S.R., to examine the fishery potential of this area in relation to oceanographic factors. In consequence, extensive fisheries for the tropical tunas have been developed, and sizeable populations of the pelagic *Sardinella*, and of various demersal species, now coming into commercial production, have been discovered.

Upwelling associated with the divergence along the equator, in both the Pacific and Atlantic Oceans, are responsible for high biological productivity. Investigations of physical and biological oceanography, in advance of and along with exploratory fishing operations, greatly accelerated the development of the equatorial pelagic fisheries for tunas and spearfishes, conducted primarily by Japanese fishermen.

A very recent example is the incipient development of large new fisheries in the Arabian Sea. Studies of this region carried out cooperatively during the Indian Ocean Expedition by oceanographers of the U.S., England, Germany, U.S.S.R., and other countries, have demonstrated that the Arabian Sea, particularly on its western side, is a region of high basic productivity associated with vertical circulation related to the monsoon winds. There are also some direct observations of sizeable populations of sardines, mackerels, tunas, and other fishes. Detailed analyses of the physical, chemical, and biological information from the Indian Ocean Expedition have made it possible to indicate in some detail the most promising areas for the development of new, large-scale commercial fisheries, which promise to be similar in productivity to those of the southeast Pacific off Peru and northern Chile.

Identification and Location of Unused Fishery Resources. In those areas of the sea that have long supported commercial fisheries, systematic scientific observations often lead to the identification of important, latent fishery resources. A well known example is the systematic study of the California Current, off the west coast of the U.S., carried on for a number of years, initially motivated by the need for investigation of the ecology and the fishery dynamics of the California sardine. One technique employed is the systematic survey of the abundance of larvae of sardines and other fishes. A dominant element in the catches of fish larvae was discovered to be the Pacific hake, from which it was inferred that there is a large latent resource of this species that might be commercially exploitable.

Based on this lead, systematic explorations by echo sounding and experimental trawling revealed large commercially exploitable concentrations along the coast of Washington, Oregon and northern California. It appears that this species moves south and somewhat offshore to spawn, and moves northerly and inshore on a feeding migration, although many of the details remain to be elucidated. That the stocks are commercially exploitable off Washington and Oregon is demonstrated by a newly developed commercial fishery, pursued primarily by Russian vessels, but to some extent also by U.S. vessels, producing on the order of 100,000 tons per year. Whether the stocks are commercially exploitable further to the south, in the vicinity of the spawning grounds, remains to be investigated.

The systematic surveys of fish larvae have also revealed that, with the decline of the sardine population, its very close competitor, the anchovy, has increased very greatly in abundance. It has been estimated by the scientists of the California Cooperative Fishery Investigations that there exists off California and Baja California a standing stock of some 4,000,000 tons of anchovies, enough to sustain a harvest of perhaps a million tons per year or more. It is believed that some reduction of the anchovy population might, simultaneously, accelerate the recovery of the heavily depleted sardine population. These investigations have also indicated that the stock of jack mackerel, of which only a few thousand tons are currently harvested near the coast, extends westward over a vast region of the Pacific and should be able to support a very much larger fishery.

Increasing demand for raw material for the manufacture of fish meal, and a decline in the supply of herring in the northeast Atlantic, stimulated investigations of other resources. This resulted, some years ago, in the development of sizeable fisheries for capelin and sand lance. Recent investigations in the northwest Atlantic, off Canada, have revealed also large, unused populations of these species.

Of greatest interest are current investigations of the Antarctic krill (*Euphausia superba*) as a possible base for a very large commercial fishery. This shrimp-like animal is the principal food of the baleen whales of the Antarctic, as well as of seals and penguins. It is estimated, on the basis of the quantities required to sustain the former large populations of whales, that the krill population could support a continuing harvest on the order of 100 million tons per year. The catch may be used for direct human consumption, since the flesh of these small crustacea is very similar to shrimp or lobster, or it might be used for the preparation of fish meal for feedstuffs for domestic animals. Studies of the feasibility of direct harvesting are currently being conducted. Another small crustacean, the red crab (*Pleuroncodes planipes*), which occurs off the coast of Mexico and is one of the principal items in the diet of the tunas, also offers an interesting possibility for direct harvest for human consumption or for fish meal.

Information for Improving Fishing Tactics. Oceanography can be helpful to fishermen in indicating those sea areas, and often the particular seasons, where abundant exploitable populations occur. But the fisherman has additional problems of locating fish shoals within a general area, and then of catching them rapidly and efficiently. Knowledge of the local distribution of harvestable species in relation to the properties of their environment, and understanding of their behavior, especially those aspects related to measurable properties of the environment, can, consequently, be useful in the fisherman's day-to-day operations. Such knowledge can often indicate measurements that the fisherman can make at sea in order to improve his scouting and catching operations, and thus cut down his cost of production. A few examples follow.

It is known that tunas, and many other species of marine fishes, tend to be more concentrated in the vicinity of seamounts, which the fishermen refer to as "banks." Charts showing the precise locations of seamounts, which are continually being discovered both by fishermen and by submarine geologists, indicate good fishing locations. Bottom topography charts, together with echo sounders for locating both the seamounts and the fish shoals and their vicinity, can be used profitably by the fishermen.

The occurrence of many fish species, and their migration paths, are determined by water temperature. Fortunately, both surface and subsurface temperatures have long been measurable by both scientists and fishermen, and, indeed, surface temperatures are routinely measured by merchant vessels as part of the marine weather reporting system. Consequently, better information is available concerning the local distributions of some kinds of fish in relation to temperature than with relation to many other aspects of their environment, and this knowledge can be of considerable tactical advantage to the fishermen. For example, the north Pacific albacore, which visit the west coast of the U.S. during the summer months, prefer water temperatures of 60°–66°F.

The U.S. Bureau of Commercial Fisheries publishes sea surface temperature charts at two-week intervals, and also forecasts the regions of expected occurrence, enabling the fishermen, by the use of these charts and their own temperature measurements while searching for albacore schools, to direct their scouting operations more efficiently. Similarly, the tropical tunas of the eastern Pacific, yellowfin and skipjack, prefer waters above 66°F and 62°F, respectively. These limiting temperatures determine the seasonal distribution of these species at the northern and southern ends of their range. This knowledge is of particular utility to distant-water vessels operating from California near the southern extreme of the range, off Peru and Chile. Knowledge of distribution and behavior of cod in relation to water temperatures in the vicinity of Bear Island, between Norway and Spitsbergen, which was elucidated by English scientists, has led to useful forecasts of the distribution of this species, and also forms the basis of the useful employment by trawler captains of bottom-water temperature measurements while searching for concentrations of cod in this area.

An interesting relationship of tuna to their environment that is of tactical value to fishermen is their distribution and behavior, in the eastern tropical Pacific, in relation to the depth of the mixed layer and the sharpness of the underlying thermocline. The tropical tuna occur in the upper mixed layer of warm, low-density water, which varies from 10–80 m depth, and is underlain by colder water. It has been shown that the percentage of successful purse seine sets on tuna schools is greatly increased when the mixed layer is shallower than the depth to which the net fishes, and when the gradient of temperature in the underlying thermocline is very sharp, presumably because the fish, under these circumstances, escape less readily under the bottom of the net before it is pursed. Consequently, by measuring the vertical distribution of temperature by bathythermographs or other means, the fishing master may select situations where the escape rate is minimized.

Local distributions of a number of fish species have been shown to be related to salinity and to oxygen. As one example, it has been shown that the distribution of hake off Chile varies seasonally with the position of the low-oxygen water of the subsurface undercurrent that flows southerly under the north-flowing Chile Current. Knowledge of such relationships will become increasingly valuable in relation to the tactical operations of commercial fishermen as they acquire instruments for measuring salinity, oxygen, and other factors, from their fishing vessels.

Another local phenomenon that often corresponds to concentrations of various kinds of pelagic fish is the occurrence of "fronts," which the Japanese call *siome*; these are boundaries between water masses. Along these boundaries, which can be often located by differences in water color, sharp temperature gradients, or floating debris, the plankton organisms are concentrated by the associated vertical circulation; this, in turn, leads to concentrations of forage fishes and to the predatory fishes that eat them. Tuna fishermen in the eastern Pacific often scout successfully along such fronts. Longline fishermen from Japan find that laying their gear along and across fronts greatly improves the fishing success for tunas and spearfishes.

A potentially powerful tool for assisting in the location of harvestable fish aggregations is the relationships of such aggregations to concentrations of their food organisms. For example, the relationship between herring and the copepod *Calanus* on which it feeds is sufficiently good so that herring fishermen in the North Sea and the Barents Sea have profitably employed their own plankton collecting, using such simple instruments as the Hardy plankton indicator.

Fishery Forecasting. Great improvements in efficiency would flow from reliable future forecasts of fishing locations and of abundance and catchability for particular species. This is yet in a very primitive state. To make such forecasts for any particular kind of fish, one needs useful estimates of the magnitude of the exploitable population, an understanding of the distribution and behavior of its members in relation to measurable physical properties of the ocean, and some means of predicting the space and time changes in the oceanic properties and processes to which the distribution and behavior are related. Some progress has been made in all of these, and some forecasts of a few to several weeks ahead have been moderately successful.

A major source of variation in abundance of the catchable sizes of fish is variation in infant survival, leading to variations in the recruitment of the young fish into the fishable stock. Through the compilation and analysis of statistics on catch and effort, and determinations of age-composition of catches, supplemented by estimates of abundance of young fish prior to their entry into the stock of commercial sizes, through sampling the larvae or juveniles, methods have been developed for forecasting fish population magnitudes in some cases. Notable examples are the New England haddock, the sockeye and pink salmon of the Frazer River, Bristol Bay red salmon, anchovies of Peru, and some stocks of North Atlantic herring.

As already noted above, we have some useful understanding of the distribution and behavior of various kinds of fish in relation to environmental factors, usually temperature, but also, in some instances, including depth of mixed layer, oxygen, salinity, and other factors. Such knowledge is sufficient, in at least a few instances, for forecasting the effects of large-scale changes in the ocean circulation; e.g., the El Niño off northern South America, which, at irregular intervals averaging seven or eight years, brings to the coast of Peru abnormally warm surface waters that cause large and predictable shifts in the populations of anchovies, tuna, and other fishes, and catastrophic effects on the guano bird population. Another example is the skipjack tuna population inhabiting the waters of the California Current Extension, which may be identified by temperature and salinity, near the Hawaiian Islands. As this current shifts northerly through the vicinity of Hawaii each summer, the "season" skipjack appear, the abundance and availability fluctuate. The extent of the shift in the boundary between the California Current Extension and the water masses to the north varies widely.

Most forecasting of ocean conditions, and consequently of the effect on the fisheries, is based on "pattern and persistence," supplemented by limited knowledge of the dynamic processes of the atmosphere and the ocean. One must rely on the facts that changes in the upper layer of the ocean, which are fundamentally due to the wind-driven circulation and to the water and heat exchanges between sea and atmosphere, tend to occur in repetitive patterns, and that anomalies tend to persist for some weeks. Since the ocean is considerably more sluggish in its changes than is the atmosphere, the ocean forecaster is somewhat better off than the local weather forecaster. It has been observed that a week in the ocean is comparable to a day in the atmosphere.

Such forecasting has enabled oceanographers, in some cases, to forecast location and success of fisheries for periods of a few weeks to a few months. For example, upwelling in the Gulf of Panama, which influences the abundance of pink shrimp in shallow waters during the winter months, can be usefully forecast. Early spring forecasts of the success of the skipjack fishery near Hawaii during the summer have been reasonably successful. Likewise, from temperature and salinity distributions and trends off the U.S. west coast, forecasts have been made each year of the expected catches of albacore and bluefin tuna, and of the areas of good albacore fishing; these forecasts have been generally useful, although they sometimes have been far from precise. The continuing monitoring of the Peru Current by the Instituto del Mar del Peru has been quite successful in enabling short-term forecasting of the anchovy fishery.

Through the prospective development of the World Weather Watch, under the auspices of the World Meteorological Organization, and with the cooperation of many nations, we appear to be on the threshold of much better ocean forecasting, through the monitoring of atmospheric circulation and heat exchange between the sea and atmosphere. These are the principal driving forces on the upper layers of the sea, and there is increasingly better understanding of the dynamic relationships between them and the ocean circulation. It should be possible, given an adequate network of stations for observing of the atmosphere over the sea and of the upper layer of the ocean, both to keep track of what the ocean is doing, in real time, and to forecast changes that will affect the fisheries. It is hoped that the next decade will witness the establishment of an observational net sufficient to describe the entire physical system, consisting of the atmosphere and the upper mixed layer of the sea, for at least the entire northern hemisphere, and, hopefully, for the whole globe. Such a description is essential to support a really large advance in meteorological and oceanographic forecasting. The cost of such a data acquisition system, and associated processing by computers will be very large. However, the same kind of forecasting that is needed by fisheries interests is also needed for other purposes, including weather forecasting, ship routing and many aspects of military ocean operations. An adequate system is now within our technical competence.

MILNER B. SCHAEFER

Cross references: *Minerals of the Ocean; Tuna Fisheries.*

OYSTERS

Oysters are found worldwide in a broad band between latitudes 64°N and 44°S. They are distributed from the intertidal zone down to approximately 130 ft (Merrill and Boss, 1966). Gunter (1950) recognized three living genera—*Ostrea, Crassostrea,* and *Pycnodonte*—which include about 100 living species. The genus *Ostrea* is larviparous, lacks a promyal chamber, and has relatively large gill ostia; *Crassostrea* is oviparous, has a promyal chamber and has relatively small gill ostia; and *Pycnodonte* has a broad hinge, is nonincubatory, and has a shell apparently filled with numerous bubbles or empty cells (Abbott, 1954). The commercial edible oysters belong to the first two genera; some of the most common species include *Ostrea edulis, O. lurida, Crassostrea virginica, C. gigas, C. angulata, C. commercialis, C. cucullata,* and *C. chilensis.*

Natural History

Shell. The soft body of an oyster is covered by two shells or valves; the right or top valve is flat and the left or bottom valve is heavier and cupped. An oyster attaches and usually rests on its left valve. The two shells are joined together at the hinge by an elastic material known as the ligament. The shape of the shell is highly variable in *Crassostrea* and subcircular in *Ostrea*. More than 95% of the shell weight is calcium carbonate. Other compounds include manganese carbonate, calcium sulfate, silica, various salts of manganese, iron, and aluminum, traces of heavy metals, and organic matter (Galtsoff, 1964). The shell has four layers: an extremely thin outer periostracum; a medium prismatic; an inner calcite ostracum, which is the major portion of the shell; and the hypostracum, a very thin layer of arogonite at the area of muscle attachment.

Growth of shell depends directly on water temperature. For example, along the east coast of North America, the growing season is longer in the south than in the north. In Canada and New England, the season of shell growth is from 4–5 months (Fig. 1); in Chesapeake Bay the season lasts 6–7 months (Fig. 2); and in the warm waters of Florida, oysters grow almost all year. Oysters in northern waters require about 5 years

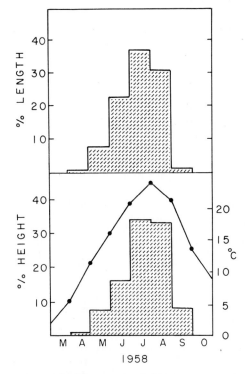

Fig. 1. Monthly increase in height and length of oysters as percentage of total year's growth, Cape Cod, Massachusetts. Temperature curve based on records taken at time of measurement.

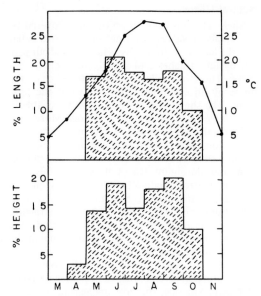

Fig 2. Monthly increases in height and length of oysters as percentage of total year's growth, Chesapeake Bay, Maryland. Temperature curve based on records taken at time of measurement.

to reach market size (4 in.); in Florida, this size is attained in some areas within 18 months.

Soft Body. The body of the oyster is almost completely covered by a membrane called the mantle (Fig. 3). Its principal function is the formation of shell. A pair of gills is on the ventral side of the oyster. Along with the mantle, the gills play a major role in respiration (Galtsoff, 1964). Cilia on the gills have a dual function—to force water through openings called ostia, and to transport particles of food to the mouth. The gills also disperse sex cells in oviparous oysters and incubate fertilized eggs in larviparous oysters.

The mouth, at the anterior end of the body, is covered by two pairs of labial palps. The ciliated palps sort the food particles before they enter the mouth. The particles entering the mouth are transported into a crescentic esophagus, which leads into the stomach (Shaw and Battle, 1957). Divisions of the stomach include the anterior chamber containing a complex caecum that extends into anteriorly and posteriorly directed spiral appendices. An extensive area of the left ventral wall of the posterior chamber of the stomach is covered by a translucent gastric shield, that fits into a corresponding elevation of the wall. Caudal to the shield is an elongated chamber that is incompletely divided by two typhlosoles into a style-sac and a mid-gut. The mid-gut leads to the ascending, median, and descending limbs of the intestine; the latter merges into the rectum. The digestive tract terminates at the anus on the dorsal surface of the adductor muscle.

The stomach is surrounded by the greenish-brown tubules of the digestive gland. Four openings connect the stomach with this gland (Chestnut, 1950). The complete digestive tract is lined by simple, ciliated, columnar epithelium except for the upper lip or fused external palps, the lower side of the gastric shield, and the tubules of the digestive diverticula.

In the European oyster, *O. edulis*, and in the American oyster, *C. virginica*, particles not only enter from the esophagus to the stomach by cilia, but probably are also pulled in by the action of the rotating crystalline style (Yonge, 1960). Mucous threads, mixed with food particles, rotate clockwise around the head of the style, and tracts of cilia carry them into the ducts of the digestive tubules. Other particles enter the food-sorting caecum and are separated by size. The large particles are passed into the intestinal groove to be added to the waste materials from the digestive gland. As the waste materials pass from the intestine to the rectum, they are consolidated into pellets, a compact form that does not foul the respiratory chamber into which the anus opens.

According to Yonge (1926) extracellular enzymes are released by the slow liquefaction of the style. Mansour-Bek (1946) has found extracellular proteolytic and lipolytic enzymes in the stomach juices of lamellibranchs. The chief enzyme liberated by the style is amylase, which converts

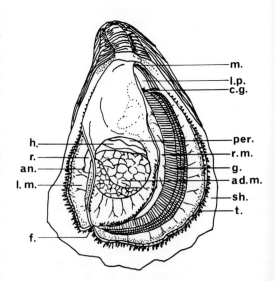

Fig. 3. Soft body of oyster, *C. virginica*, after removal of right valve: ad. m. adductor muscle; an., anus; c.g., cerebal ganglion; f., fusion of two mantle lobes and gills; g., gills; h., heart; l.m., left mantle; l.p., labial palps; m., mouth; per., pericardium; r., rectum; r.m., right mantle; sh., shell; t., tentacles. The right mantle is curled up to expose the gills. Portion of the mantle over the heart region was removed. (Drawn from Galtsoff, 1964.)

starches into simple sugars. In a more acid condition (pH 5.2) in the sac, the style is relatively firm, but in the less acid medium rising over pH 6.0 the style is absent. (The optimum pH for the action of amylase is about 5.9.) The formation and dissolution of the style are nicely balanced in a healthy oyster. In addition, the style contains a powerful oxidase system, which helps to supply the tissue with oxygen (Yonge, 1960). The role of the gastric shield, which is chitinous (Shaw and Battle, 1957), is to act as a base for the grinding action of the style and to protect the underlying epithelium from damage by the head of the rotating style.

The heart, which lies above the adductor muscle in the pericardial chamber, consists of a single ventricle and a pair of pigmented auricles. Blood pumped through the anterior aorta, flows to the organs of the visceral mass; blood pumped through the posterior aorta, flows to the adductor muscle. The ventral and dorsal circumpallial arteries supply blood to the mantle. An oyster has an open circulatory system where sinuses (open spaces) replace capillaries.

The venous system is composed of sinuses, afferent and efferent veins, and small vessels of the gills. Blood is collected in the veins, carried either to the gills or to the kidneys, and returned to the heart through the common efferent vein. The oyster also has paired tubular structures known as the accessory hearts. They lie along the inner surface of the right and left mantle fold where they form the cloacal chamber. Their primary function is the oscillation of blood in the mantle.

Oyster blood is colorless and contains neither hemoglobin nor hemocyanin; blood cells are of two types—hyaline and granular amoeboid (also called amoebocytes or phagocytes).

The adult oyster has one adductor muscle, the posterior. In *C. virginica* it is divided into a smaller, crescent-shaped, white or opaque portion and a larger translucent portion. In *O. edulis* the opaque portion is the larger (Yonge, 1960). The muscle controls the opening and closing of the shell. The larval form has an anterior adductor muscle, but it disappears after metamorphosis.

Major components of the nervous system are a pair of cerebro-pleural ganglia (Fig. 3) at the base of the labial palps, and a pair of visceral ganglia in the adductor muscle. The two pairs of ganglia are joined by the cerebro-visceral connection. Many nerves originate from the ganglia, and travel to all parts of the body. One of the longest is the circumpallial nerve, which runs along the edge of the mantle. The adult oyster has only two sense organs, the pallial organ that is attached to the anterior side of the adductor muscle and the tentacles of the mantle edge.

The larvae also have a pair of dark pigmented eyes and a pair of statocytes in the foot (Fig. 4).

The excretory system consists of a pair of tubular nephridia, one on each side of the visceral mass near the heart. The pericardial glands, wandering phagocytes, and mantle epithelium also play an excretory role in the oyster.

The reproductive organs lie between the digestive diverticula and the surface epithelium. The thickness of the gonad depends on the ripeness of the oyster; by-products are discharged through the gonoducts between the adductor muscle and gills. In *Crassostrea*, the sexes are separate and hermaphrodites are rare, but *Ostrea* is ambisexual, containing both male and female germ cells. In the former genus, the egg and sperm are discharged into open waters where fertilization takes place, in *Ostrea*, fertilization occurs inside the inhalant chamber and the larvae are incubated from a week to ten days. The length of larval life varies with temperature, food, and other environmental factors. The larvae can swim weakly by the use of a highly developed velum although they are generally transported by currents and tides.

Just before setting, the larvae develop a pair of eye spots and a foot (Fig. 4). This period is probably the most critical in an oyster's life, for unless suitable substrate is available for attachment, the oyster dies. Oyster shells are the most common cultch material, although the larvae will attach to almost any clean, hard surface.

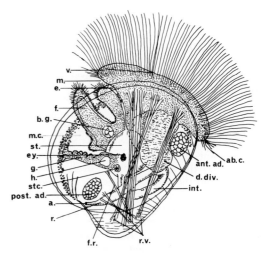

Fig. 4. Fully developed osyter larvae, *C. virginica*, view from the left side, in swimming position: a., anus; ab. c., aboral circle of cilia; ant. ad., anterior adductor muscle; b.g., byssus gland; d. div., digestive diverticula; e., esophagus; ey., eye; f., foot; f.r., foot retractor muscles; g., gill rudiment; h., heart; int., intestine; m., mouth; m.c., mantle cavity; post. ad., posterior adductor muscle; r., rectum; r.v., velar retractor muscle; st., stomach; stc., stastocysts; v., velum. (Drawn from Galtsoff, 1964.)

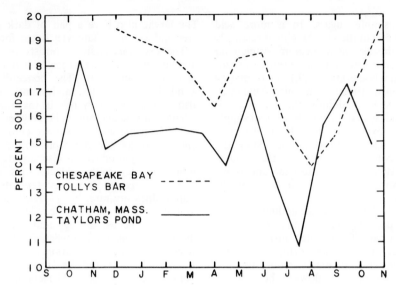

Fig. 5. Monthly averages of percentage solids for Chesapeake Bay and Chatham, Massachusetts, oysters.

Oyster Condition or Meat Quality. Oyster meats contain about every element found in seawater. Meat yield per given number of shucked animals is of importance commercially. The amount of meat varies with geographic location and the time of year. In the U.S., the meat yield is generally greater in northern waters, especially in the winter. Meat quality drops during the spawning season. After spawning, oysters build up glycogen and are in their best condition at the height of this buildup. These seasonal changes in condition are sometimes measured by calculation of "percentage solids", that is, dry weight of meat multiplied by 100 divided by the wet weight of meat (Fig. 5). In the spring, the glycogen is converted into sex products. When ripe, an oyster appears "fat" because of a thick gonadal layer.

Predators. Wherever oysters are found, there are associated pests or enemies. In high salinity (above 20 ppt salt) waters of the U.S., two of the most serious carnivorous gastropods that prey on oysters are the Atlantic oyster drill, *Urosalpinx cinerea*, and the thick-lipped drill, *Eupleura caudata*. Both species occur on the east coast of the U.S., and have been introduced to the west coast and to Europe. Other serious snail predators include the rock-shell, *Thais haemastoma*, the knobbed whelk, *Busycon carica*, and the channeled whelk, *B. canaliculatum*, on the Atlantic coast; and the dwarf triton, *Ocenebra japonica* (introduced from Japan), on the Pacific coast. Another serious oyster predator, especially in Long Island Sound, is the starfish, *Asterias forbesi*.

In lower salinity waters, below 15 ppt, the above predators are usually absent. Instead the turbellarians of the genus *Stylochus* (Fig. 6) and *Pseudostylochus* kill oysters, especially the young spat under 1 in. long. Other predators include several species of crabs (rock crab, blue crab), fish (of which the most serious are stingrays and drum), and ducks such as bluebills and scoters.

Diseases and Parasites. The oyster is vulnerable to many diseases. Several diseases have all but wiped out some oyster populations. In 1915 and 1916 at Prince Edward Island, Canada, about 90% of the oysters were killed by an unknown agent called Malpeque Bay disease. Populations in the diseased area have slowly increased from apparently resistant parents. The disease is still

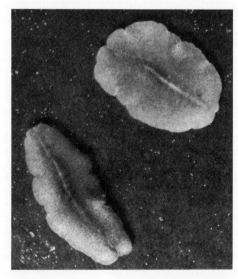

Fig. 6. The flatworm, *Stylochus ellipticus*, a serious oyster predator in the upper portion of the Chesapeake Bay.

present; oysters introduced from other areas die at the original rate. More recently a haplosporidan, *Minchinia nelsoni*, nearly wiped out the oyster population of Delaware Bay. This disease was then discovered in lower Chesapeake Bay where it seriously depleted the stocks. In Chincoteague Bay, Maryland-Virginia, both *M. nelsoni* and another haplosporidan, *M. costalis*, are causing oyster mortality. Other parasites of oysters include *Bucephalus haimeanus*, *B. cucullus*, *Mytilicola intestinalis*, *Hexamita* sp., and *Nematopsis* sp. A shell disease, thought to be a fungal infection, caused serious mortalities in Holland. In the southern waters of the U.S., a fungus, *Dermocystidium marinum*, inflicts periodic mortalities. High salinity, high temperature, and crowding on oyster beds contribute to the spread of the disease.

Commensals and Competitors. The oyster provides an excellent habitat for commensals and competitors. Many live in the shell, e.g., the boring sponge, *Cliona*, the boring clam, *Diplothyra*, and the mud worm, *Polydora*. The outer shell is also a collector for many types of fouling organisms. The most common are the barnacles, tunicates, mussels, and slipper limpet. Several species of algae grow on the oyster's shell. One species, *Codium fragile*, produces gas-filled branches that are capable of lifting an oyster off the bottom so that it is carried away with the tide. The most common commensal, considered by Stauber (1945) to be a parasite of the oyster, is the pea crab, *Pinnotheres ostreum*. This crab lives between the shell and soft body of the oyster, and has been reported to feed on the gills (Galtsoff, 1964); it also intercepts food particles collected by the oyster, thereby interfering with the bivalve's feeding.

Pollution, Recreation, and Marsh Land Developments. Pollution, domestic and industrial, has seriously affected oyster production in many areas. Galtsoff (1964) stated that domestic wastes have a threefold effect in that the sludge smothers oyster beds, reduces the oxygen content of water, ann increases the bacteria count of the water. A variety of industrial wastes enter oyster-producing waters. The most serious is probably the wastes from pulp and paper industries. The higher the waste concentration the longer the oyster remains closed. Therefore the pumping rate of oysters is reduced, and the oyster's condition declines.

A relatively new type of pollution, known as thermal pollution, comes from electric power plants, where large volumes of water are used for cooling. The water returned to the river, much warmer than before, can raise the water temperatures over oyster beds located nearby. It is not certain whether this type of pollution is beneficial or harmful to the oysters. The effects of thermal pollution will vary with the geographic location of power plant, the amount of temperature change, and the season at which it occurs.

Increased recreational use of waterways increases pollution from boat marinas and from boat traffic. In some areas this kind of pollution has forced closure of oyster beds to fishing. Also, the building marinas and waterfront developments are diminishing the productive marsh land that supplies the nutrients for aquatic plant life on which oysters feed.

Oyster Cultivation

Evidence of the first oyster cultivation dates back to the Roman era (Yonge, 1960). As far as is known, the Romans were first to grow oysters by hanging them off the bottom. As increased research has brought greater knowledge about oysters, culture techniques have become considerably refined. In France in 1857, Dr. M. Coste, an embryologist, found that by placing bundles of branches in certain areas, he could obtain excellent oyster sets (Yonge, 1960). His initial research laid the groundwork in developing France as one of the leading oyster producing areas in Europe.

The Japanese are leading oyster producers in Asia. At one time, all their oysters were grown on the bottom. In 1923, two Japanese biologists, H. Seno and J. Hori, developed the hanging method, which gave excellent results (Cahn, 1950). They found that by suspending oysters: (1) more could be grown per unit area of bottom; (2)

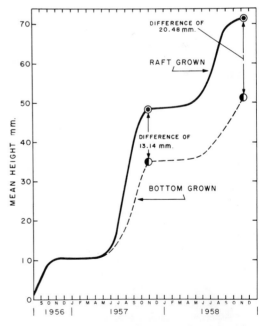

Fig. 7. Growth curves of raft-grown and bottom-grown Wareham River oysters at Chatham, Massachusetts. Symbols indicate when shell growth stopped for the year.

Fig. 8. Oyster tonger in Chesapeake Bay.

growth rate could be doubled (Fig. 7); (3) oysters could be grown where bottom was unsuitable; (4) meat quality was improved; and (5) oysters are out of reach of bottom predators. Now almost all the oysters cultivated in Japan are grown off-bottom. In Australia, stick and tray cultivation of oysters is extensive. The oysters are caught and held on sticks for two years and then scraped off and placed in wire netting trays for another year before marketing (Anon., 1966).

Little oyster culture is undertaken in the U.S.; most of the industry depends on natural maintenance of stocks. Some attempts at culture include shell planting to catch seed or improve bottom, movement of seed to growing grounds, and, frequently, to fattening areas just before harvesting. The oysters are fished by a variety of methods depending on geographical area. In Long Island Sound, Connecticut, the oysters are collected by power dredges, but in Chesapeake Bay, Maryland, hand tonging (Fig. 8), patent tonging, and sail dredging (Fig. 9) are used. Oysters on the west coast are harvested with huge hydraulic dredges.

Fig. 9. Sail dredging in Chesapeake Bay.

Methods of Seed Production. For a successful industry, seed oysters must be readily available. Methods of collection vary with each country. The Japanese catch seed oysters on shells suspended from rafts or on shells draped over wooden racks (Cahn, 1950). In France, large quantities of seed are caught on lime-coated tiles, which are laid in rows in the intertidal zone (Yonge, 1960). Australian seed oysters are caught on sticks (Anon. 1966). In the U.S., most oysters are caught on shells scattered over the bottom. At Fisher's Island, New York, spat are collected from shells suspended from rafts, and in Debob Bay, Washington, oysters are caught on shells suspended from rafts and on shells draped over wooden racks. The commercial potential of seed production with rafts is being investigated by the U.S. Bureau of Commercial Fisheries and the Maryland Department of Chesapeake Bay Affairs in Chesapeake Bay (Fig. 10).

Recently, commercial shellfish hatcheries have been established in the U.S., mainly in the Long Island Sound area. The purpose of these commercial hatcheries is to produce seed oysters in areas where natural setting is infrequent or no longer occurs.

Commercial Potential

The oyster has always been an important food resource for the world population. This importance is demonstrated by the fact that the annual world production by the 1950's was approximately 1.6 billion lb. Leading producers are Japan, France, and the U.S.; other important oyster-producing countries are Australia, Canada, Philippines, Great Britain, and Korea.

The development of off-bottom culture in Japan during the 1930's caused oyster production in that country to boom. Presently most areas in Japan, especially in the protected bays, are being utilized, and annual production has remained nearly constant between 30,000 to 35,000 metric tons of oyster meats. Production in Japan could increase with the further development of the longline technique (Cahn, 1950). This method would make less sheltered waters available for oyster culture. Production in France, by present culture methods, will probably not increase appreciably, since most areas suitable for oyster cultivation are being used.

In the U.S., oyster production has declined since 1880. The most striking decrease has been in the Chesapeake Bay region (Fig. 11) where landings dropped from 117.4 million lb of meats in 1880 to 21.2 million lb in 1965. Yet the Chesapeake Bay region probably still is the region with the greatest potential for producing oysters in the U.S. Although diseases have recently reduced populations in the Virginia portion of the Bay, the upper Bay in Maryland is relatively free from disease and predators. (Occasionally, heavy run-offs from the Susquehanna River can cause mortality in the Upper Bay). Oyster setting is excellent each year in many of Maryland's tributaries, e.g., 1.3 million bu of seed oysters were harvested in the spring of 1966 (Maryland Board of Natural Resources, 23rd annual report, 1966). By expanding the seed-producing areas in these tributaries, by making the seed available to private planters, and by increasing the number of leased

Fig. 10. Experimental rafts used by the Bureau of Commercial Fisheries Biological Laboratory, Oxford, Maryland, to catch seed oysters. Strings and bags of shells are suspended from the rafts as cultch for attachment of oyster larvae.

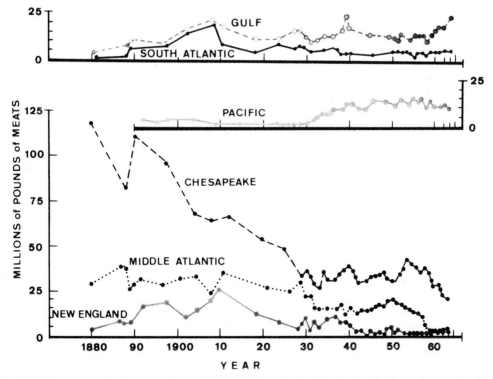

Fig. 11. U.S. oyster production by regions—Gulf of Mexico, South Atlantic, Pacific Coast, Chesapeake Bay, Middle Atlantic, and New England—from 1880 to 1964. (From Engle, 1966, after McHugh, 1963.)

acres for private farming, oyster production in the Upper Bay could be increased greatly.

The oyster industry of New England produced 26.3 million lb of meats in 1910, but in 1965 the region produced only 0.3 million lb. Many acres are still available for growing oysters, but because of the failures in natural setting, the lack of young oysters to plant on these grounds is serious. Oyster landings in the area can be increased only by substantial increase in production from commercial hatcheries; changes in methods of growing oysters—possibly through use of off-bottom techniques (Shaw, 1962, 1967); and development of methods of predator control.

Oyster production in the South Atlantic region has fluctuated only slightly over the past 25 years (Fig. 11). Development of more private oyster farming could increase production in this region.

Production in the Gulf of Mexico region is sometimes seriously affected by hurricanes and heavy rains, which can almost wipe out oyster populations. Also, the disease *Dermocystidium* can seriously deplete stocks in areas of high salinities. Recent heavy demands for oysters from northern states have caused production in the Gulf to increase. The danger in the Gulf region is that natural populations will not be able to maintain themselves under increased fishing pressure (Engle, 1966). Development of more private oyster farming would reduce the drain on the natural oyster bars.

Oyster production along the west coast of the U.S. increased rapidly during the 1930's with the introduction of seed of the Japanese oyster, *Crassostrea gigas* (Fig. 11). Since 1940, landings have fluctuated between 8 and 13 million lb of meats. Greater production in the future will depend on market demands, availability of seed, and development of new growing areas and methods.

WILLIAM N. SHAW

References

Abbott, R. T., "American Seashells," 1954, D. Van Nostrand Co., Princeton, N.J.

Anonymous, "Oyster harvesting, Georges River, N.S.W.," *Australian Fisheries Newsletter*, **25**(7), 24–25 (1966).

Cahn, A. R., "Oyster culture in Japan," 1950, U.S. Fish and Wildl. Serv. Fishery Leaflet 383, 1–80.

Chestnut, A. F., "Studies on the digestive system of the oyster," *Proc. Natl. Shellfish. Assn.*, 11–15 (1950).

Engle, J. B., "The molluscan shellfish industry—current status and trends," *Proc. Natl. Shellfish. Assn.*, **56**, 13–21 (1966).

Galtsoff, P. S., "The American oyster *Crassostrea virginica* Gmelin," *U.S. Fish and Wildl. Serv. Fish. Bull.*, **64**, 1–480 (1964).

Gunter, G., "The generic status of living oysters and the scientific name of the common American species," *American Midland Naturalist*, **43**(2), 438–449 (1950).

Mansour-Bek, J. J., "Extracellular proteolytic and lipolytic enzymes of some lamellibranchs," *Nature* (London), **158**, 378–379 (1946).

McHugh, J. L., Statement before the Sub-Committee on Fisheries and Wildlife Conservation of the Committee on Merchant Marine and Fisheries, U.S. House of Representatives, Washington, D.C. 1963, Hearings Document Serial No. 88-13, p. 138–147.

Merrill, A. S., and Boss, K. J., "Benthic ecology and faunal change relating to oysters from a deep basin in the lower Patuxent River, Maryland," *Proc. Natl. Shellfish. Assn.*, **56**, 81–87 (1966).

Shaw, B. L., and Battle, H. I., "The gross and microscopic anatomy of the digestive tract of the oyster *Crassostrea virginica* (Gmelin)," *Can. J. Zool.*, **35**, 325–347 (1957).

Shaw, W. N., "Raft culture of oysters in Massachusetts," *U.S. Fish and Wildlife Serv. Fish. Bull.*, **61**, 481–495 (1962).

Shaw, W. N., "Advances in the off-bottom culture of oysters," "Gulf and Caribbean Fish. Inst., 19th Ann. Sess. Nov. 1966," 1967.

Stauber, L. A., "*Pinnotheres ostreum*, parasitic on the American oyster, *Ostrea* (*Gryphaea*) *virginica*," *Biol. Bull.*, **88**(3), 269–291 (1945).

Yonge, C. M., "Structure and physiology of the organs of feeding and digestion in *Ostrea edulis*," *J. Mar. Biol. Assn.* (U.K.), **14**, 295–386 (1926).

Yonge, C. M., "Oysters," 1960, Willmer Brothers and Haram Ltd., Birkenhead.

Cross references: *Pearling Industry; Pollution of Seawater.*

P

PACIFIC FISHERIES

Tropical and Subtropical

The tropical and subtropical Pacific Ocean covers 23.5% of the globe (Fig. 1). Along the Equator, from the coast of Ecuador to the western shore of Sumatra, it reaches slightly more than halfway around the earth. From north to south, it covers one-third of the distance between the poles.

Land and water are most unevenly distributed. Although there are thousands of islands in the tropical and subtropical Pacific Ocean, ranging in size from a few hundred square meters to the 794,094 km^2 bulk of New Guinea and the 751,082 km^2 of Borneo, the ratio of land to water is only about 1:40.

The contribution of the tropical and subtropical Pacific Ocean to the world's marine harvest is large, accounting for about 14.7 million metric tons, or 32% of the world catch, more than half of which comes from the Southern Hemisphere. In production of marine products the region far exceeds those parts of the Indian and Atlantic Oceans that lie between latitude 30° N and 30° S, producing about four times as much per unit area as the former, about twice as much as the latter.

The numbers just given include only the commercial catch. The actual catch—all the fish taken from the sea—is a matter of conjecture, for some fish are taken at sea in commercial catches and discarded, and data on the landings from subsistence and sports fisheries are not available. The addition of these catches, particularly the subsistence fisheries of the developing nations, would probably increase these totals considerably.

Quantitative estimates of the magnitude of a fishery resource, particularly in a region as large and varied as the tropical and subtropical Pacific Ocean, can vary widely, as Schaefer (1965) pointed out. A review of the "fishery resources" of the Pacific Ocean made no more than 20 years ago could not have predicted, except in the most general terms, some of the striking recent developments. "Best estimates" are short-lived; they have been repeatedly exceeded as large and profitable new fisheries have come into being. It would have been almost impossible to forecast the development in little more than a decade of the Philippine fishery for round scad (*Decapterus russelli*) from a subsistence level to a yield of about 200,000 metric tons a year, or the rapid growth of the Peruvian catch of anchoveta, the Peruvian anchovy (*Engraulis ringens*). The term "fishery resource," then, must be largely confined to the present yield of the commercial fisheries; this cautious course is followed here.

The oceanography of the tropical and subtropical Pacific Ocean has been much studied and well described. (For a succinct treatment, see the entry "Pacific Ocean" and those for the adjacent seas in Vol. I.). Its general features are the great northward flow of cool water enriched by upwelling, along the coast of South America, a somewhat broader and less rich flow southward along Mexico and Central America, the westward turning of both these great streams as they approach the equator, a decreasing richness (as measured in terms of zooplankton, Fig. 2) as the great currents sweep westward across the open Pacific, relative poverty, as measured by standing crops of zooplankton, in most areas of the open sea, and increasing productivity in the sheltered seas off the Asian coast. For most of the year the seas in the greater part of this region have temperatures higher than 20° C. The 20° C isotherm fluctuates seasonally, reaching the southern shores of Japan in the northern summer and lying several hundred miles offshore of South America in the same season, but its average position coincides reasonably well with latitude 30° N and 30° S.

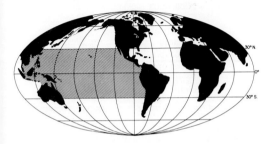

Fig. 1. The tropical and subtropical Pacific Ocean (the area between latitudes 30°N and 30°S), which covers almost one-fourth the globe, has about 29% of the world's population and supplies 32% of the world's catch of marine fish and other animals of the sea.

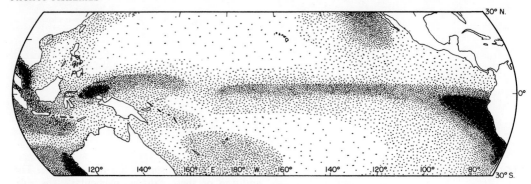

Fig. 2. Density of forage organisms (as measured in milligrams per cubic meter of zooplankton biomass in upper 300 m) of the tropical and subtropical Pacific Ocean. (Adapted from Christy and Scott, 1965; courtesy Resources for the Future, Inc. and Johns Hopkins University Press.)

The fisheries of the region are in three main categories: (1) those of the Americas; (2) those of Oceania; and (3) those of Southeast Asia. There are certain immediately apparent differences among them in size, quality, technique, and disposition of the catch:

(1) The fisheries of the Americas, in weight, are the greatest. They are highly specialized: 93% of the catch consists of fish of a single species, the Peruvian anchovy. The prevalent fishing method is seining. Most of the catch is exported as fish meal.

(2) The widespread fisheries of Oceania are small and concentrated on a single genus (*Thunnus*); almost all the catch is taken by a single nation, Japan, by longline and used for food.

(3) The fisheries of Southeast Asia, about three-fourths those of the Americas in weight, draw upon a large group of families, genera, and species, so that only one or two of the individual fisheries approach in weight even the tuna landings of Oceania. Fishing techniques are varied: some fish are seined, some are caught by pole and line, some taken in traps, some cultured. Almost the entire catch is used as food by the fishing nations.

Fisheries of the Americas. The fish catches of the tropical and subtropical Americas, i.e., the area lying between Baja California on the north and the port of Coquimbo, Chile, on the south, are dominated in weight (although not necessarily in terms of their contribution to the world economy) by the anchoveta catches of Peru and Chile (Table 1). This small (about 10 cm), short-lived herbivore is found in the relatively narrow belt of cool, enriched waters of the Peru Current. The catch is seasonal, reaching its peak in the southern summer.

The anchoveta was long known chiefly as the prey of the guano birds; it supported a bird population that numbered in the tens of millions, the excreta of which, collected from the offshore islands, for decades formed a ready source of high-grade fertilizer. Within the past 15 years, the guano bird has faced competition with fishermen, however, and the anchoveta has been put to more direct use. Schaefer (1965) said that 7.6 million tons are taken by the fishermen and 2.5 million tons by the guano birds. He pointed out that the yield of these waters is among the greatest of any on earth, exceeding 400 lb per acre (the average yield of the entire tropical and subtropical Pacific Ocean is about 1 lb per acre).

As a result of the growth of the industry, Peru is now the world's leading fishing nation, in terms of weight landed. The anchoveta catch, however, affects the human diet indirectly, for it

TABLE 1. COMMERCIAL CATCHES OF FISH AND OTHER MARINE ANIMALS IN TROPICAL AND SUBTROPICAL EASTERN PACIFIC OCEAN, 1965
(in thousands of metric tons)[a]

Nation	Catch
U. S.	120
Mexico	87
Guatemala	4
El Salvador	8
Nicaragua	2
Costa Rica	3
Honduras	—
Panama	39
Colombia	5
Ecuador	54
Peru	7,462
Chile	468
	8,252

[a] Derived from data prepared by Food and Agriculture Organization of the United Nations, 1966. Pacific catches of those nations that fish in both the Atlantic and Pacific (e.g., Mexico), were estimated on the basis of material given in Institute of Marine Resources, 1965, as was the contribution of that part of the Chilean coast lying north of latitude 30°S.

is made into fish meal to be used as food for poultry and stock. Japan, traditionally the leading fishing nation, seems to have retained this distinction, if fish for direct human use is counted first.

The great anchoveta catch is of recent origin. In 1954, Peru had a catch of only 35,400 metric tons. The landings had increased nearly 100 times, to 3.3 million metric tons, by 1960, and reached a peak of 8.9 million metric tons in 1964. In weight, the anchoveta provides not only the largest single fish catch in the tropical and subtropical Pacific Ocean, but in the entire world. In price to the fishermen, however, it ranks low—about $9 a ton, as opposed to upward of $300 for some of the food fishes, e.g., tunas.

Second both in tonnage and value to the catch of the Peruvian anchovy is that for several members of the Scombridae, of which by far the largest is the California-based fishery for the tunas, particularly the yellowfin tuna (*Thunnus albacares*) and skipjack tuna (*Katsuwonus pelamis*). This fishery, which extends from the northern limit of the subtropics almost to the southern, provides an income to the fishermen of about $30,000,000. The fishery began before World War I, and has shown almost continuous growth (Matsumoto, 1966).

Third in weight and value is the shrimp catch of Mexico and her neighbors. Wherever the habitat is suitable (relatively shallow water, muddy bottom), from Guaymas to Panama City, shrimp are caught, primarily for export to the U.S.

Many of the fish of the Americas, as in other parts of the world's ocean, are underharvested or unharvested; some may be unharvestable by present methods. Some find their way into the subsistence fisheries. Every small port along the western coast of Mexico, for example, has a variety of delectable fresh fish for sale. As to quantity, if there were one canoe or other small craft for each mile of the 7000-mile coast, and each made 100 trips a year, averaging a catch of 200 lb of fish per trip, the total amount of fish taken off the Pacific coast of the Americas for human consumption would be an additional 70,000 metric tons.

Fisheries of Oceania. The problem of estimating the importance of the subsistence fisheries (Fig. 3) becomes critical when one looks at the fishery resources of Oceania (the shores of Queensland and the Northern Territory in Australia, and the myriad of Polynesian, Melanesian, and Micronesian islands that reach from Easter Island to New Guinea). Some of the most exotic microfisheries in the world are found in this area. An example is that in American Samoa for palolos—the wormlike sacs of eggs and sperm of a sea annelid (*Eunice viridis* Kramer), released in October or November by adults whose heads remain deep in coral crevices. The wriggling freed

Fig. 3. An American Samoan netting mullet. (Photo: U.S. Bureau of Commercial Fisheries Biological Laboratory.)

sections, about $\frac{1}{16}$ in. in diameter and 12 in. long, are considered a great delicacy by the Samoans. On the whole, there is only one major fishery in Oceania: the large Japanese longline catch of tuna (Table 2). There are many accounts of this remarkable enterprise. Borgstrom (1964), is a notable example; Japanese and American scientific literature carries a wealth of material, e.g., recent reviews are offered by Yoshida, 1966, Rothschild, 1966, and Otsu and Sumida, 1966.

In longline fishing, baited hooks are strung on a mainline that may stream as far as 100 km. The hooks fish at a depth of about 100 m (Fig. 4). The desired catch is the large tunas of the middepths, that is, albacore, bigeye tuna (*Thunnus obesus*), and yellowfin tuna. Other large fish are taken also, however, and in sufficient quantities to make a substantial contribution to the Japanese catch. These fish include several species of the billfishes and sharks.

Beginning in the early 1950's, the Japanese began to expand their longline fleet and extend its operation. Today, the Japanese longliners fish all three great oceans, shore to shore. In 1964, their catch of tunas and billfishes in the Pacific Ocean was 317,300 metric tons, in the Atlantic Ocean 118,800 metric tons, tons, and in the Indian Ocean 76,500 metric tons (Federation of Japan Tuna Fishermen's Co-Operative Associations, 1966).

The U.S. has fisheries in the Pacific, too, although they are much smaller. Hawaii lands about

TABLE 2. COMMERCIAL FISH CATCHES OF OCEANIA
(in thousands of metric tons)[a]

Insular	Catch
U.S. (Hawaii)	9
French Polynesia	5
American Samoa	—
Western Samoa	—
Cook Islands	—
Tonga	—
Fiji	—
New Caledonia	1
New Hebrides	—
Australia (Queensland and Northern Territory)	8
Solomon Islands	—
New Guinea and Papua	—
Trust Territory	1
	24
Open sea—longline	
Japan	
(Japan-based)	225
(Samoa-based)	12
(New Hebrides-based)	3
(New Caledonia-based)	3
(Fiji-based)	2
Republic of Korea[b]	
(Samoa-based)	
Republic of China[b]	
(Samoa-based)	
Seine and pole and line	
Japan	
(Trust Territory)	16
	261

[a] Principal source: Food and Agriculture Organization of the United Nations, 1966; the longline statistics come from Federation of Japan Tuna Fishermen's Co-Operative Associations and Japan Tuna Fisheries Federation, 1966; Japan Fisheries Agency, 1967; and Tamio Otsu, personal communication, and are for the year 1964; data on the Japanese pole-and-line catch are for the year 1965, and are from Tohoku Regional Fisheries Research Laboratory, n.d.
[b] Statistics not available.

6000 metric tons of fish a year, mostly the skipjack tuna, which is taken by pole and line. (The statistic, 9000 metric tons, given in Table 2 was for 1965, a phenomenally good year.)

In the South Pacific, American Samoa is the site of two American canneries that have substantially bolstered the economy. The fish catch is provided by vessels from Japan, the Republic of China (Taiwan), and the Republic of Korea. The last two nations are now energetically competing with the larger Japanese fleet. At Palau in the Trust Territory of the Pacific Islands, is an American installation at which tuna are frozen for transshipment. The catch consists chiefly of tunas taken by Okinawan vessels operating under an agreement with the Trust Territory.

The U.S. interest in the fisheries of the tropical and subtropical Pacific Ocean extends beyond direct involvement in catching and processing. The nation imports fish and fish products from Oceania as well as Japan and the republics of Central and South America.

In Oceania, as in the Americas, very few of the many kinds of animals of the sea are taken. For example, in Western Samoa there are more than 600 known species of fish. Yet the Western Samoans find it more feasible to import their fish, canned, from South Africa, than to catch it themselves. These canned fish imports accounted for 11.3% of the value of the total imports in 1965. Not all the islands present so extreme a picture (in Hawaii about 60 of 600 species of fish are caught and sold on the market) but Oceania is a veritable sea of tunas and little else as far as commercial fisheries go.

Fisheries of Southeast Asia. The peoples of Southeast Asia have always turned to the sea as a source of animal protein. The Philippines, for example, are at once a fish-catching (668,000 metric tons in 1965) and fish-importing nation (Martin, 1963, estimated imports of 65,000 metric tons annually). They obtain 7 g of their daily 13 g of animal protein from fish, whereas the Mexicans obtain only 1 g of their 20 g a day (Abbott, 1966).

The Indo-Pacific fish fauna, which spreads from Indonesia eastward to Central America (although it is attenuated there) is probably the most diverse in the world. Of the thousands of species it contains, only a few, of course, are used as human food, but these few embrace a wide variety of marine animals. FAO's Yearbook, for example, lists ten broad groups of marine fisheries; Southeast Asia makes catches from eight of these (Table 3).

Fig. 4. Longline fishing is used to catch the large tunas and other fish of the middepths, a depth of about 100 m.

TABLE 3. NUMBERS OF SEPARATE MARINE FISHERIES (FAMILIES, GENERA, OR SPECIES) LISTED IN FAO'S YEARBOOK FOR ENTIRE WORLD AND FOR THE SOUTHEAST ASIAN COUNTRIES THAT BREAK DOWN THEIR CATCHES TO THESE GROUPS

Groups	World	Southeast Asia
Flounders, halibuts, soles, etc.	15	—
Cods, hakes, haddocks, etc.	17	—
Redfishes, basses, congers, etc.	29	15
Jacks, mullets, etc.	13	9
Herrings, sardines, anchovies, etc.	16	3
Tunas, bonitos, skipjacks	9	7
Mackerels, billfishes, cutlassfishes, etc.	8	7
Sharks, rays, chimaeras	7	3
Crustaceans	13	4
Mollusks	9	7

The total catch of the marine fisheries of Southeast Asia is demonstrably very great (Table 4), but these statistics do not tell the whole story. For example, the subsistence fisheries appear to be largely unreported. Applying the same conversion factors as in the Americas to the 33,612 nautical miles of shoreline in Southeast Asia would yield an additional catch of more than 300,000 metric tons a year. And the commercial catch statistics from Southeast Asia are perhaps less reliable than those from some other areas. Indonesia has reported no catches since 1963. The total catch of Mainland China from Shanghai southward, the section within the subtropics, is simply a matter of conjecture. In the prewar years, this part of the Chinese coast contributed about 60% of the marine catch (Shindo, 1964). If the last reported catches from Mainland China, which may or may not be accurate, are projected to 1965, and 60% is assigned to the section from Shanghai south, then the catch would be about 2.4 million metric tons. This rather shaky estimate has been used in Table 4, but the actual catch could be much more or much less. It seems hardly realistic to consider the fisheries of Southeast Asia without attempting some estimate, however.

In terms of food for direct human use, then, the commercial catch of Southeast Asia outweighs the catches of the other subregions of the Pacific by far. The catch is about 27 times as large as that of Oceania, and, except for Peru, about 8 times as large as that of the Americas. Southeast Asia takes its catches from a diversity of fishes and invertebrates unapproached by the other parts of the tropical and subtropical Pacific Ocean.

Research. Every nation in the tropical and subtropical Pacific region is conducting research on its fishery potentials. Even the small kingdom of Tonga (pop. 71,000), for example, has its fishery officer. On a larger scale, major research groups are supported by industry, by states, by national governments, by nations under treaties, and by the United Nations. The total number of professional scientists working on some aspect of the oceanography, biology, and fishery science of the tropical and subtropical Pacific Ocean would be easily more than 300 persons (Hiatt, 1963, listed about 250 at 29 laboratories).

This massive research effort has come into being for the most part within the past 20 years. It seems probable that within the next 20 years our knowledge of the fishery resources of the tropical and subtropical Pacific Ocean will increase and perhaps increase drastically.

Potential Yield. Several of the scientists concerned with the fishery resources of the tropical and subtropical Pacific Ocean have attempted, often necessarily on the basis of data that are sparse indeed, to estimate the yield of the region within the next few decades. Each area is discussed in turn.

Americas. A recent estimate sets the maximum sustainable yield of the anchoveta fishery of Peru at 7.5 million metric tons per year (Murphy, 1965). Some recent catches have been somewhat larger. The estimate is documented with statistics that show a declining catch per unit of effort over the past few years.

As is well known, the yellowfin tuna stock of the eastern Pacific is considered to be at a low level and the nations represented by the Inter-American Tropical Tuna Commission agreed to a quota of 77,500 tons in 1967. This catch was reached early in the season and fishing for yellowfin tuna ceased.

TABLE 4. CATCHES OF MARINE FISHERIES IN TROPICAL AND SUBTROPICAL PACIFIC OCEAN OF SOUTHEAST ASIA, 1965
(in thousands of metric tons)[a]

Nation	Catch
Ryukyus	24
Republic of China	369
Hong Kong	56
Philippines	668
Brunei	4
Indonesia	559
Malaysia	227
Singapore	12
Thailand	533
Cambodia	43
North Viet Nam	222
South Viet Nam	289
Mainland China	2,400
	5,406

[a] Food and Agriculture Organization of the United Nations, 1966 was principal source of statistics; Mainland China catch estimated as described in text.

TABLE 5. ESTIMATED MARINE FISHERY CATCHES AND HUMAN POPULATIONS, TROPICAL AND SUBTROPICAL PACIFIC OCEAN, 1965 AND 2000

Area	Catches (10^6 metric tons)				Population (10^6)		Fish/capita (kg)	
	1965	2000 (1965 × 1.5)	New	Total	1965	2000	1965	2000
Americas	8.2	12.3	—	12.3	100	200	82	62
Oceania	.3	.4	.4	.8	6	12	40	83
Southeast Asia	5.4	8.1	7.5	15.6	674	1,360	8	11
Total	13.9	20.8	7.9	28.7	780	1,572		

The shrimp catch of Mexico has reached its maximum and the sustainable harvest (for both the Atlantic and Pacific) will continue to be about 30,000 metric tons (U.S. Bureau of Commercial Fisheries, 1966), but somewhat increased catches seem to be in order for the smaller nations to the south.

Estimates for other fisheries are lacking.

Oceania. Strong evidence points to the existence of a large stock of skipjack tuna in the tropical and subtropical Pacific Ocean between the limits of the eastern Pacific fishery and the longitude of Hawaii. This evidence was summarized in Manar, 1966. If these fish can be caught—only a minuscule amount is taken at present—the maximum sustainable yield may be 400,000 metric tons per year (B. J. Rothschild, personal communication).

Little attention has been given the surface skipjack tuna resources of French Polynesia beyond a series of cruises by the U.S. Bureau of Commercial Fisheries in the late 1950's. Skipjack tuna was the most common tuna species caught. No estimates were made of the magnitude of the resource however.

On the eastern side of Oceania, substantially increased catches are seen for Queensland and the Northern Territory, especially the shrimp fisheries, which might yield as much as 20,000 additional tons a year.

In the Trust Territory, the Japanese have resumed their pre-war fishery for the skipjack tuna, reporting a catch of 16,000 metric tons in 1965. Before World War II, this fishery reached a catch of 33,000 metric tons. The fish are taken by Japan-based vessels, by the traditional method of pole and line and, with some success, by seining.

The probable future yields of the Japanese longline fishery in the tropical and subtropical Pacific Ocean are uncertain. Rothschild (1966) has found that the catch per unit of effort has been declining.

Southeast Asia. By far the largest increases are forecast for Southeast Asia. There are vast areas of shallow bottom in the South China Sea and between Indonesia and Australia. These waters have been little exploited. On the basis of the results of some exploratory fishing by Thailand, Tiews forecasts for the year 2000 a demersal catch of 6.5 million metric tons of fish and crustaceans in the trawl fisheries. He foresees an increase in the pelagic catch of 1.0 million tons. And he predicts that "the development of methods of mollusk farming might one day exceed even the whole fish and crustacean production in the region." Ignoring the mollusk harvest, the predicted total is 7.5 million metric tons.

Fisheries as Food Supply. Meanwhile, history makes its own forecasts: the world catch of fisheries has been increasing at a rate of about 7% a year. If one assumes that the rate of growth of the developed fisheries (Japanese longline, anchoveta, etc.) slows down to some extent, a total growth of 50% over the next few decades, i.e., until the year 2000, seems not unreasonable. Assuming that the new resources mentioned above come into production, then the catch of the tropical and subtropical Pacific Ocean would be 29.1 million metric tons (Table 5).

By the end of the century, the world's population is expected to have at least doubled. If all animal protein were obtained from the sea, every human being would require 100 kg of fish a year. It appears, then, that the fisheries of Southeast Asia, where it matters a great deal that supplies of animal protein be increased, might be able to supply an increasing percentage by the end of the century, although there appears little likelihood that fish would entirely support vast new increases in population. It would appear also that the Americas and Oceania may look forward to several more years of a profitable cash harvest from the sea. But again, one should emphasize that at present the true extent of the fishery resources of the tropical and subtropical Pacific Ocean is unknown; by the year 2000, when Table 5, at least, suggests things will certainly be no worse than they are now, new knowledge, new stocks, new techniques, may have altered the picture to make them much better.

THOMAS A. MANAR

References

Abbott, J. C., "Protein supplies and prospects: the problem," in Altschul, Aaron M. (Ed.), "World Protein Resources," 1966, Advances in chemistry series 57, American Chemical Society, Washington, D. C.

Borgstrom, G., "Japan's World Success in Fishing," 1964, Fishing News (Books) Ltd., London.

Christy, F. T., Jr., and Scott, A., "The Common Wealth in Ocean Fisheries," 1965, published for Resources for the Future, Inc., by Johns Hopkins Press, Baltimore, Md.

Fairbridge, R. W., (Ed.), "The Encyclopedia of Oceanography," 1966, Reinhold Publishing Corporation, New York.

Federation of Japan Tuna Fishermen's Co-Operative Associations and Japan Tuna Fisheries Federation. "Statistics of Japanese Tuna Fishery," 1966, Tokyo.

Food and Agriculture Organization of the United Nations. "Yearbook of Fishery Statistics, 1965," 1966, vol. 20.

Hiatt, R. W. (Ed.), World Directory of Hydrobiological and Fisheries Institutions, 1963, American Institute of Biological Sciences, Washington, D.C.

Institute of Marine Resources, "Review of the coastal fisheries of the west coast of Latin America," 1965, University of California, Institute of Marine Resources in collaboration with the Inter-American Tropical Tuna Commission, IMR Ref. 65-4, UCSD-34P99-7.

Japan Fisheries Agency, "Annual report of effort and catch statistics by area on Japanese tuna long line fishery, 1964," 1967, (in Japanese).

Manar, T. A. (Ed.), "Proceedings of the Governor's Conference on Central Pacific Fishery Resources," 1966, State of Hawaii, Honolulu.

Martin, C. "The fishing industry—an important food producer," in Huke, R. E. (Ed.), "Shadows on the Land: an Economic Geography of the Philippines," 1963, Bookmark, Manila.

Matsumoto, W. M., "Catch and effort statistics for the eastern Pacific tuna fishery," In Manar, T. A. (Ed.), "Proceedings of the Governor's Conference on Central Pacific Fishery Resources," 1966, State of Hawaii, Honolulu.

Murphy, G. I., "Preliminary analysis of the population dynamics of the Peruvian anchovy," 1965. mimeographed report to the Instituto del Mar del Peru.

Otsu, Tamio, and Sumida, R. F., "An estimate of total catch and effort of the Japanese longline fishery in the Pacific Ocean, 1953–63," in Manar, T. A. (Ed.), "Proceedings of the Governor's Conference on Central Pacific Fishery Resources," 1966, State of Hawaii, Honolulu.

Rothschild, B. J., "Major changes in the temporal-spatial distribution of catch and effort in the Japanese longline fleet," in Manar, T. A. (Ed.), "Proceedings of the Governor's Conference on Central Pacific Fishery Resources," 1966, State of Hawaii, Honolulu.

Schaefer, M. B., "The potential harvest of the sea," Trans. Amer. Fish. Soc., 94, 123–128 (1965).

Shindo, S., "Sea Fisheries in Communist China," 1964, Japanese Fisheries Resources Protection Association, Tokyo, (in Japanese).

Sverdrup, H. U., Johnson, M. W., and Fleming, R. H., "The Oceans; their Physics, Chemistry, and General Biology," 1942, Prentice-Hall, Inc., Englewood Cliffs, N.J.

Tiews, K., "On the possibilities for further developments of the south east Asian fisheries," Indo-Pac. Fish. Counc., Current Affairs Bulletin, No. 47, 1–13 (1966).

Tohoku Regional Fisheries Research Laboratory, "Statistics of the southern sea fishery area, fiscal year 1964–65."

U.S. Bureau of Commercial Fisheries, "Mexican Fisheries, 1965," 1966, Foreign Fish. Leafl. 7, Washington, D.C.

Yoshida, H. O., "Tuna fishing vessels, gear and techniques in the Pacific Ocean," in Manar, T. A. (Ed.), "Proceedings of the Governor's Conference on Central Pacific Fishery Resources," 1966, State of Hawaii, Honolulu.

U.S. West Coast Fisheries

Although commercial landings in the U.S. increased from 3.09 billion lb annually in the 1926–30 period to 4.94 billion lb annually in 1961–65, landings in Pacific coast states are now less than in 1926–30 (Fig. 5, and Table 6). The value of landings to Pacific coast fishermen, however, was about one-third of the national total in 1926–30 and in 1961–65. This phenomenon is explained by the increasing emphasis on Pacific coast states on high-valued species such as tunas and shellfish.

In 1961–65, California's fisheries contributed 48% of the volume and 39% of the value of landings by Pacific coast states (including Alaska). During the period 1936–45, when California's fishery for Pacific sardines was at its peak,

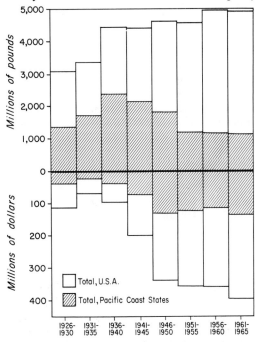

Fig. 5. Proportion of total U.S. catch and total landed value of U.S. catch contributed by the Pacific coast states of California, Oregon, Washington, and Alaska, 1926–65.

TABLE 6. AVERAGE ANNUAL COMMERCIAL LANDINGS AND VALUE OF LANDINGS TO FISHERMEN IN PACIFIC COAST STATES AND IN IN THE U.S., 1926–1965

Years	Average Annual Landings (millions of lb) In:				
	California	Oregon and Washington	Alaska	Total, Pacific Coast States	Total U.S.
1926–1930	605	141	631	1,377	3,090
1931–1935	891	157	690	1,738	3,330
1936–1940	1,434	205	800	2,439	4,433
1941–1945	1,305	244	602	2,151	4,415
1946–1950	1,012	232	557	1,801	4,607
1951–1955	656	196	346	1,198	4,585
1956–1960	607	196	372	1,175	4,974
1961–1965	581	183	444	1,208	4,941
Years	Average Annual Landed Values (millions of dollars) In:				
	California	Oregon and Washington	Alaska	Total, Pacific Coast States	Total U.S.
1926–1930	10.7	11.2	15.5	37.4	112.5
1931–1935	9.3	6.7	9.5	25.5	69.9
1936–1940	18.4	8.9	12.7	40.0	97.4
1941–1945	30.5	23.7	19.2	73.4	197.3
1946–1950	67.8	30.6	31.1	129.5	337.2
1951–1955	62.5	27.2	32.2	121.9	356.6
1956–1960	52.5	26.9	34.4	113.8	359.8
1961–1965	52.4	26.2	55.1	133.7	393.8

corresponding figures were 60% of the volume and 43% of the value. Among the Pacific coast states, only the combined landings of Oregon and Washington experienced an increase over the 1926–30 period: however, their proportion of the Pacific coast states' landed values fell from 30% in 1926–30 to 20% in 1961–65. Except during the years 1926–30 and 1961–65, the landed value of Alaska's fisheries has generally been considerably less than that of California and only slightly more than that of Oregon and Washington combined. The great increase in value of Alaskan fisheries in 1961–65 is caused mainly by a rapid growing fishery for the king crab, *Paralithodes camschatica* (Table 7). Increased landings of king crab enabled Alaska to overtake California in 1965 and remain as the leading state in value of fishery landings in 1966.

Landed values of major fisheries for Alaska and for California, Oregon, and Washington combined for the periods 1936–40 and 1961–64 are compared in Table 8. For the Pacific coast states as a whole, salmon and tuna are the most valuable species; in 1961–64 they constituted 71% of the value of all species landed. The fishery for Pacific salmon is the nation's second most valuable—exceeded only by shrimp from the Gulf of Mexico—and the Pacific fishery for tuna is the nation's third most valuable.

Largest increases between 1936–40 and 1961–64 in relative contribution to the total landed value of all species for the combined fisheries of California, Oregon, and Washington have been for tuna and for trawl-caught species. The landed value of tuna increased from about one-third of the California-Oregon-Washington total in 1936–40 to one-half the total in 1961–64; the value of trawl-caught species rose from less than 3% of the total in 1936–40 to almost 8% of the total in 1961–64. In contrast to the substantial increases in tuna and trawl-caught-species, the percentage of the total value contributed by salmon to the combined fisheries of California, Oregon, and Washington has increased only slightly, whereas that for halibut has fallen considerably. Pacific sardines, which, in the years 1936–40, contributed almost one-fourth of the total landed value, are now of insignificant value.

In Alaska, salmon made up 77% of the landed value of all species in 1936–40 and 73% in 1961–64. The percentage contribution of halibut and herring declined, whereas that of shellfish—largely king crab—increased greatly.

TABLE 7. LANDINGS OF KING CRAB IN ALASKA, 1935–1966

Year	Pounds	Year	Pounds
1935	2,000	1961	43,412,000
1940	10,000	1962	52,782,000
1945	0	1963	78,740,000
1950	1,519,000	1964	86,721,000
1955	8,163,000	1965	131,671,000
1960	28,570,000	1966	158,900,000[a]

[a] Preliminary figure.

TABLE 8. VALUE TO FISHERMEN OF CERTAIN MAJOR COMMERCIAL FISHERIES IN THE STATES OF CALIFORNIA, OREGON, AND WASHINGTON COMBINED AND IN THE STATE OF ALASKA, 1936–1940 AND 1961–1964

(thousands of dollars)

Years	California, Oregon, and Washington								Alaska				
	Salmon	Tuna[a]	Herring	Sardine	Halibut	Trawl-Caught[b] Bottom-fish	Shellfish	Total All Species	Salmon	Herring	Halibut[c]	Shellfish	Total All Species
1936	3,708	6,343	14	7,099	2,130	710	1,856	24,882	11,857	864	1,435	152	14,701
1937	5,427	8,788	17	6,814	2,177	752	2,012	28,776	11,877	1,032	1,560	219	15,272
1938	4,226	7,787	7	6,311	1,931	759	2,100	26,086	9,942	899	1,495	159	12,734
1939	4,162	8,823	7	6,856	1,978	660	1,997	27,417	9,256	928	1,620	128	12,129
1940	4,369	11,563	14	4,836	2,348	716	2,094	29,578	8,420	471	2,495	154	11,587
Average	4,378	8,661	12	6,383	2,113	719	2,012	27,348	10,270	839	1,721	162	13,285[d]
% of Total	16.0	31.7	NIL	23.3	7.7	2.6	7.4		77.3	6.3	13.0	1.2	
1961	16,286	40,226	171	1,146	3,520	5,333	8,455	81,520	35,741	559	4,888	5,117	46,470
1962	14,234	42,644	312	490	4,112	6,147	7,731	81,741	42,119	379	7,467	7,094	57,242
1963	17,713	37,455	188	299	2,722	6,634	7,601	78,402	31,298	468	4,161	9,638	45,703
1964	14,599	36,766	172	500	2,350	5,816	7,339	73,321	41,359	719	3,573	10,063	56,016
	15,708	39,273	211	609	3,176	5,982	7,781	78,746	37,629	531	5,022	7,978	51,358
Average % of Total	20.0	49.9	0.3	0.8	4.0	7.6	9.9		73.3	1.0	9.8	15.5	

[a] Albacore, bluefin, skipjack, and yellowfin.
[b] Paranzella nets contributed to value of landings in 1936–1940.
[c] Includes value of halibut caught by U.S. vessels and landed in British Columbia ports—arbitrarily assigned to Alaska.
[d] Higher than corresponding value shown in Table 6 because of different method of assigning value of halibut landed by U.S. vessels in British Columbia ports.

TABLE 9. AVERAGE ANNUAL COMMERCIAL LANDINGS IN THE STATES OF CALIFORNIA, OREGON, AND WASHINGTON COMBINED AND IN THE STATE OF ALASKA BY SPECIES AND BY SPECIES GROUPS, 1926–1964[a]

(millions of lb)

	California, Oregon, and Washington									Alaska							
	1926–1930	1931–1935	1936–1940	1941–1945	1946–1950	1951–1955	1956–1960	1961–1964		1926–1930	1931–1935	1936–1940	1941–1945	1946–1950	1951–1955	1956–1960	1961–1964
Salmon, Total	111.8	102.4	74.4	74.0	86.9	86.4	54.0	49.4		421.8[d]	489.4	560.2	445.4	353.1	246.0	213.9	269.4
Sockeye					9.3	14.3	13.3	6.2						102.7	67.5	62.5	59.4
Chinook					32.0	26.8	19.3	17.5						13.8	13.7	10.3	9.5
Coho					14.1	14.8	10.3	13.0						26.6	22.3	12.5	16.3
Pink					19.2	20.0	6.3	9.1						152.7	88.2	75.5	133.6
Chum					10.9	8.9	3.9	2.7						57.3	54.3	53.1	50.6
Steelhead Trout					1.4	1.6	0.9	0.9									
Tuna, Total	69.0	79.2	158.4	142.8	300.8	305.7	299.2	293.4		0	0	0	0	+	+	+	+
Albacore					45.5	35.7	42.6	43.2						+	+	+	0
Bluefin					11.3	10.6	18.2	26.4						0	0	0	0
Skipjack					72.3	116.2	96.5	80.2						0	0	0	0
Yellowfin					171.7	143.2	141.9	143.6						0	0	0	0
Sardine	439.0	685.2	1,181.4	1,059.4	485.8	127.0	91.0	19.5		0	0	0	0	0	0	0	0
Herring	+	+	+	+	+	+	+	+		169.0	174.2	208.4	116.8	157.2	53.0	99.8	40.8
Mackerel[b]	24.2	74.0	92.6	77.0	124.8	91.0	101.4	133.5		0	0	0	0	0	0	0	0
Halibut[c]	13.2	21.6	22.0	15.6	10.8	15.6	19.0	15.5		12.8	7.2	8.6	15.2	23.4	20.4	23.2	27.5
"Sole" and Flounders	11.2	12.0	17.4	26.6	42.6	42.2	48.2	49.2		+	+	+	+	+	+	+	0
Rockfishes	7.4	5.8	4.8	22.4	26.0	29.4	36.2	42.9		0	0	+	+	+	0	+	+
Oysters	0.7	3.6	8.8	10.1	10.3	10.4	11.6	10.2		0	0	+	+	+	0	+	0
Crab	4.5	6.0	11.9	15.4	31.5	23.1	35.8	16.6		0.2	0.8	0.9	1.1	3.2	9.3	18.8	75.0
Shrimp	2.3	2.4	1.5	0.6	0.9	1.2	5.8	6.0		3.1	2.3	2.9	1.5	2.7	2.9	7.2	13.9
Total All Species	746.0	1,048.0	1,639.0	1,549.0	1,244.0	852.0	803.0	712.0		631.0	689.8	799.6	602.0	557.0	346.0	372.0	432.0

[a] Plus symbol denotes landing of insignificant quantity.
[b] Pacific mackerel (*Scomber japonicus*) and jack mackerel (*Trachurus symmetricus*).
[c] Includes landings by Canadian vessels in U.S. ports.
[d] Average for years 1927–1930.

Volume of landings by species and species groups is shown in Table 9. For all Pacific coast states combined, salmon production averaged 318.8 million lb annually in the years 1961–64 compared to 533.6 million lb annually in 1926–30. The decline in salmon production has been more pronounced in California-Oregon-Washington than it has been in Alaska. In Alaska, pink salmon (*Oncorhynchus gorbuscha*) has contributed the greatest poundage, followed by sockeye (*O. nerka*), chum, coho (*O. kisutch*), and chinook (*O. tshawytscha*). In the California-Oregon-Washington region, chinook salmon have generally been first in poundage, followed by either coho or pink and then sockeye and chum. Landings of steelhead trout (*Salmo gairdneri*) by commercial fishermen are relatively small.

Tuna landings since World War II have averaged about twice those prevailing during the years 1936–45 and four times those in the years 1926–30. About 20% of the albacore (*Thunnus alalunga*) on the Pacific coast states are landed in Oregon and Washington; the rest of the albacore and all the bluefin (*T. thynnus*), skipjack (*Katsuwonus pelamis*), and yellowfin (*Thunnus albacares*) are landed in California.

In the 1930's and early 1940's the Pacific sardine supported the largest fishery in the Western Hemisphere, which accounted for over one billion lb taken from California waters in some years. A great decline has occurred in the sardine fishery since World War II; in 1965 the catch fell to only 2.0 million lb and the canned pack was the smallest on record.

The principal herring fisheries on the U.S. Pacific coast are in the bays and channels of southeastern and central Alaska. In contrast to the steadily increasing production of herring noted earlier for British Columbia, landings of herring in Alaska have declined substantially in recent years. Annual production of herring in Alaska in 1961–64 was less than 25% that in the years 1926–50.

Larger landings of halibut are accounted for by increased deliveries by Canadian vessels to U.S. ports. Catches by U.S. vessels have remained rather constant since the early 1930's in contrast to a substantial increase in production by Canadian vessels.

Pacific mackerel (*Scomber japonicus*) and jack mackerel (*Trachurus symmetricus*) have been the object of a steadily expanding fishery in California. As shown in Table 9, annual landings of these species reached 133.5 million lb in 1961–64 compared to only 24.2 million lb in 1926–30.

Increased production of flounders ("sole") and rockfishes in California, Oregon, and Washington mainly reflects the growth of the otter trawl

TABLE 10. LANDINGS AND VALUE OF LANDINGS TO FISHERMEN IN CALIFORNIA, OREGON, AND WASHINGTON, AND IN ALASKA IN 1964 BY TYPES OF FISHING GEAR EMPLOYED

Type of Fishing Gear	Landings (millions of lb) In:			Total, Pacific Coast States
	California	Oregon and Washington	Alaska	
Seine Nets:				
Purse, lampara, haul, etc.	376.3	10.1	241.2	627.6
Trawls:				
Otter, beam, dredges[a]	33.0	88.6	7.8	129.4
Lines:				
Set, troll, hand	72.4	32.9	38.0	143.3
Traps:				
Pots, pound, weirs, reef, floating, etc.	2.6	9.3	101.7	113.6
Gillnets:				
Drift, set, trammel	3.4	16.2	103.4	123.0
Other	13.2	1.0	0.5	14.7
Total	500.9	158.1	492.6	1,151.6

Type of Fishing Gear	Value of Landings (millions of dollars) In:			Total, Pacific Coast States
	California	Oregon and Washington	Alaska	
Seine Nets	31.3	1.6	21.3	54.2
Trawls	2.3	6.7	0.3	9.3
Lines	14.0	7.7	8.1	29.8
Traps	1.0	2.1	9.8	12.9
Gillnets	0.5	4.2	16.4	21.1
Other	1.5	0.4	0.1	2.0
Total	50.6	22.7	56.0	129.3

[a] For oysters, includes some taken by hand.

kshery, which, as in British Columbia, came into prominence during World War II. The otter trawl fishery supplies a growing volume of flounders, rockfishes, and codlike fishes to the human food market as well as to the animal food market.

Production of oysters, crab, and shrimp also has shown strong gains over the years. In California, Oregon, and Washington, Dungeness crabs (*Cancer magister*) account for almost the entire crab production, and this resource is thought to be fully utilized. A similar situation may be near for the king crab in Alaska. Production of shrimp in California, Oregon, and Washington appears to have stabilized at about 6.0 million lb; in Alaska, however, there is opportunity for considerable further expansion of shrimp production. The ocean pink shrimp—*Pandalus jordani* in California, Oregon, and Washington and *P. borealis* in Alaska—is the main object of Pacific coast shrimp fisheries.

As shown in Table 10, seine nets, mostly purse seines, account for the bulk of the fish landed in California and Alaska. In California, purse seines are used to harvest most of the tuna, mackerel, and anchovy (*Engraulis mordax*); in Alaska, they account for most of the salmon and herring. Gill nets are mainly used for taking salmon. Lines account for almost all of the Pacific halibut, most of the albacore tuna, and much of the chinook and coho salmon and sablefish. Trawls are used to harvest most of the flounders, other than Pacific halibut, and the "cods," rockfishes, and shrimp. Traps and pots find their greatest use in harvesting crabs.

A. T. Pruter

Cross references: *Crab Industry; Peru Current; Peru Fishery; Philippine Fishery; Salmon; Tuna Fisheries.*

PARASITES AND DISEASES

Marine fishery biologists have paid little attention to the role of diseases as a cause of natural mortality in fish populations, except when epizootics (diseases affecting many fish of one kind at one time) occur. However, effects of diseases on fish populations may be more subtle. For example, infections producing "castration" effects impair the reproductive potential of the species. Further, there is little doubt that in cases of nonlethal parasitemia, evidence of loss in body weight can be established if one applies the rule of *coefficient of condition* ($K = W \times 10^5/L^3$, where $W +$ weight in grams, $L +$ length in millimeters).

Even though in most instances no obvious deleterious effects are evident, a closer examination (microscopical) of the infected tissue or organ will reveal some damage. Death will invariably follow, or the fish will succumb more readily to environmental stress or become easier prey to natural predators, when the kidneys and other vital organs regulating homeostasis are seriously damaged.

Although fish have long been the source of materials for parasitological studies, only the diseases of commercially important fishes have been studied by fishery scientists mainly as they affect marketability and not for the basic understanding of their role in population dynamics. A few of the more important parasites and diseases are listed and briefly described below.

Viruses

One of the oldest and best known virus diseases in fish is called *lymphocystis*, first described from European flounders in 1874. The disease appears on the skin and fins as white or grayish nodules, or tumorlike growths that contain numerous isolated and transformed connective tissue cells so enormously enlarged that they appear to the naked eye as small "tapiocalike" beads. Fish with extensive growths on the skin are so unsightly as to make them unmarketable. These are culled and dumped overboard. This practice of recontaminating the waters is responsible for the relatively high incidence of the disease in certain regions.

Viral *lymphocystis* has been found in about 30 species of marine fishes, particularly in the European flounder *Pleuronectes flesus* (12%), the European plaice *Pleuronectes platessa* (5%), and the American plaice *Hippoglossoides platessoides* (1%). The virus is highly resistant, and there is evidence that the infection in the American plaice is the result of accidental contamination with virus brought over from Europe on commercial fishing boats.

True tumors (neoplastic diseases) of viral origin have been reported in a number of flat fishes. The viral etiology is based primarily on epizootiological evidence, histopathological evi-

Fig. 1. Lymphocystis (viral) infection on surface of an American plaice (Photo: W. Templeman, courtesy of the *Journal of the Fisheries Research Board of Canada*.)

Fig. 2. Papillomatous tumor caused by virus in lemon sole from California coast.

dence, or both. The tumors are classified as warts or papillomas of the skin and head, and appear as solitary or multiple, flat or raised, grayish or brownish growths of various sizes. They are most frequently found in young (0–1 year old) fish. Epizootics have been reported in the following flatfishes from the Pacific coast of the U.S.: Lemon or English sole, *Parophrys vetulus* (5–10%), sand sole, *Psettichthys melanosticus* (32%); Flathead sole, *Hippoglossoides elassodon* (5%); and Dover sole, *Microstomus pacificus* (6–32%). The tumors have been occasionally reported in other species of soles in the Pacific, U.S., and in European waters.

Bacterial Diseases

Surprising as it may seem, no authenticated epizootics of bacterial origin have been established for *oceanic fish*, either in natural populations or in captivity. Occasionally, and under certain unspecified conditions, fish in captivity are susceptible to bacteremia by nonspecific *Pseudomonas*, ubiquitous organisms in the marine environment.

One of the oldest and best known diseases of marine fish is the highly lethal "Red Disease" of eel (*Anguilla*), which is caused by *Vibrio anguillarum*. It is consistently reported as the cause of epizootics in eels in European waters, especially during the summer months. However, the lesions occur most commonly in eels that are kept in crowded conditions in "live" boxes. The "sores" appear as reddish patches in the skin and gradually develop into deep, sharply defined bloody ulcers.

Of interest are the reports of mycobacteriosis in several species of commercially important marine fishes. The organisms are acid-fast, resembling the forms that cause tuberculosis in higher vertebrates. The disease appears as numerous small granulomatous-like bodies of the internal organs (heart, kidney, liver, spleen, gonads, and mesenteries). Infection of the gonads has some effect on reproductive potential. In general, infected fish are more susceptible to environmental stress and in this manner the disease may be an associated cause of death.

Mycobacteriosis (tuberculosis) is found in both fresh water and marine fishes in various parts of the world. *Mycobacterium marinum* has been found in croakers (*Micropogon undulatus*), sea bass (*Centropristis striatus*), Atlantic cod (*Gadus* sp.), white perch (*Roccus americanus*), striped bass (*Roccus lineatus*), common pompano (*Trachinotus carolinus*), weakfish (*Cynoscion regalis*), spot (*Leiostomus xanthurus*), black drum (*Pogonias cromis*), tautog (*Tautog onitis*), Atlantic halibut (*Hippoglossus hippoglossus*), windowpane flounder (*Scophthalmus aquosus*), and summer flounder (*Paralichthys dentatus*). So far, the disease has been reported in over 150 species of marine and freshwater fishes, and the indications are that tuberculosis is much more widespread in marine fishes than so far indicated. It must be emphasized that the bacteria is highly fish-specific and *not pathogenic to man*, or any other higher vertebrate.

Fungal Infections

Fungal infections in feral populations of marine fish are even rarer than bacterial infections. *Ichthyophonus hoferi* (also well known as *Ichthyosporidium hoferi*) is the cause of recurrent epizootics in the herring *Clupea harengus*, from the western North Atlantic. During epizootics (acute phase of the disease), which may or may not be cyclic, mass mortalities occur that greatly diminish the herring populations for several years. The incidence of the disease ranges from 1% to as much as 80% for the herring. The fungus also occurs occasionally in winter flounders (*Pseudopleuronectes americanus*) and alewife (*Alosa pseudoharengus*) in the same locality. A chronic form of the disease has recently been discovered in large populations of adult yellowtail flounder (*Limanda ferruginea*) off the coast of Newfoundland. In both the Atlantic herring and in the yellowtail, the disease is systemic and appears as numerous whitish or yellowish, granular bodies in the flesh and all the organs of the body, including the gills, heart, and brain. Transmission of the disease is brought about by ingestion of the spores either directly or by eating diseased fish, or by the ingestion of zooplankton that have fed on infective spores.

Protozoan Parasites

Flagellates and Ciliates. With the exception of the class Sporozoa, which will be discussed below, only a relatively few forms in the classes Flagellata and Ciliata are important agents in epizootics. No authentic parasitic species of the Class Sarcodina has ever been reported for fish.

Oodinium ocellatum is a parasitic dinoflagellate responsible for recurrent epizootics in populations of puffers (*Sphaeroides maculatus*) in Sandy Hook Bay, New Jersey. However, the parasite, which occurs in the gills, is also widely distributed in tropical marine fishes and one of the primary causes of death in fish kept in captivity. *Ichthyodinium chabelardi* is parasitic in the eggs of sardines (*Sardinai pilcharus* and *Maurolicus pennanti*) off the coast of Algeria. Incidence as high as 80% (winter season) was reported; thus the infection must be considered a very important limiting factor in reproduction of the sardines in this area.

A classical example of the devastating effects of a nonparasitic dinoflagellate is represented by the "red-tide" phenomenon, which occurs frequently in tropical and semitropical waters. *Gymnodinium brevis* is the phytoplanktonlike dinoflagellate commonly found on the west coast of Florida. During episodes (blooms) in 1946–1947 more than 200,000,000 lb of dead fish accumulated on the beaches at a rate of 100 lb per linear foot. Death is caused by a poison inherent in this species of flagellate.

Sporozoans. This class of parasitic Protozoa includes four orders of which only one, the Coccidia, contains species that are important parasites of fish, mainly clupeiforms. For example, a disease of *Sardina pilcharus* is caused by the well known coccidian *Eimeria sardinae* that invades the testes. More than 50% of the sardines sampled off the coast of Portugal were found to be infected. There can be no question that such a high incidence has some effect on the reproductive potential of this species in this area. The evidence seems to indicate that intestinal and visceral coccidiosis in marine fishes may be more common than is generally suspected.

Cnidosporidia. Members of the Protozoan class Cnidosporidia are predominantly parasites of invertebrates and fishes; more than 700 species in the orders Myxosporida and Microsporida have been described from an estimated 2000 species of fishes. The myxosporidians are the most important disease producing entities in fishes, causing infections in the skin, gills, and all the internal organs, including brain, heart, cartilage, and bone. Many species are host-specific and tissue-specific. Healthy fish become infected by ingesting mature spores released in the water with intestinal and urinary wastes, or by eating infected fish. Pathological reactions vary from the development of simple connective tissue cysts to pathological manifestations characterized as cystitis, nephritis, hepatitis, enteritis, and endocarditis. Some myxosporidians will induce hypoplasia (atrophy) or hyperplasia (non-neoplastic tumors) of the infected or surrounding tissues. Hyalin degeneration of the flesh leads to conditions referred to as "mushy", "wormy," "jellied," and "milky" fish. These conditions have been reported in the halibut (*Hippoglossus stenolepis*), yellowfin tuna (*Thunnus albacares*), swordfish (*Xiphias gladius*), snoek or barracouta (*Thrysites atun*), John Dory (*Zeus faber*), and the South African stock fish (*Merluccius capensis*). The deterioration of the flesh is caused by the action of proteolytic enzymes released by the trophic (growing) stage of the parasite. The frequent occurrence of this type of disease in commercially important food fishes has been the cause of considerable concern. Although no statistics are available, the halibut fisher has been subjected to considerable annual loss due to "wormy" and "mushy" conditions. The former is caused by the myxosporidian *Unicapsula musclaris*; the latter by as yet an unidentified species of *Chloromyxum*.

Kudoa thyrsites is the cause of "milky" disease in 5% of the barracouta (a mackerel-like fish), 76% of the John Dory and 70% of the stock fish. The high incidence in the last mentioned fish was determined by the fluorescent method of detecting parasites in fish flesh. In smoked fish the lesions appear as whitish spots scattered throughout the body musculature.

"Jellied" swordfish in the North Atlantic has long been known but the exact cause of this condition has never firmly been established. One theory suggested that the "jellied" condition was due to proteolytic action of bacterial contaminants derived from the intestine of the fish during "gutting" or from unsanitary deck conditions. However, myxosporidian *Chloromyxum musculoliquefaciens* was reported by scientists to be the cause of this condition in Japanese swordfish. In one case, except for the head, the entire body of a 227.5 lb swordfish that had been stored for only 5 to 7 days after being caught was found to be "jellied." This observation suggests that spoilage in swordfish infected with myxosporidia is a postmortem phenomenon and may develop rapidly at ambient temperatures.

In marine fishes, the gall bladder is a common site of infection for myxosporidians. In such fish the liver is whitish in appearance, sometimes enlarged, and the gall bladder thickened (hypertrophied) or considerably distended with highly viscid bile.

Certain species of myxosporidians may be localized in the endocardial, myocardial, and pericardial regions of the heart. Frequently the pericardium is thickened, closely adhered to the heart wall, and comparative large areas of the myocardium may be hyalinized, much in the manner of the body musculature.

The microsporidians are small intracellular parasites of fishes (and invertebrates), with a somewhat similar life cycle as the myxosporidians. Relatively few species, mainly in the genera *Plistophora*, *Glugea*, and *Nosema*, have been reported in marine fishes. *Plistophora macrozoarcidis* causes deterioration of the flesh in 4% to 64% of the ocean pout (*Macrozoarces americanus*). There is some evidence that the deterioration of the flesh continues in fish kept in cold storage. *Plistophora gadi* produce large tumorlike masses in cod in the Barents Sea.

Of considerable interest is the microsporidiosis in European and North American flatfishes caused by *Glugea stephani*. The parasites invade the digestive tract, reproductive, and other internal organs. In the western Atlantic, the disease is quite common in young (0–1 year-old) winter flounders (*Pseudopleuronectes americanus*), often reaching epizootic proportions (25%–50%). Mortality statistics are not available, but infected fish show an apparent loss in body weight. *Glugea hertwigi* is the cause of microsporidiosis in 1.5 to 23% of the American smelt (*Osmerus mordax*). It has been suggested that this disease is responsible for the decline of smelt fishing in the North Atlantic coast. This disease is also of interest because the parasites invade the connective tissue cells that become so enlarged that they look like roe. This enlarged and transformed cell is referred to as a *Glugea-cyst*.

Nosema lophii infects the ganglion cells of the medulla, and those associated with cranial and spinal nerves in approximately 40% of relatively young angler or goose fish (*Lophius americanus*).

Helminthic Parasites

The trematodes are divided into two major groups: the Monogenea and Digenea. The Monogenea have a direct life cycle, i.e., eggs hatch into young forms and develop directly into adults. They are predominantly fish parasites found in the gills and skin. In one study, 958 species have been described, 84.1% of which occurred on one genus of fishes, 10.4% on two genera, 3.2% on three genera, and 2.4% on three or more genera. It is quite evident that the last mentioned group are highly nonspecific, and are found on a wide variety of fish. From evidence obtained on studies in captive marine fish, some species of monogenetic trematodes are highly pathogenic. It is assumed that fish with heavy parasitemia on the gills and skin are highly susceptible to the effects of abnormal environmental factors, and in some cases the lesions, e.g., destruction of gill tissue, are so severe that death is inevitable. Insofar as is known, except for one unusual instance, no epizootics of marine fish populations have been reported as caused by monogenetic trematodes. An epizootic in 1936 in the Aral Sea sturgeon, *Acipenser nudiventris*, caused by the monogenetic trematode (*Nitzschia sturionis*) depleted this fishery for more than 20 years.

The Digenea, of which about 2000 species are recognized, have an indirect and complicated life cycle. The eggs hatch in water and give rise to a free-swimming larva called the miracidium; this, in turn, invades a snail or other molluscan host and undergoes a series of changes eventually giving rise to large number of infective forms that are called cercariae. The cercariae leave the snail and eventually may penetrate directly into the fish and grow into sexually mature adults in the body tissues or in the blood stream; in most species the cercariae may encyst in the same or other species of snail, other mollusks, in marine worms, copepods, or in the skin, flesh, and other tissues of susceptible fish. The encysted stage is called metacercariae; fish become infected when they eat any of the animals harboring the metacercariae; the parasites then develop into sexually mature worms in the stomach or in the intestine. In most cases, the intestinal parasites are usually harmless except when the parasitemia load is unusually heavy, the effects of which are apparent by loss of body weight. Such fish, weakened by the infestation, are also more prone to predation and to abnormal hydrological factors.

However, more drastic effects on viability are caused by the cercariae or metacercariae, since these forms cause considerable damage during their tissue migration or affect the proper function of the skin or organ when they encyst in great numbers. A classical example of such infestations are the black cysts in the skin of Tautog (*Tautog onitis*), cunner (*Tautogolabrus adspersus*), Atlantic herring (*Clupea harengus*), menhaden (*Brevoortia tyrannus*), other North Atlantic fish,

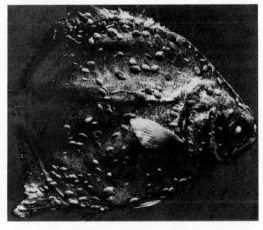

Fig. 3. Round pompano heavily infected with monogenetic trematode, *Benedenia melleni*.

and the following species of flatfish: witch flounder (*Glyptocephalus cynoglossus*), American plaice (*Hippoglossoides platessoides*), Atlantic halibut (*Hippoglossus hippoglossus*), yellowtail flounder (*Limanda ferruginea*), smooth flounder (*Liopsetta putnami*), winter flounder (*Pseudopleuronectes americanus*), and windowpane flounder (*Scophthalmus aquosus*). The incidence and degree of infestations varies considerably with the species and geographical location. Windowpane flounders are invariably heavily infested with black cysts. The cysts are metacercariae of the digenetic trematode *Cryptocotyle lingua*, the adults of which are found in gulls (*Larus argentatus*) and other sea birds. They have also been found in seals. The cercariae develop in the snail *Littorina littorea*, and after passing from the snail penetrate and encyst in the skin of fish. Massive cercarial invasions affect the metabolic function of the skin, and infestations of the eye often cause blindness. It should be emphasized that the metacercariae of this species and those of *Heterophyes heterophyes* and *Stellantchasmus falcatus* (in Pacific mullets, *Mugil cephalus*) are potential human parasites.

Marine fish harbor both the larval stage (plerocercoid) and gut-inhabiting adult tapeworms. The life cycle is indirect; eggs hatch into free-swimming coracidia, which are eaten by copepods or other crustaceans. In the copepod, the coracidium becomes transformed into a wormlike procercoid. When the infected copepod is eaten by fish, the procercoid passes from the intestine and encysts in the flesh or in or on one of the internal organs. The encysted stage is called plerocercoid. Fish infected with plerocercoid are eaten by definitive hosts, i.e., other fish or fish-eating animals, in which the parasite develops to sexual maturity in the intestine. The plerocercoid of the tapeworm *Otobothrium crenacolle* occurs in large numbers in the musculature of the butterfish (*Poronotus triacanthus*), immediately adjacent to the vertebrae. The sexually mature tapeworms develop in the intestine of hammerhead (*Sphyrna zygaena*) and other sharks.

Although tapeworms are predominantly fish parasites, the adults are less common than trematodes in marine teleosts. Many species of fish harbor the plerocercoid stage, the adults of which are invariably found in elasmobranchs. One study reported 37 species of tapeworms of different families (Diplocotylida, Ptychobothriidae, Bothriocephalidae) in the stomach and intestine of flatfish, with incidences ranging from 3% to 30%.

None of the marine fish tapeworms are pathogenic to humans.

Roundworms or nematodes are also found in fish; the larval forms usually occur in the flesh, liver, or other internal organs; sexually mature

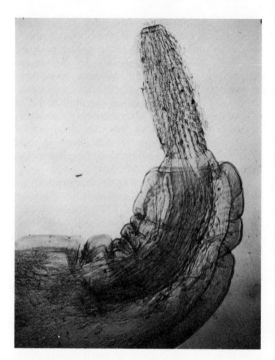

Fig. 4. Spiny-headed worm, *Echinorhynchus gadi*, common in intestine of codfish and other North Atlantic fish.

adults are intestinal inhabitants of fish or fish-eating animals. It is sufficient to mention that more than 60 species of larval and adult nematodes have been reported in flatfish and about 12 species in codfishes. Larval nematodes (*Contracaecum* sp.) of the liver of winter flounders in Long Island Sound induce tumorlike growths, while severe infestation of the liver of the Baltic Sea cod with the larvae of *Contracaecum aduncum* causes atrophy and severe changes in the coefficient of condition factor (from 0.941 to 0.629), with the fat content of the liver lowered from 57% to 14.5%. Encysted nematodes in other fish probably produce similar effects.

Of special interest to fishery biologists is the nematode *Terranova decipiens* (also called *Porrocaecum decipiens* and *Phocanema decipiens*). The larval form of this parasite is commonly called the cod-worm, because of its relatively high incidence in the flesh of various species of Gadiidae seriously affecting the commercial value of this fishery in certain areas. The adults are found in the intestine of harbor seals (*Phoca vitulina*). However, the larvae of this species or closely related forms are also found in a number of Pacific and other Atlantic fish; e.g., ocean pout (*Macrozoarces americanus*), redfish (*Sebastes marinus*), sea raven (*Hemipterus americanus*), and in several flatfish such as the witch flounder, American plaice, halibut, yellowtail flounder, smooth flounder, winter flounder, and windowpane flounder. Candling fillets with fluorescent light revealed incidences ranging from 7% to 45%. The importance of such infestations to the flat-fish fisheries remains to be determined.

Acanthocephalans or spiny-headed worms are primarily parasites of freshwater fishes. Larval, immature and adult forms occur in both marine and freshwater fish. *Echinorhynchus gadi* is circumpolar in distribution, being found in a wide variety of marine fishes. More than 60 species in the North Atlantic alone are hosts and include cod (*Gadus* sp.), striped bass (*Roccus saxitilis*), black sea bass (*Centropristis striatus*), goose fish (*Lophius americanus*), ocean pout (*Macrozoarces americanus*), and most of the species of flatfishes, especially the winter flounder. In some fish, striped bass for example, heavy infestations cause extensive ulceration and occlusion of the digestive tract. There can be no doubt that there is some effect on the "condition" factor. *Teleosentis tenuicornis* is another species of spiny-headed worms widely distributed in North Atlantic fishes, while *Pomphorhynchus laevis* is a common parasite in European flatfishes.

Parasitic Copepods

The most common group of copepod parasites belongs to the order Caligidea; more than 200 species of the genus *Caligus* alone are known. Other important genera of parasitic copepods in the skin and flesh of marine fishes are *Lepeophtheirus* (70–100 species), *Hatschekia* (75 species), *Lernanthropus* (80 species), *Lernaeeniscus* (30 species), *Pennella* (30 species), *Lerneocera* (17 species), and *Sphyrion* (6 species). Practically every commercially important fish is host for these parasites. For example, *Lernaeeniscus* are common parasites of Clupeids; *Lerneocera* infest the cod and codlike fish; *Pennella* the swordfish and other spearfish; and *Sphyrion* the redfish. These cause extensive tissue damage to the flesh, often reducing market value because of unsightly lesions.

One of the most interesting and important copepod parasites is *Lerneocera branchialis*, a relatively large species that lives in the gill cavity of cod, pollock, and whiting. The eggs hatch in water, grow for a time, and then take up a parasitic existence in the gill chamber of a flounder or lumpfish, which acts as intermediate hosts. Each fish may harbor relatively large numbers of these parasites. The females are fertilized by males, the latter dying soon after. After fertilization, the females grow to about three times the original size and then leave the intermediate host to take up a free-living existence for a time. Eventually, two or three of the females take up a second parasitic existence in the cod. If this is accomplished, a great transformation in form takes place while in the gills, changing from a characteristic crustaceanlike stage to a large, almost amorphous mass. The parasite burrows into the heart or ventral aorta, leaving only the posterior part of the body with egg sacs exposed in the gill cavity. Codfish that survive the drastic effects of these parasites are anemic and show considerable loss in weight (20–30%). Death will occur if the heart and ventral aorta is severely damaged. The incidences of this parasite in North Sea fish have been reported as follows: haddock, 10%; whiting, 80%; cod, 20%.

Sphyrion lumpi is another tissue-invading copepod parasite, and is commonly found in the redfish, *Sebastes marinus*, from certain areas in the North Atlantic. The parasites are relatively large but the "sores" are so unsightly that they affect their marketability. The incidence in endemic areas may reach 100% and in other areas the fish may be entirely free of the parasite. About 10% of the haddock in the North American coast, and a relatively small number of other fish, are also susceptible to infestation with *Sphyrion lumpi*.

Nonparasitic Diseases

Fish are subject to all kinds of teratological abnormalities such as twisted (lordosis and scoliosis) and fused spines (kyphosis), loss of fins or fin-rays, supernumerary fins, pug-headedness and other abnormalities of the skull, abnormal pigmentation (ambicolored flounders), and failure of opercular covering to develop. Fish with extreme malformations do not survive for any great length of time, since they are easy preys.

Cancers in fish are not uncommon, but fish with malignant growths are seldom seen in feral population. One analysis on 10,000 fish examined showed about 1% with some form of tumor, most of them of osteogenic origin. It is of interest that about 2% of the gray snappers (*Lutianus griseus*) in Bahamian and West Indian waters show nerve sheath tumors as one or more growths in the skin; similarly about 2.5% of the slippery dicks (*Halichoeres bivittatus*) in the Bahamas are afflicted with benign papillomas. Papillomas in flatfishes, probably caused by viruses, are discussed under viral diseases.

Ross F. Nigrelli

Cross references: *Bacteria in the Sea; Dinoflagellates; Fungi, Marine.*

PEARLING INDUSTRY

Shell Industry

The pearling industry of Australia is just over 100 years old. The main aim of the industry has always been the gathering of pearl shell, but

Fig. 1. One valve of *Pinctada maxima* showing mother-of-pearl.

recently, following the introduction of pearl culture to Australia, the use to which this shell is put has been changing.

For most of the history of the industry, shell has been collected for its "mother-of-pearl" lining, this being used for the manufacture of buttons, jewelry, knifehandles, ornaments, and other bric-a-brac. Button manufacturers have probably been the most important source of demand for pearl shell. During World War II, pearl shell was a prohibited export, and was used to make compass dials. It is particularly suited to this purpose as it is little affected by extremes of temperature and humidity.

Mother-of-pearl, or nacre, makes up almost the entire shell of the pearl oyster. It is produced by special epithelial cells in the mantle of the oyster, and is composed of finely crystalline calcium carbonate in the form of aragonite, the crystals being bound together by an organic substance, conchiolin ($C_{32}H_{98}N_2O_{11}$). Pearls, too, are composed of nacre, and the quality of a pearl is dependent on its size and shape and on the size and shape of the aragonite crystals, which are laid down in fine layers about the nucleus of the pearl. The color and luster of a pearl are the result of light diffraction by the first few layers of aragonite at the surface of the nacre. It should be noted that the epithelial cells of the mantle are the only ones capable of producing nacre, and this factor is the basis of pearl culture technique.

The industry is based on stocks of the silver-lip pearl oyster, *Pinctada maxima*. This is a much larger species than the Japanese pearl oyster, *Pinctada fucata* (=*P. martensi*). Whereas *P. maxima* can reach 30 cm across the shell, *P. fucata* seldom grows larger than 10 cm. The silver-lip pearl oyster is distributed widely in the indo-Pacific region, and in Australia occurs along the northern coast, down to about 25°S in the west and about 15°S in the east. The Arafura Sea yields 80–90% of the world's silver-lip pearl oyster and up to 80% has come from Australian operations. The main fishing grounds are in 10–50 fathoms.

The constant problem of the pearling industry has been labor. From the start, the industry has been dependent on foreign labor. The arduous work and the unfavorable climate and conditions have made it unattractive to Europeans. Divers have nearly always been Asian, and the death rate among them has been high. Causes of death ranged from beri beri to decompression sickness.

Commercial pearling in Australia began in 1861, near Nickol Bay, Western Australia. For the first few years, shell was gathered from beds exposed at low tide ("dry-shelling"), but as such stocks dwindled, shell had to be gathered by divers working in 2–3 fathoms. In Shark Bay, another species, *P. carchariarum*, was gathered solely for pearls. By 1871, 12 boats of from 12 to 50 tons were operating, and production of shell for that year was 180 tons. The industry expanded rapidly and in 1873 there were 80 boats based at Cossack.

By the 1890's, Broome had become the most important pearling centre in Western Australia, a position which it still holds. The export value of Western Australian shell in 1890 was $A252,000 and in 1905 the Broome yield of 1,394 tons constituted 85% of Australian production. The 1905 catch was taken by 25 schooners of 13–133 tons and 340 luggers of 3–14 tons. The industry employed 2,900 men—Japanese, Chinese, Arabs, Aborigines, South Sea Islanders, and a few Europeans.

The Western Australian industry prospered until about 1920, then began to decline. It was hard hit by the depression of the 1930's. In addition, in 1935, 20 boats and 141 men were lost in a cyclone. In 1936, there were 56 boats working out of Broome, and the industry employed 256 men, 205 of these being Japanese. Exports of shell between 1934 and 1939 varied between 950 and 1200 tons per annum. But prices were dropping; in 1930 the average price per ton of shell was about $A155 whereas the corresponding value for 1939 was $A87. The postwar yield of the Western Australian industry has been between 300 and 400 tons per annum.

A fishery around Thursday Island, in Torres Strait, started in 1868 when shell was worth $A600-800 per ton. Bêche-de-mer fisheries employing native divers already existed in the area. In 1886, many boats from the Thursday Island fishery moved to northwest Australia following reports of good grounds there, but most of these returned in 1890. There were 109 vessels pearling from Thursday Island in 1897. A record catch of 1,233 tons was made in 1897, but after this, catch declined up to 1914. In 1905, only 543 tons of shell were taken, the export value being $A270,000. There were 348 boats operating and 2,850 persons employed in the industry.

After World War I, the catch increased. In 1925, it was over 1,000 tons, and in 1929, following the discovery of new shell beds in 1928, 1,400 tons were taken. However, demand decreased and not all of the 1929 catch was bought. Catch was restricted to 850 tons in 1930. During the depression of the 1930's prices fell, and in 1935 the market price for Thursday Island shell was $A217 per ton. The average yearly output between 1926 and 1935 was 889 tons.

A pearl shell fishery based at Port Darwin started in 1884 when pearl oysters were discovered there. There is a strong tidal current and the water is turbid in this area and the quantity of shell taken in this fishery has seldom exceeded 300 tons per annum.

For the most part pearl shell is taken in 10–20 fathoms of water. Occasionally, depths greater than 40 fathoms have been worked, but at these depths the risk to the diver is great. At 20 fathoms a diver can work safely for up to 1 hr, and his well-being depends upon his ascent being properly "staged." Between 1909 and 1917, there were 145 deaths from diver's paralysis at Broome.

Diving dress was first used in Torres Strait in 1874, and in the 1880's in Western Australia and the northern Territory. At first, hand-pumps were used to supply air to the divers—one to each boat—but power compressors began to replace hand pumps in about 1913. After World War I, the introduction of the high-capacity compressor enabled two or more divers to go down simultaneously. Before about 1917, divers in all Australian pearling areas walked the sea-bed looking for shell. However, at Broome, the method of "working to windward" is now used. In this method the diver is towed along, a few feet from the bottom, while the lugger drifts beam on with some sail set. When the diver sights shell he signals the surface and is lowered to collect it. This method is used by Broome divers because they wear the full diving suit and this makes bottom-walking strenuous. At Thursday Island and in the Northern Territory, where the water is warmer, divers wear either the half-suit or the helmet and corselet only.

As regards the pearling lugger, the workboat of the industry, a ketch-type vessel of between 50–60 ft in length and of 15–30 tons has evolved.

The Australian Commonwealth and State Governments have played some part in the development of the industry. The Western Australian Pearling Act of 1875 has already been mentioned. The Commonwealth Government has always been concerned about both the influx of foreign labor and the fishing of Australian grounds by Japanese vessels. The Japanese have had a great influence on the development of the Australian pearling industry. They began to enter the industry in the 1890's and gradually became its principal operatives, working as divers, tenders and crew, under contract to the Australian pearling companies.

Moreover, Japan, up until 1962, maintained its own pearling fleet to fish Australian beds. In 1961 and 1962, the Japanese pearling fleet took 380 and 360 tons of shell, respectively. In earlier years, Japan's pearling fleet took much greater

Fig. 2. Thursday Island pearling luggers working the shell beds.

Fig. 3. A Torres Strait diver collecting shell.

amounts, and was reckoned by some to be responsible in part for the depression in the world price for pearl shell during the 1930's. The Japanese fleet began operations in Australian waters in 1935, and by 1940 had taken about 12,000 tons of shell, the same amount as taken by Australian pearlers over the same period. The Pearl Fisheries Act of 1952–53 was passed after discussions between the Australian and Japanese Governments and was designed to control the operations of Japanese pearling vessels in Australian waters. Under the new legislation, Japanese pearlers were required to hold licences issued by the Australian Government, and in 1954, 25 such licences were issued. Moreover, Japanese vessels were permitted to operate only in specified areas off the Northern Territory and not less than 10 miles from the Australian coast. Any catch limits imposed had to be observed as did the prescribed minimum sizes, and catch data had to be provided regularly.

Although the main emphasis of the Australian pearling industry has always been on shell collecting, natural pearls of considerable value have been taken, especially around the turn of the century. In pearl shelling operations, pearls have for the most part been regarded as a bonus and have been given to the divers collecting the shell. The annual average value of pearls taken by Broome luggers between 1900 and 1906 was $A68,690. However, the corresponding value for the years 1946–52 was $A1,980. The most valuable Australian pearl, "The Star of the West," was taken by a Broome pearler in 1917; weighed 100 grains and was valued at $A20,000. Not nearly as many pearls are found now, and it has been suggested that this is the result of operations in deeper water where turbidity and currents are reduced and there is not as much likelihood of nuclei being washed into the oysters.

It is since World War II that the greatest changes have occurred in the Australian pearling industry. Production of shell has not been as great as in prewar years. The most recent crisis in the industry, and probably the most severe that it has ever had, arose out of the introduction of plastics for button making. The result was that the world price of shell fell drastically; light grades of shell, those used in button making, became unsalable. In 1958, some shell held in the U.S. and Europe could not be sold, and between 1958 and 1962 world production of shell fell by about half. Between 1958 and 1961, the Australian Commonwealth Government spent $A102,000 in a promotion campaign to clear surplus stocks. There were 139 luggers and 1,514 men employed in Australian pearling in 1957; the corresponding statistics for 1961 were 57 and 726, respectively. Production dropped from 1,839 to 790 tons.

Pearl Culture

However, although demand for mother-of-pearl was falling, the demand for live oysters by pearl culture farms was rising. The first pearl culture farm in Australia was started in 1956, at Kuri Bay near King Sound in northern Western Australia, by an Australian company in collaboration with Japanese and U.S. interests. In 1958, round pearls of about 14 mm diameter were harvested from *P. maxima*. The company subsequently extended its operations to Torres Strait, where farms were started at Friday Island and Moa Island.

By 1962 there were eight farms in Torres Strait and these were using 80–120 tons of live shell per year. The statistics show the rapid growth of the industry. In 1963, there were 11, in 1964, 15, and in 1966, 17 pearl culture farms between Exmouth Gulf, Western Australia, and Torres Strait. In addition, four farms have been started in New Guinea waters. In 1963, 242 tons of live shell were sold to culture farms; 311,000 shells were in production and 470,000 in reserve. Of the 17 farms now operating in Australian waters, 13 are joint Australian-Japanese ventures and the others Australian-owned. In the joint ventures, Australians establish the farm and supply the live shell; the technical work, cultivation, and marketing are done by the Japanese.

By 1964, the accent in the mother-of-pearl industry was on the supply of live shell to culture farms. In Queensland, the Island Industries Board operates luggers and these provide the major part of the Queensland supply. The value of cultured pearls produced in 1964 was $A2,839,000; 51,581 round pearls were produced. A modern live shell carrier, *Kuri Pearl*, was launched for the Kuri Bay pearl farm in 1964.

The 210-ton vessel has two special holds each capable of holding 50 tons of seawater for the transport of live shell. Special inlet valves allow fresh seawater to circulate freely through the holds during transport of the shells from shell beds to the culture farms. In 1967, the same company launched the *Merindah Pearl*, which is similar to the *Kuri Pearl*, but larger and improved in the light of experience gained with the first vessel.

P. maxima is a much larger species than the Akoya oyster, *P. fucata*, which is used for pearl culture in Japan. Using *P. maxima*, pearls up to 18 mm in diameter can be produced in 2–3 years. *P. fucata* takes 4–7 years to produce an 11 mm pearl. Moreover, Australian pearls lack nothing in quality and luster, and command top prices. The technique for pearl culture in the silver-lip oyster is similar to that used with the Japanese pearl oyster. A suitable nucleus for nacre deposition is inserted in an incision made in the epithelium of the foot of the host oyster together with a small piece of mantle tissue. It is a grafting technique and if the graft is successful the mantle tissue forms a pearl sack around the nucleus and the deposition of nacre begins. The epithelial cells

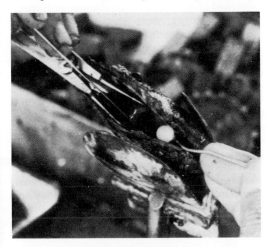

Fig. 4. Inserting the nucleus.

Fig. 5. Half-pearls on the shell.

Fig. 6. Rafts from which the pearl oysters are suspended during nacre deposition.

of the mantle are the only ones capable of producing nacre. Nuclei are manufactured in Japan, usually from the shell of a Mississippi fresh water mussel. For half-pearls, a nucleus is inserted between the mantle and shell of the oyster and hemispheres of soapstone or plastic glued to the shell are used as nuclei.

Oysters fit for nucleus insertion are 17–20 cm across the shell and the nucleus used is 11–15 mm diameter. One or two nuclei are inserted, but a particularly skillful operator can insert three. During culture, when the layers of nacre are laid down on the nucleus, the oysters are usually held in wire baskets on floating rafts. Mortality resulting from the nucleus-inserting operation is about 20%, and subsequent mortality about 10%.

Although the technique of pearl culture is known in principle, considerable skill is required of the operators. The Japanese perfected the technique in their own country at the turn of the century and have closely guarded their methods. However, the results of pearl culture farms operated solely by Australians have shown that the intricacies of the technique can, with practice, be mastered.

Pearl culture is quickly displacing the old pearling industry whose aim was the gathering of pearl shell for mother-of-pearl. However, there is still an appreciable demand for pearl shell and prices are high. In 1964–65, exports of Australian shell were worth $A850 per ton. Spent shell from culture farms is sold for its mother-of-pearl. In many respects, the Australian pearling industry has developed in a haphazard fashion, probably because of its geographic isolation and the rigors of its pursuit. It was pioneered by a few hardy individuals and would not have developed had foreign labor not been available. The Japanese have played a dominant role in the industry, and have been blamed from time to time for some of the setbacks that the Australian industry has suffered. Nevertheless, the Japanese-aided introduction of pearl culture has undoubtedly saved and revitalized the Australian pearling industry.

E. Highley

References

Brownfield, E. J., "The pearlshell industry in Western Australia," *Fish. Bull. West. Aust.*, **11**(4), 107–116 (1953).
Cahn, A. R., "Pearl culture in Japan," *Fishery Leafl. Fish Wildl. Serv. U.S.* 359, 1949.
Commonwealth of Australia, "Pearl shell, bêch-de-mer, and trochus industry of Northern Australia, 1946, Department of Commerce and Agriculture, Commonwealth Fisheries Office, Economic Report No. 1.
Commonwealth of Australia, "Australian pearling statistics," 1958, Department of Primary Industry, Canberra.
Commonwealth of Australia, "Fisheries: Australia," in "Official Year Book of the Commonwealth of Australia," 1966, Bureau of Census and Statistics, Canberra.
Commonwealth of Australia, "Australian pearling industry statistical handbook," 1967, Department of Primary Industry, Canberra.
Serventy, V., "Pearl culture in Australia," *Pacific Discovery, Calif. Acad. Sci.*, **20**(1), 22–28 (1967).
Tranter, D. J. "Pearl culture in Australia," *Aust. J. Sci.*, **19**, 230–232 (1957).
Wada, S., "Biology and Fisheries of the Silver-Lip Pearl Oyster," 1953, private publication.

Cross references: *Australian Fisheries; Oysters; Trust Territory of the Pacific Islands.*

PELAGIC DISTRIBUTION

For purposes of study, the oceans can be divided into the benthic realm or sea bottom, and the pelagic realm, the seawater itself. Benthic organisms rest on, are attached to, crawl over, or burrow into the sea bottom. Pelagic organisms float or swim freely in the seawater. With depths exceeding 10,000 m and an average depth of 3800 m (nearly $2\frac{1}{2}$ miles), there is more "living room" in the pelagic realm than in any other earth environment. Pelagic distribution refers to the location, concentration, and arrangement of life in the ocean waters.

Pelagic life is divided into three somewhat artificial categories—phytoplankton, zooplankton, and nekton. *Phytoplankton* includes the free-floating and drifting plants of the sea such as Sargassum weed, a brown alga, the more abundant and important microscopic diatoms, dinoflagellates, some blue-green algae, many groups of smaller photosynthetic flagellates, coccolithophores, and photosynthetic bacteria. The heterotrophic bacteria, viruses, and fungi are not included in this common classification (see Oppenheimer, 1963). *Zooplankton* includes all animals unable to swim effectively against the horizontal currents of the oceans. Although usually small, size is *not* the distinctive characteristic of zooplankton; swimming ability is. There are giant jellyfish that are weak swimmers and thus are members of the zooplankton. Chains of the lesser known pelagic tunicates of the genus *Salpa* may reach lengths of more than 5 m, with single individuals almost 25 cm long. The luminescent colonial tunicate, *Pyrosoma*, may grow bigger than a large watermelon. The Portuguese man-of-war, *Physalia*, may approach the size of a football.

The most abundant members of the zooplankton are the largely herbivorous crustacean copepods, the omnivorous crustacean euphausiids, and the carnivorous chaetognaths. Other animal groups such as the Foraminifera, Tunicata, Amphipoda, Mollusca, Coelenterata, Ctenophora, Pteropoda, and Radiolaria may sporadically become locally dominant, but generally make up less than 50% of the total zooplankton biomass. In neritic waters (from the shore out to depths of 200 m) the meroplankton or temporary plankton, composed of the eggs and larval stages of benthic and nektonic organisms usually forms a large part of the zooplankton.

Nekton includes all animals able to swim effectively against the horizontal ocean currents for prolonged periods of time and thus able to make long migrations. Examples are fur seals, which migrate from the Pribilof Islands to Southern California, whales and dolphins, tunas, swordfish, sharks and other fish including such small fish as herring, which are known to migrate from Iceland to Norway. Because of their large size and strongly developed muscles, the nekton, of all the organisms in the pelagic realm, are with minor exceptions, the only organisms that are commercially harvested by man. The study of the concentration, population structure, and migration of nekton has great direct economic application. The bluefin tuna of the Atlantic can migrate from the Bahamas to Norway, a distance of at least 4500 miles. The smaller Pacific albacore tuna migrates from the North Central Pacific to both Japan and California. In such cases regulation of only one of the fishing fleets may not be effective. Proper management of such populations becomes a matter of international negotiation and cooperation.

Components of Distribution

Biotic. The distribution of any species is determined by biotic and physical-chemical factors. The chief biotic factors are: food supply and nutrients, predation, reproduction, and behavior. Behavior includes the extensive migrations of the nekton to facilitate feeding and/or reproduction. For example, the California Grey Whale feeds on zooplankton in the Bering Sea in the summer and migrates in winter to the bays of Lower California (a distance of some 2000 miles) to mate and bear young. In spring, the adults and young return to the Bering Sea to feed. Thus in many nektonic species only a small part of the total possible range is occupied at any one time. The behavior of nekton leads to very uneven or patchy distribution. Although the fisherman may think that this makes his fishing operations more difficult, in reality, it is the behavioral congregation of nektonic species, as well as their size, that makes them economically harvestable.

Zooplankton species, in contrast, are distributed more uniformly throughout their range. They may also congregate for feeding or reproduction, especially in the upper 30–50 m of the sea, but only vertical congregation can be performed by behavioral activities. In some very local areas horizontal accumulation may be produced by currents, and zooplankton may become highly concentrated. Although more easily distributed by currents, many zooplankton groups can control their horizontal distribution to some extent either by vertical migration into different ocean currents or by reproductive cycles or both.

Phytoplankton species are almost completely dependent on the currents for distribution; however, reproduction cycles with resting stages on the bottom or in deeper water allow some control of distribution.

Ignoring patchiness, the average abundance of a pelagic species in a given area within its range is largely determined by available food which is ultimately dependent on the supply of mineral nutrients to the phytoplankton. Thus the highest reported concentrations of nekton and zooplankton are in regions of upwelling of mineral nutrients along the eastern temperate sides of the oceans, in regions of marked seasonal overturn of water as in the subarctic and subantarctic, or in areas of nutrient replenishment from the land, especially in bays and estuaries. Special circulation features of the equatorial regions lead to mineral replacement there with resulting standing crops larger at all levels of the food web than in the central water masses to the north or south. The lowest standing crops of phytoplankton, zooplankton and nekton are found in the warm central water masses, the deserts of the sea.

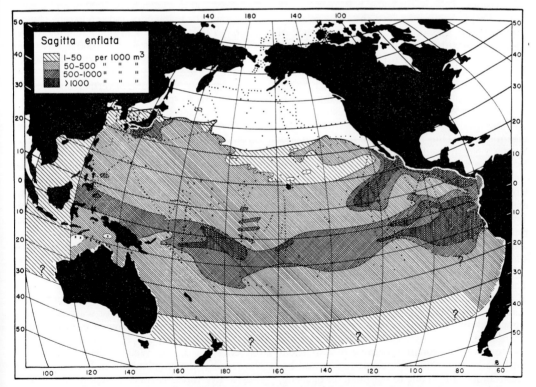

Fig. 1. Distribution of the planktonic chaetognath, *Flaccisagitta enflata* in the Pacific. This species also is common to abundant in the warm water realm of the Indian and Atlantic Oceans and adjoining seas. The distribution is typical of many species of zooplankton found in the warm water realm between the northern and southern warm water realms. (From Bieri, 1959.)

Physical-Chemical Components and Horizontal Distribution. The study of pelagic distribution in early years focused largely on physical-chemical factors limiting the ranges of species. Temperature received and continues to receive the most attention, but salinity, especially as in combination with temperature it reflects the circulation of the water, has received increasing attention. Despite the large number of works that have related distribution to temperature, salinity and circulation, there is still little agreement on what physical factors limit either horizontal or vertical ranges in the sea.

Most workers agree with Giesbrech's division of the pelagic realm into three great provinces, northern and southern cold water provinces separated by a warm water belt. On the western sides of the oceans this warm water belt extends to 45–50° N and S latitude. On the eastern margins it is much narrower extending only to about 18–20° N and S latitude (Fig. 1). It is this latitudinal distortion of the isotherms by the oceanic circulation that makes the use of terms such as "temperate," "tropical," and "subtropical" particularly inappropriate to the discussion and description of pelagic distribution. Although there is no agreement on the finer subdivisions of the warm water belt, recent work in the Pacific has emphasized the close relationship between species limits and the limits of water masses with their characteristic temperature-salinity-circulation patterns. In the Pacific there appear to be unique communities in each water mass as well as in the transition regions between water masses. These transition communities may correspond to the ecotones of terrestrial biotic communities. If so, they are of a much more gigantic scale. Thus it is possible to distinguish a Western North Pacific Central Community, with its unique fauna and flora, an Eastern North Pacific Central Community, a California Current Community, an Equatorial Community that extends into the Indian Ocean, a Northern Transition Community and a Subarctic Community (see Bieri, 1959; Bradshaw, 1959; Brinton, 1962; Ebeling, 1962; references in Brinton 1962; Banse, 1964).

Vertical Distribution and Vertical Migration. The vertical migration and vertical distribution of zooplankton and of nekton illustrate the combined action of the physical-chemical environment and the reaction of the organism to it. There is as yet little agreement on the degree and nature of vertical zonation because of insufficient work below the 200 m epipelagic region and because complications due to seasonal, ontogenetic, and daily migrations obscure zonation where it exists. Also, the depths of the water masses vary in different parts of the oceans, especially on the east and west sides, leading to different results in different areas. There is considerable evidence that many mesopelagic (approximately 200–700 m) and even bathypelagic (approximately 700–4000 m) species have horizontal range limits that correspond closely to the upper (0–500 m) water masses. This is due to the fact that the larvae and young of many of these species live in relatively narrow depth ranges near the surface (Ebling, 1962).

Bruun (in Hedgpeth, 1957) has divided pelagic life of the warm water belt vertically into the epipelagic zone, where there is sufficient light for phytoplankton production of organic matter, and the dimly lit mesopelagic zone extending below to the 10°C isotherm. Below this is the bathypelagic zone reaching to the 4°C isotherm, and below this isotherm the abyssopelagic zone extending to depths of 6000 m. In the Atlantic Ocean this boundary of 4°C between the bathypelagic and abyssopelagic zones occurs at about 2000 m while in the Indian and Pacific Oceans it would be as high as 1500–1000 m.

Vertical migration is a common phenomenon in many nekton and zooplankton species. Daily migrations from depths of 500 m and greater are known for lantern fishes and euphausiids in the deep scattering layer. Seasonal migrations from the upper 100 m to depths of 1000 m and greater are known for many copepods and some chaetognaths in the subpolar and polar regions. Ontogenetic migrations to depths of 1000 m have been demonstrated in many species of fish and invertebrates. See Cassie (1963) and Banse (1964) for comprehensive reviews of vertical distribution.

Dispersive Forces and Concentrating Mechanisms. The dispersive forces of currents and eddy diffusion tend to distribute pelagic populations throughout the world oceans. Extremes of temperature, salinity, oxygen, and mineral and organic nutrients may limit this dispersal. Behavioral patterns of the adult organisms may counteract these dispersive forces to varying degrees. But in addition to the dispersive forces and limiting factors, there are physical concentrating forces that tend to prevent completely random dispersal. Although the study of these concentrating forces is just beginning, four, perhaps five fundamental concentrating mechanisms are known.

A *vertical circulation cell* of pelagic organisms can be maintained by diurnal or seasonal vertical migration into currents going in opposite directions. For example, M. E. Johnson (1949; see references of J. W. Wells in Hedgpeth, 1957) showed that the zooplankton in Bikini lagoon is four to five times as concentrated as in the open ocean outside. The concentration is due largely to the vertical migration of the zooplankton during the day into a near-bottom countercurrent that prevents its being swept out of the lagoon into the open ocean. The retention of the zooplankton and its subsequent decay leads to an enrichment of the lagoon waters.

A *horizontal circulation cell* may be local and temporary as in short-lived eddies or oceanic in scale and permanent, lasting for thousands of years. Bieri (1959; reference in Banse, 1964) and Brinton (1962) have shown that some species of chaetognaths and euphausiids live only within the huge circulating cells of the North and South Pacific Central Waters (Fig. 2).

A *toroidal circulation cell*, a combination of horizontal and vertical circulation cells, has been postulated by Boden (1952) and Boden and Kampa (1963) as a conserving mechanism for the meroplankton around Bermuda (see references of J. S. Wells in Hedgpeth, 1957). Lagoon water, made more dense in summer by evaporation and in winter by cooling, spreads out from the island and mixes with oceanic water. This produces locally more dense water that sinks to moderate depth and may return larval forms to the vicinity of the island.

Line concentrating forces may be local or oceanic in scale. Best known are the equatorial convergences, which may concentrate phytoplankton, zooplankton, detritus, and nekton over distances of hundreds of miles as the result of the coming together of two nearly parallel ocean currents. On a smaller scale, Langmuir circulation cells are often seen at sea with long rows of debris concentrated in the zones of convergence. Fronts between cold and warm water may develop temporarily and concentrate pelagic life into lines.

Eddy diffusion, generally thought of as a dispersive force, may permit some zooplankton species to exist in local areas where the horizontal currents are weak. M. W. Johnson in 1960 suggested this in the case of the larvae of the spiny lobster, *Panulirus interruptus*. Such a mechanism may account for the existence of some neritic species of zooplankton, especially on the eastern sides of oceans where currents are more sluggish. The euphausiid, *Euphausia pacifica* found off California and the chaetognath, *Sagitta friderici* found off Northwest Africa are examples.

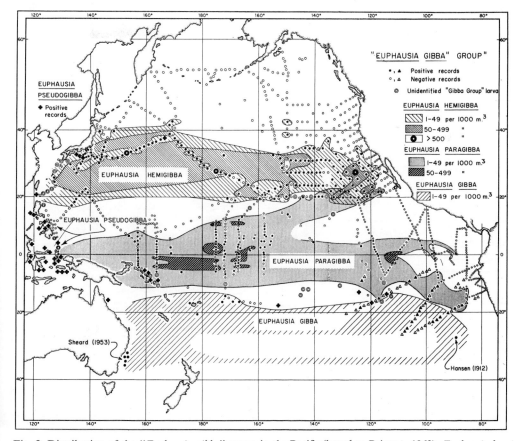

Fig. 2. Distribution of the "*Euphausia gibba*" group in the Pacific (based on Brinton, 1962). *Euphausia hemigibba* is found in the North Pacific Central Water, *E. gibba* in the South Pacific Central Water.

Future Study. Although descriptive, experimental, and theoretical methods are available for the study of pelagic distribution, the latter two methods have been largely ignored. The massive size of the oceans and the continued difficulty of working at sea would indicate that much more attention should be paid to these latter two approaches. Even with the concerted use of all three methods, it will be many years before we reach a reasonably good understanding of pelagic distribution.

ROBERT BIERI

References

Banse, K., "On the vertical distribution of zooplankton in the sea," *Progress in Oceanography*, **2**, 53–125 (1964).

Brinton, E., "The distribution of Pacific Euphausiids," *Bull. Scripps Inst. Oceanogr.*, **8**(2), 51–270 (1962).

Cassie, R. M., "Microdistribution of plankton," *Oceanogr. Mar. Biol. Ann. Rev.*, **1**, 223–252 (1963).

Fleming, R. H., and Laevastu, T., "The influence of hydrographic conditions on the behavior of fish," *FAO Fisheries Bull.*, **9**(4), 181–196 (1956).

Hedgpeth, J. W. (Ed.), "Treatise on Marine Ecology and Paleoecology," Vol. I, Ecology, 1957, Geol. Soc. Am. Mem.

Oppenheimer, C. H., (Ed.), "Symposium on Marine Microbiology, 1963, Charles C. Thomas, Springfield, Ill.

Cross references: *Phytoplankton; Plankton Resources.*

PERU CURRENT

Peru Current and Humboldt Current are names given to the ocean current or current system that flows toward the equator along the Pacific coasts of Chile and Peru. Peru Current is the more frequently used name, The current is the eastern portion of the anticyclonic gyre of surface water that is centered in the subtropical part of the South Pacific Ocean. The waters of the southern portion of the gyre, namely the transpacific West Wind Drift, approach the South American coast at about 40°S latitude, where they divide into equatorward and poleward currents. The equatorward current is the Peru Current; it follows the coast to about 5°S latitude and there turns west to join the transpacific South Equatorial Current.

A distinction is sometimes made between the Peru Coastal Current and the more offshore Peru Oceanic Current. They have a common origin in

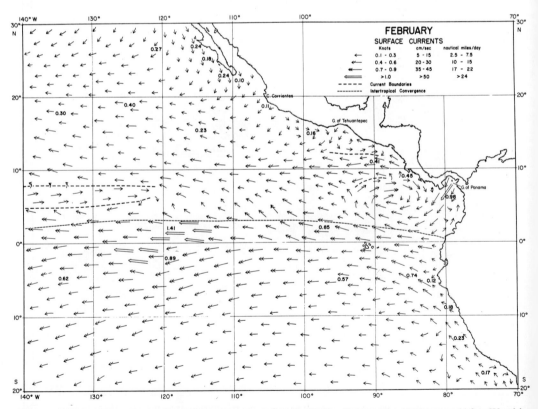

Fig. 1. Mean surface current chart of the eastern tropical Pacific for the month of February. (After Wyrtki, 1965; *courtesy Inter-American Tropical Tuna Commission*.)

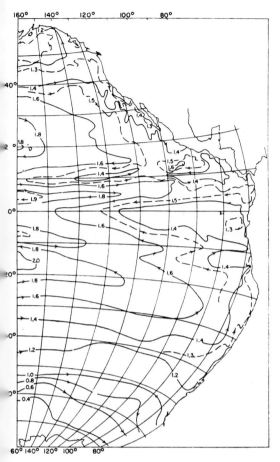

Fig. 2. Surface geostrophic circulation in the eastern Pacific: dynamic height anomalies, 0 over 1,000 db. (After Wooster and Reid, 1963; *courtesy John Wiley & Sons, Inc.*)

the West Wind Drift and a common termination in the South Equatorial Current, but they differ in properties and structure because of upwelling and other features along the coast. Between these currents, about 500 km offshore, is the weak poleward subsurface Peru Countercurrent. Another poleward current in this region, deeper and weaker than the Countercurrent, is the Peru-Chile Undercurrent, which underlies the Peru Coastal Current.

Knowledge of the direction and velocity of these currents comes mainly from charts of surface drift and dynamic topography (Figs. 1 and 2), although some direct current measurements have been made. Mean monthly charts of surface drift off Peru show the following seasonal changes which correspond to changes in the wind field. From July to October, when the southeast trade winds are strongest, the Peru Coastal Current flows at about 0.3 knot and the Peru Oceanic Current slightly faster, and there is no surface poleward flow between them. From February to April, when the southeast trades are weakest, the same two currents flow toward the equator as before, but more slowly (about 0.2 knot for the Coastal Current, slightly faster for the Oceanic Current), and poleward flow of the Peru Countercurrent appears between them at the sea surface (Fig. 1). Except in the Countercurrent, the general direction of surface flow is always equatorward, but there are coastal eddies in which local reversals of flow occur.

According to dynamic topographies and vertical profiles of properties, the Peru Oceanic Current extends to about 700 m and the Peru Coastal Current to about 200 m below the sea surface. The Peru Countercurrent, which differs from the Peru Current in being stronger at about 100 m than at the surface, extends to about 500 m below the surface. The Peru-Chile Undercurrent, which is wholly subsurface, is probably a few hundred meters thick.

Although charts (such as Fig. 1) might suggest that the Peru Currents are the principal source of water for the west-flowing South Equatorial Current, it has been estimated that their net contribution is less than one-third—about 14,000,000 m^3/sec, compared with about 49,000,000 m^3/sec transported in the South Equatorial Current. Other surface currents make up a little of this difference, but most of it is provided by the Equatorial Undercurrent, which flows to the east below the South Equatorial Current. It is thought that some of this Equatorial Undercurrent water is taken by the subsurface Peru Countercurrent into the Peru Current system, upwelled there into the northern part of the Peru Coastal Current, and thence contributed to the South Equatorial Current.

Sea surface temperatures along the Pacific coasts of Chile and Peru are generally low for the latitudes and times of year at which they occur (Fig. 3). They show a general increase from south to north, but not in an entirely regular sequence: for instance, they are generally lower at 14°S latitude than at 20°S. The coastal coolness cannot be explained entirely by the fact that the Peru Current originates at higher latitudes, therefore. Other property distributions, such as the profiles in Fig. 4, in which the upper isolines rise toward the east, indicate that upwelling occurs near the coast. In this process water ascends to the surface from below, thereby altering the temperature and other properties of the surface water.

It is generally agreed that the coastal upwelling occurs only from about 100 m to the surface, and that it is an effect of wind-induced divergence of surface waters along parts of the coast. According to theory, a wind stress parallel to the coast, which occurs for example when southeast trade winds blow along the northern and central parts

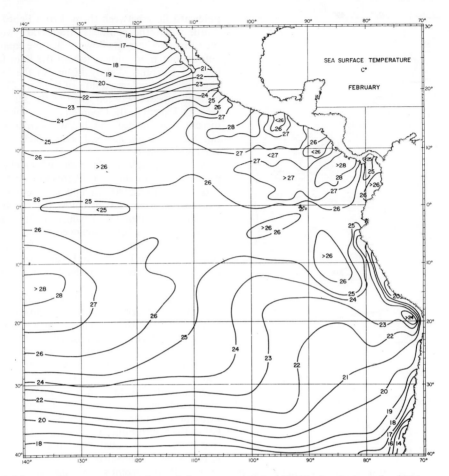

Fig. 3. Mean surface temperature chart of the eastern tropical Pacific for the month of February. (After Wyrtki, 1964; *courtesy Deutsches Hydrographisches Institute.*)

of the coast of Peru, should cause a transport (Ekman transport) of surface water away from the coast. Compensatory ascending movement of deeper water will then occur in the areas from which the surface waters were transported. Such transports have been estimated for different seasons and latitudes, from data on mean wind stress and mean orientation of the coast along the west side of South America. From these calculations, upwelling would be expected to occur to some extent all along the coast from 5 to 45°S, and to be strongest between 5 and 20° S and weakest between 20° and 25° S. It would also be expected to be strongest in winter to the north of 20° S, and strongest in spring or summer to the south of 30° S. Field observations broadly agree with this picture, but more observations are needed, especially on seasonal variations, to test the expectations.

The pattern of combined horizontal and vertical motion of water is quite complex in the Peru Current system. It was investigated from data obtained in October and November 1960, in an area broadly between 95° W longitude and the Peru coast. For the water layer between 100 m and the sea surface it was estimated that 2.6 million m^3/sec entered from the north and the same amount entered from the south, while 5.7 million m^3/sec departed to the west. The difference, outflow minus inflow within the layer, namely 0.5 million m^3/sec, was provided by ascending motion from below 100 m. For the layer between 100 and 300 m it was estimated that 3.5 and 4.4 million m^3/sec. entered from the north and south respectively, while 5.6 million m^3/sec departed to the west. The difference, inflow minus outflow within the layer, namely 2.3 million m^3/sec, went partly upwards (0.5 million m^3/sec, into the surface layer as mentioned previously) and partly downwards (1.8 million m^3/sec). Convergent horizontal flow and downward vertical motion also occurs between 300 and 700 m. It can be argued from these findings that the upwelling is basically a result of the convergent flow below 100 m (of which the Peru Countercurrent forms a part), the wind stress merely providing a mechanism for it. The water ascends at an average rate of about $20 \cdot 10^{-5}$ cm/sec, equivalent to 5 m/month; this rate is probably greatly exceeded in certain areas when suitable winds are strong.

Some oceanographers have distinguished areas of rather consistent strong upwelling from other areas along the coasts of Peru and Chile, mainly by comparing inshore and offshore surface temperatures. There is no doubt that the distribution of upwelling is patchy. Strong upwelling frequently occurs off Antofagasta in Chile (24° S), and off San Juan (15° S) and Punta Aguja (6° S) in Peru.

The average depth of the center of the permanent thermocline is from about 60 to 90 m in the Peru Coastal Current, and about twice as much in the Peru Oceanic Current. The mean temperature gradient is rather weak, less than 1°C/10 m. A summer thermocline with a stronger temperature gradient, 1.5°C/10 m or more, is formed above the permanent thermocline. There is no permanent thermocline, although there is a summer thermocline, in coastal waters south of 30°S.

Precipitation exceeds evaporation in the source region of the Peru Current, whereby surface salinity is low (about 34‰) at 40° S. It tends to increase northwards because of evaporation and mixing, becoming 35‰ or more at some latitude (depending upon the season) north of 16° S; and it tends to be higher in the Peru Oceanic Current, which is closer to the high salinity cell of the subtropical South Pacific Ocean, than in the Peru Coastal Current. Upwelling may modify surface salinity in inshore areas. At about 5° S on the coast the surface salinity drops sharply from about 35‰ to about 34‰ (Fig. 5). This front, which is recognizable as far offshore as the Galapagos Islands, separates the tropical low-salinity surface water from the subtropical high-salinity surface water of the Peru Current. The Peru Current turns westward into the South Equatorial Current to the south of the front. In the southern hemisphere winter, the front is also marked by a temperature difference, the tropical waters being then much warmer than the subtropical waters. In the northern and western parts of the Peru Current system, the salinity decreases with depth, but in the southern part (south of about 20° S) there is a salinity minimum at about 100 m and a salinity maximum at about 300 m.

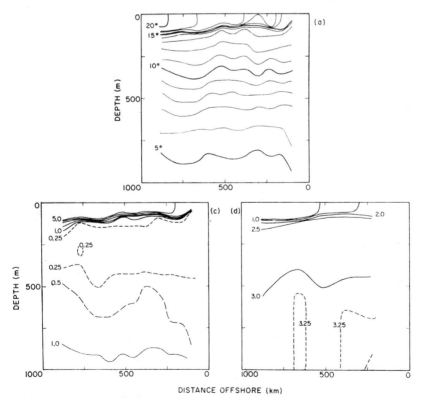

Fig. 4. Profiles of properties across the Peru Current at 14° S in July 1952: (a) temperature, °C; (c) dissolved oxygen, ml/liters; (d) phosphate-phosphorus, μg-at/liter. (After Wooster and Reid, 1963; *courtesy John Wiley & Sons, Inc.*)

The salinity minimum supplies the upwelling in the southern parts of the Peru Coastal Current.

An oxygen minimum, with concentrations of dissolved oxygen below 0.25 ml/liter, occurs throughout the Peru Current system. It commences closer to the surface and is thicker in the Coastal Current than in the Oceanic Current (Fig. 4). This probably reflects the greater biological productivity of the Coastal Current, mentioned below. Upwelling from the top of the oxygen minimum layer can result in locally low concentrations of oxygen at the surface.

The horizontal and vertical distributions of the following chemical nutrients have been studied in the Peru Current region: phosphate-phosphorus, silicate-silicon, nitrate-nitrogen, and nitrite-nitrogen. Nitrite-nitrogen has a primary maximum near the top of the thermocline and a secondary maximum in the region of the oxygen minimum, in some situations. The secondary maximum is possibly the result of reduction of nitrate by denitrifying bacteria. The other nutrients increase in concentration from the surface downwards, the gradient being particularly great in the region of the thermocline (Fig. 4). This means that high concentrations are available in the water layers that feed the upwelling, so that the upwelling process brings nutrient-rich water to the surface. Vertical eddy diffusion is probably responsible for much additional upward transport of nutrients. Surface nutrient concentrations are therefore very high (for tropical waters) in the

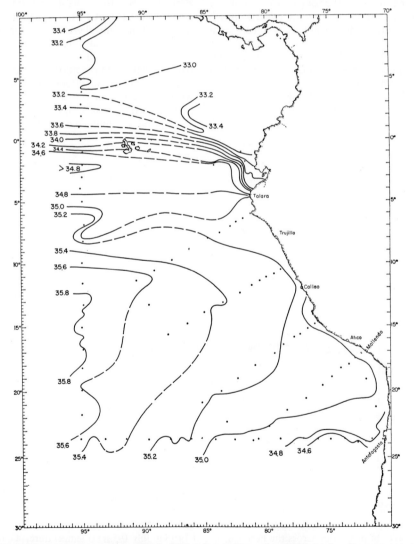

Fig. 5. Distribution of surface salinity (‰) in September to December 1960; broken lines represent interpolation. (After Wyrtki, 1963; *courtesy The Regents of the University of California.*)

region of the Peru Coastal Current, especially in the coastal upwelling areas, although the concentrations decline considerably to the west in the region of the Peru Oceanic Current. Mean surface concentrations in the Coastal Current are approximately 1.0, 10.0, and 10.0 μgm-at/liter for phosphate-phosphorus, silicate-silicon, and nitrate-nitrogen, respectively. Corresponding concentrations in the Oceanic Current are less than half as much. It is this chemical richness—a consequence of the above-mentioned physical processes and structure—that makes the Peru Coastal Current region one of the most biologically productive regions of the world ocean.

Biological Productivity

The scientific literature on the plant and animal life of the region is large and scattered. As far as human resources are concerned, the main groups of relevant organisms are the phytoplankton, the anchoveta or anchovy, which is one of the animals that feeds on the phytoplankton, and various seabirds, which feed chiefly on the anchoveta. The phytoplankton, of course, is not a human resource (or likely to be), but it is the basic foodstuff that, either in a living or a dead state, directly or indirectly supports all the animal life of the sea. The anchoveta and the birds are of direct importance to man.

Abundance of phytoplankton is frequently measured as concentration of chlorophyll a, which is one of its constituents. The surface concentration of chlorophyll a varies in physically different areas of the eastern tropical Pacific Ocean. During the period July to December, which includes the season of probable peak upwelling off Peru, the mean concentration is highest in the northern part of the Peru Coastal Current (0.70 mg/m^3), next highest in the southern part (0.30 mg/m^3), and much less elsewhere, including the Peru Oceanic Current (0.12 mg/m^3).

Such concentrations reflect a balance between accessions to the phytoplankton population and removals of various kinds from it (e.g., grazing by herbivores). There is no routinely good method for estimating removals, but accessions (production of plant material from photosynthesis) can be estimated by measuring the uptake of C^{14} in water samples. The means of such estimates for surface waters vary with areas in approximately the same way as the chlorophyll a means. This shows that the rate of accession of material to the phytoplankton population, like the concentration of the population itself, is high. This makes plenty of plant material available for grazing by animals. If there were no upward transport of nutrients, the plant population would soon deplete the surface waters and diminish. Although the rate of primary production is very high (over 200 g C/m^2 of sea surface/year, compared with less than 50 g in the centers of subtropical ocean current gyres), similar rates occur in other upwelling areas of the world (e.g., in the California Current). Phytoplankton genera that are dominant in certain areas include *Chaetoceros, Rhizosolenia, Planktoniella, Coscinodiscus, Thalassiosira, Synedra,* and *Corethron.*

The anchoveta (*Engraulis ringens*), a small clupeoid fish, was little studied until a significant commercial fishery began in Peru about 1955. The Peruvian catch increased from 59,000 tons in 1955 to 8.8 million tons in 1964, and continues to produce about 8,000,000 tons annually. It is now the world's largest fishery in terms of weight of catch of a single species, and it makes Peru the principal fishing nation of the world in terms of weight of catch of all species. The principal fishing gear is the purse-seine net. The fishery is coastal and extends into Chile. Most of the catch is reduced into fish meal, which is used for feeding domestic animals, and oil, which is used for making margarine and other products. Investigations into the population structure, life history, ecology, and population dynamics of the anchoveta are being actively pursued by scientists associated with the Instituto del Mar del Peru.

The coast and coastal islands of Peru are inhabited by very large but fluctuating numbers of seabirds (10,000,000 to 30,000,000) that are of great economic importance, and are therefore conserved by an agency of the Peruvian government, because of the guano they produce. The three principal guano-producing species are the guanay or cormorant (*Phalacrocorax bougainvillii*), piquero or booby (*Sula variegata*), and alcatráz or pelican (*Pelecanus thagus*).

Since guano is formed from anchovetas, the proper management of the anchoveta resource is of concern to man on two counts, the fish and the guano. The maximum sustainable catch of anchovetas by both men and birds has been estimated at about 10,000,000 tons/year. An estimate of the amount taken by birds, based on the size of the bird population in recent years, is about 2.5 million tons/year, leaving about 7.5 million tons/year available for man. The resource is now being managed on the basis of a catch limit by man of approximately 8,000,000 tons.

Other organisms are of interest and importance. The mean concentration of zooplankton, as captured between the surface and 300 m by a net of minimum mesh size 0.31 mm, has been estimated at 127 ml/10^{3}m^3 in the Peru Coastal Current region and 76 ml/10^{3}m^3 in the Peru Oceanic Current, for the period July to December. In the uppermost 150 m of the Coastal Current, concentrations are probably over 200 ml/10^{3}m^3. Published data are insufficient to describe adequately the seasonal variation of zooplankton standing stock, or any other biological property including chlorophyll *a* and primary production, but suitable data are now being collected by oceanographers of the U.S., Peru, and Chile. Euphausiids are an important item in the zooplankton, especially off Chile. Measurements of the standing stock of the small nekton, apart from the anchoveta and a few other small fishes and crustaceans of commercial importance, are sparsely distributed by area as well as by season, although this situation is being remedied. The available data show the expected decline westward from the Coastal Current into the Oceanic Current.

The larger nekton include cephalopods, fish, and mammals (cetaceans and pinnipeds). The commercial catch of fish (including elasmobranchs) in Peru in 1961 was 4.72 million tons, of which 4.58 million tons consisted of anchoveta; the Chilean catch in 1961 was 0.39 million tons, of which 0.26 million tons consisted of anchoveta. The catch of all fish except anchovetas was only 270,000 tons for both countries. Pelagic fish outweigh demersal fish in this catch, as might be expected from the fact that the continental shelf is narrow and subject to invasion by oxygen-poor water along the sea bottom. The pelagic bonito, *Sarda chilensis*, is the principal species taken in Peru; it also occurs in Chile and contributed 59,000 tons to the 1961 catch of both countries combined. The demersal or semidemersal merluza or hake, *Merluccius gayi*, is the principal species taken in Chile, although it is not important in the Peruvian fishery; the 1961 Chilean catch was 75,000 tons. The 1961 commercial catch of mollusks and crustaceans was 4,000 tons in Peru and 41,000 tons in Chile. The principal whale of commerce is the sperm whale, *Physeter macrocephalus*, of which almost 5,600 were caught in the waters of Peru and Chile in 1961 and again in 1962, but it is suspected that this rate of exploitation cannot be sustained. Sea lions are not now exploited commercially. A spectacular component of the nekton in the Peru Current is the large squid *Dosidicus gigas*, which is sometimes found in large aggregations, but it has no significant commercial value.

It is appropriate to mention the occurrence of important changes in physical, chemical, and biological properties that occur in parts of the Peru Current system in certain years or groups of years, and are called El Niño. The principal physical manifestation of El Niño is that sea surface temperatures become higher than normal. In consequence, anchovetas become scarce in surface waters, whereby the birds can no longer catch them, and the fledgling birds starve; another consequence is that tropical fish, including yellowfin tuna (*Thunnus albacares*) and skipjack tuna (*Euthynnus pelamis*), which are captured by U.S. fishermen along the South American coast, become more abundant than usual in Peruvian waters. Recorded El Niño periods include the years 1891, 1941, 1953, and 1957–1958. The physical situation has been variously explained by physical oceanographers and meteorologists, most of whom agree that fluctuations in the wind field are basically responsible for it, but it has not yet been possible to test these hypotheses by all the necessary observations before, during, and after a Niño.

The following important references may not convey the large volume of information available or the rate at which further information is being added. Investigators from Peru and Chile are playing a substantial role in the work currently in progress, both on the current system and its resources.

MAURICE BLACKBURN

References

Blackburn, M., "Biological oceanography of the eastern tropical Pacific: summary of existing information," 1966, U.S. Fish and Wildlife Service, Special Scientific Report, Fisheries No. 540.

Gunther, E. R., "A report, on oceanographical investigations in the Peru Coastal Current," *Discovery Reports*, **13**, 107–276 (1936).

Institute of Marine Resources, University of California, "Review of the coastal fisheries of the west coast of Latin America," 1965, IMR Reference 65–4.

Murphy, R. C., "Oceanic Birds of South America," vols. I and 2, 1936, Macmillan Company, New York.

Posner, G. S., "The Peru Current," *Bingham Oceanographic Collection, Yale University, Bulletin*, **16**(2), 106–155 (1957).

Schweigger, E., "Die Westküste Südamerikas im Bereich des Peru-Stroms," 1959, Keysersche Verlagsbuchhandlung, Heidelberg-München.

Wooster, W. S., and Gilmartin, M., "The Peru–Chile Undercurrent," *Journal of Marine Research*, **19**(3), 97–122 (1961).

Wooster, W. S., and Reid, J. L., "Eastern boundary currents," in M. N. Hill, Ed., "The Sea: Ideas and Observations on Progress in the Study of the Seas," vol. 2, 1963, Interscience Publishers, New York.

Wyrtki, K., "The horizontal and vertical field of motion in the Peru Current," Scripps Institution of Oceanography, University of California, *Bulletin*, **8**(4), 313–346 (1963).

Wyrtki, K., "The thermal structure of the eastern Pacific Ocean," *Deutschen Hydrographischen Zeitschrift*, Ergänzungsheft (A)6 (1964).

Wyrtki, K., "Surface currents of the eastern tropical Pacific Ocean," Inter-American Tropical Tuna Commission, *Bulletin*, **9**(5), 271–304 (1965).

Wyrtki, K., "Circulation and water masses in the eastern equatorial Pacific Ocean," *International Journal of Oceanology and Limnology*, **1**(2), 117–147 (1967).

PERU FISHERY

Benthonic Community

It has been a general belief that in the Peruvian Current's ecosystem, life below the thermocline down to the bottom, in the continental shelf area, was very poor. Advancing as a cause of this poverty the scarcity of dissolved oxygen, the condition was supposed to extend over the whole bottom of the shelf, except where the shore line is affected by the waves (Schweigger 1943, Popovici 1963, and Chapman, 1964).

In 1965, Del Solar et al., aboard the *Bettina*, searched for the genus Merluccius, in order to understand better its distribution; the ecological behavior of the Peruvian species of the genus was assumed to be very similar to that of others in the rest of the world. The border of the continental shelf was particularly explored and found to be the seat of a dense population of hake that occupies a strip 390 miles long between Punta Pariñas (04°41′S) and Huarmey (10°16′S) at a depth from 70–200 m on the external face of the shelf.

Merluccius gayi is the key species of a large benthonic community made up mainly of diverse species of crustaceans, mollusks, and fish that normally live in that area, so poor in oxygen that near the bottom its contents range between 0.12 and 1.09 ml O_2/liter. There is evidence indicating that the bottom occupied by this community is covered by a loose, light mud quite different from the sticky polysaprobic mud in those areas where the benthis is very poor, as is the usual case South of Huarmey. The low dissolved oxygen content of the waters close to the bottom and up to the thermocline is not a limiting factor for the development of these large bentho and benthopelagic populations.

The limiting factor for the biological productivity within the scavenger-loaded subtratum would be hydrogen sulfide; however, this would not be of importance in the water column above the mud in the inner part of the shelf. Nourishment for the population is provided mainly through an energy-transferring current in the form of migrating zooplankton and nekton from the immediate high primary productivity areas to the bottom of the shelf's edge (Del Solar, 1968).

Conservation

To conserve marine resources and promote rational exploitation, the Peruvian government and the fishing industry created, in 1967, a scientific organization now called Instituto del Mar del Peru supported mainly by the industry with some assistance from the Special Fund of the United Nations through FAO in the form of experts and materials. The institute (IMARPE) operates the R/V *Unanue* and has a modern building with ample facilities at Callao and six field laboratories at different seaports along the coast. "The work of the staff of the IMARPE in collecting data concerning catch and effort, size composition of the catches, and information concerning the biology and ecology of the anchoveta, has made it possible to monitor the development of the industrial fishery with a view to, on the one hand, insuring against the overexploitation of the anchoveta population, and on the other hand, insuring against premature curtailment of the exploitation of the anchoveta resource before such curtailment is actually needed. This is one of the few instances when adequate statistical and biological data have been collected and analyzed during the early development of an important commercial fishery. The government and industry of Peru are to be congratulated on their foresight in collecting information upon which the status of the fishery may be examined, and decisions concerning its further development made on an objective, scientific basis." (Schaefer, 1967.)

The interest of IMARPE is not directed only to anchoveta but also to other species in the present fisheries in this region. Furthermore, oceanographical investigations fall within the scope of the Institute as does its participation in various international programs for the study of the Peruvian sea.

The Living Resources of the Sea

There are 502 known marine fish species in Peru. Only those making up the most important fishery resources of Peru will be discussed here (Table 1).

Family Engraulidae (anchovies). There are ten species of this family in the waters off Peru, but *Engraulis ringens* Jenyns is the one with the largest population living in the Peruvian Current and ranging from Central Chile (37°04′S) all along the Peruvian coast to Cabo Blanco (04°15′S) in a belt close to the shore and extending 30 miles out during the summer and 120 miles during the winter.

Anchovetas spawn in both winter and summer, but with much more intensity and duration during the summer. They reach sexual maturity when approximately one year old and 12 cm long, and can produce about 9000 eggs during several spawnings in the same season. It has been indicated that *E. ringens* spawns from 94°15′S to the south, and that the young anchovetas reach 8 to 9 cm at an age of about six months being then recruited to the fishery.

The diet of this species consists of 1% zooplankton, mainly copepods, euphausids, and fish eggs, and 99% phytoplankton, with a dominance of diatoms among which the principal species are *Coscinodiscus centralis, C. curvulatus, Schrodella delicatula, Skeletonema costatum,* and *Thalassiothrix frauenfeldii* (R. de Mendiola in Jordan and C. de Vildoso, 1965).

It has been estimated, on the basis of the amounts consumed by the guano birds and the bonito only, that the annual natural mortality of anchoveta is at least 3,000,000 metric tons (Boerema & Saetersdal, 1966). The author thinks that this figure omits the predation upon the anchoveta of at least fifteen other fish species plus some cephalopods and mammals, and that if this predation is accounted for, the minimum estimate should be increased to 6.5 metric tons.

The Peruvian fishing industry caught 9,760,000 metric tons of anchoveta in 1967, operating a fleet, exclusively devoted to anchoveta, of 1,536 boats of which 58% are steel-hulled. Most of the boats, 67%, are of the 65 to 79 ft classes, while the larger classes, between 80 and 95 ft, account for 18.6% of the total. The total holding capacity of this fleet is 194,027 metric tons corresponding to a GRT of 153,557 metric tons. The average engine power for all the boats is 266 h.p. per boat, while the purse-seine nets vary between 130 and 350 fathoms length by 25 to 42 fathoms depth. Of all the boats, 98.5% are provided with echo-sounders and 82.9% with power blocks; only 76.9% of the boats have special pumps to transfer fish from the net in the water to the hold. It can be appreciated then that the anchoveta fishery uses the most sophisticated fleet in Peru.

The fishermen employed exclusively in this fishery total 18,000. It has been determined that the average anchoveta catch per vessel during the fishing season (September 1966 to June 1967) was 5,360 metric tons for vessels between 65 and 79 ft at the beam.

Almost all the anchoveta are used in the fish meal industry. A total of 144 produced in Peru 1,850,000 metric tons for animal feed (1967). Of this total, 62.1% was exported to Europe, 28.3% to the U.S. and Canada, 6% to Latin America, and 3.6% to the far East. During this same year the poultry industry of Peru used 28,500 metric tons of fish meal. At present both the Peruvian government and the industry are very much interested in using part of the anchoveta and hake catches to produce fish protein concentrate for human consumption.

Family Scombridae. There are 12 species of this family in Peruvian waters but only four of them are important. The "bonito" (striped skipjack),

TABLE 1. PERUVIAN FISHERIES PRODUCTS (1966)
In metric tons, main species

Scientific name	Common name	Total catch	Reduction	Human consumption
Engraulis ringens Jenyns	Anchoveta	8,529,820	8,529,820	
Sarda sarda chiliensis (Cuvier)	Bonito	71,430	20	71,410
Brevoortia sp.	Machete	13,419	3,833	9,586
Merluccius gagy gayi (Guichenot)	Merluza	13,359[a]	12,654	648
Pneumatophorus japonicus Jordan	Caballa	7,559	14	7,544
Katsuwonus pelamis (Linnaeus)	Barrilete	6,387		6,387
Trachurus symetricus Nichols	Jurel	4,270		4,270
Neptomenus crassus Starks	Cojinoba	5,379		5,379
Thunnus albacares Collete & Gibbs	Atun a.a.	5,461		5,461
Sciaena gilberti Abbott	Corvina	3,516		3,516
Paralonchurus spp.	Coco	3,155		3,155
Paralabrax spp.	Cabrilla	3,084		3,084
Sciaena deliciosa (Tschudi)	Lorna	2,229	68	2,161
Mustelus spp.	Tollo	6.996		6.996
Other fishes species		26,326	8,676	17,649
Peneus spp. and Trachypeneus spp.	Langostinos	357		357
Brachyura	Cangrejos	105		105
Miscellaneous crustaceans		5		5
Mytilus spp.	Choros	4,397		4,397
Pecten purpuratus	Conchas	720		720
Loligo opalescens	Calamares	554		554
Miscellaneous mollusks		424		424
Mammals	Animales	27		27
Tortoises	Tortugas	8		8
Others		3		3
Totals		8,708,990	8,555,085	153,846

[a] Corrected by the author.

Sarda chilensis (Cuvier) is the species most fished for human consumption as fresh and salted (41,980.3 metric tons) being very popular. In 1966 the canning industry used 32,450 metric tons of this species.

The "atun de aleta amarilla" (yellowfin tuna), *Thunnus albacares* Collete & Gibbs, was abundant in the external front of the Peruvian Current, but its stock is diminishing because of overfishing by foreign tuna boats. The best part of the catches are exported whole-frozen (5,441.3 metric tons in 1966).

The barrilete (skipjack tuna), *Katsuwonus pelamis* (Linnaeus) is another pelagic species occupying the same area as the yellowfin along the whole Peruvian Current, but its population is more abundant; its catches are also for export whole-frozen (5,592.6 metric tons in 1966).

The "caballa" (mackerel), *Pneumatophorus japonicus* Jordan and Hubbs, is the most abundant scombrid of the Coastal Peruvian Current, and is mainly fished in the northern part of the country mostly for salting (6,619.6 metric tons in 1966) and very little for fresh consumption. Caballa is a resource of quite a considerable volume, although little exploited as yet.

Almost all the fishing for the above species is done with purse-seines.

Family Gadidae. Although poorly represented in numbers, this family is important on account of the merluza (hake), *Merluccius gagy gayi* (Guichenot), which, after the discovery of new and extensive areas of distribution along a strip of 390 miles on the bottom next to the continental shelf, has erupted violently into the fisheries picture of Peru.

On these new bottom locations that, until recently, were considered deserts, large communities have been found, among which the merluza's is the most abundant. This species then is a new marine resource that, although its exploitation has just been initiated, has ranked as the fourth most fished species in Peru (13,359.7 metric tons in 1966).

Because this species is as yet little known to the public, only 5% of the catches are for direct consumption as whole-fresh or frozen fillets. There are good prospects of using merluza to produce high quality fish protein concentrate. The present fleet for trawl fishing is too small and not well suited to operate 50 or 60 miles from the coast where the best merluza banks are situated.

Family Sciaenidae. One of the most abundant in species, 36 now known, sciaenidae is preferred by Peruvians; of a single species of this family, the "corvina," *Sciaena gilberti* Abbot, 3516 metric tons were caught for fresh consumption in 1966, and were sold at the highest prices yet obtained. It is quite probable that this species is in danger of being overfished as an effect of the large beach-seine nets (*chinchorros*) used, especially in the southern part of the country.

The "coco", *Paralonchurus spp.* and the "lorna" *Sciaena deliciosa* (Tschudi) are fished with purseseine nets and trawls. Trawl fishing for these species, especially in the north, is substantially increasing the catches, so that quite probably the total amount of sciaenids caught at present is about 15,250 metric tons. Almost all this fish is consumed fresh and whole, although the practice of filleting has increased during the last three years.

Family Serranidae. This family has 22 species in this part of the Pacific; among them, the "cabrillas," *Paralabrax spp.* are the most important because of the strong demand for them; so strong is the demand that 3084 metric tons caught in 1966 were not enough to satisfy it. Most of the species live in rocky habitats so that trawling is difficult. The majority of catches are in the northern part of the country where they are also used for salting.

Family Triaenodontidae. These sharks, with three species in Peruvian waters, *Mustelus spp.*, are of great commercial importance and the amounts taken are now increasing considerably because of their association with merluza in the recently discovered locations on the edge of the continental shelf. They feed mainly on benthonic mollusks and crustaceans. The total catches are up to 6,996.9 metric tons (1966) and come mainly from trawl fishing in the northern part of the country.

Bottom Fishing

Fishing for demersal species by trawling started near the border with Ecuador on the areas under the influence of the Guayas river and the immediately close littoral area of Tumbes. Fishing is done there chiefly for shrimps and also for some tropical fishes. The most common shrimp species are *Peneus precipua*, *P. vannamei*, *Trachypeneus similis pacificus* and *T. Byrdi*, and *Xiphopeneus riveti*. In the area are 18 vessels not greater than 70 ft at the beam operating with shrimp trawls. Their catches for 1966 totaled 357.5 metric tons.

Farther south is Paita (05°05′S), the port with the greatest fleet for bottom fishing, i.e., 35 vessels, not larger than 80 ft at the beam, that operate in the area between Punta Pariñas and Lobos de Tierra.

Although bottom fishing is just starting, it should soon cover the locations recently discovered.

Family Otariidae. These marine mammals are represented in Peru by two littoral species, *Otaria flavescens* Shaw, the *lobo chusco de un pelo* and *Arctocephalus australis* Zimmermann, the *lobo fino de dos pelos*. The most abundant of the two species is *O. flavescens*. Males of this species may reach 100 to 135 kg weight and the females 30 to 50 kg. As for *A. australis*, it is at present very much reduced in numbers and in imminent danger of disappearing. Both species have been subject to intensive hunting on account of their hair coat and oil. Since they live on the islands and points used by the guano birds as nesting places, they benefit to a degree from some official protection. They reproduce in December and January, and the largest concentrations have been observed in the Chincha Islands. It is thought that these *lobos marinos* (sea wolves) compete with the fishermen since they damage the nets while preying on the same fish sought by the fishermen.

Order Cetacea. Several species of these big mammals are often seen in the Peruvian Current; the most abundant are the "delfines," *Paradelphinus longirostris* and *Delphinus delphis*, which are usually found in great herds. *Globicephala meleas* is seen in appreciable quantities near the coast. The greater whales such as *Balaenoptera borealis* Lesson, *B. physalus* L., and *B. musculus* L. have been the subject of intensive hunting during the last decade, but the one most sought has been the "cachalote" *Physeter catodon* L. Consequently, its numbers have declined greatly in Peruvian waters. These catches represent 90% of the mammal catches in recent years.

Marine Birds

Within the 06°20′S and the 18°40′S parallels, some millions of marine birds live as a part of the Peruvian Current's ecosystem; only ten species are endemic to the region (Murphy, 1936).

The "guanay," *Phalacrocorax bouganvillii*, the "piquero," *Sula variegata*, and the "alcatraz," *Pelecanus occidentalis thagus*, rank in that order as the more abundant species, as determined by census. They are also the more important producers of guano, that is, accumulated, dry excrement, used as fertilizer. The total of these populations varies between 5,000,000 and 36,000,000 birds. Such variations accompanied by great mortalities, coincide with El Niño years when the great alterations of the Coastal Peruvian Current extend their effect to the climate of the region.

Millions of birds of other species also frequent the current, and are anchoveta predators as well but do not produce guano because they do not nest on the islands and points especially protected by the governmental institution in charge of the guano management.

The absence of rains in the Peruvian coast allows for the accumulation of guano in the above-mentioned places. This fertilizer has been highly appreciated in the past but at present agriculture cannot depend on the scanty and fluctuating harvest of guano, which reached a maximum of 286,733 metric tons in 1955 and was produced by a bird population close to 40,000,000. The development of the anchoveta fishery has prompted discussions about priority among competitors. Schaefer, very well versed on the ecological system of the Eastern Pacific, says "the question of what share of the sustainable average harvest of anchovetas, estimated at 10,000,000 metric tons per year, should be allocated to direct capture by men and what share should be allocated to consumption by guano birds, has been the subject of great controversy, which needs to be resolved. This can only be done through objective cost-benefit analyses taking into account all the important economic and social factors. Certainly, as a minimum, everyone can agree that it is a priority matter to maintain sufficient populations of each of the bird species to prevent their being driven to extinction, because it is of overriding importance to maintain this genetic material for the future use of humanity. However, one can assert with some certainty that, for this purpose, there is required a good deal less than 16,000,000 birds." (Schaefer, 1967).

Seaweeds

In the littoral and internal sublittoral regions of Peru, there are more than 150 species of marine algae, the greater part of which is distributed from Punta Pariñas, 04°40′S, south to Chile. There are 27 species belonging to the six genera that are the main sources, in other countries, for the prime matter used in the elavoration of agar-agar such as Gelidium, Gracilaria, Gigartina, Pterocladia, Ahnfeldtia, and Hypnea. There are also other edible genera such as Ulva, Enteromorpha, Chondrus, Porphyra, and Lamminaria. (Dawson *et al.*, 1964). A small quantity of algae is used in Peru as food, but industrialization is nil. Exports of dry algae, mainly to the U.S., has been started with a total of 133.5 metric tons.

ENRIQUE M. DEL SOLAR

References

Borgo, J. L., Vasquez, A. I., and Paz Torres, A., "La Pesquería Marítima Peruana Durante 1966," 1967, IMARPE, Chucuito, Callao. Perú.

Chapman, W. M., "Fishery Aspects of the National Oceanographic Program," 1964, Van Camp Seafood Co., California.

Dawson, E. Y., Acleto, C., and Foldvik, N., "The Seaweeds of Perú," 1964, Verlag Von J. Cramer, Weinheim.

Del Solar, E. M., "Ensayo sobre Ecología de la Anchoveta" (Engraulis ringens J.). *Bol. Cia. Adm. del Guano*, **18**, Enero (1942).

Del Solar, E. M., Sanchez, J., and Piazza, A., "Exploracion de la Areas de Abundancia de la Merluza (Merluccius gayi peruanus) en la Costa Peruana, a bordo del "Bettina," 1965, Informe N° 8, IMARPE, La Punta, Callao, Perú.

Del Solar, E. M., "La Merluza (Merluccius gayi peruanus) como Indicador de la Riqueza Biótica de la Plataforma Continental del Norte del Perú," 1968, Soc. Nac. de Pesqueria, Lima, Perú.

Jordan, R. and Ch. de Vildoso, A., "La Anchoveta Engraulis ringens J.," 1965, IMARPE, Informe N° 6, La Punta, Callao, Perú.

Koepcke, W. H., "Peces Marinos Conocidos del Perú," 1962–1963, *Biota IV y V*, Lima, Peru.

Lagler, K. E., Bardach, J. E., and Miller, R. R., "Ichthyology," 1963, John Wiley & Sons Inc., New York.

Menzies, R. J., and Chin, E., "Cruise N° 11 Report R/V "Anton Brunn," 1966, Mar. Lab., Texas A & M University, Galveston, Texas.

Murphy, R. C., "Oceanic Birds of South America," vol. I, 1936, The Macmillan Co., New York.

Piazza, A. A., "Los Lobos Marinos en el Perú," 1959, Caza y Pesca N° 9, Min. de Agricultura., Lima, Perú.

Popovici, Z., "Horizontes Oceanicos de Sudamerica," *Pub. Rev. Pesca. Anuario*, 1962–1963, Lima, Perú.

Popovici, Z., and Chacón, G., "Ensayo de Oceanografia Física," 1966, IMARPE, La Punta, Callao, Perú.

Ryther, J. H., "Cruise N° 15 Report R/V "Anton Bruun," 1966, Mar. Lab., Texas A. & m. University, Galveston, Texas.

Schaefer, M. B., "Dynamics of the Fishery for the Anchoveta Engraulis ringens off Perú," (English and Spanish), 1967, IMARPE, Boletin, Chucuito, Callao, Perú.

Soc. Nacinal de Pesqueria, "Embarcaciones de la Industria de Harina de Pescado," 1967, Lima, Peru.

Cross references: *Algae, Marine; Benthonic Domain; Fish Meal; Seaweeds of Commerce.*

PESTICIDES IN THE SEA

Pesticide pollution of the land and aquatic environments is well documented, but, despite intensive research on this problem in the past two decades, it is still not clear whether these chemicals are causing irrevocable damage to the environment. Pesticides are designed to control unwanted plant and animal populations. Evidence is growing, however, that the continued use of the persistent pesticide chemicals is producing environmental changes or residues in the food web that may cause reproductive failure and lead to the extinction or genetic alteration of some species. These possibilities are of special concern in the marine environment since eventually most of man's waste products find their way to the sea.

Estuarine Pollution

Not enough is known about the kinetics and degradation of pesticides in the marine environment to permit predicting whether or not permanent changes will result from pesticide pollution. Some unfortunate examples of moderate to serious increases in the mortality of aquatic life have occurred when pesticides, in solution and adsorbed on particulate matter, have been transported relatively long distances downstream to an estuarine environment. The direct application of pesticides to the salt marsh and estuarine shore line for the control of noxious insects and aquatic vegetation, and to salt water bottoms for the control of oyster pests or predators has also, on occasion, caused mass mortalities of fish and crustaceans. There are, however, few data to support the widespread feeling that pesticides may already be causing permanent damage to plant and animal communities in the sea. This situation exists despite the fact that residues of the polychlor pesticides are demonstrably present in a wide variety of marine plants and animals and in bottom deposits. On a nationwide basis, pesticide residues in the marine environment are very small —usually in the parts per billion range.

On the other hand, industrial and domestic pollution of many estuaries has already so degraded the environment that only the most tolerant of organisms can persist. Relatively uncontaminated estuaries and coastal waters of the world are of major importance, not only from the esthetic point of view, but also as habitats for many species of commercial fish and for the general productivity of the ocean itself. In recent years, more than half the commercial fishery harvest in the U.S. was composed of six or seven species that depend directly on the estuarine environment, either during some stage of their migrations or as a permanent habitat.

Although the oceans cover nearly 75% of the earth's surface, a relatively small proportion of the total, the shallow coastal waters and estuaries, is the most productive. The abundance of nutrient salts in these waters creates the most favorable conditions for the conversion of solar energy into phytoplankton—the first and most important link in the food chain of the sea. Microscopic plants and animals flourish in estuarine waters, which are justly known as nursery areas. The juvenile stages of many commercially important species require this environment for their survival and development. Smaller fish essential to the food supply of oceanic species, such as the tuna, are produced here in countless numbers. Elimination of this environment by pollution or other factors would have catastrophic effects on the fisheries of the world.

Use and Persistence of Pesticides

Use of the synthetic organic pesticides has increased enormously since the introduction of DDT during World War II. In the period 1962–1966, average production was increasing nearly 8% yearly, and in 1966, more than 625,000 tons were produced. Roughly 200 basic pesticides are in use today, but only about 20 of these are of natural origin, i.e., inorganics and botanicals. Nearly 70,000 different registered formulations of the synthetic compounds have been developed to meet the needs for the various methods of application and for multiple-use mixtures to control two or more pests simultaneously.

The economic usefulness of herbicides has become increasingly apparent and they are now used in about the same total volume as the insecticides. Together these two pesticide types account for about 88% of the total amount of pesticides used. Lesser quantities are required as fumigants and for the control of mites, fungus, rodents, and miscellaneous pests. DDT continues to be the most widely used compound because of its toxicity to a broad array of pests. Its cost is only about 20% the average cost of 24 other important pesticides, and this economic advantage encourages its general use despite the fact that it may persist for a long period in the environment. The U.S. produces about half the world supply of DDT, and, in 1966, exported nearly 45,000 tons to other countries, primarily to control insect vectors of human disease. It is estimated that through its use the incidence of malaria has decreased one thousand- and three thousand-fold, respectively, in India and the Soviet Union in the past decade. Substantial success has been achieved also in the control of plague, typhus, and yellow fever.

The extent of pollution of the environment by a particular pesticide is primarily a function of its chemical stability, and, secondarily, of the continuity of its use or discharge in an effluent. The organophosphorus compounds tend to hydrolyze in moist environments and decay, usually to less toxic forms. They may dissipate in a few hours or days, but, depending on environmental conditions, some are long-lasting. It has been shown that as much as 50% of one chemical experimentally applied to dry soil remained active more than a year. These compounds hydrolyze rather readily in warm water, but may remain toxic for longer periods when released in the environment during the winter.

The synthetic chlorinated hydrocarbon pesticides, or polychlors, are generally much more stable than other types. Consequently, they are usually most toxic at summer temperatures because of the increased levels of metabolic activity of the affected biota. Under optimal conditions, as much as 39% of the DDT and about 30% of the aldrin and dieldrin were found in the soil 15–17 years after a single controlled application. DDT has accumulated in the soil of an extensive New York salt marsh to an average level of 13 lb per acre—presumably the result of two decades of mosquito control. Repeated applications of DDT to apple orchards have produced soil residues of 400 lb per acre after 20 years. Critical estimates suggest that worldwide residues of DDT may amount to about 1,000,000 lb at any one time.

In the following discussion of pesticide residues, DDT and its two most prevalent degradation products, DDD and DDE, are referred to collectively as DDT. Although other polychlor residues are encountered frequently in randomly collected monitor samples, DDT has been found to be almost universal.

Physical and Biological Transport

The unintentional physical transport and distribution of pesticides depend chiefly on the method of application and the nature of the site. Such factors as volatility, solubility, susceptibility to chemical decay, and photodecomposition affect their mobility. The degree to which they are absorbed physically or biologically by different soils and crops also plays an important role in determining the ultimate disposition of pesticide residues.

Meteorological conditions following a pesticide application will determine whether or not substantial amounts are washed away in surface waters to contaminate lakes and streams in the drainage basin. Adsorbed on silt, they may eventually be carried to the river delta. Mississippi River water sampled at New Orleans usually contains measurable amounts of one or more of the chlorinated pesticides; it has been estimated that the river contributes about 10 tons of pesticides a year to the Gulf of Mexico.

Land winds may carry the equivalent of 15 tons of dust per mile^2 at times. Analysis of random samples of such dust have shown DDT residues as high as 90 ppm or about $2\frac{1}{2}$ pounds per mile^2. Recent collections of dust air-borne to Barbados Island from Europe and Africa indicate that the easterly tradewinds may transport more than 1,000 lb of DDT across the ocean annually.

The evidence is clear that substantial amounts of pesticides persist on the soils and vegetation where they are applied. Variable amounts are leached into surface waters and transported away from the application site in solution or adsorbed on silt. As a consequence, this chemical burden is circulating continuously through the terrestrial and aquatic environments where sensitive species may be adversely affected.

An understanding of the full impact of pesticides has been hampered by lack of adequate data and the necessity for relying on estimates of average levels of pollution. Such estimates tend to be too high when based on biased samples collected at a time or place of known pesticide pollution. On the other hand, analyses of samples collected in systematic programs to monitor the physical environment indicate apparently insignificant levels of pollution because of the typically many-fold dilution that occurs when a pollutant is introduced into the atmosphere or into an aquatic system.

Animals do not live by averages, of course, and a brief incident of a high level of pollution may cause, as it frequently has in the past, a serious increase in fish and wildlife mortality. Such incidents become increasingly rare as knowledge of the hazards to nontarget animals grows, and the mechanics of pesticide applications are improved.

Data derived from the detection and quantification of pesticide residues in still-living systems, as opposed to the study of fish kills for example, provide a more intelligible picture of the extent to which this type of pollution now permeates the marine environment. Sophisticated analytical techniques are now available that permit the identification of residues in the parts per trillion range in water samples. Experimental evidence is ample to show that marine fauna are at least equally proficient. Not only do they detect pesticides at these and greater dilutions, but they may absorb or ingest, metabolize, and store them so that the concentration of residues in tissues may far exceed pollution levels in the ambient waters.

The oyster, for example, continuously exposed to 0.1 ppb of DDT may concentrate in its tissues up to 7.0 ppm in about a month without suffering apparent damage. This biological magnification of $70,000 \times$ indicates the extent to which minor levels of pollution may be incorporated into the food web. Fortunately this degree of magnification in one step is the exception rather than the rule.

Bacteria and phytoplankton form the broad base of marine food chains. Under experimental conditions, both absorb and metabolize DDT without demonstrable damage to the cells. Low concentrations of the pesticide may actually enhance their growth in cultures. From this initial source, residues are gradually magnified as they are transmitted through successive trophic levels to zooplankton, filter-feeding invertebrates, and fish. The fish, in turn, may be eaten by carnivorous mammals and birds. Residues may be aquired at any trophic level regardless of the feeding habit of the animal. The polychlors readily penetrate the skin or gills of crustaceans and fish so that residues are aquired even by the nonfeeding animal, although more slowly and to a lesser extent.

The food chain may be most complex and involve a large number of interacting animals, or it may be direct and short. Mullet feeding on plankton that had residues up to about 50 ppb were found to contain residues in the range of 1–15 ppm. Random observations of porpoise that feed largely on mullet in the Gulf of Mexico showed DDT residues in the blubber as high as 800 ppm. Ten large porpoises with residues of this magnitude would be carrying the equivalent of 1 lb or more of DDT in their bodies.

Sea gulls are involved in an equally direct food chain along the Gulf coast where they feed on bay scallops. These filter-feeding mollusks may contain relatively large DDT residues, and it is certain that these contents are being magnified in the gull. Numerous investigators have observed that fish-eating birds typically harbor significant amounts of DDT and other persistent pesticides.

Distribution of Residues

Evidence that estuarine fauna from several drainage basins in the U.S. contained substantial pesticide residues led to the initiation of a systematic monitoring program in 1965. The oyster and related mollusks were selected as the bioassay animals because of their widespread occurrence; their sessile habits insured a reflection of pollution at a fixed point in the estuary.

Laboratory experiments had revealed the oyster's ability to store a broad array of the persistent pesticides that might be present at very low levels in the environment. It was observed that pesticide residues were flushed out of the oyster fairly rapidly when surrounding waters were no longer polluted. It was possible, therefore, to detect when the waters became polluted with polychlors and the relative amounts present by sampling the bioassay animals periodically. As finally developed, the program involved analyses at 30-day intervals of animals from about 175 stations along the Pacific, Gulf, and Atlantic coasts.

The samples contained low levels of persistent pesticide residues throughout the year with the exception of a few locations in Washington and Maine. The residues of DDT and its metabolites were, for the most part, less than 0.1 ppm. In only one area were residues consistently as high as 0.1 to 0.5 ppm. In a few drainage basins, residues have been absent for six or more months of the year following the spring freshets. Residue levels follow a seasonal trend and are usually highest in early summer. This peak is to be expected if the residues are derived from surface waters draining agricultural land. In river basins supporting large urban populations, residues are consistently high throughout the year, although seasonal peaks may still be detected.

In one Gulf of Mexico drainage basin, intensive truck farming produces three crops annually. At least six different pesticides are applied to one or more of the major crops; a total of about 18 lb per acre is applied each year. These practices are reflected in the three seasonal peaks of residues in the monitor samples and their generally high levels as compared to other areas. We are justified in assuming that similar fluctuations of pesticide pollution are developing in coastal waters around the world, although the data remain to be collected.

Essentially random monitoring of oceanic fauna during the past few years demonstrated the worldwide occurrence of pesticide residues, but has done little to clarify their lasting significance. A few examples will illustrate the widespread dissemination of this type of pollution.

In a series of 169 fish-oil samples collected on a global basis, 157 contained pesticide residues. Residues in tunas taken off California, Ecuador, and Hawaii were sufficiently large to permit the calculation, by extrapolation, that the annual U.S. commercial tuna harvest removes about 45 lb of DDT from the ocean. Fortunately, most of this DDT is in the subcutaneous fat and organs that are discarded in processing; canned fishery products contain negligible, if any, pesticide residues.

Seabirds whose feeding is restricted to the mid-Atlantic and Antarctica contain substantial DDT residues. By implication, their diet of small fish and invertebrates has been contaminated by airborne pesticides, although water transport is not ruled out. All 90 eggs analyzed of 13 species of seabirds contained residues of DDT, as did gray seals in Scotland, a whale in California, and Weddell seals in Antarctica. This list could be expanded greatly, but the examples are sufficient to make us inquire as to the significance of the apparent global extent of this pollution.

Significance of Residues

Despite the wide occurrence of persistent pesticide residues in the world fauna, their magnitude is for the most part too small to have any known significant effects to human health. The effect of these residues on animals is a different matter. Both laboratory and field observations indicate that present levels of pollution may already be causing irreversible changes in some areas.

Oysters containing DDT residues of about 4 ppm were used experimentally as the exclusive diet for laboratory populations of shrimp and fish. The mortality of test animals was low; only one or two died each day. The longer the contaminated diet was used, however, the greater was the cumulative mortality. After four weeks, at least half the experimental animals were dead, whereas deaths in control populations were negligible. It would be impossible to detect such small daily losses under field conditions. These results suggest that continuing low levels of pesticide pollution in the estuary could cause significant but unidentified increases in the mortality of commercially important animals and of other species important in the food web.

It is equally probable that low levels of pollution can cause changes in the behavior of marine animals that might make them more susceptible to being preyed upon, or perhaps alter their physiological ability to migrate back and forth between the fresh and salt water habitats. Such derangements could be crucially important to the continued existence of many species of commercially important fish and shellfish.

Loss of ability to produce viable fry by lake trout in the state of New York has been attributed to the buildup of DDT in the un-spawned eggs. A similar situation is suspected in one species of marine fish. Circumstantial evidence is strong, too, that dietary DDT is lowering the reproductive ability of the Bermuda petrel. This bird is one of the species already endangered; only one colony of about 100 individuals is known to exist. A recent report states that nesting success in this colony has declined each year during the past decade. Analyses of dead embryos and abandoned eggs showed DDT residues to be in excess of 6 ppm. These birds feed only in the open ocean far from land, presumably in areas free from coastal pollution.

We cannot continue to beguile ourselves with the belief that the immensity of the oceans guarantees adequate dilution of our wastes. Even now, six major U.S. metropolitan areas are dumping tens of thousands of tons and millions of gallons of domestic and industrial wastes daily into near-shore waters. Just as some of the rivers and estuaries are already so saturated with wastes that the normal biota has beeen eliminated, the eventual degradation of the ocean into a biological desert may be only a question of time. One cannot know at this point what pollution load is bearable but the accumulation of pesticides in the sea may eventually be the final factor in upsetting the balance of this ecosystem. It is urgent that the further development and use of ocean resources be seen as inextricably linked with man's future. As terrestrial resources are exploited and diminish the managed harvesting of marine resources will become increasingly vital to the human economy.

Pesticides make up only one facet of the entire waste disposal problem. Their importance in ameliorating daily life, in controlling disease, and in augmenting the harvest of farm and forest products implies a steady increase in their use. Knowledge of pesticide toxicity and the registration of pesticides for public use does not guarantee their safety. Once applied, the persistent pesticides are beyond man's control and their secondary effects eventually may be disastrous. Mankind has a clearly defined responsibility to create specific pest control methods that do not alter the environment permanently.

PHILIP A. BUTLER

Cross references: *Estuarine Environment; Oysters; Pollution of Seawater.*

PHILIPPINE FISHERY

The Philippines is one of the 12 countries that reported major increases in fish catch between 1948 and 1965. Of the 50 fishing countries of the world, the Philippines ranks sixteenth in fish catch, according to 1965 figures.

Administration

Philippine Fisheries Law provides among others, a dual control of the administration of the country's fisheries by the municipal and the national governments. The municipal government has jurisdiction over inland fisheries of the municipalities, coastal areas within the three-mile limit, and fishing craft powered or non-powered of three tons gross or less. The national government, with the Philippine Fisheries Commission as the agency, has supervision over commercial fishing vessels of more than three tons operating in all Philippine territorial waters and beyond. Ordinances regulating the fisheries of the municipalities are passed for review and have to be approved by the Philippine Fisheries Commission so that these would not conflict with the policies of the National Government on fisheries, particularly those that touch on fisheries conservation.

For the purpose of administration, the Philippine Fisheries Commission, with main offices in Manila, subdivides the Philippines into eight fishery regional areas, each of which is divided into two districts. The Commission is headed by a Commissioner of Fisheries assisted by the Deputy Commissioner for Administration and Services, and the Deputy Commissioner for Research and Development. The Regional Offices are administered by Regional Directors and the District Offices by District Fishery Officers.

Inshore Fisheries

Fishing in the Philippines is done mostly inshore, i.e., within the territorial waters only. In coastal municipalities, fishermen employ small fishing craft which often use sails. In recent years, however, the fishermen have motorized their craft for better efficiency. Commercial fishing is conducted in all waters of the Archipelago with the use of large motorized commercial fishing vessels.

The fishing vessels are based mainly in Manila, Cebu, Iloilo, Bacolod, and Zamboanga. Most of the large fishing vessels are based in Manila, the biggest market for fresh fish and the largest distributing center for both fresh and processed fish. As these vessels operate in fishing grounds several hundred miles from the home port, their voyages range from 5 to 20 days. From 1961 to 1965, fisheries statistics show a gradual increase in the number of licensed commercial fishing vessels from 1560 in 1961 to 2393 in 1965, involving an investment of P34,814,082.00 in 1961 increasing to P106,625,810.00 in 1965.

The fishing vessels range in tonnage from 3 to over 100 gross tons. The major fishing gear used are the bagnet, beach seine, purse seine, otter trawl, round haul seine, hook and line, gill net, Japanese drive-in net (*muro-ami*), and fish corral. Of these, the bagnetters, purse seiners, and the otter trawlers contributed 87% of the catch reported by commercial fishing vessels in 1965. The purse seiners are of the latest vintage and are equipped with the Puretic block.

Figures show a steady increase in the production of commercial fishing vessels from 1961 to 1965 as follows: 125,626,430 kg in 1961; 150,036,540 kg in 1962; 208,747,880 kg in 1963; 258,100,120 kg in 1964; and 300,074,200 kg in 1965.

Pelagic Species. The fisheries for pelagic species of the archipelago seem to be gaining importance from year to year. To the yearly catch reported, it contributed 43.63% in 1961, 47.42% in 1962, 52.64% in 1963, 55.17% in 1964, and 59.18% in 1965.

Among the commercial pelagic species, anchovies may be considered the most important. They supply the greater portion of the materials used in the manufacture of fish paste (*bagoong*) and fish sauce (*patis*). Anchovies are caught through almost the entire year, but in different months from the northern to the southern islands, depending upon the prevailing monsoon to which the fishing ground is exposed. When they are caught during the dry season months, a major portion of the catch is sun-dried for distribution in the rural and agricultural areas in the interior localities of the provinces.

Of the 20 species of sardine and herring found locally, only 3 species are caught in commercial quantities. The catch is mostly salted, dried, or smoked. In 1935, the commercial possibilities of canning sardines was demonstrated by the Bureau of Fisheries (now Philippine Fisheries Commission) in a pilot fish canning plant at Estancia, Iloilo Province. This example was followed soon after by the establishment of a commercial sardine cannery at Madridejos on Bantayan Island.

Tuna and tuna-like fish of which there are five commercially important ones, are caught by means of indigenous fishing gear such as fish corrals (*baklad*), seines, troll lines, and hand lines. Off the coast of Antique Province, large-sized yellowfin, which appear from January to April, are caught in fish corrals. Small-sized bonito and skipjack are sometimes caught by fishing with high-powered lights and round haul or scoop seines (*sapyaw*). In the waters off the coast of Batangas Province, the frigate mackerel appear from November to May, and are caught by purse seines.

The small mackerel are usually sold fresh. A portion is preserved in the form of *daeng* or *pinakas*, where the fish are split, lightly salted, and dried.

The carangids are commercial species that include pompanos, crevalles, and scads. Most of the catch is sold fresh. Fishing for scads was not an important fishery before World War II, although some of the fish landed in Madridejose were canned sardine style. In more recent years, large schools of both round- and flat-bodied scads have appeared in the waters north and east of Palawan. Since then, this fishery has acquired considerable importance.

Demersal Species. Demersal fishing of ranking fishing countries is carried on offshore in so-called banks, which are continental shelves submerged to a depth of 100 fathoms or more. The pursuit for demersal fish in Philippine waters is carried on in areas of but a few score fathoms deep.

The most important segment of the demersal fisheries of the Philippines is the trawl fishery. The slipmouth, lizard fish, croaker, nemipterid, hairtail, and shrimp compose the predominant species sustaining this fishery. Trawl fishing is carried on in many bays and gulfs and on the narrow belt of insular coastal shelf, in some places at a depth of from 6–20 fathoms. Existing trawling grounds are estimated at about 4500 miles.[2] The principal trawling grounds are the waters of Lingayen Gulf, Manila Bay, Tayabas Bay, Ragay Gulf, San Miguel Bay, Southeastern Samar, Carigara Bay, Guimaras Strait, Eastern Visayan Sea, Asid Gulf, and northern Capiz. A recent survey made by the Philippine Fisheries Commission and trawl operators reveals more prospective inshore trawling areas in Lamon Bay, the northeastern part of Camarines Sound off eastern Luzon, Bacuit Bay, Coron Bay, Taytay Bay, Malapaya Sound in Palawan Island, and in Sibuguey Bay off Zamboanga del Sur.

The next ranking segment of demersal fisheries is the reef fishing areas of the Archipelago. A diversified group of large-sized demersal species, such as groupers and sea bass, snappers, barracuda, and runners, comprising the major portion of the first and second class fish of the fresh fish market, are generally found in the extensive shoal and fringe types of coral reefs existing in many parts of the Visayan Islands, northern Mindanao and in the islands of the Sulu group. The principal groups of commercial species but of a third class type that constitute the catch here are caesios, porgies, surgeon fish, snappers, and groupers. Commercial reef fishing was introduced in the Philippines by the Japanese from Okinawa in the 1920's with the use of the drive-in net or Japanese trap net called *muro-ami*.

According to official figures, the fisheries for demersal species contributed less to the yearly catch than the fisheries for pelagic species with the following figures: 35.33% in 1961, 45.10% in 1962, 40.48% in 1963, 62% in 1964, and 33.99% in 1965.

Estuarine Fisheries. The major part of the fish produced from the estuarine fisheries comes from the brackish water fishponds. The raising of fish in farm ponds or fish ponds is one of the most intensive methods of land use for food production in the Philippines. In this industry, marginal areas formed by the deposition of alluvial materials through daily flow and annual flooding of rivers, continually supplied with brackish water, are utilized extensively. These areas, so-called swamplands, acquire a role of great significance in the country's economy with their exploitation for brackish water fishponds. There are few regions elsewhere in the world where the utility of swamplands in the delta regions of rivers and foreshore areas is so extensively exploited as a fishery resource as in the Philippines.

The fishpond industry makes use of the milkfish (*Chanos chanos* Forsk), locally called *bañgos*, as the cultivated fish. The fry are collected in clear sandy shores. The gathering of the bañgos fry in itself is a big industry and a major source of revenue for the coastal municipalities favored with this kind of fishery. It is a common practice among many fishpond owners to buy bañgos fry and raise them in the fry ponds of the nursery of the fishpond system, which consists of nursery, fry ponds, catching pond and rearing ponds, and water supply canals.

Many other fishpond owners buy fingerlings, for stocking their rearing ponds, from large scale nursery pond operators, who make the raising of fry into fingerlings a significant commercial enterprise. This is a specialized industry in Dampalit, of the town of Malabon, located only a few miles north of Manila.

As of December 31, 1965, there were in operation in the entire Archipelago, 54,467.57 hectares of privately owned fishponds and 82,783.11 hectares of Government-leased fishponds, making a total of 137,250.68 hectares with a total annual production of 63,197,690 kg. Based on the developmental cost alone, on the average of P2,000.00 per hectare, without considering the real estate values of the areas, the investment involved in this bañgos fishpond industry amounts to P274,501,360.00.

To increase the production of bañgos, fishpond operators now resort to fertilization for enriching pond soil and water. The practice, which came to the fore in the later part of 1955, employs either the appropriate chemical fertilizer or some organic fertilizer such as various animal manures, composts, rice bran, wheat bran, corn meal, and copra meal. The systematic use of fertilizers has increased the production per hectare of the bañgos fishponds from the average of 350 kg to from 1300–1800 kg per hectare per year.

Offshore Fisheries. There is now practically no activity in the Archipelago. In the latter part of 1937, local interest in commercial offshore fishing, especially for tuna, was centered in Zamboanga City with the establishment by the Sea Foods Corporation of a tuna cannery, a complete ice plant and cold storage facilities, and a small fleet of tuna fishing boats equipped for pole and line, and longline fishing.

In the late 1930's, Davao City was also the base of a thriving offshore fishery for yellowfin tuna and skipjack. There were 8 fishing outfits of less than 12 gross tons each, most of which were tuna longliners. The field of fishing operation covered Davao Gulf and as far out in the Pacific Ocean and Celebes Sea as could be safely reached by the small craft. In 1950 tuna fishing activity offshore was tried by a company with two longline outfits, but did not prosper because of complete dependence on the local fresh fish market.

Published works on tuna fishing indicate that the waters surrounding the Archipelago are rich tuna longline fishing grounds. As an industry the exploitation of the offshore tuna fisheries has a very good prospect here.

Handling and Utilization of Fish. Commercial fishing vessels with insulated fish holds take on sufficient crushed ice to chill the fish on board from two to three weeks. Upon arrival at the port of landing, the iced fish are either sold immediately, deposited in cold storage establishments or left to wait in the boatholds until totally disposed of.

The wholesale marketing of fish is done in fish landing places, and the sale is done not in open auction but by means of secret bidding with the bidder whispering his price to the ear of the consignee. From the fish landing the wholesalers distribute the fish to retailers in the different fresh fish markets.

Before the introduction of ice and the storage of fish in cold storage plants, salting and drying were the most universal method of preserving fish and fishery products. This practice of salting fish has shown little improvement through the years despite efforts on the part of the government to raise the standard of the product by demonstrating improved methods.

The most important staple dried fish product is the *tuyo* or *uga*. These are fish of medium-sized species especially herring species, brine-salted in the round for from two to four hr and dried under the sun for three or four days.

Another form of dried fish is the *daeng* or *pinakas*, which is prepared from large mackerel, cavalla, pampano, and snapper. The dressed and split fish are brine salted and dried in the sun.

Bagoong is a salted fish paste made from small fish, usually anchovy or small shrimp, preserved in its own pickle, and allowed to ripen or age. *Patis* or fish sauce is obtained from long-standing salted fish kept in large wooden vats. After months and months of standing in the vats, the flesh of the fish gets completely autolyzed and the supernatant clear liquid is allowed to drip through a spigot, after which it is collected and bottled as *patis*. The production and bottling of patis in the modern way is growing to be a big industry especially in the towns of Malabon and Navotas, Rizal Province.

Anchovy constitutes the great bulk of dried fish in the interisland trade. They are caught in great quantities and dried in the sun without salting.

Herring, small mackerel, roundscad, gizzard shad, and *bañgos* are the usual smoked fish, called *tinapa*. The fish in the round are brined, placed in baskets, and immersed in boiling brine for cooking. Larger fish are gutted before brining. After cooking, the fish are laid in shallow bamboo trays, which are placed in smoking holes and covered with deep round baskets to confine the smoke.

Other Important Marine Resources

Salt. The production of salt by solar evaporation in specially laid out beds is a very important industry during the dry season in many coastal provinces. In other provinces salt is extracted from sea water by evaporation.

Industrial Shells. The most important of Philippine miscellaneous fishery products are the

Fig. 1. Goldlip pearl oyster shells, *Pinctada Maxima*, showing blister pearls.

Fig. 2. View of a *Chanos* or milkfish fishpond showing the water supply canal and portions of rearing ponds on both sides.

industrial shells. The important varieties are the goldlip pearl shell, blacklip pearl shell, top shell or troca, window pane shell, and green snail shell. These are gathered in sizable amounts and exported, either as raw shell or as blanks or finished buttons. These commercial shells, except the window pane shells, are found mostly in the southern waters of the Archipelago. Window pane shells, because of their translucent nature, are utilized for window panes and lampshades.

Pearl fishing in the Sulu Archipelago produces most of the shells, and sometimes pearls, for export from the Philippines. These pearl shells also yield blister pearls, which are cut from the shells and sold also as articles for jewelry.

Reptile Skins. Sea snakes (*Laticauda calubrina*), the common monitor lizard and crocodiles yield skins for export in raw forms, salted, or as cured leather for the manufacture of leather goods.

Commercial Echinoderms. The most important commercial echinoderms are the holothurians or sea cucumbers, of which about twelve species are commercially important. They are found in sandy littoral areas and reefs and, most abundantly, in the Sulu Archipelago. The common

Fig. 3. Catching marketable milkfish in rearing pond with the use of a seine net.

Fig. 4. Milkfish fingerlings enclosed in a fish corral within a nursery pond being readied for transferring to a rearing pond.

names for the smoke dried bodies of the sea cucumbers are beche-de-mer, trepang or *balatan*.

Marine Turtles. The green turtle, the loggerhead, and the hawksbill turtle are the three commercially important marine turtles. These are fairly well distributed in Philippine waters, but are more abundant in the Sulu Archipelago. The Turtle Islands, a group of several islets near Borneo, are the center of the turtle fisheries of the Philippines. The sandy beaches of the Turtle Islands serve as the breeding grounds of marine turtles. These islands supply turtle eggs for export to Jolo, Sandakan, Borneo, and seaports in the Asian mainland. The hawksbill turtle produces the tortoise shell of commerce, which is made into combs and other boudoir articles.

Seaweeds. Up and coming commercial fishery products are the marine seaweeds. In coastal towns where seaweeds occur they are ordinarily gathered for food. Some species are good sources of agar and alginates, which are utilized by other industries. The common commercial kinds are the species of *Caulerpa, Gelidiella, Gracilaria,* and *Eucheuma*. In former years the exportation of marine seaweeds was not very significant. In recent years, however, the importance of *Eucheuma* in the world market has increased. Consequently, the local production of seaweeds increased from 96,603 kg in 1961, 84,827 kg in 1962, 91,773 kg in 1963, 100,978 kg in 1964, 305,132 kg in 1965 to 1,132,600 kg in 1966.

Diginea simplex, from which a kind of vermifuge is extracted, is also gathered commercially and exported.

Cross references: *Pearling Industry; Sea Turtles; Seaweeds of Commerce.*

PHYCOCOLLOIDS AND MARINE COLLOIDS

Recognition of the importance of marine algae as a natural resource is a development of recent years. It has been brought about mainly by the progress made in the extraction, on an industrial scale, of the highly useful colloids contained in certain species of these marine algae. They range in size from the giant kelps to the microscopic phytoplankton. The marine colloids of commercial importance with which this article concerns itself are derived from the macroscopic algae, specifically the two classes known as the "brown" and the "red" seaweeds. The scientific terms "phycocolloids" and "hydrocolloids" applied to these extractives from marine algae are derived, respectively, from their origin and their character, the term "phycocolloid" (from the Greek "phycos": seaweed) denoting origin, and "hydrocolloid" denoting the water-soluble characteristic. The phycocolloids are often referred to as "hydrocolloids" since they are hydrophilic and their most common solvent is water. The majority of these colloids and their derivatives are insoluble or only slightly soluble in the common organic solvents.

The most common term of reference for all plant and seaweed polysaccharides or their derivatives in industry is "gum." This term originally referred only to plant exudates that thickened or hardened on exposure to air, but common usage now refers to those products, including the algal polysaccharides, that are dispersible in either hot or cold water to produce viscous mixtures or solutions.

Many species of marine algae have been harvested for centuries but for use as a staple in the diets of people rather than for extraction and purification of the particular hydrocolloids that the plants contained. The Chinese and Japanese are well known for their commercialization of various seaweeds or marine algae as marine vegetables in their diets. In Hawaii, as many as seventy-five species of seaweeds were used for food in the 19th century under the collective name "limu." It has been only during the 20th century that the economic significance of phycocolloids has been realized, and that colloids from marine sources have been finding increasingly wide application in industry.

Sources of Phycocolloids

The marine algae are generally divided into four classes, largely on the basis of photosynthetic pigmentation (Table 1). The classes of green algae (Chlorophyceae) and blue-green algae (Cyanophyceae) are more commonly associated with freshwater environments. While representatives of these two classes are present in salt water in both the microscopic and macroscopic forms, they provide no phycocolloid of current importance. The brown algae (Phaeophyceae) and the red algae (Rhodophyceae) are found almost exclusively in marine habitats and furnish the principal algal polysaccharides. The brown seaplants are the largest of all marine algae and are often collectively termed kelp.

The ecology and geographic distribution of the marine algae play an important role in the commercialization of a particular phycocolloid. Factors such as light intensity and penetration, water and air temperature, and the type of substratum available for the particular alga to affix itself will determine where and in what quantities any particular species will be found. These, as well as other factors, will account for seasonal variations in hydrocolloid content in the marine algae from one month to another at the same location or for consistent variations

TABLE 1. A Brief Classification Chart of Some of the Commercially Important Marine Algae

Class	Subclass	Order	Family	Genus	Extract
Phaeophyceae (brown algae)		Laminariales	Laminariaceae	Laminaria Macrocystis	algin
			Lessoriaceae	Nereocystis	
		Fucales	Fucaceae	Fucus Ascophyllum	
			Sargassaceae	Sargassum	
Rhodophyceae (red algae)	Florideae	Gelidiales	Gelidiaceae	Gelidium	agar-agar
			Gracilariaceae	Gracilaria	
		Gigartinales	Gigartinaceae	Gigartina Chondrus	carrageenan
			Solieriaceae	Eucheuma	
			Furcellariaceae	Furcellaria	furcellaran
Cyanophyceae (blue-green algae)					
Chlorophyceae (green algae)					

within the same species from one colder environment in one latitude vs a warmer environment in another latitude.

General Structure and Function

Structurally, the major polysaccharide of the algin-bearing brown marine algae (class Phaeophyceae) is a polymer of D-mannuronic and L-guluronic acids connected by β-$(1 \rightarrow 4)$ glycosidic linkages. Recent studies of the red seaweed polysaccharides (class Rhodophyceae) indicate that nearly all those studied to date are composed of galactose and substituted galactose units linked together by glycosidic linkages having an alternating $\alpha(1 \rightarrow 3), \beta(1 \rightarrow 4)$ configuration (Table 2). These phycocolloids differ from each other by having either the D or L configuration, a 3, 6-anhydro group, an ester sulfate group, or by being methylated.

Most phycocolloids are anionic in nature. These materials are often used where these charges offer specific functional advantages as opposed to some land plant colloids such as the mannogalactans (guar or locust bean gum) which are nonionic or gelatin, which is amphoteric.

The utilization of these phenomena by the selection of the proper phycocolloid system for a particular function can often result in increased efficiency for a specific use. Thus, for example, not only can a viscosity system be used to suspend ingredients, but interrupted gel systems, thixotropic gels, or highly swollen but nondissolved phycocolloid systems can be developed for a particular suspension.

The real value of such phycocolloids is in their versatility. Each phycocolloid exhibits different strength characteristics and properties in the gelled state; different flow and texture properties in the viscous state; and variations in behavior as protective colloids in a given system, depending on such influences as response to type and concentration of salts, concentration of phycocolloid and the pH of the system.

Phycocolloids of Major Commercial Importance

Agar. Agar is a dried hydrophilic, colloidal extract of certain families of the red marine algae (class Rhodophyceae). It is commercially available as thin strips or in flaked, granulated, or powder forms. Agar is white to pale yellow in color and is normally considered odorless and tasteless. Agar is insoluble in cold water, but soluble in boiling water. Agar is also known as agar-agar, seaweed Isinglass, Bar Kanten, Kobe, and Japanese gelatin.

Agar gels have been known and used by the Orientals for several hundred years. In the middle of the 17th century, the crude process of purifying and drying agar was discovered in Japan. It was introduced to Europe from the Orient and then in the 19th century to the U.S. as a replacement for gelatin in gelled desserts. In the 1880's,

TABLE 2. COMMON STRUCTURAL FEATURES OF
RED SEAWEED POLYGALACTANS

3G 1β 4G 1α 3 G 1β4 G 1α3 G 1β 4 G 1β 4 G 1α 3 G 1β 4 G 1α 3 G 1β

Polygalactan	Predominant Sugar Units
Agarose	D-galactose 3, 6-anhydro-L-galactose
Porphyran	3, 6-anhydro-L-galactose D-galactose 6-0-methyl-D-galactose L-galactose-6-sulfate 3, 6-anhydro-L-galactose
Lambda carrageenan[a]	D-galactose-2-sulfate D-galactose-2, 6 disulfate
Kappa carrageenan[b]	D-galactose-4-sulfate 3, 6-anhydro-D-galactose
Iota carrageenan	D-galactose-4-sulfate 3-6-anhydro-D-galactose-2-sulfate
Furcellaran, Hypnean	D-galactose D-galactose sulfate 3, 6-anhydro-D-galactose

[a] Major components of Lambda carrageenan
[b] Major components of Kappa carrageenan

the most sophisticated use of agar was discovered—as a bacteriological culture medium.

Agar-bearing marine algae of the class Rhodophyceae (red marine algae) are widely distributed throughout the temperate zones. The most common examples of the marine algae presently being used in commercial quantities for agar production are *Gelidium cartilaginium*, *Gelidium corneum*, *Gracilaria confervoides*, and *Pterocladia lucida*.

These algae are found in waters ranging from the intertidal zone to depths of more than 40 m, where they grow from holdfasts attached to substrata. The algae reproduce by spores and stolons and mature plants will normally range in size from 0.1 to 2.0 m in length. The agar-bearing marine algae, after harvesting, are usually dried, baled, and shipped to the manufacturing point.

Although the methods of agar production vary somewhat from region to region, agar manufacture today still follows many of the traditional steps used by the Japanese two centuries ago. The agar-bearing seaweed is cleaned and washed to remove impurities and sea salts. It is then cooked under pressure to extract the agar from the weed, after which the extractive is filtered and then poured onto trays and allowed to gel into sheets of dark-colored raw agar gel. After gelation the agar is frozen to assist in purification. The water crystallizes during freezing and, upon thawing, the water runs off carrying the impurities with it. The agar is then washed, decolorized, and dried with the product appearing as light colored flakes, which are ground to specific mesh sizes or sold in the ribbon or flake form.

Agar *in situ* is a cell-wall constituent and probably serves as a structural, ion-exchange and dialysis membrane in the plant. Agar is composed of two polysaccharides, agarose and agaropectin. Agarose, the gelling fraction of agar, is essentially an unsulfated polysaccharide while the agaropectin fraction contains various percentages (usually 5–10%) ester sulfate depending on the source of raw material. The percentage of agarose in an extractive of agar can vary from 50–90%.

The structure of agar appears to be composed of alternating D-galactopyranose units linked β-(1→4) and 3, 6-anhydro-L-galactopyranose units linked α-(1→3) with a half-ester sulfate on approximately every tenth D-galactopyranose unit.

Properties. Agar as a functional hydrocolloid has three properties of greatest importance: the ability to absorb large amounts of water; its gel strength; and its hysteresis characteristics. Although agar will swell in cold water, it is practically insoluble or only very slowly soluble in cold water and only very slightly soluble in organic solvents such as ethanolamine and is soluble in formamide. Agar will dissolve in boiling water and will exhibit gel characteristics upon cooling at sufficient concentrations. The gelation temperature of agar occurs at a point normally well below the melting temperature of the gel. Such a solution will form a thermally reversible, opaque, and brittle gel structure at approximately 32°C., which will remelt above 85°C. This high hysteresis of agar is unique among the polysaccharides and is responsible for many of its applications. Agar normally has not more than 6.5% total ash and less than 20% moisture.

Applications. A major use for agar is as a substrate or medium in bacteriological cultures. Other major uses include bakery icings, where the agar serves to control moisture penetration into the cake from the icing and also aids in controlling the melting point of the icing, and confectionery items, such as jelly candies, marshmallows, and fillings for candy bars. It is also used as the gelling agent in canned meats, fish, and poultry in various areas. Agar is commonly used as a mechanical laxative due to its ability to absorb high quantities of water.

Agar is normally used from 0.1% to 1.0% in most applications.

Algin. Algin is the generic term generally used to describe the derivatives of alginic acid, but is most commonly applied to the sodium salt, known as sodium alginate. Algin is a phycocolloid extracted from certain families of the brown marine algae (class Phaeophyceae). It is normally available in a granular or fibrous powder of white to yellow-brown color. Purified grades of algin are essentially odorless and tasteless. The sodium salt, sodium alginate, dissolves in water to form a viscous, colloidal solution. At a pH below 4, insoluble alginic acid is precipitated. Although algin is insoluble in the majority of organic solvents, aqueous solutions will tolerate certain concentrations of water-miscible organic solvents such as alcohol (up to 30% by weight) before precipitation occurs. The molecular formula for the sodium salt is generally thought to be $(C_6H_7O_6Na)_n$ with a molecular weight ranging from 32,000–250,000.

In 1883, E. C. C. Stanford, an English chemist, made the first crude extractions of the colloid from local seaweed. Previously, these macroalgae were used as a source for potash and iodine. The word "kelp" originally referred to the potash and iodine-rich ash obtained by burning the brown algae, but the term has ultimately come to apply to the large plants themselves. The colloid extracted was termed "algin," derived from the word "alga." The first successful commercial developments on a large scale were begun in the latter part of the 1920's.

Algin occurs as a cell-wall constituent of a large number, if not all, of the brown marine algae (class Phaeophyceae). Commercial production of algin is based largely on the use of the following marine algae: *Macrocystis pyrifera*, *Laminaria cloustoni*, *Laminaria digitata*, and *Ascophyllum nodosum*. Other marine algae of lesser importance as commercial sources of algin are *Nereocystis luetkeana*, *Fucus vesiculosus*, *Fucus serratus*, and *Laminaria saccharina*.

The giant kelp, *Macrocystis pyrifera*, is a principal source of algin found in the temperate zones of the Pacific Ocean. Commercially harvested beds are located off the southwestern coast of the U.S. in water ranging from 25–80 ft in depth. Each plant is secured to a rocky substrate by a holdfast. From 1–100 stipes grow out of the holdfast. Mature plants often obtain a size of from 50–200 ft in length. Several sublittoral brown marine algae appear in commercial quantities, e.g., *Laminaria cloustoni* and *Laminaria digitata*. These plants attach themselves to a rocky substrate and the former appears in waters with depths to 60 ft, while the latter is normally found in shallower water. The most common source of a littoral or "rock weed" for algin is *Ascophyllum nodosum*. It attaches itself by a disk to stones on rocky beaches and normally reaches a length of 3–5 ft. Where the wave action of the surf does not break the fronds, it might range from 1–10 ft in length.

Several commercial processes are available for the manufacture of algin. Normally, the kelp is washed and then digested with an alkali such as sodium carbonate. Following flotation to remove cellulose, and then filtration, the algin is removed from solution by precipitation as alginic acid or calcium alginate, which may then be converted to alginic acid. Neutralization with appropriate alkalies will produce various salts such as sodium or potassium alginate. After drying the algin is milled, tested, and blended for specific viscosity ranges or other functions.

The propylene glycol alginate derivative of alginic acid is prepared by the reaction of propylene oxide with partially neutralized alginic acid, when the acid is in a fibrous and swollen state due to the presence of controlled amounts of water, under moderate pressure.

Alginic acid is thought to be a linear polymer of high molecular weight consisting of β-(1→4) linked D-mannuronic and L-guluronic acid units in the pyranose ring form. The ratio of the two acid units will vary according to the source.

Properties. Sodium alginate dissolves in hot or cold water to form a viscous solution. The flow properties of the solution can be altered by the presence of certain polyvalent metallic ions (calcium is particularly effective), solution temperature, and concentration of algin. Alginate solutions are normally compatible with other water-soluble gums, proteins, starch, sugar, and most carbohydrates. Sodium alginate solutions are most stable at neutral or alkaline pH. Propylene glycol alginate is employed in acidic solutions ranging from a pH of 3 to approximately 6.8. The ester groups of propylene glycol alginate tend to hydrolize under alkaline conditions.

Algin gels are most often formed by using a polyvalent metal salt (such as a calcium salt), which results in a chemically controlled set rather than one controlled by temperature. The setting time of the gel can be further altered by the use of a phosphate or polyphosphate retarder salt, which combines with the polyvalent metallic ion and controls the release of the metallic ions necessary for gelation.

Clear, water-soluble films of sodium alginate can be formed by casting, extrusion, or evaporation and made insoluble by contact with a hardening bath of a polyvalent metal salt such as calcium chloride.

Applications. Various commercial grades of algin can be used as stabilizers in frozen confections such as ice cream to maintain texture and control formation of ice crystals. Propylene glycol alginate is widely used in sherbets in concentrations as low as 0.3% of the weight of the mix to protect the product from ice crystal growth during freezing and thawing cycles, as well as to impart texture and body to the finished dessert. Other food uses include the use of algin for bakery jellies, dessert gels, pudding mixes, and pie fillings. Propylene glycol alginate is widely used in products of low pH such as salad dressings (e.g., French dressing) to provide emulsion stability and longer shelf life. It is also used in fruit drinks as a bodying and suspending agent. Propylene glycol alginate is often added to beer at a concentration of 40–80 ppm to stabilize the foam.

Pharmaceutical applications include the use of algin as a suspending agent and a protective colloid. Sodium alginate is used as a tablet binder, and alginic acid as a tablet disintegrant. Algin is also used as a thickener for liquid shampoos and as the setting agent for dental impression compounds. The paper industry employs the filming action of algin as a surface sizing or coating to resist or control penetration of wax, oil, fat, or ink into the paper stock. Algin has been used for a number of years in the textile industry as a thickener for dyes and printing pastes in fabric printing. Other industrial applications include the use of various alginate products for thickening and creaming of latex rubber, as ceramic binders, and as thickeners in paints.

Carrageenan. Carrageenan is a hydrophilic sulfated polyanionic phycocolloid extracted from certain families of the class Rhodophyceae (red marine algae). It is normally associated with metallic ions such as sodium, potassium, calcium, etc. or combinations of these ions. Carrageenan is commercially available as a white to cream-colored powder, which is practically odorless and has a mucilaginous mouthfeel. It is soluble in water and forms clear to slightly opalescent solutions of a viscous nature that flow readily. Carrageenan is insoluble in most organic solvents, but will tolerate certain amounts of water-miscible solvents before precipitation occurs. Carrageenan displays a unique ability to alter the physical properties of fluid milk through protein interaction. Carrageenan is also known in various parts of the world as Irish moss extract, Carrageen, Chondrus extract, or carrageenin.

The gelling power of the extractive was exploited by those who lived along the French, English, and Irish coasts for hundreds of years prior to the appearance of a commercial product. The seaweed was gathered, cooked with milk, with the result that the milk gelled upon cooling. This delicacy was eaten as a dessert and known as "blancmange." The common name "Irish moss" is supposed to have derived from the association with the Irish coast dwellers and the original name, Carragheen, from the area where the greatest seaweed activity took place. Although the polysaccharide was isolated in 1837, it was not produced on a large commercial basis until the 1930's. World War II accelerated the further development of this polysaccharide in the U.S. as a replacement for other gums in short supply. Carrageenan is now well established as a major hydrocolloid in industry.

Carrageenan is extracted from a number of botanical families, principally the families *Gigartinaceae* and *Solieriaceae* of the class Rhodophyceae (red marine algae). The two principal sources for carrageenan for a number of years were *Chondrus crispus* and *Gigartina stellata*. These algae are widespread in distribution, appearing in stands or beds of commercial importance along the North Atlantic coastline of Canada and the U.S. These plants normally occur from the intertidal zone to depths of 20 ft or more. The plants vary in height but are normally less than 1 m. They are secured to rocky substrata by holdfasts. After harvesting, the carrageenan-bearing seaweeds are dried, baled, and transported to the manufacturing point. The carrageenan extracted from these two primary sources is composed of at least two fractions, designated as *kappa* and *lambda*, that occur in varying ratios and degrees of polymerization.

Other marine algae of importance in the manufacture of these types of carrageenan include *Chondrus ocellatus, Eucheuma cottonii, Gigartina acicularis, G. pistillata,* and *G. radula. Eucheuma spinosum* provides a primary source for the third type of carrageenan, *iota.*

Carrageenan is obtained by water extraction of the phycocolloid from the seaweed. Time, temperature and pH will vary according to the product being produced. The crude extract is filtered and purified to remove insolubles as well as color, odor, and flavors. The product is then either drum dried or precipitated in an organic solvent such as alcohol, and vacuum dried. The carrageenan is then ground, tested, and standardized.

The three major types of carrageenan (*kappa, lambda, iota*) are highly sulfated linear polymers containing between 20%–40% ester sulfate. They have α-(1–3), β-(1–4) linkages and differ in the amount and position of the sulfate groups as well as the amount of 3,6-anhydro-D-galactose present.

Properties. The predominantly *kappa* fraction is potassium sensitive (also to other cations such as ammonium and calcium to a lesser extent) and forms brittle gels that exhibit syneresis. The *iota* type of carrageenan is calcium sensitive and forms elastic gels with no syneresis. The predominately *lambda* fraction of carrageenan is not normally cation sensitive and forms viscous, nongelling solutions.

As a strong, negatively-charged polyelectrolyte, carrageenan forms complex structures with positive groups of other molecules or with the negative group of other molecules through the interposition of a polyvalent cation, such as calcium. This reaction is especially true in the modification of milk systems. Concentrations as low as 0.03% will affect the viscosity of milk and a concentration of 0.1% will form milk gels. This phenomenon is employed widely in the food industry.

Applications. Carrageenan products find extensive application in the food industry, particularly in dairy products, as a stabilizer, gelling agent, and viscosity control agent. The protein/carrageenan interaction is employed in products such as ice cream (to stabilize the product) chocolate milk (to provide body and suspend the cocoa), egg nogs (stabilizer), and milk-based cooked puddings (gelling agents). Carrageenan is also used as the gelling agent for many other products ranging from fish gels, baby fruit-juice gels, and dessert gels to dog food gels. A major area of use is in dietetic foods where carrageenan is used to provide the body and texture normally supplied by starch or sugar. Cosmetic applications include the use of carrageenan as a binder to give toothpastes the proper texture and consistency and in hand lotions. In Europe, carrageenan is also used in ulcer therapy and in other pharmaceutical applications. Industrial applications of carrageenan include its use in suspensions (such as graphite, abrasive or pigment suspensions), industrial gels, and emulsion stabilization systems. A growing area is in the thickening and suspension of textile dyes and pastes for printing fabrics.

Other Phycocolloids

Furcellaran, also known as Danish agar, is an extract produced in Denmark from the red marine alga, *Furcellaria fastigiata*. This extract is a sulfated (12–16% ester sulfate) polysaccharide whose structure is thought to be composed of D-galactose and 3,6-anhydrogalactose. Furcellaran gels are influenced by the presence of potassium to provide rigid gel structures. The raw material, *Furcellaria fastigiata*, is harvested commercially in waters off the North Atlantic coasts at depths ranging from the intertidal zone to 30 ft. The commercial application of furcellaran is as a gelling agent in flans and puddings and as a suspending agent in pharmaceutical suspensions. In many respects it behaves like the *kappa* fraction of carrageenan.

Another phycocolloid, funorin, is a solubilized seaweed product rather than an extract. It is found in the genus *Gloiopeltis*. Funorin is produced and used in the Orient for sizing of textiles and paper.

Porphyran is a sulfated polysaccharide found in species of the genus *Porphyra* of the red marine algae (class Rhodophyceae). To date, the plant itself has been used as a source of protein and little has been done to commercialize the extractive. It is common along the coasts of the English Isles, Japan, and the west coast of the U.S.

A number of other phycocolloids are recognized and have been investigated for physical and chemical properties, but are either not commercially available at present or are used in very limited amounts in the locality where they are produced.

Marine Colloids from Other Sources

A number of other marine sources contain potentially important colloids, although they have not been exploited or commercialized to as great a degree as the algal colloids.

Chitin is a high molecular weight, linear polymer that occurs in the cell walls of a number of fungi, as a constituent of the skeletal shield of marine invertebrates, and in the exoskeletons of many insects. The name chitin is derived from the Green work *chiton* meaning coat-of-mail because of its function as a protective coating for invertebrates.

Structurally, chitin is thought to be a linear polymer of anhydro N-acetyl-D-glucosamine. Since chitin is insoluble in water and organic solvents in the natural state, it must be deacetylated during manufacture. The accepted structure for completely deacetylated chitin is that of an anhydro-D-glucosamine polymer linked by β-(1$\to$4) glycosidic bonds. Addition of various acids will convert this to a water-soluble cationic salt. Although it is not known to be produced commercially at present, deacetylated chitin shows promise in sizing of fabrics and in adhesives where the cationic properties are of value.

W. M. REES

References

McNeely, W. H., in R. L. Whistler, Ed., "Industrial Gums," 1959, Academic Press, New York.

Anderson, N. S., Dolan, T. C. S., and Rees, D. A., *Nature*, **205**, 1060–1062 (1965).

"The Structure of Red Seaweed Polysaccharides," Technical Bulletin 3, 1967, Marine Colloids, Inc., Springfield, N.J.

O'Neill, A. N., and Stewart, D. K. R., *Can. J. Chem.* **34**, 1700 (1956).

Smith, F., and Montgomery, R., "Chemistry of Plant Gums and Mucilages," American Chemical Society Monograph No. 141, 1959, Reinhold Publishing Corp., New York.

Smith, D. B., and Cook, W. H., *Arch. Biochem. Biophys.*, **45**, 232 (1953).

Schachat, R. E., and Glicksman, M., in R. L. Whistler, Ed., "Industrial Gums," 1959, Academic Press, New York.

Tracey, M. V., *Reviews of Pure and Applied Chemistry*, 7, No. 1, 1–14 (1957).

Cross references: *Agar; Alginates from Kelp; Irish Moss Industry.*

PHOTOSYNTHESIS—*See* PRODUCTIVITY OF MARINE COMMUNITIES; PLANKTON RESOURCES

PHYTOPLANKTON

Marine phytoplankton organisms are minute algae suspended in the sea, drifting or swimming within the currents, and usually remaining near the surface of the water where many of the species derive their energy through photosynthesis. Phytoplankton organisms are extremely varied in their morphology, physiology, and the adaptations they have made to the environment. Many species are animal-like in possessing their own locomotory apparatus and in being capable of ingesting other organisms for food, and it is difficult to assign some planktonic organisms to either the plant or animal kingdoms.

A major environmental factor to which phytoplankton must adapt is the influence of gravity. The tendency for organisms to sink below the photic zone has resulted in a variety of morphological and physiological adaptations in these organisms. The adaptations generally tend to increase the buoyancy of the organisms and, as a result, there are many very bizarre forms in the phytoplankton reflecting combinations of these adaptations. Changes in cell form increasing the ratio of surface area to volume of protoplasm also increase the frictional resistance to sinking, thus giving such cells a selective advantage. Long, narrow cells and disk-shaped forms are two types commonly found in phytoplankton populations (Figs. 4, 14, 16, 18). The development of bristles and spines, particularly those that are hollow, also greatly increase the buoyancy (Figs. 9, 10, 11, 12). The size of phytoplankton cells also influences their buoyancy because small cells have a greater surface area-to-volume ratio than larger cells. Friction in such cells, therefore, retards the sinking rate more efficiently than in larger cells.

Nonmotile marine species of phytoplankton often occur in colonies of long chains or ribbons of cells (Figs. 8, 9, 11), many having long spines that may branch. The direction, hence the rate, of sinking is influenced by the configuration of these chains and the shape of individual cells. Some species of the diatom genus *Chaetoceros*, for example, form long spirals which descend in a spiral pattern. Many species of another diatom genus, *Rhizosolenia*, have solitary long narrow cells with ends tapering toward one side (Fig. 9). Such cells tend to sink in a zigzag pattern because of the sloping ends of the cells, thus reducing their rate of sinking.

Buoyancy has also been achieved by reduction in the density of the cells. In some species the cells may be imbedded in large masses of watery gelatinous material that approaches the density of sea water. With even slight turbulence in the sea, the gelatinous colonies tend to float or remain suspended in the water. In other species the cell sap may contain proportionately more ions of lower atomic weights than does sea water. For example, vacuoles of *Noctiluca* contain acidic solutions, indicating that hydrogen ions probably replace heavier ions in the cell sap (Davis, 1953). Similarly, a large-celled diatom species, *Ethmodiscus rex*, contains lower concentrations of the heavier ions than does the surrounding sea water (Beklemishev, Petrikova, and Semina, 1961). Density is also reduced by the accumulation of gas vacuoles in algae such as *Skujaella* (*Trichodesmium*), a planktonic blue-green alga.

It has been hypothesized that accumulation of fats and oils as food reserves by species of phytoplankton is a factor that contributes to their flotation. Although these materials undoubtedly tend to lower specific gravity of some cells, it has been demonstrated (Smayda and Boleyn 1965, 1966) that older cells, in which these materials usually are more abundant, tend to sink more rapidly than young cells in the same population.

Some marine phytoplankton, known as neustonic species, are capable of attaching themselves to the surface film of the sea, thus overcoming the tendency to sink. They may attach by means of certain nonwettable surfaces of the cells or by special organelles such as trichocysts and muciferous bodies or flagellum-like haptonemata, the tips of which adhere to the surface film.

The smaller phytoplankton species have advantages, in addition to the increased buoyancy mentioned earlier, in having a relatively larger surface for absorption, and in requiring a smaller amount of nutrients because of the smaller volume of protoplasm per cell. This advantage is probably especially important in regions where the concentration of nutrients may be low, such as in very slowly mixing tropical waters. The smallest phytoplankton, the species with cells measuring less than 60 μ in diameter, are not collected in most types of plankton nets and are referred to as nannoplankton.

In modes of nutrition, phytoplankton ranges from strictly autotrophic species, producing all

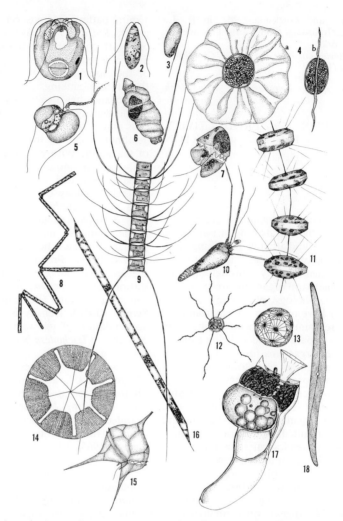

Figs. 1–18. Marine phytoplankton: (1) *Pyramimonas*, ×3700; (2) *Dunaliella*, ×4000; (3) *Micronas*, ×11,000; (4) *Pterosperma moebiusi*, a, face view, b, side view (hypothetical), ×1200; (5) *Chrysochromulina*, ×5300; (6) *Cochlodinium*, 1300; ×(7) *Gymnodinium*, ×2400; (8) *Thalassionema*, ×750; (9) *Chaeotceros*, ×650; (10) *Michaelsarsia*, ×2700; (11) *Thalassiosira*, ×1200; (12) *Meringosphaera mediterranea*, ×1750; (13) *Umbellosphaera tenuis* (after Markali & Paacshe), ×1600; (14) *Asterolampyra*, a centric diatom in valve view, ×1500; (15) *Peridinium*, ×900; (16) *Rhizosolenia* containing *Richelia*, ×200; (17) *Histioneis* containing blue-green algae within a special chamber, ×2000; (18) *Pleurosigma*, a pennate diatom in valve view, ×900.

their organic nutritional requirements from inorganic substances using energy from light, to those lacking a capacity for photosynthesis and therefore obtaining organic nutrients from external sources. The latter forms include species that absorb nutrients in solution (saprophytic or osmotrophic nutrition) and those that ingest other organisms (phagotrophic nutrition). Other phytoplankton organisms obtain their organic nutrients by various combinations of these modes of nutrition, some being able to manufacture all requirements but a few, such as one or more vitamins, and others being both photosynthetic and capable of ingesting particles of food material.

Phytoplankton species also vary in the proportion of their life histories in which they are suspended in the sea. Holoplanktonic species drift or swim during their entire life histories, whereas meroplanktonic species attach or settle on the sea floor or margin during part of their life histories. These latter species usually occur in neritic habitats.

Most of our early knowledge of marine phytoplankton was obtained from samples collected at sea and preserved in from 2–10% commercial formalin. For this reason, the best known species of phytoplankton are types that produce walls or endoskeletons of solid substances that are not subject to morphological changes when exposed to various fixatives. The protoplasm in cells fixed with formalin rarely is recognizable when the samples are examined, often many years later, and it is the species of diatoms, armored dinoflagellates, silicoflagellates and coccolithophorids that are usually described from such samples. Fortunately, species belonging to these groups often are dominant in the phytoplankton populations, and our knowledge of such groups progressed quite rapidly. Mixed in these populations, however, are certain species, often very large cells, that have no recognizable cell parts after preservation. This type of phytoplankton may contribute substantially to productivity of the sea, but is usually not found in the preserved samples made during productivity analyses. Also, some phytoplankton species that produce rigid materials may have these parts disintegrate in certain fixatives. For example, coccolithophorids produce scales or skeletons consisting of calcium carbonate (coccoliths) that are used in their identification. These scales are easily preserved in basic fixatives, but they are dissolved in acidic solutions. Coccolithophorids would not be counted as present, therefore, in samples preserved in formalin that was not previously neutralized or in other acidic fixatives.

Marine phytoplankton is composed of plants belonging to at least seven different phyla, described hereafter.

Chrysophyta. This is a large group of microscopic algae that contains three distinct classes commonly known as diatoms, haptophytes, and chrysophytes.

Diatoms (class, Bacillariophyceae). The most commonly observed types of phytoplankton in both temperate and tropical seas are diatoms, characterized by a siliceous wall that often is ornamented with pores and chambers as well as various types of protrusions. Most of the conspicuous diatoms in the sea are more or less radially symmetrical when viewed from the top or bottom of the cell (centric diatoms), but there are also many species that are bilaterally symmetrical in the same views (pennate diatoms). Diatom cell walls are composed of hydrated silica that is organized into four basic separate pieces, two valves that are the top and bottom pieces of the wall, connected by two girdle bands that often are in the form of open hoops with the ends overlapping. One valve and one girdle band are slightly smaller than the other two pieces, and the smaller girdle band fits partly inside the larger girdle band. As cells grow they often form new wall pieces, intercalary bands, similar in structure to the girdle bands and inserted between the girdle bands and the valves. The ornamentations on the wall usually are much more highly developed on valves than on girdle pieces. Pores and chambers are arranged in radiating patterns on valves of centric diatoms, whereas they are arranged bilaterally along a central line in most pennate diatoms. Taxonomy of diatoms is based primarily upon cell wall shapes and ornamentations. Some diatom species may also have spine-like projections, long setae, wing-like or keel-like protuberances, and various other structures. Spines and setae often join with those of daughter cells keeping the cells together in chains. A fibrous polysaccharide (chitan) strand joins the cells of species of *Thalassiosira* (Fig. 11). Smayda and Boleyn (1965, 1966) noted that longer chains of *Bacteriastrum hyalinum*, *Nitzschia seriata*, and *Chaetoceros lauderi* sink faster than shorter chains of the same species, whereas the sinking rate for *Skeletonema costatum* is inversely related to the number of cells per chain.

Although many species of diatoms require an external source of vitamin B_{12} and some require thiamine, they do not seem to have additional requirements for organic nutrients. As in other classes of Chrysophyta, the green color of chlorophyll pigments, chlorophylls a and c, usually is obscured by the more highly concentrated xanthophyll pigments, primarily fucoxanthin, giving a distinct brown or golden color to the chromatophores.

Haptophytes (class, Haptophyceae). This class has recently been established primarily through the research of Dr. Mary Parke and Dr. Irene Manton on marine species of *Chrysochromulina* (Fig. 5). This group has cells with two flagella and a third flagella-like structure, the haptonema, having an ultrastructure different from that of flagella. The haptonema seems to function in attaching cells to various objects, and it is often contractile. Cells of Haptophyceae usually have a covering of very small sculptured scales composed of carbohydrates.

Coccolithophorids (Figs. 10, 13) a type of phytoplankton with scales of calcium carbonate covering the cells, are placed in the Haptophyceae because some, but not all, species have been demonstrated to have haptonemata. The scales of coccolithophorids, known as coccoliths, are formed within the cell and deposited on the surface of the cell. They may be up to 50 μ in diameter, although most species have very small scales. The structure of coccoliths range from simple bodies consisting of from one to several crystals of the same size and shape, to a wide variety of coccoliths composed of variously shaped crystals arranged in complex patterns.

First described as microfossils in marine sedimentary rock, coccoliths may be very abundant in bottom mud of the sea, particularly in tropical and subtropical regions where coccolithophorids are more abundant and the number of species is greater than in other parts of the seas. As fossils they are important, along with other marine microorganisms, in identifying petroleum-bearing strata.

Coccoliths often form a loosely fitting envelope, the coccosphere, which is frequently found intact even in fossil deposits. Empty coccospheres may also be found in samples of living phytoplankton, and it has been shown that cells of some species emerge from the coccosphere at certain stages. An oceanic species of coccolithophorid, *Coccolithus pelagicus*, when grown in culture, passes through nonmotile and flagellate stages, these two phases producing coccoliths of entirely different morphologies. The nonmotile phase produces very complex coccoliths in a loosely fitting coccosphere, whereas the motile cells are naked at first and later produce very simple coccoliths that fit tightly to the cell membrane (Parke and Adams, 1960). Both of these stages in the life history of *Coccolithus pelagicus* are found in the open sea, the coccoliths of the flagellate stage having been known as *Crystallolithus hyalinus*. Other species of coccolithophorids, growing in the neritic region, may form palmelloid or filamentous stages that are attached to intertidal chalky rock (Parke, 1961).

One haptophyte species, *Prymnesium parvum*, under certain conditions may produce toxin that kills fish. This species is often abundant in brackish water ponds and has been a particular hazard to fish farming in Israel.

A common marine haptophyte, *Phaeocystis*, produces globular or irregularly shaped masses of watery gelatin in which the nonflagellate cells are imbedded.

Chrysophytes (class, Chrysophyceae). The chrysophytes are often flagellate cells with two flagella of unequal length. Cells usually are small and contain few golden-colored chromatophores. The open sea seems to contain fewer members of the Chrysophyceae than of the other classes of Chrysophyta. Marine chrysophytes generally are very fragile organisms containing no identifiable structures after treatment with preservatives. *Ochromonas*, a nannoplankton genus common in fresh-water environments, is rarely observed in samples of marine plankton, but frequently occurs abundantly in enrichment cultures from marine localities. This fact indicates that *Ochromonas* is probably more common than examination of preserved collections indicates. There is reason to believe that the genus *Pseudopedinella*, which is easily recognizable in the living condition, may be common in some regions. This genus is rarely seen, however, because it is so fragile that the cells disintegrate almost immediately when placed under the light beam of the microscope.

Colorless chrysophytes are also often abundant in marine enrichment cultures. Some of the most common types in temperate neritic habitats are choanoflagellates, characterized by having a single flagellum surrounded by a collar of fine rhizopodia. The beating of the flagellum brings currents of water to the region of the collar where food particles, such as bacteria, may be ingested by the organism. Some choanoflagellates produce a large, loosely fitting, basket-like lorica composed of pieces of silica.

Silicoflagellates are a special type of unicellular chrysophyte that produce a tubular stellate skeleton that is enveloped by the protoplasm. A single flagellum is usually present, emerging from the cell near one of the points of the skeleton. Although there are very few living species of silicoflagellates, those that exist are not uncommon. Skeletons of silicoflagellates are known in the fossil record as far back as the Cretaceous.

Chlorophyta. Primarily represented in the sea by large benthic algae such as *Ulva*, this phylum of green algae contains two classes, Chlorophyceae and Prasinophyceae.

Chlorophyceae. Marine planktonic members of the Chlorophyceae usually are flagellates with two or four flagella of equal length. Cells may be naked, but many have distinct walls surrounding the protoplast. The bright green chloroplast within the cells usually covers most of the peripheral part of the cell.

Planktonic members of this class are mostly found in neritic habitats, often being more abundant in brackish water. Species of *Chlamydomonas* and *Carteria* often occur in the phytoplankton of estuaries and tide pools that receive seepage of fresh water. *Brachiomonas*, a species with a peculiarly lobed cell, may be common in tide pools with varying salinities. *Dunaliella* (Fig. 2) often occurs as a dense bloom in many pools with water of high salinities. *D. salina* is known to change its color from green to red as the salinity of the water increases and may cause evaporating ponds for the commercial extraction of salt from sea water to become a very striking red color in later stages of evaporation. Certain coccoid species of green algae are known to occur in the sea, but they are usually associated with brackish tide pools rather than occurring in offshore localities.

Prasinophyceae. This class was recently recognized as distinct from the Chlorophyceae primarily because of the presence of organic scales that cover the flagella and the cell body. These scales are observable only with the electron microscope and are sometimes complex in their structure. A single species may have several layers of scales, each layer composed of scales of a different morphology. One of the smallest of algal cells, *Micromonas pusilla* (Fig. 3), probably belongs in the Prasinophyceae. Cells of this species are from 1–3 μ in length, possess a single flagellum, a nucleus, a chloroplast, a mitochondrion, and little else within the cell. Although capable of photosynthesis and able to live autotrophically, this species has been found in very deep water (2000 m) off the coast of California.

Other genera of Prasinophyceae usually have two or four flagella. *Pyramimonas* and *Platymonas* have deep apical cavities at the base of which are attached four flagella. *Pyramimonas* (Fig. 1) has four lobes around the apical depression, the cells being covered by scales, whereas in *Platymonas* the cells are flattened and the scales become fused together to form a theca surrounding the cell.

Large spherical green cells may be found in the plankton (size ranging up to $\frac{1}{2}$ mm in diameter). These cells contain a multinucleate protoplast surrounded by a thick stratified wall that may be ornamented with small pits. These cells, known as *Halosphaera* or *Pachysphaera*, divide to form many motile cells that resemble *Pyramimonas* in structure. The flagellate cells swim for an indefinite period before losing their flagella, secreting a wall, and increasing in size to form the large, cyst-like cells of the *Halosphaera* stage. Similar cyst-like stages may not increase in size, but produce elaborate wall structures; they are classified in the genus *Pterosperma* (Fig. 4).

All of these genera may be found in oceanic localities or in neritic situations. *Pyramimonas* and *Platymonas* often are the organisms causing blooms in tide pools or small bays. Also, these genera are sometimes established in symbiotic relationship with spumellarian radiolarians, and *Platymonas* is the algal symbiont of *Convoluta roscoffensis*, a planarian worm.

Dinoflagellates (phylum–Pyrrhophyta). These highly developed, mostly unicellular organisms are often a dominant part of the phytoplankton population. Many species are photoautotrophic, but a large number of species do not have a photosynthetic apparatus and subsist as osmotrophic or phagotrophic organisms. Most species are motile in their vegetative phase, possessing two heterodynamic flagella that are usually attached laterally on the cell (Fig. 7, 15). One flagellum is ribbon-shaped and encircles the cell in a girdle groove. Its action is an undulatory motion that seems to cause the cells to spin on their axes as they swim. The second flagellum trails behind the cell, lying in a longitudinal groove, the sulcus, in the posterior end of the cell. Some species have torsion in the middle of the cell causing the girdle groove to be displaced, or sometimes twisted several times around the cell circumference (Fig. 6).

Dinoflagellate cells often are covered by a theca (armor) composed of plates of a cellulose-like substance (Fig. 15). As the cell grows in size the plates also grow along certain of the sutures bordering the plates. Taxonomy of armored dinoflagellates is dependent upon knowledge of the arrangement, form and size of the thecal plates. There are many species of dinoflagellates, however, that do not have thecae visible with the light microscope, although recent investigators, using the electron microscope, have found a plate-like structure in the cell membrane. The unarmored dinoflagellates (Figs. 6, 7) are less well known than the armored species because they are difficult to preserve.

Protoplasm of dinoflagellates contains a single large nucleus in which the chromosomes remain as discrete structures even at interphase. This characteristic seems to be the most reliable one in recognizing members of this group. Chromatophores of dinoflagellates are variously shaped and contain a wide variety of pigments ranging from green to blue, red, and yellow-brown. Chlorophylls a and c are present in the chromatophores as well as certain xanthophyll pigments that seem to be distinctive. Marine dinoflagellates usually contain one or two deep invaginations of the cell near the points of flagellar attachment. These cavities, known as pusules, often are filled with a pinkish colored fluid and may function in the intake of water and food particles. Dinoflagellate cells often contain trichocysts and a few species have elaborate nematocysts that resemble those of Coelenterates. A few genera of unarmored dinoflagellates have a complex structure composed of a peripherally located pigment cap, usually red, that surrounds a large lamellated clear body. This structure, the ocellus, is believed to function in the orientation of the cell to other objects or to light.

Some armored dinoflagellates produce elaborate wings and keels on the armor that may aid in preventing the organisms from sinking rapidly. In *Ornithocercus* and *Histioneis* (Fig. 17) small coccoid blue-green algae are often within chambers formed by the protrusions of the armor. It is believed that there is a symbiotic relationship between the dinoflagellate and blue-green algae (Norris, 1967). Other closely related dinoflagellates contain similar blue-green algae within their protoplasts in a symbiotic relationship. Symbiosis of dinoflagellate cysts, belonging to the genus *Symbiodinium*, occurs with various coelenterates, corals, radiolarians, and foraminiferans, making these dinoflagellates extremely important in the primary productivity of many benthic, intertidal, and planktonic environments.

Dinoflagellates are known to be parasites on various planktonic animals such as tintinnids, copepods, *Oikopleura*, etc., as well as on a species of *Chaetoceros*, a diatom. Dinoflagellates that are parasitic on nuclei of other dinoflagellates are probably the most extraordinary type of parasite in this group. Some of the species commonly thought to be parasitic on copepods seem to be photosynthetic, producing large masses of brownish-colored cells within the transparent body of the animal. It is possible that the relationship between these two organisms may be one of symbiosis rather than parasitism.

Free-living, cyst-like cells are produced by the genus *Pyrocystis*, which produces zoospores that resemble unarmored dinoflagellates. The cyst-like cells are globose or lunate shaped, usually, and may be common, particularly in tropical and subtropical regions. Like several other kinds of dinoflagellates, including species of *Gonyaulax* and *Noctiluca*, *Pyrocystis* cells may be luminescent.

The cyst-like cells of *Noctiluca* often are abundant in seas throughout the temperate and tropical regions. *Noctiluca* cells are very large, filled mostly with vacuolar solution, have a single comparatively short flagellum, and a most peculiar organelle, a tentacle-like protrusion from the cell. *Noctiluca* usually subsists as a phagotrophic organism, ingesting diatoms and often other dinoflagellates. In the Bay of Bengal and the Arabian Sea, however, *Noctiluca* often contains a small green flagellate within its vacuolar solution, the flagellate most likely living in a symbiotic relationship with *Noctiluca*.

Micropaleontologists have only recently recognized that many of the species formerly placed in a group of uncertain affinity, known as hystrichospheres, actually represent cysts of dinoflagellates. Dinoflagellate cysts have been recognized in marine sediments of all types, and have even been seen in liquid petroleum. They are known to occur in deposits of the Lower Palaeozoic (400 to 600 million years ago), and have been recorded in all ages to the present time. Their frequent abundance in sediments makes them extremely valuable in determining the origin and dating of fossil materials and various strata. Several living motile dinoflagellate species have recently been demonstrated to form cyst stages in their life history (Wall and Dale, 1967).

Cryptomonads (phylum, Cryptophyta). Composed primarily of various flagellate genera, the cryptomonads contain a single class, Cryptophyceae, of quite highly organized and specialized cytology. The cells are photoautotrophic or osmotrophic, and have two flagella of approximately equal length that are attached somewhat laterally on the cell. In *Cryptomonas* there is a pouch-like invagination of the cell near the point of flagellar attachment. Outlining the invagination in regular spiral rows are highly refractive trichocysts. Cells usually contain one or two chromatophores that almost cover the peripheral part of the cell. Eyespots are rare in this group, but another type of photoreceptor system may be present. Food reserves are deposited in granules of starch-like carbohydrates. Cells do not produce a definite wall or theca, but their membrane is firm and usually does not permit the cells to change shape. Cryptomonads seem to be unique among the flagellates in possessing phycobilin pigments in their chromatophores in addition to chlorophylls and xanthophyll pigments. Phycobilin pigments occur in red and blue-green algae, groups with very different morphologies and not closely linked phylogenetically with the cryptomonads. The presence of these pigments, which usually are a shade of blue or red, undoubtedly accounts for the broad diversity of colors within species of cryptomonads. Some species may be red in a culture in log phase of growth, but they become green or brownish when growth of the culture tapers off.

Cryptomonads often are abundant in estuarine localities and in tide pools along the open coast. They are rarely seen in neritic plankton samples, but almost invariably occur in enrichment cultures inoculated from coastal regions. It is possible that cryptomonad flagellates may be useful as an indicator of neritic water masses because they are uncommon in oceanic localities.

Blue-green Algae (phylum, Cyanophyta). In the sea, blue-green algae seem to be generally far less important in phytoplankton than in fresh-water bodies of water where they often dominate the population. Marine planktonic blue-green algae seem to be restricted to a relatively few species, but they may be very important in the populations where they occur. *Skujaella* (also known as *Trichodesmium*, and sometimes placed in the genus *Oscillatoria*), for example, often is the dominant planktonic organism in tropical and subtropical localities. Because it produces gas vacuoles within its cells, the filaments float on the surface of the sea and often accumulate in wind-rows that are many miles in length. Ordinarily these filaments have little pigment, often being greyish or yellowish in color. At times, however, the filaments may have a distinct red color, and it is said that the Red Sea received its name because of the occurrence of this alga.

A species of *Nostoc* has been reported from the Mediterranean Sea and the subantarctic Indian Ocean that sometimes forms large populations at extreme depths for phytoplankton organisms (1000 m) (Bernard and Lecal, 1960).

Symbiotic associations between unicellular coccoid blue-green algae exist in the sea with dinoflagellates (*Ornithocercus*, *Histioneis*, and *Amphisolenia*), a silicoflagellate (*Dictyocha specu-*

lum), and a diatom (*Streptotheca indica*) (Norris, 1967). A very interesting symbiotic relationship seems to exist between a short, filamentous blue-green alga, *Richelia intracellularis*, and photosynthetic cells of certain species of the diatom, *Rhizosolenia* (Fig. 16). *Richelia* seems to occur only at one end of the diatom cell, usually regularly oriented in relation to the cell axis. *Richelia* occasionally has been found attached to the surface of other diatoms such as *Chaetoceros*. *Richelia* does not seem to occur in *Rhizosolenia* growing in colder waters, but is restricted to populations occurring in tropical and subtropical regions.

Of the marine blue-green algae, only *Skujaella* has been demonstrated to fix atmospheric nitrogen.

Of the remaining groups of algae, only the euglenoids and yellow-green algae have representatives in marine phytoplankton. Photosynthetic euglenoid flagellates (*Euglena*, *Eutreptia*, and *Eutreptiella* may be locally abundant in neritic situations, some species probably indicating polluted conditions. Colorless euglenoid flagellates seem to be found in most parts of the sea but we know little about these species.

Few yellow-green algae (Xanthophyceae) occur in marine environments and only a few of these are planktonic. *Meringosphaera* (Fig. 12), having naked cells with long spines radiating from them, occasionally is found in the sea. The spines of the most common species have an undulating form, and possess small, barb-like side projections. There is a strong possibility that this genus belongs in the Chrysophyceae.

Economically, marine phytoplankton has tremendous importance to humans, supporting the entire marine fisheries industry. Although species of marine phytoplankton often are successfully grown on a small scale in the laboratory under artificial conditions, there has been limited success so far in farming marine phytoplankton either in artificial or natural environments. It is possible that this important primary source of human energy may be grown on large-scale, sea-going farms in the future. Such farms would directly support populations of zooplankton and small fish such as anchovies and other animals that are important as food for marine animals that man eats. Farming of marine phytoplankton important in the nutrition of shellfish, such as oysters, has already been successful on a small scale.

RICHARD E. NORRIS

References

Beklemishev, M., Petrikova, N., and Semina, H. J., "On the cause of the buoyancy of plankton diatoms," *Akad. Nauk SSSR, Trudy Inst. Okeanol.*, **51**, 33–36 (1961).

Bernard, F., and Lecal, J., "Plancton unicellulaire récolté dans l'océan Indien par le *Charcot* (1950) et le *Norsel* (1955–56)," *Bull. Inst. océanogr. Monaco*, no. 1166, 1–59 (1960).

Davis, C. C., "Concerning the flotation mechanism of Noctiluca," *Ecology* **34**, 189–192 (1953).

Norris, R. E., "Algal consortisms in marine plankton," *in* Krishnamurthy, V., *Proc. Seminar on Sea, Salt and Plants, Bhavnagar*, 1965 (1967).

Parke, M., "Some remarks concerning the class Chrysophyceae," *Br. Phycol. Bull.*, **2**, 47–55 (1961).

Parke, M., and Adams, I., "The motile (*Crystallolithus hyalinus* Gaarder & Markali) and non-motile phases in the life history of *Coccolithus pelagicus* (Wallich) Schiller," *J. mar. biol. Ass. U.K.*, **39**, 263–274 (1960).

Smayda, T. J., and Boleyn, B. J., "Experimental observations on the flotation of marine diatoms. I. *Thalassiosira* cf. *nana*, *Thalassiosira rotula* and *Nitzschia seriata*," *Limnol. and Oceanog.*, **10**, 499–509 (1965).

Smayda, T. J., and Boleyn, B. J., "Experimental observations on the flotation of marine diatoms. II. *Skeletonema costatum* and *Rhizosolenia setigera*," *Limnol. and Oceanog.*, **11**, 18–34 (1966).

Smayda, T. J., and Boleyn, B. J., "Experimental observations on the flotation of marine diatoms. III. *Bacteriastrum hyalinum* and *Chaetoceros lauderi*," *Limnol. and Oceanog.*, **11**, 35–43 (1966).

Wall, D., and Dale, B., "The resting cysts of modern marine dinoflagellates and their paleontological significance," *Rev. Paleobot. Palynol.*, **2**, 349–354 (1967).

Cross references: *Algae, Marine; Dinoflagellates; Parasites and Diseases; Phycocolloids and Marine Colloids; Plankton Resources.*

PLANKTON RESOURCES

Plankton (from the Greek term for that which is passively drifting or wandering) consists of both plants (phytoplankton, Fig. 1), and animals (zooplankton, Fig. 2). It ranges in size from the smallest microflagellates in the nanoplankton of only 2 or 3 μ to the largest jellyfish, which may be a meter or more across. There are few free-living bacteria in the sea but they exist in abundance attached to any substrate, living or dead, including the organisms of the plankton. Plankton organisms are conveniently termed "planktonts" or "plankters." A review of knowledge of the production of marine plankton has been recently given by Raymont (1966).

Phytoplankton

The phytoplankton consists largely of three categories of plants—diatoms, dinoflagellates, and a miscellany of minute forms called the nanoplankton. These plants are responsible for almost all the primary production in the sea, as the larger attached algae occur only in the very limited coastal area where light penetration

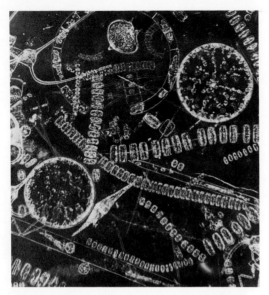

Fig. 1. Phytoplankton. (Photo: J. H. Fraser.)

adequately reaches the substrate. The floating phytoplankton, utilizing the dissolved nutrient salts and CO_2, with the action of chlorophyll in the light, produces the proteins, carbohydrates, and oils which form the food of the herbivorous animals in the sea; these, in turn, are the food of the carnivorous invertebrates and the fish, whales, etc. It is the dominant and basic link in the food chain in the sea and is thus of vital importance in the resources of the sea.

Being dependent on light, the plant production is virtually continuous throughout the year in the tropics, but there is a marked seasonal growth in the polar regions. In the temperate areas there is a spring maximum when the light increases and the fertility is high after winter mixing of the waters. This maximum then drops in the summer as the nutrients are utilized, and is followed by a further increase in the fall while there is still adequate light, but only after the reduced temperature breaks down the summer thermal layering of the water, permitting the replenishment of nutrients at the surface (Fig. 3).

Estimates of plant production can be obtained by various methods.* Where there is a marked seasonal growth, the amount of nutrient in the water, e.g., phosphate, can be measured before and after the spring season. The difference between these values can be attributed to utilization by the plants and a general assessment of the total plant production can be calculated (e.g.,

* See "Measurements of primary production in the sea," contribution to the "Plankton Symposium of the International Council for the Exploration of the Sea," *Rapp. et Procès-Verbaux*, Vol. 144, 1958.

Cooper, 1958). Such estimates cannot be accurate as no exact allowance can be made for the nutrients regenerated during the metabolic processes of the living plants and animals.

It is important to recognize the difference between phytoplankton production and standing crop, the latter being the amount of plant life present at any one time. A high production rate coupled with a high grazing rate by herbivores will result in a low standing crop. Standing crop can be measured by counting the plants occurring in known quantities of water with proper allowance made for the varying volumes of the organisms, or chemically from the total amount of chlorophyll present. This is usually done by the estimation of chlorophyll *a*. A known volume of water is filtered through a microfilter, the chlorophyll is then extracted from the filter by acetone, and the color is measured in a spectrometer. A monograph on oceanographic methodology, "Determination of photosynthetic pigments in sea-water," was issued by UNESCO in 1966.

Different areas have vastly different fertilities, depending on temperature, light penetration, and the amount of nutrient present, though the limiting factors may also be silicate, iron or other trace metals (Ryther and Guillard, 1959, Johnston, 1964), or metabolites (Lucas, 1956). The amount

Fig. 2. Zooplankton. (Photo: J. H. Fraser.)

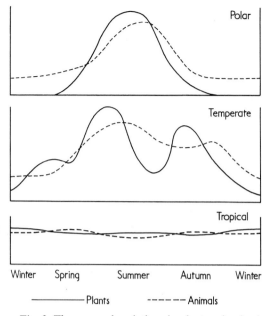

Fig. 3. The seasonal variations in plant and animal plankton in polar, temperate, and tropical zones. Note that the plants increase first and then, because food is available, the animals increase and graze down the plants. (After Fraser, 1962.)

of nutrient will largely be dependent on the degree of upwelling of nutrient-rich deep water and on land drainage. Upwelling is caused by the impinging of deep currents against the edge of a continental shelf, and by constant offshore winds that blow surface water out to sea so that deep water has to rise to replace it, e.g., the coasts of Peru and East Africa. An increase of density due to cooling of high salinity surface water causes large water masses to sink and these are replaced by upwelling of deep water in the open ocean at the "convergencies." In areas like the Sargasso Sea, the standing crop is particularly low though production is not far from normal (Ryther, 1960) because of the rapid utilization of all regenerated nutrients.

Too much light is detrimental to diatoms, causing an inhibition of growth at the actual surface, so that in bright sunlight the greatest production is subsurface, usually at about 10–20 m depth in the clear water (Steemann Nielsen, 1952). As depth increases the light is more and more attenuated until the plants lose as much by respiration as they gain by photosynthesis, i.e., the compensation depth. Below this, photosynthesis is further reduced and there is no autotrophic plant production. There is, however, evidence that some diatoms and flagellates can thrive below the compensation depth by feeding heterotrophically (Lewin, 1953, Rodhe, 1955,

Bernard, 1964). The compensation depth changes with the transparency of the water and the angle of the sun, and is about the light level of 350–500 lux, or about 0.3% of noon sunlight (Clarke, 1939). Transparency is reduced by the phytoplankton itself as well as by detritus and sediment in the water; in general deep water is clearer than shallow water where wave and tidal action stir up the bottom, and it is most turbid in areas of pollution. The compensation depth may be as much as 120 m in clear water and as little as $2\frac{1}{2}$ m in such places as the Oslo fjord (Rustad, 1946). Light can penetrate the polar icecaps sufficiently for some photosynthesis to take place (Bunt, 1964).

As neither the broad measurements nor the interpolations from pinpoint sampling can be accurate over wide areas and seasons, the total figures calculated for production in the world oceans cannot be more than a broad assessment—published figures have varied considerably but are of about the same order of magnitude (Steemann Nielsen, 1964). Kljashtorin (1961) quotes 9 mg C per m^3 daily in the central Atlantic, but says this figure could be increased by a factor of 10 in some places. Steemann Nielsen and Jensen (1957) give a total production for all seas of $1.2–1.5 \times 10^{10}$ tons of carbon annually. Using the factor of 37 (Fleming, 1940) to convert this to wet weight results in round figures of $30–40 \times 10^{10}$ tons of phytoplankton annually, which is essentially similar to that of Datsko (1959) who gives 30×10^{10}.

Standing crop is, of course, much less than total production for organisms with very short life cycles, and Bogorov (1965) gives a figure of 15×10^8 tons for the world standing crop of marine phytoplankton.

Comparing the amount of plant life in the sea with that on land, one must remember that the life span of individual terrestrial species is usually very much greater than for marine species so that standing crops on land are high. Nevertheless, productivity per unit of surface in sea water is usually assessed from quite low to about the same as on land or in places rather more. Westlake (1963) includes the following comparative figures in metric tons per hectare (1 hectare = 100 m^2 or about $2\frac{1}{2}$ acres).

Arid desert, 1
Ocean phytoplankton, 2
Coastal phytoplankton, 3 (more in polluted areas)
Temperate deciduous forest, 12
Temperate marine macrophytes, 29
Tropical marine macrophytes, 35
Tropical reed swamp, 75

Steeman Nielsen (1957) considers that terrestrial production is 60% greater than in plankton, and even in culture the most successful *Chlorella* plant produces only $\frac{1}{5}$ of the weight of terrestrial plants per unit of surface area.

Zooplankton

Zooplankton is dependent on the phytoplankton for its food supply. Herbivores feed directly on the phytoplankton and must live within reach of their food, although they are often located during daylight hours at levels below the euphotic zone, migrating towards the surface to feed at night. The most abundant marine herbivores are the copepods—though not all copepods are herbivores—and in the temperate and boreal parts of northern hemisphere the genus *Calanus* is of particular importance. Further along the food chain the carnivores feed on the herbivores and on the detritus feeders, which get their nourishment from the breakdown of dead organic matter and the bacteria that assist in the breakdown. The primary food production at the surface is therefore actively carried deeper and deeper by the living animals (Vinogradov, 1959), losing some 80–90% in energy at each stage. The food supply, which at any given time is the standing crop, not the productivity, becomes progressively reduced with depth. A small increase is seen at the bottom because it acts as a physical barrier against further sinking and so concentrates what is left to be used by detritus feeders and those that prey on them.

The level of maximum biomass of zooplankton is highly variable, but it is usually about 100–200 m, though nearer the surface in the higher latitudes; but the maximum number of different species is usually found much deeper, in the region of 1000 m. Numbers of individuals and of species fall off rapidly below 2000 m. Zooplankton is, however, present at all depths, even though numbers are greatly reduced. As no autotrophic life exists in the depths the metabolic products from living organisms and the disintegration products of the dead, regenerated into nutrients, are not utilized there by the phytoplankton, and deep waters become rich and fertile. A formula for estimating abundance with depth is given by Johnston (1962).

Zooplankton, following the development of the primary phytoplankton, is also most abundant in areas of upwelling. In polar areas the biomass of zooplankton is also high, due in part to the longer individual life cycles of the planktonts in the cold water and in part to the lack of thermal layering with the result that there are ample supplies of nutrients reaching the surface waters.

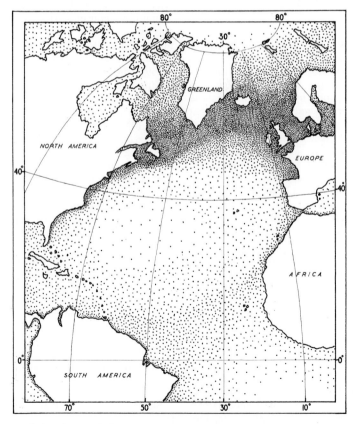

Fig. 4. The general abundance of plankton in the North Atlantic. (After Russell and Yonge, 1963.)

A chart of the abundance of plankton in the Atlantic is given in Fig. 4.

The food value of zooplankton varies greatly according to the type of organism. Crustacea have, as a rule, a high protein content and in the colder seas often a high oil content. Although there is quite a wide range according to species, figures in the region of 60% of the dry weight are given for protein, 15% lipid, 2 or 3% carbohydrate, and 20% ash, though the oil content may at times reach 70% (see details given by Raymont, Austin, and Linford, 1967). Details of the chemical composition of zooplankton are given by Vinogradov (1953) and Beers (1966). The water content of medusae and ctenophores can be of the order of 95% so that volumetric measurements are not, as a rule, very meaningful and zooplankton biomass is now usually stated as dry weight. Even so, variations arise and Lovegrove (1966) has advocated standardized techniques, which are generally accepted. If a more accurate assessment of food value is required the ash weight can be subtracted from the dry weight to give a figure for organic weight, or the albumen content can be estimated (Krey, Banse, and Hagmeier, 1957).

Where there are big seasonal differences in phytoplankton production these are reflected in the zooplankton when the peaks tend to follow those of the plants after a slight delay (Fig. 3). In some organisms egg production is known to be stimulated by food supply, e.g., in *Calanus* (Marshall and Orr, 1952). Zooplankton abundance in the coastal zone may be 20 to 40 times as great in the warm half of the year as in the colder half and 10 times as great in the summer in the slope zone at the edge of the continental shelf (Clarke, 1940). There is, however, little seasonal variation in places such as the Sargasso Sea where the dry weight of the zooplankton in the upper 100 m has a mean of 0.87 to 1.85 mg per m³ according to season, reaching about 5 mg in surface waters in June.

Arctic and Antarctic seas can be considerably richer, but there are very great variations. The zooplankton standing crop biomass (wet weight) in the Barents Sea varies from 1.5 to 3843 mg/m³ with an average of 230 mg/m³ in August (Jashnov, 1939), due mainly to one species of copepod, *Calanus* (Fig. 5). Variations occur in the dry weight obtained from the wet weight according to season and other factors; *Calanus* may be 75–85% water. Taking dry weight at 20% of wet weight, these figures, for comparison with those given above, are 0.02–55 g per 100 m³ with an average of 3.3. Manteufel (1939) says that *Calanus* can reach 9 g per m³, which would give a dry weight of about 180 g per 100 m³. Patches of the krill, *Euphausia superba* (Fig. 6) make Antarctic waters particularly rich (Foxton, 1956, Hardy, 1965), and there are more elusive patches of euphausids, *Meganyctiphanes norvegica* and *Thysanoessa longicaudata*, in the northern hemisphere.

The total world marine animal standing crop has been estimated at 32,500 metric tons of which 21,500 metric tons are zooplankton (Bogorov, 1965) but, as for phytoplankton, such estimates are extremely gross.

It is on the standing crop of zooplankton that the pelagic fish and whalebone whales feed—and, less directly, also the demersal fish. Out of the annual production of 30×10^{10} tons of wet weight phytoplankton about $40\text{--}50 \times 10^6$ tons of fish are caught, for direct consumption or for use as fish meal for animal feeding or fertilizers. This figure could probably be increased three or four times (Schaefer, 1965) by rational exploitation and by a policy of no waste, but it is still a poor yield to harvest from the original plant production. Most of the fish consumed are carnivores, often several links of the chain removed from the source. Whalebone whales and pelagic fish such as herring and anchovy are much more efficient users of the potential food supply than cod or halibut.

Economics of Plankton as Human Food Source

Can the use of ocean resources be made more efficient by adopting plankton as a source of human food? The supply of plankton is potentially available in tremendous quantities; it is the cost of extraction that makes it at present a doubtful commercial proposition. A chapter on this is given by Fraser (1962). Lacroix (1962) and others have dealt with it, and a paper by Jackson (1954) deals with costing. They suggested that the costs of extracting the plankton, as they could then be estimated, could not give profitable returns, and that it would be more economical to catch the fish, etc., which are so much easier to harvest even though there are several wasteful links farther along the food chain. The crux of the matter is in the filtration; to quote Ryther (1960) "The person who examines these data with the hope of feeding an over-populated earth on marine resources would do well to remember, when he picks a pound of beans from

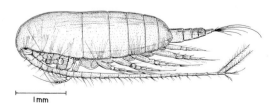

Fig. 5. The copepod *Calanus finmarchicus*. (From Sars.)

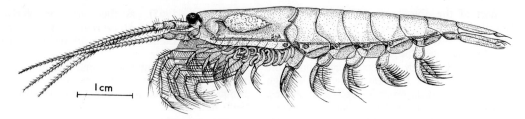

Fig. 6. The Antarctic krill, *Euphausia superba*. (From Bargmann.)

his kitchen garden, that to get the same weight of rather indigestible and unappetising plankton algae from the open sea he would need to filter some 5 million gallons of water." There is also the vicious circle of supply and demand; without knowing the demand and prices for the final product it is not easy to decide to make the capital expenditure, and without the guarantee of adequate and proved supplies these data cannot be given. As production costs would be extremely high for small bulk, this is not an industry that could gradually grow while prices and production find their own levels.

Other major factors in costing are that the richest sources of plankton are, in general, farthest from the areas where they are most required, and that, although free power is available in tidal currents, these usually contain a very high proportion of silt and detritus. Nevertheless, future engineering research may change the economics and some progress on these lines is being made (see Shropshire, 1944, and the Dietzsch patent, U.S. Patent No. 3234, 639). Some plankton, especially the larger crustacea, is being collected for food in local areas; e.g., mysids in the East Indies. The northern krill, a euphausid, *Meganyctiphanes norvegica* is caught in the Norwegian fjords by using lights to attract it into dense enough concentrations and by small meshed nets in the Norwegian Sea (Wiborg, 1966). It is used to supplement the diet used in fish farming. Where it is possible to get concentrations of fairly large species, with a fair degree of purity, then a suitable mesh can be chosen to get the maximum efficiency from the ship and the collecting gear and prospects are therefore more hopeful than a system of taking all from a wide range of sizes. Possible species for this would be the copepod *Calanus* in high northern latitudes and *Euphausia superba* in the Antarctic where the Soviets have already had experimental expeditions. Modern echo-sounders and sonar are sensitive enough to detect these and other plankton concentrations and vessels searching for dense shoals would profit by having this equipment. Many planktonts live at greater depths during the daylight hours, rising toward the surface at night, so that nightcatches are more likely to give good returns at practicable working depths.

One way to deal with bulk catches of diverse plankton, which might contain much detritus as well as plankton, is to process it in such a way that protein alone is extracted. Although this is a costly process at present, it not only gets rid of indigestible and useless bulk, but it removes the fats and oils that become rancid, and the final product is more stable. It would also utilize the protein available in the detritus and the bacteria associated with it (Krey, 1961). Improvements in techniques to make a marine protein concentrate could go a long way toward making bulk plankton collection a useful way to increase the available protein in the world's food supply. This has already been done using waste fish (U.S. Bureau of Commercial Fisheries, Fishery Leaflet 554, 1966) and in Scandinavia where at current food prices the cost of fish protein used for animal feeding is about one-thirteenth of protein derived from meat. Such marine protein concentrates could, with suitable hygiene in the preparation, be used as direct human food or in its cheaper form used to feed poultry etc. that can provide fresh animal proteins in the places where they are most needed.

The diverse collection system suffers also from the presence of unwanted species that may be distasteful or dangerously toxic, and these elements cannot always be removed during the normal drying processes. Medusae feed on prey that they paralyze with stinging cells, and concentrations of plankton containing large proportions of medusae would likely be distasteful at least, and possibly more seriously harmful. Many dinoflagellates produce toxins, and shellfish, such as mussels, feeding on these can concentrate enough paralytic poison in their own tissue to render them highly toxic for human consumption even after cooking. They occur normally in small quantities in the sea but when conditions are right they multiply rapidly and form a bloom, often known as a redtide because of the reddish tinge given to the water by the organisms themselves. Luckily they are mostly very small organisms that would escape fairly easily through most filters. Only a professional biologist could identify the dangerous organisms; his services, and the loss of plankton under his mandate when necessary, are added features of the costing.

Plankton can contribute to the supply of protein needed to relieve the world's food shortage, but supply is not the only answer. It has to be supplied at a price—including processing, transport, handling, and storage costs—that can be paid by those who most need it. There is, too, the very real danger that more food will lead not to relief but to still further increases in population and the demand for yet more food. Until the problems of controlling the world population have been solved, the sea will be looked to for more and more food, and seafood is closely linked with the plankton resources.

Costing complications do not arise when considering the value of plankton as a source of sustenance for ship-wrecked mariners. Here there are two main points in favor of using plankton—it is highly nutritious, and the liquid of the cell contents contains less salt than sea water. Against it are the risks of eating something toxic and the effects of a diet overly rich in vitamins. When the only alternative is death from thirst or starvation, there is no choice, and simple equipment for plankton catching is a wise precaution. It would be very far from an ideal food, and it is not likely that enough could be obtained to maintain life indefinitely, but it would extend the opportunity for rescue. However, this should not be regarded as an excuse for supplying an inadequate emergency ration.

JAMES H. FRASER

Cross reference: *Phytoplankton.*

POLAND FISHERIES

Before World War II, Poland had on her 70-km seacoast, only one port capable of servicing deep-sea fishing vessels, the port of Gdynia built in the 1930's. Then, there were two fishing ports servicing the Baltic fleet (Władysławovo and Hel), also put into operation at that time. Other harbors were of only local importance. Possibilities of developing sea fisheries were limited mainly because of the narrow sea coast and lack of shipyards.

Thus, the demand for fish had to be covered mainly by imports. The following is an illustration of fleet potential, catch, and consumption of fish in prewar Poland: in 1938, the Polish fishing fleet consisted of 8 steam trawlers, 20 luggers, 160 cutters (small trawlers), and 720 boats of which only 30 were engine powered. In 1936, the best of the prewar years as regards catch, Polish fishermen landed a total of 23,000 tons, mostly of Baltic sprats. Fishing for herring in the North Sea was inaugurated in 1931, the top landings (about 10,000 tons only) being attained in 1938. Imports consisted mainly of salted herring and varied from 45,000 tons (1934) to 104,000 tons (1929). There was a steady rise in fish consumption per capita: in 1924, 2.31 kg; 1927, 3.18 kg; and in 1938, 2.87 kg; of which 78% were sea fish and 22% fresh-water fish.

After 1945, based on the 500-km seacoast, consisting of the prewar coast and that of recovered western territories, Poland started developing sea fisheries on an industrial scale. This decision was based on:

(a) unlimited demand for food, especially of animal protein foodstuff;
(b) lower cost of obtaining one unit of animal protein and shorter production cycle in fisheries as compared with agriculture;
(c) shipyards capable of building vessels for the fisheries;
(d) planned economy.

The beginning was extremely hard. In 1945 there were only 3 cutters and 60 fishing boats on the entire coast. The country, exhausted by long invasion, suffered from hunger, and an immediate catch offered a quick and relatively cheap way of supplying the market with animal protein food. Thus, in the first period (1945–1949) there was a

TABLE 1. FISHING EFFORT DATA (HOURS FISHED)

Regions	Years					
	1961	1962	1963	1964	1965	1966
Baltic	209,860	70,240	363,790	360,359	413,502	404,946
High Seas						
North Sea	196,870	150,417	110,951	144,529	176,275	186,218
Atlantic North	2,488	5,744	10,919	20,321	28,910	29,865
Labrador	290	397	934	4,123	9,484	12,027
New Foundland	2,198	4,567	9,663	15,590	16,465	13,673
Georges Bank		591		566	2,843	3,750
Greenland		188	136	42	63	407
Nova Scotia		1	186		55	8
Middle Atlantic	2,536	586	7,807	10,048	14,379	21,509

POLAND FISHERIES

TABLE 2. AREAS FISHED (THOUSANDS OF METRIC TONS)

	1946	1950	1966	1970*	Main species caught (1966)	Main vessel types employed (1966)
Grand total	23.4	65.8	316.5	500.0		
Baltic catch total	22.2	58.1	104.7	103.0	cod, herring	cutters of 24 and 17 m l.o.a.
High seas catch total	1.2	7.7	211.8	398.0		
North Sea	1.2	7.7	96.0	122.0	herring	motor and steam side trawlers
North Atlantic			72.3	203.0		
Labrador			31.9		cod	factory stern trawlers
New Foundland			23.2		redfish	factory stern trawlers
Georges Bank			16.1		herring	factory stern trawlers
Greenland			0.8		cod	factory stern trawlers
Nova Scotia			0.2		herring	factory stern trawlers
Ireland grounds			0.1		herring	side trawlers
Middle Atlantic total			43.6	72.0	miscellaneous	full and semi freezer trawlers (stern and side)

* Estimates.

tendency to put quickly into operation all craft capable of fishing. At the same time long term fisheries planning began, as part of the planned national economy. A number of state-owned fishing companies were set up in the main fishing ports, cooperatives in smaller ones, private fishermen playing a supplementary role in the fisheries. The following figures illustrate the share of these three sectors in total catch:

	1949	1961	1966
state sector	35.5%	75.4%	86.5%
cooperative sector	7.0	12.8	8.3
private sector	57.5	11.8	5.2
	100.0	100.0	100.0

At the same time the three important shipbuilding centers were created by the Government. Starting with vessels of simple construction, Poland reached, in 1964, ninth place among shipbuilding countries of the world and sixth among ship-exporting ones.

Fishing Effort and Catch

Polish sea fisheries can be divided into three regions: Baltic, North Sea, and Atlantic. Tables 1, 2, and 3 illustrate the role actually played by these regions and the importance of particular species in landings.

For biological reasons (the need of conservation of fish stocks) as well as economic (low yield actually obtained per fishing unit), no growth is planned as regards catch in the Baltic and North Seas. Scientists estimate Poland's maximum landing possibilities from the Baltic grounds to be about 100,000 tons and from the North Sea about 120,000 tons per annum. Fishing on the Atlantic grounds was introduced in 1961, when the first factory trawler was put into operation. In 1961, landings from the North Atlantic totaled 4000 tons, in 1962 over 12,000 tons, in 1964 about 47,000 tons, and in 1966 over 72,000 tons.

Utilization of Catch

Fish processing has been developed parallel to the extension of landings. From the prewar small curing shops, it developed into an industrial branch.

The largest and best equipped processing plants on the coast are those in Gdynia, Gdańsk, and Swinoujście, producing both for the home market and for export. Among the inland plants, dealing with fresh and marine fish, the largest is at Giżycko, in the Mazurian Lakes district. Table 4 is an illustration of market preferences and methods of preparing fish.

Frozen cod and redfish fillets are delivered to the home market mainly on modern factory vessels operating in the northwest Atlantic. Frozen herring is landed by modern freezer trawlers fishing in the North Sea and in the ICNAF Area (Georges Bank). Salted herring is produced mainly by the conventional fleet

TABLE 3. CATCH (THOUSANDS OF METRIC TONS)

Species	1946	1956	1966
Total	23.4	122.5	316.5
Herring	1.5	64.1	117.1
Cod and cod-like	19.3	51.3	107.8
Redfish	—	—	15.1
Sprat	0.1	0.7	13.6
Mackerel	—	1.6	9.8
Other	2.5	4.8	53.1

TABLE 4. UTILIZATION OF CATCH (PROCESSED FISHERY COMMODITIES), NET PRODUCT WEIGHT, THOUSANDS OF METRIC TONS

Commodity	1958	1966	Main Commodity (1966)
Fish fillets, fresh or chilled		1.5	
Fish fillets, frozen	2.2	9.4	cod
Miscellaneous fish products, frozen	16.5	42.5	herring
Cods, hakes, haddocks, etc., salted	0.5	0.1	
Herring, salted	43.1	51.9	
Miscellaneous fish products, salted	1.4	1.5	
Herring, smoked or smoked-frozen	4.7	10.4	
Miscellaneous fish products, smoked or smoked-frozen	7.5	15.0	
Crustaceans, fresh, frozen, etc.	0.2	0.2	
Herrings, sardines, etc., in air-tight containers	3.5	7.1	herring in oil
Miscellaneous fish products, in air-tight containers	5.0	15.2	fish in tomato sauce
Fish preparations, not in air-tight containers	5.8	7.8	
Fish liver oils and similar products	0.8	3.7	
Fish meals and solubles	1.5	23.9	fish meal

operating in the North Sea. Fish meal and oils are produced aboard factory and freezer trawlers and on shore.

Distribution and Marketing

There is centralized disposition of fish landings from all Polish fishing vessels as well as of imported fish, done by a special state-owned enterprise. It is the main channel of distribution. The main duties of this firm consist in:
(a) balancing the entire fish turnover;
(b) taking over fish landed at ports;
(c) supplying fish wholesalers "Centrala Rybna" and processing plants with fish;
(d) creating fish reserves to cover the interseasonal demand; and
(e) financing the forming of fish reserves and fish transportation.

Prices paid by this firm to the owners are fixed and differentiated according to species and quality. In the fishermen's remuneration system, gross share is the main component. The selling prices are also fixed. The "Centrala Rybna" supplies its own and other retailers with fish and fish products.

Foreign Trade in Fishery Products

The actual import of fish and fish products is decreasing considerably, as compared with the prewar figures. In 1966, the main imports were fish meal, to cover the animal feeding requirements, and frozen fish for the fish processing plants.

The export figures show a constant increase. A considerable part of the export is sent to western countries, the chief buyers of fresh and sea fish being: United Kingdom, Austria, Denmark, Netherlands, German Federal Republic, Sweden, Italy, and U.S.; preserved Polish fish are imported by U.K., Austria, Belgium, Finland, Sweden, U.S., and other countries. (See Table 5.)

The Fishing Fleet

The achievement in developing the fleet is shown in Table 6.

The largest in number is the cutter fleet, which, since 1961, has consisted mainly of steel vessels (wooden ones have not been built since 1949). The cutters, built in series by the Gdynia shipyard, are suitable for Baltic operations throughout the year and North Sea operations during the summer. The annual total catch of such a cutter is around 400 tons, top vessels having caught over 800 tons of fish. Cutters of smaller size operate in the Baltic grounds only.

The conventional side trawlers, as well as drifter-trawlers are engaged in the North Sea herring fishery. The first group of vessels consists

TABLE 5. IMPORT AND EXPORT OF FISH AND FISH PRODUCTS (IN METRIC TONS)

	1950	1966
Import		
Marine fish, fresh and frozen	5,064	8,564
Salted fish	17,297	2,000
Canned fish	—	1,495
Fish meal	—	67,162
Caviar	—	10
Export		
Salmon	260	82
Marine fish, fresh or frozen	9,026	21,085
Salted fish	144	1,520
Smoked fish	—	33
Canned fish	856	4,203
Carp	1,986	328
Other fresh water fish	1,334	804
Crustaceans	—	44
Fish meal	—	2,050
Fish oils	—	215

TABLE 6. COMPOSITION OF FISHING FLEET

	At end of 1950			At end of 1966		
	number of units	000 Gross tons	Power (000 hp)	number of units	000 Gross tons	Power: (000 hp)
Factory stern trawlers	—	—	—	17	48.0	40.8
Full freezer stern trawlers	—	—	—	17	33.7	31.2
Conventional side trawlers						
motor	—	—	—	14	11.2	19.3
steam	24	7.5	15.1	53	34.6	46.2
Drifter-trawlers	3	0.8	1.0	42	7.3	13.7
Cutters	383	—	—	533	24.7	70.4
Mother and auxiliary vessels	—	—	—	5	32.7	20.2
Motor boats	—	—	—	745	—	9.4
Row boats	1,847	—	—	199	—	—

of the following types: old steam trawlers of B-10 and B-14 types (of 450 and 500 tons deadweight respectively); and modern motor side trawler of B-20 type (500 dwt tons), equipped to freeze part of the catch. The group of full freezers consists of the B-23 and B-18 types (600 and 1300 dwt tons respectively), equipped to operate in the Middle Atlantic. The factory trawlers (B-15 type built by the Gdańsk shipyard) are equipped to operate in the ICNAF area and process mainly cod and redfish into fillets, fish meal, and oils. Their dwt tonnage is 1250.

Auxiliary vessels are engaged in the North Sea herring fishery, servicing the fleet of cutters and side trawlers. A modern base-ship, put into operation in 1967, is equipped to operate in the distant Atlantic waters.

The main fishing gear used by the entire fleet is otter bottom trawl made from synthetic yarn. Midwater single and pair trawls are also used.

Organization of Fisheries

The motor and row boats are owned mainly by private fishermen, cutters by state-owned enterprises and cooperatives, and other vessels by state-owned deep sea enterprises.

The group of deep sea enterprises comprises: "Dalmor" of Gdynia; "Odra" of Swinoujście; and "Gryf" of Szczecin. To the group of enterprises exploiting Baltic grounds belong: "Koga" of Hel, "Szkuner" of Władysławovo, "Korab" of Ustka, "Kuter" of Darłowo, and "Barka" of Kołobrzeg. Then there is an enterprise (Deep Sea Fishing Bases of Szczecin) operating motherships and tender vessels, supplying the herring flotilla with barrels, salt, provisions, fresh water, and fuel, and receiving barrels of herring from the fishing vessels (of different owners). This enterprise also operates a base-processing ship, and organizes land bases for all Polish vessels at foreign ports.

All the state-owned fisheries are joined in the Fisheries Central Board (in Szczecin), this Board belonging to the Ministry of Shipping. Under this Board come also 10 fish processing plants both on the sea coast and inland, as well as 17 fish wholesale enterprises, the "Centrala Rybna" (producing also smoked fish, marinades, and ready-to-use fish products for the local market), situated in all major towns throughout the country. The "Centrala Rybna" supply the retailers with fish and fish products.

The cooperative fisheries are associated in the National Union of Fishermen's Cooperatives (in Gdynia), the private fishermen having their Private Fishermen's Association.

At the Fisheries Central Board there is a section coordinating the activities of all the three sectors.

Sea Fisheries Research

In Poland scientific research is carried out at three levels:

(1) at institutions of the Polish Academy of Science;
(2) at universities and other schools, where many chairs and scientific branches carry out fishery research, e.g., at the Academic Schools of Agriculture (Faculties of Fisheries), in Olsztyn, Cracow, and Szczecin; at the Academic School of Economics (Chair of Sea Fisheries Economics) in Sopot; at the Technical University in Gdańsk (Chair of Fish Technology, Faculties of Shipbuilding, Electrics, Mechanics); at the Technical University of Szczecin (Faculties of Electrics, Chemistry);
(3) by ministerial bodies, such as research and scientific institutions, central offices of construction and designing, central laboratories.

The main research institutions dealing with sea fisheries are:

(1) within the Ministry of Shipping:
 (a) Sea Fisheries Institute of Gdynia (SFI);
 (b) Central Laboratory of the Fish Processing Industry, Gdynia (CLFPI);
 (c) Designing Office for Marine Structures, in Gdańsk.
(2) outside the Ministry of Shipping:
 (a) Central Ship Design Office, Gdańsk;
 (b) Hydrometeorological Institute, Gdynia.

Of these, the principal ones that deal with the problems of sea fisheries are SFI and CLFPI, both being components of the Fisheries Central Board. The SFI has branches at Swinoujście, at Kołobrzeg, and at Tolkmicko. The development of SFI as well as of CLFPI is an integral element of sea fisheries growth.

In Poland, research is based on long-term and yearly plans, the research plans being adjusted to national economic plans (e.g., up to 1980, 5-year periods from the year 1961). They include the research objectives, needs of research staff, and technical requirements. Detailed projects within the objectives are specified in the successive yearly plans.

Ministerial institutions such as the SFI, as well as the CLFPI, are mainly financed from state grants and partly from their own income, e.g., the SFI has an income from selling fish caught when performing investigations at sea, this income covering about 15% of the Institute's costs.

Fisheries Education

Ensuring efficient crews has been the decisive factor in developing the fisheries. Professional training courses for fishermen and motormen were initiated by the authorities in 1947. In 1956, the Ministry of Shipping set up a training center in Gdynia with the purpose of ensuring trained men for all enterprises, for shore and shipboard employment.

There are two elementary fishery schools at Gdynia and Darłowo (training fishermen and motormen), and two officers schools in Gdynia and Szczecin. Then there is a fish processing college in Gdynia. Other specialists come to the fisheries from outside the Ministry of Shipping, from different schools both of academic and secondary level.

The number of fishermen increased from 2186 in 1945 to 10,262 in 1965. Foreign fishermen were employed aboard Polish vessels up to 1948.

To attract men to join the fisheries, a system of incentives has been introduced, the earnings of those employed aboard ship being higher than for those employed on shore. The incentives were for the purpose of ensuring that crews would stay at sea for trips lasting up to 100 days. Another problem concerned crewing of all vessel types—cutters with trips lasting a few days, as well as modern vessels with trips of around 100 days. Here, too, differentiation of earnings and differentiated living conditions aboard are the decisive factors. Thus, earnings are differentiated according to (a) place of work (shore or shipboard employment); (b) vessel type; (c) rank-rating. Effective earnings of fishermen are influenced mainly by the value of landings and the duration of the trip. Comparing for example, effective average monthly earnings of crew members on steam trawlers fishing in the North Sea with corresponding figures from factory trawlers operating in the ICNAF area, the earnings of the latter were higher by more than 50%.

Differentiation in earnings according to rank-rating is the main incentive for improving professional knowledge.

In Poland, schools and professional training are free. Monthly allowances paid to crew members during their stay at school for training purposes are much higher than the guaranteed wages. They are paid as for recreation leave, i.e., an average of their effective earnings during 12 preceding months.

The force of these incentives weakens as the general living standard rises. Thus, to ensure crews for the developing fisheries, it appears necessary to introduce methods of operating vessels that allow reducing time spent at sea (use of auxiliary vessels, exchange of crews at sea).

Poland is a member of the following international conventions:
 NEAFC (Northeast Atlantic Fishery Commission);
 ICNAF (International Commission for the North Atlantic Fisheries);
 ICES (International Council for the Exploration of the Sea);
 FAO Committee on Fisheries;
 London Fisheries Convention.

Poland is conveying and making available her statistical and research data to other members of these international bodies.

M. FORMELA

POLLUTION OF SEAWATER

Pollution is a general term associated with changes in the physical, chemical, and biological conditions of seawater as a result of discharge of polluted water or as a result of other activities of man. In a more restricted sense the term pollution is used mainly when the changes are harmful or in conflict with other uses and interests in the seawater area. As for pollution of fresh water, air, and soil, it is difficult to find a strict definition acceptable to all. In recent years, pollution of seawater has become an important topic for national and international consideration. Because of its impact on society, seawater pollution has attracted great attention from politicians, administrators, natural scientists, and technologists.

Seawater pollution is mainly due to liquid waste discharges from dwellings, factories, and farms. Pollution is mainly a result of industrialization and technological achievements, but its increase is also correlated with the rising population. In some cases the polluted material is discharged directly to the sea, but in most cases it first reaches the rivers, which then bring it to the sea. Many types of agriculture, landscaping, road building, and other large engineering tasks may lead to considerable soil erosion, and the suspended particles will usually be brought all the way to the sea. Although many rivers carry a natural load of such substances to the sea, man-initiated erosion will increase the load, and hence it seems justified to call it pollution. Some pollution is discharged at sea, mainly from ships. This includes sanitary waste, garbage, oil, and other goods transported by ships.

Polluting Material

The material polluting seawater may be classified in certain groups, according to its nature.

Solids. Usually as suspended material, solids are of inorganic (soil, silt, clay) or organic material (cellulose fibers, sewage sludge). The particles may vary greatly in size.

Organic matter. The main fraction of domestic sewage is composed of organics. Industrial and agricultural run-off also contains large amounts of organics at times. In narrow and shallow seawater areas with limited exchange of water, the load of organic matter is particularly important.

Nutrients. Most polluted water contains substances that may improve the growth of many organisms. Particularly important are certain nutrients that may stimulate the growth of algae and higher plants.

Toxic materials. Toxic materials may kill or prevent the growth of various organisms in seawater. Such materials are usually discharged from industries, but biocides (pesticides) used by agriculture may also reach seawater.

Organisms. Organisms of public health interest, such as bacteria and virus, usually come to seawater from sewage outlets, but they may also originate from food industries and agriculture.

Oil. Oil is a particularly important waste material. It may reach the sea regularly from engines and storages, but severe cases of oil pollution usually result from accidents with boats or land storage tanks. As the major part of the world's oil consumption is still carried across the sea by tankers, a threat of oil pollution exists. The situation is becoming more and more serious as the carrying capacity of each vessel is increased. Recently the threat of oil slicks from sunken tankers has assumed alarming proportions, and the damage to bird life, fish, and even human environment has been severe.

Radioactive wastes. Modern use of radioactive substances and nuclear energy for various purposes has caused a considerable problem for disposal of radioactive wastes. Usually waste occurs in the form of high activity solids in small volumes and low activity liquid wastes in large volumes. Although the enormous water masses of the oceans certainly have a capacity to receive such waste, there seems to be a general agreement that this waste should never reach the seas or be deposited at the bottom of the oceans.

Refuse. Many people use the sea as a great garbage can. Such materials as house refuse, refuse from outdoor activity and camping, septic tank solids, and the contents of chemical toilets are frequently disposed of in the sea. Because large populations live on the coasts or use the sea for recreational purposes, it is extremely difficult to control this particular misuse of the sea.

Polluted Areas

Pollution of the sea is most pronounced in shallow water areas near the coast and in bays and fjords. Estuaries of rivers are usually considered part of the marine environment in this regard. Some estuaries are among the most polluted waters in the world. Pollution is again most pronounced in the neighborhood of big cities, harbors and industrialized areas, which discharge polluting material to the sea.

In the sea itself, it is usually the surface water masses that are polluted. Certain deeper water bodies that have little exchange of water with the open sea are particularly sensitive to pollution. In such cases the sensitivity depends on the degree of stagnancy in the water masses, which again is determined by topography, current systems, and condition of adjacent waters. The bottom may be polluted directly through sedimentation of organic and inorganic particles.

The free water masses of the high seas are not yet considered as polluted or directly threatened by pollution. However, small areas may be polluted at the surface by floating materials, chiefly oil, and at the bottom by settleable matter, such as shipwrecks and solid wastes. Speculations have been made of the possibility of radioactive contamination of the high seas.

Since seawater is rich in potassium and certain naturally radioactive substances, it has a fairly high background radioactivity, and it is not easy to believe that a substantial increase in the total radioactivity can be brought about. However, there is the possibility that even when present in minute quantities, certain substances may be accumulated during biological activities, and thereby become dangerous to the organisms themselves, or to man. In radioactive pollution of seawater it is, therefore, not so much the radioactivity as such that is important, but the nature of the radioactive substances.

Effect of Pollution

To systemize the pollutional effects of different substances, it may be advantageous to talk about primary and secondary pollution. Primary pollution is the direct influence of the polluting material as it reaches the seawater in the dilution in which it will occur. Except for very narrow and stagnant seawater areas, primary pollution usually is important only in the neighborhood of the source of pollution. Primary pollution may affect the chemistry and biology of the water. The most pronounced effect of primary pollution is turbidity and coloration of the water and disappearance of dissolved oxygen. As seawater has a fairly good buffering capacity, chemical discharges will rarely affect the pH or the chemistry of the water significantly. As a result, seawater is particularly well suited to receive acid and alkaline wastes because of its buffering capacity. Organic solids that settle on the sea bottom will start to ferment. If the deposits are rich in organics, as often is the case in estuaries near big outlets, the fermentation may lead to pronounced gas fermentation. The gas formed will bring to the surface blackish flakes of the sediment which give off a strong smell of hydrogen sulfide.

The secondary effects of pollution have recently received much attention. Uusally secondary pollution is described as increased algal growth due to the supply of plant nutrients. This phenomenon is called eutrophication. Seawater as such is an excellent medium for growth of innumerable organisms. By the addition of small amounts of minor elements, it may support heavy growths of different plant organisms, mainly algae.

Eutrophication can alter the water masses in a number of ways. It may impart color to the water, reduce the transparency, and give rise to a characteristic planktonic smell. It is very important to observe that the organic matter produced by eutrophication will constitute an additional load of organic pollution in the water mass. As long as eutrophication is modest, it may not entirely disturb the biological equilibrium, and the number of plankton-eating organisms may keep up with the increased production of algae. Most frequently, however, eutrophication will throw the biological system out of balance. Some of the organisms, dead or alive, will sink to the deep and dark areas, creating an oxygen demand. For this reason, the result of eutrophication will often be lower water masses low in dissolved oxygen. The effect of increased algal growth as a result of pollution may spread over fairly wide areas.

Pollution Control

There are two ways of obtaining satisfactory control of seawater pollution: dilution and purification. Dilution may be used in order to take advantage of the large water masses that usually are available in coastal areas. Purification of the water before discharge will directly reduce the load to the receiving sea water. The general tendency in discharge philosophy today is *first* to look for dilution possibilities, and second to see what kind of purification is necessary to obtain the final desired result. Dilution is usually obtained by discharging the waste water from the pipeline along the bottom a certain distance from the coast. Except for a few rare cases of certain industrial effluents, the density of the polluted water is usually much less than that of seawater. At the point of discharge the waste water will therefore move upward and, through turbulence and diffusion, mix with sea water. In open seawater the mixture usually will reach the surface, but in cases where there is a pronounced stratification, e.g., due to brackish water in the top layer, the diluted waste water may come to a stop at a certain depth. Usually the waste water is discharged through a number of holes in the pipe; this part of a pipe usually is called a diffusor. With a diffusor it is easy to obtain dilution between 10 and 100 times only a few meters from the pipe. In the receiving water it is usual to distinguish between initial dilution, obtained in the immediate surrounding of the discharged point, and secondary dilution, which mainly is a result of movements in the general oceanographic system of the water mass.

The design of discharge arrangements requires a thorough knowledge of oceanography. In shallow areas, bays and fjords it may be difficult to get a good knowledge of water renewal time, and to evaluate the possibilities of secondary dilution. It is, however, to be expected that surveying techniques and instruments as well as theoretical treatment will improve rapidly and meet the demands. Usually all waste water must be given some treatment. For long discharge pipes it is usually enough to remove substances that may give trouble in the outlet system, or may accumulate at the bottom or surface at the point of discharge. When more treatment is necessary, it is at present usually in the form of mechanical or biological treatment. However, because of the great importance of eutrophication, which is often a pronounced result of marine pollution, technologists have focused interest on new treatment processes which may remove important nutrients. Many industrial effluents have rather special qualities and must be given particular consideration. In general these processes are composed of various mechanical, biological, and chemical methods.

Some industries, for instance mining, discharge large quantities of fine mineral particles. Such material usually will have no direct biological influence, but the quantities discharged may in themselves create certain problems. When such effluents have a density higher than the seawater, which usually is the case, the discharge will distribute itself along the bottom from the point of discharge. This gives density currents that usually creep along the bottom, and may carry the disposed material far away from the point of discharge. Density currents can lead to subwater erosion and can thus remove sediments as well as increase them.

Research

There is a great need for research in marine waste disposal. Thorough surveying of the discharged polluting water as well as the receiving water mass is necessary both as part of the planning and construction of an outfall and for later control of the system. Surveys usually will comprise a number of physical, chemical and biological determinations. As the conditions of seawater are subject to changes that only partly exhibit some sort of regularity due to tide, season, and weather, surveying must be carried out for a long period of time before it is possible to give the results a thorough examination. As surveying of marine resources requires much time, manpower, and money, a great effort must be made in planning and rational methods. The choice of parameters that will indicate the situation in question is of primary importance. Traditionally a complete surveying of the ocean requires a number of different data that can be obtained only by special instruments on research vessels.

Solution of problems of seawater pollution requires information obtained in the science of oceanography. However, the question of marine pollution raises a number of new issues and sometimes requires highly specialized information. The situation often arises that the basic sciences have not produced results directly applicable to the practical problems. Many problems associated with seawater pollution require new thinking in oceanographic research. Some of the hydraulic questions can be dealt with in models in such a way that the results are of direct use to the actual case. For chemical and biological studies it is also possible to conduct laboratory experiments that more or less simulate natural conditions. But in these cases it is fairly difficult to go from laboratory studies to field conditions. However, as more results and experience are collected, it is possible that in the future chemical-biological laboratory experiments may be of direct use in the solution of practical problems. As pollution of seawater usually occurs in estuaries and coastal areas, oceanographic research in these areas is particularly needed.

It is quite likely that the discharge of polluting material to the sea will increase significantly during the coming years. It is easy to foresee that coastal areas of considerable length may be grossly affected by pollution. Seawater pollution may in particular cause damage to harvested marine products and to all kinds of recreational life on the coast. To protect these vital aspects of modern life, it is likely that a considerable amount of money will be spent on measures against pollution. A tremendous increase in the understanding of the natural science and technological aspects of seawater pollution control is to be expected during coming decades.

A number of international organizations have adopted programs to control water pollution including seawater pollution, e.g., International Association for Water Pollution Research, the Council of Europe, OECD, and several United Nations bodies such as WHO, ECE, FAO, and UNESCO. In the international water areas certain conventions are already laid down and it is to be expected that more general conventions against pollution will be accepted. The national seawater areas are today directly under national control. However, as the water masses move freely between international and national areas, and from one national territory to another, bridges between international conventions and national regulations must be built. It is likely that one of the typical features of future society will be a controlled disposal of all kinds of waste.

Economy

It is virtually impossible to evaluate the total economic aspects of seawater pollution. A modern, technologically advanced society produces increasing amounts of polluting material that must somehow be disposed of. There are, therefore, great economic interests associated with the disposal of wastes in marine waters. The total damage and health hazards associated with the pollution of seawater can be calculated in terms of money only to a limited extent. It is primarily a question of what kind of surroundings man wants to live in. It seems quite likely that the ever-increasing standard of living will increase his desires to keep the environment in a state that will permit every citizen a maximum of freedom for all kinds of activities.

KJELL BAALSRUD

References

Pearson, E. A., "Proceedings of the First International Conference on Waste Disposal in the Marine Environment," 1960, Pergamon Press, Oxford.

Pearson, E. A., "International Conference on Advances in Water Pollution Research 1," Vol. 3, 1964, Pergamon Press, Oxford.

Pearson, E. A., "International Conference on Advances in Water Pollution Research 2," Vol. 3, 1965, Pergamon Press, Oxford.

Maroto, J. Paz, and Josa, F., "International Conference on Advances in Water Pollution Research 3," Vol. 3, 1967, Wat. Poll. Contr. Fed., Washington, D.C.

Riley, J. P., and Skirrow, G., "Chemical Oceanography," 1965, Academic Press, New York.

Olson, Th., and Burgess, Fr. J., "Pollution and Marine Ecology," 1967, Interscience Publishers, New York.

Cross reference: *Radioactivity in the Sea.*

POLYSACCHARIDE COATINGS

The film-forming properties of the methyl celluloses have been utilized in protective coatings for various food applications. A Russian article discusses the possible use of methyl celluloses in the meat industry for the coatings of various meat products (Ref. 1). Ulsenheimer (Ref. 2) used it for the preservation of eggs. The eggs were first treated with formaldehyde and then baked with methyl cellulose to decrease the evaporation of water. Eppell (Ref. 3) developed an edible film made of a mixture of methyl cellulose and low methoxyl pectin modified by a calcium salt. This film composition was designed to be used as a coating for compressed cereal bars, compressed fruit and nut bars, candy bars, jelly bars, and other foodstuffs (in either piece or bar form) that require protective coatings.

A good coating is able to protect a base material against deterioration changes either through oxidation or through moisture pickup and loss. And, it withstands effects of high and low humidities.

Some coatings are permeable to oxygen. Others lack flexibility, toughness, and brittleness. Still others rapidly deteriorate under conditions of high temperature and humidity.

An ideal edible coating should have strength, flexibility, low permeability to air and moisture, and stability to temperature and humidity variations. The edible coating must be economical to be applied at a reasonable cost, and most not result in allowing any detrimental chemical or bacteriological changes to occur during refrigerated storage, or preparation for consumption. Before such a compound can be used commercially, laboratory evidence must be accumulated to prove the technical feasibility of the process and that no health hazards exist or are created by its employment.

As fish and shellfish are among the most perishable of food products it is necessary they be handled in such a manner as to retain their "freshness" from the moment of capture to the time of consumption. There are numerous excellent methods of preserving fish, including refrigeration, freezing, canning, salting, drying, smoking, pickling, and treatment with antibiotics. However, at temperatures above 32°F (0°C), fish deteriorate so rapidly that it is necessary to pack them in ice or otherwise refrigerate them, even though they are to be held for only a short time prior to use or preservation by some other method. Yet when storage is prolonged, they become stale, with loss of flavor. The chemical reason for the loss of flavor in fishery products is generally supposed to be the evaporation of volatile constituents. In addition to loss of flavor, the refrigerated fishery products lose water by sublimation. Thus, fishery products, unless carefully protected, may rapidly lose 50 or 60% of their weight in a few months. As the fish or shellfish dries, the skin shrinks and loses its luster, the tissues become like cork, and they are unacceptable as food.

The terms "flavor," "texture," "consistency," and "lyophoresis" are often misdefined. The Institute of Food Technologists defines these terms as follows:

Flavor—Flavor is the simultaneous physiological and psychological response obtained from a substance in the mouth that culminates the senses of taste (salt, sour, bitter, sweet), smell (by which aromas are characterized as fruity, pungent, etc.), and feel. The sense of feel as related to flavor encompasses only the effect of chemical action on the mouth membranes such as heat from pepper, coolness from peppermint, etc.

Texture—Texture deals with the sense of feel or touch. Practically, it is limited to hand and mouth feel. As differentiated from flavor, texture is limited to those characteristics sensed by the force applied in mastication. From the physical standpoint it deals with the deformation or flow, but operates as a result of the applications of forces greater than gravity.

Consistency—Consistency refers to the free flow rate of a food product; it is usually applied

TABLE 1. GENERAL CHARACTERISTICS OF POLYSACCHARIDE-COATED OYSTERS, FROZEN 15 DAYS

	Organoleptic Scores (59 people)	Bacterial Count orgs/g	General Appearance	% Gain or Loss from original weight
1. Uncoated	5.2	427,000	Poor	Loss 3–12
2. With coating, fried, breading	7.2	96,000	Good	Loss 1–4
3. With coating, breading, not fried	7.0	146,000	Good	Loss 1–3

TABLE 2. GENERAL CHARACTERISTICS OF POLYSACCHARIDE-COATED OYSTERS, FROZEN 30 DAYS

	Organoleptic Scores (72 people)	Bacterial Count orgs/g	General Appearance	% Gain or Loss from original weight
1. Uncoated	2.0	635,000	Bad	Loss 3–22
2. With coating, fried, breading	5.0	249,000	Fair	Loss 3–7
3. With coating, breading, not fried	5.2	300,000	Fair	Loss 3–6

TABLE 3. ORGANOLEPTIC SCORES OF POLYSACCHARIDE-COATED FROZEN OYSTERS

Sample Treatment	Initial	Scores After Listed Storage Period[a] 1 Month	3 Months
Uncoated Oyster	9.2	6.2	4.3
Polysaccharide-coated Oyster	9.2	7.8	5.9

[a]Values are averages for participants on taste panel for the attributes of odor, appearance, flavor, and texture.

Code of scores:
  (10) No change from fresh product of highest quality;
  (8) First noticeable slight change in attributes;
  (6) Moderate degree of changed attribute. Increased in intensity and occurrence from score of 8;
  (4) Definite or strong degree of changed attribute;
  (2) Extreme degree of changed attribute.

TABLE 4. GENERAL CHARACTERISTICS OF POLYSACCHARIDE-COATED SHRIMP, FROZEN 15 DAYS

	Organoleptic Scores (59 people)	Bacterial Count orgs/g	General Appearance	% Gain or Loss from original weight
1. Without a coating and no breading	6.7	381,000	Good	Loss 2–5
2. With coating, breading, unfried	8.1	165,000	Good	Loss 1–2
3. With coating, breading, fried	7.8	143,000	Very Good	Loss 1–3
4. With breading, fried	7.3	207,000	Fair	Loss 2–6
5. With breading, unfried	7.4	216,000	Fair	Loss 2–5

TABLE 5. GENERAL CHARACTERISTICS OF POLYSACCHARIDE-COATED SHRIMP, FROZEN 30 DAYS

	Organoleptic Scores (72 people)	Bacterial Count orgs/g	General Appearance	% Gain or Loss from original weight
1. Without a coating and no breading	4.9	763,000	Poor	Loss 5–8
2. With coating, breading, unfried	6.8	208,000	Good	Loss 2–3
3. With coating, breading, fried	6.7	177,000	Good	Loss 2–4
4. With breading, fried	5.2	594,000	Poor	Loss 5–12
5. With breading, unfried	4.8	638,000	Poor	Loss 5–10

TABLE 6. ORGANOLEPTIC SCORES OF POLYSACCHARIDE-COATED FROZEN SHRIMP

Sample Treatment	Initial	Scores After Listed Storage Period[a] 1 Month	3 Months
Uncoated Shrimp	9.6	7.2	5.0
Polysaccharide-coated Shrimp	9.6	8.4	6.5

[a] Values are averages for participants on taste panel for the attributes of sweetness, odor, appearance, flavor, and texture.
Code of scores:
(10) No change from fresh product of highest quality;
(8) First noticeable slight change in attributes;
(6) Moderate degree of changed attribute. Increased in intensity and occurrence from score of 8;
(4) Definite or strong degree of changed attribute;
(2) Extreme degree of changed attribute.

to those products that are suspensions of ingredients of different particle sizes. By contrast, viscosity is used to describe the degree of fluidity of a liquid product.

Lyophoresis—Lyophoresis is a characteristic of certain food products closely related to and affecting consistency. It refers to the separation of the more liquid phase from the less liquid phase of the product such as the "weeping" of applesauce.

The Department of Food Science and Technology at Louisiana State University has been studying problems connected with fish and shellfish preservation since 1943. Their attention has recently centered on studying the value of various edible coatings that may be used to preserve fishery products. These edible coatings serve as carriers of flavors, colors, vitamins, waxing agents, and other functional additives of fish and shellfish. Various types of food coatings have been studied for increasing the storage life of fresh fishery products.

Experimental results obtained from chemical, bacteriological, histological, and organoleptic tests, showed that the use of an edible polysaccharide supplied by Food Research, Inc., Hollywood, Florida, as a coating material for the preservation of fresh fishery products resulted in superior products compared to the uncoated controls. Results of these studies also indicated that this product fulfilled most of the requirements described above for fishery products, and these products were evaluated for flavor, texture, and consistency.

Products employed in these experiments were obtained fresh, and were of known history. All the shellfish were less than 72 hours out of the water at the time the coating was applied.

Shrimp were obtained from a shrimp boat off Grand Terre, Louisiana. They were caught in nets in the early morning, separated from the trash fish, washed with seawater, and headed, shelled, and deveined. The product was packed in crushed ice in Arctic hampers and transported to the laboratory in Baton Rouge, where they arrived less than 24 hours after being caught.

Oysters were dredged during the early morning hours, and brought to New Orleans in a refrigerated truck (40°F). They were shucked in a commercial packing plant by professional shuckers, washed and drained according to F.D.A. regulations, and packed into 1-gal cans. These oysters were transported to Baton Rouge packed in crushed ice in Arctic hampers.

Fresh fish caught under supervision were purchased from fishermen, packed in crushed ice, and immediately shipped to the laboratory. They were cleaned within a few hours and cut into fillets. The fish were coated immediately.

Shrimp, oysters, and fish fillets were divided into equal portions. Half of each product was retained as the untreated control, and the other half was coated with polysaccharide, applied by dipping the products into 2-gal stainless steel pots containing the coating solutions. The coating formed a thin film around the foods, which could be controlled in thickness by adjusting the length of time in the dips.

The coated products and uncoated control samples were frozen at $-10°C$ and retained at this temperature until removed for chemical bacteriological, physical, and organoleptic testing. Examinations were made on the initial unfrozen samples and on the frozen foods at 1- and 3-month intervals.

Additional studies were made on breaded shrimp. All the shrimp were breaded according to the "Standards and Definitions" for frozen, raw, breaded, and lightly breaded shrimp, under the Federal Food, Drug and Cosmetic Act (Federal Register, July 6, 1963; 28 F.R. 6915). They included coated and uncoated, breaded and unfried; and coated and uncoated, breaded and fried. All the samples were frozen after treatment and withdrawn for testing.

Tables 1 through 6 present the tabulated results of these studies.

ARTHUR F. NOVAK
M. R. RAMACHANDRA RAO

References

1. Eppell, N. S., "Edible food coatings of a pectinate and methyl-cellulose," U.S. Patent 2,703,286, 1955.
2. Lyaskovskaya, Y., Ivanova, A., and Poletaev, T., "Polyvinyl alcohol and its possible uses in the meat industry," *Myasnaya Ind. S.S.S.R.* **25** (1), 52, 1955 (C.A. 49, 7770E).
3. Ulsenheimer, G., "Preservation of eggs," German Patent 945,063, 1956 (C.A. 53. 18333A).

Cross reference: *Breaded Fishery Products.*

PORPOISES

"Porpoise" is a broad term usually used to distinguish those relatively small toothed whales of the mammalian order Odontoceti, which, like all cetaceans, are exclusively aquatic and may be characterized by the following: a fusiform (torpedo-shaped) body; anterior limbs (flippers) that are paddle-shaped (with the joints distal to the shoulder immobile) and movable to varying degrees according to species; no external digits or claws; no external hind limbs; the tail flattened laterally and bearing horizontal flukes at the tip in all living species; only internal vestiges of the ear pinna present (the external ear only a pinhole opening); body essentially hairless (a few follicles present on the snout of adults, and these usually bearing obvious hairs in the newborn); no sebaceous (sweat) glands; and a thick subcutaneous layer of blubber. The bones of the skull are strongly telescoped, the external nares (called the blowhole) usually on top of the head, and the rostrum greatly elongated sometimes into an obvious beak, or snout, which is frequently obvious only in the skull and not in the living animal.

Members of the several families Stenidae, Phocoenidae, and Delphinidae are usually the ones referred to as porpoises, but smaller individuals of the families Ziphiidae (beaked whales), Monodontidae (narwhal and beluga), Physeteridae (sperm whales, especially the pygmy and dwarf sperm whales), and Platanistidae (long-snouted river dolphins) are frequently referred to by seamen and in the popular literature as porpoises. The many genera and species in these families have been listed by Scheffer and Rice (1963). Of the first three, the family Stenidae was only recently proposed, and contains two or three genera that are usually included with the Delphinidae by most authorities.

Technically speaking, only the members of the family Phocoenidae are "porpoises," while members of the family Delphinidae are called "ocean dolphins" or simply "dolphins." The two families have technical differences that seem to make their separation valid, but for the most part all the members of these two families are referred to popularly as porpoises. Probably the most obvious differences to a casual but competent observer would be that the "porpoises" (Phocoenidae) in general have no obvious elongated beak in life; a low and somewhat triangular-shaped dorsal fin (or none at all); and spadelike teeth with laterally-compressed, weakly two- or three-lobed crowns. "Dolphins," on the other hand, have either an obvious elongated beak or a globose, often bulging, forehead as in the pilot whales and *Grampus* in life; a high, usually falcate, dorsal fin (or none at all); and a widely-varying number of generally conically curved teeth, which may range from small and slender to thick and heavy in relative size. The terminology of "porpoise" and "dolphin" is controversial, and largely academic.

Fig. 1. Newborn captive bottlenosed dolphin swimming beside its mother. Light lines on infant's side are fetal folds and will disappear in several weeks. (Photo: Marineland of Florida.)

Probably the best known of the porpoises, actually a dolphin, is the bottlenosed dolphin, *Tursiops truncatus* (Fig. 1). This is the animal popularized to such a high degree on television, in the popular press, and as a highly-trained show animal in public and private aquariums and oceanariums. In size this species usually ranges up to 8 ft, although animals up to 9 ft are not too uncommon in captivity and in the wild they are known to reach some 12 ft in length. Much remains to be learned about this species; e.g., whether it is worldwide in distribution as this one species, as some biologists believe, or if it can be subdivided into several species or subspecies is still unknown. It is interesting to note, for example, that individuals considered to be this species commonly are found measuring 10–12 ft in length in European waters, while in the western Atlantic an animal this size is almost unheard of, although this species is studied much more extensively in that region (especially in Florida) than in Europe. One would expect a wider range of size variation to manifest itself in the many American animals taken if such a wide variation, especially on the high end of the range, was common. *Tursiops* measuring some 6–8 ft in length are so uncommon in Europe that specimens of this size, the one most suitable for exhibit and training, are regularly shipped to Europe by air. In addition to size, the coloration of *Tursiops* is very variable when one considers the extremes from almost black to piebald to buff to pure albino. However, while most individuals are shades of gray on the dorsal surface, shading to a creamy white ventrally, there is evidence that there is some geographical variation in color and pigment pattern.

In the wild, *Tursiops* usually travels in schools varying from three or four to hundreds of animals. These are usually found near the coastlines in tropical and temperate marine waters, both mainland and insular, but occasionally they occur far offshore (up to 100 miles or so), are frequently found in harbors and estuaries, and venture into pure fresh water for apparently short periods of time. (Florida's St. Johns River, for example).

Other species of porpoises travel in even larger schools than *Tursiops*, and schools estimated to number in the thousands have been reported for *Delphinus delphis*, the so-called common dolphin (Fig. 2). Some species are exclusively pelagic, and others live permanently in Arctic or Antarctic seas. Porpoises in general, therefore, can be said to be extremely cosmopolitan, probably more so than any other mammal, since they may be found in all seas of the world in a wide range of temperatures, and may occur partly or exclusively in fresh water, brackish water, and on out to the deepest seas. In addition, they eat a wide range of animal foods, and while all must surface regularly to breathe, they do not live exclusively near the surface, but instead may dive to great depths for food.

Relatively little is known of the biology of wild porpoises, but what little has been observed seems to correlate satisfactorily with extensive studies made on captives, especially the bottlenosed dolphin, in large oceanariums. It appears that the only real restriction that captivity places on the behavior of porpoises is that related to depth of the water.

Family groups, or at least communities, of *Tursiops* have been maintained for a number of years, and births are not uncommon in captivity. At Marineland of Florida, for example, three generations have been born in captivity after the original animal had been captured in local waters. Sexual maturity apparently is reached at about six years and the gestation period is about one year, in *Tursiops*, based on captive experience. Little is known of the life span of this species, but a female was maintained at Marineland of Florida for 21 years and 1 month before dying of a malady apparently other than old age, and her mother was a large adult at capture who was at least in her twenties at death. A conservative estimate of maximum age for females in this species would seem to be 25–30 years.

On the other hand, some large animals are captured whose teeth are worn down to the gums. If, as is usually supposed, this is an indication of old age when there is no obvious malformation of the jaws or unusual diet, then the maximum age attained by females of this species may be much greater, as the teeth of the 21-year old female were little worn, and her 8 ft $2\frac{1}{2}$ in. length was considerably less than that of females captured in Florida waters whose teeth were worn badly.

Gestation periods, ages at maturity, and life spans of other species of porpoises are relatively unknown, and surely vary with the species. The northern Atlantic pilot whale, for example, is believed to mature at from 6–7 (females) to 12 (males) years, have a gestation period of some 15–16 months, and live a minimum of 15 years (although much greater ages certainly must be reached) (Sergeant, 1962).

Many studies have been conducted on captive and wild porpoises relative to their behavior, social structure and activity, taxonomy and vocalizations. A cross-section of these studies, with considerable appropriate literature cited, can be found in the results of a recent generalized international symposium on the biology of cetaceans (Norris, 1966), and more details concerning the use of sounds by porpoises for echolocation (sonar) and inter- and intra-specific

Fig. 2. Captive common dolphin. (Photo: Marineland of Florida.)

transfer of information be found in Busnel (1967) and Tavolga (1967).

Because of the popularization of a supposed "speech" type of communication by porpoises in recent years, special note should be made of this aspect of their behavior. While these animals (at least *Tursiops*, *Delphinus*, and *Lagenorhynchus*) certainly communicate in part in the form of a vocal transfer of information, there is no valid evidence that they have a "language" in the sense of a human language. However, it appears from experimental evidence and observation that they do express emotions and generalized states of mind relative to certain apparent conditions of their immediate environment by use of pulsed sounds (Caldwell and Caldwell, 1967), and that they identify themselves to one another and also express emotions by the use of a puretone whistle and/or chirp (Caldwell and Caldwell, 1965, 1968). In the case of the pure-tone whistle, for example, each individual *Tursiops* and *Delphinus* appears to have its own individualized signature whistle that is almost always emitted no matter what the context. This signature whistle may be emitted more often, louder, or in a higher or lower frequency range; but the basic contour almost always remains the same no matter what the situation. In the rare cases where a secondary whistle is emitted, it too appears to be stereotyped. Contours shortened or lengthened in time of emission, or contours that are interrupted, are not considered exceptions to the signature whistle, since they are mere distortions of the basic contour.

True porpoises are rarely exhibited in aquariums and oceanariums in America, but in Europe the harbor porpoise (*Phocoena phocoena*) is sometimes seen in captivity. In America, Africa, Japan, and Australia, the porpoises (actually by definition all dolphins) most frequently seen in public establishments and research facilities, in addition to the common *Tursiops*, are: various species of pilot whales (*Globicephala*), the Pacific striped dolphin (*Lagenorhynchus obliquidens*), the killer whale (*Orcinus orca*), the false killer whale (*Pseudorca crassidens*), the spotted dolphins (*Stenella* spp.), the common dolphin (*Delphinus delphis*), and the grampus or Rissos' dolphin (*Grampus griseus*).

Some of the cetaceans not usually considered porpoises, but sometimes called by this general term, that are rather regularly exhibited are the beluga (*Delphinapterus leucas*) and the Amazonian Bufeo (*Inia geoffrensis*). Almost any of the other small toothed whales, including unusual forms of dolphins, such as the Amazonian River Dolphin (*Sotalia fluviatilis*) or generally-accepted "whales" such as the pygmy or dwarf sperm whale (*Kogia* spp.), may be on exhibit at some place and at some time when captured as accidentals, or collected when they happen to strand alive on some shore near an aquarium. Such exhibits in the past have always been short-lived.

Commercial fisheries, whereby porpoises were captured for food and oil, once existed in several areas of the world, including a large fishery for the bottlenosed dolphin at Cape Hatteras in the western Atlantic. Today such fisheries for the most part are reduced to catching pilot whales in Newfoundland and the West Indies, but occasionally some individuals of almost any species are taken anywhere and at any time (frequently by accident in a fish fishery) and are utilized for food and oil. A relatively few animals are taken for public exhibit and research, and while done by professional fishermen, this fishery presents no real problem to the population even in a local area because the animals become so wary that the collecting boats have to move farther afield.

In the old days of whaling under sail, porpoises formed an important part of the diet of the whalemen who were at sea for many months. For the most part the animals they took were for shipboard use only, rather than for commercial purposes. The meat of porpoises is said to be excellent, as it is rich in proteins. Along a different vein, however, a large fishery still exists in the Solomon Islands wherein large numbers of porposies are taken primarily for their teeth, which are used for money, and in the Arctic the narwhal is taken by the Eskimos for its long ivory unicorn-like tusk as well as for food and oil along with the beluga.

Besides being edible, the porpoise furnishes an oil that is exceptionally useful for lubricating small and delicate machinery, for example, watches and instruments of various kinds. The oil is extracted by boiling the blubber of the jaw or melon, and is sold in commercial quantities. Chemically it is a mixture of triglycerides in which isovaleric acid is prominent. The refined oil has a uniquely low pour point and high lubricity, and is resistant to oxidation and consequent gumming.

DAVID K. CALDWELL
MELBA C. CALDWELL

References

Busnel, R. G., (Ed.), "Les Systèmes Sonars Animaux, Biologie et Bionique," 1967, Jouy-en-Josas, France, 2 vols.

Caldwell, M. C., and Caldwell, D. K., "Individualized whistle contours in bottlenosed dolphins (*Tursiops truncatus*)," *Nature*, 207 (4995), 434–435 (1965).

Caldwell, M. C., and Caldwell, D. K., "Intraspecific transfer of information via the pulsed sound in captive odontocete cetaceans," *In* Busnel, R. G. (Ed.), *Les Systèmes Sonars Animaux, Biologie et Bionique*, 1967, 2, 879–936.

Caldwell, M. C., and Caldwell, D. K., "Vocalization of naive captive dolphins in small groups," *Science*, 159, 1121–1123 (1968).

Norris, K. S. (Ed.), "Whales, Dolphins, and Porpoises," 1966, University of California Press, Berkeley.

Scheffer, V. B., and Rice, D. W., "A list of the marine mammals of the world," *U.S. Fish and Wildlife Service, Special Scientific Report—Fisheries*, 1963, 431, 1–12.

Sergeant, D. E., "The biology of the pilot or pothead whale *Globicephala melaena* (Traill) in Newfoundland waters," *Bulletin of the Fisheries Research Board of Canada*, 132, i–vii + 1–84 (1962).

Tavolga, W. N. (Ed.), "Marine Bio-Acoustics," 1967, Pergamon Press, New York, Vol. 2.

Cross references: *Sounds of Marine Animals; Whales and Whaling.*

PRESERVATION — *See* **FREEZING FISH AT SEA; RADIATION PRESERVATION OF MARINE FOODS; POLYSACCHARIDE COATINGS; ANTIBIOTIC PRESERVATION**

PRODUCTIVITY OF MARINE COMMUNITIES

All the living resources in the sea depend ultimately upon the production of organic material by photosynthetic plants. In the open sea this process is carried on by microscopic organisms, the phytoplankton. Various estimates of the total quantity of organic matter produced in the oceans indicate that it may equal, or even exceed, the production of organic matter on land.

At any one time the standing crop of plant material in the oceans is very small compared to the amount found on land. This small standing crop produces organic material very effectively because of the rapid rate of turnover of the population. Under ideal conditions in the sea, the phytoplankton can double or even triple the size of the population in a day. Because of this rapid turnover rate, Ryther (1959) concluded that although the standing crop of phytoplankton of the sea constituted only 1/10 of 1% of the total plant material on earth, the annual production by this small population is 40% of the total.

The phytoplankton produced is rapidly grazed down by the herbivorous zooplankton, which, in turn, form the food for larger forms. The rate of grazing is an important factor in determining the size of the phytoplankton population at any given place and time. Thus, the plant population in the sea is more comparable to a heavily cropped pasture land ashore than it is to a field of corn or grain, where the total net production during the growing season is harvested annually or to a forest where the standing crop of plant material may represent the accumulation of 50 or more years of production.

Environmental Factors

Photosynthesis is, of course, dependent upon the energy of sunlight, which is absorbed by the chlorophyll molecule. The distribution of light in the sea is, consequently, an important factor controlling the rate of production of organic material. The amount of light reaching the surface of the ocean varies seasonally both in intensity and in the duration of daylight. The timing of the spring bloom of the phytoplankton in polar and temperate regions is closely related to the increasing light availability, not only because of its direct effect upon the rate of photosynthesis, but also because of its effect on the stability of the water column.

The amount of light available within the water column depends both upon the intensity of the incident solar radiation and also upon the transparency of the water. In the clearest ocean waters, such as those in the Sargasso Sea, sunlight to provide enough photosynthesis to balance the respiratory demands of the plant cell will penetrate no more than 100 m or so of depth. This is only about $2\frac{1}{2}\%$ of the total depth of the oceans, so that the production of organic matter is limited to a thin surface layer called the Euphotic Zone. As the plants grow, they themselves absorb light and the most highly productive areas rarely have a Euphotic Zone deeper than 50 m or so. Light can also be absorbed by nonliving suspended material or by colored compounds in solution, and the Euphotic Zone in coastal waters and harbors is further limited by these light-absorbing constituents. In heavily polluted waters, photosynthesis may be possible only to a depth of a few inches. Because the clearest water represents a paucity of living plant material, the most productive waters are found where the Euphotic Zone is shallow, and where the light is all or mostly absorbed by the plants themselves. This relationship is shown in Fig. 1.

Since the available amount of light decreases with depth, the rate of photosynthesis also

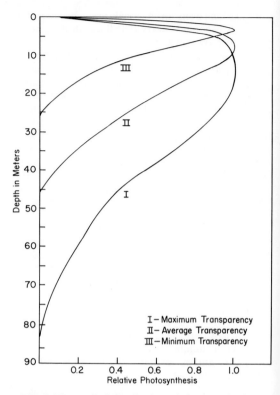

Fig. 2. The vertical distribution of photosynthesis in a homogeneously distributed population under bright sunlight in the open ocean. Data on transparency are from Jerlov (1951); on photosynthesis-light relations, from Ryther (1956). (After Yentsch, 1963.)

varies with depth in the water column. Figure 2 shows the computed rate of photosynthesis at various depths for three types of water of differing transparency. At low light intensities, photosynthesis increases linearly as light increases, but at high light intensities, such as those found near the surface in the ocean, the rate of photosynthesis by these microscopic algae is inhibited. This explains the near-surface decrease shown in Fig. 2. The relationship between photosynthesis and light intensity of various laboratory cultures of different species of phytoplankton is shown in Fig. 3.

Incident solar radiation also modifies the stability of the water column by increasing the temperature. Increased stability decreases the amount of vertical mixing, and this influences the development of phytoplankton populations in two ways. Under conditions of low stability and vigorous mixing, the plant cell may be carried to depths where light intensity is inadequate for photosynthesis. If a sufficient period of time is spent in the darker parts of the water column, the respiration of the plant cells can use up all the material produced by photosynthesis so

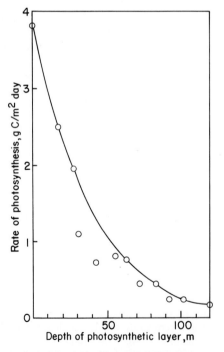

Fig. 1. Depth of photosynthetic layer and maximum rate of photosynthesis per m² surface. (After Steemann Nielsen and Jensen, 1957.)

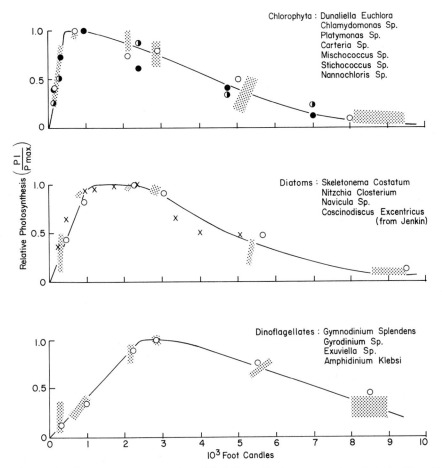

Fig. 3. The relation between photosynthesis and light intensity in some marine phytoplankton. Rectangles represent scatter of points in experiments using cultures grown at 1000 fc and measured with neutral density filters. Open circles are cultures grown under natural sunlight. Filled and half-filled circles are cultures grown at 350 and 1500 fc and measured in Woods Hole Harbor. Crosses from Jenkin (1937). (After Ryther, 1956.)

that no net growth of the population is possible. In temperate waters, the mixed layer becomes progressively shallower as the surface waters warm in the spring, and effective growth of the population can occur only when the mixed layer is shallow enough so that the phytoplankton remain in the illuminated part of the water column most of the time.

The increased stability and reduced mixing of the near-surface layers also permits the phytoplankton to assimilate nearly all the essential nutrients that can no longer be supplied at an adequate rate by vertical mixing from greater depths. It has been shown that under the stable conditions existing in the summer, populations of phytoplankton in the surface waters of the Sargasso Sea can increase their rate of photosynthesis severalfold if some water from greater depth, containing all the required plant nutrients, is mixed with the surface water before the rate of photosynthesis is measured. The nutrients that are commonly nearly exhausted in the sea and that, consequently, control the rate of photosynthesis of the population, include nitrogen compounds, phosphorus, and silicates (for the diatoms). There is evidence that certain trace metals, such as iron or manganese, may also be nearly exhausted and limit the rate of photosynthesis under some conditions, but much more extensive studies of the distribution and availability of these trace elements is needed.

The presence of organic compounds dissolved in seawater may also greatly influence the growth of phytoplankton populations. Certain growth factors, such as vitamin B_{12}, thiamine, and biotin are indispensable requirements for many marine algae. There have been some studies of the distribution of these compounds in seawater, and active studies are under way both to determine which species of phytoplankton algae require various growth factors, and also to investigate

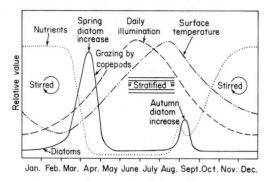

Fig. 4. Generalized diagram of seasonal cycle of diatom abundance and certain controlling factors in the temperate ocean. (After Clarke, 1954.)

which organisms in the sea produce and excrete them into the water (Provosoli, 1963). Further investigations are needed to evaluate the importance of these compounds in determining the ability of various species to survive and grow in different environmental conditions.

Thus, the environmental factors that determine the rate of production of organic matter in the sea include intensity of incident solar radiation, transparency of the seawater, stability of the water column, availability of essential nutrients, distribution and production of organic growth factors, and removal of phytoplankton by grazing zooplankton. All these environmental factors are continuously changing, and these changes must be taken into account in evaluating the dynamic process of primary production.

The seasonal cycle of phytoplankton in temperate waters shown in Fig. 4 indicates the changes of these environmental characteristics and their effects on the diatom population. During the winter period, the phytoplankton crop is small, although the nutrient concentrations are high. At this time, productivity is limited by the comparatively low light intensity and the vigorous mixing, which carries the plants out of the Euphotic Zone. As light intensity and stability increase in the spring, the phytoplankton crop increases to a maximum known as the spring bloom. This maximum population lasts only a brief time both because the available nutrients are decreased by assimilation into the living organisms and because the grazing by herbivorous zooplankton increases as the available food supply increases. During midsummer the phytoplankton standing crop is low, since it is grazed down by herbivorous zooplankton as fast as it can be produced under the low nutrient conditions. During the early fall, mixing increases enough to enrich the surface waters and produce a secondary bloom, but, as vertical mixing further increases, the population decreases to the small winter size.

Methods of Measuring Productivity

A variety of methods have been suggested for evaluating the productivity of marine communities. The earliest of these involved measuring the standing crop of phytoplankton. Samples were taken by means of a fine silk net, the species were identified, and the numbers of individuals were counted. While this method gives valuable information concerning the distribution of species of phytoplankton, it is not an adequate measure of productivity, even when the counts of individuals are corrected for the size of individuals. The standing crop is an instantaneous measure of the population and represents, at any one moment, the net balance between production and destruction. As mentioned above, grazing by herbivorous zooplankton is severe at various times of year, and the standing crop does not, therefore, give an evaluation of the rate of production of total organic matter unless the losses from the population can be adequately evaluated. Another problem is the fact that many of the phytoplankton are small enough to pass through the finest silk nets used. Analysis of total plant pigments or of the chlorophyll content of phytoplankton filtered from the water may obviate the difficulties of losing much of the population in a net sample. Comparisons of the amount of plant pigments in a net sample with the amount in the total volume of water have indicated that they may differ by a factor of 1000 or so.

The chlorophyll content of the population has been used as an index of productivity as well as a measure of the size of the standing crop. Knowing the light intensity and the amount of chlorophyll in the water, it has been postulated that one could compute the capacity of the population for phytosynthesis. It has been shown, however, that the rate of photosynthesis per unit of chlorophyll may vary both throughout the day and as a result of the nutritional conditions to which the cell is exposed, or the physiological state of the organism. Calculations of the productivity based upon the chlorophyll content of the water and light conditions are, therefore, first approximations only.

An early experimental method for the measurement of productivity involved enclosing samples of the water in bottles and measuring the changes in oxygen both under illumination and in the dark. As a result of phytsoynthesis, the oxygen content of the light bottle would increase, and as a result of respiration, the oxygen content of the dark bottle would decrease. The difference between these two oxygen contents was taken as a measurement of the gross productivity of the community. Naturally any respiration by animal forms or bacteria in the bottles would also be measured. In oceanic environments, where the phytoplankton population is sparse, the changes in the oxygen content of the sample may be too small to be measured over a short-term experiment, and the method is subject to criticism when the experiments are extended over long periods of time because of the possibilities of major changes in the character of the population being evaluated.

Another early method for the measurement of the productivity of the marine community was to evaluate the changes in the concentration of elements in the sea as a result of biological activity. The removal of carbon dioxide, phosphate, or available nitrogen as nitrate, nitrite, and ammonia, or the release of oxygen to the water can all be used as an index of the total production of organic matter by the phytoplankton. All these methods need to consider the effects of respiration by animal populations, the supply of the element by horizontal and vertical mixing and by advection, and both oxygen and CO_2 changes must be corrected for the transfer between seawater and the atmosphere. Thus, the correct application of these methods requires a complete understanding of the physical processes, and they are difficult to apply except under idealized and simplified conditions. However, for short periods of time they are valuable, and, when compared with other estimates, they permit an evaluation of the rate of delivery of nutrients to the Euphotic Zone by vertical mixing and of the recycling and regeneration of nutrients in the biological cycle.

Most of the recent measurements of productivity depend upon experimental evaluation of the rate of assimilation of radioactive carbon. This method was first described by Steemann Nielsen (1952) and has been widely used in investigations in all parts of the world oceans. The method is comparatively simple and involves adding some radioactive carbon 14 as bicarbonate to samples of water and exposing them to various light intensities or in the dark for 2–4 hours. The particulate matter is then removed by filtration, and the amount of radioactivity on the filter is measured. By knowing the total carbon dioxide in the sample and the amount of radioactive carbon dioxide added, one can compute the amount of organic carbon fixed and retained within the living cells. The appropriate light intensities can be provided by suspending the samples in the sea at the same depth from which they were collected or by exposing them on shipboard to similar light intensities artificially produced.

The experimental methods, such as the oxygen bottle or the radioactive carbon techniques, suffer from the disadvantage that they are instantaneous measures of the productivity at the time of observation and must be repeated frequently throughout the year if one is to obtain an evaluation of the total annual production. Measurements of the changes in environmental conditions, such as the nutrient concentrations, carbon dioxide, or oxygen content of the water, are most applicable during the rapid growth of the population in a plankton bloom, but are incapable of measuring productivity under steady state conditions when there is no detectable change in the concentration of the element in the environment. Using the two methods simultaneously, however, makes it possible to evaluate the rate of turnover of the population so that

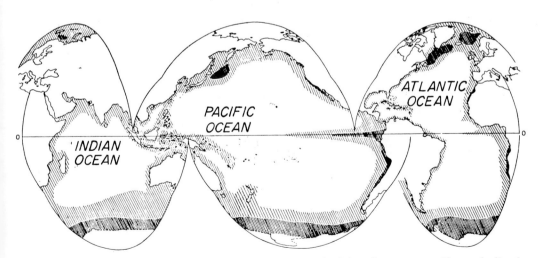

Fig. 5. Schematic representation of the probable relative productivity of ocean areas. Heavy shading indicates very productive areas; light shading indicates moderately productive regions. (After Sverdrup, 1955.)

one can evaluate the supply of nutrients by regeneration and by mixing.

Regional Variations in Productivity

The estimates of the total annual production of organic matter by marine populations have ranged from 50–1000 g of carbon per m^2 per year. The lower value is comparable to the production of organic material in deserts on land, and the higher value is comparable to the production of good farm land. The global distribution of production of organic material is indicated in Fig. 5. The highly productive areas, indicated by dark shading, all have mechanisms to enrich the surface water with water from greater depths, where the nutrient concentration is higher. For example, off the west coast of continents, such as the coast of Africa or South America, persistent trade winds move the near-surface waters off shore and permit upwelling of waters from intermediate depth. In the southern ocean there is a major current flowing completely around the earth, and associated with this current is a divergence that again brings nutrient-rich waters close to the surface. The currents in the equatorial regions are also associated with divergence of surface waters and with high rates of production

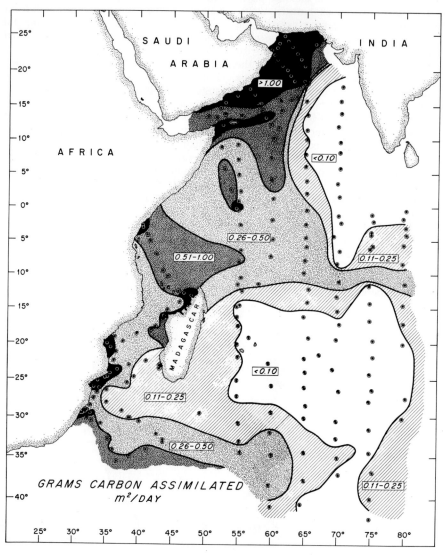

Fig. 6. Chart of the western Indian Ocean showing the positions of oceanographic stations occupied during *Anton Bruun* cruises 2, 3, 4A, 5, 6, 7, and 8 and the general level of primary organic production. (After Ryther, 1966.)

of organic material. The great fishing banks, such as Georges Bank and the Grand Banks in the North Atlantic, are comparatively shallow areas where wind stirring and tidal mixing enrich the surface waters and increase the productivity. The effect of river water carrying nutrients into the sea is important in coastal waters and in semiconfined bodies of water such as the Gulf of Mexico. However, in terms of the total oceanic production, river drainage adds only about 1% of the total nutrient requirement each year. Thus, while river drainage is very important locally, its value to the productivity of the sea has been greatly overemphasized by some.

A very detailed study of the production of organic matter by the marine community was made in the Western Indian Ocean during 1963–64, and the results of this study are summarized in Fig. 5. The values vary by a factor of more than 200, with the highest productivity being found in the Arabian Sea, where the waters were characterized by high concentrations of inorganic nutrients in the near-surface waters.

Productivity is low for large areas of the ocean, as shown in both Figs. 5 and 6. These areas are characterized by a very stable water column where the plankton has been able to extract the nutrients from the surface waters and where the replenishment of these nutrients by vertical mixing is inhibited by the stability.

Much more remains to be done in direct measurements of the productivity of marine communities before man will be able to achieve a satisfactory evaluation of the total productivity of the oceans. The extent of areas of high and low production need to be more accurately defined, and more frequent observations are required before the occasional measurements now available can be extended to a valid evaluation of annual production. Since the organic matter produced by these microscopic phytoplankton organisms is the fundamental source of organic material for all living resources of the sea, a more precise evaluation of the magnitude of the productivity is essential before it will be possible to evaluate the value of the oceans completely as a source of additional food for mankind.

BOSTWICK H. KETCHUM

References

Clarke, G. L., "Elements of Ecology," 1954, John Wiley and Sons Inc., New York.

Provasoli, L., "Organic regulation of phytoplankton fertility," in Hill, M. N., Ed. "The Sea," vol 2, 1963, Pergamon Press, New York.

Ryther, J. H., "Photosynthesis in the ocean as a function of light intensity," *Limnol. & Oceanog.* I(1), 61–70 (1956).

Ryther, J. H., "Organic production by plankton algae and its environmental control," in "Ecology of Algae, The Pymatuning Symposium in Ecology, 18–19 June 1959," Special Publ., Univ. Pittsburg Press, 1960.

Ryther, J. H., "Geographic variations in productivity," in Hill, M. N., Ed. "The Sea," Vol 2, 1963, Pergamon Press, New York.

Ryther, J. H., Hall, J. R., Pease, A. K., Bakun, A., and Jones, M. M., "Primary organic production in relation to the chemistry and hydrography of the Western Indian Ocean," *Limnol. & Oceanog.* **11** (3), 371–380 (1966).

Steemann Nielsen, E., "The use of radioactive carbon (C^{14}) for measuring organic production in the sea," *J. Cons. Explor. Mer*, **18**, 117–140 (1952).

Steemann Nielsen, E., and Aabye Jensen, E., "Primary oceanic production, the autotrophic production of organic matter in the oceans," *Galathea Report*, **1**, 49–120 (1957).

Steemann Nielsen, E., "Productivity, definition and measurement," in Hill, M. N., Ed. "The Sea," Vol 2, 1963, Pergamon Press, New York.

Sverdrup, H. U., "The place of physical oceanography in oceanographic research," *J. Mar. Res.*, **14**(4), 289–294 (1955).

Yentsch, C. S., "Primary production," *Oceanogr. Mar. Biol. Ann. Rev.* **1**, 157–175 (1963).

Cross references: *Phytoplankton; Plankton Resources.*

R

RADIATION PRESERVATION OF MARINE FOODS

The use of ionizing radiation is man's latest weapon in his arsenal against food spoilage. Like canning, irradiation preservation of food involves a transfer of energy. This method depends upon the bombardment of food by electromagnetic radiation in the form of gamma rays emitted from a radioactive isotope, such as cobalt 60, to kill spoilage micro-organisms. The radiation that speeds through the food ionizes and disrupts the electrical balance of the atoms in the food-spoilage organisms. The electrical imbalance thus generated is usually fatal to the bacteria, but the food itself is not made radioactive nor does it retain any traces of radiation. The flavor and odor of the food are not changed at the levels of radiation found best for food.

The world's first large-scale effort to determine the feasibility of using gamma radiation for seafood preservation resulted in the erection of the MPDI (Marine Products Development Irradiator) at Gloucester, Mass. The U.S. Atomic Energy Commission and the Bureau of Commercial Fisheries of the Fish and Wildlife Service of the U.S. Department of the Interior, respectively, have built and staffed the MPDI. The purpose in building the irradiator was to determine if it is commercially feasible to irradiate seafoods on a large scale and ship them to distant points in the nation and, by extension of shelf life, still retain a high degree of freshness for normal marketing at the destination.

Present investigations of the low-level irradiation of seafoods had their origin in 1960 when the U.S. Atomic Energy Commission began its support of studies into the chemistry, microbiology, and food technology of various fish and shellfish. This work was conducted at several of the Bureau of Commercial Fisheries laboratories (namely, those at Gloucester, Mass.; Ann Arbor, Mich.; and Seattle, Wash. in order to include the most important marine and freshwater fish and shellfish.

U.S. food laws demand proof of safety and effectiveness of new food additives and new food processes. Radiation processing of any food constitutes a new and very special type of food additive according to the law, which states that a food that is intentionally irradiated is considered to be adulterated unless the irradiation is performed under an exemption granted by a food regulatory agency. The purpose of this requirement is to place the burden of proof upon the food processor to show that the new food process or food additive is safe and that it accomplishes the intended effect. At present, two foods have been approved for irradiation and general consumption. They are wheat and wheat products and potatoes. As research work progresses, applications for general clearance for sale and consumption of additional foods will increase.

Methods of Food Irradiation

Irradiation of food means the subjection of food to ionizing energy in the form of gamma rays from a radioactive isotope such as cobalt 60 or cesium 137, or by electrons from particle accelerators or from particle accelerators that employ targets to produce X rays with penetrating

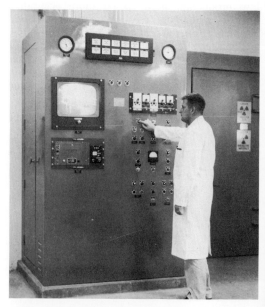

Fig. 1. Control panel for radioactive source. Closed circuit television scans conveyor line.

and ionizing characteristics very similar to gamma rays. Most irradiation facilities that process bulky packages of food use a radioisotope such as cobalt 60 because of the high penetrating power of the gamma rays. Machine sources of ionizing radiation, however, possess advantages in the irradiation of thin packages of food or of free-flowing material. Each method of irradiation possesses merits, and each must be judged in the light of penetration, ease of handling, uniformity of dose, dose rate, where it may be important, ease of dosimetry, and production rate and economics.

Regardless of the source of radiation, all methods have ionization as their objective, with the resultant death or inactivation of food-spoilage organisms. Irradiation of foods with cobalt 60 and cesium 137 does not in any way impart or induce radioactivity, because their levels of energy are not high enough to cause any material to become radioactive. Similarly, the use of electrons of 10 MEV and X rays of up to 5 MEV have been accepted as being safe for foods by the Food and Drug Administration.

The new method of food preservation employed by the MPDI relies on the use of radioactive cobalt 60, which constantly emits invisible electromagnetic gamma rays. Gamma rays are similar to X rays and possess extraordinary penetrating power. The gamma rays kill or inactivate up to 99% of the spoilage bacteria occurring on seafoods. After being passed through the irradiation process, the seafoods are still raw, moist, and cold. This condition is in contrast to that produced by heat, because the amount of radiant energy required to obtain the same degree of bacterial kill generates no appreciable heat. The viable bacteria that remain are still capable of reproduction. Therefore, irradiated seafoods, like pasteurized dairy products, must be kept refrigerated to retard the growth of the few remaining bacteria and thereby prevent subsequent spoilage of the product. Optimum radiation dose levels will vary from species to species, but most seafoods require 100,000 to 300,000 rads. (A rad is that quantity of ionizing radiation that results in the absorption of 100 ergs of energy per gram of irradiated material. Thus a rad is a measure of the amount of energy absorbed in any bombarded material regardless of the nature of the originating source of radiation.)

Two elements are necessary for the successful irradiation of fresh fishery products to obtain extended shelf life: (1) irradiation at dose levels most suitable for the species under consideration; (2) maintenance of the irradiated food at or near ice temperature in a manner calculated to prevent recontamination by micro-organisms. They are accomplished by placing the seafood in a suitable packaging material such as in approved plastic films or in the rigid metal containers commonly used in the seafood industry and then irradiating the entire container. When commercially shipped and marketed so as to maintain the food at ice temperature, the product has a shelf life extension of 1 to 2 weeks over similarly handled but non-irradiated material. Table 1 illustrates the number of fish and shellfish that have been investigated to determine their response to ionizing radiation and their subsequent shelf life when maintained at or near the temperature of melting ice.

Operation of the Irradiator

In a typical operation, fillets conventionally wrapped and packaged at the commercial seafood

TABLE 1. SHELF LIFE EXTENSION OF FISH AND SHELLFISH IRRADIATED AT OPTIMUM DOSE AND HELD AT 33°–35°F UNDER CONTROLLED CONDITIONS

Fish or Shellfish Irradiated		Optimum Irradiation Dose Levels, Rads	Extension in Shelf Life, Weeks
Common Name	Scientific Name		
Cod	Gadus morhua	150,000	4–5
Flounder	Limanda ferruginea	250,000	3–4
Haddock	Melanogrammus aeglefinus	250,000	3–4
Halibut	Hippoglossus stenolepis	200,000	2
English sole	Parophrys vetulus	300,000	4–5
Gray sole	Glyptocephalus cynoglossus	200,000	3–4
Petrale sole	Eopsetta jordani	200,000	2–3
Mackerel	Scomber scombrus	250,000	4–5
Pollock	Pollachius virens	150,000	4
Smoked chub	Coregonus artedii	100,000	6
Yellow perch	Perca flavescens	300,000	4
Lobster meat	Homarus americanus	150,000	4
Shrimp	Penaeus sp.	200,000	4
Blue crab meat	Callinectes sapidus	250,000	25
Dungeness crab meat	Cancer magister	200,000	3–6
King crab meat	Paralithodes camchatica	200,000	4–6
Soft shell clams	Mya arenaria	450,000	4
Oysters, shucked raw	Crassostrea virginica	200,000	3–4

Fig. 2. Radioactive source of six plaques at bottom of 15 ft of shielding water. In operation, the source rises along curved track.

plant are sent to the MPDI in rectangular metal containers of 10, 20, or 30-lb capacity. Upon receipt, they are immediately stored in a refrigerated storage room capable of maintaining product temperatures at 33°–35°F, and each tin is tagged with a yellow indicating disk, which changes to red when the product has received a predetermined amount of radiation. At the beginning of the irradiation process, the containers are transferred from the cold storage room by a conventional roller conveyor to the package conveyor feed station of a high-speed mechanical conveyor. This conveyor carries the packages into a vertical labyrinth to a slow-speed conveyor in the radiation cell area. It is in this area that the packages are moved through the zone of high intensity radiation.

Once inside the irradiation chamber, the packages are transferred to a slower conveyor, which carries them past the radioactive source in steps at a rate that ensures their receipt of a predetermined amount of radiation. Each package on the slow conveyor makes a round trip into and out of the radiation chamber. This round trip is accomplished by passing the package under and then over the radioactive source on one side of the center line of the source. It is returned to the starting point and again makes the same round trip on the other side of the center line of the radioactive source. The second round trip ensures the complete irradiation dose uniformity of each package. After irradiation is completed, each package is returned to the irradiator cold storage room and maintained at 33°–35°F until shipped in ice, to maintain the proper temperature during storage and subsequent transportation.

The irradiation source of 250,000 curies of cobalt 60 is made up of six replaceable subunits in a rigid stainless steel frame whose active radiation area is about 4 ft long by 1 ft wide. Each subunit contains 16 activated cobalt strips that are doubly encapsulated in welded stainless steel jackets. The double encapsulation results in a loss of about 15% of the energy emitted by the cobalt 60 source. Miller and Herbert (1964) have calculated that the MPDI system of presenting moving targets of products to be irradiated results in a utilization of about 21% of the total energy available from the radioactive source.

The original design of the MPDI called for a dose level of 250,000 rads and a production rate of at least 1 ton an hour with a maximum-to-minimum ratio of 1.3 and not greater than 1.4. At the present time, dose levels have been slightly decreased, thus increasing the production rate, which is a function of time. There is no change, however, in the maximum-to-minimum ratio. This ratio is obtained by dividing that point of maximum energy absorption received in terms of rads, by the point of minimum absorption. The resultant quotient is 1.3, which is determined by the use of phantoms of packages that are identical in shape to the food container and that contain dosimeters so scattered throughout the phantoms or units to be irradiated as to ensure the attainment of the desired dose.

Regardless of the type of food for which exemption is sought, certain requirements must be met. Among these are lack of induced radiation, assurance of full nutrition, knowledge of chemical and microbiological changes, and sound food technology, including effectiveness of the irradiation treatment.

In respect to induced radiation, studies of many irradiated foods, including those of marine origin, have been conducted under contract to the Surgeon General of the U.S. Army.

They involved high dose levels of irradiation in which the foods were subjected to 10–20 times as much irradiation as is proposed for marine foods; all studies showed that the food was not impaired. Marine fish fillets irradiated at levels higher than 2,000,000 rads have produced no adverse effects when ingested either by humans or experimental animals and when fed at high dietary levels for periods up to 2 years. One can logically infer therefore that if high

Nutrition

One index of the nutritive properties of a food lies in the protein utilization of the food by experimental animals. Studies have showed that marine fish irradiated at levels of more than 2,000,000 rads possessed the same degree of protein utilization by human volunteers and experimental animals as do nonirradiated control samples. Since animals that were fed the irradiated fish compared similarly in growth and reproduction for three generations with the control animals, which received no irradiated fish, it again is logical to infer that if high radiation doses prove acceptable, low radiation doses should also prove acceptable.

Chemistry

The Bureau of Commercial Fisheries Technological Laboratory in Gloucester, in cooperation with the Atomic Energy Commission, has placed much emphasis upon the chemical aspects of odor and flavor of irradiated marine fish. The line of attack is to try to understand why changes in odor and flavor occur when high dose levels of radiation are used. Much of the program has been devoted to collecting, separating, and identifying the volatile compounds in both irradiated and nonirradiated marine foods. More than 30 different compounds have been identified. Separation and identification of carbonyl and sulfide compounds have been made by gas chromatography and time-of-flight mass spectrometry. One of the most interesting results is that few carbonyls are developed when vacuum-packed soft-shell clam meats (*Mya arenaria*) are given massive doses of 4.5 million rads of irradiation at cryogenic temperatures and held at these temperatures for weeks. Yet, the development of carbonyls is rapid under similar treatment, but when the sample is held at room temperature during the irradiation process and subsequent storage. This observation may point the way to successful sterilization of some marine foods in the future, but no economic advantages can be forecast because of the added costs of the two processes.

Microbiology

When ionizing radiation is used with seafoods, it has as its object the suppression or kill of up to 99% of the micro-organisms initially present. Irradiation changes the microflora of seafoods (Sinnhuber and Lee, 1965; Seagran and Emerson, 1965; Slavin, Ronsivalli, Kaylor, and Carver, 1966) by suppressing some species, thus allowing other species to predominate because they possess greater innate resistance to radiation. The customary pattern is for the genus Pseudomonas, the chief spoilage bacteria of fish fillets, to be suppressed by doses as low as 100,000 rads. This low level of radiation permits another genus, Achromobacter, to become dominant up to irradiation dose levels of 300,000 rads. Above 300,000 to 500,000 rads, Achromobacter ceases activity, and yeasts then become dominant.

This shift in microflora is important when viewed in the light of toxic micro-organisms. Some microbiologists have thought that radiation might have the effect of concentrating pathogens, particularly *Clostidium botulinum* type E. Studies performed by Goldblith and Nickerson (1965, 1968) indicate that irradiated fresh haddock fillets (*Melanogrammus aeglefinus*) and cod fillets (*Gadus morhua*) are just as safe as are nonirradiated fresh haddock fillets. Fresh haddock fillets have been marketed by the millions of pounds with no recorded instance of botulism. Studies made at the Massachusetts Institute of Technology and at the Bureau of Commercial Fisheries Technological Laboratory at Gloucester indicate that normal cooking of the fish provides additional safety. Intensive studies are now underway at several university, government, and private laboratories to determine whether a botulism hazard exists with fillets irradiated at low dose levels.

Optimum Dose Level

Optimum dose levels for various seafoods, when irradiated and held under controlled laboratory conditions, do not necessarily reflect the best dose levels for the same foods when they are subjected to the stresses of commercial production, distribution, and sale. Under controlled laboratory conditions, researchers have determined that when optimum levels of radiation are exceeded by a factor of one and a half or more, odors and flavors are generated that taste panels consider different if not objectionable. When dose levels are less than optimum, product shelf life is shortened, but no appreciable reduction in shelf life occurs unless the radiation dose reduction approaches 50% or more.

Influence of Freshness

It is axiomatic that the fresher seafood is when it is irradiated, the longer it will retain its fresh attributes. The basic reason is that the tissue of freshly caught fish is almost sterile and that contamination with increasing numbers of bacteria is a function of time. Fillets cut from commercially well-handled haddock, which have not been stored in ice longer than seven days aboard trawlers, have been found suitable for irradiation. This period of seven days represents an average maximum pre-irradiation age of fillets that justifies the irradiation process for commercial shelf life extension. To determine what role freshness plays in gaining even more extended shelf life, the Bureau of Commercial Fisheries installed a 30,000-curie cobalt 60 irradiator aboard its research vessel *Delaware* at Gloucester, Massachusetts. This is the world's first irradiator installed aboard a vessel to study the effect of irradiation on fish and other seafoods immediately after capture. The commercial feasibility of irradiators installed aboard large trawlers to preserve catches at sea will be studied and evaluated in respect to land-based irradiators.

Effect of Special Attributes

All seafoods do not respond to irradiation in the same manner because of differences in their innate character. Generally, lean fish respond to the process very well, but some species of fish with a high content of fat exhibit undesirable effects. Radiation will destroy the pleasing red color of salmon (*Oncorhynchus sp.*) flesh, thereby lowering its value markedly. On the other hand, mackerel (*Scomber scombrus*), which possesses almost as high a fat content as does salmon, responds very well to irradiation treatment. Conversely, some species of fish that are relatively low in fat compared with mackerel, such as halibut (*Hippoglossus stenolepis*) and petrate sole (*Eopsetta jordani*), exhibit rancidity in a relatively short time after being irradiated. This behavior would seem to suggest that the chemical makeup of the lipids may be the governing factor rather than the amount of fat. The phenomenon of exceptionally rapid rancidity after irradiation may be minimized by excluding air from the package during irradiation, according to Dassow and Miyauchi (1965). This exclusion of air may be accomplished by vacuum packing or packing the product tightly in low oxygen-permeable materials, excluding as much air as possible. The use of approved antioxidants and the packing with inert gases to replace the naturally occurring oxygen may prove feasible with the more expensive species of seafoods despite the extra cost.

Commercialization

Successful commercialization of seafoods preserved by irradiation hinges upon three pivotal points. The first is that of maintaining a favorable temperature during the entire chain of distribution of fresh seafoods from trawler to retailer. Personnel of the MPDI have investigated the distributing aspects of the present fresh seafood industry. This investigation showed that all the common commercial methods of transporting fresh fish ensure the efficient and safe distribution of highly perishable fresh fishery products. Internal product temperatures during transporta-

Fig. 3. Loading 30-lb packages of fillets on fast conveyor for passage into irradiation area.

tion are reduced to, or maintained at, a desirably low level of 40°F or lower.

The second important point in the commercialization of irradiated fillets was to determine if an increase in distribution time and extension of shelf life of irradiated seafoods can be realized under commercial conditions of handling and shipping. It was repeatedly proved that cod and haddock fillets could be purchased on the open market, irradiated, and shipped to cities at least a thousand miles farther than the distance to which these fresh fillets are now shipped. In every instance, it was found that cod and haddock fillets irradiated at 200,000 rads and held at ice temperature had a shelf life more than double that of similarly handled nonirradiated control samples. The distances and durations involved in these shipments are unprecedented and clearly reflect, in the most practical fashion, the benefits to be gained by irradiation.

The third crucial point in the commercialization of irradiated seafoods was the market potential. To predict this potential, it was necessary to determine the degree of acceptance of irradiated fillets by seafood buyers located at the headquarters of large chain stores throughout the nation. Obviously, if there was resistance at this level in the food cycle, the consumer would never have an opportunity to express his preferences. As a consequence, commercial size shipments of

Fig. 4. Curved uppermost track guides radioactive source from bottom of pool to horizontal slot between upper and lower packages of slow conveyor.

irradiated and nonirradiated haddock fillets were sent to the buyers for their critical examination but not for sale. They tested both irradiated and nonirradiated fillets over a period of time and reported that the irradiated fillets outlasted the nonirradiated fillets from 6 days to as much as 2 weeks. They were unanimous in their acceptance of the new process.

The benefits to be realized from radiation processing of irradiated seafoods are substantial. By extending the shelf life of seafoods, irradiation will reduce the amount of spoilage now normally encountered. This will result in savings to consumers in a competitive market. Radiation preservation will provide a consistently higher level of freshness than is now obtainable with fresh seafoods. This improved quality in turn should help promote increased sale and consumption of fresh seafoods. With increased sales, some increase in profit should also follow. High among the benefits of commercialization will be the extension of fresh seafoods to areas of the nation where they are not now obtainable.

Economics

No discussion of commercialization would be complete without a consideration of the economics of the irradiation treatment of fresh seafoods. Obviously, where no competitive commercial radiation processing of seafoods exists, no precise costs are available. The MPDI, with its production rate of over a ton an hour of irradiated seafoods, serves as a model of semicommercial proportions from which we can realistically predict costs on a commercial scale. Kaylor and Slavin (1965) estimated that the cost per pound for low-level irradiation of seafoods would be about 2 cents. Hitt (1966) states that his economic study showed that the most practical estimates fell in the range of 1–2 cents per lb. All practical estimates indicate that the costs of irradiation are as low, or lower, than the cost of packaging the product. The 2-cent estimate is slightly higher than the cost for conventional freezing, but in view of the higher prices that fresh fishery products command over the products preserved by any other method, the economics of this method of preservation are indeed favorable.

No food process has ever been subjected to as much scientific and technological scrutiny prior to commercial use as has been the irradiation treatment or processing of foods. The granting of exemptions for specific irradiated foods has been slow; for a variety of reasons. Chief among these has been the lack of a precise definition of the nature of supporting evidence required. Fortunately, this situation has been corrected, and petitions for exemptions may reasonably be expected to increase.

Food irradiation presents a paradox. It is a mature, tremendously sophisticated technology that has been deterred from demonstrating what it can do; yet, at the same time, it represents an infant industry of unknown potential. It has problems that are basic in nature and wide in scope, but they are no more difficult to resolve than are those in any new technology. The equipment and facilities needed to develop the basic technology are completely new in design, but are not immoderate in cost to build and to operate. Food irradiation can be expected to develop in a relatively short time an impetus that will help to make it another valuable method of food preservation.

JOHN D. KAYLOR

References

Dassow, J. A., and Miyauchi, D. T., "Radiation Preservation of Fish and Shellfish of the Northeast Pacific and Gulf of Mexico. Radiation Preservation of Foods," 1965, Publication 1273, National Research Council, Washington, D.C.

Goldblith, S. A., and Nickerson, J. T. R., "Annual Report. The Effect of Gamma Rays on Haddock and Clams Inoculated with *Clostridium botulinum*, Type E," MIT-3325 Report number 1321, Isotopes-Industrial Technology TID 4500, 1965.

Goldblith, S. A., and Nickerson, J. T. R., "Annual Report June 1966–June 1967. The Effect of Gamma Rays on Ocean Perch Inoculated with Type E, *Clostridium botulinum*," MIT-3343-28, Isotopes-Industrial Technology (TID-4500), 1968.

Hitt, J. C., Sixth annual AEC food irradiation contractors' meeting October 3-4, 1966. Conf-661017.

Kaylor, J. D., and Slavin, J. W., "Irradiation: big advance in preserving seafood," *Fishing News International* **4**(2), 147 (1965).

Miller, P. and Herbert, R. J., *Marine Products Development Irradiator, Isotopes and Radiation Technology*, **1**, No. 4 (1964).

Seagran, H. L., Emerson, J. A., Kazanas, N., Gnaedinger, R. H., and Krzeczkowski, R., Radiation Pasteurization of Foods, Summaries of Accomplishment, Fifth Annual Contractors' Meeting Conf-651024, 1965.

Sinnhuber, R. and Lee, J. S., Progress reports Nos. 7, 17, and 18 to the U.S. Atomic Energy Commission Under Contract No. AT(04-3)-502, 1964.

Slavin, J. W., Hearings before the subcommittee on research, development, and radiation of the Joint Committee on Atomic Energy, Congress of the United States, June 9 and 10, 1965.

Slavin, J. W., Ronsivalli, L. J., and Kaylor, J. D., "Radiopasteurization of fishery products," *Activities Report* **17**, No. 2 (1965).

Slavin, J. W., Ronsivalli, L. J., Kaylor, J. D., and Carver, J. H., Study of Irradiated-Pasteurized Fishery Products. Annual Report to the U.S. Atomic Energy Commission, TID-22833, 1966.

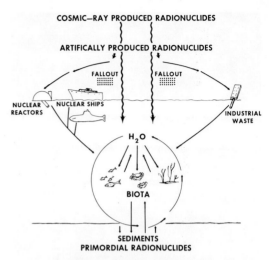

Fig. 1. Naturally occurring radioactivity, consisting of cosmic-ray produced and primordial radionuclides and artificially produced radionuclides, are continuously being cycled through the three components of the marine environment—water, biota, and sediments.

RADIOACTIVITY IN THE SEA

General

Radionuclides in the marine environment can: (1) remain in solution or in suspension; (2) precipitate and settle on the bottom; or (3) be taken up by plants and animals. Certain factors interact to dilute and disperse these materials, while other factors simultaneously tend to concentrate them. Currents, turbulent diffusion, isotopic dilution, and biological transport dilute and disperse radionuclides. Concentrating processes may be biological, chemical or physical. Radionuclides are concentrated by the biota, by uptake directly from the water, and by passage through food webs; they are concentrated chemically and physically by adsorption, ion exchange, co-precipitation, flocculation, and sedimentation. Also, radionuclides in the sea are cycled through the water, sediment, and biota (Fig. 1). This cycling can be described by the three R's: routes, rates, and reservoirs.

Each radionuclide tends to take a characteristic route and has its own rate of movement through the three components (reservoirs) of the marine environment. Radionuclides are exchanged from the water to the sediments or to the biota. In effect the sediments and biota compete for the isotopes in the water. Even though in some instances sediments initially remove large quantities of artificial radionuclides from the water and thus prevent their immediate uptake by the biota, this sediment-associated radioactivity may later affect many benthic organisms by exposing them to radiation. Also, radionuclides leach from the sediments back to the water and again become available for uptake by the biota. Even before radionuclides are leached from the sediment they may become available to the biota, due to variation in the strength of the bonds between the different radionuclides and the sediment particles. Loosely bound radionuclides can be "stripped" from particles of sediment and utilized by bottom-feeding organisms.

Classification of Radionuclides

Naturally occurring radionuclides. Naturally occurring radionuclides in the oceans, at least in part, arose from the weathering of rock. The principal categories of natural radionuclides are: (1) primordial radionuclides and their decay products (daughters); and (2) radionuclides resulting from reactions between cosmic-rays and elements in the atmosphere or in the earth.

Several long-lived naturally occurring radionuclides have been identified (Table 1). Most of the natural radiation in the sea originates from three of these: uranium 238, thorium 232, and potassium 40, the latter accounting for more than 90% of the natural radiation. The decay of potassium 40 ends with its transformation into stable calcium or argon, whereas uranium and thorium decay into a long series of radioactive "daughter" elements. Most potassium compounds are soluble in seawater and potassium ions are adsorbed on particles of clay, sediments, and other materials with an ion-exchange potential. Since uranium precipitates more slowly than thorium, its concentration reaches high levels in seawater (Ref. 1). The chemical trans-

Table 1. Principal Naturally Occurring Radionuclides in the Sea

Nuclide	Half-life (years)	Concentration (g/l)	Concentration ($\mu\mu$c/l)
Tritium	1.2×10^1	3.2×10^{-18}	3.0
Beryllium 10	2.7×10^6	1.0×10^{-13}	2.0×10^{-3}
Carbon 14	5.5×10^3	3.1×10^{-14}	1.5×10^{-1}
Silicon 32	7.1×10^2		1.2×10^{-5}
Potassium 40	1.3×10^9	4.5×10^{-5}	3.0×10^2
Rubidium 87	5.0×10^{10}	3.4×10^{-5}	2.8
Radium 226	1.6×10^3	8.0×10^{-14}	$(3.6–.25) \times 10^{-2}$
Thorium 228 (RdTh)	1.9	4.0×10^{-18}	$(0.25–1.4) \times 10^{-2}$
Radium 228 (MsTh)	6.7	1.4×10^{-17}	2.3×10^{-3}
Thorium 230 (Io)	8.0×10^4	6.0×10^{-13}	$(0.4–1.2) \times 10^{-2}$
Protactinium 231	3.2×10^4	5.0×10^{-14}	$(1.4–2.4) \times 10^{-3}$
Thorium 232	1.4×10^{10}	2.0×10^{-8}	$(0.02–1.0) \times 10^{-2}$
Uranium 235	7.1×10^8	1.4×10^{-8}	5.2×10^{-2}
Uranium 238	4.5×10^9	2.0×10^{-6}	1.15

formation of decaying uranium has some interesting effects. Salts of thorium 230 (the second long-lived daughter of uranium 238), like those of thorium 232, precipitate rapidly into the marine sediments which are being formed continuously. Precipitated thorium 230 eventually decays into radium 226, which partially dissolves and is carried upward in the water where its concentration progressively diminishes toward the surface. The concentration of radium 226 in the water thus provides an index of the mixing of the water.

Several radionuclides are produced in the upper atmosphere by cosmic rays, the most energetic radiations known. Primary cosmic rays originate outside the earth's atmosphere and secondary cosmic rays are produced by interactions between primary cosmic radiation and elements in the earth's atmosphere or in the earth itself. The nuclei of nitrogen, oxygen, and other atmospheric gases react with cosmic rays to produce radionuclides. Two of the most interesting and abundant of these are carbon 14 and tritium, which are also formed by the explosions of nuclear weapons. Other radionuclides formed by cosmic rays are beryllium 7, beryllium 10, sodium 22, silicon 32, iodine 129, and rhenium 187/osmium 187. The distribution of several of these cosmic-ray produced radionuclides in the sea can be used in studies of currents, mixing processes, and geochronology of the ocean.

Uses of naturally occurring radionuclides. The patterns of large-scale circulation and rates of water movement are often difficult to study. By correlating knowledge gained from studies of the distribution of various radioactive isotopes in the sea with knowledge obtained from classical oceanographic studies, however, it has been possible to improve our understanding of ocean circulation. The relatively new technique of using radioisotopes in this type of study makes it possible to evaluate classical models on the circulation of water in the oceans and to establish rates of movement of the water in models.

According to Broecker (Ref. 2), an isotope used in large-scale mixing studies should: (1) occur in measurable quantities in all parts of the ocean; (2) vary in concentration from one location to the next in amounts greater than the limits of measurement; (3) vary in mode and rate of introduction into the system as a known function of time and space; (4) act as an infinitely soluble salt, and thereby move with the water; and (5) have identifiable characteristics that make possible a separation into those amounts originating from natural and artificial sources.

Four naturally occurring radioisotopes—carbon 14, tritium, silicon 32, and radium 226—possess enough of the above mentioned characteristics to make them useful for circulation studies in the ocean. Of these, naturally occurring carbon 14, which has a half-life of 5600 years, has received the greatest attention. Carbon 14 reaches the ocean from the atmosphere in nearly constant amounts in both time and space. Carbon 14 measurements have a potential value in tracing water movements within oceanic reservoirs and particularly in the deep sea. The mechanism and the rate of production of tritium in the atmosphere are not understood as well as those for carbon 14.

Examination of material from the Discovery XVII satellite, which was exposed to intense solar flare radiation, has suggested that a significant fraction of tritium in the atmosphere is accreted from the sun. Because of its 12-year half-life, tritium is restricted to more rapid processes that occur in well mixed water. The application of silicon 32 to circulation studies is extremely difficult. Low levels of activity of this isotope make direct measurements in sea water time-consuming and expensive. Koczy (Ref. 5) has demonstrated that radium 226 can be used as a natural tracer. As mentioned previously, radium 226 is released from deep sea sediments into the water, and apparently this release is nearly constant with time and geographic location.

Artificially produced radionuclides. Fallout is the radioactive debris that settles to the surface of the earth after nuclear explosions. The amount and composition of fallout resulting from a specific explosion are determined by: (1) the height of the burst; (2) the size or power of the explosion; and (3) the composition of the fissionable material originally in the bomb. Since the oceans cover 71% of the earth's surface, it would be expected that about 71% of the worldwide fallout would fall into the oceans. Since about 61% of the northern hemisphere is covered by the oceans, approximately this percentage of the hemisphere's fallout would fall directly into the oceans. Also, some of the fallout on land is leached from the soil and carried by rivers to the ocean. Some of the artificial radionuclides that occur in the sea are shown in Table 2.

Although localized in distribution, another major source of artificial radioactivity is fuel processing and production plants, which discharge wastes into coastal waters or into rivers emptying into the sea. One such plant discharging radioactive effluent into the Irish Sea is the Windscale Works of the United Kingdom Atomic Energy Authority. This plant discharges about 90,000 curies of fission products per year. Over the past 10 years about 500,000 curies have been released (Ref. 6). Upon mixing with the receiving waters, radionuclides in the effluent from the Windscale plant may assume different chemical and physical states, but, in general, the distribution of radioactivity follows the known patterns of circulation and currents along the coast.

The Hanford Atomic Production Plant in the U.S. discharges radioactivity directly into the Columbia River. About 1000 curies per day of neutron-induced radionuclides are carried by the river to the Pacific Ocean 350 miles downstream (Ref. 11). Most of this radioactivity is due to radionuclides of relatively short half-lives, including neptunium 239 (half-life, 2.3 days), which has been detected at the mouth of the river. This radionuclide has little environmental significance, since it is not passed through the food chain and has such a short half-life (Ref. 4). Two other radionuclides, zinc 65 and chrom-

TABLE 2. ARTIFICIAL RADIONUCLIDES THAT HAVE BEEN IDENTIFIED IN THE SEA; CONCENTRATIONS HAVE VARIED WITH TIME AND SPACE DUE TO MAN'S ACTIVITIES

Nuclide and Daughter	Half-life
Fission Products	
Strontium 89	50.4 days
Strontium 90; yttrium 90	28 years; 64.4 hours
Yttrium 91	58 days
Zirconium 95; niobium 95	63.3 days; 35 days
Ruthenium 103; rhodium 103 m	41.0 days; 54 minutes
Ruthenium 106; rhodium 106	1.0 years; 30 seconds
Tellurium 129 m; tellurium 129	33 days; 74 minutes
iodine 129	1.6×10^7 years
Cesium 137; barium 137 m	30 years; 2.6 minutes
Cerium 141	32.5 days
Cerium 144; praseodymium 144	290 days; 17.5 minutes
neodymium 144	2.0×10^{15} years
Promethium 147; samarium 147	2.5 years; 1.3×10^{11} years
Iodine 131	8.05 days
Barium 140	12.8 days
Induced Nuclides	
Phosphorus 32	14.3 days
Sulfur 35	87.1 days
Chromium 51	27.8 days
Manganese 54	300 days
Iron 55	2.94 years
Iron 59	45.1 days
Cobalt 57	270 days
Cobalt 58	72 days
Cobalt 60	5.27 years
Zinc 65	245 days
Cadmium 113 m	14 years

ium 51, with longer half-lives, however, are of concern; both enter the Pacific Ocean and are present in detectable amounts in pelagic organisms some distance offshore. Zinc 65 is found generally in most marine organisms in the Columbia River estuary, and chromium 51 is concentrated in the lower trophic levels (Ref. 9).

It is more difficult to evaluate and control the release of radioactive wastes into the marine environment from mobile reactors, such as those in ships and submarines, than from land-based reactors. The geographical locations of these are selected with some consideration to the safety of man and his environment, and the radioactive wastes are released into a relatively restricted area. Ships and submarines, however, frequent ports and harbors where much damage would result if an accident should occur, and the wastes are dispersed over large areas of the open oceans where the possibility of danger to man and marine organisms is greatly reduced. The wastes from these mobile reactors usually consist of approximately 10 curies for each start-up and about 400 curies for a 50-day operation. These wastes become associated with the ion-exchange resins in the circuits. The radionuclides in these wastes are predominantly neutron-induced, rather than fission products. Radionuclides that may be present in the waste from nuclear-powered ship reactors are chromium 51, iron 55 and 59, cobalt 60, copper 64, zinc 65, strontium 90, zirconium 95, niobium 95, ruthenium 106, iodine 131, cesium 137, cerium 144, and tantalum 182.

Radioactivity released into the marine environment is a potential hazard to man, because he can be exposed internally by eating seafoods containing radioactivity (Table 3) and exposed externally to radioactivity that accumulates on the shore, fishing gear, and other objects that have been immersed in the water. The amount of radioactive material accumulated by seafood organisms in the vicinity of nuclear reactors is monitored to ensure the maintenance of safe levels of radioactivity. For example, the amount of radioactivity that has been disposed of at Windscale has been limited to a certain extent by the amount of radioactivity accumulated by *Porphyra* (Ref. 7), a seaweed used to make a delicacy known as laverbread. In instances such as this, the potential danger is determined by the amount of radioactivity accumulated by the seafood organisms and the quantity of the seafood eaten by an individual. Thus, an organism that concentrates less radioactivity, but is eaten in much larger quantity can be more dangerous than one that accumulates much more radioactivity, but is eaten in smaller quantities. Extensive monitoring of radioactive waste from Windscale during the past ten years, however, has shown no systematic increase in the levels of radioactivity in the environment (Ref. 6).

T. R. Rice
T. W. Duke

References

1. Arnold, J. R., and Martell, E.A., "The circulation of radioactive isotopes," *Sci. Amer.*, **201**(3), 85–93 (1959).
2. Broecker, W., "Radioisotopes and large-scale oceanic mixing," in N. M. Hill (Ed.), "The Sea," Vol. 2, Interscience Publishers, New York, 1963.
3. Chipman, W. A., "Food chains in the sea," in R. Scott Russell (Ed.), "Radioactivity and Human Diet," 1966, Pergamon Press, New York.
4. Foster, R. F., "Environmental behavior of chromium and neptunium," in V. Schultz and A. W. Klement, Jr. (Eds.), "Radioecology," 1963, Reinhold Book Corp., New York; Amer. Inst. Biol. Sci., Washington, D.C.

TABLE 3. Estimation of Accumulation in Edible Portions of Seafood Organisms of Some Important Radioactive Nuclides Relative to the Concentration in Seawater[a]

Isotope	Edible Red Algae	Mollusks	Crustaceans	Fish
Fission products				
Strontium 90	10^{-1}–1	10^{-1}–1	10^{-1}–1	10^{-1}–1
Cesium 137	1 –10	10 –10^2	10 –10^2	10 –10^2
Cerium 144	10^2	10^2	10^2	10 –10^2
Zirconium 95	10^2–10^3	10 –10^2	10^2	1 –10
Niobium 95	10^2–10^3	10^2	10^2	1 –10
Ruthenium 106	10^3	1 –10^3	1 –10^3	1 –10
Induced activities				
Zinc 65	10^2	10^3–10^5	10 –10^4	10^3–10^4
Iron 55	10^3–10^4	10^2–10^4	10^2–10^4	10^2–10^4
Cobalt 60	10^2	10 –10^3	10 –10^3	10 –10^2
Manganese 54	10^3	10^3–10^4	10^2–10^4	10^2–10^3
Chromium 51	10 –10^3	10^3	10^3	10^2–10^3

[a]From Chipman (Ref. 3).

5. Koczy, F. F., "Natural radium as a tracer in the ocean," *Proc. 2nd Int. Conf. Peaceful Uses At. Energy*, **18**, 351–357 (1958).
6. Longley, H., and Templeton, W. L., "Marine environmental monitoring in the vicinity of Windscale," in "Radioecological Monitoring of the Environment," 1965, Pergamon Press, New York.
7. Preston, A., and Jefferies, D. F., "The assessment of the principal public radiation exposure from, and the resulting control of, discharges of aqueous radioactive waste from the United Kingdom Atomic Energy Authority factory at Windscale, Cumberland," *Health Phys.*, **13**, 477–485 (1967).
8. Seymour, A. H., Held, E. E., Lowman, F. G., Donaldson, J. R., and South, D. J., "Survey of Radioactivity in the Sea and in Pelagic Marine Life West of the Marshall Islands, September 1–20, 1956," UWFL-47, 1957, University of Washington Applied Fisheries Laboratory, Seattle, Washington.
9. Seymour, A. H., and Lewis, G. B., "Radionuclides of Columbia River Origin in Marine Organisms, Sediments, and Water Collected from the Coastal and Offshore Waters of Washington and Oregon, 1961–1963," UWFL-86, 1964, University of Washington Laboratory of Radiation Biology, Seattle, Washington.
10. Strutt, R. J., "On the distribution of radium in the earth's crust. Part II—Sedimentary rocks," *Proc. Royal Soc.*, London, A**78**, 150–153 (1906).
11. U.S. Atomic Energy Commission, "Major Activities in the Atomic Energy Programs, January-December 1959," 1960, Superintendent of Documents, Government Printing Office, Washington, D.C.

Radioactive Waste in Seawater

The oceans are naturally contaminated with radioactive isotopes, although the concentrations present are very low and often difficult to discern from the background levels of natural radiation. This natural radiation mainly originates from potassium 40 whose concentration in seawater is about $300\mu\mu c$/liter. Other radioisotopes such as rubidium 87, bismuth 209, and isotopes of the uranium, thorium, and actinium decay series also occur, but in much lesser concentrations; the total contribution of all these isotopes is less than $10\mu\mu c$/liter of seawater. Other naturally occurring radioisotopes are those formed by the action of cosmic rays on the upper atmosphere of the earth and these enter the oceans in rain; belonging to this group are hydrogen 3, beryllium 7, beryllium 10, carbon 14, sodium 22, silicon 32, sulfur 35, iodine 129, rhenium 187, osmium 187. The concentrations of these isotopes in ocean waters are very low indeed although that of carbon 14 is measurable, around $3\mu\mu c$/liter, but there are other sources of this isotope, for example, atomic weapons and the burning of fossil fuels.

As far as the major areas of the oceans are concerned the most important source, so far, of radioactive wastes has been from weapons testing via the mechanisms of radioactive fallout. The explosion of an atomic weapon produces large concentrations of short-lived radioisotopes that present an immediate radiation hazard to all organisms living close to the site of the explosion. A long-term hazard is also created by the injection of debris from the explosion into the upper atmosphere where it is resident for some time and circulates so that when it is deposited as fallout it appears on the earth's surface in areas distant from that of the original explosion. The main hazard in this material arises from the presence of long-lived isotopes such as strontium 90 and cesium 137.

The major source of contamination of certain restricted sea areas is the nuclear power industry. The fuel elements in a nuclear power station are generally stored in freshwater ponds, the water acting as a shield against the radiation hazard of the elements. The metal cans of the elements are subjected to varying amounts of corrosion in the ponds with the subsequent release of small amounts of radioisotopes into the pond water. This contaminated water is periodically discharged and if the installation is on the coast or close to a river, the pond water enters the sea carrying with it small quantities of radioisotopes. These radioisotopes are of two types, neutron induced radioisotopes and fission products. The former are produced when stable isotopes present in the canning material of the fuel elements are bombarded with neutrons in a reactor.

The contribution of nuclear generating stations to the overall radioactive contamination of the seas is usually small, but much greater contributions are made by the chemical processing plants which reprocess the fuel elements from these stations.

Still another source of radioactive wastes is becoming of increasing importance, namely nuclear-powered ships and submarines. The discharging of wastes from these ships requires extremely strict control because the ships are mobile. The radioisotopes in the wastes are predominantly neutron induced and are: chromium 51, manganese 54, iron 55, iron 59, cobalt 58, cobalt 60, copper 64, zinc 65, iodine 131, and tantalum 182.

Treatment and Disposal of Wastes. In the case of radioactive fallout the only method of control is the cessation of weapons testing. Radioactive wastes from nuclear generating stations, because of their low concentration, are usually discharged to the natural environment but under strict supervision. The same controls apply to the discharging of larger amounts of wastes such as produced at Hanford and Windscale. High-active liquid wastes, with concentrations of hundreds of curies per liter, are stored in large stainless steel or concrete tanks on land. The low-active liquid wastes are treated in various ways, depending on their volume and levels of radiation. Some are evaporated and stored, others are temporarily stored to allow some of their radioisotopes to decay before discharge, while others are treated and the major portion of the radioisotopes removed by a chemical flocculation treatment, the decontaminated liquid being discharged. A certain amount of packaging of solid wastes is done, and these are dumped in deep water seaward of the continental shelves. The wastes produced in nuclear-powered ships are usually associated with the ion exchange resins in the circuits although about 10 curies of liquid wastes are occasionally discharged during start-up. These latter discharges are liable to take place in a harbor and so are carefully supervised.

The quantities of radioactive wastes discharged into a given sea area have to be strictly regulated to ensure that no radiation hazard is suffered by the human population. Radiation hazards arising on contaminated shores, affecting the recreational facilities of the beaches, and on contaminated fishing nets can be measured directly or calculated from the radioactive content of the material. The hazard, however, involved in eating contaminated seafoods, such as fish and shellfish, is often more difficult to assess, and requires a knowledge of the Recommendations of the International Commission on Radiological Protection. The Commission recommends quantities of the different radioisotopes that a person may be allowed to consume in his food or drink each day over long periods without adverse effects. Knowing these quantities, studies of the marketing of food organisms caught in the relevant sea area must be made to discover who eats it and how much; the local population may be the largest consumers but in some cases seafood may be regularly exported from one area to another so that the people likely to suffer the greatest hazard may live far from the contaminated area. Having obtained realistic values for the average daily intake of seaweed, fish, shellfish, or other material of interest, values for the maximum permissible concentrations (M.P.C.) of radioisotopes in each material can be derived:

$$\text{M.P.C. in food} = \frac{\text{permissible intake } (\mu\mu c) \text{ per day}}{\text{weight (g) of food consumed per man per day}}$$

Dispersion and Concentration of Wastes. Radioisotopes introduced into the marine environment, whether in fallout material or as wastes from an industrial site, interact with the chemical and physical environment (Fig. 2). The behavior of

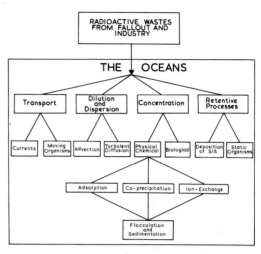

Fig. 2. Processes affecting the dispersion of radioactive wastes in the marine environment. (After Waldichuk and Mauchline and Templeton, 1964.)

radioisotopes on entering the sea depends on their chemistry and on the chemical form in which they are introduced to the environment. The factors that govern the behavior of radioisotopes and their corresponding stable isotopes in the sea are the same except that the radioisotopes are subject to radioactive decay. The isotopes are subject to the normal processes of dilution and dispersion, to concentration, and to transport. Their dispersion away from the site of introduction may be limited by retention on sand and silt or in populations of static organisms. If there is good exchange of water between the local sea area and its surrounding sea areas, then the wastes are transported away from the site of introduction and subject to dilution and dispersion by the normal processes of advection and turbulent diffusion. Radioisotopes may be transported by migrating organisms, both horizontally and vertically, but just how significant this form of transport is when compared to that originating from the physical and chemical processes of the sea is unknown.

Concentration of radioisotopes is effected by both the physical chemical environment and by the organisms; accumulation by the organisms is discussed below. Three main processes in the physical chemical environment are responsible for concentrating radioisotopes: adsorption, coprecipitation, and ion-exchange. Some radioisotopes are less liable than others to concentration by these processes and the following tend to remain in solution in sea water: phosphorus 32, chromium 51, strontium 89, strontium 90, antimony 125, iodine 131, cesium 137, radium 226, and uranium isotopes.

The great majority of fission products and other hazardous radioisotopes, however, are easily adsorbed, co-precipitated, or involved in ion exchange processes: manganese 54, iron 55, iron 59, cobalt 57, cobalt 58, cobalt 60, yttrium 90, zirconium 95, niobium 95, molybdenum 99, ruthenium 103, ruthenium 106, cerium 144, praseodymium 144, bismuth 207, actinium 227, protactinium 231, plutonium 239, and thorium isotopes.

Isotopes in this latter group, therefore, tend to contaminate the surfaces of sand and silt particles, diatom frustules, macroalgae, and marine invertebrates and fish. Adsorption on organic and inorganic particles makes them easily available to filter and deposit feeding organisms because these particles either remain for some time, in suspension or are deposited on the seabed.

The differences in the behavior of these two groups of radioisotopes are clearly seen when their distribution within the whole environment is examined, especially in coastal areas. Isotopes in the first group occur primarily in the water with much smaller proportions associated with the biomass of organisms, the suspended matter in the water, and the sediments. Isotopes in the second group are predominantly associated with suspended material in the water and with the sediments with very much smaller proportions present in the biomass and the water itself. Since isotopes in the second group are adsorbed on particulate matter, and sediment accumulations can occur in areas such as bays or river mouths where sediments are being deposited, and it has been demonstrated that washing of sediments contaminated with radioisotopes such as ruthenium 106, does not remove the isotope. These accumulations take place in sublittoral as well as tidal areas, but those in tidal areas are affected by additional factors. The sublittoral organisms are under water and so are subjected to a relatively constant radioactive environment, but this is not so in tidal areas. In tidal areas the distribution of contaminated sediments and the accumulation of radioisotopes by organisms can be affected by: (1) wind action on contaminated sediments, especially silt, lifting them from near low-water mark and depositing them near high-water mark; (2) the vertical position of organisms on the beach, and, therefore, their coverage time by contaminated water; (3) movement of organisms on the beach that alter their coverage time; and (4) fresh water seepage and rainfall, assuming that their radioactive contents are less than that of the seawater, causing some decontamination of the organisms and substratum by washing.

Accumulation by Marine Organisms. A quantitative expression of the degree of accumulation of a radioisotope is the "concentration factor" (C.F.). This is defined as:

$$\text{C.F.} = \frac{\text{concentration } (\mu\mu c/g) \text{ of isotope in fresh organism}}{\text{concentration } (\mu\mu c/ml) \text{ in seawater}}$$

The concentration factor for a radioisotope is the same, for the majority of chemical elements, as that for the corresponding stable isotopes. Consequently, a radioisotope and its stable counterpart will be present in the tissues of the organism, after equilibrium has been attained, in the same ratio of concentrations as present in the outside medium. This is so for the majority of the radioisotopes under discussion here, but the biochemical systems can discriminate in favor of lighter or heavier isotopes of some of the lighter elements, causing differences in the ratios inside and outside the tissues.

Radioisotopes (and stable isotopes) can enter the tissues of an organism by various routes (Fig. 3). Entry can be by way of contaminated food or directly from the water across semi permeable membranes or epithelial areas such as those present in the gills and certain regions of the "skin" of many organisms. The rate of absorption will depend on the species of organism, the route of entry, and the radioisotope concerned. The radioisotope will be accumulated in a tissue only if its inactive isotope or a chemically similar element is normally present. For example, if a tissue accumulates manganese and manganese 54 is present in the water, then it will be absorbed along with stable manganese and the two isotopes occur, once a steady state is attained, in the same ratio of concentrations in the tissues and the water. Calcium and strontium

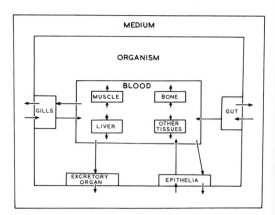

Fig. 3. Schematic presentation of the passage of an element into and out of the tissues of an organism. The arrows denote movement (flux) of the element between tissues of the organism or between the organism and the external environment. The rate of flux depends on the element and the tissue concerned, and on the relation to steady state of the concentrations of the element within the tissues and in the external environment. (After Mauchline and Templeton, 1964.)

TABLE 4. CONCENTRATION FACTORS FOR RADIOISOTOPES BY TISSUES OF MARINE ORGANISMS[a]

Isotope	Algae	Invertebrates Soft Parts	Skeletal	Soft Parts	Vertebrates Skeletal	Plankton
P^{32}	10,000	10,000	10,000	40,000	2×10^6	
Sc^{46}	1,500–2,600					
Cr^{51}	300–100,000	70–1,000				
Mn^{54}	6,500	10,000				
Fe^{55}, Fe^{59}	20,000–35,000	10,000	100,000	2,000 1,000	2,000 5,000	750 200–140,000
Co^{57}, Co^{58}, Co^{60}	450	whole 500–3,000		30–1,000 160	1,500	
Zn^{65}		whole 5,000–40,000		1,000	300–3,000	1,000
As^{76}	200–6,000	whole 20,000				
Sr^{89}, Sr^{90}	1–40	1–25	180–1,000	0.1–2.0	100–200	9
Y^{90}, Y^{91}	100–1,000	20–100	300	5		
Zr^{95}	350–1,000					1,500–3,000
Nb^{95}	450–1,000	whole 200,000				
Mo^{99}	10–100	100		20		
Ru^{103}, Ru^{106}, Rh^{106}	15–2,000	2,000		3		600–3,000
Sb^{125}		100(?)		100(?)		50
I^{131}	3,000–10,000	50–100	50	7–10		
Cs^{137}	1–100	10–100	0	10–100		0–5
Ce^{141}, Ce^{144}	300–900	300–2,000	45	12		7,500
Bi^{209}		500		50		
U^{234}, U^{235}, U^{238}				<20		
Pu^{239}	10		400	13		

[a] After Mauchline and Templeton, 1964.

90 or potassium, rubidium and cesium 137 behave similarly chemically and occur together in tissues. Here, however, the ratio of concentrations within the tissue may vary from that in the seawater because the biochemical systems distinguish between the elements. The strontium-to-calcium atom ratio in seawater is 9×10^{-3}, whereas in most tissues of marine organisms (plant and animal) it is less. One major exception is the brown seaweeds (Phaeophyceae), where in fact the tissues accumulate strontium in preference to calcium. Some fission products, e.g., ruthenium, are not accumulated within tissues to any great degree, but most of the contamination measured is on the exposed surfaces such as the skin and internal walls of the gut; and, of course, concentrations can be found in such organs as the liver of vertebrates or the digestive diverticula of invertebrates.

Some generalized concentration factors for radioisotopes are shown in Table 4, but reference should be made to Polikarpov (1966) for detailed information on many species. Fish muscle is not a notable accumulator of fission products and in consequence the concentrations of fission products in other materials in the sea usually limit the amount of wastes discharged long before levels of contamination, dangerous to man, are reached in fish. This is not so for neutron-induced radioisotopes, and the levels of phosphorus 32 or zinc 65 in fish can be the limiting factors. Shellfish are notable accumulators of neutron induced radioisotopes and where the whole shellfish is eaten, digestive tract as well, fission products present in them can also be dangerous; consequently, discharges of wastes near shellfish grounds have to be carefully monitored.

Effect on Organisms. The study of the effects of radiation on marine organisms has been neglected and little is known about the long-term effects of increasing the background levels of natural radiation. Most of the work in this field has involved very high dosages of radiation, far in excess of any likely to occur in the natural environment except in the region of a nuclear explosion. In general, algae can survive much longer dosages of radiation than animals, and simpler forms of animal life much larger dosages than higher forms. All the present information suggests that the levels of radiation allowed to accumulate in the environment by the operators of the nuclear power industries are not producing noticeable adverse effects on the marine community. This does not, however, exclude the possibility that subtle, long-term changes have been initiated by the increased levels of radiation, but only future observations and experiments will discover these.

J. Mauchline

References

Mauchline, J., and Templeton, W. L., "Artificial and natural radioisotopes in the marine environment," in Barnes, H. (Ed.), "Oceanography and Marine Biology; An Annual Review," Vol 2, 1964, George Allen and Unwin Ltd., London.

Polikarpov, G. G., "Radioecology of Aquatic Organisms," 1966, North-Holland Publishing Co., Amsterdam; Reinhold Book Corp., New York.

Effects on Fisheries

Most commercially important species of marine organisms are harvested in coastal waters where radiation from man's peaceful use of radioactive materials is most likely to reach highest levels. To date the greatest concentrations of radioactivity in aquatic environments have occurred in the open ocean in areas used as test sites for nuclear bombs. In addition to radiation from the relatively recent artificial radioactivity, marine organisms have been exposed to radiation from naturally occurring radioactivity for as long as life has existed in the oceans. Over the ages, organisms have been exposed to less radioactivity in the sea than on land because of the shielding capacity of water, and thus marine organisms possibly are more sensitive to it. In limited areas in the oceans, artificial radioactivity already has reached, on occasion, concentrations high enough to be deleterious to marine organisms, especially in the vicinity of radioactive waste disposal by sea burial.

The effects of radiation on marine organisms are very difficult to evaluate in the natural environment, e.g., near sites of nuclear bomb tests. Dead fish were observed after the detonation of nuclear devices at Bikini-Eniwetok, but it was believed that they died from blast effects (Ref. 5). Some fish in the immediate vicinity of the detonation of a large device, however, can be expected to receive lethal levels of radiation. If fish die as a result of radiation from radioactivity taken into their body or from radioactivity in the water, deaths would be expected over an extended period, as has occurred in laboratory experiments when organisms that were exposed simultaneously to one lethal level of radiation did not all die at the same time. It was also observed in the laboratory that greater differences in levels of radiation caused greater differences in time of death. The dose of radiation received by fish in the vicinity of a nuclear explosion would of course depend upon their distance from the explosion and the depth at which they were swimming. This gradient of radiation would be expected to extend the deaths over a long time and over a considerable area of the ocean. During the period between initial exposure to radiation and death, fish could migrate various distances. Although dead fish have not been seen after a nuclear test (except near the bomb site), the possibility remains that most deaths have been unobserved, especially since fish weakened by radiation might be devoured by predators or succumb to other environmental stresses, seemingly unrelated to radioactivity in the ocean.

Although it is difficult to make in situ observations on the effects of radiation on organisms in the marine environment, at least one instance of adverse effects on marine organisms has been recorded; it resulted from the uptake of radioactive isotopes. This observation was made after the detonation of a nuclear bomb in the Marshall Islands. In 1958, Gorbman and James (Ref. 10) found evidence that herbivorous fish concentrated radioactive iodine from seaweeds that had previously accumulated this radioisotope from seawater. Carnivorous fish then further concentrated the iodine to such high levels that their thyroid glands were destroyed. Although radioactive iodine is accumulated by marine organisms, it is usually not of serious concern except in areas receiving close-in fallout. Since the radioactive decay rate of iodine 131 is relatively rapid (half-life, 8 days), and since iodine 131 is diluted with stable iodine in seawater, marine organisms seldom accumulate enough of it to result in detectable injury.

Beneficial changes in marine organisms from exposure to radiation also are difficult to detect and evaluate in the natural environment. For instance, to detect an increase in growth rate, physiological vigor, or survival for organisms in the sea usually requires observations over a considerable period of time. Even though this time-consuming evaluation is usually required, Blinks (Ref. 1) has reported an effect of radiation on plant species that might be construed as beneficial. Marine algae in the most radioactive area on the Bikini Island reefs, one year after a nuclear bomb blast, had greater catalase activity than algae from a nonradioactive area. Blinks tentatively explained this phenomenon as an enzyme adaptation that protected the organism from the hydrogen peroxide produced by radiation.

Even though few observations have been made of the effects of radiation on animals and plants in the marine environment, additional knowledge has been gained from experiments conducted in the laboratory. Before considering data obtained from laboratory experiments, however, the origin of radioactivity in the marine environment, the characteristics of radiation, and the sensitivity of living material to radiation will first be explored.

Ionizing Radiation. Marine organisms are now subjected to radiation originating from radionuclides that occur naturally and from those that are artificially produced. Natural radionuclides have existed since the earth was formed and have found their way to the oceans, in part at least, from the weathering of rock. The principal categories of natural radionuclides are: (1) primordial radionuclides and their decay products (daughters); and (2) radionuclides resulting from reactions between cosmic rays and elements in the atmosphere or in the earth. More than a dozen long-lived naturally occurring radionuclides have been identified. Most of the natural radiation in the seas originates from only three of these radionuclides: potassium 40 (which alone accounts for more than 90% of the natural radiation); uranium 238, and thorium 232. Artificial or man-made radionuclides have been added to the sea in measurable amounts since 1945, primarily through the explosion of nuclear weapons and through planned or accidental discharges from nuclear installations. Whereas natural radionuclides are more or less evenly distributed throughout the oceans, concentrations of artificially produced radionuclides vary with time and location.

Radioactive substances emit one or more of the three different kinds of radiation: alpha particles, beta particles, and gamma rays. The rate at which a radioactive isotope disintegrates or decays is completely independent of external conditions. Thus, a constant proportion of the radioactive atoms present will disintegrate in a given period of time independently of the total number present. Approximately 30 different natural radioisotopes emit alpha particles and about the same number emit beta particles, whereas only about half this number emit gamma rays. Ionizing radiation is capable of penetrating matter, it possesses great energy, and most types are electrically charged. This radiation is referred to as "ionizing radiation" because of its capacity to eject planetary electrons from atoms—resulting in ions.

Response of Organisms to Radiation. A quantitative relation exists between the extent of damage to an organism and the amount of radiation it receives. Dose can be plotted against effect in two ways (Fig. 4), but opinions differ as to which curve shows the correct relation. The usual biological response to most physical and chemical factors is shown by the dotted line of Fig. 4. As can be seen, there is no effect until a certain dose is received. This is called a "threshold response." Theoretically, any dose less than threshold will cause no effect. The "nonthreshold" relation between dose and effect is illustrated by the solid line, which runs through the origin. Since the relation is linear, any dose, however small, will have a measurable effect directly proportional to the size of the dose. It is assumed now that genetic effects of radiation follow this type of dose-effect curve.

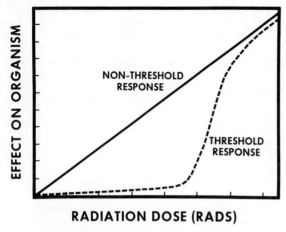

Fig. 4. Two concepts of the response of organisms to radiation. (After Comar.)

Damage from radiation takes place at the most fundamental level—in the cells of organisms. The process by which this damage occurs, however, remains more or less a mystery. Radiation affects cells in many ways: it may kill them, change their rate of growth and division, and alter their genes and chromosomes so that their progeny inherit different characteristics. Cells that are dividing rapidly are more sensitive to radiation than less active ones. Since the well-being of a cell depends on its metabolism, it is here that investigators have looked for the basic effects of radiation. The effects result from the ionizing action of radiation on atoms and molecules within the cell. Ionization involves the removal of an electron from a neutral atom, creating a positive ion. This released electron reacts with another atom to form a negative ion. Thus ion pairs result from ionization. It is believed that ionization is the major cause of injury to protoplasm and that the damage is proportional to the number of ion parts produced. Radiation ionizes atoms in its path impartially, whether contained in water, proteins, enzymes, hormones, or other biologically important substances. This change in structure of biochemically important molecules disrupts the structural components of the cell and reduces the cell's metabolic efficiency.

A comparison of data on the effects of radiation on organisms indicates that phyla differ in their response (Fig. 5). The more primitive organisms are usually more resistant than the more complex animals, such as fish. Variation between related species is considerable, however. For example, 50% of the snails, *Radix*, exposed to 10,000 roentgens (R) died after about 1 month, whereas another group of snails, *Thais*, exposed to the same dose lived for about 6 months (Ref. 2). Price (Ref. 15), who irradiated two species of shellfish, found that about 93,000 R killed 50% of the oysters, *Crassostrea virginica*, in 34 days and 50% of the clams, *Mercenaria mercenaria*, in 38.5 days. This similarity in survival time would appear to indicate a similar response to radiation for the two species. If LD-50's (the amount of radiation needed to kill 50% of the organisms) are calculated earlier or later than 30 days after irradiation, however, the tolerances of the two species are not similar (Ref. 19). Oysters have an LD-50 almost twice that of clams 20 days after irradiation and only one-fourth that of clams 40 days after irradiation.

Different stages in the life cycle of a species also differ in sensitivity to radiation (Table 5).

TABLE 5. AMOUNTS OF RADIATION REQUIRED TO KILL 50% OF THE RAINBOW TROUT, *Salmo gairdneri*, IRRADIATED AT VARIOUS STAGES IN THEIR LIFE CYCLE[a]

Stage in Life Cycle	LD-50 (R)
Gametes	50–100
1 cell	58
32 cell	313
Germ ring	454–461
Eye	415–904
Adult	1,500

[a] After Welander (Refs. 16, 18).

Gametes and eggs through the one-cell stage are very sensitive to radiation, for example, during mitosis of the single cell of silver salmon, *Oncorhynchus kisutch* the LD-50 is only 16 R (Ref. 6). The sensitivity of rainbow trout (*Salmo gairdneri*) decreases with increasing age (Refs.

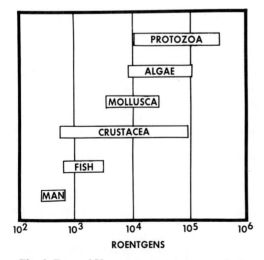

Fig. 5. Dose of X rays or gamma rays required to kill 50% of organisms. (After Donaldson.)

16, 18), and this relation appears to be the general rule for other species.

Since sensitivity to radiation is correlated with the metabolic rate of the cell, it is not surprising that the dormant eggs of aquatic invertebrates are especially resistant. The eggs of the brine shrimp, *Artemia salina*, that have been placed in water so that embryonic development is resumed are more than twice as sensitive to radiation as the dry dormant eggs (Ref. 2). Results of experiments on the effects of radiation (strontium 90-yttrium 90), on the eggs of the plaice have been inconsistent: Fedorov et al. (Ref. 7) reported very high sensitivity to low concentrations, whereas Brown and Templeton (Ref. 3) found no effect at concentrations as high as 10^{-4} c/l (curies per liter). Polikarpov and Ivanov (Ref. 14) found that a concentration of 10^{-10} c/l of strontium 90-yttrium 90 was sufficient to cause a significant number of abnormal larvae in five species of marine fish. Concentrations of radioactivity in the sea have sometimes reached or exceeded this level. Strontium 90 has been reported at concentrations of 10^{-12} and 10^{-11} c/l in the Pacific Ocean and Irish Sea (Refs. 12, 11); in 1959 the waters of the Irish Sea contained as much as 10^{-11} c/l cesium 137, 10^{-10} c/l cerium 144, 10^{-10} c/l zirconium 95, and 10^{-9} c/l ruthenium 106 (Ref. 11).

The rates of increase in both length and weight of fish can be slowed by radiation. Irradiation of rainbow trout eggs retarded the development of the young and increased the number of abnormalities (Ref. 16). Somatic damage was proportional to the amount of radiation received, and the greatest amount of damage was in tissues that were growing and dividing rapidly (Ref. 17). Also, the growth of young rainbow trout from irradiated parent stock was considerably slower than that of the young from unirradiated parents (Ref. 9).

Influence of Radiation on Populations. Populations of fishery organisms exposed to radiation may be adversely affected. If the deleterious effects of the radiation occur simultaneously with other unsatisfactory conditions, such as a shortage of food, excessive predation, or disease, the population naturally will suffer a greater reduction than it would if the radiation exposure occurred during a period of optimum conditions. Certain combinations of these deleterious factors may be sufficient to reduce the population to extinction or to the point of no recovery. Two primary considerations for evaluating effects of radiation on populations are: (1) the cumulative dose of radiation, and (2) the tolerance of the most sensitive stage. If exposure is continuous, the life span of organisms becomes critical in considering levels of radiation and their effects, because the cumulative dose received by an organism is equal to the level of radiation multiplied by length of exposure. The time of exposure of a population to acute doses of radiation is also important. For instance, if exposure to radiation occurs during spawning, the most sensitive stage in the life cycle, i.e., the egg, would be irradiated.

Fish eggs are very sensitive to radiation and may be more important in relation to radiation effects on populations than is presently believed by many investigators. For instance, the assumption has been made that when 50% or more of a population is lost, the commercial fishery for the species in the region concerned may cease to exist. The time required to lose half a fishery when different percentages of eggs are damaged, as calculated by Zaystev and Polikarpov (Ref. 20), is shown in Table 6. If radiation affects more

TABLE 6. TIME REQUIRED TO REDUCE POPULATIONS OF FISH 50%, ON THE BASIS OF THE PERCENTAGE OF EGGS DESTROYED[a]

Eggs Destroyed, %	Time (years)		
	Mullet	Horse-mackerel	Anchovy
5	55	68	100
10	29	34	50
20	15	18	25
30	11	13	17
40	8	10	13
50	7	8	10

[a]After Zaystev and Polikarpov (Ref. 20).

than 10% of the eggs, commercial catches of fish will decrease appreciably. These calculations, however, were based on the assumption that the percentage of eggs damaged each year would remain constant. There is no way to predict the effect at any given time because the amount of radioactivity in the water and other factors which influence the size of the population fluctuate irregularly. High levels of radioactivity in conjunction with low spawning success could result in a large reduction.

Radiation can affect not only a population of fishery organisms receiving the radiation but also the descendants of the population. Somatic changes to the individuals of a population, if they are severe enough and if they occur in sufficient numbers, can reduce the population immediately. Establishing the occurrence of a genetic change in a population is more difficult than detecting somatic damage. Most mutations are not advantageous to the animal and are eliminated because the animals with the mutations cannot compete with the existing population. For instance, if the genetic change results in slower movement, the affected animals would probably be the first to fall prey to predators, and the chances of the mutations becoming established in the population would be reduced. The probability of a mutation becoming established in a breeding population is highest in populations isolated from other similar breeding populations. Possibly populations of sedentary organisms (i.e., animals that only move over short distances), or highly motile animals with restricted breeding groups (such as the salmon, which spawns in the same stream in which it was reared), would be types of population in which new mutations would have a better chance of becoming established.

In addition, radiation damage can affect individual organisms—and thus populations—by reducing the rate of growth, shortening the life span, reducing fecundity, and changing behavior. How radiation affects marine organisms in these ways is not well understood. There is little doubt, however, that levels of radioactivity in the seas, particularly in coastal waters, will continue to increase as man increases the use of nuclear energy. That the maximum levels of radionuclides now allowed in natural waters may occasionally result in damage to aquatic organisms already has been demonstrated (Ref. 8). Also, catastrophic or emergency conditions may occur at any time and would definitely result in damage to marine organisms. For these reasons it is necessary that more research be carried out on the effects of radiation on marine organisms.

T. R. Rice
J. W. Angelovic

References

1. Blinks, L. R., "Effects of radiation on marine algae," *J. Cell. Comp. Physiol.*, **39** (suppl. 1–2), 11–18 (1952).
2. Bonham, K., and Palumbo, R. F., "Effects of X-rays on snails, crustacea, and algae," *Growth*, **15**, 155–188 (1951).
3. Brown, V. M., and Templeton, W. L., "Resistance of fish embryos to chronic irradiation," *Nature*, **203** (4951), 1257–1259 (1964).
4. Comar, C. L., "Fallout from nuclear tests," 1963, Understanding the Atom Series. U.S. Atomic Energy Commission, Oak Ridge, Tennessee.
5. Donaldson, L. R., "Evaluation of radioactivity in the marine environment of the Pacific Proving Ground," in S. H. Small (Ed.), "Nuclear Detonations and Marine Radioactivity," 1963, Kjeller, Norway.
6. Donaldson, L. R., and Foster, R. F., "Effects of radiation on aquatic organisms," in "The Effects of Atomic Radiation on Oceanography and Fisheries," 1957, Nat. Acad. Sci.-Nat. Res. Counc., Publ. 551, Washington, D.C.
7. Fedorov, A. F., Podymakhin, V. N., Kilezhenko, V. P., Buyanov, N. I., and Goloskova, E. M., "Radiation conditions in the fishery regions of the North Atlantic (June-August 1961)," in "Radiochemical and Ecological Studies of the Sea, 1964, U.S. Dept. Commerce, JPRS: 25, 966, Washington, D.C.
8. Foster, R. F., and Davis, J. J., "The accumulation of radioactive substances in aquatic forms," *Proc. 2nd Int. Conf. Peaceful Uses At. Energy*, **13**, 364–367 (1955).
9. Foster, R. F., Donaldson, L. R., Welander, A. D., Bonham, K., and Seymour, A. H., "The effect on embryos and young of rainbow trout from exposing the parent fish to X-rays," *Growth*, **13**, 119–142 (1949).
10. Gorbman, A., and James, M. S., "An exploratory study of radiation damage in the thyroids of coral reef fishes from the Eniwetok Atoll," in V. Schultz and A. W. Klement, Jr. (Eds.), "Radioecology," 1963, Reinhold Book Corp., New York; Amer. Inst. Biol. Sci., Washington, D.C.
11. Mauchline, J., "The biological and geographical distribution in the Irish Sea of radioactive effluent from Windscale Works 1959 to 1960," Report AHSB(RP)-R-27, United Kingdom Atomic Energy Authority, Production Group, Windscale (From Polikarpov.)
12. Miyake, Y., and Saruhashi, K., "Vertical and horizontal mixing rates of radioactive material in the ocean," in "Disposal of Radioactive Wastes," Vol. II, 1960, Int. At. Energy Agency, Vienna.
13. Polikarpov, G. G., "Radioecology of Aquatic Organisms," 1966, Reinhold Book Corp., New York.
14. Polikarpov, G. G., and Ivanov, V. N., "Injurious effect of strontium-90–yttrium-90 on early development of mullet, wrasse, horse mackerel, and anchovy," a translation of *Doklady Akad. Nauk SSSR*, **144**, 491–494 (1961).
15. Price, T. J., "Accumulation of radionuclides and the effects of radiation on mollusks," in "Biological Problems in Water Pollution," Trans. 3d Seminar, Publ. No. 999-WP-25, 1965, Robert A. Taft Sanitary Engineering Center, Cincinnati, Ohio.
16. Welander, A. D., "Some effects of X-irradiation of different embryonic stages of the trout (*Salmo gairdnerii*)," *Growth*, **18**, 227–255 (1954).
17. Welander, A. D., Donaldson, L. R., Foster, R. F., Bonham, K., and Seymour, A. H., "The effects of roentgen rays on the embryos and larvae of the chinook salmon," *Growth*, **12**, 203–242 (1948).
18. Welander, A. D., Donaldson, L. R., Foster, R. F., Bonham, K., Seymour, A. H., and Lowman, F. G., "The effects of Roentgen Rays on Adult Rainbow Trout," UWFL-17, 1949, University of Washington Laboratory of Radiation Biology, Seattle, Washington.
19. White, J. C., Jr., and Angelovic, J. W., "Tolerances of several marine species to Co 60 irradiation," *Chesapeake Sci.*, **7**(1), 36–39 (1966).
20. Zaystev, Yu. P., and Polikarpov, G. G., "The problems of radioecology of hyponeuston," in "Radiochemical and Ecological Studies of the Sea," U.S. Dept. Commerce, JPRS: 25,966, Washington, D.C.

REDFISH FISHERY

Deep, cold water seems to be the preferred habitat of the redfish for they are mostly taken in depths from 50 to 150 fathoms, although catches have been obtained in 200 fathoms. Its range in the northern regions appears limited only by depth, for it has been captured almost everywhere where hook and line or otter trawl fishing is carried on in waters deeper than 20 fathoms. In July 1912, the research vessel *Grampus* took redfish in 6 out of 7 hauls in 25 to 60 fathoms of water between Cape Ann and Penobscot Bay. Finding redfish in depths less than 40 fathoms is the exception rather than the rule.

Throughout the entire Gulf of Maine, along all the Nova Scotian shore, in the Gulf of St. Lawrence, on the famous Grand Bank of Newfoundland, off Labrador, and around Iceland redfish have schooled in dense populations undisturbed through passing ages. Both sides of the Atlantic, northward to Spitzenbergen, know this little fish, which even occurs off New Jersey, in deep water.

In some regions along the New England coast redfish are known as red perch. Unlike the perch, its color varies from a bright orange to flame red, with some individuals tending toward a grayish or brownish red. Black eyes contrast vividly with the brightly colored body. The large head makes up about a third of the entire body length.

Not all marine biologists are in agreement with the present classification of the *Sebastes* group. Because of its wide range there appear to be some racial variations. Color differences in fish taken from various grounds have been observed and loosely associated with depth. When exploitation of redfish extended into the Gulf of St. Lawrence the most striking variation was in eye diameter. This appeared to be associated with depth. Sampling has not been intensive enough to arrive at any definite conclusions. Landings from newly exploited grounds between 1940 and 1946 disclosed a variability in body structure, particularly in the diameter of the eye orbit and the chin appurtenance, or schnabel.

For many years the two names *Sebastes marinus* and *S. viviparous* included the known populations of North Atlantic redfish. Those inhabiting deep water were considered *S. marinus* while the shallow water form of the European coast was named *S. viviparous*. After the expansion of the commercial fishery into new grounds, and the subsequent observations of a wider variation in body form, Travin (Ref. 1) described the large-eyed, long-chinned form as a new species, *Sebastes mentella*. This was a first step toward clarifying the status of the many Sebastes groups. Andriiashev (Ref. 2) questioned the validity of *S. mentella* as a full species claiming insufficient knowledge of the variability of characteristics used to describe it. Racial studies of the entire Sebastes group continue in many research centers. At the request of the Committee on Research and Statistics of the International Commission for the Northwestern Atlantic Fisheries an important review of meristic and morphometric characteristics was begun and results are available (Ref. 3).

Many changes have occurred in Sebastes classification and are summarized in the following citations:

Perca marina (Linnaeus)—1758
Sebastes norvegicus (Cuvier and Valenciennes) —1829
Sebastes viviparous (Kroyer)—1844–45
Sebastes marinus (White)—1851
Sebastes fasciatus (Storer)—1854
Sebastes marinus viviparous (Jordan and Gilbert)—1883
Sebastes mentella (Travin)—1951
Sebastes marinus mentella (Andriiashev)—1954.

The diet of the redfish includes a wide variety of crustaceans, especially mysids, euphausiids, and decapod shrimps. Small mollusks and other invertebrates and small fishes round out its epicurean menu. In certain seasons, captured redfish have been taken gorged with small, bright red shrimp. Most individuals reaching the filleting tables are under 14 in. in length, for redfish are among the smaller foodfishes marketed in quantity. They contribute hardly more than ½ lb of flesh per individual and frequently less, yet nature has been so bountiful that millions of pounds of these fillets appear on American tables each year. Age studies have shown that the average growth is not more than 1 in. a year, diminishing as the fish reaches its later years. This slow growth rate is not a factor in its favor if overexploitation without some form of protection is applied to this fishery, especially as most fishes under 7–8 in. have been found to be immature.

Various methods for determining the age of redfish have been explored, and the use of otoliths transcends other means in reliability. Of several early redfish age studies, one of the first applied to western Atlantic redfish was completed in 1949 (Ref. 4) using scales from a selected area of the body. This was referred to as the "pectoral patch." Scales taken from this area had the largest and most uniformly shaped scales containing the highest circuli count.

Redfish scales are ctenoid on the outer edge and somewhat square to slightly oblong. Radii extend from the focus to the anterior margin. The initial western Atlantic age study indicated that redfish are extremely slow growing, averaging

Fig. 1. Redfish: *Sebastes mentella*, above, and *Sebastes marinus*, below.

less than a 25-mm increase in length per annulus to the ninth annulus. This was the same for fish taken in both the Gulf of Maine and off western Nova Scotia. Marketable redfish, those over 200 mm long, are mostly more than 10 years of age. A superficial examination of scales from larger fish indicate many to be upwards of 20 years old.

Because of the difficulty in separating the open and closed phases of annuli bands in scale reading, this method was superseded by an intensification in the use of otoliths. In 1959 another age study of Gulf of Maine redfish was made (Ref. 5), this time using the otoliths' opague and hyaline bands. The opague band begins to form in April, somewhat similar to scales, and the hyaline band in September. Together these form one annulus. Results of this study do not conflict with scale determinations, but do indicate that otoliths are more reliable in aging older fish. Other more recent studies do not disprove that Sebastes is a slow-growing, long-lived fish and if undisturbed might attain 40 or even more years.

Spawning time for redfish is during the spring months and into midsummer, depending upon the area. Perfunctory observations have indicated that in the grounds far to the east and north, such as the Gulf of St. Lawrence, redfish spawn before spring has really arrived and is completed before many of the grounds in the Gulf of Maine have even begun to spawn. During the period of spawning the decks of vessels bringing in heavy catches will be covered with immature fry or larvae, pressed from the parents' bodies. Many of these will live for hours when placed in a bucket of seawater, but survival is unsuccessful for these premature young and tiny fish. Protection during the spawning season has been considered, but its value is doubted.

By actual count, the female redfish produces up to 135,000 young each year, but 25,000–40,000 is more the average. This species is ovoviviparous, the offspring are extruded alive and, although there is some motility they are still at the whim of wind and wave for some time. When compared with the oviparous varieties of fishes, such as the cod, which may produce as many as 4 million ova, these may seem few indeed, but being born already hatched or liberated from the egg capsule, they have had the advantage of the protection afforded by the mother's body until able to move about under their own volition, to some extent. Adverse currents, however, may sweep masses of helpless fry offshore into deep waters where they cannot survive. Countless numbers become food for predators of many kinds, even the parents themselves.

Conservation planning for the maintenance of redfish stocks at a commercially beneficial level begins with a study of its life history. Because they do prefer deep water, research has been confronted with many barriers. The type of gear presently in use limits exploitation to the areas of the continental shelf less than 200 fathoms. Redfish captured in their deep-water habitat arrive at the surface with internal organs dislodged and eyeballs forced outward due to changes in pressure. As a result those that escape from the surfacing net or are culled because of small size are unable to descend again and soon die. This fact has prevented the successful tagging of redfish everywhere except in the Passamaquoddy Bay.

In the region between Campobello Island and Eastport, Maine, but principally along the shores of the latter, a unique population of redfish has provided an opportunity for a visual study of its habits. These fish appear along certain wharves during the hours of darkness, especially when the immense tides influenced by the Bay of Fundy are falling. By working throughout these descending tides and taking advantage of this local population's tendency to rise close to the surface in the nightime, redfish captured on barbless hooks have been marked with identifying tags.

During a period of about five years (1956–61) over 6,000 redfish have been tagged and, from those recaptured, the amount of growth increment has been recorded. Some fish have been retaken as many as five times. More than 2000 of those marked with the Petersen disk tag have been recaptured. After measurement recordings, these were returned to the water unharmed.

The growth increment of Eastport tagged fish was considerably less than for fish taken in the Gulf of Maine (Ref. 5). The average increase was only 1 mm per year. Otolith examinations of tagged fish exhibited a broad hyaline zone at the edge, which was not present in otoliths from untagged fish. This hyaline formation appeared to coincide with time of tagging (Ref. 6). After $2\frac{1}{2}$ years of diminished growth, the rate gradually increased to match that of the untagged. One explanation that appears plausible suggests that the bright new tag startles their preferred live food, the euphausiid shrimp, *Meganyctiphanes norvegica*. As the tag becomes less visible because of fouling by marine organisms, the euphausiids may be more readily captured.

Another interesting phenomenon was disclosed during this study at Eastport. The taggings were made at three different wharves and recaptured fish were taken only at the wharves where they had been tagged.

There is some reason to believe that this may be an isolated population and not entirely representative of the schools inhabiting so much of the northern seas. Also, there is no evidence that the Eastport fish move to offshore grounds at any time of the year.

In the entire Gulf of Maine the redfish is host to an ectoparasitic copepod, *Sphyrion lumpi*, which is rarely found on redfish in other waters. Since the inception of the redfish fishery these parasites have been a source of annoyance to the processors, who soon became aware of the percentages of infestation in catches from the various grounds.

Since it was impossible to use a manufactured tag for the study of redfish movements an effort was made to enlist the crustacean parasite. Redfish sampling of the commercial catch included the recording of the number and location of parasites per fish. This was done only when the catch was known to be "pure," or all from one narrow area. *Sphyrion lumpi*, then, served as a natural tag (Ref. 7).

From the beginning it was obvious that the percentage of infestation differed for each of the several Gulf of Maine grounds. As this degree of infection remained fairly constant, apparently the amount of intermixing is slight, even though some of the grounds are sufficiently close. Here, again, the Eastport study results are indicative of self-sustaining individual stocks.

In the earlier years of this fishery it was possible to run offshore from Gloucester a few hours and catch as much as 60,000 lb and even twice that amount in a single day. But over the years more vessels entered the fishery, the type of gear improved, and the empirical knowledge of the fisherman increased to such an extent that today, the catch-per-unit-of-effort has fallen to 8,000–10,000 lb for an average day's work in the same local waters. It is understandable that an untapped, virgin fishery will yield immense returns for a day of effort in its first years. As time goes on, production will tend to level off to a point where it can sustain itself through annual recruitment of new year classes, if the fishery is not exploited to the point of extinction. In the levelling-off process, and as a natural but unplanned "pruning" measure, larger boats requiring greater returns for a day's work are forced out of the fishery, or move on to other more productive grounds. By the same token, the inefficient fisherman who was carried along in the tide of plenty, must turn elsewhere for a livelihood.

New and ingenious instruments and devices have become a "must" on redfish vessels. Radar has given the fisherman a third eye by which he finds his way in fog or dark. Loran provides a beam upon which to ride to and from his favorite fishing grounds. Fathometers contribute a printed record of the bottom contour over which he is traveling, as well as the depths, and reveals the preferred haunts of the redfish. The screen of his fishscope in the boat's pilothouse informs him of the kind of fish he is moving over and an indication of the school's size. All these, while helping the fisherman, have also increased the need for wisdom in conservation practices.

Each year has found the fisherman seeking new grounds, for the Gulf of Maine alone cannot keep the redfish fleets of Gloucester, Rockland, and Portland fully occupied. Year by year the fishermen have moved eastward in their search. They have scoured the entire lengthy shore of the Nova Scotian coast and reduced those once-abundant populations. They have exploited the Grand Banks of Newfoundland and then worked up through the Gulf of St. Lawrence with its two populations of large- and small-eyed redfish. They have moved on to Labrador. All these locations yielded abundantly at first from their untouched fisheries and then fell to a self-sustaining level. These northern waters are locked in ice part of the year, forcing the fisherman to limit his efforts to grounds nearer home. Not only are such long trips expensive, but the time away from home has increased from the few days of years ago to two weeks and more, depending upon the weather.

Many of the fishing vessels concentrating on redfish, in the early days of this fishery were of the renowned schooner type. All these schooners were converted sailing vessels in which diesel engines had been installed. During the years before mechanical power, schooners were designed along yacht lines for speed. Keen competition had stimulated the incentive to arrive first in port and reap the benefit of highest prices. Steam-driven vessels began to replace sail early in the 20th century and with them came experimentation and radical changes in design aimed at greater capacity as well as speed. After World

War I, oil-fueled diesel engines superseded steam, and greater impetus was given to improving design and speed in each successive vessel constructed. From all this experimentation the present dragger-type of vessel evolved. While not as seaworthy, dragging trawlers have tremendous capacity, and trips of 250,000 lb and greater are not uncommon.

These draggers do just that—they drag fishing gear along the very bottom of the sea. The net opens on the ocean's floor into a wedge-shaped fishing device. A heavy implement called a "door," because it resembles a huge door, attached at either side of the net, fans out through the water when pulled by the vessel, much as a kite rises in the air when the string is pulled by a small boy. Gliding on edge along the bottom, these doors keep the net wide open. Connected to the doors are the wings, which form the larger part of the funnel and lead or guide the fish into the cod-end, or bag. It is in this cod-end that they are gathered together and later hauled to the surface. Along the top of the net are set, at intervals, a number of floats or cans, as they are sometimes called, that help to keep the net open by their buoyancy. Along the net bottom are heavy wooden rollers, actually slices cut from large trees, through which runs a heavy chain. These rollers guide or carry the net over rocks and gear-damaging obstructions.

Tows of 2 hr are generally made before hauling in the net and its catch. It is a marvel of ingenuity that this mixture of netting and heavy doors, rollers and cables, dumped over the vessel's side en masse, lands on the bottom in proper position and functions perfectly, for the bottom may be at 100 or even 200 fathoms. Most redfish effort is confined to daylight, for this species is seldom available in quantity after dark. Apparently they rise from the bottom at night, out of reach of present fishing gear. While this fact lengthens each voyage, the work is not as demanding as that required for cod, pollock, and haddock, where fishing is round the clock.

Many processing companies have standing agreements with certain boats to take all the fish in each trip. On days of light landings one company owning such a trip will frequently share it with other companies, the custom being reciprocal. Unloading of fish begins early each morning, often before 7 am. An association of what are known locally as "lumpers" usually takes over the discharging of the catch. They agree to take out all the fish for a set price calculated on the thousands of pounds in the fare.

As soon as the vessel has been moored to the purchaser's wharf, the boom has been adjusted to the mast, and the weighing scale has been put in place, unloading gets underway. Lumpers down in the hold of the vessel fork the fish from pens into canvas baskets, which are raised by a man on deck operating a donkey engine. By the aid of the boom, high above on the mast, the filled basket is swung onto the wharf where it is weighed by a company man and checked by a fisherman, also known as "purser," from the vessel unloading. Most plants now have what is called a de-icer set beside the weighing scale. This cylindrical device, covered with a wide-mesh wire, tumbles the fish around and allows bits of ice and unusable small fish to fall through onto the wharf. The fish are then carried by a moving belt or an escalator, into the scaling machine. This is also cylindrical but covered with wire of a smaller mesh, something like poultry fencing. Tumbling continues and the scales are rubbed off either against the wire or against other fish.

Strong streams of water directed into the mass of tumbling fish washes them and also carries away all the loosened scales. A powerful, endless belt conveyor moves the freshly-scaled fish into the plant where they are stored in a pen room and iced until needed. As the fish are required for processing they are shovelled from the pen room onto another endless belt, which passes between rows of filleters. Men exceptionally skilled, with sharp knives rapidly remove the fillets from each side of the fish. These are collected in pans, and each filleter's accumulation is weighed and recorded. A minimum per man is required. An ingenious filleting machine has been developed in recent years; although its use

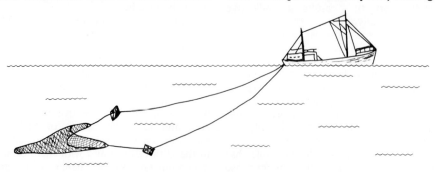

Fig. 2. Redfish dragger towing.

eliminates many filleters, human attention is a necessity in its operation. As this machine is extremely expensive it has not become popular.

The amount of fish flesh suitable for human food as taken from the redfish averages somewhere between 27 and 30% of the whole carcass. Body frames, including the heads, are moved away on another belt and deposited in a "gurry" hopper and later removed by truck to a dehydrating plant for reduction into meal. This meal in turn becomes a supplement in swine and poultry feeds and eventually reappears on the market as pork, eggs, turkeys, or chickens.

At other tables in the processing plant, girls and women "candle" the fillets over glass plates under which light is thrown upward by electric bulbs. Each fillet is carefully inspected, and bits of bone, dark stomach lining, or copepod remnants are removed with the aid of specially designed scissors. Moving farther along on still another belt, the fillets pass through a tank of prepared brine, which washes and also firms the flesh, at the same time enhancing the flavor. Following the bath in the brine tank, the fillets are packed in cartons according to the weight desired to meet current market demand; perhaps in 1, 5, or 10 lb packages. A considerable amount of redfish frames, the remains after the fillets have been removed, is always in demand by local lobstermen, who have found it to be excellent lobster bait.

In recent years a newly developed product has appealed to the housewife for its convenience and ease of preparation. Breaded and precooked fish sticks made from rectangular strips of redfish meat have increased the market for this species. An additional field of employment has been created for women by this new form of fish product from the sea, for they tend the various cookers as well as operate the packaging machines.

It is apparent that new industrial technology, which includes every step from the fisherman to the retailer, must place a greater urgency on conservation, if this species is to survive. Discovery of new grounds may delay the final reckoning, but the redfish fishery is limited to the comparatively narrow confines of the continental shelf, much of which has been scoured time and again. Wise use of such a bountiful natural resource is required if future generations are to share this beneficial and delicious source of food for mankind.

G. M. CLARKE

References

1. Travin, V. I., "A new species of sea perch in the Barents Sea (Sebastes mentella Travin sp. nov.)", *Dokl. Akad. Nauk SSSR*, **77** (4), 741–44, 1951 (translated by C. Richard Robins).
2. Andriiashev, A. P., "XLII. Family Scorpaenidae," in "Fishes of the Northern Seas of Russia," 1954, Guides to the fauna of the U.S.S.R. published by the Zoological Institute of the Academy of Sciences of the U.S.S.R., Moscow and Leningrad, No. 53, 329–339 (translated by C. Richard Robins).
3. Kelly, G. F., Barker, A. M., and Clarke, G. M., "Racial Comparisons of Redfish from the Western North Atlantic and the Barents Sea," U.S. Dept. of the Interior, Fish and Wildlife Service, Washington, D.C.
4. Perlmutter, A., and Clarke, G. M., 1949, "Age and Growth of Immature Rosefish (Sebastes marinus) in the Gulf of Maine and off Western Nova Scotia," U.S. Dept. of the Interior, Fish and Wildlife Service, Washington, D.C.
5. Kelly, G. F., and Wolf, R. S., 1959, "Age and Growth of the Redfish (Sebastes marinus) in the Gulf of Maine," U.S. Dept. of the Interior, Fish and Wildlife Service, Washington, D.C.
6. Kelly, G. F., and Barker, A. M., "Effect of Tagging on Redfish Growth Rate at Eastport, Maine," U.S. Dept. of the Interior, Fish and Wildlife Service, Washington, D.C.
7. Sindermann, C. J. (1961-ICES/ICNAF REDFISH SYMPOSIUM) Special Publication No. 3, "Parasitological Tags for Redfish of the Western North Atlantic," U.S. Dept. of the Interior, Fish and Wildlife Service, Washington, D.C.

RED TIDE—See **DINOFLAGELLATES; MORTALITY IN THE SEA**

RYUKYUAN FISHERIES

The 73 islands of the Ryukyus chain extend for 374 statute miles between Japan and Taiwan; 47 are inhabited; only 26 are considered important. Of the 850 mile^2 land area, only 28% is arable; 86% of the nearly 1,000,000 total population live on the island of Okinawa. The Ryukyuan population of Okinawa averages more than 2000 per mile^2, compared to 679 in Japan and 401 in India. The population density continues to increase; in the past 20 years, the Ryukyuan population has increased 62%. The scarcity of productive agricultural land for food and livelihood necessitates dependence on the marine resources surrounding the islands.

The Islands lie along the easterly margin of the main stream of the Kuroshio (black) current. Due to the warm waters of the Kuroshio, a variety of marine fauna, largely tropical, is found in the vicinity of the Ryukyus. The Kuroshio water area is not considered highly productive, except where the Kuroshio may mix with colder and more nutrient rich water masses along the Asian continental shelf or off Japan in the region of the Kuroshio-Oyashio interaction.

Ryukyuan fisheries are administered by the Government of the Ryukyu Islands (GRI) Fisheries Division in the Agriculture and Forestry

Department. The organization and much of the legislation regulating fisheries are patterned after the Japanese system. The Japanese method of tonnage rights is in use, tonnage rights being administered by the GRI Fisheries Division. The licensing system is a means of regulating and controlling each section of the Ryukyuan fishing industry. The fishing right or license to operate a vessel in a fishery is not generally negotiable (i.e., traded, sold, or used as security for a loan) in the Ryukyus as in Japan; nor are fishing vessels licensed to fish in designated areas as in Japan.

For statistical and licensing purposes the GRI classifies fishing operations as inshore, offshore, and deep-sea fisheries. Inshore fishing is performed by canoes and vessels less than five gross tons. Offshore fishing is performed by vessels 5 to 50 gross tons, and deep-sea operations are performed by vessels 50 gross tons and larger.

Two fishery high schools, one at Naha, with approximately 800 students, and one at Miyako, with 450 students, provide a three-year course of basic vocational training and experience for young Ryukyuan men who plan to enter the fishing or maritime industry. In addition, a two-year postgraduate course for deck and engineer officers is given to approximately twelve students each year. The high schools operate two longline tuna fishing vessels for the practical training of the students: the *Shonan Maru*, 295 gross tons, and the *Tonan Maru*, 159 gross tons. Students major in one of the following branches: marine products technology; mechanical engineering; radio telegraphy; or navigation and fisheries. The demand for trained fishermen and experienced and licensed personnel aboard Ryukyuan deep-sea fishing vessels, together with the yearly increase in deep-sea tuna vessel operational tonnage, has caused a personnel shortage in the key positions aboard the vessels.

A variety of fishing gear is employed in the Ryukyuan fisheries. The deep-sea vessels, besides two 750-ton otter trawlers and two sets of paired trawlers, use only longline tuna gear. Among the gear used by inshore and offshore vessels are:

(a) Tuna longlines;
(b) Skipjack poles and lines (live-bait);
(c) Bottom longlines;
(d) Hand lines;
(e) Drive-in nets;
(f) Fish traps;
(g) Lift nets with night lights;
(h) Gill nets (surface and bottom);
(i) Beach seines;
(j) Spears and harpoons (for bill fish).

Spiny lobsters are taken at night by divers in shoal water areas throughout the Ryukyus from May to November. Traps are not used.

Catch and Landings

The Deep-Sea Fishery. The Ryukyuan fishing industry experienced its greatest relative growth in the early 1960's due to the increase in Ryukyuan deep-sea tuna vessel operational tonnage. Total metric tons of Ryukyuan fish landings, local and foreign port, increased from 16,460 in 1962 to 25,608 in 1966, a 55% increase. During this same period, the landings of tuna, swordfish, and shark (included in the total landings), obtained almost entirely from deep-sea vessels, increased 206%, from 4937 in 1962 to 15,105 in 1966.

Fig. 1. Typical Ryukyuan deep-sea tuna fishing vessel *Yashima Maru* #8. This 213-metric ton vessel, built in Japan in 1965, is powered with a 620-hp diesel; it is 40.25 m long, 6.96 m wide, and has 212.62 m^3 of fish hold space. Its fishing rounds range from the Indian Ocean off East Africa to the New Guinea area.

Fig. 2. Ryukyuan deep-sea trawler. One of the two 750-ton Ryukyuan trawlers operating in the South Atlantic, based at Abidjan.

As late as 1960, the deep-sea fishing fleet consisted of a few old, outmoded longline tuna vessels of Japanese registry operated by Ryukyuans under charter. By 1964 the Ryukyuan tuna fleet was Ryukyuan-owned and registered; older, inefficient vessels were being replaced by newly-constructed vessels and used vessels purchased in good condition. As of April 1967, the Ryukyuan deep-sea commercial fishing tonnage consisted of:

(a) 48 longline tuna vessels   9742 gross tons
(b) 4 trawlers (2 pair)   378 gross tons
(c) 2 trawlers (side)   1488 gross tons
                        Total   11,608 gross tons

The operation of trawlers by Ryukyuans is a comparatively recent innovation. Two sets of paired trawlers (one set towed by two vessels of approximately 100 gross tons each) have recently been operated by Ryukyuan fishing companies fishing off the Continental Shelf of the China coast. Except for a brief unsuccessful trial period of local marketing, the catch of these vessels has been landed and marketed in Japan (due to the relatively low local market value of trawl-caught fish). Two additional Ryukyuan trawlers of 750 gross tons each have been operating for approximately one year in the South Atlantic, and their catch has been sold to Japanese transport vessels and marketed in Japan. Only a relatively small proportion of the deep-sea fishing vessel catch is landed in the Ryukyus. This is due to a number of reasons: higher market value for the catch in the Japanese market; prompt payment for the catch in Japanese markets; more rapid unloading and sale of the catch in Japan compared to the Ryukyus; unfamiliarity of Ryukyuans with trawl-caught fish, and consequent low local market value; and the fact that a number of Ryukyuan tuna vessels operating from foreign bases in the Indian Ocean, the Atlantic, and Fiji, transship their catches to Japan in Japanese refrigerated transport vessels.

The Offshore Fishery. In past years, the skipjack live-bait fishery, an offshore fishery, has been the most important facet of the Ryukyuan fishing industry, both in quantity and value of landings. Since 1962 when 7283 metric tons of skipjack were landed locally, the skipjack catch has declined each year; 1966 skipjack landings were reported as 3522 metric tons, taken by 61 Ryukyuan skipjack vessels with a combined gross tonnage of some 1500 tons. The majority of the skipjack catch is caught in the Miyako-Yaeyama area, and is smoked dried into katsuobushi (skipjack sticks). Each skipjack vessel usually maintains a smoking-drying unit at its home port.

By present methods, this fishery is limited by the availability of live bait, the range and capability of the Ryukyuan fishing vessels, and the abundance of seasonally migrating skipjack in the fishing area. The skipjack fishing season usually starts in late April or May. By mid-October the season is over, and the majority of the skipjack vessel owners then suspend fishing operations until the next spring.

In a recent year, several owners made arrange-

Fig. 3. Concrete blocks for fish shelters. Twenty sites, totaling 2,080 blocks, were constructed by Ryukyuan fisheries cooperatives in 1965. The blocks provide shelter for fish that would not ordinarily be attracted to open, sandy bottom areas.

ments to keep their vessels employed for the greater part of the year by fishing off Taiwan after termination of the local season, purchasing live bait from Taiwanese vessels. There has also been interest in developing local skipjack purse seining operations; to date skipjack purse seining has not been tried in the Ryukyu area.

The Inshore Fishery. The inshore fishery supplies a substantial share of the fish catch for local consumption. Many inshore fishermen are part-time operators. The majority of the inshore vessels are Ryukyuan canoes—long, narrow, graceful, open craft without stabilizers or outriggers. The canoe engines are detached from the engine bed and propeller shaft when the canoe is beached, much as an outboard is removed from a small craft when not in use. Outboard motors are seldom used with Ryukyuan fishing vessels or canoes. The GRI reported 2400 registered offshore fishing craft in 1966; approximately 75% of these are mechanized with gasoline or diesel engines usually of 3–10 hp. Very few inshore vessels are equipped with such mechanical fishing aids as winches or line haulers.

Fisheries Processing

The principal fish processing industry in the Ryukyus is the preparation of smoke-dried skipjack sticks (katsuobushi). The GRI reported that 141 metric tons valued at $192,000 were exported to Japan in 1966. During the same period, 770 metric tons valued at $1.2 million were imported into the Ryukyus from Japan. Fishcake (kamabuko) is prepared for local consumption and, in 1965, amounted to approximately 1800 metric tons valued at $892,000. There are no fish canneries in the Ryukyus; canned fish products constituted 42% of the $6.6 million value of 1966 Ryukyuan marine product imports. No frozen fish packaging industry has been established, although packaged frozen fish products are imported by local military clubs and messes.

Problem Areas

The following conditions have inhibited the growth of the Ryukyuan fishing industry:

(1) Limited capital among Ryukyuan investors to finance the initial cost of large vessels and plant equipment;

(2) Lack of trained, licensed, and experienced personnel required for key positions aboard large fishing vessels;

(3) Scarcity of local bait supply (frozen tuna bait is imported from Japan; live bait for skipjack is found in the Ryukyus in relatively limited quantity);

(4) Lack of repair and construction facilities, with a logistic support of spare parts and construction materials, to effect low cost vessel construction and repair in the Ryukyus;

(5) Lack of an up-to-date, well-trained and qualified fishery extension and demonstration section within the local government, which is attentive to the needs and problems of the local fishing industry.

Potential Areas of Growth

Indications are that the major growth in the Ryukyuan fishing industry during the coming years will continue as a result of expansion and increased efficiency within the deep-sea fishing fleet—from the catch of the longline tuna vessels and trawlers, rather than from the catch of vessels fishing in the inshore and coastal waters of the Ryukyus. Other potentials for development as yet untried are:

(1) Purse seining—tentative purse seine trials have been scheduled in the Miyako area for the summer of 1968;

(2) Trawling—there is a potential for a few small vessels operating beam trawls or other trawls catching bottom fish or possibly shrimp in local waters;

(3) Sale of local fish to military commissaries, messes, and local clubs, and packaging frozen fish for local consumption—a large potential market exists.

Ole J. Heggem

References

Shapiro, S., "Aquatic Resources of the Ryukyu Area," Natural Resources Section Report No. 117, 1948, U.S. Dept. of Interior, Fish and Wildlife Service, Fishery Leaflet 333.

Hiyama, Y., "Systematic List of Fishes of the Ryukyu Islands," Natural Resources Section Report No. 150, 1951, General Headquarters Supreme Commander for the Allied Powers (SCAP).

"Ryukyu Islands Facts Book," 1966, compiled by Comptroller Department, United States Civil Administration of the Ryukyu Islands.

Statistical data from various publications of the Fisheries Division, Agriculture and Forestry Department, Government of the Ryukyu Islands.

Cross references: *Artificial Reefs; Education for the Commercial Fisheries.*

S

SALMON, PACIFIC

The Pacific salmon are in the family Salmonidae, which also includes the trout, the chars, and the Atlantic salmon. This family is characterized by fish having an ellipsoid-shaped body covered with small cycloid scales, a prominent adipose fin, a large fleshy appendage at the base of each pelvic fin, and a lateral line that is well developed. The Pacific salmon are in the genus *Oncorhynchus*, and can be distinguished from other genera in the family by the number of rays in the anal fin. Pacific salmon have 13 to 19 rays in this fin, while the Atlantic salmon, trout, and chars have 8 to 12 anal fin rays. The North American species of Pacific salmon are the chinook (*Oncorhynchus tshawytscha*), the coho (*Oncorhynchus kisutch*), the sockeye (*Oncorhynchus nerka*), the chum (*Oncorhynchus keta*), and the pink (*Oncorhynchus gorbuscha*). There is an additional Asian species (*Oncorhynchus Masou*) whose common name is simply masu salmon.

The various species of salmon vary greatly in size reached at maturity. The chinook is the largest of the Pacific salmon with a record weight of 126 lb and an average weight of about 20 lb. Coho salmon range in size up to 30 lb with most fish being in the 8–12 lb category. Chum salmon weigh as much as 33 lb and the average weight is about 8 lb. Sockeye salmon are usually between 5–7 lb, but weights up to 15 lb have been recorded. Pink salmon usually weigh between 3–5 lb and specimens up to 12 lb have been taken. The masu salmon has an average weight of about 10 lb and a maximum weight of 20 lb.

The chinook salmon can generally be distinguished from the other species by the heavy black spotting on the back, the dorsal fins, and both lobes of the caudal fin, as well as by the black pigmented skin along the bases of the teeth. Young chinook salmon can be recognized in fresh water by strongly developed parr marks. The black spotting of the adult coho salmon is confined to the back and upper lobe of the caudal fin and is not as pronounced as on the chinook salmon. This species also differs from the chinook salmon in not having black pigment along the bases of the teeth. Young coho salmon in fresh water can usually be recognized by the elongated anterior rays of the anal fin, parr marks that are more elongated than those of chinook salmon, and frequently an orange tinge to the lower fins. The adult chum salmon has no large black spots on the body or fins. The narrow part of the body just in front of the tail is slenderer than in other species, and as the adults approach fresh water for the spawning journey, a series of dusky streaks or bars forms on the sides of the body. The young in fresh water can be recognized by slender parr marks, which barely extend below the lateral line.

The mature sockeye salmon male in fresh water becomes a brilliant red and the female is a dark red. This species is distinguished from others by having 28 to 40 long, slender, closely set gill rakers on the first gill arch. There are no black spots. Young sockeye salmon have oval parr marks, which extend only slightly below the lateral line. The adult pink salmon can readily be distinguished by the heavy, oval black blotches on the caudal fin, and the young are also easily recognized because they have no parr marks. The masu salmon is closely related to the coho and replaces it to the south in Asia. The young have 8–11 large transverse bands, which are persistent even in adult fish.

Chinook salmon enter the Sacramento-San Joaquin system of California to spawn, but are found only rarely in streams south of San Francisco Bay. Spawning occurs in rivers north to the Bering Sea and on the Asiatic side south rarely as far as the Amur River, and in Japan in the rivers of northern Hokkaido. In the North Pacific Ocean chinook salmon are found generally to the north of 46° NL in the eastern half of the ocean, but west of the 180th meridian they occur almost as far south as 42° NL.

In North America coho salmon are present in rivers and streams from Monterey Beach, California, on the south to the Chukchi Sea on the north. In Asia they are rarely found in the Anadyr River but occur in large quantities in Kamchatka and south almost to the Amur River. This species is also present on Sakhalin Island and in Hokkaido. During their fast-growing period in the ocean, coho salmon are distributed across the northern North Pacific Ocean and in the Bering Sea. They are found as far south as California waters in the eastern half of the northern Pacific and, in the western North Pacific, are found as far south as 42° NL.

Chum salmon are widely distributed along the North American Coast from the Klamath River in California, north to the Arctic Coast of Alaska and even in the MacKenzie River. In Asia chum salmon are abundant in the Amur River and are found on the island of Sakhalin and in Hokkaido streams and south almost to Tokyo. Chum salmon are in the Pacific Ocean north of 45° NL in the eastern part of the ocean, and south almost to 36° NL in the western Pacific, as well as in the Bering Sea.

Sockeye salmon occur only occasionally in North American coastal streams south of the Columbia River, but they are present north to the Yukon River of Alaska and along the Asian Coast from the Anadyr River, in the rivers of Kamchatka, where it is the principal species in the Kamchatka River, and south to northern Hokkaido where it is very rare. Sockeye salmon are distributed in the ocean throughout the northern North Pacific and the Bering Sea.

Pink salmon are found only occasionally in streams south of Puget Sound and from there northward into Arctic Ocean streams of Alaska. Pink salmon are also found in the MacKenzie River. On the Asiatic side of the Pacific, pink salmon have been reported from the Lena river and south into the streams of Sea of Okhotsk, the Kurile Islands, Sakhalin, Hokkaido, and on the northeastern coast of Hondo. Pink salmon are found usually north of the 40th parallel across the entire north Pacific Ocean.

Masu salmon are found only in streams on the Asiatic side of the Pacific Ocean from the Amur to the Pusan River, in Sakhalin streams, and on the island of Hokkaido and Hondo. They are also occasionally taken on the western coast of Kamchatka.

Life History of Pacific Salmon

The life history of the various species of Pacific salmon is similar in several respects. All salmon spawn only once and die shortly after spawning. Spawning takes place almost entirely in fresh water, although pink and chum salmon may spawn in streams that are inundated with brackish water at high tide. Salmon migrate to the sea after a period in streams or rivers, put on their greatest growth in the ocean, and then return to the stream of their birth to spawn. All the species of salmon prepare the nest in about the same way. The female does almost all the digging of the nest, called the redd, by turning on her side and forming a depression by sweeping up sand and gravel with violent movements of her body and tail. The redd is an oval depression about twice the length of the fish and several inches deep. At the downstream end is a semicircular mound of sand and gravel that has been washed from the redd. The female settles into the redd to deposit her eggs and the male moves quickly alongside and curves his body against hers. The female releases a group of eggs into the nest at the same time the male emits a milky cloud of sperm, which fertilizes the eggs. She then covers this group of eggs by moving forward a short distance and, with movement of her body and tail, stirring up gravel to cover the eggs. This spawning act is repeated several times until all eggs have been released and covered with gravel. The female remains nearby to guard the nest as long as she is physically able. In a period of a few days to a few weeks all males and females die.

Chinook salmon generally ascend the larger streams to spawn and they are abundant in such rivers as the Sacramento, Columbia, Fraser, and Yukon. The center of abundance of this species is the Columbia River, in which adults return to spawn during every month of the year. Here the principal runs are designated as the spring run, summer run, and fall run. Chinook salmon in the spring and summer runs usually spawn in the upper tributaries from 200–500 miles from the ocean. Fall-run fish usually enter coastal streams and the lower tributaries of larger rivers from August to mid-November. Farther to the north British Columbia and Alaska, chinook salmon generally enter the streams from May to July.

Each female chinook salmon carries from 2000–13,000 eggs (average about 5000). After the eggs are deposited they are protected by a cover of from 8–14 in. of gravel. Hatching may take as long as four months, depending on water temperature. After hatching, the young salmon remain within the gravel for a few more weeks until the egg sac is nearly absorbed. The young then push their way up through the gravel into the open water where they are free-swimming and begin to feed on aquatic and terrestrial insects. In fresh water young salmon are eaten by other fish such as coho salmon, char, trout, squawfish, and by such water birds as cormorants and mergansers. The young of fall-run chinook salmon generally migrate to the ocean during the summer and fall months whereas the young of spring- and summer-run fish usually migrate in the spring of their second year. In the ocean chinook salmon feed largely on herring, anchovy, and sardines. In the rich waters of the North Pacific chinook salmon grow rapidly. In their second year in the ocean, chinook salmon usually weigh from $1\frac{1}{2}$–$4\frac{1}{2}$ lb, in the third year from 3–14 lb, in the fourth year from 6–30 lb, and in the fifth year 15–43 lb. In the ocean some sharks feed on salmon, as do fur seals, sea lions, and some whales. After several years of heavy feeding and rapid growth in the ocean, the fish approach maturity. At this stage they begin the ocean migration that leads them with great precision to the river in which they originated. They ascend the parent stream, spawn, and die to complete the life cycle.

Adult coho salmon begin the freshwater migration between September and December, often coincident with a freshet. This species enters the larger rivers, but also is common in the very small coastal streams throughout its range. Spawning takes place often in tiny tributaries only 3 or 4 ft wide. The young emerge from the gravel in the early spring. Coho salmon usually remain in fresh water for about a year after fry emergence and then begin their downstream journey to the sea. In Alaskan streams coho commonly remain in fresh water for two years. Young coho move downstream primarily at night and often during periods of higher runoff. In fresh water they are preyed upon by other fish, especially trout, char, and sculpins, and by birds. Young coho in fresh water and the estuaries are themselves frequently predators of smaller pink and sockeye salmon. Most coho salmon enter salt water by the end of July, in their second summer, where they feed on crustaceans and small fish. Larger coho salmon feed principally on squid, small fish, and euphausiids. Adult coho salmon return to spawn late in the year following that in which they entered the ocean. In their last summer of ocean life coho salmon grow very rapidly and commonly double their weight during this period.

Chum salmon enter fresh water to spawn from July to January. To the north, in Alaska and northern British Columbia, the runs are primarily between July and early September, while south of Vancouver Island the runs are from October through January. Chum salmon usually spawn in smaller coastal streams or in the lower portions, of larger rivers. However, they ascend several hundred miles up the Yukon to spawn and also spawn in the headwaters of some Asian rivers. Chum salmon frequently dig their nests in the same area used by pink salmon. The young emerge from the gravel in March, April, and May. Newly emerged chum salmon are a little over 1 in. long. In most North America streams young chum salmon almost immediately move downstream to salt water, but in some Asian streams and the Yukon River it may be several weeks before the young reach the ocean.

Chum salmon usually migrate downstream at night and upon entering salt water remain in inshore waters until late July or August. During this period of their life they are fed upon by other fish such as trout, pollack, and cod. After midsummer chum salmon gradually disperse throughout the offshore waters of the North Pacific Ocean and Bering Sea where they intermix with Asian fish. The central North Pacific is an important feeding area for mature chum salmon from western Alaska and much of Asia. Asian fish are found to the east at least as far as 160° WL. Chum salmon usually mature at 4 years but 3- and 5-year-old fish are common.

Sockeye salmon are typically lake-dwelling during the brief fresh water part of their life. After migrating from the lake, they usually spend 2 or 3 years at sea. Adult sockeye may begin the upstream migration as early as May and in some areas the migration may extend into October. Sockeye salmon spawn in outlet and inlet streams of lakes and also along some lake shores. After leaving the protection of the gravel the young move into lakes where they remain for 1–3 years before migrating to the ocean at a length of 3–6 in. The downstream migration takes place from April to June, usually under the protection of darkness.

In the North Pacific sockeye salmon feed heavily on amphipods, copepods, euphausiids, pterepods, fish, squid, and similar animals. In the ocean growth is rapid and those fish that return to fresh water after spending two years in the ocean are about 21 in. long.

In the ocean there is a considerable overlap of sockeye salmon from Kamchatka and North America. Immature Kamchatkan sockeye salmon range eastward in the North Pacific Ocean to approximately 170° WL immediately south of the Aleutian Islands, and in the Bering Sea to at least 175° WL and as far north as about 60° NL in July and August. Maturing Kamchatkan sockeye salmon occur in the North Pacific Ocean in June at least as far east as 175° WL in the vicinity of the Aleutian Islands and possibly in early May as far east as 160° WL. Within the area of overlap of maturing Bristol Bay and Kamchatkan sockeye salmon, maturing Bristol Bay salmon usually dominated in the North Pacific Ocean and Bering Sea eastward to approximately 180° and maturing Kamchatkan sockeye salmon dominated westward to about 173° EL.

Maturing sockeye salmon in the high seas begin moving shoreward in May and June. The very important Bristol Bay runs travel an average of about 24 miles per day when heading toward their home stream.

Pink salmon move from the ocean into the streams where they spawn from July to November. Adults usually migrate only a short distance from salt water and sometimes spawn in streams in areas that are affected by the tide. In larger rivers pink salmon may move considerable distances upstream. The young migrate to salt water as soon as the yolk sac is absorbed, in March through early June, but principally in April. In many of the smaller coastal streams of Alaska the young salmon emerge from the gravel and migrate to the sea in one night. Downstream movement begins as darkness approaches and ceases before morning. In the estuaries, schools of young fish migrate along the shore near the surface where the currents gradually carry them toward the ocean. Here their food consists of euphausiids, amphipods, pteropods, small fish, crustaceans, larval squid, and copepods. Pink salmon migrate from the stream to the estuaries at about 1½ in. in length and when they return to spawn after 15–17 months of life in the ocean their average length is about 20 in.

Fishing and Processing Methods

Salmon fishing in North American waters takes place primarily along the coast and most fish are taken by trolling gear, gill nets, and seiners. An ocean commercial salmon trolling vessel is usually from 25–60 ft long. Trolling is a hook-and-line method, which uses four or more lines attached to long trolling poles. From 100–150 fathoms of line are wound on a power reel. Metal spoons, wooden or plastic plugs, and herring are used as lures. Heavy weights carry these lures to the proper depth.

Purse seining is usually a two-boat operation with one boat serving as an auxiliary to the seiner. Purse seine vessels are from 30–90 ft long. The seine is 200–300 fathoms long, 20–40 fathoms deep at the ends, and 25–80 fathoms deep at the center. A series of rings is attached to the bottom of the seine through which a line is threaded. After the net is set in a circle, pulling on this threaded line causes the bottom of the net to pucker and the catch is concentrated in a bowl-like bag. The size of the bowl is reduced by bringing the ends of the net aboard through an elevated power block, which concentrates the fish, and they are then loaded into the vessel by power operated dip nets.

A typical gill net vessel is 22–40 ft long. The length of the net in North American waters ranges from 100–300 fathoms and the size of the meshes varies depending on the species sought. Salmon are captured by swimming into the net where they become entangled. The gill net usually drifts with the current, although there are some areas in Alaska where they are attached to the shore.

Salmon are sold as fresh fish, and are also canned, smoked, and fresh-frozen. Salmon eggs are used as bait for sport fishermen, or eaten as red caviar, or salted as food. Salmon for the fresh market are usually troll caught in the ocean within a few miles of the port of landing. The fish are stored in ice for the return to the dock. Coho salmon is the principal species used in the fresh market, although chinook and pink salmon also are sold fresh. Some gill net and troll caught fish are quick-frozen and processed into fillets or steaks. By far the greatest percentage of sockeye, pink, and chum salmon are processed in tall 1-lb cans; ½-lb cans are also used.

The beginning of the commercial salmon fishing industry in North America is credited to the Hume Brothers and their business partner, A. S. Hapsgood. These men brought fishing gear and canning equipment from their home in Maine and in 1864 processed the first commercial pack at Sacramento, California. The firm then moved to the Columbia River and packed the first salmon there in 1866. As the sale of salmon increased, new canneries opened on the Fraser River in British Columbia in 1867, on Puget Sound in 1877, and on Prince of Wales Island in Alaska in 1878. From these early beginnings commercial processing spread throughout the Pacific Northwest from California to northern Alaska. The catch of salmon in North American waters gradually increased and reached a peak in the 1930's. Subsequently, the take sharply decreased, although in recent years there have been signs of an upturn, and the prospects for a gradual increase in the catch of Alaskan salmon are good. The highest and lowest catches (millions of fish) of the various species in North America in 1922–1961 are as follows:

Sockeye salmon, 42.1 in 1938 and 13.2 in 1959;
Pink salmon, 90.0 in 1936 and 20.2 in 1960;
Chum salmon, 20.0 in 1928 and 5.4 in 1955;
Coho salmon, 10.1 in 1951 and 3.8 in 1960;
Chinook salmon, 4.2 in 1925 and 2.1 in 1944.

Fishermen of Japan and the U.S.S.R. also operate major commercial fisheries for salmon. The Japanese fishery began in 1952 and has developed rapidly. The Japanese mother ships are accompanied by catcher boats, which use drifting surface gill nets to capture salmon. Land based salmon fishing vessels working out of ports in northern Japan employ either drift gill nets or floating longlines. Japanese coastal salmon fishermen use traps and fixed gear.

Fishermen of the U.S.S.R. take salmon only in coastal waters. They use traps, beach seines, and weirs to capture maturing salmon en route to their spawning grounds. Offshore or high seas salmon fishing with nets is prohibited in Canada and the U.S.

In 1961 the catch (thousands of pounds) by species of salmon for Canada, Japan, U.S., and U.S.S.R. was as follows:

Country	All species	Sockeye	Pink	Chum	Coho	Chinook
All four countries	964,554	228,833	394,295	234,354	67,513	39,559
Canada	125,112	26,622	50,059	14,609	24,729	9,093
Japan	349,036	80,903	167,295	90,375	9,433	1,030
U.S.A.	314,749	104,051	110,226	49,079	23,389	28,004
U.S.S.R.	175,657	17,257	66,715	80,291	9,962	1,432

This compilation shows, surprisingly, that the U.S. was second in production of salmon to Japan. By far the greatest part of the catch in the U.S. comes from Alaskan waters. The Japanese catch is mostly from the high seas of the North Pacific Ocean.

In Lake Superior some 192,000 yearling coho salmon were planted in 1966, but the size of the spawning adults and percentage return in 1967 was considerably smaller than in Lake Michigan.

The salmon stocking program in the Great Lakes has grown steadily since the first plantings in 1966. Yearling chinook salmon were introduced in 1967 and coho and chinook salmon were planted in Lake Huron in 1968. In 1968, sportsmen caught about 91,000 coho salmon in Lake Michigan and about 12,000 in Lake Superior.

After World War II the Japanese made known their intentions to prosecute a high seas gill net fishery for salmon in the North Pacific. In order to prevent overexploitation of salmon, Canada, Japan, and the U.S. in 1953, entered into an agreement called the International Convention for the High Seas Fisheries of the North Pacific Ocean; it applies primarily to salmon but also to other species of concern to these nations. Under its terms, Japan agreed to abstain from fishing in the area east of 175° WL exclusive of the Bering Sea. Canada and the U.S. agreed to carry out necessary conservation measures to obtain maximum sustained production, and to carry out extensive scientific research. This Convention has been successful in protecting most North American salmon from Japanese nets.

From San Francisco in the south to the Bristol Bay area of Alaska well over a million sport anglers fish for salmon each year. In 1966, approximately 700,000 coho salmon and 444,000 chinook salmon were caught by sport fishermen in California, Oregon, Washington, and Idaho. Anglers in British Columbia took an average of about 350,000 salmon each year in 1958–1961; Alaska also has an important sport fishery for salmon.

One of the principal salmon sport fishing areas along the Pacific Coast is in the estuary and several miles off the mouth of the Columbia River. The reliable catch records from this area demonstrate the phenomenal increase in sport fishing between 1958 and 1967. In 1958, less than 70,000 salmon were taken in the Columbia River estuary by anglers, but the sport catch increased each year until 1967 when an all-time record was set of about 300,000 coho and 77,000 chinook salmon. On peak fishing days in 1967 more than 2,000 sport fishing boats were concentrated at the mouth of the Columbia River. Anglers took almost 50,000 coho salmon in one week of intensive fishing during the last week of August 1967 and the first few days of September. Many other ports in northern California, Oregon, Washington British Columbia, and Alaska are centers of similar intensive salmon angling activity. Despite the record catches in 1967, more than 250,000 coho salmon returned to the lower Columbia River and a record 100,000 coho salmon were counted over Bonneville Dam, which is a little over 100 miles from the ocean.

Sport fishermen usually take salmon in the ocean by trolling a herring or artificial lure. Lures for coho salmon are fished near the surface where the fish feed most frequently. Chinook salmon are often caught on lures fished as deep as 50 ft. Salmon can frequently be located by observing flocks of gulls that feed on small fish such as herring and anchovies near the surface. The salmon are often feeding on the same schools of small fish.

Salmon stop heavy feeding when they enter fresh water, but they can frequently be induced to strike at a lure cast from shore or trolled or cast from a boat. Although their digestive organs can no longer assimilate food, they occasionally will take bait such as salmon eggs or worms. Salmon are taken by anglers in most of the Pacific Coast rivers from Northern California to the Yukon River of Alaska. In Idaho there is a substantial fishery for chinook salmon in the upper tributaries of the Columbia River several hundred miles from the ocean.

An important sport fishery for salmon has developed in the Great Lakes. Of 658,000 yearling coho salmon planted by the Michigan Department of Natural Resources in Lake Michigan in the spring of 1966, some 33,000 were caught as adults by anglers in 1967. The average weight of the fish was about 12 lb. It was estimated that $\frac{1}{3}$ of the coho salmon planted in Lake Michigan in 1966 survived to maturity. Salmon not taken by fishermen spawned in streams or were captured at weirs by the Michigan Department of Natural Resources. Eggs stripped from fish at the weirs in 1967 provided fish for later plantings in the Great Lakes.

Research

Major salmon research efforts are being undertaken by the U.S. Fish and Wildlife Service, The International Pacific Salmon Fisheries Commission, the fishery agencies of the various Pacific Coast states, the Fisheries Research Board of Canada, and major West Coast universities such as Oregon State University, the University of Washington, and the University of British Columbia. The High Seas Salmon Research Program of the U.S., Canada, and Japan is being coordinated by the International North Pacific Fisheries Commission. When this Commission established a North Pacific research program in 1955, knowledge of the distribution, abundance, origin, and movements of salmon on the high seas was almost nonexistent. The research program first sponsored by the Commission devoted itself to studies of offshore distribution of salmon. Because this was an international program and knowledge was needed for managing salmon on an international scale, much of the first research was directed toward the identification of the continent of origin of salmon on the high seas. Studies included morphological and meristic studies, biochemical and physiological investigations, age and growth studies, and studies of the parasites of salmon. Salmon were tagged to study their movements on the high seas. The ocean environment was studied to provide a background of the understanding of the characteristics and variation of salmon distribution on the high seas. From Kotzebue Sound, Alaska, south 2000 miles to the Sacramento River of California, biologists are seeking answers that will assist in maintaining the salmon resource. To the north, research is mainly concerned with sockeye and pink salmon, and to the south, chinook and coho salmon. Chum salmon receive less attention. In Alaska much of the salmon research is carried out at wilderness field stations. One such remote station is maintained by the U.S. Bureau of Commercial Fisheries at Olsen Creek to study the important pink and chum salmon runs of Prince Williams Sound, which spawn mostly in gravel areas affected by the tide. One of the principal objectives of the study at Olsen Creek is to determine how many female pink salmon are needed for optimum utilization of the gravel. After several years of study in this stream biologists have been able to calculate that under natural conditions at Olsen Creek about two females per yard^2 of stream gravel result in optimum production of adult pink salmon. In most pink salmon streams there are many fewer salmon than necessary to give this optimum density of spawners.

The year or so spent in the stream before migrating to the ocean is critical for survival of coho and chinook salmon. For these two species, the size of the stream and the amount of food available often limit production. Thus for coho and chinook salmon, the problem of the biologists is to determine how many young fish a stream can support and how many adults are needed to yield this number of young. Scientists at Oregon State University have determined that there is a rapid decline in the biomass of coho salmon from the time of emergence from the gravel in March and April to early June. Many young salmon may be swept downstream in periods of flood water. In periods of low water the amount of water and food available for these small coho salmon have considerable effect on the carrying capacity of a stream.

Sockeye salmon after emerging from the gravel nest live usually 1–2 years in a lake before migrating to the sea. For this species the number of young that can be reared in the nursery lake often limits the maximum size of the run. Biologists, therefore, are directing much effort toward finding how many young sockeye salmon can be produced in lakes and what the natural factors are which limit production.

After salmon have spent 1–3 years in the ocean, some internal biological mechanism modifies their constant wandering search for food and directs them toward the stream in which they were hatched. A salmon may be over 1000 miles from his natal stream when he starts the journey back. The ocean navigating system that directs salmon to precise landfalls is not understood. Scientists are experimenting with fish to find out how they are able to navigate so exactly. At the University of Wisconsin scientists have found evidence that some species of fish can perceive the altitude of the sun, much as the sailor does when he measures the height of this body above the horizon with his sextant. Whether salmon in the North Pacific use such a natural directional system is not known.

After salmon arrive in an area within the influence of waters from their own river, some may still be 2000 miles from the riffle or tributary where they will spawn. Salmon are able to find this small tributary and there is strong evidence pointing to the internal mechanism that guides the fish once they have reached fresh water. Many animals have an acute sense of smell, but this sense reaches an astounding degree in some fish. Experiments with eels have demonstrated their ability to detect an alcohol in a dilution equal to a few drops in a lake several miles in diameter. Eels may be able to detect only two or three molecules of the alcohol in the olfactory sac at one time. The amazing ability of fish to detect odors has suggested to scientists that salmon may be able to smell their way home. Such a theory implies that each stream has a permanent, characteristic odor that salmon can detect in minute concentrations and use as a guide to reach the home tributary. Tests have proved that fish can indeed distinguish between the waters of different streams, and it is generally believed that this is the method by which salmon are able to find their way so accurately to the spawning area whence they came several years before.

In the area from British Columbia to California, and to some extent in Alaska also, an increasing percentage of adult salmon find passage hindered or blocked by power, irrigation, or flood control dams. The Columbia River and its wild tributary, the Snake River, are rapidly becoming a series of placid pools tamed by walls of concrete, polluted by human and industrial wastes, and with water temperatures raised to critical levels by atomic reactors and heat absorption from the sun.

Fishways have been constructed over the low dams of the main Columbia and Snake Rivers, and these have been generally successful in passing adult salmon. Yet there have been unexplained losses of upstream migrating salmon that seem to be associated with dams. Downstream migrating salmon pass through the turbines of the low dams or over the spillways with usually only a small percentage lost. Each dam may take only a small toll of adults moving upstream and young salmon moving downstream, but the cumulative effects of all dams may result in disaster to the Columbia River salmon runs. Scientists of the U.S. Bureau of Commercial Fisheries Engineering Research Laboratory at Bonneville Dam are working with adult salmon in experimental fishways to get information that will allow more efficient fish passage facilities to be designed. Fish passage problems at low dams are serious but rather insignificant when compared with the difficulty of getting downstream migrating young salmon safely past a dam that may be 400 ft high and may form a lake 15 miles long. If natural runs are to be maintained in situations such as this, young salmon must be collected at the head of a reservoir or in streams above the reservoir and transported below the dam. Successful collection of fingerling salmon requires that they be guided and concentrated for subsequent downstream transport. Scientists are attempting to develop practical means of guiding and collecting fingerling salmon so they may be passed around high dams.

In recent years fish passage facilities have been provided at most dams, but at some high dams fishway construction has not seemed feasible and other methods have been provided for maintaining the runs. Chief among these methods has been the production from salmon and steelhead hatcheries. In recent years research efforts have resulted in a great improvement in the success of rearing salmon in hatcheries. Many hatchery diseases can now be controlled and nutritious diets have been developed for young salmon. Experiments have pointed the way toward the best size, time, and area of release for fingerling salmon. A start has been made toward increasing production through selective breeding. We can now have considerable confidence that a modern hatchery can successfully rear most races of salmon and steelhead and produce good returns.

In recent years scientists have turned their attention to improving survival of salmon eggs and fry by constructing spawning channels. An artificial spawning channel generally consists of a dam at the head to control the flow of water and a man-made channel of appropriate width and slope with gravel of optimum size for best egg and fry survival. Many of the conditions detrimental to the survival of salmon eggs and fry in nature can be controlled in such artificial spawning areas. The optimum number of adults can be allowed to spawn. Flood waters, which wash eggs and fry from the gravel, can be prevented. Predators can be controlled. Gravel can be selected to allow optimum circulation of water with its life giving oxygen and the amount of water needed to ensure best production of fry can be maintained. Artificial channels show considerable potential as another tool for maintaining salmon runs. For example, at Jones Creek, a tributary of the Fraser River in British Columbia, one of the first artificial spawning channels for salmon was constructed in 1954, and here survival of pink salmon eggs and fry has been four to six times that of natural streams.

These are only a few samples of the many salmon research projects. Equally interesting research is being undertaken by universities and fishery agencies in such fields as salmon behavior, physiology, population dynamics, genetics, nutrition, and disease.

GEORGE Y. HARRY, JR.

Cross references: *Biological Clocks; Breeding in Marine Animals.*

SALT, SOLAR

The oceans of the world contain enormous reserves of sodium chloride (salt), estimated at 5×10^{16} tons (50 million billion tons) (Ref. 1). Recovery of salt from seawater by solar evaporation has been an important source of salt, and its importance is increasing. Where conditions permit, solar salt can be produced and transported at low cost. In the U.S. about 5% of the total salt production of 25,000,000 tons (1965) was solar salt. However, of the 94,000,000 tons of salt produced in the rest of the world, the contribution of solar salt approaches 50%.

Seawater and its Concentration

The composition of seawater is given in Table 1. Only the major compounds that relate to the production of solar salt are given. Some 50 elements have been found in seawater in trace quantities.

TABLE 1. COMPOSITION OF SEAWATER

Compound	Weight Percent
Sodium chloride	2.68
Magnesium chloride	0.32
Magnesium sulfate	0.22
Calcium sulfate	0.12
Potassium chloride	0.07
Sodium bromide	0.008

The total solids content in the waters of the oceans is about 3.5 weight percent. Some of the smaller bodies of water, such as the Red Sea, contain about 4.0% solids. The Dead Sea and Great Salt Lake are special cases where the concentration is close to saturation—23-27% solids. A remarkable characteristic of seawater is that regardless of geographical location the relative proportions of the various ions are almost independent of salinity. Therefore, the physical chemistry of the concentration of seawater is also independent of salinity.

Sodium chloride constitutes about 75% of the total solids in seawater. It is quite apparent that the oceans represent a tremendous reserve of salt. Just one mile³ of seawater contains about 130,000,000 tons of salt, which is the present total annual world production. However, seawater is 96.5% water, and the practical recovery of salt is dependent upon an economic process to remove vast quantities of water.

Salt is recovered from seawater by solar evaporation. This process uses solar energy, so heat is obtained at no cost. Figure 1 shows the changes in composition and volume that occur as seawater is concentrated. Usiglio (Ref. 2) was the first to study the concentration of seawater. His data are still considered to be among the best available.

Seawater is first passed through a series of concentrating ponds where it is evaporated to the salt point, which is approximately 26° Baumé (Bé) (1.22 specific gravity). Brine at this point is sometimes referred to as "pickle" liquor. As shown in Fig. 1, a large part of the calcium sulfate is crystallized in the concentrating ponds where it is deposited. This serves to improve the purity of the salt produced substantially. It is also apparent that in evaporating from seawater concentration to the salt point, the volume is reduced by 90%. At the salt point, the brine is transferred into crystallizing, or "garden," pans, where salt is crystallized and deposited on the floor of the pan during evaporation. (Throughout this article, the crystallizing areas will be called pans to distinguish them from the concentrating areas, which will be called ponds.) During salt crystallization some calcium sulfate also deposits, but all the magnesium and potassium salts remain in solution. The practical limit of concentration is 29° Bé (1.25 sp. gr.). At this point the magnesium salt concentration is so high that the evaporation rate is reduced substantially. Also, above 30° Bé, there is danger of crystallization of magnesium salts, which would reduce the salt purity. The 29° Bé brine, called bittern, is discarded, carrying with it practically all the magnesium and potassium salts. The bittern may be processed in a chemical plant to recover magnesium salts, potassium salts, sodium sulfate, and bromine. The crystallized salt is recovered and processed as described later.

Successful operation of a seawater solar evaporation process to produce salt is dependent on three major factors:
(1) availability of large, level land area with suitable soil conditions;
(2) favorable meteorological conditions;
(3) available markets for the product.

Under favorable conditions the yield of salt is about 40 tons per acre of total concentrating and crystallizing area. Thus a large plant, which may well produce 500,000-1,000,000 tons of salt per year, requires many thousands of acres. The land must be reasonably level and close to sea level elevation. Another extremely important requirement is a relatively impervious soil, which will not allow excessive brine loss due to seepage. Although the theoretical ratio of concentrating to crystallizing area is 10:1, brine seepage losses may increase this ratio to as high as 15:1.

Concentrating ponds, which have natural mud bottoms, are built by forming mud dikes to contain the brine. The dikes must be substantial enough to withstand wave action and to carry harvesting vehicle traffic. Concentrating ponds, which generally follow the natural terrain contours, may be irregular in shape and from several hundred to several thousand acres in size. Crystallizing pans are also constructed by building dikes with the existing soil, and should be regular in shape to allow economical harvesting. They are generally rectangular, and up to 50 acres. It is desirable to establish a salt floor of 6-12 in. in crystallizing pans. This serves as a harvesting

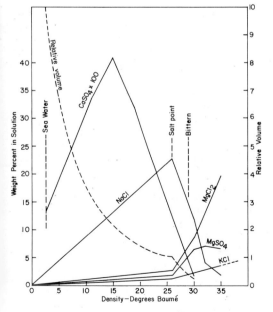

Fig. 1. Seawater concentration curves.

base and also keeps the new salt crop clean. In some locations, it is not possible to maintain salt floors because of heavy rains that occur during a rainy season. In these cases the salt must be harvested right down to the mud base.

In order to produce solar salt the meteorological conditions must be such that there is sufficient evaporation to produce economic quantities of salt. In some areas there may be evaporation throughout the year. In others, there may be a seasonal evaporation period followed by a rainy season. The important factors are incident radiant energy, air temperature, humidity, wind, and rainfall (Ref. 3). These factors must combine in such a fashion that they effect a net evaporation that will support a solar plant. The theoretical amount of salt deposited during the crystallizing step is about 600 lb per short ton of water evaporated. Since a solar plant is so dependent upon weather conditions, one year's production of salt is generally stockpiled as a protection against a year when little or no evaporation may occur.

One artificial method for promoting evaporation is the use of dyes (Ref. 4) to increase absorption of radiant energy. One such dye, Naphthol Green B is reported to increase production 15–20% when used at 5 ppm in the crystallizing pan brine (Ref. 5).

The concentrating area consists of a series of separate ponds through which brine flows in series to be concentrated to the salt point. Crystallizing pans may be operated either in batch or series type of operation. In batch operation fresh brine is added periodically as evaporation occurs until the desired salt crop is obtained. It may be necessary to bleed some brine if the mother liquor concentration becomes too high. In series operation brine flows continuously through a series of pans at a rate that permits it to enter at the salt point and leave at the bittern point. Brine depth in crystallizers is generally kept at 6–12 in. A depth of less than 6 in. may allow some loss of radiant energy due to excessive reflection off the salt crop. Usually, a salt crop of 4–6 in. in thickness is formed per season. Where possible, transfer of brine in the crystallizing pan system is best accomplished by gravity. Where transfer must be done by pumping, the pumps will salt up because the brine is usually slightly supersaturated. It is necessary either to add a small amount of water continuously to the pump intake or to periodically wash out the pumps.

Some rather interesting micro-organisms occur in seawater and its concentrates. The organisms that grow in media above 2% salt concentration are referred to as halophilic, or salt-loving. Two types of algae are commonly found. Dunaliella viridis is green in color, and is a good food source for brine shrimp, which are quite common. The green algae thrive best at a salinity somewhat below saturation. Dunaliella saline is a red algae that grows in saturated brine. Probably the most striking organisms are the red bacteria that often thrive in the crystallizing pans, coloring the brine pink to deep red. The classification of these bacteria is still somewhat unsettled. The scientific names now generally given cover two main classes of the red bacteria, Halobacterium and Sarcina-micrococcus (Ref. 6). It is commonly believed that the red color imparted to the crystallizing pan brines improves evaporation rates. However, it has been pointed out that the bacteria are discrete particles that may cause loss of radiant energy by reflection (Ref. 5).

Harvesting and Processing

The harvesting operation consists of recovering the salt from the crystallizer pan floors and transporting it to the stock pile. Usually, the salt is washed before stockpiling, to improve purity and cleanliness. The frequency of harvesting depends generally upon the climate; if there is a wet season, it may be necessary to harvest annually. In locations where continuous crystallization is possible, it is usually advantageous to harvest continuously.

The harvesting methods, of which there are a variety, depend upon climate, ground conditions, and size of the operation. Three general methods are widely used. Where the climate or ground conditions are not favorable, only a thin layer of salt is accumulated, but it must be harvested as frequently as every week. In these cases, the salt is harvested manually. Where climate and ground conditions permit, a permanent salt floor is maintained of a thickness of one or more harvests. The current crop is recovered down to this floor. This method of crystallizer operation is preferred, as there is least chance of harvested salt contamination with clay or sand, and the salt floor provides structural strength sufficient to support heavy harvesting equipment. In many other operations where there is a rainy season and it is not possible to maintain a salt floor, or soil conditions are poor, all harvesting equipment must be supported by the current crop. Special harvesting equipment has been developed for this condition.

The harvesting machines are of many designs but, in general, they operate on the principle of inserting a blade underneath the salt to be harvested, to lift the salt off the floor where it can be broken if necessary, picked up, and elevated into the transport system. These machines can be independently driven, pushed, or pulled, mounted on rollers, caterpillars, wheels, or on floats.

Figure 2 shows salt recovery equipment used where a salt base is maintained in the pond. If

Fig. 2. A salt harvesting machine.

this is not the case, the equipment must operate on the bed to be recovered, but otherwise it is operated entirely on the salt base. The pans are drained prior to salt recovery. In some locations, draining the pans is not practical, in which case harvesting is done either from barges pulled by cables or by partially submerged tractors loading boats, as in Fig. 3. In still other locations, where a brine only is required, the salt is dissolved in situ by flooding the drained pans with water.

Transporting the salt to the stock pile in modern plants is done by using rail cars, trucks, carryalls, belts, lorries, or pumping as a slurry. The choice is largely determined by economics, as all methods are widely used. Where permanent salt beds are maintained, heavy trucks can be used, and have replaced rail cars in many plants. Pipelines are perhaps economical for short distances of less than 1 mile, but are being used for distances of 10 miles. Slurry transport is well adapted to washing operations.

Salt is usually washed in an inclined screw with a countercurrent flow of saturated brine. This washing reduces both insoluble and soluble impurities in the salt. The wash brine is recycled through a settling basin to settle out insolubles, and some of the brine is bled off to keep the soluble impurities within desired limits. If the salt is coarse, some grinding of the salt prior to washing may be desirable.

The washed salt is drained before stockpiling. Draining may be done in elevators or in centrifugals. Salt is frequently piled in long windrows, using slingers or more elaborate equipment.

Reclaiming salt from the stockpile may be carried out with specialized equipment, with mechanical shovels, or with simple bulldozers that push the salt into elevators, which load the salt onto trucks, or lorries. Stockpiled salt usually contains 3–4% moisture.

If a screened salt is required, the salt is further processed by drying, crushing, and screening. A product meeting requirements of many types of uses for industry, farm, and food processing may be obtained from solar salt.

Solar salt, as produced in modern plants, is of good quality. It is a clean, white product containing approximately 99.7% sodium chloride. The principal impurities are calcium sulfate and magnesium salts. Solar salt grows in the shape of "hopper" crystals, which generally knit tightly in the crystallizing pans. In processing, these break up into irregularly shaped crystals characteristic of solar salt. The crystals are generally quite coarse, much of the salt being larger than $\frac{1}{4}$ in. mesh size.

As with all products, solar salt must have markets that it can serve. Most solar plants have the natural advantage of low cost water transportation to move the product to its markets.

HOWARD W. FIEDELMAN
HORACE W. DIAMOND

Fig. 3. Salt harvesting in Torrerieja, Spain.

References

1. De Fler, P., Solar salt production, 1967, unpublished.
2. Usiglio, J., *Annales Chim. Phys.*, **27** (3), 92–107 (1849).
3. Kohler, M. A., et al., "Evaporation from Pans and Lakes," 1955, U.S. Weather Bureau Research Paper No. 38.
4. Bloch, M. R., et al., "Solar evaporation of salt brines," *Ind. Eng. Chem.*, **43**, 1544–53 (1951).
5. Bonython, C. W., "Factors determining the rate of evaporation in the production of salt," "Second Symposium on Salt," Vol 2, 1966, Northern Ohio Geological Society, Cleveland, Ohio.
6. Gunsalaus, et al., "The Bacteria," Vol IV, 1962, Academic Press, New York.
7. Kerns, W. H., "U.S. Bureau of Mines Yearbook —Salt," 1965.
8. Ver Planck, W. E., "Salt in California," Division of Mines Bulletin 175, 1957.
9. Sverdrup, H. U., et al., "The Oceans," 1942, Prentice-Hall Inc., Englewood Cliffs, New Jersey.
10. Kaufmann, D. W., "Sodium Chloride, A.C.S. Monograph No. 145," 1960, Reinhold Publishing Corp., New York.
11. Office of Saline Water, Research and Development Report No. 25, 1959.
12. Garrett, D. E., "Factors in the design of solar salt plants," Part I & II, "Second Symposium on Salt," Vol 2, 1966.

Cross references: *Chemistry of Seawater: Minerals of the Ocean.*

SARDINES—See ECOLOGY, MARINE

SATELLITE SENSING OF MARINE PHENOMENA

Operating communication satellites are already relaying telephone calls and television programs to Europe and Asia. Observational satellites are tracking hurricanes. Some research satellites are being tested and others are being designed to determine what kinds of valuable information can be sensed, or "seen." Methods of sensing include:

(1) photography;
(2) passive radiometry;
(3) active methods;
(4) combinations of several techniques.

Satellites carry special kinds of receivers called remote sensors. These instruments detect various wavelengths of electromagnetic radiations. Electromagnetic waves include gamma rays, X rays, ultraviolet light, visible light, infrared light, microwaves (including radar), and radio waves. To make it easier to describe all the electromagnetic waves, they have been arranged in a series according to their wavelength, called the electromagnetic spectrum. Gamma rays are the shortest and radio waves the longest. Different types of receivers pick up radiations from various parts of this spectrum.

Every known substance radiates and reflects electromagnetic energy according to its nature and condition. The differences in wavelength and intensity tell much about distant objects. Certain very narrow bandwidths of the electromagnetic spectrum will carry information on a single feature without the interference of other signals above and below it. These are called "windows." For instance, sea surface temperature can be measured through one of these "windows" in the infrared part of the spectrum. Satellites can carry instruments that measure the intensity of radiations from the ocean over very restricted bandwidths, much the same way radios can tune in on a single station. Some of these radiations originate from the ocean itself and some are reflected.

In the early 20th century, global weather predictions came into being for a variety of purposes. Now, global oceanographic predictions for shipping, fishing, and military purposes are maturing. Many seafaring men are contributing. The total system is developing as rapidly as communication and sensing equipments are made practical and available. Results are advancing from historical oceanographic atlases to daily and, eventually, hourly synoptic charts transmitted by radio. But, even if all vessels at sea, all offshore towers, all unmanned stations, and all islands were eventually equipped to contribute, large gaps would persist. It is logical that the technological breakthrough of the last decade be utilized to increase our capability to provide timely and global coverage. Satellite technology is already providing continuous communication and observational services. It now remains to determine and develop the kind of capability required to make the sea more useful to mankind.

The usefulness of certain oceanographic information has already been clearly demonstrated; however, much remains to be done. As an illustration of usefulness, the prediction of thermocline depth, at which cod feed in the Lofoten Islands, saves fishermen from Oslo the expense of running north to find that fish are out from the Islands in waters too dangerous for small craft. When the thermocline is high, cod are in the sheltered waters. The area between Iceland and Europe, where the basic information for these predictions is gathered, has been intensively studied.

As an illustration of work that needs to be done, the anchovy fishermen of Chile must first go out to determine if the sea is cool enough for fishing. Unseasonably warm waters are believed to push the fish beyond the operating range of the fleet, or below the reach of their nets. Significantly, data from this area and the South Pacific, in general, are sparse. Advance knowledge of surface temperatures would increase the efficiency of vessel and reduction plant operations. There are many other examples of the need for improved prediction.

Data obtained from a limited but growing cooperative surface network of ships and buoys in the eastern North Pacific are collected and collated by the U.S. Naval Fleet Numerical Weather Facility, Monterey, California. Part of the data contributed by the Bureau of Commercial Fisheries (BCF) comes from commercial fishermen on a voluntary basis. The Bureau takes this material, interprets it, and makes it available to fishermen at frequent intervals. The relation between temperature and fish has been described in restricted areas. Appropriate reporting of other

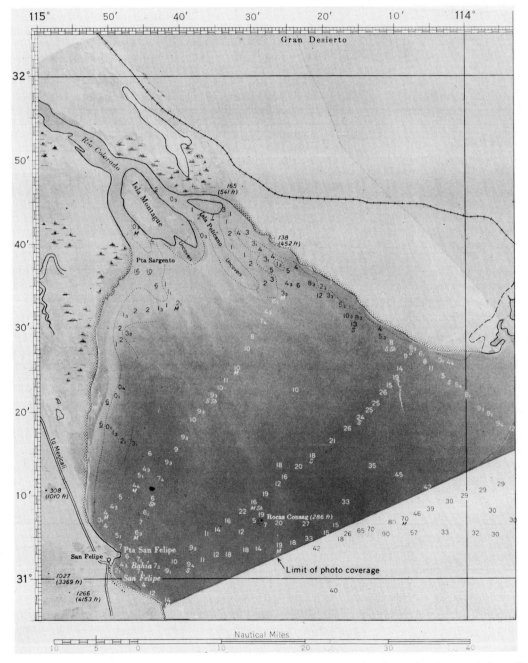

Fig. 1. Nautical chart overlay on Gemini photo of mouth of the Colorado River. The photograph fits well and illustrates the changes in the Colorado delta since the chart was made.

areas might facilitate such scientific descriptions.

The U.S. Naval Oceanographic Office (NAVOCEANO), for instance, gathers, for military purposes, and distributes by fascimile, data on surface temperature, layer depth isopleth (depth of thermocline), and sea state and direction for the North Atlantic; similar information on weather is available. Such information might assist certain fisheries in locating fish and inform them that fishing is feasible at a given time and place.

There is little room for doubt that a wider and more adequate reporting of certain ocean phenomena would be beneficial. The usefulness of data on temperature and sea state has been clearly established and demonstrated in specific situations. These are the only types of data, in addition to weather information, now available.

Satellites provide a means of obtaining both a global coverage and a new look at the sea. In polar orbit, a single satellite may scan every part of the earth at least twice a day. The new, direct look, remote sensing, may measure many characteristics of the surface and near-surface waters.

This work is sponsored by the National Aeronautics and Space Administration (NASA). One of its missions is "the establishment of long-range studies of the potential benefits to be gained from the opportunities for and problems involved in the utilization of aeronautical and space activities for peaceful and scientific purposes." The objectives of NASA's Earth Resources Program are to identify user requirements and to determine experimentally the feasibility of employing orbiting spacecraft for peaceful applications in geoscience.

The Spacecraft Oceanography Project (SPOC), one of four potential application areas, was established to develop and coordinate applications for Oceanography-Marine Technology. It evolved from an agreement in 1965 between NASA and the U.S. Navy. NAVOCEANO was designated to carry out the agreement. Through informal cooperative working arrangements, the Environmental Science Services Administration (ESSA) and BCF are working with NAVOCEANO to assist in the provision of meteorological and biological requirements. In fact, all government agencies with an interest in oceanography participate in this effort through a SPOC Advisory Group.

A major application of satellite relay stations as data relay links between buoys and ships and shore stations is entirely possible. Feasibility studies are in progress to determine if information collected and recorded by a buoy or ship can, on interrogation, be transmitted to a relay satellite. Under a separate program, NASA is developing the Interrogation, Recording, and Locating System (IRLS) experiment for this purpose. IRLS differs from other data-relay systems, such as Relay, Telstar, and Syncom, in that it is also designed to locate the surface transmitter.

Photography, the first remote sensing techniques appears potentially useful in many ways. Pictures of bottom structure, such as the Gemini shot of the mouth of the Colorado River (Fig. 1) indicates that application may help locate thousands of "doubtful shoals" and verify charts.

The determination of the best film and filter combinations will help bring out features of interest. A set of nine cameras, capable of simultaneously photographing a single area with three different films and nine filters, provides an excellent tool for studying the response of films and filters (Fig. 2). With a multifilter and multifilm approach, such as this, it is hoped to determine whether significant observations of plankton colonies can be made. Differences in color have already been related to biological or chemical characteristics in agriculture; possibly this will afford a new analytical tool to chart red tide and other plankton colonies. Many other aspects of photography are being explored.

Passive radiometry can be designed to provide images in the same manner as television scanning provides a picture. An infrared scanner on an aircraft made the "picture" of a warm current (Fig. 3). This information, recorded electronically, may be expressed in numbers. It is anticipated that high-flying satellites may rapidly scan large areas and provide accurate average temperature readings for incremental squares. Infrared (IR) investigations have concentrated on temperature, but there are other applications. It has been shown with laboratory absorption spectrometers that fish oils make a unique "signature," different from that of mineral oils, in a very narrow portion of the IR spectrum. Fish oils appear on the ocean surface where larger fish are feeding on smaller ones. How long do these films last? How do they differ from those caused by fish feeding on plankton? Can these oils be "seen" by a satellite? The answers to such questions may lead to useful forecasting and reporting.

In active methods, as opposed to passive radiometry, satellites also scan incoming radiations from the area beneath it. Here, the radiation is an echo of a signal produced by the satellite. It may be possible to obtain an index of sea state from measurements of radar backscatter. The relationship between the amount of backscatter reaching the satellite and the angle at which the radar beam hits the sea surface becomes less pronounced as roughness increases. By scanning ahead and back along the path of flight, a cross section of return signal strength related to surface roughness might be obtained. Laser altimeters also may have application to sea state measurements.

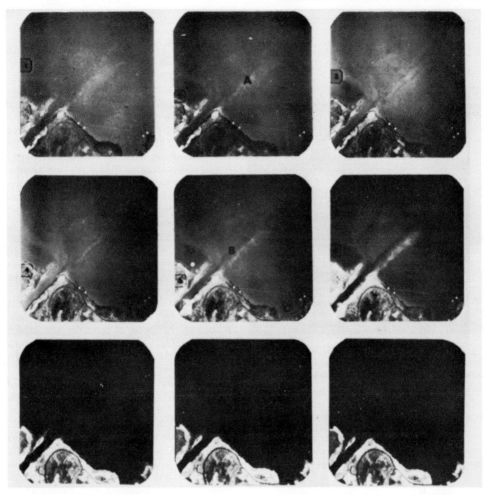

Fig. 2. Nine-lens camera photograph. Approximate spectral range of each filter: 1, 400–450; 2, 450–500; 3, 500–550; 4, 550–600; 5, 600–650; 6, 650–700; 7, 700–770; 8, 770–820; 9, 820–900 mμ. Further experiments are revealing usefulness of different films and filters. (Photo: ARPA.)

Combinations of various techniques can be used to provide corrections for interfering factors. Temperature measurements by microwave will have to be corrected for sea state or vice versa.

Some windows may provide consistent readings that are above or below the true values. Isolated surface sensors that communicate observations to the satellite, may provide corrections for larger nearby areas.

Experiments to develop these methods require that many greatly different groups work together smoothly. When instruments are ready to "fly," they are first placed aboard aircraft that pass at varying altitudes over areas where intense surface observations are in progress. Ultimately, field activity will be scheduled to coincide with the flights of experimental satellites so that comparisons can be made.

As scientific and engineering principles prove useful, NASA will turn the information over to operating agencies. Applications developed may have a profound effect on fishing operations. If they should result only in global reporting of adequately measured sea surface temperature twice daily, then the entire effort will have been well spent. More than this is expected, however. Knowledge of the sea has advanced through the ability of people to work together well, and through the introduction of new instruments that extend human perception. This particular effort is already providing the cooperation of many skills and new methods of observation.

This article is limited to those physical measurements that directly relate to biology and fishing operations. Most studies of sea color caused by plankton, detection of fish oils, tracks of fish schools caused by temperature

Fig. 3. Infrared scan of edge of Gulf Stream taken from an aircraft. The area shown is about 1 mile². In this instance the scanning rate was synchronized with the speed of the aircraft.

disturbance, bioluminescence, red tide studies, pollution tracing, and other marine biological experiments are so recently initiated that only their existence can be mentioned at this time.

JULIUS ROCKWELL, JR.

SCALLOPS—*See* **SEA SCALLOP FISHERY**

SCHOOLING BEHAVIOR

Gregarious organisms occur in groups; these groups have been called many things ranging from the prosaic "herds" to the poetic "murmurs," but the novelty in terminology has far outstripped any real differences in the nature of the groups. Among fish, the more popular terms for groups have been shoals, schools, aggregations, and pods. Though some of these terms have been carefully defined, the usage has been less precise in the literature, reflecting a common viewpoint that a school or a shoal is any group of fish. The imprecise usage has not been a totally arbitrary preference of terms, but in part a practical necessity because, in nature, it is frequently impossible to analyze groups in sufficient detail to determine whether they would meet the special qualifications of schools, aggregations, or pods.

Definition

In Breder's usage, a school is a group in which all individuals are headed in the same direction, are uniformly spaced in close rank, and are swimming at the same speed. He recognized the superficial resemblance schools bear to groups oriented into a stream or current and stationary with respect to the substrate, but he reserved judgment in including these as schools because the individuals may be coordinated with non-living objects rather than with one another. He applied the term "pod" to groups of individuals in contact with one another and recognized two kinds: the polarized pod, with all fish headed in the same direction; and the unpolarized pod of irregularly oriented individuals. Groups with individuals irregularly spaced, or spaced but irregularly oriented were assigned to the collective category of aggregations. In this scheme, schools and pods are ideal representations of typological concepts, and aggregations are all groups that deviate from the ideals. Keenleyside defined a school as any group of mutually attracted individuals regardless of their orientation, spacing, and swimming speeds, and thus included Breder's three categories under one heading.

Though both investigators based their definitions on concepts of spatial design and mutual interaction, they were not using the same definitive criteria for schools, and failure to realize this has led to some unnecessary confusion in the literature. Keenleyside was concerned specifically with the question of group motivational behavior (mutual attraction), but Breder was concerned specifically with mechanical models and, excluding the implications for fish oriented in streams, mutual attraction was not more than the cause for existence of the groups. Superficially there would appear to be no conflict involved by restricting the term school to a specific arrangement of fish, but conflict does arise if it is asserted that only schools show schooling behavior because this implies something more than mutual attraction. To a degree the confusion stems from the use of the term "schooling" to define both form and behavior, but there are more serious problems. A close analysis of the typological ideals reveals that the stipulation the group must be swimming becomes, unwittingly, a definitive specification of the school, because translational motion of a group and orientation of its component members are cause and effect relationships. Moreover, the concepts are unclearly related to the dimension of time. They represent neither instantaneous group configurations nor continuous group configurations, but rather some mean configuration over an arbitrarily designated time interval. In stressing the ideal mean configurations, the important parameters of variation are excluded.

Because they are based on divergent conceptual schemes, the definitions of Breder and Keenleyside cannot be unified without major concessions in one way or another. A solution is most readily obtained by expanding on Keenleyside's definition.

Characteristics of Schooling

A logical analysis of schooling behavior begins with the recognition that regimented behavior of individuals is essential to the unity of the school. The school remains a unit when the individuals are coordinated with one another either collectively or unilaterally. Schooling behavior is not defined by the spatial orientation of individuals, but rather, the spatial orientation is defined by the occupation of the school; that is, whether it is stationary, feeding, or swimming off to some other locality. The independent variable is the intact unit and the variables of spacing, direction, and speed of individuals are dependent upon the occupation of the school.

Mutual attraction between individuals in a school is an ideal concept that defies proof in reality, though it gains logical support from the fact that most fish school with their own kind. But attraction may be unilateral in schools of mixed species composition in which some individuals would merely tolerate their use by others as objects for coordination. Is there mutual or unilateral attraction when the gray snapper follows the school master?

Provided that individuals in a school remain coordinated with one another, there is considerable latitude for variations in spacing, orientation, and speed, and these variations are indeed prominent features of schools in nature. The distance between individuals may extend over a range determined by visual conditions, and should decrease as the swimming speed increases in order to minimize the tactical problems in maintaining the unity of the school. The directional headings of individuals must vary minimally as the school swims at high speeds on a straight course and may become highly irregular when the school is stationary or feeding. Individual speeds will vary in a similar manner with the important distinction that the least variation may occur at zero speeds as well as at high speeds; the greatest variation will occur when schools are disrupted as they are when feeding. Individuals must continually compensate for deviations from the mean spacing, orientation, and speed, and the tolerance time for compensation is minimal in fast-swimming schools and maximal in stationary schools. When a school is analyzed over the course of a day, it can be expected to exhibit a wide and ever-changing variety of mean configurations.

To give a definition that integrates the independent and dependent variables, a school is a coordinated group of individuals, all of which continually compensate in speed, orientation, and spacing in order to preserve the unity of the school. Rather than seeking mean arbitrary types within a continuum of configurations, a description of schooling behavior should be based upon a continuous record of individual speeds, spacings and headings such as has been obtained from the analysis of time-lapse cinematographs by Hunter. Although his method excludes the vertical dimension of a school, and thus the more difficult measurements, it has the distinct advantages of objectivity and inclusiveness. A complete characterization of a school might include indices of species composition, size distribution of individuals, the size of the school, and its occupation at the time of the observations. It should be noted that schooling behavior is not determined by the size of the school. Two fish will exhibit the essential characteristics of schooling behavior, and some of the behavior can be obtained from a solitary fish.

Time Factor

The dimension of time must be included in a description of schooling behavior not only on a daily basis, but also throughout the life of a species, because schooling behavior is regulated by many time-dependent variables. Some of the sharpest changes are associated with reproductive seasons. Schooling ceases among adults of a wide range of species such as sunfish, bass, and mouthbreeders when one or both sexes retreat to nesting sites, and is resumed when care of the young is completed. The exact opposite occurs among solitary fish such as pike and trout. Their migration to a common spawning ground results in the occurrence of groups. In still other species, such as perch, whitefish, and North Sea herring, migration to specific spawning sites occurs but it does not effect drastic changes because these species exhibit strong schooling tendencies throughout the year.

Spawning

Whether spawning groups should be considered schools depends on the rigidity of the criteria, particularly of motivation. Overtly, the structure of a spawning group may seem to be indistinguishable from a school, but spawning behavior is a classical example of a special kind of unilateral attraction, that between opposite sexes. A general policy for the disposition of spawning groups would be difficult to formulate because there is a wide range of hierarchical structures with some species showing considerable antagonism and others showing very little. If antagonism is equated with repulsion, the complexity of the problem increases because repulsion is a functional component of schooling behavior. By countering the force of attraction, repulsion establishes the minimal distance of separation between individuals in a school. When fighting occurs within spawning groups, it signals a loss of mutual respect for individual separation. The decision whether to include spawning groups in schools, would be arrived at arbitrarily, at best.

Effect of Age

Schooling behavior changes with age in some species. The young of the black bullhead, for example, form dense stationary schools with disoriented individuals in contact with one another, or moving schools of regularly spaced individuals. The intensity of schooling declines in older fish. In most species, the schooling response persists once it has developed among the juveniles, but juveniles rarely coexist in schools of larger fish either because they occupy exclusive habitats as, for example, the occupancy of shallow marginal waters by juvenile minnows or suckers, or the surface as opposed to depths by juvenile North Sea herring. In a dynamic sense, juveniles may be segregated because they cannot swim fast enough to stay with the larger fish, or because the fish establish dominance hierarchies and drive away others of different sizes. The segregation of schools according to size, and thus of age, of the fish continues throughout the life span of the species and the age-exclusiveness of schools is most clearly defined in the first few years while the age-length correlations are most distinct. This does not mean that the individuals within a school are of uniform length—they may vary by a factor of 2 or 3 depending on the mean—but that the mean lengths between schools differ significantly.

Environmental Factors

Some changes in schooling behavior are correlated with either short-term or long-term environmental variations, which are generally interpreted as sources of physiological stress. In the laboratory, increased stress as caused by low temperatures, elevated carbon dioxide or chlorine content, or stimulants, such as strychnine or caffeine, may intensify the schooling response, as do noises, sudden motion of objects, a strange object in the aquarium, or novel surroundings, which might be created merely by transferring the fish to a different aquarium. The Schreckstoff (fright substance) released by injury to the skin in certain fish, notably the freshwater ostariophysids, initiates a flight reaction that is rapidly transmitted as a visual sign and causes fish to form tight schools or to seek cover, depending on species. In the absence of Schreckstoff, a flight reaction is transmitted by visual means. A reduction in stress as might be represented by the monotony of conditions in the laboratory aquarium induces a relaxation in the regimentation of schools, to different degrees in different species, and the response may vary with the size of the aquarium.

Shape

In nature, schools are generally spindle-shaped, and if only two or three fish are involved, they usually resort to following one another. Compact pods or balls are frequently seen, and a common formation is the mill in which the school rotates in one spot like a large wheel on a fixed axis. The significance of such formations is generally unknown, although some are thought to represent reproductive or protective behavior and others may represent a response to physical conditions. Many species become quiescent during the cold seasons, and some, such as the bullheads, are known to form dense congregations in specific areas suggestive of denning among snakes. The overall form of a school may depend upon the number of individuals comprising it. Changes in the structure of a large school of black mullet were correlated, along the axis of progression, with metabolic reduction of dissolved oxygen. This is an interesting example of an aquatic group establishing its own limiting conditions, and is undoubtedly important in any very large school of fish because a school changes many conditions in its environment. It not only reduces dissolved oxygen, but also the density of food while increasing the content of metabolic wastes. The form and distribution of schools is related to currents and bottom topography. Stationary schools head into a current, and, since currents are modified by topographical features, schools often occupy characteristic sites. North Sea herring are frequently located near the substrate over local depressions in the bottom contours, and the specific distribution of schools is readily seen in the transparent tropical waters. In shallow waters, the form of a school and its course may be determined by objects along the way and by the color of the substrate especially where patches of vegetation form a very dark contrast against an area of white sands.

In nature, as in the laboratory, fish flee from fright situations such as Schreckstoff or intrusion of strange objects and again, as in the laboratory, there appears to be a graded fright reaction. Moderate fright induces compact schooling, but excessive fright may cause fish to seek seclusion among plants or under objects. The latter behavior would be particularly characteristic of some shallow-water species. It is a general hypothesis that schooling and escape behavior are graded fright responses, and one sees an interesting parallel in plains mammals which converge into a pack when alarmed by an intruder and then turn to flee with the continued advance of the intruder.

Orientation

Schooling is a form of behavior and should not be considered an adjective modifying species. There is no dichotomy between schooling and nonschooling fishes. The two categories lie at opposite ends of a continuous scale expressing time devoted to schooling behavior. Occupying the top of the scale would be species of whitefish, herring, sardine, mackerel, and tuna, and, at the bottom, solitary predators such as trout, pike, and barracuda, or demersal or seclusive species such as the sculpins and flatfish.

An interesting clue to schooling motivation is the optomotor response, which is characteristic of animals in general. An organism is oriented in its environment with points of visual fixation, and when either the organism or the points are moved relative to the other, the organism adjusts its position to conserve the original orientation. By means of this response, fish hold their positions in streams or currents, or follow a moving object when other points of fixation are lacking. This might account for the generally lower exhibition of schooling by shallow-water or inshore species than by species in open water, but it would have little bearing on the behavior of solitary species. The shallow-water species, in addition to other fish, have many stationary objects upon which to fixate, while in the relatively unbounded reaches of the pelagic zone of oceans and deep lakes, fish must serve one another as points of optical fixation. Reports of pelagic fishes holding pace with drifting flotsam indicate the strength of the drive for optical fixation though there may be other motivational factors involved. That the optomotor drive is, indeed, persistent is further illustrated by studies on visual thresholds. Some fish will follow a rotating drum of alternating black and white bars at a level of brightness comparable to a very dark night. Operationally, schooling behavior satisfies the optomotor drive, but the drive is not an exclusive explanation of schooling motivation.

Attractions

In experimental situations, given a choice, a solitary fish usually swims to the larger of two schools, but whether it remains with the school depends on the species involved. It would, of course, school persistently with individuals of its own kind and size. A single fish of one species will school with a group of a different species if no choice exists, and in nature an occasional straggler may end up with a school of a different species. As a result of conspecific preference, schools in nature are predominantly monotypic, and, although exceptions are not uncommon, they are probably short-lived anomalies. Mixtures of the young of different species are frequent occurrences in the shallow marginal water of lakes, but they are not sharing one another mutually. When alarmed, they swim to deeper water in predominantly segregated schools. No doubt, a big factor in the occurrence of mixed schools is the accidental result of occupancy of the same habitat by several species.

Apparently, vision plays the dominant role, not only in bringing a solitary fish into a school, but also in recognition of species; this is shown by studies that either remove characteristic marks from individuals or substitute models of varying shapes and patterns of color. It is noteworthy that many species appear to exhibit no contrast in color other than the dark eyes and that everything but the eye tends to blend into the background of diffuse brightness of the water. It is equally noteworthy that these fish have very silvery sides that flash brilliantly when they catch the sun. Other sense organs are only relatively well developed in the fish and their roles in species recognition are not well understood. It is known that fish respond preferentially to the odors of their own species and it has been shown that individuals respond to sounds made during the breeding seasons, but whether a school of nonbreeding fishes makes specific recognition noises remains debatable.

Feeding

Ichthyologists have long sought evidence of the survival value or the evolutionary significance of schooling behavior, and although it must exist, it remains elusive or speculative. Schools may assure that fish find mates during the breeding season, but if schools migrate to particular spawning grounds, as do whitefish and herring, the argument resolves into one of redundancy or double assurance. It has been suggested that schools improve the feeding efficiency of individuals in pursuit of planktonic organisms, that a plankter might readily dart away from one fish only to fall into the grasp of another, but this explanation is questionable in view of the efficiency with which solitary fish capture plankton and also the fact that in a feeding school, individuals dart about independently selecting one prey organism after the other. Schools have been considered protective devices of various sorts. Clumping of individuals into schools would reduce the chance of their discovery by wandering predators, but the validity of this reasoning is not assured because we know so little about the ways of the predators. Schools have been interpreted as a mass of confusion and distraction that lessens the capturing efficiency of a pursuing predator and also, when in compact form, as mimics of larger animals, and thus a threat symbol. Perhaps the survival value of schools will become better known through analogies when the survival value of gregariousness in the more approachable terrestrial organisms is better understood.

Effect of Light

The foregoing discussions pertain to fish under good visual conditions in daylight. The fate of schools at night is another matter. Blinded fish and fish in laboratory darkness (below the visual threshold) do not school. Observations in the laboratory, by infrared image conversion, on groups of fresh-water fish by John and on the jack mackerel by Hunter, using time-lapse infrared cinematography, show that schooling ceases entirely when the visible illumination falls to the order of 1×10^{-6} foot candles, which is probably near the visual threshold of fish as it is for man. The schooling under decreasing illumination does not change abruptly. As the light falls, the school fragments into progressively smaller schools that persist for progressively shorter periods until, near the end, they seldom consist of more than two fish or persist for more than a few seconds. As implied in the transformation of the school, there is one threshold of illumination for maximal schooling and a second threshold for minimal schooling. The upper threshold and the rate of decline of schooling to the lower threshold vary with the species, at least under these laboratory conditions. The upper threshold of the jack mackerel is in the order of 1×10^{-5} foot candles, and for the black bullhead, 1×10^{-2} foot candles. For most species studied, the upper threshold lies between 1×10^{-5} and 1×10^{-3} foot candles, and the rate of decline in schooling is gradual.

For fish near the surface in clear water, there would generally be sufficient light for visual coordination and schooling, but the limited observations in nature reveal the first surprise, that disorder sets in among schools near the time of sunset when illumination is more than adequate for visual coordination. The dispersal of schools under adequate illumination means that schooling behavior is not dependent solely on visual conditions, and this is supported by the varying effects of fright and monotony and by changes associated with seasons. Very little is known about the nocturnal behavior of schools, but at least two patterns occur. Some species, such as the perch, settle to the substrate and others, such as sardines and North Sea herring, rise to the surface and disperse. There have been a number of claims that schools remain unified at depth in sub-threshold illumination at night, but verification of such claims must await better methods of detection and analysis of schools and corresponding measurements of illumination.

Hearing

Much discussion has centered around the possibility that schools might remain intact without visual coordination (the dark night) by using some combination of auditory, olfactory, or lateral line information. If specific sounds are made by fish in darkness, they would be potentially useful, especially when representing the combined output of hundreds or thousands of individuals, but there is no evidence that schools remain unified through auditory information, or that isolated fish at much distance can localize a sound source. Specific odors are likewise a potential to unity of the school, but unlike sound, which disseminates in all directions from the source, odors are generally a unidirectional gradient extending downstream from the source, and thus are of limited utility. Other problems arise in a large water mass in subthreshold illumination because the fish have no points of optical fixation and would most probably drift with the water, in which case the odor would not disseminate from the source. However, there would be an odor gradient diminishing from the center of the school. Because fish are known to have a keen olfactory sense, one could postulate that the unity of a school is assured if the fish but show a positive response to the odor gradient. The function of the lateral line system in fish is not totally understood. It is known as a detector of mechanical disturbances in the water. At short range, it responds to turbulence. It has been postulated that it serves at somewhat greater range to localize a sound source. The lateral line system could undoubtedly assist in regulating the minimum distance between fish, but it would be purely speculative to assert that it serves to bring them together or to keep them from drifting apart.

There is no doubt that highly regulated schools require visual coordination, but what happens to pelagic schools under varying conditions of "darkness" in nature may depend on the size of the school and the locomotory activity of individuals. Fish are essentially weightless in their aquatic medium and can maintain a suspended position with small effort. If the individuals in a large school simply stop swimming, they will remain grouped though they may drift with the water mass. The act of dispersing would require a positive effort, but, even so, if the school consisted of hundreds or thousands of individuals, dispersal would scarcely result in solitary isolation of all individuals. Moreover, the volume of water to accommodate dispersal would be greatly restricted for those species that rise to the surface at night. The ocean is not a homogeneous mass of water. It is bounded by thermal gradients and strata of varying salinities, and these strata maintain their identity over long periods. In other words, though the area over which a species might disperse is relatively large, it is limited to a species range and the range is subdivided by environmental gradients. The area is in any case a mere fraction of the whole ocean.

It is impossible, in so short an article, to give due recognition to the many scientists who have contributed to our understanding of schooling behavior. The comprehensive analysis of social groups by Breder (1959) includes an extensive bibliography. Most of the other references have been selected because they contain pertinent contributions made since 1959.

KENNETH R. JOHN

References

Blaxter, J. H. S., and Parrish, B. B., "The importance of light in shoaling, avoidance of nets and vertical migration by herring," *J. Cons. perm. int. Explor. Mer.,* **30**, 40–57 (1965).

Breder, C. M., Jr., "On the survival value of fish schools," *Zoologica,* **52**, 25–40 (1967).

Breder, C. M., Jr., "Studies on social groupings in fishes," *Bull. Amer. Mus. Nat. Hist.,* **117**, 393–482 (1959).

Dijkgraaf, S., "The functioning and significance of the lateral-line organs," *Biol. Rev.,* **38**, 51–105 (1963).

Hunter, J. R., "Procedure for analysis of schooling behavior," *J. Fish. Res. Bd. Canada,* **23**, 547–562 (1966).

John, K. R., "Illumination, vision, and schooling of *Astyanax mexicanus* (Fillipi)," *J. Fish. Res. Bd. Canada,* **21**, 1453–1473 (1964).

Keenleyside, M. H. A., "Some aspects of the schooling behavior of fish," *Behaviour,* **8**, 183–248 (1955).

McFarland, W. N., and Moss, S. A., "Internal behavior in fish schools," *Science,* **156**, 260–262 (1967).

Pfieffer, W., "The fright reaction of fishes," *Biol. Rev.,* **37**, 495–511 (1962).

Shaw, E., "Schooling in fishes: critique and review," in "The Development and Evolution of Behavior," L. R. Aronson et al., editors, 1969, W. H. Freeman and Co., San Francisco.

Wynne-Edwards, V. C., "Animal Dispersion in Relation to Social Behavior," 1962, Hafner Publishing Co., New York.

Cross references: *Breeding in Marine Animals; Salmon, Pacific.*

SEA FARMING

Human survival depends upon an adequate supply of food and water. Although science, technology, and money are now devoted to trying to reach and live on the moon, a more immediate problem is to find out how to survive on this planet. Twelve species of plants stand between mankind and starvation. We rely on a few cereals and root crops, two sugar plants, legumes, and tree crops for food. At least half the world's population gets 60% of its energy from one plant species—the rice plant. The present population of the world is 3.3 billion; it is increasing at a rate of 3% per year. If present trends continue, by 1985 food requirements will be at least twice what they were in 1965.

The food problem is linked to the problem of economic growth of a developing country. Increasing the food production depends on increasing the financial demand for food and on creating a modern agricultural system. An annual growth rate of 4% in food production and of 5.5% in gross national income must be achieved by the developing nations to meet minimum food requirements during the period of 1965–1985.

Currently these countries are increasing their food production by only 2.7% annually and their gross incomes by 4.5%. To achieve the necessary growth in food production and economic activity, annual capital investment in the developing nations will have to increase from 15% of gross national product to 19%. For the 1965 base year this 4% increment would have amounted to $12 billion. To achieve such a feat will require capital and technical involvement of the developed and developing nations alike on a scale unparalleled in the peacetime history of man.

The bulk of the increase in food supplies must come from farm crops, principally through higher yields. The current agricultural production is about 2 tons of dry organic material per acre per year. On a small scale it is possible to produce 15 to 20 times more organic matter than contemporary agriculture achieves. Such a "laboratory farm" is a problem in engineering (how to expand laboratory experiments into large scale production); a problem in economics (whether the cost of such an undertaking is prohibitive); or a problem in psychology (how to make people adjust to a revolutionary change in their eating habits). Thirty per cent of the earth's land surface is tillable, of this 30%, 80% is cultivated. If the remaining 20% of tillable soil were utilized it could support only the present world population, never an increased population. What is needed is a widespread, low cost production of food supplement rich in high quality protein.

The least tapped source of protein resides in the oceans. Fish seem to be an almost inexhaustible source of animal protein. The annual U.S. catch is 5 billion lb. Ten million tons go into the production of fish oils and fish meal. The oils include 523,000 tons of fish body oils, from menhaden and herring-like fishes; 386,000 tons of whale oil; 103,000 tons of sperm oil and 62,000 tons of fish liver oils. Marine oils are the cheapest of all fats and oils, They are used in margarine and cooking oils, in paints, lubricants, and mineral flotation. Whale oils are used in fat liquoring of leather; and fish liver oils are processed as a source of vitamins A and D. Estimates are that the seas could supply 500 billion lb of fish annually without becoming depleted.

Equally important are plant products. Red and brown algae are harvested for animal food and for preparing extracts usable to industry. Seaweed extracts, agar, algin, and carrageenin, are used as thickeners in food products. About $44,000,000 worth of seaplant products are produced each year. The products are prepared by washing and bleaching the sundried, baled, raw weed followed by steam extraction for 6–12 hours. The aqueous extract is filtered and dried, often decolorized, and again bleached.

Canada and the U.S. are agricultural countries. This tends to reduce fish consumption and interest in sea farming. The subsidizing of scientific land farming likewise is a deterrent. Another drawback is that Americans have fastidious tastes and rarely eat snails, mussels, and other seafoods popular in many parts of the world.

Other parts of the world have developed their sea resources in coastal regions. These estuarine regions have supported the culture of fish and invertebrates in tremendous numbers as well as great fisheries. In pound yield per surface area these estuaries are much richer in food resources than the open sea.

There are extensive culture fisheries in the estuarine areas of the Indian Ocean coastal zone. These brackish water swamps are banked with dikes to take in water at high tide, and the suspended silt settles on the bottom slowly raising its level. Within several years, the land is raised sufficiently to be used as rice paddies. When the land level is high enough, intake of tidal water is discontinued and rainwater leaches the salt from the soil. After two to three seasons, the paddy may be planted. The plot will have a canal system inside the dike; these canals are used for culturing brackish water fishes. But during this silting period, the area is used for growing fish. The tidal water brings in the fry of commercially important fish including prawns. Screens prevent the escape of fish or the ingress of extraneous fish, and sluice gates control the tidal flow.

The incidental use of swampy land for raising fish promoted a system of intensive brackish water fish culture in India and Pakistan. Fish ponds of different design and shape began to be constructed for the sole purpose of commercial fish culture.

On the islands of Java and Madura, ponds or "tambaks" produce nearly 33,000,000 lb of fish yearly. The industry is based on the milk fish *Chanos chanos*. Post larvae and juveniles are collected from inshore areas from September to December and from April to May and transported to the fish pond areas in flat watertight bamboo baskets. The annual requirements for the tambaks of Java and Madura is 190–200,000,000 fry.

In India grey mullet, pearl spot, prawns, or milk fish are seined from nearby areas and transported to Government-owned fish ponds called "porong" ponds. Each porong type farm has fry ponds of 90–900 m^2, and rearing ponds of 900–4500 m^2. The irregularly shaped sections are connected by secondary sluice gates and the whole complex is controlled by a main gate located in a deep portion having a channel in the middle. The ponds are drained and dried to eradicate predatory and weed fish and to hasten the decomposition of organic matter. Then 3–5 cm of tidal water is taken in and allowed to stand. Within 3–7 days a brownish, greenish, yellowish layer of microorganisms (principally bacteria, unicellular and filamentous blue-green algae, and diatoms) develops on the bottom. Growth and production of fish in the ponds depends on the growth of algae.

The product is marketable in 6–10 months. The ponds are drained at low tide to capture the fish. The tendency of the fish to swim against the current is used for partial fishing. When tidal waters are let into the ponds, the fish swim against the current and can then be led into a catching pond.

In Java and Madura annual production is 19,759,000 lb of milk fish, 6,482,000 lb of penaeid prawns, and 3,300,000 lb of other fish.

The addition of inorganic or organic matter is a valuable way of improving the yield of fish in a pond or lake. In Europe the yield of fish has been improved by adding phosphate and potassium or organic matter to fish ponds, but results with nitrogen are conflicting. For example, base deficient ponds must be treated with lime before other substances are added. The results so far relate the yield of fish and the chemical treatment directly ignoring the intermediate links. Until these have been studied, no final word on the best treatment for specific ponds or species of fish will be possible.

In protein-short Africa a useful source of additional protein food might be the tiny fish used to clear mosquito larvae from ponds. *Nothobranchuis taenipygus* will hatch from eggs and grow to full size (4.6 cm) in 3–4 weeks in the right conditions. The fish will live in a few inches of water. Areas from 6–12 in. deep allow the maximum number to be bred in the minimum volume of water. Young *Nothobranchius* feed in algae and protozoa, older fish prefer insect larvae. Apart from some chemical fertilizers to increase the growth of algae in the feeding pools, little has to be brought into the area to provide supplies of fish.

In Northern America, sea farming is limited. Pacific oysters are raised in Washington and there is some oyster farming on the east coast. In the northwest coastal areas of the U.S. there are some low cost fish farms that use marine fish requiring fresh water environment for propagation.

In Italy, the Volturno River discharges into Lake Patria. The communication of the estuary with the sea consists of a very narrow shallow watercourse. Nutrients enter the lake from volcanic subsoil. Lake Patria teems with mullet and eels. The soft muddy subsoil makes mussel farming on the bottom impossible but by hanging cultures above the bottom the mussels grow and fatten rapidly.

The Oosterschelde, in the southwestern section of the Netherlands, is an estuary with a different pattern. The Oosterschelde is an embayment penetrating far into the land that receives little influx from the surrounding fertile arable land. Yet river and seawater meet under the influence of tides. The Rhine, the Meuse, and Scheldt Rivers discharge fresh water rich in nutrients and organic materials into the North Sea. Tides and winds bring about a thorough mixing and microorganisms mineralize the remains of fresh water organisms and other organic material. It is this mixed coastal water that pours into the Oosterschelde with the tides. The high and constant salinity together with the discharge of nutrients conducive to rich plankton development makes good oyster water. Under natural conditions the oyster population was limited by an unprepared bottom and by failure of the oyster larvae to settle on beds where growth and fattening would be optimal. An exploitation of the natural resource by applying techniques of cultivation has resulted in a 30 fold increase in production over what nature formerly yielded.

The Galician bays of Spain are examples of estuaries in which shelter is a more important factor than discharge of nutrient rich fresh water, since the seawater is sufficiently rich through upwelling. In the early 1950's, experiments were carried out to take advantage of the favorable conditions in these estuaries by growing mussels in hanging cultures following experiences gained in the Mediterranean Sea. Instead of racks, which are unsuitable where the tidal range is great, rafts were used. At first these were old boats equipped with outriggers. The success was promising and now about 1500 installations consisting mainly of specially constructed rafts are in use in Galician bays. Each raft carries 800 ropes; each rope is about 6 m long and carries 100 lb of marketable mussel. It takes eight months to rear a mussel seed to marketable size. The total production is greater than 80,000,000 lb.

The Japanese utilize a similar technique to cultivate oysters. By growing oysters on long ropes hanging from simple rafts, the plankton of the whole water column is available to the oyster, and there is a 50 fold increase in yield.

Another type of estuary, with extensive tidal flats and channels of varying depth through which tidal currents run, is located behind the Frisian Islands of Texel, Vlieland, and Terschelling, the Dutch Wadden Zee (western section). These flats are productive with mussels, *Mytilus edulis*, and cockles, *Cardium edule*, brown shrimp, *Crangon crangon*, and polychaete worms of various species. Young flatfish and sole use this area as nursing grounds. Man fishes for shrimp, and cultivates mussels under the low water line. The production of marketable mussels surpasses 150,000,000 lb per year.

The productivity of the Wadden Zee is partially mechanical. Suspended matter is carried into the Wadden Zee by the flood tide and only part of this material is returned at ebb tide. There is the intensive CO_2 assimilation, and filter feeders (cockles, mussels) filter off enormous volumes of water brought in on the flood tide. The digestible material is transformed into molluscan flesh: the remainder is deposited as fecal pellets and is available to the bottom organisms. The large numbers of benthic organisms per unit of surface area can survive only through the continuous accumulation of organic matter brought in from the North Sea by the flood tides. Annual protein production per meter^2 is among the highest in the world.

The North Sea (southern portion) is a type of estuary. Rich in plankton, it supports an abundance of fish and bottom invertebrates. The important source of its fertility is the supply of ocean water around the Shetland Islands. Vast quantities of protein and other organic matter accumulate in the bodies of invertebrates living on and in the bottom, and many fish are supported by this supply.

The bottom fish of the North Sea and the waters surrounding the British Isles were fished vigorously up to 1914, when catches started to drop off and the average fish size decreased. Between 1914–1918 the war imposed a closed season in the North Sea, and stocks were replenished. Then fishing intensity increased again and by 1938 the fishing curve had dropped. World War II imposed another rest and the stocks improved. So many nations depend on the North Sea fisheries that the mesh size of nets is regulated, but uniformity and compliance is difficult to achieve.

Man has not learned how to intelligently regulate private entry into a public resource. The only conservation and control mechanisms that have been generally used to prevent overfishing have been the legal imposition of primitive methods and other constraints into fishery and shellfish acquisition techniques. Fishermen are therefore faced with the anomaly of being legally constrained to use some of the most primitive fishing methods in the world, while berated for failing to compete successfully.

The extremely effective methods of fish capture already "on the shelf" are outlawed. The Bristol Bay salmon fishery of the Northwest, for example, has outlawed fishwheels, fish traps, and beach seines, which could readily capture the entire legally allowable catch with great reduction in expense. Yet even today the picture of men in small boats, laboriously catching a few fish in the middle of the night in the freezing and dangerous rivers of the Northwest, is an accurate portrayal of the situation. And in most other near-shore and open-sea fisheries, there are similar constraints on net length, areas they may be fished, boat size, and seasons permitted.

A conspicuous exception to this generally bleak picture is offered by the pelagic tuna fishery of the Pacific coast. Because this fishery slowly advanced its operations farther offshore, it has been able to escape serious constraint and hence has been able to develop its own technology freely and effectively. It is thus one of the few, if not the only, domestic fisheries that competes successfully with foreign fisheries.

In contrast to fishery methods in the U.S., fishery research is quite sophisticated. The emphasis has been on studying life histories of pelagic fish. Pelagic fish spawn their eggs freely into the open waters, unlike many near-shore fish, which spawn on some substrate. These developing pelagic larvae therefore become temporary members of the plankton population, and they can be caught and thoroughly sampled with simple gear such as plankton nets. Thus, one can learn much more about the worldwide distribution of pelagic fish, at the most critical stage of their life history.

In the laboratory many pelagic fishes can now be raised in tanks from the egg stage, permitting greater knowledge about their life histories, metabolism, behavior, and growth. The blood type of commercially important fish is an extremely valuable tool for tracing specific populations and differentiating them from others that closely resemble them. Fish tags of various sophisticated kinds, including tagging based on acoustic principles, permit studies of migrations of population.

Fish scales and otoliths (ear bones) help to determine the exact spawning ground from which pelagic salmon originate, and to assess the ages of populations of various pelagic fish in sufficient detail so that their increases and decreases—the dynamics and statistics of the population—can be far better understood. An equally powerful entree into understanding natural changes in the abundance of species has come from studies at Scripps of fossil fish and plankton excellently preserved in certain rare kinds of sediments found in a few basins on the sea floor along the Pacific coast. From these fossils and sediments one can now reconstruct the events of each 3- or 4-year period over the last several thousand years, appraise the success of a species from its abundant variations, and better understand the nature of the relationships between it and its competitors and associates.

Scientific understanding is necessary to predict fishery stocks from year to year. Less than a hundred years ago man believed the sea so vast and the populations of fish so immense that his efforts, however intense, could have no effect on the population. An estimate of the inexhaustible resource of the sea was based on lack of understanding. Now there is evidence of the extraordinary vulnerability of hatching and larval fish to changes in the environment—a drop of a degree in temperature may delay hatching several days, so that the egg hatches at a time when it is most subject to predation.

The decline of the sardine industry along the California and Oregon coast is an example. At the peak of the industry in 1936–39, the catch was used for fish meal for livestock food and oil for industrial purposes.

It is generally believed that the sardines declined because of changes in the temperature of the ocean, brought about by shifts in the ocean currents. However, the very heavy fishing at a period of unfavorable environmental change may also have contributed to the decline of the fish stocks. On the other hand, there is some evidence that a fairly heavy fishery of adults during favorable years might have the reverse effect—that is, removing the mature large fish makes it possible for the young fish to grow faster and replace the older ones that have been removed.

The peak production of California sardines has now been equalled or surpassed by the menhaden fishery of the South Atlantic and Gulf States, which in 1961, accounted for about 45% of the entire fish catch of the U.S., including Alaska. Menhaden are used for fish meal and it is obvious that they have replaced sardines in the economy.

Today's fishing is following essentially the same route as early land hunting: eliminating the higher predators, the bigger fish that prey on preferred marine fish harvest. The adult tuna, mackerel, salmon, halibut, plaice, and cod eat the smaller anchovy, sardine, and shad which offer far greater potential for increasing food production by reason of their lower position in the food chain and consequent greater abundance. By reducing the population of these predators one can expect to increase the yield of smaller fish six to ten times over the population of the higher predators eliminated.

The idea of farming the broad expanse of open sea by using automated ships to gather and process plankton for use directly as food is not practical. The concentration of plankton in the richest surface waters averages only a few parts per million. To harvest this 20 billion tons of diffuse product from the 40,000,000 billion tons of surface water in the sea does not seem economically feasible.

Another constraint on productivity of the open oceans results from trapping of nutrients in deep water below the stable density layers that separate deep water from the surface waters where photosynthesis takes place. To raise this nutrient-rich deep water to the surface and keep it at the surface requires tremendous energy. The only way to overcome this energy barrier is to artificially stimulate overturning and to physically restrain the mechanically raised water by introducing it into a shallow lagoon or somehow impounding it. Otherwise it will promptly return to its own density level.

Finally, we would have to control predation and encourage only carefully bred, efficient and effective domestic fish, which might be easily caught, to forage on the phytoplankton.

The use of fish in its natural state presents certain problems of transportation and preservation since spoilage and decay are rapid and dangerous. Fish flour and fish powder are odorless and tasteless. The protein content exceeds 80%. Neither is designed to be consumed in its basic form as fish flour or powder, but rather as a component of standard cereal grains products. In concentrations of up to 10% they will not affect flavor or odor. These concentrates recently have been added to flour for breads and pastries, used as ingredients in noodles and other farinaceous foods, to rice and corn, and dairy products to provide even greater amounts of protein. It has even been added to high-quality vegetable products to upgrade them even further.

Only hake or hake-like fish are used for these concentrates in the U.S. The addition of $\frac{1}{10}$ lb of concentrate to $\frac{9}{10}$ lb of rice provides a total protein quantity and quality that equals nine eggs. The concentrates are cheap and easy to produce, and could be effective in preventing the protein malnutrition endemic in almost half the world's population. There is a word of caution about these concentrates. Under certain conditions fish become poisonous by concentrating certain chemical compounds in the environment. In only a few cases can these poisonous compounds be identified. In the preparation of fish protein concentrates, undesirable and toxic compounds may be chemically concentrated.

The small size of most of the food in the sea is one of the most serious constraints imposed by nature. In the ocean, unlike the land, most of the food is cycled through all but the final portions of the food chain in microscopic steps. The reasons for this are obscure, but they probably relate to the absence over most of the ocean of a substrate in which large plants can attach at depths shallow enough for them to receive the light needed for photosynthesis.

Each of the microscopic steps up the marine food chain is perhaps only 10 to 15% efficient, so of that great amount of potential food initially fixed by photosynthesis in the phytoplankton, only a very small portion emerges in the form of fish to be caught.

The codfish, for example, eats predatory crustaceans and gastropods and small fish, which in turn eat small herbivorous invertebrates. Thus, each million pounds of cod production requires 10,000,000 lb of smaller fish and crustaceans, which in turn, requires 100,000,000 lb of small herbivorous invertebrates. These in turn consume ten times their own weight in plant matter so that the cod requires 1000 times its weight in plant matter and a vast amount of bottom acreage.

About 70% of the total annual food requirements of the adult winter flounder are met by detritus and phytoplankton-eating invertebrates. Not more than 80g/m^2/year are available in estuaries. This is equivalent to 320 kg per acre per year, and will support 45.6 kg of flounder. Thus the maximum flounder production would be 100 lb per acre per year assuming no competition from other species and no mortality. A realistic production might be 25–30 lb per acre per year.

The direct requirements to develop 100 lb of flounder could be 300 lb of algae. The remainder of the diet, the 80 g per m^2 of bottom feeding invertebrates require 7000 lb of feed per acre per year. The total direct and indirect requirements of vegetable matter for the production of 180 lb of winter flounder is 7300 lb.

For this reason, large-scale controlled fish farms can be achieved only in restricted lagoons and estuaries, where the migratory propensities of genetically improved fish can be curbed, and where fertilization of the water with essential nutrients is accomplished by artificially induced upwelling or by direct introduction, and where both nutrients and fish can be retained.

In shallow waters, rice paddies, or estuaries the autotrophic (plants) and heterotrophic (animal) layers are in close contact. In the sea the autotrophs are small and the heterotrophs are large, whereas on land the autotrophs are large (trees) and heterotrophs are small. Organic detritus is the chief link between these levels of primary and secondary productivity, rather than a grazing food chain as on land. Although these productive areas may be less stable than the land, the advantage in such a detritus food chain is that microbial manipulations can be used to produce protein, which is available as food. Thus an estuary has food all the time for its populations.

Certain steps in the food chain of the sea can be circumvented increasing the ultimate yield by a factor of six to ten for each step circumvented. A filter feeding creature like a mussel, clam, or oyster can aggregate phytoplankton in one step into protein. Instead of moving uneconomically large masses of water to extract the plankton, man can, at low cost, use such efficient creatures to concentrate the plankton wherever the natural movement of the water is sufficient to keep replenishing the plankton food supply. The concentration of plankton by natural advection is important in many places in the sea—at boundaries between current systems and around islands and shoals—and this can be a highly efficient concentrating mechanism. It can increase the local supply of plankton by a factor of 50 or more over the basic productivity of the contributing water.

The coral ring atolls of the Pacific, many of them enclosing shallow lagoons hundreds of square miles in area, are natural sites for fish farming. In some of these atolls phosphate rock, deposits of potential on-site fertilizer, can be introduced directly into the lagoon. Then small power plants can fix the added nitrogen needed to maintain a sustained level of fertility in these enclosed waters.

Before man can use these sites for large-scale experiments, he has much to learn about fish communities, the development of genetic strains with desirable characteristics, control of predators and the inevitable diseases, and the control of unwanted plants and animals. This intensive kind of fish farming is eventually practical and realizable. It is being done now on a small scale in many restricted bays, estuaries, ponds, and lagoons in China, Japan, India, Scotland, and Israel.

For such farming attempts, cold-blooded marine creatures other than fish may prove to be especially advantageous. Experimental culture of sea turtles now going on in Malaya and the Caribbean, for example, look quite promising. In fact, the great and often overlooked range of genetic material among the cold-blooded creatures of the sea may prove to be the greatest exploitable aspect of its living resources.

GUY C. MCLEOD

References

"Chemistry and the oceans," *Chemical & Engineering News*, June 1, 1964.
Lauff, G. H. (Ed.), "Estuaries," No. 83, 1967, American Assn. for the Advancement of Science, Washington, D.C.
Terry, R. D. (Ed.), "Ocean Engineering," Vols. 1–8, 1966, North American Aviation, Inc., El Segundo, California.

"The ocean," *International Science and Technology*, February, 1966.
"The World Food Problem," Vols. I and II, 1966, U.S. Government Printing Office, Washington, D.C.

Cross references: *Coral Reefs*; *Estuarine Environment*; *Fish Protein Concentrate*; *Tuna Fisheries*.

SEA LAMPREY

Distribution

The sea lamprey, *Petromyzon marinus* (family Petromyzonidae), is found in the Atlantic Ocean from Iceland and northern Europe to northwestern Africa and in North America from the Maritime Provinces of Canada to Florida. Although usually found in marine waters, the lamprey has adapted to life in the fresh waters of the Great Lakes and in several lakes of New York. Construction of canals enabled lampreys which had deserted the Atlantic Ocean for Lake Ontario to move farther west. For a time Niagara Falls blocked their passage from Lake Ontario, but completion of the Welland Canal in 1829 gave them a way into Lake Erie. This passageway enabled them in about 135 years to destroy the 10,000,000-lb lake trout fishery of the Great Lakes.

Even after the Welland Canal was completed, the lampreys seem to have been slow in establishing themselves in Lake Erie. The first lamprey was caught there in 1921. Because the waters of Lake Erie were too warm and spawning conditions poor, the lampreys did not thrive there. By the 1930's, however, they reached Lakes Huron and Michigan, where food, cold waters, and clear, gravel-bottomed tributary streams were ideal for growth and survival. Then they moved toward Lake Superior, but the dam and the navigation locks at the head of St. Mary's River slowed the rate of their invasion into this lake; but enough arrived in the lake to establish a rapidly growing population. The first lampreys were taken in 1946 off Isle Royale and Whitefish Point.

Sea lampreys have been abundant for centuries in several lakes of New York, including Cayuga Lake. They may have entered by way of the "Champlain Sea" or the Hudson-Champlain estuary and Mohawk outlet.

Natural History

The life history of the sea lamprey comprises three distinct phases—spawning, larval, and adult.

Adult lampreys congregate in late winter in bays and estuaries. Before beginning the spawning migration, which may last 20 weeks, their body changes. The sex glands grow enormously, the muscles, skin, and eyes degenerate, and the digestive tract shrinks. The lampreys do not feed, but live on stored fats and body tissues. When streams warm to about 40°F, lampreys ascend those that contain gravel, clear water, and a moderately strong current.

A male chooses a satisfactory spawning site in the stream and starts building a nest. A female soon joins him. They clear a small area, picking up stones with their mouths and piling them in a crescent-shaped mound on the downstream side of the nest. If the water temperature is over 50°F when they finish the nest, they begin to spawn. The female, averaging 61,500 eggs, lays a few eggs at first, and the male immediately fertilizes them. The current carries the eggs to the rim of the nest where they lodge in the spaces among the stones. The female lays more eggs, and the process is repeated. The pair continues laying and fertilizing eggs until they are spent. Spawning may take 1 to 3 days and then both die within a few hours and decompose rapidly. Those lampreys that do not have an opportunity to spawn die also.

Depending on the water temperature, hatching occurs in 2 to 3 weeks. Less than 1% of the heavy, small, shell-less eggs hatch. The newly hatched larvae remain in the nest until about the 20th day. Then about ¼ in. long, they drift downstream to quiet waters. In the soft bottom each larva digs a burrow that will be its home for about 5 years unless erosion washes it away. Protruding the head slightly from the burrow, a larva sucks food, mainly microscopic organisms, from the water passing the mouth of the burrow. A filtering apparatus in the throat passes food organisms to the digestive tract and keeps out debris. Throughout their larval life the young lampreys, celled ammocetes (Fig. 1), are blind and harmless.

After about 5 years a larva becomes adult. It develops large, prominent eyes, a round mouth lined with horny teeth, a filelike tongue, and enlarged and unpaired fins. Its slim body, with a soft skeleton of cartilage rather than bone, becomes dark blue above and silvery white below. Now 4–7 in. long, it may emerge from its burrow when late autumn rains raise the stream level, but usually it waits until the spring ice breakup and high water before migrating downstream to the lakes to begin the parasitic phase of its life.

The adult lamprey is a jawless predator that does not school and has no known enemies; it attaches its suckerlike mouth (Fig. 2) to almost any part of a fish, which thrashes about violently, but rarely shakes off the lamprey. Sometimes several lampreys feed upon one fish at the same time. The strong teeth and the rasping tongue soon penetrate the fish's scales and skin. Lamphredin, a substance in the lamprey's saliva, dissolves the torn flesh or other parts of the fish—eggs from the ovaries or stomach and intestinal

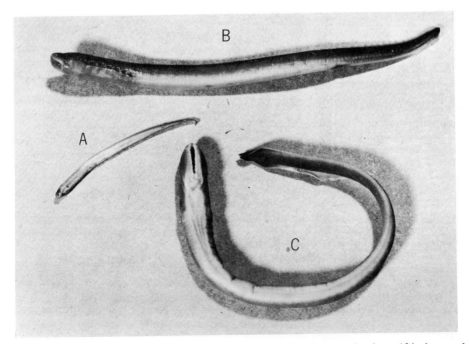

Fig. 1. Stages of development in the sea lamprey of the Great Lakes. A, an eyeless larva 1¾ in. long and about 1 year old. B, a 3-year old larva 4½ in. long in the nonparasitic life history stage. C, a juvenile parasitic lamprey that recently metamorphosed from the nonparasitic form. (Photo: U.S. Bureau of Commercial Fisheries.)

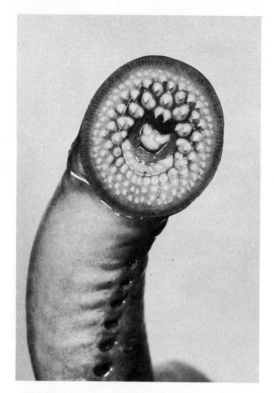

Fig. 2. The mouth of the sea lamprey is lined with horny teeth. The rasping tongue is in the center of the mouth. (Photo: U.S. Bureau of Commercial Fisheries.)

contents—and keeps the blood from clotting. Feeding stops when the fish dies or the lamprey becomes glutted. A lamprey may remain attached to a fish for weeks, but some fish may die in as few as 4 hours. If a fish escapes, it is so badly scarred that it is often unmarketable.

The lamprey, which destroys no less than 20 lb of fish in its life, grows rapidly. It becomes 12–24 in. long and weighs about 8 oz; rarely does it weigh more than 1 lb.

The effect of the lamprey attacks on the lake trout, the most prized food fish because of its fine, delicate flavor and the mainstay of the fishing industry of the Great Lakes, was devastating. In 1944, commercial fishermen caught 10,604,000 lb of lake trout in the Great Lakes, valued at $3,792,000. In 1964, only two decades later, their catch declined to 102,000 lb, with a value of $67,000. Lampreys preyed heavily not only on the large lake trout of Lakes Huron, Michigan, and Superior, but also on the larger chubs and whitefish. Like the lake trout, chubs and whitefish inhabit the deep-water environment, the domain also of the lampreys. Losses to other channels in the trade and to sport fisheries caused by the lamprey cannot be estimated.

Great Lakes Fishery Commission

Because the catches of lake trout continued to decrease each year, the Great Lakes States, the Province of Ontario, and the Federal Governments of Canada and the U.S. joined efforts to eradicate or control the lamprey. In 1956, however, their objective became one of the responsibilities of the Great Lakes Fishery Commission, established in 1955 by the Convention on Great Lakes Fisheries between Canada and the U.S. The Department of Fisheries of Canada and the Bureau of Commercial Fisheries of the U.S. Fish and Wildlife Service are under contract to the Commission to conduct research on the lampreys.

Research

To control the lamprey, the fishery scientists of Canada and the U.S. tried to find (1) commercial uses for the lamprey and (2) vulnerable periods in its life. They were unsuccessful in the first effort. The people of the U.S. do not find them palatable, although many people of Asia, Australia, Europe, and New Zealand relish them. Analyses indicate that vitamin A potency and oil yield of the lamprey are too low for commercial exploitation. Biological supply houses require only a few lampreys as study specimens.

Greater success attended the second research enterprise. It was found that the most vulnerable periods in the lampreys' life occur when they are in streams, either as adults migrating from the Great Lakes upstream to spawn and die, as juveniles migrating downstream to prey on lake trout and other fish, or as larvae burrowing in the soft mud bottom.

Control Measures

To control the lampreys, the fishery scientists designed and installed mechanical weirs and electric weirs in streams. At the same time they were seeking other control measures and began testing chemicals. Because the chemicals proved more effective in lamprey control than weirs, the weirs are now used only for assessing the size of lamprey populations.

Mechanical weirs were installed to prevent adult lampreys from entering streams to spawn, but they proved expensive and undependable and flash floods washed them out.

The fishery scientists then developed a combination of mechanical and electric weirs. An electric field produced in a stream by electric weirs blocked the upstream migration of lampreys. These weirs were superior to the mechanical variety because they were less likely to be washed out by floods, did not clog with debris, and were cheaper to install and maintain. Fish and some lampreys enter traps placed at each end of the electric weir, while others penetrate the electric field and are killed. A different type of weir, which is energized by pulsed, direct-current electricity, is used in streams where movements of important food fish coincide with the lamprey migration. This weir (Fig. 3) guides the food fish and most of the lampreys into traps where they can be sorted and separated. The lampreys are destroyed and the other fish are passed upstream.

Attempts were made to stop the downstream migration of juvenile lampreys. Mechanical, inclined-plane screens that strain all the water of a stream were installed. These are extremely vulnerable to floods and accumulations of debris, which occur during the peak of the downstream migration of lampreys. Replacing these screens with devices designed to electrocute immediately all downstream migrants is not economically feasible. Voltages that kill fish only stun young lampreys and enough electricity to kill lampreys at this stage would cost an exorbitant amount.

The fishery scientists next considered using selective poisons to kill juvenile lampreys. After 3 years of testing some 5000 chemicals they found that halogenated nitrophenols successfully kill larvae in streams but do not harm fish. One of these chemical lampricides, 3-trifluormethyl-4-nitrophenol, was used to destroy lamprey larvae in streams tributary to Lake Superior. Addition of this chemical to lamprey-infested streams requires great skill and precision by research teams. The procedure first involves a stream survey with portable electric shocking devices that drive buried lamprey larvae out of the bottom for capture and counting. A two-man crew makes the survey, and a record is made of the abundance and distribution of larvae. If larvae are present, other fishery scientists map the stream, measure water flows at a number of sites, run analyses of the chemical properties of the water, and determine the points where the chemical should be introduced to provide complete coverage.

Just before treatment, a test is run in which lamprey larvae and game fish are placed in a series of jars containing aerated stream water at the prevailing temperature. Into the jars varying amounts of lampricide are added to determine the concentration and time of exposure that should be used to obtain a complete kill of lamprey larvae and a minimum loss of game fish.

When this pretreatment information is collected and analyzed, lampricide is introduced by proportioning pumps into an infested stream at a rate to achieve the desired concentration throughout the water course. Chemical tests are made at numerous sites during the period of treatment to ensure an adequate concentration of lampricide.

Fig. 3. Electric barrier (weir) in use on Pere Marquette River near Ludington, Michigan. On the left the alternating current electrodes are suspended upstream to prevent the passage of fish and lampreys. Downstream on the right is a suspended direct current electrode array used to guide fish and lampreys migrating upstream into two traps. (Photo: U.S. Bureau of Commercial Fisheries.)

After the treatment, fishery scientists traverse the stream with electric shockers to check on the presence of live larvae. Seldom are live larvae found because of the care taken in treating the stream. If many live larvae are found, the stream is retreated.

Chemical control operations by Canadian and U.S. research teams began on streams tributary to Lake Superior in 1958. By the end of 1965 all lamprey streams, except a small one at the northern end, had been treated once and 74 streams twice or more to eliminate re-established populations. As a result of this control effort the size of the adult lamprey population in 1962 declined 82% from the 1957–1961 level.

During 1963 to 1965 the population of adults increased slightly but remained 77% below that of the 1957–61 level; however, another decrease occurred in 1966 and 1967, which was 92% of the mean high abundance. Adult lampreys that congregate in the autumn below the locks connecting Lake Superior and Lake Huron and at other locations in northwestern Lake Huron are being tagged. Recaptured lampreys will reveal their contribution to lamprey populations in Lake Superior and Lake Michigan.

In Lake Michigan all 99 lamprey streams have been treated once and 42 streams twice since 1960. The catch of spawning lampreys at three assessment weirs has declined to about 30% of the average catch for the period 1958–1962.

During 1965, fishery scientists continued surveys in Lakes Huron, Erie, and Ontario to locate lamprey streams and gather information for planning chemical operations. They are continuing research also on chemicals or combinations of chemicals that may be effective lampricides.

After their chemical treatments of the tributary streams of Lake Superior and Lake Michigan had substantially reduced the lamprey populations, the fishery scientists of the U.S. and Canada further advanced lake trout rehabilitation. In 1958 they began planting hatchery-reared yearling lake trout, produced in State, Federal, and Provincial hatcheries, in Lake Superior. Their plantings in 1967 brought the number of trout planted in Lake Superior to 19.1 million. As a result of these efforts, the lake trout population in Lake Superior has been increasing at an annual rate of about 25% in recent years. In 1965 they made their first effort to restore the trout fisheries in Lake Michigan by planting nearly 1.3 million lake trout. The total plant of lake trout was 4.8 million by the end of 1967.

LOLA T. DEES

SEALS—*See* SOUTH AFRICA, MARINE RESOURCES; GOVERNMENT DEVELOPMENT OF MARINE RESOURCES

SEA SCALLOP FISHERY

Populations of the largest American scallop (*Placopecten magellanicus* Gmelin) occur from Labrador to New Jersey. Individual animals with an estimated age of 18–20 years and a diameter of nearly 9 in. (225 mm in height and 210 mm in width) have been recorded in Maine.

Posgay (1957) has stated that the greatest known sea scallop grounds are found between the 20- and the 50-fathom curves on Georges Bank. The fishery had its beginning in Maine waters where populations occur discretely in major estuaries and embayments from the Piscataqua River, separating Maine and New Hampshire, to the St. Croix River, which forms the international boundary between Maine and New Brunswick. Vertical distribution in Maine ranges from mean low water in some areas to depths of several hundred feet in others. Concentrations of commercial importance are generally limited to the area from Penobscot Bay eastward. Sporadically, commercial fishing has been carried on in western Maine waters. Frequently scuba divers are able to gather scallops in commercial quantities from rocky bottoms

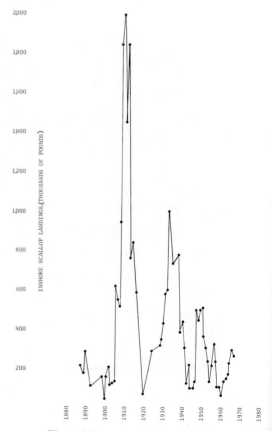

Fig. 1. Maine sea scallop production.

that are impossible to drag with conventional scallop gear.

In its publications *Fishery Statistics of the United States*, the Bureau of Commercial Fisheries does not differentiate between landings of bay and sea scallops except in New England. Table 1 shows the historic catch of sea scallops in New England only, as reported by the Bureau from what prior to 1939 consisted of sporadic sampling of catch. These data show considerable departure in some years from those reported by the Department of Sea and Shore Fisheries for the State of Maine alone (Fig. 1). Landings of sea scallops have also been made for many years at New York and New Jersey ports, but data cannot be included in Table 1 for lack of statistical separation from bay scallop catches.

In 1929 it was reported that only one vessel from Boston made a trip to Georges Bank for scallops in 1928, and in 1930 only Boston scallop boats were reported to have fished on Georges Bank. It is not clear when Maine fishermen commenced scalloping on Georges. In 1936 the Commissioner of Sea and Shore Fisheries reported that "several boats sail each summer from Maine ports to Georges Bank after scallops." No offshore scalloping by Maine boats was done prior to 1931.

Several sizes of dredges have evolved in the sea scallop fishery. Those used inshore differ greatly in size from those used on Georges Bank. In Maine a single 4-ft drag is generally used although tandem 3-ft drags have been popular. Only on occasion have drags as wide as 6 ft been used, and then only when it was presumed the scallop population was sufficiently great to make the larger size economically practical. Royce (1947) described the 11-ft dredge that, he stated, had become fairly standard in the fishery, although smaller boats used dredges from 8–10 ft in width. Larger vessels have experimented with 13-ft gear.

The Maine Fishery

The fishery appears to have had its beginning about 1880 but records of landings are extant only since 1887 and then with several gaps, principally in the 1890's and between the end of World War I and the 1930's. Consecutive annual landings data are limited to the period since 1938.

The open season for fishing the very limited inshore stocks from November 1 through the following March 31 has also been affected by two other considerations: (1) the low water temperatures of the winter make the scallops less active and, therefore, less able to escape the small inefficient drags used in the fishery; and (2) it provides off-season employment for part-time fishermen. Since the catch is reported by calendar year, each year's landings include portions of two fishing seasons.

Records of commercial landings in Maine (Fig. 1) date from 1887; since then, production and value have fluctuated widely from year to year as well as over periods of years. Production gradually increased from the 1880's to an all-time peak of more than 2,000,000 lb in 1910. A general decline in abundance followed, with a 75% reduction in catch by 1916.

During World War I, landings dropped rapidly and in 1919 amounted to 73,000 lb, the second lowest production on record since 1899. During the 1920's landings gradually increased and reached a second peak of slightly more than 1,000,000 lb in 1933. Although production remained relatively high during the 1930's, there was a continuous decline that terminated in a new annual low of approximately 100,000 lb in 1944. Landings gradually increased during the late 1940's to a new cyclic peak of 500,000 lb. in 1950.

Landings and Abundance

Production has fairly closely followed cyclic trends, with alternating peaks and lows at about 10-year intervals (Table 2). Peaks of production occurred in 1889, 1910, 1933, and 1950. Low points in landings came in 1899, 1919, 1944, and 1960. The next abundance peak will probably occur in the early 1970's. In the 40-year period since 1910, each of the two cyclic peaks has declined 50% from the preceding high. The

TABLE 1. NEW ENGLAND SEA SCALLOP CATCH 1887–1965

Year	Thousands of Pounds (Meats)	Year	Thousands of Pounds (Meats)
1887	247	1944	4,263
1888	201	1945	3,994
1889	311	1946	9,578
1898	289	1947	13,039
1902	134	1948	12,483
1905	440	1949	13,980
1908	1409	1950	13,753
1919	196	1951	14,444
1924	435	1952	15,392
1928	475	1953	19,987
1929	824	1954	15,594
1930	947	1955	16,848
1931	1081	1956	16,881
1932	1572	1957	18,781
1933	2158	1958	16,410
1935	1670	1959	20,259
1937	5730	1960	22,462
1938	5850	1961	23,775
1939	7178	1962	21,724
1940	5391	1963	17,794
1942	6164	1964	14,536
1943	4842	1965	12,335

TABLE 2. MAINE INSHORE SCALLOP LANDINGS SELECTED YEARS 1889–1960

Year	Weight of Meats (000 lbs)
1889	295
1899	53
1910	2027
1919	73
1933	1073
1944	101
1950	512
1960	72

consistency with which periodic highs have alternated with lows in landings at approximately decade intervals suggests, with what else is known of the fishery, that scallop abundance has likewise fluctuated in the inshore growing areas of Maine.

This assumption has been supported by the biological sampling of M. S. Chrysler (1920) in 1917, and of the Department of Sea and Shore Fisheries from 1949–1953 and in 1957–1958, which forecast the general trends landings were to take during immediately subsequent years.

At intervals since the early 1930's the scallop fishery in Maine has been divided into an inshore open season, small boat fishery, and a year-round, offshore, large boat fishery. The inshore fishery is carried on during a fishing season covering parts of two calendar years. The present open season was established in 1947. Prior to 1947 the open season had been from the first of December through April 14 of the year following. During World War II the closure was suspended and fishing was permitted throughout the year. Data of the several fisheries are shown in Table 3 for the first decade after World War II.

Sampling of the catch from 1949–1953 and again during 1957–1958 by the Department of Sea and Shore Fisheries indicated that toward the end of each fishing season (January–March) an increasing number of scallops enter the fishery after completing their sixth growing season. The first major contribution of a year class population to the fishery occurs in the sixth year after spawning. By this time the population of older and larger scallops has been so reduced by fishing that fishermen depend upon the new crop for a continuing source of supply. The sixth year population contributes from 25 to 40% of the calendar year catch. Each year class is fished twice during the calendar year. The January–March catch represents some 12 to 15% of the landings for the year of the six-year-old stock. Following the seventh growing season the same year class adds another 13 to 25% to the year's catch; the total contribution varying with the relative importance of the year class to the total available population. Landed value for the shucked meats, which has ranged from six cents in 1887 to 70 cents in 1965, with a median of 37 cents per pound, has continuously attracted intensive commercial activity; for example, studies have shown (Dow 1956) a correlation of .74 between average annual price paid fishermen and offshore landings during the post-World War II period. By contrast, an inverse correlation ranging from −.4 for the 1899–1955 period to −.7 for the post-World War II period suggests

TABLE 3. MAINE SCALLOP FISHERIES

Calendar Year	Fishing Season	Calendar Year Landings	Inshore Landings	Winter Fishery Landings
1946		136,531	136,531	
	1946–47			178,507
1947		507,032	507,032	
	1947–48			254,379
1948		453,686	453,686	
	1948–49			453,600
1949		508,916	508,916	
	1949–50			469,741
1950		524,824	511,783	
	1950–51			543,039
1951		676,803	377,548	
	1951–52			490,925
1952		1,495,754	313,890	
	1952–53			679,655
1953		1,697,172	242,958	
	1953–54			272,537
1954		707,758	143,741	
	1954–55			386,647
1955		1,113,564	219,313	
	1955–56			513,196

Fig. 2. Scallops of three different year classes of approximately equal diameter.

that catch generally influences price in the inshore fishery.

The beginning of a new inshore fishing season, with the accumulated scallop growth of the preceding growing season, a seasonally undisturbed population, and comparatively better fishing weather, account for the higher average landings during November when an average 32% of the winter catch is made. Catch declines rapidly in December to 21%, to 17% in January, and to a low of 14.8% for the season in February. In March average production increases slightly to 15.2%. During the winter, average landed price increases steadily to an average seasonal peak in March, indicating that as catch declines seasonally, a steady market forces price up.

Fishing Efficiency

Dragging experiments indicate that the commercial scallop dredge does not have a high degree of efficiency (Fig. 3). Tag returns (Baird, 1954) likewise support these observations. Estimates of efficiency range from less than 5% to not more than 15%. Even if a dredge is operating properly on smooth bottom, the rapid short range mobility of the scallop permits the animal to evade the gear and frequently to escape. The demonstrated decrease in scallop activity when water temperatures are low suggests that the scallop is less likely to escape during the winter months. This advantage may be more than offset by unfavorable weather conditions during the winter. It has been estimated that only about one-third of the average open season is utilized by commercial fishermen because of unfavorable weather.

Biological Research

Not only did the commercial fishery for sea scallop have its beginning in Maine, but biological research on the scallop was begun there also. Although research in Maine has been intermittent, the location of the inshore populations and relatively easy access to growing areas has

Fig. 3. A large scallop dredge used in Penobscot Bay.

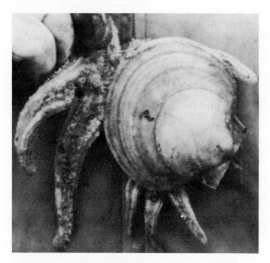

Fig. 4. Scallop being attacked by two starfish.

generally made it possible to obtain biological information not easily attainable elsewhere. Unfortunately no research program has been sufficiently long-lived to produce all the information necessary to a reasonably thorough understanding of the resource.

Extensive scallop mortalities have been reported from time to time. A short-term investigation was carried on in 1917 by M. A. Chrysler of the University of Maine Biology Department as a result of mass mortalities in Penobscot Bay, reported by commercial fishermen during the preceding winter. Although Chrysler admitted the information was inadequate to justify conclusions, he was of the opinion that predation by starfish could account for the decline in population. He wrote: "The only piece of evidence against this view is the apparent greater mortality among mature scallops. In the absence of other evidence this might imply that some disease was working among the older scallops . . .".

Not until 1947 did the Department of Sea and Shore Fisheries again become involved in biological research on the sea scallop. At that time a graduate student, Walter Welch, now a Bureau fishery biologist, was employed on a fellowship basis to study scallop growth rates and spawning behavior. This program was continued until 1953 by Frederick T. Baird, Jr., a department biologist. Both Welch and Baird found no measurable migration of scallops. Posgay has reported essentially the same tagging results on Georges Bank.

Another phase of the research program was a study of the early life of the scallop. Spawning was induced in the scallops being held at the laboratory, but survival did not extend beyond the trocophore stage.

Dragging operations aboard a commercial fishing boat in March 1951 brought up large numbers of young scallops attached to a bryozoon of the genus *Gemellaria*, probably of the species *loricata*. The bryozoon was attached to living and empty shells of large scallops. The smallest of the young scallops was about 1 mm in diameter.

Samples of the bryozoon with attached larvae were transferred to the laboratory where they could be observed. The scallops lived and grew well. They detached from the bryozoon when they had reached some 4–5 mm in diameter. The small scallops then attached themselves to the shells of larger scallops and to the sides and bottom of the holding tanks.

Dragging operations during the following summer by research personnel indicated that attachment to scallop shells by the young scallops (from the previous year's spawning) lasted from about the time they were 4–5 mm in diameter until they had grown to about 10 mm in diameter. Beyond this size, the young scallops apparently did not rely upon byssal attachment, for they were picked up from the bottom when a small mesh liner was used in the drag. During the shell attachment period, young scallops were

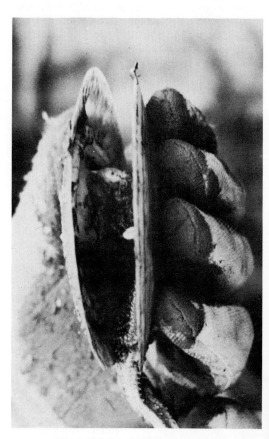

Fig. 5. Starfish feeding on scallop.

found on other marine mollusk shells and on bottom debris.

The habit young scallops have of attaching themselves to shells and other anchors by means of their byssus has led commercial fishermen to return shucked shells to the grounds from whence they came. Since the very small scallops apparently attach themselves to bryozoa at some time during the larval period, it would appear that this process may be an essential part of the scallop's early life history.

Observations made on scallops held at the research laboratory in Boothbay Harbor indicate that scallops up to about 3 in. in diameter are active for about 6 months each year; except during the spawning season (August–October) and during the winter (January–March). As the spawning season approached, they attached themselves to the bottom of the holding tanks by means of their byssus.

Growth Rates

In Maine inshore waters, sea scallops grow rapidly. Attained sizes at various ages have been determined by Baird as follows:

Growing Season	Age	Size
1	½ yr	2 mm
2	1½ yrs	5–12 mm
3	2½ yrs	2.2 in.
4	3½ yrs	2.9 in.
5	4½ yrs	3.5 in.
6	5½ yrs	4.1 in.
7	6½ yrs	4.4 in.
8	7½ yrs	4.7 in.
9	8½ yrs	4.9 in.
10	9½ yrs	5.1 in.

and meat yields (adductor muscle only) are related to shell size:

Size	Meat Yield (Oz)
2.0 in.	.092
2.4 in.	.15
2.8 in.	.25
3.1 in.	.36
3.5 in.	.62
3.9 in.	.67
4.3 in.	1.02
4.7 in.	1.23
5.1 in.	1.50
5.5 in.	1.90
5.9 in.	2.07
6.2 in.	2.38

Natural Mortality

Assessment of natural mortality, other than those obvious cases of mass mortality, has been extremely difficult. Mass mortalities, such as those reported by Chrysler in 1920 and by Dickie

Fig. 6. Close-up of tagged scallop.

and Medcof in 1963, can frequently be identified as occurring for the individual scallop at about the same time. Equally frequently the cause cannot be adequately identified.

In commercial catches or biological sampling varying numbers of empty paired scallop valves will be recovered together with living animals, single valves, other species, and an assortment of undifferentiated debris including cobbles and boulders. It has been tempting for biologists to use paired valves in which the hinge and ligament are intact as evidence of recent mortality. How recent the mortality is may be as much as several years elapsed time. In the early 1950's when Maine biologists were studying the sea scallop population, several experiments were set up in an effort to evaluate the use of paired valves as a basis for natural mortality assessment.

Intact empty shells or "clappers" from commercial fishing operations, from research sampling, and from freshly shucked animals were held under various natural and simulated conditions: on sea bottom below mean low tide, intertidally, in suspended mesh containers, in running seawater tanks, and in aquaria with living scallops. Periodic examination of the shells for more than a year (in some experiments up to two years) did not indicate any reliable means of estimating "recent" mortalities by this method.

Seawater Temperature

The only data that indicate why abundance fluctuations have taken place are records of seawater temperature taken at Boothbay Harbor by the U.S. Fish and Wildlife Service or its predecessor agencies since March 1905. Since the offspring of any year's spawning—August–October—becomes of major importance to the fishery 6 years later, it appears from a study of

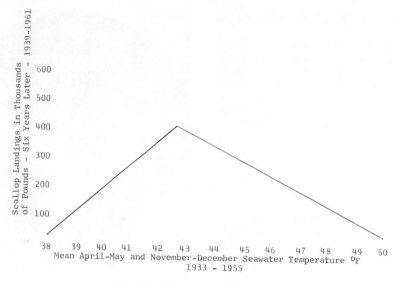

Fig. 7. Maine inshore scallop fishery.

TABLE 4. RELATIONSHIP BETWEEN SEAWATER TEMPERATURES SIX YEARS EARLIER AND HIGHS AND LOWS IN MAINE'S SCALLOP LANDINGS

Seawater Temperature		Scallop Landings (Meats only)	
Year	°F	Year	000 lbs
1913	47.4	1919	73
1927	46.2	1933	1073
1938	45.1	1944	101
1944	46.5	1950	512
1954	50.2	1960	72

temperature and production records that an association exists between seawater temperature 6 years earlier and highs and lows of scallop landings. The values in Table 4 suggest an optimum temperature of 46.0°F to 46.5°F for peak landings. Less favorable high and low temperatures yield the lowest landings.

Inshore scallop populations in Maine declined steadily in Maine after 1950, and by 1960 had reached the lowest production on record since 1899. During the period 1947–1950, annual landings from the inshore grounds had averaged 500,000 lb. A consistent decline occurred after 1950, and in 1960 the catch was only 72,000 lb.

Studies of the apparent relationship between seawater temperature before and after the August–October spawning period suggest that this is an important factor in determining the survival of each year class. On the basis of this apparent relationship, predictions were made in 1962 of the probable future abundance of inshore scallop stocks in Maine, based on seawater temperature only. Table 5 shows these forecasts and actual landings to 1967.

These forecasts were based on monthly mean temperatures, including some temperature estimates reported by the Bureau of Commercial Fisheries. Recently Bureau personnel have corrected mean temperature records (Welch, 1967), and, because of lack of daily observations, have deleted several previously reported monthly means. For this reason, a new base (April–May and November–December) temperature has been calculated. Although predictions derived from this curve do not differ substantially from those based on average values for the period, Fig. 7 graphically shows the apparent relationship between seawater temperature and the relative abundance and landings of inshore sea scallops. With a linear correlation value of .8 between the two series, factors other than seawater

TABLE 5. FORECASTS OF SCALLOP ABUNDANCE

March–April and Oct.–Nov. Mean Temp. °F		Predicted Landings (000 lbs)		Landings (000 lbs)
45.6	1956	240	1962	154
45.3	1957	270	1963	175
44.2	1958	350	1964	233
43.7	1959	370	1965	297
43.9	1960	350	1966	271

temperature fluctuations appear to account for less than 40% of the variance in production.

The lack of positive correlation between Maine and New England (i.e., Georges Bank) scallop landings, and in fact a generally inverse relationship between the two, suggests that seawater temperature is not as important a factor in influencing abundance on Georges Bank as it has been with respect to the inshore growing areas of Maine (Dow, 1962) and Canada (Dickie, 1958). Or if temperature is as important, conditions on Georges Bank and on the inshore grounds have been quite different.

ROBERT L. DOW

References

Baird, F. T., Jr., "Observations on the early life history of the giant scallop (*Pecten magellanicus*)," 1953, Maine Dept. of Sea & Shore Fisheries, Research Bull. No. 14.

Baird, F. T., Jr., "Migration of the deep sea scallop (*Pecten magellanicus*)," 1954, Maine Dept. of Sea & Shore Fisheries. Fisheries Circular No. 14.

Baird, F. T., Jr., "Meat yield of Maine scallops," 1954, Maine Dept. of Sea & Shore Fisheries. Research Bull. No. 16.

Dickie, L. M., "Effects of high temperature on survival of the giant scallop," *J. Fish. Res. Bd. Canada*, **15**, 1189–1211 (1958).

Dow, R. L., "The Maine sea scallop fishery," 1956, Maine Dept. of Sea & Shore Fisheries. Fisheries Circular No. 19.

Dow, R. L., and Baird, F. T., Jr., "Scallop resource of the United States Passamaquoddy area," 1960, U.S. Fish & Wildlife Serv. Special Scientific Report. Fisheries No. 367.

Medcof, J. C., and Bourne, N. "Causes of mortality of the sea scallop, *Placopecten magellanicus*," 1964, Fish. Res. Bd. Canada.

Merrill, A. S., Posgay, J. A., and Nichy, F. E., "Annual marks on shell and ligament of sea scallop (*Placopecten magellanicus*)," U.S. Fish & Wildlife Serv. *Fishery Bulletin*, **65**, No. 2 (1961).

Premetz, E. D., and Snow, G. W., "Status of New England sea-scallop fishery," *Comm. Fish. Rev.*, **15**, No. 5, 348 (1953).

Royce, W. F., "Gear used in the sea scallop fishery," 1947, U.S. Fish & Wildlife Serv., Fishery leaflet 225.

Welch, W. R., "Growth and spawning characteristics of the sea scallop, *Placopecten magellanicus* (Gemlin), in Maine waters, Univ. of Maine, unpublished manuscript.

Welch, W. R., "Monthly and annual means of surface seawater temperature, Boothbay Harbor, Maine, 1905 through 1966," May, 1967.

SEA TURTLES

Five species of sea turtles, each belonging to a different genus, occur throughout most of the temperate and tropical seas of the world. Although one may occasionally come into a bay or estuary, for the most part all species are restricted to the open sea. Sea turtles, and only sea turtles, have paddle-like legs or flippers (Fig. 1).

Fig. 1. Hatchling loggerhead sea turtle. Note paddle-like flippers. (Photo: Marineland of Florida.)

Little is known of the biology of sea turtles at sea, but considerable information on their habits has been gained by tagging animals as they come ashore to nest and by their subsequent capture in commercial fisheries.

Nesting by sea turtles in the U.S. is mainly concentrated along the southeastern Atlantic coast and in the Gulf of Mexico. All five species are known to nest in that area, but only the loggerhead is common. The leatherback and the green turtle nest occasionally in southeastern Florida, but their principal nesting grounds lie farther south in the Caribbean and West Indies. The ridley occasionally nests in southern Texas, but for the most part nests farther south in Mexico. The hawksbill, primarily a tropical species, may still nest occasionally in Florida, but there are no recent records for this. Pacific sea turtles may occasionally nest in southern California, but they are apparently uncommon there and generally nest only in Lower California, or on still more southern mainland beaches of the eastern Pacific. Green turtles, at least, may nest on some of the Hawiian islands.

Nesting takes place during late spring and summer, and based on the results of tagging, a number of qualified statements can be made. An individual turtle may lay eggs several times a season at approximately 2-week intervals (there are records at present of three nestings in one season by an individual female). There is evidence that a female may not nest every year, but that she does return to the same place when her time comes again and this may be within a matter of a few yards and certainly with a mile or so.

Although individuals of a species may nest anywhere throughout their nesting range, most of the laying is accomplished in very restricted areas, known as rookeries, which are used year after year. The turtles may come from long distances, as much as 1000 miles or more to nest on a rookery, but the mechanisms they use for navigation are unknown. Groups of turtles apparently stay together and come ashore to lay on the same or closely approximated nights; thus the same turtles nesting in a group one night will probably repeat the process about 2 weeks later. Presumably the individuals remain together in the interim and may migrate together to the rookery area from some common feeding ground in groups or flotillas. Probably all movement, at the rookery and away from it, is in these same flotillas which, when seen at sea, may number in the hundreds of individuals. Whether the adults return to nest on the beach where they were hatched is unknown, but it is presumed that they do.

Only the females (recognizable by their short tails extending little if at all beyond the hind edge of the carapace) come ashore during nesting. The males (long-tailed, extending far beyond the hind edge of the carapace) lie in the water just off the beach, and mating for later layings seems to take place in the area at this time. In some areas of the Pacific, males may haul out of the water to sun themselves (especially green turtles), but this is unrelated to nesting.

There appears to be no clear-cut correlation between nesting and the phase of the moon or stage of the tide, although speculation concerning this remains a controversial subject. Detailed studies on the rookeries indicate that some nesting takes place every night during the season and at every level of the tide. Daily or weekly variation in numbers of turtles nesting on a given night (or day, as in the case of the ridley frequently) or stage of the tide seem to be more closely related to the arrival of flotillas of turtles off the beach than to moon and tide. Arrival of the flotillas appears to occur randomly. The fact that more people often are searching for nesting turtles on clear, brightly moonlit nights may be the chief basis for this widespread belief in a correlation with the full phase of the moon in addition to a slight positive tendency toward this phenomenon.

The turtles come ashore to lay almost exclusively at night, with the exception of the ridley, which lays almost exclusively during the day. Nests are usually placed near the bases of prominent sand dunes. After selecting a site, the turtle uses her hind and fore flippers to dig a preliminary pit or depression so that her entire body is lowered several inches below the original surface of the beach. This done, she uses her hind flippers to dig a neat, nearly cylindrical hole, somewhat larger at the bottom, in the sand (Fig. 2). Depending on the kind of turtle and the

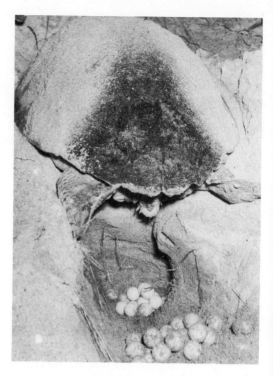

Fig. 2. Nesting loggerhead sea turtle. One wall of nest removed to show eggs in position. (Photo: Marineland of Florida.)

size of the individual, a hole 18–36 in. deep is dug, after which the eggs are deposited—nearly filling it.

The almost spherical eggs have a leathery shell that helps protect them as they fall into the nest in ones, twos, and threes. After the eggs are deposited (about 75–200 of them, usually about 100–150), she uses the hind flippers to drag sand back into the hole to cover the eggs. After the eggs are covered, the hind flippers pack the sand firmly. This done, the turtle usually flings sand about the nest site, using all four flippers again, apparently in an effort to disguise the exact location of the nest itself. The entire process usually lasts 1–2 hours. After the nest has been disguised to her satisfaction, the turtle returns to the surf, often stopping several times to rest.

On land, in addition to man's depredations on the nesting females and the eggs, the eggs and young are eaten by raccoons, birds, sand crabs, dogs, hogs, etc. In the water, the baby turtles must face carnivorous fish and sea birds. Consequently, relatively few turtles reach maturity, and many young must hatch to insure the survival of a few.

Growth and Feeding Habits

After the nesting female has returned to the water, the sun will provide the heat to hatch her eggs in about 2 months. The tiny hatchlings (only 1–4 in. long, depending on the species) mostly come out of the nest at one time and usually facilitate this by clawing at the sand above them and then packing it down as it falls below them, so that in a sense they actually cause themselves to rise as a group by removing material from above and thus opening the way as the floor of the nest rises below them. Hatching or to be more correct escaping from the nest usually occurs in the cool of the night when the killing heat of the sun is reduced and when many of the predators (especially birds) are less active. The young turtles seldom err in making their way directly to the sea, and experimental evidence indicates that this is a matter of vision and the lighter horizon provided by the open water or the reflections of the sky off the open water even if the turtle is behind some obstruction.

Nothing is known of the whereabouts of the young turtles after they enter the ocean. Only rarely are hatchlings captured at sea, and then usually on the surface in the general vicinity of a nesting ground. Young captive sea turtles grow rapidly, and though captive conditions may be more beneficial than natural ones, at least from the aspect of food supply, such growth indicates that sea turtles mature much earlier than formerly believed. Some of the nesting individuals may be only 15–20 years old at most, rather than the hundreds often attributed to them (Fig. 3). This does not rule out the possibility of great age, and many of the adults may be very old, although recent research indicates that such estimates should be modified.

The scarcity of records of small turtles seen in the ocean as compared with adult or subadult individuals is a most intriguing and puzzling circumstance. While the apparent rarity may result from not looking in the proper places, it may be that the lack of numbers is real. Considering the high mortality of the very young, the rapid growth of those that survive, and long life after reaching maturity, it is quite possible for the population to consist mostly of larger turtles.

Although most species of sea turtles tend to be either carnivorous or herbivorous at a particular life stage (for example, young green turtles are carnivorous when very young and herbivorous as adults, while loggerheads are carnivorous at all ages), all actually are omnivorous when necessity dictates. This is particularly true of young turtles and those kept in captivity. The ridley and hawksbill, like the loggerhead, show preferences for animal food at all ages. The leatherback, as far as is known on the other hand, seems to be a true omnivore.

Fisheries

The green turtle and its eggs are the most important of the sea turtles as a source of human food. Regrettably, it has been overfished throughout its range, but conservation efforts are being made to revive it (Parsons, 1962).

Limited fisheries are carried out on the other four species of sea turtles in various parts of the world, primarily for food and/or on the eggs. The hawksbill in addition provides the raw material for tortoise shell used in jewelry. The leatherback, although sometimes used, has the most regularly pelagic habits of any of the sea turtles, and, because it is thus a rarity to most fisheries, it is little used and its flesh is not especially esteemed. Its eggs, on the other hand, are utilized by man along with those of the other species. The heavy, tough folds of skin of the neck and at the point where the flippers join the body have use as special-application leather products, after proper tanning.

Summary

While most of the biological observations made have been based on studies of American sea turtles, they generally apply to all sea turtles. For more specific details of sea turtle biology, for characters used to distinguish the various subspecies of the five currently recognized species, and for additional general material on sea turtles, the reader is referred to Carr (1952, 1956, 1967); and to Caldwell (1960, 1962).

DAVID K. CALDWELL
MELBA C. CALDWELL

Fig. 3. Adult loggerhead sea turtle with a hatchling several weeks old to show comparative sizes of essentially newborn and mature animals. A minimum age differential in this case might be expected to be some 15–20 years. (Photo: Marineland of Florida.)

References

Caldwell, D. K., "Sea turtles of the United States," 1960, *U.S. Fish and Wildlife Serv., Fish. Leaflet*, 492.

Caldwell, D. K., "Sea turtles in Baja Californian waters (with special reference to those of the Gulf of California), and the description of a new subspecies of Northeastern Pacific green turtle," *Los Angeles County Mus., Cont. in Sci.*, **61**, 1–31 (1962).

Carr, A., "Handbook of turtles; the turtles of the United States, Canada, and Baja California, 1952, Comstock Publ. Assoc., Ithaca, New York.

Carr, A., "The Windward Road," 1956, Alfred A. Knopf, New York.

Carr, A., "So Excellent a Fish, A Natural History of Sea Turtles," 1967, The Natural History Press, Garden City, New York.

Parsons, J. J., "The Green Turtle and Man," 1962, University of Florida Press, Gainesville.

Cross references: *Biological Clocks in Marine Animals; Salmon, Pacific.*

SEAWEEDS OF COMMERCE

Seaweed is now the raw material for a number of industries and an important item in the diet of many Orientals. Only the red and brown seaweeds are used commercially; the green species will not be considered here, although several edible species are cultivated.

The Japanese agar industry started over 300 years ago, and an industry based on brown seaweeds spread throughout Europe in the 18th century. The manufacture of agar remained a Japanese monopoly until 1929 when a small production unit opened in California; several factories started production in the 1940's, but these were all small and had little effect on the general situation. At least seven factories, with an estimated annual production of 3000 tons of agar, are now being built in Chile, Portugal, and Russia, and the manufacture of this product will soon be distributed throughout the world. These new factories will use seaweed that is now processed in Japan, and their output will make little difference to the present world shortage of agar.

The industry based on the brown seaweeds, unlike the agar industry, has been most unstable. Initially the seaweed was burnt and the ash (known as kelp in Europe) was a valuable source of alkali carbonates and iodine. The discovery of iodine in the Chile nitrate deposits later caused a steady decline in the fortunes of the kelp industry. A little iodine may still be made from kelp in Japan, but manufacture in Scotland ceased in 1935 and in France by 1960.

Alginic acid was discovered in the brown seaweeds in 1881, but the early manufacturers were unsuccessful and the first viable unit, the Kelco Company, started in 1929. This was followed some five years later by Alginate Industries Ltd. The manufacture of alginates expanded rapidly in the postwar years, and it is now the biggest outlet for seaweed in the western countries. This section of the industry is still expanding, and two large units are now being built in Canada.

The production of seaweed meal started at Boothbay, Maine, in 1870, but made little progress until the 1950's, and the demand has expanded rapidly in the last decade. The manufacture of liquid fertilizers from brown seaweeds was started in Britain in 1950; this section of the industry expanded tenfold in the last decade and shows considerable potential.

An industry based on the red seaweed, *Chondrus crispus*, started in the mid-19th century when this seaweed was first used in refining beer. Again, growth was relatively slow until the 1950's when a rapid expansion took place. Certain other red seaweeds, notably *Gigartina* spp. and *Eucheuma* spp., yield a somewhat similar product, and furcellaran, or Danish agar, has been manufactured in increasing quantities in postwar years from *Furcellaria fastigiata*. The current demand for these seaweed extracts exceeds the supply of the raw materials.

In addition to these industrial uses of seaweed, the production of edible seaweeds is a major industry in the Orient. It is particularly well

Fig. 1.

established in Japan where seaweed "farming" (mariculture) is practised on a large scale; Korea and China are also notable producers of edible seaweeds, and improved hybrids have been reported from China.

Only a few years ago, there was no untoward demand for the widely scattered plants used by the seaweed industry; the supply of brown seaweeds is still adequate but all the red seaweeds are in short supply and only the development of new collecting areas can supply the needs of this expanding industry. The increased demand for seaweeds has now focused attention on methods of cultivating selected species, and the Atlantic Regional Laboratory of the National Research Council of Canada has already announced its intention of establishing a field station to apply the principles of scientific agriculture to marine plants. This step is not as revolutionary as it sounds since, as recorded above, some of the edible seaweeds are already cultivated and "Fucus" farms were once common on the west coast of Ireland. Attempts to transplant a temperature-resistant strain of *Macrocystis pyrifera* from Lower California to La Jolla and recent reports from Russia that *Ahnfeltia plicata* has been successfully introduced near Vladivostock indicate widespread interest in these new ideas.

Distribution

Until the early years of this century, the seaweed industry existed as small units in areas where there was an abundance of seaweed. This position changed around 1910, when the U.S. Department of Agriculture, in an attempt to locate an indigenous supply of potassic fertilizer, organized a survey of the eastern Pacific ranging from Alaska to Southern California. There was little change in the situation until 1940, when wartime conditions prompted surveys in many parts of the world. At this time, there was a notable shortage of agar and the idea of making a jute substitute from alginic acid was fostered by the British Government. These demands caused the development of new survey techniques and expanded knowledge of the world's resources.

The Scottish Seaweed Research Association, now the Institute of Seaweed Research, formed in 1944, has carried out extensive surveys in recent years. In the immediate postwar period, surveys were carried out in Ireland, Norway, and Canada, but it was not until the late 1950's that the demand for seaweed began to exceed the supply of readily available material.

Around 1958, the Japanese *Gelidium* beds, which had proved adequate for centuries, went into a mysterious decline and forced a worldwide search for new sources of supply. This resulted in the large-scale exploitation of supplies of suitable seaweeds for the manufacture of agar in Chile and Portugal. Both countries will manufacture substantial quantities of agar in the near future. By the early 1960's, the demand for carrageenin, and to a lesser extent for furcellaran, caused a shortage of other red seaweeds; the main producers of these products built drying units in Nova Scotia and Prince Edward Island to encourage the collection of seaweed. Other developments in Canada suggest a considerable

TABLE 1. ANNUAL SEAWEED HARVEST (THOUSAND METRIC TONS; FRESH WEIGHT)

	1938	1948	1958	1961	1962	1963	1964	1965
Morocco	—	—	—	3.4	1.9	2.1	2.2	2.6
Sudan	—	—	0.6	0.7	0.9	—	—	—
Tanzania (Tanganika)	—	—	—	0.1	0.1	0.1	—	—
Tanzania (Zanzibar)	—	—	—	0.2	0.2	0.2	0.4	0.2
Canada	0.6	4.4	15.4	18.5	23.5	22.5	19.8	25.0
Mexico	—	—	6.5	15.6	21.3	19.5	23.3	17.0
U.S.	—	1.1	1.2	3.2	2.5	3.3	2.2	2.4
Argentina	—	—	2.0	10.1	9.1	8.3	10.2	19.9
Taiwan	—	—	0.7	0.9	1.0	0.8	0.8	1.1
Japan	442.7	169.9	339.7	424.8	501.7	425.7	361.0	395.1
Korea								
(aquaculture)	—	—	2.1	3.1	6.1	5.1	11.7	—
(marine fisheries)	44.0	6.6	28.8	36.9	45.6	37.8	42.8	—
Philippines	—	—	—	0.1	0.1	0.1	0.1	0.3
Ryukyu Islands	4.2	0.1	0.3	0.1	0.1	0.1	0.1	0.1
Norway	—	2.5	82.7	72.1	71.0	56.9	63.6	74.2
Portugal	—	—	—	—	3.1	4.2	4.8	2.7
Spain	—	—	—	4.4	4.9	4.3	5.3	—
Scotland	—	—	—	18.4	18.4	14.5	18.4	21.2
France[a]	(411.5)	(39.0)	(29.0)	(171.9)	(18.3)	(75.4)	(12.9)	(14.8)

[a] Data not included in the total as the products are not clearly defined.

development of the seaweed industry in the next year or two.

Until a decade ago, it was widely felt that the supply of seaweed was inexhaustible, but it must now be recognized that the supply will fluctuate unless special precautions are taken.

An opportunity to map the world's seaweed resources occurred at the First International Seaweed Symposium in 1952. Some quantitative data followed at the subsequent Symposium and F.A.O. has published (Table 1) data on the production of seaweed over the last few years. The importance of seaweed in Japan is immediately obvious from the table, but the discrepancies are less obvious and the wide differences in monetary value of the various seaweeds is not considered. Notable omissions from the table are Chile and Ireland and the data for the U.S. is at least 100,000 tons lower than the known annual harvest.

Seaweed Surveys

The distribution of the seaweeds is largely controlled by water temperatures; the brown seaweeds are more prolific in the cooler seas and the red seaweeds prefer warmer waters. Water temperature, in turn, is influenced by latitude and the circulation pattern of the major oceanic currents. These two factors combine to give the main seaweeds of commerce a wide geographical range within which there are slight differences in growth rates and the times at which the plants spore. Salinity is another factor that affects the growth of seaweed. Only a few species, such as *Ulva* and *Enteromorpha*, flourish at low salinities, and it is usual to find smaller plants as the salinity decreases.

The economically important seaweeds fall into two main groups: one grows in the intertidal fringe, and the other is permanently submerged. This apparently simple distinction is, however, complicated by many factors. The extent of the intertidal zone, for instance, is partly controlled by the gradient of the beach and it also increases with latitude and tidal range. There are considerable local variations, e.g., the tidal range in the British Isles varies from less than 10 ft in Shetland to almost 30 ft on parts of the west coast. One prominent seaweed of the intertidal zone, *Chondrus crispus*, extends into the sublittoral and a substantial part of the Canadian harvest grows sublittorally. There are also anomalies in the seaweeds of the sublittoral zone. Some of these become detached and may exist as free-floating plants, while others are cast on the shore and collected. The important free-floating seaweeds are *Furcellaria fastigiata*, harvested in the Central Kattegat, and *Gracilaria* spp., found off the coast of Chile, Portugal, etc.; these forms are usually harvested by netting. Cast seaweeds are important in some areas, e.g., the cast red seaweeds of Prince Edward Island and the cast stipes of *Laminaria hyperborea* of the Scottish coast. Other members of this group, e.g., *Macrociptis pyrifera*, *M. integrifolia*, and *Nerocystis leutleana*, are large plants that extend to the water surface. These diverse factors complicate survey operations.

The earliest seaweed survey was carried out before World War I and covered the whole west coast of North America. Practically all the subsequent surveys were carried out since 1940. The earlier surveys used boats and were obviously time-consuming, but aerial surveys were introduced about 1930 and echo-sounding was used in the 1940's. The introduction of aqualung diving techniques has extended knowledge of seaweeds in the last few years without adding much quantitative data. Despite pressure from the industry for increased supplies, the world's seaweed resources are still little known. In fact, only the coasts of British Columbia, California, New Zealand, Norway, Nova Scotia, Scotland, and Tasmania have been surveyed quantitatively (Table 2). There is little known of the quantity of seaweed on other coasts despite the fact that exports from some areas, e.g., Chile and Portugal, are now valued at over 1 million dollars.

There is no doubt of the extent of the market for certain seaweeds and the need for further surveys, but progress is negligible. This is partly due to the shortage of trained personnel and the need, despite the inherent possibilities of aerial photography and echo sounding, to walk laboriously over vast areas of the shore or to sample even larger areas of the seabed from small boats.

Harvesting

The use of seaweed is inextricably bound to the need to harvest and conserve the crop, but very few countries exert any control over seaweed harvesting and the need for conservation is often

TABLE 2. SURVEYED SEAWEED RESOURCES

Country	Millions of Tons (Fresh Weight)	
	Littoral	Sublittoral
Alaska		19.2
British Columbia		1.5
Puget Sound–San Diego, U.S.		13.8
San Diego–Cedros Is., Mexico		8.5
Nova Scotia	1.2	0.9
Saragasso Sea		4–11
Norway		ca. 20
Scotland	0.2	ca. 10
Russia–White Sea		1.5
–Black Sea		1–2
New Zealand		0.8
Tasmania		0.35

overlooked. There are, in fact, two conflicting views on the availability of seaweed; one considers the supply is inexhaustible while the other sees a need for conservation and "farming" the available resources. Only a decade ago, the resources seemed more than adequate, but there is now a shortage of the red seaweeds. The rapid expansion of the industry now calls for conservation, and the decline of the Japanese *Gelidium* beds and fungal disease in *Porphyra* beds add further emphasis to this need.

A most comprehensive study of seaweed growth and harvesting has been made on the *Macrocystis* beds off California; these beds had been studied continuously since 1910 and the seaweed companies have actively cooperated. In many instances, a licence is needed to collect seaweed and a royalty is exacted. The companies in California proposed 15-year leases at an annual fee of $100 per square mile for leasing the beds with a tax of 10 cents per wet ton harvested; the income is used to aid administrative costs and research by the Fish and Game commission. The regulations in Nova Scotia prescribe a licence of not less than $50 or more than $500 and a royalty of 10 cents per wet ton harvested, but if the royalty exceeds the licence, the latter is deductible from the total payment. In Britain, the owner of the foreshore exacts payment for the collection of seaweed, the State claims a royalty equivalent to approximately 20 cents per wet ton, but private landowners usually negotiate a rental for exclusive collecting rights over a long period.

The need for research into algal growth and reproduction becomes more evident as the industry expands into new areas. The newly developed collecting areas tend to seek advice from elsewhere and tacitly assume the experience in another country will prove adequate; such an assumption may serve as a guide but each seaweed and every environment must be separately considered. The need for quantitative surveys and controlled experimental harvesting should be based on a scientific assessment of the local situation.

The actual harvesting of seaweed is often based on collecting by hand but, nevertheless, it can be a major industry in some areas. For instance, the Scottish alginate industry provides employment for about 600 collectors, and even this effort is insufficient to supply all the seaweed for the alginate factories in Scotland. Most of the *Chondrus crispus* collected in Nova Scotia is also collected by hand using a long-handled rake on the sea bed at low water. Many tools have been used to collect seaweed, and these are adequately illustrated in the standard texts. Girl divers are a well know feature of the Japanese industry, which is wholly dependent on hand collecting, and the aqualung is now used by collectors in Europe and probably elsewhere.

Mechanical harvesting of seaweed is best developed in California where it has been used on the *Macrocystis* beds for over 50 years. The plant is cropped with a mower, mounted on a power-driven barge, and cutting about 3 ft below the surface; regrowth is rapid and over 100,000 tons are harvested annually. The earlier harvesters carried about 300 tons of seaweed and had the mower in the bow, but the newer craft are longer and have the mower at the stern. This type of harvester is also used in Chile and Tasmania and probably by the newly established industry in British Columbia.

While vast quantities of *Laminaria* spp. grow in many parts of the world, the harvesting of these seaweeds has not been developed commercially. The Institute of Seaweed Research developed two types of harvester, the so-called "cut and suck" method and the continuous grapnel. A grapnel harvester, with a belt 18 in. wide mounted on a 50-ft fishing boat, had an average harvesting rate of 3 tons per hour. A somewhat similar harvester has been patented (U.S. 2,941,334) for harvesting *Chondrus crispus*, but it does not appear to have been successful. Other experimental harvesters have appeared in the last few years.

Other forms of harvesting have been reported. Chapman mentions the use of the Agassiz trawl for the collection of *Gracilaria* in New South Wales and claims it harvest 600 lb per hour. A dredge is used to collect *Chondrus crispus* from the south coast of the Northumberland Straits and yet only cast seaweed is collected on Prince Edward Island and raking is practiced elsewhere in Nova Scotia.

Only in a few instances is the seaweed processed in the wet state and drying is usually necessary. Under favorable conditions, seaweed can be air-dried, but heated driers are needed in most parts of the world. Many different types of dryer are used and care must be taken to avoid decomposition through overheating.

Utilization

Apart from edible uses, seaweed is essentially the raw material for the extraction of a range of carbohydrates, e.g,. agar, carrageenin, and furcellaran from the red seaweeds and sodium alginate from brown seaweeds. In addition, some brown seaweed is used in animal feeds to provide vitamins and trace elements, and a smaller amount is sold as manure or processed to give liquid fertilizers. In every instance, there is remarkably little information available on the total production of these products.

The Japanese production of agar in 1958 involved the use of 16,559 metric tons of *Gelidium*;

imports amounted to 3569 metric tons and the agar production amounted to 2026 metric tons, a yield of only 10.1%, which appears low. Since that time, the crop from the Japanese *Gelidium* beds has declined, but considerable quantities of seaweed have been imported from Chile and Portugal and some *Anfeltia plicata* is imported from Russia. Since 1960, the Japanese have been unable to import seaweed and crude agar from South Korea, where 800 metric tons are said to be produced annually. Notwithstanding these adverse conditions, the Japanese production came to 2600 metric tons in 1961 and now probably exceeds 3000 metric tons.

Agar is also produced in Chile, Portugal, New Zealand, South Africa, Spain, and the U.S. but Spain, with three major producers, is the only country with a substantial production of agar. The total production of agar from all sources is probably about 6000 metric tons, and this will use about 50,000 tons (airdry weight) of seaweed. This commodity is still in short supply, and since 1958 prices have risen from $1.25 to the current price of around $4.00.

The production of furcellaran is limited to the 15,000 tons (wet weight) of *Furcellaria fastigiata* collected in the Kattegat and a relatively small amount collected in Prince Edward Island; one of the major Danish producers is now building a factory on Prince Edward Island and it must be assumed that the area has a large supply of this seaweed.

The use of brown seaweeds in animal foodstuffs, etc., is not known, but most of the Norwegian production of *Ascophyllum nodosum* is used for this purpose and a similar amount of this seaweed is produced in Ireland. The Norwegian production of seaweed is well documented; it increased from 160 tons in 1950 to 14,500 tons in 1965.

There are no data on the production of liquid fertilizers from the brown seaweeds, but the three British Producers must make about 250,000 gal, and smaller quantities are produced in Ireland and Norway. The volume of sales of the major producer increased twelvefold between 1956 and 1965, and this section of the industry appears to have considerable potential.

The Oriental use of edible seaweeds probably accounts for more seaweed than all the industrial outlets, but only the Japanese data are available (Table 3).

The production of Amanori now exceeds 6000 sheets and another 1000 sheets are manufactured in South Korea, about half of which is exported to Japan; thus the production of this seaweed now amounts to 260,000 tons. The production of edible seaweeds in China is believed to be considerable, and seaweed cultivation has been practiced there since the 1930's. While edible seaweeds are used in other parts of the Far East, there are no other records of their cultivation by modern methods.

E. Booth

References

Chapman, V. J., "Seaweeds and Their Uses," 1950, Methuen, London.
Davison, E. F., "Marine Botany," 1966, Holt, Rinehart and Winston, Inc. New York.
Glicksman, M., "The utilisation of natural polysaccharide gums," in *Advances in Food Research*, **11** (1962).
Kirby, R. H., "Seaweeds in Commerce," 1953, H.M. Stationery Office, London.
Newton, L., "Seaweed Utilisation," 1951, Sampson Low, London.
Schofield, W. L., "History of kelp harvesting in California," *California Fish and Game*, **45** (3), 135–157 (1959).
Braarud, T., and Sørensen, N. S., Eds., "Proceeding, of the Second International Seaweed Symposiums Trondheim, 1955," Pergamon Press, London, 1956,
de Virville, A. D., and Feldmann, J., Eds., "Proceedings of the Fourth International Seaweed Symposium, Biarritz, 1961," 1964, Pergamon Press, London.
Young, E. G., and McLachlan, J. H., Eds., "Proceedings of the Fifth International Seaweed Symposium, Halifax, 1965," 1966, Pergamon Press, London.

Cross references: *Agar; Alginates from Kelp; Irish Moss Industry; Phycocolloids and Marine Colloids.*

SHARKS

Biological Characteristics

Sharks and their evolutionary derivatives, the sawfish, torpedoes, guitarfish, skates, and rays (Elasmobranchii), are called cartilaginous fish because their skeletal elements lack the architecture of true compact bone with included living bone cells. Their skeletons usually do have heavy mineral depositions, however, and the difference between a fragment of calcified cartilage and one of bone is not always apparent without microscopic examination. The elasmobranch fish are further characterized by the presence of five, six, or seven external gill slits, skin armed with placoid scales or dermal denticles, well developed vertebral centra with the notochord constricted

TABLE 3. PRODUCTION OF EDIBLE SEAWEEDS IN JAPAN

Japanese Name	Seaweed	Production (Wet Tons: Av. 1955–60)
Konbu	*Laminaria* spp.	142,000
Wakame	*Undaria pinnatifida*	50,200
Amanori	*Porphyra* spp.	78,000 (2118 sheets)
Aonori	*Monostroma* spp. *Enteromorpha* spp.	9,000 (255 sheets)

segmentally, and upper jaws free from fusion with the chondrocranium. The sharks (Selachii) differ from the other elasmobranchs (Batoidea) chiefly in body shape. They are usually fusiform and always have their gill slits either more or less above the plane of the pectoral fins or in advance of them, whereas the batoids are flattened forms with ventral gill slits and pectoral fins attached at their forward ends to the sides of the head.

Shark jaws are armed with teeth that are replaced frequently. The teeth are not set in sockets but are attached only to ligamentous bands that cover the biting surface of each jaw. They originate as buds in a protected germinal area inside the mouth and move forward into their functional positions on the ligamentous band, to which they become attached. The outer functional teeth are shed frequently. In some rapidly growing sharks the functional life of a tooth may be as little as two weeks, but the rate of tooth shedding varies greatly. The common requiem sharks—tiger sharks, dusky sharks, and their near relatives—usually have only one series of teeth functional at one time and these may be lifted into a more or less erect position by red muscles at the corners of the jaw that apply tension to the ligamentous band on which the teeth rest. Most sharks also protrude their jaws, upper and lower jaws moving as a unit. The upper jaws slide forward and outward while supported and held rigid by processes that slide along ligament attachment areas of the chondrocranium. The jaws of sharks are thus remarkably strong and are functionally more efficient and flexible than is suggested by the external appearance of the shark's mouth under an overhanging snout.

Sharks have no swimbladder and tend to sink to the bottom if not supported by the hydrodynamic lift given to the swimming shark by fins and body surfaces that may act as planes. The energy required of the shark to maintain a position above the bottom is a major biological disadvantage. The tendency to sink is offset to a varying extent in sharks by the presence of a large amount of fat stored as oil in the liver. Oils with a specific gravity of 0.92 to 0.94, which occur in most sharks living near the surface or in shallow water, no doubt reduce the energy required to maintain their positions. Some deepwater and midwater species have hydrocarbon oils, chiefly squalene, in the livers, the proportions increasing with the age of the shark. One kind of deepwater spiny dogfish, *Centrophorus*, may have as much as 90% squalene in its liver oil, which has a specific gravity of about 0.86. It is probable that *Centrophorus* and some other sharks that contain a substantial amount of hydrocarbon oils in their livers attain nearly neutral buoyancy.

A few sharks lay eggs in leathery cases, which sometimes become attached to objects on the bottom by tendrils. The developing embryos are nourished by large yolks, and the young emerge from the cases in substantially the form of adults. Egg laying is restricted to representatives of a few families: the bullhead sharks (Heterodontidae), which produce a peculiar spiral form of egg case; most of the cat sharks (Scyliorhinidae); many of the smaller species related to the nurse shark (Orectolobidae); and the giant whale shark (Rhincodontidae). The rest of the sharks retain developing eggs, either with or without shells or shell membranes, in the oviducts of the mother where various arrangements provide nourishment to supplement that supplied to the embryo by the egg yolk.

One evolutionary trend in sharks is to produce few young, large in relation to the size of the mother, and thus increase probabilities of survival. This trend may hold for small as well as large species. *Squaliolus*, a midwater species that may be the smallest of all sharks, reaches sexual maturity at a length of about 15 cm (6 in.) for males and 20 cm (8 in.) for females. In the only known example of a gravid female, the 22-cm (8.7-in.) mother contained 3 embryos 9 cm (3.6 in.) long. Females of the common sand shark, *Odontaspis taurus*, have young when they are from about 275 cm (9 ft) to 335 cm (11 ft) long, and produce a maximum of two young in a litter, each about 100 cm (39 in.) long at birth.

All sharks are carnivorous and although the two largest—the whale shark and the basking shark—are primarily plankton feeders, the rest are predators on prey that is large in relation to their size. Many species may eat carrion and, like predators in general, they are especially attracted to partially disabled fish and other marine animals. Although sharks have a reputation as scavengers, the large voracious species prefer fresh fish whenever obtainable. Some sharks have alternate feeding habits to which they may turn when circumstances require. Thus the silky shark, a species that reaches a length of 3 m (10 ft) or more, may at times pack its stomach with fish less than 10 cm long, even though its teeth and jaws are strong and adapted to cut pieces from large prey.

Sharks are responsive to changes in water temperature and in some areas, such as in continental shelf waters of the Atlantic coast of the U.S., are strongly migratory. To reach waters of suitable temperature, sharks may need only to move to a different depth. An alternative, however, is to travel great distances; on the Atlantic coast of the U.S., migrations of some of the larger species may extend more than a thousand miles.

Temperature is not the only factor governing shark migration. Sharks may have to migrate to find a suitable area to give birth to their young. Some shark species or populations have very restricted nursery areas, and the need of the mature females to return there periodically influences the extent and direction of seasonal migrations. Among most species some segregation by size occurs when sharks gather in schools, and migratory schools of adults are often segregated by sex. The geographical ranges of adult males, adult females, and young may be appreciably different in some migratory populations. Ranges then overlap completely only for short periods in limited areas.

About 300 species of sharks are known. More than half are either restricted to depths greater than 200 m (109 fathoms) or enter lesser depths rarely and briefly. Furthermore, many of the sharks—particularly the larger species that are well known from near-surface or inshore waters—spend a substantial proportion of their lives at depths of 200–500 m (109–273 fathoms) or more. The deeper limits of the ranges of large species are not well known. A few small species are known to occur to depths of about 2500 m (1367 fathoms), but little information is available on shark distribution at depths greater than 500 m (273 fathoms).

The abundance of sharks is difficult to estimate. A few species are conspicuous and a nuisance to fishermen, who may overestimate their numbers. The relative abundance of sharks may also be underestimated because, as a group, their greatest numbers occur in habitats that are out of sight and beyond the reach of most fishermen. As predators their numbers are necessarily held within vague but real limits by the food supply, but at least a few sharks are normally present in all marine situations within a depth range from 200–500 m, where temperatures are about 9 to 23°C. The various species differ greatly in their tolerance of the extremes of water temperature and other rapidly shifting environmental factors common to near-shore situations. Except for a few species, the nurse shark for example, they enter near-shore waters as invaders when conditions of temperature, light, and food supply are favorable, but do not remain as permanent residents.

Some sharks, such as the spiny dogfish, gather in large schools to feed. Schools of hundreds of blue sharks are commonly encountered at the surface well offshore or near deep water, and the migratory requiem sharks may travel in aggregations. Some deepwater varieties, *Etmopterus* for instance, are gregarious. Russian observers have noted the presence of the small and little known midwater sharks, *Isistius*, attacking their fragile plankton nets in swarms. Groups of sharks, sometimes containing more than one species, may assemble in response to the presence of a superior food supply. As a rule, however, sharks scatter widely within their preferred ranges and feed singly or in small aggregations of less than a dozen individuals.

Landings of sharks, skates, rays, and chimaeras by fisheries throughout the world in 1966 are estimated at about 860,000,000 lb (live weight). Sharks probably account for at least 500,000,000 lb of this total, but this is only a little more than a half of 1% of the total landings of marine fish. Fishery statistics do not furnish reliable evidence of the abundance of sharks relative to the abundance of other fish.

The demand and marketability of sharks and shark products varies greatly in different parts of the world, as does the availability of reliable estimates of fishery landings. Landings of sharks in fisheries of some European countries may exceed 1% of their total landings and yield a lesser percentage, but still a substantial return, in value. Shark landings in some non-European countries without either an advanced technology or a well developed marketing system sometimes contribute more than 1% of total landings. In either kind of situation only a few kinds of sharks are selected for landing, and probably more sharks are discarded at sea than are landed.

The present utilization of sharks for food is paradoxical. A part of the production goes to the gourmet trade and to somewhat restricted markets at a price relatively high in relation to the price of other fishery products. The other parts of the production, much of it prepared as a dried salted product, is consumed as a very low-cost protein food.

The optimum utilization of sharks requires not only an advanced technology but also a considerable amount of skilled labor. Sharks and marine skates and rays all retain urea in their body fluids and tissues as a part of their physiological arrangement for the maintenance of salt and water balance. Only the freshwater rays that are permanent residents of some South American rivers are free from appreciable quantities of urea. Urea itself is not harmful when sharks are used as food, and substantial portions of it drain off and are removed by good processing and handling. Decomposition of shark flesh, however, is accompanied by the release of free ammonia through the action of enzymes on urea, which greatly intensifies the unpleasant consequences. Putrefactive decomposition may not be faster for sharks than for bony fish, but it is much more easily detectable. It follows that successful utilization of sharks for food requires rapid handling and good processing from the moment of capture, nearly always with procedures that are not readily mechanized.

As might be expected, sharks of some species find greater acceptance as food items than others.

In addition, they appear to receive greater acceptance in areas where culinary arts are highly developed and where the shark meat is used as one of several ingredients in final food preparation.

Fibers from the fins of large requiem sharks and sawfish are used to make a soup of superlative quality. The demand for dried shark fins for the preparation of the soup, usually by Chinese cooks, has always exceeded the supply and kept the price for suitable dried fins remarkably high. The preparation of the soup, however, is an art that has not been widely mastered even among Chinese cooks.

Industrial uses of shark products have been important from time to time in the past and some products are still in demand and are used to the extent that they are available. One U.S. company has prepared and marketed shark leather for more than 30 years. The hides of the larger requiem sharks and a few other large species make a long-wearing leather that resists scuffing; it is currently used chiefly for some high quality shoes. The leather is comparatively expensive and the hides require special treatment before tanning. The supply of shark hides has never been large enough to meet the full potential of the market, in spite of high prices paid for salted green shark hides.

From about 1935 to 1950 the principal source of vitamin A was from shark-liver oil. In the U.S. the demand was greatly expanded by improved methods in animal feeding. During this period the poultry industry became the largest consumer, and shark fishing by U.S. fishermen became a well established activity. The development of an inexpensive method for making vitamin A from vegetable sources caused the market for the more expensive shark liver oil in the U.S. to disappear almost completely, and landings of sharks have since been negligible. Shark livers remain as an alternate source and in parts of the world where production costs are sufficiently low, shark-liver oils are still utilized.

Shark bile is used for biochemical research to some extent due to its high content of sterols and bile acids.

Although the use of sharks as a food and industrial resource varies greatly in different parts of the world, they are economically important because of the severe economic losses they inflict on some fisheries by their attacks on fishing gear and catches. The extent of annual loss from shark depredation is impossible to assess with precision. A part of it is in damaged nets and gear, but losses of catch may be greater. Sharks compete with man in the harvesting of some of the more desirable food resources of the sea, but the extent that shark predation affects stocks of desirable food fish is unknown. It is possible that fishery losses to sharks the world over and the gain to fisheries through utilization of sharks are in the same order of magnitude. Sharks seem to be a more severe damaging factor to fisheries, however, in areas where they are not utilized. Good management of ocean fisheries may eventually require the encouragement of commercial utilization of some shark species in some areas as a method to reduce damage to other fisheries.

STEWART SPRINGER

Behavior and Attack Patterns

Sensory Receptors. Although the shark has very limited intelligence, it has an elaborate sensory apparatus. In addition to the familiar sense receptors of eyes, ears, and nose, the shark has a lateralis system, for the detection of low frequency vibrations, on the head and sides; sensory crypts that function as mechanoreceptors liberally peppered over the back; and on the head sensory cells, called ampullae of Lorenzini, that may detect vibrations, heat, pressure changes, and electrical impulses. Sharks are very alert to vibrations in the water. The erratic movements of a wounded fish, or the crash of a plane can be detected from considerable distances by the lateralis system.

Of all the shark's sensory systems, the most acute is the sense of smell. The nostrils are located on the ventral side of the head and do not open into the mouth. They have incurrent and excurrent openings through which seawater flows continuously, bathing large sacs lined with folds of tissue supplied with abundant olfactory cells. In the hammerhead (family *Sphyrnidae*) the nostrils are at the outer ends of the head, and it swims swinging its head from side to side, permitting this shark to sample a relatively wide olfactory path of water.

Contrary to popular opinion, the eyes of sharks are efficient and well-adapted to the shark's needs. They are highly sensitive to contrasts of light and shadow and quickly discern moving objects, thanks to a rod-rich retina. When a shark is in the dark the *tapetum lucidum*, a mirror-like layer of cells located behind the retina, reflects the light back through the retina and restimulates the light-sensitive rods. In some species of sharks a further refinement increases the efficiency of the eye. A tapetal curtain composed of expansible pigment-filled cells occludes the tapetum reflexly when the shark is in bright light and thus prevents damaging overdoses of light on the sensitive retinal photoreceptors.

Feeding and Attack Patterns. All sharks are carnivores. Anatomically they are admirably adapted for the perception, pursuit, and seizure of living food. The lateralis system of the shark first detects the presence of food through vibrations in the water. As the shark swims towards the disturbance, olfaction becomes the primary sensory tool. Close in, vision takes over and the shark usually circles the prey before attacking it. When captive sharks were fitted with eye occlusors that obstructed their vision they had great difficulty in homing in on the bait.

Rarely do sharks roll on their sides when attacking even a vertically oriented bait. As the shark approaches a bait it raises its head and sinks the sharply pointed teeth of the lower jaw into the flesh. The upper jaw, not fused to the brain case, is then protruded and the myriad teeth imbed in the prey and the head shakes violently from side to side until a large chunk is bitten off. Many sharks swallow their prey whole depending upon its size. Some, like the great white shark, *Carcharodon carcharias*, can bite a porpoise or a sea turtle in two.

Sharks can and do fast for long periods of time during which they subsist on oil stored in the liver. The liver consequently varies greatly in size. Competition is a potent stimulus to feeding. Captive sharks held in a large pen may ignore bait offered to them until one more adventurous and aggressive shark finally approaches and bites it. Others then advance and consume the bait voraciously. Sometimes when large numbers of sharks congregate around food a fantastic delirium obsesses them and they bite at anything in sight, even each other, and continue to feed, completely oblivious to wounds. Appropriately this activity is called a "feeding frenzy."

The problem of shark attack is now under intensive and extensive study. The Shark Research Panel of the American Institute of Biological Sciences, sponsored by the Office of Naval Research, has gathered and catalogued the records of nearly 1500 shark attacks. When an attack occurs anywhere in the world, complete and verified data are collected and added to the file that is housed in the Mote Marine Laboratory at Sarasota, Florida. For the first time attacks are being analyzed scientifically.

Many antishark measures have been developed and tested but none has been found to be effective against all species of sharks; there is no deterrent to sharks in a "feeding frenzy." U.S. Navy fliers are currently issued a 6-oz packet, known as "Shark Chaser" or "Shark Repellent Compound," that consists of a cake of copper acetate and nigrosine dye. It is presumed that the copper acetate is repugnant to the shark's sense of smell, but this is doubtful. The black nigrosine dye, however, does function on many species of sharks as a good visual deterrent. Under investigation is the Johnson "Shark Screen"—a plastic bag with an inflatable collar into which a man climbs after he has filled it with water. Dark-colored shark screens are unattractive to sharks and silver plastic bags seem to attract sharks and have frequently been bitten. The dark-colored shark screens, in addition to deterring a shark from attacking, serve to confine human effluent, including blood, that might be attractive to sharks.

Various electrical devices have been tested and some hold promise, but the danger of utilizing any electrical system in seawater must be emphasized. Moreover, it has been found that when a particular electrical pulse frequency is employed to frighten away lemon sharks, the same frequency, under identical test conditions, serves as a strong attractant to tiger sharks. Thus, before any electrical device is placed on the market, it should be tested on a wide variety of sharks under field conditions as well as controlled conditions within shark enclosures.

One of the most effective methods of protecting bathers from sharks is the practice of "meshing" as currently employed in Australian and South African waters. Gill nets are placed parallel to the shore and sharks are caught as they attempt to swim through them. By means of meshing, the population of sharks is quickly reduced to the point where the ratio of shark to swimmer is highly in favor of the swimmer. No attack on a bather has taken place at any meshed beach in Australia or South Africa since this program began.

In response to requests from bathers, skin and scuba divers, and U.S. Navy and Air Force service personnel, the AIBS Shark Research Panel provides the following advice.

To swimmers and bathers: (1) Always swim with a companion. Do not become a lone target for attack by swimming away from the general area occupied by a group of swimmers and bathers. (2) If dangerous sharks are known to be in the area, stay out of the water. (3) Because blood attracts and excites sharks, do not enter the water or remain in it with a bleeding wound. (4) Avoid swimming in extremely turbid or dirty water where underwater visibility is very poor.

To skin and scuba divers: (1) Always dive with a companion. (2) Do not spear, ride, or hang on to the tail of any shark. To provoke a shark, even a small and seemingly harmless one, is to invite possibly severe injury. (3) Remove all speared fish from the water immediately; do not tow them in a bag or on a line cinched to the waist. (4) As a rule a shark will circle its intended victim several times; get into a boat or out of the water as quickly as possible after sighting a circling shark, before it has time to make an aggressive "pass." Use a rhythmic beat with the feet and do not make an undue disturbance in the water as you move toward the boat or the shore. If wearing scuba apparatus, remain submerged until you have reached the boat. (5) If a shark moves in and there is not time to leave the water, try not to panic and keep the shark in view. A shark can often be discouraged by releasing bubbles or, at close range, by deliberately charging it and hitting it on the snout with a club or "shark billy." Since the hide of a shark is very rough and may cause serious skin abrasions, hit the shark with your bare hands only as a last resort. Shouting underwater may or may not discourage a shark.

To survivors of air and sea disasters: (1) Do not abandon your clothing when entering the water. Clothing, especially on the feet and legs, is your only protection against the rough skin of a shark. (2) Place wounded survivors in a life raft; all should use the raft if there is room. (3) Remain quiet—conserve energy. (4) If you must swim, use regular strokes, either strong or lazy, but keep them rhythmic. (5) Do not trail arms or legs over the side of the raft. (6) Do not jettison blood or garbage, for this attracts sharks. (7) Do not fish from a life raft when sharks are nearby. Abandon hooked fish if a shark approaches. (8) When a shark is at close range, use "Shark Chaser" (U.S. Navy repellant) if available; the black dye will repel many species of sharks. (9) If your group is threatened by a shark while in the water, form a tight circle and face outward; if approached, hit the shark on the snout with any instrument at hand, preferably a heavy one; hit a shark with your bare hand only as a last resort.

General advice: (1) Always swim with a companion. (2) Avoid swimming at night, or in extremely turbid or dirty water, where underwater visibility is very poor. (3) Remain calm when a shark is sighted; leave the water as quickly as possible. (4) If an attack does occur, all effort should be made to control hemorrhage as quickly as possible—even before the victim reaches shore. If the wound is serious, the victim should be treated by a physician as soon as possible. (5) Adopt a sensible attitude toward sharks. Remember that the likelihood of attack is less than that of being struck by lightning. Attack is almost assured, however, when one deliberately grabs, injures, or in some other way provokes even a small and seemingly harmless shark.

PERRY W. GILBERT

Cross reference: *Salmon, Pacific.*

SHELLS—*See* TRUST TERRITORY OF THE PACIFIC ISLANDS; PEARLING INDUSTRY

SHRIMP FISHERIES

World Resume*

The fisheries for shrimps and prawns produce an average of about 60,000 metric tons each year. This represents only about 1% of the weight of seafood landed, but in terms of value these fisheries are of greater importance, representing about 5% of the world total. More nations land shrimp than nearly any other kind of marine product, and in at least 20 countries shrimp fishing is a substantial industry.

Shrimp landings have increased greatly in recent years. Many countries have gone from subsistence fishing, where most of the production was consumed locally, to a much more intense fishery, using modern gear. The stimulus for this has been a substantial rise in demand for shrimp, and over a third of the production has entered the export market. Most of this increased activity has taken place since World War II, and the trend has been even more marked since the mid-1950's.

Table 1 shows landings by country for 1963–1965. The data are supplied by the countries involved, and not all nations that land shrimps are listed, nor are the data complete or comparable. Production by some other countries is given in the text.

Species Exploited. There are no generally accepted meanings for the terms "shrimps" and "prawns." In many European countries and elsewhere the name shrimp is reserved for small individuals, and prawn means large ones, but in most parts of the U.S. all sizes are called shrimp. Shrimps and prawns belong to the sub-order Natantia of the crustacean order Decapoda. Species in the family Penaeidae are of greatest importance, while three families under the section Caridea, the Palaemonidae, the Pandalidae, and the Crangonidae, support fisheries of smaller magnitude, especially in northern waters.

A great many species support from large to insignificant shrimp fisheries in various parts of the world. These are listed by Holthuis and Rosa (1965). The genus *Penaeus* is the most important commercially, with at least a dozen species supporting major commercial fisheries, and additional members being caught in considerable numbers. Other genera in the family Penaeidae that support significant to large commercial fisheries include *Metapenaeopsis*, *Metapenaeus*, *Parapenaeopsis*, and *Xiphopeneus*. Species of this family are more important than other kinds of shrimp because they are much more abundant and ordinarily grow to a larger size.

Fishing Grounds. Fishing for shrimp usually

---

* Contribution No. 1017 from the Institute of Marine Sciences, University of Miami.

TABLE 1. LANDINGS OF SHRIMPS AND PRAWNS BY COUNTRIES, 1963–1965*

Country	Thousand Metric Tons		
	1963	1964	1965
Africa			
Algeria	0.9	0.9	1.1
Morocco	1.2	1.2	0.7
Senegal	0.4	0.4	0.8
North America			
Canada	0.8	0.5	0.8
Cuba	1.8	3.5	1.9
El Salvador	3.6	3.4	4.0
Greenland	3.3	3.8	5.1
Mexico	72.0	69.0	59.1
U.S.	109.0	96.1	109.6
South America			
Argentina	0.8	0.7	0.7
Brazil	31.9	30.3	—
Chile	3.6	5.9	5.9
Colombia	1.8	2.6	2.7
Ecuador	5.2	5.0	5.7
Peru	0.8	0.9	0.4
Venezuela	3.9	4.3	7.5
Asia			
China	9.0	9.8	14.2
Hong Kong	1.0	0.5	0.6
India	81.6	94.9	77.3
Japan	88.7	80.2	69.0
Korea	14.2	18.1	17.2
Malaya	19.0	19.3	20.8
Thailand	26.7	33.3	39.2
Turkey	0.3	0.3	0.2
Europe			
Belgium	1.1	1.1	1.0
Denmark	4.8	3.7	5.2
France	2.3	2.6	2.3
Fed. Rep. of Germany	41.1	27.5	27.3
Iceland	0.6	0.5	0.9
Italy	4.3	4.7	4.7
Netherlands	20.9	18.6	14.7
Norway	11.7	11.0	10.4
Portugal	0.2	0.2	0.2
Spain	13.0	12.7	11.8
Sweden	4.8	4.3	3.5
United Kingdom	1.7	1.7	1.4
Oceania			
Australia	5.7	6.1	5.5

* From *Bulletin of Fishery Statistics,* 13, Food and Agriculture Organization, 1965.

takes place close to the coastline. In some regions much of the production comes from shallow waters, in estuaries, brackish lakes and near the beach. In the largest fisheries most of the production comes from offshore, out to about 45 fathoms in depth, and the trend is in this direction, since larger shrimp are caught some miles off the coast. In a few areas, particularly in northern waters, shrimp are caught in greater depths.

For penaeid shrimp there are two kinds of environmental conditions necessary for a substantial commercial industry to exist: (1) extensive coastal areas of estuaries with shallow waters of low salinity, to serve as nursery areas for juvenile shrimp. The size of nursery grounds can thus be used as one approximate criterion of the potential for commercial stocks; (2) extensive, relatively shallow (10–45 fathoms usually) areas of muddy bottom offshore. Shrimp fisheries do not occur in areas of clean sand or rock, or at any considerable distance from brackish water estuaries.

Major Fisheries. The state of advancement of the world shrimp fisheries varies greatly from country to country. Many of the areas that formerly exploited their resources only with primitive gear have adopted modern methods. Even yet, however, the amount of fishing has been so limited in many countries that the extent of available populations is only vaguely known. In some places where large shrimp stocks are known to exist, lack of skilled labor and inadequate shore transportation and storage facilities have prevented the establishment of large fisheries.

About half the world's supplies of shrimp come from Asia. The importance of this area has been increasing rapidly, largely because of the growth of the fisheries in India and Pakistan. North and South America produce the next largest quantities, with large catches coming from the Gulf of Mexico and the Caribbean Sea. Although the fisheries of Europe are of some local importance, landings are relatively small. New fisheries for shrimp off West Africa may develop important industries for several countries there.

The leading shrimp producing nation of the world is the U.S. India is the second largest shrimp producing country with Japan, Mexico, Thailand, Pakistan, and Chile following in that order. China may exceed India in landings, but the volume is uncertain.

Asia. Prawns form the most valuable marine product of India. An average catch of about 75,000 tons has been made in recent years, representing about 10% of the total marine landings of the country. Another estimated 25,000 tons of prawns of marine origin are landed from estuaries and brackish areas along the coast, for a total of about 100,000 tons. Beginning in the 1950's the industry took a sudden spurt. The small catches, which had been sundried and sold locally, were replaced by much larger catches, the product being canned and frozen and most of it being exported. About 80% of the Indian marine catches of shrimps and prawns come from the west coast. Catches are higher in the northern areas, but the southern coast supports fisheries for the larger species so that the processing plants are concentrated there.

Two of the more important areas are the province of Travancore Cochin and the coastal area around Bombay.

Most of the prawn fishing in the sea is still conducted by relatively primitive methods. Dugout canoes are used on the southwest coast and catamarans on the southeast coast. Other small boats operate "boat seines," these being simple gears with two forward-projecting wings, operated from two boats. "Stake nets," fixed gears of various designs, are used in estuaries and creeks. There are a few larger vessels of American type operating in deeper waters.

Most of the fishing is carried out in depths to 20 fathoms. It appears possible that some resources are available to be exploited beyond this depth. There seem to be considerable areas of potential shrimp grounds, especially on the east coast off the larger rivers.

Pakistan has greatly expanded her shrimp fisheries in recent years. In West Pakistan 12 species enter the commercial catches in largest numbers. The total catch in 1965 was 22,000 metric tons, heads-on. The shrimp season in West Pakistan is from October to the end of April, with the best period from November to the middle of March. Fishing fleets of West Pakistan consisted in 1967 of about 5500 boats, including trawlers, launches, sail boats, and a few row boats. The catch in 1965 along this coast was about 18,000 metric tons, heads-on. This was mostly from around Karachi and on the Sind Coast. Very little of this is locally consumed; most is exported frozen to the U.S. Dried and canned prawns are also exported. West Pakistan had about 21 fish processing plants in 1967 but not all of these are in operation. It is estimated that the maximum utilization is only about 30% of capacity. The fishery in East Pakistan is of small importance. The coastline is about 200 miles long. Shrimp grounds exist on the delta of the Ganges-Brahmaputra, and expansion of the industry seems possible.

The catch of "shrimps and lobsters" in Japan is given by FAO statistics as 63,400 metric tons in 1965, but it is unclear how much of this total is shrimp. The most valuable species, "kuruma-ebi," *Penaeus japonicus*, was landed in annual amounts of 3100 to 3200 metric tons in 1963–1965. *Penaeus semisulcatus* and *Metapenaeus joyneri* are other important species in southern Japan, while *Pandalus kessleri* and *P. borealis* are the major species exploited in the north.

The principal Japanese fishing grounds are in Uchera Bay, Hokkaido, the waters of northeastern Hokkaido, the Inland Sea, and Osaka and Ise and Ariake Bays. The main fishing season in the southern fishery is from June to September. Most of the catch is boiled and dried, but some freezing and canning is done by plants handling other marine products. Of interest is the small quantity of shrimp shipped live to market, chilled, and packed in sawdust. In recent years the Japanese shrimp fishery has expanded to the Bering Sea and off the Pribilov Islands, exploiting *Pandalus*. These are canned and frozen aboard ship. Japanese shrimp boats are also operating out of British Guiana and Surinam, South America, catching *Penaeus aztecus* and other species. Japanese shrimp catches have declined in recent years. In 1965 their catch in the Bering Sea fell nearly 65% from 1964. Japanese purchases of shrimp aboard are expected to increase.

Shrimp are abundant along the mainland coast of China. Figures of landings are not available but the annual catch has been estimated at over 200,000,000 lb, caught mostly by beam trawls. Perhaps 90% of the catch is consumed by the local inhabitants in the coastal areas. Dried and salted shrimp are used throughout the country in various dishes. Most important commercial shrimp are *Penaeus carinatus, Metapenaeus monoceros,* and *P. japonicus*. In Hong Kong about 600 junks using small beam trawls supply the colony with shrimp. Korea, Thailand, Ceylon, the Philippines, Formosa, Indonesia, South Vietnam, Malaysia, and Singapore all have small shrimp fisheries, landing between 1,000,000 and 40,000,000 lb per year. Fishing methods are largely primitive, ranging from beach seines and cast nets to traps. Most of the catch is dried or consumed fresh.

In several countries of Asia, especially Japan, the Philippines and Indonesia, shrimp culture operations are conducted. These range from simple entrapment of larval stages with subsequent harvesting of adults from rice paddies and other enclosed brackish impoundments, to the complex farming of *Penaeus japonicus* in southern Japan.

In the mid-1960's countries bordering the Persian Gulf, including Iran, Saudi Arabia, Kuwait, Bahrein and others, had become major factors in the world shrimp market. Their combined shipments to the U.S. were surpassed only by those of Mexico (Farber, 1968). The first significant shipment from this area was in 1963, and the volume doubled in each succeeding two years. Persian Gulf shrimp resource has been described as enormous, and many more boats are expected in the later part of the 1960's. Most of the shrimp were shipped to the U.S. and Japan.

Australia. The principal commercial species in New South Wales are *Metapenaeus macleayi, M. mastersii,* and *Penaeus plebejus*; in Queensland, *Metapenaeus mastersii, Penaeus plebejus,* and *P. merguiensis*; in Western Australia, *Penaeus latisulcatus* and *P. esculentus*. The major fishing grounds are off the east coast of the continent. In western Australia shrimp are fished mainly in

Shark Bay, this industry having started in 1962.

Until the mid-1920's shrimp fishing in Australia was done on a limited local basis. From that time to 1948 the industry gradually developed, at which time offshore fishing (i.e., out to about 25 miles) began to increase. From annual catches of about 300 tons 50 years ago, total production has risen to about 6000 tons in the early 1960's. Offshore fishing is done with otter trawls.

There is probably room for considerable expansion of the shrimp fisheries of Australia.

North America. The U.S. is the principal shrimp nation in the world in two respects: it produces more shrimp than any other country; and it absorbs a substantial amount of the catch of many other nations—sometimes virtually their whole production.

The U.S. shrimp fishery consists of two parts, one of much greater importance than the other. The southern fishery, which extends through eight states from North Carolina to Texas, produces over 200,000,000 lb of shrimp per year. The Pacific coast fishery produces about 18,000,000 lb annually. The shrimp fishery is the most valuable in the U.S., in 1966 accounting for 21% of the value of all marine species landed. The species of principal importance in the southern fishery are brown shrimp, *Penaeus aztecus,* pink shrimp, *P. duorarum,* and white shrimp *P. setiferus.*

Fishing methods used in a large part of the world to catch penaeids are patterned on the methods used in the southern U.S. fishery. At the present time virtually the whole American catch of southern shrimp is made by shrimp (otter) trawls. The commonest vessel is the so-called "Florida-type" boat, up to 80 ft or more in length, of wood or steel, and powered by 150–200 h.p. diesel motors. A second, smaller vessel employed in the Gulf of Mexico area is the "Biloxi-type," used in the inshore areas.

A single otter trawl was used by these vessels up to the late 1950's, when most vessels adopted a "double rig," involving two smaller trawls about 45 ft across the mouth. Most of the vessels employ a try-net, a miniature trawl about 10 ft across the mouth. This is used when fishing begins to determine whether there are sufficient numbers of shrimp to justify shooting the main net. Echo sounders have not been especially useful in shrimp fisheries except to give accurate indications of depth and of smoothness and type of bottom. Attempts to design echo sounders that would discriminate shrimp from the bottom have met with only partial success. Shrimp react to an electrical stimulus by darting out of the bottom. Attempts have been made to take advantage of this behavior by attaching electrodes to the trawl to increase the efficiency of the gear. Experiments are promising but so far no use is being made of this on a commercial basis.

Shrimp from warm waters are usually caught along with a great variety of other animals, including small fish, crabs and other crustaceans, sponges, and mollusks. On deck the shrimp are sorted from the "trash," and washed and stored in the hold in ice. Usually the "heads" (the cephalothorax, which includes most of the viscera) are removed before storage. This is desirable since spoilage takes place much more rapidly in unheaded shrimp. Some fishermen do not use ice, and spoilage is more rapid. In an increasing number of cases shrimp are frozen on the boats in large blocks, to be thawed and processed on shore.

Traditionally shrimp have been consumed fresh near their point of capture, or have been dried or salted. In technologically advanced countries in recent years the trend has been away from this, and the great bulk of shrimp are now frozen or (to a lesser extent) canned. In the U.S. freezing greatly outweighs any other method of preparing shrimp, and in other countries where shrimp are intended for the U.S. market freezing is also the principal method of preparation.

In many of the countries where the shrimp fisheries have expanded rapidly, fishermen have begun to operate in deeper waters where larger individuals congregate. Most of the shrimp fishing at the present time is done out to a maximum of about 30 fathoms. It is clear that there are shrimp stocks beyond this depth. For example, exploratory fishing by U.S. research vessels and others have shown that large stocks of penaeid shrimp exist in deep water. Royal red shrimp, *Hymenopenaeus robustus,* occurs to depths of 300 fathoms or more, and some sporadic attempts have been made by U.S. boats to harvest this resource; the cost of capture is so great, however, that there is no sustained fishery. A potential exists for this species and for a considerable number of others if it becomes economical to catch them.

A far greater potential exists for small species of shrimp. In the Americas for example, the seabobs, *Xiphopeneus kroyeri* and other species, exist in extremely large numbers. There is an increasing market for these, and catches are rising rapidly. Within the next few years there is no doubt that production of these small shrimps will increase.

On the west coast of North America several genera and species of the family Pandalidae are caught. *Pandalus jordani* is fished from northern California through Washington and Oregon. This species is replaced off British Columbia by *Pandalus borealis,* which is predominant in catches there and in Alaskan waters.

The fishery for *Pandalus jordani* in California,

Oregon, and Washington is carried out largely by Gulf of Mexico type shrimp trawls. West coast landings were about 17,500,000 lb in 1966 compared to 13,000,000 lb in 1965, and most of this total (about 14,500,000 lb) came from Alaska.

In Alaska 80–90% of the shrimp catch consists of *Pandalus borealis*. Approximately 15% of the catch is *Pandalopsis dispar* with smaller quantities of *Pandalus goniurus, P. hypsinotus,* and *P. platyceros*. Catches are made near shore in southeastern Alaska, the Gulf of Alaska and the Bering Sea (Harry, 1964).

Alaskan fishermen have exploited this resource for 35 years or more, but it is only in the present decade that catches have been substantial. Beginning in 1961, Japanese and, in 1963, Soviet fishermen entered this fishery, and catches by these boats have greatly exceeded those of the Americans in recent years. Table 2 shows catches by the U.S. and other fishermen from 1960 to the present.

In southeastern Alaska beam trawls are used by American fishermen. The beam is 32–52 ft long. In the Gulf of Alaska most of the gear consists of otter trawls, about 70 ft across the mouth.

The Japanese and the Soviets are using large factory ships supplied by catcher vessels employing otter trawls. The catch is canned or frozen in blocks.

The shrimp fishery in Maine has been increasing since 1961. Detailed information will be found later in this article.

The commercial shrimp fishery in Canada is limited to the Pacific Province of British Columbia (Butler, 1967). In 1946 landings were about 119,000 lb, in 1966 about 764.5 metric tons. The value of this catch, some $300,000, is only half of 1% of the total catch value of the Province's fisheries.

There are six species of importance, including *Pandalus jordani, P. borealis, P. platyceros, Pandalopsis dispar, P. hypsinotus,* and *P. danae*. Most of the catch is taken by beam trawls but about 7% is caught in small traps or pots. This trap has a single chamber and is variable in design. Typically it is oblong with a conical entrance in each end made of netting. The frame is usually of welded iron rods covered with netting and wire mesh. The traps vary in size from 61 × 30 × 30 to 91 × 45 × 45 cm. Traps are fished on the bottom by means of a ground line. Up to 40 traps may be put on a ground line spaced 9–18 m apart. Traps are lifted once or twice daily and bait is renewed each time. Codfish, shark, herring, and scrap fish are used as bait. The areas of most importance are the lower mainland of British Columbia and Vancouver Island. Some shrimp are also taken in Chatham Sound and in the Queen Charlotte Islands.

The modern shrimp fishery of Mexico started in the 1930's when U.S. boats fished on the west coast and Japanese exploratory vessels made observations on the east coast. Offshore exploitation of the large populations in the Gulf of Mexico and the Caribbean Sea, principally off the Campeche Banks, began by U.S. trawlers in 1946. American-owned freezing plants were established at Ciudad del Carmen, and by 1954 some 200 Mexican trawlers were operating on the Campeche Bank.

Mexican shrimp grounds extend the entire length of the coast of the country. Guaymas and Mazatlan are the two principal Pacific ports, between them handling about half the west coast landings. East coast fisheries center at Ciudad del Carmen, Campeche and Tampico.

About 75% of Mexico's shrimp production is from the west coast. Here *Penaeus stylirostris, P. vannamei, P. californiensis,* and *P. brevirostris* are the important species. The latter is taken in deeper water, out to 45 fathoms. On the east coast, *Penaeus duorarum, P. aztecus,* and *P. setiferus* are caught in greatest numbers. In inshore waters fishery cooperatives are the only groups legally allowed to catch shrimp commercially. They exploit juveniles leaving the estuaries. The gear used includes cast nets, traps, and weirs. Offshore otter trawls are used, of the same types employed in the U.S. Mexico has been the chief foreign supplier of shrimp to the U.S., and most of its production goes there. In 1967 this amounted to 39,000,000 lb.

Central America and the Caribbean. The shrimp fishery in the Pacific started in northern Mexico in 1941 and in Panama in 1950. The Costa Rican fishery started in 1952 and that in El Salvador in 1958; Guatemala and Nicaragua were later. The fishery now operates south to Ecuador.

On the Pacific coast white shrimps (*Penaeus stylirostris, P. vannamei,* and *P. occidentalis*) are the most important. The latter two were sometimes called brown shrimp along with *P. californiensis,* the true brown shrimp. The pink shrimp *P. brevirostris* is of considerable importance in El Salvador and Panama. There are in

TABLE 2. CATCHES BY U.S. AND OTHER FISHERMEN IN S. E. ALASKA, THE GULF OF ALASKA, AND THE BERING SEA (IN MILLIONS OF LB)

Year	U.S.	Foreign	Total
1960	7.4	—	7.4
1961	16.0	22.5	38.5
1962	16.9	46.3	63.2
1963	15.2	69.7	84.9
1964	7.7	46.3	54.0
1965	16.8	30.2	47.0
1966	24.0	35.2	59.2

addition large quantities of four species of smaller shrimp, *Xiphopeneus riveti, Protrachytene precipua,* and the seabobs and *Trachypenaeus byrdi* and *T. faoea.*

The shrimp fishery in the Caribbean Sea developed first in Cuba in 1953, followed by production in Venezuela in 1957. Honduras and Nicaragua began sizeable operations in 1958. The Caribbean shrimp fisheries are based principally on four species: white shrimp, *Penaeus schmitti,* pink shrimp, *P. duorarum,* spotted pink shrimp, *P. brasiliensis,* and the brown shrimp, *P. aztecus.*

Shrimp are fished on the entire Pacific coastline of Guatemala, which is 155 miles long. In 1962 the Government set a limit of 50 vessels, but in 1965 there were only 20 fishing. Guatemala exported between 1–2,300,000 lb of heads-off shrimp between 1962 and 1965. The record catch was made in 1964 when almost 3,000,000 lb were taken. There is no sustained fishery on the small Caribbean coast.

More shrimp are produced in the 175 miles coastline of El Salvador than any comparable area between Mexico and Panama. Before 1955 shrimp were exploited only by canoe fishermen, taking juveniles. In 1962 the Government limited the number of vessels to 73. In the 1960's between 7,500,000 and 8,500,000 lb were exported.

Shrimp are found along the entire 190 mile Pacific coastline of Nicaragua, but rocky conditions inshore limit the boats to depths greater than about 10 fathoms. Nicaragua has a Caribbean coastline of about 285 miles, including an excellent series of brackish water estuaries. A large-scale fishery began there in 1958. Nearly all the production is exported to the U.S. Landings rose from 266,000 lb in 1960 to 3,100,000 lb in 1965, but these figures do not include catches made off the coast by U.S. vessels based in Florida. As many as 30 trawlers may fish offshore some part of the year. They carry each others catches home, transferring the catch at sea.

Despite a long Pacific coastline of 360 miles in Costa Rica the fishing areas are relatively small. The fishery here started in 1952 and in 1965 the fleet consisted of 52 trawlers. Catches of heads-off shrimp from 1960 to 1965 ranged from 1–2,700,000 lb. In 1960 the fleet commenced to take the pink shrimp (*Penaeus brevirostris*) in considerable quantities. Seabobs there appear to be very abundant and are taken to the domestic market.

Panama has the largest shrimp fishery south of Mexico on the Pacific. Shrimp are found along most of the 700-mile coastline. The principal fishing grounds are in the Gulf of Panama and Gulf of Chiriqui. Modern trawling began in 1946; in 1965 about 200 vessels were in operation. Originally the fishery exploited the abundant white shrimp, but later the pink shrimp was caught in considerable quantities. An especially large run of pinks occurred in 1957, causing the industry to increase its capacity greatly. It is believed that white shrimp are being exploited to their capacity at the moment while the pink shrimp are sporadic in their appearance. Seabobs find a ready market in Panama, with catches from 1960 to 1965 ranging from 4,400,000–7,100,000 lb heads-off. These are packed, peeled and deveined. Total catches climbed from 3,600,000 lb in 1954 to 15,500,000 lb in 1964, but dropped back to 12,800,000 lb in 1965. There is no shrimp fishery on the Caribbean coast of Panama, although there may be stocks that could be exploited.

The Honduras fishery began in the Caribbean in 1958. It has been very unstable because shrimp are found in quantity only along a 100-mile stretch between Punta Patuca and the Nicaraguan border. The land opposite the shrimp grounds is uninhabited and the grounds are inaccessible. In 1966 three freezer plants were in operation. In 1964 about 50 vessels were fishing off Nicaragua, but by 1965 this had dropped to 30. Nearly all of the production is exported to the U.S. In the 1960's catches varied from 227,000 lb to 1,600,000 lb, but this does not include the catch made by U.S. vessels fishing offshore, which is included in the statistics of the U.S. fishery. In British Honduras small quantities are caught in the southern part of the shelf area.

Shrimp fisheries of most island countries of the Caribbean are sporadic, and are largely small scale operations for local markets. This is because favorable ecological areas and suitable trawling grounds are missing. In the Bahamas, Jamaica, Haiti, Dominican Republic, Puerto Rico, and Barbados small quantities of shrimp are caught. A shrimp fishery began in Cuba in 1953 on the south side of the island; this expanded until 1957, when stocks fell sharply. This fishery seems not to have been revived although Cuba's fin fish industry has expanded greatly with U.S.S.R. help.

South America. Shrimp are found only in the northern part of the 500-mile Pacific coastline of Colombia. The trawl fishery began operating in 1943. The biggest fleet fished in 1960 when 90 boats operated; by 1963 this had fallen to 62 boats. About 90% of the production is exported; in the 1960's between 1,900,000–2,800,000 lb were exported to the U.S. The white shrimp, *Penaeus occidentalis,* constitutes 99% of the landings. There appear to be only sparse shrimp resources off Columbia's Caribbean coast, although small local fisheries are based on immature shrimp in some areas.

Ecuador has about 500 miles of coastline but fishing is confined to the 100 miles on the northern edge, and to the Gulf of Guayaquil on the extreme south. The fishery began in 1954; exports averaged about 5,000,000 lb in the early 1960's. About 68 boats were in operation in the mid-1960's. Nearly all the catch is exported.

Peru is on the extreme southern edge of the range of tropical species of shrimp and has only a small fishery. This is centered in a 30 mile area next to the border of Ecuador. After a small beginning in 1952 the fishery has made some progress in recent years. About 80% of the production is exported to the U.S.; this was 256,000 lb in 1960, and 446,000 lb in 1965.

The shrimp fishery of Chile has increased from an estimated 100 tons in 1954 to in excess of 10,000 tons in 1966. More than half the landings come from the Port of Valparaiso, where five companies receive the catches of 16 boats. An additional 20 boats work from other ports of the country. The same vessels often catch fish, particularly hake, and shrimp landings are more variable than if the fleet were pursuing only shrimp.

Three species of shrimp are of at least potential importance from Chilean waters. These include *Heterocarpus reedi,* which account for about 95% of present landings, *Rhynchocinetes typus,* a small shallow water species, and *Hymenopenaeus diomedeae,* which is landed in very small quantities. *Heterocarpus* is a deep sea species, caught at about 180 m. This fishery began in 1953. Prior to that time most of the landings were *Rhynchocinetes.*

The normal gear is the otter trawl, using heavy doors. Most of the catch is cooked on shore by boiling or under steam, and then frozen. A small proportion of the cooked shrimp are canned. Most of the production is exported.

Shrimp is by far the most valuable fishery of Venezuela. From a small local fishery, the industry expanded at a meteoric rate beginning in 1959, until Venezuela produced more shrimp in 1965 than any other country south of Mexico. The large production is almost exclusively from a small area in the western part of the country, the Gulf of Venezuela and Lake Maracaibo. There are three separate fisheries in the Maracaibo area. In the lake about 3200 fishermen operated an estimated 636 canoes and other small boats in 1965. They fish with modified beach seines and throw nets and catch mostly white shrimp (*Penaeus schmitti*), which accounts for about 75% of the total Venezuelan production. A fleet of about 55 trawlers operated mostly by Italian fishermen worked in the eastern Gulf of Venezuela, in 1965. A second trawl fleet, of 14 Florida-type boats in 1965, worked in the western Gulf. The trawlers catch pink shrimp (*P. duorarum*), spotted pink shrimp (*P. braziliensis*), and brown shrimp (*P. aztecus*) in that order of importance. Few large shrimp have been found east of the Maracaibo area. A small quantity is landed around Margarita Island, and seabobs (*Xiphopeneus kroyeri*) exist in abundance in some areas, including the Gulf of Paria.

The waters off the northeast coast of South America between the mouth of the Orinoco and the mouth of the Amazon are the richest shrimp fishing grounds of the western hemisphere. They were first exploited on a large scale in 1959, and by 1965 they supplied the U.S. with over 50,000,000 lb of shrimp. Additional quantities are exported to Japan, France, the United Kingdom and other countries. These grounds comprise about 75,000 miles.[2]

Fishing is carried on from 50–100 miles offshore, in depths of 15–34 fathoms. The sea bottom has a gently sloping shelf out to about 32–34 fathoms, after which it shelves off abruptly. This dropoff is the most productive area. The fishermen recognize five grounds: the West Grounds off the coast of Guyana; the Middle Grounds off the coast of Surinam; the East Grounds off the coast of French Guiana; the Rock offshore from Cayenne, and the Gullies in the Amazon area. Strong current and heavy seas make fishing difficult and the successful vessels are larger than those operating in Florida. Most are 72–75 ft long with high-powered engines.

The most important shrimp is the spotted pink shrimp, *Penaeus brasiliensis,* with lesser quantities of the northern pink shrimp *P. duorarum,* brown shrimp, *P. aztecus,* and white shrimp, *P. schmitti.* The latter occurs only in shallow water and is lightly fished. Seabobs, *Xiphopeneus kroyeri,* are extremely abundant and constitute a reserve of enormous size.

Most of the vessels now entering the fishery are made of steel and have their own freezing systems. They use brine refrigeration and not ice. All the vessels operate under the U.S. flag except for one Japanese fleet in Paramaribo, and British units in Georgetown and Port of Spain. Many of the boats transfer their early catch to another vessel in the fleet, which delivers it to the home port so that fishing can be continued.

In Trinidad a local fishery is carried out from dozens of fishing villages around the west coast facing the Gulf of Paria. A reported 1200 small trawlers of less than 40 ft fish close to shore. These land about 50 tons of white and brown shrimp a month for the local market. An export fishery for shrimp began in 1965. Most of the fishing is off the Guianas and Brazil. The industry is growing rapidly and nearly 1,000,000 lb were exported to the U.S. in the first nine months of 1966. About 60 vessels landed in Port of Spain in 1966.

Guyana has a local shrimp fishery conducted in the rivers, by means of Chinese trap nets. The shrimp caught are very small and are used for local consumption. Landings in 1956 were approximately 800,000 lb, heads-on. It has been estimated that small shrimp could support a fishery of 4–5,000,000 lb a year if the market were available. About 100 vessels, American and British, based in Georgetown in the mid-60's, were fishing on the Guiana grounds. The West Grounds and the Gullies are 9–80 hours from Georgetown. This fishery began in 1959. U.S. imports from Guyana were nearly 8,000,000 lb in 1965.

Surinam has had an important local fishery for seabobs and other small species for many years. Here (as in many other parts of the world) the local shrimp fisheries are conducted in shallow waters, ordinarily on juvenile stocks. The gear consists of traps, beach seines, throw nets, and other simple devices. In Surinam the Chinese shrimp net and the pin seine are operated. These are long conical nets held in place by poles or pilings driven in the bottom. They operate by tidal action. These nets are attended by canoe. Small shrimp are boiled, sun-dried and peeled. In 1956 the annual catch of seabobs was estimated to be in excess of 1,000,000 lb, heads-on. The seabob resource could probably produce 5–6,000,000 lb a year.

The export fishery started in 1956. An American fleet of 15 vessels and a Japanese fleet of 10 vessels fishing the Guiana grounds land their catch at Paramaribo. The U.S. imports from Surinam were between about 1,000,000–1,500,000 lb in 1962 to 1965. Perhaps an equal quantity is exported to Japan.

The export fishery in French Guiana began in 1963. Plants exist in two cities, St. Laurent near the mouth of the Mornoni River, and Cayenne, near the mouth of the Cayenne River. Between 50 and 60 trawlers are supplying these plants. In addition to the large shrimp, a considerable quantity of seabobs are caught for the French market. U.S. imports of shrimp from French Guiana range from 2,800,000 lb in 1963 to nearly 4,000,000 lb in 1965.

Shrimp are abundant along the entire coast of Brazil, and, while local fisheries exist, no large-scale export industry had developed by 1968. Unlike the countries to the north, Brazil has not encouraged foreign participation in the shrimp industry, and as a consequence the great growth of the industry in the Guianas and some of the Caribbean islands has not taken place there.

The annual Brazilian shrimp catch was estimated at 20–25,000,000 lb, live weight. Three separate fisheries account for this. The northern fishery, which contributes about half the catch, is carried on east of the mouth of the Amazon to near Sao Luiz. This is conducted by traps, nets, and seines in the estuaries and along the beaches. About 60% of the catch is seabobs, the rest being juvenile white and brown shrimp. The second fishery is conducted by finfish trawlers from Rio de Janeiro to Santos which take incidental quantities of shrimp along the south and central coast. The large southern fishery is near Rio Grande and is based on young spotted pink shrimp. This fishery is seasonal and variable. It has been estimated that Brazil could produce more shrimp than all the rest of Latin America combined, meaning in excess of the 120,166,000 lb produced in these countries in 1965.

In Argentina two species, *Artemesia longinaris* ("camaron") and *Hymenopenaeus mulleri* ("langostino") constitute most of the catch. These are taken from Rio de Janeiro, Brazil, to Puerto Deseado, Argentina. Their main concentration appears to be south of 35° S. They are fished in Argentina south of 37° S.

From 1934 to 1963 the combined landings of these species fluctuated between 200 and 3240 metric tons; average annual landings were 780 metric tons for langostino and 543 metric tons for camaron. These landings are small compared to those of fish.

Since 1947 the grounds south of Rawson have been the most important. After 1956 langostino landings fell very seriously (1,935.2 metric tons in 1956; 362.3 metric tons in 1963). Camaron landings have varied much less (236.4 to 677.0 metric tons).

Uruguay has had only a very small shrimp fishery, the largest landing reported being 103 metric tons in 1961. These came from coastal lagoons. The species caught is *Penaeus aztecus* (Mistakitis, 1964).

Europe. The species landed in the largest quantities in Europe are the common shrimp *Crangon crangon* and the pink shrimp, *Pandalus borealis*. Species of lesser importance include *Pandalus montagui, Leander adspersus* and *L. serratus*.

About half the shrimp produced in Europe are caught by West German boats. The species is principally *Crangon crangon,* caught along the North Sea coast in a 15-mile strip and in the estuaries of the Ems, Elbe, and Weser Rivers. Depths fished range from 11–85 fm. The main gear is the beam trawl, about 45 ft in length and 22 ft wide. They are operated by 30–45 diesel powered cutters of shallow draft. In Germany most of the shrimp landed are used to produce fish meal; a considerably smaller quantity is sold for food.

The German shrimp fisheries operate from the end of March or beginning of April to about the end of November. Maximum catches of small shrimp are obtained from July to October, when 70 to 80% of their catches are taken. Catches of large shrimp used for human consumption have two peaks, April–May and October; the fishery for large shrimp operates throughout the year. Fishing is done on sandy or muddy bottoms. On the German coast fishing is done from near the tide mark out to 15 or 20 m in depth.

The type of shrimp used for poultry feed runs between 1000 to 1400 to the pound; shrimp for human consumption average between 200 to 250 to the pound. Shrimp are graded on board for size in many cases. Those to be sold for food are cooked aboard the vessel in salted brine. These shrimp are of small size and removal of the shells poses a problem. In some parts of the coast of Germany shrimp are parcelled out to families and peeling is a cottage industry. Peeling machines of American design have been tried with varying success in recent years. Much of the production is canned. Shrimp to be made into meal are landed raw.

The principal shrimp fishing areas in the Netherlands are the estuaries of the Ems, Rhine, Maas, and Schelde Rivers. Shrimp are caught by otter trawls and beam trawls, or in the estuaries by elongated fixed nets which trap the shrimp in the rising tide. Threequarters of the Dutch catch are of small shrimp, which are made into fish meal. Catches have ranged from about 11,000 to 20,000 tons in recent years. Those sold for human food are cooked on the boats.

In Belgium fishing is done along a coastal area of the North Sea. Half the catch is made from April to July, but fishing is done all year. Heads-on catches in recent years have been about 1000 metric tons annually.

The principal species caught in Britain are *Crangon crangon* and *Pandalus montagui*. The Thames estuary, Morecambe Bay, and Liverpool Bay in the Irish Sea, the North Sea coast, and the area of the Wash produce most of the shrimp. The offshore fishery is conducted by 30–40 ft diesel-powered smacks using beam trawls of 12–20 ft wide. On the east coast otter trawls fish to a small extent. In recent years the catch has ranged from about 1000 to 2000 metric tons. Most is sold fresh but increasing quantities are frozen or canned.

The catch of shrimp from Spain consists mostly of *Parapenaeus longirostris* and *Crangon crangon*, although several species are caught. The annual catch is about 20–26,000,000 lb, and is marketed mostly in Spain. The main fishing grounds are on the south coast off Barcelona and Valencia.

Shrimp fishing in Norway extends from the Skagerak in the south to the Varanger Fjord in the north. The species caught is *Pandalus borealis*. It is captured in deep fjords off the coast where the sea bottom is soft. The gear used is otter trawls 80–100 ft wide. The fishing is conducted in waters from 50 to 200 fathoms deep. Most of the shrimp are frozen and packaged for export to the United Kingdom and Sweden. Annual production is about 20,000,000 lb.

In Sweden about 5,000,000 lb of *Pandalus borealis* are taken around Goteborg.

In Denmark most of the catch (3000–5000 metric tons a year) is of *Pandalus borealis*, and is made outside Danish territorial waters, in depths of 75–200 fathoms, north of Cape Skagen. Catches are best from March through October, especially in June and July. The catch is landed iced or boiled and most of it is canned.

The catch in Greece consists of about half of *Penaeus trisulcalus* and about a third *P. longirostris*. Annual catches may be about 250–400 tons.

Most shrimp landed in Italy are caught in the Adriatic Sea and south of Sicily. They are caught by trawlers and are sold fresh. Some shrimp are landed in France, Portugal, and other European countries.

A Greenland fishery for *Pandalus borealis* began to expand about 1948. Fishing is mostly off the west coast at a depth of about 200 fathoms in very cold water. Catches of over 3000 tons have been recorded in the 1960's.

Africa. Shrimp are taken commercially along most of the southern shores of the Mediterranean and as far south on the west coast of Africa as the Canary Islands. Morocco, Egypt, Algeria, Libya, Turkey and Tunisia all produce some shrimp. Egypt produces some shrimp from the Red Sea.

Shrimp fisheries along the coast of East Africa are inconsequential. Only three species, *Penaeus indicus, P. monodon,* and *P. semisulcatus* (and possibly *Metapenaeus monoceros*) are abundant enough to support commercial exploitation. Only Mozambique and the United Arab Republic land enough shrimp to be recorded in the FAO statistics. In 1965 Mozambique exported 200 tons and the United Arab Republic 900 tons of fresh, frozen, dried, and salted crustaceans and mollusks, an unknown fraction of which were shrimp.

From Dar es Salaam, Kenya, northwards, the steeply sloping coastline, the lack of extensive brackish and fresh water areas suitable to the developmental stages of shrimp, and the lack of trawlable offshore areas greatly limit the abundance of shrimp populations. South of Dar es Salaam shrimp occur as far as Algoa Bay. The potential of these resources is impossible to estimate at present.

In recent years it has been discovered that shrimp occur off the coast of west Africa in the regions of Nigeria, Ghana, and Angola. One of the principal species, *Penaeus duorarum*, is apparently identical with the pink shrimp of the Gulf of Mexico (Hoestlandt, 1967). In Africa this species occurs from Cap Blanc, Mauritania to Luanda, Angola. Two other species, *P. kerathurus* and *Parapenaeopsis atlantica*, occur in parts of the coast.

U.S.S.R. The Soviet Union began fishing in 1965 in the Gulf of Alaska for *Pandalus borealis*. In that year between 2500 and 3000 tons were landed, which increased in 1966 to 10,700 tons. Vessels used are SRTM freezer-trawlers of 30 m length. Their average catch was about 500 lb per 24 hour day; the biggest catch was 10 tons per hour, taken near the Pribilov Islands in the Bering Sea, and 7 tons per hour off the Shumagin Islands in the Gulf of Alaska. Most of the catch is sent to Japan.

C. P. IDYLL

Maine Shrimp Fishery

That Northern shrimp is an important food of cod has been known for many years, ever since the presence of *Pandalus borealis* in Maine waters was observed, during colonial times, by fishermen who found cod stomachs filled with shrimp in several areas of the western Gulf of Maine. After the lath lobster trap was developed in 1870 to replace the hoop-net, fishermen frequently caught larger specimens of *Pandalus borealis* in their lobster pots.

Populations of the Northern shrimp (*Pandalus borealis*) support a variable fishery in Maine and Massachusetts during the winter. This species occurs from the Gulf of Maine northward, ranging both sides of the North Atlantic, into the Arctic Ocean, Bering Sea, and the Gulf of Alaska. The Gulf of Maine is generally considered to be the southern limit of the species in the Northwestern Atlantic, and Maine and Massachusetts are the only Atlantic coastal states with a commercial fishery.

Catches are made during the fall-to-spring period on grounds within a 50-mile radius of Portland by small draggers operating from nearby ports. Landings have fluctuated widely from year to year, and appear to be related primarily to abundance and availability. Recent increased catches also reflect more boats, greater effort, more efficient methods, and more reliable information on availability based on forecasts of continued greater abundance.

Biology. *Pandalus borealis* is a sex reversal species, as are many shrimp. During initial development, most of the young mature as males. Later they pass through a transitional stage to mature as females. Spawning occurs in the winter, with February and March the peak months for this activity. Until 1964 no commercial fishery for shrimp was conducted except on the inshore spawning grounds. In more recent years, fishermen have operated farther offshore before the bulk of the spawning population moves into the spawning grounds.

Fig. 1. Top, several ages of *Pandalus borealis* up to mature female; bottom, mature female.

Since shrimp is the only species of commercial interest that is naturally available in greater concentrations during the winter, the species has attracted a considerable amount of fishing effort in Maine throughout the history of the fishery.

Maine Landings. Beginning with 1939, records of monthly landings have been made by the Bureau of Commercial Fisheries and the Maine Department of Sea and Shore Fisheries. Some of the annual data of the fishery are shown in Fig. 2.

The principal fishing ports in Maine are located in Cumberland and Lincoln Counties, where nearly 85% of the annual catch is landed. Landings in Sagadahoc County average 15% of the catch, while only occasional catches are landed at Knox, York, and Hancock County ports. Landed value has ranged from 4 to more than 30 cents per pound; with 13 cents being the average for the greater landings of recent years.

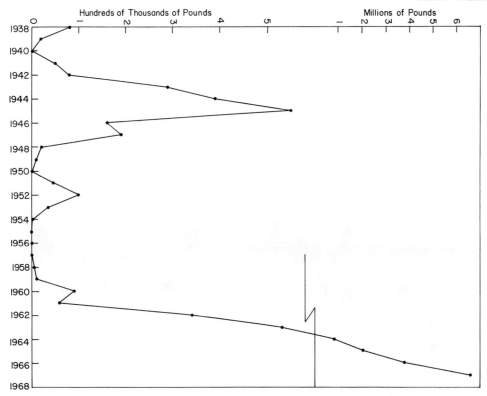

Fig. 2. Total landings of northern shrimp at Maine ports by fishing seasons.

Early Surveys. The first survey effort was made in 1927 by General Seafoods Corporation. Then in 1936 the vessel *New Dawn* supported by the Fisherman's Relief Corporation of Portland and the Federated Fishing Boats of New England and New York, Inc. collaborated on a further survey. Woods Hole Oceanographic Institution, individual fishermen, and others conducted exploratory fishing in the Gulf of Maine. All these efforts were summarized by Walford in 1936 and Fig. 3 is an adaptation of the summary he made of sampling locations where shrimp were found. Figure 3 clearly shows the difference in distribution of the summer-fall population and the winter population, with the winter population moving inshore to more shallow spawning areas. In 1937, General Seafoods again carried on experimental fishing in the Gulf of Maine. In 1938, with the cooperation of the Fish and Wildlife Service and the Department of Sea and Shore Fisheries, a survey for shrimp was made from York County to Mount Desert Island.

Small scale commercial operations were commenced the following year in Casco Bay. During the next several years the fishery expanded rapidly and landings increased to the 1945 peak of more than 500,000 lb. Thereafter, catches declined to a low of 7000 lb in 1950. Another cycle of

* Summer and Fall Population

• Winter Population

Fig. 3.

Fig. 4. Shrimp dragger.

increasing landings commenced in 1951 but was short lived, and after 1953 no shrimp were landed until 1958 when less than 5000 lb were taken. Since then, production has increased rapidly and amounted to more than 6,500,000 lb during the 1966–1967 season.

In addition to an apparent cyclic variation in abundance, as indicated by landings (Fig. 2), there is evidence that seawater temperature is an important factor and that monthly, seasonal, and annual mean temperatures, as measured at Boothbay Harbor by the Bureau of Commercial Fisheries, are related to the magnitude of the shrimp catch two years later.

Seawater Temperature. Twenty-eight complete years of temperature records have been maintained at Boothbay Harbor by the Bureau of Commercial Fisheries since commercial shrimp

Fig. 5. Northern shrimp in peeling machine.

SHRIMP FISHERIES

TABLE 3. ANNUAL SEAWATER TEMPERATURE—SHRIMP LANDINGS

Annual Seawater Temperature °F	Shrimp Landings, Thousands of Pounds	Temperature Year	Annual Seawater Temperature °F	Shrimp Landings, Thousands of Pounds	Temperature Year	Annual Seawater Temperature °F	Shrimp Landings, Thousands of Pounds	Temperature Year	Annual Seawater Temperature °F	Shrimp Landings, Thousands of Pounds	Temperature Year
52.0	0	1953	49.3	104	1950	47.9	340	1960	45.2	8	1938
51.4	38	1951	48.8	11	1957	47.9	2068	1963	44.6	79	1940
50.3	0	1954	48.6	5	1956	47.4	90	1958	43.5	54	1939
50.2	0	1952	48.6	10	1947	47.3	27	1946			
50.0	0	1955	48.2	18	1937	47.3	529	1961			
						47.1	194	1945			
						47.0	64	1959			
						46.9	3691	1964			
						46.6	898	1962			
						46.6	389	1942			
						46.4	162	1944			
						46.1	292	1941			
						45.8	6412	1965			
						45.5	83	1936			
						45.3	554	1943			

fishing commenced. During five years in which annual mean temperature averaged 50° F or higher, shrimp landings averaged less than 8000 lb two years later. Annual average landings of approximately 30,000 lb are associated with temperatures between 48° and 50°F. Annual average landings of 47,000 lb are associated with temperatures of less than 45.3°F. In 15 years temperatures ranged between 45.3° and 47.9°F and annual shrimp landings two years later

TABLE 4. JUNE SEAWATER TEMPERATURE AND NORTHERN SHRIMP LANDINGS TWO YEARS LATER

June Temperature °F	Shrimp Landings, Thousands of Pounds	Temperature Year	June Temperature °F	Shrimp Landings, Thousands of Pounds	Temperature Year	June Temperature °F	Shrimp Landings, Thousands of Pounds	Temperature Year
58.0	11	1957	56.3	389	1942	52.6	64	1959
57.7	0	1953	56.0	8	1938	52.3	90	1958
57.7	38	1951	56.0	2068	1963	51.9	79	1940
57.7	45	1949	55.7	104	1950	49.8	54	1939
57.1	0	1954	55.3	340	1960			
56.6	18	1937	55.0	529	1961			
56.5	0	1955	54.4	898	1962			
56.4	0	1952	54.1	5	1956			
			54.0	10	1947			
			53.6	7	1948			
			53.6	27	1946			
			53.5	194	1945			
			53.2	292	1941			
			53.2	3691	1964			
			53.1	83	1936			
			53.1	6412	1965			
			53.0	554	1943			
			52.8	162	1944			

TABLE 5. NOVEMBER-FEBRUARY SEAWATER TEMPERATURE AND NORTHERN SHRIMP LANDINGS TWO YEARS LATER

November-February Temperature °F	Shrimp Landings Thousands of Pounds	Temperature, Year	November-February Temperature °F	Shrimp Landings, Thousands of Pounds	Temperature Year	November-February Temperature °F	Shrimp Landings, Thousands of Pounds	Temperature Year	November-February Temperature °F	Shrimp Landings, Thousands of Pounds	Temperature Year
44.7	38	1951	43.1	104	1950	41.0	529	1961	37.0	83	1936
44.6	0	1952	43.0	11	1957	41.0	3691	1964	36.8	54	1939
44.4	0	1954	41.8	90	1958	40.9	340	1960	36.8	18	1937
44.4	0	1955	41.6	5	1956	40.4	2068	1963	33.8	79	1940
43.2	0	1953	41.1	45	1949	40.3	10	1947			
						39.9	64	1959			
						39.7	898	1962			
						38.9	7	1948			
						38.8	194	1945			
						38.7	162	1944			
						38.6	292	1941			
						38.2	8	1938			
						37.7	389	1942			
						37.7	6412	1965			
						37.5	27	1946			
						37.0	554	1943			

averaged more than 1,000,000 lb, suggesting that there is a definite optimum temperature range for the species in the Gulf of Maine.

The association of different levels of mean annual temperature with shrimp landings two years later is shown in Table 3.

Seasonal temperatures or those of selected months rather than annual means appear to be more closely associated with fluctuations in shrimp landings. Table 4 illustrates the observation that the least productive years are associated with consistently high, very low, or greatly fluctuating temperatures from 1⅔ to 2⅓ years preceding, while years of landings in excess of 100,000 are associated with moderate fluctuations in temperature.

TABLE 6. OPTIMUM TEMPERATURES BY MONTHS

Month	Optimum Temperature °F
October	49.2–52.8
November	44.8–48.5
December	37.4–42.2
January	32.8–39.0
February	32.0–37.0
March	33.6–38.3
April	37.0–43.5
May	46.1–50.0
June	52.8–56.3
July	54.9–60.8

Mean temperatures during the winter and the magnitude of shrimp catches two years later suggest that winter temperatures have greater influence on the abundance and availability of shrimp two years later than do annual or selected monthly temperature means. This relationship is shown in Table 5.

Table 6 shows the monthly temperature ranges associated with those years of greatest shrimp landings. No year in which shrimp production exceeded 150,000 lb is associated with temperatures outside these limits, and no year in which shrimp landings were less than 150,000 lb is associated with temperatures within all these ranges.

Conclusions

(1) Very high temperatures have been associated with virtually no shrimp landings (Tables 3 and 5).

(2) Highly variable temperatures have been associated with landings of 10,000 lb or less (Tables 3, 4 and 5).

(3) High temperatures have been associated with landings ranging from 10,000 to less than 50,000 lb (Table 4).

(4) Medium high and low temperatures have been associated with landings ranging from approximately 25,000 to 100,000 lb (Tables 4 and 5).

(5) Landings greater than 150,000 lb have been consistently associated (during the entire October–July period) with the optimum temperatures shown in Table 6.

If the assumption that a causal relationship exists between seawater temperature and shrimp abundance is valid, landings should continue at a relatively high level for the foreseeable future. Current and recent temperatures have fallen within the optimum range. If the predicted long-term downward trend of seawater temperature continues, conditions favorable for the Northern shrimp should continue for the next two decades.

ROBERT L. DOW

References

Dow, R. L., "Fluctuations in maine shrimp landings," *Commercial Fisheries Review*, 25, No. 4 (1963).
Rathbun, R., "The shrimp and prawn fisheries," "The Fisheries and Fishery Industries of the United States," Section V, "History and Methods of the Fisheries," Vol. II, 1887, Government Printing Office. Washington, D.C.
Scattergood, L. W., "The northern shrimp fishery of Maine," *Commercial Fisheries Review*, 14, No. 1 (1952).
Walford, L. A., "Notes on shrimp fishing along the New England coast," ms., Bureau of Fisheries, Department of Commerce, Washington, D.C.
Fishery Statistics of the United States, 1924 to 1965.

Cross references: *Canadian Fisheries, British Columbia; Pacific Fisheries; U.S. West Coast; Peru Fishery; U.S.S.R. Fisheries.*

SOUNDS OF MARINE ANIMALS

That marine animals make sounds has probably been known since man first went to sea, for some of the sounds are so loud that no electronic equipment is needed to hear them. That fish could hear, however, was not generally acknowledged until the early 20th century when more sophisticated instruments and techniques proved that all species studied had at least some hearing ability. Behavioral response to sound is less well understood, only a few investigators within the past 15 years having succeeded in influencing fish behavior to a significant degree. Recent evidence on the biological function of sound suggests its use in communication for some species, particularly in environments where vision is limited.

Sound production in marine mammals, particularly cetaceans, has developed much more than in either fishes or invertebrates. Aristotle (350 BC) stated that captured dolphins made moaning sounds above water. Many whalers reported similar findings, particularly from harpooned or wounded individuals, but not until the 1940's, when hydrophones and sonar systems came into increasing use after World War II was it confirmed that cetaceans produced sounds underwater, both for communication and orientation. While the significance of the latter is known for several species, little is known about the function of the so-called communication whistles.

Mechanisms of hearing in cetaceans are more complex than in fish. Directional hearing, a necessity for sound orientation, is well developed, but the exact acoustic pathway is still under debate. Frequency hearing thresholds have been studied in detail for only one species, *Tursiops truncatus*, the bottlenose porpoise, and the results indicate that frequencies well over 100 kHz (1 kilohertz = 1000 cycles per second) may be detected; lower limit frequency data are scarce for all species.

Marine invertebrate bio-acoustics has developed even less than fish or mammal bio-acoustics, partly because of the nature of the sounds. About all that can be said is that many invertebrates make sounds, a fact known for centuries. Some of the noisier species, such as snapping shrimp, have been studied in more detail because of their relevance to underwater warfare during World War II, yet the functional significance of their sounds is still uncertain. More is known about sound-producing mechanisms and the sounds themselves, since they can readily be described, than about sound receptor mechanisms, which can generally be studied only by physiological techniques or behavioral conditioning.

Fish

Sound Production. Although there are more than 10,000 species of shallow-water fish in the world, less than 20% of these are known sound producers. For deep-sea forms, Marshall (1967) has shown that sound-producing mechanisms occur only in slope-dwelling benthopelagic fish, those that swim near the dimly lit or dark bottom, but do not rest there habitually. Here population densities are high enough so that sound may be useful in assembling breeding populations.

Variations in the way fish produce sound are numerous, but most of the mechanisms may be grouped under either stridulatory or swim bladder types, neither of which has evolved solely for sound production. Stridulatory mechanisms involve the movement of hard, rough body parts against each other, resulting in a scratchy or rasping sound. Pharyngeal teeth, which grind food in the throat, are the most common example of this type. Some fish such as the puffer (*Spheroides maculatus*) have their teeth modified to form beak-like structures that produce grating sounds when the upper and lower plates are moved against each other (Fish, M. P., 1954). Generally pharyngeal sounds have their energy distributed quite evenly over a wide frequency range, resulting in a lack of tonal or resonance quality. Jacks (*Caranx* sp.) emit sounds like this with durations of less than 100 msec and frequencies extending to near 9000 Hz. Still other species make low-frequency vibrations by moving modified skeletal parts, such as the pelvic girdle in the longhorn sculpin, (*Myoxocephalus octodecimspinosus*) or special fin rays in some of the catfishes.

Swim bladder mechanisms, the second major type, are made up of the swim bladders themselves plus the muscles that control their vibratory movements (Fig. 1). These muscles may be either intrinsic, actually forming a part of the bladder wall as in the toadfish (*Opsanus tau*), or extrinsic, connecting the bladder to the body wall and ribs, as in the particularly noisy croaker family (Sciaenidae). Swim bladder sounds, possessing a vibrant or tonal quality, generally contain several harmonics with a fundamental often below 200 Hz; occasionally a signal may be nearly a pure tone. The actual number of harmonics, as well as the duration of the call, is highly variable from species to species but is rather consistent within a species. Some intraspecific variation, however, is due to the size of the individual, larger specimens producing lower pitched sounds, and to the type of bottom. The spontaneously produced boatwhistle sounds of the toadfish, sound pressure levels often exceeding 45 db re 1 microbar with the fish 2 ft from the hydrophone, have most of their energy concentrated below 300 Hz.

Some noisy fish combine stridulatory and swim bladder mechanisms, the sounds being produced by pharyngeal teeth and enhanced by the air bladder. These sounds generally extend from 100 to 8000 Hz with most of their energy below 1000 Hz (Tavolga, 1964).

Some of the families of fish that are particularly noted for sound production are the toadfish (Batrachoididae), croakers and drums (Sciaenidae), squirrelfish (Holocentridae), sea robins (Triglidae), triggerfish (Balistidae), sea bass (Serranidae), and several families of catfish.

Sound Reception. Parker (1918) gave conclusive evidence that fish could hear. Little was known about frequency range, however, until the early 1930's, when several German investigators made quantitative threshold studies. The hearing literature has been thoroughly reviewed by Kleerekoper and Chagnon (1954) and Schwartzkopff (1962). Fish lack both a true outer and middle ear, but since their flesh is nearly acoustically transparent these organs are not essential for conducting sound to the site of auditory reception, the inner ear. A delicate membranous labyrinth filled with fluid (endolymph), the inner ear, is made up of three pockets, the utriculus sacculus, and lagena, each containing a calcareous stone called an otolith. The pars inferior, composed of the lagena and sacculus with their otoliths, is the primary site of sound reception. Here the otoliths, held vertically over the acoustic sensory hair cells, are displaced by the incoming signal, causing movements of the hair cells which are detected by branches of the auditory nerve. In addition to the inner ear, the lateral line may function as a receptor for low-frequency vibration, particularly in the near field (Harris and van Bergeijk, 1962). The lateral-line system has many components similar to the ear and, together with the ear, is designated as the acousticolateralis system.

Because fish flesh immersed in water is transparent to sound, one of the few discontinuities in a fish is the swim bladder. A sound wave striking this reflector causes it to vibrate. In some major groups of fish (including the mostly fresh water Ostariophysi) specialized structures have

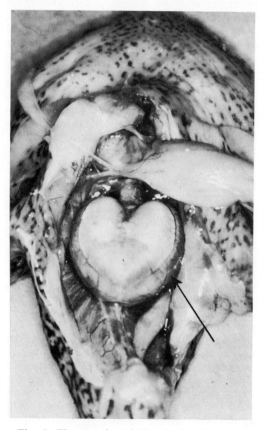

Fig. 1. The sound-producing mechanism of the toadfish, *Opsanus tau*, is the heart-shaped swim bladder, shown by an arrow in this photograph of a dissected specimen.

evolved which transmit this vibration efficiently to the inner ear, the entire mechanisms acting as a middle ear. Hearing is greatly improved in these species, with upper frequency limits being higher, auditory thresholds lower and pitch discrimination better than in nonspecialized forms.

The range of detectable frequencies is narrow in the nonspecialized fish, generally not extending below 50 or above 1000 Hz, while for the specialized forms it may extend up to 3000 Hz (Jacobs and Tavolga, 1967). The thresholds of most nonspecialized species rarely go below −20 db re 1 microbar compared to −45 db re 1 microbar for some of the specialized fish. In the latter, the connecting mechanism between the swim bladder and ear, known as the Weberian apparatus, is composed of four small bones that are modified vertebrae. Removal of the apparatus results in a marked depression of hearing sensitivity. Figure 2 compares several audiograms including one for a marine grunt (*Haemulon* sp.), and one for the goldfish (*Carassius auratus*), an Ostariophysine fish with much better hearing.

Biological Significance. Even though some hearing ability has been shown for all species studied, in most cases a function has not been demonstrated for the sounds that fish make or receive either from other fish or artificial sources. This is not to say, however, that the sounds are useless, for it is unlikely that fish hearing and sound-producing mechanisms would have evolved to their present level if this were so. Location of a sound source by a fish is said by many investigators to be anatomically impossible, except perhaps for low frequencies (and consequently large displacements) in the near field.

Demonstrating biological significance usually involves controlling the behavior of a fish under experimental conditions by bringing about an appropriate reaction to a signal. These signals may be synthesized electronically or previously recorded while a fish is engaged in a particular behavior, such as aggression, feeding, or breeding. The sounds are then played to the experimental fish and its behavior is observed. Only recently has anyone succeeded in influencing the calling of fish in the ocean. Moulton (1956) found that he could both elicit and suppress the calling of sea robins (*Carolinus* sp.) in Woods Hole Harbor by playing back appropriate signals. Winn (1967), playing back the boatwhistle sounds of male toadfish (*Opsanus tau*) to individual male toadfish on their breeding nests, succeeded in changing the rate of natural calling. The change varied with the playback rate, suggesting that temporal coding may be one means of sound communication. Spontaneous calling is often associated with breeding activities, many fish making only certain sounds during this season.

Sound may also affect movement, as Stout (1963) has shown for the satinfin shiner (*Notropis analostanus*). Some playback signals caused changes in aggressive behavior while others altered courtship behavior. Moulton (1956) modified the movements of menhaden and butterfish in a sound field; the fish tended to orient in the region of greatest sound intensity.

Certainly it has been established that fish produce characteristic sounds under specific circumstances; examples are: sounds produced during competitive feeding by sea robins; territorial defense grunts and predator staccatos of squirrelfish; spontaneous yelps of marine catfish in aggregations; escape sounds of codfish; spontaneous male sounds of toadfish during reproduction; courtship sounds of male satinfin shiners; and disturbance sounds of many fish (Tavolga, 1965; Winn, 1964).

Practical Applications. Several aspects of fish bio-acoustics may offer either immediate or future practical applications, including locating populations of soniferous fish with underwater listening gear, attracting commercially valuable and repelling undesirable species with sound playbacks, and guiding small fish through unsafe waters.

Japanese and Russian fishermen have used sonic devices for centuries to drive fish into their nets, but almost all these mechanical devices are operated either at the surface or the bottom, where they also cause visual stimuli such as splashing and turbulence. Electronic underwater sound projectors recently have attracted fish over short distances. But the possible lack of directional hearing, except in the near field, reduces the chance of future large-scale operations. Hashimoto and Maniwa (1967) were able to

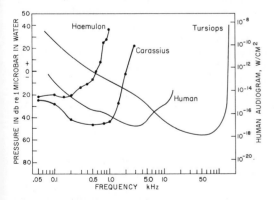

Fig. 2. Comparison of audiograms for: a marine fish without a specialized ear-swim bladder connection (*Haemulon* sp.), a fresh-water fish with a specialized connection, the Weberian apparatus (*Carassius* sp.); the bottlenose porpoise (*Tursiops truncatus*); a human. (Adapted and modified from Jacobs and Tavolga, 1967 and Johnson, 1966.)

attract carp in artificial ponds by playing back carp feeding sounds. They also found that several marine fish responded positively when their own feeding sounds were played back. Nelson and Gruber (1963) attracted large sharks in their natural environment by playing low-frequency (20–60 Hz) pulsed sounds; continuous low-frequency sounds produced no results. Van-Derwalker (1967) demonstrated that low-frequency vibrations up to 280 Hz at pressure levels up to 86 db re 1 microbar could be effective in guiding migrating salmon downstream around hazardous obstructions such as dams.

That fish produce low-frequency sounds underwater with reasonably good efficiency certainly warrants study by underwater acousticians interested in the design of sound projectors. An understanding of hearing mechanisms may likewise prove useful for designing hydrophones, particularly if directional hearing should be demonstrated.

Invertebrates

Sound Production. Some marine invertebrates make sound and probably perceive it, but the significance of the sounds and mechanisms of reception are mostly unknown. Part of the problem results from the nature of the sounds, which are short transients (wide-band clicks spaced many times their own time duration apart). This type of sound is efficient from the animals' standpoint for, as Frings (1964) points out, the marine organism lives in a relatively dense medium requiring much more energy to move than air, and it must produce a relatively loud sound if it is to be useful over the high background sound level. The sharp click, with durations of about 1 msec and energy distributed over a wide frequency range, does not require the expenditure of so much energy as a continuous sound. Unfortunately for the underwater acoustician, this type of sound is the most difficult to study, since listening and analyzing equipment works best with longer sounds that start and stop gradually.

Nearly all noise-making marine invertebrates are crustaceans that make sound by rubbing together parts of their hard chitinous exoskeleton. The clicking and rasping sounds can be quite loud, levels as high as 50 db re 1 microbar having been recorded 1 m from individual snapping shrimp (*Alpheus* sp.) and 40 db re 1 microbar 1 km from the shrimp beds. These small organisms have a specialized claw with a short movable "finger," bearing a large lobe near the base. The "finger" closes against an immovable "thumb" possessing a depression into which the lobe fits (Fig. 3). Most evidence indicates that the pop-like sound is produced when the claw snaps closed, but some sound may result when

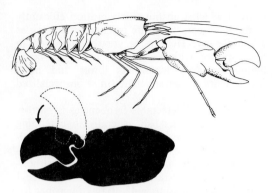

Fig. 3. Snapping shrimp showing the enlarged claw responsible for the loud pop produced when the movable "finger" snaps closed against the immovable "thumb." (After Dumortier, 1963.)

the plunger leaves the depression. The individual snap is a short pulse with frequencies extending to over 20 kHz. A chorus of thousands of these organisms has been likened to the crackling of a brush fire.

The spiny lobster, (*Panulirus argus*) also has a specialized sound-producing mechanism (Moulton, 1957). A toothed ridge on the carapace, extending anteriorly from beneath each stalked eye and lying against the basal segment of the antenna, is rubbed by a rough membrane when the base of the antenna is raised. Two types of sounds are produced by this organism, a rasp made up of 100 msec pulses with frequencies extending from 40–9000 Hz, and a slow rattle of 5 or 6 pulses with frequencies from 500–3300 Hz.

The American lobster (*Homarus americanus*) is unique among invertebrates, making a low-frequency growl that is nearly a pure tone with a fundamental around 120–150 Hz. The sound, sometimes a second long, is produced internally, the suggested mechanism being the vibration of a taut proventriculus, or stomach. The vibration is caused by contraction of the second antennal muscles which lie adjacent to the stomach wall (Fish, J. F., 1966).

Numerous other invertebrates, including gastropods, make cracking sounds. Mussels snap their byssal threads; oysters and clams clap their shells together (Fish, M. P., 1964).

Sound Reception. Most invertebrates possess mechano-receptors that are basically gravity and equilibrium organs, but may also receive sounds. Crustaceans have specialized vibration organs called phonoreceptors that may function similarly. But, to date, there is no conclusive evidence indicating that the organs are used for this purpose.

Biological Significance. The presence of sound-producing mechanisms is not evidence for biological significance, but it is likely that at least some invertebrates use this means of communication. Many make sounds when seized by an enemy, suggesting a defensive use. An American lobster in shallow water, after being aggravated by hand, can then be made to growl simply by waving a hand near its head. A territorial function has been suggested for the sounds of the spiny lobster. A reproductive function of sounds, common among fish, has not yet been shown for a marine invertebrate, except the fiddler crab when on land (Salmon, 1965).

Practical Applications. Once a sound has been identified with its source, local distribution of the organism can be learned with a hydrophone, a task much simpler than trawling. Of course this is only possible if the animal is consistently sonic, at least for a particular time such as breeding season, and occurs in sufficient abundance to be heard for some distance. M. W. Johnson (1948) made a survey of this type to determine snapping shrimp distribution off Point Loma, California, and from this was able to identify a particular habitat that also supported a commercially important, but silent, sponge. Even estimates of population size can be made if enough data are accumulated to correlate sound levels from the species with size of catch.

Mammals

Sound Production. Sounds have now been recorded underwater from many species of cetaceans (whales), including both odontocetes (toothed whales) and mysticetes (baleen whales), and recently from pinnipeds (seals and walruses). Two major classes of signals have been described: (1) narrow-band, whistle-like (communication sounds, which have frequencies extending from 1 to over 15 kHz, often frequency modulated; and (2) wide-band echolocation clicks with frequencies to nearly 200 kHz. Many of the sounds have been described on a phonograph record by Schevill and Watkins (1962). The sounds of odontocetes are much better known, particularly those of smaller porpoises, than mysticete signals. *Tursiops truncatus,* the Atlantic bottlenose porpoise, probably studied more than any other cetacean, emits loud whistles with frequencies from 4–15 kHz and durations occasionally as long as 3 sec, but generally under 0.5 sec.

Low-frequency (below 1 kHz) moans, groans, and growls were long thought the only sounds that mysticetes made. The low-frequency groaning of the humpback whale (*Megaptera novaeangliae*), for example, seldom exceeds 600 Hz. Recently, however, Perkins (1966), has reported communication whistles and weak echolocation pulses for a baleen whale, *Balaenoptera physalus,* the finback. Its whistles or chirps span a range from 1500–2500 Hz and vary in duration from 50 to 600 msec.

Echolocation sounds cover a wide range of frequencies with most of their energy in the sonic range below 20 kHz but with components extending to over 150 kHz. The sounds, essentially white-noise clicks, take on a tonal quality when the rate exceeds about 20 per sec (Kellogg, 1961) and the human ear can no longer distinguish the component clicks. When the rate increases still further to several hundred per second, the signals sound like barks or groans. The rate is extremely variable, apparently increasing as a navigating animal nears its target. Backus and Schevill (1966) observed in the case of the sperm whale that a click may in turn be composed of a single pulse or as many as nine short pulses from 0.1–2 msec long.

The pinnipeds make underwater clicks that are apparently useful for target discrimination and navigation, but recent experiments by Schusterman (1966) suggest that the system is much less developed than in the cetaceans. In addition to these clicks, subsurface barking of sea lions has been described several times.

The odontocetes have been studied to determine possible mechanisms for the two types of sounds, which may on occasion even be produced simultaneously. While true vocal cords are lacking, a larynx is present with projections and thin membranous parts that can be made to vibrate when air is artificially passed through the region. This has been demonstrated with dead animals; the resulting sounds have some properties in common with those produced naturally but are not identical with them. Apparently sounds can be emitted without loss of air through the blowhole; a series of nasal air sacs seem to be involved (Fig. 4), air passing from one to another through tightly compressed valves, the sacs acting as resonating chambers (Norris, 1964).

Sound Reception. If sounds are to have any value they must be heard. From the few data available on cetacean hearing, work having been performed only on *Tursiops truncatus,* it appears hearing is highly sensitive over a wide frequency range. Schevill and Lawrence (1953) studied the auditory response of this animal, particularly the upper threshold, using a conditioned response experimental technique, and found it to extend as high as 120 kHz with a sharp cut-off above this point. C. S. Johnson (1966) obtained the first detailed audiogram for this animal by training it to respond to pure-tone signals by pushing a lever-operated switch. The audiogram (from 75 Hz–150 kHz) shows the lowest threshold, −55 db re 1 microbar, at about 50 kHz (Fig. 2).

The exact hearing mechanism, particularly the acoustic pathway to the inner ears, is not fully

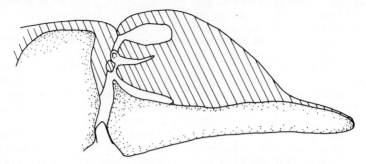

Fig. 4. Diagram of a cetacean head showing the nasal passages and sacs where sounds may be produced. (After Norris, 1964.)

understood. Several pathways have been hypothesized, including the "vestigial" ear canals, which could enhance directional hearing, and the lower jaw, which has a hollow oily core and close connection to the inner ears. The inner ears are acoustically isolated by dense bone and mucous foam.

Biological Significance. Echo navigation is obviously a functional use of sound by marine mammals. But to demonstrate with certainty a significance for the various communication-type signals is more difficult. It does appear, however, that particular whistle contours—time-frequency excursions of narrow-band whistles—occur repeatedly during similar behavior patterns, even with different species. Dreher and Evans (1964) noted 32 different contours established for four types of cetaceans, five of the 32 being used by all the animals involved in the study. Sexual functions have been attributed to the high-frequency yelping sounds of captive male *Tursiops* when the courting female deserts him. To study *Tursiops truncatus* whistles, Lang and Smith (1965) provided two isolated specimens with a two-way telephone. During the alternate 2-min periods when the line was open, the porpoises repeatedly communicated with whistles in a tight sequence, a particular sound produced by one often being answered by a characteristic sound from the other.

Practical Applications. Tuffy, a porpoise used on the Sealab II project in 1965, was trained to home on a sonar signal, acting as a courier between the undersea habitat and the support ship (Pauli and Clapper, 1967). The animal proved so valuable that several are now being trained for Sealab III to assist in searching for lost divers. They will be trained to dive to the lab in response to a signal from a small transducer carried by the diver, pick up an emergency breathing hose or rescue line, and carry it to the man.

Species identification on the basis of sound alone may be possible in the future, but the variety of sounds produced by any one form complicates the problem. Even behavioral patterns may be classified on the basis of characteristic signals. It remains, however, to collect enough sound contours from the same species under the same "known" behavioral situations before anything such as this can be accomplished.

JAMES F. FISH
HOWARD E. WINN

References

Backus, R. H., and Schevill, W. E., "Physeter clicks," in K. E. Norris, (Ed.), "Whales, Dolphins, and Porpoises," 1966, Univ. of Calif. Press, Berkeley.

Dreher, J. J., and Evans, W. E., "Cetacean communication," in W. N. Tavolga, (Ed.), "Marine Bio-acoustics," 1964, Pergamon Press, Oxford.

Dumortier, B., "Morphology of sound emission apparatus in Arthropoda," in R. -G. Busnel (Ed.), "Acoustic Behavior of Animals," 1963, Elsevier, Amsterdam.

Fish, J. F., "Sound production in the American lobster, *Homarus americanus* H. Milne Edwards (*Decapoda reptantia*)," *Crustaceana*, **11**, 105–106 (1966).

Fish, M. P., "The character and significance of sound production among fishes of the western North Atlantic," *Bull. Bingham Oceanogr. Coll.*, **14**, 1–109 (1954).

Fish, M. P., "Biological sources of sustained ambient sea noise," in W. N. Tavolga (Ed.), "Marine Bio-acoustics," 1964, Pergamon Press, Oxford.

Frings, H., "Problems and prospects in research on marine invertebrate sound production and reception," in W. N. Tavolga (Ed.), "Marine Bio-acoustics," 1964, Pergamon Press, Oxford.

Harris, G. G., and vanBergeijk, W. A., "Evidence that the lateral-line organ responds to near-field displacements of sound sources in water," *Jour. Acoust. Soc. Amer.*, **34**, 1831–1841 (1962).

Hashimoto, T., and Maniwa, Y., "Research on the luring of fish shoals by utilizing underwater acoustical equipment," in W. N. Tavolga (Ed.), "Marine Bio-acoustics," Vol. 2, 1967, Pergamon Press, Oxford.

Jacobs, D. W., and Tavolga, W. N., "Acoustic

intensity limens in the goldfish," *Animal Behaviour*, **15**, 324–335 (1967).

Johnson, C. S., "Auditory thresholds of the bottlenosed porpoise (*Tursiops truncatus*, Montagu)," 1966, U.S. Naval Ordnance Test Station, China Lake, Calif., Tech. Publ. 4178.

Johnson, M. W., "Sound as a tool in marine ecology, from data on biological noises and the deep scattering layer," *Jour. Mar. Res.*, **7**, 443–458 (1948).

Kellogg, W. N., "Porpoises and Sonar," 1961, University of Chicago Press.

Kleerekoper, H., and Chagnon, E. C., "Hearing in fish, with special reference to *Semotilus atromaculatus atromaculatus* (Mitchill), "*Jour. Fish Res. Bd. Canada*, **11**, 130–152 (1954).

Lang, T. G., and Smith, H. A. P., "Communication between dolphins in separate tanks by way of an electronic acoustic link," *Science*, **150**, 1839–1944 (1965).

Marshall, N. B., "Sound-producing mechanisms and the biology of deep-sea fishes," in W. N. Tavolga (Ed.), "Marine Bio-acoustics," Vol. 2, 1967, Pergamon Press, Oxford.

Moulton, J. M., "Influencing the calling of sea robins (*Prionotus* spp.) with sound," *Biol. Bull.*, **111**, 393–398 (1956).

Moulton, J. M., "The movements of menhaden and butterfish in a sound field," *Anat. Rec.*, **125**, 592 (1956).

Moulton, J. M., "Sound production in the spiny lobster *Panulirus argus* (Latreille)," *Biol. Bull.*, **113**, 286–295 (1957).

Nelson, D. R., and Gruber, S. H., "Sharks: attraction by low-frequency sounds," *Science*, **142**, 975–977 (1963).

Norris, K. S., "Some problems of echolocation in cetaceans," in W. N. Tavolga (Ed.), "Marine Bio-acoustics," 1964, Pergamon Press, Oxford.

Parker, G. H., "Hearing in fishes," *Copeia*, **1918**, 11–12 (1918).

Pauli, D. C., and Clapper, G. P., "Sealab Report," An experimental 45-day undersea saturation dive at 205 feet, Sealab II Project Group, 1967, ONR REPORT ACR-125.

Perkins, P. J., "Communication sounds of finback whales," *Norsk Hvalfangst-Tidende*, **10**, 199–200 (1966).

Salmon, M., "Waving display and sound production in the courtship behavior of *Uca pugilator*, with comparisons to *U. minax* and *U. pugnax*," *Zoologica*, **50**, 123–150 (1965).

Schevill, W. E., and Lawrence, B., "Auditory response of a bottlenosed porpoise, (*Tursiops truncatus*), to frequencies above 100 kc," *Jour. Exp. Zool.*, **124**, 147–165 (1953).

Schevill, W. E., and Watkins, W. A., "Whale and porpoise voices," 1962, Woods Hole Oceanographic Inst. Publication, and phonograph disc.

Schusterman, R. J., "Underwater click vocalizations by a California sea lion: effects of visibility," *Psych. Rec.*, **16**, 129–136 (1966).

Schwartzkopff, J., "Vergleichende Physiologie des Gehörs und der Lautäusserungen," *Fortschr. Zool.*, **15**, 214–336 (1962).

Stout, J. F., "The significance of sound production during the reproductive behavior of *Notropis analostanus* (family Cyprinidae)," *Animal Behavior*, **11**, 83–92 (1963).

Tavolga, W. N., "Sonic characteristics and mechanisms in marine fishes," in W. N. Tavolga (Ed.), "Marine Bio-acoustics," 1964, Pergamon Press, Oxford.

Tavolga, W. N., "Review of marine bio-acoustics. State of the Art: 1964," 1965, U.S. Naval Training Device Center, Port Washington, N.Y., Tech. Rep. 1212-1.

VanDerwalker, J. G., "Response of salmonids to low-frequency sound," in W. N. Tavolga (Ed.), "Marine Bio-acoustics," Vol. 2, 1967, Pergamon Press, Oxford.

Winn, H. E., "The biological significance of fish sounds," in W. N. Tavolga (Ed.), "Marine Bio-acoustics," 1964, Pergamon Press, Oxford.

Winn, H. E., "Vocal facilitation and the biological significance of toadfish sounds," in W. N. Tavolga (Ed.), "Marine Bio-acoustics," Vol. 2, 1967, Pergamon Press, Oxford.

Cross references: *Lobsters; Porpoises.*

SOUTH AFRICA, MARINE RESOURCES

The coastline of South Africa (including South West Africa) is well over 2000 miles. The southern tip of the continent divides two major current systems, the Agulhas Current off the east coast and the Benguela Current off the west coast. The Agulhas Current, a typical western boundary current, is a warm, swiftly flowing body of water. Surface temperatures at the core of this current are usually of the order of 22°–26°C, and velocities are normally between 2–4 knots. The Agulhas continues on its southwesterly course until it encounters the edge of the Agulhas Bank, which deflects the main body of the current to the south. Close inshore off the east coast, moderately strong, cold countercurrents have often been observed.

The Benguela Current is a cold, sluggishly moving current. Benguela surface water is typically below 16°C, and close inshore values as low as 9°C have been recorded. The current is variable in both direction and velocity. As a rule it flows at a rate of $\frac{1}{2}$ to $1\frac{1}{2}$ knots in a north to northwesterly direction, but periodically a complete reversal of flow has been known.

The presence of the two different current systems along the South African coast is also associated with widely divergent climatic conditions, which are in turn reflected in the marine resources of these regions. The east coast is characterized by a relatively diverse fauna, each species being represented by a comparatively limited number of individuals. The west coast fauna, however, is noted for its limited number of species with numerically strong populations.

It is thus clear why the South African fishing industry is centered on the west coast.

The prolific fauna of the west coast is the direct result of high primary productivity rendered possible by an abundance of plant nutrients in the upper layers of the sea. The Benguela Current region is a typical example of surface water enrichment brought about by upwelling. The prevailing southerly and southeasterly winds along the west coast induce an offshore transport of surface water, which is replaced by cold, nutrient-rich (phosphates, nitrates, silicates, and trace elements) water welling up from below the surface. This also explains why surface temperatures in the region are so very much lower than on the east coast. Upwelling varies not only according to locality, but also seasonally. Between Cape Town and the Orange River it is most marked during summer, but further north it is most intense during spring. Normally the vertical displacement of subsurface water is of the order of 150–200 m, but cases have been noted where it has exceeded 350 m. The process of upwelling, therefore, is primarily responsible for the high fertility of the waters of the west coast of South Africa; in fact, the Benguela Current is one of the most productive areas of the world. Estimations of primary productivity by the carbon-14 method have yielded figures of up to 3.8 g of carbon per m^2 per day.

Major developments have occurred since World War II, resulting in much fuller utilization of the marine wealth of the west coast. The biggest impetus was not given by the development of traditional fisheries, e.g., hake and rock lobster, but by the intensive exploitation of pelagic shoal fish, e.g., the pilchard and anchovy. Within just over a decade South Africa has risen from obscurity to a major fishing nation. Catch statistics released by the FAO indicate that during 1965, South and South West Africa, with an aggregate fish catch of 1.34 million metric tons, ranked seventh among world fishing powers. Table 1 gives an indication of the rate at which total fish landings have increased since 1948.

The rise in landings has been paralleled by a concomitant increase in the value of marine products; it is estimated that the country's

Fig. 1. A modern South African stern trawler.

production is presently worth approximately R100 million (U.S. $140 million) a year. Although the pelagic fish catch has mainly been responsible for the increased landings, the trawl fish catch has also shown a threefold increase since 1948.

The past decade has evidenced vast technological advances in the harvesting and processing of raw materials from the sea. While the sealing, whaling, line-fishing, and other branches of industry still employ traditional gear and methods, the pelagic and trawling industries are undergoing a major metamorphosis. South Africa today possesses highly efficient fishing fleets, modern factories for processing the catch, and alert sales and marketing organizations for disposing of the final product.

South Africa has always adopted a somewhat conservative policy in respect of the harvesting of her marine resources. A variety of measures are enforced by law to ensure rational exploitation. The State has created a Division of Sea Fisheries under the Department of Industries for the purpose of conducting research into the living resources of the sea to provide a scientific basis for management and control.

South Africa has already taken the initiative in mining the sea floor, although at this stage operations are limited to recovering diamonds. As new techniques are perfected it is expected that the practicability of mining other valuable deposits will receive serious consideration. The present keen interest in offshore drilling for oil is probably an indication that South Africa with her advancing technology is approaching a fuller appreciation of the potentials of her waters.

TABLE 1

Year	Landings (metric tons)
1948	169,000
1952	629,000
1956	518,000
1960	868,000
1965	1,342,000

Fig. 2. Fishing boat with a load of pilchards, about to discharge its catch at a factory.

Pelagic Fish

The most important pelagic fish off the shores of South Africa are the pilchard (*Sardinops ocellata*), anchovy (*Engraulis capensis*), maasbanker or jack/horse mackerel (*Trachurus trachurus*), true mackerel (*Scomber japonicus*), round-herring or red-eye sardine (*Etrumeus micropus*), snoek (*Thyrsites atun*), and tuna (four species).

The pilchard has a very wide geographical distribution and is known from St. Lucia Bay (north of Durban) to Bahia dos Tigres on the Angolan coast. The main commercial concentrations, however, are limited to the Walvis Bay region, the waters off St. Helena Bay and the area between Cape Point and Cape Agulhas. The species is normally found within 25 miles of the coastline, but occasionally shoals have been reported up to 80 miles offshore.

The South African pilchard is a fast growing fish and reaches sexual maturity at the age of approximately 2½ years, by which time it attains a length of 21 cm. The main spawning seasons are spring and early summer. Spawning occurs offshore, and three main grounds have been identified namely, those off Walvis Bay, near St. Helena Bay, and east of Cape Point. The pilchard is a filter-feeder, its diet consisting of both phytoplankton and zooplankton. Tagging experiments have established that there is periodically an influx of pilchards from the Walvis region into Cape waters.

The geographical distribution of the South African anchovy is not as well known as that of the pilchard. It is probably limited more or less to the region between Cape Agulhas and Walvis Bay, the highest concentrations apparently being found in Cape waters. The first indications of the presence of this species were obtained in the course of experimental sampling of juvenile fish, when vast quantities of anchovies were caught.

Reaching sexual maturity at a length of about 11 cm, the anchovy has a very limited lifespan, probably about two years. It spawns mainly during late spring and early summer, the highest egg concentrations occurring within 50 miles of the coast, east of Cape Point. Like the pilchard, the anchovy is a filter-feeder with a mixed diet of phytoplankton and zooplankton.

Little is known about the general biology of the maasbanker, mackerel, and round-herring. These species are caught mainly on the Cape west coast.

The exploitation of the above species really gained momentum only after 1950. Increased fishing effort coupled with a gradual relaxation of catch restrictions led to an increase in landings as will be seen from Table 2.

It is significant that pilchard catches off the Cape declined sharply after 1963, and the anchovy (which was exploited on a commercial basis for the first time in 1964) has risen to a dominant position in the annual catch.

As can be expected, the pelagic fishing fleet has undergone striking changes during the past decade or two. In Cape waters there has been a sharp reduction in the number of boats, but a steep rise in the average size of the fishing unit. Today the pelagic fish harvest of the west coast is being gathered by a fleet of over 200 vessels equipped with synthetic fiber nets, echo sounders, power blocks, etc., and the use of sonar for fish-finding is gradually gaining ground. Two large factory ships with catcher boats have recently entered the fishery.

The entire pelagic catch is converted to fish

TABLE 2. CATCH IN METRIC TONS

Species	1950	1954	1958	1962	1966
Pilchard	133,001	338,960	423,739	806,322	767,351
Anchovy	—	—	—	—	159,444
Maasbanker	50,352	118,143	56,419	67,226	26,827
Mackerel	—	4,043	20,077	21,224	55,598
Round-herring	—	—	—	—	4,518
Total	183,353	461,146	500,235	894,772	1,013,738

meal, oil, and canned fish. There is strict quality control and South African marine products enjoy a keen demand and high prices on critical markets. The principal markets are the United Kingdom, West Germany, Israel, Japan, and the U.S.

A variety of legal control measures are currently applied to the capture of pelagic fish. In South West Africa fishing is regulated by an over-all catch quota. In the Cape the most important measures are the specification of permissible net mesh-sizes, a closed season of four months' duration, and the limitation of the aggregate boat hold capacity of the fleet.

The tunas enjoy a world-wide distribution and the most common species off South African shores are the yellow-fin tuna (*Thunnus albacares*), long-fin tuna (*T. alalunga*), big-eye tuna (*T. obesus*), and blue-fin tuna (*T. thynnus orientalis*). The different species show very definite temperature preferences, being normally encountered off the continental shelf at temperatures between 15° and 23°C. Test fishing has shown that on the west coast tuna are most abundant during winter. Observations to date have shown that the tunas, which are migratory fish, do not spawn in South African waters. They are voracious feeders and show a preference for fish, crustaceans and cephalopods.

Very promising results obtained from test fishing during 1960–1961 led to an upsurge of interest in tuna fishing. Soon a small fleet of vessels was operating successfully, using the Japanese long-line method, and attractive prices were being obtained for the frozen product. Unfortunately, though, the catch contained a very small proportion of yellow-fin tuna, the most sought-after species, and in the face of dwindling catches, this fishery has virtually ended. It is nevertheless considered highly likely that investigations into methods of capture and the exploration of new areas may cause a revival of tuna fishing.

The snoek (*Thyrsites atun*) is also limited principally to the cold inshore waters of the west coast. The species is especially abundant off the Cape coast in winter and off Walvis Bay in summer and early autumn. It has been possible

Fig. 3. Typical scene at a Cape west coast fishing factory.

to establish by tagging that snoek migrate between South West African and Cape waters. Snoek spawn during spring in the warmer offshore waters. Their diet is varied, but the main food item is always small pelagic fish. Snoek is still caught in the traditional manner, namely by handlines with baited hooks or lures, trailed behind drifting vessels. A variety of craft varying from small dinghies with outboard motors to 70–80 ft diesel-powered craft, is employed. The bulk of the catch is salted and dried and is disposed of in South and Central Africa.

Demersal Fish

The rich waters of the west coast sustain vast populations of bottom fish. The most important among these is the hake (*Merluccius capensis*), which constitutes 85–90% by weight of trawl catches. Far less abundant, yet of great commercial importance, are the kingklip (*Genypterus capensis*), sole (*Austroglossus spp.*), kob (*Johnius hololepidotus*), and panga (*Pterogymnus laniarius*). The highest concentrations of demersal fish have been recorded to the northwest of Cape Town.

The Cape hake is a whitefish and it appears that there are two distinct populations on the trawling grounds. Fifty per cent of the population attains sexual maturity at an age of 40 cm, which coincides with an age of 3–4 years. Peak spawning occurs during spring and early summer. Preliminary studies seem to indicate that in the case of the Cape hake, diurnal vertical migration is much less pronounced than for example in the European cod. The diet of the adult hake is comprised of rattails, maasbanker and squid, cannibalism being quite common.

Bottom trawling constitutes the oldest sector of importance of the South African fishing industry, dating back to the start of the century. For many years trawling operations were limited to the traditional grounds northwest of Cape Town, but various factors have led to a partial shift of activities to the Lüderitz ground. This was among others due to the decline of catches in the south, the high density of fish to the north and the advent of large, modern, long-range stern trawlers. The landings of bottom fish have fluctuated somewhat, but a marked increase was recorded in 1960 and again in 1966, as shown in Table 3.

TABLE 3

Year	Landings (metric tons)
1956	89,700
1958	75,100
1960	117,400
1962	106,900
1964	106,800
1966	127,300

Trawling is normally conducted in depths of 100–400 fathoms.

In spite of a gradual process of modernization, the majority of South African trawlers still utilize ice for storing their catch. Fresh hake is in keen demand on the home market, but a good export market has been established, notably in Australia. Various fancy packs, e.g., fish sticks, smoked fish, etc., are also produced, while dried, salted hake has for many years been supplied to noncritical markets.

The regulation of the trawl fishery is achieved by prescribing minimum mesh-sizes for the cod ends of trawl nets. On the east coast cod-ends should be at least 76 mm (stretched) and on the west coast 102 mm. While these measures may have been adequate when hake was exploited at a low level, it is likely that the mesh-size regulation will be reviewed in the light of increased fishing pressure from foreign trawlers.

Crustaceans

Rock Lobster. The Cape spiny lobster *Jasus lalandii* is the most important of the three species of rock lobster occurring off South Africa. It is most abundant between Cape Point and Hollams Bird Island (north of Lüderitz), and is found in depths down to 80 fathoms. East of Cape Point there is a rapid decline in population density.

A slow-growing crustacean, the spiny lobster takes about seven years to attain a carapace length of 9 cm. Males attain sexual maturity at carapace lengths between 6.0 and 6.5 cm, while the entire female population becomes sexually mature only upon reaching 7.0 cm. Spawning occurs principally from June to August and the species spawns only once per year. In the case of males, moulting occurs in spring and summer, and in females, in May and June, i.e., shortly before spawning. Tagging experiments have demonstrated that the species does not undertake major migrations and all marked specimens were recaptured within a radius of 5 miles of their point of release. The spiny lobster feeds mainly on black mussels, crabs, and kelp.

Rock lobster fishing has been practised by South African fishermen for many decades, yet methods of capture have undergone little change. Fishing craft vary from 12-ft dinghies powered by outboard motors to 70-ft motor vessels, acting as a mother ship to as many as 12 dinghies. Baited hoop-nets are still used, although some interest in traps is being shown.

Whereas a large proportion of the catch was formerly canned, the bulk of the raw material is today converted into frozen tails, for which there appears to be an insatiable demand in the U.S. An interesting development has been the export of live rock lobster to Europe. Rock lobster offal is converted to meal for use as fertilizer.

Being a very valuable asset, the rock lobster has always been afforded special protection. This is achieved among others by the imposition of an annual export quota, the declaration of closed seasons, size limits, prohibition on the landing of soft-shelled lobsters or females in berry, and the banning of trawling in certain inshore waters where rock lobsters are abundant. Commercial diving for rock lobster is not allowed.

Panulirus homarus, which is locally known as the east coast rock lobster, has a very wide distribution in the Indian Ocean. The east coast of South Africa represents a fringe of the distribution of *P. homarus* and hence it is not surprising that the species is not particularly abundant in those waters nor does it attain a large size. In appears to increase in size and numbers northeast of Port Elizabeth and occurs on a rocky substratum in the immediate subtidal zone. Sexual maturity is attained at carapace lengths in excess of 5 cm. The male displays peak sexual activity in summer, the female spawning during the same season. The main food items of the species are brown mussels and barnacles, cannibalism being rather pronounced.

On account of its low numerical abundance and restriction to a zone of the sea where traditional methods of capture can not be employed, the east coast rock lobster has not been exploited successfully on a commercial basis.

The presence of the Natal rock lobster, *Palinurus gilchristi,* was established by bottom trawling many years ago. It occurs in deep water off Natal and northward in the Mozambique Channel. To date no detailed study of the species has been undertaken.

Limited trawling for Natal rock lobster has been conducted by South African ships since the 1950's. In recent years, however, a host of foreign vessels has started operations in the area, systematically stripping the grounds. As a result trawlers have been compelled to go farther afield and currently most of the fleets are fishing in the Mozambique Channel.

Shrimps and Prawns. Preliminary investigations regarding the occurrence and abundance of shrimps and prawns have shown that these crustaceans are concentrated mainly in three areas, namely, off the Natal coast, off the Agulhas Bank, and along the South West African coast.

Off Natal, the King prawn (*Nephrops andamanica*), which is actually a lobster, Knife prawn (*Hymenopenaeus triarthrus*), Brown prawn (*Penaeus indicus*), and Tiger prawn (*Penaeus monodon*) are fairly abundant, especially in autumn and spring. The Red shrimp (*Solenocera africanum*) and the White shrimp (*Macropetasma africanum*) predominated in catches made off the Agulhas Bank and South West Africa. Depth distribution of the various crustaceans was found to vary between 3–240 fathoms, depending on the species.

Experimental fishing for prawns on a semi-commercial basis has demonstrated that the Natal grounds are capable of sustained but limited fishing operations. A variety of shrimp trawls has been employed with promising results. It is certainly possible to produce a high quality frozen product for home consumption and for the export market.

Mollusks. South Africa possesses a diverse molluscan fauna, but the most important members of this phylum, from a commercial point of view, are the abalone (*Haliotis midae*), squid (*Loligo reynaudii*), oyster (*Crassostrea sp.*), giant periwinkle or turban (*Turbo sarmaticus*), white (*Donax serra*), and black mussel (*Chloromytilus meridionalis*).

Although found over a considerable length of coastline east and west of Cape Point, the South African abalone or perlemoen is most abundant between Cape Agulhas and Saldanha Bay. It is a slow-growing animal that reaches sexual maturity at a shell-breadth of 10.5 cm, corresponding to an age of about 12 years. Spawning occurs in spring and autumn, millions of eggs being released by each individual at a time. Movement is limited in the species, but it has been established that individuals are capable of moving over 200 m in a month. The abalone is vegetarian.

Abalone are collected commercially by divers operating from small motor-powered vessels and using compressors and air-hoses. During the past years annual landings have varied between 700,000 and 1,400,000 lb. While the entire catch used to be exported in canned form, there has recently been a tendency toward exporting the frozen product. Consumer demand in the Far East remains keen.

At present the most important conservational measure applied to the abalone is a minimum size limit of $4\frac{1}{2}$ in. shell breadth. It is nevertheless felt that this resource should be afforded more drastic protection, for example through the imposition of an appropriate catch quota.

To date no assessment has been made of the abundance of the cephalopod, *Loligo reynaudii*, in South African waters. This species, commonly known as squid, appears in considerable quantities in the catches of trawlers. Landings in recent years have shown a sharp rise, namely from 122,000 lb in 1956 to 527,000 lb in 1966. The squid is a valuable bait item for anglers in this country.

Oysters (*Crassostrea* sp.) occur in vast quantities at certain localities along the south and east coasts, usually on reefs in the subtidal zone. This resource has never been exploited on an appreciable scale and warrants fuller investigation. Currently commercial diving for oysters is prohibited, a minimum size limit of 2 in. is applied and the collection of this shellfish is not permitted during the period December to February. Oysters have been reared successfully in the Knysna Lagoon and their artificial culture holds great promise.

The South African turban or giant periwinkle, *Turbo sarmaticus*, is an inhabitant of the intertidal and subtidal zone and is very abundant east of Cape Point. A minimum size limit of 2½ in. is enforced and the collection of limited quantities for private use is permitted; commercial harvesting is, however, not allowed, pending an investigation into the extent of the resource.

Another important bait organism in South Africa is the redbait, *Pyura stolonifera*. Extensive colonies of this ascidian inhabit the intertidal and especially the subtidal zones, showing a preference for warmer water. They are considered as a possible source of protein food. The collection of limited quantities of redbait is permitted for private use and for commercial purposes.

The bivalve, *Donax serra* (white mussel), which is also utilized as bait by anglers, occurs in isolated patches between Walvis Bay and East London. It is found in the intertidal and immediate subtidal sand. This resource is most vulnerable to commercial exploitation and the ravages of "red tide" and accordingly only limited collection by private individuals and bait suppliers is allowed. A minimum size limit of 1¼ in. is in force.

Vast beds of black mussel (*Chloromytilus meridionalis*) are found around the coast of South Africa, notably in the subtidal zone. Currently no commercial harvesting of this species is allowed, since it forms a significant proportion of the diet of a variety of fish and crustaceans of economic importance and as it is prone to poisoning by toxic dinoflagellates.

Sharks. The history of commercial shark fishing in South Africa has been one of varying fortune. At one stage vitamin oils extracted from shark livers commanded good prices, but it has since become uneconomic to produce this commodity, probably because of synthetically produced vitamins. The demand for salted, dried shark is still keen, but better prices are fetched by frozen meat at present.

Of the various sharks occurring off the coasts of South Africa, only two species, namely the tope or vaalhaai (*Galeorhinus galeus*) and the Joseph (*Callorhynchus capensis*) are exploited commercially. Fishing is conducted by handlining or trawling, mainly in the Cape agulhas region and northward as far as St. Helena Bay. The only commodities produced, frozen meat and shark fins, are exported, the former to Australia and Italy and the latter to the Far East.

Marine Mammals

Seals. The Cape fur-seal (*Arctocephalus pusillus*) is the only seal species breeding and feeding in the waters around Southern Africa, the total population being estimated at roughly 1,000,000 animals. Breeding places are scattered around the South and South West African coasts from Cape Cross in the north to Algoa Bay (Port Elizabeth) in the east. With the exception of three places on the more or less deserted mainland along the west coast, rookeries are situated on numerous small rocky islands close inshore. As a rule, the Cape fur-seal does not undertake distant migrations. Rookeries are inhabited throughout the year and most seals forage only short distances from their rookeries.

The pupping and mating season occurs (more or less concurrently) during the early southern summer months (November/December) during which period the active bulls establish, defend and maintain "harems," containing from 1–12 cows, plus their pups, within the geographical confines of the rookery.

All animals moult each year in the summer. Pups lose their black coats after 4 months, changing these for olive-grey yearling coats, which form the much sought seal skins of fashion, which are lost through moulting some 12 months later.

The Cape fur-seal has been exploited for over 300 years, not only for the fur skins of the 7- to 10-month old pups, but also for the thick subcutaneous layer of fat or blubber of both the pup and adult seals, which yields oil for the manufacture of soap and balanced livestock feeds. The skins of adult seals can also be used for leather, but are not much in demand as they are inferior to the skins of hair seals. On account of the general inaccessibility of and lack of facilities near the rookeries where the seals are killed, carcasses are not generally processed, but upon removal of the skin and blubber, are returned to the sea. Approximately 60,000 pups and 5000 adult seals are killed annually for commercial purposes in South Africa and South West Africa, from which approximately 60,000 pup furskins, 5000 adult skins, and 500 adult skins, and 500 (short) tons of oil are obtained. All Cape fur-seals are protected under Government regulations, sealing being undertaken either by Government employees or private persons to whom concessions have been granted.

Whales. Whaling off South Africa today is based chiefly on the sperm whale (*Physeter catodon*), with smaller catches of fin (*Balaenoptera physalus*), sei (*Balaenoptera borealis*), and Bryde

Fig. 4. Cape fur-seals basking in the sun.

whales (*Balaenoptera brudei*). Minke whales (*B. acutorostrata*) are also taken periodically.

The headquarters of the sperm whale is in the tropics, and only the larger males penetrate into the Antarctic Ocean, so that most of the population avoids the intensive Antarctic pelagic fishery. Consequently, this species seems to have a greater potential for exploitation by local industry than most of the baleen whales. The sperm whale population as a whole shifts away from the Equator in spring and back again in autumn. It is on the northward leg of this migration in February to April that South African whaling stations make their best catches of sperm whales, though the larger males arrive later and depart earlier. Most conceptions occur in spring and summer, and the calves are born 14–15 months later. Nursing continues for 2 years, so that a female usually produces only one young every 4 years, and the level of recruitment is correspondingly low. Females reach sexual maturity at a body length of about 28 ft, and males at about 45–46 ft, although spermatozoa are produced from a length of 39–40 ft onwards.

Both sei and fin whales undertake migrations between the rich feeding grounds of the Antarctic in summer and subtropical calving and breeding grounds in winter. As a consequence they are exploited in both areas, a fact that has undoubtedly led to the sharp decrease in fin whales off South Africa after 1961. Fin whales are most prevalent off South Africa in midwinter (July), when they are moving north, whereas sei whales are most abundant in spring (August to October) on the southward leg of their migration. The full effects of the recent increased exploitation of sei whales in the Antarctic have still to be felt, but it is possible that their future availability off South Africa may decrease significantly.

Two land stations undertake whaling operations

Fig. 5. Whale-catcher and whales alongside a jetty.

in South African waters, one at Durban, Natal, and the other at Donkergat in Saldanha Bay, Cape Province. Whale-catchers of about 500 tons are in use at each station, 10–12 being based at Durban and 4–6 at Donkergat. Two spotter-planes assist the catchers at Durban and one at Donkergat. The species of whales taken are the same at both stations, the bulk of the catch consisting of sperm whales. Products of the industry include whale and sperm oil, meat meal, meat extract, and frozen meat, for all of which a ready export market exists.

As South Africa is a member of the International Whaling Commission, the whaling industry is administered in conformity with the regulations of the Commission's schedule. An 8-month open season for sperm whales usually extends from February–March to September–October, during the last 6 months of which baleen whales may also be taken. Since 1966 both stations have operated on a quota basis, the baleen whale catch being limited to 236.8 Blue Whale Units at Durban and 162.7 B.W.U. at Donkergat, and the sperm whale catch to 2846 individuals of this species at Durban and 798 whales at Donkergat.

Guano - Producing Seabirds. Approximately 6000 (short) tons of guano are collected annually and sold for fertilizer in South Africa and South West Africa. Of this 4000 tons are collected on State-controlled islands, while the balance is obtained from privately owned bird-platforms erected in the shallow salt water lagoons near Swakopmund and Cape Cross in South West Africa. Some thirty small islands, stretching from Port Elizabeth on the south east coast to approximately 130 miles south of Walvis Bay, were proclaimed as sanctuaries for guano-producing seabirds in 1890. Since then guano has been collected annually under the auspices of a State department for sale to local farmers.

South African guano is obtained from three species of seabirds, i.e., gannets (*Morus capensis*), cormorants (mainly the Cape cormorant, *Pahalacrocorax capensis*) and the Cape or Blackfooted penguins (*Spheniscus demersus*).

Most of the guano collected on the islands (approximately 80% is obtained from gannets, while the platforms rely almost exclusively on cormorants for their guano harvests.

The birds are protected by law and are permitted to breed undisturbed on the islands and platforms, most of the guano being produced in the summer when the chicks are still confined to their nests. During the late summer, after the chicks have left their nests, the guano is collected and shipped to Cape Town.

An average analysis of the plant-food value of guano follows:

Nitrogen (as N)	12.5%
Total phosphorus (as P)	4.7%
Total calcium (as Ca)	1.7%
Total potash (as K)	1.6%

The collection of penguin eggs on Dassen Island during autumn has been conducted under State supervision for many years. Vast quantities were gathered in former years, but now it has dwindled to about 5000 dozen per annum. Eggs are sold to the public at R1.50 (U.S. $2.20) per dozen.

Marine Algae

Seaweeds and sea-kelp are important as sources of alginic acid, agar, fertilizer, fodder, and agaroids. Especially alginic acid and its salts find wide industrial usage. Of the economically important kelp or sea-bamboos *Macrocystis angustifolia* has the most limited distribution, occurring patchily between Cape Point and Dassen Island. *Laminaria pallida* and *Ecklonia maxima* are found from Cape Algulhas westward and northward as far as Lüderitz. *Ecklonia*, being very abundant, is harvested between Cape Agulhas and St. Helena Bay, but legislation prohibits the cutting of all these species at sea and only plants cast ashore may be collected.

Sources of agar are the seaweeds *Gelidium pristoides*, *Gelidium cartilagineum*, *Gelidium amansii*, and *Gracilaria confervoides*. The various species of *Gelidium* occur east of Cape Point, attaining their highest concentration east of Port Elizabeth and as far as Port St. Johns. The Gelidiums are not cast up and have to be removed from their rocky substratum. Harvesting is confined to the intertidal zone and as yet little is known about the distribution of the genus below the water line. *Gracilaria confervoides* is a very interesting weed in the sense that it

Fig. 6. Gannets roosting on an island off the Cape west coast.

requires very specialized conditions for growth, e.g., a silty bottom and comparatively little disturbance. These requirements are met in Saldanha Bay, where it is abundant and probably to a lesser extent in Hout Bay near Cape Town. Apart from their collecting weed washed up on beaches, processors are permitted to cut weed during periods when this raw material is scarce.

The primary source of agaroids is *Gigartina radula,* which is abundant along rocky parts of the west coast. A variety of red and brown algae, suitable for use in fertilizer or fodder, is encountered along the entire South African coast. *Porphyra capensis* is plentiful along the west coast and could be considered as a source of food.

The harvesting of marine algae has steadily progressed, and it is certain that the available raw material will still permit a considerable increase in production. Table 4 shows the quantities collected during 1966.

South Africa is steadily building up an export market for seaweeds. Japan is the principal buyer of agar weeds and *Ecklonia maxima* is imported by several European countries.

Minerals

Marine geological research is gradually leading to a better knowledge of the mineral wealth of the continental shelf of South Africa. Scientific surveys are making favourable progress, but it will necessarily take a considerable time before the picture can be completed.

The presence of phosphorite nodules on the Agulhas Bank has been known for almost a century. Especially rich deposits, however, have been discovered beyond the 100 fathom line off the south-western Cape with phosphate (P_2O_5) contents ranging between 20–32%.

Another mineral is glauconite, which contains 2 to 9% K_2O and is a possible source of potassium for agricultural fertilizers; it frequently occurs in association with phosphorite. It is widely distributed throughout the surface sediments of the South African continental shelf. The presence of barytes deposits consisting of 75–77% $BaSO_4$ has also been established, while manganese nodules are found in certain areas of the shelf with notably high concentrations on the abyssal sea floor.

The most interesting development of exploiting submarine deposits is the successful underwater recovery of diamonds off the coast of South West Africa. The sediments being mined at present are believed to be an extension of the alluvial deposits found on the adjoining continent. Gems have generally been small, but of high quality. Most of the work is by suction dredging and is conducted very close inshore where the operations are attended by numerous hazards, often with the loss of expensive equipment. A considerable overburden of sediment has to be removed before the diamondiferous deposits are reached, which contain economic quantities of gem and industrial stones. Research and development will in due course certainly lead to more advanced and efficient methods of diamond recovery from the sea.

Future Prospects

These depend to a very large extent upon the expansion rate of knowledge of the sea (i.e., research effort), the demand for and the degree of utilization of marine products, and various technological and engineering advances, to mention but a few. That the world's population will become more dependent on the sea for supplying food and raw material is beyond dispute. Likewise, there seems little doubt that there will be an increasing tendency to farm the sea as opposed to the present techniques of random exploitation.

The coast east of Cape Agulhas, although representing roughly 40% of the South African coastline, probably does not produce more than 5% of its marine products. The only major activity in this region is pelagic whaling off Durban. Notwithstanding the lower primary productivity of the east coast, this vast discrepancy in terms of yields seems unrealistic and emphasizes the need for thorough investigation. Certain mollusks are reasonably abundant along the east coast and while the general feeling is that no single species warrants large-scale exploitation, it is considered that a multiple-species enterprise could well succeed. The exact location and extent of prawn-fishing grounds has still to be established, especially in view of the excellent quality of the various species occurring off Durban. The nature of the sea floor along the East Coast is such that very restricted areas are suitable for bottom trawling; there is of course no reason why successful fishing could not be done with midwater trawls. It also seems highly desirable to do a thorough stock assessment off Durban to establish if the whale populations are capable of sustaining a higher annual yield.

TABLE 4

Species	Weight (lbs)
Ecklonia maxima	4,269,327
Gracilaria confervoides	2,619,049
Gelidium pristoides	134,002
Gelidium cartilagineum	121,617
Gigartina radula	720
Porphyra capensis	18,100
Total	7,162,815 lb

The successful exploitation of tunas by foreign vessels off the east coast render a proper examination of this migratory fish of paramount importance. The general impression gained, therefore, is that the area east of Cape Agulhas is undoubtedly capable of making a more significant contribution towards South Africa's earnings from the sea.

On the west coast it is doubtful whether the current inshore resources, such as pelagic fish and rock lobster, will permit any appreciable increase in intensity of fishing. It is rather thought that the future lies in bottom trawling. At the same time the increasing activities of foreign fleets are being viewed with growing concern in South African circles, and ultimately some form of joint control will have to be developed. In this area, also, midwater trawling may prove remunerative.

The abundance and low utilization of sea weeds along South African shores involuntarily cause one to reflect on the future potential of this resource. In this world of diminishing raw materials the time is not far off that this material will be used profitably.

In spite of her terrestrial mineral resources, South Africa cannot disregard the possibilities the sea offers in this field. Once the major engineering difficulties have been overcome, there is no reason why marine mining should not become commonplace, provided that it can compete favorably with land operations. Geologists are optimistic about the possibility of finding oil below the continental shelf in certain regions of South Africa and enormous interest is being shown by foreign and local enterprise.

Considering the intensification of South Africa's marine research effort and the increasing participation by private firms and universities, there can be no doubt that the sea will progressively play a more active part in the future of South Africa.

G. H. STANDER

Cross references: *Abalone; Agar; Alginates from Kelp; Lobsters; Mussels; Oysters; Peru Current; Sharks; Shrimp Fisheries; Tuna Fisheries.*

SPONGE INDUSTRY

General Biology

Sponges (Porifera) are aquatic, predominantly marine animals distinct from Protozoa in having cellular construction and from other Metazoa in lacking true tissues. Sponges are sessile organisms with the larva, formed as a result of sexual reproduction, having a short, free-living existence.

The production and maintenance of a water current is the chief activity of a sponge. The organism acts essentially as a pump forcing through its body a current of water of considerable volume but at low pressure. By means of this current, most exchanges between the sponge and its environment are effected. The elaborate filter feeding apparatus is based upon an extensive system of pores and canals and powered by the action of peculiar flagellated cells, choanocytes. The inhalant current is drawn in through the dermal pores to inhalant canals that open ultimately by one or several branches to small chambers, $40-60\mu$ in diameter lined by choanocytes. The flagellae of the choanocytes beat in uncoordinated fashion, but the orientation of the collars is such that each cell directs its current toward the single exhalant opening the apopyle. From the apopyles the current passes to exhalant canals, which open by way of an osculum. This system depends for its functioning upon the fact that, in each flagellated chamber, the diameter of the apopyle is much greater than the total diameter of all inhalant apertures entering the chamber. Consequently, water is sucked into the chambers from the inhalant system and an inward current is set up at the sponge surface.

The food of sponges is organic detritus, bacteria, and small zoo- and phytoplankton. These particles are incorporated either by choanocytes, where they adhere to the collar and are passed to the cell body, or by amoebocytic cells termed archaeocytes. Archaeocytes ingest larger particles directly through the lining membrane of inhalant canals as well as receiving particles from adjacent choanocytes. Digestion is entirely intracellular and takes place in vacuoles within archaeocytes. Waste products are voided from thin walled vacuoles at the surface or through the lining of an exhalant canal.

Support for the soft strictures of the sponge body is provided by calcareous or siliceous spicules, by spongin fibers, or by a combination of spongin and spicules. Spongin, so important in commercial species, where it makes up the entire skeleton, is a halogenated scleroprotein with chemical characteristics very similar to those of collagen.

Sponges are capable of only very basic responses to environmental stimuli. The cells lining the surface, inhalant and exhalant apertures, and larger canals are capable of slow contraction in response to chemical, mechanical, light, or heat stimulation. Stimulus leads to closure of external apertures and overall contraction of the body. Despite recent claims based upon histological evidence, sponges do not appear to possess nerve cells, and excitation, insofar as it occurs, is a matter of transmission from cell to cell (neuroid transmission) at a rate of a few millimetres per minute.

Classification

The division of the phylum into three classes is based upon the chemical constitution and growth form of the skeletal components.

  I. *Calcarea* with skeleton of calcium carbonate.
  II. *Hexactinellida* with skeleton of silica and spicules having a 6-rayed (triaxon) or modified 6-rayed pattern.
  III. *Demospongiae* with skeletal elements of silica, spongin fibers, or both. Spicules never triaxon.

The Demospongiae is by far the largest and most widely dispersed group of sponges, and all species of economic significance belong in this group.

Boring Sponges

Clionid sponges are found in tidal and shallow waters throughout the world. In tropical reef habitats they are found in great profusion and are a significant agency in reducing coral substrate to fine detritus. It is in temperate regions, however, where oyster and scallop fishing is of importance, that boring sponges pose an economic problem.

The presence of the sponge in oyster shells is apparent from the bright yellow to orange spots scattered over the shell. These are the inhalant and exhalant apertures of the sponge; the remainder of the tissue is lodged in galleries excavated in the shell.

The mechanism of boring remains to be demonstrated conclusively, but it has been observed by various workers that protoplasmic extensions of sponge cells etch fine lines in calcareous and in proteinaceous conchiolin layers of the shell. The shape proscribed by the etched lines corresponds with the shape of the fine particles of shell, which are extruded from the sponge oscules. It seems likely that etching is effected by localized acid secretion perhaps in association with an enzyme. Removal of the chips could be effected by the extension of protoplasmic filaments into the substrate along the line of weakness thus prising the chips out of position. It is unlikely that mechanical adhesion coupled with cell contractility could account for the removal of fragments.

Shells that are infested by boring sponges tend to flake while being opened. This coupled with the fact that the sponge decomposes soon after removal from the water, means that these oysters cannot be opened without contamination. This reduces their consumer appeal and hence their market value. There is good evidence that heavy sponge infestation leads to a loss of condition in oysters.

The annual loss to the oyster culture industry in eastern Canada from sponge action has been estimated as "many thousands of dollars." No actual estimates can be located for the eastern U.S., but whole commercial enterprises in the South Atlantic States have been ruined as a result of sponge infestation.

Commercial Sponges

Species of *Spongia* and *Hippospongia* have no mineral skeleton, they are variable in shape and surface characteristics and thus have proved difficult to classify with certainty. The names in common usage for these sponges reflect only characteristics of the dried spongin skeleton and give no clue as to the appearance of the living

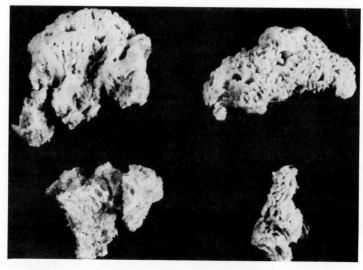

Fig. 1. *Tedania toxicalis*, deLaubenfels. This sponge produces a poison that is lethal to many other organisms.

sponge. Since all statistics on the sponge industry are compiled using common names, it is necessary to retain these and to attempt to assign a valid scientific name to each major category.

Many varietal and subspecific names have been assigned to commercial sponges. These are of dubious scientific merit and can be ignored.

In Table 1, the area in which the species is commercially significant is given. The actual distribution area of the species will always exceed this. It is however most improbable that there are any species of *Spongia* and *Hippospongia* common to the Mediterranean and the West Indian/Gulf of Mexico region. It is also certain that species of *Spongia* from the west central Pacific, which are of good commercial quality, do not belong to *Spongia officinalis* as suggested by deLaubenfels. Much more information is required before the systematic status of these obviously very closely related sponges can be clarified.

The Sponge Industry

Until about 1840, the world sponge supply derived entirely from southern Europe and North Africa. In 1841 a shipwrecked French sponge merchant, observing the fine quality of sponges in the Bahama Islands, sent a trial shipment to Paris. Eight years later, sponge exports from the Bahamas were valued at $10,000. In 1849 the first shipment of native sponges from Key West, Florida arrived in New York. After some initial difficulties, the American industry became established around the Florida Keys, later to spread to the Gulf of Mexico. About the turn of the century Tarpon Springs became the principal market and also, in 1905, the headquarters of the immigrant Greek divers. The Tarpon Springs Sponge Exchange—a non-profit organization established in 1908—still handles all but a fraction of the Florida trade.

There are two basic methods of sponge harvesting: diving and hooking. Both methods are employed in all major sponge producing areas. In the western Mediterranean near Sicily sponges are taken in dredges that operate at depths from 450–600 ft. This method is destructive, as the smaller sponges are taken or else smothered by the heavy gear. The earliest and most primitive method of collection, still practiced in parts of Tunisia, is wading in the shallows and collecting sponges with the toes. As sponges became less accessible in deeper waters, nude diving logically followed. The Greek divers developed particular skill in this method. Although most Greek divers today use modified lightweight diving gear, in some places the traditional diving techniques are still employed. A heavy stone is carried to enable the diver to reach the bottom quickly, and a rope attached to his waist enables him to signal his boatmen to pull him up fast.

In America there are three distinct groups of spongers employing two basic methods. The hooker in the Tarpon Springs area uses a glass-bottomed bucket for viewing and a long, 4-pronged rake to tear the sponge from its base. As shallow water sponges tend to be soft and have an open texture this technique frequently tears the sponge and, although it may leave a new growing center, results in damaged specimens. As this method is confined to shallow waters, the beds tend to be worked more often, resulting in smaller sponges of poor quality.

In the Florida Keys area a system is employed in which a pole is used to hook the sponge by means of a rope loop. This procedure is regarded as more satisfactory from a conservation viewpoint, as smaller and poorly formed specimens

TABLE 1.

Scientific Name		Common Name	Region of Commercial Importance
Spongia	*graminea*	grass, glove	Gulf of Mexico Caribbean
	barbara	yellow	Gulf of Mexico Caribbean
	zimocca	Zimocca, chimousse Fine dure	Mediterranean
	officinalis	Turkey cup or solid Fine Levante	Mediterranean
Hippospongia	*lachne*	wool, sheepswool	Gulf of Mexico Caribbean
	communis	Honeycomb	Mediterranean
	gossypina	velvet	Caribbean
Axinella polycapella		finger	Gulf of Mexico Caribbean

cannot be taken. No diving is permitted in this area.

Diving boats working out of Tarpon Springs are still designed after the Greek sponging boats of the Mediterranean. They are heavy, seaworthy boats, diesel-powered, with the air pump worked off the main engine. Most divers use modified deep-water equipment with weighted rubber suit and helmet, and carry a short, 2-ft long hooking claw. Cheaper lightweight equipment, such as synthetic foam suits and air-supplied masks, has been successfully used by individual fishermen, and as most diving in the Gulf of Mexico is done in 20–30 ft of water, it would seem that adoption of this method could greatly increase the return from the same fleet of boats.

When collected, the sponges are first crushed to hasten the decay of organic matter, then are piled base down in the well of the boat and covered with wet sacks. In hot weather the sponge decays rapidly, and within about 2–3 days is ready to be scraped, washed, and strung out to dry. In the Bahamas, sponges are left to decay in "kraals," shallow ponds of salt water.

In the dealers' hands the sponges are further cleaned to remove sand, grit, and shell; they may also be bleached and dyed. Greek sponges for export to Great Britain undergo a special treatment involving immersion in hydrochloric acid, rinsing in seawater, and a further immersion in a mixture of potassium permanganate and oxalic acid. This results in a dark stain, and the yellow color so valued on the sponge market is obtained by a second bath of acid.

The best grade of wool sponges is the "forms." These are well shaped sponges of good texture, without tears and imperfections. Very large sponges, or those with blemishes, must be divided into smaller pieces and trimmed. These are then called "cuts" or "seconds" and bring a lower price. In all producer countries there is a complex system of commercial classification indicating geographical origin, method of collection, quality, etc.

The production of sponges in the U.S. reached its peak in 1936–1937. Over 600,000 lb of sponges were taken valued at more than $1.2 million. At this time the industry employed 72 diving outfits, 256 sponge hooks, and gave livelihood to nearly 1000 fishermen. Sale value of landed sponges reached a peak in 1945–1946, when only about 160,000 lb of sponges returned almost $3,000,000. Of the 191 boats in this fleet, 75 were diving boats. This large number of boats caused serious overfishing of the beds. By 1951 only 32 boats remained in the industry, and of these only 2 were diving outfits. A catch of under 12,000 lb in that year returned only $82,000.

A number of factors have contributed to this decline, the most significant being disease, lack of conservation methods, and a decline in consumer demand. The sponge industry was set back in 1938–1939, and again in 1947–1948, when disease swept through the commercial beds of North America. In the earlier epidemic it has been estimated that between 90–95% of the commercial sponges were destroyed in most areas of the Caribbean and the Gulf of Mexico. Velvet sponges, the most highly valued, disappeared entirely from the Bahamas, and only a few have been since taken from the Florida coast. This disease was believed to have been caused by a parasitic fungus, *Spongiophaga communis*. If this organism was also responsible for the heavy sponge losses reported in 1878 and 1898, then its incidence would appear to be rare but disastrous.

The 1947 attack of disease was investigated by scientists from the Marine Laboratory, University of Miami, but no evidence of fungal infection was detected. The oceanographic conditions were not outside the range normal for the region, nor were there any extraordinary oceanic conditions present, such as the planktonic or fungal blooms reported before previous epidemics.

Since 1951 there has been a steady increase in the landings of sponges, and the fishing fleet has again increased in size. There have been fewer wild fluctuations in price or catch and scientific surveys have been undertaken to assist the industry. The need for scientific management of sponge resources and acceptance of conservation methods is still acute, and younger men are needed in the fleets. More detailed research into the biology of commercial sponges could provide answers to many problems that beset the industry and perhaps lead to well organized and profitable farming procedures.

Statistics

While documentation and statistics are most accessible for the sponge industry of the U.S., it must be remembered that the Mediterranean trade still takes first place in production figures. However, in the Mediterranean, fishing is conducted by so many nationalities, overlapping in their areas of operation and their markets, that accurate statistics are difficult to obtain. In recent years Tunisia has consistently headed the list of production figures, followed by Greece, Turkey and the U.S. A report from Libya indicates that "foreign enterprises" account for the largest part of that country's production, and that for this reason the industry contributes little to the Libyan economy. Greek fishermen farm the coasts in Greek, Libyan, Tunisian, and Cypriot waters, and it is evident that the same sponges may be reported in part both for the country owning the beds and by that whose

TABLE 2. SPONGES: CATCH BY COUNTRIES, 1958–1966 (KILOGRAMS)

Country	1958	1961	1962	1963	1964	1965	1966
GRAND TOTAL	268,600	216,000	264,900	279,500	245,000	239,200	210,500
Libya	6,900	19,102	5,263	—	—	—	—
Tunisia	159,000	85,948	119,319	120,743	73,071	102,266	66,217
U.S.	13,154	16,783	21,772	24,948	19,958	20,865	(21,000)
Cyprus	—	—	—	800	—	70	—
Philippines	1,453	(3,500)	(3,500)	(3,000)	(3,500)	6,000	(6,000)
Turkey	23,000	22,000	45,000	50,000	48,000	41,000	(45,000)
Spain	(65,000)	68,700	(70,000)	(80,000)	98,000	69,000	(69,000)
Greece	—	—	—	—	2,500	—	3,300

nationals do the fishing. Table 2 has been supplied by the Fishery Statistics and Economic Data section of FAO.

In no country does sponge production make a major contribution to the national economy. It is, however, of major importance to the local population wherever fishing takes place. During World War II the sponge fishing industry of Florida, based on collection, preparation, and marketing, was one of the major industries of the state. Entire populations of many Greek islands depend on sponge fishing for their livelihood. In Kalymnos 80% of the working force are sponge fishers, as are about 20% of the men of Aegena, Lymnos, and Volos. Both Libya and Egypt have taken steps to prohibit foreign sponge fishers and to expand their local industries.

Cultivation

Because of their high regenerative capacity sponges are admirably suited to cultivation and farming procedures. There are records of experiments in sponge culture during the 18th and 19th centuries in Mediterranean waters. At Andros Island in the Bahamas a sponge farm has been in existence for over 30 years. Greece has several beds that are being farmed on an experimental basis, and for some years before World War II the Japanese were experimenting with sponge culture in the Marshall and Caroline Islands. Several attempts have been made about the coasts of Florida to grow sponges from cuttings. Failures have occurred for a variety of reasons, many of them other than biological or technological. Poor localities, an influx of fresh water, disease in the grounds, storms, poaching, and deliberate sabotage by fishermen are some of the causes of failure. The practicability of sponge farming has been demonstrated by the success both of the Bahamas and British Honduras government sponsored programs and of the Japanese experiments, which were interrupted by the war in the Pacific.

The technique of sponge cultivation consists essentially of cutting a sponge into pieces, attaching these to some type of submerged anchor and then allowing regeneration to proceed. The sponge must be cut cleanly to avoid crushing and the pieces quickly returned to the water, as exposure to sunlight and drying rapidly kills the animal. The Japanese experimenters believed that cutting and fixing should be carried out under water. They tried several different techniques, the most successful being a method in which the sponge cuttings were strung at 4 in. intervals on a wire, suspended on a bottle float $1\frac{1}{2}$–2 ft below the surface, and anchored by a concrete disk. The experimenters at Ailinglapalap between 1940 and 1943 used a sponge later identified by de Laubenfels as *Spongia officinalis mollissima,* a species known otherwise only from the Mediterranean. This identification is obviously erroneous. Although sponges were transported between islands in Micronesia, Japanese authorities emphasized that all were native to the region. Indeed, in the circumstances of 1940 the likelihood of Mediterranean sponges being successfully transported to the Pacific is extremely remote. When examined in 1946 these sponge cuttings were found to have grown to between 4–$6\frac{1}{2}$ in. in diameter. This method produced most rapid growth with highest percentage survival. At Palau and Truk both set disk and hanging raft methods were tried, the latter proving incomparably better.

Cultivation in both Florida and the Bahamas has consisted of attaching cuttings to submerged rocks or concrete disks. The sliced piece rapidly adheres to its base, the raw surfaces healing over within about 2 weeks. The living material withdraws from the extremities of the cutting, and at the end of about 6 months the sponge has rounded out by new growth. A recently completed study of the sponge fisheries of the Gulf of Mexico conducted by the Marine Laboratory of the University of Miami provides valuable data concerning growth rates of sponges both from larvae and from cuttings.

The advantages of sponge cultivation are fairly obvious in view of the fluctuating returns from natural fishing grounds and the great difficulties encountered in trying to bring about an understanding of the need for sound conservation practices. The planting method is simple, and the materials and techniques are basic. Concentration of sponges of a similar size could facilitate harvesting and replanting; these could be a single operation if sponges were cut rather than torn from the base. Under good scientific management some control should be possible over quality, size, and shape.

However, there are a number of difficulties to be considered. The choice of a bottom or growing area is of primary importance, and it might take some years to find suitable areas and obtain long-term leasing rights. Returns could not be expected in under 4 years, and protection of the areas from vandalism and theft could be expensive, with the ever-present threat of complete loss of the crop by storms or disease. Any attempt at scientific sponge farming would need to be preceded by a type of investigation in depth similar to that carried out in Florida, Gulf of Mexico, by Dr. Storr and the Marine Laboratory of the University of Miami. Sponge growth rates are affected by temperature, salinity, turbidity, water currents, bottom slope, sediment, and competition for space. Sponge cultivation would not bring any quick return for the money invested, but would require a fairly large investment of both time and effort over a period of 4–5 years. Whether, in view of the declining market for natural sponges, this is a justifiable economic proposition is debatable.

Future of the Industry

The greatest threat to the sponge industry lies in the low cost and increasing versatility of synthetic sponges. Before 1938 the sponge market was divided, according to the U.S. Sponge and Chamois Industry trade reports, as follows:

Amateur cleaners and housewives	25%
Pottery, tile, shoe, and miscellaneous manufacturers	25%
Professional painters, decorators, and wall washers	50%

During World War II, increased prices and short supply coincided with the advent of the cheap synthetic product. High prices forced industrial users to try synthetic substitutes, and radical changes in methods and materials in the painting and decorating trades further reduced the need for sponges.

For many purposes the natural sponge has advantages not yet available in the synthetic substitute. The plastic fiber is not as strong as the natural spongin fiber, and the foamlike mesh of the synthetic sponge tends to retain dirt and odors. However, while there is still a market for natural sponges for toilet purposes and in a few trades, a whole generation of housewives is unaware of the uses and advantages, and often even of the existence, of the natural product. Unless the sponge producer can keep the price of the natural sponge down to a level where it can compete with the synthetic product, it would appear unlikely that the industry will be able to regain its share of this market.

PATRICIA R. BERGQUIST
CATHERINE A. TIZARD

References

Cahn, A. R., "Sponge production and International sponge trade of the United States," 1946, U.S. Fish and Wildlife Serv., Fishery Leaflet 170.

Cahn, A. R., "Japanese sponge culture experiments in the South Pacific Islands," 1948, U.S. Fish and Wildlife Serv., Fishery Leaflet 309.

Galtsoff, P. S., "Sponges," 1963, U.S. Fish and Wildlife Serv. Bureau Comm. Fisheries, Fishery Leaflet 490.

Hartman, W. D., and Goreau, T., "Boring sponges as controlling factors in the formation and maintenance of coral reefs," *AAAS Publ.*, **75**, 25–54 (1963).

Laubenfels, M. W., de, "The Order Keratosa of the Phylum Porifera," *Occ. Pap. Allan Hancock Found.*, **3**, 1–217 (1948).

Laubenfels, M. W., de, "Sponges from the Gulf of Mexico," *Bull. Mar. Sci. Gulf and Caribbean*, **2**(3), 511–557 (1953).

Laubenfels, M. W., de, and Storr, J. F., "The taxonomy of American commercial sponges," *Bull. Mar. Sci. Gulf and Caribbean*, **8**(2), 99–117 (1958).

Moore, H. F., "The commercial sponges and the sponge fisheries," *Bull. U.S. Bureau of Fisheries*, **28**(1), 399–511 (1910).

Storr, J. F., "The sponge industry of Florida," *State of Florida Board Conservation Educational Series*, **9**, 1–29 (1957).

Storr, J. F., "Ecology of the Gulf of Mexico Commercial Sponges and its relation to the Fishery," 1964, U.S. Fish and Wildlife Serv. Special Scientific Report 466.

Stuart, A. H., "World Trade in Sponges," 1948, U.S. Dept. Commerce Industrial Ser. 82.

Tressler, D. K., "Marine Products of Commerce," 1948, Ch. 36 "Commercial sponges," by H. F. Moore, Rev. P. S. Galtshoff, Reinhold Publishing Corp., New York.

Vacelet, I., "Répartition générale des Eponges et Systématique des éponges cornée de la région de Marseille et de quelques Stations Mediterranéenes," *Rec. Trav. Stat. Mar. de Endoume*, **26**(16), 39–107 (1959).

Warburton, F. E., "Control of the boring sponges on oyster beds," *Prog. Rept. Atl. Coast Sta. F.R.B. Canada*, **69**, 7–11 (1958).

Warburton, F. E., "Effects of boring sponges on oysters," *Prog. Rept. Atl. Coast. Sta. F.R.B. Canada*, **68**, 3–8 (1958).

Cross reference: *Toxicology of Marine Animals.*

SPORT FISHING

Sport fishing, or angling, is the art of catching fish for fun or sport, usually employing a line, with baited hook or lure attached, either held in the hand or attached to a hand-held pole or rod. The rod, in turn, may or may not be equipped with a reel to facilitate casting and/or retrieval and storage of the line. A reel-equipped, hand-held rod appears in a Chinese painting now hanging in the National Museum of Tokyo, said to date from the period 1190–1230 A.D. Even older in origin, some forms of net fishing and spear fishing are also regarded as sport fishing under certain circumstances. Sport fishing has achieved its greatest recreational significance since 1950. Especially in the U.S. and Canada, widespread participation in recreational fishing has become both sociologically and economically significant. People fish primarily for the fun of catching fish, refreshing the mind and renewing the spirit in the process, and for the opportunity to get away from the normal pattern of life and contemplate.

Little information is available or verifiable concerning the extent of angling outside the western countries. With respect to the Soviet Union, for example, Dr. Donald E. Bevan, Fisheries Research Institute, University of Washington College of Fisheries (Seattle), advised in 1965, that President Kuznetzov of the Federation of Sport Fishing of the U.S.S.R. has estimated that there were about 1,500,000 organized sport fishermen in the Soviet Union on January 1, 1964. He also estimated there were 6,000,000 more outside the organizations who were also more or less systematically occupied with sport fishing. Elsewhere, it may at least be surmised that primitive forms of sport fishing are in current use wherever fish occur in some abundance throughout the world. Because of the obvious rapid growth and widespread participation in all kinds of outdoor recreation in the U.S., Congress established a temporary (1959–1962) Outdoor Recreation Resources Review Commission (ORRRC) to assess the matter and recommend a national program to accommodate future needs. Within the scope of its work, a special study was undertaken of outdoor recreation opportunities that would be available for U.S. tourists in several foreign countries. Findings suggested that sport fishing ranks universally high as a leading form of outdoor recreation.

Denmark's sport fishing by her nationals emphasizes salt water, as does that of the Netherlands, while both countries offer excellent "coarse fishing" for tourists. Fishing is virtually the national pastime in France, with over 2,500,000 registered anglers in 1957. In Germany, where fishing licenses are also required, the holder of the right to fish is responsible for fish stocking and protection. Sport fishing is an ancient and widely indulged form of recreation in Japan; fishing with well trained cormorants and various kinds of net fishing are also forms of sport fishing there. In Great Britain, fishing was reported to be the most popular of the country sports. Since the ORRRC made its report on outdoor recreation abroad, a national survey by Research Services, Ltd. (London) set the number of anglers in Britain at 2,200,000 male adults (16 years and over) in 1964; few ladies fish for sport in Great Britain.

An economic survey of fishing and hunting in Canada revealed that 10.8% of Canadians 14 years of age or older fished in 1961, and that men anglers outnumbered women 6 to 1.* Fresh water sport fishermen (about 1,257,000—with only 40% required to buy licenses) outnumbered salt water anglers (about 149,000—of whom some 64,000 apparently fished exclusively in salt water) by more than 8 to 1. Pacific salt water anglers (about 92,000) were found to outnumber Atlantic salt water anglers (about 57,000) somewhat less than 2 to 1. All in all, sport fishing provided 20,000,000 man-days of recreation for Canadians in 1961, who spent $188 million on their sport that year, averaging $143 per fisherman. Capital items, i.e., such fishing equipment as boats, motors, trailers, fishing tackle, camping gear used primarily for fishing, etc., accounted for 51% of the expenditures. Expendable items, including accommodations where utilized, food, supplies of various nature, consumed something under 27%. Costs of operating private vehicles (averaging $7\frac{1}{2}$ cents per mile), made up the 22% remaining.

The greatest participation in sport fishing in the western world unquestionably occurs within the U.S. In 1965, the total number (including those of any age who fish only occasionally and/or spend little or virtually nothing in the process) was estimated by the Sport Fishing Institute to have been about 55,000,000. In 1966, the Fish and Wildlife Service published the third in a series of national surveys of fishing and hunting. It reported that 28,348,000 habitual ("real" or "substantial") anglers—those 12 years or older who participated at least three times or who spent at least $5.00 during the year—devoted 522,759,000 recreational days to fishing and spent $2,925,304,000 on their preferred means of outdoor recreation.† In addition, 3,241,000

---

* Benson, D. A., "Fishing and Hunting in Canada," 1963, Can. Wildl. Serv., Ottawa, Ont.

† 1965 National Survey of Fishing and Hunting, U.S. Bureau of Sport Fisheries and Wildlife, Resource Publication 27. U.S. Gov. Printing Office, Washington, D.C.

youngsters, aged 9–11, also fished during some portion of 28,265,000 recreational days in 1965. Thus, well over 31,589,000 habitual anglers 9 years or older fished during a total of more than 550,974,000 recreational days. These data exclude the many millions of angling youngsters under 9 years old and the many added millions of incidental anglers of all ages (those fishing less frequently than some part of three days or spending less than $5.00 during 1965) who collectively complete the over-all total of angling participants estimated by the Sport Fishing Institute. One among every ten females fish, compared with one among every three males. Primarily because of various legal exclusions (too old, too young, property owners, disabled servicemen, aborigines, specified waters, etc.), only 59% of all habitual anglers 12 years or older are licensed. These include those various categories of salt water fishermen who are now licensed in some manner in eight coastal states (Alabama, Alaska, California, Louisiana, North Carolina, Oregon, Texas, Washington). An estimated 601,000 U.S. anglers fished in Canada (up by 37% since 1960), some 138,000 fished in Mexico, and 57,000 fished in other foreign countries in addition to the U.S.

Fresh waters supported 426,922,000 recreational fishing days by 23,962,000 habitual anglers. They spent an average of $89 per person during the year or $4.98 per day. Salt water supported 95,837,000 recreational fishing days by 8,305,000 habitual anglers. They spent an average of $96 per person in 1965 or $8.34 per day. Fresh water fishing generated $2,125,652,000 of gross business activity, compared with $799,656,000 generated by salt water fishing. The over-all total of angler expenditures for necessary goods and services (nearly 3 billion dollars) was comprised of expenditures for: (1) primary fishing equipment—rods, reels, lures, tackle boxes, lines, landing nets, etc. (11.0%); (2) auxiliary equipment—boats, motors, boat trailers, camping gear, coolers, cameras, etc. (26.9%); (3) food and lodging, in excess of what would normally have been spent staying at home (15.2%); (4) transportation—largely out-of-pocket costs of operating motor vehicles plus some travel on common carriers (14.7%); (5) fishing licenses and privilege fees (4.5%); and (6) bait, guides, and other miscellaneous expenses (27.7%).

Among the 8,305,000 habitual salt water anglers, some 4,486,000 fished exclusively in coastal marine waters; the remaining 3,919,000 fished extensively in inland fresh waters as well. Over half of all saltwater anglers (4,178,000) and of total marine angling activity (55,950,000 recreational days) occurred on the Atlantic coast where annual angling costs, averaging $79.27 (about $5.92 per angler-day), were lowest. Roughly ¼ of all salt water anglers (2,084,000) and marine angling activity (22,390,000 recreational days) occurred on the Gulf coast where annual angling costs, averaging $84.50 (about $7.87 per angler-day), were intermediate between east and west coasts. The nearly ¼ of salt water anglers (2,043,000) and ⅕ of marine angling activity (17,497,000 recreational days) remaining were found on the Pacific coast. There, angling costs averaged $143.11 for the year (about $16.71 per angler-day).

A comparison of data for three national surveys of fishing shows a strong growth trend for fishing participation in the U.S. that greatly exceeds that of additions to the general population (Fig. 1). Long-term data for annual sales of fishing licenses show clearly that the growth trend of angling participation has been one of continuous strong acceleration since World War II. It should be borne in mind that many anglers (41%, possibly more) are not required to purchase licenses to fish, depending upon a variety of circumstances of age, nature or method, of fishing, peculiarities of residence etc.

Such statistics, though useful for predicting future demand, do not indicate possible influences of other forms of outdoor recreation on fishing activity, nor the motivating effects of current trends toward more leisure time, paid vacations, and higher incomes. To provide some of those data, which are vital to sound planning for the future, the temporary Outdoor Recreation Resources Review Commission sponsored a special study of demand factors among American adults (18 years or older) by the University of Michigan Survey Research Center. The study, which furnished the hard core data for the basic ORRRC report, revealed that 20% of all those interviewed had fished "often" (5 or more times), and 18% "a few times" (1–4 times) during 1959.* On this basis, nearly 43,000,000 adults fished in some degree during the year.

The report authors also ranked outdoor activities according to "involvement," based separately on *prompted* and *spontaneous* mentions of activities of the interviewees, with fishing found to be one of the *most* "involving" activities. To determine potential needs for facilities, the survey also attempted to define the extent to which people would like to increase their participation in various activities in the future, viz.:

* Mueller, E., Gurin, G., and Wood, M., "Participation in Outdoor Recreation: Factors Affecting Demand Among American Adults," 1962, ORRRC Study Report 20, Washington, D.C.

Activity	Per cent who would like to do more often or take up	
	Total	Did not do at all previous year
Fishing	21	8
Swimming, or going to beach	14	5
Auto riding for sightseeing	13	3
Camping	13	9
Horseback riding	12	10
Boating and canoeing	11	6
Hunting	10	5
Picnics	10	3
Skiing, other winter sports	5	4
Hiking	5	3
Nature or bird walks	4	2

From these data, it would appear that facilities will have to be increased not only in accordance with population growth, and to relieve present overcrowding, but also to allow for some rise in participation rates. The interviewers first asked the people, "How do you usually spend most of your leisure time?" in order to focus on the regular day-to-day leisure patterns. Fishing was the *most* frequently mentioned of all active sports; conclusion: ". . . swimming, hunting, and especially fishing seem to be of the greatest importance and salience . . . That one out of six Americans spontaneously mentions fishing in the context of questions asking about activities they engage in 'quite a lot,' would seem to attest to the importance of this recreational activity."

A nationwide detailed survey of the habits and preferences of Americans engaged in outdoor recreation was conducted in 1960–1961 by the U.S. Bureau of the Census. The survey determined that about 35% of the U.S. population 12 years of age and over (133,000,000) fished at least once during the period June, 1960—May 1961. These 47,000,000 anglers fished an average of 11.9 days per year, participation rates being highest in the south, and lowest in the northeast. The estimates are comparable to those computed for the ORRRC by the Michigan Survey Research Center. In the Census Bureau survey, fishing was found to be the preferred outdoor activity among 33% of the population. From the tabulated material, it was determined that about 26% of all fishing activity (man-days) occurred in the spring, 47% in summer, 18% in the fall, and

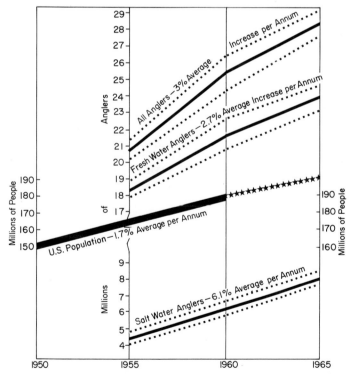

Fig. 1. Ten-year trends of increasing participation by Americans in recreational fishing, compared with the trend for growth of the entire U.S. population, based on the findings of three national surveys of fishing and hunting. Dotted lines paralleling the angler-trend lines indicate the boundaries of statistical reliability of the survey findings, based on 95% confidence limits of the data secured in the sampling procedure. These are the limits between which would fall the figure derived from a complete enumeration of the population with a 19 out of 20 probability.

9% in winter. In the south, fishing participation during the winter was 16% compared to 4% in the northeast, 6% in the northcentral, and 7% in the west. There were about 3,700,000 persons who fished during the December-January-February period in the northcentral and northeast regions; undoubtedly, the great majority were ice fishermen. The study estimated that total annual man-days of fishing during 1960 equalled about 560,000,000 (more recent data indicated that they had increased to 678,000,000 by 1965).

A special study of angling was prepared by the U.S. Bureau of Sport Fisheries and Wildlife for the Outdoor Recreation Resources Review Commission with advice and assistance from the Sport Fishing Institute. A major conclusion was that U.S. anglers would probably increase 50% by 1976 and 150% by 2000; by comparison, the U.S. population was estimated to increase by only 30% and by 98%, respectively, over 1960 figures. By century's end, the number of fisherman-days were expected to be about three times what they were in 1960. In round figures, then-current trends indicated that 63,000,000 anglers would fish during 1,300,000,000 days in the year 2000. Among its many other important findings the angling study team determined that the average daily catch per angler is about 1.3 lb of fish in fresh water and about 7 lb of fish in salt water (5 lb on the Atlantic and Gulf coasts; 15 lb on the Pacific Coast.)* Using these data, counting the fishing of all anglers regardless of their ages and frequency of fishing, and projecting 1965 fishing estimates in accordance with the rates of annual increases shown in Fig. 1, it is possible to calculate that total catch of fish by anglers from all U.S. waters was a little over 1.5 billion lb in 1967. This was made up of about 659 million lb of fresh water fish and about 843 million lb of salt water fish, the vast majority of which are eaten by their captors. By comparison, the U.S. commercial catch of edible food finfish fluctuates annually around 1.7 billion lb, thereby indicating the substantial, if secondary, contribution made by sport-caught fish to the national diet.

The report concluded that the resulting heavy fishing demand can be met with only slight reductions in the present average catch. New reservoirs are expected to add 10 million surface acres by 2000, doubling present impounded waters; these new waters will supply about $\frac{1}{3}$ of the expected increase in fishermen-trips. Improvement of existing waters through better management and the capacity of existing waters to absorb more fishing effort is anticipated to meet an additional $\frac{1}{3}$ of increased demand. Marine waters can absorb the remaining $\frac{1}{3}$, provided that estuaries are not damaged or rendered unfit as spawning and nursery grounds, and anadromous species are not blocked by more man-made structures. Essential requisites are that siltation and pollution must be prevented and controlled more effectively, fishery research must be expanded to provide bases for improved management, and the problem of getting adequate funds for these conservation programs must be solved. At least 75% of the angling activity in 1960 occurred on public waters, and this percentage was expected to rise in the future.

RICHARD H. STROUD

* King, W., Swartz, A., Hemphill, J., and Stutzman, K., "Sport Fishing—Today and Tomorrow," 1962, Outdoor Recreation Resources Review Commission. Study Report 7, Washington, D.C.

STANDING CROPS—*See* **PRODUCTIVITY OF MARINE COMMUNITIES; BENTHONIC DOMAIN; PLANKTON RESOURCES**

T

TAGGING—See MARKING OF FISH

TEREDO—See MARINE BORERS

TOXICOLOGY OF MARINE ANIMALS

Poisonous and venomous marine animals are of three major types: (1) those that are *poisonous* to eat; (2) those that produce their poisons by means of specialized poison glands, but lack a traumagenic apparatus (the *crinotoxic* organisms); and (3) those that produce their poisons by means of specialized venom glands, and possess a traumagenic organ (the *venomous* marine animals). Marine zootoxins are also sometimes grouped into two major categories, the oral poisons and the parenteral poisons, or venoms. Crinotoxins are thought to be largely parenteral poisons, but this has not been determined, and some of these crinotoxins may also be effective oral intoxicants. The term "poisonous" may be used in the generic sense, referring to both oral and parenteral poisons, but it is more commonly used in the specific sense to designate oral poisons. Thus, all venoms are poisons, but not all poisons are venoms. Generally speaking, knowledge of the chemistry of crinotoxins is too meager to permit a very intelligent classification at this time.

As man attempts to harness ocean resources, toxic marine organisms will become increasingly important since they are an integral part of the biological economy of the sea. They are not only a health hazard, but also a vast supply of untapped biodynamic agents that are beginning to be utilized as a valuable source of new pharmaceuticals.

Toxic Invertebrates

Protozoans—One-Celled Animals. Poisonous dinoflagellates are best known because of their role in causing paralytic shellfish poisoning. The species of dinoflagellates identified include *Gonyaulax catenella* Whedon and Kofoid, *G. tamarensis* Lebour, and *Pyrodinium phoneus* Woloszynska and Conrad. Paralytic shellfish poisoning occurs along the Pacific coast of North America, northeastern coast of North America, Europe, South Africa, and rarely elsewhere. The toxic dinoflagellates are ingested by a variety of

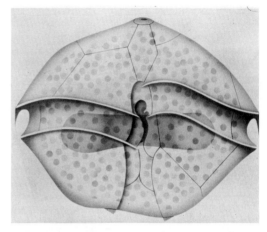

Fig. 1. *Gonyaulax catenella* Whedon and Kofoid. This dinoflagellate is the causative agent of paralytic shellfish poisoning (approx. 5000 ×). (Drawing by R. Kreuzinger.)

mussels, clams, scallops, etc. The poison accumulates in the digestive glands of mussels and clams. In some of the clams the gills may be quite toxic, whereas in *Saxidomus*, the butter clam, the poison is in the siphons. The distribution of the poison in the body of the animal appears to vary somewhat with the species of shellfish and with the season of the year. Man is poisoned by eating shellfish that have been feeding on toxic dinoflagellates.

Paralytic shellfish poisoning may be diagnosed readily by the presence of pathognomonic symptoms that usually appear within 30 min. Initially, there is a tingling or burning sensation of the mouth and face, with gradual progression to the neck, arms, and extremities. The paresthesia later changes to numbness, so that voluntary movements are made with difficulty. In severe cases, ataxia and general motor incoordination are accompanied in most instances by a peculiar feeling of lightness. Constrictive sensations of the throat, incoherence of speech, and aphonia are prominent symptoms in severe cases. Weakness, dizziness, malaise, prostration, headache, salivation, rapid pulse, intense thirst, dysphagia, perspiration, anuria, and myalgia may be present. Gastrointestinal symptoms of nausea, vomiting, diarrhea, and abdominal pain are less common. As a rule, the reflexes are not affected. Pupillary changes are variable, and there may be an impairment of vision or even temporary blindness. Mental symptoms vary, but most victims are calm and conscious of their condition throughout their illness. Occasionally, patients complain that their teeth feel "loose or set on edge." Muscular twitchings and convulsions are rare.

The toxic principle is termed saxitoxin or paralytic shellfish poison, and has the molecular formula of $C_{10}H_{17}N_7O_4 \cdot 2HCl$. The principal action of the toxin is on the central nervous system, i.e., respiratory and vasomotor centers, and on the peripheral nervous system, i.e., neuromuscular junction, cutaneous tactile endings, and muscle spindles. Absorption occurs through the gastrointestinal tract, and rapid excretion of the active toxin occurs through the kidneys. The treatment is largely symptomatic.

Dinoflagellates have also been identified as the causative agent of venerupin shellfish poisoning. The species involved is *Exuviaella mariaelebouriae* Parke and Ballantine. Man is poisoned by eating shellfish that have been feeding on toxic dinoflagellates. The symptoms usually develop within 24–48 hours after ingestion of the toxic mollusks, but an incubation period is believed to extend up to 7 days. The intial symptoms are: anorexia, gastric pain, nausea, vomiting, constipation, headache, and malaise. Body temperature usually remains normal. Within 2–3 days nervousness, hematemesis, and bleeding from the mucous membranes of the nose, mouth, and gums develops. Halitosis is a dominant part of the clinical picture. Jaundice, petechial hemorrhages, and ecchymoses of the skin are generally present, particularly about the chest, neck, and upper portion of the arms and legs. Leucocytosis, anemia, retardation of blood clotting time, and evidence of disturbances in liver function have been noted. The liver is generally enlarged, but painless. In fatal cases the victim usually becomes extremely excitable, delirious, and comatose. There is no evidence of paralysis or other neurotoxic effects usually observed in paralytic shellfish poisoning. In the outbreaks that have been reported, there was found to be an average case fatality rate of 33.5%. In severe cases, death occurs within one week; in mild cases, recovery is slow, with the victim showing extreme weakness. The development of ascites is a frequent complication.

Treatment is symptomatic: bed rest, injections of intravenous glucose, and the administration of vitamins B, C, D, and insulin.

There is no information available concerning the chemistry and pharmacology of these poisons.

Porifera—Sponges. There is very little information available concerning toxic sponges, although it is known that some species do produce poisonous substances. Sponges of the genera *Fibulia, Hemectyon, Tedania, Microciona,* and others are capable of inflicting a dermatitis that, in some instances, is believed to be due to a chemical irritant. Alcoholic extracts from the marine sponge *Suberites domunculus* Olivi produce vomiting, diarrhea, prostration, intestinal hemorrhages, and respiratory distress when injected intravenously into laboratory animals. The chemistry and pharmacology of the poisons involved are unknown.

Coelenterata—Hydroids, Jellyfish, Sea Anemones, Corals. Coelenterates inflict their injurious effects upon man by the use of their nematocyst apparatus. The venom is conveyed from the capsule of the nematocyst through the tubule into the tissues of the victim. Theoretically, any coelenterate equipped with a nematocyst apparatus is a potential stinger. The injurious effects may range from a mild dermatitis to almost instant death. The severity of the stinging is modified by the species of coelenterate, the type of nematocyst that it possesses, the penetrating power of the nematocyst, the area of exposed skin of the victim, and the sensitivity of the person to the venom.

Some of the sea anemones have been found to be poisonous to eat, particularly when raw. It is not known whether oral actinian intoxications are caused by their nematocyst poisons, or are the result of other noxious chemical substances contained in the tissues of their tentacles.

Stinging hydroids include certain species of the genera *Sarsia, Liriope, Halecium, Millepora, Gonionemus, Olindias, Olindioides, Pennaria, Physalia, Aglaophenia, Lytocarpus,* and *Rhizophysa*. The stings produced by these organisms vary from a mild stinging sensation to an extremely painful one (from *Olindias* and *Physalia*). There may be redness of the skin, urticarial rash, hemorrhagic zosteriform, or a generalized morbilliform rash, vesicle and pustule formation, and desquamation of the skin, abdominal pain, chills, fever, malaise, diarrhea, etc. These signs

Fig. 2. *Mytilus californianus* Conrad. Poisonous dinoflagellates are ingested by shellfish such as this species, which thereby become toxic (1.5 ×). (Photo: R. Kreuzinger.)

and symptoms may be accompanied by headache, malaise, primary shock, collapse, faintness, pallor, weakness, cyanosis, nervousness, hysteria, chills, fever, muscular cramps, abdominal rigidity, etc. Death may result in rare instances.

Stinging jellyfish include members of the genera *Aurelia, Carybdea, Cassiopea, Catostylus, Chironex, Chirodropus, Chiropsalmus, Chrysaora, Cyanea, Lobonema, Linuche, Lychnorhiza, Nausithoe, Pelagia, Rhizostoma, Sanderia,* and *Tamoya.* The stings from most of these jellyfish are usually relatively mild. *Cassiopea, Catostylus, Chrysaora, Cyanea, Rhizostoma,* and *Sanderia* may be moderate to severe, but *Chironex* and *Chiropsalmus* can be fatal. The effects in severe cases may consist of extremely painful localized areas of whealing, edema, and vesiculation, which later result in necrosis involving the full thickness of the skin. The initial lesions, caused by the structural pattern of the tentacles, are multiple linear wheals with transverse barring. The purple or brown tentacle arks form a whiplike skin lesion. Painful muscular spasms, respiratory distress, a rapid weak pulse, prostration, pulmonary edema, vasomotor and respiratory failure, or death may result. The pain is said to be excruciating, the victim frequently screaming and becoming irrational. Death may take place within 30 sec to 3 hr, but the usual time is less than 15 min. The cough and mucoid expectoration that are present in some of the other forms of jellyfish attacks are generally absent in *Chironex* and *Chiropsalmus* stings.

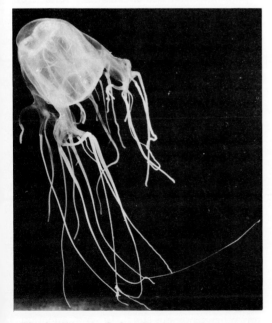

Fig. 3. *Chironex fleckeri* Southcott. This sea wasp is one of the most dangerous stingers in the sea (0.5 ×). (Photo: K. Gillett.)

Sea anemone stings tend to be more localized in their effects. There may be itching and a burning sensation at the sting site, accompanied by swelling and erythema, ultimately followed by local necrosis and ulceration. Severe sloughing of the tissues may occur, with a prolonged period of purulent discharge. Multiple abscesses have been reported. Localized symptoms may be accompanied by such generalized effects as fever, chills, malaise, abdominal pain, nausea, vomiting, headaches, a feeling of extreme thirst, and prostration. Sea anemone ulcers tend to be resistant to treatment and are slow to heal. As in the case of other types of coelenterate stings, they can be quite mild with little or no ill effects to the victim.

Coral cuts and stings are ill-defined problems. Although such cuts and ulcers are well known to most individuals working in tropical waters, the actual stinging ability of scleractinian or stony hexacorals is not well defined. The scleractinian corals are generally considered to be of minor significance among venomous coelenterates. However, there are a few genera that have members reputedly capable of stinging human beings, viz., *Acropora, Astreopora, Goniopora,* and *Plesiastrea.* The severity of coral lesions is probably due to a combination of factors: laceration of tissues by the razor-sharp exoskeleton of the coral, effects of the nematocyst venom, introduction of foreign materials into the wounds, secondary bacterial infection, and adverse climatic and living conditions. Coral ulcers are slow to heal.

Coral cuts should receive the following prompt treatment: cleansing the wound, removing foreign particles, debriding if necessary, and applying antiseptic agents. Considerable difficulty can be prevented by promptly painting coral abrasions with an antiseptic solution such as 2% tincture of iodine, etc. In severe cases, it may be necessary to give the patient bed rest with elevation of the limbs, kaolin poultices, magnesium sulfate in glycerin solution dressings, and antibiotics.

Jellyfish tentacles that are adhering to the skin of the victim should be immediately removed with the use of sand, clothing, bathing towel, seaweed, gunny sacks, or other available materials. This is one of the most important steps, because as long as the tentacles are on the victim's skin, they continue to discharge their venom. Alcohol, sun lotion, oil, or other readily available materials should be applied promptly to the wheals or skin lesions to inhibit the further activity of adherent microscopic nematocysts. Numerous local remedies have been advocated in various parts of the world: sugar, soap, vinegar, lemon juice, papaya latex, ammonia solution, sodium bicarbonate, plant juices, boric acid solution, etc. These have been used with varying degrees of success. Topical or oral cortisone preparations are sometimes useful. Oral antihistamines and topical antihistaminic creams alleviate urticarial lesions and symptoms. Opiates may be required to alleviate pain. Severe stings may require epinephrine (7 minims) subcutaneously, repeated as necessary. Intravenous hypertonic glucose solutions may also be useful. Muscular spasms can be relieved with the use of intravenous injections of 10 ml of 10% calcium gluconate or sodium amytal intravenously. Artificial respiration and oxygen may be required. A number of fatalities have occurred as a result of stings on the lower extremities and the immediate use of a tight tourniquet might save the life. Cardiac and respiratory stimulants and other supportive measures may be required. There are no known specific antidotes for coelenterate venoms.

The chemistry and pharmacology of coelenterate toxins are largely unknown. It has now been determined that the rapid death in *Chironex* stings is not due to anaphylactic shock, but rather to the direct action of the venom on the heart muscle (Endean, 1967).

Echinodermata—Starfish, Sea Urchins, Sea Cucumbers. Toxic substances have been reported as present in certain species of the asteroid genera *Aphelasterias, Asterias, Marthasterias, Asterina, Astropecten, Echinaster, Solaster,* and *Pycnopodia*. Poisonous starfish are believed to be toxic to eat. Contact with the slime of some species of asteroids may result in a contact dermatitis. In both cases the poison is thought to be produced by the glandular cells present in abundance in the epidermis of starfish. The chemical and pharmacological properties of these asterotoxins have not been fully determined, but at least one of these poisons resembles holothurin, which is found in certain species of toxic sea cucumbers. The treatment of biotoxications from starfishes is symptomatic.

Fig. 4. *Acanthaster planci* (Linnaeus). This starfish is equipped with venomous spines; Great Barrier Reef, Australia; diameter 15 in. (Photo: K. Gillett.)

Fig. 5. *Diadema antillarum* Philippi. The spines of this sea urchin are reputed to contain a poison and can inflict very painful wounds; West Indies; length of spines 12 in. (Photo: R. Straughan.)

Acanthaster planci is the only known venomous asteroid. The spines of this starfish are elongate, pungent, and covered by a venom-producing integument. The nature of the poison is unknown. Contact with the spines of *A. planci* may produce an extremely painful wound, redness, swelling, protracted vomiting, numbness, and paralysis.

Poisoning from sea urchins may result from ingestion of their gonads, as is the case in *Paracentrotus lividus. Tripneustes ventricosus,* and *Centrechinus antillarum.* However, in most instances sea urchin poisonings are due to stings from either their spines or pedicellariae.

The hollow, elongate, fluid-filled spines of echinothruid and diadematid sea urchins are particularly dangerous to handle. Their sharp, needlelike points are able to penetrate the flesh with ease, and these produce an immediate and intense burning sensation. As the spines penetrate, they release a violet-colored fluid, which causes discoloration of the wound. Intense pain is soon followed by redness, swelling, and aching sensations. Partial motor paralysis of the legs, slight anesthesia, edema of the face, and irregularities of the pulse have been reported. Secondary infection is a frequent complication with some species. The pain usually subsides after several hours, but the discoloration may continue for three to four days.

Insofar as the venom is concerned, sea urchin stings should be handled in a manner similar to any other venomous sting. However, attention is directed to the need for prompt removal of the pedicellariae from the wound, for when they are detached from the parent animal, they frequently continue to be active for several hours. During this time they will introduce venom into the wound.

The extreme brittleness and retrorse barbs of some sea urchin spines present an added mechanical problem. Nielly (1881) recommended that grease be applied, stating that this would allow the spines to be scraped off quite easily. Cleland (1912), Earle (1940), and others are of the opinion that some sea urchin spines need not be removed, as they are readily absorbed. Absorption of the spines is said to be complete within 24–48 hr. However, the spines of some sea urchins are not readily absorbed, and months later roentgenological examination may reveal them in the wound. It is recommended that the spines of *Diadema* be removed surgically.

Intoxications can reputedly result from ingestion of toxic sea cucumbers. The poison of sea cucumbers is termed holothurin, which is said to be concentrated in the organs of Cuvier. It is believed to be a steroidal glycoside having an empirical formula of $C_{50}H_{82}O_{26}S$. It appears to have a direct contractural effect on muscle. It also has a nerve-blocking effect similar to that of cocaine, procaine, and physostigmine in laboratory animals, but its effects on humans have not been fully determined.

Little information is available regarding the clinical effects of holothurin in humans. Reported symptoms of dermal contact with sea cucumber poison are burning pain, redness, and a violent inflammatory reaction. If the fluid contacts the eyes of the victim, blindness may result. Ingestion of sea cucumber poison may be fatal. Treatment is symptomatic, but pharmacological studies suggest that anticholinesterase agents may be effective in the event of ingestion of holothurin (Friess, 1963).

Mollusca—Snails, Bivalves, and Cephalopods. Whelk poisoning is caused by the ingestion of toxic univalves of the genus *Neptunea* and some of their close relatives. The poison is restricted to the salivary glands of the shellfish. The poison is believed to be tetramine, which is an autonomic ganglionic blocking agent. The symptoms consist of nausea, vomiting, anorexia, weakness, fatigue, faintness, dizziness, photophobia, impaired vision, and dryness of the mouth. Treatment is symptomatic.

Cone shell envenomations are caused by stings produced by univalves of the genus *Conus*. They inflict their stings by means of venomous radular teeth. The radular teeth originate in the radular sheath where they reside until used. When needed, a single tooth passes from the sheath through the pharynx where it is charged with venom that is produced in the venom duct and purveyed to the hollow supervoluted radular

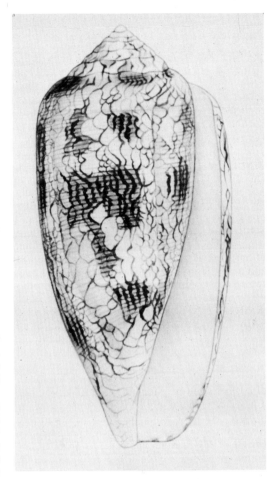

Fig. 6. *Conus textile* Linnaeus. Cone shells are equipped with venomous radular teeth (2 ×). (From Hiyama.)

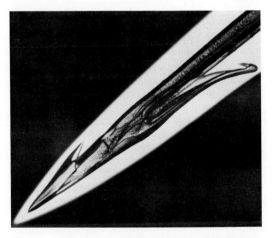

Fig. 7. Photograph showing enlarged view of tip of venomous radular tooth of *Conus striatus* Linnaeus. The tip of tooth is charged with venom (80 ×). (Photo: R. Kreuzinger.)

tooth under pressure by the muscular venom bulb. The tooth then passes from the pharynx into the anterior opening of the proboscis where it is held ready to be plunged into the flesh of the victim. Most cone stings result from the careless handling of the mollusks by curious shell collectors. The chemistry and pharmacology of cone shell venom has not been fully determined, but there is evidence that cone shell venoms may vary from one species to the next.

Stings produced by *Conus* are of the puncture wound variety. Localized ischemia, cyanosis, and numbness in the area about the wound, or a sharp stinging or burning sensation are usually the initial symptoms. The presence and intensity of the pain varies considerably with the individual. Some state that the pain is similar to a wasp sting, whereas others find it excruciating. Swelling of the affected part usually occurs. The numbness and parasthesias begin at the wound site and may spread rapidly involving the entire body, particularly about the lips and mouth. In severe cases paralysis of the voluntary muscles is initiated early, first by motor incoordination and followed by a complete generalized muscular paralysis. Knee jerks are generally absent. Aphonia and dysphagia may become very marked and distressing to the victim. Some patients complain of a generalized pruritus. Blurring of vision and diplopia are commonly present. Nausea may be present, but gastrointestinal and genitourinary symptoms are usually absent. The recovery period in less serious cases varies from a few hours to several weeks. Until fully recovered, victims complain of extreme weakness and tiring easily with the least amount of physical exertion. Coma may ensue, and death is said to be the result of cardiac failure. Treatment is symptomatic.

Mollusks of the genus *Murex* contain a poison in their purple gland that has been termed murexine. Little is known about murexine poisoning in man. Treatment is symptomatic.

Poisonings have been caused by the ingestion of cephalopods of the genera *Ommastrephes* and *Octopus* in certain areas of Japan (Kawabata, Halstead, and Judefind, 1957). The nature of the poisons involved is unknown, but there is no evidence that bacterial contaminants are involved. The predominant symptoms consist of nausea, vomiting, abdominal pain, diarrhea, fever, headache, chills, weakness, and severe dehydration. Paralysis and convulsions are sometimes present, but death is rare. Most victims recover within a period of 48 hr. Treatment is symptomatic.

Cephalopods inflict their envenomations with the use of a well-developed apparatus, the beak and salivary glands. The sharp parrotlike beak produces the initial wound into which is introduced the toxic saliva or venom, cephalotoxin.

Cephalopod lesions usually consist of two small puncture wounds produced by the sharp, parrotlike, chitinous jaws of the mollusk. Usually the pain is immediate and consists of a sharp burning or stinging sensation. It is sometimes described as similar to a bee sting, which at first is localized, but may later radiate to include the entire appendage. Within a few minutes a tingling or pulsating sensation develops in the area about the wound. There is some indication that coagulation time is retarded since bleeding is profuse and prolonged in most cephalopod bites. Swelling, redness, and heat usually develop about the wound. Some victims complain of an intense itching sensation about the affected area. Motor and severe sensory disturbances are generally absent. In severe cases there may be numbness of the mouth and tongue, blurring of vision, difficulty in speech and swallowing, loss of tactile sensation, floating sensation of the hands, etc. Muscular paralysis, loss of equilibrium, and deaths have been reported. Treatment is symptomatic.

The chemical composition of cephalotoxin is unknown, and the poisons may vary from one species to the next.

Toxic Vertebrates—Phylum Chordata

Agnatha—Lampreys and Hagfish. The slime and flesh of certain lampreys and hagfish are reported to produce a gastrointestinal upset, nausea, vomiting, and dysenteric diarrhea. The slime and skin is said to contain a poison that is not destroyed by either gastric juices or heat. Nothing else appears to be known regarding the nature of cyclostome poisons.

Condrichthyes—Sharks, Rays, Skates, and Chimeras. The musculature of some sharks, such as the Greenland shark *Somniosus microcephalus* (Bloch and Schneider), is said to be poisonous to eat (Jensen, 1914, 1948), but the livers of several species of tropical sharks may cause severe intoxication. The musculature may cause symptoms of a mild gastroenteritis. Ingestion of toxic shark livers may be very severe, with the onset of symptoms within a period of less than 30 minutes. Nausea, vomiting, diarrhea, abdominal pain, headache, weak pulse, malaise, cold sweats, oral paresthesia, burning sensation of the tongue, throat, and esophagus may be present. The neurological symptoms develop at a later time, consisting of extreme weakness, trismus, muscular cramps, sensation of heaviness of the limbs, loss of superficial reflexes, ataxia, delirium incontinence, respiratory distress, visual disturbances, convulsions, and death. The recovery period, if the victim recovers, varies from several days to several weeks. The mortality rate is not known. The severity of the symptoms varies with the amount of shark liver

TOXICOLOGY OF MARINE ANIMALS

Fig. 8. *Somniosus microcephalus* Bloch and Schneider. The flesh of this shark is sometimes poisonous to eat. Length is 3 ft or more. (From Halstead.)

eaten, the species of shark, physical condition of the victim, and other factors not yet clearly understood. The nature of the poison is unknown; it is not destroyed by heat or gastric juices.

Horn sharks inflict their envenomations by means of two dorsal stings adjacent to the anterior margins of each of the two dorsal fins. The nature of the venom is unknown. Symptoms consist of immediate intense stabbing pain, which may continue for several hours. Swelling and redness of the affected parts are usually present. Lesions are of the puncture wound variety. Deaths have been reported. See recommended treatment at end of this article.

Stingrays constitute the most important single group of venomous fish since they cause the largest number of serious stings. Most marine stingrays inhabit shallow coastal waters, bays,

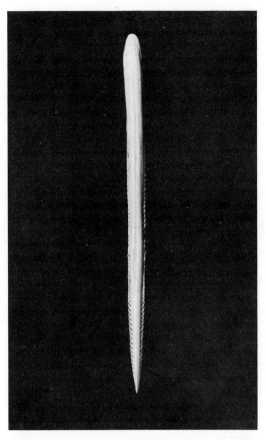

Fig. 10. Sting from a dasyatid stingray. The venomous spines of these rays can inflict a serious envenomation. (Photo: R. Kreuzinger.)

brackish water, lagoons, and river mouths. Venom is secreted and introduced into the body of the victim by the sting or venom apparatus located on the tail of the ray. The chemical nature of stingray venom has not been fully determined. The venom appears to be a toxic protein, which affects the cardiovascular, respiratory, and central nervous systems of mammals. Death is the result of cardiac standstill. The venom also produces respiratory depression (Russell, 1965).

The musculature and viscera of some of the chimeras or ratfish have been found to be toxic. However, aside from some vague references to the toxicity of these fish in humans, nothing is known of the nature of the poisons or the symptoms they produce. The reproductive organs have been reported to be poisonous. Chimeras are also capable of inflicting stings with their venomous dorsal spines, but the nature of the venom is unknown.

Osteichthyes—Ichthyotoxic Bony Fish. *Ciguatoxic Fish.* Ciguatera poisoning is one of

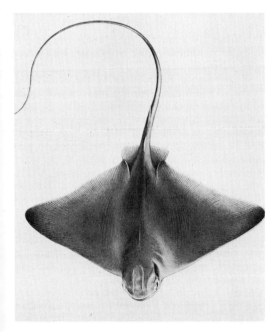

Fig. 9. *Myliobatis californicus* Gill. The base of the tail of this stingray is equipped with a venomous spine. Width is 36 in. (Photo: M. Shirao.)

681

the most treacherous and common forms of fish poisoning in tropical waters because it is usually caused by fish that appear to be edible and in most parts of the world are considered to be valuable food fish. About 300 species of marine fish have been included in this form of ichthyosarcotoxism; the most common are seabass, grouper, barracuda, snapper, parrotfish, wrasse, surgeonfish, and various other types of shore fish.

The symptoms consist of tingling about the lips, tongue, and throat, followed by numbness, which may develop immediately or at any time within a period of 30 hr after ingestion. The usual time interval for the development of symptoms is from 1–6 hr. The initial symptoms in some instances consist of nausea, vomiting, metallic taste, dryness of the mouth, abdominal cramps, tenesmus, and diarrhea followed by perioral tingling and numbness. The muscles of the mouth, cheeks, and jaws may become drawn and spastic with a feeling of numbness. Generalized symptoms of headache, anxiety, malaise, prostration, dizziness, pallor, cyanosis, insomnia, chilly sensations, fever, profuse sweating, rapid weak pulse, weight loss, myalgia, and joint aches are frequently present. Victims usually complain of a feeling of profound exhaustion and weakness. The feeling of weakness may become progressively worse until the patient is unable to walk. Muscle pains are generally described as a dull, heavy ache or cramping sensation, but on occasion may be sharp, shooting, and affect particularly the arms and legs. Victims complain of their teeth feeling loose and painful.

Visual disturbances consisting of blurring, temporary blindness, photophobia, and scotoma are common. Pupils are usually dilated and the reflexes diminished. Frequently reported are skin disorders, which are generally initiated by an intense generalized pruritus, followed by erythema, maculopapular eruptions, blisters, extensive areas of desquamation—particularly of the hands and feet—and occasionally ulceration. There may also be a loss of hair and nails.

In severe intoxications, the neurotoxic components are especially pronounced. Paresthesias involve the extremities, and paradoxical sensory disturbances may be present in which the victim interprets cold as a "tingling, burning, dry-ice or electric-shock sensation," or hot objects may give a feeling of cold. Ataxia and generalized motor incoordination become progressively worse. The reflexes are diminished and muscular paralyses develop. There may be clonic and tonic convulsions, muscular twitchings, tremors, dysphonia, dysphagia, coma, and death by respiratory paralysis. The limited morbidity statistics show a case fatality rate of about 7%. Death may occur within 10 min, but generally requires several days.

When the victim survives, recovery is slow and convalescence may be prolonged, extreme weakness, sensory disturbances, and excessive weight loss being the last symptoms to disappear. Complete recovery may require several years, as long as 25 years in some cases. Individuals who have been severely intoxicated have stated that during periods of stress, fatigue, exposure, or poor nutrition, there is a recurrence of the myalgia and joint aches similar to those suffered during the original acute period of the disease (Halstead, 1958, 1959, 1967).

Ciguatoxin is a complex poison that appears to have several fractions. There is a fat-soluble fraction, which is a light yellow, viscous, fatlike oil with an empirical formula approximately $C_{28}H_{52}NO_5Cl$. It is said to be a quarternary ammonium compound, and a positive ninhydrin test can be obtained on hydrolysis (Mosher, 1966). There is also a water-soluble fraction, which has been obtained from some species of ciguatoxic fish, but it is not known which fraction is responsible for the ciguatera syndrome in humans. One or more fractions of ciguatoxin may occur within a single species of fish. Unfortunately the pharmacology of ciguatoxin is not fully understood. There is evidence that at least one of the fractions of ciguatoxin is an irreversible anticholinesterase. However, other fractions showed no evidence of anticholinesterase. The chain of events leading to death by ciguatera fish poisoning are: (a) inhibition of cholinesterase, (b) acetylcholine accumulation, (c) disruption of nerve function—centrally, peripherally, or both (d) respiratory failure, and (e) death by asphyxia (Li, 1965). For recommended treatment of fish poisoning, see the end of this article.

Scombrotoxic Fish. Some of the scombroids (tuna, skipjack, and bonito) may on rare occasions cause ciguatera, but usually they produce an entirely different form of intoxication termed scombroid poisoning. This is the only type of fish poisoning in which bacteria appear to play an etiological role in the formation of the toxin. If scombroids are inadequately preserved, a toxic "histamine-like" substance is formed, possibly from the decarboxylation of histidine, a normal constituent of fish flesh. Victims complain of fish having a "sharp or peppery" taste. The symptoms most often present are nausea, vomiting, flushing of the face, intense headache, epigastric pain, burning of the throat, difficulty in swallowing, thirst, pruritus, swelling of the lips and urticaria, which are typical of a histamine reaction. Symptoms generally subside within 12 hr. For some unknown reason, scombroid fish appear to be more prone to producing intoxications of this type than other fish. The mortality rate is unknown. It is an established fact that when scombroid and certain other types of fish are left to stand at room temperature (about 15°C) for several hours, the histidine present in the musculature of the fish, due to the action of bacteria, undergoes decarboxylation and rapidly converts to histamine. The presence of histamine is used routinely in determining fish spoilage. For many years it was assumed that the toxic effects of scombroid poisoning were due primarily to histamine poisoning. However, experimental evidence has shown that the effects are due primarily to some other substance (Geiger, 1955) since histamine is generally ineffective when taken by mouth. Scombrotoxin is apparently produced by certain little-known strains of marine bacteria, possibly acting on the musculature of the fish. The biogenesis of the poison and its chemical structure are unknown. This poison has recently been termed "saurine" by Japanese investigators. It is insoluble in ether, acetone, benzene or chloroform, and possibly alcohol. Saurine is extracted with 80% methanol at room temperature. It is stable to boiling for 1 hr in 60% methanol with 1% concentrated HCl. Studies of the chemistry of this substance are being continued by the National Institute of Health in Japan.

Hallucinogenic Fish. This form of poisoning is caused by the ingestion of certain types of fish inhabiting the tropical Pacific, chiefly members of the genera *Mugil, Neomyxus, Mulloidichthys, Upeneus, Acanthurus,* or *Kyphosus.* The poison is heat-stable and is not destroyed by gastric juices. The symptoms may develop within minutes to 2 hr after ingestion, persisting for up to 24 hr. Symptoms consist of dizziness, loss of equilibrium, lack of motor coordination, hallucinations, and mental depression. A common complaint is that "someone is sitting on my chest." The conviction that they are going to die, or other terrible nightmares is a consistent part of the clinical picture. Other complaints consist of itching, burning of the throat, muscular weakness, and rarely abdominal distress. According to the reports received to date the intoxication is generally mild and nonfatal (Helfrich, 1961). The nature of the poison is unknown. Treatment is symptomatic.

Tetrodotoxic Fish. Tetraodon poisoning is one of the most violent forms of fish poisoning. It is produced by tetraodontoid or pufferlike fish. The disease is characterized by rapidly developing symptoms. The onset and symptoms in puffer poisoning vary according to the person and the amount of poison ingested. However, malaise, pallor, dizziness, paresthesias of the lips and tongue, and ataxia most frequently develop within 10–45 min after ingestion of the fish; but cases have been reported in which the symptoms did not develop for 3 hr or more. The paresthesias, which the victim usually

Fig. 11. *Sphaeroides annulatus* (Jenyns). This species is representative of a large number of puffers that contain violently toxic viscera. Length is 10 in. (Photo: Hiyama.)

describes as a "tingling or prickling sensation," may subsequently involve the fingers and toes, then spread to other portions of the extremities, and gradually develop into severe numbness, which may involve the entire body; patients have stated that it felt as though their bodies were "floating." Hypersalivation, profuse sweating, extreme weakness, precordial pain, headache, subnormal temperatures, decreased blood pressure, and a rapid, weak pulse usually appear early in the succession of symptoms.

Gastrointestinal symptoms of nausea, vomiting, diarrhea, and epigastric pain are sometimes present early in the disease, whereas in other cases they are totally lacking. Contradictory statements appear in the literature relative to pupillary changes, but these differences can probably be resolved on the basis of the time at which the examination is made. Apparently the pupils are constricted during the initial stage and later become dilated. As the disease progresses the eyes become fixed and the pupillary and corneal reflexes are lost.

Shortly after the development of the paresthesias, respiratory symptoms become a prominent part of the clinical picture. Respiratory distress, increased rate of respiration, movements of the nostrils, and diminution in depth of respiration are generally observed. Respiratory distress later becomes very pronounced, and the lips, extremities, and body become intensely cyanotic. Petechial hemorrhages involving extensive areas of the body, blistering, and subsequent desquamation have been reported. Severe hematemesis has also been known to occur. Muscular twitching, tremor, and incoordination become progressively worse and finally terminate in an extensive muscular paralysis. The first areas to become paralyzed are usually the throat and larynx, resulting in aphonia, dysphagia, and later complete aphagia. The muscles of the extremities become paralyzed and the patient is unable to move.

As the end approaches, the eyes of the patient become fixed and glassy, and convulsions may occur. The victims may become comatose, but in most instances they retain consciousness and their mental faculties remain acute until shortly before death. Death results from a progressive ascending paralysis involving the respiratory muscles. On the basis of Japanese statistics, the case fatality rate is 61.5%. If death occurs, it generally takes place within the first 6 hr, or within 24 hr at the latest. The prognosis is said to be good if the patient survives for 24 hr.

Puffer poison has been given greater attention by pharmacologists than any other fish poison. Studies have shown that the primary action of puffer poison or tetrodotoxin is on the nervous system, producing both central and peripheral effects. Comparatively low doses of the poison will readily inhibit neuromuscular function. Major effects include respiratory failure and hypotension. It is believed that puffer poison has a direct action on respiratory centers. There are little or no effects on gut activity. The retching and vomiting observed in tetradon poisoning are believed to be elicited by action of the compound on the chemoreceptive trigger zone of the area postrema.

Puffer poison has no major direct cardiotoxic effects, but the poison still causes significant decreases in cardiac contractile force associated with hypotension. Japanese workers have reported that tetrodotoxin depresses the vasomotor centers. The finding that blockage of peripheral sympathetic fibers may contribute to the hypotension suggests that the negative inotropic effect observed in the intact animal is caused largely by indirect actions of puffer poison. It is believed that the neuromuscular paralysis may be due to inhibition of conductivity (Katagi, 1927). Murtha (1960) gives the intraperitoneal LD_{50} for mice as 0.02 µg/g, killing the mice within 3.4 to 3.7 min after injection.

The chemistry of tetrodotoxin has been under study by Japanese chemists for several decades. This subject has been completely reviewed by Halstead (1965). Chemists have isolated and identified a decomposition product of the toxin believed to be of significance in establishing the formula of the poison. The product obtained was by alkaline hydrolysis of puffer poison and was shown to have the structure of 2-amino-6-hydroxymethyl-8-quinazolinol, with oxalic acid as a byproduct. The poison reacts with water under mild conditions to yield a substance with the empirical formula $C_{11}H_{19}O_9N_3$. This derivative has acid properties and has been termed tetrodonic acid. This acid yields a hydrobromide salt on treatment with hydrobromic acid, a molecule of water being released.

Ichthyootoxic Fish. Most ichthyootoxic fish are fresh water species and do not come within the scope of this presentation. One of the few marine fish offenders is the cabezone *Scorpaenichthys marmoratus,* found along the California coast. Symptoms develop soon after ingestion of the roe, and consist of abdominal pain, nausea, vomiting, diarrhea, bitter taste, dryness of the mouth, intense thirst, cold sweats, rapid irregular weak pulse, pupillary dilatation, syncope, chest pain, pallor, dysphagia, and tinnitus. In severe cases there may be muscular cramps, convulsions, and coma. The victim usually recovers within a period of 5 days. No deaths have been reported from eating marine ichthyootoxic fish, but deaths have been caused by some of the fresh water species. The nature of most ichthyootoxins is unknown.

Ichthyocrinotoxic Fish. The subject of ichthyocrinotoxic fish is one of the least known areas of marine zootoxicology. Ichthyocrinotoxins are produced by specialized glandular structures, but there are no traumagenic organs present for injecting the poisons. Ichthyocrinotoxins are generally excreted by skin glands and thereby released into the environment. The victim is intoxicated by ingesting the fish, by coming in contact with the slime of the fish, or by ingestion of water containing the poison. Fish that have been identified as ichthyocrinotoxic include such groups as lampreys, hagfish, moray eels, seabass, triggerfish, puffers, trunkfish, etc. It should be kept in mind that ichthyocrinotoxic fish may also contain one or several other different types of ichthyotoxins.

Very little has been reported regarding the actual mechanism by which these poisons are encountered by man. The slime of some species of cyclostomes is toxic to ingest and may produce an inflammatory reaction if brought in contact

Fig. 12. *Scorpaena guttata* Girard. The dorsal spines of the California scorpionfish are venomous. Length is 12 in. (Photo: M. Shirao.)

with the mucous membranes of humans. The skin of certain species of moray eels and puffers is poisonous to eat. Dermal contact with the slime of *Rypticus saponaceus* and probably other species of ichthyocrinotoxic fish may produce a dermatitis. Whether the poisons produced in the skin of these fish are chemically identical with those found in the flesh of these same fish is not known. The skin secretions of only two ichthyocrinotoxic fish have been chemically identified thus far. One of these is ostracitoxin (pahutoxin) from *Ostracion lentiginosus*. The poison has been identified as choline chloride ester of 3-acetoxyhexadecanoic acid. The biological activity of ostracitoxin resembles that of the steroidal saponins isolated from echinoderms. A second ichthyocrinotoxin that has been studied to some extent is derived from the toxic skin secretions of the soapfish *Rypticus saponaceus*. The skin of this fish releases a foamy toxic secretion when it is disturbed. The poison is believed to be a protein or polypeptide.

Tetrodotoxin has been studied at length, but it is not known whether tetrodotoxin obtained from the viscera is chemically identical with the poison produced by the skin of these fish. For a complete review of this subject, see Halstead (1967).

Treatment of Fish Poisoning

The treatment of oral fish poisonings is largely symptomatic. There are no specific antidotes and an attack does not impart immunity. Gastric lavage and catharsis should be instituted as soon as possible. In many instances 10% calcium gluconate given intravenously has given prompt relief, whereas in others it has been ineffective. Paraldehyde and ether inhalations have been reported useful in controlling the convulsions. Nikethamide or one of the other respiratory stimulants is advisable in cases of respiratory depression. Where excessive mucous production is a factor, aspiration and constant turning are essential. Atropine has been found to make the mucus more viscid and difficult to aspirate, and is not recommended. Oxygen by inhalation and intravenous administration of fluids supplemented with vitamins given parenterally are usually beneficial. If laryngeal spasm is present, intubation and tracheotomy may be necessary. In case of severe pain, opiates such as morphine, given in small divided doses will probably be required. Cool showers have been found to be effective in relieving severe itching.

Patients suffering from the paradoxical sensory disturbance should be given fluids slightly warm or at room temperature, as well as vitamin B complex supplements. Antihistaminic drugs will be found to be useful in the treatment of scombroid poisoning. Banner et al. (1965) have recommended the use of 2-PAM (2-pyridine aldoxime methochloride) for the treatment of ciguatoxications, but caution is advisable in employing any anticholinesterase drug in the routine therapy of ciguatera fish poisoning since ciguatoxin appears to be a complex of poisons having multiple actions.

Efforts in treating venomous fish stings should be directed toward achieving three objectives: (1) alleviating pain, (2) combating effects of the venom, and (3) preventing secondary infection. The pain results from the effects of the trauma produced by the fish spine, venom, and the introduction of slime and other irritating foreign substances into the wound. In the case of stingray and catfish stings, the retrorse barbs of the spine may produce severe lacerations with considerable trauma to the soft tissues. Wounds of this type should be promptly irrigated, or washed out with cold salt water or sterile saline if such is available. Fish stings of the puncture-wound variety are usually small in size, and removal of the poison is more difficult. It may be necessary to make a small incision across the wound and then apply immediate suction, and possibly irrigation. At any rate, the wound should be sucked promptly in order to remove as much of the venom as possible. However, it should be kept in mind that fish do not inject their venom in the manner employed by venomous snakes, so at best results from suction will not be too satisfactory.

There is a division of opinion as to the advisability and efficacy of using a ligature in the treatment of fish stings. If used, the ligature should be placed at once between the site of the sting, and the body, but as near the wound as possible. The ligature should be released every few minutes to maintain adequate circulation. Most doctors recommend soaking the injured member in hot water for 30–60 min. The water should be maintained at as high a temperature as the patient can tolerate without injury, and the treatment should be instituted as soon as possible. If the wound is on the face or body, hot moist compresses should be used. The heat may have an attenuating effect on the venom since boiling readily destroys stingray venom *in vitro*. The addition of magnesium sulfate or epsom salts to the water is believed to be useful. Infiltration of the wound area with 0.5–2% procaine has been used with good results. If local measures fail to prove satisfactory, intramuscular or intravenous demerol will generally be efficacious. Following the soaking procedure, debridement and further cleansing of the wound may be desirable. Lacerated wounds should be closed with dermal sutures. If the wound is large, a small drain should be left in it for 1–2 days. The injured area should be covered with an antiseptic and sterile dressing.

Prompt institution of the recommended treatment usually eliminates the necessity of antibiotic therapy. If delay has resulted to any extent, the administration of antibiotics may be desirable. A course of tetanus antitoxin is advisable as a precautionary measure.

The primary shock which follows immediately after the stinging generally responds to simple supportive measures. However, secondary shock resulting from the action of stingray venom on the cardiovascular system requires immediate and vigorous therapy. Treatment should be directed toward maintaining cardiovascular tone and the prevention of any further complications. Respiratory stimulants may also be required. The Commonwealth Serum Laboratories, Melbourne, New South Wales, Australia, have recently developed an antivenin for the treatment of stonefish *Synanceja* stings (Wiener, 1959).

Toxic Reptiles

Five species of marine turtles have been identified as toxic to humans. They are members of the genera *Caretta, Chelonia, Eretmochelys, Dermochelys,* and *Pelochelys*. The symptoms of chelonitoxication vary with the amount of flesh ingested and the person. Symptoms generally develop within a few hours to several days after eating the turtle. The initial symptoms usually consist of nausea, vomiting, diarrhea, facial tachycardia, pallor, severe epigastric pain, sweating, coldness of the extremities, and vertigo. There is frequently reported an acute stomatitis, consisting of a dry burning sensation of the lips, tongue, lining of the mouth, and throat. Some victims complained of a sensation of tightness of the chest. The victim frequently becomes lethargic and unresponsive. Swallowing becomes very difficult, and hypersalivation is pronounced. The oral symptoms may be slow to develop but become increasingly severe after several days. The tongue develops a white coating, the breath becomes foul, and later the tongue may become covered with multiple pinhead-sized, reddened papules. The pustules may persist for several months, whereas in some instances they break down into ulcers. Desquamation of the skin over most of the body has been reported (Cooper, 1964). Some victims develop a severe hepatomegaly with right upper quadrant tenderness. The conjunctivae become icteric. Headaches and a feeling of "heaviness of the head" are frequently reported. Deep reflexes may be diminished. Somnolence is one of the more pronounced symptoms of severe intoxications and usually indicates an unfavorable prognosis. At first the victim is difficult to awaken and then gradually becomes comatose and soon dies. The symptoms are typical of a hepatorenal death. The over-all case fatality rate on reported outbreaks is about 28%. Treatment is symptomatic. The nature of chelonitoxins is unknown.

There are more than 50 species of venomous sea snakes, but only 14 species have been identified as toxic to humans. Those most frequently involved are *Enhydrina schistosa* (Dandin), *Hydrophis cyanocinctus* Daudin, *H. spiralis* (Shaw), *Kerilia jerdoni* (Gray), and *Pelamis platurus* (Linnaeus). Sea snakes inflict their wounds with the use of fangs that are reduced in size but are of the Elapine or cobra type. There is no clinical evidence of direct cardiovascular involvement. Failing vision is considered to be a terminal sign. The generalized pains resulting from muscle movements and the myoglobinuria are said to be the oustanding clinical signs of sea snake envenomation. In fatal cases, respiratory paralysis with terminal hypertension and cyanosis precede death. Little or no peripheral paralysis may be evident, and the victim dies from bulbar paralysis. Death may take place within several hours or several days after the bite.

Treatment. Sea snake poisoning is a medical emergency requiring immediate attention and the exercise of considerable judgment. Tragic consequences may result from delayed or inadequate treatment. Before any first aid or therapeutic measures are instituted, it is always important to determine if envenomation has occurred. Needless treatment can cause discomfort and deleterious results. A sea snake may bite without injecting venom. The fangs and teeth of a sea snake are generally small and may not have penetrated the skin sufficiently to have resulted in envenomation.

Absorption of sea snake venom is rapid. In most instances the venom is absorbed before first aid can be administered. Suction is of value

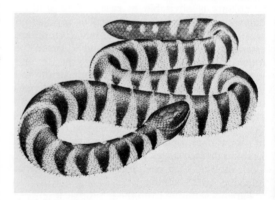

Fig. 13. *Lapemis hardwicki* Gray. Most sea snakes are docile, but some of the species are aggressive and can inflict fatal envenomations. Length is 30 in. (Photo: Maki.)

only if it can be applied within the first few minutes following the bite. Incision and suction are said to be of little value in sea snake envenomations. It is generally advisable to leave the bite alone. The affected limb should be immediately immobilized and all exertion must be avoided. The patient should lie down and keep the immobilized part below the level of the heart. A tourniquet should be applied tight enough to occlude the superficial venous and lymphatic return. Apply the tourniquet to the thigh in leg bites or to the arm above the elbow in upper limb bites. It should be released for 90 sec every 10 min. A tourniquet is of little value if applied later than 30 min following the bite, and it should not be used for more than 4 hr. The tourniquet should be removed as soon as antivenin therapy has been started. Some workers believe that the tourniquet is of little or no value to sea snake bites. If sea snake antivenin or a polyvalent antivenin containing a krait (Elapidae) fraction is available, it should be administered intramuscularly either in the buttocks or at some other site distant from the bite. The antivenin should be given only after the appropriate skin or conjunctival test has been made. Usually one unit (vial or ampule) is sufficient until the patient can be transported to a physician. Keep the patient warm. He should not be given alcoholic beverages of any kind. Transport the patient to a physician. For further information see Halstead (1959, 1968).

BRUCE W. HALSTEAD, M.D.

References

Akiba, T., and Hattori, Y., "Food poisoning caused by eating asari (*Venerupis semidecussata*) and oyster (*Ostrea gigas*) and studies on the toxic substance venerupin," *Japan. J. Exp. Med.*, 20, 271–284 (1949).

Boylan, D. B., and Scheuer, P. J., "Pahutoxin: a fish poison," *Science*, 155(3758), 52–56 (1967).

Cooper, M. J., "Ciguatera and other marine poisonings in the Gilbert Islands," *Pacific Sci.*, 18(4), 411–440 (1964).

Fish, C. J., and Cobb, M. C., "Noxious marine animals of the central and western Pacific," 1954, U.S. Fish Wildlife Serv., Res. Rept. No. 36.

Geiger, E., "On the specificity of bacterium-decarboxylase," *Proc. Soc. Exp. Biol. Med.*, 55, 11–13, (1944).

Gudger, E. W., "A new purgative, the oil of the 'castor oil fish,'" *Boston Med. Surg. J.*, 192, 107–111 (1925).

Gudger, E. W., "Poisonous fishes and fish poisoning, with special reference to ciguatera in the West Indies," *Am. J. Trop. Med.*, 10, 43–55 (1930).

Halstead, B. W., "Animal phyla known to contain poisonous marine animals," in E. E. Buckley and N. Porges (Eds), Venoms, 1956, Am. Assoc. Adv. Sci., Washington, D.C.

Halstead, B. W., "Poisonous fishes," *Public Health Repts.*, 73(4), 302–312 (1958).

Halstead, B. W., "Dangerous Marine Animals," 1959, Cornell Maritime Press, Cambridge.

Halstead, B. W., "Fish poisonings—their diagnosis, pharmacology, and treatment," *Clin. Pharmacol. Therap.*, 5(5), 615–627 (1964).

Halstead, B. W., "Poisonous and Venomous Marine Animals of the World," Vol I, "Invertebrates," 1965, U.S. Government Printing Office, Washington, D.C.

Halstead, B. W., "Poisonous and Venomous Marine Animals of the World," Vol. II, "Vertebrates," Part I, 1968, U.S. Government Printing Office, Washington, D.C.

Halstead, B. W., "Poisonous and Venomous Marine Animals of the World," Vol. III, "Vertebrates," Part II, in press, U.S. Government Printing Office, Washington, D.C.

Maretzki, A., and del Castillo, J., The toxin of soapfish (*Rypticus saponaceus* Bloch and Schneider), 1965, unpublished.

Murtha, E. F., "Pharmacological study of poisons from shellfish and puffer fish," *Ann. N.Y. Acad. Sci.*, 90(3), 820–836 (1960).

Rodahl, K., "Toxicity of polar bear liver," *Nature*, 164(4169), 530–531 (1949).

Saunders, P. R., "Venoms of scorpionfishes," *Proc. West. Pharm. Soc.*, 2, 47–54 (1959).

Saunders, P. R., "Pharmacological and chemical studies of the venom of the stonefish (genus *Synanceja*) and other scorpionfishes," *Ann. N.Y. Acad. Sci.*, 90(3), 784–804 (1960).

Cross references: *Biotoxins, Marine; Dinoflagellates; Sea Turtles.*

TRUST TERRITORY OF THE PACIFIC ISLANDS

Occupying 3,000,000 miles^2 of the western Pacific in an area that lies just above the equator is the Trust Territory of the Pacific Islands. Its western boundary is west of Tokyo and its eastern boundary is 3,000 miles to the east. Within this area lies what is known as Micronesia, which includes 2141 islands, excluding the Gilbert Islands, which are under British control and thus are not covered in this article. Most of these are small atolls only a few feet above the sea, but there are also large islands that rise as high as 2595 ft. The largest island in the Trust Territory, Babelthuap in Palau, is 25 miles long and about 8 miles wide, but the total land area of the Trust Territory is only 700 miles.^{2} Only 96 islands are inhabited by the 93,000 Micronesians living in this ocean area, which is approximately the size of the continental U.S. The Trust Territory is comprised of several groups of islands, each group having a large number of islands, usually scattered over a vast area. These groups include

TABLE 1. JAPANESE SKIPJACK IN THE MANDATED ISLANDS 1930–1949 (METRIC TONS)*

Year	Saipan	Yap	Truk	Ponape	Jaluit	Palau	Total
1930	258.0	.9	913.4	6.4	N/A	157.1	1,335.8
1931	564.3	.4	1,097.1	525.2	81.3	548.1	2,816.4
1932	1,309.7	N/A	810.3	534.2	614.8	1,592.3	4,861.3
1933	1,762.3	N/A	1,883.4	926.9	172.4	2,144.5	6,889.5
1934	2,516.0	4.2	1,200.0	1,202.5	255.1	3,778.7	8,956.5
1935	1,786.0	N/A	3,002.4	1,313.1	229.8	5,391.0	11,722.3
1936	1,696.0	N/A	5,870.2	2,695.8	167.7	3,836.0	14,265.7
1937	2,697.3	N/A	12,433.5	4,064.0	91.3	13,774.7	33,060.8
1938	2,392.0	149.3	5,294.8	1,495.6	6.7	3,420.2	12,758.6
1939	2,087.0	36.1	7,640.0	3,707.8	N/A	3,584.2	17,019.7
1940	3,379.1	3.6	7,217.1	1,586.3	1.5	6,047.4	18,235.0

* Source: "The Japanese Tuna Fisheries," 1948, U.S. Department of the Interior, Fish and Wildlife Service, Fishery Leaflet No. 297.

the Marshalls, the Carolines, and the Marianas. Truk lies in the approximate center of the Trust Territory and during the war was considered Japan's strongest Pacific base. Guam, in the Marianas, has long been a territory of the U.S. and is the most modern of all the islands in the western Pacific.

Considering the Trust Territory islands as a whole, several similar characteristics are found. These include large, relatively shallow lagoons (150 ft), small island masses within the lagoons or on the reef perimeter, as is the case with the atolls, and abrupt drop-offs from the edges of the fringing reefs into the abyssal depths.

A number of these lagoons are very large in total area. Truk, for example, as an enclosed lagoon area of over 1200 miles^2, and Kwajalein's lagoon is approximately the same size. Normally lagoons are not considered assets, but with growing knowledge of how to live in and harvest the seas, it is probable that these large shoal areas, adjoining the abyssal depths, will soon become one of the greatest resources the islands have to offer.

The Japanese Period

The Japanese recognized the potential of the Trust Territory and started developing it shortly after seizing the islands from Germany at the beginning of World War I. During this period it was known as the Mandated Islands. One of the first moves made by Japan in the 1920's was to start an investigation of the skipjack and longline fisheries. Commercial skipjack fishing at Palau was started in 1925, but production did not become sizable until the 1930's; in 1937 production reached its peak when over 33,000 metric tons of skipjack were landed in the Trust Territory. Because of the great quantity of fish produced in 1937, the market for katsuobushi (dried fish stick) dropped and the number of operating boats declined. It is reported that there were a total of 147 skipjack boats licensed in the Trust Territory.

It should be noted that 1937 was unquestionably an unusual year, as the entire Pacific basin had unusually good catches that year. Decline in production has been attributed to the return of normal fishing conditions, and to the Japanese-Chinese war, which resulted in scarcity of fuel, engine parts, labor, etc. (Table 1).

Longline fishing operations were also carried on from various bases in the Trust Territory during the prewar years. Most of these catches were landed in Japan because of the higher price and greater demand. Table 2 shows the number of longliners based in the islands during 1940.

Table 3 shows the landings of tuna, excluding skipjack, in the Trust Territory from 1930 to 1940.

Current Offshore Tuna Fisheries

In 1964 Van Camp Sea Food Co., with head offices in Long Beach, California, set up the first full-scale commercial fishing operation based in the Trust Territory since the Japanese were forced out at the end of the war. Van Camp brought in 12 new Okinawan skipjack boats manned with 12 Okinawans on each boat, many

TABLE 2. JAPANESE LONGLINERS BASED IN THE TRUST TERRITORY, 1940[a]

Palau	14[b]
Ponape	5
Saipan	2
Truk	1[c]

[a] Data compiled by Japanese Tuna Fishermen's Association, in "The Japanese Tuna Fisheries," 1948, U.S. Dept. of Interior, Fish & Wildlife Service Fishery Leaflet No. 297.
[b] All longliners reported about 40 tons.
[c] Possible errors, Trukese report a greater number working from Dublon.

TABLE 3. TUNAS, EXCLUDING SKIPJACK LANDED IN THE TRUST TERRITORY, 1930–1940 (METRIC TONS)*

Year	Saipan	Palau	Truk	Ponape	Jaluit	Total
1930	4.53	92.26	8.53	3.54	2.37	111.23
1931	16.73	156.61	29.43	4.83	3.85	211.45
1932	48.24	137.62	5.18	34.69	135.72	361.45
1933	.31	242.23	55.39	41.42	25.87	365.22
1934	27.26	278.88	55.39	26.49	31.36	419.40
1935	42.92	301.18	98.50	23.50	13.91	480.01
1936	151.02	213.26	178.02	29.96	14.85	587.11
1937	88.88	189.78	342.18	56.37	3.96	681.17
1938	33.94	73.13	101.44	60.21	ND	268.72
1939	34.88	188.94	93.60	31.58	55.14	354.14
1940	84.51	686.57	46.62	17.31	7.97	842.98

* Source: "The Japanese Tuna Fisheries," 1948, U.S. Dept. of Interior, Fish & Wildlife Service, Fishery Leaflet No. 297.

of whom had fished in the islands prior to the war. Eight Micronesians were also placed on each boat to learn from the Okinawans how to fish skipjack.

In addition, the government sent a number of selected Micronesians to Hawaii to work on the skipjack boats operating there. During the initial training period a boatyard was put up in Palau, and a Hawaiian style tuna boat was built to be fished by the Micronesians returning from Hawaii at the end of their training period. The objective of this program was to see if by providing Micronesians with a vessel designed to be operated by half the number of fishermen (10 vs. 20) it might not be possible to encourage more Micronesians to take up commercial fishing as a full-time occupation. This program is still in operation, and while promising results are being obtained, it is still too early to draw any definite conclusions.

While no other fishing bases have been opened elsewhere in the Trust Territory, the results of Van Camp's operations in Palau are promising. Their annual catches since starting operations are shown in Table 4.

As the world demand for canned tuna increases, it seems probable that Van Camp will increase their effort in Palau and that other packing companies will follow their lead and set up operations in Truk, Ponape, and possibly Saipan and the Marshall Islands.

While the Japanese were not able to re-establish their shore-side fishing bases in the Trust Territory, they rapidly developed the longline fishery in the Trust Territory, to where, by 1962, they were producing approximately 50,000 tons of

Fig. 1. Six of the 12 Okinawan skipjack boats now fishing for Van Camp in Palau.

TABLE 4. ANNUAL SKIPJACK LANDINGS—PALAU
1964–1966 LB

1964 (5 months)	2,664,071
1965	6,007,618
1966	6,181,281

tuna from Trust Territory waters. These fish are landed in Japan and either consumed there or exported to the U.S. where they are canned for domestic or export use.

Early Research

The Fourth Session of the Pacific Science Congress recommended the establishment of a science laboratory in the Palau Islands. This was done, and in 1931 the Palau Tropical Biological Station began operation. The laboratory had a fishing section, a processing technology section, and an aquaculture section. It operated 3 boats, 2 10-tonners and the 180-ton steel vessel with a 360 hp diesel engine. This laboratory carried out a varied research program during its existence, which included experimental longlining for yellowfin and other tuna, marketing research, skipjack fishing in local waters, pearling ground surveys, guidance to native fishermen, katsuobushi processing experiments and production surveys, skipjack canning experiments, transplantation of trochus and sea cucumbers, basic biological studies, and oceanographic surveys.

Culture of Pearls and Mother of Pearl

While skipjack fishing was the biggest Japanese marine resource, other specialized fisheries were of importance, as many of them were in the developmental stages and have during recent years achieved considerable economic importance to Japan and other countries where production techniques have been constantly studied and improved. The pearl culture industry is an excellent example of this. In 1920 Mikimoto set up an experimental pearl culture station in Palau. He used the locally available *Pinctada margaratifera* or black lip pearl shell for culturing the pearls and later he extended his work to include the yellow-lipped pearl oyster, *Pinctada maxima*, which he had to import by pearl luggers from the Arafura Sea. In 1935 and 1936 Mikimoto successfully transported *Pinctada martensii*, the Japanese pearl oyster, to Palau. These experiments proved successful and excellent pearls of fine luster and color were produced.

Mikomoto's operations apparently induced imitators as two other companies began operations in Palau, these being the South Seas Pearl Co. in 1936 and the Horiguchi Pearl Trading Company (Horiguchi Shinju Boeki KK) in 1937. These companies remained in business until 1941, till the outbreak of World War II.

The Palau pearls were reported to reach a marketable size in a period of 3 years; as many as 5 years are required in the colder northern waters of Japan where the growth rates are slower. The Japanese apparently believe that *P. martensii* produces pearls of higher quality than does *P. maxima* or *P. margaratifera,* but that they are not comparable in size. Also, Palau is reportedly one of the few places in the world that can produce a "black pearl." This color and type is much sought after and is still, apparently, not generally available.

Whether the inshore waters of Palau are suited for large-scale pearl production is still to be determined. Certain reports indicate that Palau is an ideal location, another states that the prewar pearl enterprises were not highly successful commercially and that this was apparently due to unfavorable environmental conditions. On the island of Ebon in the Marshalls, Shinju Kabushiki Kaisha of Tokyo started a pearl culture experimental operation that carried on from 1935 until 1942 when the plantings were abandoned. The results of this operation are unknown, but it was patterned after the Palau operation. *P. margaratifera* and *P. maxima* approximately 6 in. in diameter were used for culturing the pearls.

During the pre-war years, the mother of pearl fishery for trochus (*Trochus niloticus*), was a most important fishery. These shells were used for manufacturing high quality buttons in Japan. Investigations started by the Japanese in the early 1920's showed that trochus was established only in Palau and Yap. Because of the value of this shell, the Japanese started a program of introductions and 6724 shells were transplanted in 5 years, during 1927 to 1931, from Palau to Truk. Additional transplants were made to Saipan and Puluwat in 1937 and 1938 and to Ponape in 1939 from Truk. Jaluit also received 1540 animals from Truk this year. Available records do not show how trochus came to be established elsewhere in the Marshalls, but it was harvested there in quantity until the price began to drop in 1956.

Japanese production figures show that approximately 100–200 tons of trochus shells was produced annually during the prewar years. Since 1953 annual production has varied between 100–400 tons. Before the market collapsed, trochus provided the Micronesians of the Trust Territory a very important source of income. The abrupt decline in value after 1956 was a result of the gains made by the plastics industry in manufacturing buttons at a much lower price and of nearly comparable quality. The drop in value from $358,342 in 1956 to $34,323 in 1965 shows just how much this fishery has been hurt. The future of trochus in the Trust Territory is dim and it is probable that during the coming years its use will be confined to the manufacture of handicraft items and jewelry from the polished shell.

Although the black lip pearl oyster *Pinctada margaratifera* is widely distributed throughout the Trust Territory and has been taken at Saipan, the Palau islands, Ponape, Kapingamarangi, Nukuoro, and Likiep, it does not appear to be abundant in any of these areas. Since the war, there has been very little work done with this species and at present, very little is known about its distribution and abundance. According to one report, the Japanese believed that limited commercial production might be possible in Palau, Truk, Ponape, and elsewhere.

The black lip oyster occurs normally in waters from 2–30 m in depth. Where abundant, they can be found anywhere on the reef flats exposed at low tide, but they also occur below low water, attached to coral or rubble. The gold lip pearl shell was reported to be abundant in Nukuoro before the war by divers who used to free dive as deep as 30 fathoms to get them. These shells were reportedly sent to Yap where they had considerable value as local money. Now, however, there are reportedly no longer available.

Sea Cucumbers (Holothurians)

Holothurians formed a small fishery of considerable importance during the prewar years. These animals are elongate and cylindrical and are very common on reefs and sand flats. It is difficult to tell the head from the tail and many species have no really noticeable appendages. In feeding, the sea cucumber ingests a large volume of sand. This passes through its body and out the anus as a column, edible particles being digested out of the sand during its passage through the body, which is flexible, tough, and often thick.

The larger holothurians occur in greatest abundance in shallow tropical reef areas and for centuries have been the object of profitable fisheries for food. In the Indo-Pacific area they are often called "trepang," while in French Oceania they are referred to as "beche-de-mer."

It is estimated that there are over 50 different species in the Trust Territory. Of these only about a dozen are used in the manufacture of trepang. Because of the abundance of these species, the great number of islands in the Trust Territory, and the ease of processing the live animals into a marketable product, there is very likely a potential commercial fishery for this species on every island with a moderately extensive lagoon and surrounding reef flat. It is hoped that it will be possible to revive it during the next few years.

Prices paid for trepang, which is eaten by shaving off thin slices for flavoring in soup, are based on quality, species, and size. A Japanese purchaser reports that prices vary between $0.25 and $0.50 per pound. Unconfirmed market reports also indicate that the price may now go as high as $0.75 per pound for top quality species.

A review of the reported exports of trepang from the Trust Territory during 1941 gives a good idea of the future potential of this resource, which is now lying dormant.

TABLE 5. TREPANG EXPORTS—1941*

Saipan	119,673 lb	Truk	1,142,779
Yap	68,952	Ponape	201,784
Palau	341,244		
	Total Trepang Export 1941		1,874,432 lb
	Estimated price, 1967		$.030/lb
	Estimated Value		$562,329.60

* Source: "Survey of the Fisheries of the Former Japanese Mandated Islands," 1947, U.S. Department of Interior, Fish and Wildlife Service Fishery Leaflet 273.

Marine Turtles

The two most common sea turtles in the Trust Territory are the Green Sea Turtle (*Chelonia mydas*) and the Hawksbill (*Eretmochelys imbricata*). These turtles are easily distinguishable by the plates or scutes on the shell. The Hawksbill has overlapping scutes and the Green Turtle has none. The scutes of the Hawksbill, along with the plastron or ventral surface of both turtles, is sought after by the Japanese for the manufacture of such items as combs, jewelry boxes, earings, rings, cuff links, tie pins, etc. The Japanese market is most interested in obtaining the yellow shell with only a very few red spots as this type is considered the best quality. Within the Trust Territory the people price the turtles as a source of food and eggs, and they use the shell to make handicraft items such as fans, purses, local money, etc.

In Palau it has been noted that the egg-laying habits of the Green Turtle and the Hawksbill differ. The Hawksbill will lay its eggs on the beaches of deserted islands within the lagoon, but Green Turtles are seldom found laying in similar areas. Palauans say that the Green Turtles lay most of their eggs in the Helens Reef area, about 300 miles southwest of Palau, and also on a small island north of Palau and just off the island of Kayangel.

Since the end of World War II, there has been no enforcement of the existing regulations governing the taking of turtles or their eggs. Consequently, relatively few turtles are now taken by the local fishermen.

The Trust Territory has recently started a Marine Conservation Program and one of its major objectives is to protect the marine turtles.

Educational programs are now in progress to explain to the people of the islands what conservation really is, what it can do for the people, and what will happen in a few years if a meaningful conservation program is not put into effect.

Attempts are also being made to protect newly hatched turtles and raise them in captivity for a few months until their chances for survival are considerably increased. As personnel and funds become available, this work will be extended to other islands in the Trust Territory, in a serious attempt to conserve the existing stocks of marine turtles and do everything possible to increase their numbers.

Marine Crocodiles

The marine crocodile, *Crodylus porosus* is common in the Palau Islands, but not established elsewhere in the Trust Territory. No fishery exists for them at this time despite high prices paid for their skins in Australia and Japan. As a consequence, these beasts have been allowed to multiply and they now present a formidable hazard to the underwater fisherman. Palauans often take crocodiles with throwing spears. The largest recorded specimen landed in Palau in the last 15 years measured 14 ft 10 in., and there are reports of one monster landed many years ago that measured 25 ft.

The Japanese prefer skins that measure from 5–18 in. across the hide when skinned out. The Australian tannery, A. C. Galstaun and Company will take skins with a minimal size of 4 in. and no maximum size. Prices vary according to the size and condition of the skin. Top-quality skins will bring as much as $9.00 an inch measured across the hide.

As these beasts are wary of people, they are difficult to approach. They are most easily fished at night with spotlights and shotguns loaded with slugs. They will normally remain fixed in one spot as long as the spotlight is held steady and the boat's motor does not change pitch. If not killed with the first shot, (difficult to do at night from a moving boat), the crocodile will quickly disappear into the water where he will be lost. If he struggles during capture, his hide can be damaged and its value reduced considerably.

The Palauans report that the Japanese used to have a crocodile "farm" where crocodiles were raised to a marketable size. Reptiles usually have a rapid growth rate during their juvenile period; Schmidt and Inger (1962) reported that well-fed crocodiles kept in zoological gardens grow as much as 1 ft a year up to the length of 6–8 ft, after which the rate of growth became progressively slower.

The female usually lays 30–50 eggs in muddy or sandy areas along the river banks. The nest is somewhat loosely guarded by the mother, who opens it when she hears the peeping of the young

Fig. 2. The salt water crocodile, *Crocodylus porosus,* is considered one of the most dangerous of all crocodiles. A study is now underway to determine the abundance and distribution of this species in Palau. This specimen measures 14 ft 10 in., and was taken by the fishermen pictured. Despite being speared seven times, it survived and lived for several years in captivity.

about to hatch. The marine crocodile has spawned in captivity in the zoo at the Entomology Laboratory in Palau but the eggs did not hatch. Since the zoo has a concrete floor and is not designed to encourage reproduction, it would seem possible, with a better designed facility, to establish a commercially valuable crocodile farm.

Sponges

Sponges formerly represented one of the more important fisheries in world-wide tropical waters. With the rapid growth of the synthetic sponge industry, the demand for natural sponge dropped until today there is little if any left. However, new uses are continually arising for out-of-date products and in the case of the sponge, it appears that one of the potential uses might be in the pharmacological field. If so, this may represent an important new resource, as hundreds of kinds of sponges occur in the Trust Territory. Many of these species are minute, living in small patches on dead coral, rocks, and other substrata. Several species form a thin layer a few millimeters thick over the reefs. Many are brightly colored, adding striking colors to the natural underwater beauty of corals and brightly colored fish.

In the lagoons of the high islands, large brown barrel sponges occur, as do the long purple organ-pipe sponges. Ponape, Likiep, and Kusaie have a species of sponge used locally for bathing. Palau's sponges were reported to be so abundant that they were used as a source of cultural material in sponge experiments carried out by the Japanese.

These experiments were carried out in Truk, Alinglaplap, Ponape, and other, unrecorded, areas. The results indicated that commercial sponge culture is feasible in the Trust Territory, particularly at Namorik where the Japanese believed the physical, chemical and oceanographic conditions of this atoll would be the best natural area for it.

In addition, Namorik has many areas where neither wind nor wave action would strike the culture. The Japanese estimated that 18,000,000 sponges could be produced annually in this atoll. They also thought Palau would be an excellent sponge culture area as they found that even during the poorest of conditions, in 18 months to 2 years, the cultured sponges were large enough to be of commercial value.

Shortly after the end of World War II samples of the sponges being cultured in Alinglaplap were collected and sent to Dr. Lewis Radcliffe, Executive Secretary of the Sponge Institute, Washington, who sent a sample on to Dr. M. W. de Laubenfels for identification. He reported as follows:

"The specimen is *Spongia officinalis*, subspecies *mollissima*, known as Fine Levant or Turkey Solid. One expects to find this exclusively in the eastern Mediterranean, and it is absent or rare elsewhere in the world. The specimen is one of the finest I have seen. Its fibers are a little bit weak, perhaps as a result of chemical bleaching, but in general it is worthy of enthusiasm."

Until such time as a new use for sponges can be developed, it seems unlikely that this group of animals will constitute an important resource.

Clams

The largest and one of the most popular shells found in the islands is *Tridacna gigas*. It is commonly called the "Man-Eating Clam" or the "Giant Clam," or simply "Tridacna" after its generic name. There are several species of *Tridacna* common in Trust Territory waters. The Palauans probably export more of the large *T. gigas* shell than the rest of the Trust Territory combined. In addition, the people are very fond of the meat, and often, rather than bother taking the shell, a fisherman will simply cut out the meat.

The distribution of these shells varies through the Trust Territory. Exact records relating to their relative abundance are not available, but they have been observed in most of the islands. The Palaus and the Marshalls seem to have the greatest abundance of *T. gigas*. Because of their relatively small inshore areas on Yap, and Ponape, large specimens have not been located here recently. Possibly they can still be found on the outer reef fringes as the local people seldom dive in these areas.

Because of its food value and because the larger shells can be sold to visitors as well as to the Trading Companies for export, a considerable number of Tridacna shells are harvested in the islands today. This is leading to a scarcity of shells and a growing concern that some sort of conservation practice should be initiated. The Trust Territory is presently studying a plan that will restock depleted areas by setting up underwater sanctuaries where brood stock can be concentrated and where the animals can, in effect, be "farmed." This should not prove too difficult as this group requires only planktonic food, sunlight, and the unicellular symbiotic algae called zooxanthellae, which lives in its mantle, for nourishment.

Hopefully, with the help of the island inhabitants and an organized program of research, enough knowledge of the life habits of this interesting group can be obtained to make it possible not only to conserve the group more effectively, but also to help it increase substantially in abundance.

Miscellaneous Shells

The colorful marine sea shells so popular with collectors are seldom regarded as a resource. However, the growing popularity of this hobby has made sea shells a thriving business in many parts of the world. The dedicated hobbyist is growing rapidly in numbers, and being determined to add to his collection through purchasing, trading, and collecting, he can be found in increasing numbers in the more remote parts of the world looking for new species to add to his collection. The Trust Territory Islands are recognized as being one of the best sources of sea shells and this minor resource should be developed further in order to attract more visitors as well as to bring in outside capital through the export sales of shells.

The local people do not presently realize the potential value of such shells, which are often picked up by children and fishermen who simply use them as playthings or give them away to visitors. The shells are generally improperly cared for and are usually chipped or broken. Attempts have been made to teach those who are interested in selling sea shells the proper methods of cleaning and preserving them; however, after a few weeks of doing the job properly, they soon tire of the extra work and go back to the old habits, with the result that buyers refuse their shells and the project is dropped.

Some thought should also be given to the conservation of certain species of shells, particularly the rarer species. Because of the growing interest in shells, a wealth of information has been accumulated on their life habits. A good deal of this information could be put to use in setting up some general guide lines for conserving the more popular and rare species.

It does seem doubtful that shell collecting will ever provide an exclusive source of income for any great number of people; however, it does seem probable that shells will constitute a source of extra income for islanders, particularly if they can be encouraged to handle them properly.

Crustaceans

The two most important types of crustaceans presently known in the islands are the mangrove crab (*Scylla serata*) and the spiny lobster (*Pannulirus spp.*) of which there are two or more species. Perhaps nearly as important are the coconut crab (*Birgus latro*) and the common land crab (*Cardisoma sp.*) both of which live on land but spend their juvenile stages in the sea. The Kona Crab or Red Frog Crab (*Ranina ranina*), the Mantis Shrimps (*Squilla spp.*) and the fresh water shrimps are all considered delicacies and are very much sought after, particularly by U.S. personnel living in the islands.

The mangrove crab is most abundant in Palau, Yap, Truk, and Ponape where mangrove areas exist around high islands. The spiny lobster is found in the reefs of all the islands, but it it does not appear to be present in sufficient abundance to warrant the establishment of an export lobster tail fishery. Little is known about the distribution, abundance, and life history of these animals in the Trust Territory. Because of their potential value, an intensive investigation of this important group should be initiated as soon as is possible.

Marine Algae

There are now a number of various species of marine algae being harvested in the tropical seas, which occur in sufficient abundance to create an above average income for those who have learned to harvest them. One of the more important species is *Eucheuma spinosum*, an alga that as a rule grows between the levels of the highest low tide and the lowest low tides, but is generally more widely spread geographically at any locality in the lower reaches of that vertical spread. It grows mixed in with *Fungia* coral and synoptid holothurians (sea cucumbers) in the shoal water fronting deeper inshore water. The Trust Territory falls within the zone that supports commercial operation elsewhere. *Eucheuma* is common in the Philippines, Indonesia, New Guinea, etc. Prior to the war, Tiyota Kanda of the Palau Tropical Biological Station, collected *Eucheuma* in at least two areas in Koror. Recent algae collections have yet to turn up this species, but these collections have not been carried on very intensively.

The best market for *Eucheuma* is presently Marine Colloids Co. in Rockland, Maine. This company has been in operation over 30 years and processes up to 20,000,000 lb of dried weed per year. Marine Colloids has recently developed a new large market for the extract of *Eucheuma* and as a result there is now an unlimited market demand for dried *E. spinosum* at a price of approximately $200 per ton.

There is presently some research work being carried out in Mindanao that is aimed at determining if *Eucheuma* can be introduced into a suitable environment where it does not occur naturally and literally be "farmed." This appears to be possible as *Eucheuma* reproduces rapidly and can be taken from the same area in approximately 1 month to 45 days if the gatherers leave a portion of the plant to allow it to grow again.

Aside from *Eucheuma* other tropical algae in demand are *Gilidilla acerosa* and *Gracilaria sp.* They bring approximately $0.30 per lb in the Philippines where they are sold to the local agar factories. The women and children who gather marine algae there reportedly earn as much as eight to ten pesos (U.S. $2.00 to $2.60) per day.

While little is now known of the distribution and abundance of the commercially important marine algae in the Trust Territory, the potential value of this undeveloped resource warrants further intensive investigation. In addition, experimental studies should also be carried out in several places in the Trust Territory to determine if it will be possible to introduce *Eucheuma* and other valuable species to those islands where it is not presently established, and once successfully introduced to start "farming" it.

Mining

The shore side mineral deposits of the Trust Territory are considered to be minimal. The possibility exists that underwater deposits of phosphate and bauxite exist, particularly in Palau where the Japanese carried on mining operations for these minerals prior to the war. Undoubtedly manganese nodules are also present.

Future Development

To determine the course of the future development of the marine resources of the Trust Territory we should look at what has been accomplished by Japan. Japan is a relatively small island complex that at one time faced the same problems as the Trust Territory does today. Few raw materials are indigenous to its shores and none exist in sufficient quantities to meet the minimal requirements of the people. For centuries Japan has been more dependent on the productivity of the sea than any other country. It seems likely that no nation in the world consumes as great a variety of seafood as Japan and, in addition, it is one of the world's leaders in underwater mining and aquaculture. Eighty-eight percent of its products are produced in domestic waters which include approximately 220,000 tons of oysters, seaweeds, and other products cultivated in the shallow inshore waters. There are 760 eel farms, which raise 2200 tons annually to supply over 1000 eel restaurants in Tokyo. The first shrimp farms were started in 1954 by Dr. Motosaku Fujinaga and by 1963 he was producing over 1000 tons annually.

In 1963 Yukimasa Kuwatani succeeded in raising oysters in tanks, feeding them on diatoms and keeping them alive in water clear and clean because of continuous circulation and aeration. Large artificial reefs have been created out of concrete blocks, thus providing new fishing grounds in areas where there was formerly no fishery.

In the offshore fisheries, reports indicate that the Japanese are now seining skipjack in the tropical seas of the western Pacific. If as successful as initial reports indicate, this could open up the largest undeveloped tuna resource in the world—the tropical Pacific skipjack. The "Governor's Conference on Central Pacific Fishery Resources" held in Hilo, Hawaii in 1966 concluded that minimal potential additional yields of 100,000–200,000 metric tons of skipjack in the Central Pacific (excluding the Western Pacific) would be possible. Adequate live bait resources have been one of the limiting factors during past years, and new techniques of harvesting must be developed if it is going to be possible to utilize this tremendous marine resource to its ultimate capacity. Should a method be found that will permit the exploitation of this potential resource, the Trust Territory stands to benefit, as it is located in the centre of this extensive fishery.

While the yellowfin resource of the Pacific does not have the potential of the skipjack resource, the May 9, 1967 issue of *Suisan Keizai Shimbun* stated that a recent survey to the Central-West Pacific off the Marshall and Caroline Islands by a government-owned research vessel exceeded expectations and indicated the possibility that the area could support a year-round fishery.

By providing receiving stations in these areas, it may be possible to purchase fish from the long-liners fishing this area while also selling fuel, ice, water, and other supplies the boats would require to remain on the fishing grounds for longer periods of time.

Peter T. Wilson

Cross references: Agar; Algae, Marine; Artificial Reefs; Clams; Irish Moss Industry; Pearling Industry; Phycocolloids; Sea Turtles; Sponge Industry; Tuna Fisheries.

TUNA FISHERIES

From 1958–1963, the world catch of tuna has varied from about 700,000–900,000 metric tons. Of this amount, Japan has consistently caught about $\frac{2}{3}$, and her catch has originated in all tropical and temperate oceans of the world.

Although catch statistics for tuna on a world basis are not very complete, they are improving. Even with the available figures, it might prove useful to look at the world catch for the past few years from a variety of points of view. First, what species are caught and how much of each species? Second, where did the catch come from? Third, who did the catching? And fourth, what effect did these large catches have on the fishery and on the resource?

Figure 1 shows the world catches for the years 1958–1963, by species. The data used are all derived from FAO's Fisheries Yearbook, supplemented by the Statistics and Survey Division of the Japanese Fisheries Agency. It is evident that catches increased regularly from 722,000 metric tons in 1958 to 905,000 tons in 1962 and then fell

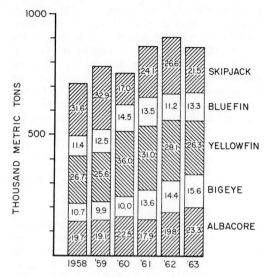

Fig. 1. World landings of tuna by species, 1958–1963. Numerals represent species composition in percent.

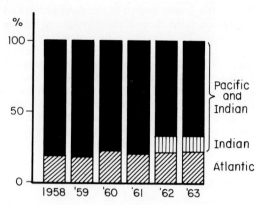

Fig. 2. World landings of tuna by oceans, 1958–1963.

slightly to 859,000 tons in 1963. Yellowfin catches over this period fluctuated around 30% of the total catch, skipjack around 25%, albacore around 20%, bluefin around 13% and bigeye around 12%.

To answer the second question as to where this fish came from is not easy, since the area of origin even by oceans was not available until very recent years. Figure 2 indicates that the Pacific and Indian Oceans together have supplied between 70–80% of the catch, of which the Indian Ocean's contribution was about 10%, at least in 1962 and 1963. The Atlantic contribution has remained very near to 20%.

Figure 3 shows that Japan has consistently caught tuna in all three oceans, and that her catch approaches ⅔ of the world total. In the Pacific, the U.S. catch is quite impressive, amounting to about ⅓ of the Japanese Pacific catch in 1962 and 1963 and between 15–20% of the world catch during the 6-year period from 1958–1963. In the Atlantic, Spain, France, and Portugal catch appreciable amounts and Peru, China, and the Ryukyus in the Pacific appear in the chart.

What effect have these substantial catches had on the fishery and on the world tuna resources? Only an approximate answer can be given, and this approximation can best be arrived at from Japanese records. That the size of some stocks has been affected, appears certain, but the effect seems to vary according to boat size and hence with the area of operation. This is shown roughly by comparing the catch per day-out fishing and the average length of trip for the Japan-based fleet for five different size categories of fishing boats for the five years 1959–1963.

From this tabulation it can be seen, at least for the vessels 200 tons and over, that the catch per day shows signs of decreasing, and the length of trip has increased proportionately even over this short period. In the largest size class this tendency is the most pronounced. This change in catch rate can make the difference between success and failure in a fishing operation that takes fishermen so far from the home port, with accompanying increase in cost. It is particularly impressive to Japanese fishermen, whose larger boats travel for thousands of miles to make their catch.

Size of Vessels			1959	1960	1961	1962	1963
30–50	GT[a]	C/D[b]	0.5	0.4	0.5	0.5	0.4
		LT[c]	17.0	19.9	21.1	20.9	20.9
50–100	GT	C/D	0.6	0.6	0.6	0.7	0.7
		LT	29.7	31.1	34.1	30.0	33.6
100–200	GT	C/D	1.0	1.0	0.9	1.9	0.8
		LT	45.3	47.4	52.0	52.8	65.1
200–500	GT	C/D	1.9	1.9	1.9	1.8	1.8
		LT	81.4	84.1	87.6	98.4	106.0
500+	GT	C/D	5.2	4.3	4.1	3.9	3.7
		LT	128.6	128.1	125.8	151.0	156.0

[a] GT = gross tons. [b] C/D = average catch per day. [c] LT = length of trip.

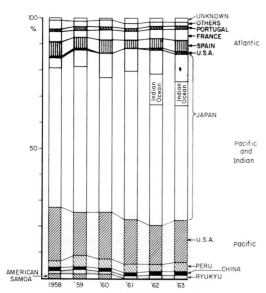

Fig. 3. World tuna catch by countries and oceans.

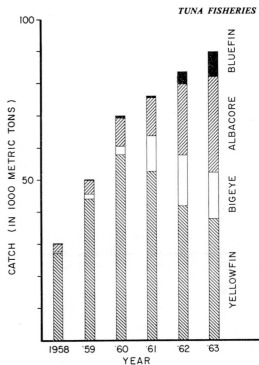

Fig. 4. Catches by species (in weight) by Japanese Atlantic tuna fishery.

Although a few large American tuna seiners now move from their traditional eastern Pacific fishing grounds to the Atlantic during restricted seasons in search of better catches, the effective history of tuna fishing in the Atlantic is, for the most part, a Japanese story.

In 1952 the restrictions on fishing were removed, and Japan quickly moved into offshore and distant water areas for its principal tuna catch. This move was accelerated and made effective by government assistance in the building of larger vessels, so that by 1956 large (300 GT and more) Japanese vessels could be found fishing successfully in all the western half of the Pacific and the Indian Ocean. About this time too (1955), "H" bomb tests in the Bikini Island area contaminated waters of the central Pacific, which in turn affected markets and fishing to the point that brought Japanese tuna fishing to a near-crisis situation.

TABLE 1. Estimated Effort (In Number of Hooks) and Catches by Species (In Number of Fish) of the Japanese Atlantic Longline Fishery, 1956–1963[a]

	1956	1957	1958	1959	1960	1961	1962	1963[b]
Effort	130,000	3,370,000	7,980,000	15,230,000	20,900,000	26,400,000	54,100,000	?
Bluefin	0	200	500	3,400	6,900	4,300	52,900	111,600
Albacore	1,100	31,600	99,300	354,800	456,000	425,200	1,086,700	1,467,100
Bigeye	200	8,600	14,900	44,600	71,200	251,400	370,900	342,300
Yellowfin	12,000	258,300	744,700	1,092,000	1,168,000	971,700	976,300	879,000
Total catch	13,300	298,700	859,400	1,494,800	1,702,100	1,652,600	2,486,800	2,800,000

[a] After Shiohama (1965) and Bureau of Statistics (1965).
[b] Number of fish were converted from weight.

TABLE 2. Yearly Change in Hook-Rates by Species in the Whole Atlantic Ocean[a]

	1956	1957	1958	1959	1960	1961	1962	1963[b]	1964[b]
Bluefin	—	+	+	+	+	+	0.1	0.1	0.1
Albacore	0.8	0.9	1.2	2.3	2.2	1.6	2.0	2.5	2.7
Bigeye	0.2	0.3	0.2	0.3	0.3	1.0	0.7	0.3	0.4
Yellowfin	9.2	7.7	9.3	7.2	5.6	3.7	1.8	1.5	1.0
Total	10.2	8.9	10.7	9.8	8.1	6.3	4.6	4.4	4.2

[a] After Shiohama (1965) and *Tuna Fishing*, Nos. 43–75.
[b] Hook-rates for 1963 and 1964 are estimated from data in *Tuna Fishing*; the rate of coverage is unknown.

Japanese fishermen, in order to survive, were forced to look quickly for new but remote productive fishing grounds. The captains of the larger vessels moved from the central Pacific and Indian Ocean to the eastern Pacific and the Atlantic. They found good fishing in both places. This quick maneuver really saved the Japanese tuna industry from economic collapse. It also marked the beginning of the eastern Pacific and the Japanese Atlantic tuna fishery.

The first Japanese longliner to fish the Atlantic was a research vessel from Chiba Prefecture; it fished experimentally for tuna in the Caribbean under a cooperative agreement with Venezuela in 1955. In 1956 a commercial longliner fished for tuna off the mouth of the Amazon River under a cooperative agreement with Brazil. In the same year, several independent vessels fished on a trial basis in various parts of the Atlantic and exported their catches to Italy. By 1957, as many as 26 large commercial vessels fished the Atlantic successfully and a large research vessel conducted extended exploratory and survey cruises during 1956 and 1957. By the end of 1957, the principal tuna fishing grounds of the Atlantic had already been mapped out.

Up to and including 1961, the majority of the tuna fleet consisted of independent vessels from 300 to 500 GT (Table 3). After that date, due to the rapid decline in tuna abundance in other oceans, and also because of the increasing number of overseas bases for operations in the Atlantic (these were acquired by agreement with strategically located countries), a number of smaller vessels were attracted into the area so that the average size of vessel was reduced from 569.4 GT in 1960 to 330.2 GT in 1962, and by this latter year there were 117 vessels of this average tonnage operating in the Atlantic.

The conventional mothership-type of operation, which had proved effective in the early years of the central and western Pacific fishery, was not permitted by the Japanese government to operate in the Atlantic, but the modified version of it, the deckboat-carrying mothership, which had first been tested experimentally about 1955 and found good, was permitted to operate there. The first operation of this kind entered the Atlantic in 1961. It was found so effective that by 1963 there were 28 such units fishing in many parts of the Atlantic. As each of these deckboat-carrying motherships (of 2,000 tons or less) fished effectively itself and in addition had 1–4 portable ships of approximately 20 GT on board, which could be dropped to fish anywhere, each such mothership unit introduced a lot of new gear into the Atlantic.

Because of its remoteness from the home islands, the Japanese Atlantic tuna fishery has developed some other characteristics that increase efficiency and reduce costs. One such distinction is the development of new and closer markets for their catches in Europe and America. Introduction of tuna to these countries started in a small way, but both the trading bonds and volume of trade are growing. To facilitate getting the fish to these new and growing markets, Japan has made special arrangements with several foreign governments, such as Brazil, Argentina, and Israel to use their land bases for transshipping tuna to such importing countries as the U.S., whose domestic laws prohibit direct landing of cargoes from foreign ships. Japanese ships also deliver tuna directly to such new tuna importing countries as Italy,

TABLE 3. SUMMARY OF ATLANTIC OCEAN FLEET OPERATION[a]

Year	Number of boats	Number of trips	Catch[b] (MT)	No. trips per boat	Catch per trip (MT)	Catch per boat (MT)	Average size of boats (GT)
1960	88(50,103 GT)	243	72,946	2.8	300.2	828.9	569.4
1961	86(48,014 GT)	258	82,251	2.9	318.8	934.7	558.3
1962	117(38,630 GT)	306	60,369	2.6	197.3	516.0	330.2
1963	100(37,476 GT)	303	59,407	3.0	196.1	594.1	374.8

[a] Mothership type operations are not included. After Bureau of Statistics (1964).
[b] Most probably includes the catches of billfish.

TABLE 4. SUMMARY OF PORTABLE-BOAT-CARRYING-MOTHERSHIP OPERATION*

Year	Number of boats	Average size of boat	No. of trips	Total days out fishing	Catch (MT)	Days at sea per boat	Catch per day out fishing	Catch per boat	Average length of trip
1962	11(11,178 GT)	1,016 GT	24	3048	18,285	277	6.0	1662	127 days
1963	28(27,502 GT)	982 GT	65	8185	41,823	292	5.1	1494	126 days

* This type of operation started in the Atlantic Ocean in 1961 for which year data are not available.

TABLE 5. YEARLY FLUCTUATION IN YELLOWFIN HOOK-RATES*

	1956	1957	1958	1959	1960	1961	1962
5°S–10°N	9.4	7.7	9.8	8.2	7.3	5.4	3.1
All Atlantic	9.2	7.6	9.3	7.2	5.6	3.7	1.8

* After Shiohama, 1965.

Yugoslavia, Greece, France, Libya, etc. This type of operation is growing. It is not necessarily a popular development as it keeps vessel operators and crews away from home for long periods of time and all fishing crews cannot be flown home at the end of each trip. This increased efficiency is required, however, to stay in business. It is dictated by the relentless march of falling catch rates, by great distances from home and by progressively increasing general operating costs.

The progressive and intensive development of tuna fishing in the Atlantic has also dictated changes in the species of tuna caught, and in the seasons and areas the various species could be caught.

In the early years of fishing in the Atlantic, fishing effort was principally directed toward yellowfin tuna off Guinea, the Ivory Coast, and Brazil. As the yellowfin fishery developed, the combined effects of the constantly increasing number of boats, the progressively and relentlessly dropping catch rate for yellowfin, the finding of good albacore, bigeye, and bluefin grounds and the developing taste and demand for the latter species by Europeans as well as in the U.S., resulted in a great expansion of fishing grounds throughout the Atlantic. This expansion is shown graphically in Fig. 5. From a modest beginning off the coast of South America in 1956, the Japanese tuna fishery in the Atlantic had spread to every part of this ocean by 1959 and fishing intensities continued to increase over the whole area at least until 1962.

The erratic seasonal pattern of the Atlantic tuna fishery is shown in Fig. 6. Fishing grounds and fishing seasons for each species seem to vary even more in the Atlantic than in the other known tuna fishing areas. Yellowfin catches occur in two peaks, one in February–March and the second in August. This catch is mostly made in equatorial waters and is relatively stable, except that the hook-rate has fallen precipitously. Albacore catches are made in December in the southern hemisphere and in July in the northern hemisphere. In general, albacore are more abundant in the western half of the Atlantic and bigeye in the eastern half. Bigeye catches also show two seasonal peaks, one in the summer months in the northern hemisphere and the other in the fall months between 5° and 20°S latitude. Bluefin are most abundant in March and April and again in September to November in waters off Brazil.

The complaints of fishermen fishing the area of the highly variable nature of the fishery seem to be well borne out.

As can be seen from the foregoing, the development of the Japanese tuna fishery in the Atlantic has been dramatic, ingenious, and arduous. What effect has this fast expansion of activity and fishing effort had on fishing and on the fish stocks?

To arrive at a more precise picture of the growth of the fishery and the changing species composition, refer to Table 2 and Fig. 7. The remarkable rise in the Atlantic tuna fishery, and the changing species composition are clearly shown. The most remarkable thing about the latter is the progressive reduction in the catch of yellowfin after 1960 and the equally progressive increase in species other than yellowfin, particularly albacore.

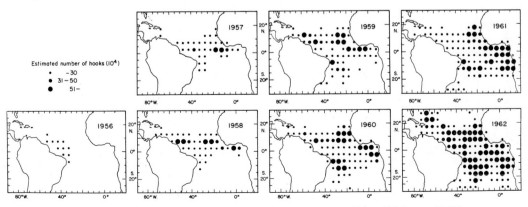

Fig. 5. Estimated total fishing effort by year and 5° square. (After Shiohama, 1965.)

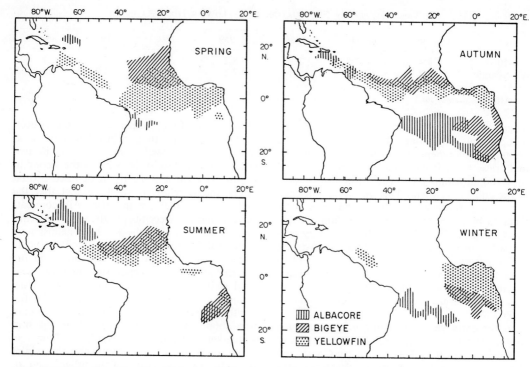

Fig. 6. Distribution of longline fishing ground for each species and season. Charts are prepared considering the average of monthly distributions of hook-rate, 1957–1962. (After Shiohama, 1965.)

Figure 7 is designed to show how the effort, as measured in millions of hooks fished, is related to the amount (in this case numbers of fish) caught and its species composition. Again it becomes broadly apparent that effort after 1959 and 1960 was increasing faster than catch and that this was particularly true of the yellowfin component of the catch, with the proportionate increase in the other three species.

Another look at the hook-rate is given for yellowfin alone in Fig. 8, prepared by Kurogane and reported at the Tuna Conference in 1965. The actual hook-rate has been smoothed out by the author by eliminating seasonal influences, to better indicate the trend. This again shows unmistakeably that yellowfin of the Atlantic, at least that component of the stock taken by longlines, has been fished to a new low level. A fall in hook-rate from a high of 20 yellowfin per 100 hooks, then to 10, and now to 3–5, even in the best yellowfin territory, cannot readily be ignored, either operationally or from a conservation point of view.

Figure 9 reveals that, although the hook-rate for yellowfin has fallen rather precipitously, as has that for the fishery as a whole (influenced heavily by the yellowfin), the hook-rates for albacore, bluefin, and even bigeye have not. In fact the hook-rate for albacore during the past few years, when new grounds have been discovered, has actually gone up. Recent good albacore catches have helped keep the tuna fishermen in business, but even here the particularly good catches of albacore temporarily loaded the market and once more brought the twin dilemma, so

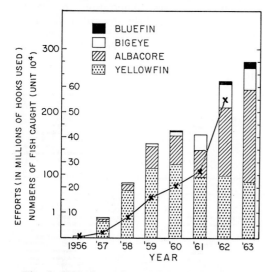

Fig. 7. Annual number of fish caught shown by series. (After Shiohama, 1965, and Statistics and Surveys Division, 1965.)

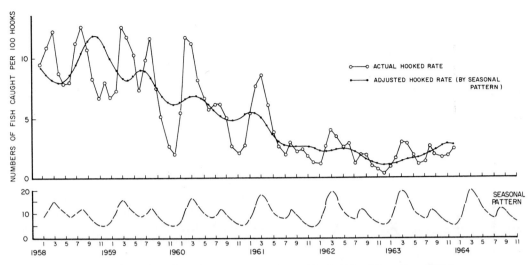

Fig. 8. Hooked-rate of yellowfin, Atlantic combined. (After Kurogane, 1965.)

familiar to fishermen, of dwindling resources on the one hand and temporarily oversupplied markets on the other.

The Japanese tuna fishery in the Atlantic, though only 15 years old, has already run the full gamut of problems. With many other countries building tuna fishing fleets, it looks very much like Japan will soon have others with whom to share her tuna problems, not only in the Atlantic but in the world ocean.

It would appear that Japan has been the principal tuna fishing country in the Atlantic during the past several years. But by no means has she been the only tuna fishing country there. In 1963, the most recent year for which statistics are available, Japan caught the most albacore, 29.7

TABLE 6. WORLD ATLANTIC TUNA CATCHES (IN THOUSANDS OF METRIC TONS)

	1958	1959	1960	1961	1962	1963
Albacore						
Japan	2.0	3.6	8.5	11.8[d]	22.0	29.7
Spain	28.9	26.7	25.9	19.9	22.3	28.3
France	17.4	16.8	17.2	15.2	18.0	14.4
Brazil[a]	—	5.1	4.7	4.4	1.4	—
Bigeye						
Japan	+	2.0	3.3	11.5[d]	15.7	14.5
Bluefin[b]						
Japan	+	0.3	0.7	1.4[d]	3.7	7.8
Spain	13.8	9.0	9.8	6.8	9.3	7.9
France	1.7	2.4	0.8	1.7	1.4	1.2
Portugal	6.1	8.4	8.1	7.8	8.3	8.7
Italy	3.0	2.1	1.4	1.6	1.6	1.8
Norway	3.0	2.5	3.3	6.7	6.8	0.1
Morocco	10.5	5.0	6.0	2.3	0.3	1.5
Angola	3.5	1.5	2.8	2.5	2.4	—
U.S.	1.1	1.3	0.6	1.1	3.2	3.9
Yellowfin						
Japan	27.2	44.1	57.8	52.6[d]	41.9	37.7
Spain					2.5	0.8
France	9.6	3.5	13.8	13.1	14.8	22.7
Venezuela	0.9	1.6	2.1	2.0	3.6	3.1
U.S.	0.3	0.1	—	—	+	0.2

[a] Catches other than albacore are included.
[b] Catches in the Mediterranean Sea are included.
[c] Catches of albacore are included.
[d] Catches by mothership are not included but believed to be minor.

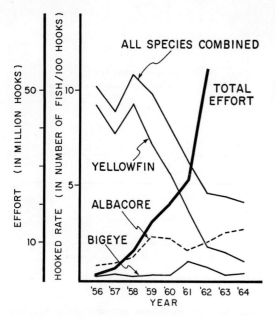

Fig. 9. Yearly change in hook-rates, by species and total efforts (in number of hooks used).

thousand metric tons, but Spain caught nearly as much and France caught half as much. Japan caught all the bigeye and 7.8 thousand tons of bluefin, but Spain caught 7.9 thousand tons and Portugal caught 8.7 of this species; of course these countries make most of their catches in the Mediterranean. In the catch of yellowfin, however, Japan has been supreme, having caught 41.9 thousand tons in 1962 and 37.7 thousand in 1963. (See Table 6). It should be noted, however, that France, Japan's chief competitor here, caught 14.8 thousand tons in 1962 and 22.7 thousand in 1963—a gain of 8 thousand tons in a year when fishing hook-rates reduced Japan's catch.

The Japanese tuna fisheries of the Atlantic have had good days and bad. But some trends appear certain. Even the present stocks with their falling hook-rates will have to be shared with newcomers, and more information on stocks and their better utilization will be needed if all, newcomers and present fishermen, are going to be able to continue a profitable harvest of this unique ocean crop.

J. L. KASK

Cross reference: *Japanese Fisheries.*

U

UNITED STATES AND WORLD FISHERY STATISTICS

The U.S. is favorably situated as a major fish-producing nation. The general tidal shoreline of the Atlantic is 6370 miles in length; 4097 miles on the Gulf of Mexico; 17,542 miles on the Pacific, and 900 miles around Hawaii. With such a vast shoreline distance and a wide range in temperature of the land and oceans, it is not surprising to find a large number of species harvested in U.S. territorial waters or on the high seas adjacent to the coast.

More than 60 major river systems empty their nutrients into the coastal waters from the cold Arctic to the tepid waters of the Gulf of Mexico. The Mississippi River, for example, stretches the entire length of the U.S. draining vast plains and mountains on either side. On the Atlantic and Gulf coasts, extensive sunken coastal plains provide numerous bays, inlets, and estuarine areas as habitats and sanctuaries for numerous species. Into this immense paradise of ichthyological fauna the early settlers began their colonization of the North American continent. It is not surprising that fishing played such an important part in the development of the colonies and later the economy of the U.S. Much of the early competition between European governments for territory on the North American continent was prompted by the desire for fishing rights. Capital for the developing industries in the U.S. was largely obtained from fishing and whaling. Naval skill, so important in protecting the young nation, was acquired in fishing and whaling, and nations participating in the fisheries along the North Atlantic coast did so not only to obtain a product of commerce but also to train sailors. Today's fishing industry in the U.S. is, however, a far cry from the number 1 position it occupied in the early colonies. For example, in 1966, the U.S. produced 4.4 billion lb of fishery products (round weight), slightly over 5% of the world's catch, but used more than 11% of the world supply. While the U.S. has declined as a major fish-producing nation it has become the world's largest importer and user of fishery products. Prior to 1956, the U.S. maintained the number 2 position among the fish-producing nations, following Japan—the world leader. A static U.S. fleet and rapid expansion by the fishing fleets in Peru and the U.S.S.R. had relegated the U.S. to 5th place by 1960. Thus, in less than 20 years, despite some of the richest coastal waters of the world, landings and relative position have substantially declined. While world data place the U.S. behind Red China, this author does not believe the data from China are valid for comparative purposes for the following reasons: (1) Red China has no major fleets on the high seas; the Japanese and the U.S.S.R. have extensive operations utilizing catch, processing, and supply craft, (2) there is no observable concentration of fishing effort along her shores as is the case in Peru; (3) the U.S. has a catch by sportsmen,

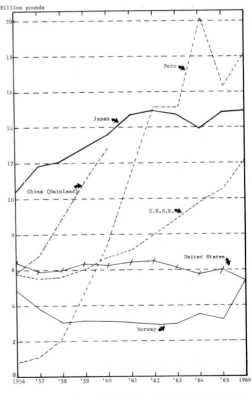

Fig. 1. World catch of fish and shellfish by leading countries, 1956–1966 (live weight basis).

Table 1. U.S. Landings of Fish and Shellfish, 1950–1965

Year	Landings used for human food	Landings used for industrial products[a]	Total		Average price per pound
	Million pounds	Million pounds	Million pounds	Million dollars	Cents
1950	3307	1594	4901	347	7.09
1951	3048	1385	4433	365	8.23
1952	2778	1654	4432	364	8.20
1953	2519	1968	4487	356	7.94
1954	2705	2057	4762	359	7.55
1955	2579	2230	4809	339	7.05
1956	2690	2578	5268	372	7.06
1957	2475	2314	4789	354	7.39
1958	2651	2096	4747	373	7.86
1959	2369	2753	5122	346	6.76
1960	2498	2444	4942	354	7.15
1961	2490	2697	5187	362	6.98
1962	2540	2814	5354	396	7.40
1963	2556	2291	4847	377	7.78
1964	2497	2044	4541	389	8.57
1965	2587	2190	4777	446	9.34

[a] Manufactured into meal, oil fish solubles, homogenized condensed fish, shell products, and used as bait and animal food.

Table 2. U.S. and World Landings, 1950–1965 (Live Weight)

Year	World Landings, thousand metric tons	U.S. Landings, million pounds	U.S. Landings, million metric tons
1950	21,100.0	[a]6344	[a]2878
1951	23,600.0	[a]5731	[a]2600
1952	25,200.0	[a]5842	[a]2650
1953	25,900.0	[a]5897	[a]2675
1954	27,600.0	[a]6131	[a]2781
1955	28,900.0	[a]6151	[a]2790
1956	30,500.0	[a]6590	[a]2989
1957	31,500.0	[a]6085	[a]2760
1958	32,800.0	[a]5971	[a]2709
1959	36,400.0	6373	2891
1960	39,500.0	6205	2815
1961	43,000.0	6464	2932
1962	46,400.0	6554	2973
1963	47,600.0	6122	2777
1964	52,000.0	5836	2647
1965	52,400.0	6006	2724

[a] Revised to improve the accuracy of the mollusk data. All data include Hawaii and exclude whale products. Live weight includes the weight of shells for mollusks.

whereas Red China has no sport catch and all fish taken are considered as food or commercial fish—hence, the higher catch figure in Red China. Nevertheless, there are substantial data to demonstrate clearly that total world production increased from 21.1 million metric tons in 1950 to 52.4 million metric tons in 1965. Meanwhile the U.S. total landings, number of fishermen, and boats registered a decline.

The U.S. landings of all fishery products on a round weight basis for the years 1950–1965 are shown in Table 1.

Examination of data in Tables 1 and 2 shows very clearly the steady increases in world landings of fish and shellfish from 1950 to 1965, while the U.S. landings declined. Only the quantity of landings used for industrial purposes was greater in 1965 than in 1950, and even this reached a peak three years earlier, and has since shown a steady decline.

TABLE 3. WORLD PRODUCTION BY CONTINENTS (LIVE WEIGHT)
Thousand Metric Tons

Year	North America	South America	Asia	Europe	Africa	Oceania	U.S.S.R.
1950	3780	520	7670	6170	1200	90	1627
1951	3530	570	9020	6970	1390	90	1977
1952	3480	620	10,320	7080	1670	100	1888
1953	3750	610	10,550	7150	1750	100	1983
1954	3980	670	11,140	7680	1780	100	2258
1955	3950	830	11,900	7840	1820	100	2495
1956	4340	910	12,280	8270	1940	100	2616
1957	3980	1170	13,720	7880	2070	110	2531
1958	3990	1630	14,590	7750	2130	110	2621
1959	4260	2950	15,870	8170	2250	120	2756
1960	4080	4430	17,450	8090	2290	130	3051.0
1961	4330	6300	18,200	8360	2480	130	3250.0
1962	4490	8290	18,630	8640	2630	140	3616.5
1963	4380	8420	18,970	8980	2750	140	3977.2
1964	4300	11,010	19,290	9740	3020	150	4475.8
1965	4430	8980	19,950	10,810	3060	150	4979.5

World landings increased chiefly as a result of rapid expansion in certain continents and in specific countries. Those continents showing the greatest gains were South America, U.S.S.R., and Asia. In South America the chief contributor to this gain was Peru, where the exploitation of the anchoveta fishery for processing into fish meal was responsible for the increase. This fishery now provides one of the chief sources of exports for Peru's expanding foreign trade. Increased use of fish meal as a poultry supplement in the U.S. and Western Europe provides most of the outlet for this product. Peru's fishery has expanded faster than that of any other country.

The need for high quality protein led the Soviet Union to expand its ocean going factory, fishing, and supply craft. In contrast to Peru, where the fishery is concentrated near its own shores, the Soviets have dispatched their fleets all over the world to fish on known concentrations of usable fish. Soviet ships now fish off the Atlantic and Pacific coasts of the U.S. and have conducted extensive explorations in the Gulf of Mexico.

Japan, also in need of protein and jobs for her expanding population following World War II, increased her fishing fleet and steadily expanded her sphere of operation. Japanese fleets now operate in every major ocean of the world, off the Atlantic and Pacific coasts of the U.S. and in the Gulf of Mexico. Japan is the chief foreign source for this country's tuna, both raw and canned. More than ⅔ of the U.S. supply of tuna is received from foreign sources.

While South America, the U.S.S.R., and Asia were registering significant gains, landings in North America, including the U.S., increased only 17%. Since the U.S. registered a decline during this period, gains were principally the result of increases in Canada and Mexico, two countries that send substantial supplies of fishery products to the U.S.

About 128,000 fishermen, use approximately 12,000 vessels of 5 net tons or greater, and over 65,000 boats of less than 5 net tons in making the U.S. catch. However, about 7,500 fishermen operating less than 1000 purse seine and otter trawl vessels account for more than 70% of the U.S. landings. The U.S. operates no large ocean going factory ships, in striking contrast to the Japanese and Russians. Over 90% of the U.S. vesssels are less than 70 ft in length and 95% are less than 100 gross tons.

With a small antiquated fishing fleet it is not surprising that all the growth in the U.S. fishing industry has occurred largely in the taking of shellfish. Since 1950 landings of shellfish have increased 47%, from 558,000,000 lb (11% of the total U.S. landings) to 821,000,000 or 17% of the total in 1965. The value of these crustaceans and mollusks has likewise increased from $114 million in 1950 (33% of the value of U.S. landings) to $204 million in 1965, 46% of the total value.

Landings of finfish on the other hand have declined from 4.3 billion lb in 1950 to 3.9 billion in 1965, a decline of 9%. Meanwhile the value of finfish increased from $230 million in 1950 to $241 million in 1965, an increase of only 5%. A large part of the shellfish catch is taken by small vessels and boats whereas the finfish catch is taken mostly by the larger craft of the U.S. fishing fleet. Poor returns, among other causes, obviously contributed to declining U.S. finfish landings.

While world catch was forging ahead, and specific continents and countries were making

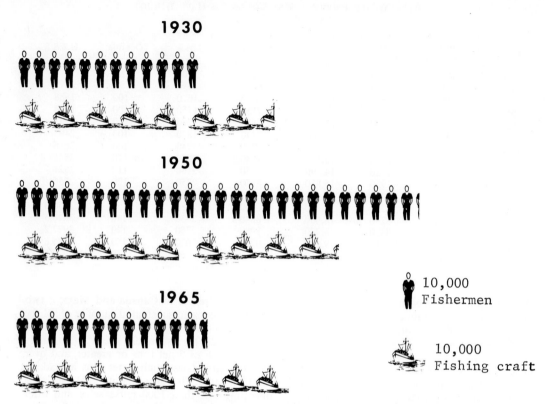

Fig. 2. Number of U.S. fishermen and fishing craft, 1930, 1950, and 1965.

spectacular gains, U.S. landings reached a peak and began a decline. Despite some year-to-year fluctuations in the number of vessels engaged in fishing, the general trend was for more, slightly larger and more powerful craft. However, the number of fishermen declined from 161,463 in 1950 to 128,565 in 1965. Likewise, the number of boats declined from 80,814 in 1950 to 67,221 in 1965, a decline of 13,593 boats and 32,898 fishermen. Some of the decline is accounted for by the trend to larger and better fishing craft, thus reducing the number of smaller older units. Furthermore, many of the less productive craft were forced out of business by the more efficient units.

However, not all segments of the fishing industry declined or for that matter were in a weakened economic condition. The shrimp and menhaden industries of the Gulf of Mexico experienced a phenomenal growth during this period. For example, in 1940 the states bordering on the Gulf of Mexico accounted for only 6% (250,000,000 lb) of the volume and 11% ($10.6 million) of the value of all U.S. landings. In 1965 this same area accounted for 31% of the volume and 26% of the value of all fish and shellfish landed in the U.S. The increase in volume in the Gulf states was the result of a rapidly expanding menhaden industry and to a lesser extent discovery of new shrimping grounds. The increase in value was largely the result of the increasing popularity of shrimp and a corresponding price increase.

The greatest decline in volume occurred in the Pacific coast states and was largely the result of the almost complete disappearance of pilchards, used in the preparation of canned sardines and for reduction into fish meal and oil. The pilchard or Pacific sardine fishery is an example of the great fluctuations that occur in some major fisheries. In the relatively short span of less than 40 years, this species moved from a position of minor importance in Pacific Coast landings to that of the leading species taken by U.S. fishermen and then, in an even shorter period, the catch declined to less than 1% of peak production. The catch increased to such proportions in San Francisco that it became the nation's leading fishing port in 1936, 1938, and 1939. The disappearance of the fish, for causes not fully understood, caused San Francisco to disappear as a major fishing port and dropped California from its position as the leading fish producing state in the nation. It was replaced by Louisiana, whose volume increased remarkably as a result of the phenomenal growth of the menhaden fishery in

Table 4. U.S. Landings of Fish and Shellfish, 1950–1965
(Thousands of Pounds and Thousands of Dollars)

Year	Fish		Shellfish	
	Quantity	Value	Quantity	Value
1950	4327	$230	558	$114
1951	3823	236	590	124
1952	3850	226	569	134
1953	3845	199	623	153
1954	4122	220	620	136
1955	4190	196	605	140
1956	4664	217	583	152
1957	4182	197	589	154
1958	4158	216	567	154
1959	4484	199	626	146
1960	4250	197	682	156
1961	4550	215	625	146
1962	4709	230	635	165
1963	4121	210	718	166
1964	3829	213	701	175
1965	3948	241	821	204

that state. The New England states, the Great Lakes, and Mississippi River sections also registered declines, while the Chesapeake Bay states increased slightly and the South Atlantic states remained unchanged.

The U.S. fishing fleet is composed mostly of small independent producing units without legal ties to large companies, an arrangement that permits fishing craft to land their fares at any port of choice or convenience. However, certain ports stand out as major landing ports and are important because of their contribution to the history of commercial fishing. During the 40-year period from 1927–1966, San Pedro, California, led all other domestic fishing ports in volume of landings 26 times; Boston, Massachusetts, 5 times; Monterey and San Francisco, California, 3 times; and Lewes, Delaware, and Reedville, Virginia, once each. In addition to being the leading fishing port in most of the past 40 years, San Pedro holds the all-time record for landings at a domestic fishing port, with a total of 848,000,000 lb in 1950. This catch included 547,000,000 lb of sardines, 162,000,000 lb of tuna, and 129,000,000 lb of jack and Pacific mackerel. San Francisco ranks second in total volume on the basis of the nearly 815,000,000 lb received in

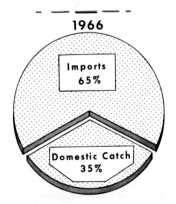

Fig. 3. U.S. supply of fishery products, 1957–1966 (round weight basis).

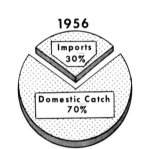

Fig. 4. Domestic supply of fishery products increased 64% since 1956.

TABLE 5. VOLUME OF THE U.S. CATCH BY REGIONS, VARIOUS YEARS, 1940–1966 (MILLION POUNDS)

Region	1940	1950	1960	1965	1966
New England and Middle Atlantic					
Quantity	982	1498	1636	1058	852
Per cent	24	31	33	22	20
Chesapeake					
Quantity	321	381	435	591	502
Per cent	8	8	9	12	12
South Atlantic					
Quantity	325	261	379	357	368
Per cent	8	5	8	8	8
Gulf					
Quantity	250	571	1266	1463	1196
Per cent	6	12	26	31	27
Pacific Coast					
Quantity	2020	1997	1061	1147	1254
Per cent	50	41	21	24	29
Great Lakes and Mississippi River					
Quantity	162	177	154	141	181
Per cent	4	3	3	3	4
Hawaii					
Quantity	a	16	11	20	13
Per cent	—	—	—	—	—
Total quantity	4060	4901	4942	4777	4366

[a] Data not available.

TABLE 6. VALUE OF THE U.S. CATCH BY REGIONS, VARIOUS YEARS, 1940–1966 (MILLION DOLLARS)

Region	1940	1950	1960	1965	1966
New England and Middle Atlantic					
Value	28.1	89.4	83.2	99.5	99.8
Per cent	28	26	23	22	21
Chesapeake					
Value	7.5	25.0	34.9	40.2	35.3
Per cent	8	7	10	9	7
South Atlantic					
Value	4.1	18.9	20.2	26.8	7.1
Per cent	4	5	6	6	6
Gulf					
Value	10.6	50.4	85.5	113.5	122.6
Per cent	11	15	24	26	26
Pacific Coast					
Value	40.2[a]	139.2[a]	112.1	148.4	168.1
Per cent	40	40	32	33	36
Great Lakes and Mississippi River					
Value	8.5	21.0	15.0	13.7	6.4
Per cent	9	6	4	3	3
Hawaii					
Value	[b]	3.5	2.7	3.6	3.1
Per cent	—	1	1	1	1
Total value	99.0	347.4	353.6	445.7	472.4

[a] Does not include types of compensation (food, use of boats, gear, etc.) that were added to the value in later years.
[b] Not available.

1936. Included in this catch were 789,000,000 lb of sardines landed at shore plants in the bay area and on factory ships anchored in the vicinity of San Francisco, beyond territorial waters. Lewes, Delaware, and Reedville, Virginia—the only Atlantic Coast ports, other than Boston, that have led in volume—are both almost exclusively menhaden ports.

During the 40-year period, San Pedro led in value of landings 20 times and Boston 17. In 1950, the catch landed at San Pedro yielded an all-time high of $38.1 million. Of this amount, tuna accounted for $24.6 million; and sardines, $9.4 million. San Diego, California, was the only other port to lead in value of landings, having done so in 1948 and 1949. At a time when the major nations were expanding their fleets, the U.S. fleet was static.

More than 250 species of fish and shellfish are landed and sold by U.S. fishermen. However, 10 species or groups of species (menhaden, crabs, salmon, tuna, shrimp, flounders, haddock, herring, ocean perch, and whiting) account for approximately 3/4 of the total U.S. landings. Turning to value of the species we find that 10 species (not necessarily the same species) account for more than 76% of the total value of all U.S. landings.

Strong competition from other animal protein foods, despite a rapidly increasing population, kept the annual per capita consumption of commercially-caught fish and shellfish in the U.S. at an average 10–11 lb (edible weight) except for the periods 1931–1934 and 1942–1944 when consumption declined sharply due to the depression of the 1930's and World War II. Although the static domestic industry has failed to increase its catch, the U.S. is one of the leading consumers of fish and shellfish and perhaps the world's

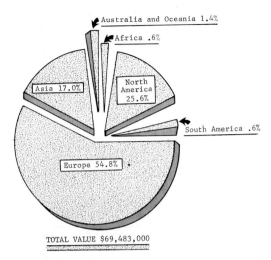

Fig. 5. Value of U.S. imports to continent of destination, 1965.

largest market for aquatic products. Americans are eating as much fish as ever, but more of what they are consuming comes from foreign fisheries.

In 1947, the U.S. was near the forefront as a great fishing power of the world: Japan had not yet recovered from the ravages of World War II; the U.S.S.R. had no high-seas fishery; and Peru had not yet begun exploiting the anchoveta resource. During this period the U.S. supply of all fishery products totaled 5.0 billion lb and the domestic fleet contributed all but 14%; imports on a live weight basis totaled only 710,000,000 lb. The domestic industry accounted for nearly 80% of this country's total supply of fish meal; 91% of the needed groundfish and Atlantic ocean perch fillets; 81% of the supply of other fillets and steaks; 94% of the domestic canned tuna was packed from domestically caught fish; and about 80% of the supply of shrimp was of domestic origin. The U.S. produced 83% of its total supply of edible fishery products (3.6 billion lb) and 94% of its total supply of industrial products (1.4 billion lb).

The rate of consumption of edible fish in the U.S. remains about static but the use of industrial fishery products is increasing. With a rapidly increasing population and a failure of the fishing fleets to increase landings, the U.S. has turned to products of foreign fisheries to meet domestic demands.

In 1966, the U.S. total supply of fishery products was 12.5 billion lb—146% larger than in 1947. The domestic catch was at about the same level as in 1947 (4.4 billion lb) but imports of 8.1 billion lb were over 1,000% larger, accounting for 65% of the total 1966 U.S. supply of fishery

TABLE 7. U.S. FISHERMEN AND FISHING CRAFT, 1950–1965

Year	Fishermen	Vessels	Boats
1950	161,463	11,496	80,814
1951	155,403	11,242	78,549
1952	151,559	11,065	77,071
1953	152,907	10,621	76,060
1954	144,645	11,179	70,911
1955	144,359	11,796	71,496
1956	141,547	11,458	70,061
1957	138,171	11,671	66,299
1958	128,960	11,496	63,795
1959	128,985	12,109	63,192
1960	130,431	12,018	65,039
1961	129,693	11,964	66,724
1962	126,333	11,511	64,222
1963	128,470	11,928	66,045
1964	127,875	11,808	64,604
1965	128,565	12,311	67,221

TABLE 8. RELATIVE VOLUME OF U.S. LANDINGS BY SPECIES, 1965

Species	Quantity Thousand lb	Percent of Total	Record Catch, Thousand lb, and Year	
Menhaden	1,726,089	36.1	1962	2,347,944
Crabs	335,407	7.0	1965	335,407
Salmon	326,871	6.8	1936	790,884
Tuna	318,895	6.7	1950	391,454
Shrimp	243,645	5.1	1954	268,334
Flounders	180,121	3.8	1965	180,121
Haddock	133,892	2.8	1929	293,809
Herring	110,293	2.3	—	—
Ocean Perch, Atlantic	83,608	1.8	1951	258,320
Whiting	82,574	1.7	1957	133,041
Other	1,231,371	25.9	—	—
Total	4,776,766	100.0	—	—

TABLE 9. RELATIVE VALUE OF U.S. LANDINGS BY SPECIES, 1965

Species	Value, Thousand dollars	Percent of Total	Record Value, Thousand dollars, and Year	
Shrimp	82,409	18.5	1965	82,409
Salmon	65,159	14.6	1965	65,159
Tuna	41,734	9.4	1950	61,342
Crabs	30,792	6.9	1965	30,792
Oysters	27,868	6.3	1961	33,204
Menhaden	27,073	6.1	1956	28,425
Lobsters, Northern	21,957	4.9	1965	21,957
Flounders	17,948	4.0	1965	17,948
Clams	16,733	3.8	1965	16,733
Haddock	13,630	3.1	1965	13,630
Other	100,376	22.4	—	—
Total	445,679	100.0	—	—

products. U.S. fishing fleets were able to supply only 47% (2.6 billion lb) of the 5.4 billion lb of the supply of edible fishery products and only 25% (1.8 billion lb) of the supply of industrial products (7.0 billion lb). The domestic fishery accounted for only 33% of the total domestic supply of fish meal; 19% of the groundfish and Atlantic ocean perch fillets; 50% of the supply of other fillets and steaks; 33% of the domestic canned tuna was packed from domestically caught fish; and only 43% of the supply of shrimp was of domestic origin.

The domestic fishery processing industry, unlike that segment concerned with the harvesting of the oceans, has made substantial gains during the past 20 years and now prepares products worth over $1 billion at the point of production. This billion dollar business uses foreign and domestic raw material (fish) and several methods of preserving seafood products to insure a wholesome supply to the American public. These methods are refrigeration, freezing, canning, and curing.

Refrigeration is perhaps the most important and because of this method, large quantities of frozen products are available, kept in good condition for further processing, or available to meet changing market demands. The freezing and holding of seafood products has substantially increased its use, and few food processors depend

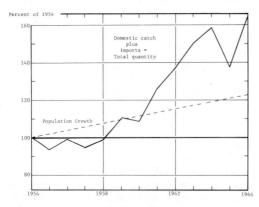

Fig. 6. Total quantity of fishery products (round weight) compared with population growth.

TABLE 10. VALUE OF PROCESSED FISHERY PRODUCTS VARIOUS YEARS

Year	Thousand dollars
1950	578,692
1955	663,890
1960	812,847
1961	876,659
1962	958,825
1963	914,492
1964	964,720
1965	1,117,958

TABLE 11. MONTH OF LARGEST AMOUNT OF HOLDING OF FROZEN FISH, VARIOUS YEARS (THOUSANDS OF POUNDS)

Year	Month and Day	Quantity
1950	November 1	166,105
1955	January 1	194,338
1960	November 30	237,163
1965	November 30	232,107
1966	November 30	281,728

more on refrigeration than the fish industry. Because of increasing demand for fillets and steaks, sticks and portions, sea scallops, shrimp products, crab products, fish and shellfish speciality dinners, and other items, the marketing of frozen seafood products has increased from less than one-half billion pounds in 1950 to well over a billion pounds at present.

The canning of seafood products in hermetically sealed containers processed with heat is an important segment of the fishery processing industry. In 1966 over 500 packers in nearly every state produced over 100 different canned seafood products from many different species of fish and shellfish. Over 1 billion lb of seafood products were canned in 1966 and the production was valued at $500,000,000 f.o.b. point of production. Modern cannery methods provided the means to develop the salmon, tuna, sardine, shrimp, clam, crab, and pet food industries of the U.S. Canned fish and shellfish products are very important in the general food canning industry and each year exceed canned meat and meat products both in volume and value to the producer.

The curing of fishery products by drying, smoking, salting, or pickling is a very ancient process developed years before the discovery and practical use of refrigeration and canning. Very important in the early development of the U.S., better ways of keeping fishery products fresh and wholesome led to a decline in the production and marketing of cured seafood products.

U.S. catch and imports of fishery products include not only products for human food, but also large quantities used in the industrial fishery processing industry. Whole or waste material from human preparations are used for bait and animal food, and there is a very large fishery (menhaden and other species) for material used in the manufacture of fish meal and oil. Fish meal indirectly is very important because it is used in feeding animals (swine and poultry) that ultimately end as food for humans. Other products produced by the industrial fishery processing industry include poultry grit, lime, and buttons from shells, seaweed products, fish foods, pearl products, leathers, and fertilizers. All industrial products were valued at $73 million in 1966.

The U.S. with only 6% of the world's population, catches and imports for consumption about 11% of the world's supply of fishery products. In the years immediately following World War II, the domestic fleet was able to supply the country with nearly all its demand for fishery products. Imports were small, accounting for only 14% of the needed total supply. Failure of the U.S. to maintain a modern fishing fleet; a rapidly increasing population that required additional supplies of seafood; and economic factors in foreign countries such as subsidy programs, lower wages, and other production costs, has drastically changed this balance in less than 20 years. Development of foreign trade is U.S. policy. The growing demand for fishery products in the U.S. provided foreign countries with sufficient incentive to send an ever increasing volume of fishery products to U.S. markets. In 1947, about 14% of the domestic supply of fishery products was imported. The U.S. accounted for 82% of the supply of edible products and 94% of industrial fishery products. In 1966, nearly 53% of the supply of edible products and 75% of industrial products were imported.

TABLE 12. VOLUME AND VALUE OF CANNED FISHERY PRODUCTS, VARIOUS YEARS (THOUSANDS OF POUNDS AND THOUSANDS OF DOLLARS)

Year	For Human Consumption		Bait and Animal Food		Total	
	Quantity	Value	Quantity	Value	Quantity	Value
1950	835,812	317,082	131,303	14,253	967,115	331,335
1955	588,307	275,382	257,073	28,193	845,380	303,575
1960	666,277	342,747	422,490	44,848	1,088,767	387,595
1965	749,518	446,833	372,111	48,398	1,021,629	495,231

TABLE 13. CURED FISHERY PRODUCTS, VARIOUS YEARS
(THOUSANDS OF POUNDS AND THOUSANDS OF DOLLARS)

Year	Quantity	Value
1950	85,429	35,022
1955	80,501	37,684
1960	68,319	42,845
1965	74,448	54,166

TABLE 14. VALUE OF INDUSTRIAL PRODUCTS
VARIOUS YEARS

Year	Meal	Oils	Other Products	Total
1950	29,254	17,473	30,461	77,188
1955	34,730	14,805	32,891	82,426
1960	25,282	13,386	27,927	66,595
1965	35,707	14,852	32,271	82,830

Imports for consumption in 1966 were valued at $723,000,000. About 120 different seafood items were imported in 1966 from over 125 countries through about 40 U.S. Customs Districts.

In contrast to the import situation, exports of fishery products from the U.S. to foreign countries are small although relatively important. In 1966, exports of domestic fishery products were valued at $85 million and exports of foreign-produced fishery products, $14.8 million. Although exports were sent to over 115 countries in 1966, about 52% of the shipments were sent to Europe. Canada and a few countries in Asia also received considerable quantities from the U.S. and are important. Trade with other countries is rather small and relatively unimportant. Recently, the Bureau of Commercial Fisheries, cooperating with the domestic industry, has been promoting U.S. seafood products in food fairs and shows in Europe in an effort to increase exports to world markets and create a more equal balance in total U.S. trade in fishery products.

Trend by Countries and Areas

In 1966, U.S. fishery products were exported to 114 countries. Of total exports, 72% was shipped to five countries: Canada, the United Kingdom, Japan, Sweden, and the Netherlands (Table 16). Shipments to Canada increased substantially. Trade with the United Kingdom and Japan was on the same level as in 1964, while the value of exports to Sweden and the Netherlands was less than in the previous year. Exports to Europe accounted for 55% of this trade; to North America, 26%; Asia, 17%; and the remaining to South America, Africa, and Oceania.

The future of the U.S. commercial fishery industry is not bright. Per capita consumption has not increased in many years. Landings are declining and the chief source of all fishery products is presently from foreign nations. Such major resources as menhaden, pilchards, and New England groundfish are not being maintained.

TABLE 15. SUMMARY OF IMPORTS AND EXPORTS OF FISHERY PRODUCTS, 1950–1966
(THOUSANDS OF POUNDS AND THOUSANDS OF DOLLARS)

Year	Imports				Exports			
	Edible Products		Nonedible Products	Total	Edible Products		Nonedible Products	Total
	Quantity	Value	Value	Value	Quantity	Value	Value	Value
1950	639,725	158,414	39,882	198,296	121,623	18,856	8,618	27,474
1951	646,668	158,363	54,094	212,457	165,624	27,072	8,659	35,731
1952	705,118	183,121	57,308	240,429	62,056	15,511	6,436	21,947
1953	726,195	195,869	49,611	245,480	69,308	17,084	10,794	27,878
1954	804,054	203,722	48,687	252,409	62,724	16,238	15,289	31,527
1955	780,185	208,973	49,896	258,869	109,750	24,923	15,054	39,977
1956	801,655	234,699	48,031	282,730	101,918	22,939	16,564	39,503
1957	900,227	252,788	46,487	299,275	85,221	20,549	15,403	35,952
1958	1,020,326	283,822	46,959	330,781	65,468	19,440	11,564	31,004
1959	1,141,114	314,650	55,467	370,117	80,688	26,747	17,495	44,242
1960	1,095,014	310,596	52,685	363,281	61,454	25,622	18,543	44,165
1961	1,087,175	339,318	61,301	400,619	40,137	19,594	15,116	34,710
1962	1,255,532	405,832	83,975	489,807	56,530	22,470	13,258	35,728
1963	1,196,977	399,928	100,784	500,712	64,745	30,376	26,229	56,605
1964	1,318,099	433,674	130,569	564,243	94,835	42,878	21,326	64,204
1965	1,398,778	479,412	121,492	600,904	96,444	49,308	20,175	69,483
1966	1,593,359	568,216	152,231	720,447	109,604	62,882	21,931	84,813

Table 16. U.S. Exports of Fishery Products by Selected Countries of Destination, 1961–1966 (Thousands of Dollars)

Country	1961	1962	1963	1964	1965	1966
Canada	10,265	8,846	11,156	10,434	15,542	14,886
United Kingdom	4,554	8,249	13,081	15,102	15,530	19,004
Japan	2,984	939	7,819	9,200	9,940	17,994
Sweden	1,665	1,076	4,473	6,425	4,875	2,653
Netherlands	2,385	2,273	2,593	4,879	4,127	4,380
France	1,007	1,073	1,889	2,325	3,177	5,814
West Germany	1,555	1,467	3,638	3,146	2,616	2,566
Belgium	351	547	445	1,115	2,076	2,494
Italy	423	869	1,643	656	1,760	1,571
Switzerland	738	1,712	2,229	1,284	980	1,673
Greece	364	487	566	471	912	761
Australia	458	198	203	426	736	969
Hong Kong	368	383	388	603	677	841
Norway	2,390	403	1,539	1,064	420	1,278
Philippines	582	320	403	1,043	363	542
Venezuela	360	274	183	238	250	293
Mexico	459	375	263	417	204	300
Ecuador	82	171	1	38	15	9
Cuba	—	243	—	—	—	—
Other	3,720	5,823	4,094	5,338	5,283	6,785
Total	34,710	35,728	56,606	64,204	69,483	84,813

Despite vessel construction subsidies, fleet improvement is slow. Modern scientific methods of harvesting are yet to be applied to catching fish. Little research has been directed toward understanding the structure of the industry, and demand studies are almost unheard of. There is considerable reluctance to supplying even basic data so necessary for much of this research.

Charles H. Lyles

Cross references: *Freezing Fish at Sea; Japanese Fisheries; Peru Fishery; Sport Fishing; U.S.S.R. Fisheries.*

U.S.S.R. FISHERIES

The highly developed Soviet fishing industry, now in fourth place behind Peru, Japan, and Communist China, deserves a hard look by U.S. fisheries, now lagging in sixth place. The Soviets are modernizing and automating their fishing vessels, at the same time gaining both oceanographic intelligence data and world prestige. Fisheries are a basic industry in the Soviet economy, providing about 1/3 of the annual total animal protein consumed in the U.S.S.R.

During the last 25 years its catch quadrupled from 1.4 million metric tons landed in 1940 to over 6.7 million metric tons landed in 1968. In 1969, Soviet fishery landings will reportedly further increase to over 7.0 million metric tons. The annual rate of increase has been fairly consistent during recent years, although there is reason to believe that the Soviets may find it difficult to maintain. The preliminary 1968 catch estimates indicate that the Soviets barely achieved their catch plan. Major factors behind these increases were: generous investments in the distant-water fleet; the introduction of flotilla fishing where the trawlers are accompanied by freezing and processing vessels; the creation of a large marine research organization; and the expansion of operations into all the world's oceans. During 1946–1965, Soviet fishery investment amounted to an estimated $4 billion, most of which was used for domestic construction or purchase abroad of new fishing vessels.

In 1967 the Soviet catch of fish and shellfish amounted to 5.8 million tons. Catches of whales, other sea mammals and the harvest of aquatic products totalled about 900,000 tons. In 1965 the Soviet Union caught about 11% of all fishery resources landed throughout the entire world, although her population amounted to only about 7% of the world total.

Soviet vs U.S. Fisheries

In contrast to Soviet fishing activities is the decline of U.S. fisheries, which in 1966 were the smallest since 1953. During 1966, the U.S. imported 65% of its fish or almost 3.7 million tons including some Soviet fishery products. In effect, more than every other fish in a U.S. frying pan is imported. While the Soviet fish catch has quadrupled, that of the U.S. has actually declined.

The U.S. has traditionally been one of the world's major producers of fishery products. Before World War II and until 1959, the U.S. ranked second only to Japan in size of catch. It dropped to third place in 1959, behind the expanding fisheries of mainland China. Following closely, the fisheries of Peru and the Soviet Union surged ahead, and in 1960 the U.S. dropped to fifth place among the fishing nations of the world. By 1967, the U.S. was relegated to sixth place by Norway.

Several factors have contributed to this loss of stature as a world leader in the fishing industry. For some U.S. fishermen it has not been possible to maintain their competitive position against foreign imports. As a result, productivity has not improved in certain segments of the industry. The U.S. presently has 12,300 fishing vessels above 5 gross tons and the average age of these is about 20 years. Only about 600-odd vessels entered the fisheries in 1966, most of which were wooden shrimp boats built for the relatively prosperous shrimp fishing industry in the Gulf of Mexico.

Another factor is the high cost of boat-building. A 1792 law still on the books forbids U.S. fishermen to land catches in a foreign-built boat.

Organization of Soviet Fisheries Activities

By comparison, the Soviets attempt to organize their fishing activities as a science. Because of the centralized nature of the Soviet state, the U.S.S.R. Ministry of Fisheries controls the funds and coordinates a general plan for all stages of the fishing industry from landings by fishing vessels to the work of fish-processing plants and fishery research institutes. The Government can and does dictate the diversion of capital investments into fisheries; it also determines the prices for fishery products as well as the salaries of the fishermen.

The U.S.S.R. Ministry of Fisheries is headquartered in Moscow whence it supervises the various Territorial Administrations, which are divided functionally (one for marine fisheries, one for fresh water fisheries, and one for fisheries conservation and reproduction) but closely cooperate with each other.

In marine fisheries, the executive authority is delegated to five so-called "Main Administrations" each of which is responsible for coastal, offshore and high seas fisheries in various geographical areas:

Northern (White Sea, Barents Sea, Norwegian Sea, Northwest Atlantic, and North Atlantic);
Western (Baltic Sea, North Sea, North Atlantic, including Northwest Atlantic, Gulf of Mexico, Southern Atlantic);
Black Sea (Black and Azov Seas, Mediterranean, Red Sea, and Indian Ocean);
Caspian Sea;
Far Eastern (Sea of Okhotsk, Bering Sea, North and South Pacific, and Indian Ocean). These administrations are controlled directly by the Fisheries Ministry, and in turn control smaller administrative units, called Regional Administrations. Inland (fresh water) Fisheries and Fisheries Conservation Administrations are also divided into smaller units.

Soviet trade unions bargain with the Fisheries Ministry on employee production norms, and also have a voice in hiring and dismissal policies. Reportedly a joint "labor disputes committee" also exists.

Emphasis on Research

Research is the foundation of Soviet fishery development plans. The Soviet Institute of Marine Fisheries and Oceanography (VNIRO) coordinates the work of 22 research laboratories employing over 900 scientists and 3000 technicians and workers. Leningrad's Research Institute of Lake and River Fisheries handles fresh water research. The All-Union Research Institute of Pond Fisheries in Moscow handles fresh water and fish farming research. Altogether 135 research laboratories with over 2000 scientists are engaged in fishery research. The Soviets are using a converted submarine, the *Severyanka*, for fishery research.

The Soviets are working on ways to forecast locations and migrations of fish schools from oceanographic and meteorological data. The procedure, which is also applied in the U.S. as a kind of fish-forecasting service, is similar to weather forecasting. Other research is aimed at causing fish to school and to be fished easily. Lights have been used to attract fish to the intake of pumps placed aboard fishing vessels. The fish are subsequently pumped aboard the vessels; no nets are used.

Cod-fishing ships have been experimenting recently with a conductor cable based on sonar principles, using a transducer fixed to the trawl. The transducer gives information on vertical and horizontal spread of the trawl to determine the catch efficiency of the net and the number of fish entering the net.

The Leningrad State Institute for designing fishing vessels has plans to develop an underwater laboratory to work at depths of up to 300 m. It is equipped with movie, TV, and still cameras, and has a lock for egress of divers and a section that can surface independently for rescue purposes.

The Soviets are fish farming by modifying the sea environment and by artificial breeding. A number of artificial breeding plants in the Caspian and Ural Seas are experimenting on changing the spawning cycle of fish. The work of these plants is being expanded.

Related to the artificial breeding facilities are a growing number of "incubating" stations, which serve as a new home for transplanted species of fish. Several examples of fish transplants are pink salmon eggs, king crab eggs, baby king crabs. Also, 300 mature crabs were successfully transported from the far eastern Sea of Okhotsk and the Bering Sea to the Barents and White Seas (near Norway and northern Russia).

The final success of these programs is not yet certain. Many salmon, which reproduced surprisingly well in the new habitat, "defected" into Norwegian rivers; some were found as far away as the United Kingdom. There is no evidence that the king crab transplantation program is successful but, in all fairness, it must be said that it is too early to make a final appraisal.

Two new Soviet vessels, *Akvarium I* and *Akvarium II*, carry live fresh water fish from the lower Volga River to Moscow, a 10-day journey. Live fish transportation is increasing yearly. The fish is brought part of the way in rail tankers and is then transferred to trucks. Part of the shipment is kept in a live box in the Moscow River port cold-storage area, which holds up to 500 metric tons of live fish. In 1965 more than 5000 tons of live fish were shipped to Moscow. In 1966 more than 6000 tons were shipped, and by 1970 up to 10,000 tons will be shipped. Several additional aquarium ships are planned.

The Soviet Union is engaged in a comparatively small experimental operation on its southern Pacific coast for the systematic growth of seaweed. In the same area they grow experimental patches of "sea cabbage," which they mix with tomatoes in a stewed concoction and sell for domestic consumption.

The Soviets produce fish meal from "trash" fish plus the residue (heads, guts, fins, and bones) of certain edible species. The meal is used as fertilizer and livestock feed.

At present, the Soviets appear to have done less than the U.S. in terms of research and development of fish-based dehydrated food suitable for human consumption. If they should get into large-scale production of such an inexpensive protein-rich food as FPC before the U.S. does, they could conceivably gain considerable leverage in international diplomacy.

Problems of Soviet Fisheries

As the industrialization of the Soviet economy continues, power plants and pollution are causing trouble (as in the U.S.). Due to the pollution problems in the Caspian Sea, a few shipments of caviar to foreign purchasers have been returned to the Soviet Union with the complaint that the sturgeon roe had a bad taste. For that reason a special institute for conducting sturgeon research at Astrakhan, located on the Volga River, close to its entrance into the Caspian Sea, has been investigating these complaints.

The Soviet Government is also beginning to crack down on industrial facilities that contribute to the increased pollution of water. It was recently reported that the manager of a large chemical plant was jailed for dumping improperly processed pollutants into a river. Wastes from his factory apparently had killed hundreds of thousands of sturgeon fingerlings.

Another problem facing Soviet fisheries has been the influence of hydroelectric power dams on anadromous fish—fish such as sturgeon that go upstream to spawn. The obstacle presented by the dams for these fish has been almost solved through the construction of "fish ladders." During 1963–1965, the dam at Volgograd has been used for an interesting device, known as a "fish elevator." The fish are attracted to swim against an artificial current at the bottom of the dam. When a sufficient number move into position, a lock-like slat is lowered creating a wall that isolates the fish inside from others and releases them to swim upstream. As long as the fish attempt to move upstream the process is repeated. After the fingerlings develop and begin to move downstream, it is said that they pass through the turbines at the dam without harm. This is not consistent with experience in the U.S. and it is probable that the Soviet Union also suffers losses as fingerlings pass through turbines.

Foreign Policy with Regard to Fisheries

An important area of Soviet combined foreign policy and "business" is their program of assistance to the newly developing nations in many parts of the world. For example: the U.S.S.R. has offered large fishery development projects to India, Ceylon, and Tanzania, and is currently proposing to provide Senegal with a modern tuna fleet and processing industry. In turn, Soviet fishing, merchant, and oceanographic ships can resupply or be repaired at Senegal's port of Dakar.

Another country, Ghana, had one of the most rapidly developing domestic fishing industries in Africa largely due to exploitation of available fishery resources with aid from the U.S.S.R. After Nkhrumah's overthrow, however, Soviet influence rapidly decreased, although frozen fish are still being landed in Ghana from Soviet fishing vessels.

Although most of the fish caught by the Soviet fishing fleets is for domestic consumption, a significant amount is landed in underdeveloped countries. Soviet vessels unload more than 2000 tons of fresh frozen fish every month in Nigeria. The same thing is happening in the Congo (Brazzaville), Liberia, Sierra Leone, Guinea, and other countries.

The U.S.S.R. is building a shipyard for Egypt at Alexandria on the Mediterranean Sea and a fishing port at Ras Banas on the Red Sea. She has aided in the construction of a modern, highly sophisticated fishing port and terminal at Havana, Cuba.

The Soviets are using their strengthened maritime position to further Communist political objectives. The fruits of heavy Soviet exploitation of the oceans—food rich in protein—can be offered to hungry countries, in exchange for sympathy and support in international forums. Further efforts are probable as Soviet ocean-ranging fleets move into waters off South America where population growth could explode into political and economic upheaval.

Thus the Soviet fishing fleet is an effective arm for extending the U.S.S.R.'s influence throughout the world, particularly to the developing nations.

Soviet Vessels off U.S. Coasts

The Soviets have maintained a 12-mile territorial sea limit off their shores, but Soviet fishing vessels were free to come within 3 miles of the U.S. coast until Congress extended the U.S. fisheries jurisdiction to 12 nautical miles in late 1966.

In 1966 over 340 Soviet fishing vessels and 20,000 fishermen operated on Georges Bank fishing grounds and in the mid-Atlantic Bight. Of that number, 124 vessels were identified as large stern factory trawlers, over 150 were medium side trawlers, and 63 were support and processing vessels. The total number of Soviet fishing vessels sighted off the U.S. Atlantic coast in 1966 was below the 460 sighted in 1965, but because more large stern factory trawlers were fishing in 1966 (whose capacity is 5–7 times that of a medium trawler), the fishing effort probably remained the same. In 1965, the U.S.S.R. landed over 500,000 metric tons of fish on Georges Bank and an estimated 35,000 tons in the mid-Atlantic Bight.

Of course not all the vessels sighted fished off U.S. Atlantic coasts the entire year. The largest concentration of Soviet fishing vessels occurred in late June 1966, when a total of almost 160 fishing and support vessels operated on Georges Bank (Soviet fishing in mid-Atlantic only lasted from January to early June). Of these, 65 were large factory stern trawlers, 77 were medium trawlers, and 15 were support and processing vessels. Some experts feel that the giant trawlers of the Soviet Union are sweeping up immature fish before they can spawn, but it is difficult to document such allegations scientifically.

The Soviets are not limiting their fishing to the U.S. east coast; recently west coast fishermen were distressed by Soviet fishing operations close to shore.

Soviet-U.S. Agreements

In 1964, the expanding Alaska king crab industry was seriously threatened by the proliferation of Soviet crab-fishing fleets into the Gulf of Alaska. The U.S. Government, however, requested that the U.S.S.R. abstain from fishing for king crab in the Gulf of Alaska and in 1965 concluded an agreement with the Soviet Government under which the exploitation of king crab resources in the Bering Sea also is effectively policed by the U.S. As a result, the Soviet catches of adult male crabs in the eastern Bering Sea decreased from an average annual catch of 3.0 million crabs during 1960–1964 to 2.4 million crabs in 1965–1966. During the negotiations for the 2-year renewal of the king crab agreement, the U.S. obtained a 15% decrease in the U.S.S.R. catch quota applicable to the 1967–1968 fishing seasons.

In July 1966 fishery experts of the U.S. and the Soviet Union concluded a week of technical discussions in Moscow on problems relating to the conservation and use of fishery resources off the U.S. coast. The two delegations agreed to recommend to their respective Governments measures to alleviate the short-term problems and to establish procedures looking to long-term solutions.

Among the recommendations were the following proposals:

(1) that scientists and technical experts of the two countries meet periodically;

(2) that exchanges of fisheries personnel aboard fishing and research vessels of the two countries in both the Atlantic and Pacific areas be initiated;

(3) that the Soviet Government take action to ease problems arising out of concentrations of vessels on fishing grounds customarily used by American fishermen;

(4) that there be no Soviet fishing within 12 miles of the Washington-Oregon Coast, except for research vessels;

(5) that special instructions be issued to the Soviet fleets in the Pacific Northwest reiterating earlier instructions not to fish for salmon.

Senator Warren G. Magnuson discussed the impact of Soviet fisheries with U.S.S.R. officials in the fall of 1966, during his visit to Moscow. During the same year, the scientists of both countries discussed the state of fishery stocks of mutual interest in the northeastern Pacific along the coast of the states of Washington and Oregon, and in the mid-Atlantic Bight. Such discussions were necessary to establish some common basis for estimating the size of the standing stocks and the maximum sustainable yield of certain species of fish exploited jointly by U.S. and Soviet fishermen. The talks also dealt with navigational and other technical problems that have arisen because of the appearance on the traditional fishing grounds of U.S. fishermen of large groups of Soviet vessels using different fishing tactics. The U.S. and the U.S.S.R. delegations identified problem areas and reviewed data presented by both sides and laid foundations on which a future fishery agreement could be reached.

The final negotiations on problems arising out of greatly increased Soviet fishing operations off the U.S. coasts took place in Washington during January-February 1967 and were preceded by a renegotiation of the U.S.-U.S.S.R. King Crab Agreement.

The Agreement, of one year's duration, specifies several areas seaward of 12 miles from the Oregon-Washington coast in which Soviet vessels would either refrain from fishing or from concentrating their efforts. In certain other areas off the Oregon-Washington coast, measures would be taken, jointly and separately, to protect stocks of fish. Additional protection would be provided for the fishing gear of U.S. halibut fishermen in areas near Kodiak Island, Alaska, in the halibut season. Under the Agreement, Soviet vessels would transfer cargoes in several designated areas off Washington and Oregon, and off Alaska in the 9-mile zone contiguous to the U.S. territorial sea. Soviet vessels would also continue to fish within the 9-mile exclusive contiguous zone for the duration of the agreement in two limited areas of the central and western Aleutians and a smaller area in the northern Gulf of Alaska.

The Agreement also provides for cooperation in scientific research, exchange of scientific data and personnel, exchanges of fishermen or their representatives aboard vessels of the two countries, and general procedures for reducing conflicts between vessels and gear of the two countries.

The two delegations also reviewed certain fishery problems of the U.S. Atlantic Coast, and agreed that these matters should be considered further at a meeting to be held in late May 1967 just prior to the annual meeting of the International Commission for the Northwest Atlantic Fisheries. This meeting resulted in full-scale negotiations in Moscow during which an agreement was reached on fishery problems in the mid-Atlantic Bight. This agreement was signed in Moscow in November 1967 and renegotiated with some small changes in December 1968 for the next 2 years (1969–1970).

In 1964, the Department of State discussed gear conflict off Kodiak with Soviet officials, and got agreement that lessened crab pot losses in the area. The agreement provides for the establishment of a number of areas in the vicinity of Kodiak Island, where U.S. king crab pots are concentrated, in which Soviet mobile bottom trawl gear will not operate during July-October, inclusive.

Future Plans

The Soviets are rapidly expanding their fishing industry and plan to build more and more large ocean-going trawlers and factory ships. With these, they will spread their fleets farther south. The Soviets plan to depend more and more on large, ocean-going factory and mother ships. Fish can be processed on board, and ships can stay at sea a year or more, while the crews are rotated about every 3–5 months, often by passenger liners.

The 5-year Soviet plan for fishing industry development provides for a 50% increase over the 1965 fishery landings by 1970. By then, total fishery production should reach about 8.5 million tons. Of this, 7.8 million tons will be fish and shellfish catches, and the rest whales, other marine animals, and aquatic products. Up to 90% of the Soviet fish will be caught on the high seas. Whether or not the U.S.S.R. will be able to fulfill this ambitious plan remains to be seen; at the time of this writing it seems almost certain that the 1970 catch quota will not be reached.

HASTINGS KEITH

V

VITAMINS FROM MARINE SOURCES

Vitamins from marine sources may be considered in two categories: (1) vitamins that are extracted and concentrated (oil-soluble vitamins A and D); and (2) vitamins that are not extracted, but are valuable nutritional components when marine products are used as human or animal food (these include water-soluble vitamins, such as those in the B complex).

Vitamins A and D

Cod-liver oil was used for its medicinal properties, for example, as a cure for rickets long before vitamins A and D were established as the effective factors by McCollum (Refs. 7, 8) and Zucker (Ref. 10). Although the physiological modes of action of these vitamins are still unknown, their chemistry is now understood. In fact, both vitamins are made synthetically. Before synthetic vitamin A became available, marine sources were the most important natural sources of vitamins A and D, and the oils and concentrates used to supply these vitamins for medicinal or food uses or for animal feeds were processed from fish livers or viscera.

In the U.S., the fish-liver oil industry was an important branch of the fishing industry from 1930 to 1950. The volume of production reached a peak in 1941 and then gradually decreased. This reduction apparently was due to a shortage in the supply of raw material, and attempts were made to find new sources of vitamin-containing oils. However, in 1949, synthetic vitamin A became available in large quantities and liver oils decreased in price abruptly and dramatically. Thereafter, U.S. production of vitamin oils from fish livers became negligible.

Despite this situation in the U.S., the production of liver oils for vitamin concentrates is still a valuable and important industry in other countries. In 1963, the world production of liver oils was 65,000 metric tons. The principal producing countries in order of production were United Kingdom, Norway, Iceland, Japan, France, and Poland.

Vitamin concentrates from fish livers are imported into the U.S. from the producing countries and used both as a medicinal grade in pharmaceuticals and for incorporation into animal feed mixtures. In 1964, the U.S. imported 1,908,000 gal of liver oils valued at $1,631,000 (Ref. 6). This amount is small compared with that in 1948 when the U.S. produced 740,137 gal of liver oils valued at $12,508,000 and imported 2,653,000 gal valued at $14,146,000.

The livers of a number of species of fish contain relatively large amounts of vitamins A and D. Since these vitamins are oil-soluble, they are associated with the lipid fraction. The livers are categorized on the basis of (1) high oil content-low vitamin A potency, (2) low oil content-high vitamin A potency, and (3) high oil content-high vitamin A potency. Examples of the three categories are livers from (1) cod and dogfish shark, (2) halibut, tuna, and whale, and (3) soupfin and miscellaneous species of shark. The most important source of raw material is the cod family, which is the principal source for medicinal grade oils.

A number of factors other than species of fish affect the potency of the oils. These include age, size, sex, nutritional condition, and spawning stage of the fish as well as season and geographical source. The value of the oils is related to vitamin A potency rather than to vitamin D potency. If a particular vitamin D potency is required, the necessary amount of synthetic vitamin D is blended into the oil. Ranges of potency of vitamins A and D in some of the important commercial species are given in Table 1.

The methods of handling and processing the raw material have been developed to obtain high-quality oils that are light colored with little or no flavor and odor and with a minimum of free fatty acids.

Treatment of raw material. High-quality oils can be obtained only if the fish are eviscerated rapidly and if the processing of the livers is started immediately after the fish are caught. In some areas the fish livers are rendered on board ship to effect a crude separation and the oils are then further refined at shore plants. If the processing is delayed for any reason, a number of problems will arise, and their severity will be directly related to the amount of decomposition that has occurred in the livers. Processing livers in which decomposition has started is complicated

TABLE 1. AMOUNTS OF VITAMINS A AND D IN SOME FISH LIVER OILS USED TO PROCESS VITAMIN CONCENTRATES

Category and Species	Scientific Name	Oil Content, %	Vitamin A Content, USP Units per g of oil	Vitamin D Content I.U. per g of oil
High oil-low potency				
Cod	Gadus sp.	20–60	1,000–30,000	85–500
Dogfish shark	Squalus acanthias	50–72	2,000–25,000	5–25
Low oil-high potency				
Halibut	Hippoglossus hippoglossus	8–27	20,000–300,000	1,000–5,000
Tuna				
Albacore	Germo alalunga	7–20	10,000–60,000	25,000–250,000
Bluefin	Thunnus thynnus	4–6	25,000–100,000	20,000–70,000
Skipjack	Katsuwonus pelamis	4–6	30,000–60,000	25,000–250,000
Yellowfin	Neothunnus macropterus	3–5	35,000–90,000	10,000–45,000
Bonita	Sarda chiliensis	4–12	15,000–60,000	35,000–50,000
High oil-high potency				
Hammerhead shark	Sphyrnidae sp.	30–75	5,000–150,000	Negligible
Soupfin shark	Galeorhinus zyopterus	55–72	15,000–200,000	5–25

by problems such as formation of emulsions and increased difficulties in extracting the oil. These result in a lower yield of vitamins and sometimes a lower potency oil. In some instances, additional operations and refining are necessary, thus increasing the cost of operation. The final product, even with refining, often will be of lower quality and thus bring a lower price because of undesirable characteristics such as greater amounts of free fatty acids and odors and flavors that may be carried over with the oil.

If processing of the oil cannot start immediately the livers can be removed from the fish and frozen at once for future processing. Processing then must be done as soon as the livers are thawed. When freezing facilities are not available, chemical preservatives such as salt can be utilized. Clean, fresh livers cut into 2–3 in. sections and packed tightly into containers with 10% by weight of good quality salt and a minimum access to air can be preserved for several months before processing.

Methods of processing. A number of methods are used for processing (Refs. 1, 3, 9) with the method of choice being determined by the amount of oil present, by the potency, by the use to be made of the extracted oil, by the amount of yield desired, and by the size of the plant, which, in turn, determines the amount and the sophistication of the equipment.

Livers with high oil content and low vitamin A potency. Livers with a high oil content are processed with steam to free the oil. Both direct and indirect contact with steam are used, direct contact usually being preferred because the equipment is simpler and the livers are heated faster. The Melbu Cooker, developed in Norway, is an example of comparatively inexpensive, simplified equipment. The cooker, with a capacity of about 100 gal, has a waterjacket and is heated from below. Steam from the top of the jacket is directed to the bottom of the liver cooker. The livers are heated rapidly because of the dual effect of the waterjacket and the direct contact with live steam. The liver mass reaches the desired temperature of 185°–192°F (85°–90°C) in 1 hr. The cooked material is allowed to stand for a few minutes to permit the oil to separate. The oil is then decanted with a skimming device. The residue still contains substantial amounts of oil, most of which is recovered by pressing or by centrifuging while the residue is still warm. The residue still contains some oil, which is released by autolysis. The oil recovered from this fermentation process is dark, so it is used as an industrial oil. The residual press cake can be dried to a liver meal and used for animal feed as a source for protein and for water-soluble vitamins.

Shipboard processing. Shipboard processing is done out of Great Britain, where trawlers often remain at sea for several weeks and have equipment on board to render oil for vitamin production. Commercial equipment is available for this purpose. On these vessels, livers are removed from the fish, and steam cooking is started immediately. The livers are brought to the boiling point, the amount of steam is reduced, and the livers are simmered for 15 min. If too much steam is used at this point, poor separation of the oil results. After being cooked, the liver mass is allowed to stand to permit the oil to separate, a stage requiring about 20 min. The oil is then filtered and placed in storage tanks. The residue is stored separately. When the boat returns to the plant, the oil is centrifuged to remove water and suspended material; it then is dried under vacuum, cooled, and stored. The residue is treated to remove remaining oil, which is of lower grade and is used for technical purposes.

Livers with a high vitamin A potency. Livers with a high vitamin A potency require special processing methods to obtain maximum yield of oil and vitamins. These methods also are used for fish viscera with a high vitamin A potency. Originally, solvent extraction was used, especially for livers with a low oil content. These methods result in dark, often viscous oils with a high free fatty acid content and often result in destruction of vitamin A from oxidation. Because of these major disadvantages, they have not been used to any extent since the development of digestion procedures.

A mild alkali digestion is one of the simplest and most effective methods for the extraction of high-potency oils. The livers are usually ground, placed in a digestion tank, and mixed with ½–1 part of water to 1 part of liver. Alkali, either 1–2% by weight of sodium hydroxide or 2–5% by weight of sodium carbonate, is added; the mixture then is cooked with live steam and with agitation. The digestion is carried out at 180°–190°F (82°–88°C) and should continue until the liver particles are semicolloidal or have been liquefied. The average liver mass at pH 8–9 will digest at this temperature in 1 hr. The time, however, varies with a number of factors such as species, size of liver particles, pH, and amount of agitation. The processing techniques need to be varied to obtain maximum recovery of the oil and vitamins.

After digestion of the livers, the liquor is passed through a three-phase centrifuge. The oil, which is in the form of an emulsion that is broken by adjustment of the pH, is separated in a purifier-type centrifuge. The water phase sometimes is given a wash-oil treatment to remove any residual oil. An oil of low vitamin A potency is added, the mixture is heated to 175°F (80°C), and the oil is separated by centrifuging. This procedure is repeated as many times as is justified by further recovery of the vitamins. The wash-oil method can also be used in a counter-current system to increase the potency of oils of a low vitamin A value.

Refining of oil. When the oils are ready to market, they contain less than 0.3% moisture and no suspended protein or nitrogenous material. If the livers were fresh and if the oils have been properly processed, the oils will meet these conditions and do not require further refinement. The oils, however, are often "winterized" to remove stearin. The stearin is removed by slowly cooling the oils to 35°F (2°C) to cause the stearin to precipitate, which is then removed by filtration with a pressure filter. Medicinal grade cod liver oils are usually treated by steam stripping under vacuum to deodorize them. These oils are very light in color and nearly odorless and tasteless.

Vitamin Concentrates

Low-potency oils may be further processed by molecular distillation or by saponification to prepare vitamin concentrates or to obtain fraction containing specific vitamins. The oils are molecularly distilled under high vacuum to obtain a concentrate of vitamin A and to separate the glycerides, vitamin D, vitamin E, sterols, and free fatty acids. The conditions of molecular distillation are quite mild in that the oil is subjected to heat for too short a time to damage the oil and the vitamins. The process was developed for preparation of concentrates of vitamin A. Other fractions such as vitamin D, vitamin E, and sterols can be separated whenever economics and need warrant their recovery. Vitamin D was never recovered from this source in the U.S. because cheaper sources were available by the time molecular distillation was a commercial process. Vitamin D is produced in other countries, by this process, however.

Saponification is a much older process, and a number of methods have been patented. Basically, the triglycerides are separated into glycerol and fatty acids. The material not affected by this procedure is known as unsaponifiable matter, and the vitamin A is included in this portion. Thus the unsaponifiable matter is removed from the oil and purified to obtain the vitamin A concentrate.

Liver oils in Japan are molecularly distilled to fractionate them into vitamins A, E, and glycerides (Ref. 9). The vitamin A fraction may be used directly. Alternatively, the liver oil may be saponified and esterified to make a concentrate of vitamin A acetate or palmitate. Cholesterol is also produced from liver oils and squid oil and is used as raw material for the production of vitamin D_3.

Vitamins in Fishery Products

In general fish and fishery products are good sources of the water-soluble vitamins, including thiamine (B_1), riboflavin (B_2), nicotinic acid, pantothenic acid (B_6), vitamin B_{12}, folic acid, choline, and vitamin C. Although many gaps exist in our knowledge about the amount of these vitamins found in fish and shellfish, the information available has been compiled and reviewed (Refs. 2, 4, 5).

The B vitamins are usually considered as a group, for they seem to follow similar patterns in distribution and occurrence. Their distribution in different parts of the fish is important because the parts are utilized for various products. These vitamins are more generally distributed throughout the fish than vitamins A and D, and are found in the flesh as well as in the organs. The lateral dark flesh is a good source of these vitamins; in many species, their concentration in these areas is as great as in the liver. The light flesh also contains enough of the vitamins that they make a significant contribution to human nutrition. Very little work has been done on cooked fish despite the interest from the nutritional standpoint because of the number of variables involved in each method of cooking. In general, however, the losses will not be great if the fish is not cooked in too much water, or for too long, or at too high a temperature. The canning process seems to cause little or no decrease in most of the vitamins. The effect of the salting process on vitamin content varies from a negligible loss if the fish are lightly salted to as much as 50% if they are heavily salted.

The vitamin content of fish meals is dependent on the raw material used and on the method of processing, with both temperature and length of time at elevated temperatures affecting the vitamin content.

The interest in the vitamins present in fishery products is primarily from a nutritional standpoint because the prices of the fishery products are not directly affected by the amounts present, as they are in the case of vitamin oils, and because the marine sources are not used to manufacture preparations of these vitamins. Nevertheless, these vitamins do play a role in the market and in the demand for products such as fish meals and fish solubles. The liver meals, which are the dried residues from the extraction of the oils containing vitamins A and D, contain all the water-soluble vitamins found in the livers and consequently are good sources of the vitamin B complex. Fish meals made from whole fish or waste contain a fair amount of the water-soluble vitamins, but about half of the vitamins present in the original raw material will go into the stickwater during the cooking-pressing operation. These vitamins are not lost however. The stickwater is either put back into the meal to process a "whole" meal or is condensed to 50% solids and used as a supplement for animal feeds. The vitamins present in the fish meal and fish solubles are included in the computations made on mixed feed formulations to obtain vitamin requirements in the feeds. This is true for processed pet foods as well as for feeds for poultry, swine, and cattle.

NEVA L. KARRICK

References

1. Aure, L., "Manufacture of fish liver oil," in "Fish Oils," M. E. Stansby, Ed., 1967, Avi Publishing Company, Westport, Connecticut.
2. Braekkan, O. R., "B-Vitamins in fish and shellfish" in "Fish in Nutrition," E. Heen and R. Kreuzer, Eds., 1962, Fishing News (Books), London.
3. Butler, C., "The Fish Liver Oil Industry," Fishery Leaflet 233, 1955, U.S. Fish and Wildlife Service, Washington, D.C.
4. Higashi, H., "Vitamins in fish," in "Fish as Food," G. Borgstrom, Ed., Vol. 1, 1961, Academic Press, New York.
5. Love, R. M., Lovern, J. A., and Jones, N. R., "The Chemical Composition of Fish Tissues," Food Investigation Special Report No. 69, 1959, Dept. of Scientific and Industrial Research, London.
6. Lyles, C. H., "Fishery statistics of the United States 1964," "Statistical Digest," No. 58, 1966, Bureau of Commercial Fisheries, Washington.
7. McCollum, E. V., and Davis, M., "Necessity of certain lipids in diet during growth," *J. Biol. Chem.*, **19**, 245 (1914).
8. McCollum, E. V., Simmonds, N., Becker, J. E., and Shipley, P. G., "Experimental rickets," *J. Biol. Chem.*, **53**, 252–312 (1922).
9. Sone, H., "Fish oils and the fish oil industry in Japan," in "Fish Oils," M. E. Stansby, Ed., 1967, Avi Publishing Company, Westport, Connecticut.
10. Zucker, T. F., Pappenheimer, A. M., and Barnett, M., "Observations on cod liver oil and rickets," *Proc. Soc. Exptl. Biol. Med.*, **19**, 167–169 (1922).

Cross reference: *Fish Meal.*

W

WHALES AND WHALING

Whales belong to the mammalian order Cetacea, a wholly marine group that also includes the dolphins and porpoises. The Cetacea are divided into three suborders: Archaeoceti, exclusively fossil; Mysticeti, baleen or whalebone whales, which includes most of the commercially important species; and Odontoceti, toothed whales, comprising sperm whales, beaked whales, dolphins, and porpoises. The term "whale" does not indicate a natural division of the order and sometimes it is used to mean the whole species of the cetacea, irrespective of size. In practice, however, it is generally used for any species of the order that attains a body length exceeding about 12 ft. The living species in this category are listed below together with their approximate maximum body lengths. In the following list the distributions of each species are also shown briefly.

Order Cetacea

Suborder Mysticeti (baleen or whalebone whales)
  Family Balaenidae (right whales)
    Balaena mysticetus (Greenland right whale or bowhead), 18m, Arctic Ocean.
    Eubalaena glacialis (black right whale), 18m, temperate waters of the North Atlantic, North Pacific, and southern hemisphere. Several names are given, but probably these are local races of one species.
    Caperea marginata (pigmy right whale), 6m, Southern hemisphere. Uncommon.
  Family Eschrichtidae
    Eschrichtius gibbosus (gray whale), 13.7m, North Pacific Ocean, but subfossil representatives occur also in the North Atlantic.
  Family Balaenopteridae (rorquals and humpback whale)
    Balaenoptera musculus (blue whale), 30m, all oceans. A smaller kind, known as the pigmy blue, has recently been found in the waters around Kerguelen Island (Ichihara, 1961) and was named as a subspecies, *B. m. brevicauda* (Zemsky and Boronin, 1964).
    B. P-Physalus (fin whale), 26m, all oceans.
    B. borealis (sei whale), 18m, all oceans, but seems to avoid coldest waters.
    B. edeni (Bryde's whale), all tropical and subtropical seas.
    B. acutorostrata (minke or little piked whale), 10m, all oceans, though rare in tropical waters. Certain varieties have been reported and the southern form is named *B. bonaerensis*.
    Megaptera novaeangliae (humpback whale), 15m, all oceans.
Suborder Odontoceti (toothed whales)
  Family physeteridae
    Physeter catadon (sperm whale), 18m, worldwide in tropical and temperate waters, though some of the males are found in cold waters.
    Kogia breviceps (pigmy sperm whale), 4m, worldwide in tropical and temperate waters. *K. simus* may represent a separate species. Both relatively rare.
  Family Ziphiidae (beaked whales)
    Hyperoodon ampullatus (bottle-nosed whale), 9m, North Atlantic. In the southern hemisphere its congener *H. planifrons* occurs, but has no commercial importance.
    Ziphius cavirostris (Cuvier's beaked whale), 8m, all oceans.
    Berardius bairdii (Baird beaked whale), 12m, North Pacific Ocean. In the southern hemisphere its congener *B. arnuxii* occurs, but has no commercial importance.
    Mesoplodon, 5m, the genus has a worldwide distribution. About 10 species, though most of them are apparently rare.
    Tasmacetus shepherdi, 6m, known only from New Zealand, rare.
  Family Monodontidae
    Monodon monoceros (narwhal), 4.8m, Arctic Ocean. Formerly of economic importance, on account of its tusk and oil.
    Delphinapterus leucas (white whale or beluga) 5.5m, Arctic Ocean and adjacent seas.
  Family Delphinidae
    Orcinus orca (killer whale), 9m, all oceans.
    Pseudorca crassidens (false killer whale), 5.5m worldwide in tropical and temperate waters.
    Globicephala melaena (pilot whale or blackfish), 8.5m, A boreal species in the North

Atlantic Ocean, though a number of different forms have worldwide distribution. *G. scammoni* occurs in the North Pacific and *G. macrorhyncha* in the tropical and subtropical waters of the world.

Of the Mysticeti the Greenland right, black right, and gray whales had been hunted almost to extinction in the old days of whaling with open boats and hand harpoons by the close of the 19th century. Under protection of about 30 years the gray whale is regaining rapidly its numbers in recent years, though still there is no scientific evidence for the recovery of the right whales.

In the modern whaling the species belonging to the family Balaenopteridae, excluding the minke whale, and the sperm whale out of the Odontoceti are only of major importance. Other small species are the object of local industries or are taken occasionally among catches of large whales. The minke and bottle-nosed whales are taken mainly from Norway and the Baird beaked whale from Japan. The pilot whale is seen in schools, sometimes amounting to several hundreds, and taken by driving from Faroe Island and also from Newfoundland in considerable numbers. The white whale enters the mouths of rivers and is taken by nets. It is hunted not only for its oil but also for its skin which provides valuable leather, generally called "porpoise-hide." The killer whale is taken in rather small numbers but is of some significance as a predator on other marine animals. Damages on longline catches of tunas by killer whale (and possibly also by the false killer whale) are reported from the Pacific and Indian Oceans. It also attacks the carcasses of larger whales before they are hauled out of the water. Dolphins and porpoises are also sometimes deemed a nuisance by fishermen when they destroy the fishing gear and rob the catch in some localities.

Most baleen whales are distributed widely both in the northern and southern hemispheres. There is, however, an equatorial gap between the northern and southern populations and no very free interchange between the hemispheres is considered. Their lives are dominated by the seasons, and they migrate over long distances between the summer feeding grounds in high latitudes and the winter breeding grounds in low latitudes. They are so-called "filter-feeders" and live on planktonic crustaceans, which are produced abundantly in cold waters in high latitudes. In the Antarctic the krill, *Euphausia superba,* is the staple food of all the baleen whales, though some other kinds of euphausiid and amphipod are reported from the stomachs of whales near the Antarctic convergence.

Sperm whales differ in many respects from the baleen whales. They feed on squid, though fish also have occasionally been found in their stomachs. Sperm whales are polygamous and the male is much larger than the female whereas baleen whales are monogamous and the female is slightly bigger than the male. These are reflected in the distribution of the sperm whales. Most of them, including virtually all the females and probably the immature of both sexes, remain in tropical and warm temperate waters, and only some of males, probably surplus to the breeding schools, wander into higher latitudes. Seasonal movement of main herds also takes place, but there is no apparent segregation into northern and southern populations. They are divided rather into Atlantic, Pacific, and Indian Ocean stocks. Figure 1 shows the essential form of the migrations.

Abundance

Modern whaling began in 1864 when the Norwegian, Svend Foyn, invented a new technique using harpoon guns and steam catcher boats. This permitted the taking of blue and fin whales, which had been too large and fast for the old whaling methods. Modern whaling was at first carried out from small land stations on the north coast of Norway, but soon it shifted to Iceland, the Faroe, Shetland, and Hebrides Islands, and Ireland. Later it spread to Newfoundland, and to Korea, Japan, British Columbia, and Alaska in the north Pacific.

In 1904 Norwegian C. A. Larsen established a land station at South Georgia with Argentinian capital, which was the start of the Antarctic whaling. Antarctic whaling too was at first conducted from land stations on South Georgia as well as by factory ships moored in harbors at South Shetland and other islands. In the 1925–1926 season, however, a factory ship, with a slipway

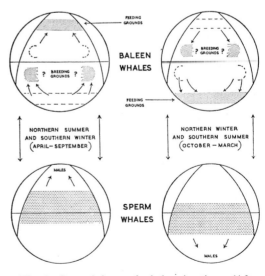

Fig. 1. General form of whale migrations. (After Mackintosh, 1965.)

to take the carcasses out of the water, operated on the high seas, independently of land bases or anchorages. The invention of the slipway led to a great expansion of the Antarctic whaling and in the 1930–1931 season 41 factory ships operated with more than 200 catcher boats. In this season the catch by pelagic factories alone amounted to 37,465 whales, including 28,325 blue whales, and the production of whale oil in the total Antarctic exceeded 600,000 tons. The catch of blue whales as well as oil production are both the highest ever recorded in the Antarctic.

In the following season most of the pelagic factories were laid up due to the overproduction in the previous season, but since 1932 the Antarctic whaling has been subject to restrictions, first by production agreement between whaling companies and later by agreement or convention among countries concerned.

Postwar whaling in the Antarctic began in 1945–1946, but an over-all limit was agreed upon in 1944. It was first set at 16,000 blue whale units (1 blue = 2 fin = 2.5 humpbacks = 6 sei), which was about ⅔ of the average prewar catch in respect of the mass of whales taken. Major species taken in the prewar whaling were humpbacks (in Atlantic sector by land-based operation), blue and then fin whales, but the catch of blue whales declined year by year. In the postwar season, too, this decline has continued, and reciprocally the whaling had been supported by increased catch of the fin whale until 1961–1962 season. Since then also the catch of fin whale has decreased and in recent seasons the dominant baleen whale taken in the Antarctic is the sei whale (Fig. 2).

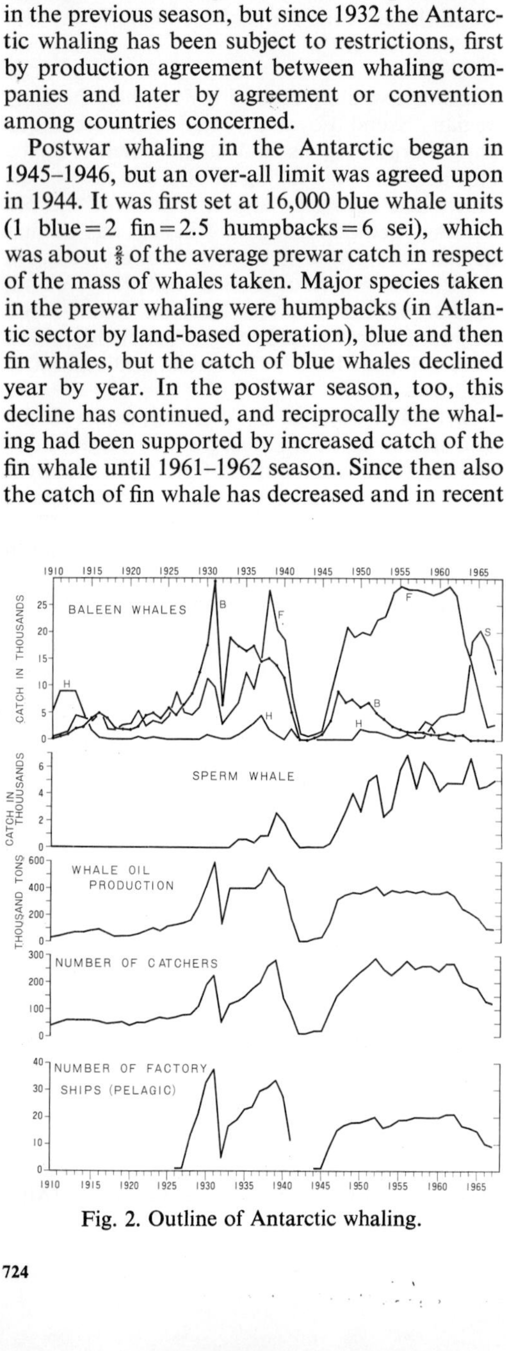

Outside the Antarctic, whaling has been carried on from land stations in various localities. In the north Pacific, however, several factory ships are operating and the catch has increased considerably in recent years. Here the industry depends on sperm whales rather than baleen whales, and the number of sperm whales taken in the total north Pacific is about half the world catch. In Table 1 are shown the numbers of whales taken in the various areas of the world in recent 10 years and in Table 2 by species. It will be seen from Table 2 that the catches of blue, fin, and humpback whales have been decreased remarkably, and by 1967 blue and humpback whales were protected throughout the world by the International Whaling Convention. On the other hand catches of sei and sperm whales rose, due mainly to the increased catch in the Antarctic (sei) and in the north Pacific (sperm).

Treaty Regulations

As early as 1931 an International Convention for the Regulation of Whaling was drawn up in Geneva. It provided for the protection of right whales and females with calves, but had little effect on Antarctic whaling. In 1937 an International Conference was held in London and the International Agreement for the Regulation of Whaling was signed by representatives of 9 countries; it provided for (1) the protection of gray and right whales, (2) minimum length set for each species, (3) protection of calves or females accompanied by calves, (4) opening and closing date for factory ships in the Antarctic, (5) limitation of season within 6 months for land station and (6) prohibition of factory ships in the waters north of 40° SL, except part of the north Pacific. It also provided for the appointment of at least one inspector on each factory ship and complete utilization of the carcass. In 1938 another International Conference was held; it provided a sanctuary in the Pacific sector of the Antarctic and temporal protection of humpback whales in the Antarctic; some amendments to the Principal Agreement were also made. Thus in these two years a big advance was made for the protection of whale stocks. But the most important one is the over-all limit to the Antarctic catch, agreed upon in 1944, prior to the reopening of the postwar Antarctic whaling, though eventually this did not prevent further reduction of the blue whale stock and failed to maintain the fin whale stock at its optimum size.

In 1946 an International Conference took place in Washington, D.C. and the International Convention for the Regulation of Whaling was concluded. This new Convention came into effect in 1948 and the International Whaling Commission,

Fig. 2. Outline of Antarctic whaling.

WHALES AND WHALING

TABLE 1. WHALES KILLED IN THE VARIOUS MAIN AREAS IN RECENT 10 YEARS

Years	Antartic		Africa		N. Atlantic[a]		N. Pacific		Chile and Peru		Australia and N. Zealand		Others	
	Number	%	Number	%	Number	%	Number	%	Number	%	Number	%	Number	%
1956–1957	36,069	61.1	2,536	4.3	979	1.7	10,838	18.4	4,893	8.3	2,286	3.9	1,389	2.3
1957–1958	39,403	61.5	3,026	4.7	886	1.4	12,585	19.6	4,871	7.6	2,278	3.6	1,026	1.6
1958–1959	38,787	60.3	3,609	5.6	606	0.9	12,469	19.3	5,633	8.8	2,131	3.3	1,138	1.8
1959–1960	38,688	61.0	3,523	5.5	596	0.9	11,327	17.9	5,535	8.7	2,170	3.4	1,650	2.6
1960–1961	41,127	62.7	3,352	5.1	608	0.9	10,697	16.3	5,936	9.0	2,018	3.1	1,903	2.9
1961–1962	38,552	58.3	3,947	6.0	738	1.1	13,429	20.3	5,639	8.5	1,355	2.1	2,430	3.7
1962–1963	30,159	47.4	4,054	6.4	600	1.0	18,292	28.8	4,784	7.5	843	1.3	4,847	7.6
1963–1964	29,942	47.5	4,210	6.7	613	1.0	18,407	29.1	3,574	5.7	849	1.4	5,406	8.6
1964–1965	32,563	50.3	5,398	8.4	726	1.1	19,497	30.2	2,637	4.1	668	1.0	3,191	4.9
1965–1966[b]	24,680	44.0	4,148	7.4	971	1.7	22,296	39.4	886	1.6	606	1.1	2,563	4.5

[a] Excluding Spain, Azores, and Madeira.
[b] Excluding whaling from Brazil, Peru, and one Chilean shore station.

TABLE 2. WHALES KILLED IN RECENT 10 YEARS, BY SPECIES

Years[a]	Blue	Fin	Humpback	Sei	Sperm	Others	Total
1956–57	1,775	31,626	3,196	3,138	19,156	99	58,990
1957–58	1,995	31,587	2,923	5,670	21,846	54	64,075
1958–59	1,442	30,942	5,055	5,539	21,298	97	64,373
1959–60	1,465	30,985	3,576	7,035	20,344	84	63,489
1960–61	1,987	31,790	2,840	7,785	21,130	109	65,641
1961–62	1,255	30,178	2,436	8,804	23,316	101	66,090
1962–63	1,429	21,916	2,758	9,549	27,858	69	63,579
1963–64	372	19,182	318	13,690	29,255	184	63,001
1964–65[b]	513	12,317	452	25,453	25,548	297	64,680
1965–66[c]	132	6,535	8	22,992	26,222	261	56,150

[a] 1956–1957 = Antarctic season 1956–1957 and summer 1957.
[b] Excluding one Chilean shore station.
[c] Excluding whaling from Brazil, Peru, and one Chilean shore station.

an executive body, was founded. All the regulatory measures are contained in the schedule attached to the Convention and they are subject to amendment by the Commission. The Commission is composed of one member from each Contracting Government, who attends the meetings, normally held once a year, with experts and advisers. Most of the regulatory measures have been inherited from the Agreement and the work of the Commission has been in varying and extending them rather than in making new ones.

In May 1965 a special meeting of the Commission was held at the request of several Contracting Governments in order to obtain some agreed total quota that would allow the restoration of the Antarctic whale resources. It was agreed that Commissioners recommend to their Governments a quota for the 1965–1966 Antarctic season of 4,500 blue whale units and that further reduction should be made in the 1966–1967 and 1967–1968 seasons so that the quota for the 1967–1968 season would be less than the combined sustainable yields of the fin and sei whale stocks as determined on the basis of more scientific evidence. During the 17th Meeting of the Commission (1965) this recommendation was implemented by an amendment of the schedule. For the subsequent two seasons quota of 3500 and 3200 blue whale units were agreed at the 18th (1966) and 19th (1967) meetings, respectively. At the 19th meeting the size of the stock of fin whales was estimated 39,800 whales with a sustainable yield of 4800. The corresponding figures for sei whales are 70,000–90,000 and 4400–7000. This means the combined sustainable yield in terms of blue whale units is 3100 to 3600.

For the North Pacific whaling a working group, set up among scientists from Canada, Japan, U.S., and U.S.S.R., has been working on the stock assessment of whales in the north Pacific. According to its recommendations the Commission prohibited the taking of blue and humpback whales in the north Pacific at its 17th meeting in 1965. The protection of these species in the north Atlantic has been carried on since 1954 and that of blue whales in the waters between the Equator and 40° SL was formally decided at the 19th meeting in 1967. Thus both species are totally protected throughout the world, except for a small group of South American nations, which have a separate Permanent Commission. The north Pacific working group had also recommended in 1966 that the catch of fin whales in the entire north Pacific, which exceeded 3000 in 1965, should be held below the estimated sustainable yield of 1800.

Apart from the baleen whale the sperm whale is the only one of major importance today. The species was also hunted almost everywhere in warm and temperate waters in the old whaling days when fleets based on Nantucket and New Bedford, Mass., dominated the activity. In modern whaling it has been taken in much smaller numbers than baleen whales, but the catch increased in the postwar years, with the reduction of the stocks of baleen whales in the Antarctic. The most protective measure for this species in the Convention is the minimum length (38 ft for factory ships and 35 ft for land stations). The regulation was set in order to protect the breeding stocks. In practice, however, this regulation has not worked satisfactorily. Arrangements have already been made by the Commission and the stock assessment of this species in the world oceans will be carried out in the near future.

Other important points to note are the International Observer Scheme, which has not yet come into operation, and the national quotas, which are concluded, outside the Convention, among the countries concerned.

Methods of Capture

Since 1864 the harpoon method has been used in principle, but the efficiency of whale catchers has increased considerably and whaling itself has expanded from land-based local industry to pelagic operations in the Antarctic. The earliest catchers were small steam vessels of about 100 gross tons, but the largest at present exceeds 900 gross tons, and has a speed of up to 18 knots.

A swivel harpoon gun, with a pistol grip and a sighting-bar along the top, is mounted on the gun platform at the bow, which is joined with the wheel bridge by an elevated walk-way, the flying bridge. The harpoon, 90 mm in diameter and 1.5 m long, has a grenade of about 40 cm in length at its head and four hinged barbs. The grenade contains about 1 kg of blackpowder, fused to explode 3 sec after the harpoon is lodged in the whale body. In former times its head was pointed, but now the flat-topped grenade is in use, in order to prevent ricochet when it strikes the water surface at a flat angle.

When loaded, the harpoon is about 1.85 m long in total and weighs 65–70 kg. Attached to it is the forerunner, a fine manilla or nylon rope, 100–120 m long, coiled on a tray in front of the gun. It flies out with the harpoon when the gun is shot. The forerunner is connected with the whale line, made of manilla, and it passes under the gun platform, running up to the accumulator on the mast, round the drum of the whale winch, and then down to the hold. Total line is about 1000 m long and each catcher has two sets. The accumulator is a pulley hung on the mast and connected with the shock-absorber installed on the bottom of the ship as shown in Fig. 3. On the mast there is also a crow's nest or a "barrel," a look-out for finding whales.

When a whale is sighted, at first by the blow,

which can be seen at a distance of several miles, the catcher chases it at full speed. In former times whales were stalked with low speed, but this method of hunting has been changed, because of the use of the diesel catcher instead of the steamer.

Whales must break surface to breathe, but this happens very quickly and with minimum exposure of the body. Large baleen whales usually blow a few times at intervals of about 20 sec and then dive for 5–10 min. Sperm whales dive to greater depths, blow 10–15 times at short intervals, and remain submerged for 20 min or more. Moreover, a whale changes its swimming direction during the dive when chased. Much skill is needed, therefore, in maneuvering the catcher to within gunshot range. In most catchers sonar or some other whale-finder device is installed. The harpoon kills quickly, but the gun is reloaded without delay and the second harpoon is fired when the whale is not killed by the first.

In connection with the humane killing of whales an electric harpoon gun was tried extensively in recent years, with considerable success. However, recent advances in the design and technique of explosive harpooning have enabled this to kill as quickly as the electrical method.

The dead whale is drawn alongside the catcher and air is pumped into the body cavity through a spear-like pipe to make the carcass float. It is buoyed, with a mast carrying a flag, or sometimes with a radio transmitter or radar reflector, and the catcher deserts the carcass to seek another whale. Later the carcasses are collected by the catcher or by a buoy boat and towed back to the factory ship.

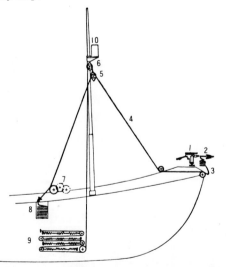

Fig. 3. Shock absorbing mechanism of whale catcher: 1, harpoon gun; 2, harpoon; 3, forerunner; 4, whale line; 5, accumulator; 6, pulley; 7, whale winch; 8, hold; 9, shock absorber; 10, crow's nest. (After Schubert, 1955.)

The factory ship is of about 20,000 gross tons and is equipped with a slipway, an opening of about 6 × 6 m, at its stern. The upper deck is the flensing deck, where the carcasses are dismembered with the help of steam winches. The lower is the factory deck, where oil-extracting and other equipment is installed. Below this deck there are many tanks that carry the fuel oil of the whole expedition at the beginning of a season's work and whale oil at the end.

The whale carcasses are hauled up on the flensing deck through the slipway by means of 2 40-ton winches. An apparatus called a claw, with scissors-like shape and about 2.5 tons in weight, is used to clasp the tail flukes so that the carcass is drawn up easily onto the deck. First the blubber is peeled off from the carcass, cut into small pieces, and sent down into the oil-extracting apparatus through its openings on the flensing deck. Meat is refrigerated for human consumption as well as pet food or used for the production of extract, meal, and oil, etc. Bones are cut by steam-driven bone-saws and thrown into the apparatus for extraction of oil. Among viscera the liver is used for the production of vitamins, but most of the remaining parts are discarded.

There are several types of oil-extracting apparatus; in the modern factory ships the Hartmann and Kvaerner types are generally used. The former was invented in Germany and the latter in Norway. In both cases the raw material is cooked by steam in a rotating drum, and the oils thus produced are separated from glue water in a separator, connected with the extractor, purified by a centrifugal machine, and then stored in tanks.

A recent Antarctic expedition consisted of 1 factory ship, 10 or more catcher boats, buoy boat, scouting boat, refrigeration ship, and transport vessel. The number and kind of vessels is different according to the production plan of each expedition.

Future Food Potential

The lives of most baleen whales are dominated by the seasons. They feed in the summer months in the cold waters of high latitudes and in autumn they migrate to the winter breeding grounds in warmer latitudes. It is possible that they eat little or even abstain from food while they stay in the breeding grounds. They must live on the energy stored in the fat which they build up in the feeding grounds.

About 90% of the world catch of fin, blue, and humpback whales is normally from the southern hemisphere, and they must have obtained nearly all their nourishment in the Antarctic or sub-Antarctic. Recent heavy reduction of stocks of these species is due to overfishing. Since 1960 quantitative assessments of the Antarctic stocks

TABLE 3. STOCKS OF ANTARCTIC BALEEN WHALES

Species	Original Stock Size	Present Size (1967)	Sustainable Yield (1967)	Maximum Sustainable Yield
Blue	More than 150,000	930–2,790	Less than 200	6,000
Fin	More than 250,000	39,800	4,800	20,000
Humpback*	About 20,000	2,000	Less than 100	Less than 1,000
Sei	—	70,000–90,000	4,400–7,000	5,400–6,300

* Figures for humpbacks are only for 2 groups out of 6. Present sizes and sustainable yields for blue and hump backs are figures in 1962.

of baleen whales have been carried out, results of which are summarized in Table 3, which indicates clearly how the stocks of baleen whales in the Antarctic, excepting that of the sei whale, have been declined. The stock of the sei whale was exploited quite recently and the present stock size is still a little above the level at which the maximum sustainable yield is produced. The last column of Table 3, the maximum sustainable yield, is important. Assessments for humpbacks were made only for two groups out of six, since there has been little or no catching from other groups for many years. This species concentrates in the warm coastal waters when breeding, whereas other species remain in oceanic waters. Thus this species is hunted in breeding season from land stations, and the sustainable yield, shown in the table, should include catches in temperate waters.

It can be assumed provisionally that the maximum sustainable yield of the humpback whale in the whole southern hemisphere may be around 3000 and a half or 1500 can be taken in the Antarctic. Then the industry can take 6000 blue, 20,000 fin, 1500 humpback, and 6000 sei whales, i.e., 17,600 blue whale units in the Antarctic. This is a little higher than 16,000, the blue whale unit limit set at first for Antarctic pelagic whaling in the postwar seasons, and is the highest number that could be taken from the stock without either causing it to decline or allowing it to grow. This could only be attained when all of the stocks of baleen whales are permitted to rebuild. It was also calculated that the years to reach the level of maximum sustainable yield if no catch after 1961–1962 season is 50 or more for both blue and humpback whales.

The average body weights of blue, fin, humpback, and sei whales taken in the Antarctic are about 84, 50, 33, and 22 metric tons respectively. Therefore the maximum sustainable yield of

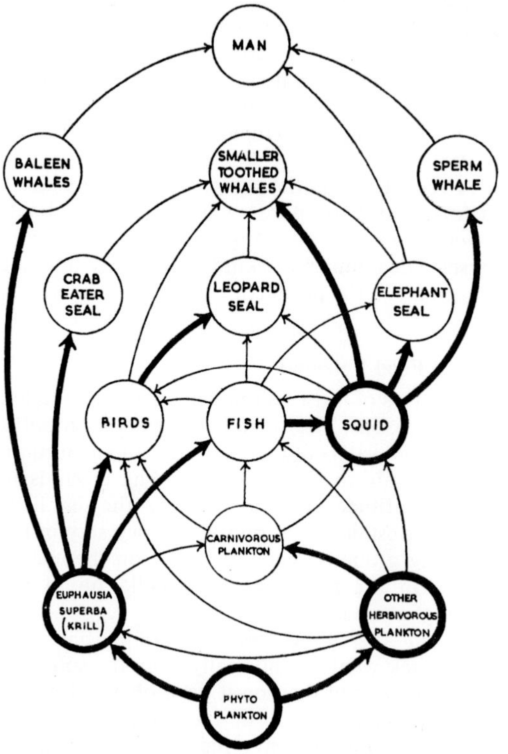

Fig. 4. Food chains in the Antarctic. (After Mackintosh.)

Fig. 5. Tying finback whale flukes to the bow off Eureka, California. (Photo: U.S. Bureau of Commercial Fisheries.)

baleen whales in the Antarctic is about 1.85 million metric tons in term of biomass, from which about 352,000 tons of whale oil and over 800,000 tons of whale meat can be produced.

For the sperm whale no precise stock assessment has been determined, but the catches in the Antarctic were usually around 5000, most of them adult males. Since the mean body weight is about 33 metric tons the annual production is 165,000 tons in terms of biomass, from which about 36,000 tons of sperm oil is produced. The oil of sperm whales is, however, used for industrial purposes, due to its chemically different nature, whereas the baleen whale oil is converted into edible fats. The meat of sperm whales is not used for food in general, due to its lower quality. In the schedule of the Convention there is a proviso excepting the meat of sperm whales from complete utilization of carcass.

The next matter to be considered is the food chains in the Antarctic. As shown in Fig. 4 the food chain leading to the baleen whales is very short, whereas the sperm and other toothed whales depend on a more complex series. There are several competitors of food for baleen whales —crabeater seal (*Lobodon carcinophagus*), birds including several species of penguins, fish, and other animals. In the Antarctic the continental shelf is relatively narrow and there is no positive evidence of extensive concentrations of fish in shelf waters. Recently some dense concentrations of an edible fish, *Micromesistius australis*, were discovered in oceanic waters, but it is not known yet whether it is in widespread or long-standing populations. All available evidence at present suggests that the baleen whales are the most suitable for harvesting in the Antarctic, except for possible exploitation of the krill, *Euphausia superba*.

HIDEO OMURA

References

Budker, P., "Baleines et Baleiniers," 1957, Paris. (English edition "Whales and Whaling," 1958, London.)
Fraser, F. C., "Whales and dolphins," in Norman, J. R., and Fraser, F. C., "Giant Fishes, Whales and Dolphins," London (new edition 1948).
International Whaling Commission, Reports 1–17, 1950–67, Office of the Commission, London.
International Whaling Statistics, 1–58, 1930–66, Oslo.
Mackintosh, N. A., "The Stocks of Whales," 1965, London.
Norris, K. S., Ed., "Whales, Dolphins, and Porpoises," 1966, University of California Press, Berkeley and Los Angeles.
Schubert, K., "Der walfang der gegenwart," "Handbuch der Seefischerei Nordeuropas," XI, 6, 1955, Stuttgart.
Slijper, E. J., "Walvissen," 1958 (English edition, "Whales," 1962, London.)

Cross reference: *Porpoises.*

Fig. 6. Humpback whale (*Megaptera novae angliae*), 37 ft female. Note barnacles on anterior end. (Photo: U.S. Bureau of Commercial Fisheries.)

ZOOPLANKTON—See PELAGIC DISTRIBUTION; PLANKTON RESOURCES

INDEX

Abalone, 1, 2, 4, 53, 333
  divers for, 3, 45
  juice, 32
  South Africa, 660
Abyssal zone, 4, 53, 184
  environmental factors in, 5
  fauna in, 6, 7, 9
  reproduction in, 7
Acoustics
  fish detection by, 225, 226, 229, 235
Agar, 9
  from algae, 18
  composition, 15, 523
  harvesting, 10, 11
  production, 523
  properties, 522, 523
  sources of, 12
  South Africa, 663
  uses of, 12, 523, 629
Agaroids, 11, 12, 14, 663, 664
Agaroses, 14
Air-blast freezing, 279
Alaska fishery
  crabs, 150, 155, 159
  salmon, 589
Albacore, 696, 699, 700. *See also* Tuna
Albatross, 410, 463
Algae. *See also* Phytoplankton
  agar from, 18
  antibacterial action of, 18, 31
  artificial reefs for, 35
  in benthonic domain, 52
  blue-green, 417, 531, 532
  brown, 18, 20, 522, 608
  as food source, 16, 20, 608
  and fungi, 284
  green, 31, 530
  halophilic, 596
  lipids in, 17
  nutritional value of, 17
  nutrients for, 555
  photosynthesis by, 16, 18
  in polysaccharides, 522
  proteins in, 17
  red, 18, 31, 326
  reef-forming, 147
  South Africa, 663
  toxic, 65
  Trust Territory, 694
  types of, 16
Algin, 19, 20
  extraction of, 21, 524
  from kelp, 19, 524
  manufacture of, 21, 524
  in polysaccharides, 19

properties of, 522
uses of, 21, 22, 524
Alginates, 19, 22, 524, 626
Alginic acid, 18, 21, 626, 663
  chemical nature, 524
  properties, 21
Alvin, 463
Ambergris, 23
Amino acids
  in algae, 17
  in fish, 55, 56, 450
Anchovy, 178, 216, 313, 478, 507
  California, 466
  Peru, 509, 510
  Philippine, 517
  Russia, 228
  South Africa, 657
Animal food, 18, 24, 25, 102, 291, 452. *See also* Pet foods
Antibiotics, 62, 63
  in algae, 18
  preservation by, 27
Anticholinesterase, 65, 682
Antioxidants, for fish oils, 219, 264
Aquaculture. *See* Sea farming
Ascomycetes, 283
Ascophyllum, 20
Atlantic Northwest fishery, 39, 91, 92
  clams, soft-shell, 112
  clams, surf, 122–124
  cod, 129, 130, 343
  menhaden, 395
  redfish, 579
Atlantic States Marine Fisheries Commission, 381
Atlantis, 463
Australian fishery
  abalone, 2, 45
  crayfish, 44
  estuarine, 43
  mackerel, 42
  oysters, 46
  pearls, 47
  prawns, 45, 47
  salmon, 43
  scallops, 45
  shark, 41
  shrimp, 637, 638
  snapper, 43
  tuna, 46
  whales, 46
  yellow jacket, 43
Automation in dead reckoning, 430

Bacteria, 6, 48, 398, 399
  effect of environment on, 400

functions of, 50, 402
  toxic, 402
  types of, 400, 402
Bahamas
  conch fishery, 137
  sponges, 667, 668, 669
Baleen whales, 466, 723, 725, 726
Bait, 4
  clams as, 113, 119, 123
  for cod, 130
  sardines as, 262
  snails as, 137
  squid as, 103
  for squid, 103
  for tuna, 46, 256
  worms as, 372
Bangos, 518
Barents Sea, 95
  capelin in, 98, 99
Barnacles as food, 325
Baselines, legal, 350, 351
Bathysphere, 462
Batter mixes, 68, 69
Bay, definition of, 350
Beagle, 460
Benthic. *See* Benthonic
Benthonic fish, 7, 8
  and food chain, 51, 53
  in Peru current, 509
Benthonic zone, 182, 183, 185
  foraminifera in, 275
Bermuda petrel, 516
Biological clocks, 57
Bioluminescence, 144
Biomass, 8
  bacterial, 50
  and production, 51, 52, 53
  of zooplankton, 536
Biotic environmental factors, 181, 185
Biotoxins, 61–66, 418
Birds, marine, 190, 478, 507, 512, 515, 516, 663
Bloodworms, 372, 373, 375
Bloom, flagellate, 18, 20, 173, 222, 417, 418, 465
Blue-green algae, 417, 531, 532
Blue whales, 189, 723, 724, 726
Bombay duck, 317, 318
Borers, marine, 377
  control of, 379
  damage by, 378
  distribution of, 378
  protective methods, 380
  types of, 377
Botulism, 563
Breading, automatic, 68, 69, 70

731

INDEX

Breeding behavior, 57, 71. *See also* Reproduction; Spawning
  capelin, 95
  environmental factors, 72, 74, 76
  migrations, 75
  sexual cycles in, 73, 74
British Columbia fishery, 92, 93, 94, 589, 639. *See also* Canadian fishery
British fishery
  administration, 85
  cod, 78, 83
  demersal, 79, 81
  equipment, 81, 82
  haddock, 78, 83
  localities, 78
  methods, 79, 80
  plaice, 78, 83
  processing, 84, 85
  shrimp, 643
  size, 78
  statistics, 79–83
Bromine, 86, 87, 89, 369
Brown algae, 18, 608
  chemical nature of, 522
  as food, 20
Buffer capacity, seawater, 111, 112

Calories, in seafood, 449, 450
Canadian fishery, 91, 92. *See also* British Columbia fishery; Labrador fishery
  cephalopods in, 102
  inshore, 345
  lobsters in, 352
Cannery waste, 216
Canning
  abalone, 4
  crabmeat, 153, 154, 156
  miscellaneous products, 711
Capelin
  Barents Sea, 98, 99, 100
  breeding of, 95
  Greenland, 99, 100
  distribution, 95
  Iceland, 99, 305
  morphology, 95
  Norway, 100
  Newfoundland, 100
  Pacific, 100
  spawning, 97, 98, 419
Carbon, radioactive
  assimilation of, 557
  in oceans, 567, 570
Carrageenin, 18, 92, 326, 608. *See also* Seaweeds; Chondrus crispus
  chemistry, 328, 525
  harvesting, 326
  processing, 327, 525
  properties, 328, 525
  uses, 328, 526
Cast nets, 247
Cells, effect of radiation on, 575, 577

Cephalopods, 101. *See also* Octopus; Squid
  behavior of, 104
  benthonic, 53
  catching of, 103
  eyes of, 105
  as fertilizer, 102
  as food, 102
  reproduction of, 104
  toxicity of, 680
CERES Fish Finder, 201
Cetaceans, sounds produced by, 649. *See also* Porpoises; Whales
Challenger, 5, 410, 460, 463
Chemical composition
  of fish, 54, 55, 56
  of seawater, 106, 107
Chesapeake Bay
  clams, 114, 116, 117
  conch fishery, 139
  menhaden fishery, 394, 398
  oysters, 474, 475
Chimeras, 680
Chitin, 526
Chlorella, 17, 18
Chlorophyll
  in algae, 18
  in dinoflagellates, 531
  in phytoplankton, 534, 556
Chlorophyta, 530
Cholesterol, 56
Chondrus crispus, 18, 326, 525, 626, 628. *See also* Carrageenin
  harvesting, 629
Chrysophyta, 529, 530
Ciguatera poisoning, 65, 681, 682
Clams
  farming of, 54
  soft-shell, 112–118
    anatomy, 114
    diseases of, 116
    harvesting of, 117
    life cycle, 115
  surf, 119
    anatomy, 119
    catch (U.S.), 118
    harvesting, 122
    life cycle, 121
    reproduction, 120
  Trust Territory, 693
Cnidaria, 62, 63, 322
Coccolithophorids, 529
Coastline, legal aspects, 350, 351
Cobalt, radioactive, 561, 562, 570
Cod
  British fishery, 78, 83
  fish meal from, 133
  in fish sticks, 132
  food required by, 611
  Greenland fishery, 100
  Iceland fishery, 304, 307
  Labrador fishery, 130, 343
  liver oil from, 56, 133, 718, 720
  migration, 75

  New England, 129, 130
  Newfoundland, 435, 437
  Northwest Atlantic, 129, 343
  Pacific, 94
  vessels for, 130
Cod-liver oil, 56, 133, 718, 720
Coelenterates, toxic, 676
Collagen, 230
Colloid, hydrophilic, 21
Columbia River, 589, 593
Commensalism, 182, 473
Communities, 187, 189
Conchs, 135
  as food, 137
  reproduction, 138, 139
  shells, 136, 137
Contact plate freezing, 278
Contiguous zones, 347, 348
Continental shelves
  as legal zones, 349
  minerals in, 215, 407, 408
Copepod
  anatomy, 141
  eye of, 144
  feeding of, 144
  as food, 146
  life cycle, 145
  migration, 146
  parasite, 493
  reproduction, 143, 145
Coral, 147
  Mediterranean, 53
  uses of, 149
  toxicity, 677
Crab, 150. *See also* King crab; Dungeness crab; Red crab
Crab meat, processing of, 153, 156, 161
Crayfish, Australian fishery, 44, 45
Crocodiles, 692
Crustaceans, 8, 9, 54, 120, 152. *See also* Crabs; Lobsters; Shrimp; Copepods
  British fishery, 79
  interstitial, 324
  reproduction in, 71, 74, 76, 153
  sound emission by, 652
  South Africa fishery, 659
  toxicity of, 64
Crystallization, of salt, 595, 596
Currents, 6, 299, 330
  Benguela, 655
  and distribution, 499
  and mortality, 420
  Peru. *See* Peru current

DDT
  oysters affected by, 515, 516
  pollution by, 514, 515
  residues, 515, 516
Dead reckoning, 428, 429
  automation in, 430
  equipment for, 431
Deep sea bottom, 349
  exploration, 413, 414

732

INDEX

minerals in, 409, 410
   sampling, 414
Dehydration, preservation by, 266
Demersal fish
   capelin, 98
   farming of, 54
   in Peru fishery, 511
   in Philippine fishery, 518
   South Africa, 659
Desalination
   electrodialysis, 168
   flash evaporation, 163, 164
   freezing, 166, 167
   ionic processes, 168
   reverse osmosis, 167
Detection, electronic, 200–203, 225, 229
Deuteromycetes, 284
Diamonds, from sediment, 664
Diatoms, 6, 18, 529
   in benthonic domain, 52
   as food foraminifera, 275
   and light, 535
Dinoflagellates, 18, 531, 532. *See also* Phytoplankton; Red tide
   biology of, 530, 531
   and blue-green algae, 417, 532
   cell structure, 531
   life cycle, 171
   as parasites, 531
   poisoning by, 51, 61, 173, 175, 675
   and red tide, 173, 174
   reproduction, 171
Direction, determination of, 429, 431, 432
   by salmon, 593
Diseases, 488, 493
   bacterial, 489
   fungal, 489
   of oysters, 472
   viral, 488
Distribution
   biotic factors, 498, 499
   effect of currents on, 499
   of fish, 312
   horizontal, 500
   lobsters, 352, 357, 361
   pelagic, 498
   and tagging, 390
   vertical, 500
Divers
   for abalone, 3, 45
   for agar, 11
   in mineral exploration, 416
   for pearls, 495
   protection from sharks, 634
   for sponges, 667, 668
   for spiny lobsters, 360
Dolphins. *See* Porpoises
Doppler method
   in geodesy, 380
   in navigation, 430
Dredges
   harvesting by, 248

for oysters, 290
   for scallops, 619
Dungeness crabs, 151
   harvesting, 152
   reproduction, 153

Echinoderms, 8, 9
   predation by, 423
   spawning of, 74
   toxicity of, 64, 678
Echo sounders, 200, 201, 225
Ecology, 178
   ecosystem, 182, 189
   of estuaries, 206
   history, 184, 185
   and productivity, 221, 222
   of teleost fish, 178–180
Economics of marine resources, 210–213, 235
Ectyonin, 62
Eddy diffusion, in Peru current, 501, 506
Education, 191
   administration, 195
   ocean engineering, 458
   organization, 194
Eels
   conger, 197
   deep-water, 197
   as food, 199
   fresh-water, 196, 199
   garden, 198
   migration, 75, 313
   moray, 197, 198
Electricity
   for harvesting, 228
   from tides, 214, 288
Electrodialysis, desalination by, 168
Electronic equipment, 199, 225
   fish detection by, 200, 201
   mineral exploration, 413, 414
   for navigation, 433
   radar, 235
Endocrinology, in reproduction, 73, 74, 75
Estuaries
   ecology, 206
   environment in, 206–209
   and fish farming, 609, 611
   government study of, 288, 289, 381
   oyster farming in, 207
   oxygen in, 208
   pesticides in, 513
   Philippine, 518
   pollution of, 207–209, 473, 513, 544
   salinity of, 206, 207, 209
   shellfish in, 206, 207, 636
   silt in, 207, 208
   temperature in, 207
   yeasts in, 284
Euphotic zone. *See* Light
Eutrophication, 545
Evaporation
   desalination by, 162, 164

flash, 163, 164
   salt produced by, 595
Exploration, biological, 459, 460, 461, 463
   of sea floor, 413
Explosives, sea-floor exploration by, 414
Eye
   of cephalopods, 105
   of copepods, 144
   of pelagic fish, 180
   of sharks, 633

Factory ships, 40, 252
   British, 81, 234
   freezing on, 276, 277, 278
   U.S., 705
   whale, 260, 723, 724, 727
Fallout, radioactive, 568, 570
Fatty acids, 56, 218, 219
Fertilizer, 102, 630, 663
Fiddler crabs, biological clock in, 59, 60
Fillets, breaded, 66
Filleting, 131, 582
   at sea, 276, 277
Fisheries. *See also* specific country
   conventions, 286, 344
   economic management, 213, 235
   in Japan, 333, 335
   explorations for, 466
   forecasting of, 468
   improved procedures, 467
   new areas, 465, 466
   and oceanography, 464
   unused resources, 466
   U.S. commissions, 380
Fisherman's asdic, 202
Fishery engineering, 225
   new methods in, 228, 229, 250
   production increased by, 227
Fish farming. *See* Sea farming
Fish glue, 229
   chemistry, 230
   from cod skins, 133
   manufacture, 230
   properties, 231
   uses, 231, 232
Fish-liver oils, 56, 216, 217, 218, 633, 718. *See also* Cod-liver oil
Fish meal, in animal feed, 24, 25, 216, 261
   from anchovies, 510
   from cod, 133
   curing of, 264
   drying of, 263
   on factory ships, 260
   fish used for, 261
   for FPC, 268
   manufacture of, 24, 262, 263
   from menhaden, 398, 452
Fish oils. *See* Oils
Fish pastes and sauces, 267, 517, 519
Fish ponds, 518, 608

733

INDEX

Fish protein concentrate, 91, 135
  acceptability of, 270
  in food products, 270
  government research, 288
  processing of, 266, 267
  proteins in, 266, 271, 611
  solvent extraction, 268, 269, 270
Fish sticks, breaded, 66
  cod, 132
  redfish, 583
  tuna, 586
Fixing (navigation), 429, 431
Flash evaporation, 163, 164, 301
Flavor, definition, 547
  and irradiation, 560, 563
Flounder, food requirement of, 611
Flying fish, 311, 312, 319
Food chain, 16, 49, 51, 220, 222
  in benthonic domain, 52
  efficiency of, 611
  and foraminifera, 275
  effect of pesticides, 515
  in pelagic domain, 52
  phytoplankton in, 176, 515
  and seaweeds, 49, 50
Foraminifera
  and food chain, 275
  morphology, 273, 274
  planktonic, 275
  reproduction, 275
  in sediment, 6, 276
  shells, 273, 276
Forecasting, 468, 469
Freezing
  desalination by, 166, 167
  methods of, 277, 279
  preservation by, 131, 234, 236, 260, 276
Fungi, 280, 284, 399
  infections from, 489
Furcelleran, 11, 12, 526, 626, 630

Galactose, 328
Gamma rays, 560, 561
Gas bladders. *See* Swim bladders
Gastropods, 1
  toxic, 63
Gelatin, 230
Geodesy, 383. *See also* Navigation
  networks in, 384, 386, 388
  systems for, 386
George's Bank, 559, 616, 617, 620, 623
Gillnets, 244, 255, 307, 316, 345, 346
  for salmon, 591
  for sharks, 634
Gills, function of, 179
Glauconite, 408, 412
Grand Bank, 95, 96, 100, 113, 437, 559. *See also* Newfoundland fishery
  cod, 130
  halibut, 439
  redfish, 439, 579, 581
Gravity measurement, 387

Great Barrier Reef, 148
Green algae, 31, 530
Greenland fishery
  capelin in, 95, 99, 100, 101
  cod in, 100
  halibut in, 100
Gribbles, 377
Groundfish
  fish meal from, 261
  Newfoundland, 437
Grunions, 57
  spawning of, 74
Guano, 149
  and anchovy fishery, 478, 507, 512
  South Africa, 663
Guillotine cutters, 67, 68
Gulf Marine Fisheries Commission, 382
Gulf of Mexico
  blue crab, 291
  industrial fish, 291
  menhaden, 291, 395, 706
  oysters, 290
  redfish, 290
  red snapper, 290
  Spanish mackerel, 290
  shrimp, 292, 636, 706
  sponge industry, 667
  sports fishery, 290
  stone crab, 292

Hadal zone, 5, 53, 184, 186
Haddock, 278, 279
  British fishery, 78
  Iceland fishery, 83, 304
  Newfoundland, 438
  size of, 40
  spawning of, 77
Hagfish, 179, 180, 311, 680
Hake, 659
Halibut
  British Columbia fishery, 93
  freezing of, 277
  Greenland, 100
  Japan, 334
  liver oil, 718
  Newfoundland, 437, 439
Haliotids. *See* Abalone
Haptophytes, 529
Harvesting
  agar, 10, 11
  British methods, 232, 235
  carrageenin, 326
  crabs, 152, 156
  by dredges, 248
  by electricity, 228
  kelp, 19, 20
  nets and gear, 237–247
  new area location, 465
  production, 224
  seaweed, 628, 629
  solar salt, 596
Hawaii fishery, 246
Heat exchanges, 302
Hermit crabs, 139

Herring. *See also* Sardine
  Atlantic fishery, 91, 92, 436
  British Columbia, 93
  British fishery, 78, 84
  fish meal from, 261
  gillnets for, 255
  Iceland fishery, 303, 307, 308
  survival of, 313
  West Coast, 487
High seas, 347
Holothurians. *See* Sea cucumber
Holothurin, 679
Humber Fish Detection System, 202, 203
"Hydrate" freezing process, 167
Hydrocolloid. *See* Colloid, hydrophilic
Hydrogenation, of fish oils, 218, 220
Hydrographic surveys, 388
Hydroids, toxic, 676

Ice, preservation with, 27, 29, 131, 236, 519
Iceland fisheries, 303
  capelin, 99, 305
  cod, 304, 307
  haddock, 304
  herring, 303, 307, 308
  redfish, 305
  saithe, 305
  whaling, 308
Icthyology, 311
  schooling research, 605
Immersion freezing, 276, 278
Indian fishery, 315
  mechanization of, 317
  problems of, 318
  processing in, 318
Industrial fish, 262, 291, 293
Inertial guidance, 430
Internal waters, 347
International Commission for Northwest Atlantic Fisheries, 344
International fishery conventions, 286, 344
Interstitial fauna, 319
  collection of, 324
  locomotion, 321
  metazoa, 322–324
  morphology, 320
  protozoa, 322
  reproduction, 321
Iodine, from kelp, 626
Irish moss. *See* Carrageenin; Chondrus crispus
Irradiation. *See* Radiation preservation
Isinglass, 133

Japanese fishery, 330
  abalone, 2, 333
  administration, 336, 338, 339
  agar, 10
  economics, 333

INDEX

king crab, 156, 331, 334
  legal management, 336, 339, 340
  licensing system, 335, 338, 340
  mariculture, 333
  organization, 335
  oysters, 333, 473
  pearls, 333, 495
  rights system, 337, 339
  treaties affecting, 331, 333, 334
  tuna, 479, 688, 696
  whaling, 334
Jasus lobster, 362
Jellyfish
  as food, 324
  toxic, 63, 677

Kelp
  algin from, 19, 524
  harvesting, 19, 20, 629
  iodine from, 626
  South Africa, 663
King crab, 54, 155
  Alaskan, 155, 159, 484
  biology, 155
  harvesting, 156
  Japan, 156, 331, 334
  processing, 156
  production, 157, 158
  regulations, 158
  U.S.S.R., 716, 717
Krill, 189, 466, 538

Labrador fishery, 342
  cod, 130, 343, 345
  offshore international, 344
  salmon, 345
  seals, 439
Laminaria, 18, 20, 629, 663
Lamprey, 612
  behavior, 613
  control of, 614, 615
  in Great Lakes, 614, 616
  population, 616
  spawning, 613
  toxicity, 680
Lateral line, 180, 183, 606
Leather
  from marine crocodiles, 692
  from seals, 661
  from sharks, 633
  from turtles, 625
Light
  and diatoms, 535
  as ecological factor, 180, 185, 186, 222
  and photosynthesis, 553, 556
  schooling affected by, 606
  variation with depth, 183, 221, 554
Limnoria, 377, 378
Lipids
  in algae, 17
  in fish, 56
  removal of, 266, 267
Littoral zones, 184, 186

Lobsters
  Atlantic Coast, 91, 352, 357
  giant, 353
  larvae, 356, 363
  markets for (spiny), 361
  migrations, 358
  morphometric studies, 356, 359
  Newfoundland, 436
  Norwegian, 352
  offshore, 353, 354, 356, 357, 358
  parasites, 356, 358
  populations, 356, 357, 358
  preservation, 361
  research program, 357
  rock, 659
  size of, 353, 354, 355
  spiny, 359, 361, 362, 652, 659
  tagging of, 356, 357, 358
  tails, 361
  transportation of, 353, 359, 361
  traps for, 352, 360
  trawls for, 353, 354, 57
Locomotion, 180
  of interstitial fauna, 321
Long Island, surf clam fishery, 122, 124
Longline method, 247, 255
  Australia, 41, 46
  cod, 130
  Iceland, 306
  swordfish, 363
  tuna, 297, 479, 698

Mackerel
  Australian fishery, 42
  Gulf fishery, 290
  Indian fishery, 316
  Northwest Atlantic fishery, 365
  populations of, 366
  preservation of, 29
  South African, 657
  spawning of, 366
  survival of, 313, 366, 367
Magnesium
  from seawater, 368
  uses of, 372
Maine
  conchs, 139
  herring, 262
  lobsters, 352, 355
  mackerel, 367
  marine worms, 372
  redfish, 579, 581
  sardines, 262
  scallops, 617, 619
  seaweed meal, 626
  shrimp, 644
Management, Japanese fishery, 335, 336, 339
Manganese
  nodules, 214, 410, 463
  and photosynthesis, 555
Mariculture. See Sea farming
Marine control networks, 384, 386, 388
Marine ecosystem, 182, 184, 220

Marine technical institutes, 192, 193
  administration of, 195
  courses in, 194, 195
Marine technology, education, 192
Marine worms, 372
  abundance, 375
  harvesting, 373, 374
  production of, 375, 376
  species of, 373, 375
Maritime Provinces, fisheries, 91, 230. See also Atlantic Northwest Fisheries
Marking. See Tagging
Mechanization, in fishery practice, 226, 250, 317
Menhaden
  abundance of, 394
  geographic distribution, 395
  harvesting, 397
  meal from, 261, 398
  migration of, 394
  oil from, 216, 398
  processing of, 398
  size of, 393
Mercenene, 64
Mesh size, of nets, 40, 41, 235, 239
Mexican fishery
  abalone, 2
  shrimp, 479, 639
Microbiology, 398
  sample collection for, 402, 403
Micro-organisms, marine. See Microbiology; Bacteria
Migration, 313
  of copepods, 146
  of eels, 75, 313
  interstitial fauna, 320
  of menhaden, 394
  for reproduction, 75
  of salmon, 314
  and tagging, 314, 390
  vertical, 500
Mineral content, fish, 55, 451
Mineralization, by bacteria, 402
Minerals
  authigenic, 412
  in beach deposits, 405
  in continental shelves, 407, 408
  in deep-sea floor, 409
  exploration for, 411, 412
  placer, 412
  in seawater, 107, 109, 214, 463
  South Africa, 664
  in sub-seafloor rocks, 409
Mink, feeding of, 25, 291, 452
Mollusks, 63, 324. See also Clams
  toxic, 679
Morphology, fish, 311
  capelin, 95
  foraminifera, 273, 274
  interstitial fauna, 320
Mortality, 416
  and currents, 420
  and disease, 418
  factors in, 416, 417
  Peruvian anchovy, 313

INDEX

Mortality (cont.)
  and pesticides, 513
  and red tide, 174, 175
  salmon, 419
  and salinity, 420
  and tagging, 390
  and temperature, 419
Mother-of-pearl. *See* Nacre
Mother ships, 260, 276
Murexine, 680
Mussels
  biomass of, 53
  European fisheries, 421
  as food, 421, 422, 424
  growth of, 422, 423
  and parasites, 423
  poisoning by, 417, 418
  predation of, 423, 424, 425
  South Africa, 660, 661
  and water pollution, 424
Myxosporidians, 490

Nacre, 494, 496, 690
National Sea Grant Program, 427
National Science Foundation, 427
Navigation
  celestial, 432
  electronic aids, 225
  by fish, 593
  science of, 428
Nekton, 498, 499, 500
  in Peru current, 508
Nematocysts, 62, 676
Nematodes, 492
Neritic province, 183, 184, 185
Nets
  cast, 247
  dip, 247
  gill. *See* Gillnets
  manufacture, 237, 239
  materials for, 130, 226, 307, 308, 333
  mesh structure, 40, 41, 235, 239
  miscellaneous, 247
  reef, 247
  ring (crabs), 152
  in seines, 253, 255
  trammel, 245
New England fisheries, 39, 40. *See also* Atlantic Northwest Fisheries; Maine
  clams, 113, 114, 116
  cod, 129
  conchs, 139
  lobsters, 352, 357
  mackerel, 365, 367
  menhaden, 395
  oysters, 476
  scallops, 617
Newfoundland fishery. *See also* Grand Bank
  capelin, 95, 96, 100, 436
  cephalopods, 102, 103
  cod, 435, 437
  groundfish, 437
  haddock, 438
  halibut, 437
  herring, 91, 436
  history of, 39
  lobsters, 436
  mackerel, 356
  redfish, 439
  salmon, 436
  seals, 439
  squid, 436
  whales, 437, 440
New Zealand fishery, 440
  by-products, 443
  crawfish, 443
  shellfish, 443
Norwegian fishery, 445
  capelin, 99, 100
  exports, 446
  lobster, 352
  regulations, 448
  seals, 449
  whales, 448
Nutrients. *See also* Proteins; Vitamins
  in seafood, 449, 452
  supply of, 222, 223
  vitamins, 720
Nylon
  as net material, 226, 238, 244, 307, 333
  as tag material, 391

Ocean Engineering, 192, 455
  education for, 458
  government programs, 459
  operations and projects, 455, 456
  technical societies, 458, 461
Oceania fisheries, 479, 480. *See also* Pacific fisheries
Oceanic province, 183, 184, 185
Oceanography, 192, 459
  and fisheries, 464, 465
  government study of, 289
  history of, 459
  laboratories, 461
  organizations, 461, 462
  and pollution control, 545, 546
  and weather control, 453
Ocean space
  boundaries of, 351
  legal aspects, 347
  zones (legal), 347
Octopus
  benthonic, 53
  bite of, 64, 105
  eye of, 105
  nervous system, 105
Odor detection
  by fish, 593, 606
  by sharks, 633
Oils
  fish, 55, 56, 216, 450, 607
  composition, 218
  hydrogenation, 218, 220
  manufacture, 217, 263, 398
  oxidation of, 264
  removal of, 263, 264, 266, 268
  uses of, 220, 607
  whale, 727
fish-liver, 56, 133, 216, 217, 218, 451, 633
  composition, 218
  processing, 718, 719
  production, 718
  refining, 720
Ooze, 6, 276, 410
Otter trawl, 130, 241, 243, 291
  for crabs, 156, 158
  for lobsters, 354, 357
  for shrimp, 638, 639, 641
Oxygen
  in estuarine waters, 208, 209
  in seawater, 106, 107
Oysters
  biology of, 469
  farming of, 207, 473
  mantle of, 470, 471, 494
  mortality of, 419
  parasites, 472
  pearl, 333, 496, 497, 691. *See also* Pearls
  predators, 472
  polysaccharide-coated, 549
  shells, 149, 370, 469, 493, 496, 497
  South Africa, 660
  spawning of, 76
  sponge infestation of, 666
  and water pollution, 473, 515, 516

Pacific fisheries
  abalone, 1
  Americas, 478, 481
  crab, 150, 484
  Oceania, 479, 482
  research in, 481
  salmon, 484
  Southeast Asia, 480, 481, 482
  subtropical, 477, 478
  tropical, 477, 478
  West Coast, 150, 483
Pacific Marine Fisheries Commission, 381, 382
Palau Islands, 690, 691, 693
Panulirus lobster, 361, 362
Parasites, 488
  copepod, 493
  dinoflagellates as, 531
  helminthic, 491, 492
  on mussels, 424
  on oysters, 472
  protozoan, 489, 490
Pearls
  Australia, 47, 493, 494, 496
  culture of, 496, 497
  Japan, 333, 495
  Trust Territory, 690
  Palau, 690
Pelagic zone, 52, 182, 183, 185, 186
  distribution in, 498
  divisions of, 500

INDEX

Pelicans, 507
Penguins, 190
Peru current, 502
  benthonic community in, 509
  eddy diffusion in, 506
  productivity in, 507
Peruvian fishery, 478, 509
  anchovy, 178, 216, 313, 479, 509
  other species, 510–512
Pesticides
  DDT, 514
  in estuaries, 513
  persistence of, 513
  polychlors, 514
  residual, 515, 516
  transport of, 514
  types of, 514
Pet foods, 25, 291, 452, 721
Petroleum
  offshore drilling, 38, 287, 455
  pollution by, 544
  resources, 214, 412
Philippine fishery
  administration, 517
  anchovies, 517
  demersal fish, 518
  estuarine fishery, 518
  fish ponds, 518
  marketing, 519
  preservation, 519
  tuna, 517, 519
Pholads, 377, 378
Phosphorescence, 171
Phosphorite nodules, 215, 408, 664
Phosphorus
  in bacteria, 50
  in photosynthesis, 555
  in seawater, 571
Photography
  fish glue in, 231
  submarine, 413, 463
Photosynthesis, 220
  and algae, 16, 554
  and phytoplankton, 534, 535, 553
  nutrients available for, 555
Phycocolloids, 521, 522, 523. See also Seaweed; Polysaccharides
Phycomycetes, 280
Phytoplankton, 49, 50, 498, 527, 533. See also Algae
  abyssal zone, 6
  bloom, 222, 465
  buoyancy of, 527
  cells, 527, 529, 530, 531
  communities of, 187, 553, 555
  as human food, 537
  and light, 180, 186, 222, 556
  nutrition of, 528, 534
  Peru current, 507
  production due to, 176, 177, 533
  types of, 529, 530, 532
Piccard, A., 462
Piloting, 431
Pigmentation, 57, 144, 181
Pilchard, 657, 706. See also Sardine

Plaice, American, 492
  British fisheries, 78, 83
  Newfoundland fishery, 438
Plankton. See also Phytoplankton; Zooplankton
  economics of, 537
  as food source, 537, 538
  resources, 533
Poisoning. See also Venomous marine animals
  ciguatera, 681, 682
  by dinoflagellates, 51, 61, 173, 175, 417
  hallucinogenic, 683
  scombroid, 682
  shellfish, 61, 118, 175, 402, 417, 675
  by teleost fish, 681, 682
  tetraodon, 683, 685
  treatment of, 685, 686
  by tuna, 682
Poland fisheries, 539–543
Pollution, 543
  causes of, 545, 546
  and clam culture, 118
  control of, 545, 546
  effects of, 545
  estuarine, 207, 208, 209, 544
  legislation, 285
  marine, 190, 544, 545
  and oyster culture, 473
  by pesticides, 513, 515
  radioactive, 544
  research on, 546
  thermal, 207, 473
Polychlors, 514
Polyethylene, 226, 308
Polygalactans, 523
Polypropylene, 130, 307
Polysaccharides, 521. See also Phycocolloids; Seaweed
  and algae, 19, 522
  chemical nature of, 14, 522
  edible coatings of, 547, 548, 549
Porifera, 665. See also Sponges
  toxic, 676
Porpoises, 550
  biology of, 551
  communication by, 552
  fisheries, 552
  oil from, 552, 553
  schooling of, 551
Potassium, radioactive, 566, 573
Pots, for crabs, 152, 156. See also Traps
Prawns, 45, 47, 608, 635
  Indian fishery, 636
  South Africa, 660
Predation
  mortality caused by, 417
  of oysters, 472
  by parasites, 424
  by starfish, 423
Preservation
  by antibiotics, 27–30
  of crabmeat, 154

  by dehydration, 266
  by drying, 132, 310, 318
  by freezing,
    air-blast, 279
    brine, 277, 278, 641
    contact-plate, 278
    on factory ships, 234, 236, 260, 276, 277, 278
  of halibut, 277
  by ice, 27, 29, 131, 236, 519
  of lobsters, 361
  by pickling, 318
  by polysaccharide coatings, 547
  by radiation. See Radiation preservation
  by salting, 132, 310, 318, 519
  of shellfish, 28, 29, 638, 641
  by smoking, 132
  by superchilling, 236
  U.S. methods, 638
Production
  and biomass, 51, 52, 53
  harvesting, 225
  of marine communities, 553
  measurement of, 556, 557
  of mussels, 425
  of Peru current, 507
  of phytoplankton, 533, 535
  primary, 221
  secondary, 222
  variations in, 558
  and vertical mixing, 224, 554
Proteins
  in algae, 17
  in FPC, 271
  in fish, 55, 450
  in mussels, 425
  as ocean resource, 265, 266, 538, 607
  solubilization of, 266, 267
Proximate composition, fish, 55
Pseudopodia, 272, 274
Puffers, toxicity of, 65, 684
Purse seines. See Seines

Quahogs
  biology, 126
  distribution, 125, 128
  mortality, 127
  predators, 127
  Rhode Island fishery, 128
  spawning, 126
  spoilage, 128

Radiation, ionizing, 575
  effect on organisms, 575, 576
  effect on populations, 577
Radiation preservation
  chemical changes by, 563
  dosage, 561, 562, 563
  economics of, 565
  history of, 560
  methods of, 560
  microbiology of, 563
  and shelf life, 565
  technology of, 561, 562

INDEX

Radioactive waste, 286, 568, 569, 570
  dispersion of, 570
  as pollutant, 544, 570
  treatment and disposal, 570–572
Radioactivity, 566
  accumulation by organisms, 572
  cells affected by, 576
  concentration of, 571, 572
  fisheries affected by, 574, 575
  hazards of, 569, 571
Radiometry, 600
Radionuclides
  classification, 566
  uses of, 567
Radium, in oceans, 567
Red algae, 18, 31, 326, 523
Red crab fishery, 159–161
Redfish, 579
  biology of, 579, 580
  food of, 579
  Gulf of Mexico, 290
  harvesting, 582
  Iceland, 305, 579
  Labrador, 343, 344, 579, 581
  new areas for, 581
  New England, 579
  Newfoundland, 439, 579, 581
  processing, 582
  spawning, 580
  tagging, 580
Red Sea, 407
  minerals from, 409
Red tide, 18, 173, 174, 417, 418, 490
Reef nets, 247
Reefs
  artificial, 35
  barrier, 147, 148
  coral, 147
  economic aspects, 148
  fishing in, 149
Rendering, 262, 264
Reproduction. *See also* Breeding Behavior; Spawning
  abyssal zone, 7
  copepods, 143, 145
  crabs, 153, 155
  dinoflagellates, 171
  endocrinology in, 72, 73, 75
  fish, 71, 74, 181, 312, 313
Research ships, 460, 463
Respiration, of fish, 179
Reverse osmosis, 167
Ruthenium, radioactive, 569, 572, 574
Ryukyuan fishery, 583
  deep-sea, 584
  inshore, 586
  offshore, 585

Saithe, 305
Salinity
  as ecological factor, 179, 185, 467, 500
  effect on mussels, 423
  in estuaries, 206, 207, 209
  and mortality, 420
  of Peru current, 505
Salmon
  Australia, 43, 44
  British Columbia, 93, 589
  chinook, 588, 589
  chum, 588, 589, 590
  coho, 588, 590
  fishery methods, 591
  Labrador, 345
  marketing, 591
  migration, 314, 589, 590
  mortality, 419
  Newfoundland, 436
  Pacific, 588
  pink, 589, 590
  processing, 591
  reproduction, 589
  research, 592
  sockeye, 91, 588, 589, 590, 593
  spawning, 588, 589
  tagging, 389
  West Coast, 487, 588, 589
Salt, 594
  harvesting, 596
  as preservative, 132, 310, 318, 519
  processing, 597
  solar evaporation, 299, 411, 412, 519, 595
Sampling
  for biological research, 402–404
  of sea bottom, 414, 415
  of sediments, 402, 414, 415
Sand
  as mineral source, 405, 408
  organisms in, 319, 320
  sampling of, 415
Sandworms, 372, 373, 375
Sardine fishery
  California, 189, 487, 610
  India, 316
  Maine, 262
  South Africa, 657
Sargasso Sea, 75, 178, 314, 535, 554
Satellites, 598, 600
  in geodesy, 386
  navigation by, 433
  oceanographic uses of, 598
Saxitoxin, 676
Scallops, 616
  Australian fishery, 45
  dragging for, 619, 620
  Georges Bank, 616, 617, 623
  Maine, 617, 618, 619, 622
  mortality, 620, 621
  New England, 617
  research, 620
  and sea temperature, 621, 622
Schooling
  characteristics, 603
  definition, 602, 603
  environmental factors, 604, 605
  formations in, 604
  optomotor response in, 605
  and spawning, 603
Sea anemones, 9, 325, 677
Sea birds. *See* Birds, marine
Sea cucumber, 7, 9, 32, 36, 64, 520
  toxicity, 678, 679
  Trust Territory, 691
Sea farming, 24, 51, 454, 607
  clams, 54
  demersal fish, 54
  estuarine, 609, 611
  in India, 608
  mussels, 609
  oysters, 207, 473, 609
  pelagic fish, 610
  shellfish, 117
  sponges, 669
  U.S.S.R., 714
Sea floor. *See* Deep sea bottom
Sea foam, 284
Sea Grant Colleges, 289, 427
Sea gulls, 515
Seals, 285
  Labrador fishery, 439
  Newfoundland fishery, 439
  Norwegian fishery, 449
  South Africa, 661
Sea snakes, 65, 520
  toxicity, 686
Sea star, 9
  toxicity of, 64, 678
Sea urchin, 36, 50
  toxicity, 64, 678
Seawater
  bromine from, 86–89
  chemistry of, 106, 594
  desalination, 162, 287
  discoloration, 173, 176
  elements in, 107, 108, 368, 405, 406
  equilibrium models, 111
  heat and power from, 298
  magnesium from, 368
  minerals in, 404
  pollution of. *See* Pollution
  radioactive waste in, 569, 570
  salinity, 179, 185, 594
  temperature, 298, 299, 300
Seaweeds, 19, 20, 626. *See also* Agar; Kelp; Carrageenin; Phycocolloid
  antibacterial action of, 31
  biomass of, 53
  distribution of, 628
  and food chain, 49, 50
  Japan, 628, 630
  Peru, 512
  Philippines, 521
  and phycocolloids, 521
  South Africa, 663
  surveys of, 628
Sediment, 319, 320
  abyssal, 6
  bacteria in, 399
  diamonds in, 664
  foraminiferal, 6, 276

738

INDEX

radioactive isotopes in, 572
sampling of, 402, 403, 413, 414, 415
Seines
 British, 82
 Danish, 253
 drag, 239
 Japanese, 333
 mackerel, 367
 purse, 94, 227, 241, 253, 307, 333, 367, 395, 488, 591
Set net fishery right, 337
Sharks, 630
 attack patterns, 633, 634
 Australian fishery, 41, 42
 biology of, 631
 commensalism, 182
 eyes of, 633
 feeding habits, 633, 634
 habitat, 178
 Indian fishery, 316, 317
 leather, 633
 liver oil, 633
 and odor detection, 633
 protection from, 634, 635
 South Africa, 661
 squalene in, 56, 631
 toxicity, 64, 680
 uses, 633
Shellfish. *See also* Crabs; Lobsters; Shrimp
 antibacterial agents in, 32
 breaded, 66
 British Columbia fishery, 93
 in estuaries, 206
 farming of, 117
 New Zealand, 443
 poisoning by, 61, 118, 175, 402, 417, 675
 preservation of, 28, 29
Shells
 abalone, 4
 clam, 693
 conch, 136, 137
 diatoms, 6
 foraminifera, 273, 276
 oyster, 149, 370, 469, 493, 494
 pteropod, 6
 Philippines, 519
 as sand component, 412
 scallop, 620
 trochus, 690
 Trust Territory, 694
Shipworms, 377
Shrimp, 635
 African, 643, 660
 Australian, 637
 in benthonic domain, 53
 breaded, 67
 Caribbean fishery, 639, 640
 Europe, 642, 643
 frozen, 638, 641
 Gulf fishery, 636, 638
 Indian fishery, 316, 318, 636
 Japan, 637
 Maine, 644

Mexican, 479, 639
Pacific coast, 638, 639
Pakistan, 637
polysaccharide-coated, 549
South America, 640–642
species, 635, 639
staining of, 391, 393
world fisheries, 635, 636
Silicon, radioactive, 567, 570
Skins, cod, 133
Skipjack. *See* Tuna
Snails, 1, 50, 138, 139. *See also* Abalone
 toxic, 679
Snoek, 658
Sodium chloride. *See* Salt
Solar evaporation
 desalination by, 162
 salt produced by, 299, 412, 519, 595, 596
 technology of, 595, 596, 597
Solvent extraction, 268, 269, 270
Sonar, in ocean surveys, 413, 414
Sound detection
 biological effects of, 651
 by fish, 650, 651
 by invertebrates, 652
 by mammals, 653
Sound emission, 649
 in depth recording, 413
 by electric arc, 414
 by fish, 180, 649, 651
 by invertebrates, 652
 by mammals, 653
 by porpoises, 552, 641
 and schooling, 606
South Africa fisheries, 655
 abalone, 660
 algae, 663
 anchovy, 657
 demersal fish, 659
 guano, 663
 harvesting, 656, 657
 lobster, 659, 660
 minerals, 664
 mussels, 661
 oysters, 660
 pilchard, 657
 seals, 661
 sharks, 661
 shrimp, 660
 squid, 660
 technology of, 656, 657
 tuna, 658
 whales, 661, 662
Spain
 cephalopod fishery, 102
 shrimp fishery, 643
Spawning, 72, 74. *See also* Reproduction; Breeding behavior
 capelin, 97
 death caused by, 419
 migrations in, 75, 76, 390
 redfish, 580
 of salmon, 419, 588
 and schooling, 603

stimuli for, 77
and tagging, 389
Sperm whales, 616
Spiny lobster, 359, 360
 markets for, 361, 362
 sound emission by, 652
 South Africa, 659
 Trust Territory, 694
Spoilage, 27, 28. *See also* Preservation; Radiation Preservation
 of fish products, 547
 of quahogs, 128
 of whales, 30
Sponges, 665
 antimicrobial agents from, 32
 Bahamas, 667, 668, 669
 biology of, 665
 boring, 666
 commercial, 666
 farming of, 669
 Gulf of Mexico, 667
 harvesting, 667
 Mediterranean, 667, 668, 669
 processing, 668
 toxic, 61, 676
 Trust Territory, 693
Sport fishery, 671
 Canada, 671
 Gulf of Mexico, 290, 672
 Europe, 671
 U.S., 671, 672, 673
 U.S.S.R., 671
Squalene, 56, 631
Squid. *See also* Cephalopods
 as bait, 103
 behavior of, 104
 food value of, 103
 Japanese fishery, 102
 Newfoundland fishery, 436
 as prey, 106
 seines for, 103
 South Africa, 660
 toxicity of, 105
Standing crop, 51, 220, 499
 of phytoplankton, 534, 553, 556
 of zooplankton, 535, 536, 537
Steered Narrow Beam System, 203, 204
Stingrays, 64, 65, 681
Stone crab, 292
Sturgeon, 715
Strontium, radioactive, 571, 572, 577
Support ships, 260
Swim bladder, 179, 180, 650
Swordfish
 "jellied," 490
 longline fishing, 363
Symbiosis, 182

Tagging, 314, 389
 king crabs, 155
 lobsters, 356, 357, 358
 by mutilation, 390
 purposes of, 389, 390

739

INDEX

Tagging (cont.)
  of redfish, 580, 581
  selection of method, 392
  by staining, 391
  by tattooing, 391
  of turtles, 623
Tags
  Atkins, 392
  color of, 392
  dart, 393
  materials for, 391
  Petersen, 392, 580
  recovery of, 392
  types of, 392
Tangle net, 156
Tanner crabs, 150
Tapeworms, in fish, 492
Taxonomy, fish, 311
Telemetry, 229
Teleost fish
  ecology, 178, 179
  reproduction, 71, 74, 181
  toxicity, 65
Temperature, seawater, 298, 299, 300
  as ecological factor, 178, 185, 187, 375, 467
  in estuaries, 207, 473
  of interstitial water, 320
  and mortality, 419, 420
  effect on mussels, 422
  effect on oysters, 467
  of Peru current, 503
  effect on scallops, 621
  effect on shrimp, 646–648
Teredos, 377, 378
Territorial sea, 347, 348, 351
Tetrodotoxin poisoning, 65, 683, 684
Thawing, of sea-frozen fish, 279
Tidal rhythm, 59, 60
Tides, power from, 214, 288
Time, in navigation, 429, 431
Toxic fish. *See* Venomous marine animals; Poisoning
Training vessels, 196
Traps, 245, 260
  for crabs, 152, 156, 639
  for lobsters, 352, 360
Trawl Warp Tensiometer, 204
Trawlers
  British, 80, 81, 82, 234, 235, 278
  dragging, 582
  drifter, 255
  factory, 40, 250, 276
  freezing on, 234, 236, 276, 277, 278, 540
  Icelandic, 306, 308
  lobster, 353, 354
  otter, 130, 354
  shrimp, 252
  side, 250, 542
  South Africa, 659
  stern, 250

Trawling, mid-water, 228
Trematodes, 491
Tritium, in seawater, 567
Trochus shells, 690
Trust Territory fisheries, 687
  abalone, 1
  algae, 694
  clams, 693
  crocodiles, 692
  mother of pearl, 690
  pearls, 690
  sea cucumbers, 691
  sponges, 693
  tuna, 688
  turtles, 691
Tuna
  Canadian fishery, 92
  fish meal from, 261
  freezing of, 276, 277, 278
  Hawaiian fishery, 296
  Indian fishery, 316
  Japanese fishery, 480, 696, 698
  location of, 467
  mechanization of fishery, 226, 227
  Oceania, 479, 517
  South Africa, 658
  Trust Territory, 688, 695
  West Coast, 487
Turtles, 623
  behavior of, 624, 625
  fisheries, 625
  leather from, 625
  nesting habits, 624
  Philippines, 521
  tagging of, 623
  toxic, 65, 686
  Trust Territory, 691

United States fisheries
  fishery products, 711
  fishing fleet, 707
  preservation methods, 710
  statistics, 704–709
  vs. U.S.S.R., 713, 716
U.S. Food and Drug Administration, 30
U.S. Government, marine resources research, 285
U.S.S.R. fisheries, 713
  agreements with U.S., 716
  organization, 714
  research, 714
  off U.S. coast, 716

Vapor compression, desalination by, 162
Vegetable oils, breaded fish, 70
Venomous marine animals. *See also* Poisoning
  cephalopods, 680
  gastropods, 63
  jellyfish, 63, 676, 677
  mollusks, 679, 680

puffers, 65, 684
sea cucumber, 679
sea star, 64, 678
sea snakes, 65, 686
sea urchin, 64, 678
sharks, 64, 680
snails, 679
sponges, 61, 676
sting rays, 64, 65, 681
treatment of victims, 685
turtles, 65, 686
Vitamins, 216, 718
  in algae, 17
  concentrates, 720
  in fish, 56, 451
  liver oils, 718, 719, 720
Viruses, in fish, 488

Waste materials
  and pollution, 544, 545, 546
  radioactive. *See* Radioactive waste
Water pollution. *See* Pollution
Weather
  control of, 453, 454
  forecasting, 468, 469
Weirs, 245
  for lampreys, 614
Whales, 722
  ambergris from, 23
  Australian fishery, 46
  baleen, 466, 723, 725, 726
  blue, 189, 723, 724, 726
  biology, 722
  factory ships, 260, 723, 724, 727
  Greenland, 723
  harpooning, 726
  humpback, 728
  Icelandic fishery, 308
  Japanese fishery, 334
  Newfoundland fishery, 437, 440
  Norway, 448
  South Africa, 661
  sperm, 616
  spoilage, 30
  squid as food for, 106, 723
  treaty control, 725, 726
Whiting, 78, 83
White Line System, 202, 203
Woods Hole Oceanographic Institution, 463
World statistics, 704, 705, 707
Worms. *See* Marine worms

Yeasts, 284, 398
Yellow fin, 696, 699, 700. *See also* Tuna

Zooplankton, 498, 499, 536
  biomass of, 536, 537
  food value of, 537
  in Peru current, 507
  standing crop, 537

740